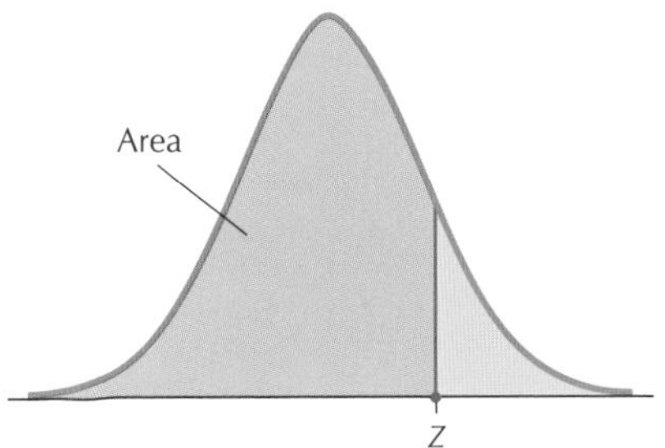


Table C Standard normal distribution (*continued*)

| z | 0.00 | 0.01 | 0.02 | 0.03 | 0.04 | 0.05 | 0.06 | 0.07 | 0.08 | 0.09 |
|---|---|---|---|---|---|---|---|---|---|---|
| **0.0** | 0.5000 | 0.5040 | 0.5080 | 0.5120 | 0.5160 | 0.5199 | 0.5239 | 0.5279 | 0.5319 | 0.5359 |
| **0.1** | 0.5398 | 0.5438 | 0.5478 | 0.5517 | 0.5557 | 0.5596 | 0.5636 | 0.5675 | 0.5714 | 0.5753 |
| **0.2** | 0.5793 | 0.5832 | 0.5871 | 0.5910 | 0.5948 | 0.5987 | 0.6026 | 0.6064 | 0.6103 | 0.6141 |
| **0.3** | 0.6179 | 0.6217 | 0.6255 | 0.6293 | 0.6331 | 0.6368 | 0.6406 | 0.6443 | 0.6480 | 0.6517 |
| **0.4** | 0.6554 | 0.6591 | 0.6628 | 0.6664 | 0.6700 | 0.6736 | 0.6772 | 0.6808 | 0.6844 | 0.6879 |
| **0.5** | 0.6915 | 0.6950 | 0.6985 | 0.7019 | 0.7054 | 0.7088 | 0.7123 | 0.7157 | 0.7190 | 0.7224 |
| **0.6** | 0.7257 | 0.7291 | 0.7324 | 0.7357 | 0.7389 | 0.7422 | 0.7454 | 0.7486 | 0.7517 | 0.7549 |
| **0.7** | 0.7580 | 0.7611 | 0.7642 | 0.7673 | 0.7704 | 0.7734 | 0.7764 | 0.7794 | 0.7823 | 0.7852 |
| **0.8** | 0.7881 | 0.7910 | 0.7939 | 0.7967 | 0.7995 | 0.8023 | 0.8051 | 0.8078 | 0.8106 | 0.8133 |
| **0.9** | 0.8159 | 0.8186 | 0.8212 | 0.8238 | 0.8264 | 0.8289 | 0.8315 | 0.8340 | 0.8365 | 0.8389 |
| **1.0** | 0.8413 | 0.8438 | 0.8461 | 0.8485 | 0.8508 | 0.8531 | 0.8554 | 0.8577 | 0.8599 | 0.8621 |
| **1.1** | 0.8643 | 0.8665 | 0.8686 | 0.8708 | 0.8729 | 0.8749 | 0.8770 | 0.8790 | 0.8810 | 0.8830 |
| **1.2** | 0.8849 | 0.8869 | 0.8888 | 0.8907 | 0.8925 | 0.8944 | 0.8962 | 0.8980 | 0.8997 | 0.9015 |
| **1.3** | 0.9032 | 0.9049 | 0.9066 | 0.9082 | 0.9099 | 0.9115 | 0.9131 | 0.9147 | 0.9162 | 0.9177 |
| **1.4** | 0.9192 | 0.9207 | 0.9222 | 0.9236 | 0.9251 | 0.9265 | 0.9279 | 0.9292 | 0.9306 | 0.9319 |
| **1.5** | 0.9332 | 0.9345 | 0.9357 | 0.9370 | 0.9382 | 0.9394 | 0.9406 | 0.9418 | 0.9429 | 0.9441 |
| **1.6** | 0.9452 | 0.9463 | 0.9474 | 0.9484 | 0.9495 | 0.9505 | 0.9515 | 0.9525 | 0.9535 | 0.9545 |
| **1.7** | 0.9554 | 0.9564 | 0.9573 | 0.9582 | 0.9591 | 0.9599 | 0.9608 | 0.9616 | 0.9625 | 0.9633 |
| **1.8** | 0.9641 | 0.9649 | 0.9656 | 0.9664 | 0.9671 | 0.9678 | 0.9686 | 0.9693 | 0.9699 | 0.9706 |
| **1.9** | 0.9713 | 0.9719 | 0.9726 | 0.9732 | 0.9738 | 0.9744 | 0.9750 | 0.9756 | 0.9761 | 0.9767 |
| **2.0** | 0.9772 | 0.9778 | 0.9783 | 0.9788 | 0.9793 | 0.9798 | 0.9803 | 0.9808 | 0.9812 | 0.9817 |
| **2.1** | 0.9821 | 0.9826 | 0.9830 | 0.9834 | 0.9838 | 0.9842 | 0.9846 | 0.9850 | 0.9854 | 0.9857 |
| **2.2** | 0.9861 | 0.9864 | 0.9868 | 0.9871 | 0.9875 | 0.9878 | 0.9881 | 0.9884 | 0.9887 | 0.9890 |
| **2.3** | 0.9893 | 0.9896 | 0.9898 | 0.9901 | 0.9904 | 0.9906 | 0.9909 | 0.9911 | 0.9913 | 0.9916 |
| **2.4** | 0.9918 | 0.9920 | 0.9922 | 0.9925 | 0.9927 | 0.9929 | 0.9931 | 0.9932 | 0.9934 | 0.9936 |
| **2.5** | 0.9938 | 0.9940 | 0.9941 | 0.9943 | 0.9945 | 0.9946 | 0.9948 | 0.9949 | 0.9951 | 0.9952 |
| **2.6** | 0.9953 | 0.9955 | 0.9956 | 0.9957 | 0.9959 | 0.9960 | 0.9961 | 0.9962 | 0.9963 | 0.9964 |
| **2.7** | 0.9965 | 0.9966 | 0.9967 | 0.9968 | 0.9969 | 0.9970 | 0.9971 | 0.9972 | 0.9973 | 0.9974 |
| **2.8** | 0.9974 | 0.9975 | 0.9976 | 0.9977 | 0.9977 | 0.9978 | 0.9979 | 0.9979 | 0.9980 | 0.9981 |
| **2.9** | 0.9981 | 0.9982 | 0.9982 | 0.9983 | 0.9984 | 0.9984 | 0.9985 | 0.9985 | 0.9986 | 0.9986 |
| **3.0** | 0.9987 | 0.9987 | 0.9987 | 0.9988 | 0.9988 | 0.9989 | 0.9989 | 0.9989 | 0.9990 | 0.9990 |
| **3.1** | 0.9990 | 0.9991 | 0.9991 | 0.9991 | 0.9992 | 0.9992 | 0.9992 | 0.9992 | 0.9993 | 0.9993 |
| **3.2** | 0.9993 | 0.9993 | 0.9994 | 0.9994 | 0.9994 | 0.9994 | 0.9994 | 0.9995 | 0.9995 | 0.9995 |
| **3.3** | 0.9995 | 0.9995 | 0.9995 | 0.9996 | 0.9996 | 0.9996 | 0.9996 | 0.9996 | 0.9996 | 0.9997 |
| **3.4** | 0.9997 | 0.9997 | 0.9997 | 0.9997 | 0.9997 | 0.9997 | 0.9997 | 0.9997 | 0.9997 | 0.9998 |

The cover image for *Discovering Statistics* is taken from one of the book's case studies, "Trial of the Pyx: How Much Gold Is in Your Gold Coins?" (from Chapter 7, Sampling Distributions). The Case Study describes a procedure held annually in London since the Middle Ages to ensure that the coins of the realm contain the proper amount of gold. Using the sampling distribution of the sample mean, the probability is calculated that the Master of the Mint would have been caught and punished had he cheated the throne by debasing the gold content in the coins.

CASE STUDY

### Trial of the Pyx: How Much Gold Is in Your Gold Coins?

The kings of bygone England had a problem: How much gold should they put into their gold coins? After all, the very commerce of the kingdom depended on the purity of the currency. How did the lords of the realm ensure that the coins floating around the kingdom contained reliable amounts of gold?

From the year 1282, the Trial of the Pyx has been held annually in London to ensure that newly minted coins adhere to the standards of the realm. It is the responsibility of the presiding judge to ensure that the trial proceeds lawfully and to inform Her Majesty's Treasury of the verdict. Six members of the Company of Goldsmiths compose the jury, who are given two months to test the coins. It works like this: A ceremonial boxwood chest, called the Pyx, is brought forth, and a sample of 100 of the coins cast that year at the mint is put into it. The Pyx is then weighed. In times past, each gold coin, called a guinea, had an expected weight of 128 grams, so the total weight of the guineas in the Pyx was expected to be 12,800 grams.

If the weight of the coins in the Pyx was much less than 12,800 grams, the jury concluded that the Master of the Mint was cheating the crown by pocketing the excess gold, and he was severely punished. On the other hand, if the coins in the Pyx weighed much more than 12,800 grams, that wasn't good either, since it cut down on the profits produced by the kings' coin-minting monopoly.

By how much could the Master of the Mint debase the coinage before getting caught? We shall see in the chapter's Case Study, The Trial of the Pyx, which unfolds in Section 7.2.

Learn more as this Case Study unfolds in Chapter 7, pages 359–362.

CASE STUDY

### Trial of the Pyx: How Much Gold Is in Your Gold Coins?

Medieval English kings devised a procedure to ensure that the coins of the realm contained the proper amount of gold. A sample of 100 of the gold coins that were cast each year was placed in a ceremonial box called the Pyx. At the chosen time, the Company of Goldsmiths jury weighed the gold coins. The weight of the entire sample of coins was supposed to be 12,800 grams. If the weight was much less than 12,800 grams, the jury concluded that the Master of the Mint was cheating the crown by pocketing the excess gold, and he was severely punished.

**Problem 1**

How did the jury determine what was "much less than 12,800 grams"?

**Solution**

If the weight of the coins was within 32 grams of the expected 12,800 grams, the jury accepted the year's gold coins as pure. Thus, the *sum* of the weights of the gold coins had to lie between 12,768 grams and 12,832 grams. However, the methods we have

# Discovering Statistics

DANIEL T. LAROSE
Central Connecticut State University

W. H. FREEMAN AND COMPANY / NEW YORK

*Senior Publisher:* Craig Bleyer
*Publisher:* Ruth Baruth
*Development Editors:* Shona Burke, Tony Palermino
*Executive Marketing Manager:* Jennifer Somerville
*Market Development Manager:* Kirsten Watrud
*Senior Media Editor:* Roland Cheyney
*Assistant Editor:* Brian Tedesco
*Editorial Assistant:* Katrina Wilhelm
*Marketing Assistant:* Eileen Rothschild
*Photo Editor:* Cecilia Varas
*Photo Researcher:* Patty Cateura
*Cover and Text Designer:* Blake Logan
*Senior Project Editor:* Mary Louise Byrd
*Illustrations:* Macmillan Publishing Solutions
*Production Manager:* Susan Wein
*Composition:* Macmillan Publishing Solutions
*Printing and Binding:* RR Donnelley

Chapter Opener photo: Roz Woodward/Photodisc/Getty Images


Library of Congress Control Number: 2008932351

Casebound ISBN-13: 978-1-4292-2798-8
ISBN-10: 1-4292-2798-2

Paperback ISBN-13: 978-1-4292-2808-4
ISBN-10: 1-4292-2808-3

Instructor's Edition ISBN-13: 978-1-4292-2809-1
ISBN-10: 1-4292-2809-1



Printed in the United States of America

First printing

W. H. Freeman and Company
41 Madison Avenue
New York, NY 10010
Houndmills, Basingstoke RG21 6XS, England
**www.whfreeman.com**

# Brief Contents

# Contents

## Chapter 13 Regression Analysis 665

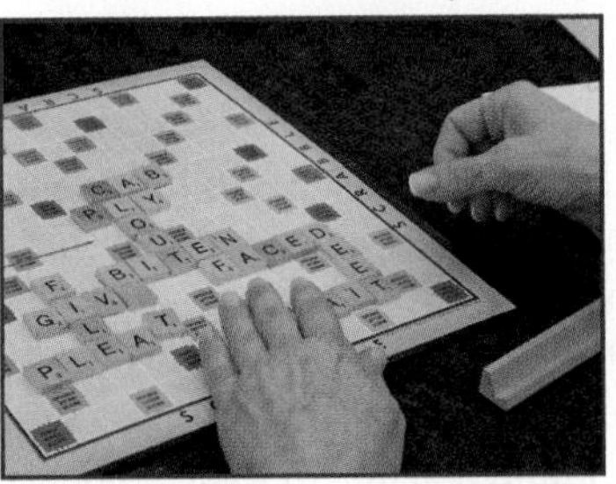

# Preface to the Student

Our twenty-first-century world is flooded with data. Stock market returns and sports results snake across our TV screens in a nonstop stream. Grocery purchases are beep-beeped into data warehouses that enable the retailer to analyze the purchases and recommend individualized offers to their customers. Political candidates recite statistical facts and figures often massaged to support their positions on the issues. To develop a deeper sense of meaning and comprehension of data, students today need to study **statistics**: the art and science of collecting, analyzing, presenting, and interpreting data. *Discovering Statistics* will help you develop the quantitative and analytical tools needed to understand statistics in today's data-saturated world.

## Introductory Statistics Course

*Discovering Statistics* is intended for an algebra-based, undergraduate, one-semester course in general introductory statistics for non-majors. The only prerequisite is intermediate algebra. Introductory statistics will prepare you to work with data in fields such as: psychology, business, nursing, education, and liberal arts, to name a few.

The text was written with the following goals in mind:

- Emphasize student motivation.
- Emphasize statistical literacy and develop statistical thinking.
- Use real data.
- Stress conceptual understanding in addition to knowledge of procedures.
- Foster active learning in the classroom.
- Use technology for developing conceptual understanding and analyzing data.
- Use assessments to improve and evaluate student learning.
- Demonstrate, through Case Studies, how newly acquired analytic tools may be applied to a familiar problem.

These goals follow the Guidelines for Assessment and Instruction in Statistics Education (GAISE) model, endorsed by the American Statistical Association.

## Approach of *Discovering Statistics*

**Balanced analytical and computational coverage.** The text integrates data interpretation and discovery-based methods with complete computational coverage of introductory statistics topics. Through unique and careful use of pedagogy, the text helps you develop your "statistical sense"—understanding the meaning behind the numbers. Equally, the text includes integrated and comprehensive computational coverage, including step-by-step solutions within examples. Select examples include screen shots and computer output from TI-83/84, Excel, and Minitab, with keystroke instructions located in the Step-by-Step Technology Guides at the end of sections.

**Communication of results.** *Discovering Statistics* emphasizes how, in the real world, to explain statistical results to others who have never taken a statistics course. The text emphasizes the importance of keeping in mind how to interpret these results to non-specialists.

**Emphasis on variability.** The importance of variability in the introductory statistics curriculum cannot be overstated. Without a solid appreciation of how statistics may vary, there is little chance that you will be able to understand the crucial topic of sampling distributions.

**Emphasis on sampling distributions and the Central Limit Theorem.** All statistical inference in this text depends on sampling distributions. All test statistics, all confidence interval formulas depend on sampling distributions. It is vital that you have a grasp of sampling distributions and the Central Limit Theorem to completely understand classical statistical inference.

**Use of powerful, current examples with real data.** The terrorist attacks on September 11, 2001, the bird flu pandemic, and the popularity of online dating represent the variety of examples included in *Discovering Statistics*. Example and exercise topics reflect real-world problems and engage your interest in their solution. Real data (with sources cited) are frequently used to further demonstrate relevance of topics.

## Features of *Discovering Statistics*

**Case Studies.** A Case Study begins each chapter and is developed throughout the section examples, using the new set of tools that the section provides.

**The Big Picture.** Brief paragraphs look at "where we are coming from and where we are headed . . . ". (Chapter 2, page 34)

*The Big Picture* Where we are coming from, and where we are headed...

In Chapter 1, we became acquainted with the building blocks of data analysis, such as the concepts of population, sample, types of variables, and the importance of a random sample. In Chapter 2, we apply the adage "A picture is worth a thousand words." The human mind can assess information presented pictorially, better than it can through words and numbers alone. We can recognize the entire scope of the interrelationships within a picture and somehow formulate a synthesis. Psychologists sometimes call this innate pattern recognition ability the "gestalt" experience. Statistical graphics take advantage of these abilities to quickly summarize data.

**Matched Objectives.** Each section begins with a list of numbered objectives headed "By the end of this section, I will be able to . . . ". The objective numbers are matched with the numbered topics within each section as well as with the end-of-section summary. (Chapter 9, pages 455, 461)

### 9.1 Introduction to Hypothesis Testing

*Objectives:* By the end of this section, I will be able to...

1 Construct the null hypothesis and the alternative hypothesis from the statement of the problem.

2 Explain the role of chance variation in establishing reasonable doubt.

3 State the two types of errors made in hypothesis tests: the Type I error, made with probability $\alpha$, and the Type II error, made with probability $\beta$.

#### 1 Constructing the Hypotheses

Recall that inferential statistics refers to methods for learning about the unknown value of a population parameter by studying the information in a sample. The basic idea of hypothesis testing is the following:

### Section 9.1 SUMMARY

1 Statistical hypothesis testing is a way of formalizing the decision-making process so that a decision can be rendered about the unknown value of the parameter. The status quo hypothesis that represents what has been tentatively assumed about the value of the parameter is called the null hypothesis and is denoted as $H_0$. The alternative hypothesis, or research hypothesis, denoted as $H_a$, represents an alternative conjecture about the value of the parameter.

2 In a hypothesis test, we compare the sample mean $\bar{x}$ with the value $\mu_0$ of the population mean used in the $H_0$ hypothesis. If the difference is large, then $H_0$ is rejected. If the difference is not large, then $H_0$ is not rejected.

3 When performing a hypothesis test, there are two ways of making a correct decision: to not reject $H_0$ when $H_0$ is true and to reject $H_0$ when $H_0$ is false. Also, there are two types of error: a Type I error is to reject $H_0$ when $H_0$ is true, and a Type II error is to not reject $H_0$ when $H_0$ is false. The probability of a Type I error is denoted as $\alpha$ (alpha). The probability of a Type II error is denoted as $\beta$ (beta).

A Decision Is Not Proof

*It is important to understand that the decision to reject or not reject $H_0$ does not prove anything.* The decision represents whether or not there is sufficient evidence against the null hypothesis. This is our best judgment given the data available. You cannot claim to have *proven* anything about the value of a population parameter unless you elicit information from the entire population, which in most cases is not possible. ■

**Developing Your Statistical Sense.** This feature will empower you with useful perspectives that data analysts need to know. You will learn to think like a real-world statistical analyst. (Chapter 9, page 460)

**What Does This Mean?** Feature boxes foster an intuitive approach and interpretation of results. Whenever a new formula or statistic is being introduced, the emphasis is on "What does this really mean?" Developing this understanding is just as important as getting the right answer, especially when the software can do the calculations. In the workplace, you may need to explain to your managers what the statistical results really mean. (Chapter 4, page 191)

Interpreting the Slope

What is the meaning of the slope coefficient $b_1$? In data analysis, we interpret the slope of the regression line as the *estimated change in y per unit increase in x*. In our temperature example, the units are degrees Fahrenheit, and the value of $b_1$ is 0.9865. Therefore, $b_1 = 0.9865$ means the following: "For each increase of 1°F in low temperature, the estimated high temperature increases by 0.9865°F." Also note that the slope is positive, which concurs with our assessment of the overall relationship between high and low temperatures. Further, is it a coincidence that both the slope and the correlation coefficient are positive? Not at all.

**What If Scenarios.** The scenarios help you focus on statistical thinking rather than rote computation. Because of the availability of powerful statistical computer packages, statistical analysis is easy to do badly. The wrong analysis is worse than useless. It can cost companies lots of money, may convince lawmakers to pass legislation affecting millions of people, can incorrectly determine effects of pharmaceuticals or environmental pollution, and can have many other serious ramifications. The What If scenarios are extensions of examples or exercises aimed at honing your critical-thinking skills. In What If exercises, the original problem set-up is altered in a specific but non-quantifiable way. You are then asked to think about how that change would percolate through the results, without recourse to calculations. The exercises as well as the scenarios are marked with the What If icon. (Chapter 3, page 94)

Give the Calculator a Rest

What If Scenarios offer you a chance to reflect on how changes in the initial conditions will percolate through the various aspects of a problem. The only requirement is to put your calculator down and think through the problem. No data are given. You are asked to find the answers by using your knowledge of what the statistics *represent*.

Consider Example 3.5 once again. Now imagine: *what if* there was a data entry error (a typo), and the subscription cost for the most expensive journal, the *Journal of Pure and Applied Algebra,* was not \$3631 but actually some *unspecified* dollar amount greater than \$3631. Describe how and why this change would have affected the following, if at all:

**a.** The mean subscription cost

**b.** The median subscription cost

**c.** The mode subscription cost

**Solution**

**a.** Consider Figure 3.3, a dotplot of the subscription costs, with the triangle indicating the mean cost of \$1185.10. Recall that this represents the balance point of the data. As the cost of the most expensive journal increases (blue arrow), the point at which the data balance (that is, the mean) also moves somewhat to the right. Thus, the

**Stepped Example Solutions.** In selected examples, you are guided through the key steps needed to work the calculations and find the solution. (Chapter 9, page 522)

**Step 1** **State the hypotheses.** Our hypotheses are

$$H_0: p = 0.21 \quad \text{versus} \quad H_a: p \neq 0.21$$

where $p$ represents the population proportion of Americans who smoke tobacco.

**Step 2** **Find $Z_{\text{crit}}$ and state the rejection rule.** We have a two-tailed test, with $\alpha = 0.10$. This gives us our critical value $Z_{\text{crit}} = 1.645$ from Table 9.14 and the rejection rule from Table 9.15: Reject $H_0$ if $Z_{\text{data}} > 1.645$ or $Z_{\text{data}} < -1.645$.

**Step 3** **Find $Z_{\text{data}}$.** Our sample proportion is $\hat{p} = 78/400 = 0.195$. Since $p_0 = 0.21$, the standard deviation of the sampling distribution of $\hat{p}$ is

$$\sigma_{\hat{p}} = \sqrt{\frac{p_0(1-p_0)}{n}} = \sqrt{\frac{(0.21)(0.79)}{400}} \approx 0.020365$$

Thus, our test statistic is

$$Z_{\text{data}} = \frac{\hat{p} - p_0}{\sigma_{\hat{p}}} = \frac{\hat{p} - p_0}{\sqrt{\frac{p_0(1-p_0)}{n}}} = \frac{0.195 - 0.21}{\sqrt{\frac{(0.21)(0.79)}{400}}} \approx -0.74$$

*What Results Might We Expect?*

$Z_{\text{data}} = -0.74$ denotes that the sample proportion $\hat{p} = 0.195$ lies 0.74 standard deviations below the hypothesized proportion $p_0 = 0.21$. This is illustrated in Figure 9.64. Clearly, the sample proportion $\hat{p} = 0.195$ is not extreme, and so, by the essential idea about hypothesis testing for the proportion, we would not expect to reject $H_0$. ■

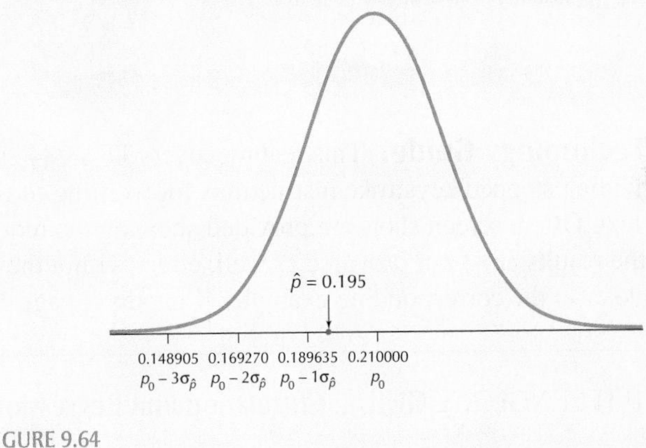


FIGURE 9.64

**What Results Might We Expect?** This feature, located in example solutions, challenges you to predict what the results of a particular problem will be. You are presented with a graphical view of the situation, and, before performing any calculations, you are asked to bring your intuition and common sense to bear on the problem and to state what results we might expect once we do the number crunching. (Chapter 9, page 522)

**Definitions and Formulas.** Easily located in highlighted boxes, key definitions and formulas are important to understand when working examples and exercises. Important vocabulary and formulas are also listed (with page references) at the end of each chapter. (Chapter 1, page 6)

**What Is Statistics?**
**Statistics** is the *art* and *science* of

- collecting,
- analyzing,
- presenting, and
- interpreting data.

**Exercises.** *Discovering Statistics* contains a rich and varied collection of section and chapter exercises:

- Clarifying the Concepts (conceptual)
- Practicing the Techniques (skill-based)
- Applying the Concepts (real-world applications)

At the end of each chapter, **Review Exercises** and a **Chapter Quiz** help to test your overall understanding of each chapter's concepts and to practice for exams. The answers to odd-numbered exercises and all chapter quiz exercises are located in the back of the book.

CHAPTER 5 *Quiz*

**TRUE OR FALSE**

**1.** True or false: An outcome is a collection of a series of events from the sample space of an experiment.

**2.** True or false: It is fine for an event to consist of only one outcome.

**3.** True or false: For any event *A* (even events like *A*: the moon is made of green cheese) the probability of *A* plus the probability of $A^C$ always add up to 1.

**FILL IN THE BLANK**

**4.** The minimum value that a probability can take is ______ and the maximum value is ______.

**5.** The union of two events is associated with the English word ______, and the intersection of two events is associated with the English word ______.

**6.** Someone has told you that there is a 50-50 chance of rain tomorrow. This means that the probability of rain tomorrow equals ______.

**Step-by-Step Technology Guide.** This feature covers TI-83/84 calculators, Excel, and Minitab, providing stepped keystroke instructions for working through selected examples in the text. Often, screen shots are provided showing the required steps. Screen shots of the results are often provided as well, either within the Step-by-Step Technology Guide or in the corresponding example. (Chapter 4, page 193)

## STEP-BY-STEP TECHNOLOGY GUIDE: Correlation and Regression

We illustrate using Example 4.5, the temperature data (page 178).

### TI-83/84

*Step 1* Turn diagnostics on as follows. Press **2nd 0**. Scroll down and select **DiagnosticOn** (Figure 4.23). Press **ENTER** twice to turn diagnostics on.
*Step 2* **Enter** the **X (Low Temp)** data in **L1**, and the **Y (High Temp)** data in **L2**.
*Step 3* Press **STAT** and highlight **CALC**.
*Step 4* Press **8** (not **4**) to choose **LinReg(a+bx)**, as shown in Figure 4.24.
*Step 5* On the home screen, **LinReg(a+bx)** appears. Press **ENTER**.

```
CATALOG
 DependAsk
 DependAuto
 det(
 DiagnosticOff
▸DiagnosticOn
 dim(
 Disp
```

Figure 4.23

```
EDIT CALC TESTS
5↑QuadReg
6:CubicReg
7:QuartReg
8:LinReg(a+bx)
9:LnReg
0:ExpReg
A↓PwrReg
```

Figure 4.24

### EXCEL

*Step 1* Enter the *x* variable in column **A** and the *y* variable in column **B**.
*Step 2* Click on **Data > Data Analysis > Regression** and click **OK**.
*Step 3* For **Input Y Range**, select cells **B1–B10**. For **Input X Range**, select cells **A1–A10**. Click **OK**.

### MINITAB

**Correlation Coefficient *r***

*Step 1* Select **Stat > Basic Statistics > Correlation...**
*Step 2* Select the variables you want to analyze and click **OK**.

**Regression**

*Step 1* Enter the *x* variable variable in **C1** and the *y* variable in **C2**.
*Step 2* Click on **Stat > Regression > Regression**.
*Step 3* Select the *y* variable for the **Response Variable** and the *x* variable for the **Predictor Variable**. Click **OK**.

**Try This in Class!** In-class activities promote active and cooperative learning. Each chapter ends with a set of class-based activities, Try This in Class!, in which you can engage with other students during class time using manipulatives such as dice. (Chapter 9, page 537)

*Try This in Class!* *In-class activities to enhance your understanding of statistics*

TESTING FOR THE POPULATION MEAN OF THE SUM OF TWO FAIR DICE

Is the population mean $\mu$ of the sum of two fair dice equal to 7? The population standard deviation is $\sigma = 2.4152$. Each student will generate sample data to perform the following hypothesis test with level of significance $\alpha = 0.10$:

$$H_0: \mu = 7 \quad \text{versus} \quad H_a: \mu \neq 7$$

where $\mu$ represents the population mean sum of two fair dice. Assume that the distribution is normal. The instructor should prepare a number line on the board to display the collection of test statistic $Z_{\text{data}}$ values. For $\alpha = 0.10$ and a two-tailed test, the value of $Z_{\text{crit}}$ is 1.645. Thus, the instructor should draw lines separating the critical regions $Z_{\text{data}} < -1.645$ and $Z_{\text{data}} > 1.645$ from the noncritical region $-1.645 < Z_{\text{data}} < 1.645$.

1. Each student should toss two dice 10 times, recording the sum each time, and then calculate the mean sum $\bar{x}$ of this sample of size 10.
2. Using $n = 10$, $\sigma = 2.4152$, and $\mu_0 = 7$, each student should calculate his or her own value for the test statistic $Z_{\text{data}}$ using his or her own value of $\bar{x}$ generated above.
3. Give each student a Post-It Note. Each student should record the following on the note: his or her name and the sample mean $\bar{x}$ and test statistic $Z_{\text{data}}$ for his or her sample.
4. Each student should place the Post-It Note on the number line that the instructor prepared earlier.
5. Identify any Post-It Notes that indicate evidence against the null hypothesis that the population mean is 7. What is the value of $Z_{\text{data}}$? Of the sample mean $\bar{x}$? Do you agree that these sample means are unusual?
6. Approximately what percentage of sample means indicate evidence against $H_0: \mu_0 = 7$? Approximately what percentage of sample means indicate evidence consistent with $H_0: \mu_0 = 7$?
7. Overall, what is your personal conclusion about whether or not the population mean of the dice equals 7? What evidence can you point to in support of your conclusion?

APPLET

**Applets.** Interactive statistical applets are located on your CD and the book companion Web site: www.whfreeman.com/discostat. Applet icons in the text mark the related chapter material and exercises.

CAUTION

**Caution notes.** Signaled by the Caution icon, these warnings in the text help you to avoid common errors and misconceptions.

## Supplements

The following electronic and print supplements are available with *Discovering Statistics.*

STATSPORTAL

courses.bfwpub.com/discostat (access code required; available packaged with *Discovering Statistics* or for purchase online). StatsPortal is the digital gateway to *Discovering Statistics,* designed to enrich the course and enhance your study skills through a collection of Web-based tools. StatsPortal integrates a suite of diagnostic, assessment, tutorial, and enrichment features, enabling you to master statistics at your own pace. It is organized around the following learning components:

**Interactive eBook** offers a complete and customizable online version of the text, fully integrated with all the media resources available with *Discovering Statistics.* The eBook allows you to quickly search the text, highlight key areas, and add notes about what you're reading.

**Resources** organizes all the resources for *Discovering Statistics* into one location for ease of use. The resources include the following:

- **StatTutor Tutorials** offer more than 150 audio-multimedia tutorials tied directly to the textbook, including video, applets, and animations.
- **Stats@Work Simulations** put you in the role of a statistical consultant, helping you to better understand statistics interactively in the context of real-life scenarios. You are asked to interpret and analyze data presented in report form, as well as to interpret current events.
- **Statistical Applets** help you master key statistical concepts and work exercises from the text. There are 16 interactive applets.
- **EESEE Case Studies**
- **Data Sets** are available in ASCII, Excel, TI, Minitab, SPSS, and JMP formats.
- **Statistical Software Manuals** for TI-83/84, Excel, Minitab, SPSS, and JMP are provided.
- **WHFStat macros for Excel**
- **Student Solutions Manual**

**Assignments** (for instructor use only) organizes assignments and grades through an easy-to-create assignment process providing access to questions from the Test Bank, Web Quizzes, and Exercises from *Discovering Statistics*.

**Online Study Center:** www.whfreeman.com/osc/discostat (access code required; available for purchase online) offers all the resources available in StatsPortal except the eBook and Assignments Center.

**Companion Web Site:** www.whfreeman.com/discostat is an open-access Web site that includes statistical applets, data sets, and self-quizzes.

**Interactive Student CD-ROM** is included with every new copy of *Discovering Statistics;* it contains access to data sets and applets (also found on the companion Web site).

**Printed Student Solutions Manual** offers detailed solutions for key exercises from each section of *Discovering Statistics*. ISBN: 1-4292-2753-2

**EESEE (Electronic Encyclopedia of Statistical Examples and Exercises) Case Studies.** Developed by The Ohio State University Statistics Department, these electronic case studies provide a wide variety of timely, real examples with real data. EESEE case studies are available via an access code–protected Web site. Access codes are included with new copies of *Discovering Statistics* (printed behind the CD in the back of the book), or subscriptions can be purchased online.

## Acknowledgments

I would like to join W. H. Freeman and Company in thanking the many instructors from across the United States and Canada who offered comments that assisted in the development and refinement of this book. Their contributions included class testing, manuscript reviewing, and participating in surveys about the book and general course needs.

ARKANSAS George Bratton, *University of Central Arkansas*
ARIZONA Cheryl Ossenfort, *Coconino Community College*
CALIFORNIA Christine Cole, *Moorpark College*; Carol Curtis, *Fresno City College*; Kevin Fox, *Shasta College*; Kristin M. Hartford, *Long Beach City College*; Elizabeth Hamman, *Cypress College*; Sara Jones, *Santa Rosa Junior College*; Wendy Miao, *El Camino College*; Michael A. Nasab, *Long Beach City College*; Keith Oberlander, *Pasadena City College*; Greg Perkins, *Hartnell College*; Zika Perovic, *MiraCosta College*; Ladera Rosenburg, *Long Beach City College*; Sherman Sowby, *California State*

*University, Fresno*; James Wan, *Long Beach City College*; Michael Zeitzew, *El Camino College*
**CANADA** Susan Chen, *Camosun College*; Shaun Fallat, *University of Regina*; Dorothy Levay, *Brock University*
**COLORADO** Holly Ashton, *Pikes Peak Community College*; Dean Barchers, *Red Rocks Community College*; Nels Grevstad, *Metropolitan State College of Denver*; Jay Schaffer, *University of Northern Colorado*
**DELAWARE** Derald E. Wentzien, *Wesley College*
**FLORIDA** Abraham Biggs, *Broward Community College*; Lisa M. Borzewski, *St. Petersburg College*; Janette H. Campbell, *Palm Beach Community College*; Zhao Chen, *Florida Gulf Coast University*; Lani Kempner, *Broward Community College*; Nancy Liu, *Miami Dade College*; Panagiotis Nikolopoulos, *Nova Southeastern University*; William Radulovich, *Florida Community College at Jacksonville*; Traci M. Reed, *St. Johns River Community College*; Pali Sen, *University of North Florida*; Jerry Shawver, *Florida Community College at Jacksonville*; Deanna Voehl, *Indian River State College*
**GEORGIA** Donna Brouillette, *Georgia Perimeter College*; Ayona Chatterjee, *University of West Georgia*; Wanda Eanes, *Macon State College*; Todd Hendricks, *Georgia Perimeter College*; Shahryar Heydari, *Piedmont College*; Barry J. Monk, *Macon State College*; Chandler Pike, *University of Georgia*; Kim Robinson, *Clayton State University*; Howard L. Sanders, *Georgia Perimeter College*; Karen H. Smith, *University of West Georgia*; Martha Tapia, *Berry College*
**HAWAII** David Ching, *University of Hawai'i at Mānoa*; Eric Matsuoka, *Leeward Community College*
**ILLINOIS** Virginia Coil, *College of Lake County*; James Cicarelli, *Roosevelt University*; Faye Dang, *Joliet Junior College*; Linda Hoffman, *McKendree University*; Glenn Jablonski, *Triton College*; Julius Nadas, *Wilbur Wright College*; Stephen G. Zuro, *Joliet Junior College*
**INDIANA** Ewa Misiolek, *Saint Mary's College*
**IOWA** Russell Campbell, *University of Northern Iowa*
**KANSAS** Linda Herndon, *Benedictine College*; James Leininger, *MidAmerica Nazarene University*; Leesa Pohl, *Donnelly College*
**KENTUCKY** Brooke Buckley, *Northern Kentucky University;* Lloyd Jaisingh, *Morehead State University*; Christopher Schroeder, *Morehead State University*; Marlene Will, *Spalding University*
**LOUISIANA** Arun K. Agarwal, *Grambling State University*; David Busekist, *Southeastern Louisiana University*; Julien Doucet, *Louisiana State University at Alexandria*; Diane Fisher, *University of Louisiana at Lafayette*; David Gurney, *Southeastern Louisiana University*; Nabendu Pal, *University of Louisiana at Lafayette*; Victor S. Swaim, *Southeastern Louisiana University*
**MARYLAND** Cathy Hess, *Anne Arundel Community College*; Steven Hundert, *College of Southern Maryland*; Annette Noble, *University of Maryland Eastern Shore*; Steve Prehoda, *Frederick Community College*; Fary Sami, *Harford Community College*; Kim Sheppard, *Cecil College*
**MASSACHUSETTS** Mary Fowler, *Worcester State College*; LeRoy P. Hammerstrom, *Eastern Nazarene College*; Bonnie Wicklund, *Mount Wachusett Community College*
**MICHIGAN** Jennifer Borrello, *Grand Rapids Community College*; Lorraine Gregory, *Lake Superior State University*; Linda Reist, *Macomb Community College*; Kathy Zhong, *University of Detroit Mercy*
**MINNESOTA** Ken Grace, *Anoka-Ramsey Community College*; Mezbahur Rahman, *Minnesota State University, Mankato*
**MISSOURI** Kathy Carroll, *Drury University*; Christina Morian, *Lincoln University of Missouri*
**MONTANA** Debra Wiens, *Rocky Mountain College*
**NEBRASKA** Polly Amstutz, *University of Nebraska at Kearney*; Kathy Woitaszewski, *Central Community College*
**NEW JERSEY** Robert Thurston, *Rowan University*; Cathleen Zucco-Teveloff, *Rowan University*
**NEW YORK** David Bernklau, *Long Island University*; Jadwiga Domino, *Medaille College*; Reva Fish, *University at Buffalo*; Maryann Justinger, *Erie Community College*; Michael Kent, *Borough of Manhattan Community College*; William Price, *North Country Community College*; Sharon Testone, *Onondaga Community College*; Nicholas Zaino, *University of Rochester*
**NORTH CAROLINA** Emma B. Borynski, *Durham Technical Community College*; Ayesha Delpish, *Elon University*; Jackie MacLaughlin, *Central Piedmont Community College*; Jeanette Szwec, *Cape Fear Community College*; John Russell Taylor, *The University of North Carolina at Charlotte*; James Truesdell, *Chowan University*
**OHIO** G. Andy Chang, *Youngstown State University*; Don Davis, *Lakeland Community College*; Arjun Gupta, *Bowling Green State University*; William Huepenbecker, *BGSU Firelands*; Gaurab Mahapatra, *The University of Akron*; Mahbobeh Vezvaei, *Kent State University*
**OKLAHOMA** Mickle Duggan, *East Central University*; John Nichols, *Oklahoma Baptist University*; William Warde, *Oklahoma State University*
**OREGON** Jong Sung Kim, *Portland State University*; Carrie Kyser, *Clackamas Community College*
**PENNSYLVANIA** Linda M. Myers, *Harrisburg Area Community College*; Sandra Nypaver, *Mount Aloysius College*
**SOUTH CAROLINA** Diana J. Asmus, *Greenville Technical College*; Thomas Fitzkee, *Francis Marion University*; Erwin Walker, *Clemson University*
**TENNESSEE** Aniekan Ebiefung, *University of Tennessee at Chattanooga*; Frankie E. Harris, *Southwest Tennessee Community College*; Marc Loizeaux, *University of Tennessee at Chattanooga*; Mary Ella Poteat, *Northeast State Technical Community College*
**TEXAS** Ananda Bandulasiri, *Sam Houston State University*; Ferry Butar Butar, *Sam Houston State University*; Ola Disu, *Tarrant County College*; Emmett Elam, *Texas Tech University*; Maggie Foster, *Tarrant County College*; Grady Grizzle, *North Lake College*; Jada P. Hill, *Richland College*; Melinda Holt, *Sam Houston State University*; Jianguo Liu, *University of North Texas*; David D. Marshall, *Texas Woman's University*; Melissa Reeves, *East Texas Baptist University*; Ricardo Rodriguez, *Eastfield College*; Daniela

Stoevska-Kojouharov, *Tarrant County College*; Jo Tucker, *Tarrant County College*
**UTAH** Kari Arnoldsen, *Snow College*; Joe Gallegos, *Salt Lake Community College*; Ruth Trygstad, *Salt Lake Community College*
**VIRGINIA** John Avioli, *Christopher Newport University*; Robert May, *Virginia Highlands Community College*; Mike Shirazi, *Germanna Community College*; Glenn Weber, *Christopher Newport University*; Ken Wissmann, *Shenandoah University*
**WASHINGTON** Margaret Balachowski, *Everett Community College*; Kelly Brooks, *Pierce College*; Abel Gage, *Skagit Valley College*; John Kellermeier, *Tacoma Community College*
**WISCONSIN** William K. Applebaugh, *University of Wisconsin – Eau Claire*; David M. Reineke, *University of Wisconsin – La Crosse*; Vicki Whitledge, *University of Wisconsin – Eau Claire*

I would also like to acknowledge the following people for their help in class testing chapters of the text: Ferry Butar Butar, *Sam Houston State University*; Barry Monk, *Macon State College*; Michael A. Nasab, *Long Beach City College*; Traci M. Reed, *St. Johns River Community College*; Ricardo Rodriguez, *Eastfield College*; Karen H. Smith, *University of West Georgia*; Martha Tapia, *Berry College*; Mahbobeh Vezvaei, *Kent State University*; James Wan, *Long Beach City College.*

*Discovering Statistics* owes much to the untiring efforts of the team of professionals at W. H. Freeman and Company. I would like to thank Craig Bleyer, Mary Louise Byrd, Shona Burke, Kirsten Watrud, and Jennifer Somerville for contributing their talents to the creation of the book. A big thank you goes to Tony Palermino, whose development editing greatly improved the text, as well as to Pam Bruton, Penny Hull, Ann Cannon, Christina Morian, Sivanandan Balakumar, and Jackie Miller. Most especially, I would like to thank Ruth Baruth, Mathematics and Statistics Publisher, who recognized the need for a book like *Discovering Statistics* and helped make it a reality.

I also wish to thank Dr. Timothy Craine and Dr. Chun Jin, Chair and Assistant Chair of the Department of Mathematical Sciences at Central Connecticut State University, Dr. Dipak K. Dey, Chair of the Department of Statistics at the University of Connecticut, and Dr. John Judge, Chair of the Department of Mathematics at Westfield State College. Thanks to my daughter and statistician-in-training Chantal Danielle (20), for carrying on the love of statistics to the next generation, and to my twin children, Tristan Spring and Ravel Renaissance (9), for demonstrating that there is life beyond the computer screen. Above all, I extend my deepest gratitude to my darling wife of 24 years, Debra J. Larose, for her love, understanding, and true belief in the project.

# About the Author

Since his days of collecting baseball cards as a youngster and scrutinizing the statistics of his favorite players, Dan Larose has loved statistics. He also loved language and writing, so when Dan went to college he majored in French, then philosophy, and finally linguistics and computer science. This background in the liberal arts honed his writing ability. However, his love of statistics never left him, so he went on to earn an M.S. (1993) and a Ph.D. in statistics (1996) from the University of Connecticut. Today, Dan is Professor of Statistics in the Department of Mathematical Sciences at Central Connecticut State University.

At CCSU, Dan designed, developed, and now directs the world's first online Master of Science degree and Graduate Certificate program in data mining. He has published three books on data mining and one book on SAS programming. *Discovering Knowledge in Data: An Introduction to Data Mining* has been translated into French and Polish. He is the founder of DataMiningConsultant.com, and his consulting clients include *The Economist* magazine, the CIT Group, KPMG International, Sonalysts, Inc., Booz Allen and Hamilton, the Hospital for Special Care, and Microsoft. His consulting work includes a $750,000 Phase II grant from the Air Force Office of Research, Storage Efficient Data Mining of High Speed Data Streams. He is the Series Editor for the new Wiley Series on Methods and Applications in Data Mining.

However, Dan's favorite work is imparting a love of statistics to a new generation, and he trusts that *Discovering Statistics* will help to do so.

Dan lives in Tolland, Connecticut, with his wife and three children.

Chapter 1

# The Nature of Statistics

CASE STUDY

## Does Friday the 13th Change Human Behavior?

Superstitions affect most of us. Some people will never walk under a ladder, while others will alter their path to avoid a black cat. Do you think that people change their behavior on Friday the 13th? Perhaps, suspecting that it may be unlucky, some people might elect to stay home and watch television rather than venture outdoors or drive on the highway.

But how does one go about researching such a question? How would *you* do it? In this chapter, we will learn about a British study that considered this question.

## OVERVIEW



## The Big Picture

### Where we are coming from, and where we are headed...

We begin Chapter 1 by sharing some stories with you, data stories. Hidden within many data sets lie undiscovered items of real human interest. For example:

- Why did many of the survivors of Hurricane Katrina choose not to evacuate?
- How did you feel right after the events of September 11, 2001? Do you think that men and women experienced different emotions?

Our goal is to demonstrate that the field of statistics, often said to be too dry, actually provides us with the tools to understand and appreciate the emotions and experiences of the people behind the numbers.

In Section 1.2, we will learn about the building blocks of data analysis, including many basic but crucial concepts that the rest of the book depends on, such as *population, sample, variable, element,* and *observation*. In Section 1.3, we will investigate ways of gathering data, including random sampling, questionnaires, and statistical studies.

Where do we go from here? Once we have gathered the data, we will need to find ways to describe the characteristics of the data set. Humans, especially "right-brain people," understand graphical representations of information very well, and so in Chapter 2 we will investigate methods of *describing data using graphs and tables.* Later, in Chapter 3, we will look into *describing data numerically,* which is often characterized as a left-brain function. ■

## 1.1 Data Stories: The People Behind the Numbers

**Objective: By the end of this section, I will be able to...**

1 Realize that behind each data set lies a story about real people undergoing real-life experiences.

We begin *Discovering Statistics* by sharing some data stories. The field of statistics is often regarded as dry, too antiseptic and unfeeling. We hope to dispel this misapprehension by sharing these data stories with you. We hope that these stories will kindle a response in you, be it sympathy or curiosity or concern, for behind every data set lies a story about the lives of real people. Individual people are speaking to us from behind the numbers.

### Example 1.1 The reasons Katrina survivors did not evacuate

Hurricane Katrina was the costliest and one of the deadliest hurricanes in American history. Damages exceeded $50 billion and fatalities exceeded 1300, according to the National Oceanic and Atmospheric Administration. In September 2005, a survey was

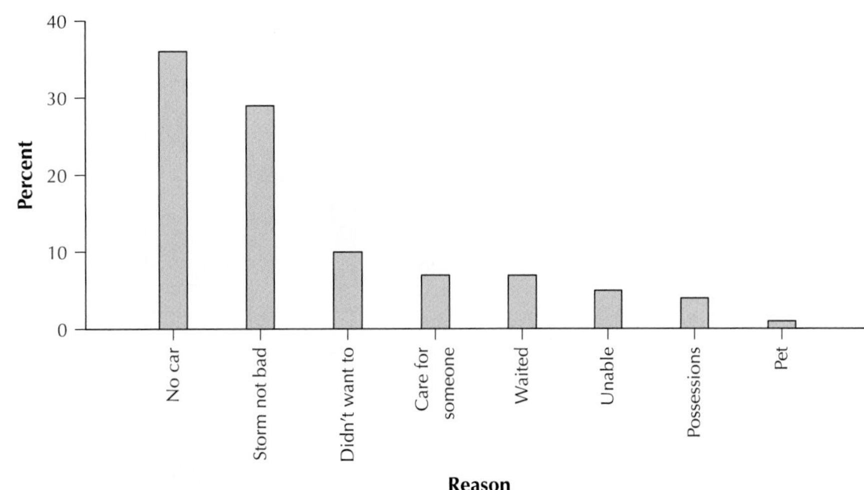


FIGURE 1.1 Bar graph of Katrina survivors' reasons for not evacuating.

conducted of a group of hurricane survivors who had later been moved to shelters in the Greater Houston area. The respondents who did not evacuate were asked what was their most important reason for not evacuating. Figure 1.1 provides a bar graph of the responses, with Table 1.1 supplying more detailed information.

Table 1.1 Katrina survivors' most important reasons for not evacuating

| Reason | Percent |
|---|---|
| I did not have a car or a way to leave | 36 |
| I thought the storm and its aftermath would not be as bad as they were | 29 |
| I just didn't want to leave | 10 |
| I had to care for someone who was physically unable to leave | 7 |
| I waited too long | 7 |
| I was physically unable to leave | 5 |
| I worried that my possessions would be stolen or damaged if I left | 4 |
| I didn't want to leave my pet | 1 |

## Example 1.2 Were there gender differences in the emotions experienced immediately after September 11, 2001?

What was your strongest emotion in the days following the September 11, 2001, terrorist attacks in New York City and Washington, DC? Do you think that men and women felt the same emotions, or did the reactions of men and women differ? In an NBC News Terrorism Poll conducted the day after the tragic events, the following question was asked: "Which one of the following emotions do you feel the most strongly in response to these terrorist attacks: sadness, fear, anger, disbelief, vulnerability?" Figure 1.2 is called a *clustered bar graph* and shows the results. The dominant emotion felt by the men was anger, while the women tended to feel either sadness, anger, or disbelief. Note how the bar graph makes these findings—that there were indeed systematic differences in the emotions felt by men and women regarding the events of September 11, 2001—crystal clear. We will learn how to construct bar graphs in Chapter 2, "Describing Data Using Graphs and Tables."

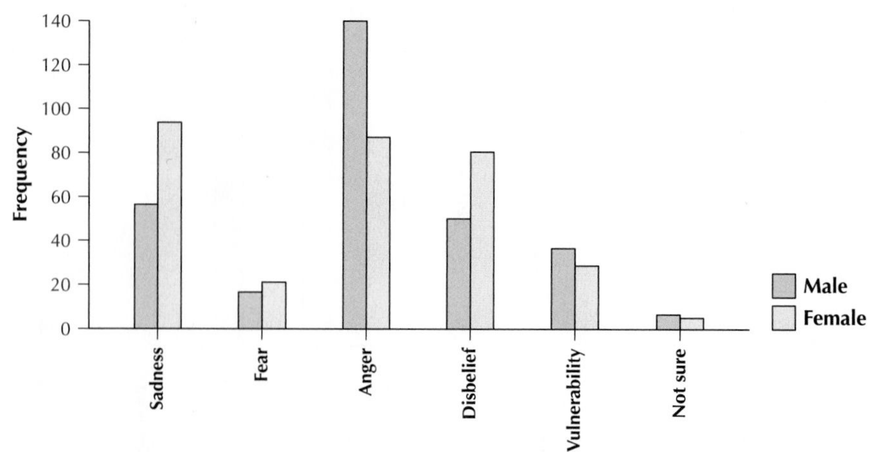


FIGURE 1.2 Clustered bar graph of strongest emotions felt regarding the September 11, 2001, attacks (by gender).

## Example 1.3 The ballot that changed history

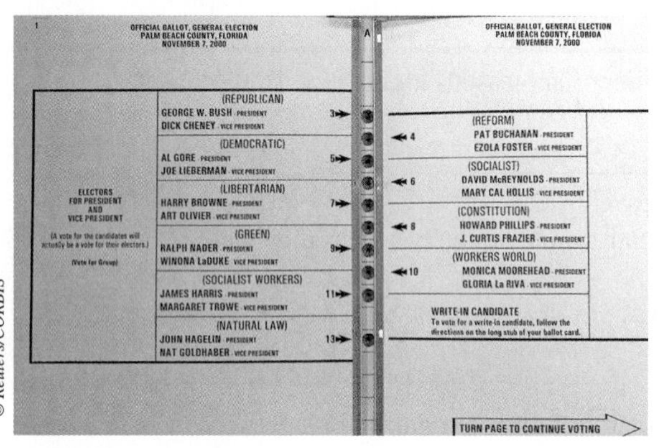




Look at the accompanying illustration. Do you find this 2000 presidential election ballot confusing? There is evidence that many Palm Beach County, Florida, residents did find the ballot confusing. According to a review of discarded ballots by the *Palm Beach Post,* confused voters marked more than one choice on the county's "butterfly ballot," thereby presumably costing Gore 6607 votes.[1] Bush officially won Florida by 537 votes—and with it the presidency.

There may also be evidence that many confused Palm Beach County voters chose Reform Party candidate Pat Buchanan by mistake as well. The scatterplot in Figure 1.3 shows, for each of Florida's counties, the number of votes for Buchanan versus the total number of votes. There is a clear trend of a linear (straight-line) relationship except for one glaring point that lies outside the trend, which is Palm Beach County. Do you think that this confusing ballot had an effect on the results of the election? We will learn how to construct and interpret scatterplots in Chapter 4, "Describing the Relationship Between Two Variables," and we will learn just how unusual the Palm Beach County results were in Chapter 13, "Regression Analysis." ■

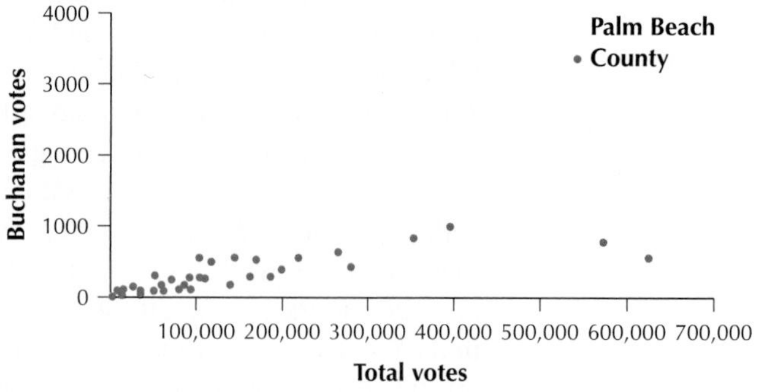


FIGURE 1.3 Scatterplot of Buchanan votes versus total votes in Florida counties, 2000.

## Section 1.1 EXERCISES

Refer to Example 1.1 for Exercises 1–4.

**1.** If you had been in New Orleans during Hurricane Katrina, which of the reasons in Table 1.1 might have been most powerful for convincing you not to evacuate? Which of the reasons would not have played a part in convincing you not to evacuate?

**2.** Refer to Figure 1.1.

- **a.** What does the graph say was the most common reason why the Katrina survivors did not evacuate?
- **b.** What does Table 1.1 say was the most common reason?
- **c.** Which is more descriptive, the table or the figure?
- **d.** Why do you think the text in Figure 1.1 has been shortened?

**3.** If you were writing a news story that sought to display the Katrina survivors in the most sympathetic light, which reasons from Table 1.1 might you emphasize?

**4.** If you were writing a news story that sought to display the Katrina survivors in a less favorable light, which reasons from Table 1.1 might you emphasize?

Refer to Example 1.2 for Exercises 5–8.

**5.** What was your strongest emotion in the days following the September 11, 2001, terrorist attacks in New York City and Washington, DC?

**6.** Can you find the bar in Figure 1.2 that reflects the emotion you felt? Was this the most common emotion felt by others of your gender?

**7.** Do you think the emotions felt were different for men and women? If so, then what evidence from Figure 1.2 would you offer in support of such a view?

**8.** Suppose you did not believe that the emotions felt were different for men and women. What evidence from Figure 1.2 could be offered in support of that position?

Refer to Example 1.3 for Exercises 9–12.

**9.** Do you think that the Palm Beach County votes for Buchanan were unusually high? If so, what is it about Figure 1.3 that leads you to think so?

**10.** If the number of votes for Buchanan followed the pattern of the other counties in Florida, about how many votes would Buchanan have gotten in Palm Beach County? (*Hint:* Look at the vertical axis, and give a rough estimate.)

**11.** About how many votes did Buchanan actually receive in Palm Beach County?

**12.** Do you think that Figure 1.3 gives convincing evidence that what occurred in Palm Beach County was unusual?

# 1.2 An Introduction to Statistics

*Objectives:* **By the end of this section, I will be able to…**

1. Describe what statistics is.
2. State the meaning of descriptive statistics.
3. Understand that inferential statistics refers to learning about a population by studying a sample from that population.

## 1 What Is Statistics?

Do you believe in aliens? According to a recent survey, 54% of the men surveyed responded that they believed in aliens, and 33% of the women did (Figure 1.4). These numbers are examples of *statistics*. Think about these numbers. Here are some questions we could ask about this survey:

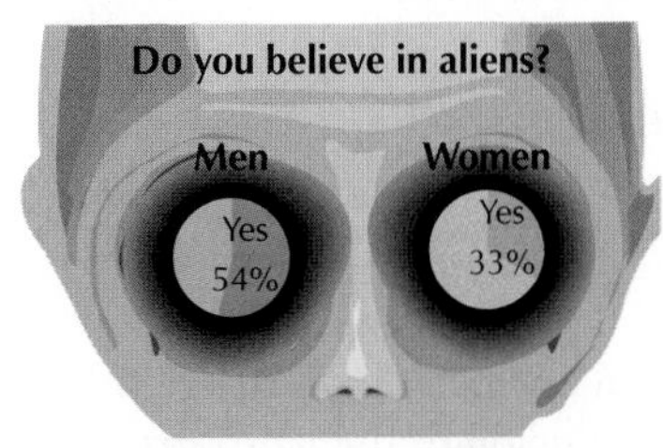


FIGURE 1.4 Graphs comparing percentages of men and women who believe in aliens. (© *USA Today*)

- How did the pollsters arrive at these figures?
- Are the figures accurate? Could they be inaccurate?
- Why do pollsters never ask me my opinion about aliens?
- This survey found that more men than women believed in aliens. But is this difference meaningful or just a product of random chance (like getting 5 heads in a row when tossing a coin)?

These are some of the types of questions we shall be investigating throughout this book.

## Examples of Statistics

Many people, including the author, first became interested in statistics as children collecting baseball cards. The back of each card contains the player's statistics season by season. You can always tell diehard baseball fans by the number and quality of the statistics that they are able to recite about their favorite teams. Television networks routinely employ sports statisticians to collect and report statistics about sports figures. Table 1.2, for example, contains batting averages of the league-leading hitters from 2002 to 2006.

Table 1.2 **Batting-average leaders, Major League Baseball, 2002–2006**

| Season | Player | Team | Batting average |
|---|---|---|---|
| 2006 | Joe Mauer | Minnesota Twins | .347 |
| 2005 | Derrek Lee | Chicago Cubs | .340 |
| 2004 | Ichiro Suzuki | Seattle Mariners | .372 |
| 2003 | Albert Pujols | St. Louis Cardinals | .359 |
| 2002 | Barry Bonds | San Francisco Giants | .370 |

The informal meaning of the term *statistic* refers to a number that describes a person, a group, or a set of items. (On page 12, we provide a more precise definition of statistic.) For example, Joe Mauer's batting average of .347 is a statistic, because it is a number that describes his batting performance for the entire 2006 season. Apart from sports, most people become familiar with statistics through exposure to media reports or advertising, such as

- "Polls indicate a majority of Democrats support stem cell research."
- "The median home sales price in Connecticut has climbed in recent months to $250,000."
- "Three out of four dentists surveyed recommend sugarless gum for their patients who chew gum."

You may have noticed that the section title, "What Is Statistics?" refers to **statistics** in the singular. Why? Because the *field of statistics* involves much more than just collecting and reporting numerical facts. The field of statistics may be defined as follows.

**What Is Statistics?**
**Statistics** is the *art* and *science* of

- collecting,
- analyzing,
- presenting, and
- interpreting data.

A statistician, then, is not simply a sports analyst but any person trained in the art and science of statistics. You may be surprised at the inclusion of the word *art* in the definition of statistics. But there is no question that judgment, experience, and even a little intuition are indispensable tools for any statistician's portfolio.

For today's college student, the field of statistics is especially relevant and useful.

- For example, a *business major* may be interested in whether she should consider diversifying her portfolio to tech stocks, based on their price/earnings ratio.
- A *psychology major* may be interested in determining whether there are differences in therapeutic outcomes between traditional counseling methods and a new cognitive approach.
- An *education major* may be interested in whether listening to a Mozart sonata before taking an exam can significantly improve your grade.

The field of statistics can help solve each of these puzzles.

CASE STUDY

## Does Friday the 13th Change Human Behavior?

How would researchers go about studying whether superstitions change the way people behave? What kind of evidence would support the hypothesis that Friday the 13th causes a change in human behavior? T. J. Scanlon and his coresearchers thought that if there were fewer vehicles on the road on Friday the 13th than on the previous Friday, this would be evidence that some people were playing it safe on Friday the 13th and staying off the roads.[2] Note that the researchers didn't simply argue about the validity of the Friday the 13th superstition. Such discussions are interesting but largely subjective. What they deemed important is the effect of such a superstition on human behavior and how to measure such an effect as a change in behavior.

***Phase 1*** **Data collection.** The first phase of a statistical study, as in the definition of statistics, is to *collect* the data. The researchers obtained data kept by the British Department of Transport on the traffic flow through certain junctions of the M25 motorway in England.

***Phase 2*** **Data analysis.** Next comes the analysis of the data. The authors compared the number of vehicles passing through certain junctions on the M25 motorway on Friday the 13th and the previous Friday during 1990, 1991, and 1992.

Table 1.3 Traffic through M25 junctions, 1990–1992

| Friday the 6th | Friday the 13th | Difference |
|---|---|---|
| 139,246 | 138,548 | 698 |
| 134,012 | 132,908 | 1104 |
| 137,055 | 136,018 | 1037 |
| 133,732 | 131,843 | 1889 |
| 123,552 | 121,641 | 1911 |
| 121,139 | 118,723 | 2416 |
| 128,293 | 125,532 | 2761 |
| 124,631 | 120,249 | 4382 |
| 124,609 | 122,770 | 1839 |
| 117,584 | 117,263 | 321 |

Table 1.3 shows that, in every instance, the number of vehicles passing through these junctions on Friday the 13th was less than on the preceding Friday. Now, let's

examine the data graphically. The clustered bar graph in Figure 1.5 illustrates the difference in the number of vehicles traveling on the M25 motorway on Friday the 6th (in green) and the subsequent Friday the 13th (in yellow) for 10 pairs (clusters) of dates. Note that, *in every instance,* the green bar is longer than its partner yellow bar. This indicates that the number of vehicles on the motorway *decreased* on Friday the 13th when compared with the previous Friday in every instance.

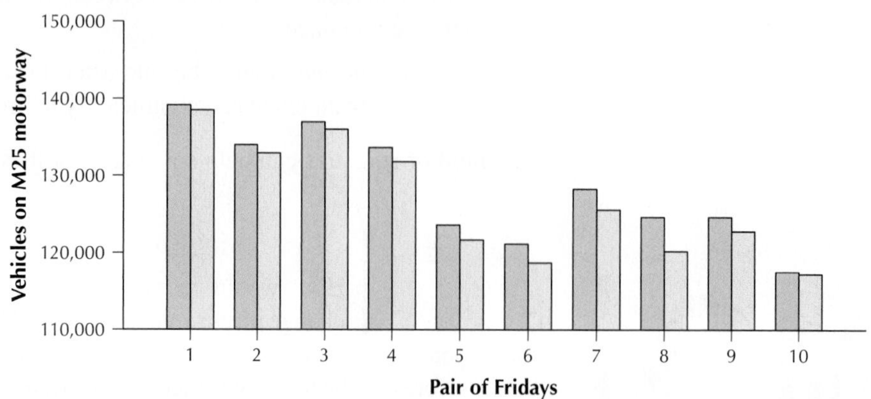


FIGURE 1.5
Clustered bar graph of motorway traffic.

***Phase 3*** **Data presentation.** The *presentation* of the results is important, and the researchers found a highly respectable journal, the *British Medical Journal,* in which to publish their findings. Other avenues for presentation are delivering a talk at a conference, writing up a report for one's supervisor, or presenting a class project.

*What Do These Numbers Mean?*

***Phase 4*** **Data interpretation.** Finally, the last facet in our definition of statistics is *interpretation.* It is crucial for those who are performing a statistical study to make their results understandable to nonstatisticians. It is not sufficient for the statistician alone to understand the results. Rather, the statistician must communicate the results clearly, whether in writing or orally. In this case, the researchers chose decrease in number of vehicles as the criterion on which to base support for their hypothesis that people changed their behavior on Friday the 13th. Their finding of a consistent decrease in traffic on Friday the 13th supports their hypothesis. ■

## 2 Descriptive Statistics: The Building Blocks of Data Analysis

Every data set holds within it a story waiting to be told, as we saw in the Friday the 13th Case Study. To provide us with the tools to uncover these stories we need to learn some simple concepts, *the building blocks of data analysis.*

> **Descriptive statistics** refers to methods for summarizing and organizing the information in a data set.

In **descriptive statistics** we use numbers (such as counts and percents) and graphics to describe the data set, as a first step in data analysis. In Chapters 2 to 4, we will examine descriptive methods much more closely. First, we need to introduce a few terms. Suppose a data analyst for a health maintenance organization (HMO) is collecting data about the patients in a particular hospital, including the diagnosis, length of stay, gender, and total cost. The sources of the information (the patients) are called the **elements**. The patients' characteristics (for example, diagnosis, length of stay) are called the **variables**. Finally, the complete set of characteristics for a particular patient is called an **observation**.

**Elements, Variables, and Observations**

An **element** is a specific entity for which information is collected.

A **variable** is a characteristic of an element, which can assume different values for different elements.

An **observation** is the set of values of the variables for a given element.

When data are presented in tables and spreadsheets, it is typical practice to have the columns indicate the variables, and the rows to indicate the elements. So, for the hospital patients, the observation (specific values for the set of all the variables) for each element (patient) would appear as a row in the table.

## Example 1.4 A remarkable student named Maria

Information was collected on four students from two area colleges and is presented in Table 1.4. What are the elements? What are the variables? Provide the observation for Maria.

Table 1.4 Data set of four elements and seven variables

| Student | Age | Gender | Ethnicity | No. of children | Marital status | GPA | College |
|---|---|---|---|---|---|---|---|
| Jamal | 19 | Male | African American | 0 | Single | 4.00 | Western CC |
| Maria | 25 | Female | Hispanic | 2 | Married | 3.95 | Northern State Univ. |
| Chang | 20 | Female | Asian | 0 | Single | 3.90 | Northern State Univ. |
| Michael | 47 | Male | European American | 3 | Divorced | 3.75 | Western CC |

**Solution**

The elements are the students Jamal, Maria, Chang, and Michael. The seven variables are

- age
- gender
- ethnicity
- number of children
- marital status
- GPA
- college

Suppose we are especially interested in Maria, a 25-year-old married Hispanic mother of two, who attends Northern State University and has a grade point average (GPA) of 3.95. Since the observation for Maria consists of the values for the variables in Maria's entire row, her observation is ■

| Student | Age | Gender | Ethnicity | No. of children | Marital status | GPA | College |
|---|---|---|---|---|---|---|---|
| Maria | 25 | Female | Hispanic | 2 | Married | 3.95 | Northern State Univ. |

Notice that we have variables that can take on various types of values, some of which are numbers and some of which are categories. For example, Maria is 25 years old, has two children, and has a GPA of 3.95, each of which is numeric. On the other hand, Maria is Hispanic, married, and enrolled at Northern State University, characteristics that do not have numeric values but instead are categories. This leads us to define two types of variables: **qualitative** and **quantitative**.

A **qualitative variable** is a variable that does not have a numeric value but is classified into categories. A **quantitative variable** is a variable that takes numeric values.

Qualitative variables are also called *categorical variables*, because they can be grouped into categories. Maria's qualitative variables include her gender, ethnicity, marital status, and college. In contrast, Maria's grade point average is an example of a *quantitative variable*. Other quantitative variables include age and number of children.

Quantitative variables can be classified as either **discrete** or **continuous**.

A **discrete variable** can take either a finite or a countable number of values. Each value can be graphed as a separate point on a number line, with space between each point. A **continuous variable** can take infinitely many values, forming an interval on the number line, with no space between the points.

### Example 1.5 Discrete or continuous?

In Table 1.4, determine whether the following variables are discrete or continuous: (a) number of children and (b) GPA.

**Solution**

**a.** Since the number of children per student is finite, the variable "number of children" is discrete.

**b.** Since GPA can take an infinite number of possible values, for example in the interval 0.0 to 4.0, the variable "GPA" is continuous. ■

### Example 1.6 Most active stocks, June 30, 2007

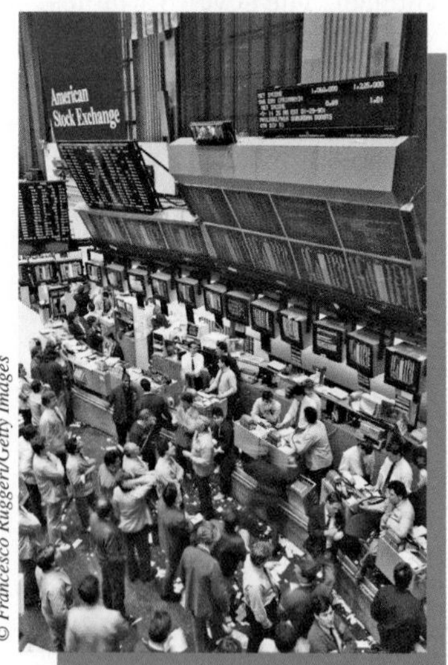

© Francesco Ruggeri/Getty Images

One of the most widespread applications of statistical analysis occurs in the business world. Managers examine patterns and trends in data, thereby hoping to increase profitability. Table 1.5 shows the five most active stocks on the New York Stock Exchange (NYSE), as reported by *USA Today* for June 30, 2007. What are the elements and the variables of this data set? Which variables are qualitative? Which are quantitative? Provide the observation for Pfizer, Inc.

Table 1.5 Most active stocks on NYSE, June 30, 2007

| Stock | Symbol | Last | Volume | Change |
|---|---|---|---|---|
| Ford Motor Company | F | $ 9.42 | 50,308,258 | −0.07 |
| General Electric Company | GE | $38.28 | 44,692,276 | +0.16 |
| Pfizer, Inc. | PFE | $25.57 | 44,346,322 | −0.06 |
| General Growth Properties, Inc. | GGP | $52.95 | 40,926,559 | −0.05 |
| AT&T, Inc. | T | $41.50 | 37,579,025 | +0.76 |

**Solution**

The *elements* are the five most active stocks traded on the NYSE on this day in 2007. The *variables* are as follows:

- Symbol: The Ticker symbol used to represent the stock.
- Last: The most recent trading price for the stock.
- Volume: How many shares of the stock were traded that day.
- Change: The change in share price (in dollars) between the opening price and the closing price that day.

The Ticker symbol, since it is a label for the stock, is qualitative. The other variables are quantitative. The observation for Pfizer, Inc., includes the Ticker symbol and the set of the day's stock data for that company. That stock has a Ticker symbol of PFE, its last share price was \$25.57 per share, 44,346,322 shares of its stock were traded, and the price decreased by \$0.06 per share. ■

| Stock | Symbol | Last | Volume | Change |
|---|---|---|---|---|
| Pfizer, Inc. | PFE | \$25.57 | 44,346,322 | –0.06 |

## Levels of Measurement

Data may be classified according to the following four *levels of measurement.*

- *Nominal data* consist of names, labels, or categories. There is no natural or obvious ordering of nominal data (such as high to low). Arithmetic cannot be carried out on nominal data.
- *Ordinal data* can be arranged in a particular order. However, no arithmetic can be performed on ordinal data.
- *Interval data* are similar to ordinal data, with the extra property that subtraction may be carried out on interval data. There is no *natural zero* for interval data.
- *Ratio data* are similar to interval data, with the extra property that division may be carried out on ratio data. There does exist a natural zero for ratio data.

### Example 1.7 Levels of measurement

Identify which level of measurement is represented by the following data.

**a.** Years covered in European History 101: 1066–1492
**b.** Annual income of students in Statistics 101 class: \$0–\$15,000
**c.** Course grades in English 101: A, B, C, D, F
**d.** Student gender: male, female

**Solution**

**a.** The years 1066 to 1492 represent interval data. There is no natural zero (no "year zero"; the calendar goes from 1 B.C. to A.D. 1). Also, division (1492/1066) does not make sense in terms of years, so that the data are not ratio data. However, subtraction does make sense, in that the course covers 1492 – 1066 = 426 years.

**b.** Student income represents ratio data. Here division does make sense. That is, someone who made \$4000 last year made twice as much as someone who made \$2000 last year. Also, some students probably had no income last year, so that \$0, the natural zero, also makes sense.

**c.** Course grades represent ordinal data, since (a) they may be arranged in a particular order, and (b) arithmetic cannot be performed on them. The quantity A – B makes no sense.

**d.** Student gender represents nominal data, since there is no natural or obvious way that the data may be ordered. Also, no arithmetic can be carried out on student gender. ■

## 3 Inferential Statistics: How Do We Get There from Here?

Descriptive methods of data analysis are widespread and quite informative. However, the modern field of statistics involves much more than simply summarizing a data set. For example, suppose a sociologist claims that one-third of American teenagers have been

© Somos Images/CORBIS

the targets of cyberbullying, that is, have received a threatening message or have had their emails or text messages forwarded without their consent, an embarrassing picture posted without permission, or rumors spread about them online. How should the sociologist go about collecting evidence to support her claim? One method would be to ask each and every person in the population of all American teenagers. In general, a **population** is the collection of *all* elements (persons, items, or data) of interest in a particular study.

However, to ask every teenager in America about his or her online experiences is a daunting task that is expensive, time-consuming, and, in the end, simply impossible. So, unfortunately, the population proportion of American teenagers who have been the targets of cyberbullying remains *unknown.* This proportion who have been targets of cyberbullying is one characteristic of the population of American teenagers. A characteristic of a population is called a **parameter**. The actual value of a population parameter is often unknown.

A **sample** is a subset of the population from which information is collected. For example, from a sample of 100 teenagers at a local mall, 18 said they had been the targets of cyberbullying. That is, the sample proportion of students who had been targets is $18/100 = 18\%$. This proportion is a characteristic of the sample and is called a **statistic**. The advantage here is that, since the sample is relatively small, the characteristics of the sample can be determined.

**Populations, Parameters, Samples, and Statistics**

A **population** is the collection of *all* elements (persons, items, or data) of interest in a particular study. A **parameter** is a characteristic of a population.

A **sample** is a subset of the population from which information is collected. A **statistic** is a characteristic of a sample.

A sample is a subset of a population.

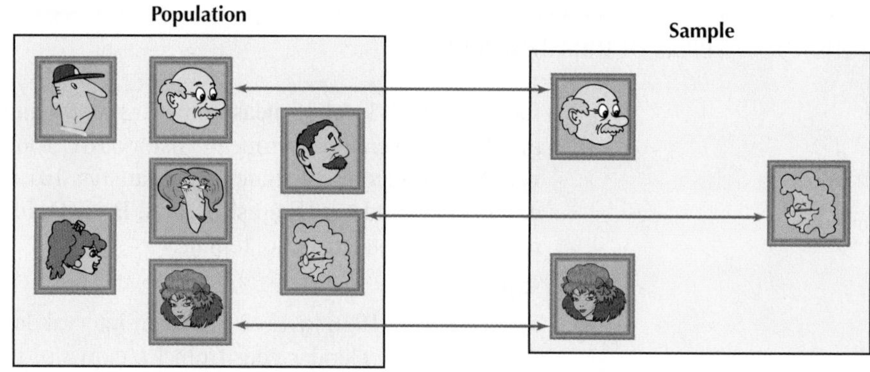


The U.S. Constitution requires that a census be conducted every 10 years. A **census** is the collection of data from every element in the population. As you can imagine, such a task is very difficult and very expensive. In fact, the Census Bureau estimates that the 2000 U.S. census "undercounted the actual U.S. population by over three million individuals."[3]

Because the population you are interested in may be too large to allow you to elicit information from every element, it is often best to gather data from a sample, a subset of that population. Also, time and money often constrain the researcher to choosing a sample rather than studying the entire population. Also, in some experiments, the resource is exhausted when testing is done, for example, in estimating the mean lifetime of light bulbs. Finally, it may be simply impossible to gather information from the entire population, such as when studying the quality of water in Lake Erie.

To estimate the proportion of all American teenagers who have been subjected to cyberbullying, we can use statistical inference. **Statistical inference** refers to learning about the characteristics of a population by studying those characteristics in a subset of the population (that is, in a *sample*). The Pew Internet and American

Life Project conducted a survey of 886 teenagers and found that 284 of them (32%) said they had been the targets of cyberbullying.[4] These 886 teenagers represent a sample, and their characteristics can be known. For example, we know that 284 of the 886 teenagers in the sample said they have been subjected to cyberbullying. At this point, the sociologist can make the *inference* that the proportion of *all* American teenagers who have been subjected to cyberbullying is 32%, because this is the proportion in the sample. In doing this, the sociologist is performing *statistical inference.*

**Statistical inference** consists of methods for estimating and drawing conclusions about population characteristics based on the information contained in a subset (sample) of that population.

"Now wait just a minute," you might object. "How can you say that the proportion of *all* American teenagers who have been subjected to cyberbullying is 32% just because your *sample* proportion is 32%?" Actually, you have a point. We *are* generalizing. We are taking what we know about a portion of the whole (a sample) and using it to draw a conclusion about the whole (the population). But even though the true proportion of American teenagers who have been the targets of cyberbullying is probably not exactly 32%, it is most likely not very far from 32%. The 32% is an *estimate,* an approximation based on sample data. Later we will learn how we can get the estimate as close as we wish to the actual value just by taking a large enough sample.

Finally, we need to point out one further attribute of parameters and statistics. The value of a parameter, even though it is unknown, is a fixed constant. For example, the average age of all persons in your home state (population) at noon today is unknown, but it still exists, and it is a single number. On the other hand, the value of a statistic depends on the sample. For example, a sample of 100 people in your hometown may produce an average age of 31. The average age of a sample of 100 people in a neighboring town may be 32. Later, we will learn that this is because a statistic is a *random variable.*

Of course, to deliver a valid estimate, the sample needs to be *representative* of the population. The sample should not differ systematically in any major characteristic from the population. We will learn more about this in Section 1.3, when we study sampling methods. Table 1.6 summarizes the attributes of a population and a sample. Isn't it convenient that the characteristic associated with the population starts with the same letter, and the same is true for sample?

Table 1.6 Summary of attributes of population and sample

| | **Population** | **Sample** |
|---|---|---|
| **Thumbnail definition** | All elements | Subset of population |
| **Characteristic** | Parameter | Statistic |
| **Value** | Usually unknown | Usually known |
| **Status** | Constant | Depends on sample |

## Example 1.8 Descriptive statistics or statistical inference?

State whether the following situations illustrate the use of descriptive statistics or statistical inference.

**a.** In Baltimore County, Maryland, the average amount spent per week on gasoline consumption in a sample of 500 commuters was $75. The county government infers that the average amount spent weekly by all Baltimore County commuters is $75.

**b.** A sample of 100 residents of Broward County, Florida, yielded 27 residents who work for the government at the local, state, or federal level. Thus, 27% of these 100 residents work for the government.

**c.** The average age of a sample of 200 residents of Garden City, New York, was 34 years old.

**d.** In a survey of 1000 citizens in the Seattle, Washington, metropolitan area, 570 said they would pay higher prices in order to reduce greenhouse emissions. City planners conclude that 57% of all Seattle citizens would do so.

### Solution

**a. Statistical inference.** A sample was taken, and a sample statistic ($75 per week) was calculated. Then, the County government used this statistic to make the *statistical inference* that this was the average amount spent by all Baltimore County commuters.

**b. Descriptive statistics.** Though a sample was taken, there was no attempt to make an inference from this sample of 100 workers to the entire population of Broward County, Florida. So, there is no statistical inference here.

**c. Descriptive statistics.** The average age of 34 years old is a descriptive statistic, since it describes the sample. But no inference is made regarding a larger population.

**d. Statistical inference.** The survey found that 57% of the sample of 1000 citizens would pay higher prices in order to reduce greenhouse emissions. This 57% is a statistic. Then the city planners used this statistic in order to perform statistical inference about the population of all Seattle citizens. ■

### A Statistical Literacy Quiz

Regardless of major, every student in America (indeed, every citizen) needs to become *statistically literate* in order to survive in today's wired society. Why not take this quiz to find out if you are statistically literate? Answer each question true or false.

1. A fair coin is tossed five times and comes up heads each time. That means that tails is "due" and the chances of tails on the next toss is increased.
2. One politician says that the mean income is rising, while another politician says that the median income is falling. One of them has to be lying.
3. Jim is tested for HIV and the test comes back positive. Thus, Jim is HIV-positive.

The correct answer to each question is *false*. Question 1 deals with something called "the gambler's fallacy," and we will cover this, along with the explanation for Question 3, in Chapter 5. We will deal with Question 2, the quirks of means and medians, in Chapter 3. ■

## Section 1.2 SUMMARY

1 The field of statistics is the art and science of collecting, analyzing, presenting, and interpreting data.

2 Descriptive statistics refers to methods for summarizing and organizing the information in a data set. Data sets include information collected on elements (hospital patients or homes in Florida). The types of information collected are called variables (diagnosis of patient, home sales price). Variables may be either quantitative (numerical) or qualitative (categorical). A discrete variable is a quantitative variable that can take either a finite or a countable number of possible values. A continuous variable is a quantitative variable that can take an infinite number of possible values. A population is a collection of all elements of interest, while a sample is a subset of the population.

3 Inferential statistics consists of methods for estimating and drawing conclusions about population characteristics based on the information in the sample. The characteristics for a population are called parameters, while the characteristics for a sample are called statistics. The value of the parameter is usually unknown but constant, while the value of the statistic varies from sample to sample.

## Section 1.2 EXERCISES

### CLARIFYING THE CONCEPTS

**1.** Write a sentence describing in your own words the field of statistics.
**2.** Write a sentence or two about how the field of statistics relates to your own life experience. For example, do you follow Major League Baseball or figure skating? Do you use statistics at work?
**3.** True or false: Statistical inference refers to methods for summarizing and organizing the information in a data set.
**4.** What do we call the entities from which the data are collected?
**5.** Describe the difference between a qualitative and a quantitative variable.
**6.** What is another term for a qualitative variable?
**7.** True or false: The actual value of a population parameter is usually unknown.
**8.** What is the difference between a sample and a population?
**9.** Explain what a statistic is.
**10.** Describe one difference between a statistic and a parameter.
**11.** What do we call the process by which we learn about the characteristics of a population by studying those characteristics in a sample?
**12.** Name at least one problem with drawing conclusions about a population using the information in a sample.
**13.** Since the researcher knows that a sample statistic will probably not equal the value of the population parameter, why would he or she use a sample rather than a population?
**14.** What is a census?

### PRACTICING THE TECHNIQUES

For Exercises 15–28:

- **a.** State whether the variable is qualitative or quantitative.
- **b.** Identify the level of measurement represented by the data.

**15.** The temperature (in degrees Fahrenheit) in the room you are in right now
**16.** Whether you own an iPOD or not
**17.** The price of tea in China
**18.** The SAT Math score of the person sitting next to you (scores range from 200 to 800)
**19.** The winning score in next year's Super Bowl
**20.** The winning team in next year's Super Bowl
**21.** The rank of the winning Super Bowl team in their division
**22.** The number of friends on a student's FaceBook page
**23.** Your favorite video game
**24.** How many contacts you have on your cell phone
**25.** Your favorite topping on pizza
**26.** Your credit card balance
**27.** How old your car is
**28.** What model your car is

For Exercises 29–34, identify the population and the sample.
**29.** A researcher is interested in the median home sales price in Tarrant County, Texas. He collects sales data on 100 home sales.
**30.** A psychologist is concerned about the health of veterans returning from war. She examines 20 veterans and assesses whether they show signs of post-traumatic stress disorder.
**31.** A sociologist wants to learn about the number of meetings per year of the 4-H clubs in Utah. He collects information from 10 different 4-H clubs in various parts of Utah.
**32.** A physical therapist would like to determine whether a new exercise method can delay the onset of osteoporosis in older women. She chooses 10 of her patients to use the new method.
**33.** An educator asks a sample of students at Portland Community College whether they would be interested in taking a course online.
**34.** A financial adviser would like to assess the effect of mergers on price/earnings ratio. She collects data on 50 companies that recently underwent a merger.

For Exercises 35–42, state whether descriptive statistics or statistical inference was used, and explain why.
**35.** The average price in a sample of 15 homes sold in Newington, Connecticut, for the week of April 21 was $253,200.
**36.** According to the Department of Transportation, 60% of all automobile passengers wear seat belts. This is based on a survey of 1000 automobile passengers, of whom 600 wore seat belts.
**37.** In a sample of 500 subjects, it was found that daily exercise lowered the average cholesterol level by 10%. A medical spokesperson then stated that daily exercise can lower everyone's cholesterol level by 10%.
**38.** In a sample of 140 traffic fatalities in New York, 75 involved alcohol.
**39.** The goals-against average for the Charlestown Chiefs hockey team in a sample of 20 games was 3.57 goals per game.
**40.** The Department of Health and Human Services conducted a survey, where it was found that the percentage of 15- to 18-year-olds using illicit drugs has dropped in the last two years. The department concluded that illicit drug use has fallen among all 15- to 18-year-olds.

**41.** The average on the first statistics exam for a sample of 10 students in Ms. Reynolds' class was 70.

**42.** A Department of Homeland Security survey of 750 Americans found that more than half of those surveyed feel less safe than they did 10 years ago. The department concluded that more than half of all Americans feel less safe than they did 10 years ago.

## APPLYING THE CONCEPTS

**Endangered Species.** For Exercises 43 and 44, refer to the following table, which lists 4 of the 11 animal species in Minnesota identified by the U.S. Fish and Wildlife Service as threatened (T) or endangered (E).

| Species | Status | Type | Scientific name |
|---|---|---|---|
| Karner blue butterfly | E | Insect | *Lycaeides melissa samuelis* |
| Canada lynx | T | Mammal | *Lynx canadensis* |
| Piping plover | E | Bird | *Charadrius melodus* |
| Gray wolf | E | Mammal | *Canis lupus* |

**43.** Do these 4 species represent a sample or a population?

**44.** Refer to the table of endangered species.
- **a.** List the elements.
- **b.** List the variables.
- **c.** Identify the qualitative variables. Identify the quantitative variables.
- **d.** Provide the observation for the Canada lynx.

**45. Top Five Employers in Santa Monica, CA.** Refer to the following table.
- **a.** List the elements.
- **b.** List the variables.
- **c.** Which of the variables is quantitative?
- **d.** Which of the variables is qualitative?
- **e.** Provide the observation for St. John's Health Center.

| Company | Employees | Industry |
|---|---|---|
| City of Santa Monica | 1892 | Government |
| St. John's Health Center | 1755 | Health services |
| The Macerich Company | 1605 | Real estate investment trust |
| Fremont General Corporation | 1600 | Property and casualty insurance |
| Entravision Communications Corp. | 1206 | Diversified media company |

*Source:* Santa Monica Chamber of Commerce.

**46. Genetically Engineered Crops.** Genetically engineered (GE) crops are now planted on the majority of acreage in many states around the country. There are three varieties of GE corn: insect-resistant, herbicide-tolerant, and stacked genes. The following table contains the proportion of the corn grown in each of five states, that is GE, along with the GE type most prevalent in each state, for 2007.[5]

| State | Proportion of GE corn | Most prevalent type |
|---|---|---|
| Texas | 79% | Herbicide-tolerant |
| Missouri | 62% | Insect-resistant |
| Minnesota | 86% | Herbicide-tolerant |
| Ohio | 41% | Stacked genes |
| South Dakota | 93% | Stacked genes |

- **a.** List the elements.
- **b.** List the variables.
- **c.** Which of the variables is quantitative?
- **d.** Which of the variables is qualitative?
- **e.** Provide the observation for Texas.

**47. Crime Statistics for Stillwater, OK.** Refer to the following table.
- **a.** List the elements.
- **b.** List the variables.
- **c.** Which of the variables is quantitative?
- **d.** Which of the variables is qualitative?
- **e.** Provide the observation for motor vehicle thefts.

| Crime type | 2005 Total | Per 100,000 people | National per 100,000 people | Compared to national average |
|---|---|---|---|---|
| Robberies | 10 | 24.4 | 195.4 | Better |
| Aggravated assaults | 83 | 202.4 | 340.1 | Better |
| Burglaries | 317 | 772.9 | 814.5 | Better |
| Larceny/thefts | 1147 | 2796.7 | 2734.7 | Worse |
| Motor vehicle thefts | 55 | 134.1 | 526.5 | Better |

**48. Web Page Statistics for Kirksville Rocks.** Refer to the following table, which contains statistics for the Web site Kirksville Rocks! (which contains live-music listings for Kirksville, Missouri), logged from April 27, 2005, to February 16, 2007.
- **a.** List the elements.
- **b.** List the variables.
- **c.** Which of the variables are quantitative?
- **d.** Which of the variables are qualitative?
- **e.** Provide the observation for **www.kvrocks.com/venues/**.

| Web page | Requests | Percent | Title of page |
|---|---|---|---|
| www.kvrocks.com/ | 35,012 | 44.71% | Home Page |
| www.kvrocks.com/musicians/ | 6,069 | 7.75% | Musicians Page |
| www.kvrocks.com/events/ | 3,867 | 4.94% | Events Page |
| www.kvrocks.com/venues/ | 3,332 | 4.26% | Venues Page |

**49. Hospitals near Jackson, MS.** Refer to the following table.
- **a.** List the elements.
- **b.** List the variables.

c. Which of the variables is quantitative?
d. Which of the variables is qualitative?
e. Provide the observation for Montfort Jones Memorial Hospital.

| Hospital name | Beds | City | Zip |
|---|---|---|---|
| Hardy Wilson Memorial Hospital | 49 | Hazlehurst | 39083 |
| Humphreys County Memorial Hospital | 34 | Belzoni | 39038 |
| Jefferson County Hospital | 30 | Fayette | 39069 |
| Lackey Memorial Hospital | 15 | Forest | 39074 |
| Leake Memorial Hospital | 25 | Carthage | 39051 |
| Madison County Medical Center | 67 | Canton | 39046 |
| Montfort Jones Memorial Hospital | 72 | Kosciusko | 39090 |
| Rankin Medical Center | 134 | Brandon | 39042 |
| University of Mississippi Medical Center—Holmes County | 25 | Lexington | 39095 |

**2006 American League Team Pitching Statistics.** Use Table 1.7 for Exercises 50–53. The abbreviations mean the following:

- ERA: earned-run average (earned runs allowed per nine innings)
- HR: total number of home runs given up by the team's pitchers
- BB: walks given up by the team's pitchers
- SO: total number of strikeouts made by the team's pitchers

**TABLE 1.7 Final team pitching statistics, American League, 2006**

| Team | ERA | HR | BB | SO |
|---|---|---|---|---|
| Los Angeles Angels | 4.04 | 158 | 471 | 1164 |
| Oakland Athletics | 4.21 | 162 | 529 | 1003 |
| Chicago White Sox | 4.61 | 200 | 433 | 1012 |
| Detroit Tigers | 3.84 | 160 | 489 | 1003 |
| Seattle Mariners | 4.60 | 183 | 560 | 1067 |
| New York Yankees | 4.41 | 170 | 496 | 1019 |
| Boston Red Sox | 4.83 | 181 | 509 | 1070 |
| Minnesota Twins | 3.95 | 182 | 356 | 1164 |
| Texas Rangers | 4.60 | 162 | 496 | 972 |
| Toronto Blue Jays | 4.37 | 185 | 504 | 1076 |
| Kansas City Royals | 5.65 | 213 | 637 | 904 |
| Cleveland Indians | 4.41 | 166 | 429 | 948 |
| Tampa Bay Devil Rays | 4.96 | 180 | 606 | 979 |
| Baltimore Orioles | 5.35 | 216 | 613 | 1016 |

*Source:* MLB.com.

**50.** How many elements are there? Identify three of them.
**51.** How many variables? Identify three of them. Are they qualitative or quantitative?
**52.** Which team had the lowest earned-run average? The most home runs? The most walks per game?
**53.** Provide the observation for the New York Yankees.

**Births and Maternal Age in Westchester County, NY.** The following table represents the number of births and the average maternal age for 10 hospitals in northwest Westchester County, New York, in 2002. Refer to the table for Exercises 54–56.

| Hospital | Births | Average maternal age |
|---|---|---|
| Briarcliff Manor | 71 | 34.1 |
| Buchanan | 25 | 31.6 |
| Cortlandt | 348 | 32.2 |
| Croton-on-Hudson | 93 | 33.5 |
| Mount Pleasant | 277 | 32.8 |
| Ossining 1 | 80 | 32.1 |
| Ossining 2 | 371 | 29.2 |
| Peekskill | 365 | 29.0 |
| Pleasantville | 79 | 32.9 |
| Sleepy Hollow | 134 | 29.2 |

**54.** What are the elements?
**55.** List the variables. Are they quantitative or qualitative?
**56.** Give the observation for Sleepy Hollow Hospital.
**57. Radio Ratings.** The ratings company Arbitron (**www.arbitron.com**) reports that, in the winter of 2004, 14% of all radio listeners listened to adult contemporary (AC) music. The breakdown within the AC category was Mainstream AC (7.8%), Hot AC (3.7%), Mod AC (0.8%), and Soft AC (1.7%).

a. What are the elements? Can they be listed? Why or why not?
b. What are the variables? Are they qualitative or quantitative?
c. What proportion of AC listeners are Mainstream AC listeners?

**58. Commodity Prices.** The financial company Bloomberg (**www.bloomberg.com**) reported that, on July 30, 2004, the prices in dollars for the following commodities were oil ($43.80, up 1.05%), gold ($393.70, up 3.90%), and coffee ($66.45, down 1.20%).

a. What are the elements?
b. What are the variables? Are they quantitative or qualitative?
c. Explain the meaning of the value of percent increase for oil.

**Light Bulb Lifetime.** Use the following information for Exercises 59 and 60. An electrical company has developed a new form of light bulb that it claims lasts longer than current models. The company has 1 million bulbs in its inventory.

**59.** How do you think the company found evidence for its claim?
**60.** Suppose you take a representative sample of 100 of the new light bulbs and find the average lifetime to be 2000 hours.

a. Is this a statistic or a parameter?
b. Write a sentence that estimates the average lifetime of all the new light bulbs.

**61. Tornado Deaths.** The Tornado Project (**www .tornadoproject.com**) reported the following list of the 10 years with the fewest tornado deaths.

| Year | Deaths | Year | Deaths |
|---|---|---|---|
| 1910 | 12 | 1996 | 26 |
| 1986 | 15 | 1972 | 26 |
| 2004 | 16 | 1980 | 27 |
| 1981 | 24 | 1963 | 27 |
| 1962 | 25 | 1951 | 29 |

**a.** What are the elements?
**b.** What is the variable? Is it quantitative or qualitative?
**c.** Could this be considered a representative sample of the number of annual tornado deaths for all years? Explain why or why not.

## 1.3 Gathering Data

*Objectives:* **By the end of this section, I will be able to...**

1 Explain what a random sample is, and why we need one.

2 Identify other sampling methods.

3 Explain the main factors that go into designing a good questionnaire.

4 Understand the difference between an observational study and an experiment.

### 1 Random Sampling

We can use the information gathered from a sample to generalize about the population when it is impractical or impossible to take a census of the entire population. However, if we get a "bad" sample, the information gleaned from the sample will be misleading, with potentially catastrophic consequences. This section introduces a method of sampling that minimizes many potential biases, which could lead to incorrect generalizations about the population. This sampling method is called *random sampling*. Everyday examples of random sampling include (a) randomly selecting lottery numbers from a basket which continuously churns the number-balls, (b) randomly choosing one card from a deck of playing cards that has been well shuffled, and (c) randomly pulling a name out of hat, after the names have been well stirred. Since random samples are not always practicable or desirable, this section also discusses some of the many alternative sampling methods available, including stratified sampling and cluster sampling.

#### What Is a Random Sample, and Why Do We Need It?

Survey sampling, or polling, has now become so widespread that hardly a day goes by without the results of some new poll or survey making the headlines. Political polls are a good example of statistical sampling at work. The pollsters canvass about 1000 or so respondents, analyze the sample results, and then report their statistical inference that, for example, "54% of Minnesotans support Garrison Keillor for president."

Today many polls are conducted quite scientifically, and their results are usually very accurate. However, such was not always the case. In 1936, the *Literary Digest* had correctly predicted the past three presidential elections and went to work to predict the winner of the contest between Republican Alf Landon and Democrat Franklin Roosevelt. The magazine sent ballots to 10 million citizens. The results ran strongly in favor of Landon, leading the *Literary Digest* to predict Landon to win the election. About 25% of the ballots were returned, giving the newsweekly a sample size of 2.5 million. George Gallup, on the other hand, was working with a sample size that was about 1000 times *smaller* than the *Literacy Digest*'s. However, Gallup predicted a victory for Roosevelt. Clearly, with more data, the *Literary Digest* should have been able to give a more accurate prediction, right? Not necessarily. Roosevelt won in a landslide, and the embarrassed *Literary Digest* later declared bankruptcy.

The problem stemmed from the way that the *Literary Digest* identified its sample. It used lists of people who owned cars and had telephones, which in the 1930s excluded millions of poor and underprivileged people, who overwhelmingly supported Roosevelt. Its sample of 2.5 million therefore was highly biased toward the richer folks, who were less likely to have any great fondness for Roosevelt and his New Deal policies. Gallup, on the other hand, chose his sample more scientifically, and even though his sample size was smaller, it was more representative of the population as a whole.

One inexpensive way of eliminating many types of bias is to make sure your sample is a **random sample**.

A **random sample** (also known as a **simple random sample**) is a sample for which every element has an equal chance of being selected.

The *Random Sample* applet allows you to produce a random sample of up to 100 elements, in the form of a lotto.

*Note:* When we take a sample, we usually discard any repeated elements because we already have their information.

**How the Gallup Organization Obtains a Random Sample**

The Gallup Organization (**www.gallup.com**) has been conducting polls since the 1930s. People often wonder how a random sample of 1000 adults can represent the sentiments of the more than 300 million American adults. How does Gallup obtain a random sample in the first place? Gallup's objective is to make sure that every American has an *equal probability of selection,* that is, an equal chance of being selected, for their poll. Now, where is the place that most Americans are likely to be found? The answer for most Americans is: at home. (Gallup is aware that those who reside on college campuses, on military bases, and in other institutions are thereby not represented, but they usually accept this compromise due to the practical problems of including these institutions.)

In the early days, Gallup conducted interviews in person, going house to house. However, today it is much less expensive to conduct telephone interviews. How does Gallup help to ensure that its telephone sample is truly random? What about the Americans whose phone number is unlisted? The first step is to construct a table of all the telephone exchanges in America, along with an estimate of the proportion of Americans living in that exchange area and the broad characteristics of that population in terms of income, age, ethnicity, education, and so on. Gallup then uses *random digit dialing,* a computer program that generates random four-digit numbers, which are then appended to the telephone exchanges. Thus, each household phone number in America has an equal chance of being included in the sample, regardless of whether it is listed or unlisted. Finally, as of January 1, 2008, Gallup added a data base of cell phone numbers, in order to contact those who can more readily be reached via cell phone.

## Example 1.9 Do you prefer watching the Super Bowl or the commercials?

© Punchstock/Image 100

In February 2007, the Gallup Organization used random digit dialing in a poll of Americans who planned to watch the Super Bowl (Indianapolis Colts versus Chicago Bears). One question they asked was whether the subjects preferred to watch the game or the commercials. Does this represent a random sample?

### Solution

Since random digit dialing ensures that each household phone number in America has an equal chance of being included in the sample, the sample is random.

A perhaps surprising 33% of respondents reported that they preferred watching the commercials, compared with 66% who preferred watching the game. Gender and age seemed to affect how one responded to this question. Twice the proportion of female viewers (44%) as male viewers (22%) preferred watching the commercials. Among females only, more than twice as many younger (aged 18 to 49) women preferred watching the commercials (56%) as older (aged 50 and over) women (26%). ■

## Example 1.10 Online polls

An online newspaper reports that, in an online poll of its readership, 60% say that they get most of their news from online sources. Does this number accurately reflect the proportion of all Americans who get most of their news from online sources?

### Solution

No, the sample is not random. Only those Americans who are online already (and already using an online news source) can respond to this online poll. Therefore, the sample is not random, and it is biased. It overestimates the proportion of Americans who get their news from online sources. Further, there is no mechanism to guard against a single person responding repeatedly and getting their vote counted multiple times. Online polls are not scientific, and their results should not be considered a true reflection of the sentiments of all Americans. ■

Random samples may be generated using technology, using the *Random Sample* applet, or using the random number table provided in Table A in the Appendix (page T-2). At the end of this section, we demonstrate how to generate random samples using the TI-83/84 graphing calculator, Excel, and Minitab.

## Example 1.11 Generating a random sample using technology

In 2004, *Inc. Magazine* published a list of the top 25 cities for doing business, shown in Table 1.8. Use the TI-83/84, Excel, or Minitab to generate a random sample of 7 cities from this list.

**Table 1.8** Top 25 cities for doing business, according to *Inc. Magazine*

| | | |
|---|---|---|
| 1. Atlanta, GA | 10. Suburban Maryland/DC | 19. Austin, TX |
| 2. Riverside, CA | 11. Orlando, FL | 20. Northern Virginia |
| 3. Las Vegas, NV | 12. Phoenix, AZ | 21. Middlesex, NJ |
| 4. San Antonio, TX | 13. Washington, DC, metro area | 22. Miami–Hialeah, FL |
| 5. West Palm Beach, FL | 14. Tampa–St. Petersburg, FL | 23. Orange County, CA |
| 6. Southern New Jersey | 15. San Diego, CA | 24. Oklahoma City, OK |
| 7. Fort Lauderdale, FL | 16. Nassau–Suffolk, NY | 25. Albany, NY |
| 8. Jacksonville, FL | 17. Richmond–Petersburg, VA | |
| 9. Newark, NJ | 18. New Orleans, LA | |

**Solution**

We used the instructions provided in the Step-by-Step Technology Guide at the end of this section to create three random samples, listed below. Note that each random sample is different, as yours will be. ■

| Random sample 1 using the TI-83/84 | Random sample 2 using Excel | Random sample 3 using Minitab |
|---|---|---|
| 9. Newark, NJ | 6. Southern New Jersey | 3. Las Vegas, NV |
| 25. Albany, NY | 23. Orange County, CA | 21. Middlesex, NJ |
| 6. Southern New Jersey | 11. Orlando, FL | 18. New Orleans, LA |
| 20. Northern Virginia | 14. Tampa–St. Petersburg, FL | 7. Fort Lauderdale, FL |
| 24. Oklahoma City, OK | 25. Albany, NY | 2. Riverside, CA |
| 10. Suburban Maryland/DC | 7. Fort Lauderdale, FL | 25. Albany, NY |
| 1. Atlanta, GA | 17. Richmond–Petersburg, VA | 10. Suburban Maryland/DC |

## 2 More Sampling Methods

The idea behind random sampling is quite intuitive: choosing names at random out of a hat, for instance. However, in certain circumstances, simple random sampling can also have shortcomings. A simple random sample may not provide sufficient information about subgroups within the population. For example, suppose you are interested in knowing the proportion of those of Hispanic or Latino descent in Walnut, California, who are registered Democrats. A random sample of size 100 of all the voters in Walnut may yield only 20 of Hispanic or Latino descent, which may be too small a sample to be useful for statistical inference (and which undercounts the actual proportion of registered voters who are of Hispanic or Latino descent, which is about 35%). Therefore, the researcher needs other methods for obtaining samples, depending on the situation and the research question.

### Stratified Sampling

Often, researchers are interested in investigating characteristics of a certain subgroup of a population, such as those of Hispanic or Latino descent in Walnut, California. In cases like this, the researcher divides the population into subgroups, or *strata,* according to some characteristic, such as race or gender. Then a random sample is taken from each stratum. In this way, the researcher knows that a sample will be obtained from each stratum and that it will be large enough to provide reliable statistical inference for each stratum.

### Systematic Sampling

Perhaps the easiest method of sampling is systematic sampling, which is used when a random sample is either not needed or unobtainable. In systematic sampling, each element of the population is numbered, and the sample is obtained by selecting every $k$th element, where $k$ is some whole number. The ancient Romans understood well how to use systematic sampling. When a Roman legion showed cowardice in battle or mutinied, every 10th member was selected and summarily executed before his comrades. Literally, the legion was *decimated,* from the Latin *decem,* meaning "ten."

### Cluster Sampling

Cluster sampling is used when the population is widely scattered geographically or poses other logistical difficulties. For example, if we were interested in estimating the mean income of Manhattan residents, it would be time-consuming and expensive to visit 1000 different locations in Manhattan to elicit sample information. In cluster sampling, the population is divided into *clusters,* such as precincts or city blocks. Then several clusters are chosen at random, and all of the elements within the chosen

*Note:* All of the sampling methods mentioned here involve randomness. However, only the simple random sample is used throughout the text. Therefore, whenever you see the phrase *random sample*, it should be understood as *simple random sample*.

clusters are selected for the sample. One disadvantage of cluster sampling is that the respondents from within a certain cluster will tend to be more similar to each other than the elements of a random sample would be. For example, if one of the clusters in the Manhattan income survey was a Fifth Avenue block, the mean income of residents there would be at the higher end of the income scale.

### Convenience Sampling

In convenience sampling, subjects are chosen based on what is convenient for the survey personnel. If you were to estimate the true proportion of females taking an introductory statistics course using only the people in your class, this would be considered a convenience sample. This method is to be contrasted with *probability sampling,* where each member of the population has a known probability of selection. As we shall see below, convenience sampling usually does not result in a representative sample.

## 3 Surveys and Questionnaires

When the information cannot be found from existing sources, a researcher can commission a survey or questionnaire. Because surveys and questionnaires are so widely used, it is important to know how to design them correctly. A badly designed survey can yield misleading results. Here we learn about some common pitfalls in the design and implementation of a survey or questionnaire, including the wording of the questions.

The **target population** is the complete collection of all elements that we are interested in studying.

The **potential population** is the collection of elements from the target population that had a chance of being sampled.

**Selection bias** occurs when the population from which the actual sample is drawn is not representative of the target population, due to an inappropriate sampling method.

### Example 1.12 Selection bias

Suppose Ashley would like to estimate the proportion of American voters who would favor abandoning the present system of Social Security in favor of a system where retirement funds would be invested in the stock market. Ashley goes to the mall with her clipboard, and canvasses as many people as she can on Monday between 9 A.M. and 5 P.M. To each person, she asked the question "Do you favor or oppose abandoning the present Social Security system in favor of a system that invests retirement funds in the stock market?"

**a.** Identify Ashley's target population.
**b.** Identify Ashley's potential population.
**c.** Discuss any possible problems.

**Solution**

**a.** Ashley's **target population** is the population of all American voters.
**b.** The collection of all the American voters who visited the mall on Monday between 9 A.M. and 5 P.M. represent her **potential population**.
**c.** It appears that Ashley's survey may suffer from **selection bias**. The population of people who went to the mall on Monday between 9 A.M. and 5 P.M. is not representative of the target population of all American voters. Since many American voters work on Mondays between 9 A.M. and 5 P.M., they are not elements of the sampled population. Further, the proportion of retirees at the mall during that time was larger than in the target population of all American voters. These retirees tend to oppose strongly any tampering with the Social Security system and probably overwhelmingly responded in the negative to the survey question. ■

### Five Factors for Good Questionnaire Design

You may have heard of the aphorism "Be careful what you ask for; you may get it." This warning is certainly relevant to the issue of questionnaire design. The wording of questions can greatly affect the responses. Here are several factors to consider when designing a questionnaire.

1. **Remember: simplicity and clarity.** Do not use four-syllable words when one-syllable words will do. Respondents will be shy about asking you to clarify the question. The result will be confused responses and muddled data.
2. **When reporting results, include the actual question asked.** Be careful about drawing generalizations. The conclusions you draw may not have been what your respondents had in mind when they answered the questions.
3. **Avoid leading questions.** The respondent is often eager to please and will try to tell you what he or she thinks you want to hear. For example, a researcher is interested in determining the proportion of Americans who favor preserving the welfare system. A leading question would be "A child growing up poor in America faces more than his fair share of crime and negligence. Do you support preserving the welfare safety net to help ensure that children are given a fair chance?"
4. **Avoid asking two questions in one.** Avoid questions like "Have you argued with your friends or family in the last month?" This is really two questions in one, and you will not know which question the respondents are answering.
5. **Words mean different things to different people.** Avoid using terminology like "often" or "sometimes." Instead, try to use specific terms such as "three times a week."

If you use ambiguous terms, the data you collect will be ambiguous, and any conclusions you draw will probably not be valid.

## 4 Experimental Studies and Observational Studies

Two major types of statistical studies are **experimental studies** and **observational studies**. We have seen that researchers can gather data by consulting existing sources, by distributing a questionnaire, or by taking a sample. However, you may not be able to obtain the information you require by using survey or sampling methods. In this case, you may have to conduct an experimental study.

**Experimental Studies**

In an **experimental study**, researchers investigate how varying the explanatory variable affects the response variable.

A **predictor variable** (also called an **explanatory variable**) is a characteristic purported to explain differences in the response variable.

A predictor variable that takes the form of a purposeful intervention is called a **treatment**.

A **response variable** is an outcome, a characteristic of the subjects of the experiment presumably brought about by differences in the predictor variable or treatment.

The **subjects** in a statistical study represent the elements from which the data are drawn.

*Note:* What is the difference between an element and a subject? *Subject* is a term usually reserved for statistical studies, while the term *element* can be used for any data set.

We illustrate experimental studies using the following example.

## Example 1.13 Newborn babies and a heartbeat: an experimental study

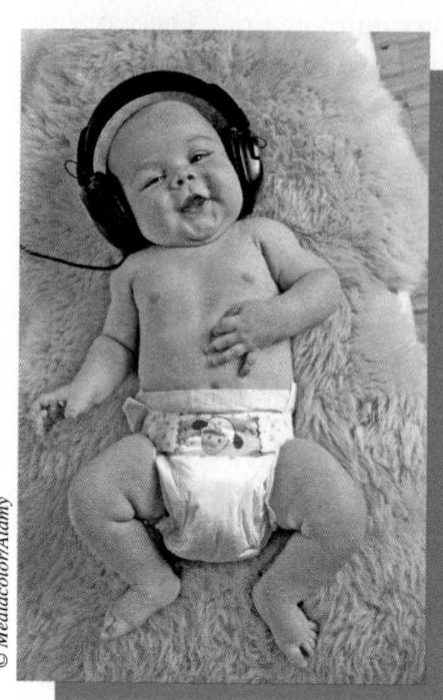
© Mediacolor/Alamy

A psychologist wanted to test whether the sound of a human heartbeat would help newborn babies grow. A baby nursery at a hospital was set up so that the sound of a human heartbeat could be heard throughout the nursery. The heartbeat sound was played in the nursery for a large batch of newborn children, who were then weighed to determine their weight gain after four days in the nursery. Later, a second batch of children occupied the nursery, but no heartbeat sound was played. These children were also weighed after four days in the nursery. Babies were randomly placed into the two groups. Identify the following:

**a.** The subjects
**b.** The predictor variable
**c.** The treatment
**d.** The response variable

### Solution

**a.** The babies were the **subjects** of this experimental study.
**b.** The **predictor variable** is whether or not the heartbeat sound was played in the nursery.
**c.** The **treatment** is the sound of the human heartbeat.
**d.** The **response variable** is the baby's weight gain, which is the outcome of the study.

The results were consistent with the psychologist's conjecture; the babies who listened to the heartbeat sound had a greater average weight gain than the babies for whom no heartbeat sound was played. ■

There are three main factors that should be considered when designing an experimental study: *control, randomization,* and *replication.*

***Control.*** A control group is necessary to compare against the treatment group, if we wish the results of our experiment to be useful. The control group in the above example is the group of babies for whom the heartbeat sound was not played. Had the psychologist omitted this control group, there would have been nothing to compare his results against. In some experiments, especially in medicine, members of the control group receive a placebo, a nonfunctioning simulated treatment. Sometimes, the symptoms of the members of the control group improve simply by taking the placebo, a phenomenon known as the *placebo effect.*

***Randomization.*** Many biases can be introduced into an experiment. For example, a well-meaning doctor may want to place underweight high-risk babies in the group with the heartbeat, in the hope that such babies will flourish. To eliminate biases like these, the placement of the subjects into the treatment and control groups should be done randomly.

***Replication.*** One major theme of statistical investigation is that larger samples are usually better, because they allow more precise inference. In a statistical study, the treatment and the control groups each must contain a large enough number of subjects to allow detection of meaningful differences between the treatment and control. For example, if a researcher examined only three babies with the heartbeat sound and three babies without the heartbeat sound, this would not be a sufficient number of replications. In Chapter 8, we will learn how large a sample size is sufficient for the needs of a particular study.

## Observational Studies

There are circumstances where it is either impossible, impractical, or unethical for the researcher to place subjects into treatment and control groups. For example, suppose we are interested in whether women who work outside the home suffer less depression than women who remain at home with the children. The explanatory variable here is whether or not a woman works outside the home. However, it is not possible for the researcher to take women and randomly separate them into groups that either work outside the home or do not work outside the home. Sometimes an experimental study is not possible for ethical reasons.

Further, suppose you are interested in whether babies born to chemically dependent mothers display differences in cognitive skills from babies born to mothers who are not chemically dependent. It is clearly not ethical to randomly assign half of the mothers in the study to become chemically dependent during their pregnancy. Therefore, researchers need another type of statistical study: the observational study. In an observational study, the researcher observes whether the subjects' differences in the predictor variable are associated with differences in the response variable. No attempt is made to create differences in the predictor variable.

A sample survey is an example of an observational study. Data about a response variable may be obtained through the survey, along with information about possible predictor variables. No attempt is made to manipulate the variables. The researcher analyzes the information to determine whether differences in the predictor variable are associated with differences in the response variable.

### Example 1.14 Is Ecstasy toxic to your neurons?

According to the British medical journal *The Lancet,* experimental studies carried out on animals (nonhuman primates, squirrel monkeys, and rodents) have revealed that large doses of the drug Ecstasy (methylene-dioxy-methamphetamine, or MDMA) produce "large and possibly permanent damage" to neural axons in the brain. Explain why the researchers did not carry out their experiment on humans.

**Solution**

It is not ethical to randomly assign half of the human subjects to receive large doses of the drug Ecstasy, especially in view of its effect on animals. The difficulty of performing experimental studies on humans concerning the effects of controlled substances is addressed by the authors of the *Lancet* study:

> *Only a prospective[experimental] study . . . could definitively show that recreational MDMA use was neurotoxic in human beings. For ethical, political, and legal reasons such a study is unlikely to ever be done. Instead, we have to rely upon evidence from observational studies of recreational MDMA users.*[6] ■

### Example 1.15 Secondhand smoking and illness in children

A 2006 Surgeon General's report found that "the evidence is sufficient to infer a causal relationship" between secondhand tobacco smoke exposure from parental smoking and respiratory illnesses in infants and children.[7] Was this report based on an experimental study or an observational study?

**Solution**

It would have been unethical to force the parents of a treatment group to smoke tobacco, given the health risks associated with tobacco use. Therefore, the study must have been an observational one. ■

## Section 1.3 SUMMARY

1 A random sample is a sample for which every element has an equal chance of being included. A random sample can minimize many potential biases, which could lead to incorrect generalizations about the population.

2 Other sampling methods include stratified sampling, systematic sampling, cluster sampling, and convenience sampling.

3 When constructing a survey, follow the five factors for good questionnaire design.

4 There are two types of statistical studies: experimental studies and observational studies. In an experimental study, researchers investigate how varying the predictor variable affects the response variable. It is not always possible to conduct an experimental study, however, and sometimes an observational study is used instead.

## STEP-BY-STEP TECHNOLOGY GUIDE: Generating a Random Sample

We illustrate using Example 1.11 (page 20).

### TI-83/84

***Step 1*** Enter a "seed," which can be any nonzero number.
***Step 2*** Press **STO⇒** .
***Step 3*** Press **MATH**, highlight **PRB**, select **1: rand**, and press **ENTER** (see Figure 1.6, which uses 1776 for the seed). Your seed number is now in the calculator's memory.
***Step 4*** Press **MATH**, highlight **PRB** and select **5: randInt(**.
***Step 5*** Enter **1, N, *two times* n**, where **N** = population size and **n** = sample size. We enter twice the sample size in case there are repeats. For Example 1.11, since **n** = 7, we enter **randInt(1, 25, 14)** and press **ENTER** (Figure 1.7).
***Step 6*** Store the random sample in list **L1** as follows: press **STO⇒** , then **2$^{ND}$**, then **L1** (Figure 1.7). Then press **ENTER**.
***Step 7*** View the random sample by pressing **STAT**, highlighting **EDIT**, and pressing **ENTER** (Figure 1.8). Note that there is a repeat (**6**). We therefore select the next number, **10**, to round out our sample. The random sample for Example 1.11 is therefore **9, 25, 6, 20, 24, 10, 1** (Figure 1.9).

Figure 1.6

```
randInt(1,25,14)
{9 25 6 20 24 6...
Ans→L1
{9 25 6 20 24 6...
```
Figure 1.7

```
L1   L2     L3     1
9    ------ ------
25
6
20
24
6
10
L1 ={9,25,6,20,2...
```
Figure 1.8

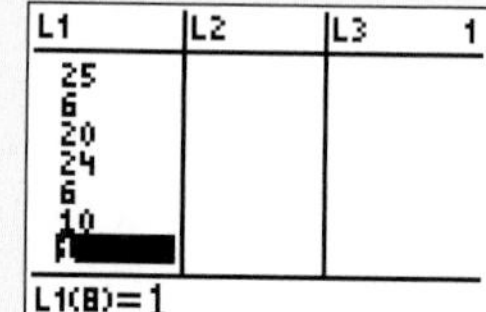

Figure 1.9

### EXCEL

***Step 1*** Select cell **A1**. Click the **Insert Function** icon $f_x$.
***Step 2*** For "Search for a function," enter **randbetween**. Click **Go**, then **OK**.
***Step 3*** For **Bottom**, enter **1**. For **Top**, enter population size **N**. For Example 1.11, **N** = 25. Click **OK**.
***Step 4*** Cell **A1** now contains a random integer between **1** and **N**. Copy and paste cell **A1** into *twice* as many cells as needed for the sample size **n**, just in case there are repeats. For Example 1.11, copy and paste into cells **A2** to **A14**. The results are shown in Figure 1.10. Note that **8** is repeated, so that our random sample is **8, 2, 20, 16, 23, 7, 22**.

A1 $f_x$ =RANDBETWEEN(1,25)

| | A | B | C | D | E | F |
|---|---|---|---|---|---|---|
| 1 | 8 | | | | | |
| 2 | 2 | | | | | |
| 3 | 20 | | | | | |
| 4 | 8 | | | | | |
| 5 | 16 | | | | | |
| 6 | 23 | | | | | |
| 7 | 7 | | | | | |
| 8 | 22 | | | | | |
| 9 | 1 | | | | | |
| 10 | 13 | | | | | |
| 11 | 19 | | | | | |
| 12 | 1 | | | | | |
| 13 | 14 | | | | | |
| 14 | 22 | | | | | |

Figure 1.10

## MINITAB

*Step 1* Click on **Calc > Random Data > Integer . . .**
*Step 2* In the **Generate _ rows of data** section, enter *twice* your desired sample size, just in case there are repeats. For example, if your desired sample size is **7**, enter **14**.
*Step 3* In the **Store in column _** section, enter whichever column is convenient for you, such as **C1**.
*Step 4* For **Minimum value**, enter **1**. For **Maximum value**, enter your population size, **N**. Click **OK**.
*Step 5* The random integers appear in column **C1**. Start from the top and go down the list, omitting any repeats, until you have your sample of size **n**. Our random sample (Figure 1.11) is therefore **3, 18, 2, 11, 21, 7, 25.**

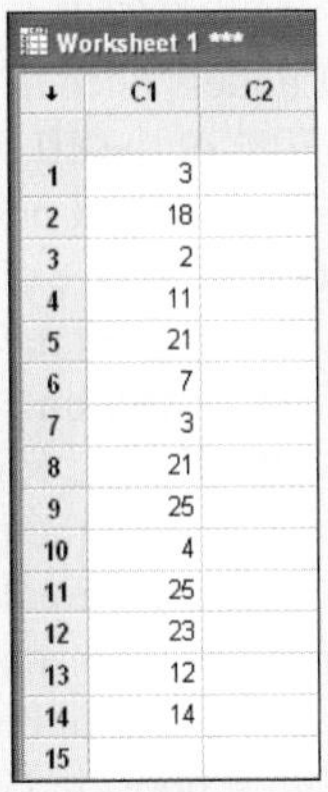

Worksheet 1 ***

| ↓ | C1 | C2 |
|---|---|---|
| 1 | 3 | |
| 2 | 18 | |
| 3 | 2 | |
| 4 | 11 | |
| 5 | 21 | |
| 6 | 7 | |
| 7 | 3 | |
| 8 | 21 | |
| 9 | 25 | |
| 10 | 4 | |
| 11 | 25 | |
| 12 | 23 | |
| 13 | 12 | |
| 14 | 14 | |
| 15 | | |

Figure 1.11

Excel and Minitab base the seed on the current time, so that you need not set it yourself.

## Section 1.3 EXERCISES

### CLARIFYING THE CONCEPTS

1. Describe a situation where the information from a badly chosen sample could lead to negative consequences.
2. What type of bias did the *Literary Digest* poll (pages 18–19) exhibit? How did it affect the results?
3. How could the *Literary Digest* have decreased the bias in its poll?
4. Was the *Literary Digest* poll a random sample?
5. Describe in your own words what a random sample is.
6. What is the difference between the target population and the potential population?
7. Give an example of the target population for a research question you are interested in. Now give an example of the potential population for this research question.
8. Describe in your own words what is meant by selection bias.
9. Describe the difference between an observational study and an experimental study.
10. Why would a researcher conduct an observational study rather than an experimental study?

### PRACTICING THE TECHNIQUES

For Exercises 11–15, state which type of sampling is represented.

11. Students in your class are divided into males and females. A random sample of size 3 is then drawn from each of the groups.
12. You are interested in estimating the average number of hours dormitory residents spend studying. Three dormitories are chosen at random. In each dormitory, one floor is chosen at random and all the students on that floor are interviewed.
13. You are researching the proportion of college students who prefer hip-hop music to other forms of music. You obtain a listing of all the students at your college and contact every 20th student on the list.
14. An instructor in a large lecture course of 300 students would like to get a student sample, and he selects every 10th name from the class roster.
15. Your campus statistical consulting center uses random digit dialing to locate potential subjects for a political survey.
16. Provide an example of a survey question that lacks simplicity and clarity. Alter it so that it is simpler and clearer, as Factor 1 recommends.
17. You have probably heard of the warning "Be careful what you ask for; you may get it." Which of the five factors best describes this danger?

For Exercises 18–20, state which type of study is involved, experimental or observational. Also, identify the response variable and the predictor variable.

18. A sociologist would be interested in whether large families (at least four children) attend religious services more often than smaller families do.
19. A financial researcher would be interested in whether companies that give large bonuses to their chief executive officers (at least $1 million per year) have a higher stock price.
20. A manufacturer would be interested in whether a wireless adapter will improve the performance of its electronics equipment.
21. Select a random sample of size 7 from Table 1.8 (page 20). Why is your sample different from the one shown in the text?

## APPLYING THE CONCEPTS

**22. Contradicting Ann Landers.** "If you had to do it over again, would you have children?" This is the question that advice columnist Ann Landers once asked her readers. It turns out that nearly 70% of the 10,000 responses she received were "No." A professional poll by *Newsday* found that 91% of respondents would have children again. Explain the apparent contradiction between these two surveys using what you have learned in this section.

**23. High School Dropouts.** For the following survey, describe the target population and the potential population, and discuss the potential for selection bias. Researchers are interested in the proportion of high school students in New England who drop out (leave school before graduating). A survey is made of 15 high schools in Greater Boston.

**24. Living Below the Poverty Level.** For the following survey, describe the target population and the potential population, and discuss the potential for selection bias. A sociologist is interested in the proportion of people living below the poverty level in Chicago. He takes a random sample of phone numbers from the Chicago phone directory and asks each respondent his or her annual household income.

**25. Rap or Hip-Hop.** Describe what is wrong, if anything, with the following survey question. "Do you enjoy listening to rap or hip-hop music?"

**26. Financial Ruin.** Describe what is wrong, if anything, with the following survey question: "Do you think that we should tax and spend our way into financial ruin?"

**27. Abortion.** Suppose 67% of female respondents respond affirmatively to the question "Do you support the right of a woman to terminate a pregnancy when her life is in danger?" Would the researcher be justified in reporting, "Two-thirds of women support abortion?"

**28. Cholesterol-Lowering Medication.** Cholesterol researchers are investigating whether there is any difference between a new medication and a placebo (inactive pill) in lowering LDL cholesterol levels in the bloodstream.

- **a.** Describe what the response variable and the predictor variable are.
- **b.** Describe how the cholesterol researchers would go about collecting their data. Would they use an experimental study or an observational study?
- **c.** What is the treatment? What is the control?

**29. Santa Monica Employers.** Refer to Table 1.9 for the following.

- **a.** We are about to select a random sample of the companies listed in Table 1.9 and determine how many employees the largest employer in that sample has. Do we know how many employees this will be before we select the sample? Why or why not?
- **b.** Select a random sample of size 3 from the table.
- **c.** If you take another sample of size 3, is it likely to comprise the same three employers? Why or why not?
- **d.** Of the employers in your random sample, which has the most employees? How many employees does it have?

**TABLE 1.9 Top 10 Employers in Santa Monica, CA**

| Employer | Employees |
|---|---|
| 1. City of Santa Monica | 1892 |
| 2. St. John's Health Center | 1755 |
| 3. The Macerich Company | 1605 |
| 4. Fremont General Corporation | 1600 |
| 5. Entravision Communications Corporation | 1206 |
| 6. Santa Monica/UCLA Hospital | 1165 |
| 7. Santa Monica College | 1050 |
| 8. Metro-Goldwyn Mayer, Inc. | 1050 |
| 9. The Rand Corporation | 1038 |
| 10. Santa Monica/Malibu School District | 1008 |

**30. Santa Monica Employers.** Refer to Table 1.9 for the following.

- **a.** We are about to select another random sample and determine how many employees the largest employer in that sample has. Do we know how many employees this will be before we select the sample? Do we know whether it will be the same as in the previous exercise? Why or why not?
- **b.** Select another random sample of size 3 from the table.
- **c.** Which employer in your new sample has the most employees? How many employees does it have?
- **d.** Compare your answers in (**c**) with those from Exercise 29(**d**). What can we say about a quantity like "the largest number of employees in a random sample of employers?"

**31. Most Active Stocks.** Here is a list of the five most active stocks on the NYSE on March 27, 2008.

| Stock | Price |
|---|---|
| Citigroup | $20.98 |
| Merrill Lynch | $40.93 |
| Lehman Brothers | $38.86 |
| Washington Mutual | $10.04 |
| Ford | $ 5.63 |

- **a.** We are about to select a random sample and determine the lowest price in the sample. Do we know what this price will be before we select the sample? Why or why not?
- **b.** Select a random sample of size 2 from the table.
- **c.** If you take another sample of size 2, is it likely to comprise the same two companies? Why or why not?
- **d.** Which stock in your sample has the lowest price? What is that price?

**32. Most Active Stocks.** Refer to the list of stocks from the previous exercise.

- **a.** We are about to select another random sample and determine the lowest price in the sample. Do we know what this price will be before we select the sample? Do we know whether it will be the same as in the previous exercise? Why or why not?
- **b.** Select another random sample of size 2.
- **c.** Which stock in your new sample has the lowest price? What is that price?

**d.** Compare your answers in (**c**) with those from (**d**) in Exercise 31. What can we say about a quantity like "the lowest price in a random sample of stocks"?

**33. Mediterranean Diet.** The American Heart Association reported the results of an experimental study.[8] Patients who ate a Mediterranean diet had a significantly lower risk of having a second heart attack than did patients who ate a Western diet. Identify the response variable and the predictor variable in this experimental study.

**Evidence for an Alternative Therapy?** Use the following information for Exercises 34–36. A company called QT, Inc. sells "ionized bracelets," called Q-Ray Bracelets, that it claims help to ease pain by balancing the body's flow of "electromagnetic energy." QT, Inc. claims that Q-Ray Bracelets can ease pain caused by cancer, restore well-being, and provide many other health benefits. The Mayo Clinic decided to conduct a statistical study to determine whether the extravagant claims for Q-Ray Bracelets were justified.[9] In the study, 305 subjects wore the Q-Ray "ionized" bracelet and 305 wore a placebo bracelet (identical to the ionized bracelet except for the ionization) for four weeks, at the end of which certain measures of pain were evaluated and compared between the treatments. The subjects, upon entry to the study, were randomly assigned to receive either the ionized bracelet or the placebo bracelet.

**34.** Identify the following aspects of this study.

**a.** The control

**b.** The randomization

**c.** The replication

**35.** Identify the following aspects of this study.

**a.** The predictor variable

**b.** The treatment

**c.** The response variable

**36.** Does this statistical study represent an experimental study or an observational study? Explain why.

Use the Random Sample applet for Exercises 37–39.

**37.** Generate a random sample of 7 cities from Table 1.8 (page 20).

**38.** Generate another random sample of 7 cities from Table 1.8. Are all the cities in the two samples the same?

**39.** Before we generate a third sample of 7 cities, choose a city from Table 1.8.

**a.** Will this city appear in the random sample?

**b.** Is there any way of telling for certain in advance whether this city will appear in the random sample?

**c.** Now go ahead and generate the third random sample of 7 cities. Is your city in the sample?

## *Try This in Class!* In-class activities to enhance your understanding of statistics

Throughout this course we will work together on various activities in class. In this way, we will share in each other's experience with statistics. Also, simply talking about the problem can help put it into perspective.

### GENERATING RANDOM DATA

**1.** Here is a fun way to generate random samples of students from your class by rolling a pair of dice! (For large lecture classes, it is probably easier to use technology or the Random Sample applet to generate a random sample. See the Step-by-Step Technology Guide at the end of Section 1.3 for instructions.) Make a table constructed as follows:

| | 1 | 2 | 3 | 4 | 5 | 6 |
|---|---|---|---|---|---|---|
| 1 | | | | | | |
| 2 | | | | | | |
| 3 | | | | | | |
| 4 | | | | | | |
| 5 | | | | | | |
| 6 | | | | | | |

Print this table on a sheet of paper and pass the paper around the class, having each student choose one cell and write his or her name in the cell. The numbers across the top are for the results of tossing a red die, and the numbers down the left are for the results of tossing a black. When a certain combination of red and black numbers is tossed, the student whose cell that combination refers to becomes part of the sample. In Chapter 5 we will learn that all the cells are equally likely to be picked if the dice are fair. If there are fewer than 36 students in your class, simply write *Roll Again* in the remaining empty cells after all students have chosen their cells. If there are more than 36 students in your class, just make more than one of these tables, and use a third colored die (for example, a green die) to choose which table is to be sampled. If a student is chosen more than once, simply roll again.

Do you think this arrangement will produce random samples? Can you suggest any other arrangements that will produce random samples from the population of students in your class?

**2.** Now let's take samples of the class.

**a.** Let the instructor call the first 10 names on an alphabetical list of those present for class today. Is this a random sample? Why or why not? Find the proportion of females in this sample.

**b.** Let the instructor ask those students who watched ESPN this week to raise their hands. Is this a random sample? Why or why not? Find the proportion of females in this group. Is this proportion the same as in (**a**)? Why doesn't it have to be the same?

**c.** Get the dice out and generate a sample using the table(s) you constructed. Is this a random sample? Why or why not? Find the sample proportion of females. Is this the same as in (**a**) or (**b**)?

**d.** Choose another sample as in (**c**). Is this a random sample? Find the sample proportion of females. Is it the same as in (**a**) or (**b**) or (**c**)?

## CHAPTER 1 VOCABULARY

**SECTION 1.2**

- **Census** (p. 12). The collection of data from every element in the population.
- **Continuous variable** (p. 10). A quantitative variable that can take an infinite number of possible values.
- **Descriptive statistics** (p. 8). Methods for summarizing and organizing the information in a data set.
- **Discrete variable** (p. 10). A quantitative variable that can take either a finite or a countable number of possible values.
- **Element** (p. 8). A specific entity on which information is collected.
- **Observation** (p. 8). The set of values of all variables for a given element.
- **Parameter** (p. 12). A characteristic of a population.
- **Population** (p. 12). The collection of *all* elements (persons, items, or data) of interest in a particular study.
- **Qualitative variable** (p. 9). A variable that does not assume a numerical value but is usually classified into categories. Also called a **categorical variable**.
- **Quantitative variable** (p. 9). A variable that takes numerical values.
- **Sample** (p. 12). A subset of the population from which information is collected.
- **Statistic** (p. 12). A characteristic of a sample.
- **Statistical inference** (p. 12). Methods for estimating and drawing conclusions about population characteristics based on the information contained in a subset (sample) of that population.
- **Statistics** (p. 6). The art and science of collecting, analyzing, presenting, and interpreting data.
- **Variable** (p. 8). A characteristic of an element, which can assume different values for different elements.

**SECTION 1.3**

- **Experimental study** (p. 23). A statistical study where researchers investigate how varying the predictor variable affects the response variable.
- **Observational study** (p. 23). A statistical study where the researcher observes whether the subjects' differences in the predictor variable are associated with differences in the response variable. No attempt is made to create differences in the predictor variable.
- **Potential population** (p. 22). The collection of elements from the target population that had a chance of being sampled.
- **Predictor variable (explanatory variable)** (p. 24). A characteristic purported to explain differences in the response variable.
- **Random sample** (p. 19). A sample in which every element has an equal chance of being included. Also called **simple random sample**.
- **Response variable** (p. 24). An outcome. A characteristic in the subjects of the experiment that may be caused by differences in the predictor variable or treatment.
- **Selection bias** (p. 22). Occurs when the population from which the actual sample is drawn is not representative of the target population.
- **Subjects** (p. 24). The elements in a statistical study from which the data are drawn.
- **Target population** (p. 22). The complete collection of all elements that we are interested in studying.
- **Treatment** (p. 24). A predictor variable that takes the form of a purposeful intervention.

## CHAPTER 1 REVIEW EXERCISES

**SECTION 1.2**

Refer to Table 1.10 for Exercises 1–3.

**TABLE 1.10 World water usage**

| Country | Continent | Climate | Water use (per capita gallons per day) | Main use |
|---|---|---|---|---|
| Iraq | Asia | Arid | 3311 | Irrigation |
| United States | North America | Temperate | 1565 | Industry |
| Pakistan | Asia | Arid | 1486 | Irrigation |
| Canada | North America | Temperate | 1268 | Industry |
| Madagascar | Africa | Tropical | 1212 | Irrigation |
| North Korea | Asia | Temperate | 1194 | Not reported |
| Chile | South America | Arid | 1176 | Irrigation |
| Bulgaria | Europe | Temperate | 1158 | Not reported |
| Afganistan | Asia | Arid | 1039 | Irrigation |
| Iran | Asia | Arid | 986 | Irrigation |

1. Use the table to find each of the following.
   **a.** How many elements are there? List a few.
   **b.** What are the variables? List them.
   **c.** What is the observation for Canada?
   **d.** Which country uses the most water per capita?
   **e.** What does "per capita" mean?

2. Identify the type of each of the following variables. What are the possible values?
   **a.** Continent.
   **b.** Climate.
   **c.** Water use.
   **d.** Main use. Is "not reported" a possible value of this variable?

3. Which two temperate countries share the same continent, climate, and main use?
   **a.** Is their water use the same?
   **b.** Is it likely that their water use would be exactly the same?

4. The following table contains population figures for the five most populous states.

| State | Pop. (1960, in 1000s) | Pop. (2005, in 1000s) | Increase |
|---|---|---|---|
| California | 15,717 | 36,132 | 20,415 |
| Texas | 9,580 | 22,860 | 13,280 |
| New York | 16,782 | 19,255 | 2,473 |
| Florida | 4,952 | 17,790 | 12,838 |
| Illinois | 10,081 | 12,763 | 2,682 |

   **a.** Identify the elements and the variables.
   **b.** Are the variables qualitative or quantitative?
   **c.** Provide the observation for the state of Florida.
   **d.** Which three states had the largest population increases? Which two states had the smallest population increases?

5. An electrical company has developed a new form of light bulb that it claims lasts longer than current models. The company has 1 million bulbs in its inventory. Consider the population average lifetime.
   **a.** What is the only way to find out the population average lifetime of the 1 million bulbs in the inventory?
   **b.** Suppose someone who worked for you wrote you a memo suggesting that it was crucial to know the exact value of the population average lifetime of all 1 million new light bulbs. How would you respond? What might you suggest instead?

6. Consider the chapter Case Study, "Does Friday the 13th Change Human Behavior?" (pages 1 and 7). There are two ways of making a decision error: (i) There is nothing to the superstition, but you believe in it and stay home. (ii) There is something to the superstition, but you venture out anyway.
   **a.** Which of (i) or (ii) do you think is the more serious error? Explain.
   **b.** Which of the two situations is more likely?

7. Suppose we are interested in the proportion of left-handed statistics students, and we take a sample to estimate the percent of students in our class who are left-handed.
   **a.** What is the population?
   **b.** What is the sample?
   **c.** What is the variable? Is it quantitative or qualitative?
   **d.** Is the sample proportion likely to be exactly the same as the population proportion? Is it likely to be very far away from the population proportion? Explain.

**SECTION 1.3**

8. Refer to the *Literary Digest* poll discussed in Section 1.3.
   **a.** What was the population? Be specific.
   **b.** What was the sample?
   **c.** Discuss whether the sample was similar to the population in all important characteristics.

9. Suppose you are interested in finding out how the statistics grades for your class compare with those of the college as a whole.
   **a.** Would you use an experimental study or an observational study?
   **b.** Discuss how this study situation would preclude effective randomization.

10. A long-running television advertisement claimed that "3 out of 4 dentists surveyed recommend sugarless gum for their patients who chew gum."
   **a.** If in fact only 4 dentists were surveyed, which of the study factors were violated?
   **b.** Use this situation to discuss why replication is important.

11. In the study situation described in Exercise 28 (page 28), there is a patient with severe hypercholesterolemia (very high LDL cholesterol levels), and so the doctor assigns this patient to the group of patients who receive the new medication rather than the placebo.
   **a.** Which of the experimental factors did the doctor violate?
   **b.** Use this situation to discuss why randomization is important.

12. Suppose we are interested in determining whether differences exist in the cognitive levels of children from single-parent families and those from two-parent families. Would we use an observational study or an experimental study? Clearly describe why.

13. Referring to the study in the previous exercise, suppose the children from single-parent families showed lower average cognitive skills than children from two-parent families. Does this mean that living in a one-parent family causes lower levels of cognitive skills? Why or why not?

# CHAPTER 1 Quiz

### TRUE OR FALSE

**1.** True or false: Statistical inference consists of methods for estimating and drawing conclusions about sample characteristics based on the information contained in the population.

**2.** True or false: Selection bias occurs when the population from which the actual sample is drawn is not representative of the target population.

**3.** True or false: A parameter is a characteristic of a sample.

### FILL IN THE BLANK

**4.** Statistics is the art and science of __________, analyzing, presenting, and interpreting data.

**5.** An __________ is the set of values of all variables for a given element.

**6.** A random sample is a sample for which every element has an ________ ________ [two words] of being included in the sample.

**7.** A statistic is a characteristic of a __________.

### SHORT ANSWER

**8.** Is a sample survey examining the effects of secondhand smoke an example of an experimental study or an observational study?

**9.** Which type of sampling is represented by the following: You would like to estimate the average number of hours per week Americans watch television. You interview all the members of your family and your best friends' families.

**10.** State which type of statistical study is involved in the following. A large pharmaceutical company is interested in whether a new drug will reduce Alzheimer's disease symptoms in elderly patients.

**11.** For the study in the previous exercise, identify the predictor variable and the response variable.

### CALCULATIONS AND INTERPRETATIONS

**Information Technology Economy.** In 2004, the Bureau of Economic Analysis of the United States published a table (Table 1.11) of the gross domestic product of three sectors of the information technology economy (the entries are billions of dollars). Refer to this table for Exercises 12–14.

**TABLE 1.11 2004 gross domestic product in three technology sectors**

| Sector | 1996 | 1997 | 1998 | 1999 | 2000 | 2001 | 2002 | 2003 |
|---|---|---|---|---|---|---|---|---|
| Hardware | 201 | 232 | 242 | 252 | 244 | 190 | 189 | 208 |
| Software | 166 | 194 | 238 | 278 | 317 | 320 | 324 | 329 |
| Communi-cations | 183 | 188 | 206 | 231 | 250 | 264 | 272 | 292 |

**12.** List the elements.

**13.** What are the variables? Be specific. Are they quantitative or qualitative?

**14.** A slump in hardware production may be a signal for the beginning of a recession. For which year did hardware production fall sharply?

**Crop Damage from Pesticides.** Use the following information for Exercises 15–17. Agricultural researchers are investigating whether a new form of pesticide will lead to lower levels of insect damage to crops than the traditional pesticide.

**15.** Describe the response variable and the predictor variable.

**16.** Would the researchers use an experimental study or an observational study?

**17.** What is the treatment? What is the control?

**18.** Describe what is wrong, if anything, with the following survey question. "How often would you say that you attend the movie theater: often, occasionally, sometimes, seldom, or never?"

**Cigarette Preference by Race.** Use the following information for Exercises 19–21. The National Household Survey on Drug Abuse states that the most popular cigarette brand among whites and Hispanics is Marlboro, while the most popular cigarette among blacks is Newport.

**19.** Do you think that the researchers contacted every single smoker in the country?

**20.** Should the favorite brands be considered statistics or parameters?

**21.** The survey reported that 60.6% of African American respondents living in the Northeast preferred Kool cigarettes. Use this statistic in a sentence, as an estimate of its associated parameter.

Chapter 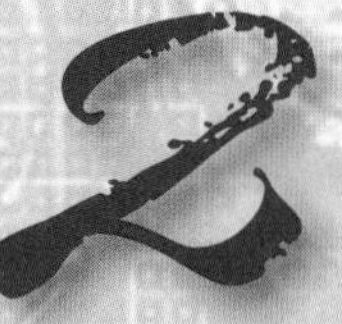

# Describing Data Using Graphs and Tables

CASE STUDY

## The Caesar Cipher

Over two thousand years ago, Julius Caesar developed the Caesar Cipher, which was a means of encoding his messages so that enemy generals would not be able to understand them if they were intercepted. He did this quite simply by shifting each letter in the message a certain number of places. For example, if each letter is shifted one place to the right, then:

| The message | Would be encoded as |
|---|---|
| **DEAR BRUTUS,**<br>**ALL GAUL IS HERE,**<br>**WISH YOU WERE BEAUTIFUL** | **EFBS CSVUVT,**<br>**BMM HBVM JT IFSF,**<br>**XJTI ZPV XFSF CFBVUJGVM** |

Where does statistics come in? Well, what if you were an enemy general and you intercepted a message from Caesar to one of his generals? You would not know which shift was being used, so *how could you use statistics to decode the message?* The answer is to make use of your knowledge of modern English letter frequencies (for simplicity, we assume that Caesar was fluent in English, a language that wouldn't develop until hundreds of years later). Now, in a typical English text 1000 letters long, the frequencies (counts) of the letters in the English alphabet are approximately the following:

| A | B | C | D | E | F | G | H | I | J | K | L | M |
|---|---|---|---|---|---|---|---|---|---|---|---|---|
| 73 | 9 | 30 | 44 | 130 | 28 | 16 | 35 | 74 | 2 | 3 | 35 | 25 |

| N | O | P | Q | R | S | T | U | V | W | X | Y | Z |
|---|---|---|---|---|---|---|---|---|---|---|---|---|
| 78 | 74 | 27 | 3 | 77 | 63 | 93 | 27 | 13 | 16 | 5 | 19 | 1 |

Armed with this knowledge, you as the enemy general could decide on the most likely shift of a given encoded message. For example, suppose you had intercepted the following message:

**LI ZH ZLQ, SLCCD IRU HYHUBRQH (HAWUD SHSSHURQL).**

Your statistical knowledge of the frequency of letters would help you to decode the message and perhaps change history (or perhaps not). For now, however, we will wait until the Case Study on page 39 to decode the secret message, at which time we will discuss how to solve this problem.

## OVERVIEW



## The Big Picture

### Where we are coming from, and where we are headed...

In Chapter 1, we became acquainted with the building blocks of data analysis, such as the concepts of population, sample, types of variables, and the importance of a random sample. In Chapter 2, we apply the adage "A picture is worth a thousand words." The human mind can assess information presented pictorially, better than it can through words and numbers alone. We can recognize the entire scope of the interrelationships within a picture and somehow formulate a synthesis. Psychologists sometimes call this innate pattern recognition ability the "gestalt" experience. Statistical graphics take advantage of these abilities to quickly summarize data.

Before a researcher can draw conclusions from a data set, he or she must first *organize* the data, usually into some sort of a table or graph. The purpose of this chapter is to introduce graphical and tabular methods for summarizing data. We construct and interpret graphical descriptive tools such as bar graphs and tabular descriptive tools such as frequency distributions. Later, in Chapter 3, we learn how to describe a data set using statistics rather than tables and graphs.

# 2.1 Graphs and Tables for Categorical Data

## Objectives: By the end of this section, I will be able to...

1. Construct and interpret a frequency distribution and a relative frequency distribution for qualitative data.
2. Construct and interpret bar graphs and Pareto charts.
3. Construct and interpret pie charts.

## 1 Frequency Distributions and Relative Frequency Distributions

### Frequency Distributions

Recall from Chapter 1 that categorical (qualitative) data take values that are non-numeric and are usually classified into categories. In this section we learn graphical and tabular methods for handling categorical data. Let us begin with an example.

The January 2004 issue of *Trends and Tudes,* the newsletter from Harris Interactive (**www.harrisinteractive.com**), described a survey that asked teenagers which career they considered to have "very great prestige." Doctors led the list of results, followed by member of Congress, military officer, firefighter, and scientist. Suppose that, as part of her thesis project, a college student conducted a survey of the students in her sister's high school homeroom, asking, "Which of the following careers do you think has the highest prestige: Actor, Athlete, Doctor, Entertainer, Firefighter, Lawyer, Member of Congress, Military Officer, Police Officer, Scientist?" Each student's chosen career was recorded. Note that *career* is a qualitative variable, not quantitative. The results from a random sample of 20 students are presented in Table 2.1.

Table 2.1 Prestigious career survey data set

| Student | Career | Student | Career | Student | Career |
|---|---|---|---|---|---|
| 1 | Doctor | 8 | Athlete | 15 | Lawyer |
| 2 | Scientist | 9 | Doctor | 16 | Military Officer |
| 3 | Military Officer | 10 | Scientist | 17 | Doctor |
| 4 | Military Officer | 11 | Doctor | 18 | Scientist |
| 5 | Doctor | 12 | Military Officer | 19 | Doctor |
| 6 | Scientist | 13 | Scientist | 20 | Lawyer |
| 7 | Military Officer | 14 | Lawyer | | |

It is not immediately clear from this data set which career is the most popular choice among the 20 students in the sample. That is why we need ways to summarize the values in a data set. One popular method used to summarize the values in a data set is the **frequency distribution** (or *frequency table*).

The **frequency,** or **count,** of a category refers to the number of observations in each category. A **frequency distribution** for a qualitative variable is a listing of all the values (for example, categories) that the variable can take, together with the frequencies for each value.

## Example 2.1 What careers do teenagers admire?

Create a frequency distribution for the variable *career* from Table 2.1.

**Solution**

For each career choice, we compute the **frequency**; that is, we **count** how many students preferred that particular career. Table 2.2 shows the frequency distribution. For example, five students chose Scientist. The frequency distribution summarizes the data set so that quick observations can be made, such as "Doctor was preferred by the greatest number of students in the survey."

Table 2.2 Frequency distribution of career

| Variable: career | Frequency |
|---|---|
| Doctor | 6 |
| Scientist | 5 |
| Military Officer | 5 |
| Lawyer | 3 |
| Athlete | 1 |

As the data set gets larger, the need for summarization gets more and more acute. (Imagine if our survey consisted of 20,000 students rather than 20.) Take a moment to add up the frequencies in Table 2.2. What do they add up to? This number is the sample size: $n = 20$. Now, is this just a coincidence, or does this happen every time? Actually, this happens every time: the sum of the frequencies equals the sample size, $n$. One way to check if you made a mistake in forming your frequency distribution table is to add up the frequencies and see if the sum equals the sample size.

### Relative Frequency Distributions

Next, suppose you didn't know the size of the sample in the survey. Suppose you were told only that 6 students preferred Doctor. What would your response be? The logical question is "Is that a lot?" If our sample size was 10 students, then 6 students preferring Doctor is certainly a lot. However, if our sample size was 1000 students, then only 6 students preferring Doctor is *not* a lot. So, the number's significance depends on what you compare the 6 students to—that is, "relative to what?" or "compared to what?" In statistics, we compare the frequency of a category with the total sample size to get the **relative frequency**.

The **relative frequency** of a particular category of a qualitative variable is its frequency divided by the sample size. A **relative frequency distribution** for a qualitative variable is a listing of all values that the variable can take, together with the relative frequencies for each value.

## Example 2.2 Relative frequency of career preferences

Create a **relative frequency distribution** for the variable *career* using the information in Table 2.2.

**Solution**

The relative frequency of the students who preferred Doctor is the frequency 6 divided by the sample size 20:

$$\text{relative frequency of Doctor} = \frac{\text{frequency}}{\text{sample size}} = \frac{6}{20} = 0.30$$

The relative frequency of the students who preferred Doctor is 0.30, or 30%. So, if someone told you that 30% of the students preferred Doctor, without telling you the sample size, you would have a better idea of the relative strength of preference for that group than if you were given the frequency alone.

To construct the relative frequency distribution, shown in Table 2.3, for the career survey data, divide each frequency in the frequency distribution by the sample size 20. Note that the relative frequencies always add up to 1.0, which represents 100%. ■

Table 2.3 Relative frequency distribution of career preference by students

| Variable: career | Relative frequency |
|---|---|
| Doctor | 6/20 = 0.30 |
| Scientist | 5/20 = 0.25 |
| Military Officer | 5/20 = 0.25 |
| Lawyer | 3/20 = 0.15 |
| Athlete | 1/20 = 0.05 |
| Total | 20/20 = 1.00 |

## 2 Bar Graphs and Pareto Charts

Frequency distributions and relative frequency distributions are tabular, and thus useful for summarizing data sets. The graphical equivalent of a frequency distribution or a relative frequency distribution is called a **bar graph** (or **bar chart**).

A **bar graph** is used to represent the frequencies or relative frequencies for categorical data. It is constructed as follows:

1. On the horizontal axis, provide a label for each category.
2. Draw rectangles (bars) of equal width for each category. The height of each rectangle represents the frequency or relative frequency for that category. Ensure that the bars are not touching each other.

### Example 2.3 Bar graphs of career preferences

Construct a frequency bar graph and a relative frequency bar graph for the career preference distributions in Tables 2.2 and 2.3.

**Solution**

The bar graphs are provided in Figures 2.1a and 2.1b. Across the horizontal axis are the five career categories. Next, draw rectangles, each of whose height represents either the frequency or the relative frequency for that category, represented on the vertical axis. For example, in Figure 2.1a, the first rectangle (Doctor) reaches a height of 6, while the second rectangle reaches only to 5. Note that the rectangles are of equal width, and none of them touch each other. Also notice that the two bar graphs are exactly alike except for the scale indicated on the vertical axis. This is because we divide each frequency by the same number, the sample size, to get the relative frequency. ■

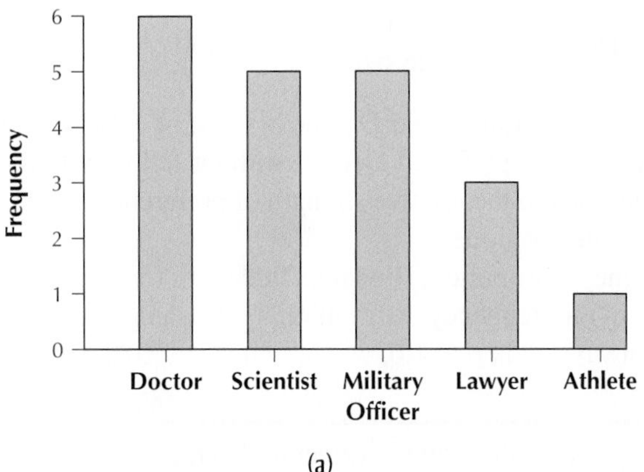


(a)

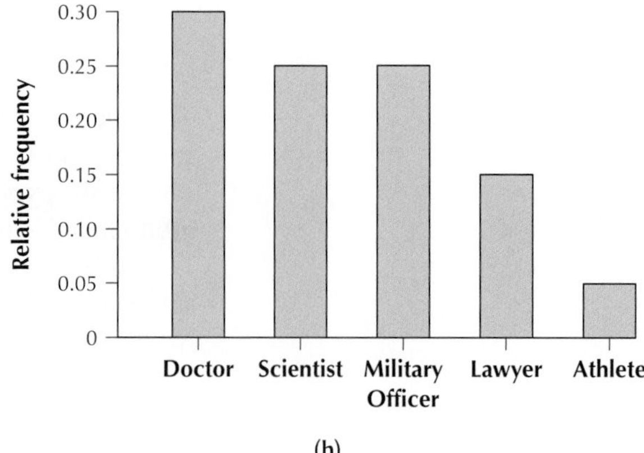


(b)

FIGURE 2.1 (a) Frequency bar graph; (b) relative frequency bar graph.

Both Figure 2.1a and Figure 2.1b are examples of **Pareto charts**. Figures 2.3a and 2.3b (page 40) are examples of bar graphs that are not Pareto charts.

A **Pareto chart** is a bar graph in which the rectangles are presented in decreasing order from left to right.

## 3 Pie Charts

**Pie charts** are a common graphical device for displaying the relative frequencies of a categorical variable.

A **pie chart** is a circle divided into sections (that is, slices or wedges), with each section representing a particular category. The size of the section is proportional to the relative frequency of the category.

Pie charts are typically made using technology. However, one can construct a pie chart using a protractor and a compass. Since a circle contains 360 degrees, we need to multiply the relative frequency for each category by 360°. This will tell us how large a slice to make for each category, in terms of degrees.

### Example 2.4 Pie chart for the careers data

Construct a pie chart for the careers data from Example 2.2.

**Solution**

The relative frequencies from Example 2.2 are shown in Table 2.4. We multiply each relative frequency by 360° to get the number of degrees for that section (slice) of the pie chart.

Our pie chart will have five slices, one for each career category. Use the compass to draw a circle. Then use a protractor to construct the appropriate angles for each section. From the center of the circle, draw a line to the top of the circle. Measure your first angle using this line. For Doctor, we need an angle of 108°. This angle is shown in Figure 2.2. Then, from there, measure your second angle—in this case, the 90° right angle for Scientist. Continue until your circle is complete. ■

Table 2.4 Finding the number of degrees for each slice of the pie chart

| Variable: career | Relative frequency | Multiply by 360° | Degrees for that section |
|---|---|---|---|
| Doctor | 6/20 = 0.30 | 0.30 × 360° = | 108° |
| Scientist | 5/20 = 0.25 | 0.25 × 360° = | 90° |
| Military Officer | 5/20 = 0.25 | 0.25 × 360° = | 90° |
| Lawyer | 3/20 = 0.15 | 0.15 × 360° = | 54° |
| Athlete | 1/20 = 0.05 | 0.05 × 360° = | 18° |
| Total | 20/20 = 1.00 | | 360° |

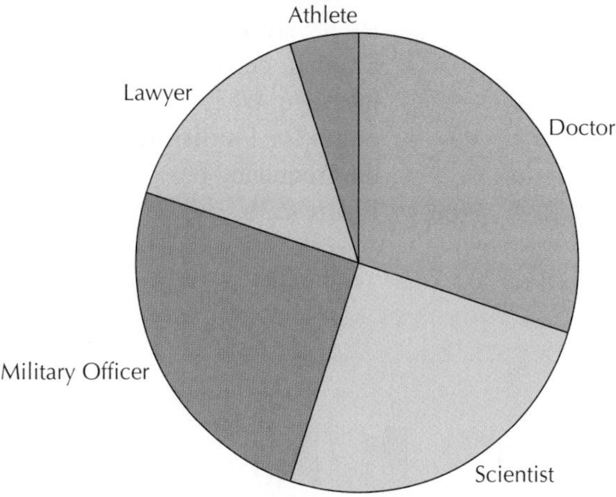


FIGURE 2.2 Pie chart of the careers data.

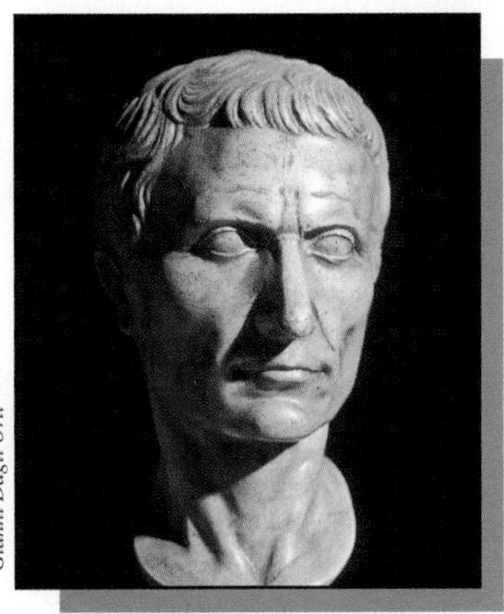



CASE STUDY

## The Caesar Cipher

Recall the Caesar Cipher from the chapter introduction. It's time to decipher that secret message from Caesar to one of his generals that we intercepted:

**LI ZH ZLQ, SLCCD IRU HYHUBRQH (HAWUD SHSSHURQL).**

We will make a frequency distribution and bar graph of the letters in the message and then compare them with the frequency distribution and bar graph of the letters in the English language. The letter frequencies in a sample of modern English text that is 1000 letters long are given in Table 2.5.

Table 2.5 Frequency distribution of English letters in a sample of 1000 letters

| A | B | C | D | E | F | G | H | I | J | K | L | M |
|---|---|---|---|---|---|---|---|---|---|---|---|---|
| 73 | 9 | 30 | 44 | 130 | 28 | 16 | 35 | 74 | 2 | 3 | 35 | 25 |

| N | O | P | Q | R | S | T | U | V | W | X | Y | Z |
|---|---|---|---|---|---|---|---|---|---|---|---|---|
| 78 | 74 | 27 | 3 | 77 | 63 | 93 | 27 | 13 | 16 | 5 | 19 | 1 |

Using this frequency distribution, we can observe that the letter **E** far outstrips all other letters in the alphabet in frequency. Other high-frequency letters are **A, I, N, O, R, S,** and **T.** Compare this with the frequency distribution of the letters in the coded

Table 2.6 Frequency distribution of letters in coded message

| A | B | C | D | E | F | G | H | I | J | K | L | M |
|---|---|---|---|---|---|---|---|---|---|---|---|---|
| 1 | 1 | 0 | 2 | 0 | 0 | 0 | 7 | 2 | 0 | 0 | 4 | 0 |
| **N** | **O** | **P** | **Q** | **R** | **S** | **T** | **U** | **V** | **W** | **X** | **Y** | **Z** |
| 0 | 0 | 0 | 3 | 3 | 4 | 0 | 4 | 0 | 3 | 0 | 1 | 2 |

message, shown in Table 2.6. From this frequency distribution, we can see that **H** is the most frequently occurring letter in the coded message. Other frequently occurring letters are **L, Q, R, S, U,** and **W.** Since **E** is the most frequently occurring letter in English, perhaps this means that **E** is encoded as **H,** the most common letter in our message. We turn to bar graphs to provide us with further evidence. The frequency bar graph for English letters, constructed from Table 2.5, appears in Figure 2.3a, while the frequency bar graph of letters in the coded message, from Table 2.6, is shown in Figure 2.3b.

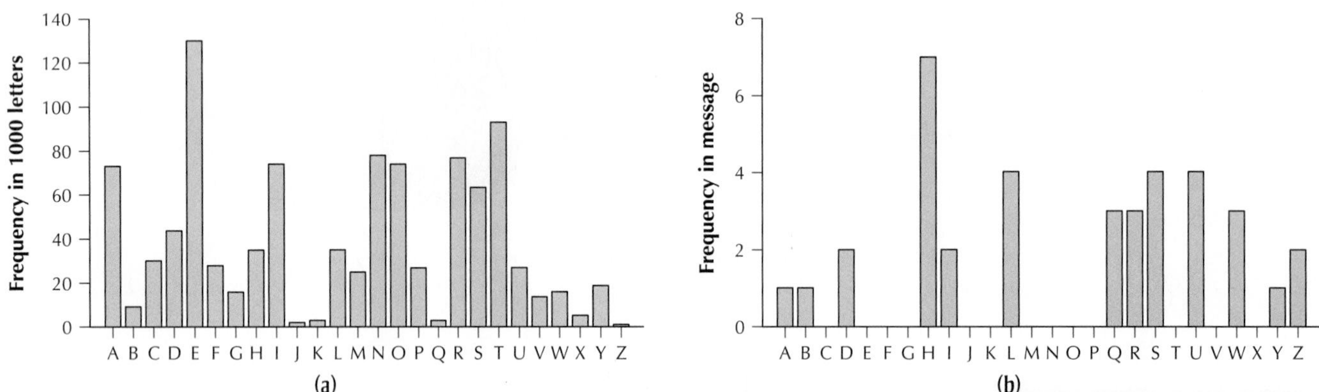


FIGURE 2.3 (a) Frequency bar graph of English letters; (b) frequency bar graph of letters in coded message.

Caesar used a simple shift of the letters for his code. If we substitute **H** for **E**, then the original letters have been shifted three places to the right (**E** → **F** → **G** → **H**). But this may just be an aberration. Is there further evidence for a "right shift of 3"? Let's see if this "right shift of 3" makes sense for the other high-frequency letters in the coded message. To undo a "right shift of 3," we would need to shift the letters in the coded message back three to the left to get the original letters. If the letter **L** is shifted back three places to the left, you get **I**, one of the high-frequency letters in English. Shift the letter **Q** three places, and you get **N**, another letter of high frequency in English. Shift the other letters of high frequency in our coded message, and you get **O**, **P**, **R**, and **T**, respectively, all high-frequency letters. There is a strong probability that we have found the correct decoding mechanism.

Let us now proceed to decode the message by shifting every letter in the coded message three places to the left (for example, **L** → **K** → **J** → **I**). It turns out that the decoded message reads

**IF WE WIN, PIZZA FOR EVERYONE (EXTRA PEPPERONI).**

Small wonder that Caesar went on to win an empire! We have gotten a taste of how the analysis of frequency distributions and bar graphs can be useful for solving problems.

## STEP-BY-STEP TECHNOLOGY GUIDE: Frequency Distributions, Bar Graphs, and Pie Charts

We use the data set in Table 2.7 to demonstrate how to use technology to construct a frequency distribution, relative frequency distribution, bar graph, and pie chart. Table 2.7 lists the declared majors of 25 randomly selected students at a local business school. (MIS stands for management information systems.)

**TABLE 2.7 Declared majors of business school students**

| | | | | |
|---|---|---|---|---|
| Management | MIS | Management | MIS | Marketing |
| Marketing | Marketing | Management | Finance | Accounting |
| Accounting | Accounting | MIS | Management | MIS |
| Management | MIS | Management | Economics | Accounting |
| Finance | Management | Economics | Marketing | Finance |

## EXCEL

### Frequency Distributions

***Step 1*** Enter the data in Column A, with the topmost cell indicating the variable name, *Major*.

***Step 2*** Select cells A1–A26, click **Insert > PivotTable**, and click **OK**.

***Step 3*** Under **Choose fields to add to report**, select ***Major***.

***Step 4*** Click on ***Major*** and drag to the **Values** box at the lower right of the screen. The resulting frequency distribution is shown in Figure 2.4. In Excel, this takes the form of a *pivot table*, which is an interactive tabular format.

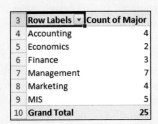

| 3 | Row Labels | Count of Major |
|---|---|---|
| 4 | Accounting | 4 |
| 5 | Economics | 2 |
| 6 | Finance | 3 |
| 7 | Management | 7 |
| 8 | Marketing | 4 |
| 9 | MIS | 5 |
| 10 | **Grand Total** | **25** |

Figure 2.4

### Bar Graphs and Pie Charts

**Note:** Excel can make bar graphs or pie charts using frequency distributions but not from the raw data.

***Step 1*** Enter the frequency distribution as shown in Figure 2.5.

***Step 2*** Select cells A1 to B7. For a bar graph, click **Insert > Column**. For a pie chart, click **Insert > Pie**.

***Step 3*** The resulting frequency bar graph and pie chart are shown in Figures 2.6 and 2.7.

| | A | B |
|---|---|---|
| 1 | Major | Count |
| 2 | Accounting | 4 |
| 3 | Economics | 2 |
| 4 | Finance | 3 |
| 5 | Management | 7 |
| 6 | Marketing | 4 |
| 7 | MIS | 5 |

Figure 2.5

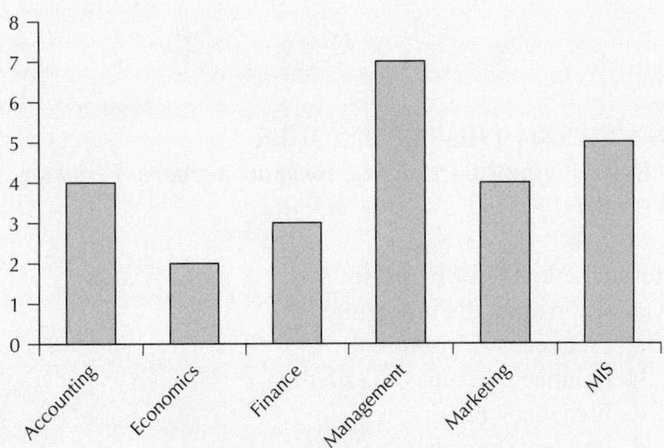


Figure 2.6 Excel frequency bar graph.

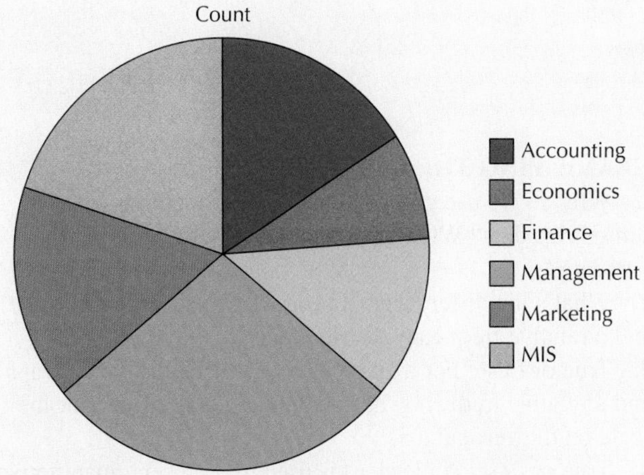


Figure 2.7 Excel pie chart.

*(Continued)*

## MINITAB

### Frequency Distributions

***Step 1*** Name your variable *Major* and enter the data into the **C1** column.
***Step 2*** Click **Stat > Tables > Tally**.
***Step 3*** Under **Display**, select **Counts** and **Percents**.
***Step 4*** Click inside the **Variables** box until you see your variable *major* listed. Select the variable **C1 *Major***, and click **Select**. Then click **OK**.

### Bar Graphs

***Step 1*** Name your variable *Major* and enter the data into the **C1** column.
***Step 2*** Click **Graph > Bar Chart**. For raw data select **Bars Represent: Counts of Unique Values**, select **Simple**, and click **OK**. (For summarized data such as a frequency distribution, select **Bars Represent: Values from a Table**, and select **Simple**. Then click **OK**.)
***Step 3*** In the **listing of variables** box, click on the ***Major*** variable to select it for analysis. Then click **OK**.

### Pie Charts

***Step 1*** Name your variable *Major* and enter the data into the **C1** column.
***Step 2*** Click **Graph > Pie Chart**. For raw data select **Chart Counts of unique values**. Then click in the **Variables** box to select the variable *Major*, and click **OK**. (For summarized data such as a frequency distribution, select **Chart Data from a Table**. Then select the category variable for **Categorical variable**, and select the variable with the frequencies or relative frequencies for the **Summary variable**. Then click **OK**.)

## Section 2.1 SUMMARY

In this section, we learned about tabular and graphical methods for summarizing qualitative (categorical) data.

1 Frequency distributions and relative frequency distributions list all the values that a qualitative variable can take, along with the frequencies (counts) or relative frequencies for each value.

2 A bar graph is the graphical equivalent of a frequency distribution or a relative frequency distribution. When the rectangles are presented in decreasing order from left to right, the result is a Pareto chart.

3 Pie charts are a common graphical device for displaying the relative frequencies of a categorical variable. A pie chart is a circle divided into sections (that is, slices or wedges), with each section representing a particular category. The size of the section is proportional to the relative frequency of the category.

## Section 2.1 EXERCISES

### CLARIFYING THE CONCEPTS

**1.** Why do we use graphical and tabular methods to summarize data? What's wrong with simply reporting the raw data?

**2.** What's the difference between a frequency distribution and a relative frequency distribution?

**3.** True or false: For a given data set, a frequency bar graph and a relative frequency bar graph look alike except for the scale on the vertical axis.

**4.** True or false: A pie chart is used to represent quantitative data.

### PRACTICING THE TECHNIQUES

**5.** Using Figure 2.6 (page 41), construct a relative frequency bar graph.

Refer to Table 2.8 for Exercises 6–8. For the indicated variable, construct the following:

- **a.** Frequency distribution
- **b.** Relative frequency distribution
- **c.** Frequency bar graph
- **d.** Relative frequency bar graph
- **e.** Pareto chart, using relative frequencies
- **f.** Pie chart

**TABLE 2.8 World water usage**

| Country | Continent | Climate | Main use |
|---|---|---|---|
| Iraq | Asia | Arid | Irrigation |
| United States | North America | Temperate | Industry |
| Pakistan | Asia | Arid | Irrigation |
| Canada | North America | Temperate | Industry |
| Madagascar | Africa | Tropical | Irrigation |
| North Korea | Asia | Temperate | Not reported |
| Chile | South America | Arid | Irrigation |
| Bulgaria | Europe | Temperate | Not reported |
| Afghanistan | Asia | Arid | Irrigation |
| Iran | Asia | Arid | Irrigation |

6. The variable *continent*.
7. The variable *climate*.
8. The variable *main use*.

9. The table below shows energy-related carbon dioxide emissions (in millions of metric tons) for 2004, by end-use sector, as reported by the U.S. Energy Information Administration.

| Sector | Emissions |
|---|---|
| Residential | 1213.9 |
| Commercial | 1034.1 |
| Industrial | 1736.0 |
| Transportation | 1939.2 |

Use the table to construct the following:
**a.** Relative frequency distribution
**b.** Frequency bar graph
**c.** Relative frequency bar graph
**d.** Pareto chart, using relative frequencies
**e.** Pie chart of the relative frequencies

10. Table 2.9 shows the numbers of children enrolled in Head Start programs, as reported by the U.S. Department of Health and Human Services in 2000, by ethnic group.

**TABLE 2.9 Ethnicity of children enrolled in Head Start**

| Census category | No. of Head Start children |
|---|---|
| White | 235,945 |
| Black | 259,004 |
| Hispanic | 169,909 |
| American Indian/Alaskan Native | 27,128 |
| Asian/Pacific Islander | 21,917 |
| Total | 713,903 |

Use the table to construct the following:
**a.** Relative frequency distribution
**b.** Frequency bar graph
**c.** Relative frequency bar graph
**d.** Pareto chart, using frequencies
**e.** Pie chart of the relative frequencies

For Exercises 11–13, use the information in Table 2.10. The table contains the Top 10 Best Sellers List from *USA Today* for the week of March 3, 2007. Among the information listed is the rank, title, first author, whether the work is fiction (F) or nonfiction (NF), whether the book is hardcover (H) or paperback (P), and whether the first author is female or male. For the indicated variable, construct the following:
**a.** Frequency distribution
**b.** Relative frequency distribution
**c.** Frequency bar graph
**d.** Relative frequency bar graph
**e.** Pie chart

11. The variable *type*.
12. The variable *book*.
13. The variable *gender*.

## APPLYING THE CONCEPTS

**Musical Activities.** Use the following information for Exercises 14–16. Every year, *USA Weekend* conducts a survey of the nation's teenagers, asking them various questions about lifestyle and music. In 2002, nearly 60,000 teenagers responded to the poll, conducted in part through *USA Weekend's* Web site. One question was, "Do you listen

**TABLE 2.10 Top 10 best sellers from *USA Today* (March 3, 2007)**

| Rank | Title | Author | Type | Book | Gender |
|---|---|---|---|---|---|
| 1 | The Secret | Rhonda Byrne | NF | H | Female |
| 2 | Innocent in Death | Nora Roberts | F | H | Female |
| 3 | Step on a Crack | James Patterson | F | H | Male |
| 4 | Bridge to Terabithia | Katherine Paterson | F | P | Female |
| 5 | Sisters | Danielle Steel | F | H | Female |
| 6 | The Measure of a Man: A Spiritual Autobiography | Sidney Poitier | NF | P | Male |
| 7 | The Memory Keeper's Daughter | Kim Edwards | F | P | Female |
| 8 | The Audacity of Hope | Barack Obama | NF | H | Male |
| 9 | You: On a Diet | Michael Roizen | NF | H | Male |
| 10 | The Best Life Diet | Bob Greene | NF | H | Male |

to music while you are . . . ?" and listed several options. Respondents were asked to select all that apply. The most common responses are shown in the table.

| | |
|---|---|
| Doing chores | 79% |
| On the computer | 73% |
| Doing homework | 72% |
| Eating meals at home | 33% |
| In the classroom | 18% |

**14.** Do you think that the sample is representative of all U.S. teenagers? How might the sample systematically introduce bias?
**15.** Assuming that the sample size is 6000, construct a *frequency* bar graph of the five activities listed in the table.
**16.** Can you construct a pie chart of the five activities listed in the table? Explain precisely why or why not.

**Music and Violence.** Another poll question asked by *USA Weekend* was "Do you think shock rock and gangsta rap are partly to blame for violence such as school shootings or physical abuse?" The results are shown in the table. Use this information to answer Exercises 17–20.

| | |
|---|---|
| Yes | 31% |
| No | 45% |
| I've never thought about it | 24% |

**17.** Do you think that there may have been other possible responses besides the three above? List at least two other possible responses.
**18.** Assuming the sample size is 6000, construct a frequency distribution of the responses.
**19.** Construct a frequency bar graph of the responses.
**20.** Construct a relative frequency pie chart of the responses.

**Astrological Signs.** Use the following information for Exercises 21 and 22. The General Social Survey collects data on social aspects of life in America. Here, 1464 respondents reported their astrological sign. A pie chart of the results is shown here.

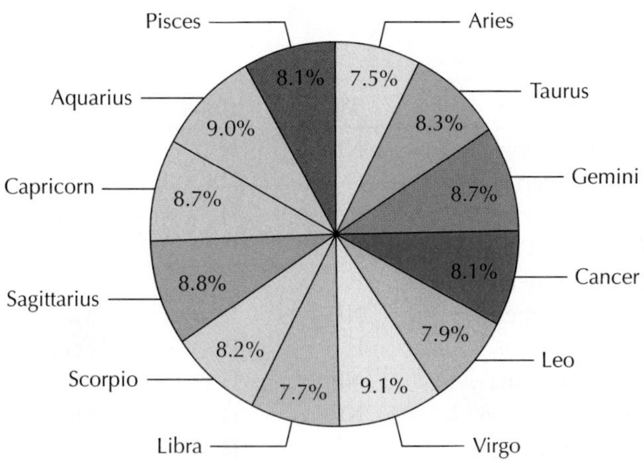

Pie chart of astrological signs.

**21.** Use the percentages in the pie chart to do the following:
  **a.** Construct a relative frequency bar graph of the astrological signs.
  **b.** Construct a relative frequency bar graph, but this time have the *y* axis begin at 7% instead of zero. Describe the difference between the two bar graphs. When would this one be used as opposed to the earlier bar graph?
**22.** Construct a frequency distribution of the astrological signs. Which sign occurs the least? The most?

**Satisfaction with Family Life.** Use the following information for Exercises 23–25. The General Social Survey also asked respondents about their amount of satisfaction with family life. Here, 1002 respondents reported their satisfaction levels with family life, as shown in the frequency distribution below.

| Satisfaction | Frequency |
|---|---|
| Very great deal | 415 |
| Great deal | 329 |
| Quite a bit | 99 |
| A fair amount | 90 |
| Some | 27 |
| A little | 25 |
| None | 17 |
| Total | 1002 |

**23.** What is your opinion of the categories? Do you think that everyone interprets these phrases in the same way? For example, what is the difference between "quite a bit" and a "great deal"?
**24.** Construct a Pareto graph of the categories. Do you think that, overall, people are happy with their family lives?
**25.** Construct a frequency or a relative frequency bar graph of the categories. Do you think that this bar graph reinforces the positive message in the data?

**Venture Capital Deals.** Use the following information for Exercises 26–28. **TheDeal.com** reports that the most active venture capital investors, for the 12 months ending July 17, 2007, are as shown in the table.

| Investor | No. of deals |
|---|---|
| Intel Capital Corp. | 41 |
| New Enterprise Associates | 22 |
| Sequoia Capital | 19 |
| Venrock Associates | 19 |
| Canaan Partners | 18 |
| Draper Fisher Jurvetson | 18 |

**26.** Construct a relative frequency distribution of the number of deals.
**27.** Construct the following:
  **a.** Frequency bar graph of the number of deals
  **b.** Relative frequency bar graph of the number of deals
**28.** Construct a pie chart of the number of deals.

**Health Insurance Coverage.** Use the following information for Exercises 29–33. The health insurance coverage of Americans in 2005 was reported by the Centers for Disease Control, as shown in the following tables.[1]

**Under Age 65 Years**

| Type of insurance | Count (in millions) |
|---|---|
| Private | 126.8 |
| Medicaid | 13.7 |
| Other | 5.7 |
| Uninsured | 35.8 |

**Age 65 Years and Over**

| Type of insurance | Count (in millions) |
|---|---|
| Private | 20.8 |
| Medicaid and/or Medicare | 11.7 |
| Other | 2.1 |
| Uninsured | 0.2 |

**29.** For those under 65 years, construct a relative frequency distribution.
**30.** For those under 65 years, construct a relative frequency bar graph.
**31.** For those 65 years and over, construct a relative frequency distribution.
**32.** For those 65 years and over, construct a relative frequency bar graph.
**33.** Comment on a comparison of the proportions of uninsured in the two groups.

**Abortion and Rape.** Use the following information for Exercises 34 and 35. Another question asked by the General Social Survey was "Please tell me whether or not you think it should be possible for a pregnant woman to obtain a legal abortion if she became pregnant as a result of rape?" Here, 934 respondents reported their responses as shown in the bar graph below.

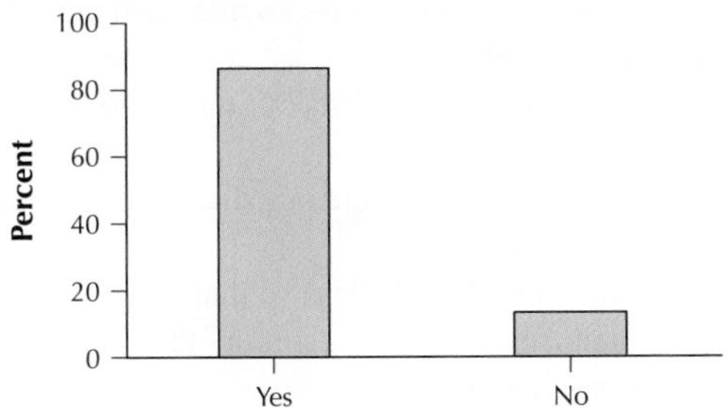


Bar graph of responses: possible for abortion if result of rape?

**34.** Estimating the percentages as best you can from the bar graph, construct a frequency or relative frequency distribution of the responses.
**35.** Estimating the percentages as best you can from the bar graph, construct a frequency or relative frequency pie chart of the responses.

**9/11 Commemorative Web Sites.** In the wake of the September 11 events, many people reached out through the Internet to find solace. The Pew Internet and American Life Project collected data on how people used the Internet to communicate and learn more about the tragic terrorist attacks. One question from their survey, taken from October 19 to November 18, 2001, asked: "Did you visit a memorial or commemorative web site as a result of the terrorist attacks in New York City and Washington?" Use the following table of the responses and their frequency to answer Exercises 36–40.

| Response | Frequency |
|---|---|
| Yes, visited site because of attacks | 475 |
| No, have not visited this kind of site | 2003 |
| Don't know/Refused to answer | 6 |
| Total | 2484 |

**36.** Construct a relative frequency distribution of the responses.
**37.** Construct a relative frequency bar graph of the responses.
**38.** If we wanted to change our relative frequency bar graph to a frequency bar graph, would we have to redraw the bars? Explain why or why not.
**39.** Construct a pie chart of the responses.
WHAT IF? **40.** *What if* we doubled the frequencies of each response. How would that affect the following?
- **a.** Relative frequency distribution of the responses
- **b.** Relative frequency bar graph of the responses
- **c.** Pie chart of the responses

**Types of Food Store.** According to the U.S. Census Bureau, there were 190,514 food stores of various types in the United States in 2000, categorized as shown in Table 2.11. Use the table to answer Exercises 41 and 42.

**TABLE 2.11 Types of food stores in the United States in 2000**

| Type of food store | Frequency |
|---|---|
| Supermarkets | 73,357 |
| Convenience food stores | 30,748 |
| Convenience food/gasoline | 23,035 |
| Delicatessens | 6,123 |
| Meat and fish markets | 8,941 |
| Retail bakeries | 20,418 |
| Fruit and vegetable markets | 2,971 |
| Candy, nut, confectionery | 5,029 |
| Dairy products stores | 2,340 |
| Other food stores | 25,552 |
| Total | 198,514 |

**41.** Construct a relative frequency distribution of the types of food store.
**42.** Construct a relative frequency bar graph of the types of food store.

**Causes of Death.** Refer to Table 2.12 for Exercises 43–45.

**TABLE 2.12 Causes of death**

| Cause of death | Deaths | Percent |
|---|---|---|
| Diseases of heart | 654,092 | 27.3 |
| Malignant neoplasms | 550,270 | 22.9 |
| All other causes | 435,769 | 18.2 |
| Cerebrovascular diseases | 150,147 | 6.3 |
| Chronic lower respiratory diseases | 123,884 | 5.2 |
| Accidents | 108,694 | 4.5 |
| Diabetes mellitus | 72,815 | 3.0 |
| Alzheimer's disease | 65,829 | 2.7 |
| Influenza and pneumonia | 61,472 | 2.6 |
| Nephritis | 42,762 | 1.8 |
| Septicemia | 33,464 | 1.4 |
| Intentional self-harm | 31,647 | 1.3 |
| Chronic liver disease | 26,549 | 1.1 |
| Hypertension | 22,953 | 1.0 |
| Parkinson's disease | 18,018 | 0.8 |
| Total | 2,389,365 | 100.0 |

*Source:* U.S. Centers for Disease Control and Prevention.

**43.** Construct a relative frequency bar graph of the cause of death.

**44.** Would a Pareto chart of the cause of death differ from your graph in Exercise 43? Explain.

**45.** Construct a pie chart of the cause of death. Consider whether it would be helpful to combine some of the smaller categories.

**Brisbane Babies.** Use the following information for Exercises 46–49. The *Brisbane Sunday Mail* newspaper reported that on December 18, 1997, 44 babies were born at the Brisbane Hospital in Australia. Their genders are given below.

| | | | | | | | | |
|---|---|---|---|---|---|---|---|---|
| Girl | Girl | Boy | Boy | Boy | Girl | Girl | Boy | Boy |
| Boy | Boy | Boy | Girl | Girl | Boy | Girl | Girl | Boy |
| Boy | Boy | Boy | Girl | Girl | Girl | Girl | Boy | Boy |
| Boy | Girl | Boy | Girl | Boy | Boy | Boy | Boy | Boy |
| Girl | Boy | Boy | Boy | Boy | Girl | Girl | Girl | |

**46.** Construct a pie chart of the numbers of boys and girls born that day.

**47.** Construct a frequency distribution and a relative frequency distribution for the gender of the babies born that day.

**48.** Construct a relative frequency pie chart for the gender of the babies born that day.

**49.** Construct a relative frequency bar graph of gender. If we wanted to change our relative frequency bar graph to a frequency bar graph, would we have to redraw the bars? Explain why or why not.

**Educational Goals in Sports.** Use your knowledge of Excel or Minitab to solve Exercises 50 and 51. Open the **Goals** data set. The subjects are students in grades four, five, and six from three school districts in Michigan. The students were asked which of the following was most important to them: good grades, sports, or popularity. Information about the students' age, gender, race, and grade was also gathered, as well as whether their school was in an urban, suburban, or rural setting.[2]

**50.** Generate bar graphs for the following variables.

- **a.** *Gender.* Estimate the relative frequency of girls in the sample. Of boys.
- **b.** *Goals.* About what percentage of the students chose "grades" as most important? About what percentage chose "popular"? About what percentage chose "sports"?

**51.** Generate relative frequency distributions for the following variables.

- **a.** *Gender.* How close were your estimates in the previous exercise?
- **b.** *Goals.* How close were your estimates in the previous exercise?

**Analysis of Households.** For Exercises 52–54, use your knowledge of Excel or Minitab. Open the data set **Household.**

**52.** How many observations are in this data set? How many variables?

**53.** Which of the variables are qualitative? Which of the variables are quantitative?

**54.** What would a relative frequency distribution of the variable *state* look like? A pie chart? A bar graph?

## CONSTRUCT YOUR OWN DATA SETS

**Environmental Club.** Use the following information for Exercises 55–57. You are the president of the College Environmental Club, which has members among all four classes, freshmen, sophomores, juniors, and seniors. The total number of members in the club is 20.

**55.** Set the frequency of each class so that each class has an equal number of members.

- **a.** Construct a frequency distribution of the variable *class.*
- **b.** Construct a relative frequency distribution of the variable *class.*

**56.** Set the frequency of each class so that there are more sophomores than freshmen, more juniors than sophomores, and more seniors than juniors.

- **a.** Construct a Pareto chart of the variable *class.*
- **b.** Construct a pie chart of the variable *class.*

**57.** Set the frequency of each class so that there are more seniors than any other class while the other three classes have equal numbers.

- **a.** Construct a frequency bar graph of the variable *class.*
- **b.** Construct a relative frequency bar graph of the variable *class.*

## 2.2 Graphs and Tables for Quantitative Data

**Objectives: By the end of this section, I will be able to…**

1. Construct and interpret a frequency distribution and a relative frequency distribution for quantitative data.
2. Use histograms and frequency polygons to summarize quantitative data.
3. Construct and interpret stem-and-leaf displays and dotplots.
4. Recognize distribution shape, symmetry, and skewness.

### 1 Frequency Distributions and Relative Frequency Distributions

In Section 2.1, we introduced tables and graphs for summarizing qualitative data. However, most of the data sets that we will encounter in this book are quantitative rather than qualitative. Recall from Chapter 1 that quantitative data take on numerical values. We can apply frequency and relative frequency distributions to quantitative data, just as we did for the qualitative data in Section 2.1.

### Example 2.5 Ages of missing children in California

The National Center for Missing and Exploited Children (**www.missingkids.com**) keeps an online searchable data base of missing children nationwide. Table 2.13 contains a listing of the 50 children who have gone missing from California and who would have been between 1 and 9 years of age as of March 4, 2007. Suppose we are interested in analyzing the ages of these missing children. Use the data to construct a **frequency distribution** and a **relative frequency distribution** of the variable *age*.

Table 2.13 Missing children and their ages

| Child | Age | Child | Age | Child | Age | Child | Age |
|---|---|---|---|---|---|---|---|
| Amir | 5 | Carlos | 7 | Octavio | 8 | Christian | 8 |
| Yamile | 5 | Ulisses | 6 | Keoni | 6 | Mario | 8 |
| Kevin | 5 | Alexander | 7 | Lance | 5 | Reya | 5 |
| Hilary | 8 | Adam | 4 | Mason | 5 | Elias | 1 |
| Zitlalit | 7 | Sultan | 6 | Joaquin | 6 | Maurice | 4 |
| Aleida | 8 | Abril | 6 | Adriana | 6 | Samantha | 7 |
| Alexia | 2 | Ramon | 6 | Christopher | 3 | Michael | 9 |
| Juan | 9 | Amari | 4 | Johan | 6 | Carlos | 2 |
| Kevin | 2 | Joliet | 1 | Kassandra | 4 | Lukas | 4 |
| Hazel | 5 | Christopher | 4 | Hiroki | 6 | Kayla | 4 |
| Melissa | 1 | Jonathan | 8 | Kimberly | 5 | Aiko | 3 |
| Kayleen | 6 | Emil | 7 | Diondre | 4 | Lorenzo | 9 |
| Mirynda | 7 | Benjamin | 5 | | | | |

**Solution**

We can construct the frequency distribution for the variable *age* and can construct the relative frequency distribution by dividing the frequency by the total number of observations, 50. See Table 2.14. ■

Table 2.14 Frequency distribution and relative frequency distribution of *age*

| Age | Tally | Frequency | Relative frequency |
|---|---|---|---|
| 1 | III | 3 | 0.06 |
| 2 | III | 3 | 0.06 |
| 3 | II | 2 | 0.04 |
| 4 | ~~IIII~~ III | 8 | 0.16 |
| 5 | ~~IIII~~ IIII | 9 | 0.18 |
| 6 | ~~IIII~~ ~~IIII~~ | 10 | 0.20 |
| 7 | ~~IIII~~ I | 6 | 0.12 |
| 8 | ~~IIII~~ I | 6 | 0.12 |
| 9 | III | 3 | 0.06 |
| Total | | 50 | 1.00 |

We can combine several ages together into "classes," in order to produce a more concise distribution. **Classes** represent a range of data values and are used to group the elements in a data set.

## Example 2.6 Frequency and relative frequency distributions using classes

Combine the age data from Example 2.5 into three classes, and construct frequency and relative frequency distributions.

### Solution

Let us define the following classes for the age data: 1–3 years old, 4–6 years old, and 7–9 years old. For each class, we group together all the ages in the class. Table 2.15 provides the frequency distribution and relative frequency distribution for these three age classes. ■

Table 2.15 Distributions for the variable *age*, after combining into three classes

| Class | Frequency | Relative frequency |
|---|---|---|
| 1–3 | 8 | 0.16 |
| 4–6 | 27 | 0.54 |
| 7–9 | 15 | 0.30 |
| Total | 50 | 1.00 |

### Choosing Which Distribution to Use

So which frequency distribution is the "right" one, Table 2.14 or Table 2.15? There is no absolute answer. It depends on the goals of the analysis, as well as other factors. For example, from Table 2.15, we can see that the majority (0.54 = 54%) of missing children are aged 4–6, an observation that was not immediately apparent from Table 2.14. So, combining data values into classes can lead to interesting overall findings. However, whenever data values are combined into classes, some information is lost. For example, it is not possible, using Table 2.15 alone, to determine that age 6 has the highest proportion of missing children. ■

We use the following to construct frequency distributions and histograms (for a discussion of histograms, see pages 50–51).

The **lower class limit** of a class equals the smallest value within that class.
The **upper class limit** of a class equals the largest value within that class.
The **class width** equals the difference between the lower class limits of two successive classes.

## Example 2.7 Class limits and class widths for the age data

Find the **lower class limits**, the **upper class limits**, and the **class width** for the classes in Example 2.6.

### Solution

The first class consists of all children aged 1, 2, and 3 years. The smallest value in this class is 1 year old. Therefore, the lower class limit for this class equals 1. The upper class limit for this class equals 3, the largest value in the class. The second class consists of all children aged 4, 5, and 6 years. So the lower class limit for this class is 4 years old, and the upper class limit is 6 years old. Similarly, for the class containing 7-, 8-, and 9-year-olds, the lower class limit is 7 and the upper class limit is 9. The class width is the difference in lower class limits between successive classes. Since our lower class limits are 1, 4, and 7, the class width of each class is 3 because the lower class limits differ by 3. For example, 4 – 1 = 3. ■

Next, we show how to construct frequency distributions for quantitative data.

To construct a frequency distribution for quantitative data:

1. Determine how many classes you will use.
2. Determine the class width. It is best (though not required) to use the same width for all classes.
3. Determine the upper and lower class limits. Make sure the classes are nonoverlapping.

## Example 2.8 Constructing a frequency distribution: the management aptitude test

Twenty management students, in preparation for graduation, took a course to prepare them for a management aptitude test. A simulated test provided the following scores:

| | | | | | | | | | |
|---|---|---|---|---|---|---|---|---|---|
| 77 | 89 | 84 | 83 | 80 | 80 | 83 | 82 | 85 | 92 |
| 87 | 88 | 87 | 86 | 99 | 93 | 79 | 83 | 81 | 78 |

Construct a frequency distribution of these management aptitude test scores.

### Solution

***Step 1*** **Choose the number of classes.** It is generally recommended that between 5 and 20 classes be used, with the number of classes increasing with the sample size. A small data set such as this will do just fine with 5 classes. In general, choose the number of classes to be large enough to show the variability in the data set, but not so large that many classes are nearly empty.

***Step 2*** **Determine the class widths.** First, find the *range* of the data, that is, the difference between the largest and smallest data points. Then, divide this range by the number of classes you chose in Step 1. This gives an estimate of the class width. Here, our largest data value is 99 and our smallest is 77, giving us a *range* of 99 – 77 = 22.

In Step 1, we chose 5 classes, so that our estimated class width is 22/5 = 4.4. We will use a convenient class width of 5. It is recommended that each class have the same width.

***Step 3*** **Determine the upper and lower class limits.** Choose limits so that each data point belongs to only one class. For example, suppose we chose one class to be 75–80 and the next class to be 80–85. Then, to which class would a data value of 80 belong? The classes should not overlap. Therefore, we define the following classes:

75–79 80–84 85–89 90–94 95–100

*Note:* In this example, we have data values that are integers. If the data values, instead, had decimal values, then we would choose the class limits accordingly. For example, if the data values ranged from 75 to 100 but were of the form 75.6, we could choose the class limits of the first class to be 75.0–79.9, the second class to be 80.0–84.9, and so on.

Note that the lower class limit of the first class, 75, is slightly below that of the smallest value in the data set, 77. Also note that the class width equals 80 – 75 = 5, as desired.

Using these five classes, we now proceed to construct the frequency and relative frequency distributions for the management aptitude test scores (see Table 2.16). We count the number of data values that fall into each class, and we divide each frequency by the sample size (20) to obtain the relative frequency. We see that the majority of the students (0.40 + 0.30 = 0.70) received scores between 80 and 89 and that only one received a score above 94. ■

Table 2.16 Distributions for the management aptitude test scores

| Class | Tally | Frequency | Relative frequency |
|---|---|---|---|
| 75–79 | \|\|\| | 3 | 0.15 |
| 80–84 | ~~\|\|\|\|~~ \|\|\| | 8 | 0.40 |
| 85–89 | ~~\|\|\|\|~~ \| | 6 | 0.30 |
| 90–94 | \|\| | 2 | 0.10 |
| 95–100 | \| | 1 | 0.05 |
| Total | | 20 | 1.00 |

## 2 Histograms and Frequency Polygons

### Histograms

There are many different methods of summarizing numeric data graphically. One example of a graphical summary for quantitative data is a **histogram**.

A **histogram** is constructed using rectangles for each class of data. The heights of the rectangles represent the frequencies or relative frequencies of the class. The widths of the rectangles represent the class widths of the corresponding frequency distribution. The lower class limits are placed on the horizontal axis (along with the upper class limit of the rightmost class), so that the rectangles are touching each other.

### Example 2.9 Histogram of management aptitude test scores

Construct a histogram of the frequency of the management aptitude test scores from Example 2.8.

**Solution**

***Step 1*** **Find the class limits and draw the horizontal axis.** Note that the class limits for these data were found in Example 2.8 and are given in Table 2.16. The lower class limits are 75, 80, 85, 90, and 95. The upper class limit of the rightmost class is 100.

Draw the horizontal axis, with the numbers 75, 80, 85, 90, 95, and 100 equally spaced along it. The numbers indicate where the rectangles will touch each other.

***Step 2*** **Determine the frequencies and draw the vertical axis.** Use the frequencies given in Table 2.16. These will indicate the heights of the five rectangles along the vertical axis. Find the largest frequency, which is 8. It is a good idea to provide a little bit of extra vertical space above the tallest rectangle, so make 9 your highest label along the vertical axis. Then provide equally spaced labels along the vertical axis between 0 and 9.

***Step 3*** **Draw the rectangles.** Draw your first rectangle. Its width is from 75 to 80, and its height is 3, the first frequency. Draw the remaining rectangles similarly.

The resulting frequency histogram is shown in Figure 2.8a. The relative frequency histogram is shown in Figure 2.8b. Note that the two histograms have identical shapes and differ only in the labeling along the vertical axis. ■

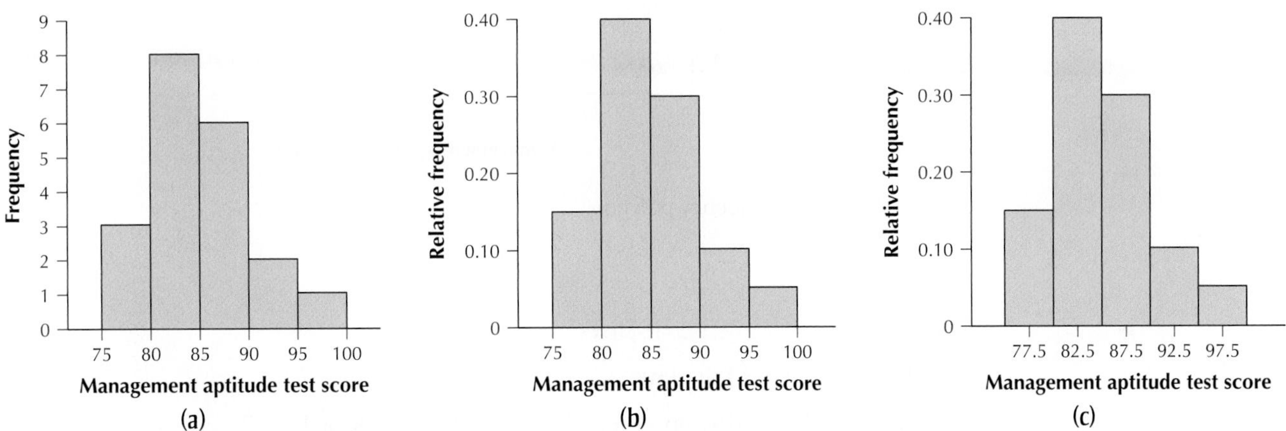


FIGURE 2.8 (a) Frequency histogram; (b) relative frequency histogram; (c) histogram using midpoints.

*Note:* Histograms are often presented using class midpoints rather than class limits. The class limits can be inferred by splitting the difference between the class midpoints. In Figure 2.8c, the upper class limit for the leftmost class is halfway between 77.5 and 82.5, that is, 80. Otherwise, Figure 2.8c is equivalent to Figure 2.8b.

Note that the histogram, unlike the frequency distribution, provides us with an overall impression of the data distribution. This characteristic will be crucial later on, when we evaluate the fitness of data sets to undergo certain data analysis methods. Also, notice that the rectangles are contiguous (touching), unlike the rectangles of the bar graphs in Section 2.1. Since the data are numeric, the horizontal axis in a histogram should be considered as the number line. A **class midpoint** is the average of two consecutive lower class limits. For example, the class midpoint for the leftmost class in Figure 2.8c is $(75 + 80) / 2 = 77.5$.

The *One Variable Statistics and Graphs* applet can display histograms for a selection of data sets in *Discovering Statistics,* including the management aptitude test scores. The applet allows you to experiment with different class widths.

## Frequency Polygons

**Frequency polygons** provide the same information as histograms, but in a slightly different format.

> A **frequency polygon** is constructed as follows. For each class, plot a point at the class midpoint, at a height equal to the frequency for that class. Then join each consecutive pair of points with a line segment.

## Example 2.10 Constructing a frequency polygon

Construct a frequency polygon for the management aptitude test data in Example 2.8.

**Solution**

The midpoints for the classes were calculated for Figure 2.8c. Plot a point for each frequency above each midpoint, and join consecutive points. The result is the frequency polygon in Figure 2.9. ■

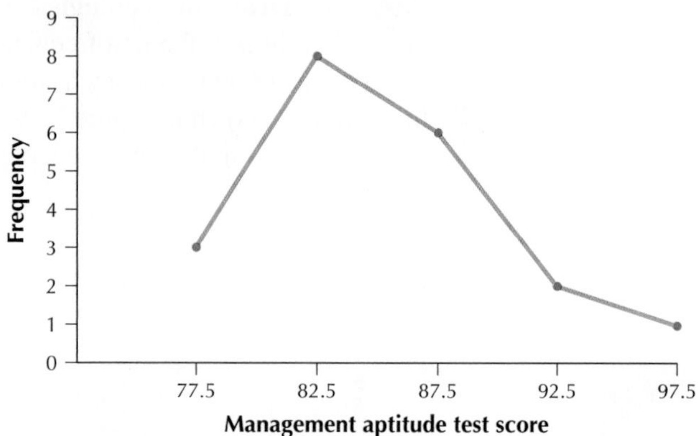


FIGURE 2.9 Frequency polygon.

# 3 Stem-and-Leaf Displays and Dotplots

## Stem-and-Leaf Displays

**Stem-and-leaf displays** were developed by Professor John Tukey of Princeton University in the late 1960s. This type of display generally contains more information than either a frequency distribution or a histogram. We will demonstrate how to construct a stem-and-leaf display in Example 2.11.

## Example 2.11 Stem-and-leaf display of psychology final-exam scores

Construct a stem-and-leaf display for the final-exam scores of 20 psychology students, given below.

| | | | | | | | | | |
|---|---|---|---|---|---|---|---|---|---|
| 75 | 81 | 82 | 70 | 60 | 59 | 94 | 77 | 68 | 98 |
| 86 | 68 | 85 | 72 | 70 | 91 | 78 | 86 | 51 | 67 |

**Solution**

First, find the leading digits of the numbers. Each number has one of the following as its leading digit: 5, 6, 7, 8, 9. Place these five numbers, called the **stems,** in a column:

5
6
7
8
9

Each number represents the tens place of the test scores. For example, 5 represents 5 tens. Now consider the ones place of each data value. For example, the first score, 75, has 5 in the ones place. Place this number, called the **leaf,** next to its stem:

| | |
|---|---|
| 5 | |
| 6 | |
| 7 | 5 |
| 8 | |
| 9 | |

The second score, 81, has 1 in the ones place, and the third score, 82, has 2 in the ones place. Write the leaves 1 and 2 next to the stem 8:

| | |
|---|---|
| 5 | |
| 6 | |
| 7 | 5 |
| 8 | 12 |
| 9 | |

Continue this process with the remaining data, placing each ones value next to its stem. Then, for each stem, order the leaves from left to right in increasing order. This produces the stem-and-leaf display:

| | |
|---|---|
| 5 | 19 |
| 6 | 0788 |
| 7 | 002578 |
| 8 | 12566 |
| 9 | 148 |

Notice that the two 68s refer to two different students who happened to get the same grade on the exam. In general, the leaf units represent the smallest decimal place represented in the data values. Then the stem unit consists of the remainder of the number. For example, suppose we have a data value of 127. Then the 7 is the leaf unit, and the 12 is the stem. Or else, suppose our data value is 0.146. Then our leaf unit is the 6 and the stem is the 14. Note that the stem-and-leaf display contains all the information that a histogram turned on its side does. But it also contains more information than a histogram, because the stem-and-leaf display shows the original values.

*Split stems* may sometimes be used in a stem-and-leaf display to provide a clearer idea of the data distribution when too many data points fall on just a few stems. When using split stems, each stem appears twice, with the leaves 0 to 4 on the upper stem and the leaves 5 to 9 on the lower stem. The above stem-and-leaf display of psychology scores would appear as follows when using splits stems:

| | |
|---|---|
| 5 | 1 |
| 5 | 9 |
| 6 | 0 |
| 6 | 788 |
| 7 | 002 |
| 7 | 578 |
| 8 | 12 |
| 8 | 566 |
| 9 | 14 |
| 9 | 8 |

The *One Variable Statistics and Graphs* applet can display stem-and-leaf displays for a selection of data sets in *Discovering Statistics,* including the psychology final-exam scores. The applet allows you to experiment with split stems if you like. ■

## Dotplots

A simple but effective graphical display is a **dotplot**. In a dotplot, each data point is represented by a dot above the number line. When the sample size is large, each dot may represent more than one data point. Figure 2.10 is a dotplot of the 20 management aptitude test scores.

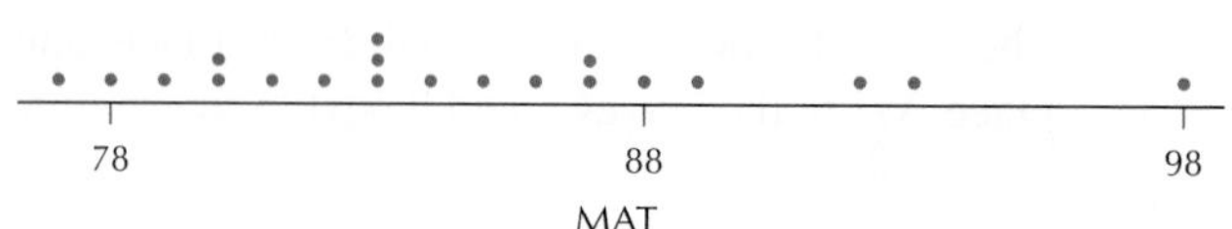


FIGURE 2.10 Dotplot of the managerial aptitude test scores. The two dots above 87 indicate that two tests had the same score of 87. Which test score was the most common?

Dotplots are useful for comparing two variables. For example, suppose that an instructor taught two different sections of a management course and gave a simulated management aptitude exam in each section (MAT-1 and MAT-2). The instructor could then compare these two groups of scores directly, using a Minitab comparison dotplot, as in Figure 2.11. Although there is much overlap, Section 1 had the highest score, while Section 2 had the three lowest scores. Therefore, it looks as if Section 1 did better, though we will have to wait until Chapter 3 to decide for sure.

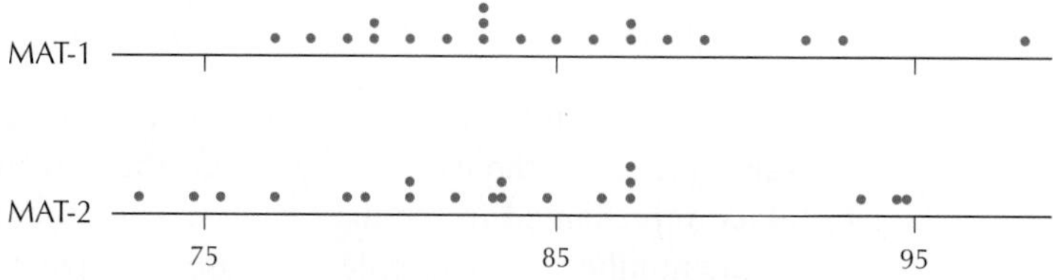


FIGURE 2.11 Comparison dotplot of MAT test scores for the two sections. Note that the two sections are graphed using the same number line, which makes comparison easier.

## 4 Distribution Shape, Symmetry, and Skewness

Frequency distributions are tabular summaries of the set of values that a variable takes. We now generalize the concept of **distribution**.

> The **distribution** of a variable is a table, graph, or formula that identifies the variable values and frequencies for all elements in the data set.

For example, a frequency distribution is a distribution since it is a table that specifies each of the values that a variable can take, along with the frequencies. However, our definition of "distribution" also includes histograms, stem-and-leaf displays, dotplots, and other graphical summaries. (In Chapter 6, we will introduce distributions defined by formulas.) These graphical distributions invite us to consider the shape of a distribution. The *shape* of a distribution is the overall form of a graphical summary, approximated by a smooth curve.

### The Bell-Shaped Curve

Figure 2.12 contains the relative frequency histogram of the heights of 1000 college women. Note that there are relatively fewer women in both the left-hand tail (shorter women) and the right-hand tail (taller women). Instead, as height increases from left to right, the relative frequency gradually increases until it reaches a peak near 65 inches tall and then gradually decreases. Thus, the distribution of heights is said to be *bell-shaped*.

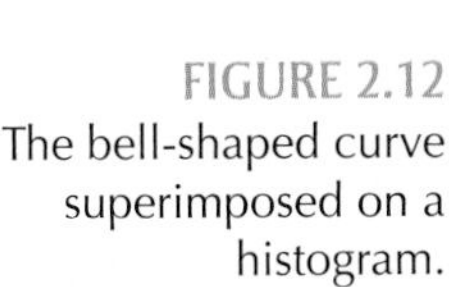

FIGURE 2.12
The bell-shaped curve superimposed on a histogram.

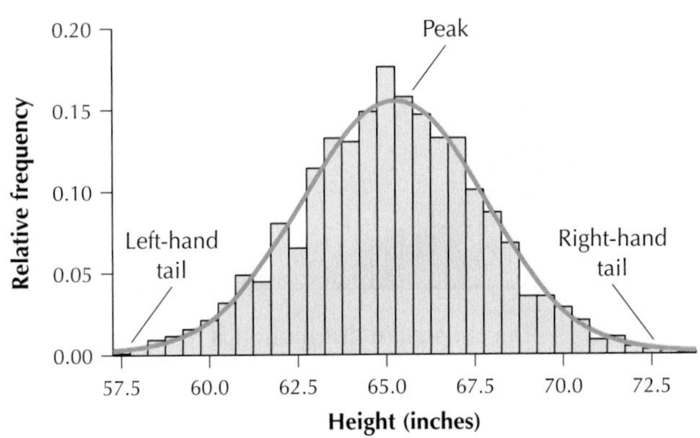


The rectangles represent the actual data. However, the smoothed curve represents an approximation of the overall form of the distribution, and thus the smoothed curve represents the shape of the distribution, which is bell-shaped. The formal name of the bell-shaped distribution is the *normal* distribution. In Chapter 6 we will learn much more about this important distribution. In Chapter 7, we will learn how to assess whether or not a particular distribution is normal (bell-shaped). Starting in Chapter 8, many of the methods for statistical inference we will learn depend on this assessment.

## Analyzing the Shape of a Distribution

We next learn some tools for analyzing the shape of a distribution. An image has *symmetry* (or is **symmetric**) if there is a line (axis of symmetry) that splits the image in half so that one side is the mirror image of the other. For example, the butterfly in Figure 2.13 has symmetry, since a line drawn down the middle of the butterfly would create two mirror images of each other. Symmetry is important in the world of statistics as well, and it is important to develop the talent for recognizing which distribution shapes are symmetric.

For example, is the distribution in Figure 2.12 symmetric? The smoothed curve in Figure 2.12 is in fact perfectly symmetric. However, the histogram rectangles reflecting the actual data are only nearly symmetric, since a vertical line drawn down the middle of the distribution would not result in two perfect mirror images. *Due to random variation, data from the real world rarely exhibit perfect symmetry.* With this in mind, the data analyst is usually content with the approximate symmetry exhibited by the data in Figure 2.12.

FIGURE 2.13
A butterfly is symmetric.
© Burke/ Triolo/Jupiterimages

However, not all distributions are symmetric, or even nearly symmetric. In Chapter 8 we will discuss a distribution called the chi-square distribution, which is not symmetric but is **skewed**. It often has a longer "tail" on the right than on the left (see Figure 2.14). Since the right-hand tail is longer, we say that this distribution is *right-skewed.* Examples of right-skewed data are usually found whenever one deals with money. For example, if we graph the incomes of the families in your home state, the graph will probably be right-skewed. Most of us will lie somewhere in the middle or left with the bulk of the data, while the incomes of folks like Donald Trump and Bill Gates lie far out on the right of the graph, in the right-hand tail. Figure 2.15 shows a *left-skewed* distribution. Good examples of left-skewed data are retirement ages or death ages. Often, exam grade data can be left-skewed, as several students bump up against the 100% boundary on the right, most students are somewhere in the middle, and a few students stagger in with 40s and 50s in the left-hand tail.

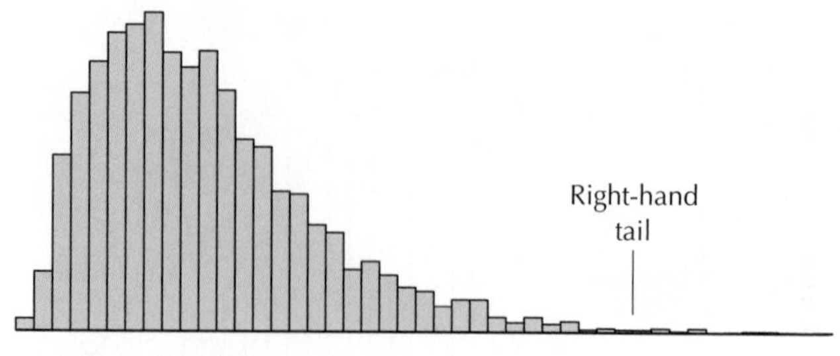


FIGURE 2.14 The chi-square distribution is right-skewed.

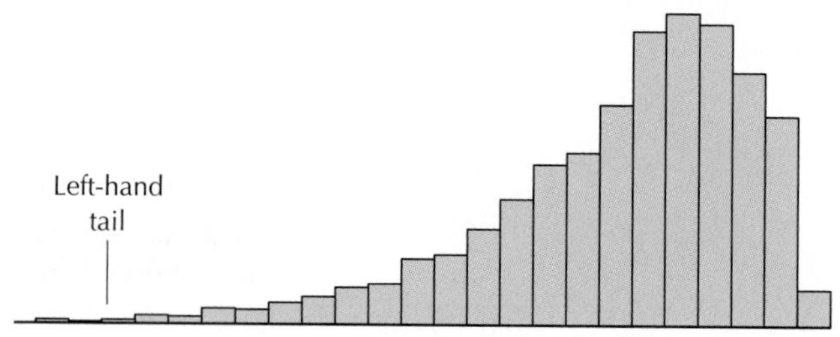


FIGURE 2.15 Some distributions are left-skewed.

## Example 2.12 Choosing the appropriate graphical summary

Statistically literate citizens recognize that one may select different graphical summaries, depending on the intention of the presenter. Figures 2.16a, 2.16b, and 2.16c contain a dotplot, a histogram, and a stem-and-leaf display of the average size of households in the 50 states and the District of Columbia. Which graphical summary—the dotplot, the histogram, or the stem-and-leaf display—is most useful if our primary objective is

**a.** to assess symmetry and skewness?
**b.** to be able to construct it quickly using paper and pencil?
**c.** to retain complete knowledge of the original data set?
**d.** to give a presentation to people who have never had a stats course before?
**e.** to be given maximum freedom in choosing how the data are to be interpreted?

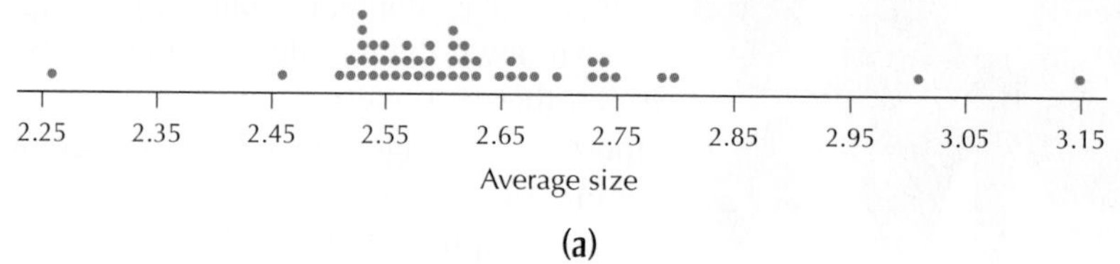


(a)

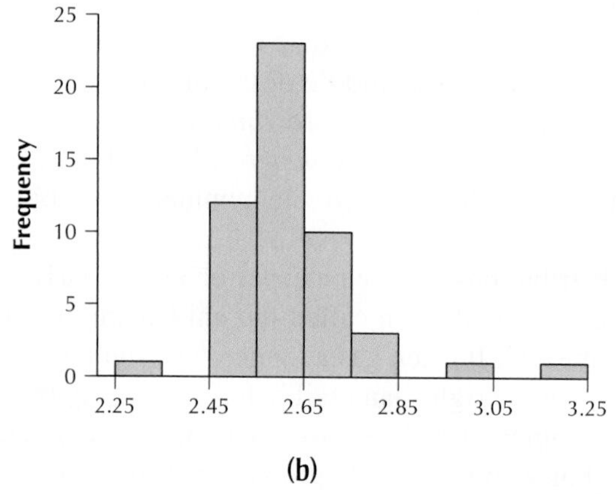


(b)

```
Stem-and-leaf of Average  N=51
Leaf Unit = 0.010

        22 6
        23
        24 6
        25 1223333344455556677788999
        26 011112223356678
        27 0334459
        28 0
        29
        30 1
        31 5
```

(c)

FIGURE 2.16 (a) Dotplot; (b) histogram; (c) stem-and-leaf display.

### Solution

**a.** All three graphics are good at assessing symmetry and skewness.
**b.** The dotplot's great asset is its simplicity. It can be quickly drawn, with minimal preparation, in contrast to the other two summaries, which require some organization or calculation.

**c.** The stem-and-leaf display was invented in order to retain complete knowledge of the data set. Histograms are the least effective in this regard.

**d.** The histogram is widely used in the real world and is probably the best choice for a presentation in front of those who have not had a stats course before.

**e.** We will learn in Section 2.4, "Graphical Misrepresentations of Data," that two histograms representing the same data set can look completely different, depending on the number of classes and class widths. Histograms therefore provide maximum freedom for the presenter to choose how the data are to be presented. Statistically literate citizens should be ready to examine the data more closely, in order to debunk possible misrepresentation. ■

## Example 2.13 Exploring the calorie content of 961 items of food



The data set **Nutrition** (on your CD and the book companion Web site) contains 22 variables' worth of nutrition information on 961 food items, courtesy of the U.S. Department of Agriculture (USDA). The variable *calories* contains the number of calories for each food item. Construct a histogram of *calories* using Minitab. Explore the data set to see if any interesting patterns emerge.

### Solution

Figure 2.17 shows the histogram of the data. Again, as with many variables, the data are right-skewed. Most of the calorie counts are less than 200. However, if you look carefully, there is one food item with a calorie count of over 6000! This amount is equivalent to everything a typical person consumes in two or three days. What could this food item be? According to the data set, this is carrot cake with cream cheese frosting. But 6000 calories in a piece of cake? How could this be? The weight in grams (variable *wt_grams*) of the serving is 1536 grams, which is more than three pounds. That's a lot of carrot cake! ■

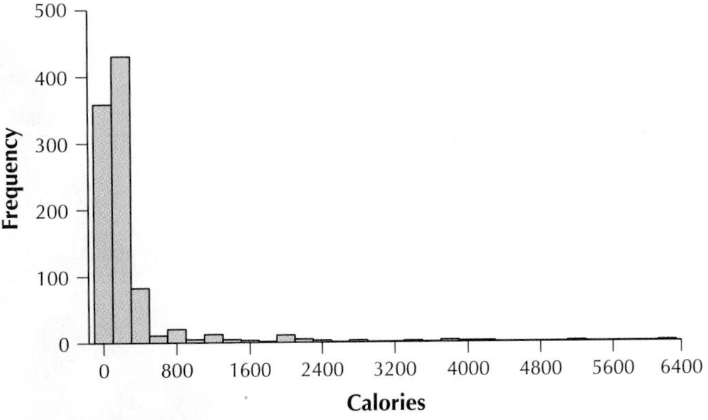


FIGURE 2.17 Histogram of calories of 961 food items.

### Exploring the Data

Let's investigate the data of Example 2.13 further and compute the *number of calories per gram* by taking the variable *calories* and dividing by *wt_grams*. This will give us the food item that has the greatest number of calories per unit weight (see Figure 2.18). Note that the histogram is much less skewed than previously. This is because we have accounted for the differences in calories that resulted from differences in weight. Most of the food items have less than 4 calories per gram, with various other food items going up to about 7 calories per gram. Note that one group of food items has a distinctly higher ratio: about 9 calories per gram. The food items in this group include lard (9.02 calories per gram), corn oil (8.93 calories per gram), and vegetable shortening (8.83 calories per gram). ■

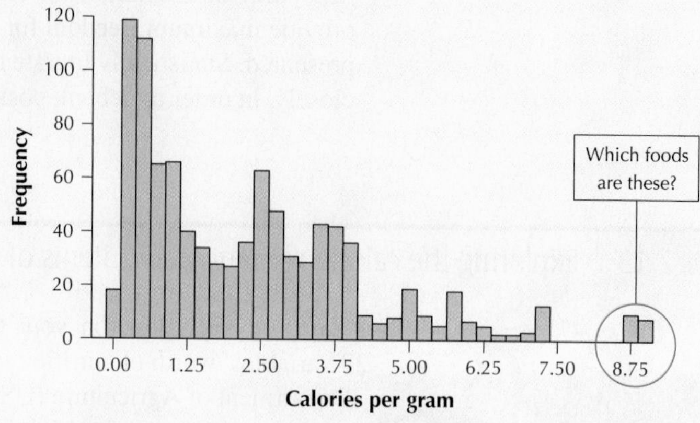


FIGURE 2.18 Histogram of number of calories per gram for 961 food items.

Do the high-calorie food items have anything in common? As it turns out, we have uncovered a well-defined group of food items that the USDA terms "fats and oils." In fact, the food item with the highest ratio of calories to weight in grams is lard. Small wonder, then, that the USDA advises us to consume fats and oils only sparingly in our daily food regimen. Fats and oils, in fact, reside at the top of the so-called food pyramid, as shown in Figure 2.19.

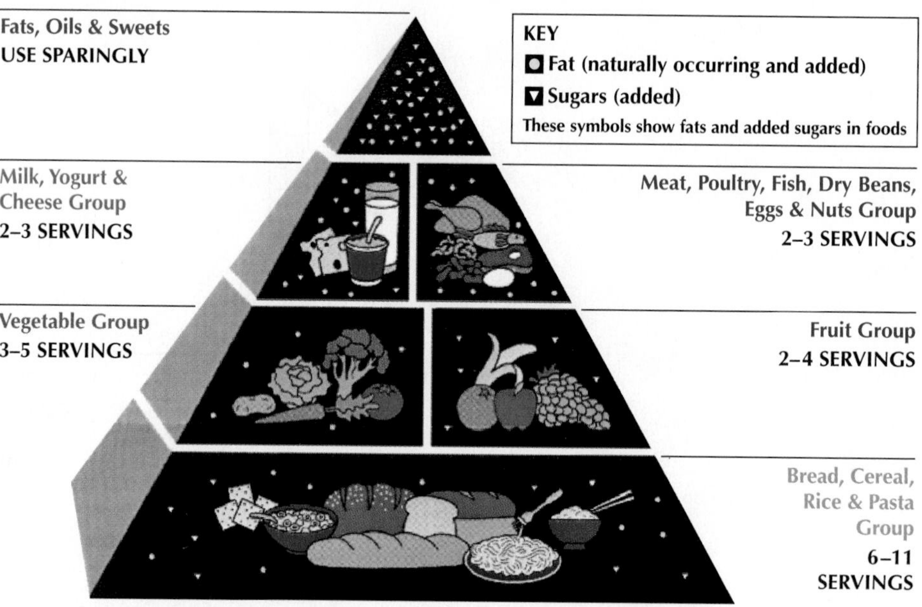


FIGURE 2.19 Food pyramid.
*© USDA and the U.S. Department of Health and Human Services*

What is remarkable here is that, in our exploration of this data set, we have uncovered this well-defined group of foods, "fats and oils," by examining some histograms, checking out some carrot cake, and having our interest piqued. This is an example of *exploratory data analysis*.

## STEP-BY-STEP TECHNOLOGY GUIDE: Quantitative Data

Suppose we would like to produce a histogram of the management aptitude test scores from Example 2.8 (page 49).

### TI-83/84

#### Entering a Data Set

***Step 1*** Press **STAT**, then press **ENTER**. Highlight the **L1** list.
***Step 2*** Clear out any old data in **L1**. Press the **up arrow** key, then **CLEAR**, then **ENTER**.
***Step 3*** Enter the first data value **77** and press **ENTER**.
***Step 4*** Continue entering data until the entire data set is in **L1** (Figure 2.20).

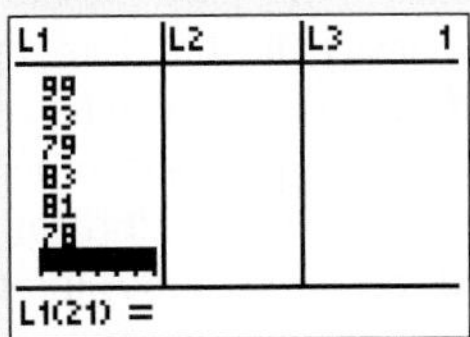


Figure 2.20 All data entered.

#### Constructing a Histogram

***Step 1*** Press **2nd**, then **Y=**. In the STAT PLOTS menu, select **1**, and press **ENTER**.
***Step 2*** Select **ON**, and press **ENTER**. Select the histogram icon (Figure 2.21), and press **ENTER**.
***Step 3*** Press **WINDOW**. Set the following values:

**a.** **Xmin:** lower class limit of leftmost class (75)
**b.** **Xmax:** upper class limit of rightmost class (100)
**c.** **Xscl:** class width (5)
**d.** **Ymin:** 0
**e.** **Ymax:** Some value slightly larger than the highest expected frequency (10).

***Step 4*** Press **GRAPH** (shows histogram without any details).
***Step 5*** Press **TRACE**. Selecting each class in turn provides class limits and class frequency. The histogram is given in Figure 2.22.

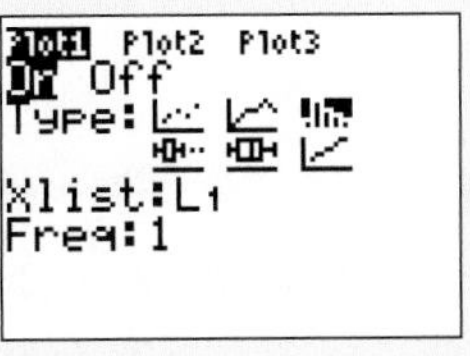


Figure 2.21 Selecting the histogram icon.

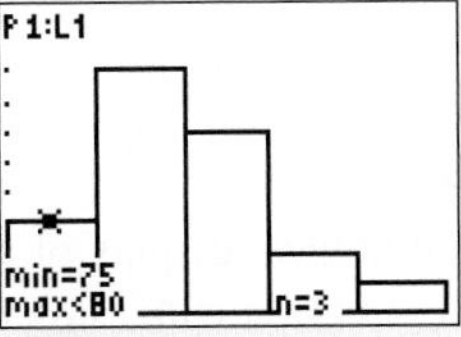


Figure 2.22 Histogram with leftmost class selected.

### EXCEL

#### Constructing a Histogram

Make sure the Data Analysis package has been installed on your version of Excel.
***Step 1*** Click **Data > Data Analysis**.
***Step 2*** Select **Histogram** and click **OK**.
***Step 3*** For the *input range*, select the cells in which the data set resides. Then click **OK**.

### MINITAB

#### Constructing a Histogram

***Step 1*** Enter the management aptitude test scores into column **C1**.
***Step 2*** Click **Graph > Histogram**.
***Step 3*** In the **Graph Variables** section, choose **Simple** and click **OK**. Select **C1** ***Scores***, and click **Select**. Then click **OK**.
***Step 4*** The histogram is shown in Figure 2.23. Note that by default Minitab uses midpoints rather than class limits to define the classes. Double-clicking anywhere on the midpoint values (78, 81, . . .) brings up a dialog box providing a wide range of options for changing the number of classes, class limits, etc.

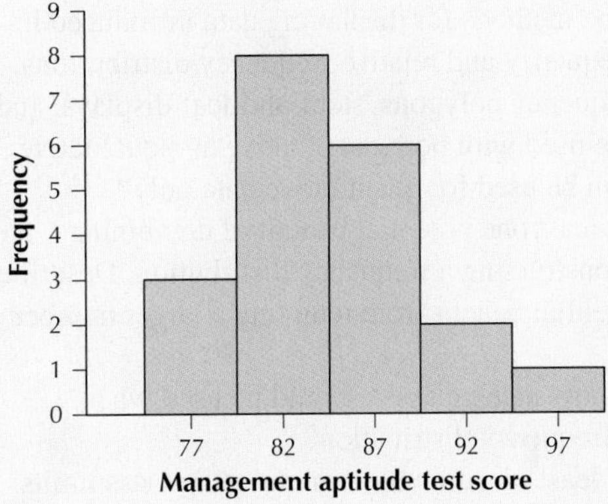


Figure 2.23

*(Continued)*

### Constructing a Stem-and-Leaf Display

***Step 1*** Enter the management aptitude test scores into column **C1**.
***Step 2*** Click **Graph > Stem-and-Leaf**.
***Step 3*** Click inside the space indicated **Variables**, select **C2 Scores**, and click **Select**. Then click **OK**.
***Step 4*** The output shown in Figure 2.24 tells us that the leaf unit is defined to be ones (1.0). Therefore, the stem unit is tens. (Ignore the leftmost column, which simply provides a cumulative count of the data points from the minimum and maximum.) The first row shows 7 7, indicating a single data point, 77. The second row shows 7 89, indicating two data points, 78 and 79.

```
Stem-and-leaf of MAT
Leaf Unit = 1.0
  1   7  7
  3   7  89
  6   8  001
 10   8  2333
 10   8  45
  8   8  677
  5   8  89
  3   9
  3   9  23
  1   9
  1   9
  1   9  9
```

Figure 2.24

## Section 2.2 SUMMARY

In this section, we learned about using graphs and tables for summarizing quantitative (numerical) data.

Quantitative variables can be summarized using frequency and relative frequency distributions.

Histograms are a graphical display of a frequency or a relative frequency distribution with class intervals on the horizontal axis and the frequencies or relative frequencies on the vertical axis. A frequency polygon is constructed as follows: for each class, plot a point at the class midpoint, at a height equal to the frequency for that class; then join each consecutive pair of points with a line segment.

Stem-and-leaf displays contain more information than either a frequency distribution or a histogram, since they retain the original data values in the display. In a dotplot, each data point is represented by a dot above the number line.

An image has symmetry (or is symmetric) if there is a line (axis of symmetry) that splits the image in half so that one side is the mirror image of the other. Nonsymmetric distributions with a long right-hand tail are called right-skewed, while those with a long left-hand tail are called left-skewed.

## Section 2.2 EXERCISES

### CLARIFYING THE CONCEPTS

**1.** Which of the methods for displaying data introduced in this section (frequency and relative frequency distributions, histograms, frequency polygons, stem-and-leaf displays, and dotplots) can be used with both quantitative and qualitative data? Which can be used for quantitative data only?

**2.** Describe at least one potential benefit of combining classes when constructing a frequency distribution. Describe at least one potential benefit from retaining a larger number of classes.

**3.** In general, how many classes should be used when constructing a frequency distribution?

**4.** Describe at least one drawback of choosing class limits that overlap.

**5.** Describe at least one way that a dotplot may be useful.

**6.** In your own words, describe what is meant by "symmetry." Provide an example of a shape that is symmetric and an example of a shape that is not symmetric.

**7.** What are some examples of data sets that are often right-skewed? Left-skewed?

**8.** For a bar graph (not specifically a Pareto chart), does it matter which order the bars are in? What does this mean for the relevance of symmetry and skewness for summaries of categorical data (such as we studied in Section 2.1)?

### PRACTICING THE TECHNIQUES

**9.** A fair die was thrown 100 times, and the values were recorded. The accompanying histogram shows the results.

- **a.** Which value occurred most frequently?
- **b.** Which values occurred least frequently?
- **c.** How often was a 3 observed?
- **d.** What percentage of times was a 3 observed?

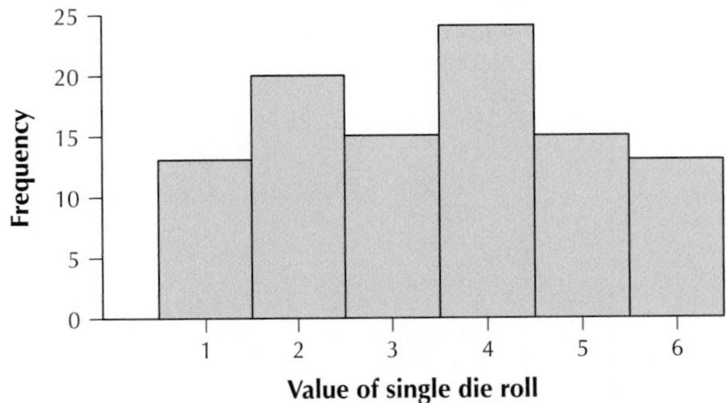


10. A random sample of 1000 police officers was taken, and the number of motor vehicle citations each handed out in a particular week was recorded. The results are shown in the accompanying histogram.

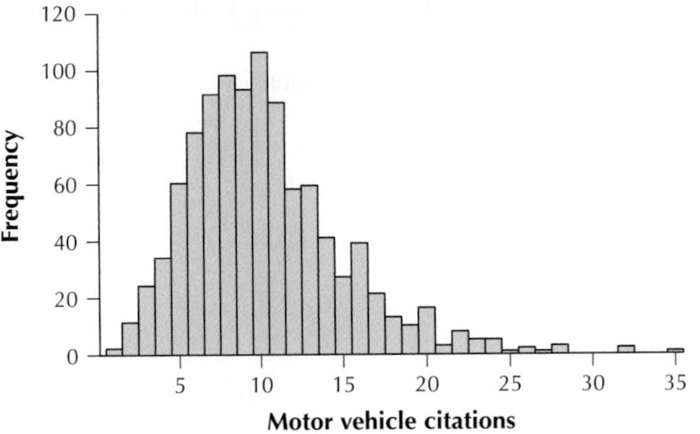


a. What was the highest number of citations issued?
b. What was the lowest number of citations issued?
c. What is the most frequent number of citations issued? About how many police officers issued this many citations?
d. Describe the shape of the distribution.

11. A campus-wide statistics midterm worth 50 points resulted in the scores provided in the histogram below.

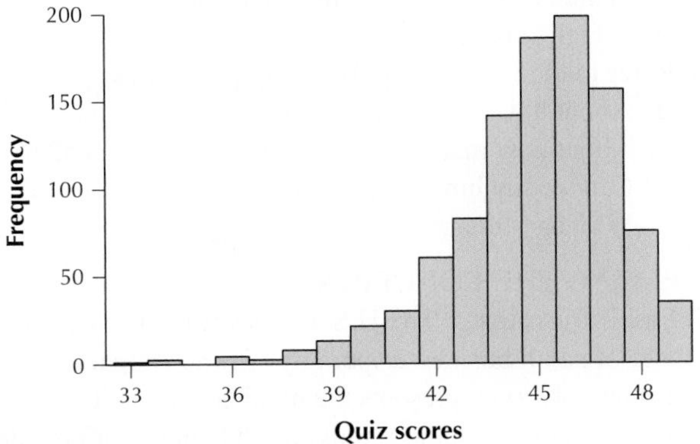


a. Which score occurred with the greatest frequency?
b. Which score occurred with the lowest frequency?
c. What is the highest score? Lowest score?
d. Describe the shape of the distribution.

12. The heights of 1000 12th-grade males are shown in the histogram.

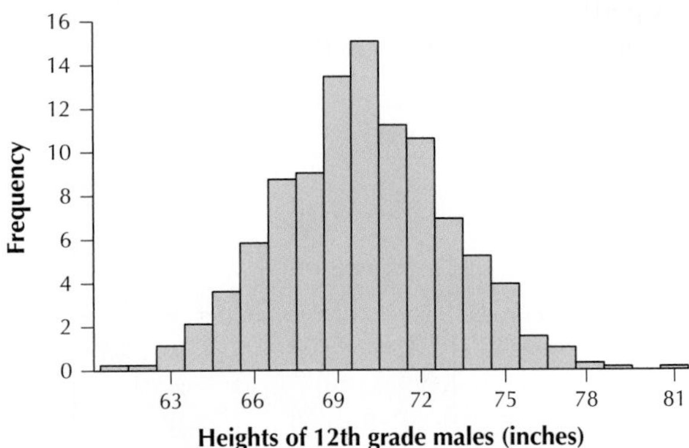


a. How tall is the tallest male? The shortest male?
b. Which height occurs with the greatest frequency?
c. Describe the shape of the distribution.

13. A middle school soccer league held a fundraiser. The number of items sold by the children is shown in the histogram below.

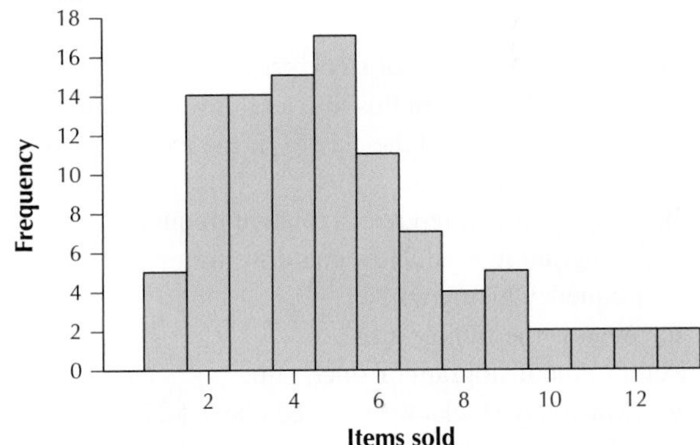


a. Two children sold the most items. How many items was that?
b. What was the smallest number of items sold? How many children sold that many?
c. What number of items sold occurred with the greatest frequency?
d. Describe the shape of the distribution.

14. Refer to the histogram of items sold at the fundraiser.
a. How could we turn this into a relative frequency histogram? Would the classes or the rectangles be affected?
b. Suppose we were given a relative frequency histogram instead. How could we turn it into a frequency histogram?
c. What is the total number of children?

15. Refer to the histogram of items sold at the fundraiser.
a. How many children sold 10 or more items?
b. What proportion of children sold 10 or more items?

**c.** How many children sold more than 10 items?

**d.** What proportion of children sold more than 10 items?

**16.** A portfolio contains stocks of 19 technology firms. The stock prices are shown in the accompanying histogram.

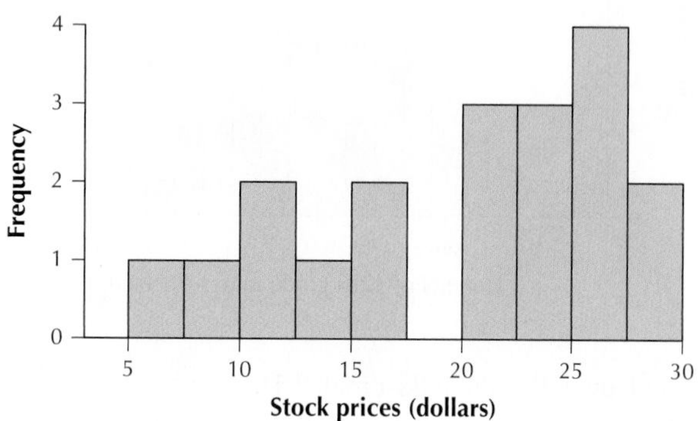


**a.** How many classes are there?

**b.** What is the class width? Is it the same for each class?

**c.** Is this a frequency histogram or a relative frequency histogram?

**17.** Refer to the histogram of stock prices.

**a.** How could we turn this into a relative frequency histogram? Would the classes or the rectangles be affected?

**b.** Suppose we were given a relative frequency histogram instead. How could we turn it into a frequency histogram?

**c.** What is the sample size?

**18.** Refer to the histogram of stock prices.

**a.** How many stocks were priced above $27.50?

**b.** What is the relative frequency of stocks priced above $27.50?

**c.** How many stocks had a price below $15?

**d.** What is the relative frequency of stocks with a price below $15?

**19.** Refer to the histogram of stock prices.

**a.** How many stocks are priced between $17.50 and $20?

**b.** What is the relative frequency of stocks priced below $5?

**c.** Which class has the largest relative frequency? Calculate this relative frequency.

**d.** What is the frequency of stocks priced between $10 and $15?

**e.** How many stocks had a price of $40?

**20.** Would you characterize the shape of the stock prices distribution as (a) tending to be symmetric, (b) tending to be right-skewed, (c) tending to be left-skewed, or (d) too close to call?

**21.** Refer to the accompanying stem-and-leaf display. Reconstruct the data set.

```
Stem-and-leaf of Data  N  = 20
Leaf Unit = 1.0

    2  3
    2  45
    2  67
    2  889
    3  011
    3  2223
    3  5
    3  67
    3  9
    4  0
```

**22.** Refer to the stem-and-leaf display. Construct a relative frequency distribution, using appropriate values for the class width and the lower class limit of the leftmost class.

**23.** Refer to the stem-and-leaf display. Construct a frequency histogram.

**24.** Refer to the stem-and-leaf display. Construct a dotplot.

**25.** The frequency polygon below represents the quiz scores for a course in introductory statistics.

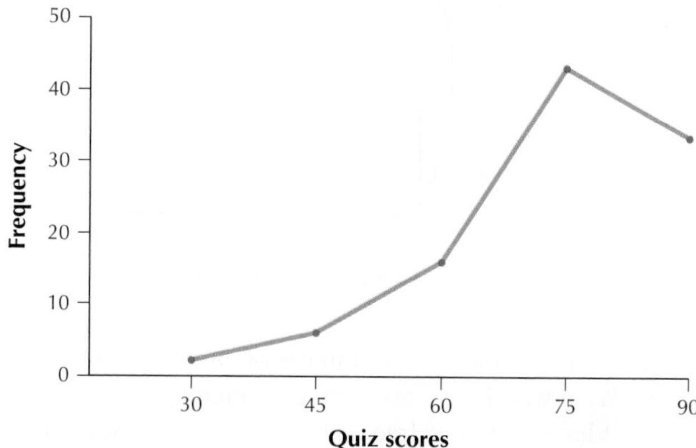


**a.** What is the class width?

**b.** What is the lower class limit of the class that has 45 as its midpoint?

**c.** What is the upper class limit of the class that has 45 as its midpoint?

**d.** Which class has the highest frequency?

**e.** Which class has the lowest frequency?

**26.** Refer to the frequency polygon of quiz scores.

**a.** About how many students scored higher than 82.5?

**b.** About how many students scored lower than 52.5?

**c.** Can we say how many students scored in the 90s? Why or why not?

## APPLYING THE CONCEPTS

**27. Small Businesses.** The U.S. Census Bureau tracks the number of small businesses per city. The accompanying frequency polygon represents the numbers of small businesses per city (in thousands) for 266 cities nationwide.

**a.** What is the class width?

**b.** What is the lower class limit of the leftmost class? (*Hint:* Don't forget about the units.)

**c.** Which class has the highest frequency?

**d.** Which class has the lowest frequency?

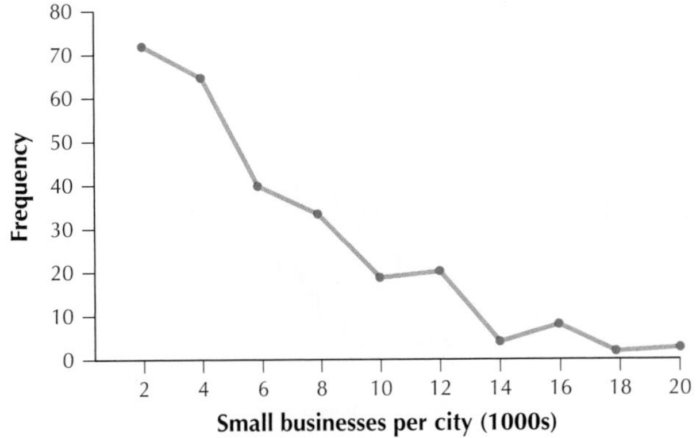


28. Refer to the frequency polygon of small businesses per city.
   a. About how many cities have between 1000 and 3000 small businesses?
   b. About how many cities have more than 19,000 small businesses?
   c. About how many cities have between 9000 and 11,000 small businesses?

29. **Countries and Continents.** Suppose we are interested in analyzing the variable *continent* for the ten countries in Table 2.17. Construct each of the following tabular or graphical summaries. If not appropriate, explain clearly why we can't use that method.
   a. Frequency distribution
   b. Relative frequency distribution
   c. Frequency histogram
   d. Dotplot
   e. Stem-and-leaf display

**TABLE 2.17 Countries and continents**

| Country | Continent |
|---|---|
| Iraq | Asia |
| United States | North America |
| Pakistan | Asia |
| Canada | North America |
| Madagascar | Africa |
| North Korea | Asia |
| Chile | South America |
| Bulgaria | Europe |
| Afghanistan | Asia |
| Iran | Asia |

**Hospitals near Jackson, Mississippi.** Answer Exercises 30–34 using the information in Table 2.18.

30. Construct a relative frequency distribution of the number of hospital beds. Use a class width of 20 beds, with the lower class limit of the leftmost class equal to 0.

31. Construct a frequency histogram and a relative frequency histogram, using the same classes as in the previous exercise.
   a. Which class or classes have the highest frequency? Lowest?
   b. Would you say that the data distribution is left-skewed, right-skewed, or symmetric?
   c. If all we had was the frequency histogram, could we reconstruct the data set?

**TABLE 2.18 Hospitals near Jackson, Mississippi**

| Hospital | Beds | City |
|---|---|---|
| Hardy Wilson Memorial Hospital | 49 | Hazlehurst |
| Humphreys County Memorial Hospital | 34 | Belzoni |
| Jefferson County Hospital | 30 | Fayette |
| Lackey Memorial Hospital | 15 | Forest |
| Leake Memorial Hospital | 25 | Carthage |
| Madison County Medical Center | 67 | Canton |
| Montfort Jones Memorial Hospital | 72 | Kosciusko |
| Rankin Medical Center | 134 | Brandon |
| University of Mississippi Medical Center—Holmes County | 25 | Lexington |

32. Construct a dotplot.

33. Construct a stem-and-leaf display, using ones as leaf units.

34. Compare the information in the stem-and-leaf display with that in the histogram. If all we had was the stem-and-leaf display, could we reconstruct the data set?

**Santa Monica Employers.** Answer Exercises 35–38 using the information in Table 2.19, which lists the number of employees for the ten largest employers in Santa Monica, California.

**TABLE 2.19 Santa Monica employers**

| Company | No. of employees | Company | No. of employees |
|---|---|---|---|
| City of Santa Monica | 1892 | Santa Monica Hospital | 1165 |
| St. John's Health Center | 1755 | Santa Monica College | 1050 |
| Macerich Co. | 1605 | Metro-Goldwyn Mayer | 1050 |
| Fremont General Corp. | 1600 | Rand Corp. | 1038 |
| Entravision Communications Corp. | 1206 | Santa Monica School District | 1008 |

35. Construct a relative frequency distribution of the number of employees. Use class width of 200 employees, with the lower class limit of the leftmost class equal to 900.

36. Construct a frequency histogram, using the same classes as in the previous exercise.

37. Construct a relative frequency histogram. What is the only difference between the frequency and relative frequency histograms?

38. Construct a dotplot.

**Miami Arrests.** Answer Exercises 39–41 using the information in the following table. The table gives the monthly number of arrests made for the year 2005 by the Miami-Dade Police Department.

| | | | | | |
|---|---|---|---|---|---|
| Jan. | 751 | May | 919 | Sept. | 802 |
| Feb. | 650 | June | 800 | Oct. | 636 |
| Mar. | 909 | July | 834 | Nov. | 579 |
| Apr. | 881 | Aug. | 789 | Dec. | 777 |

**39.** Construct a relative frequency distribution of the monthly number of arrests. Use class width of 50 arrests, with the lower class limit of the leftmost class equal to 550.
**40.** Construct a frequency histogram and relative frequency histogram, using the same classes as in the previous exercise. Which class or classes have the highest frequency? Lowest?
**41.** Construct a dotplot.

**American League Shutouts.** Use the information in Table 2.20 to answer Exercises 42–45. The table gives the number of shutouts (SO) pitched by the 14 American League baseball teams in the 2007 season.

**TABLE 2.20 Shutouts pitched in 2007 season**

| Team | SO |
|---|---|
| Boston Red Sox | 13 |
| Seattle Mariners | 12 |
| Toronto Blue Jays | 9 |
| Los Angeles Angels | 9 |
| Detroit Tigers | 9 |
| Baltimore Orioles | 9 |
| Cleveland Indians | 9 |
| Chicago White Sox | 9 |
| Oakland Athletics | 9 |
| Minnesota Twins | 8 |
| Kansas City Royals | 6 |
| Texas Rangers | 6 |
| New York Yankees | 5 |
| Tampa Bay Devil Rays | 2 |

**42.** Construct a stem-and-leaf display.
**43.** Construct a relative frequency distribution, with a class width of 2, and the lower class limit equal to 2. Can we tell, using this relative frequency distribution only, what is the most common number of shutouts? Explain.
**44.** Construct a frequency histogram and relative frequency histogram. Would you characterize the data set as left-skewed, right-skewed, or symmetric?
**45.** Using your choice of tabular or graphical summaries, answer the following questions.

- **a.** What proportion of teams have at least 8 shutouts?
- **b.** What proportion of teams have fewer than 2 shutouts?
- **c.** What proportion of teams have between 12 and 14 shutouts, inclusive?
- **d.** What proportion of teams have at most 10 shutouts?
- **e.** What proportion of teams have more than 10 shutouts?

**New York Townspeople.** Use the following information for Exercises 46–48. For towns in New York State, the following histogram provides information on the percentage of the townspeople who are between 18 and 65 years old (data set **New York**).

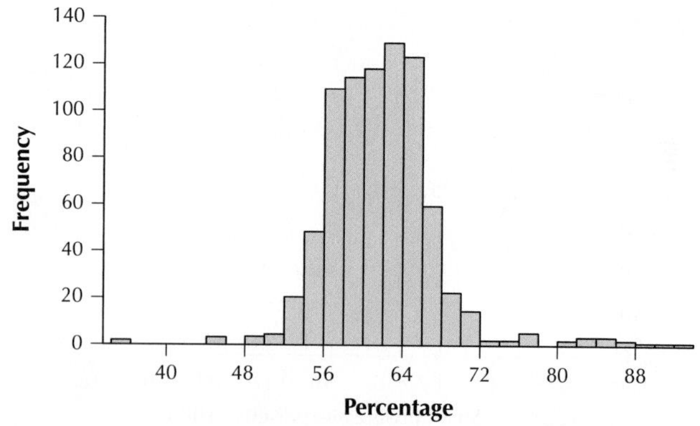


**46.** Would you characterize the distribution as left-skewed, right-skewed, or fairly symmetrical?
**47.** Provide an estimate of the "typical" percentage of townspeople who are between 18 and 65 years old. Is this typical value near the middle or near one of the "tails" of the distribution?
**48.** Provide a rough estimate of the sample size.

**Psychology Exam Data.** Use the following psychology exam data set from Example 2.11 for Exercises 49–52.

| | | | | | | | | | |
|---|---|---|---|---|---|---|---|---|---|
| 75 | 81 | 82 | 70 | 60 | 59 | 94 | 77 | 68 | 98 |
| 86 | 68 | 85 | 72 | 70 | 91 | 78 | 86 | 51 | 67 |

**49.** Without using the computer, construct the following:

- **a.** A frequency distribution
- **b.** A relative frequency distribution
- **c.** A relative frequency histogram

**50.** Without using a computer, construct a dotplot.
**51.** Compare and contrast the relative usefulness of each of three graphical presentation methods—dotplot, histogram, and stem-and-leaf display—if our primary objective is

- **a.** to assess symmetry and skewness.
- **b.** to be able to construct it quickly using paper and pencil.
- **c.** to give a presentation to people who have never had a stats course before.
- **d.** to be given maximum leeway in choosing how the data are to be presented.

WHAT IF? **52.** *What if* we subtract the same amount (say, 10) from each psychology exam score. Explain how this would affect the following. What would change? What would stay the same?

- **a.** Relative frequency histogram
- **b.** Dotplot
- **c.** Stem-and-leaf display

**Analysis of Households.** For Exercises 53–58, use your knowledge of Excel or Minitab. Open the data set **Household.**
**53.** How many observations are in this data set? How many variables? Which of the variables are qualitative? Which of the variables are quantitative?

**54.** The variable *tot_hhld* is the total number of households in each state and the District of Columbia. Construct a histogram of *tot_hhld*. Comment on the symmetry or the skewness of the histogram. What is the average number of households?
**55.** Is there one state whose total number of households is unusually large? Which state is this? How did you find out?
**56.** The variable *ave_size* gives the average household size for the state. Construct a histogram of *ave_size*. Comment on the symmetry or skewness of the histogram.
**57.** Is there one observation where the average household size is unusually small? Which observation is this? Which two states have unusually large average household sizes? How did you find out?
**58.** Try constructing a histogram of the variable *state*. Is there a problem? Why won't the software let you construct a histogram of *state*?

**Fats and Cholesterol.** For Exercises 59–63, use your knowledge of Excel or Minitab. Open the **Nutrition** data set.
**59.** How many observations are there in the data set? How many variables?
**60.** The variable *fat* contains the fat content in grams for each food. Construct a histogram of *fat*. Comment on the symmetry or the skewness of the histogram.
**61.** Is there a particular type of food whose fat content is particularly large? Which type of food item is this (actually, a set of similar food items)?
**62.** The variable *cholesterol* contains the cholesterol content in milligrams for each food. Construct a histogram of *cholesterol*. Comment on the symmetry or the skewness of the histogram.
**63.** Which food item is highest in cholesterol?

APPLET Use the *One Variable Statistics and Graphs* applet for Exercises 64–69. Work with the **Earthquakes** data set, which shows the magnitude on the Richter scale of 57 earthquakes that occurred during the week of October 15–22, 2007.
**64.** Click on the **Histogram** tab.
- **a.** How many classes are there in the histogram?
- **b.** What is the class width?

**65.** Click on the leftmost rectangle in the histogram.
- **a.** What is the frequency for this class?
- **b.** What are the lower and upper class limits?

**66.** Click on the number line and drag slowly all the way to the left.
- **a.** What happens to the number of classes as you drag to the left?
- **b.** What happens to the class widths as you drag to the left?

**67.** Click on the number line and drag slowly all the way to the right.
- **a.** What happens to the number of classes as you drag to the right?
- **b.** What happens to the class widths as you drag to the right?

**68.** Click on the **Stem-and-Leaf** tab.
- **a.** How many stems are there?
- **b.** Without counting, state how many leaves there are. How do we know this?

**69.** Select **Split Stems**.
- **a.** Now how many stems are there?
- **b.** How many leaves are there?
- **c.** Which stem-and-leaf display is preferable for the **Earthquakes** data, regular or split stems?

### CONSTRUCT YOUR OWN DATA SETS

**70.** Construct your own right-skewed data set of about 20 values. Just make up the data points, but be sure you know what the data represent (income, housing costs, etc.).
- **a.** Construct a stem-and-leaf display of your data set.
- **b.** Construct a dotplot of your data set.

**71.** Construct your own symmetric data set of about 20 values. Just make up the data points, but be sure you know what the data represent (for example, runs in a baseball game, number of right answers on a quiz).
- **a.** Construct a stem-and-leaf display of your data set.
- **b.** Construct a dotplot of your data set.

## 2.3 Further Graphs and Tables for Quantitative Data

**Objectives:** By the end of this section, I will be able to…

1. Build cumulative frequency distributions and cumulative relative frequency distributions.
2. Create frequency ogives and relative frequency ogives.
3. Construct and interpret time series graphs.

## 1 Cumulative Frequency Distributions and Cumulative Relative Frequency Distributions

Since quantitative data can be put in ascending order, we can keep track of the accumulated counts at or below a certain value using a **cumulative frequency distribution** or **cumulative relative frequency distribution.** For example, if we list the prices of homes for sale in a neighborhood, a cumulative frequency distribution tells us how many homes are priced at $300,000 or less.

For a discrete variable, a **cumulative frequency distribution** shows the total number of observations *less than or equal to* the category value. For a continuous variable, a **cumulative frequency distribution** shows the total number of observations *less than or equal to* the upper class limit.

A **cumulative relative frequency distribution** shows the proportion of observations less than or equal to the category value (for a discrete variable) or the proportion of observations less than or equal to the upper class limit (for a continuous variable).

### Example 2.14 Constructing cumulative frequency and relative frequency distributions

The first three columns in Table 2.21 contain the frequency distribution and relative frequency distribution for the total 2007 attendance for 25 Major League Baseball teams. Construct a cumulative frequency distribution and a cumulative relative frequency distribution for the attendance figures.

**Solution**

To find the cumulative frequency for a class, add the frequencies of the classes equal to or below the upper class limit of that class. For example, the cumulative frequency for the class 2.70–3.09 is the sum of the frequency for this class and the frequencies for the classes 1.90–2.29 and 2.30–2.69. The procedure for the cumulative relative frequencies is similar. The results are shown in the last two columns of Table 2.21, where we can see that more than two-thirds (0.68) of these teams had attendance of 3.09 million or less. ■

Table 2.21 Cumulative frequency distribution and cumulative relative frequency distribution

| Attendance (millions) | Frequency | Relative frequency | Cumulative frequency | Cumulative relative frequency |
|---|---|---|---|---|
| 1.90–2.29 | 5 | 0.20 | 5 | 0.20 |
| 2.30–2.69 | 6 | 0.24 | $5 + 6 = 11$ | $0.20 + 0.24 = 0.44$ |
| 2.70–3.09 | 6 | 0.24 | $5 + 6 + 6 = 17$ | $0.44 + 0.24 = 0.68$ |
| 3.10–3.49 | 4 | 0.16 | $5 + 6 + 6 + 4 = 21$ | $0.68 + 0.16 = 0.84$ |
| 3.50–3.89 | 3 | 0.12 | $5 + 6 + 6 + 4 + 3 = 24$ | $0.84 + 0.12 = 0.96$ |
| 3.90–4.29 | 1 | 0.04 | $5 + 6 + 6 + 4 + 3 + 1 = 25$ | $0.96 + 0.04 = 1.00$ |
| Total | 25 | 1.00 | | |

## 2 Ogives

Just as histograms and frequency polygons are the graphical equivalent of frequency distributions, we have the following graphical equivalent of a cumulative frequency distribution.

An **ogive** (pronounced "oh jive") is the graphical equivalent of a cumulative frequency distribution or a cumulative relative frequency distribution. Like a frequency polygon, an ogive consists of a set of plotted points connected by line segments. The *x* coordinates of these points are the upper class limits; the *y* coordinates are the cumulative frequencies or cumulative relative frequencies.

## Example 2.15 Constructing an ogive

Construct a relative frequency **ogive** for the attendance data in Table 2.21.

### Solution

For the *x* coordinates, we use the upper class limits for attendance, and for the *y* coordinates we use the cumulative relative frequencies. The result is shown in Figure 2.25.

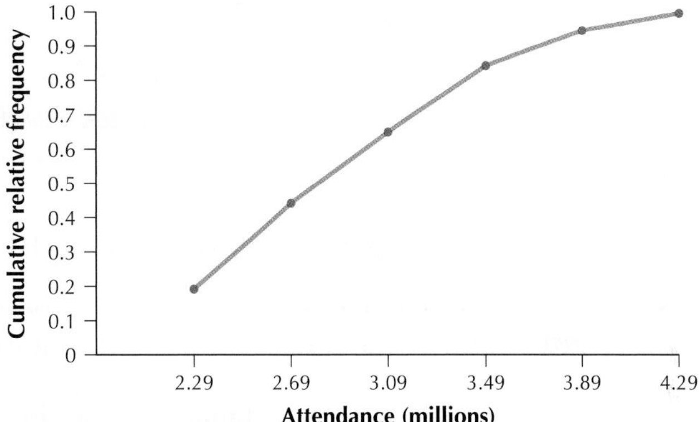


FIGURE 2.25 Ogive for baseball attendance.

## 3 Time Series Graphs

Data analysts are often interested in how the value of a variable changes over time. Data that are analyzed with respect to time are called *time series data.*

A graph of time series data is called a **time series plot.** The horizontal axis of a time series plot represents time (for example, hours, days, months, years). The values of the time series data are plotted on the vertical axis, and line segments are drawn to connect the points.

## Example 2.16 Constructing a time series plot

Table 2.22 contains the amount of carbon dioxide in parts per million (ppm) found in the atmosphere above Mauna Loa in Hawaii, measured monthly from October 2006 to September 2007. Construct a **time series plot** of these data.

Table 2.22 Atmospheric carbon dioxide at Mauna Loa, October 2006 to September 2007

| Month | Carbon dioxide (ppm) | Month | Carbon dioxide (ppm) |
|---|---|---|---|
| Oct. | 379.03 | Apr. | 386.37 |
| Nov. | 380.17 | May | 386.54 |
| Dec. | 381.85 | June | 385.98 |
| Jan. | 382.94 | July | 384.35 |
| Feb. | 383.86 | Aug. | 381.85 |
| Mar. | 384.49 | Sept. | 380.58 |

*Source:* Dr. Pieter Tans, Earth System Research Laboratory, National Oceanic and Atmospheric Administration, www.esrl.noaa.gov/gmd/ccgg/trends.

### Solution

We indicate the twelve months October through September on the horizontal axis of the time series plot (Figure 2.26). Then, for each month, we plot the amount of carbon dioxide. Finally, we join the points using line segments. Note that the carbon dioxide level increases from the fall through the winter and peaks in the spring. It then decreases through the summer. In the Step-by-Step Technology Guide, we illustrate how to construct this time series graph using technology. ■

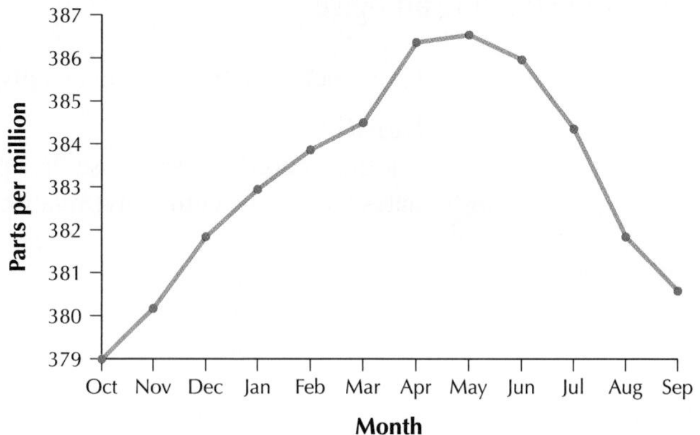


FIGURE 2.26 Time series plot. Carbon dioxide levels at Mauna Loa, Hawaii.

## Example 2.17 Constructing a time series plot using technology

The data set **Mauna Loa** contains the carbon dioxide levels at Mauna Loa from September 1999 to September 2007. Use technology to construct a time series plot of the data.

### Solution

We use the instructions provided in the Step-by-Step Technology Guide at the end of this section. The resulting time series plot is shown in Figure 2.27. (The year on the horizontal axis indicates September of each year. For example "1999" refers to September 1999.) We observe both a seasonal pattern and a long-term trend. Every autumn and winter, the carbon dioxide level increases, and every summer it decreases. In autumn and winter, leaves and other deciduous vegetation decays, releasing its store of carbon back into the atmosphere. In the spring and summer, the new year's leaves require carbon to grow and extract it from the atmosphere, thereby reducing the atmosphere's carbon dioxide level. Thus, the Earth "inhales" carbon each summer and "exhales" it each winter. However, the carbon dioxide level of each successive September does not quite reach the low level of the previous September. This leads to an overall increasing trend in the amount of carbon dioxide in the atmosphere as we move from 1999 to 2007. ■

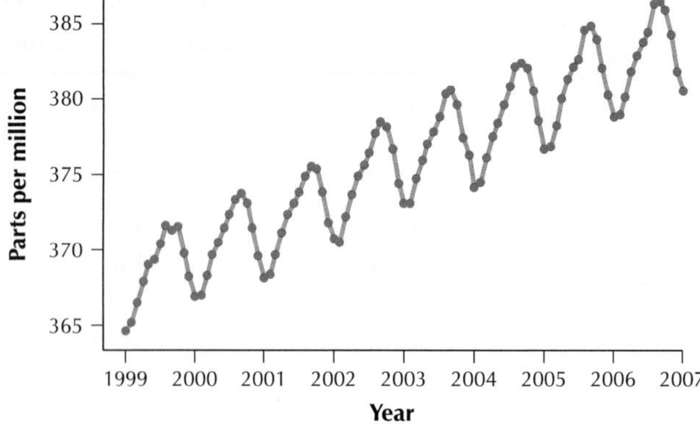


FIGURE 2.27 Watching the Earth breathe. Carbon dioxide levels at Mauna Loa, Hawaii.

## STEP-BY-STEP TECHNOLOGY GUIDE: Time Series Plots

We illustrate how to construct a time series plot using Example 2.16 (page 67).

### TI-83/84

***Step 1*** Enter your time index (integers 1, 2, . . .) into list **L1**.
***Step 2*** Enter the values of your time series variable into list **L2**.
***Step 3*** Press **2nd**, then **Y=**. In the STAT PLOTS menu, select **1**, and press **ENTER**.
***Step 4*** Select **ON**, and press **ENTER**. Select the time series icon (Figure 2.28), and press **ENTER**.
***Step 5*** Press **WINDOW**. Set the following values:

a. **Xmin: 0**
b. **Xmax: 13**
c. **Xscl: 1**
d. **Ymin: 375**
e. **Ymax: 390**

***Step 6*** Press **ZOOM > 9:ZOOMSTAT** and press **ENTER**. The time series plot is shown in Figure 2.29.

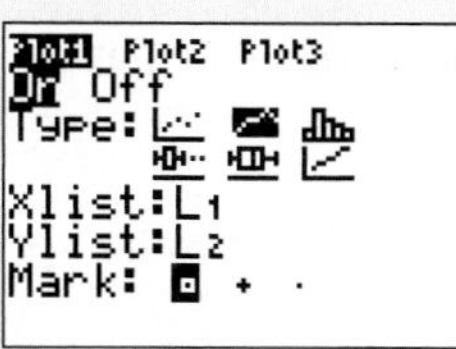

Figure 2.28 Selecting the time series icon.

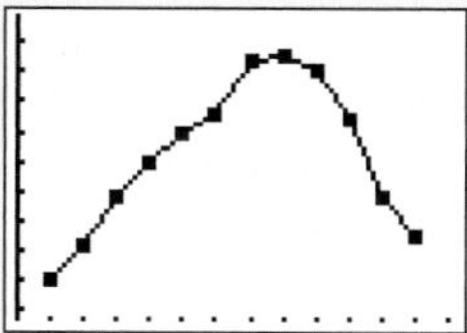
Figure 2.29 TI-83/84 time series plot.

### EXCEL

***Step 1*** Enter the month data into column **A** (see Figure 2.30).
***Step 2*** Enter the values of your time series variable into column **B** (see Figure 2.30).
***Step 3*** Select cells A1–B12 and click **Insert > Line** (in the **Chart** section).
***Step 4*** Choose the type labeled "Line with markers."

| | A | B |
|---|---|---|
| 1 | Oct | 379.03 |
| 2 | Nov | 380.17 |
| 3 | Dec | 381.85 |
| 4 | Jan | 382.94 |
| 5 | Feb | 383.86 |
| 6 | Mar | 384.49 |
| 7 | Apr | 386.37 |
| 8 | May | 386.54 |
| 9 | Jun | 385.98 |
| 10 | Jul | 384.35 |
| 11 | Aug | 381.85 |
| 12 | Sep | 380.58 |

Figure 2.30

### MINITAB

***Step 1*** Enter the values of your time series variable into column **C1**.
***Step 2*** Click **Graph > Time Series Plot . . .**
***Step 3*** Select **Simple** and click **OK**.
***Step 4*** For **Series**, double-click on **C1**.
***Step 5*** Click **Time/Scale**. Select **Calendar > Month**.
***Step 6*** For **Start value**, enter 10 (for October). For **Increment**, enter **1**.
***Step 7*** Click **OK** and **OK**.

## Section 2.3 SUMMARY

1 A cumulative frequency distribution shows the total number of observations less than or equal to the category value (for a discrete variable) or the upper class limit (for a continuous variable). A cumulative relative frequency distribution shows the proportion of observations less than or equal to the category value (for a discrete variable) or the upper class limit (for a continuous variable).

2 An ogive is the graphical equivalent of a cumulative frequency distribution or a cumulative relative frequency distribution. The $x$ coordinates of the points are the upper class limits; the $y$ coordinates are the cumulative frequencies or cumulative relative frequencies.

3 Data that are analyzed with respect to time are called time series data. A graph of time series data is called a time series plot. The horizontal axis of a time series plot represents time (for example, hours, days, months, years). The values of the time series data are plotted on the vertical axis, and line segments are drawn to connect the points.

## Section 2.3 EXERCISES

### CLARIFYING THE CONCEPTS

1. Explain the difference between a frequency distribution and a cumulative frequency distribution.
2. Explain the difference between a cumulative frequency distribution and a cumulative relative frequency distribution.
3. What is the graphical equivalent of a cumulative frequency distribution?
4. Explain how to construct an ogive.
5. What do we call data that are analyzed with respect to time?
6. Explain how to construct a time series plot.

### PRACTICING THE TECHNIQUES

For Exercises 7–9, use the histograms for the indicated exercises from Section 2.2 to

- **a.** construct a frequency distribution.
- **b.** construct a cumulative frequency distribution.
- **c.** construct a cumulative relative frequency distribution.

7. Exercise 9.
8. Exercise 13.
9. Exercise 16.

For Exercises 10–12, use the frequency distributions for the indicated exercises from Section 2.2 to

- **a.** construct a frequency ogive.
- **b.** construct a relative frequency ogive.

10. Exercise 35.
11. Exercise 29.
12. Exercise 49.
13. The frequency ogive below represents the unemployment rate (in percentages) for 367 cities nationwide.[3]

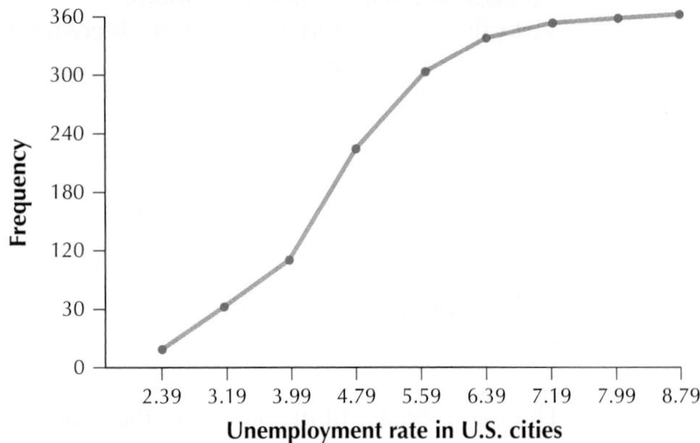


- **a.** What is the class width?
- **b.** What is the upper class limit of the leftmost class?
- **c.** What is the class midpoint of the leftmost class?

14. Refer to the frequency ogive of unemployment rates.
- **a.** About how many cities have unemployment rates 3.99 and below?
- **b.** About how many cities have unemployment rates 5.59 and below?
- **c.** About how many cities have unemployment rates 5.6 and above?

15. Describe what changes you would make to convert the frequency ogive of unemployment rates to a relative frequency ogive of unemployment rates. Would the points and line segments change at all?
16. The relative frequency ogive below represents the gasoline consumption per day (in millions of gallons) in August 2007 for the fifty states.[4]

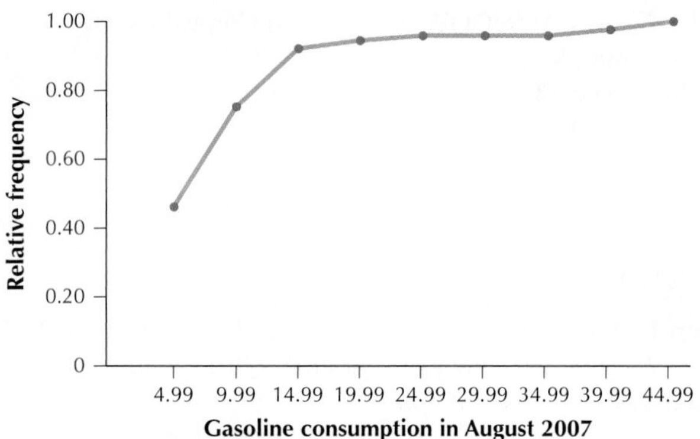


- **a.** What is the class width?
- **b.** What is the upper class limit of the rightmost class?
- **c.** What is the class midpoint of the rightmost class?

17. Refer to the relative frequency ogive of gasoline consumption.
- **a.** About what proportion of states consumed at most 4.99 million gallons per day?
- **b.** About what proportion of states consumed at most 9.99 million gallons per day?
- **c.** Based on your answers to **(a)** and **(b)**, about what proportion of states consumed between 5 million and 9.99 million gallons per day?

18. The following graph is a time series plot of the number of HIV/AIDS cases in children under 13 in the United States from 1992 to 2004.[5]

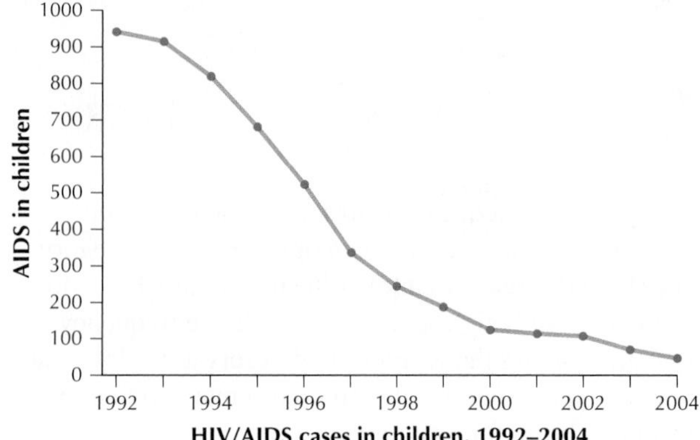


- **a.** About how many children had HIV/AIDS in 1992?
- **b.** About how many children had HIV/AIDS in 2004?

**c.** Between which two years do we see the steepest decline in the number of cases?

**d.** Overall, does this graph represent encouraging or discouraging news?

## APPLYING THE CONCEPTS

**Agricultural Exports.** For Exercises 19–21, refer to Table 2.23. The table gives the value of agricultural exports (in billions of dollars) from the top 20 U.S. states in 2006.

**TABLE 2.23 Agricultural exports (in billions of dollars)**

| State | Exports | State | Exports |
|---|---|---|---|
| California | 10.5 | Arkansas | 1.9 |
| Iowa | 4.2 | North Dakota | 1.9 |
| Texas | 3.8 | Ohio | 1.7 |
| Illinois | 3.8 | Florida | 1.7 |
| Nebraska | 3.3 | Wisconsin | 1.5 |
| Kansas | 3.2 | Missouri | 1.5 |
| Minnesota | 3.0 | Georgia | 1.4 |
| Washington | 2.2 | Pennsylvania | 1.4 |
| North Carolina | 2.1 | Michigan | 1.2 |
| Indiana | 2.0 | South Dakota | 1.2 |

*Source*: U.S. Department of Agriculture.

**19.** Construct a cumulative frequency distribution of agricultural exports. Start at $0 and use class widths of $1.5 billion.

**a.** Which class has the highest frequency? How many states belong to this class?

**b.** Of the classes that have nonzero frequency, which class has the lowest frequency? Which state does this represent?

**c.** List the states that belong to the leftmost class.

**20.** Construct a cumulative relative frequency distribution of agricultural exports. Start at $0 and use class widths of $1.5 billion.

**a.** What proportion of states belong to the class with the highest frequency?

**b.** What proportion of states have agricultural exports between $4.5 billion and $8.9 billion?

**21.** Use your cumulative relative frequency distribution to construct a relative frequency ogive of agricultural exports.

**Earthquake Magnitudes.** The U.S. Geological Survey tracks the occurrence of earthquakes around the world. The following data are the magnitudes of 57 earthquakes that occurred during the week of October 15–22, 2007, and registered 4.0 or higher on the Richter scale. For Exercises 22–25, use the following classes: 3.7–4.09, 4.1–4.49, 4.5–4.89, 4.9–5.29, 5.3–5.69, 5.7–6.09, 6.1–6.49, 6.5–6.89.

| | | |
|---|---|---|
| 5.1 | 4.5 | 4.8 |
| 6.2 | 4.5 | 5.1 |
| 5.0 | 4.6 | 5.3 |
| 4.8 | 4.8 | 4.9 |
| 4.7 | 4.6 | 4.9 |
| 4.9 | 4.9 | 5.2 |
| 5.0 | 5.1 | 4.5 |
| 5.0 | 5.2 | 6.1 |
| 5.2 | 4.6 | 5.5 |
| 5.2 | 4.8 | 4.5 |
| 5.8 | 5.0 | 6.6 |
| 5.1 | 5.0 | 4.3 |
| 5.1 | 4.6 | 4.8 |
| 4.7 | 4.6 | 4.8 |
| 5.3 | 4.9 | 4.6 |
| 4.6 | 4.7 | 5.2 |
| 4.0 | 5.0 | 4.9 |
| 5.5 | 4.6 | 5.1 |
| 5.4 | 5.5 | 4.2 |

**22.** Construct a cumulative frequency distribution of the earthquake data.

**a.** How many earthquakes were of magnitude 4.49 or lower?

**b.** How many earthquakes were of magnitude 4.5 or higher?

**23.** Construct a cumulative relative frequency distribution of the earthquake data.

**a.** What proportion of earthquakes were of magnitude 4.49 or lower?

**b.** What proportion of earthquakes were of magnitude 4.5 or higher?

**24.** Construct a frequency ogive of the earthquake data.

**a.** How many earthquakes were of magnitude 5.29 or lower?

**b.** How many earthquakes were of magnitude 5.3 or higher?

**25.** Construct a relative frequency ogive of the earthquake data.

**a.** What proportion of earthquakes were of magnitude 5.29 or lower?

**b.** What proportion of earthquakes were of magnitude 5.3 or higher?

**26. Interest Rates.** The following data represent the prime lending rate of interest, as reported by the Federal Reserve, every six months from January 2003 to July 2007.

| | |
|---|---|
| Jan. 2003 | 4.25 |
| July 2003 | 4.00 |
| Jan. 2004 | 4.00 |
| July 2004 | 4.25 |
| Jan. 2005 | 5.25 |
| July 2005 | 6.25 |
| Jan. 2006 | 7.26 |
| July 2006 | 8.25 |
| Jan. 2007 | 8.25 |
| July 2007 | 8.25 |

**a.** Construct a time series plot of the prime lending rate of interest.

**b.** What trend do you see?

**27. Rainfall in Fort Lauderdale.** The following data represent the total monthly rainfall (in inches) in 2005 in Fort Lauderdale, Florida, as reported by the U.S. Historical Climatology Network.

| | |
|---|---|
| Jan. | 1.65 |
| Feb. | 0.52 |
| Mar. | 4.84 |
| Apr. | 1.64 |
| May | 3.94 |
| June | 15.64 |
| July | 7.13 |
| Aug. | 8.67 |
| Sept. | 10.40 |
| Oct. | 10.36 |
| Nov. | 5.20 |
| Dec. | 2.72 |

**a.** Construct a time series plot of the data.

**b.** Is it wetter in summer or winter in Fort Lauderdale?

WHAT IF? **28.** In Exercise 27, *what if* we add 3 inches to each month's rainfall amount. Describe how this would affect the time series plot. What would change? What would stay the same?

**29. Cigarette Use Among 12th-Graders.** Table 2.24 presents the percentages of 12th-graders who smoke cigarettes, for the years 1975–2004.

**a.** Construct a time series plot of the data.

**b.** Describe any trends that you see.

**30. Trade Deficit.** Table 2.25 presents the annual trade deficits (imports minus exports) for the United States from 1991 to 2005, in billions of dollars.

**a.** Construct a time series plot of the data.

**b.** Describe any trends that you see.

**TABLE 2.24 12th-graders who smoke**

| Year | Percent | Year | Percent |
|---|---|---|---|
| 1975 | 36.7 | 1990 | 29.4 |
| 1976 | 38.8 | 1991 | 28.3 |
| 1977 | 38.4 | 1992 | 27.8 |
| 1978 | 36.7 | 1993 | 29.9 |
| 1979 | 34.4 | 1994 | 31.2 |
| 1980 | 30.5 | 1995 | 33.5 |
| 1981 | 29.4 | 1996 | 34.0 |
| 1982 | 30.0 | 1997 | 36.5 |
| 1983 | 30.3 | 1998 | 35.1 |
| 1984 | 29.3 | 1999 | 34.6 |
| 1985 | 30.1 | 2000 | 31.4 |
| 1986 | 29.6 | 2001 | 29.5 |
| 1987 | 29.4 | 2002 | 26.7 |
| 1988 | 28.7 | 2003 | 24.4 |
| 1989 | 28.6 | 2004 | 25.0 |

*Source:* Monitoring the Future Study, University of Michigan.

**TABLE 2.25 U.S. trade deficits**

| Year | Trade deficit in $ billions |
|---|---|
| 1991 | 7,767 |
| 1992 | 9,796 |
| 1993 | 17,641 |
| 1994 | 24,460 |
| 1995 | 23,355 |
| 1996 | 25,614 |
| 1997 | 27,380 |
| 1998 | 43,387 |
| 1999 | 69,250 |
| 2000 | 96,605 |
| 2001 | 93,988 |
| 2002 | 109,975 |
| 2003 | 125,734 |
| 2004 | 150,519 |
| 2005 | 168,650 |

*Source:* U.S. Census Bureau.

## 2.4 Graphical Misrepresentations of Data

### Objective: By the end of this section, I will be able to…

1 Understand what can make a graph misleading, confusing, or deceptive.

In the Information Age, when our world is awash in data, it is important for citizens to understand how graphics may be made misleading, confusing, or deceptive. Such an understanding enhances our statistical literacy and makes us less prone to being deceived by misleading graphics.

**Eight Common Methods for Making a Graph Misleading**

1. Graphing/selecting an inappropriate statistic.
2. Omitting the zero on the relevant scale.
3. Manipulating the scale.
4. Using two dimensions (area) to emphasize a one-dimensional difference.
5. Careless combination of categories in a bar graph.
6. Inaccuracy in relative lengths of bars in a bar graph.
7. Biased distortion or embellishment.
8. Unclear labeling.

## Example 2.18 Inappropriate choice of statistic

The United Nations Office on Drugs and Crime reports the statistics, given in Table 2.26, on the top 5 nations in the world ranked by numbers of cars stolen in 2000. The car thieves seem to be preying on cars in the United States, which has endured nearly as many cars stolen as the next four highest countries put together. (See also the bar graph in Figure 2.31.) However, the United States has a much greater population than these other countries. Is it possible that, *per capita* (per person), the car theft rate in the United States is not so bad?

Table 2.26 Top five nations for total number of cars stolen in 2000

| Country | Cars stolen |
|---|---|
| 1. United States | 1,147,300 |
| 2. United Kingdom | 338,796 |
| 3. Japan | 309,638 |
| 4. France | 301,539 |
| 5. Italy | 243,890 |

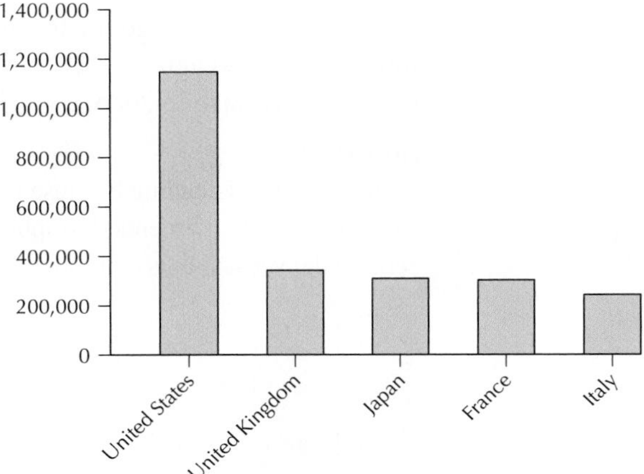


FIGURE 2.31 Bar graph of the top five nations for number of cars stolen in 2000.

### Solution

In this case, the total number of cars stolen is an *inappropriate statistic* since the population of the United States is greater than the populations of the other countries. To find the per capita car theft rate, divide the number of cars stolen in a country by that country's population. The resulting list in Table 2.27 of the top 5 countries for per capita car theft contains a few surprises. Note that the United States has disappeared from the list. It is found in ninth place, with 0.00409 car thefts per capita. Also, let's face it, Denmark and Norway are not exactly hotbeds of crime, are they? Do Danes and Norwegians individually have a higher risk of having their cars stolen than Americans? Most likely not. Another variable not factored in here is the number of cars in the country, which we would need for such an assessment of risk. ■

Table 2.27 Top five nations for total number of cars stolen per capita in 2000

| Country | Cars stolen per capita |
|---|---|
| 1. Australia | 0.00712 |
| 2. Denmark | 0.00600 |
| 3. United Kingdom | 0.00567 |
| 4. New Zealand | 0.00563 |
| 5. Norway | 0.00516 |

### Choose the Appropriate Statistic

The bottom line is that *we need to be careful how we use statistics*. Put in an extreme form, "Figures don't lie, but liars figure." One table of statistics tells us the car theft epidemic is striking the United States with special vehemence. The other table asserts the contrary. An American insurance company looking to increase car insurance rates could point to the first table to support its rate request. A citizens group opposing the request could cite the second table. Which table of statistics is true? They both are! We need to be careful how we phrase our research questions and how we choose the type of statistical evidence to bring to bear on the research question. ■

## Example 2.19 Omitting the zero

MediaMatters.com reported that CNN.com used a misleading graph, reproduced here as Figure 2.32, to exaggerate the difference between the percentages of Democrats and Republicans who agreed with the Florida court's decision to remove the feeding tube from Terri Schiavo in 2005. Explain how Figure 2.32 is misleading.

### Solution

Figure 2.32 is misleading because the vertical scale does not begin at zero. MediaMatters .com published an amended graphic, reproduced here as Figure 2.33, which includes the zero on the vertical axis and much reduces the difference among the political parties. ■

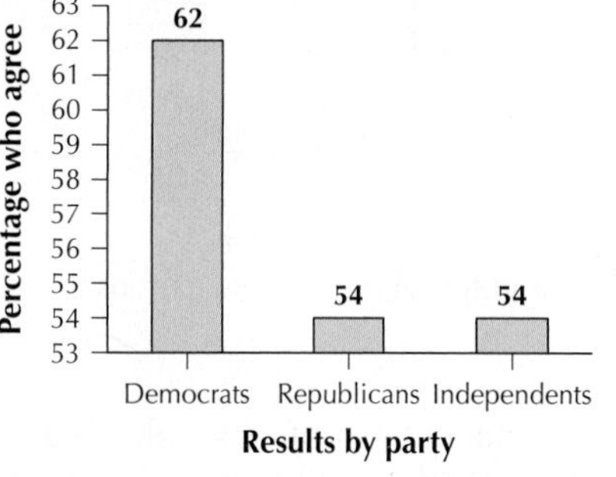


FIGURE 2.32

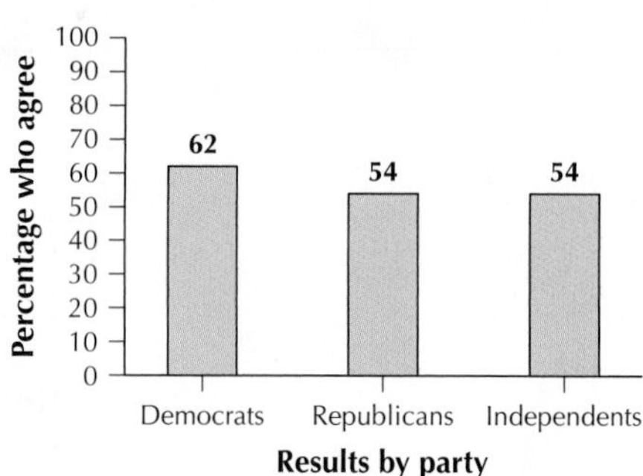


FIGURE 2.33

## Example 2.20 Manipulating the scale

Figure 2.34 shows a Minitab relative frequency bar graph of the majors chosen by 25 business school students. Explain how we could manipulate the scale to de-emphasize the differences.

**Solution**

If we wanted to de-emphasize the differences, we could extend the vertical scale up to its maximum, 1.0 = 100%, to produce the graph in Figure 2.35.

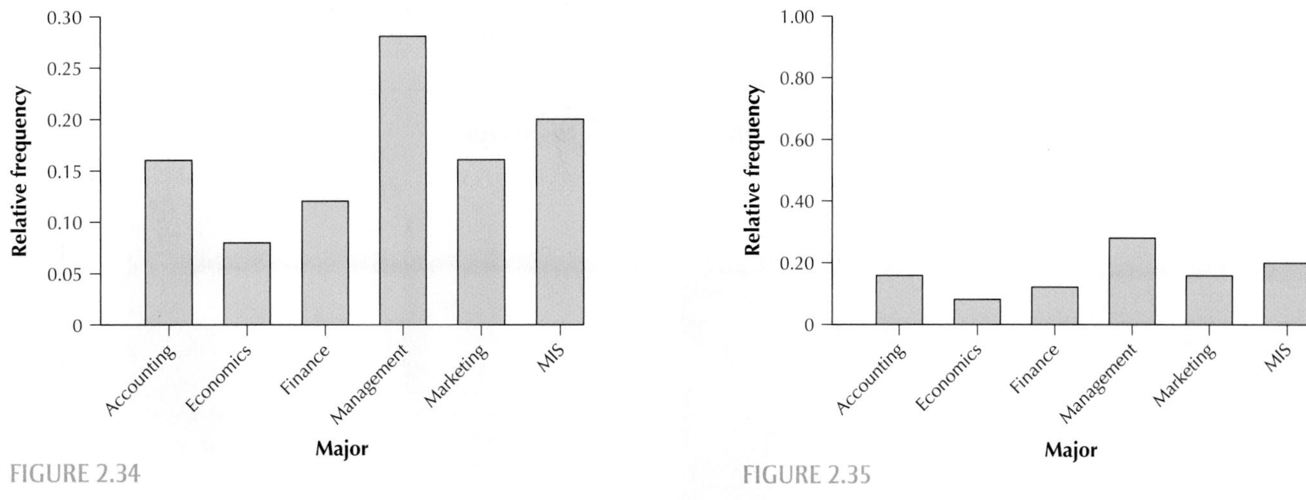


FIGURE 2.34

FIGURE 2.35

## Example 2.21 Using two dimensions for a one-dimensional difference and unclear labeling

Figure 2.36 compares the leaders in career points scored in the NBA All-Star Game among players active in 2007. Explain how this graphic may be misleading.

**Solution**

The height of the players is supposed to represent the total points, but this is not clearly labeled. Points should be indicated using a vertical axis, but there is no vertical axis at all. Further, note that Shaquille O'Neal dominates the graphic, because his body *area* is larger than the body areas of the other players. This is misleading. All four players should have the same body width, just as all bars in a bar graph have the same width. This graph uses two dimensions (height and width) to emphasize a one-dimensional (points) difference.

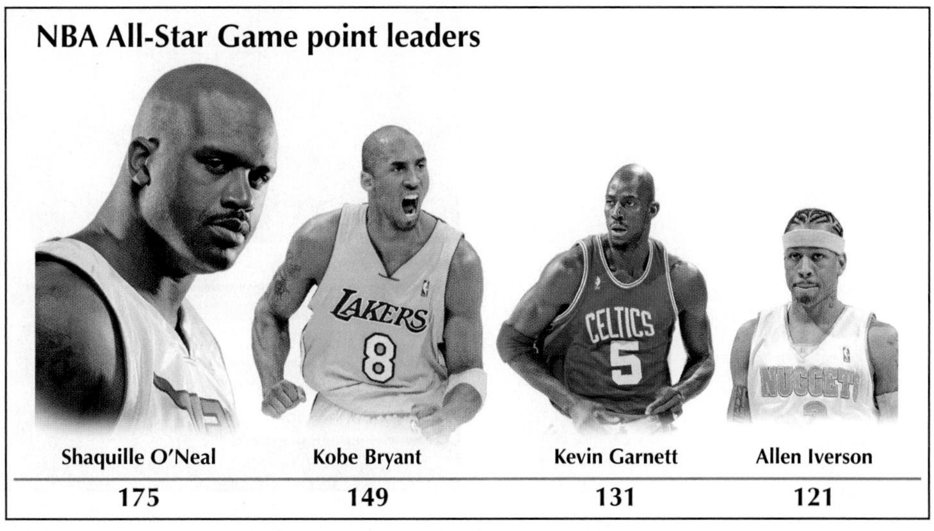


FIGURE 2.36 *(O'Neal: AP Photo/Alan Diaz. Bryant: AP Photo/Mark J. Terrill. Garnett: AP Photo/David Zalubowski. Iverson: AP Photo/David Zalubowski.)*

## Example 2.22 Careless combination of categories in a bar graph and biased embellishment

Figure 2.37 shows a graphic of how often people have observed drivers running red lights. Explain how this graphic may be considered both confusing and biased.

### Solution

One problem with this graphic is that the categories of *seldom* and *never* have been combined, which may not be appropriate. Also, as we learned in Chapter 1, what is "seldom" to one person may not be "seldom" to someone else. A third problem is that the red light of the "Seldom/never" category is lit up, which may be evidence of bias on the part of the designer of the graphic. ■

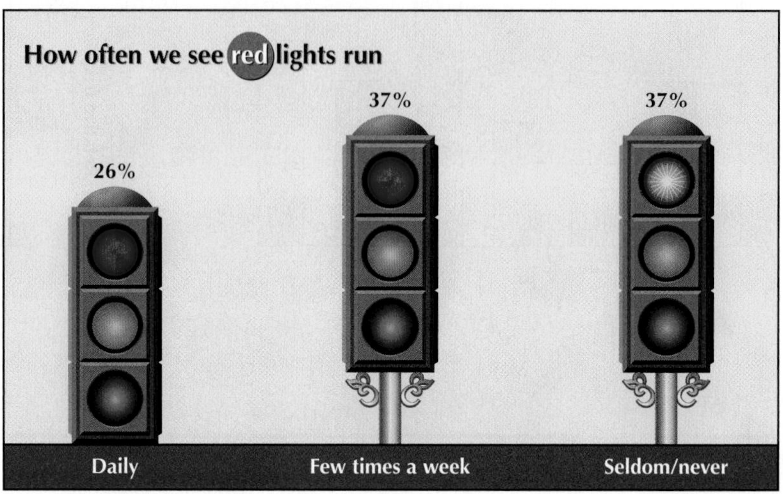


FIGURE 2.37

## Example 2.23 Inaccuracy in relative lengths of bars in a bar graph and unclear labeling

Figure 2.38 is a horizontal bar graph of the three teams with the most World Series victories in baseball history. Explain what is unclear or misleading about this graph.

### Solution

Note that 127 is more than twice as many as 52, and so the Yankees' bar should be more than twice as long as the Cardinals' bar, which it is not. ■

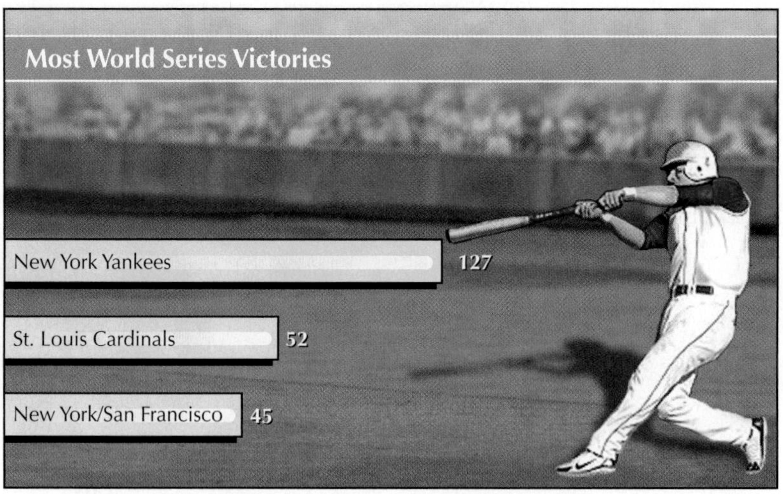


FIGURE 2.38

When constructing a histogram, changing the number of classes or the width of the interval can sometimes lead to a completely different-looking distribution. Thus, we need to exercise care when someone shows us a histogram, since it presents, not the data themselves, but one of many ways of classifying the data.

## Example 2.24 Presenting the same data set as both symmetric and left-skewed

The National Center for Education Statistics sponsors the *Trends in International Mathematics and Science Study (TIMSS)*. In 2003, science tests were administered to eighth-grade students in countries around the world (see Table 2.28). Construct two different histograms, one that shows the data as almost symmetric and one that shows the data as left-skewed.

Table 2.28 Science test scores

| Country | Score | Country | Score | Country | Score |
|---|---|---|---|---|---|
| Singapore | 578 | New Zealand | 520 | Bulgaria | 479 |
| Taiwan | 571 | Lithuania | 519 | Jordan | 475 |
| South Korea | 558 | Slovak Republic | 517 | Moldova | 472 |
| Hong Kong | 556 | Belgium | 516 | Romania | 470 |
| Japan | 552 | Russian Federation | 514 | Iran | 453 |
| Hungary | 543 | Latvia | 513 | Macedonia | 449 |
| Netherlands | 536 | Scotland | 512 | Cyprus | 441 |
| United States | 527 | Malaysia | 510 | Indonesia | 420 |
| Australia | 527 | Norway | 494 | Chile | 413 |
| Sweden | 524 | Italy | 491 | Tunisia | 404 |
| Slovenia | 520 | Israel | 488 | Philippines | 377 |

### Solution

Figure 2.39 is nearly symmetric. But Figure 2.40 is clearly left-skewed. It is important to realize that *both figures are histograms of the very same data set*. Clever choices for the number of classes and the class limits can affect how a histogram presents the data. The reader must therefore beware! The histogram represents a summarization of the data set, and not the data set itself. Analysts may wish to supplement the histogram with other graphical methods, such as dotplots and stem-and-leaf displays, in order to gain a better understanding of the distribution of the data. ■

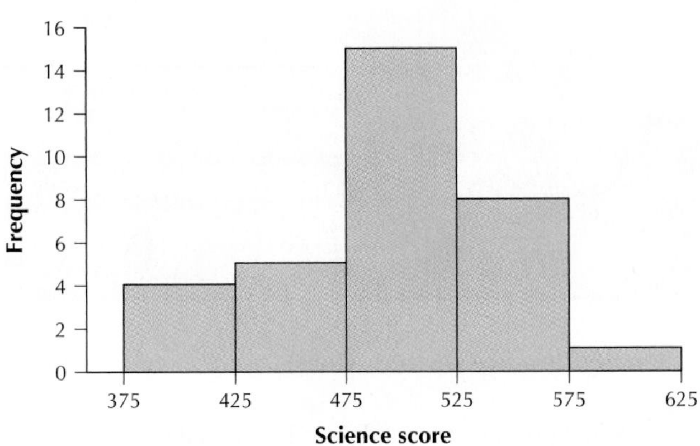


FIGURE 2.39 Nearly symmetric histogram.

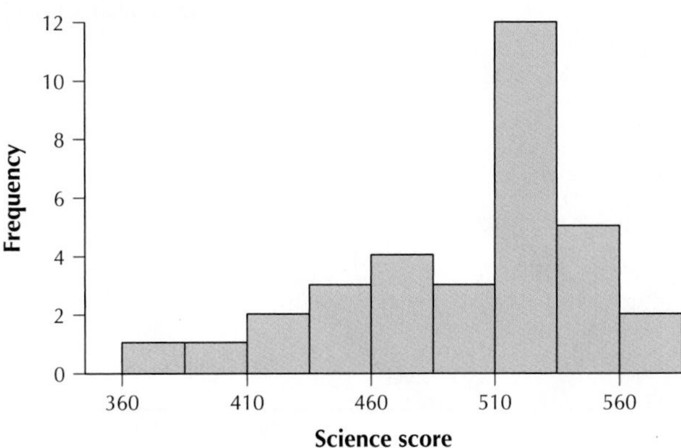


FIGURE 2.40 Left-skewed histogram.

The *One Variable Statistics and Graphs* applet allows you to experiment with the class width and number of classes when constructing a histogram.

## Section 2.4 SUMMARY

Understanding how graphics are constructed will help you avoid being deceived by misleading graphics. Some common methods for making a graph misleading include manipulating the scale, omitting the zero on the relevant scale, and biased distortion or embellishment. The savvy reader should understand that the graphics represent a summary of the data set, and not the data set itself.

## Section 2.4 EXERCISES

### CLARIFYING THE CONCEPTS

1. Explain in your own words why it is important to be aware of the methods that can be used to make graphics misleading.
2. True or false: What we have learned in this chapter proves that all statistics are misleading.

### PRACTICING THE TECHNIQUES

Refer to Example 2.18 for the following exercises.

3. Which do you think is more effective at convincing the American public that a problem exists, Table 2.27 or Figure 2.31?
4. How would factoring in the *number of cars per country* affect the rankings, in your view?
5. If you were an insurance claims adjuster arguing for higher car insurance rates, would you prefer Table 2.26 or Table 2.27? Why?

### APPLYING THE CONCEPTS

6. **Eating Bread.** Consider the accompanying graphic (similar to one found in *USA Today*) of the types of bread people eat.
   a. What type of graph is it supposed to represent, among the graphs that we have learned in this chapter?
   b. Consider how the *wheat* category dominates the graph. Which of the eight common methods for misrepresenting data is present here?
   c. Construct a graphic that is not misleading in this way.

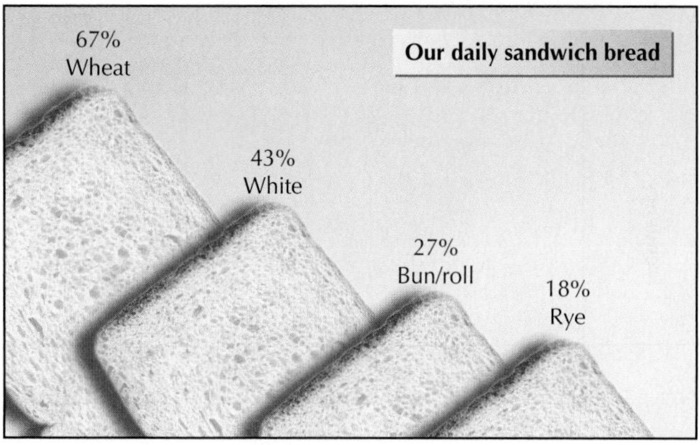


7. **Child-Rearing Costs.** Consider the accompanying graphic (similar to one found in *USA Today*) of child-rearing costs by type of cost.
   a. Identify one problem with the graphic that makes it misleading.
   b. Construct a graphic that is not misleading in this way.

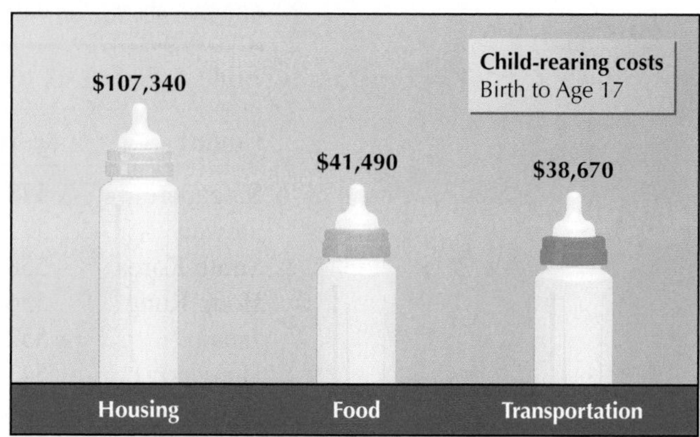


8. **Going to the Game.** Consider the accompanying graphic (similar to one found in *USA Today*) of the proportions of people who go to see professional sports events.
   a. Identify two problems with the graphic that make it misleading.
   b. Construct a graphic that is not misleading in these ways.

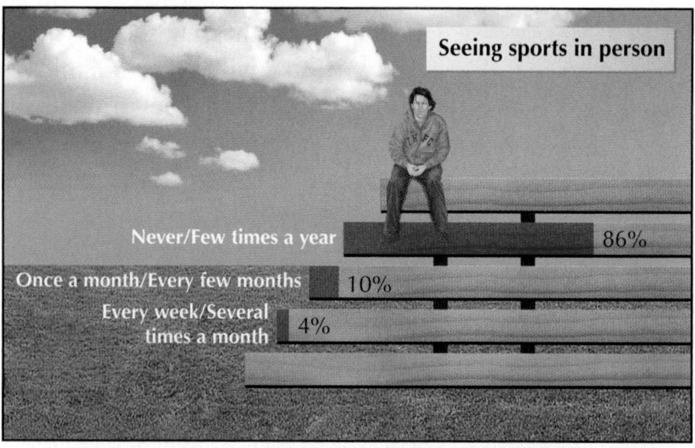


9. **Family Practice Doctors.** Consider the accompanying graphic.
   a. What point is the graphic trying to make?
   b. Which of the eight common problems is most obviously present here?
   c. Construct a graphic that is not misleading in this way.

*LA Times*

**10. Fuel Economy Standards.** Consider Figure 2.41, which compares fuel economy standards from 1978 to 1985.

**a.** What point is the graphic trying to make?

**b.** Which of the eight common problems is most obviously present here?

**c.** Construct a graphic that is not misleading in these ways.

**11. What's Your Sign?** The General Social Survey collects data on social aspects of life in America. Consider the accompanying bar graph of the results of asking 1464 people what their astrological sign is.

**a.** Which of the eight common problems is most obviously present here?

**b.** Construct a graphic that is not misleading in this way.

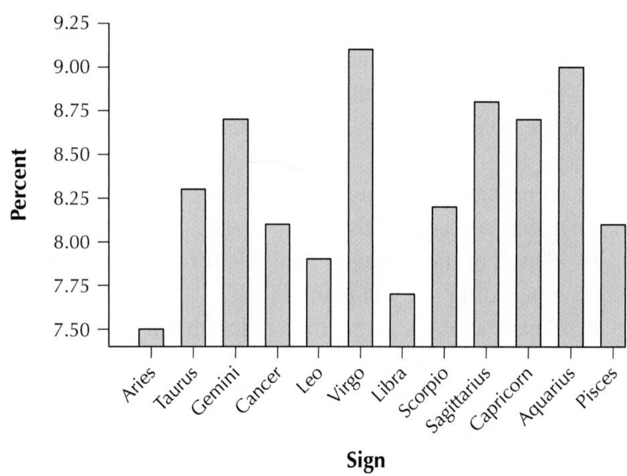


**12. American League Shutouts.** The bar graph below shows the number of team shutouts in the American League in the 2002 season.

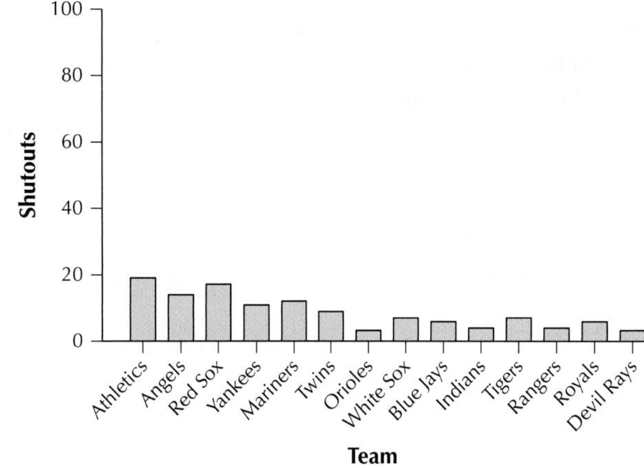


**a.** Which of the eight common problems is most obviously present here?

**b.** Construct a graphic that is not misleading in this way.

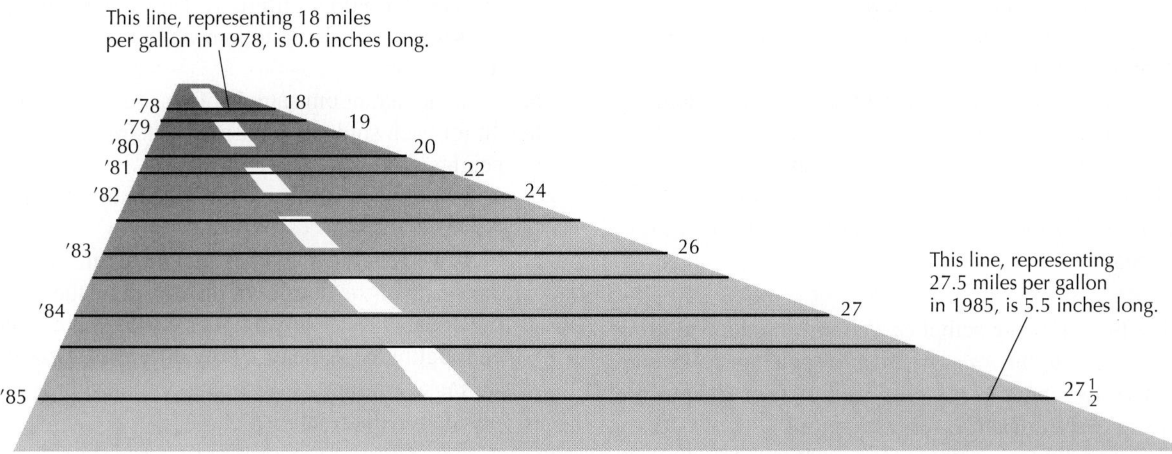


Figure 2.41 Fuel economy standards in miles per gallon.

**13. Cancer Pamphlet Readability.** The readability of 30 cancer pamphlets was examined and recorded in terms of school grade level. The following table displays the grade level of the 30 pamphlets. Let each grade level represent a category.

**a.** Construct a bar graph that overemphasizes the difference among the grade levels.
**b.** Which of the common methods for making graphics misleading are you using in **(a)**?
**c.** Construct a bar graph that underemphasizes the difference among the grade levels.
**d.** Which of the common methods for making graphics misleading are you using in **(c)**?
**e.** Construct a bar graph that fairly represents the data.

| | | | | | | | | | |
|---|---|---|---|---|---|---|---|---|---|
| 6 | 6 | 6 | 7 | 7 | 7 | 8 | 8 | 8 | 8 |
| 8 | 8 | 8 | 8 | 9 | 9 | 9 | 9 | 10 | 11 |
| 12 | 12 | 12 | 12 | 13 | 13 | 14 | 15 | 15 | 16 |

**14. Music and Violence.** *USA Weekend* conducted a poll that asked, "Do you think shock rock and gangsta rap are partly to blame for violence such as school shootings or physical abuse?" The results are shown in the following table.

| | |
|---|---|
| Yes | 31% |
| No | 45% |
| I've never thought about it | 24% |

**a.** Construct a bar graph that overemphasizes the difference among the responses.
**b.** Construct a bar graph that underemphasizes the difference among the responses.
**c.** Construct a bar graph that fairly represents the data.

APPLET Use the *One Variable Statistics and Graphs* applet for Exercises 15–17. Work with the TIMSS scores from Example 2.24.

**15.** Click on the **Histogram** tab. Experiment with the class widths by clicking and dragging on the number line. Produce a histogram that is nearly symmetric, like Figure 2.39.

**16.** Produce a histogram that is somewhat left-skewed, like Figure 2.40.

**17. Click** on the **Stem-and-Leaf** tab. The previous two exercises left us with two different ideas as to the shape of the distribution.

**a.** Now produce a stem-and-leaf display of the TIMSS scores.
**b.** Compare the regular stem-and-leaf display with the split-stem stem-and-leaf display. Which is preferable for this data set?
**c.** Use your preferred stem-and-leaf display from **(b)** to describe the shape of the distribution.
**d.** Which of the two histograms does your description in **(c)** support?

## *Try This in Class!* In-class activities to enhance your understanding of statistics

**1.** Construct a frequency distribution and a relative frequency distribution of the genders of the students in your classroom.

**2.** Where do you think your favorite sport would fall in the distribution of favorite sports? Construct a frequency distribution and a relative frequency distribution of the favorite sports of the students in your classroom. Where do you fall in this distribution?

**3.** Where do you think your height would fall in a stem-and-leaf display of the heights of students in your class? Construct such a stem-and-leaf-display, using 1s as leaf units. Did you fall about where you thought you would in this display? How would you describe the shape of the distribution?

APPLET For the following activities, you may randomize using (a) technology, (b) the *Random Sample* applet, or (c) two dice and a 6 by 6 table of students' names. (See Chapter 1's Try This in Class!, page 29, for instructions.)

**4.** Generate a random sample of ten students, recording the gender for each student.

**a.** Is the distribution of genders the same as for the entire class (viewed as a population)?
**b.** If we generated another random sample of ten students, would the distribution of genders be the same? Why or why not? Would you expect it to be roughly the same or wildly different? How could we make the sample distribution more similar to the population distribution?

**5.** Generate a random sample of ten students, recording the height for each student. Construct a dotplot or histogram of the heights.

**a.** Is the shape similar to the distribution for the entire class?
**b.** If we generated another sample of ten students, would the shape of the dotplot or histogram be the same as previously? Why or why not? Would you expect it to be roughly the same or wildly different? How could we make the sample distribution more similar to the population distribution?

# CHAPTER 2 VOCABULARY

## SECTION 2.1

- **Bar graph (bar chart)** (p. 37). Used to represent the frequencies or relative frequencies for categorical data. It is constructed as follows: (1) On the horizontal axis, provide a label for each category, and (2) draw rectangles (bars) of equal width for each category. The height of each rectangle represents the frequency or relative frequency for that category. Ensure that the bars do not touch each other.
- **Frequency (count)** (p. 36). The number of observations in each category.
- **Frequency distribution** (for a qualitative variable) (p. 35). A listing of all values that the variable can take and the frequencies for each value.
- **Pareto chart** (p. 38). A bar graph in which the rectangles are presented in decreasing order from left to right.
- **Pie chart** (p. 38). Used for categorical data, a pie chart is a circle, divided into sections (that is, slices or wedges), with each section representing a particular category. The size of the section is proportional to the relative frequency of the category.
- **Relative frequency** (for a qualitative variable) (p. 36). The frequency of a class or category, divided by the sample size.
- **Relative frequency distribution** (for a qualitative variable) (p. 36). A listing of all values that the variable can take and the relative frequencies for each value.

## SECTION 2.2

- **Class** (p. 48). A range of data values used to group the elements in a data set.
- **Class limit (lower)** (p. 49). The smallest value within that class.
- **Class limit (upper)** (p. 49). The largest value within that class.
- **Class midpoint** (p. 51). Average of two consecutive lower class limits.
- **Class width** (p. 49). The difference between the lower class limits of two successive classes.
- **Dotplot** (p. 54). A simple graph in which each data point is represented by a dot above the number line. When the sample size is large, each dot may represent more than one data point.
- **Frequency distribution** (for quantitative data) (p. 47). A listing of the frequencies for a set of classes for a quantitative variable. Constructed as follows: (1) determine how many classes you will use, (2) determine the class width, using equal widths, and (3) determine the upper and lower class limits, so that all classes are non-overlapping.
- **Frequency polygon** (p. 51). Constructed as follows: (1) for each class, plot a point at the class midpoint, at a height equal to the frequency for that class, and (2) join each consecutive pair of points with a line segment.
- **Histogram** (p. 50). Constructed using rectangles for each class of data. The heights of the rectangles represent the frequencies or relative frequencies of the class. The widths of the rectangles are all the same and correspond to the class width of the corresponding frequency distribution. The lower class limits are placed on the horizontal axis (along with the upper class limit of the rightmost class), so that the rectangles are touching each other.
- **Relative frequency distribution** (for quantitative data) (p. 47). Similar to a frequency distribution except that the relative frequencies instead of the frequencies are provided.
- **Skewed distribution** (p. 55). A right-skewed distribution has a longer tail on the right side than on the left. A left-skewed distribution has a longer tail on the left side.
- **Stem-and-leaf display** (p. 52). A graphical display for numerical data. The stem unit is ten times the leaf unit, whose value is defined in the display. For each stem, the leaf units are arranged in increasing order.
- **Symmetric** (p. 55). An image is symmetric if there is a line (axis of symmetry) that splits the image in half so that one side is the mirror image of the other.

## SECTION 2.3

- **Cumulative frequency distribution** (p. 66). For a discrete variable, shows the total number of observations *less than or equal to* the category value. For a continuous variable, shows the total number of observations *less than or equal to* the upper class limit.
- **Cumulative relative frequency distribution** (p. 66). For a discrete variable, shows the proportion of observations less than or equal to the category value. For a continuous variable, shows the proportion of observations less than or equal to the upper class limit.
- **Ogive** (pronounced "oh jive") (p. 67). The graphical equivalent of a cumulative frequency distribution or a cumulative relative frequency distribution. Like a frequency polygon, an ogive consists of a set of plotted points connected by line segments. The $x$ coordinates of these points are the upper class limits; the $y$ coordinates are the cumulative frequencies or cumulative relative frequencies.
- **Time series plot (time series graph)** (p. 67). A graph of time series data. The horizontal axis represents time (for example, hours, days, months, years). The values of the time series data are plotted on the vertical axis, and line segments are drawn to connect the points.

## SECTION 2.4

- **Eight Common Methods for Making a Graph Misleading** (p. 73)
  1. Graphing/selecting an inappropriate statistic.
  2. Omitting the zero on the relevant scale.
  3. Manipulating the scale.
  4. Using two dimensions (area) to emphasize a one-dimensional difference.
  5. Careless combination of categories in a bar graph.
  6. Inaccuracy in relative lengths of bars in a bar graph.
  7. Biased distortion or embellishment.
  8. Unclear labeling.

## CHAPTER 2 REVIEW EXERCISES

### SECTION 2.1

1. **Parts of Speech.** The accompanying bar graph summarizes the frequencies for the various parts of speech. Should we be interested in determining whether this graph is symmetric or skewed? Clearly explain why or why not.

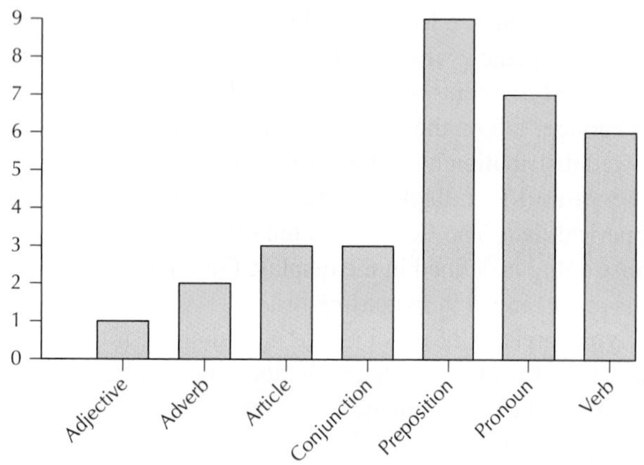


**Best Sellers.** Refer to Table 2.29 for Exercises 2–5.

2. Suppose you categorize the price of books in this way: books less than \$7 are inexpensive, books between \$7.01 and \$24.00 are moderately priced, and books more than \$24 are expensive.
   - **a.** Construct a frequency distribution for the books that are inexpensive, moderately priced, and expensive.
   - **b.** Using this categorization, construct a relative frequency distribution for the books that are inexpensive, moderately priced, and expensive.

3. Construct a relative frequency bar graph using these categories:
   - **a.** Fiction whose author is female
   - **b.** Fiction whose author is male
   - **c.** Nonfiction whose author is female
   - **d.** Nonfiction whose author is male
   - **e.** Hardcover whose author is female
   - **f.** Paperback whose author is male

4. Refer to the previous exercise. Construct pie charts using the same categories.

5. Can you reconstruct the original data set showing the list price of the books just by using the frequency distribution? Why or why not?

### SECTION 2.2

**Management Aptitude Test Scores.** Use Figures 2.8a and 2.8b, reproduced from Section 2.2, for Exercises 6 and 7.

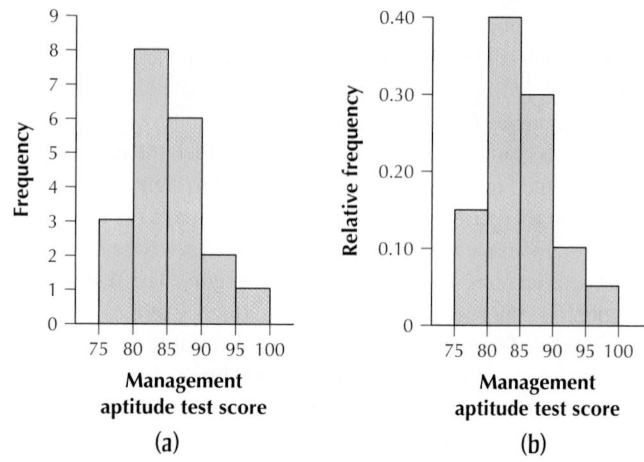


FIGURE 2.8 (a) Frequency histogram; (b) relative frequency histogram.

6. Suppose we were given the histogram in Figure 2.8a but did not have the original data set. Would we have enough information to construct the histogram in Figure 2.8b? Explain why or why not.

7. Suppose we were given the histogram in Figure 2.8b but did not have the original data set. Would we have enough information to construct the histogram in Figure 2.8a? Explain why or why not. If not, what piece of information is needed?

**New York Townspeople.** For towns in New York State, the accompanying histogram provides information on the percentage of the townspeople who are between 18 and 65 years old (data set **New York**). Refer to the histogram for Exercises 8–10.

**TABLE 2.29 Top 10 best sellers from *USA Today* (March 3, 2007)**

| Rank | Title | Author | Type | Book | Gender | Price (\$) |
|---|---|---|---|---|---|---|
| 1 | The Secret | Rhonda Byrne | NF | H | Female | 23.95 |
| 2 | Innocent in Death | Nora Roberts | F | H | Female | 25.95 |
| 3 | Step on a Crack | James Patterson | F | H | Male | 27.99 |
| 4 | Bridge to Terabithia | Katherine Paterson | F | P | Female | 6.99 |
| 5 | Sisters | Danielle Steel | F | H | Female | 27.00 |
| 6 | The Measure of a Man: A Spiritual Autobiography | Sidney Poitier | NF | P | Male | 14.95 |
| 7 | The Memory Keeper's Daughter | Kim Edwards | F | P | Female | 14.00 |
| 8 | The Audacity of Hope | Barack Obama | NF | H | Male | 25.00 |
| 9 | You: On a Diet | Michael Roizen | NF | H | Male | 25.00 |
| 10 | The Best Life Diet | Bob Greene | NF | H | Male | 26.00 |

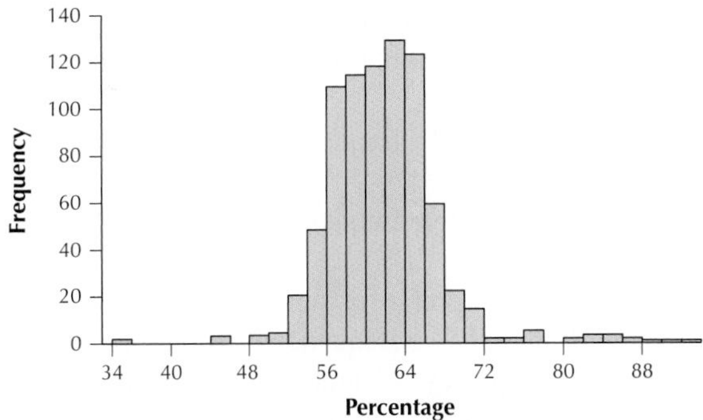


8. Using only the histogram, construct a frequency distribution and a relative frequency distribution of the proportion of townspeople between the ages of 18 and 65. Estimate the frequencies from the histogram.

9. If you had to place a fulcrum under the histogram so that it would balance perfectly like a seesaw, where would you place it?

10. Provide a rough estimate of the relative frequency of towns having 80% or more between 18 and 65 years old.

**Households.** Use the following information for Exercises 11–13. The data set **Household** contains eight variables' worth of information about the households in all 50 states plus the District of Columbia. The average size of the households is plotted in the accompanying dotplot, reproduced from Figure 2.16a.

2.25 2.35 2.45 2.55 2.65 2.75 2.85 2.95 3.05 3.15
Average size

Dotplot of average household size.

11. Would you characterize the dotplot as (a) tending to be symmetric, (b) tending to be right-skewed, (c) tending to be left-skewed, or (d) too difficult to say?

12. Refer to the dotplot of average household size.
  **a.** What is the range of the data set?
  **b.** How many states have an average household size of 2.55?
  **c.** Which average household size is the most common?
  **d.** How many states seem to have an unusually large average household size? How many states seem to have an unusually small average household size? What is your evidence for saying so?

WHAT IF? 13. *What if* the data were faulty, and each data point should have had 0.5 added to it. How would that affect the shape of the distribution?

## SECTION 2.3

14. **Countries and Continents.** Use the frequency distribution you created in Exercise 29 in Section 2.2 (page 63) to
  **a.** construct a cumulative frequency distribution.
  **b.** construct a cumulative relative frequency distribution.

15. **Hospitals near Jackson, Mississippi.** Use the frequency distribution of hospital beds from Table 2.18 (page 63) to
  **a.** construct a cumulative frequency distribution.
  **b.** construct a cumulative relative frequency distribution.

16. **American League Shutouts.** Use the frequency distribution in Table 2.20 (page 64) to
  **a.** construct a frequency ogive.
  **b.** construct a relative frequency ogive.

17. **Miami Arrests.** The Miami-Dade Police Department published the monthly number of arrests made for the year 2005, given in the following table. Construct a time series graph of the data.

| | | | | | |
|---|---|---|---|---|---|
| Jan. | 751 | May | 919 | Sept. | 802 |
| Feb. | 650 | June | 800 | Oct. | 636 |
| Mar. | 909 | July | 834 | Nov. | 579 |
| Apr. | 881 | Aug. | 789 | Dec. | 777 |

## SECTION 2.4

18. **Sports Clothing.** Consider the accompanying graphic of the types of sports clothing that children own.
  **a.** What type of graph does it represent, among the graphs that we have learned about in this chapter?
  **b.** Describe the difference between the representation of the NFL category versus the other categories.
  **c.** Which of the eight common methods for misrepresenting data is present here?
  **d.** Construct a graphic that is not misleading in this way.

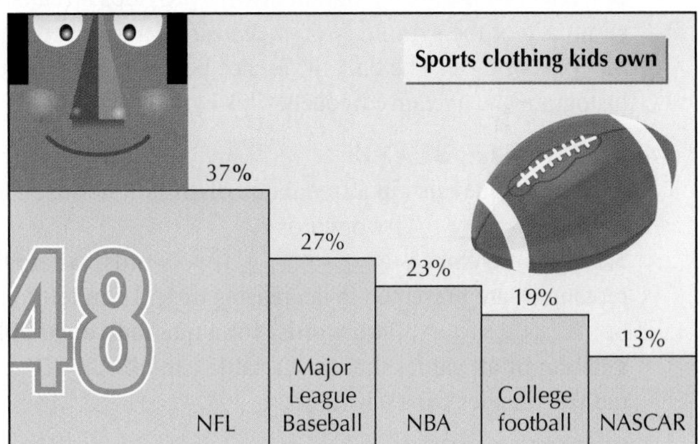


19. **Carbon Dioxide Emissions.** The following table contains the energy-related carbon dioxide emissions (in millions of metric tons) for 2004, by end-use sector, as reported by the U.S. Energy Information Administration.

| Sector | Emissions |
|---|---|
| Residential | 1213.9 |
| Commercial | 1034.1 |
| Industrial | 1736.0 |
| Transportation | 1939.2 |

  **a.** Construct a bar graph that overemphasizes the differences among the sectors.
  **b.** Which of the common methods for making misleading graphics are you using in **(a)**?
  **c.** Construct a bar graph that underemphasizes the differences among the sectors.
  **d.** Which of the common methods for making graphics misleading are you using in **(c)**?
  **e.** Construct a bar graph that fairly represents the data.

**Discovery Using the Computer.** For Exercises 20 and 21, use your knowledge of Minitab or Excel. Open the **Goals** data set. Students in grades four, five, and six from three school districts in Michigan were asked which of the following was most important to them: good grades, sports, or popularity. Information about the students' age, gender, race, and grade was also gathered, as well as whether their school was in an urban, suburban, or rural setting.[6]

**20.** Observations and variables.

**a.** How many observations are in the data set?

**b.** How many variables?

**c.** List the numeric variables.

**d.** List the categorical variables.

**21.** Consider the variable *school.*

**a.** Generate a relative frequency distribution of the variable *school.*

**b.** From which school do more students come from than any other school? Which school has the fewest students?

**c.** Generate a relative frequency bar graph of the variable *school.*

**d.** Generate a relative frequency pie chart of the variable *school.*

**e.** Which do you prefer for summarizing the variable *school:* the relative frequency distribution, the relative frequency bar graph, or the relative frequency pie chart? Why? Which contains the most information? Which is more intuitive?

## CHAPTER 2 *Quiz*

### TRUE OR FALSE

**1.** True or false: Histograms are superior to stem-and-leaf displays because histograms retain the information contained in the data set.

**2.** True or false: A histogram always provides a realistic summary of the symmetry or skewness of a data set.

**3.** True or false: The only difference between a frequency histogram and a relative frequency histogram is the vertical axis.

### FILL IN THE BLANK

**4.** The frequencies in a frequency distribution must add up to the _____ _____ [two words].

**5.** A Pareto chart is a _____ _____ [two words] where the rectangles are presented in decreasing order from left to right.

**6.** A _____ _____ [two words] for a qualitative variable is a listing of all values that the variable can take, together with the frequencies for each value.

### SHORT ANSWER

**7.** What must the relative frequencies in a relative frequency distribution add up to?

**8.** If there is a line that splits an image in half so that one side is the mirror image of the other, we say that the image is what?

**9.** If the right tail of a distribution is longer than the left tail, we say that the distribution is what?

### CALCULATIONS AND INTERPRETATIONS

For the following exercises, refer to Table 2.5 from the Case Study (page 39). Ignore the consonants and consider the vowels (A, E, I, O, U) only. Construct the following:

**10.** Frequency distribution

**11.** Relative frequency distribution

**12.** Cumulative frequency distribution

**13.** Cumulative relative frequency distribution

**14.** Frequency bar graph

**15.** Relative frequency bar graph

**16.** Pie chart of the relative frequencies

**17.** Ogive of the frequencies

**18.** Relative frequency ogive of the frequencies

**9/11 and Pearl Harbor.** What were the feelings of Americans in the days immediately following the events of September 11, 2001? The terrorist attacks on New York City and Washington, D.C., on September 11, 2001, were often compared to the Japanese attack on Pearl Harbor on December 7, 1941. In an NBC News Terrorism Poll, the following question was asked: Would you say that Tuesday's attacks are more serious than, equal to, or not as serious as the Japanese attack on Pearl Harbor? This poll was conducted on September 12, 2001, and the sample size was 618. Use the results given in the relative frequency distribution below to answer Exercises 19–21.

| Response | Relative frequency |
|---|---|
| More serious than Pearl Harbor | 0.6667 |
| Equal to Pearl Harbor | 0.2492 |
| Not as serious as Pearl Harbor | 0.0469 |
| Not sure | 0.0372 |
| Total | 1.0000 |

**19.** Construct the frequency distribution of responses.

**20.** Construct a frequency bar graph of the responses.

**21.** Construct a relative frequency bar graph of the responses. If we wanted to change our relative frequency bar graph to a frequency bar graph, would we have to redraw the bars? Explain why or why not.

**Favorite Music.** *USA Weekend* conducted a poll that asked, "If you had to choose just one type of music to listen to exclusively, which would it be?" The results are shown in the table below. Use this information for Exercises 22–25.

| | | | |
|---|---|---|---|
| Hip-hop/rap | 27% | R&B | 6% |
| Pop | 23% | Country | 5% |
| Rock/punk | 17% | Techno/house | 4% |
| Alternative | 7% | Jazz | 1% |
| Christian/gospel | 6% | Other | 4% |

**22.** Assuming that the sample size is 6000, construct a frequency distribution of the most popular types of music.

**23.** Construct a relative frequency bar graph of the most popular types of music.

**24.** Construct a pie chart of the most popular types of music.

**25.** Describe how a frequency bar graph and a relative frequency bar graph differ. How are they similar?

Chapter 3

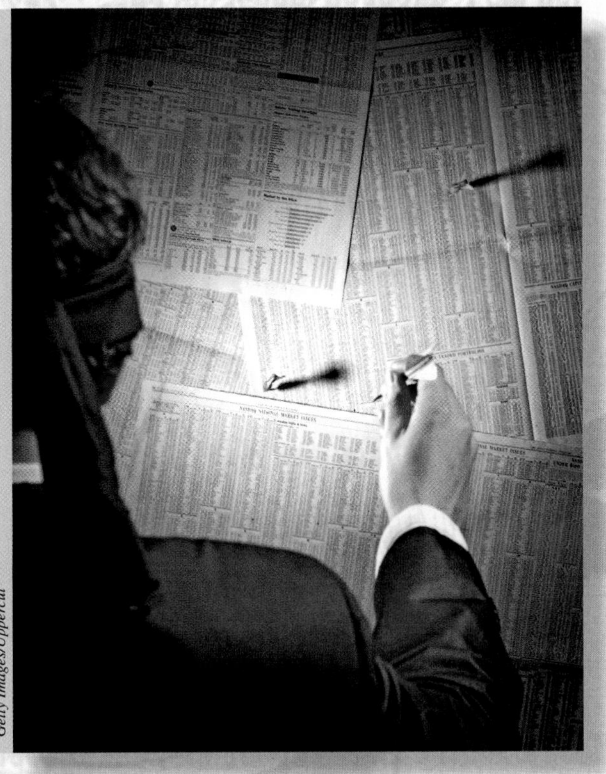

Getty Images/Uppercut

# Describing Data Numerically

CASE STUDY

## Can the Financial Experts Beat the Darts?

Have you ever wondered whether a bunch of monkeys throwing darts to choose stocks could select a portfolio that performed as well as the stocks carefully chosen by Wall Street experts? The *Wall Street Journal* (**www.wsj.com**) apparently believes that the comparison is worth a look. Starting in 1988, the *Journal* has run a contest between stocks chosen randomly by *Journal* staff members (rather than monkeys) throwing darts at the *Journal* stock pages (mounted on a board) and stocks chosen by a team of four professional financial experts.

At the end of six months, the *Journal* compares the percentage change in the price of the experts' stocks and the dartboard's stocks. So, who do you think did better? Did the six-figure-salary financial experts put the random dart selections to shame? We examine the results in the Case Study on pages 95 and 113.

## OVERVIEW



## The Big Picture

### Where we are coming from, and where we are headed...

The first three chapters of the book prepare you to find out all you ever really wanted to know (and probably more) about a particular data set. Chapter 1 introduced the language of descriptive statistics. Chapter 2 showed ways to summarize data sets using tables and graphs. In Chapter 3, we will "crunch the numbers." That is, we will develop some numerical summaries to help us discover important characteristics about the data set. We expand our role as data analyst, to dig in and uncover the numerical side of the "vital information" about the data set. This vital information includes questions such as

- Where is the center of the data?
- How spread out are the data?
- How can we determine whether a particular result is unusual or not?

Here in Chapter 3, we also become acquainted with some powerful and widespread methodologies for applying the tools of descriptive statistics. These applications of descriptive statistics enable us to extract more information from a particular data set, in order to arrive at a deeper meaning of the story behind the data. These tools include percentiles, $z$-scores for detecting outliers, and Chebyshev's Rule and the Empirical Rule for estimating the proportion of data contained in an interval.

In Chapter 4, we will learn ways of describing the relationship between two variables. However, as we learned in Chapter 1, making inferences from a sample to the general population carries with it a measure of uncertainty. That is why we will need to study *probability* in Chapters 5 and 6. Probability is the bridge between descriptive statistics (Chapters 1–4) and inferential statistics (Chapters 7–13).

So, in this chapter, we learn how to summarize an entire data set with a few numbers. For example, one summary statistic in baseball is a player's batting average (the ratio of hits to at-bats). Suppose we want to know how good a hitter Derek Jeter of the New York Yankees is. Instead of examining thousands of individual at-bats over Jeter's long career, we simply consider one number, his lifetime batting average. This one number powerfully summarizes a very large data set, the many thousands of at-bats over Jeter's career. Since Jeter's lifetime batting average is well over .300, we can see that the Yankee shortstop is a very good hitter. Using summary statistics like lifetime batting average, we can compare his performance with those of some of the greatest players who ever played the game. For more fun statistics about Derek Jeter and all of Major League Baseball, check out **www.baseball-reference.com**.

# 3.1 Measures of Center

## Objectives: By the end of this section, I will be able to...

1. Calculate the mean for a given data set.
2. Find the median, and describe why the median is sometimes preferable to the mean.
3. Find the mode of a data set.
4. Describe how skewness affects these measures of center.

As data analysts, we often encounter unfamiliar data sets. We can examine a data set using the tabular and graphical methods we learned in Chapter 2. In addition, we can use *numerical summaries* of the data set to help us uncover characteristics of the data. We would first like to know: Where is the middle, or *center,* of the data set? In this section, we will learn about three statistics that can tell us where the center of the data set lies: the mean, the median, and the mode.

## 1 The Mean

The most well known and widely used **measure of center** is the **mean.** In everyday usage, the word *average* is often used to denote the mean.

> To find the **mean** of the values in a data set, simply add up all the numbers and divide by how many numbers you have.

### Example 3.1 Bird flu cases, 2006

Health officials are carefully monitoring the number of humans who contract avian influenza (bird flu) type A (H5N1) worldwide. Table 3.1 contains the number of cases of bird flu for the five countries with the highest number of cases in 2006.[1] Find the mean number of cases of bird flu for these five countries.

Table 3.1 Bird flu cases by country, 2006

| Country | Cases |
|---|---|
| Vietnam | 93 |
| Indonesia | 26 |
| Thailand | 22 |
| China | 12 |
| Turkey | 12 |

**Solution**

To find the mean, we add up the number of cases for all the countries and divide by the number of countries, 5:

$$\text{mean number of cases of bird flu} = \frac{93 + 26 + 22 + 12 + 12}{5} = \frac{165}{5} = 33$$

That is, these five countries had a mean of 33 bird flu cases in 2006.

## Notation

Statisticians like to use specialized notation. The notation is worth learning because it saves a lot of writing and because certain concepts can best be understood using this special notation.

- The **sample size,** which refers to how many observations you have in your sample data set, is always denoted by $n$. Here, the five countries from Table 3.1 can be considered a sample taken from the population (which in this case is all the countries in the world). Thus, here, $n = 5$.
- We denote the $i$th data value by $x_i$, where $i$ is simply an index or counter indicating which data point we are specifying. For example, in Table 3.1, $x_1 = 93$, $x_2 = 26$, $x_3 = 22$, and $x_4 = 12$. The last data value is $x_n = x_5 = 12$.
- The notation for "add them together" is $\Sigma$ (capital sigma), the Greek letter for "S," because it stands for "Summation." To add up the number of cases for all five countries, we could write out 93 + 26 + 22 + 12 + 12, or we could simply represent this sum as $\Sigma x_i$ or, even more simply, as $\Sigma x$.
- The **sample mean** is called $\bar{x}$ (pronounced "x-bar"). You should try to commit this to long-term memory, since $\bar{x}$ may be the most important symbol used in this book and will return again and again in nearly every chapter.

The **sample mean** can be written as $\bar{x} = \Sigma x/n$. In plain English, this just means that, in order to find the mean $\bar{x}$, we

1. Add up all the data values, giving us $\Sigma x$.
2. And divide by how many observations are in the data set, giving us $\Sigma x/n$.

So, for example, the sample mean number of cases of bird flu can be written as

$$\bar{x} = \frac{\Sigma x}{n} = \frac{93 + 26 + 22 + 12 + 12}{5} = \frac{165}{5} = 33$$

*What Does This Number Mean?*

### The Mean as the Balance Point of the Data

Let's explore the bird flu data a bit further. Consider the dotplot of the bird flu cases for each country, given in Figure 3.1. To find out where the mean of the number of cases of bird flu lies on this number line, imagine that the dots are little blocks on a ruler or a seesaw and that you must decide where to place the fulcrum so that the ruler balances perfectly. *The place where the data set balances perfectly is the location of the mean.* Placing the fulcrum too far to the right or left would create an imbalance. This data set balances precisely at the sample mean, $\bar{x} = 33$. ■

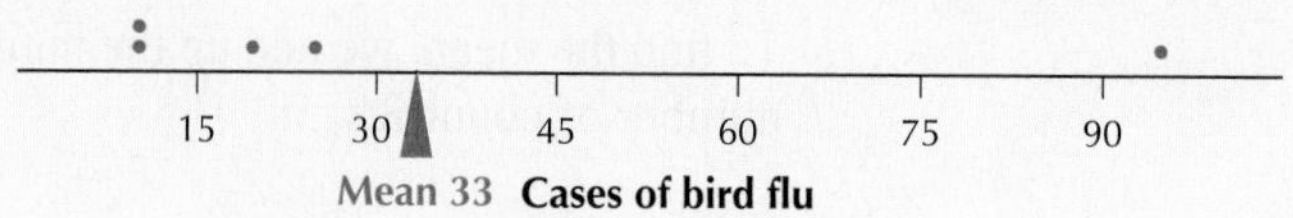


FIGURE 3.1 Where is the balance point for this dotplot?

Your Statistical Sense

## Checking Your Results Against Experience and Common Sense

*When you have found the balance point, you have found the mean.* Keep this idea in mind throughout your study of statistics, as a reality check against calculations of the mean. When you calculate the mean, or have a computer or calculator do it for you, don't just naively accept whatever value pops out. Since the mean always indicates the place where the data values are in balance, the mean is often near the center of the data. If the value you have calculated lies nowhere near the center of the data, then you may want to check your calculations.

For example, suppose we were finding the mean of the bird flu data, and we accidentally entered 930 for the first value instead of 93 for the number of cases for Vietnam. Then our value for the mean resulting from this *incorrect* calculation would be

$$\bar{x} = \frac{\sum x}{n} = \frac{930 + 26 + 22 + 12 + 12}{5} = \frac{1002}{5} = 200.4$$

The mean number of cases cannot equal 200.4 because all the values in the data set are less than 200.4. The mean can never be larger or smaller than all the values in the data set.

Don't automatically assume that the result you get from a computer or calculator is right. Remember GIGO: Garbage In Garbage Out. If you enter the wrong data, the calculator or computer will not bail you out. Human error is one reason for the explosion of faulty statistical analysis in the newspapers and on the Internet. Now more than ever data analysts must use good judgment. When you calculate a mean, always have an idea of what you *expect* the sample mean to be, that is, at least a ballpark figure. ■

## The Population Mean $\mu$

The mean value of a population is usually unknown. For example, we cannot know the mean systolic blood pressure of *all* the residents in your hometown at noon today. Instead, data analysts use *estimation*. We could select a random sample of, say, 30 residents, find the mean systolic blood pressure $\bar{x}$ of this sample, and use this $\bar{x}$ as an estimate of the unknown population mean systolic blood pressure. Generally, Greek letters are used to represent the (usually unknown) population parameters (such as the population mean). We denote the **population mean** with $\mu$ (mu), which is the Greek letter for "m." The **population size** is denoted by $N$. When all the values of the population are known, the population mean is calculated as

$$\mu = \frac{\sum x}{N}$$

## The Mean Is Sensitive to Extreme Values

One drawback of using the mean to measure the center of the data is that the mean is sensitive to the presence of extreme values in the data set. We illustrate this phenomenon with the following example.

### Example 3.2 Mean of home sales prices

Table 3.2 contains a sample of 6 home sales prices for Broward County, Florida, as listed in the *New York Times* Online Real Estate section (**realestate.nytimes.com**) for December 14, 2006. We would like to get an idea of the typical home sales price in Broward County. Find the mean sales price of the homes in this sample.

Table 3.2 Home sales prices in Broward County, Florida

| Location | Price |
|---|---|
| Coral Springs | $415,000 |
| Weston | $449,000 |
| Coconut Creek | $472,000 |
| Davie | $649,000 |
| Hallandale | $889,000 |
| Fort Lauderdale | $975,000 |

**Solution**

$$\bar{x} = \frac{\sum x}{n} = \frac{415{,}000 + 449{,}000 + 472{,}000 + 649{,}000 + 889{,}000 + 975{,}000}{6}$$

$$= \frac{3{,}849{,}000}{6} = \$641{,}500$$

So the sample mean home sales price is $641,500. Now, suppose that we append a seventh home to our sample, a home in Hillsboro Beach listed for $5,900,000, which is much more expensive than any of the other homes in the sample. Recalculating the mean, we get

$$\bar{x} = \frac{\sum x}{n}$$

$$= \frac{415{,}000 + 449{,}000 + 472{,}000 + 649{,}000 + 889{,}000 + 975{,}000 + 5{,}900{,}000}{7}$$

$$= \frac{9{,}749{,}000}{7} \approx \$1{,}392{,}714$$

Note that the mean sales price more than doubled from $641,500 to $1,392,714 when we added this extreme value. Also, this new mean is much higher than every price in the original sample. Thus, it is highly unlikely that this new mean of about $1.4 million is representative of the *typical* sales price of homes in Broward County.

This example shows why we say that the mean is sensitive to the presence of extreme values. For situations like this, we prefer a measure of center that is not so sensitive to extreme values. Fortunately, the *median* is just such a measure.

## 2 The Median

Recall that the median strip on a highway is the slice of land in the *middle* of the two lanes of the highway. In statistics, the **median** of a data set is the *middle data value* when the data are put into ascending order. There are two cases, depending on whether the sample size is odd or even.

**The Median**

The **median** of a data set is the *middle data value* when the data are put into ascending order. Half of the data values lie below the median, and half lie above.

- If the sample size $n$ is odd, then the median is a unique middle value. That is, the median is the $\left(\frac{n+1}{2}\right)^{\text{th}}$ observation when the data are put in ascending order.
- If the sample size $n$ is even, then the median is the mean of the two data values in the middle. That is, the median is the mean of the two data values that lie on either side of the $\left(\frac{n+1}{2}\right)^{\text{th}}$ position.

The case when the sample size is even is clear if you hold up four fingers on one hand. Notice that there is no unique finger in the middle. Since there is no middle value when the sample size is even, we take the two data values in the middle and split the difference.

## The Median Is Not Sensitive to Extreme Values

News reports prefer the median to the mean when reporting home sales prices. For example, the *San Jose Mercury News* reported on December 14, 2006, that the median California home sales price was $469,000. Unlike the mean, the median is not sensitive to extreme values. For example, if someone purchases a very expensive house this month, the mean home sales price will jump, but the median home sales price will be less affected. Let's look at an example of how this would occur.

## Example 3.3 Median of home sales prices

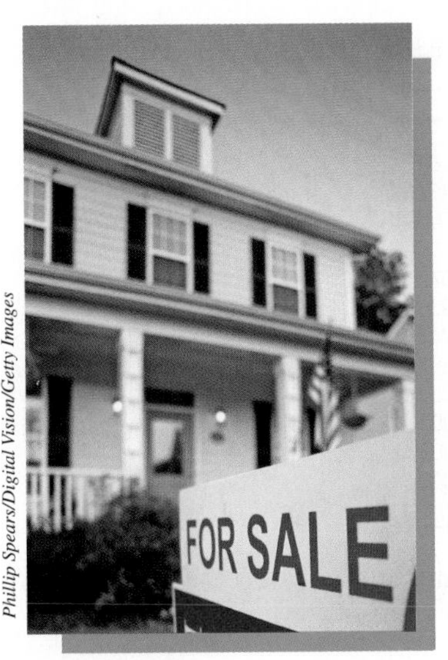


Phillip Spears/Digital Vision/Getty Images

Find the median home sales price for the Broward County data from Table 3.2.

### Solution

Fortunately, the data are already presented in ascending order in the table. Since $n = 6$ is even, the median is the mean of the two data values that lie on either side of the $\left(\frac{n+1}{2}\right)^{\text{th}} = \left(\frac{6+1}{2}\right)^{\text{th}} = 3.5$th position. That is, the median is the mean of the 3rd and 4th data values, $472,000 and $649,000. Splitting the difference between these two, we get

$$\text{median price} = \frac{472{,}000 + 649{,}000}{2} = \$560{,}500$$

We note that in Table 3.2 there are exactly as many homes with prices lower than $560,500 as there are homes with prices higher than $560,500.

Now, what happens to the median when we add in the $5,900,000 home from Hillsboro Beach? Since $n = 7$ is odd, the median is the unique $\left(\frac{n+1}{2}\right)^{\text{th}} = \left(\frac{7+1}{2}\right)^{\text{th}} = 4$th observation, given by the home in Davie for $649,000. The extreme value increased the median only from $560,500 to $649,000. Recall that the mean more than doubled to almost $1.4 million. Thus, the median home sales price is a better measure of center because it more accurately reflects the typical sales prices of homes in Broward County. Figure 3.2 shows how the mean (red triangles) changes significantly with the addition of the extreme value, while the median (green triangles) changes relatively little. ■

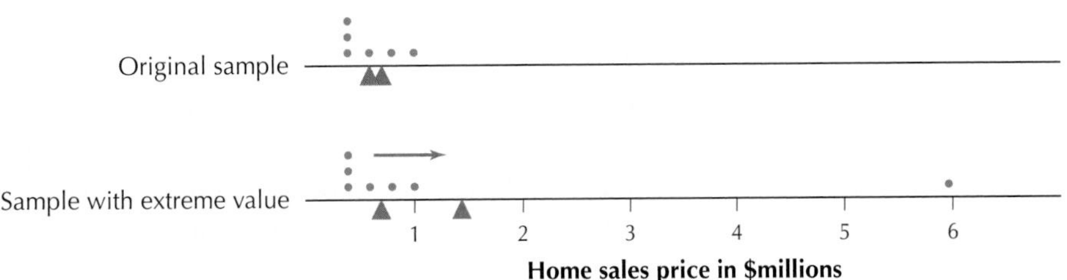


FIGURE 3.2 The mean (red triangles) is sensitive to extreme values, but the median (green triangles) is not.

The *Mean and Median* applet allows you to insert your own data values and see how changes in these values affect both the mean and the median.

Note that the formula $\frac{n+1}{2}$ gives the *position,* not the *value,* of the median. For example, the median home sales price for Table 3.2 is *not* $\frac{n+1}{2} = \frac{6+1}{2} = 3.5$.

## Example 3.4 Using technology to find the mean and median

Find the mean and median of the home sales prices in Table 3.2, using (a) the TI-83/84, (b) Excel, and (c) Minitab.

### Solution

Using the instructions in the Step-by-Step Technology Guide on page 97, we get the following output.

**a.** The first TI-83/84 screen shows $\bar{x} = 641{,}500$ and $n = 6$. The second screen shows the median Med = 560,500.

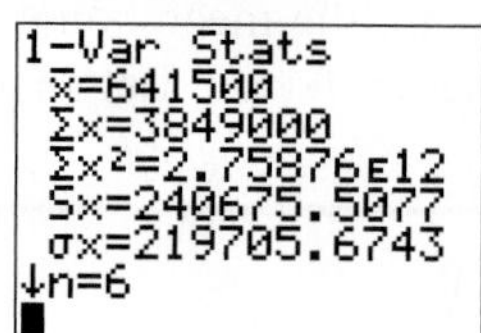


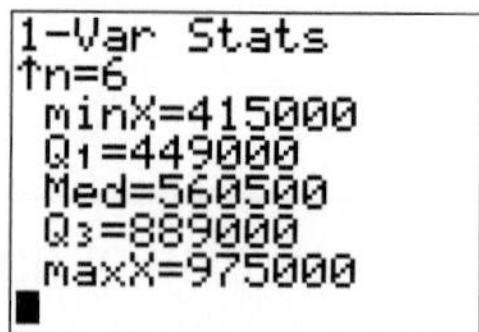


**b.** The mean, median, and sample size ("Count") $n$ are highlighted in the Excel output.

| *Home Price* | |
|---|---|
| Mean | 641500 |
| Standard Error | 98255.36457 |
| Median | 560500 |
| Mode | #N/A |
| Standard Deviation | 240675.5077 |
| Sample Variance | 57924700000 |
| Kurtosis | -1.878590247 |
| Skewness | 0.616113252 |
| Range | 560000 |
| Minimum | 415000 |
| Maximum | 975000 |
| Sum | 3849000 |
| Count | 6 |

**c.** Minitab gives us N = 6. (This is actually $n$ the sample size, not $N$ the population size.) The mean and median are highlighted. ■

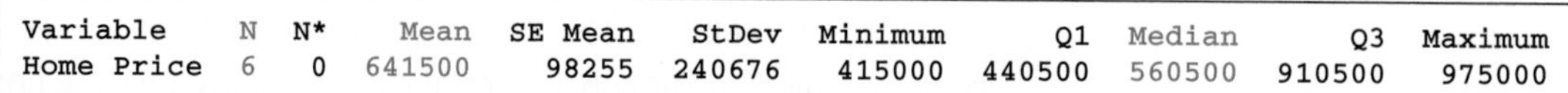

| Variable | N | N* | Mean | SE Mean | StDev | Minimum | Q1 | Median | Q3 | Maximum |
|---|---|---|---|---|---|---|---|---|---|---|
| Home Price | 6 | 0 | 641500 | 98255 | 240676 | 415000 | 440500 | 560500 | 910500 | 975000 |

## 3 The Mode

A third measure of center is called the **mode.** French speakers will recognize that the term *mode* in French refers to *fashion*. The popularity of clothing, cosmetics, music, and even basketball shoes often depends on just which style is in fashion. In a data set, the value that is most "in fashion" is the value that occurs the most.

> The **mode** of a data set is the data value that occurs with the greatest frequency.

## Example 3.5 Cost of mathematical journals

The rising cost of research journals has been taking an increasing bite out of library and research budgets. Table 3.3 contains the annual subscription cost of ten research journals in mathematics and statistics for 2006. Find the following.

**a.** The mean journal subscription cost
**b.** The median journal subscription cost
**c.** The mode journal subscription cost

Table 3.3 Annual subscription cost for ten research journals

| Journal | Annual cost (dollars) |
|---|---|
| Annals of Probability | 250 |
| Annals of Statistics | 250 |
| Bulletin of the American Mathematical Society | 402 |
| Mathematical Computation | 467 |
| Annals of Applied Probability | 850 |
| Proceedings of the American Mathematical Society | 1022 |
| Topology | 1582 |
| Journal of Mathematical Economics | 1653 |
| Transactions of the American Mathematical Society | 1744 |
| Journal of Pure and Applied Algebra | 3631 |

*Source:* American Mathematical Society Journal Survey, **www.ams.org/membership/journal-survey.html**.

## Solution

**a.** The sample mean journal cost is

$$\bar{x} = \frac{\sum x}{n} = \frac{250 + 250 + 402 + 467 + 850 + 1022 + 1582 + 1653 + 1744 + 3631}{10}$$

$$= \frac{11{,}851}{10} = \$1185.10$$

**b.** Since we have $n = 10$ journals, the median is the mean of the two data values that lie on either side of the $\left(\frac{n+1}{2}\right)^{\text{th}} = \left(\frac{10+1}{2}\right)^{\text{th}} = 5.5$th position. That is, the median is the mean of the 5th and 6th data values, \$850 and \$1022. The mean of these two journal costs is

$$\text{median journal cost} = \frac{850 + 1022}{2} = \$936$$

**c.** The mode is the data value that occurs with the greatest frequency. There are two journals that cost \$250 each. No other cost occurs more than once. Therefore, the mode cost is \$250. Note that this value of \$250 for the mode is not a very good measure of center for this data set. It is the minimum value. This illustrates a weakness of using the mode as a measure of center. ■

One of the strengths of using the mode is that it can be used with categorical, or qualitative, data. Suppose you asked your friends what their favorite flower was. Six of them answered "rose," five answered "iris," three answered "lily," and one answered "daffodil." Note that these data are categorical, not numerical. What is the most popular flower among your friends? That is, what is the mode of the variable *favorite flower*? The most frequently occurring answer is "rose," with six friends preferring roses. Since this is the highest frequency, the rose represents the mode of the variable *favorite flower.* Now think about the following question for a moment: What is the mean of the variable favorite flower? Actually, because the data values are qualitative (non-numeric), the idea of a mean or median for the variable *favorite flower* is essentially meaningless. We cannot use arithmetic with categorical variables, and so means and medians cannot be found.

Finally, it may happen that no data value occurs more than once, in which case we say there is *no mode*. Or more than one data value could occur with the greatest frequency; in which case we would say there is more than one mode. Data sets with one mode are *unimodal;* data sets with more than one mode are *multimodal*.

## What If Scenario ? Give the Calculator a Rest

What If Scenarios offer you a chance to reflect on how changes in the initial conditions will percolate through the various aspects of a problem. The only requirement is to put your calculator down and think through the problem. No data are given. You are asked to find the answers by using your knowledge of what the statistics *represent*.

Consider Example 3.5 once again. Now imagine: *what if* there was a data entry error (a typo), and the subscription cost for the most expensive journal, the *Journal of Pure and Applied Algebra,* was not \$3631 but actually some *unspecified* dollar amount greater than \$3631. Describe how and why this change would have affected the following, if at all:

**a.** The mean subscription cost

**b.** The median subscription cost

**c.** The mode subscription cost

### Solution

**a.** Consider Figure 3.3, a dotplot of the subscription costs, with the triangle indicating the mean cost of \$1185.10. Recall that this represents the balance point of the data. As the cost of the most expensive journal increases (blue arrow), the point at which the data balance (that is, the mean) also moves somewhat to the right. Thus, the mean cost will increase.

**b.** Recall from Example 3.5 that the median is the mean of the 5th and 6th data values, \$850 and \$1022. In other words, the median *ignores* most of the data values, including the most expensive journal, the one whose cost has increased. Therefore, the median will remain unchanged.

**c.** The mode also remains unchanged, since the only data value that occurs more than once is the original mode, \$250.

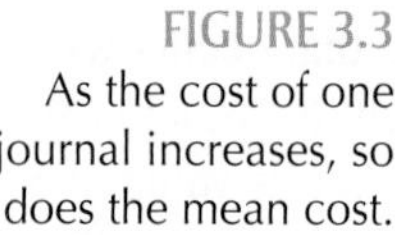
FIGURE 3.3
As the cost of one journal increases, so does the mean cost.

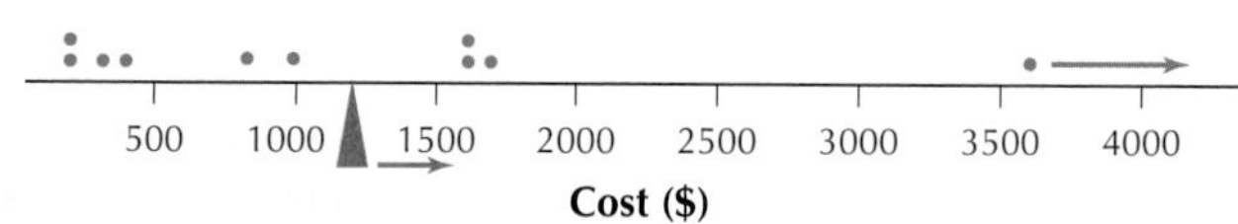


## 4 Skewness and Measures of Center

The skewness of a distribution can often tell us something about the relative values of the mean and the median.

### Example 3.6 Calories per gram

In Example 2.13 (page 57) we found that the distribution of calories per gram of 961 food items (from the data set *Nutrition*) was right-skewed, as shown in Figure 3.4. Use Excel to calculate the mean and median of these data.

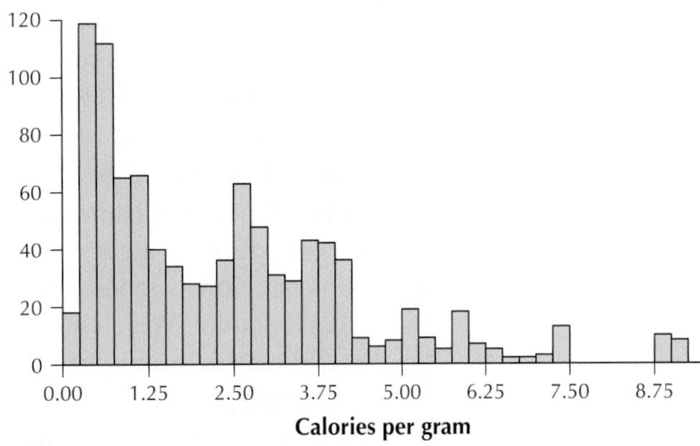


FIGURE 3.4 Histogram of number of calories per gram for 961 food items.

### Solution

The Excel descriptive statistics are shown here. Note that the mean calories per gram is greater than the median.

| *Calories per Gram* | |
|---|---|
| Mean | 2.253382 |
| Median | 1.84 |

For right-skewed data, the mean is usually greater than the median. We can generalize this as follows (Figure 3.5). ■

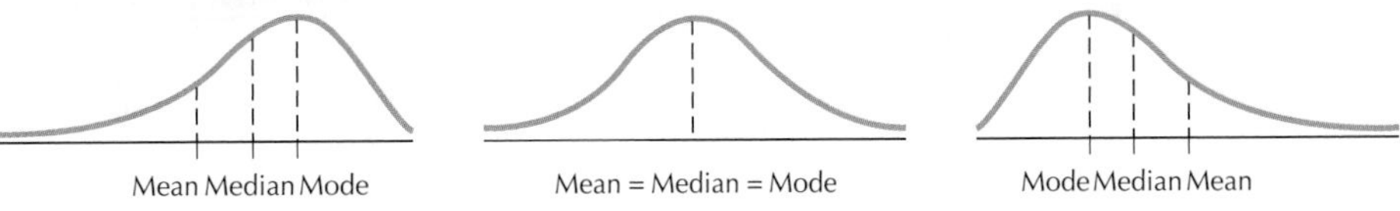


FIGURE 3.5 How skewness affects the mean and median.

**How Skewness Affects the Mean and Median**

- For a right-skewed distribution, the mean is larger than the median.
- For a left-skewed distribution, the median is larger than the mean.
- For a symmetric unimodal distribution, the mean, median, and mode are fairly close to one another.

Getty Images/Uppercut

CASE STUDY

## Can the Financial Experts Beat the Darts?

Recall the contest held by the *Wall Street Journal* to compare the performance of stock portfolios chosen by financial experts and stocks chosen at random by throwing darts at the *Journal* stock pages. By October 1998, 100 such contests had been held. We will examine the results of these 100 contests in various ways, using the methods we have learned thus far, and will return to examine them further as we acquire more analysis tools. Let's start by reporting the raw result data. The percentage increase or decrease in stock prices was calculated for the portfolios chosen by the professional financial advisers and by the randomly thrown darts and was compared with the percentage net change in the Dow Jones Industrial Average (DJIA).

### Exploratory Data Analysis

Figure 3.6 shows comparative dotplots of the percentage net change in price for the professionally selected portfolio, the randomly selected darts portfolio, and the DJIA, over the course of the 100 contests. First, estimate where the center of each distribution is located by choosing one spot to place a fulcrum that will perfectly balance the data. This balance spot is the *mean*. For fun, write down your guess for the mean for the professionals so you can see how close you were when we provide the descriptive statistics later. Now compare this with where you would find the balance spot (mean) for the darts dotplot. Which numerical value is larger, the balance spot for the pros or the darts? Please take a moment to think about what it is you are doing. You are comparing the mean portfolio performances for the professionals and the darts, and you are not even close to using a formula or a calculator. This is *exploratory data analysis*. You are using *graphical* methods to compare *numerical* statistics.

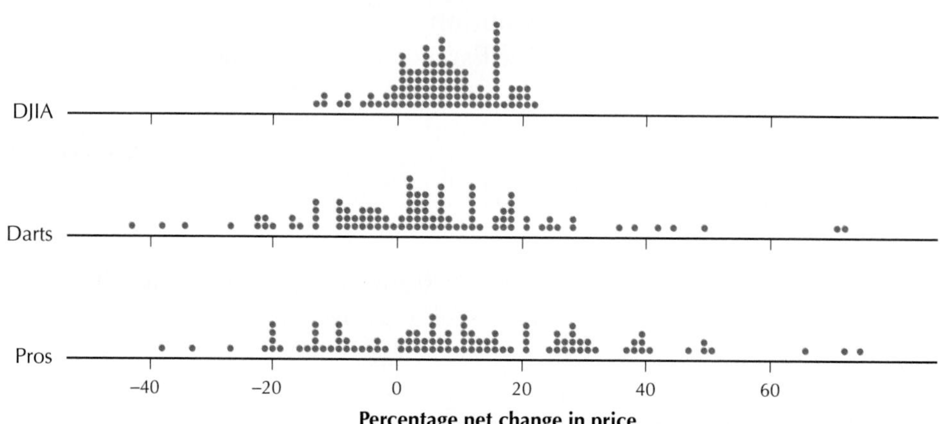


FIGURE 3.6
Dotplots of the percentage net price change for the Dow Jones Industrial Average, the randomly selected darts portfolio, and the professionally selected portfolio.

Hopefully, you discovered that the estimated mean for the pros is greater than the estimated mean for the darts. This is not particularly surprising, is it? Next, where would you place your fulcrum to find the mean for the DJIA dotplot? Compare the numerical value for the DJIA balance spot to the mean you found for the dotplot for the pros. Write down your estimate of the means for the DJIA and darts dotplots, so you can see how close you were later. Again, hopefully, you found that the estimated professionals' mean was higher than that of the DJIA. Now, a tougher comparison is to compare the estimated DJIA mean with that of the darts. Which of these two do you think is higher?

Finally, Minitab provides us with the mean percentage net price changes, as shown in Figure 3.7. Over the course of 100 contests, the mean price for the portfolios chosen by the professional financial advisers increased by 10.95%, by 6.793% for the DJIA, and by 4.52% for the random darts portfolio.

FIGURE 3.7
Mean percentage net price change for the professionals, darts, and DJIA.

| Variable | N | Mean |
|---|---|---|
| Pros | 100 | 10.95 |
| Darts | 100 | 4.52 |
| DJIA | 100 | 6.793 |

This is evidence in support of the view that financial experts can consistently outperform the market. We return to this Case Study in Section 3.2 (page 113).

## STEP-BY-STEP TECHNOLOGY GUIDE: Descriptive Statistics

### TI-83/84

***Step 1*** Enter the data in **L1** using the instructions **(STAT > 1: Edit)** found in the Step-by-Step Technology Guide in Section 2.2.
***Step 2*** Press **STAT**. Use the **right arrow** button to move the cursor so that **CALC** is highlighted.
***Step 3*** Select **1-Var Stats**, and press **ENTER**.
***Step 4*** On the home screen, the command **1-Var Stats** is shown. Press **2nd**, then **L1** (above the **1** key) and press **ENTER**.

### EXCEL

***Step 1*** Enter the data in column **A**.
***Step 2*** Select **Data > Data Analysis**.
***Step 3*** Select **Descriptive Statistics** and click **OK**.
***Step 4*** For the **Input Range**, click and drag to select the data in column A.
***Step 5*** Check **Summary Statistics** and click **OK**.

### MINITAB

***Step 1*** Enter the data in column C1.
***Step 2*** Select **Stat > Basic Statistics > Display Descriptive Statistics...**
***Step 3*** The variable selection dialog box appears. Select the variable you want to summarize by double-clicking on it until it appears in the **Variables** box.
***Step 4*** Click **STATISTICS**.
***Step 5*** Select the desired statistics and click **OK**.

## Section 3.1 SUMMARY

1 Measures of center are introduced in Section 3.1. The sample mean ($\bar{x}$) represents the sum of the data values in the sample divided by the sample size ($n$). The population mean ($\mu$) represents the sum of the data values in the population divided by the population size ($N$). The mean is sensitive to the presence of extreme values.

2 The median occupies the middle position when the data are put in ascending order and is not sensitive to extreme values.

3 The mode is the data value that occurs with the greatest frequency. Modes can be applied to categorical data as well as numerical data but are not always reliable as measures of center.

4 The skewness of a distribution can often tell us something about the relative values of the mean and the median.

## Section 3.1 EXERCISES

### CLARIFYING THE CONCEPTS

**1.** Explain what a measure of center is.

**2.** Which measure may be used as the balance point of the data set? Explain how this works.

**3.** Explain what we mean when we say that the mean is sensitive to the presence of extreme values. Explain whether the median is sensitive to extreme values.

### PRACTICING THE TECHNIQUES

**4.** Clickstream analysis is the study of how humans behave on the Internet.[2] One measure is the number of new page requests (clicks) that the visitor makes. A sample of the visitors to a particular Web site had the following total numbers of clicks.

1 5 3 4 3 2 3 7

Use this information to find the following measures of center.

**a.** Mean **b.** Median **c.** Mode

**5.** The table below contains the miles per gallon (city driving) for six 2004 vehicles. Use this information to find the following measures of center.

**a.** Mean **b.** Median **c.** Mode

| Vehicle | City mpg |
|---|---|
| Honda Civic | 36 |
| Toyota Camry | 24 |
| Ford Taurus | 20 |
| Pontiac Grand Prix | 20 |
| Jaguar X-Type | 18 |
| Lincoln Town Car | 17 |

**6.** Which states have the youngest overall populations? The table below shows the 8 states with the lowest median age.[3] Use this information to find the following measures of center. (The data points themselves are medians, but you can still find summary statistics for a set of medians.)

**a.** Mean **b.** Median

| State | Median age |
|---|---|
| Utah | 28.5 |
| Texas | 33.2 |
| Alaska | 33.9 |
| Georgia | 34.3 |
| California | 34.4 |
| Arizona | 34.5 |
| Idaho | 34.6 |
| Colorado | 34.7 |

Use the information in Table 3.4 to answer Exercises 7–10.

**TABLE 3.4 Top ten NASCAR winners in the modern era**

| Rank | Driver | Total | Super speedways | Short tracks | Road courses |
|---|---|---|---|---|---|
| 1 | Darrell Waltrip | 84 | 18 | 47 | 9 |
| 2 | Dale Earnhardt | 76 | 29 | 27 | 3 |
| 3 | Jeff Gordon | 75 | 15 | 15 | 10 |
| 4 | Cale Yarborough | 69 | 15 | 29 | 9 |
| 5 | Richard Petty | 60 | 19 | 23 | 8 |
| 6 | Bobby Allison | 55 | 24 | 12 | 7 |
| 7 | Rusty Wallace | 55 | 5 | 25 | 11 |
| 8 | David Pearson | 45 | 20 | 1 | 4 |
| 9 | Bill Elliott | 44 | 16 | 2 | 4 |
| 10 | Mark Martin | 35 | 5 | 7 | 5 |

*Source:* Nascar.com.

**7.** Refer to the super speedways data. Find the following measures of center.

**a.** Mean **b.** Median **c.** Mode

**8.** Refer to the short tracks data. Find the following measures of center.

**a.** Mean **b.** Median **c.** Mode

**9.** Refer to the road courses data. Find the following measures of center.

**a.** Mean **b.** Median **c.** Mode

**10.** Refer to the totals data. Find the following measures of center.

**a.** Mean **b.** Median **c.** Mode

**11.** Use the information in Table 3.5 to find the following measures of center for the variable *age*.

**a.** Mean **b.** Median **c.** Mode

**TABLE 3.5 Best actress Oscar winners, 1996–2005**

| Year | Actress | Film | Age |
|---|---|---|---|
| 1996 | Susan Sarandon | Dead Man Walking | 49 |
| 1997 | Frances McDormand | Fargo | 39 |
| 1998 | Helen Hunt | As Good As It Gets | 34 |
| 1999 | Gwyneth Paltrow | Shakespeare in Love | 26 |
| 2000 | Hilary Swank | Boys Don't Cry | 25 |
| 2001 | Julia Roberts | Erin Brockovich | 33 |
| 2002 | Halle Berry | Monster's Ball | 35 |
| 2003 | Nicole Kidman | The Hours | 35 |
| 2004 | Charlize Theron | Monster | 28 |
| 2005 | Hilary Swank | Million Dollar Baby | 30 |

For Exercises 12–14, use the data in Table 3.6, on the 2004 SAT I results, for the eight states with the highest participation rates. The SAT results are the means for each particular state. (The SAT scores are means, but you can still find summary statistics for sets of means.)

**TABLE 3.6 2004 SAT scores for the states with the highest participation rates**

| State | Participation rate | SAT I Verbal | SAT I Math |
|---|---|---|---|
| New York | 87% | 497 | 510 |
| Connecticut | 85% | 515 | 515 |
| Massachusetts | 85% | 518 | 523 |
| New Jersey | 83% | 501 | 514 |
| New Hampshire | 80% | 522 | 521 |
| Maine | 76% | 505 | 501 |
| Pennsylvania | 74% | 501 | 502 |
| Delaware | 73% | 500 | 499 |

*Source:* College Board.

**12.** For the participation rate, find the following measures of center.

**a.** Mean **b.** Median **c.** Mode

**13.** For SAT I Verbal, find the following measures of center.

**a.** Mean **b.** Median **c.** Mode

**14.** For SAT I Math, find the following measures of center.

**a.** Mean **b.** Median **c.** Mode

## APPLYING THE CONCEPTS

**15. Liberal Arts Majors.** Here are the declared liberal arts majors for a sample of students at a local college:

| | | | | |
|---|---|---|---|---|
| English | History | Spanish | Art | Theater |
| Theater | Philosophy | English | Music | Communication |
| Political science | Communication | History | English | Art |
| English | History | Spanish | Economics | Communication |
| Music | English | Economics | Theater | Music |

**a.** What is the mode of this data set? Does this mean that most students at the college are majoring in this subject?

**b.** Does the idea of the mean or median of this data set make any sense? Explain clearly why not.

**c.** How would you respond to someone who claimed that economics was the most popular major?

**Fiction Best Sellers.** For Exercises 16 and 17, refer to Table 3.7, which lists the top five hardcover fiction best sellers from the *New York Times* best-seller list for January 30, 2007.

**TABLE 3.7** ***New York Times*** **hardcover fiction best sellers**

| Rank | Title | Author | Price |
|---|---|---|---|
| 1 | Plum Lovin' | Janet Evanovich | $16.95 |
| 2 | For One More Day | Mitch Albom | $21.95 |
| 3 | Cross | James Patterson | $27.99 |
| 4 | The Hunters | W. E. B. Griffin | $26.95 |
| 5 | Exile | Richard North Patterson | $26.00 |

**16.** Find the mean, median, and mode for the price of these five books on the best-seller list. Suppose a salesperson claimed that the price of a typical book on the best-seller list is less than $20. How would you use these statistics to respond to this claim?

**17.** Multiply the price of each book by 5.

**a.** Now find the mean of these new prices.

**b.** How does this new mean relate to the original mean?

**c.** Construct a rule to describe this situation in general.

**Car Model Years.** Refer to Figure 3.8 for Exercises 18–20. The data represent the model year for a sample of cars in a used car lot.

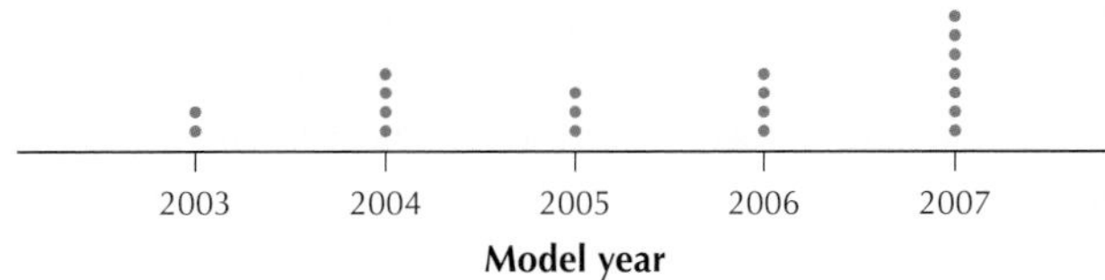


Figure 3.8 Dotplot of model year.

**18.** What are the mean, median, and mode of the model year?

**19.** Calculate a new statistic "age of the car in 2009" as follows: take the model year and subtract it from 2009.

**a.** Find the mode of the car ages.

**b.** Find the mean and median of the car ages.

**20.** What will be the mean, median, and mode of the car ages in 2012?

**21. Skewness and Symmetry.** Consider the accompanying distributions. What can we say about the values of the mean, median, and mode in relation to one another?

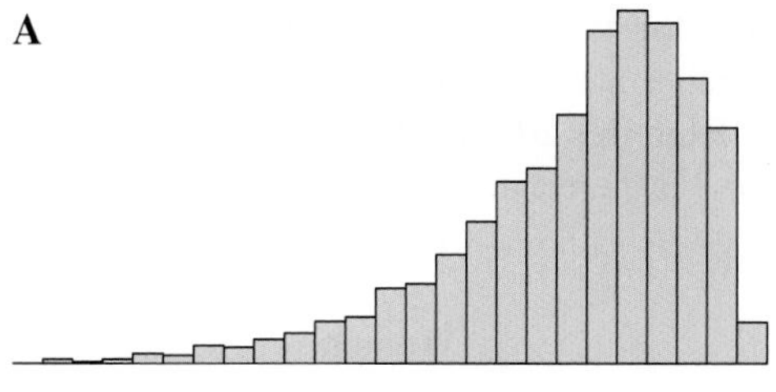


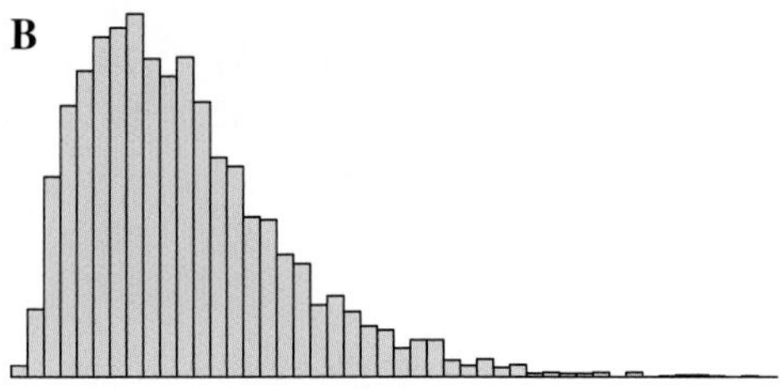


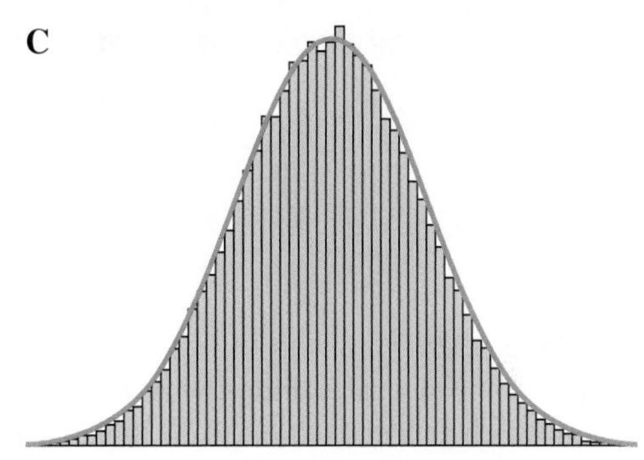


**a.** The distribution in A

**b.** The distribution in B

**c.** The distribution in C

**Nutrition Ratings of Breakfast Cereals.** Refer to the following information for Exercises 22–25.

**Descriptive Statistics: Rating**

| Variable | N | Mean | Median | TrMean | StDev | SE Mean |
|---|---|---|---|---|---|---|
| Rating | 59 | 45.46 | 42.00 | 44.76 | 14.39 | 1.87 |

| Variable | Minimum | Maximum | Q1 | Q3 |
|---|---|---|---|---|
| Rating | 22.40 | 93.70 | 35.25 | 53.37 |

**22.** Find the following sample statistics.

**a.** The sample size

**b.** The sample mean

**c.** The sample median

**d.** The highest and lowest ratings in the sample

**23.** What do these statistics tell us about the skewness of the distribution?

**24.** If we subtracted five points from each cereal's rating, how would that affect the mean, median, and mode? Would it affect each of the measures equally?

**25.** If we cut each of the cereals' ratings in half, how would that affect the mean, median, and mode? Would it affect each of the measures equally?

**Pulse Rates for Men and Women.** Use the following data set of the pulse rates for a sample of 65 women and 65 men to answer Exercises 26–29.[4] What is the mean pulse rate for humans? Is there evidence that it differs for females and males?

**Women**

| | | | | | | | | | |
|---|---|---|---|---|---|---|---|---|---|
| 69 | 62 | 75 | 66 | 68 | 57 | 61 | 84 | 61 | 77 |
| 62 | 71 | 68 | 69 | 79 | 76 | 87 | 78 | 73 | 89 |
| 81 | 73 | 64 | 65 | 73 | 69 | 57 | 79 | 78 | 80 |
| 79 | 81 | 73 | 74 | 84 | 83 | 82 | 85 | 86 | 77 |
| 72 | 79 | 59 | 64 | 65 | 82 | 64 | 70 | 83 | 89 |
| 69 | 73 | 84 | 76 | 79 | 81 | 80 | 74 | 77 | 66 |
| 68 | 77 | 79 | 78 | 77 | | | | | |

**Men**

| | | | | | | | | | |
|---|---|---|---|---|---|---|---|---|---|
| 70 | 71 | 74 | 80 | 73 | 75 | 82 | 64 | 69 | 70 |
| 68 | 72 | 78 | 70 | 75 | 74 | 69 | 73 | 77 | 58 |
| 73 | 65 | 74 | 76 | 72 | 78 | 71 | 74 | 67 | 64 |
| 78 | 73 | 67 | 66 | 64 | 71 | 72 | 86 | 72 | 68 |
| 70 | 82 | 84 | 68 | 71 | 77 | 78 | 83 | 66 | 70 |
| 82 | 73 | 78 | 78 | 81 | 78 | 80 | 75 | 79 | 81 |
| 71 | 83 | 63 | 70 | 75 | | | | | |

**26.** Examine Figure 3.9.
- **a.** Without doing any calculations, what is your impression of which gender, if any, has the higher overall pulse rate?
- **b.** Find the mean pulse rate for the males by estimating the location of the balance point.
- **c.** Find the mean pulse rate for the females by estimating the location of the balance point.
- **d.** Which gender has the higher mean pulse rate? Does this agree with your earlier impression?

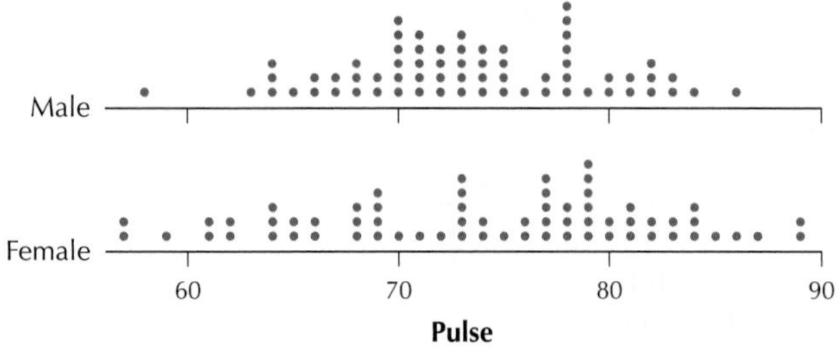


Figure 3.9 Comparative dotplots of pulse rates, by gender.

**27.** Find the following medians.
- **a.** The median pulse rate for the males
- **b.** The median pulse rate for the females
- **c.** Which gender has the higher median pulse rate? Does this agree with your findings for the mean earlier?

**28.** Find the following modes.
- **a.** The mode pulse rate for the males
- **b.** The mode pulse rate for the females
- **c.** Which gender has the higher mode pulse rate? Does this agree with your findings for the mean earlier?

WHAT IF? **29.** *What if* the fastest pulse rate for the men was a typo and should have been an unspecified lower pulse rate. Describe how and why this change would have affected the following, if at all. Would they increase, decrease, remain unchanged? Or is there insufficient information to tell what would happen? Explain your answers.
- **a.** The mean men's pulse rate
- **b.** The median men's pulse rate
- **c.** The mode men's pulse rate

**Baseball Player Salaries.** Refer to Table 3.8 for Exercises 30 and 31.

**TABLE 3. 8 Top five highest paid Major League Baseball players, 2004 season**

| Player | Team | Salary |
|---|---|---|
| 1. Alex Rodriguez | New York Yankees | $21,726,881 |
| 2. Manny Ramirez | Boston Red Sox | $20,409,542 |
| 3. Carlos Delgado | Toronto Blue Jays | $19,700,000 |
| 4. Derek Jeter | New York Yankees | $18,600,000 |
| 5. Barry Bonds | San Francisco Giants | $18,000,000 |

**30.** Find the mean and median of the salaries.

**31.** Suppose a new college graduate can expect to earn a mean of $60,000 a year for an entire career of forty years. How much money will he or she earn over his or her lifetime? What percentage is this of what Alex Rodriguez made in 2004?

**32.** Refer to Table 3.5 (page 98).
- **a.** Find the mode for the variable *actress*.
- **b.** Discuss the meaning of the mean or the median for this variable.

**33.** Refer to Table 3.6 (page 98). Would you expect your measures of center for the participation rate to be representative of all 50 states? Explain clearly why or why not.

**Body Temperature of Women and Men.** Use the following information for Exercises 34–37. We are often told that the mean human body temperature is 98.6 degrees Fahrenheit. Do you think that there is evidence that the mean body temperature differs between women and men? The following data set shows the body temperatures for a group of women and men.[5]

**Women**

| | | | | | | | | | | | |
|---|---|---|---|---|---|---|---|---|---|---|---|
| 96.4 | 96.7 | 96.8 | 97.2 | 97.2 | 97.4 | 97.6 | 97.7 | 97.7 | 97.8 | 97.8 | 97.8 |
| 97.9 | 97.9 | 97.9 | 98.0 | 98.0 | 98.0 | 98.0 | 98.0 | 98.1 | 98.2 | 98.2 | 98.2 |
| 98.2 | 98.2 | 98.2 | 98.3 | 98.3 | 98.3 | 98.4 | 98.4 | 98.4 | 98.4 | 98.4 | 98.5 |
| 98.6 | 98.6 | 98.6 | 98.6 | 98.7 | 98.7 | 98.7 | 98.7 | 98.7 | 98.7 | 98.8 | 98.8 |
| 98.8 | 98.8 | 98.8 | 98.8 | 98.8 | 98.9 | 99.0 | 99.0 | 99.1 | 99.1 | 99.2 | 99.2 |
| 99.3 | 99.4 | 99.9 | 100.0 | 100.8 | | | | | | | |

**Men**

96.3 96.7 96.9 97.0 97.1 97.1 97.1 97.2 97.3 97.4 97.4 97.4
97.4 97.5 97.5 97.6 97.6 97.6 97.7 97.8 97.8 97.8 97.8 97.9
97.9 98.0 98.0 98.0 98.0 98.0 98.0 98.1 98.1 98.2 98.2 98.2
98.2 98.3 98.3 98.4 98.4 98.4 98.4 98.5 98.5 98.6 98.6 98.6
98.6 98.6 98.6 98.7 98.7 98.8 98.8 98.8 98.9 99.0 99.0 99.0
99.1 99.2 99.3 99.4 99.5

**34.** Refer to Figure 3.10.

- **a.** Without doing any calculations, what is your impression of which gender has the higher body temperature?
- **b.** Find the mean body temperature for the males by estimating where the fulcrum would be placed that would balance the male body temperatures.
- **c.** Find the mean body temperature for the females by estimating where the fulcrum would be placed that would balance the female body temperatures.
- **d.** Which gender do you think has the higher mean body temperature, using your estimates from **(b)** and **(c)**?

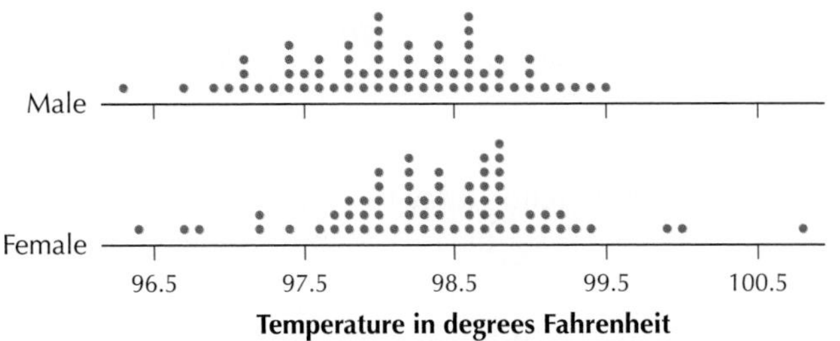


Figure 3.10 Comparative dotplots of body temperatures, by gender.

**35.** Find the following medians.

- **a.** The median body temperature for the males
- **b.** The median body temperature for the females
- **c.** Which gender has the higher median body temperature? Does this agree with your findings for the mean?

**36.** Find the following modes.

- **a.** The mode body temperature for the males
- **b.** The mode body temperature for the females
- **c.** Which gender has the higher mode(s)? Does this agree with your findings for the mean and median?

WHAT IF? **37.** *What if* the lowest temperature reading for the women was a typo and should have been an unspecified temperature lower than the original 96.4 degrees. Describe how and why this change would have affected the following, if at all. Would they increase, decrease, remain unchanged? Or is there insufficient information to tell what would happen? Explain your answers.

- **a.** The mean women's body temperature
- **b.** The median women's body temperature
- **c.** The mode women's body temperature

**CONSTRUCT YOUR OWN DATA SETS**

**38.** Construct your own data set with $n = 10$, where the mean, the median, and the mode are all the same. Yes, just make up your own list of numbers, as long as the mean, median, and mode are all the same. Draw a dotplot. Comment on the skewness of the distribution.

**39.** Construct your own data set with $n = 10$, where the mean is greater than the median, which is greater than the mode. Draw a dotplot. Comment on the skewness of the distribution.

**40.** Construct your own data set with $n = 10$, where the mode is greater than the median, which is greater than the mean. Draw a dotplot. Comment on the skewness of the distribution.

**41.** Construct your own data set with $n = 3$. Let the mean and median be equal. Now, alter the three data values so that the mean of the altered data set has increased while the median of the altered data set has decreased.

Use the *Mean and Median* applet for Exercises 42 and 43.

**42.** Insert three points on the line by clicking just below it, two near the left side and one near the middle.

- **a.** Click and drag the rightmost point to the right.
- **b.** Describe what happens to the mean when you do this.
- **c.** Describe what happens to the median when you do this.

**43.** Explain why each of the measures behaves the way it does in the previous exercise.

## 3.2 Measures of Variability

*Objectives:* **By the end of this section, I will be able to…**

1. Understand and calculate the range of a data set.
2. Explain in my own words what a deviation is.
3. Calculate the variance and the standard deviation for a population or a sample.

## The Range

Section 3.1 asked the question "Where is the center of the data set?" Is that all there is to know about a data set? Definitely not! Two data sets can have exactly the same mean, median, and mode and yet be quite dissimilar. We need measures that summarize the data set in a different dimension, namely, the variation or variability of the data. Section 3.2 will tackle how we can further refine our description of a data set by asking the question "How spread out is the data set?"

### Example 3.7 Different data sets with the same measures of center

AP Photo/Martin Meissner

Table 3.9 contains the heights in inches of the players on two volleyball teams.

Table 3.9 Women's volleyball team heights (in inches)

| Western Massachusetts University | Northern Connecticut University |
|---|---|
| 60 | 66 |
| 70 | 67 |
| 70 | 70 |
| 70 | 70 |
| 75 | 72 |

**a.** Describe in words the variability of the heights of the two teams.
**b.** Verify that the means, medians, and modes for the two teams are equal.

**Solution**

**a.** There are some distinct differences between the teams. The Western Massachusetts (WMU) team has a player who is relatively short (only 60 inches; 5 feet tall) and a player who is very tall (75 inches; 6 feet, 3 inches tall). The Northern Connecticut (NCU) team has players whose heights are all within 6 inches of each other. But despite these differences, the mean, median, and mode of the heights for the two teams are precisely the same.

**b.** As illustrated in Figure 3.11, the mean height (red triangle) for each team is 69 inches, the median height (green triangle) for each team is 70 inches, and the mode height (yellow triangle) for each team is 70 inches.

$$\bar{x}_{\text{WMU}} = \frac{60 + 70 + 70 + 70 + 75}{5} = \frac{345}{5} = 69$$

$$\bar{x}_{\text{NCU}} = \frac{66 + 67 + 70 + 70 + 72}{5} = \frac{345}{5} = 69$$

Clearly, these measures of location do not give us the whole picture. We need **measures of variability** (or **measures of spread**) that will describe how spread out the data values are. Figure 3.11 illustrates that the heights of the WMU team are *more spread out* than the heights of the NCU team.

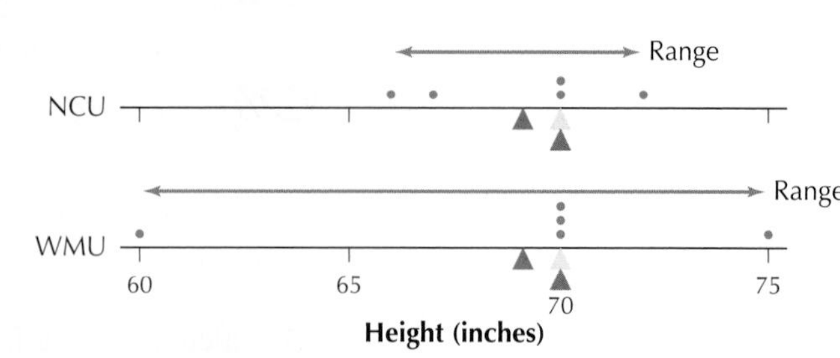


FIGURE 3.11 Comparative dotplots of the heights of two volleyball teams.

Just as there were several measures of the center of a data set, there are also a variety of ways to measure how spread out a data set is. The simplest measure of variability is the **range**.

> The **range** of a data set is the difference between the largest value and the smallest value in the data set:
>
> range = largest value – smallest value

A larger range is an indication of greater variability, or greater spread, in the data set.

### Example 3.8 Range of the volleyball teams' heights

Calculate the range of player heights for each of the WMU and NCU teams.

**Solution**

From Figure 3.11, it is intuitively clear that the heights of the WMU team are more spread out than the heights of the NCU team. This should be reflected in the ranges. The range of the WMU team should be larger than the range of the NCU team, reflecting its greater variability.

$$\text{range}_{\text{WMU}} = \text{largest value} - \text{smallest value} = 75 - 60 = 15 \text{ inches}$$

$$\text{range}_{\text{NCU}} = \text{largest value} - \text{smallest value} = 72 - 66 = 6 \text{ inches}$$

The range for WMU is indeed larger than the range for NCU, reflecting WMU's greater variability in height. ■

The range is quite simple to calculate. However, it does have its drawbacks. For example, the range is quite sensitive to extreme values, since it is calculated from the difference of the two most extreme values in the data set. *It completely ignores all of the other data values in the data set.* We would prefer our measure of variability to quantify spread with respect to the center, as well as to actually use all of the available data values. Such a measure is the **standard deviation.**

## 2 What Is a Deviation?

Before we learn about the standard deviation, we need to get a firm understanding of what a **deviation** means, in the statistical sense. We shall use an example to do this.

### Example 3.9 Amanda's and Bethany's delivery service

Amanda and Bethany have contracted with a local document dispatch service to deliver documents in midtown Manhattan. Their clients would like to be able to count on a consistent delivery time schedule. Not only do the documents have to arrive in a timely fashion, but the delivery times should not vary a lot. One particularly important client kept careful track of the delivery times (in minutes) for documents delivered by Amanda and Bethany. See Table 3.10 and Figure 3.12. Describe the variability in Amanda's and Bethany's delivery times.

Table 3.10 Delivery times (in minutes) for Amanda and Bethany

| | | | | | | | | | | |
|---|---|---|---|---|---|---|---|---|---|---|
| **Amanda** | 15 | 17 | 12 | 15 | 10 | 16 | 14 | 20 | 16 | 15 |
| **Bethany** | 5 | 22 | 15 | 25 | 4 | 18 | 10 | 29 | 17 | 5 |

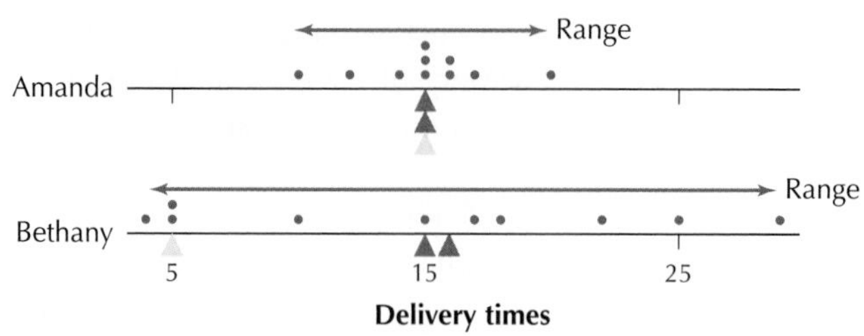


FIGURE 3.12 Comparative dotplots of Amanda's and Bethany's delivery times.

**Solution**

Note that Bethany's delivery times vary more than Amanda's. Bethany has some long delivery times, some short, and some of moderate length. On the other hand, Amanda's delivery times are mostly of moderate length. We can say that Bethany's delivery times

- are *more spread out,*
- show *greater variability,* and
- have *more variation.*

**Develop Your Intuitive Sense**

For each new statistic that you learn, try to get an intuitive feel for what it means. That is just as important as being able to calculate it—even more so nowadays because we have calculators and computers to do the calculations. For example, suppose you had miscalculated Amanda's range as being larger than Bethany's range. A glance at Figure 3.12, combined with your intuitive grasp of the range as a measure of spread, would have helped you to catch your error.

In this section, we will learn three different ways of measuring the spread of a data set (further methods await later sections): the range, the variance, and the standard deviation. Now, we know that the greater the spread of a data set, the greater the range. This holds true for all of the measures of spread that we will discuss in this book. In other words, the greater the spread of a data set, the greater the value of the range, the standard deviation, and the variance.

## Example 3.10 Bethany's measures of spread will be larger

*Without calculating* the range, the variance, and the standard deviation for Amanda's and Bethany's delivery times, can we say which of the two women will have the larger measures of spread?

**Solution**

It is clear from Figure 3.12 that Bethany's data set is more spread out than Amanda's. Thus, Bethany's data will have the larger measure of spread, whether we use the range, variance, or standard deviation.

Notice that even though these last two measures have not yet been defined, you can state with confidence which data set has the larger, say, standard deviation. You might think that, if the variation in Bethany's delivery times is larger, doesn't that mean that the mean length of her delivery times is longer as well? Not at all. The two sample means are equal, as shown by the red triangle in Figure 3.12.

$$\bar{x}_{\text{Amanda}} = \frac{\sum x}{n} = \frac{15 + 17 + 12 + 15 + 10 + 16 + 14 + 20 + 16 + 15}{10} = \frac{150}{10} = 15$$

$$\bar{x}_{\text{Bethany}} = \frac{\sum x}{n} = \frac{5 + 22 + 15 + 25 + 4 + 18 + 10 + 29 + 17 + 5}{10} = \frac{150}{10} = 15$$

Some measures of spread, such as the variance or standard deviation, rely on the concept of *deviation*.

> **Deviation**
>
> A **deviation** for a given data value $x$ is the difference between the data value and the mean of the data set. For a sample, the deviation equals $x - x$. For a population, the deviation equals $x - \mu$.
>
> - If the data value is larger than the mean, the deviation will be positive.
> - If the data value is smaller than the mean, the deviation will be negative.
> - If the data value equals the mean, the deviation will be zero.
>
> The deviation can roughly be thought of as the distance between a data value and the mean, except that the deviation can be negative while distance is always positive.

## Example 3.11 Calculating the deviations

Calculate the deviations for Amanda's and Bethany's delivery times.

**Solution**

- Amanda's mean delivery time was $\bar{x} = 15$ minutes. Her first delivery (first data value) also happened to be $x = 15$ minutes long. So the deviation for this first delivery is $x - \bar{x} = 15 - 15 = 0$. A deviation of zero occurs when a data value has the same value as the mean.
- Amanda's second delivery was $x = 17$ minutes long. So the deviation for this data value is $x - \bar{x} = 17 - 15 = 2$. A deviation of 2 implies that this data value lies two units (in this case, minutes) above the mean.
- Amanda's third delivery was 12 minutes long, resulting in a deviation of $x - \bar{x} = 12 - 15 = -3$.

Continuing in this way, we find the deviations for all of Amanda's and Bethany's delivery times, as recorded in Table 3.11. ■

Table 3.11 Delivery times and their deviations

| | Delivery | | | | | | | | | |
|---|---|---|---|---|---|---|---|---|---|---|
| | **1** | **2** | **3** | **4** | **5** | **6** | **7** | **8** | **9** | **10** |
| Amanda's times | 15 | 17 | 12 | 15 | 10 | 16 | 14 | 20 | 16 | 15 |
| Amanda's deviations | 0 | 2 | −3 | 0 | −5 | 1 | −1 | 5 | 1 | 0 |
| Bethany's times | 5 | 22 | 15 | 25 | 4 | 18 | 10 | 29 | 17 | 5 |
| Bethany's deviations | −10 | 7 | 0 | 10 | −11 | 3 | −5 | 14 | 2 | −10 |

Figure 3.13 illustrates these deviations in comparative dotplots. The green arrows indicate positive deviations, and the red arrows indicate negative deviations from the mean $\bar{x} = 15$. For example, the longest green arrow for Amanda is for her eighth delivery, which took 20 minutes. The deviation for this delivery is $x - \bar{x} = 20 - 15 = 5$. The shortest red arrow for Amanda is for her seventh delivery, which took 14 minutes. The deviation for this delivery is $x - \bar{x} = 14 - 15 = -1$. The lower dotplot shows Bethany's delivery times, with positive and negative deviations from the mean

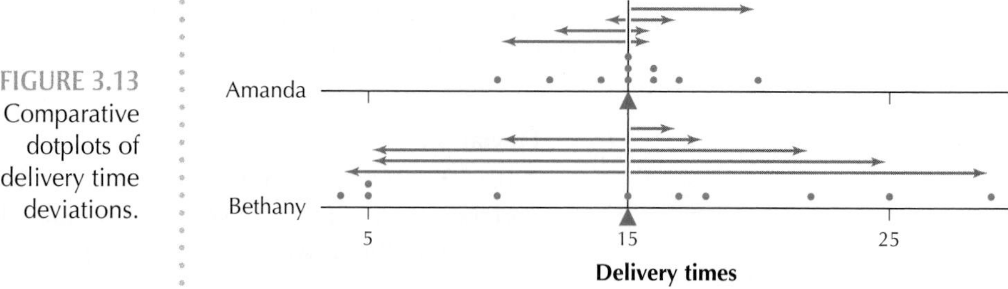


FIGURE 3.13 Comparative dotplots of delivery time deviations.

$\bar{x} = 15$. (Three of Amanda's deviations and one of Bethany's deviations are not indicated in the figure. Which deviations are these, and why are they not shown?) Bethany's arrows do seem longer, on average. Let us consider using the "mean deviation" as a way of measuring the spread. The mean of Amanda's deviations is

$$\text{Amanda's mean deviation} = \frac{0 + 2 + (-3) + 0 + (-5) + 1 + (-1) + 5 + 1 + 0}{10}$$
$$= \frac{0}{10} = 0$$

*Note:* Another way to avoid the negative deviations is to take their absolute value. The *mean absolute deviation* (*MAD*) is a measure of spread that looks at the average of the absolute values of the deviations:

$$\text{MAD} = \sum\left(\frac{|x-\bar{x}|}{n}\right).$$

You can confirm that Bethany's mean deviation is zero as well. It is not a coincidence that the mean deviation for both Amanda and Bethany turned out to be zero. In fact, *for any conceivable data set, the mean deviation is zero,* since the mean is the unique point on the number line where the negative deviations (the red arrows in Figure 3.13) precisely balance the positive deviations (the green arrows). Unfortunately, then, the mean deviation is not a very useful measure of spread. To avoid this problem of the positive and negative deviations canceling each other out, statisticians work with the squared deviations. The next two measures of spread use the squared deviations.

## 3 The Standard Deviation and the Variance

By far the most common measure of variability is the *standard deviation.* To compute the standard deviation and variance, we consider the squared deviations instead of the absolute deviations. Table 3.12 shows the squared deviations for Amanda and Bethany. Note that Bethany's squared deviations are on average larger than Amanda's, reflecting the greater spread in Bethany's delivery times. It is therefore logical to build our measure of spread using the *mean squared deviation.*

Table 3.12 Squared deviations of delivery times for Amanda and Bethany

| | Delivery | | | | | | | | | |
|---|---|---|---|---|---|---|---|---|---|---|
| | 1 | 2 | 3 | 4 | 5 | 6 | 7 | 8 | 9 | 10 |
| Amanda's times | 15 | 17 | 12 | 15 | 10 | 16 | 14 | 20 | 16 | 15 |
| Amanda's deviations | 0 | 2 | −3 | 0 | −5 | 1 | −1 | 5 | 1 | 0 |
| Amanda's squared deviations | $0^2 = 0$ | $2^2 = 4$ | $(-3)^2 = 9$ | $0^2 = 0$ | $(-5)^2 = 25$ | $1^2 = 1$ | $(-1)^2 = 1$ | $5^2 = 25$ | $1^2 = 1$ | $0^2 = 0$ |
| Bethany's times | 5 | 22 | 15 | 25 | 4 | 18 | 10 | 29 | 17 | 5 |
| Bethany's deviations | −10 | 7 | 0 | 10 | −11 | 3 | −5 | 14 | 2 | −10 |
| Bethany's squared deviations | $(-10)^2 = 100$ | $7^2 = 49$ | $0^2 = 0$ | $10^2 = 100$ | $(-11)^2 = 121$ | $3^2 = 9$ | $(-5)^2 = 25$ | $14^2 = 196$ | $2^2 = 4$ | $(-10)^2 = 100$ |

## The Population Variance $\sigma^2$

Suppose that we consider Amanda's and Bethany's delivery times to be populations rather than samples. Then the mean of each population is a population mean ($\mu = 15$) rather than a sample mean ($\bar{x} = 15$). Each deviation is $(x - \mu)$, and the mean squared deviation is called the **population variance** and is symbolized by $\boldsymbol{\sigma^2}$. This is the lowercase Greek letter sigma, not to be confused with the uppercase sigma ($\Sigma$) used for summation.

**Population Variance $\sigma^2$**

The **population variance $\boldsymbol{\sigma^2}$** is the mean of the squared deviations in the population and is found by

$$\sigma^2 = \frac{\sum(x - \mu)^2}{N}$$

Examine this formula for a moment and try to see what it is trying to tell you, piece by piece.

- $(x - \mu)$ is the deviation, the signed distance from the data value $x$ to the population mean $\mu$.
- Then we square each deviation to get $(x - \mu)^2$.
- Finally, we add all the squared deviations and divide by the population size $N$.

## Example 3.12 Calculating the population variances for Amanda and Bethany

David Greedy/Getty Images

Calculate the population variances of the delivery times for Amanda and Bethany.

### Solution

Using the squared deviations from Table 3.12, we have

$$\sigma^2 = \frac{\sum(x - \mu)^2}{N} = \frac{0 + 4 + 9 + 0 + 25 + 1 + 1 + 25 + 1 + 0}{10} = \frac{66}{10} = 6.6$$

for Amanda, and

$$\sigma^2 = \frac{\sum(x - \mu)^2}{N} = \frac{100 + 49 + 0 + 100 + 121 + 9 + 25 + 196 + 4 + 100}{10} = \frac{704}{10} = 70.4$$

for Bethany. Clearly, the population variance of the delivery times for Bethany is greater than that for Amanda, thus indicating that Bethany's delivery times are more variable than Amanda's.

However, what is the *meaning* of the values we got for $\sigma^2$, 6.6 and 70.4, apart from their comparative value? How would you explain these results in layman's terms to someone who has never taken a statistics course? The problem is that the units of these values are *minutes squared,* certainly not a useful measure. Unfortunately, the intuitive meaning of the variance is not self-evident. ■

## The Population Standard Deviation $\sigma$

Working with the variance is quite useful in the realm of mathematical statistics, but the lack of a clear-cut interpretation of its meaning leads us to de-emphasize the variance (whether population variance or sample variance) in this text. Instead, we shall focus on the *standard deviation,* which has a much clearer interpretation in practice. The standard deviation is simply the square root of the variance. By taking the square root, we return the units of measure back to the original data unit (for example, "minutes" rather than "minutes squared"). The symbol for the **population standard deviation** is $\boldsymbol{\sigma}$: conveniently, $\sqrt{\sigma^2} = \sigma$.

**Population Standard Deviation $\sigma$**

The **population standard deviation $\sigma$** is the square root of the population variance and is found by

$$\sigma = \sqrt{\frac{\sum(x-\mu)^2}{N}}$$

## Example 3.13 Calculating the population standard deviations for Amanda and Bethany

Calculate the population standard deviations of the delivery times for Amanda and Bethany.

**Solution**

Since Bethany's population variance of 70.4 is larger than Amanda's population variance of 6.6, Bethany's population standard deviation will also be larger, since we are simply taking the square root. We have

$$\sigma = \sqrt{\frac{\sum(x-\mu)^2}{N}} = \sqrt{\frac{66}{10}} = \sqrt{6.6} \approx 2.6$$

for Amanda, and

$$\sigma = \sqrt{\frac{\sum(x-\mu)^2}{N}} = \sqrt{\frac{704}{10}} = \sqrt{70.4} \approx 8.4$$

for Bethany.

The population standard deviation of Bethany's delivery times is 8.4 minutes, which is larger than Amanda's 2.6 minutes. As expected, the greater variability in Bethany's delivery times leads to a larger value for her population standard deviation $\sigma$. ■

### The Standard Deviation

So how do we interpret these values for $\sigma$? One quick thumbnail interpretation of the standard deviation is that it represents a "typical" deviation. That is, the value of $\sigma$ *represents a distance from the mean that is representative for that data set.* Remember that it is not the mean deviation, which is always zero. ■

### Communicating the Results

As you study statistics, keep in mind that during your career you will likely need to explain your results to others who have never taken a statistics course. Therefore, you should always keep in mind *how to interpret your results to nonspecialists.* Communication and interpretation of your results can be as important as the results themselves. ■

## The Sample Variance $s^2$ and the Sample Standard Deviation $s$

In the real world, we usually cannot determine the exact value of the population mean or the population standard deviation. We use the sample mean and **sample standard deviation** to estimate the population parameters. In this text, we will work with sample statistics unless the data set is identified as a population. The **sample variance** also depends on the concept of the mean squared deviation. If the sample mean is $\bar{x}$, and the sample size is $n$, it would be reasonable, working by analogy with the population variance, to expect that the formula for the sample variance would be

$$\frac{\sum(x-\bar{x})^2}{n}$$

However, data analysts would like this statistic (the sample variance) to be an unbiased estimator of the parameter (the population variance). An *unbiased estimator* is a statistic whose mean value equals the parameter it is trying to estimate. The above formula has been found to underestimate the population variance, so that we need to replace the $n$ in the denominator with $n-1$. We therefore have the following.

**Sample Variance $s^2$**

The **sample variance $s^2$** is approximately the mean of the squared deviations in the sample and is found by

$$s^2 = \frac{\sum(x-\bar{x})^2}{n-1}$$

The sample standard deviation is perhaps the second most important statistic you will encounter in this book (after the sample mean $\bar{x}$). It is the most commonly used measure of spread. The sample standard deviation is simply the square root of the sample variance and takes as its symbol the letter $s$, which is the Roman letter for the Greek $\sigma$. Again, $s = \sqrt{s^2}$.

**Sample Standard Deviation $s$**

The **sample standard deviation $s$** is the square root of the sample variance $s^2$:

$$s = \sqrt{s^2} = \sqrt{\frac{\sum(x-\bar{x})^2}{n-1}}$$

The value of $s$ may be interpreted as the typical difference between a data value and the sample mean, for a given data set.

## Example 3.14 Calculating the sample variance and the sample standard deviation

Calculate the sample variances and sample standard deviations of Amanda's and Bethany's delivery times.

**Solution**

Using the squared deviations from Table 3.12, we have the sample variance

$$s^2 = \frac{\sum(x-\bar{x})^2}{n-1} = \frac{0+4+9+0+25+1+1+25+1+0}{9} = \frac{66}{9} \approx 7.3$$

for Amanda, and the sample variance

$$s^2 = \frac{\sum(x-\bar{x})^2}{n-1} = \frac{100+49+0+100+121+9+25+196+4+100}{9} = \frac{704}{9} \approx 78.2$$

for Bethany.

We find the sample standard deviations for Amanda's and Bethany's delivery times, as follows. For Amanda, we have

$$s = \sqrt{\frac{\sum(x-\bar{x})^2}{n-1}} = \sqrt{\frac{0+4+9+0+25+1+1+25+1+0}{9}} = \sqrt{\frac{66}{9}} \approx 2.7$$

and for Bethany, we have

$$s = \sqrt{\frac{\sum(x-\bar{x})^2}{n-1}} = \sqrt{\frac{100+49+0+100+121+9+25+196+4+100}{9}} = \sqrt{\frac{704}{9}} \approx 8.8$$

Amanda's delivery times typically differ from the mean $\bar{x} = 15$ by only 2.7 minutes, while Bethany's delivery times typically differ from the mean by 8.8 minutes. ■

### Less Variation Is Better

Amanda's sample standard deviation of $s = 2.7$ shows a remarkable consistency in her delivery times. Amanda's clients would certainly appreciate this consistency. In most real-world applications, consistency is a great advantage. In statistical data analysis, less variation is often better. Throughout the text, you will find that smaller variability will lead to

- more precise estimates and
- higher confidence in conclusions. ■

*A Note on Rounding: Discovering Statistics* typically rounds to one or two decimal places when *presenting* results. However, when results are to be used for later calculations, students should not round until the last calculation.

The above formulas for the variance and standard deviation can sometimes be time-consuming, since each individual deviation needs to be calculated. You may use the following *computational formulas* to simplify your calculations. The computational formulas are equivalent to the above definition formulas.

**Computational Formulas for the Variance and Standard Deviation**

**Population variance:**

$$\sigma^2 = \frac{\sum x^2 - \left(\sum x\right)^2/N}{N}$$

**Population standard deviation:**

$$\sigma = \sqrt{\frac{\sum x^2 - \left(\sum x\right)^2/N}{N}}$$

**Sample variance:**

$$s^2 = \frac{\sum x^2 - \left(\sum x\right)^2/n}{n-1}$$

**Sample standard deviation:**

$$s = \sqrt{\frac{\sum x^2 - \left(\sum x\right)^2/n}{n-1}}$$

where $\Sigma x^2$ means that you square each data value and then add up the squared data values, and $(\Sigma x)^2$ means that you add up all the data values and then square the sum.

## Example 3.15 Calculating the population variance and population standard deviation using the computational formulas

Table 3.13 lists the amount of farmland (in 1000s of acres) in each county in the state of Connecticut. Since the data set contains *all* $N = 8$ counties in Connecticut, it can be considered a population. Calculate the population variance and population standard deviation using the computational formulas.

Table 3.13 Farmland in Connecticut

| County | Farmland (1000s of acres) |
|---|---|
| Fairfield | 12.8 |
| Hartford | 50.2 |
| Litchfield | 93.6 |
| Middlesex | 17.9 |
| New Haven | 26.0 |
| New London | 58.9 |
| Tolland | 36.8 |
| Windham | 61.1 |

*Source:* U.S. Bureau of Economic Analysis.

**Solution**

Construct a table similar to Table 3.14. The data values (farmland = $x$) go in the first column, and the squared values go in the second column. Notice that we don't have to find the deviations. Add up the columns to get $\sum x = 357.3$ and $\sum x^2 = 20{,}997.91$, respectively. Then substitute these values into the computational formula for the population variance:

$$\sigma^2 = \frac{\sum x^2 - \left(\sum x\right)^2/N}{N} = \frac{20{,}997.91 - (357.3)^2/8}{8} = 629.9998438 \approx 630.0$$

The population standard deviation is therefore

$$\sigma = \sqrt{\sigma^2} = \sqrt{629.9998438} \approx 25.1$$

The standard deviation of farmland for all counties in Connecticut is almost 25,100 acres. ■

Table 3.14 Calculating $\sum x$ and $\sum x^2$

| $x$ | $x^2$ |
|---|---|
| 12.8 | 163.84 |
| 50.2 | 2520.04 |
| 93.6 | 8760.96 |
| 17.9 | 320.41 |
| 26.0 | 676.00 |
| 58.9 | 3469.21 |
| 36.8 | 1354.24 |
| 61.1 | 3733.21 |
| $\sum x = 357.3$ | $\sum x^2 = 20{,}997.91$ |

## Example 3.16 Calculating the sample variance and sample standard deviation using the computational formulas

Suppose we take a sample of the three counties from Table 3.13 that have the most farmland. Calculate the sample variance and sample standard deviation using the computational formulas.

**Solution**

Table 3.15 shows the calculations required to get $\sum x = 213.6$ and $\sum x^2 = 15{,}963.38$.

Table 3.15 Calculating $\sum x$ and $\sum x^2$

| $x$ | $x^2$ |
|---|---|
| 93.6 | 8760.96 |
| 61.1 | 3733.21 |
| 58.9 | 3469.21 |
| $\sum x = 213.6$ | $\sum x^2 = 15{,}963.38$ |

Then, substituting these values into the computational formula for the sample variance gives us

$$s^2 = \frac{\sum x^2 - \left(\sum x\right)^2/n}{n-1} = \frac{15{,}963.38 - (213.6)^2/3}{3-1} \approx 377.53$$

The sample standard deviation is therefore

$$s = \sqrt{s^2} = \sqrt{377.53} \approx 19.4$$

The standard deviation of farmland for this sample of three counties in Connecticut is almost 19,400 acres. ■

## Example 3.17 Using technology to find the variance and standard deviation

Find the sample standard deviation and sample variance of the city gas mileage for the six 2004 cars shown in the table below. Use (a) the TI-83/84, (b) Excel, and (c) Minitab.

| Vehicle | City mpg |
|---|---|
| Honda Civic | 36 |
| Toyota Camry | 24 |
| Ford Taurus | 20 |
| Pontiac Grand Prix | 20 |
| Jaguar X-Type | 18 |
| Lincoln Town Car | 17 |

### Solution

Using the instructions in the Step-by-Step Technology Guide on page 97, we get the following output. The TI-83/84 does not report the sample variance. Instead, use the following: variance = (standard deviation)$^2$.

**a.** The TI-83/84 output is shown below. The sample standard deviation $s$ is given as $Sx = 7.03562364$.

Do not confuse $Sx$ with $\sigma x$, which the TI-83/84 uses to label the population standard deviation.

The sample variance is

$$s^2 = (7.03562364)^2 \approx 49.5$$

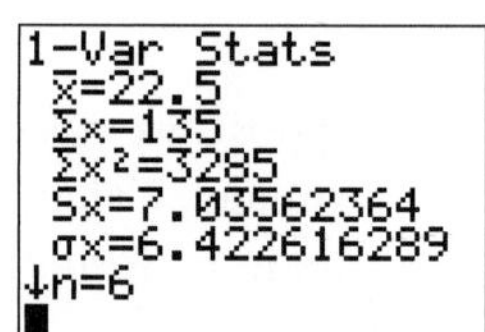
1-Var Stats
x̄=22.5
Σx=135
Σx²=3285
Sx=7.03562364
σx=6.422616289
↓n=6

**b.** The Excel output is shown below. The sample standard deviation and sample variance are highlighted.

| City MPG | |
|---|---|
| Mean | 22.5 |
| Standard Error | 2.872281 |
| Median | 20 |
| Mode | 20 |
| Standard Deviation | 7.035624 |
| Sample Variance | 49.5 |
| Kurtosis | 3.641384 |
| Skewness | 1.873582 |
| Range | 19 |
| Minimum | 17 |
| Maximum | 36 |
| Sum | 135 |
| Count | 6 |

**c.** The Minitab output below gives us the sample standard deviation "StDev," highlighted here. (Minitab rounds $s$ to 7.04.) We then find the sample variance using $s^2 = 7.04^2 = 49.5616$. (Because of the rounding of 7.04, this value of $s$ is less accurate than the value of 49.5 found in **(b)**).

```
Variable  N  N*   Mean  SE Mean  StDev  Minimum     Q1  Median     Q3  Maximum
City MPG  6   0  22.50     2.87   7.04    17.00  17.75   20.00  27.00    36.00
```

## What If Scenario — Give the Calculator a Rest

In Example 3.14, *what if* there was a typo, and Bethany's second-longest delivery time (25 minutes) was actually some unspecified number between 15 minutes (the mean) and 25 minutes. Describe how and why this change would have affected the following, if at all. Would they increase, decrease, remain unchanged? Or is there insufficient information to tell what would happen? Explain your answers.

**a.** Range

**b.** Variance

**c.** Standard deviation

### Solution

**a.** Since the range looks only at the maximum and minimum delivery times, the range would remain unaffected by a change in the second-longest delivery time.

**b.** The original deviation for this delivery was $x - \bar{x} = 25 - 15 = 10$, so that the squared deviation was 100. Now, since this delivery time is smaller than 25 (but still larger than the mean 15), the deviation $x - \bar{x}$ is also smaller. Thus, the squared deviation is also smaller than the original. Therefore, the mean squared deviation is smaller. Since *the variance is a measure of the mean squared deviation,* we conclude that the variance is smaller.

**c.** Since the variance is smaller, the standard deviation, which is the square root of the variance, is also smaller.

## CASE STUDY — Can the Financial Experts Beat the Darts?

Recall from Section 3.1 the *Wall Street Journal* competition between stocks chosen randomly by *Journal* staff members throwing darts and stocks chosen by a team of four financial experts. Note from Figure 3.14 that the DJIA exhibits less variability than the other two portfolios. This smaller variability is due to the fact that the DJIA is made up of 29 component stocks, whereas each portfolio is made up of only 4 stocks.

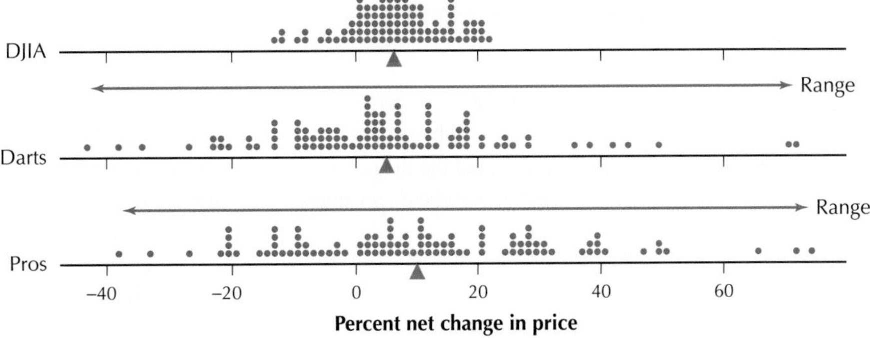


FIGURE 3.14 Comparative dotplots of the net change in prices.

Smaller sample sizes can be associated with increased variability, since an unusual result in one value has a relatively strong effect on the mean when it is not offset by a large sample. We will learn more about this in Chapter 7, "Sampling Distributions."

Which of the portfolios, pros or darts, shows greater variability? It is difficult to determine, just by examining Figure 3.14, which has the greater range or standard deviation. We therefore turn to the Minitab descriptive statistics in Figure 3.15. The Max for the pros was a 75% increase, while the Min was a 37.8% *decrease*. Thus, the range for the pros is $75-(-37.8) = 112.8$. The Max for the darts was a 72.9% increase, while the Min was a 43.0% decrease. So the range for the darts is $72.9-(-43.0) = 115.9$. This means that the range for the darts is greater than that for the pros.

From Figure 3.15, the standard deviations are 22.25, 19.39, and 8.031 for the pros, the darts, and the DJIA, respectively. Thus, the standard deviation for the pros is greater than for the darts, while the range for the darts is greater. *Measures of spread may disagree about which data set is more variable.* However, as we stated earlier, as a measure of spread, the range cannot give us the whole picture, since it takes into account only the two most extreme data values. Our conclusion, therefore, is that the returns for the professionals exhibit the greater variability.

FIGURE 3.15 Descriptive statistics for the portfolios.

| Variable | N | Mean | Median | TrMean | StDev | SE Mean |
|---|---|---|---|---|---|---|
| Pros | 100 | 10.95 | 9.60 | 10.19 | 22.25 | 2.22 |
| Darts | 100 | 4.52 | 3.25 | 3.73 | 19.39 | 1.94 |
| DJIA | 100 | 6.793 | 7.000 | 7.026 | 8.031 | 0.803 |

| Variable | Minimum | Maximum | Q1 | Q3 |
|---|---|---|---|---|
| Pros | -37.80 | 75.00 | -6.23 | 26.63 |
| Darts | -43.00 | 72.90 | -6.53 | 14.88 |
| DJIA | -13.100 | 22.500 | 1.525 | 13.250 |

Why all the effort to find which portfolio has the most variation? Because in finance, as in most other fields, high variability is not necessarily advantageous. Would you want to put your money into a fund that had high variability in its returns? Perhaps not, since high variability in finance is associated with greater *risk*. The professionals evidently chose higher-risk stocks with greater potential for high returns—but also greater potential for losing money.

## Section 3.2 SUMMARY

1 The simplest measure of variability, or measure of spread, is the range. The range is simply the difference between the maximum and minimum values in a data set, but the range has drawbacks because it relies on the two most extreme data values.

2 A deviation is the difference between a data value and the mean of the data values.

3 The variance and standard deviation are measures of spread that utilize all available data values. The population variance can be thought of as the mean squared deviation. The standard deviation is the square root of the variance. We interpret the value of the standard deviation as the typical deviation, that is, the typical difference between a data value and the mean.

## Section 3.2 EXERCISES

Unless a data set is identified as a population, you can assume that it is a sample.

### CLARIFYING THE CONCEPTS

**1.** Explain what a deviation is.
**2.** What is the interpretation of the value of the standard deviation?
**3.** Which is better, less variability or more variability? Why?

### PRACTICING THE TECHNIQUES

Use the following data set for Exercises 4–6.

6 9 3 6

**4.** Find the range.
**5.** Find the variance.
**6.** Find the standard deviation.

Use the following data set for Exercises 7–9.

10 25 0 15 10

**7.** Find the range.
**8.** Find the variance.
**9.** Find the standard deviation.

Use the following data set for Exercises 10–12.

−5 −10 −15 −20

**10.** Find the range.
**11.** Find the variance.
**12.** Find the standard deviation.

For the following exercises, make sure to state your answers in the proper units, such as "years" or "years squared."

Refer to Table 3.5 (page 98) for Exercises 13 and 14.

**13.** Find the range for *age*.
**14.** Find the variance and standard deviation of *age*. Which measure do you find to be more easily understood and interpreted?

Use the following information for Exercises 15 and 16.

| Vehicle | City mpg |
|---|---|
| Honda Civic | 36 |
| Toyota Camry | 24 |
| Ford Taurus | 20 |
| Pontiac Grand Prix | 20 |
| Jaguar X-Type | 18 |
| Lincoln Town Car | 17 |

**15.** Find the variance and standard deviation of the city mpg for the six automobiles.
**16.** Find the range for the city mpg for the six automobiles.

Use the following information for Exercises 17 and 18.

| State | Median age |
|---|---|
| Utah | 28.5 |
| Texas | 33.2 |
| Alaska | 33.9 |
| Georgia | 34.3 |
| California | 34.4 |
| Arizona | 34.5 |
| Idaho | 34.6 |
| Colorado | 34.7 |

**17.** Find the variance and standard deviation of median age. Which measure do you find to be more easily understood and interpreted?
**18.** Find the range for median age.

### APPLYING THE CONCEPTS

**NASCAR Winners.** Refer to Table 3.4 (page 98) for Exercises 19–22.

**19.** Construct separate dotplots of the short tracks winners and the road courses winners, using the same scale. Which data set would you say has the greater variability?
**20.** Construct separate dotplots of the super speedways winners and the road courses winners, using the same scale. Which data set would you say has the greater variability?
**21.** Now find the range of each of the two data sets. Which data set has the greater variability? Do these results agree with your judgment above?
**22.** Now find the variance and standard deviation of each of the two data sets. Which data set has the greater variability? Do these results agree with your judgment above?

**Zooplankton and Phytoplankton.** Refer to the table below for Exercises 23 and 24. *Meta-analysis* refers to the statistical analysis of a set of similar research studies. In a meta-analysis, each data value represents an effect size calculated from the results of a particular study. The table contains effect sizes calculated in a meta-analysis for zooplankton and phytoplankton.[6]

| Zooplankton | | Phytoplankton | |
|---|---|---|---|
| −2.37 | −3.00 | 10.61 | 3.04 |
| −0.64 | −0.68 | 2.97 | 0.65 |
| −2.05 | −1.39 | 1.58 | 2.55 |
| −1.54 | −0.64 | 2.55 | 1.05 |
| −6.60 | −3.88 | 5.67 | 2.11 |
| 0.26 | | 1.57 | |

**23.** Calculate the ranges for the zooplankton and the phytoplankton.
- **a.** Which has the greater range?
- **b.** Which plankton group has the greater variability according to the range?

**24.** Calculate the standard deviations for the zooplankton and the phytoplankton.

**a.** Which has the greater standard deviation?

**b.** Which plankton group has the greater variability according to the standard deviation? Does this concur with your answer from the previous exercise?

**c.** Without calculating the variances, say which group has the greater variance. How do you know this?

**Top-Selling Soft Drinks.** Refer to Table 3.16 for Exercises 25–27.

**TABLE 3.16 Top-Selling Soft-Drink Brands**

| Observation | Brand | Millions of cases sold |
|---|---|---|
| 1 | Coke Classic | 1929 |
| 2 | Pepsi-Cola | 1385 |
| 3 | Diet Coke | 811 |
| 4 | Sprite | 541 |
| 5 | Dr. Pepper | 537 |
| 6 | Mountain Dew | 536 |
| 7 | Diet Pepsi | 530 |
| 8 | 7UP | 220 |
| 9 | Caffeine-Free Diet Coke | 180 |
| 10 | Caffeine-Free Diet Pepsi | 97 |

*Source: Wall Street Journal Almanac.*

**25.** Find the mean number of cases sold. Calculate the deviations.

**26.** Find the range, variance, and standard deviation for the number of cases sold. Explain what these numbers mean.

WHAT IF? **27.** Suppose Coke Classic were removed from the data set.

**a.** *Without recalculating them,* describe how this would affect the values of the measures of spread you found above.

**b.** Now recalculate the four measures of spread with Coke Classic removed. Was your judgment in **(a)** supported?

**Ant Size.** Use the following information for Exercises 28 and 29. A study compared the size of ants from different colonies. The masses (in milligrams) of samples of ants from two different colonies are shown in the accompanying table.[7]

| Colony A | | Colony B | |
|---|---|---|---|
| 109 | 134 | 148 | 115 |
| 120 | 94 | 110 | 101 |
| 94 | 113 | 110 | 158 |
| 61 | 111 | 97 | 67 |
| 72 | 106 | 136 | 114 |

**28.** Calculate the range for each ant colony.

**a.** Which has the greater range?

**b.** Which colony has the greater variability according to the range?

**29.** Calculate the variance for each colony.

**a.** Which has the greater variance?

**b.** Which colony has the greater variability according to the variance? Does this concur with your answer from the previous exercise?

**c.** Without calculating the standard deviations, say which colony has the greater standard deviation? How do you know this?

**SAT Scores.** Refer to Table 3.6 (page 98) for Exercises 30–32.

**30.** Construct dotplots of the SAT I Verbal and the SAT I Math scores. Would you say that the two data sets have similar variability or that one data set varies much more than the other?

**31.** Find the range, variance, and standard deviation for both the SAT I Verbal and the SAT I Math scores. Do your findings agree with your judgment from the previous exercise? Explain.

WHAT IF? **32.** Now suppose we omit New Hampshire from the calculations.

**a.** *Without recalculating them,* describe how this would affect the values of the measures of spread you found for the SAT I Verbal scores above. How about the SAT I Math scores?

**b.** Now recalculate the three measures of spread for the SAT I Verbal scores with the New Hampshire data removed. Was your judgment in **(a)** supported?

**c.** Recalculate the three measures of spread for the SAT I Math scores with the New Hampshire data removed. Was your judgment in **(a)** supported?

**Volleyball Team Heights.** Refer to Table 3.9 (page 102) for Exercises 33–35.

**33.** Consider the deviations for each team.

**a.** Would you say that a typical distance from the mean for the WMU team is about 70 inches, about 6 inches, or about 3 inches?

**b.** What would you say a typical distance from the mean is for the NCU team?

**34.** Consider the standard deviations.

**a.** Find the standard deviation for each team.

**b.** Interpret the meaning of the values you obtained for the standard deviations so that someone who has never studied statistics would understand it.

**c.** Which team has the greater variability according to the standard deviation?

WHAT IF? **35.** Suppose that the shortest player on the NCU team was not 66 inches but only 62 inches tall.

**a.** *Without recalculating them,* what effect would this change have on the three measures of spread for the NCU data?

**b.** Now recalculate the three measures of spread for the NCU data, with the shortest player now 62 inches instead of 66 inches tall. Was your judgment in **(a)** supported?

**Common Syllables in English.** Refer to Table 3.17 for Exercises 36–39.

**TABLE 3.17 Some common syllables in English**

| Syllable | Frequency |
|---|---|
| *an* | 462 |
| *bi* | 621 |
| *sit* | 104 |
| *ed* | 907 |
| *its* | 293 |
| *est* | 186 |
| *wil* | 470 |
| *tiv* | 136 |
| *en* | 675 |
| *biz* | 114 |

**36.** Find the mean and the range of the syllable frequencies.

**37.** Would you say that a typical distance from the mean for the frequencies is about 900, about 500, about 300, or about 100?

**38.** What is your estimate of the value of a typical distance from the mean for the syllable frequencies?

**39.** Find the standard deviation of syllable frequencies.

**a.** How far is it from your estimate of the typical deviation earlier?

**b.** Interpret the meaning of this value for the standard deviation so that someone who has never studied statistics would understand it.

**40. Pulse Rates for Males and Females.** The comparative dotplots in Figure 3.16 show pulse rates for a sample of males and females.

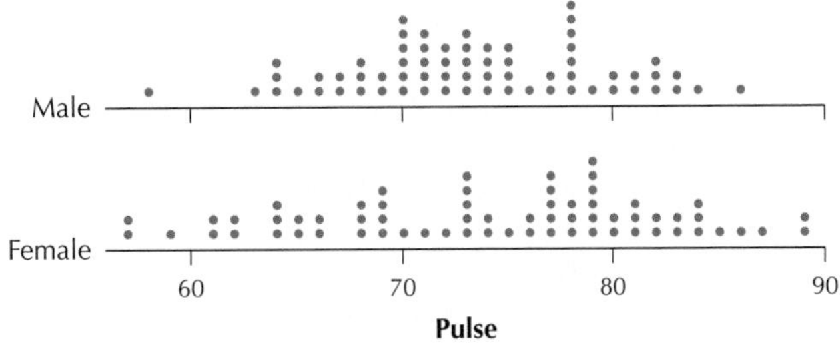


Figure 3.16 Comparative dotplots of pulse rates, by gender.

**a.** What is your impression of which gender has the greater amount of variability in pulse rates?

**b.** Estimate the maximum pulse rate, minimum pulse rate, and range of pulse rates for males.

**c.** Estimate the maximum pulse rate, minimum pulse rate, and range of pulse rates for females. According to the range, which gender has more variability in pulse rates.

**d.** Would the typical distance from the mean for females be about 78, about 28, about 18, about 8, or about 0.8?

**e.** What is your estimate of the typical distance from the mean for females? For males? Is your estimate larger for females or males, and why?

**41.** Figure 3.17 gives descriptive statistics for the pulse rates in Figure 3.16.

| Variable | Gender | N | Mean | Median | TrMean | StDev |
|---|---|---|---|---|---|---|
| Pulse | female | 65 | 74.15 | 76.00 | 74.27 | 8.11 |
| | male | 65 | 73.369 | 73.000 | 73.407 | 5.875 |

| Variable | Gender | SE Mean | Minimum | Maximum | Q1 | Q3 |
|---|---|---|---|---|---|---|
| Pulse | female | 1.01 | 57.00 | 89.00 | 68.00 | 80.00 |
| | male | 0.729 | 58.000 | 86.000 | 70.000 | 78.000 |

Figure 3.17 Descriptive statistics for the pulse rates, by gender.

**a.** How close (in terms of beats per minute) were your earlier estimates of the maximum, minimum, and range of the pulse rates for each gender?

**b.** What is the standard deviation of the pulse rates for the females?

**c.** How close (in terms of beats per minute) is this value of the standard deviation to your earlier estimate of the typical distance from the mean for the females?

**d.** What is the standard deviation of the pulse rates for the males?

**e.** How close (in terms of beats per minute) is this value of the standard deviation to your earlier estimate of the typical distance from the mean for the males?

**f.** For which gender is the standard deviation greater? Does this agree with your earlier surmise as to the relative sizes of the standard deviations?

**g.** Overall, which gender has greater variability in pulse rates?

**Body Temperature for Males and Females.** Use the following information for Exercises 43 and 44. Refer to the human body temperature data from Section 3.1 and Figure 3.18. Is there evidence that the variability in body temperatures differs by gender?

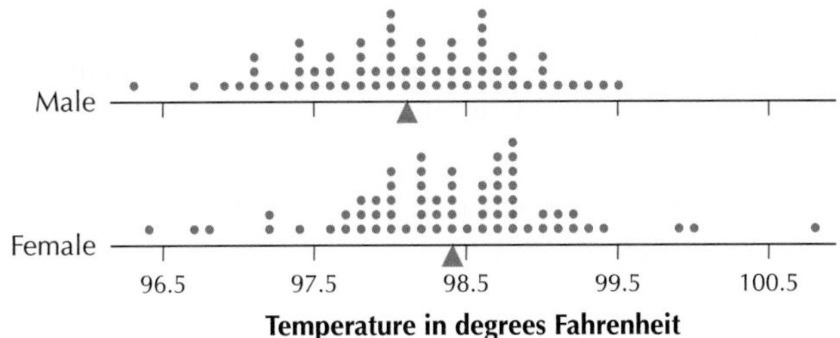


Figure 3.18 Comparative dotplots of body temperatures, by gender.

**42.** Refer to Figure 3.18.

**a.** Which gender's body temperature do you think shows more variability?

**b.** Estimate the maximum and minimum body temperatures for the males.

c. Estimate the maximum and minimum body temperatures for the females.
d. Estimate the range of body temperatures for the males and for the females. Which gender has the greater range of body temperatures? According to the range, which gender has more variability in body temperatures?
e. The mean body temperatures are indicated by the red triangles in Figure 3.18. What is the estimated mean body temperature for the males? For the females?

**43.** Figure 3.19 gives descriptive statistics for body temperature, by gender.

| Variable | Gender | N | Mean | Median | TrMean | StDev |
|---|---|---|---|---|---|---|
| Temp | female | 65 | 98.394 | 98.400 | 98.390 | 0.743 |
| | male | 65 | 98.105 | 98.100 | 98.114 | 0.699 |

| Variable | Gender | SE Mean | Minimum | Maximum | Q1 | Q3 |
|---|---|---|---|---|---|---|
| Temp | female | 0.092 | 96.400 | 100.800 | 98.000 | 98.800 |
| | male | 0.087 | 96.300 | 99.500 | 97.600 | 98.600 |

Figure 3.19 Descriptive statistics for body temperatures, by gender.

a. Compare the values for the mean body temperatures with those that you estimated above. How close were your estimates?
b. Compare the values for the minimum, maximum, and range of body temperatures with those that you estimated above. How close were your estimates?
c. What is the standard deviation of body temperature for females? For males?
d. According to the standard deviation, which gender has more variability in body temperatures?

**The 1970 Military Draft.** The 1970 military draft has come under criticism for not being truly random. We will examine elsewhere whether there were differences in the means among the mean selection ranks per month. However, what about the variability? Here we will examine whether June or December had more variability in the selection ranks. Table 3.18 shows the selection rankings for each day in both months. For example, June 1 was the 249th date selected, and December 31st was the 100th date selected.

**TABLE 3.18 Selection rankings for the 1970 military draft, June and December**

**June**

| | | | | | | | | | |
|---|---|---|---|---|---|---|---|---|---|
| 249 | 228 | 301 | 20 | 28 | 110 | 85 | 366 | 335 | 206 |
| 134 | 272 | 69 | 356 | 180 | 274 | 73 | 341 | 104 | 360 |
| 60 | 247 | 109 | 358 | 137 | 22 | 64 | 222 | 353 | 209 |

**December**

| | | | | | | | | | |
|---|---|---|---|---|---|---|---|---|---|
| 129 | 328 | 157 | 165 | 56 | 10 | 12 | 105 | 43 | 41 |
| 39 | 314 | 163 | 26 | 320 | 96 | 304 | 128 | 240 | 135 |
| 70 | 53 | 162 | 95 | 84 | 173 | 78 | 123 | 16 | 3 |
| 100 | | | | | | | | | |

**44.** Refer to Table 3.18.
a. Find the range of the June selection rankings.
b. Find the range of the December selection rankings.
c. What does the range say about which month had the greater variability in selection rankings?

**45.** Refer to Table 3.18.
a. Using a computer or by hand, construct any plot you like, such as a histogram, dotplot, or stem-and-leaf display, of each data set.
b. Based on your graphics, which month would you say has the greater overall variability?

**46.** Refer to Table 3.18.
a. Which month would you estimate has the larger standard deviation?
b. Would you estimate that a typical distance from the mean for the month of December is about 300, about 200, about 100, or about 50?
c. Find the standard deviation of the June selection rankings.
d. Find the standard deviation of the December selection rankings.
e. Which month has the larger standard deviation? Does this agree with your estimate above?
f. Do your findings regarding the standard deviation agree with your findings regarding the range, in terms of which month had more variability?

WHAT IF? **47.** *What if* the 180 selection ranking (data value) in June was not 180 but actually 360. Describe how and why this change would have affected the following, if at all. Would they increase, decrease, remain unchanged? Or is there insufficient information to tell what would happen? Explain your answers.
a. Range
b. Variance
c. Standard deviation

## CONSTRUCT YOUR OWN DATA SETS

**48.** Construct two data sets, A and B, that you make up on your own, so that the range of A is greater than the range of B. Verify this.

**49.** Construct two data sets, A and B, that you make up on your own, so that the standard deviation of A is greater than the range of B. Verify this.

**50.** Construct two data sets, A and B, that you make up on your own, so that the mean of A is greater than the mean of B, but the standard deviation of B is greater than that of A. Verify this.

**51.** Construct two data sets, A and B, that you make up on your own, so that the mean of A is greater than the mean of B, and the standard deviation of A is greater than that of B. Verify this.

**52.** Construct two data sets, A and B, that you make up on your own, so that the range of A is greater than the range of B, but the standard deviation of B is greater than that of A. Verify this. (*Hint:* Remember the sensitivity of the standard deviation to extreme values.)

# 3.3 Working with Grouped Data

**Objectives: By the end of this section, I will be able to…**

1. Calculate the weighted mean.
2. Estimate the mean for grouped data.
3. Estimate the variance and standard deviation for grouped data.

## 1 The Weighted Mean

*Note:* Before tackling this section, you may wish to review Section 2.2, "Graphs and Tables for Quantitative Data" (page 47).

Sometimes not all the data values are of equal importance. Rather, certain data values are assigned greater weight than other data values when calculating the mean. For example, have you ever figured out what your final grade for a course was based on the percentages listed in the syllabus? What you actually found was the **weighted mean** of your grades.

**Weighted Mean**

To find the **weighted mean:**

1. Multiply each data point $x_i$ by its respective weight $w_i$.
2. Sum these products.
3. Divide the result by the sum of the weights:

$$\bar{x}_w = \frac{\sum w_i x_i}{\sum w_i} = \frac{w_1 x_1 + w_2 x_2 + \cdots + w_n x_n}{w_1 + w_2 + \cdots + w_n}$$

### Example 3.18 Weighted mean of course grades

The syllabus for the Introduction to Management course at a local college specifies that the midterm exam is worth 30%, the term paper is worth 20%, and the final exam is worth 50% of your course grade. Now, say you did not get serious about the course until Halloween, so that you got a 20 on the midterm. You started working harder, and got a 70 on the term paper. Finally, you remembered that you had to pay for the course again if you flunked and had to retake it, and so you worked hard and smart for the last month of the course and got a 90 on the final exam. Calculate your course average, that is, the weighted mean of your grades.

**Solution**

*Note:* The weights $w_i$ do not have to be percentages that add up to 1.

The data values are $x_1 = 20$, $x_2 = 70$, and $x_3 = 90$. The weights are $w_1 = 0.30$, $w_2 = 0.20$, and $w_3 = 0.50$. Your course weighted mean is then calculated as follows:

$$\bar{x}_w = \frac{\sum w_i x_i}{\sum w_i} = \frac{w_1 x_1 + w_2 x_2 + w_3 x_3}{w_1 + w_2 + w_3} = \frac{(0.30)(20) + (0.20)(70) + (0.50)(90)}{0.30 + 0.20 + 0.50} = \frac{65}{1.0} = 65$$

Despite your failing the midterm miserably, you got serious and barely managed to pass the course with a 65 average, which is a D. ■

## 2 Estimating the Mean for Grouped Data

Thus far in Chapter 3, we have computed measures of center and spread from a raw data set. However, data are often reported using frequency distributions. Without the original data, we cannot calculate the exact values of the measures of center and spread. The remainder of this section examines methods for approximating the mean, variance, and standard deviation of *grouped data*—that is, data summarized using frequency distributions.

For each class in the frequency distribution, we estimate the class mean using the class *midpoint*. The class midpoint is defined as the mean of two adjoining lower class limits. The midpoint for the $i$th class is denoted $m_i$.

## Example 3.19 Finding the class midpoints

There were 1150 children adopted in the state of Georgia in 2006, according to the Administration for Children and Families.[8] The frequency distribution of the ages of the children at adoption is shown in Table 3.19. Find the class midpoints.

Table 3.19 Frequency distribution of children adopted in Georgia, by age

| Class: age | Midpoint $m_i$ | Frequency $f_i$ |
|---|---|---|
| 0–0.99 | 0.5 | 12 |
| 1–5.99 | 3.5 | 611 |
| 6–10.99 | 8.5 | 320 |
| 11–15.99 | 13.5 | 161 |
| 16–17.99 | 17.0 | 46 |

### Solution

The midpoint for the first class (ages 0–0.99) is the mean of the lower class limits for this class (0) and the adjoining class (1). That is, the midpoint is $(0 + 1)/2 = 0.5$.

Similarly, the midpoint for the second class (ages 1–5.99) is $(1 + 6)/2 = 3.5$. The remainder of the class midpoints are shown in Table 3.19. ■

The product of the class frequency $f_i$ and class midpoint $m_i$ is used as an estimate of the sum of the data values within that class. Summing these products across all classes and dividing by the total sample size thus provides us with an **estimated mean for data grouped into a frequency distribution.**

*Note:* Notice the difference between $\hat{\mu}$ and $\bar{x}$. The estimated mean $\hat{\mu}$ is used only when the data are grouped into a frequency distribution. The sample mean $\bar{x}$ is used when all the data values in the sample are known. Both $\hat{\mu}$ and $\bar{x}$ are used to estimate the population mean $\mu$.

**Estimated Mean for Data Grouped into a Frequency Distribution**

Given a frequency distribution with $k$ classes, the **estimated mean** for the variable is given by

$$\hat{\mu} = \frac{\sum m_i f_i}{N} = \frac{m_1 f_1 + m_2 f_2 + \cdots + m_k f_k}{N}$$

where $m_i$ and $f_i$ represent the midpoint and frequency of the $i$th class, respectively, and $N$ represents the total population size, which is the sum of the class frequencies: $N = \sum f_i$. The "hat" notation for $\hat{\mu}$, called "mu-hat," indicates that $\hat{\mu}$ is an estimate and not the exact value of the mean.

## Example 3.20 Calculating the estimated mean for grouped data

Calculate the estimated mean age of the adopted children in Table 3.19.

### Solution

The midpoints $m_i$ and frequencies $f_i$ are provided in Table 3.19. For example, $m_1 = 0.5$ and $f_1 = 12$. We thus calculate the sum of the products as follows:

$$\sum m_i f_i = (0.5)(12) + (3.5)(611) + (8.5)(320) + (13.5)(161) + (17)(46)$$
$$= 6 + 2138.5 + 2720 + 2173.5 + 782 = 7820$$

Next we calculate the total population size:

$$N = \sum f_i = 12 + 611 + 320 + 161 + 46 = 1150$$

The estimated mean is therefore

$$\hat{\mu} = \frac{\sum m_i f_i}{N} = \frac{7820}{1150} = 6.8$$

The estimated mean age of the children adopted in Georgia in 2006 is 6.8 years.

## 3 Estimating the Variance and Standard Deviation for Grouped Data

We also use class midpoints and class frequencies to calculate the **estimated variance for data grouped into a frequency distribution** and the **estimated standard deviation for data grouped into a frequency distribution.**

**Estimated Variance and Standard Deviation for Data Grouped into a Frequency Distribution**

Given a frequency distribution with $k$ classes, the estimated variance for the variable is given by

$$\hat{\sigma}^2 = \frac{\sum (m_i - \hat{\mu})^2 \cdot f_i}{N}$$

and the estimated standard deviation is given by

$$\hat{\sigma} = \sqrt{\hat{\sigma}^2} = \sqrt{\frac{\sum (m_i - \hat{\mu})^2 \cdot f_i}{N}}$$

where $m_i$ and $f_i$ represent the midpoint and frequency of the $i$th class, respectively, and $N$ represents the total population size, which is the sum of the class frequencies: $N = \sum f_i$. The "hat" notation for $\hat{\sigma}$, called "sigma-hat," indicates that $\hat{\sigma}$ is an estimate and not the exact value of the standard deviation.

*Note:* Notice the difference between $\hat{\sigma}$ and $s$. The estimated standard deviation $\hat{\sigma}$ is used only when the data are grouped into a frequency distribution. The sample standard deviation $s$ is used when all the data values in the sample are known. Both $\hat{\sigma}$ and $s$ are used to estimate the population standard deviation $\sigma$.

You should carry as many decimal places as you can for the value of $\hat{\mu}$ when calculating $\hat{\sigma}^2$, and for $\hat{\sigma}^2$ when calculating $\hat{\sigma}$.

### Example 3.21 Calculating the estimated variance and standard deviation for grouped data

Calculate the estimated variance and standard deviation of the ages of the adopted children in Table 3.19.

**Solution**

Table 3.20 contains the calculations required for finding $\sum (m_i - \hat{\mu})^2 \cdot f_i = 20{,}068$. The variance is therefore estimated as

$$\hat{\sigma}^2 = \frac{\sum (m_i - \hat{\mu})^2 \cdot f_i}{N} = \frac{20{,}068}{1150}$$

and the standard deviation is estimated as

$$\hat{\sigma} = \sqrt{\hat{\sigma}^2} = \sqrt{\frac{20{,}068}{1150}} \approx 4.177371755 \approx 4.2$$

Table 3.20 Calculating $\Sigma(m_i - \hat{\mu})^2 \cdot f_i$

| Class: age | Midpoint $m_i$ | Frequency $f_i$ | $\hat{\mu}$ | $m_i - \hat{\mu}$ | $(m_i - \hat{\mu})^2 \cdot f_i$ |
|---|---|---|---|---|---|
| 0–0.99 | 0.5 | 12 | 6.8 | –6.3 | 476.28 |
| 1–5.99 | 3.5 | 611 | 6.8 | –3.3 | 6653.79 |
| 6–10.99 | 8.5 | 320 | 6.8 | 1.7 | 924.8 |
| 11–15.99 | 13.5 | 161 | 6.8 | 6.7 | 7227.29 |
| 16–17.99 | 17.0 | 46 | 6.8 | 10.2 | 4785.84 |

$$\sum(m_i - \hat{\mu})^2 \cdot f_i = 20{,}068$$

In other words, the age of the adopted children typically differs from the mean age of 6.8 years by about 4.2 years.

## Example 3.22 Using technology to find the estimated mean, variance, and standard deviation for grouped data

Use the TI-83/84 calculator to find the estimated mean, variance, and standard deviation for the frequency distribution in Table 3.20.

**Solution**

Following the instructions in the Step-by-Step Technology Guide, we get the estimated mean $\hat{\mu}$ (shown in the output as $\bar{x}$) = 6.8 and the estimated standard deviation $\hat{\sigma}$ (shown in the output as $\sigma x$) = 4.177371755.

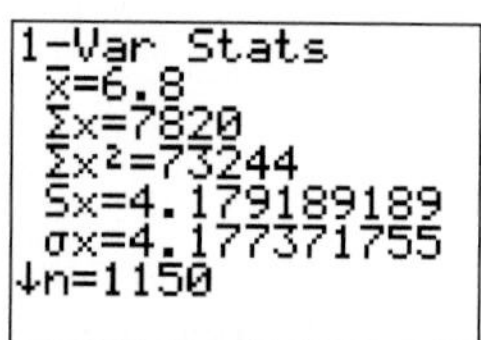


## STEP-BY-STEP TECHNOLOGY GUIDE: Estimating the Mean, Variance, and Standard Deviation for Grouped Data

### TI-83/84

***Step 1*** Press **STAT** and select **1:Edit**. Enter the class midpoints in **L1** and the frequencies or relative frequencies in **L2**.

***Step 2*** Press **STAT**, select the **CALC** menu, and choose **1: 1-Var Stats**.

***Step 3*** Press **2nd 1 Comma 2nd 2**, so that the following appears on the home screen: **1-Var Stats L1, L2**

***Step 4*** Press **ENTER**.

## Section 3.3 SUMMARY

1 The weighted mean is the sum of the products of the data points with their respective weights, divided by the sum of the weights.

2 Since we do not have access to the original raw data, it is not possible to find exact values for the mean, variance, and standard deviation of data that have been grouped into a frequency distribution. The estimated mean $\hat{\mu}$ in this case is the sum of the products of the class frequencies $f_i$ and class midpoints $m_i$, divided by the total sample size $N$.

3 Class midpoints and class frequencies are also used to find the estimated variance $\hat{\sigma}^2$ and estimated standard deviation $\hat{\sigma}$ of grouped data.

## Section 3.3 EXERCISES

### CLARIFYING THE CONCEPTS

1. Explain why the formulas for the mean, variance, and standard deviation of grouped data will provide estimates only and not the exact values for these measures.
2. Explain what the "hat" notation indicates for $\hat{\mu}$ and $\hat{\sigma}$.
3. What quantity do both $\hat{\mu}$ and $\bar{x}$ estimate?

### PRACTICING THE TECHNIQUES

For Exercises 4–9, the data values and weights are provided. Find the weighted mean.

4. $x_1 = 50, x_2 = 60; x_3 = 70; w_1 = 0.25, w_2 = 0.50, w_3 = 0.25.$
5. $x_1 = 50, x_2 = 80, x_3 = 70; w_1 = 0.25, w_2 = 0.40, w_3 = 0.35.$
6. $x_1 = 100, x_2 = 120, x_3 = 150; w_1 = 10, w_2 = 20, w_3 = 5.$
7. $x_1 = 3.0, x_2 = 2.5, x_3 = 3.5, x_4 = 4.0, x_5 = 3.0; w_1 = w_2 = w_3 = w_4 = 3, w_5 = 4.$
8. $x_1 = 70, x_2 = 80, x_3 = 85, x_4 = 95; w_1 = 0.20, w_2 = 0.30, w_3 = 0.25, w_4 = 0.25.$
9. $x_1 = 1.0, x_2 = 1.5, x_3 = 2.5, x_4 = 3.0, x_5 = 3.5; w_1 = w_2 = 14, w_3 = w_4 = 15, w_5 = 16.$

For Exercises 10–12, the class limits are provided. Find the class midpoints.

**10.**

| | |
|---|---|
| 0–1.99 | 6–7.99 |
| 2–3.99 | 8–9.99 |
| 4–5.99 | |

**11.**

| | |
|---|---|
| 10–12.49 | 15–17.49 |
| 12.5–14.99 | 17.5–19.99 |

**12.**

| | |
|---|---|
| 0–4.99 | 20–29.99 |
| 5–9.99 | 30–49.99 |
| 10–14.99 | 50–99.99 |
| 15–19.99 | 100–199.99 |

For Exercises 13–15, find the estimated mean for the frequency distribution.

**13.**

| Midpoint $m_i$ | Frequency $f_i$ |
|---|---|
| 5 | 10 |
| 10 | 20 |
| 15 | 20 |
| 20 | 10 |
| 25 | 10 |

**14.**

| Midpoint $m_i$ | Frequency $f_i$ |
|---|---|
| −10 | 3 |
| −5 | 2 |
| 0 | 5 |
| 5 | 12 |
| 10 | 8 |
| 15 | 10 |

**15.**

| Midpoint $m_i$ | Frequency $f_i$ |
|---|---|
| 50 | 20 |
| 150 | 30 |
| 250 | 30 |
| 450 | 40 |
| 750 | 30 |
| 1150 | 20 |

For Exercises 16–18, find the estimated variance and standard deviation

**16.** For the frequency distribution in Exercise 13,
**17.** For the frequency distribution in Exercise 14,
**18.** For the frequency distribution in Exercise 15.

### APPLYING THE CONCEPTS

**19. Dupage County Age Groups.** The Census Bureau reports the following 2006 frequency distribution of population by age group for Dupage County, Illinois, residents less than 65 years old.

| Age | Residents |
|---|---|
| 0–4.99 | 63,422 |
| 5–17.99 | 240,629 |
| 18–64.99 | 540,949 |

a. Find the class midpoints.
b. Find the estimated mean age of residents of Dupage County.
c. Find the estimated variance and standard deviation of ages.

**20. Capacity of Lakes and Reservoirs.** The National Water and Climate Center tracks the percentage capacity of lakes and reservoirs (how full the lake or reservoir is). The table below contains the percentage capacity for a random sample of lakes and reservoirs in Arizona, California, and Colorado.[9] Find the weighted mean percentage capacity.

| Arizona | California | Colorado |
|---|---|---|
| $n = 4$ | $n = 6$ | $n = 5$ |
| $\bar{x} = 48.5$ | $\bar{x} = 81$ | $\bar{x} = 47$ |

**21. Broward County House Values.** Table 3.21 gives the frequency distribution of the dollar value of the owner-occupied housing units in Broward County, Florida.

**a.** Find the class midpoints.

**b.** Find the estimated mean dollar value for housing units in Broward County.

**c.** Find the estimated variance and standard deviation of the dollar value.

**TABLE 3.21 Broward County house values**

| Dollar value | Housing units |
|---|---|
| 0–$49,999 | 5,430 |
| 50,000–$99,999 | 90,605 |
| 100,000–$149,999 | 90,620 |
| 150,000–$199,999 | 54,295 |
| 200,000–$299,999 | 34,835 |
| 300,000–$499,999 | 15,770 |
| 500,000–$999,999 | 5,595 |

*Source:* Census Bureau.

**22. Lightning Deaths.** Table 3.22 gives the frequency distribution of the number of deaths due to lightning nationwide over a 67-year period. Find the estimated mean and standard deviation of the number of lightning deaths per year.

**TABLE 3.22 Lightning deaths**

| Deaths | Years |
|---|---|
| 20–59.99 | 13 |
| 60–99.99 | 21 |
| 100–139.99 | 10 |
| 140–179.99 | 6 |
| 180–259.99 | 10 |
| 260–459.99 | 7 |

*Source:* National Oceanic and Atmospheric Administration.

**23. Calculating a Course Grade.** An introductory statistics syllabus has the following grading system. The weekly quizzes are worth a total of 25% toward the final course grade. The midterm exam is worth 32%; the final exam is worth 33%; and attendance/participation is worth 10% toward the final course grade. Anthony's weekly quiz average is 70. He got an 80 on the midterm and a 90 on the final exam. He got 100 for attendance/participation. Calculate Anthony's final course grade.

**24. Wages for Computer Managers.** The U.S. Bureau of Labor Statistics (BLS) publishes wage information for various occupations. For the occupation "computer and information systems management," Table 3.23 gives the wages reported by the BLS for the top-paying states in May 2006. Find the weighted mean wage across all five states, using the employment figures as weights.

**TABLE 3.23 Wages for computer managers**

| State | Employment | Hourly mean wage |
|---|---|---|
| New Jersey | 12,380 | $60.32 |
| New York | 18,580 | $60.25 |
| Virginia | 9,540 | $59.39 |
| California | 35,550 | $57.98 |
| Massachusetts | 10,130 | $55.95 |

**25. Ant Size.** A study compared the size of ants from different colonies.[10] The mass (in milligrams) was measured for ants from three different ant colonies, with the descriptive statistics shown here. Find the weighted mean size of all the ants.

| Colony | Frequency | Mean size (mass in mg) |
|---|---|---|
| A | 130 | 89.66 |
| B | 111 | 100.00 |
| C | 120 | 92.16 |

**26. Salaries of Scientists and Engineers.** The National Science Foundation compiles statistics on the annual salaries of full-time employed doctoral scientists and engineers in universities and four-year colleges. The mean 2004 annual salary for the fields of science, engineering, and health are $67,000, $82,200, and $70,000, respectively. Suppose we have a sample of 10 professors, 5 of whom are in science, 2 in engineering, and 3 in health, and each of whom is making the mean salary for his or her field. Find the weighted mean salary of these 10 professors.

**27. Challenger Exercise.** Assign the weights $w_i$ to show that the formula for the sample mean from Section 3.1 $\bar{x} = \sum x_i/n$ is a special case of the formula for the weighted mean $\bar{x}_w = \sum w_i x_i / \sum w_i$.

## 3.4 Measures of Position

**Objectives:** By the end of this section, I will be able to…

1. Find percentiles for both small and large data sets.
2. Calculate $z$-scores, and explain why we use them.
3. Use $z$-scores to detect outliers.

## Percentiles

In this section we learn about *measures of position,* which tell us the position that a particular data value has relative to the rest of the data set. For example, a prestigious nursing school may grant admission to only the top 10% of applicants. How high a score would you need to enter? This is one type of question we will answer in this section. The first measure of position we consider is the percentile. A **percentile** gives the percentage of data values that are less than or equal to a certain value.

> **Percentile** Let $p$ be any integer between 0 and 100. The $p$th **percentile** of a data set is the data value at which $p$ percent of the values in the data set are less than or equal to this value.

### Example 3.23 Jasmine's Math SAT percentile

Many students take the Scholastic Aptitude Test (SAT) when applying to college. The test results that you receive include your score and the percentile that your score represents. Suppose Jasmine's Math SAT score was 650, which represented the 90th percentile. What does "90th percentile" mean?

**Solution**

The 90th percentile simply means that 90% of all scores on the Math SAT fell at or below Jasmine's score of 650. We call the percentile a *measure of position* since it indicates the position of Jasmine's Math SAT score relative to all other Math SAT scores. Clearly, Jasmine is good at math. Figure 3.20 indicates the position of Jasmine's score relative to the rest of the test takers.

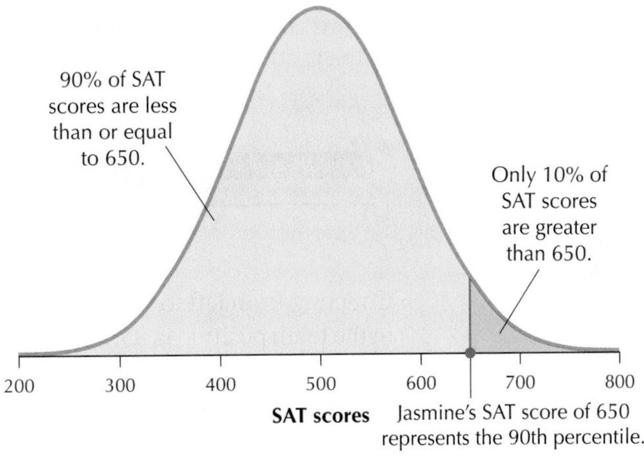
90% of SAT scores are less than or equal to 650.
Only 10% of SAT scores are greater than 650.
200
300
400
500
600
700
800
SAT scores
Jasmine's SAT score of 650 represents the 90th percentile.

FIGURE 3.20 The 90th percentile is the score with 90% of the data values at or below its value.

For large data sets, calculation of the percentiles is best left to computers. However, for small data sets, we can use the following step-by-step method to calculate the position of any percentile.

**Step 1** Sort the data into ascending order (from smallest to largest).

**Step 2** Calculate

$$i = \left(\frac{p}{100}\right)n$$

where $p$ is the particular percentile you wish to calculate, and $n$ is the sample size.

*Note:* These steps do not give the value of the $p$th percentile itself, but rather the *position* of the $p$th percentile in the data set when the data set is in ascending order.

***Step 3*** **a.** If $i$ is an integer (a whole number with no decimal part), the $p$th percentile is the mean of the data values in positions $i$ and $i + 1$. **b.** If $i$ is not an integer, round up and use the value in this position.

## Example 3.24 Finding percentiles of a small data set

Yolanda would like to go to a prestigious graduate school of the arts. She knows that this school accepts only those students who score at the 75th percentile or higher in a grueling dance audition. The following data represent the dance audition scores of Yolanda's group. Yolanda scored 85. Find the 75th percentile of the data set. Will Yolanda be accepted at the prestigious graduate school of the arts?

78 56 89 44 65 94 81 62 75 85 30 68

### Solution

***Step 1*** Sort the data into ascending order:

30 44 56 62 65 68 75 78 81 85 89 94

***Step 2*** The particular percentile we wish to calculate is the 75th percentile, so $p = 75$. There are 12 scores in our data set, so $n = 12$. Calculate

$$i = \left(\frac{p}{100}\right)n = \left(\frac{75}{100}\right)12 = 9$$

So, $i = 9$.

***Step 3*** Here, since $i$ is an integer, the 75th percentile is the mean of the data values in positions 9 and 10.

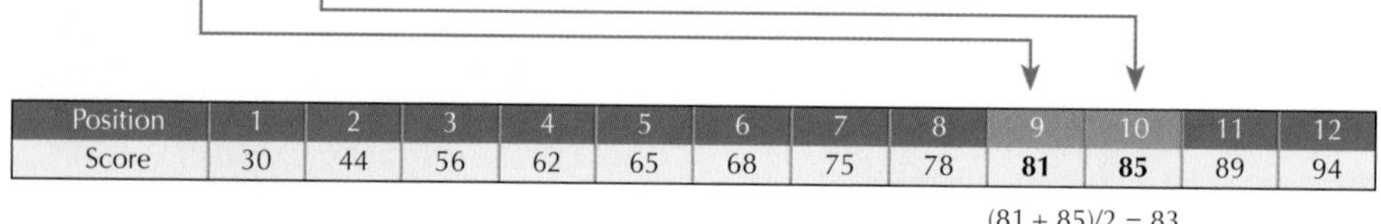

| Position | 1 | 2 | 3 | 4 | 5 | 6 | 7 | 8 | 9 | 10 | 11 | 12 |
|---|---|---|---|---|---|---|---|---|---|---|---|---|
| Score | 30 | 44 | 56 | 62 | 65 | 68 | 75 | 78 | **81** | **85** | 89 | 94 |



Counting from left to right, the data value in the ninth position is 81, and the data value in the tenth position is 85. The mean of these two values is 83. Thus, the 75th percentile is 83. Yolanda's dance score of 85 is therefore above the 75th percentile. She will be accepted to the prestigious graduate school. ■

## Example 3.25 Finding percentiles of a large data set

How old are the oldest Major League Baseball players? The data set **Baseball2007** is found on the CD. It contains data on the 516 American League baseball players in the 2007 season, including the age of each player. Find the 95th percentile of the ages of the American League baseball players in 2007.

### Solution

Minitab provides a frequency distribution for the age variable, along with the cumulative counts and percentages (Figure 3.21). The sample size $n$ (N in Minitab) is 516.

The particular percentile we wish to calculate is the 95th percentile, so we use $p = 95$ in the percentile formula:

$$i = \left(\frac{p}{100}\right)n = \left(\frac{95}{100}\right)516 = 490.2$$

Since $i = 490.2$ is not an integer, we round up to the next integer. The 95th percentile is in the 491st position, when the data are in ascending order. Now, in the "CumCnt" (cumulative count) column in Figure 3.21, note that there are 488 players who are 36 years old or less, and there are 498 players who are 37 years old or less. We are looking for the 491st. Since there are 10 players that are 37 years old, the 491st must be one of these 10 guys who are 37 years old. Hence, the 95th percentile age of American League baseball players in 2007 is 37 years old. The oldest player is Roger Clemens of the New York Yankees, at 44 years old. ■

| Age | Count | CumCnt | CumPct |
|---|---|---|---|
| 20 | 1 | 1 | 0.19 |
| 21 | 8 | 9 | 1.74 |
| 22 | 16 | 25 | 4.84 |
| 23 | 27 | 52 | 10.08 |
| 24 | 34 | 86 | 16.67 |
| 25 | 54 | 140 | 27.13 |
| 26 | 51 | 191 | 37.02 |
| 27 | 47 | 238 | 46.12 |
| 28 | 33 | 271 | 52.52 |
| 29 | 40 | 311 | 60.27 |
| 30 | 44 | 355 | 68.80 |
| 31 | 38 | 393 | 76.16 |
| 32 | 24 | 417 | 80.81 |
| 33 | 25 | 442 | 85.66 |
| 34 | 9 | 451 | 87.40 |
| 35 | 21 | 472 | 91.47 |
| 36 | 16 | 488 | 94.57 |
| 37 | 10 | 498 | 96.51 |
| 38 | 5 | 503 | 97.48 |
| 39 | 4 | 507 | 98.26 |
| 40 | 5 | 512 | 99.22 |
| 41 | 1 | 513 | 99.42 |
| 42 | 2 | 515 | 99.81 |
| 44 | 1 | 516 | 100.00 |
| N= | 516 | | |

The oldest player: Roger Clemens

FIGURE 3.21 Finding the 95th percentile of player ages.

## 2 z-Scores

Recall that the standard deviation is a common measure of the variability, or spread, of a data set. The value of the standard deviation is interpreted as a typical deviation from the mean. The SAT is designed so that the distribution of scores is bell-shaped with a mean of 500 and a standard deviation of 100. Note in Figure 3.22 that we can measure the distance from a particular SAT score to the mean in terms of standard deviations. For example, an SAT score of 600 lies one standard deviation above the mean, while an SAT score of 300 lies two standard deviations below the mean.

The term ***z*-score** indicates how many standard deviations a particular data value is from the mean. If the $z$-score is positive, then the data value is above the mean. If the $z$-score is negative, then the data value is below the mean.

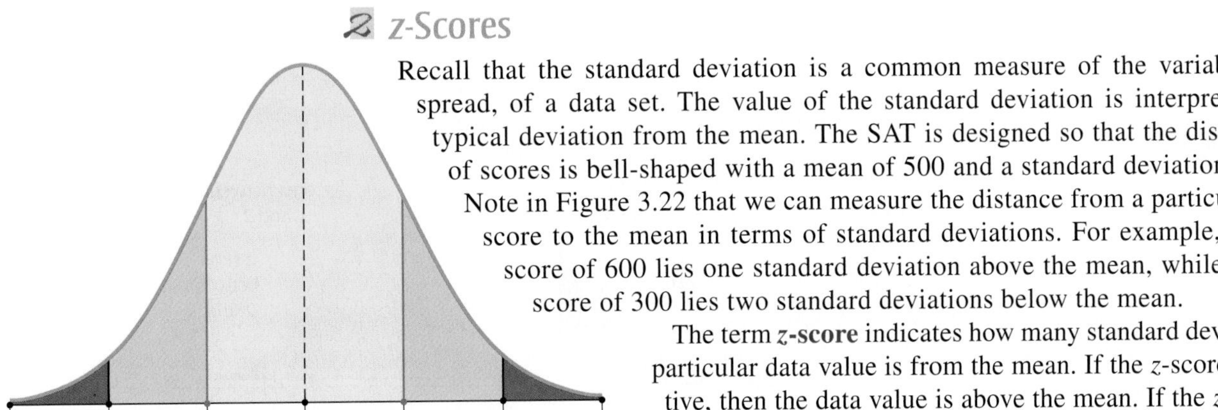


FIGURE 3.22 The distribution of SAT scores.

**z-Score**

The **z-score** for a particular data value from a *sample* is

$$z\text{-score} = \frac{\text{data value} - \text{mean}}{\text{standard deviation}} = \frac{x - \bar{x}}{s}$$

where $\bar{x}$ is the sample mean, and $s$ is the sample standard deviation.

The z-score for a particular data value from a *population* is

$$z\text{-score} = \frac{\text{data value} - \text{mean}}{\text{standard deviation}} = \frac{x - \mu}{\sigma}$$

where $\mu$ is the population mean, and $\sigma$ is the population standard deviation.

In this section, we will use the sample $z$-score unless otherwise indicated.

## Example 3.26 Jasmine's Math SAT z-score

Suppose the mean score on the Math SAT is $\mu = 500$, with a standard deviation of $\sigma = 100$ points. How many standard deviations is Jasmine's score from the mean?

### Solution

Here $\mu = 500$, $\sigma = 100$, and Jasmine's score is $x = 650$. Her $z$-score is

$$z\text{-score} = \frac{\text{data value} - \text{mean}}{\text{standard deviation}} = \frac{x - \mu}{\sigma} = \frac{650 - 500}{100} = 1.5$$

Jasmine's $z$-score of 1.5 indicates that her Math SAT is 1.5 standard deviations from the mean of 500. $Z$-scores can be positive or negative. Jasmine's $z$-score is positive (1.5), which means that her Math SAT score falls above the mean. Bright lady! Consider Figure 3.23, which shows the distribution of SAT scores, with a mean of 500 and a standard deviation of 100. The black arrows represent "units" of one standard deviation each, that is, each arrow is 100 SAT points long. Counting the arrows as you go above or below the mean is thus the same as counting the number of standard deviations above or below the mean. Jasmine's SAT score lies between 600 and 700, an area with $z$-scores ranging from 1 to 2. ■

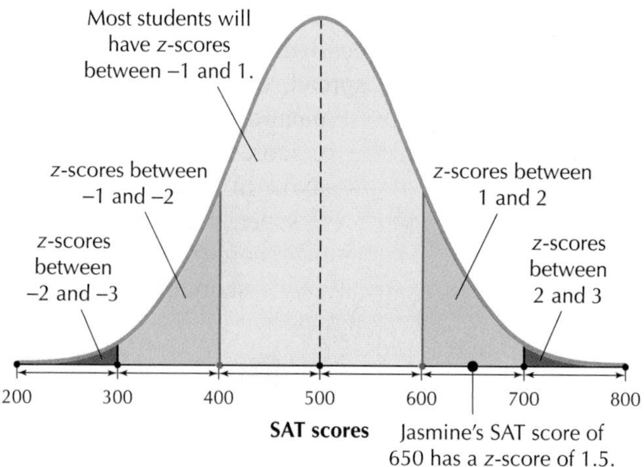


FIGURE 3.23 Jasmine's z-score of 1.5 places her 1.5 standard deviations above the mean.

Your Statistical Sense

### Why Use z-Scores?

One wonderful advantage of using $z$-scores is that they can provide us with information about data values *even when we do not fully understand the original data set.* For example, suppose that your friend just discovered a new galaxy, and she told you that this galaxy held 50 billion stars. If you are not interested in astronomy, then you probably don't know whether this is a large or small galaxy.

But, if she told you that the $z$-score for the number of stars in this new galaxy was $-3$, then you would understand that this galaxy is much smaller than the average galaxy, even if you had never so much as looked through a telescope! You would understand that the number of stars in the newly discovered galaxy is 3 standard deviations below the mean. ■

In Example 3.26, since the standard deviation equals 100, the $z$-score represents units of 100. That is, a $z$-score of 1 represents 1 standard deviation above the mean, which is 100 points above the mean. Thus, the *scale* of the $z$-scores for the SAT scores in Figure 3.23 is in units of 100, since the standard deviation equals 100. However, if the standard deviation was, say, $\sigma = 14.2$, then the scale would be different.

## Example 3.27 Lead levels in blood

A study of workers who were exposed to lead at their jobs found that their mean blood lead level was 31.4 $\mu$g/dl (micrograms per deciliter) with a standard deviation of 14.2 $\mu$g/dl.[11]

**a.** If we calculate $z$-scores, what is the scale?

**b.** Calculate the $z$-scores for the following workers:

**i.** Ryan, with a blood lead level of 78.26 $\mu$g/dl.

**ii.** Megan, with a blood lead level of 1.58 $\mu$g/dl.

**iii.** Kyle, with a blood lead level of 55.54 $\mu$g/dl.

**c.** For each worker, interpret the value of the $z$-score.

**Solution**

**a.** If we calculate $z$-scores for the workers' lead levels, the scale of the $z$-scores will be 14.2 $\mu$g/dl, since that is the value of the standard deviation.

**b.** Here are the workers' lead levels.

**i.** Ryan:

$$z\text{-score} = \frac{x - \bar{x}}{s} = \frac{78.26 - 31.4}{14.2} = \frac{46.86}{14.2} = 3.3$$

**ii.** Megan:

$$z\text{-score} = \frac{x - \bar{x}}{s} = \frac{1.58 - 31.4}{14.2} = \frac{-29.82}{14.2} = -2.1$$

**iii.** Kyle:

$$z\text{-score} = \frac{x - \bar{x}}{s} = \frac{55.54 - 31.4}{14.2} = \frac{24.14}{14.2} = 1.7$$

**c.** Ryan's lead level lies 3.3 standard deviations above the mean; Megan's lead level lies 2.1 standard deviations below the mean; and Kyle's lead level lies 1.7 standard deviations above the mean. ■

## Example 3.28 Using the $z$-score to compare data from different data sets

Andrew is bragging to his friend Brittany that he did better than she did on the last statistics test. Andrew got a 90 while Brittany got an 80. Andrew's class mean was 80 with a standard deviation of 10. Brittany's class mean was 60 with a standard deviation of 10. The professors in both classes grade "on a curve" using $z$-scores. Who did better relative to his or her class?

### Solution

Brittany can use $z$-scores to show that she did better *relative to her class*. Figure 3.24 shows comparative dotplots of the scores in the two classes. The blue dots represent Brittany's and Andrew's scores. Brittany found her $z$-score by subtracting her class mean from her score of 80 and then dividing by the standard deviation $s = 10$:

$$z\text{-score}_{\text{Brittany}} = \frac{x - \bar{x}}{s} = \frac{80 - 60}{10} = 2$$

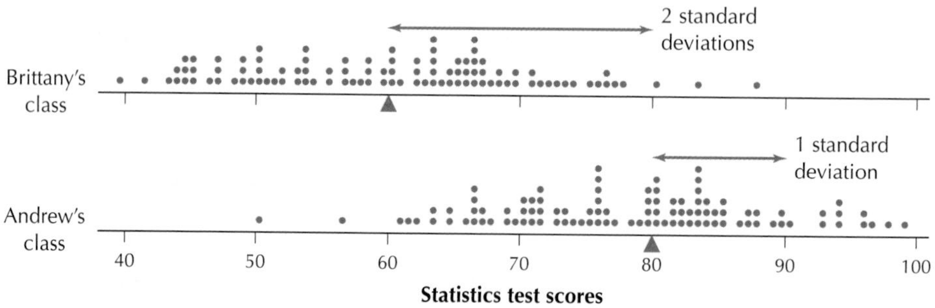

FIGURE 3.24 Brittany actually did better relative to her class.

Brittany's $z$-score is 2. What does that mean? It means that *Brittany scored 2 standard deviations above the mean* of 60. Brittany then found the $z$-score for Andrew:

$$z\text{-score}_{\text{Andrew}} = \frac{x - \bar{x}}{s} = \frac{90 - 80}{10} = 1$$

Andrew's $z$-score was 1, which means that Andrew scored one standard deviation above the mean. From Figure 3.24 we can observe that Andrew's exam score of 90 lies closer to the mean exam score of 80 for his class. That is, the blue arrow is shorter for Andrew than for Brittany. Finally, note that 10 of the 100 students who took the exam in his class did better than he did, whereas only 2 did better than Brittany in her class. So Brittany is right. Relative to her class, she did better than Andrew, even though Andrew got a higher score. The $z$-scores allowed her to compare their grades, even though they were in different classes. ■

## 3 Detecting Outliers Using $z$-Scores

An **outlier** is an extremely large or extremely small data value relative to the rest of the data set. It may represent a data entry error, or it may be genuine data. Usually an outlier can be identified because it is much farther than three standard deviations from the mean, that is, its $z$-score is less than $-3$ or greater than 3. How can we use this information? Although we will investigate this question more later on in this chapter, we can give the following rough guidelines here. Look at Figure 3.25.

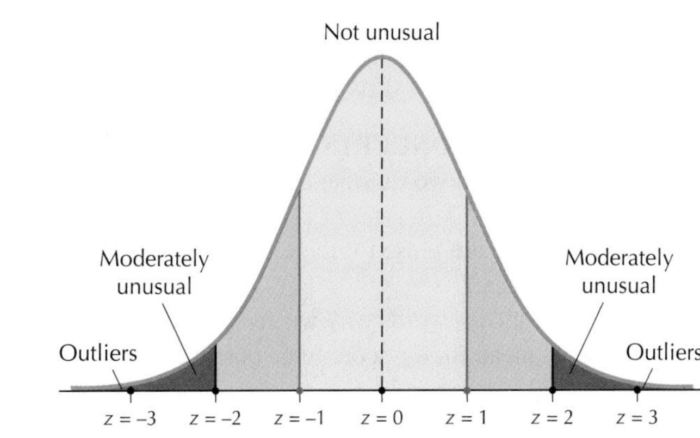


FIGURE 3.25
$z$-Scores help to identify outliers.

*Note:* If an outlier is detected, it does not automatically follow that it should be discarded. Outliers often indicate the presence of something interesting going on in the data that would call for further investigation. On the other hand, it could simply be a typo. The analyst should check with the data source. Finally, the analyst may prefer to use the *robust method of detecting outliers* that we will learn in Section 3.6.

## Guidelines for Identifying Outliers

1. A data value whose $z$-score lies in the following range is not considered to be unusual:

$$-2 < z\text{-score} < 2$$

2. A data value whose $z$-score lies in either of the following ranges may be considered moderately unusual:

$$-3 < z\text{-score} \leq -2 \quad \text{or} \quad 2 \leq z\text{-score} < 3$$

3. A data value whose $z$-score lies in either of the following ranges may be considered an outlier:

$$z\text{-score} \leq -3 \quad \text{or} \quad z\text{-score} \geq 3$$

## Example 3.29 Detecting outliers

For the three workers in Example 3.27, determine whether each of their blood lead levels represents an outlier.

**Solution**

Ryan's $z$-score is 3.3, which is greater than 3. Thus, Ryan's lead level of 78.26 $\mu$g/dl represents an outlier. Megan's $z$-score is $-2.1$, which lies between $-3$ and $-2$. Hence, Megan's lead level of 1.58 $\mu$g/dl may be considered moderately unusual but is not an outlier. Kyle's $z$-score is 1.7, which lies between $-2$ and 2. Thus, Kyle's lead level of 55.54 is not considered unusual. ■

## Section 3.4 SUMMARY

1 In this section we learned about measures of position, which tell us the position that a particular data value holds relative to the rest of the data set. The $p$th percentile of a data set is the value at which $p$ percent of the values in the data set are less than or equal to this value.

2 The $z$-score indicates how many standard deviations a particular data value is from the mean. The $z$-score equals the data value minus the mean, divided by the standard deviation.

3 An outlier is an extremely large or extremely small data value relative to the rest of the data set. An outlier can be identified when its $z$-score is less than $-3$ or greater than 3.

## Section 3.4 EXERCISES

### CLARIFYING THE CONCEPTS

1. Explain in your own words what the 95th percentile of a data set means.
2. Why doesn't it make sense for there to be a 120th percentile of a data set?
3. Explain in your own words why we need *z*-scores.
4. What does it mean for a *z*-score to be positive? Negative? Zero?
5. Is it possible for the 1st percentile to equal the 99th percentile? Explain.
6. Refer to the previous exercise. Describe the data set where this might happen. Describe the spread of the data set using any measure you like.

### PRACTICING THE TECHNIQUES

Refer to the following table for Exercises 7–11.

| Vehicle | City mpg |
|---|---|
| Honda Civic | 36 |
| Toyota Camry | 24 |
| Ford Taurus | 20 |
| Pontiac Grand Prix | 20 |
| Jaguar X-Type | 18 |
| Lincoln Town Car | 17 |

7. Find the following percentiles:
   a. 50th
   b. 75th
   c. 25th
8. Find the following percentiles:
   a. 10th
   b. 95th
   c. 5th
9. Find the *z*-scores for the city mpg for the following automobiles:
   a. Toyota Camry
   b. Lincoln Town Car
   c. Jaguar X-Type
10. Find the *z*-scores for the city mpg for the following automobiles:
   a. Honda Civic
   b. Ford Taurus
   c. Pontiac Grand Prix
11. Does the mileage for the Honda Civic represent an outlier? Explain.

### APPLYING THE CONCEPTS

Refer to Table 3.24 for Exercises 12–15.

**TABLE 3.24 Calories in 12 breakfast cereals**

| Cereal | Calories |
|---|---|
| Apple Jacks | 110 |
| Basic 4 | 130 |
| Bran Chex | 90 |
| Bran Flakes | 90 |
| Cap'n Crunch | 120 |
| Cheerios | 110 |
| Cinnamon Toast Crunch | 120 |
| Cocoa Puffs | 110 |
| Corn Chex | 110 |
| Corn Flakes | 100 |
| Corn Pops | 110 |
| Count Chocula | 110 |

12. Find the following percentiles:
   a. 50th
   b. 75th
13. Find the following percentiles:
   a. 25th
   b. 95th
14. Find the *z*-scores for the calories for the following cereals:
   a. Corn Flakes
   b. Basic 4
15. Find the *z*-scores for the calories for the following cereals:
   a. Bran Flakes
   b. Cap'n Crunch

Refer to Table 3.25 for Exercises 16–20. The table gives the number of American adults who have used the indicated "nonvitamin, nonmineral, natural products."

**TABLE 3.25 Use of dietary supplements**

| Product | Usage (in millions) | Product | Usage (in millions) |
|---|---|---|---|
| Echinacea | 14.7 | Ginger supplements | 3.8 |
| Ginseng | 8.8 | Soy supplements | 3.5 |
| Ginkgo biloba | 7.7 | Ragweed/chamomile | 3.1 |
| Garlic supplements | 7.1 | Bee pollen or royal jelly | 2.8 |
| Glucosamine | 5.2 | Kava kava | 2.4 |
| Saint-John's-wort | 4.4 | Valerian | 2.1 |
| Peppermint | 4.3 | Saw palmetto | 2.0 |
| Fish oil | 4.2 | | |

*Source:* Centers for Disease Control and Prevention, Vital and Health Statistics, 2004.

16. Find the following percentiles:
  a. 5th
  b. 95th

17. Find the following percentiles:
  a. 50th
  b. 75th

18. Find the $z$-scores for usage for the following products:
  a. Fish oil
  b. Valerian

19. Find the $z$-scores for usage for the following products:
  a. Ginseng
  b. Garlic supplements

20. Does the usage for echinacea represent an outlier? Explain.

21. **Females SAT Performance.** Females tend to perform well on the Verbal SAT. The College Board reported that the mean score on the Verbal SAT for females in 2003 was 503. Assume that the standard deviation is 100 and that we do not know the distribution. Brandy scored in the 50th percentile on the Verbal SAT.
  a. Draw a distribution curve where the 50th percentile is above the mean.
  b. Draw a distribution curve where the 50th percentile is below the mean.
  c. If the data distribution is bell-shaped, what was Brandy's score?

**Household Median Income.** Refer to Table 3.26 of state household median incomes for Exercises 22–24.

22. Find the following percentiles for state median household income:
  a. 25th
  b. 50th
  c. 75th

23. Find the following percentiles for state median household income:
  a. 5th
  b. 95th
  c. 1st
  d. 99th

24. The mean for the 50 states is about \$62,000 with a standard deviation of about \$8500. Use these figures to find the $z$-scores for the following states:
  a. West Virginia
  b. New Jersey
  c. Kansas

25. **Expenditure per Pupil.** The 5th percentile expenditures per pupil nationwide in 2005 was \$6381, the 50th percentile was \$8998, and the 95th percentile was \$17,188.[12]
  a. Determine whether the distribution of expenditures is symmetric, left-skewed, or right-skewed.
  b. Would we expect the mean expenditure per pupil to be less than, equal to, or greater than \$8998? Explain.
  c. Draw a distribution curve that matches this information.

**TABLE 3.26 Household median incomes**

| State | Income ($) | State | Income ($) | State | Income ($) |
|---|---|---|---|---|---|
| West Virginia | 47,550 | Florida | 57,473 | Washington | 66,531 |
| Mississippi | 47,847 | North Carolina | 58,227 | Virginia | 66,889 |
| New Mexico | 48,422 | Maine | 58,802 | Wisconsin | 66,988 |
| Arkansas | 49,551 | Nevada | 59,588 | Hawaii | 67,564 |
| Oklahoma | 51,377 | Missouri | 59,764 | Rhode Island | 67,646 |
| Montana | 51,791 | Utah | 59,864 | Michigan | 67,995 |
| Louisiana | 52,299 | Nebraska | 60,129 | Colorado | 68,089 |
| Alabama | 53,754 | Oregon | 60,262 | Illinois | 69,168 |
| Kentucky | 54,030 | Georgia | 60,676 | Delaware | 69,469 |
| Idaho | 54,279 | Iowa | 61,238 | Alaska | 69,868 |
| South Dakota | 55,359 | Kansas | 61,926 | New Hampshire | 72,369 |
| Tennessee | 55,605 | Vermont | 62,331 | Minnesota | 72,379 |
| South Carolina | 56,110 | Indiana | 63,022 | Maryland | 77,938 |
| Texas | 56,278 | Ohio | 63,934 | Massachusetts | 78,312 |
| Arizona | 56,857 | Pennsylvania | 64,310 | Connecticut | 81,891 |
| North Dakota | 57,070 | New York | 65,461 | New Jersey | 82,406 |
| Wyoming | 57,148 | California | 65,766 | | |

*Source:* National Bureau of Economic Research, 2005.

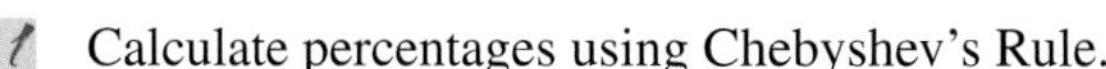

# 3.5 Chebyshev's Rule and the Empirical Rule

*Objectives:* **By the end of this section, I will be able to…**

1. Calculate percentages using Chebyshev's Rule.
2. Find percentages and data values using the Empirical Rule.

*Portrait of Pafnuty Tchebyshev-Yaroslav Sergeyevich (1899–1978)/State Central Artillery Museum, St. Petersburg, Russia/The Bridgeman Art Library*

## 1 Chebyshev's Rule

P. L. Chebyshev (1821–94, Russia) derived a result, called **Chebyshev's Rule,** that can be applied to any continuous data set whatsoever.

> **Chebyshev's Rule**
>
> The proportion of values from a data set that will fall within $k$ standard deviations of the mean will be *at least*
>
> $$\left(1 - \frac{1}{k^2}\right)100\%,$$
>
> where $k > 1$. Chebyshev's Rule may be applied to either samples or populations.

**What Does This Formula Mean?**

### Chebyshev's Rule

Let's dissect Chebyshev's Rule. The word "within" describes an *interval* of values along the number line. Say you go to a fair, and a gypsy offers to guess your age *within* 2 years of your true age. Suppose you are 20 years old. Then the gypsy could guess from 18 to 22 years old and be right. That is, "within 2 years of your true age" refers to the interval of values between $20 - 2 = 18$ and $20 + 2 = 22$, as shown in Figure 3.26. ■

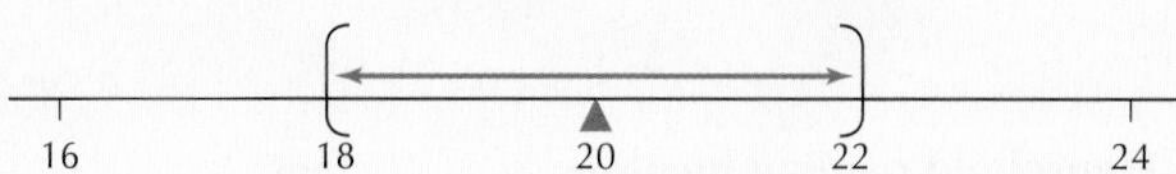


FIGURE 3.26 Within two years of 20 years old.

## Example 3.30 Finding the scores within 3 standard deviations of the mean

An instructor giving an exam expects that the mean will be 75 and the standard deviation will be 5. What exam scores represent the scores within 3 standard deviations of the mean exam score?

### Solution

We have $\bar{x} = 75$ and $s = 5$. The instructor is interested in estimating the proportion *within 3 standard deviations,* so that $k = 3$. In general, "within 3 standard deviations of the mean" means

**a.** the mean minus 3 standard deviations ($\bar{x} - 3s$), up to

**b.** the mean plus 3 standard deviations ($\bar{x} + 3s$), as shown in Figure 3.27.

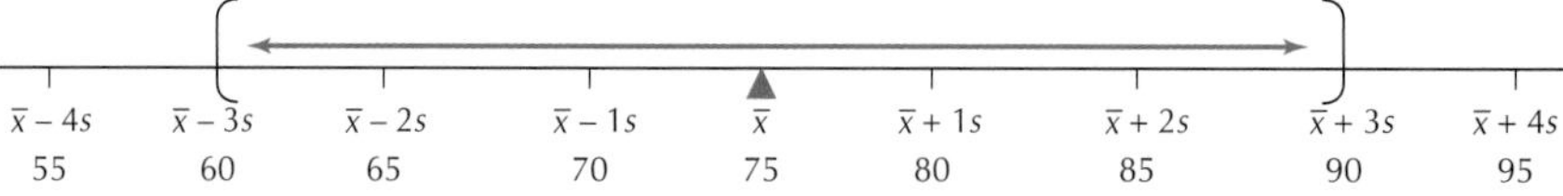


FIGURE 3.27 Within three standard deviations of the mean.

Since $\bar{x} = 75$ and $s = 5$, "within 3 standard deviations of the mean" means from $75 - 3(5) = 60$ to $75 + 3(5) = 90$. ■

### The "At Least" Is Not Optional

Chebyshev's Rule states that at least a certain percentage of the distribution lies between these two numbers. Data analysts say that Chebyshev's Rule gives a lower bound to this percentage, since it uses the phrase "at least." When giving percentages using Chebyshev, make sure you use the words "at least." There is a big difference between "about 84%" and "at least 84%." "At least 84%" could mean 100%, 95%, 84%, or anything in between. It is rather imprecise. ■

A key to solving a Chebyshev-type problem lies in finding the value of $k$, the number of standard deviations that a data value lies from the mean. The following example demonstrates how to find the value of $k$, as well as how to calculate the lower bound.

## Example 3.31 Finding $k$ and calculating the lower bound

Continuing the previous example, find a lower bound on the percentage of students who will score between 60 and 90 on the upcoming exam.

**Solution**

First, we need to find $k$. This is done as follows:

$$k = \left| \frac{\text{numerical value} - \text{mean}}{\text{standard deviation}} \right| = \left| \frac{60 - 75}{5} \right| = |-3.0| = 3$$

$$k = \left| \frac{\text{numerical value} - \text{mean}}{\text{standard deviation}} \right| = \left| \frac{90 - 75}{5} \right| = |3.0| = 3$$

Thus, $k = 3$. You may have noticed that the $k$ is found in a manner similar to finding the $z$-score. What is the difference, then, between $k$ and a $z$-score? The difference is that $z$-scores are always associated with an existing data value, whereas $k$ need not be.

Once you have found the value of $k$, plug it into the *at least* $\left(1 - \frac{1}{k^2}\right)$ 100% formula from Chebyshev's Rule to solve the problem, as follows:

$$\textit{at least} \left(1 - \frac{1}{k^2}\right)100\% = \textit{at least} \left(1 - \frac{1}{3^2}\right)100\% = \textit{at least} \left(1 - \frac{1}{9}\right)100\%$$

$$= \textit{at least} \left(\frac{8}{9}\right)100\% \approx \textit{at least } 88.8889\% \approx \textit{at least } 88.9\%$$

That is, at least 88.9% of the students will score between 60 and 90 on the upcoming exam. Note that you were told nothing about the shape of the distribution. Chebyshev's Rule can be applied to distributions of any shape, even unknown distributions, as long as the data are continuous. ■

## 2 The Empirical Rule

Metaphorically, the **Empirical Rule** is a high-performance Porsche compared to Chebyshev's Rule, which is a go-anywhere ATV (all-terrain vehicle). If your path calls for dirt roads and mud, you wouldn't use the Porsche but would hop in the ATV. Same here. If you have a skewed data distribution or an unknown distribution, use Chebyshev's Rule. However, in the right circumstances, you should use the Empirical Rule. What are those circumstances? *Use the Empirical Rule when you know that your distribution is bell-shaped (normal).* Of course, on the autobahn, the Porsche would leave the ATV in the dust. Same here. The Empirical Rule, when the data distribution is bell-shaped, outperforms Chebyshev. The more information you have, the more precise and powerful your statistical analysis will be.

**The Empirical Rule**

If the data distribution is bell-shaped:

- About 68% of the data values will fall within one standard deviation of the mean.
- About 95% of the data values will fall within two standard deviations of the mean.
- About 99.7% of the data values will fall within three standard deviations of the mean.

Stated in terms of *z*-scores:

- About 68% of the data values will have *z*-scores between 1 and –1.
- About 95% of the data values will have *z*-scores between 2 and –2.
- About 99.7% of the data values will have *z*-scores between 3 and –3.

How do we arrive at the percentages shown in Figure 3.28? Subtracting the 68% from the 95% leaves 27% with *z*-scores between 1 and 2 and between –1 and –2, combined. By symmetry, this leaves about 13.5% (half of 27%) of *z*-scores between 1 and 2. Similar arithmetic leads to the other percentages shown in Figure 3.28. Although Figure 3.28 is for *z*-scores, the percentages are approximately correct for any data set that has a bell-shaped distribution. The Empirical Rule applies only to the integers 1, 2, and 3. For example, we won't learn until Chapter 6 how to calculate the percentage of values that fall within 1.5 standard deviations of the mean.

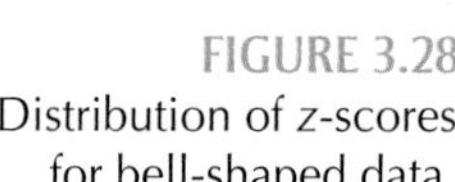
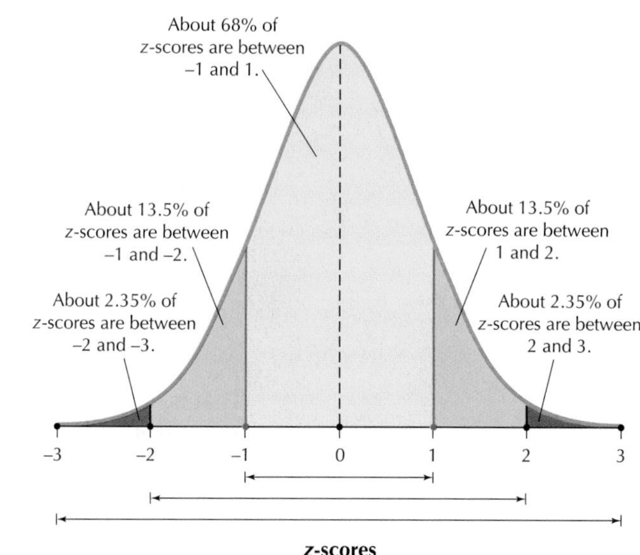


FIGURE 3.28
Distribution of *z*-scores for bell-shaped data.

You should always draw a picture of the data distribution when you have an Empirical Rule problem.

When trying to find percentages or proportions in an Empirical Rule problem, use the following steps:

1. Draw a bell-shaped curve.
2. On the number line, mark the mean right in the middle, and your lower and upper data values to the left and right of the mean.
3. Indicate the proportion of the area that lies between these values.

## Example 3.32 Using the Empirical Rule to find percentages

Suppose we know that student grade point averages (GPAs) are bell-shaped with a mean of 2.5 and a standard deviation of 0.5.

**a.** Find the proportion of GPAs between 2.0 and 3.0.

**b.** Find the proportion of GPAs that lie either above 3.5 or below 1.5.

### Solution

**a.** Again, the key is finding $k$, just as we did for Chebyshev's Rule:

$$k = \left|\frac{\text{numerical value} - \text{mean}}{\text{standard deviation}}\right| = \left|\frac{2.0 - 2.5}{0.5}\right| = |-1| = 1$$

$$k = \left|\frac{\text{numerical value} - \text{mean}}{\text{standard deviation}}\right| = \left|\frac{3.0 - 2.5}{0.5}\right| = |1| = 1$$

So, $k = 1$. We are interested in finding the proportion of GPAs within 1 standard deviation of the mean. We are told that the data are bell-shaped; therefore, we may use the Empirical Rule. Since $k$ represents the number of standard deviations between the data value and the mean, $k = 1$ represents "within one standard deviation of the mean." Thus, according to the Empirical Rule, about 68% of the GPAs lie between 2.0 and 3.0. That English word "about" is not optional; it is required. The Empirical Rule is an approximation of normal distribution probabilities that we will examine more closely in Chapter 6.

**b.** Again start by finding $k$:

$$k = \left|\frac{\text{numerical value} - \text{mean}}{\text{standard deviation}}\right| = \left|\frac{1.5 - 2.5}{0.5}\right| = |-2| = 2$$

$$k = \left|\frac{\text{numerical value} - \text{mean}}{\text{standard deviation}}\right| = \left|\frac{3.5 - 2.5}{0.5}\right| = |2| = 2$$

So, $k = 2$. At this point we are faced with the temptation to simply look up "2 standard deviations" in the Empirical Rule and respond with "about 95%" as our answer. Please do not give in to this temptation. The question is not asking for the proportion of GPAs *within* 2 standard deviations of the mean but rather the proportion *outside* this range. We are interested in what proportion are above 3.5 or below 1.5, that is, what proportion lie 2 or more standard deviations away from the mean. Consider Figure 3.29. We know from the Empirical Rule that about 95% of the GPAs lie within 2 standard deviations of the mean, so that about 95% of the GPAs lie between 1.5 and 3.5.

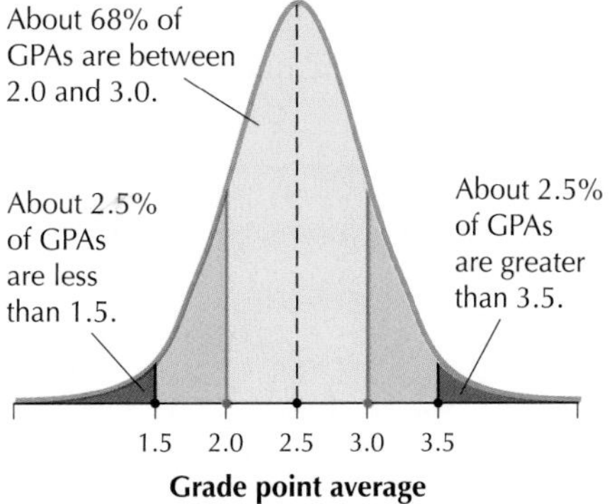


FIGURE 3.29 Distribution of GPAs.

The left-over area in the two tails in Figure 3.29 is the proportion of GPAs above 3.5 or below 1.5. To find the percentage in the left-over area, we subtract 100% − 95% = 5%. Since the bell-shaped curve is symmetrical about the mean, half of the left-over area is above 3.5 and half is below 1.5. Thus, *about 5%* of the GPAs lie either above 3.5 or below 1.5. ■

## Example 3.33 Using the Empirical Rule to find data values

So far we have given values for the GPAs and asked for relevant proportions. We can turn this around and ask which GPAs are associated with various proportions. Suppose the college is interested in identifying extreme outlier GPAs. Assuming the distribution of GPAs is bell-shaped, find the range of GPAs that define the middle 99.7% (approximately) of all GPAs.

### Solution

Here we have to work backward. According to the Empirical Rule, 99.7% is associated with $k = 3$. So, about 99.7% of the GPAs lie within $k = 3$ standard deviations of the mean. The mean plus 3 standard deviations is $2.5 + 3(0.5) = 4.0$, and the mean minus 3 standard deviations is $2.5 - 3(0.5) = 1.0$. Therefore, 99.7% of all GPAs lie within 1.0 and 4.0. ■

If we know that the distribution is bell-shaped, may we still apply Chebyshev's Rule? Yes, Chebyshev applies to any distribution. We will wrap up this section with an example that brings together the notions of *percentile, z-score, Chebyshev's Rule,* and the *Empirical Rule.*

## Example 3.34 Putting it all together

Matthew and Ashley just completed their English test. Matthew's score was at the 80th percentile. Ashley's $z$-score was 1. Who did better and why?

### Solution

Well, Matthew's score was at the 80th percentile, so 80% of the test scores were at or below Matthew's test score. If more than 80% of scores were below Ashley's test score, she wins. However, we are not told that the scores are bell-shaped, so we can't use the Empirical Rule and must use Chebyshev. Ashley's $z$-score was 1, which is $k = 1$ standard deviation above the mean. But Chebyshev's Rule does not apply either, because $k$ must be greater than 1. So the answer is, *we don't know who did better.*

Now, say we add the extra knowledge that the English test scores were *bell-shaped.* Bell-shaped means that we can use the Empirical Rule. Again, 80% of the scores were at or below Matthew's score. This time we know that Ashley's $z$-score of 1 indicates that, not only is she above the "middle" 68% (from the Empirical Rule), but she is also above half of the remaining 32% (Figure 3.30). So Ashley's test score beat about 68% plus about 16% = about 84%. Therefore, Ashley did better than Matthew. ■

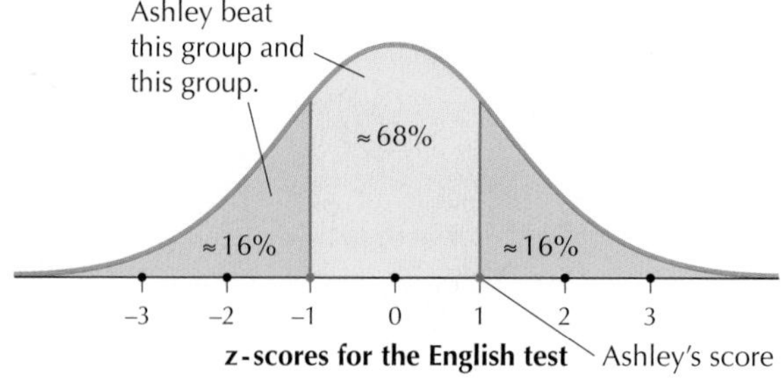


FIGURE 3.30 Sketch of the Matthew and Ashley problem, assuming a bell-shaped distribution.

## Section 3.5 SUMMARY

Chebyshev's Rule allows us to set a lower bound on the proportion of data values that lie within a certain interval. Chebyshev's Rule states that the proportion of values from a data set that will fall within $k$ standard deviations of the mean will be at least $[1 - (1/k)^2]$ 100%, where $k > 1$.

For bell-shaped distributions, the Empirical Rule may be applied, which is more precise than Chebyshev's Rule. The Empirical Rule states that, for bell-shaped distributions, about 68%, 95%, and 99.7% of the data values will fall within one, two, and three standard deviations of the mean, respectively.

## Section 3.5 EXERCISES

### CLARIFYING THE CONCEPTS

1. When may Chebyshev's Rule be used?
2. When may Chebyshev's Rule not be used? (*Hint:* Use $k$ in your answer.)
3. Explain what the following phrase means: "within two standard deviations of the mean."
4. Explain the difference between "about 75%" and "at least 75%."
5. True or false: The Empirical Rule is more precise than Chebyshev's Rule.
6. When may the Empirical Rule be used?
7. May Chebyshev's Rule be used for a bell-shaped distribution?
8. True or false: If a distribution is bell-shaped, then at least 95% of the data values will fall within two standard deviations of the mean.

### PRACTICING THE TECHNIQUES

9. A data distribution has a mean of 100 and a standard deviation of 10. Suppose we do not know whether the distribution is bell-shaped.
   - **a.** Estimate the proportion of the data that falls between 80 and 120.
   - **b.** Estimate the proportion of the data that falls between 70 and 130.

10. A data distribution has a mean of 100 and a standard deviation of 10. Assume that the distribution is bell-shaped.
    - **a.** Estimate the proportion of the data that falls between 80 and 120.
    - **b.** Estimate the proportion of the data that falls between 70 and 130.

11. A data distribution has a mean of 500 and a standard deviation of 100. Suppose we do not know whether the distribution is bell-shaped.
    - **a.** Estimate the proportion of the data that falls between 300 and 700.
    - **b.** Estimate the proportion of the data that falls between 100 and 900.

12. A data distribution has a mean of 500 and a standard deviation of 100. Assume that the distribution is bell-shaped.
    - **a.** Estimate the proportion of the data that falls between 300 and 700.
    - **b.** Estimate the proportion of the data that falls between 100 and 900.

13. A data distribution has a mean of 100 and a standard deviation of 10. Assume that the distribution is bell-shaped.
    - **a.** Find the range of data values that define the middle 95% of the data. (*Hint:* See Example 3.33.)
    - **b.** Find the data value that lies above the upper 97.5% of the data. (*Hint:* Draw a picture.)
    - **c.** Find the data value that lies below the upper 97.5% of the data.

14. A data distribution has a mean of 500 and a standard deviation of 100. Assume that the distribution is bell-shaped.
    - **a.** Find the range of data values that define the middle 68% of the data. (*Hint:* See Example 3.33.)
    - **b.** Find the data value that lies above 84% of the data. (*Hint:* Draw a picture.)
    - **c.** Find the data value that lies below 84% of the data.

### APPLYING THE CONCEPTS

15. **Heating Degree-Days.** The National Climate Data Center reports that the mean annual heating degree-days (an index of energy usage) for the period 1949–2006 was 4500 with a standard deviation of 200. Suppose we do not know the shape of the data distribution. If possible, find the percentage of years with heating degree-days within the following ranges. If not possible, explain why.
    - **a.** Between 4100 and 4900 heating degree-days
    - **b.** Between 3900 and 5100 heating degree-days
    - **c.** Between 4300 and 4700 heating degree-days

16. **Solar Power Production.** The U.S. Department of Energy reports that the mean annual production of solar power in the United States for the years 1989–2006 was 66 trillion Btu (British thermal units) with a standard deviation of 4 trillion Btu. Suppose we do not know the shape of the data distribution. If possible, find the percentage of years with solar power production within the following ranges. If not possible, explain why.
    - **a.** Between 62 trillion and 70 trillion Btu
    - **b.** Between 60 trillion and 72 trillion Btu
    - **c.** Above 72 trillion Btu

17. Refer to Exercise 15. Suppose that we know now that the distribution of heating degree-days is bell-shaped. If

possible, recalculate the three percentages in Exercise 15, with this added knowledge.

**18.** Refer to Exercise 16. Suppose that we know now that the distribution of solar power production is bell-shaped. If possible, recalculate the three percentages in Exercise 16, with this added knowledge.

**19. Shares Traded on the NYSE.** The *Statistical Abstract of the United States* reports that the mean daily number of shares traded on the New York Stock Exchange in 2005 was 1602 (in millions). Let the standard deviation equal 500 million shares.

- **a.** A broker said that one particular day was very slow on the NYSE. He said it was so slow that the $z$-score for the number of shares traded was $-3$.
    - **i.** Find the number of shares traded that has a $z$-score of $-3$.
    - **ii.** Comment on the veracity of the broker.
- **b.** Assume that the number of shares traded is bell-shaped. Which day had more shares traded: a day with a $z$-score of $-1$ or a day that was at the 25th percentile?
- **c.** Assume we don't know the distribution of shares traded. What is the percentage of days in which between 602 million and 2602 million shares are traded?

**20. Life Expectancy.** The U.S. National Center for Health Statistics reported that the mean life expectancy for 20-year-old males in 2001 was 75.5 years. Assume that the standard deviation is 5 years.

- **a.** Assuming that the distribution is bell-shaped, what age can about 95% of all 20-year-old males expect to live to? Provide an upper and lower bound.
- **b.** Suppose we don't know the shape of the distribution. Redo **(a)** without the benefit of the bell-shaped assumption. Comment on the relative precision of the two methods.

**Wind Speed in San Francisco.** Use the following information for Exercises 21–26. Windy day in San Francisco? It must be summer. The mean wind speed in San Francisco is 13.6 mph in July and 7.2 mph in January, according to the National Oceanic and Atmospheric Administration. For wind speed in July, assume we have $\mu = 13.6$ mph and standard deviation $\sigma = 6$ mph, and for wind speed in January, assume we have $\mu = 7.2$ mph and $\sigma = 7.2$ mph.

**21.** Suppose we do not know the shape of the distribution of wind speed in July.

- **a.** Estimate the proportion of times that wind speed in July is between 7.6 mph and 19.6 mph.
- **b.** Estimate the proportion of times that wind speed in July is either less than 7.6 mph or greater than 19.6 mph.
- **c.** Discuss whether we can estimate the proportion of times that wind speed is greater than 19.6 mph.

**22.** Suppose we do not know the shape of the distribution of wind speed in January.

- **a.** Estimate the proportion of times that wind speed in January is between 0 mph and 14.4 mph.
- **b.** Estimate the proportion of times that wind speed in July is either less than 0 mph or greater than 14.4 mph.
- **c.** Discuss whether we can estimate the proportion of times that wind speed is greater than 14.4 mph.

**23.** Assume that the distribution of wind speed in July is bell-shaped.

- **a.** Estimate the proportion of times that wind speed in July is between 1.6 mph and 25.6 mph.
- **b.** Estimate the proportion of times that wind speed in July is less than 1.6 mph.

**24.** Assume that the distribution of wind speed in January is bell-shaped.

- **a.** Estimate the proportion of times that wind speed in January is between 1.2 mph and 13.2 mph.
- **b.** Estimate the proportion of times that wind speed in January is less than 1.2 mph.

**25.** Refer to the previous two exercises. Compare the proportion of calm days in each month, where it is a calm day if wind speed $<1.6$ mph in July or $<1.2$ in January.

**26.** Assume that the distributions of wind speed in January and in July are bell-shaped.

- **a.** Wind speed on a certain day in July reached 30 mph. Determine whether this is an outlier or moderately unusual or not unusual.
- **b.** Wind speed on a certain day in January reached 30 mph. Determine whether this is an outlier or moderately unusual or not unusual.

**Grade Inflation.** Use the following information for Exercises 27–31. Many educators are concerned about grade inflation. A study found that one low-performing high school (with mean combined SAT score of 750) had a higher mean grade point average (mean GPA = 3.6) than a high-performing school (with mean combined SAT score of 1050), which had mean GPA = 2.6.[13] Assume that GPA at the high-performing school has standard deviation $\sigma = 0.5$.

**27.** Suppose we do not know the shape of the distribution of GPA at the high-performing school.

- **a.** Find the proportion of students with a GPA between 2.1 and 3.1.
- **b.** Find the proportion of students with a GPA between 1.6 and 3.6.
- **c.** If possible, find the proportion of students with a GPA greater than 3.6. If not possible, clearly explain why not.

**28.** Now assume that the distribution of GPA at the high-performing school is bell-shaped.

- **a.** Find the proportion of students with a GPA between 2.1 and 3.1.
- **b.** Find the proportion of students with a GPA between 1.6 and 3.6.
- **c.** If possible, find the proportion of students with a GPA greater than 3.6. If not possible, clearly explain why not.

29. Compare your responses to **(a)** and **(b)** in the previous two exercises. Which answer is more precise? What extra information accounts for this increased precision?

30. Suppose the mean GPA at the low-performing school was $\mu = 3.6$ with standard deviation $\sigma = 0.5$, and that the maximum GPA is 4.0. Does it make sense to assume that the distribution of GPA at this school is bell-shaped? Draw a graph to support your finding.

31. The maximum GPA at these schools is 4.0. Compare whether it is unusual for a student to get a 4.0 at each school.

**Cholesterol Levels in Food.** For Exercises 32–34, refer to Table 3.27 of 20 food items from the **Nutrition** data set that have the highest levels of cholesterol per gram of weight.

32. Working with Chebyshev's Rule.
   - **a.** What proportion of the food items would Chebyshev's Rule say were within 1.5 standard deviations of the mean cholesterol per gram of weight?
   - **b.** Compare your result from **(a)** with the actual proportion of food items in Table 3.27 that are within 1.5 standard deviations of the mean.
   - **c.** What proportion of the food items would Chebyshev's Rule say were within 2.5 standard deviations of the mean?
   - **d.** Compare your result from **(c)** to the actual proportion of food items in Table 3.27 that are within 2.5 standard deviations of the mean.
   - **e.** In view of your results from **(a)–(d)**, comment on the precision of Chebyshev's Rule.

33. Working with the Empirical Rule.
   - **a.** By hand or using a computer, construct a dotplot of the cholesterol levels. Indicate the mean using a triangle.
   - **b.** Evaluate whether the dotplot from **(a)** shows evidence that the distribution of food items is bell-shaped.
   - **c.** Based on **(b)**, should we apply the Empirical Rule? Why or why not?

**TABLE 3.27 Cholesterol levels of 20 food items**

| Food item | Milligrams of cholesterol per gram weight |
|---|---|
| Yolk of a raw egg | 12.5294 |
| Chicken liver | 6.3000 |
| Beef liver, fried | 4.8235 |
| Fried eggs | 4.5870 |
| Scrambled eggs | 3.5246 |
| Pound cake | 2.2069 |
| Butter | 2.2000 |
| Fried shrimp | 1.9765 |
| Braised beef heart | 1.9294 |
| Cheesecake | 1.8478 |
| French toast | 1.7231 |
| Quiche Lorraine | 1.6193 |
| Braunschweiger | 1.5614 |
| Egg, cheese, bacon sandwich on English muffin | 1.5435 |
| Canned shrimp | 1.5059 |
| Whipped cream | 1.4000 |
| Waffles | 1.3600 |
| Spinach soufflé | 1.3529 |
| Roasted veal rib | 1.2824 |
| Braised lean lamb chops | 1.2292 |

34. Working with $z$-scores.
   - **a.** Find the $z$-score for the yolk of a raw egg.
   - **b.** Find the $z$-score for chicken liver.
   - **c.** Determine whether the cholesterol level for the yolk of a raw egg could be considered an outlier.
   - **d.** Determine whether the cholesterol level for chicken liver could be considered an outlier.

## 3.6 Robust Measures

*Objectives:* **By the end of this section, I will be able to…**

1. Find quartiles and the interquartile range.
2. Calculate the five-number summary of a data set.
3. Construct a boxplot for a given data set.
4. Apply robust detection of outliers.

## Quartiles and the Interquartile Range

In Sections 3.1 and 3.2 we learned that both the mean and the standard deviation are sensitive to the presence of extreme values (outliers). Data analysts sometimes prefer more **robust measures**, that is, statistics that can summarize a data set while being less sensitive to the presence of outliers. For example, the median is a robust measure of center because it is less sensitive than the mean to the presence of outliers. An alternative method of summarizing a data set, which is robust and has come into widespread use, is the five-number summary, which we examine in this section. Before we examine the five-number summary, we first need to define the **quartiles**. Just as the median divides the data set into halves, the quartiles divide the data set into quarters (Figure 3.31).

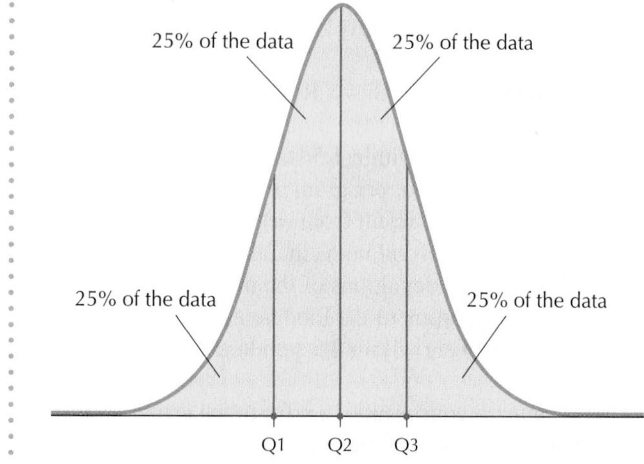


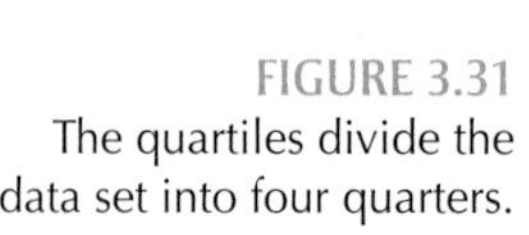
FIGURE 3.31
The quartiles divide the data set into four quarters.

> **The Quartiles**
>
> The **quartiles** of a data set divide the data set into four parts, each containing 25% of the data.
>
> - The *first quartile* (Q1) is the 25th percentile.
> - The *second quartile* (Q2) is the 50th percentile, that is, the median.
> - The *third quartile* (Q3) is the 75th percentile.
>
> For small data sets, the division may be into four parts of only approximately equal size.

It may be helpful to note that the phrase *third quartile* is akin to the phrase *three quarters,* which is 75%, representing the 75th percentile. Also, the phrase *first quartile* is akin to the phrase *one quarter,* which is 25%, representing the 25th percentile.

### Example 3.35 Finding the quartiles for a small data set: the dance audition scores

In Example 3.24 (page 126) we examined the dance scores of 12 students auditioning for admission into a prestigious graduate school of the arts. Recall that we found the 75th percentile of the dance audition scores to be 83. By definition, the 75th percentile is the third quartile Q3. Therefore, this score of 83 is also the third quartile (Q3) of the audition scores. Now we will find the first quartile and the median (second quartile).

## Solution

To find the quartiles, we use the steps for finding percentiles from Section 3.4. First, arrange the data set in ascending order, as follows:

30 44 56 62 65 68 75 78 81 85 89 94

Here, $n = 12$. To find Q1, plug $p = 25$ into the equation $i = \left(\frac{p}{100}\right)n$, where $n = 12$. We get $i = \left(\frac{p}{100}\right)n = \left(\frac{25}{100}\right)12 = 3$. Since 3 is an integer, we know that the 25th percentile is the mean of the dance scores in the 3rd and 4th positions. The score of 56 is in the 3rd position, while 62 is in the 4th position. Since $(56 + 62) / 2 = 59$, we get the 25th percentile of the dance scores to be 59 (Figure 3.32).

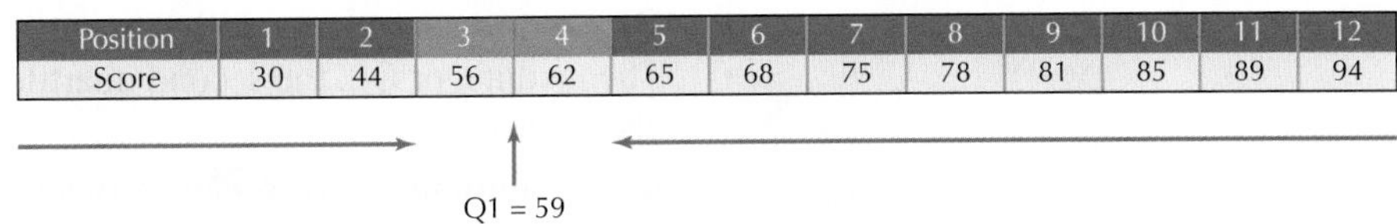

| Position | 1 | 2 | 3 | 4 | 5 | 6 | 7 | 8 | 9 | 10 | 11 | 12 |
|---|---|---|---|---|---|---|---|---|---|---|---|---|
| Score | 30 | 44 | 56 | 62 | 65 | 68 | 75 | 78 | 81 | 85 | 89 | 94 |



FIGURE 3.32 The 25th percentile splits the difference between 56 and 62.

To find the median (the second quartile, Q2), plug $p = 50$ into your steps for finding the percentiles: $i = \left(\frac{p}{100}\right)n = \left(\frac{50}{100}\right)12 = 6$. Since 6 is an integer, we know that the 50th percentile is the mean of the dance scores in the 6th and 7th positions, that is, 68 and 75. Since $(68 + 75) / 2 = 71.5$, the 50th percentile of the dance scores is 71.5 (Figure 3.33). This agrees with the method we learned for finding the median, on page 90.

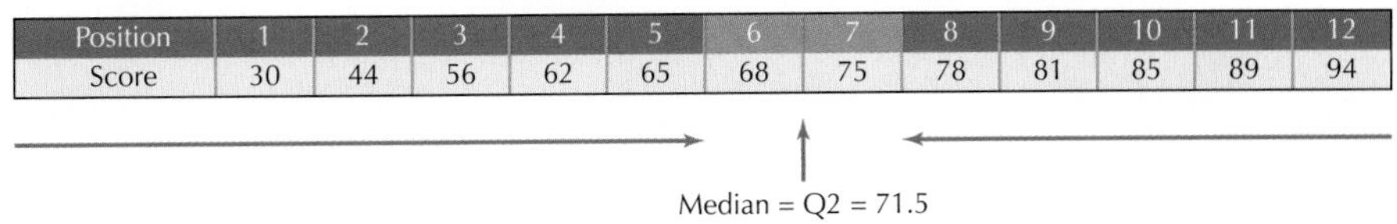

| Position | 1 | 2 | 3 | 4 | 5 | 6 | 7 | 8 | 9 | 10 | 11 | 12 |
|---|---|---|---|---|---|---|---|---|---|---|---|---|
| Score | 30 | 44 | 56 | 62 | 65 | 68 | 75 | 78 | 81 | 85 | 89 | 94 |



FIGURE 3.33 The 50th percentile splits the difference between 68 and 75.

In Example 3.24, we determined that the 75th percentile was 83. Therefore, the quartiles for the dance score data set are Q1 = 59, median = Q2 = 71.5, and Q3 = 83. Note that these quartiles divide the data set into four equal sections, of 3 observations each (Figure 3.34). ■

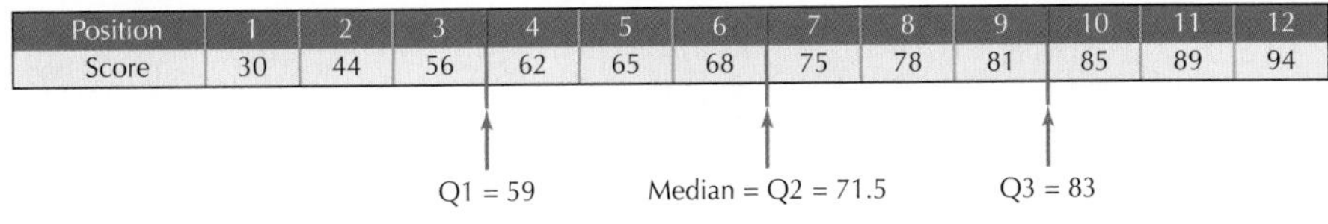

| Position | 1 | 2 | 3 | 4 | 5 | 6 | 7 | 8 | 9 | 10 | 11 | 12 |
|---|---|---|---|---|---|---|---|---|---|---|---|---|
| Score | 30 | 44 | 56 | 62 | 65 | 68 | 75 | 78 | 81 | 85 | 89 | 94 |



FIGURE 3.34 The quartiles for the dance audition data.

Of course, for small data sets, the division into quarters is not always exact. For example, what if one dancer had sprained her ankle that morning and could not make the audition? Then there would have been only 11 dance scores, which cannot be divided equally into four quarters. In this case, therefore, the quartiles would divide the data set up into four sections of approximately equal size. However, for large data sets, which the data analyst most often encounters, this becomes less of an issue.

## Example 3.36 Quartiles of a large data set: cholesterol levels in food

The U.S. Department of Agriculture recommends a diet low in cholesterol, to reduce the risk of heart disease. The data set **Nutrition** contains information on the cholesterol content (in milligrams) of 961 different foods. Find the mean, standard deviation, and quartiles.

### Solution

The Minitab descriptive statistics for the cholesterol data are shown in Figure 3.35. Note that the mean cholesterol content is 32.55 mg and that the standard deviation is about 120 mg. Recall that a standard deviation that is much larger than the mean may be associated with strongly skewed distributions. Compare the value for the mean with the values for the quartiles.

- Q1, the first quartile, or 25th percentile, is 0 mg of cholesterol.
- The median, or Q2, the second quartile (50th percentile), is also 0 mg of cholesterol.
- Q3, the third quartile, or 75th percentile, is 20 mg of cholesterol. ■

| Variable | N | Mean | StDev | Min | Q1 | Median | Q3 | Max |
|---|---|---|---|---|---|---|---|---|
| Cholesterol | 961 | 32.55 | 119.96 | 0 | 0 | 0 | 20 | 2053 |

FIGURE 3.35 Descriptive statistics for the cholesterol data.

*Note:* Minitab uses a different way to calculate the quartiles than the way we have learned, which results in different values than our hand-calculation methods. However, for large data sets, the difference is minimal.

Figure 3.36 shows that the data distribution is extremely right-skewed. There are only a few foods with over 1000 mg cholesterol, and another handful with over 500 (see data on disk). Therefore, it appears that we have outliers in this data set. What is the effect of these outliers on the mean and standard deviation? Does the mean represent a truly typical cholesterol content level for the data set, or is its value unduly increased by the outliers? Let's find out.

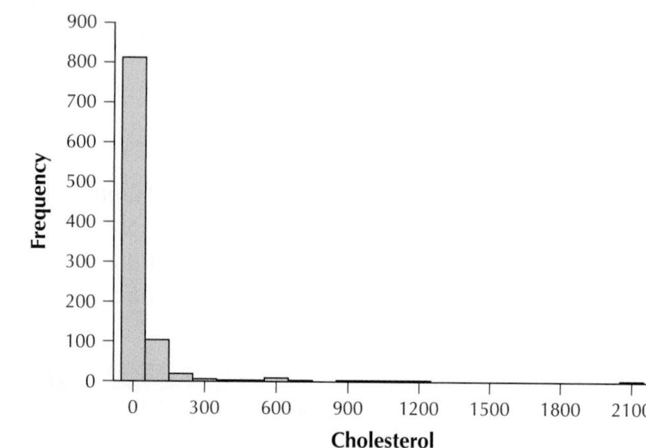


FIGURE 3.36 Cholesterol content (mg) of 961 foods.

*Developing Your Statistical Sense*

### The Mean Is Not Always Representative

Note that the median is 0 mg of cholesterol, meaning that at least half of the food items tested by the USDA in this data set had no cholesterol at all. We are intrigued by this result and ask Minitab to provide us with a frequency distribution for the cholesterol content, along with the cumulative percentages ("CumPct"). Figure 3.37 provides a portion of this frequency distribution, with the following results:

- 61.91% of the food items have no cholesterol at all, which explains why Q1 and the median are both zero.

61.91% of food items had zero cholesterol. Thus, Q1 = 0 and median = 0.

| Chol | Count | CumPct | Chol | Count | CumPct |
|---|---|---|---|---|---|
| 0 | 595 | 61.91 | 20 | 5 | 75.23 |
| 1 | 12 | 63.16 | 21 | 5 | 75.75 |
| 2 | 8 | 64.00 | 22 | 6 | 76.38 |
| 3 | 7 | 64.72 | 23 | 3 | 76.69 |
| 4 | 11 | 65.87 | 24 | 4 | 77.11 |
| 5 | 10 | 66.91 | 25 | 3 | 77.42 |
| 6 | 6 | 67.53 | 26 | 4 | 77.84 |
| 7 | 5 | 68.05 | 27 | 9 | 78.77 |
| 8 | 7 | 68.78 | 28 | 3 | 79.08 |
| 9 | 4 | 69.20 | 29 | 4 | 79.50 |
| 10 | 8 | 70.03 | 30 | 3 | 79.81 |
| 11 | 3 | 70.34 | 31 | 6 | 80.44 |
| 12 | 4 | 70.76 | 32 | 5 | 80.96 |
| 13 | 4 | 71.18 | 33 | 3 | 81.27 |
| 14 | 5 | 71.70 | 34 | 4 | 81.69 |
| 15 | 7 | 72.42 | 35 | 2 | 81.89 |
| 16 | 5 | 72.94 | 36 | 1 | 82.00 |
| 17 | 6 | 73.57 | 37 | 4 | 82.41 |
| 18 | 7 | 74.30 | 38 | 1 | 82.52 |
| 19 | 4 | 74.71 | 40 | 1 | 82.62 |

75th percentile (Q3) = 20 mg cholesterol

81st percentile is 32 mg. The mean is 32.55 mg!

FIGURE 3.37 Partial frequency distribution of cholesterol content.

- The 75th percentile, Q3, is verified to be 20 mg cholesterol.
- The 81st percentile of the data set is 32 mg cholesterol.

Think about these results for a moment. We found that the 81st percentile is 32 mg cholesterol. In other words, 81% of the food items have a cholesterol content of 32 mg or less. And yet, this 32 mg is still *less than the mean* cholesterol content, reported by Minitab to be 32.55 mg. In other words, the mean of this data set is larger than 81% of the data values in the data set. More than four out of five food items have less cholesterol than the mean.

It seems clear, then, that *the mean 32.55 mg cannot be considered as typical or representative* of the data set. Its value has been exaggerated by the presence of the outliers, to such an extent that it is now larger than 81% of the data. We need another, more *robust* measure of center, one that is *resistant* to the undue influence of outliers, such as the median. Here, the value of the median is 0 mg cholesterol. An argument may certainly be made that this is indeed typical and representative of the data set, since 61.91% of the food items have no cholesterol content at all. ■

A robust measure of variability is the **interquartile range**, or **IQR**.

**Interquartile Range**

The **interquartile range (IQR)** is a robust measure of variability. It is calculated as

$$IQR = Q3 - Q1$$

The interquartile range is interpreted to be the spread of the middle 50% of the data.

The Latin word *inter* means "between," so the *inter*quartile range is the *difference between* the quartiles Q3 and Q1. "The IQR" represents how spread out the "middle half" of the data set is. A larger IQR implies a greater degree of variability, or spread, in the data set. Since the IQR ignores both the highest 25% and the lowest 25% of the data set, it is completely unaffected by outliers and is thus quite robust.

## Example 3.37 Finding the interquartile range for the dance audition scores

In Example 3.35, we found that, for the dance audition score data, Q1 = 59 and Q3 = 83. Find the IQR for the dance score data and explain what it means.

**Solution**

Since Q1 = 59 and Q3 = 83, the IQR is IQR = Q3 − Q1 = 83 − 59 = 24.

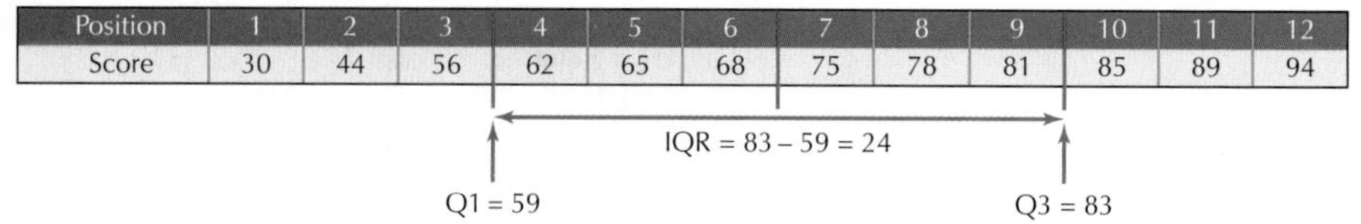

| Position | 1 | 2 | 3 | 4 | 5 | 6 | 7 | 8 | 9 | 10 | 11 | 12 |
|---|---|---|---|---|---|---|---|---|---|---|---|---|
| Score | 30 | 44 | 56 | 62 | 65 | 68 | 75 | 78 | 81 | 85 | 89 | 94 |



FIGURE 3.38 The interquartile range for the dance audition data.

We would say that the middle 50%, or middle half, of the dance audition scores ranged over 24 points (see Figure 3.38). What would happen if we introduced an outlier into this data set? For example, what if we changed the lowest score from 30 to 3? The IQR would remain completely unaffected, as it would even if we changed the 44 to a 4. However, if we changed the 56, then the IQR would be affected, since Q1 would then change. ■

## 2 The Five-Number Summary

Because the mean and the standard deviation are sensitive to the presence of outliers, data analysts sometimes prefer a less sensitive set of statistics to summarize a data set. One robust method of summarizing data that is used widely is called the **five-number summary**. The set consists of five measures we have already seen.

**The Five-Number Summary**

The **five-number summary** consists of the following set of statistics, which together constitute a robust summarization of a data set:

1. Smallest value in the data set (minimum)
2. First quartile, Q1
3. Median, Q2
4. Third quartile, Q3
5. Largest value in the data set (maximum)

### Example 3.38 The five-number summary for the dance audition scores

Find the five-number summary for the dance audition data.

**Solution**

Examining Figure 3.39, we can without difficulty find the five-number summary for the dance audition data.

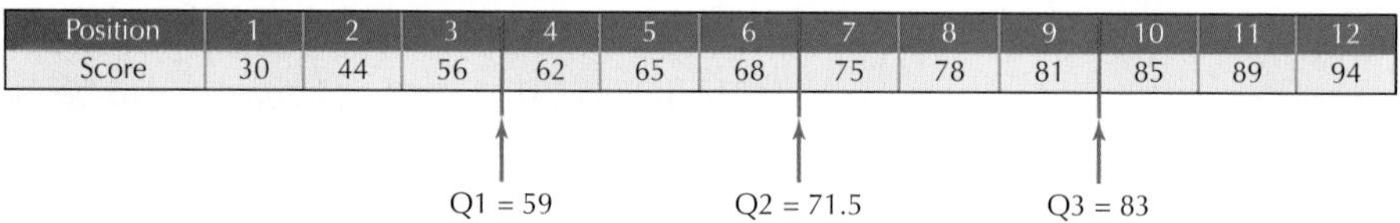

| Position | 1 | 2 | 3 | 4 | 5 | 6 | 7 | 8 | 9 | 10 | 11 | 12 |
|---|---|---|---|---|---|---|---|---|---|---|---|---|
| Score | 30 | 44 | 56 | 62 | 65 | 68 | 75 | 78 | 81 | 85 | 89 | 94 |



FIGURE 3.39 The quartiles for the dance audition data.

1. Minimum = 30
2. First quartile, Q1 = 59
3. Median = Q2 = 71.5
4. Third quartile, Q3 = 83
5. Maximum = 94

More succinctly, the five-number summary is often reported as Min = 30, Q1 = 59, Med = 71.5, Q3 = 83, Max = 94. Which parts of the five-number summary are less robust than others? Since the minimum and maximum are the most extreme values, these are clearly very sensitive to outliers. However, Q1, the median, and Q3 are very resistant to the influence of outliers. ■

## Example 3.39 The five-number summary for the cholesterol data

Find the five-number summary for the cholesterol data.

**Solution**

Minitab's reporting of the descriptive statistics makes it particularly straightforward to report the five-number summary, as here in Figure 3.35 (repeated from page 144) for the cholesterol data.

| Variable | N | Mean | StDev | Min | Q1 | Median | Q3 | Max |
|---|---|---|---|---|---|---|---|---|
| Cholesterol | 961 | 32.55 | 119.96 | 0 | 0 | 0 | 20 | 2053 |

FIGURE 3.35 Descriptive statistics for the cholesterol data.

The five-number summary for the cholesterol data set is

1. Smallest value in the data set = Min = 0
2. First quartile, Q1 = 0
3. Median = 0
4. Third quartile, Q3 = 20
5. Largest value in the data set = Max = 2053

Or, simply, Min = 0, Q1 = 0, Med = 0, Q3 = 20, Max = 2053. ■

The five-number summary is gaining increasing use as a robust alternative to the more widespread mean and standard deviation as summary measures of data sets. The five-number summary is strongly associated with a certain type of graphical summary of data, called a *boxplot*, which we examine next.

## 3 The Boxplot

The **boxplot** is a convenient graphical display of the five-number summary of a data set. The boxplot allows the data analyst to evaluate the symmetry or skewness of a data set.

## Example 3.40 The boxplot of the dance audition scores

Interpret the boxplot for the audition scores in Figure 3.40.

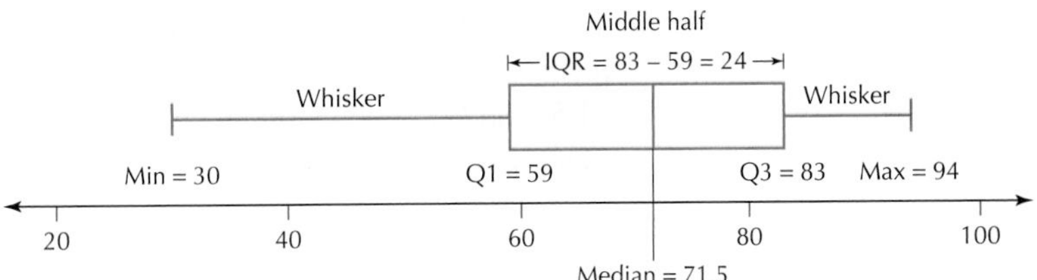


FIGURE 3.40 Boxplot of the dance score data.

**Solution**

Let's examine this boxplot carefully. The horizontal axis represents the dance scores. The red box itself represents the middle half of the data set. The right-hand side of the box, called the *upper hinge,* is located at Q3, which is 83. The left-hand side of the box, called the *lower hinge,* is located at Q1, which is 59. The solid vertical line inside the box is located at the median, which is 71.5. The horizontal lines emanating from the left and right of the box are called the *whiskers*. If there are no outliers, the whiskers extend as far as the maximum and minimum values of the data set, which are represented by the vertical lines at Max = 94 and Min = 30.

At the end of this section (page 154), we demonstrate how to create a boxplot using technology. ■

**Constructing a Boxplot by Hand**

1. Determine the lower and upper fences:
   a. Lower fence = Q1 − 1.5(IQR)
   b. Upper fence = Q3 + 1.5(IQR), where IQR = Q3 − Q1
2. Draw a horizontal number line that encompasses the range of your data, including the fences. Above the number line, draw vertical lines at Q1, the median, and Q3. Connect the lines for Q1 and Q3 to each other so as to form a box.
3. Temporarily indicate the fences as brackets ([ and ]) above the number line.
4. Draw a horizontal line from Q1 to the smallest data value greater than the lower fence. This is the lower whisker. Draw a horizontal line from Q3 to the largest data value smaller than the upper fence. This is the upper whisker.
5. Indicate any data values smaller than the lower fence or larger than the upper fence using an asterisk (*). These data values are outliers. Remove the temporary brackets.

## Example 3.41 Constructing a Boxplot by hand

Construct a boxplot for the dance score data.

### Solution

From Example 3.38, the five-number summary for the dance score data is Min = 30, Q1 = 59, Med = 71.5, Q3 = 83, Max = 94. From Example 3.37, the interquartile range for the dance score data is IQR = Q3 − Q1 = 83 − 59 = 24.

***Step 1*** Determine the lower and upper fences:
a. Lower fence = Q1 − 1.5(IQR) = 59 − 1.5(24) = 59 − 36 = 23
b. Upper fence = Q3 + 1.5(IQR) = 83 + 1.5(24) = 83 + 36 = 119

***Step 2*** Draw a horizontal number line that encompasses the range of your data, including the fences. Above the number line, draw vertical lines at Q1 = 59, median = 71.5, and Q3 = 83. Connect the lines for Q1 and Q3 to each other so as to form a box, as shown in Figure 3.41a.

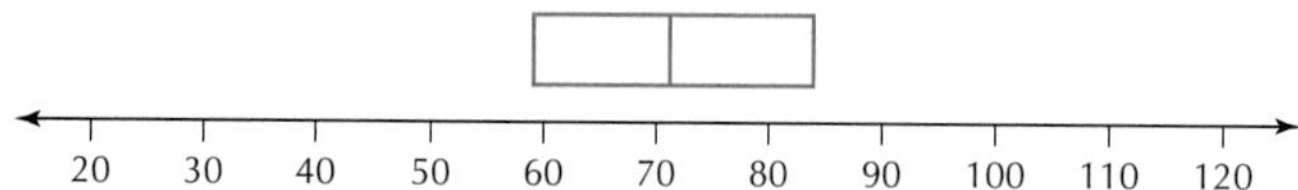


FIGURE 3.41a Constructing a boxplot by hand: Steps 1 and 2.

***Step 3*** Temporarily indicate the fences (lower fence = 23 and upper fence = 119) as brackets above the number line. (See Figure 3.41b.)

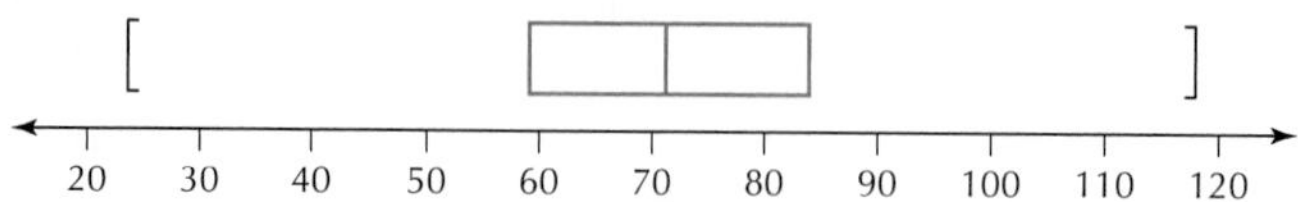


FIGURE 3.41b Constructing a boxplot by hand: Step 3.

**Step 4** Draw a horizontal line from Q1 = 59 to the smallest data value greater than the lower fence. The lowest data value is Min = 30. This is greater than the lower fence = 23. So draw the line from 59 to 23. Draw a horizontal line from Q3 = 83 to the largest data value smaller than the upper fence. The largest data value is Max = 94, which is smaller than the upper fence. So draw the line from 83 to 94. (See Figure 3.41c.)

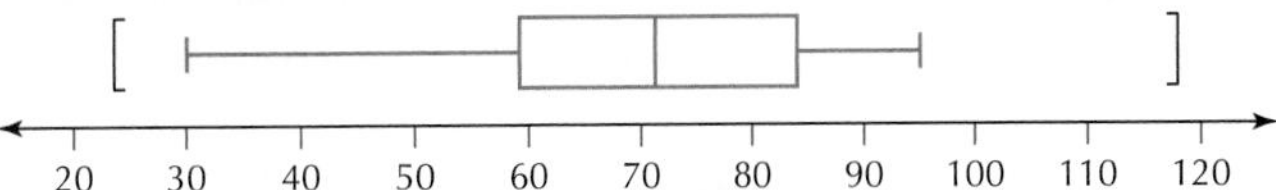


FIGURE 3.41c Constructing a boxplot by hand: Step 4.

**Step 5** There are no data values lower than the lower fence or greater than the upper fence. Thus, there are no outliers in this data set. Therefore, simply remove the temporary brackets, and the boxplot is complete, as shown in Figure 3.41d.

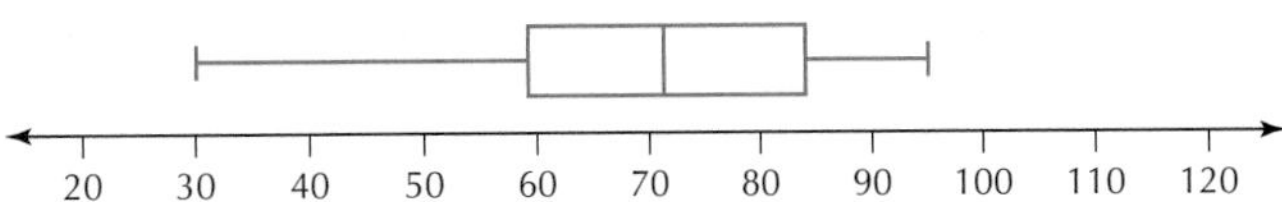


FIGURE 3.41d The completed boxplot.

The next examples show how to recognize when boxplots indicate that a data set is right-skewed, left-skewed, or symmetric.

## Example 3.42 Boxplot for right-skewed data: strikeouts per player in the 2002 baseball season

The number of strikeouts per player in the 2002 American League season is a right-skewed distribution, as shown in histogram of the data in Figure 3.42. The five-number summary is Min = 0, Q1 = 8, Med = 32, Q3 = 66, and Max = 176. How is this skewness reflected in a boxplot (Figure 3.43)? Well, in right-skewed data, the median is closer to Q1 than to Q3, and the lowest non-outlier is closer to Q1 than the highest non-outlier is to Q3. This means that the median is closer to the lower hinge than the upper hinge, and the upper whisker is much longer than the lower whisker. This combination of characteristics indicates a right-skewed data set.

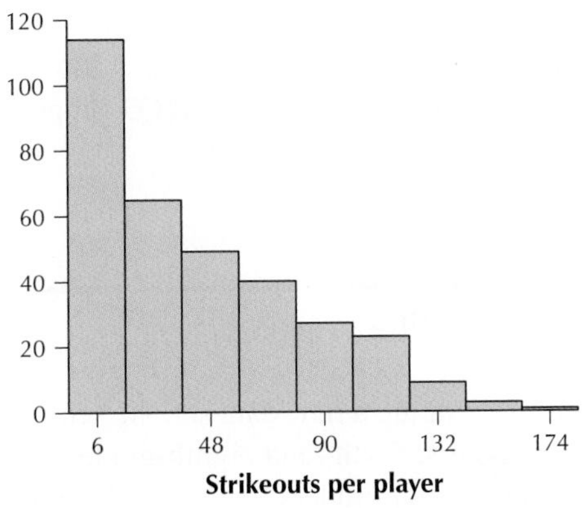


FIGURE 3.42 Strikeouts are right-skewed.

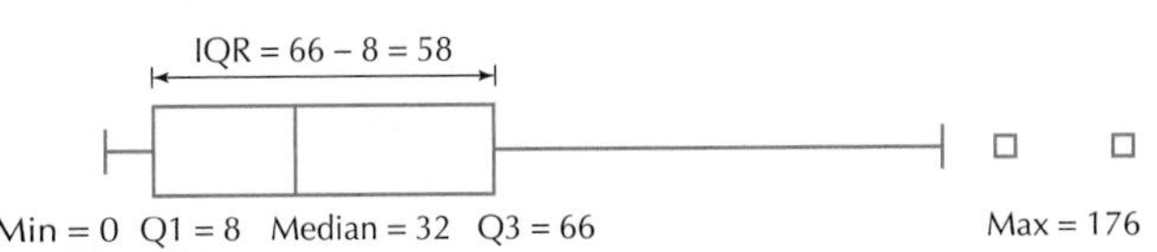


FIGURE 3.43 TI-83/84 boxplot of strikeouts: right-skewed.

The two little boxes at the right represent outliers. (The TI-83/84 uses little boxes rather than asterisks.) These players are Mike Cameron of the Seattle Mariners, who led the league that year with 176 strikeouts, and Alfonso Soriano of the New York Yankees, with 157 strikeouts. When there are no outliers, the whiskers extend as far as the minimum and maximum values. However, when there are outliers, the whiskers extend only as far as the most extreme data value that is not an outlier. ■

## Example 3.43 Boxplot for left-skewed data: exam scores

Figure 3.44 is a histogram of 650 exam scores. Some instructors suggested that the exam may have been a bit on the easy side. Clearly, the data are left-skewed, with many students getting scores in the 90s, and fewer getting grades in the 70s or 80s. Now, with right-skewed data, remember that the median was closer to Q1 than to Q3. What do you think will happen for left-skewed data?

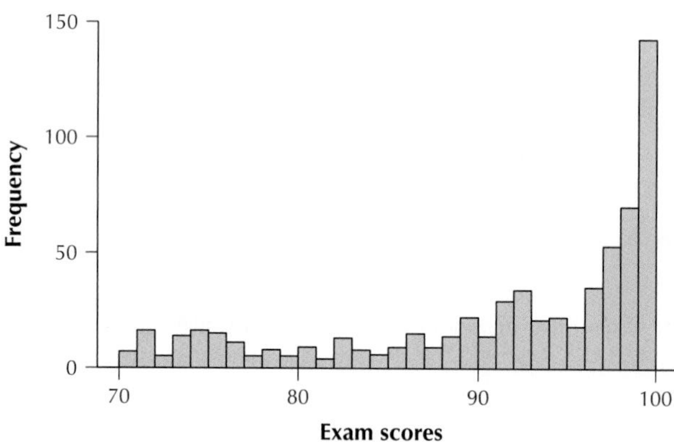


FIGURE 3.44 Histogram of exam scores.

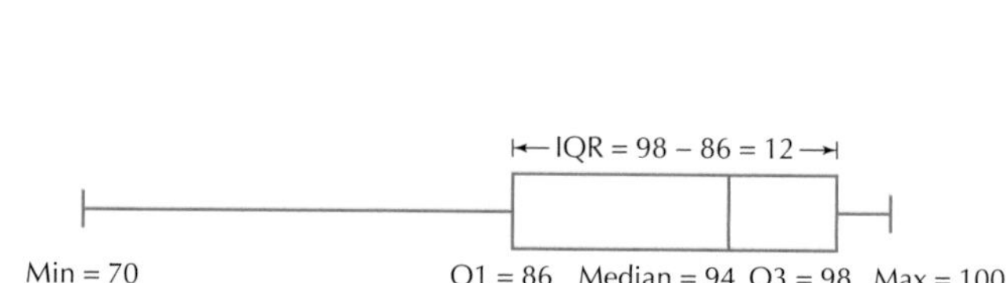


FIGURE 3.45 TI-83/84 boxplot of the exam scores.

### Solution

The five-number summary is Min = 70, Q1 = 86, Med = 94, Q3 = 98, and Max = 100. So, this time, with left-skewed data, the median is closer to Q3 than to Q1. Bet you guessed it!

In the boxplot (Figure 3.45), notice that the median (94) is closer to the upper hinge (Q3, 98) than to the lower hinge (Q1, 86), and the lower whisker is much longer than the upper whisker. This combination of characteristics indicates a left-skewed data set. ■

*What Result Might We Expect?*

**Symmetric Data and Boxplots**

So, can you now predict how a boxplot of *symmetric* data will look? The median will be about the same distance from Q1 (lower hinge) and Q3 (upper hinge). And the upper and lower whiskers will be about the same length. ■

## Example 3.44 Boxplot for symmetric data: middle school physical fitness scores

Let's check these characteristics by examining a new data set, the scores on a physical fitness test given to 2000 middle-schoolers. Sixth- and eighth-grade students (1000 of each) took the same test. Many of the eighth-graders did better, leading to a bimodal distribution, as shown in Figure 3.46. Notice that, despite the bimodality, the data set is still basically symmetric, since one may draw an axis of rough symmetry through the 150 score or thereabouts. We predicted that the median would be roughly the same distance from the other quartiles. Is this true?

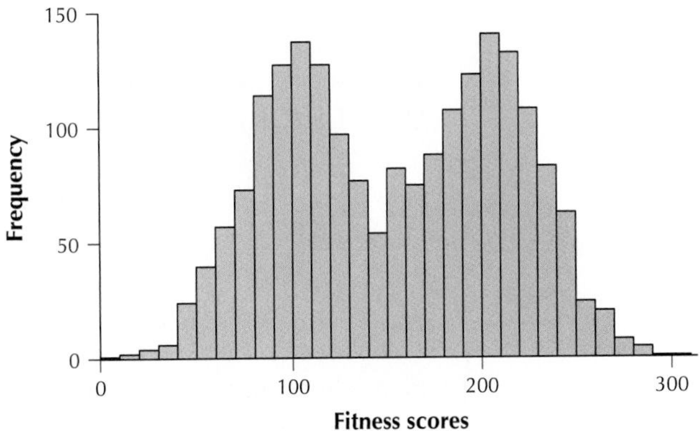


FIGURE 3.46 Histogram of fitness scores.

*Note:* Since a bell-shaped curve is symmetric, its boxplot will reflect this. The boxplot for a bell-shaped curve will have roughly the same distance from the median to each quartile, and it will have upper and lower whiskers of about the same length. On the other hand, just because a boxplot is symmetric does not mean the data set is bell-shaped, as shown by Figures 3.46 and 3.47.

### Solution

The five-number summary is Min = 1, Q1 = 99, Med = 153, Q3 = 200, Max = 297. The median is 54 points away from Q1 and 48 points away from Q3. These distances are not exactly the same, but they are still fairly close, indicating that the data set is fairly symmetric.

The boxplot for this nearly symmetric data set looks nearly symmetric itself (Figure 3.47). You would need to look closely to recognize that the median line (153) is slightly closer to the upper hinge (Q3 = 200) than to the lower hinge (Q1 = 99). Also, the upper and lower whiskers are each about the same length. ■

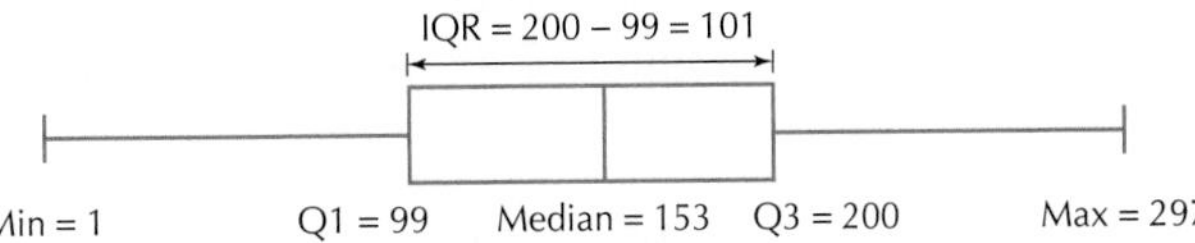


FIGURE 3.47 Boxplot of the fitness scores.

## 4 Robust Detection of Outliers

When using the mean and standard deviation as your summary measures, in most cases outliers occur more than three standard deviations from the mean. However, due to the sensitivity of these measures to the outliers themselves, we often use a more **robust method of detecting outliers**. Earlier we mentioned that, when constructing a boxplot, data values lower than the lower fence and higher than the upper fence are considered outliers. We can use this method to detect outliers without constructing a boxplot.

**Robust Detection of Outliers**

Use a five-number summary or a boxplot to detect outliers, as follows:

A data value is an outlier if

a. it is located 1.5(IQR) or more below Q1, or

b. it is located 1.5(IQR) or more above Q3.

## Example 3.45 Robust detection of outliers for the dance audition data

Determine if there are any outliers in the dance score data.

**Solution**

Recall for the dance score data set that IQR = 24, Q1 = 59, and Q3 = 83. So we have 1.5(IQR) = 1.5(24) = 36. The first step is to find the two quantities Q1 – 1.5(IQR) and Q3 + 1.5(IQR):

$$Q1 - 1.5(IQR) = Q1 - 36 = 59 - 36 = 23$$

$$Q3 + 1.5(IQR) = Q3 + 36 = 83 + 36 = 119$$

Thus, for this data set, a data value would be an outlier if it were 23 or less or 119 or more. Since there are no data values that are 23 or less or 119 or more in the data set, no outliers are identified by the robust method. ■

## What If Scenario: Robust Detection of an Outlier

*What if* the minimum dance score of 30 is changed to 23. Based on Example 3.45, this new value should be detected as an outlier. Note that changing the minimum value does not affect the calculation of Q1, Q3, the IQR, or the thresholds for outlier detection.

Figure 3.48 shows that the box, hinges, and whiskers are all located at precisely the same spots as in the boxplot of the original dance score data. However, the software has calculated, using the robust detection method, that the new data value of 23 is an outlier and indicates it as such with a blue dot. Comparing this boxplot to the earlier one (see Figure 3.40), we notice that the lower whisker is shorter. In Figure 13.48, the whisker terminates at the dance score of 44 instead of 30.

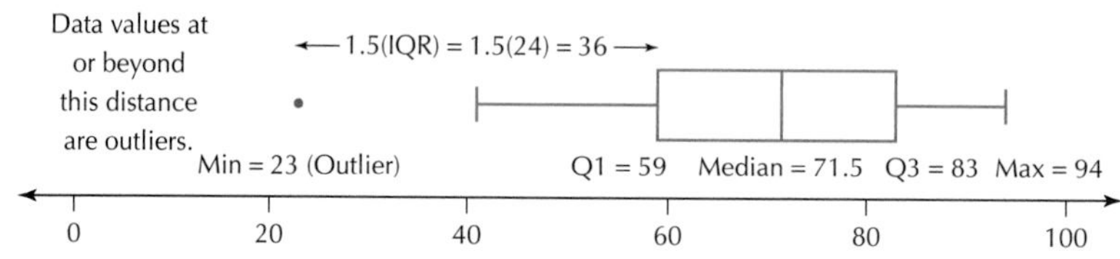


FIGURE 3.48 Boxplot of dance score data showing presence of outlier, after change.

The next example shows how comparison boxplots may be used to compare two data sets side-by-side.

## Example 3.46 Comparison boxplots: comparing body temperatures for women and men

Determine whether the body temperatures of women or men exhibit greater variability.

**Solution**

Consider the comparison boxplots in Figure 3.49. The box for females (on top) lies slightly to the right of that for the males, meaning that the first quartile, the median, and the third quartile are each higher for the women than the men. Therefore, the middle 50% of the body temperatures is higher for women than men.

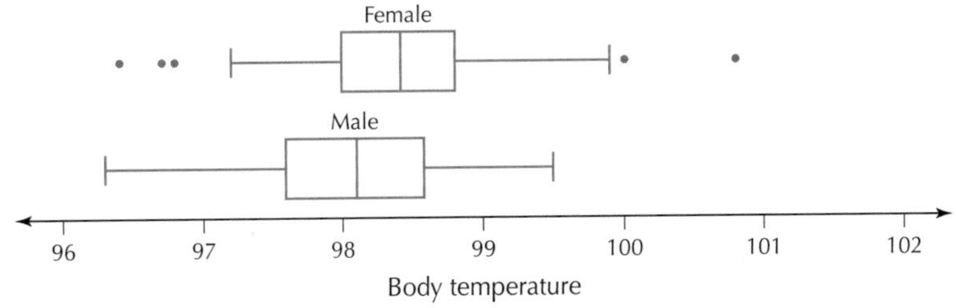


FIGURE 3.49 Comparison boxplots of female and male body temperatures.

This figure seems to offer some evidence that the mean body temperature for women may be higher than that for men. We will formally test whether there is a difference in the true mean body temperature between women and men in Chapter 10. The location of the box is an indication of the center of the data. But where would we look for a difference in the variability of body temperatures between women and men? From Figure 3.50, for the females we have

$$\text{IQR} = \text{Q3} - \text{Q1} = 98.8 - 98.0 = 0.8$$

For the males we have

$$\text{IQR} = \text{Q3} - \text{Q1} = 98.6 - 97.6 = 1.0$$

So the IQR for males is greater.

Let's determine which data set has greater variability based on the three different measures of spread that we have learned: the range, the standard deviation, and the IQR.

| Gender | N | Mean | Median | StDev | Min | Max | Q1 | Q3 |
|---|---|---|---|---|---|---|---|---|
| female | 65 | 98.394 | 98.4 | 0.743 | 96.4 | 100.8 | 98.0 | 98.8 |
| male | 65 | 98.105 | 98.1 | 0.699 | 96.3 | 99.5 | 97.6 | 98.6 |

FIGURE 3.50 Descriptive statistics for body temperature, by gender.

Range for women = 100.8 – 96.4 = 4.4 Range for men = 99.5 – 96.3 = 3.2
Standard deviation for women = 0.743 Standard deviation for men = 0.699
IQR for women = 0.8 IQR for men = 1.0 ■

### When Measures of Spread Disagree

Two measures of spread that are sensitive to the presence of extreme values—range and standard deviation—find that the female body temperatures are more variable. The measure of spread that is resistant to the effects of extreme values—IQR—finds that the male body temperatures are more variable. How do we resolve this apparent inconsistency? What appears to be happening is that, for the middle 50% of each data set, the men are more variable, but as we move toward the tails, the women are more spread out.

Note that there are outliers for the women but not for the men. In part, this may be because the IQR for the women is smaller, and thus the distance 1.5(IQR) is smaller as well. For example, the woman whose body temperature is 100 degrees is identified as an outlier because 100 is the same as the outlier threshold Q3 + 1.5(IQR) = 98.8 + 1.5(0.8) = 100. *The same temperature in a man would not be classified as an outlier, even though the male temperatures are lower overall* (and Q3, specifically, is lower). This is because the temperature of 100 is not higher than Q3 + 1.5(IQR) = 98.6 + 1.5(1.0) = 100.1, the male outlier threshold. Thus, the measures of spread that are sensitive to outliers indicate that women have greater variability, while the measure of spread that is not sensitive to outliers indicates that men have greater variability. ■

## STEP-BY-STEP TECHNOLOGY GUIDE: Boxplots

We will make boxplots for the data in Example 3.38 (page 146). Use the Step-by-Step Technology Guide at the end of Section 3.1 (page 97), and choose the appropriate statistics. Note that Minitab's values for Q1 and Q3 differ from those we calculated earlier because Minitab uses a different way of calculating the quartiles. If you wish to check your hand calculations, do so with the TI-83/84, which uses the method we learned here.

### TI-83/84

***Step 1*** Enter the data in list **L1**.
***Step 2*** Press **2nd Y=**, and choose 1: **Plot 1**.
***Step 3*** Turn plots **On**. Highlight the boxplot icon, as shown in Figure 3.51.
***Step 4*** Press **ZOOM**, and choose **9: ZoomStat**. A boxplot similar to Figure 3.43 is then produced.

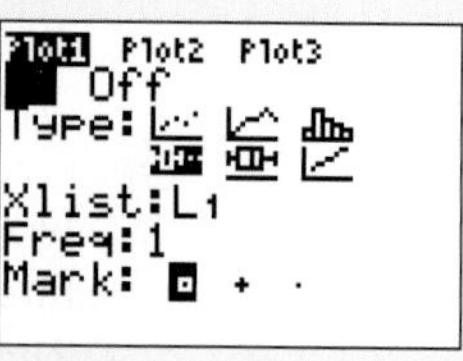


Figure 3.51

### MINITAB

***Step 1*** Enter the data in column **C1**, and name your data *Scores*.
***Step 2*** Click **Graph > Boxplot.**
***Step 3*** Select **Simple** and click **OK.**
***Step 4*** Select the variable *Scores,* and click **OK**, as shown in Figure 3.52. A boxplot similar to Figure 3.40 in Example 3.40 is then produced.

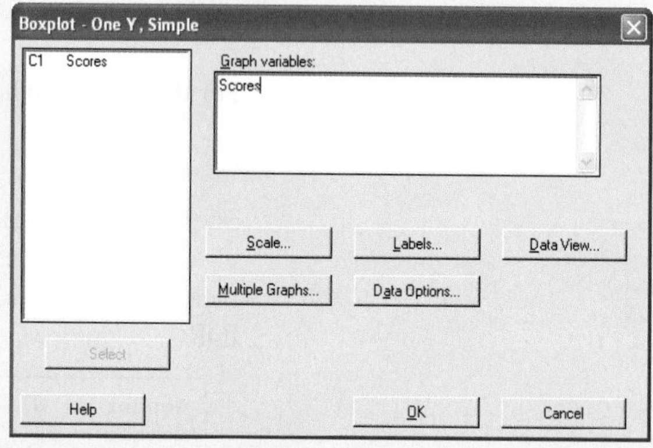


Figure 3.52

## Section 3.6 SUMMARY

1 Section 3.6 presents robust measures and methods, which are not sensitive to the presence of outliers. Quartiles divide the data set into approximately equal quarters. The interquartile range is a measure of variability found by taking the difference between the third and first quartiles.

2 The five-number summary is a robust alternative to the usual mean-and-standard-deviation method of summarizing a data set. It consists of simply reporting the minimum, first quartile, median, third quartile, and maximum of the data set.

3 A boxplot is a graphical representation of the five-number summary and is useful for investigating skewness and the presence of outliers.

4 The robust method of detecting outliers is to consider a data value an outlier if it is located 1.5(IQR) or more below Q1, or it is located 1.5(IQR) or more above Q3.

## Section 3.6 EXERCISES

### CLARIFYING THE CONCEPTS

**1.** Why do we need robust measures? Why can't we just use the mean and standard deviation all the time?

**2.** Describe in your own words what we mean by robust measures.

**3.** What is the difference between the third quartile and the third percentile?

**4.** The median divides the data into halves. What do the quartiles divide the data into?

**5.** Consider whether the following scenarios are possible. If it is possible, then clearly describe what the data set would look like. What would the boxplot for such a data set look like? If it is not possible, why not?

- **a.** A scenario where the first and second quartiles of a data set are equal.

**b.** A scenario where all five numbers in the five-number summary are equal.
**c.** A scenario where the mean of a data set is larger than Q3.
**d.** A scenario where the mean of a data set is smaller than Q1.
**e.** A scenario where the median of a data set is smaller than Q1.
**f.** A scenario where Q1 is smaller than Q3.

**6.** Explain what the IQR actually means, so that a nonspecialist could understand it.

**7.** True or false: The five-number summary consists of the following: Minimum, Q1, Mean, Q3, Maximum.

**8.** True or false: The IQR is sensitive to the presence of outliers.

## PRACTICING THE TECHNIQUES

Refer to Table 3.5 (page 98) for Exercises 9–13.

**9.** Find the following quartiles for *age:*
**a.** Q1
**b.** Q2
**c.** Q3

**10.** Find the interquartile range of *age*.

**11.** Find the five-number summary for *age*.

**12.** Use the robust method to investigate the presence of outliers.

**13.** Construct a boxplot for *age*.

Here are the final-exam scores for 20 psychology students:

| | | | | | | | | | |
|---|---|---|---|---|---|---|---|---|---|
| 75 | 81 | 82 | 70 | 60 | 59 | 94 | 77 | 68 | 98 |
| 86 | 68 | 85 | 72 | 70 | 91 | 78 | 86 | 51 | 67 |

Use this information to answer Exercises 14–18.

**14.** Find the following quartiles for *final-exam score:*
**a.** Q1
**b.** Q2
**c.** Q3

**15.** Find the interquartile range of *final-exam score*.

**16.** Find the five-number summary for *final-exam score*.

**17.** Use the robust method to investigate the presence of outliers.

**18.** Construct a boxplot for *final-exam score*.

## APPLYING THE CONCEPTS

**Women's Volleyball Team Heights.** Refer to Table 3.9 (page 102) for Exercises 19–21.

**19.** Find five-number summaries for the volleyball team heights for WMU and NCU. Comment on whether one should use a five-number summary for data sets of size 5.

**20.** Find the interquartile range for each team. Interpret what this value actually means, so that a nonspecialist could understand it.

**21.** By hand, construct a boxplot for each team. Comment on why your boxplots look unusual.

**Most Active Stocks.** Use Table 3.28 for Exercises 22–29. These companies represent the 11 most actively traded stocks on the New York Stock Exchange for February 5, 2007. Variables include the stock price and the net change in stock price, with both variables in dollars.

**TABLE 3.28 The most active stocks on the NYSE**

| Company | Price | Change |
|---|---|---|
| Pfizer Inc. | 26.88 | +0.08 |
| Slm Corp. | 42.37 | −4.09 |
| Ford Motor Co. | 8.33 | +0.10 |
| General Electric Co. | 36.37 | +0.10 |
| Equity Office Properties Trust | 55.46 | +0.08 |
| AT&T Inc. | 37.79 | −0.33 |
| EMC Corp. | 13.69 | −0.05 |
| LSI Logic Corp. | 9.14 | −0.14 |
| Motorola, Inc. | 19.87 | +0.02 |
| Advanced Micro Devices | 15.60 | −0.09 |

**22.** Find the five-number summary for *price*.

**23.** Find the interquartile range for *price*. Interpret what this value actually means, so that a nonspecialist could understand it.

**24.** Use the robust method to investigate the presence of outliers in *price*.

**25.** Construct a boxplot for *price*.

**26.** Find the five-number summary for *change*.

**27.** Find the interquartile range for *change*. Interpret what this value actually means, so that a nonspecialist could understand it.

**28.** Use the robust method to investigate the presence of outliers in *change*.

**29.** Construct a boxplot for *change*.

**Breakfast Cereals.** Refer to Table 3.24 (page 132) for Exercises 30–35.

**30.** Find the five-number summary for *calories*.

**31.** Find the interquartile range for *calories*. Interpret what this value actually means, so that a nonspecialist could understand it.

**32.** Use the robust method to investigate the presence of outliers in *calories*.

**33.** Construct a boxplot for *calories*.

**34.** Calculate the mean and standard deviation of *calories*.

**35.** Find the $z$-score for Basic 4, and use it to determine whether the cereal is an outlier. Compare the result with that from the robust method.

**Dietary Supplements.** Refer to Table 3.25 (page 132) for Exercises 36–41.

**36.** Find the five-number summary for *usage*.

**37.** Find the interquartile range for *usage*. Interpret what this value actually means, so that a nonspecialist could understand it.

**38.** Use the robust method to investigate the presence of outliers in *usage*.

**39.** Construct a boxplot for *usage*.

**40.** Calculate the mean and standard deviation of *usage*.

**41.** Find the $z$-score for echinacea, and use it to determine whether the product is an outlier. Compare the result with that from the robust method.

**Household Median Incomes.** Refer to the table of state household median incomes for Exercises 42–47. The data are expressed in thousands of dollars, and the states included are the 16 with the highest state household median income, plus the state with the lowest.

**TABLE 3.29 Household median incomes (in thousands of dollars)**

| State | Median income |
|---|---|
| West Virginia | 47.6 |
| Washington | 66.5 |
| Virginia | 66.9 |
| Wisconsin | 67.0 |
| Hawaii | 67.6 |
| Rhode Island | 67.6 |
| Michigan | 68.0 |
| Colorado | 68.1 |
| Illinois | 69.2 |
| Delaware | 69.5 |
| Alaska | 69.9 |
| New Hampshire | 72.4 |
| Minnesota | 72.4 |
| Maryland | 77.9 |
| Massachusetts | 78.3 |
| Connecticut | 81.9 |
| New Jersey | 82.4 |

*Source:* National Bureau of Economic Research, 2005.

**42.** Find the five-number summary for *income*.
**43.** Find the interquartile range for *income*. Interpret what this value actually means, so that a nonspecialist could understand it.
**44.** Use the robust method to investigate the presence of outliers in *income*.
**45.** Construct a boxplot for *income*.
**46.** Calculate the mean and standard deviation of *income*.
**47.** Find the $z$-score for West Virginia, and use it to determine whether the state is an outlier. Compare the result with that from the robust method.

**Nutrition.** Use the data set Nutrition for Exercises 48–51.
**48.** Open the data set **Nutrition.**
   **a.** How many observations are in the data set?
   **b.** How many variables?
**49.** Use a statistical computing package (like Minitab) to explore the variable *iron*.
   **a.** Find the mean and standard deviation for the amount of iron in the food.
   **b.** Find the five-number summary, the range, and the interquartile range.
**50.** Which food item has the maximum amount of iron? Does this surprise you?
**51.** Use the computer to generate a boxplot. Also, comment on the symmetry or the skewness of the boxplot.

## *Try This in Class!* In-class activities to enhance your understanding of statistics

### MEASURES OF CENTER

**1.** What is your guess of the typical height of all students in your class?
**2.** Make a dotplot of the heights of the students in your class.
**3.** Discuss where to place the center of this distribution of student heights. Without crunching any numbers, form a consensus on the location of the center.
**4.** Calculate the mean, median, and mode of the student heights.
**5.** Which measure (mean, median, or mode) comes closest to the consensus of where the center is located in Activity 3?
**6.** What is the relation between these measures and your guess of the typical height in Activity 1?
**7.** Which measure (mean, median, mode, class consensus, your guess) do you think is the best measure of the center of student heights?

### MEASURES OF SPREAD

**8.** Do you think that the distribution of the heights of all students in your class is more spread out or less spread out than the distribution of the heights of only the females in your class?
**9.** Would the values of our measures of spread (range, standard deviation) be larger for the entire class or for only the females?
**10.** Make a dotplot of the heights of only the females in the class. Make sure it uses the same scale as the dotplot for the heights of all the students in the class.
**11.** Use the two dotplots to assess which group has greater variability.
**12.** Back up your intuition by calculating and comparing our measures of spread (range, standard deviation) for the two groups.

## CHAPTER 3 FORMULAS AND VOCABULARY

### SECTION 3.1

- **Mean** (p. 87). A measure of center calculated by adding all the data values and dividing by the number of data values.
- **Measure of center** (p. 87). Indicates where the middle of the data set tends to be.
- **Median** (p. 90). The data value that occupies the middle position when the data are put in ascending order. If the sample size is even, then the median is the mean of the two data values in the middle positions.

- **Mode** (p. 92). The data value that occurs with the greatest frequency.
- **Population mean** (p. 89). Calculated as $\mu = \sum x/N$.
- **Population size** (p. 89). Denoted by $N$.
- **Sample mean** (p. 88). The mean equals the sum of the data values divided by the sample size: $\bar{x} = \sum x/n$.
- **Sample size** (p. 88). Denoted by $n$.

## SECTION 3.2

- **Deviation** (p. 103). $x - \bar{x}$, the difference between a data value and the mean of the data set.
- **Measure of variability (measure of spread)** (p. 102). Quantifies how spread out the data values are. All quantities learned in this section are measures of variability.
- **Population standard deviation** (p. 108). The square root of the population variance.

$$\text{Definition formula: } \sigma = \sqrt{\frac{\sum(x-\mu)^2}{N}}$$

$$\text{Computational formula: } \sigma = \sqrt{\frac{\sum x^2 - \left(\sum x\right)^2/N}{N}}$$

- **Population variance** (p. 107). The mean of the squared deviations.

$$\text{Definition formula: } \sigma^2 = \frac{\sum(x-\mu)^2}{N}$$

$$\text{Computational formula: } \sigma^2 = \frac{\sum x^2 - \left(\sum x\right)^2/N}{N}$$

- **Range** (p. 103). The difference between the largest value and the smallest value in the data set.
- **Sample standard deviation** (p. 109). The square root of the sample variance.

$$\text{Definition formula: } s = \sqrt{\frac{\sum(x-\bar{x})^2}{n-1}}$$

$$\text{Computational formula: } s = \sqrt{\frac{\sum x^2 - \left(\sum x\right)^2/n}{n-1}}$$

- **Sample variance** (p. 109).

$$\text{Definition formula: } s^2 = \frac{\sum(x-\bar{x})^2}{n-1}$$

$$\text{Computational formula: } s^2 = \frac{\sum x^2 - \left(\sum x\right)^2/n}{n-1}$$

- **Standard deviation** (p. 103). Measures the amount of variability in a data set. The greater the amount of variability, the larger the value of the standard deviation. A rough interpretation of the standard deviation is that it indicates a typical distance from the mean for the data values in the data set.

## SECTION 3.3

- **Estimated mean for data grouped into a frequency distribution** (p. 120).

$$\hat{\mu} = \frac{\sum m_i f_i}{N} = \frac{m_1 f_1 + m_2 f_2 + \cdots + m_k f_k}{N}$$

- **Estimated standard deviation for data grouped into a frequency distribution** (p. 121).

$$\hat{\sigma} = \sqrt{\hat{\sigma}^2} = \sqrt{\frac{\sum(m_i - \hat{\mu})^2 \cdot f_i}{N}}$$

- **Estimated variance for data grouped into a frequency distribution** (p. 121).

$$\hat{\sigma} = \frac{\sum(m_i - \hat{\mu})^2 \cdot f_i}{N}$$

- **Weighted mean** (p. 119).

$$\bar{x}_w = \frac{\sum w_i x_i}{\sum w_i} = \frac{w_1 x_1 + w_2 x_2 + \cdots + w_n x_n}{w_1 + w_2 + \cdots + w_n}$$

## SECTION 3.4

- **Outlier** (p. 130). An extremely large or extremely small data value relative to the rest of the data set.
- **Percentile** (p. 125). Let $p$ be any integer between 0 and 100. The $p$th percentile of a data set is the value at which $p$ percent of the values in the data set are less than or equal to this value.

**Calculating the Position of the $p$th Percentile**

***Step 1*** Sort the data into ascending order (from smallest to largest).

***Step 2*** Calculate $i = (p/100)n$, where $p$ is the particular percentile you wish to calculate, and $n$ is the sample size.

***Step 3*** **a.** If $i$ is an integer (a whole number), the $p$th percentile is the mean of the data values in positions $i$ and $i + 1$. **b.** If $i$ is not an integer, round up and use the value in this position.

- ***z*-score** (p. 127). Indicates how many standard deviations a particular data value is from the mean.

**a.** The $z$-score for a particular data value from a *sample* is

$$z\text{-score} = \frac{\text{data value} - \text{mean}}{\text{standard deviation}} = \frac{x-\bar{x}}{s}$$

**b.** The $z$-score for a particular data value from a *population* is

$$z\text{-score} = \frac{\text{data value} - \text{mean}}{\text{standard deviation}} = \frac{x-\mu}{\sigma}$$

## SECTION 3.5

- **Chebyshev's Rule** (p. 134). The proportion of values from a data set that will fall within $k$ standard deviations of the mean will be *at least* $\left(1 - \frac{1}{k^2}\right)100\%$, where $k > 1$.
- **Empirical Rule** (p. 135). If the data distribution is bell-shaped:

About 68% of the data values will fall within one standard deviation of the mean.

About 95% of the data values will fall within two standard deviations of the mean.

About 99.7% of the data values will fall within three standard deviations of the mean.

SECTION 3.6

- **Boxplot** (p. 147). A convenient graphical display of the five-number summary of a data set, useful for investigating skewness and the presence of outliers.
- **Five-number summary** (p. 146). Provides the following five measures: smallest value in the data set (min), first quartile, median, third quartile, largest value in the data set (max).
- **Interquartile range (IQR)** (p. 145). A robust measure of variability, calculated as IQR = Q3 − Q1.
- **Quartiles** (p. 142). Can be used to divide the data set into four parts, each containing approximately 25% of the data. The *first quartile* (Q1) is the 25th percentile. The *second quartile* (Q2) is the 50th percentile, that is, the median. The *third quartile* (Q3) is the 75th percentile.
- **Robust measures** (p. 142). Summary descriptive measures that are not sensitive to the presence of outliers. Such measures are less affected by the presence or absence of extreme values in the data set.
- **Robust method of detecting outliers** (p. 151). Use a five-number summary or a boxplot to detect outliers, as follows. A data value is an outlier if

  **a.** it is located 1.5(IQR) or more below Q1, or
  **b.** it is located 1.5(IQR) or more above Q3.

## CHAPTER 3 REVIEW EXERCISES

SECTION 3.1

**Soft Drink Sales.** Refer to Table 3.16 (page 116) for Exercises 1–3.

**1.** Find the mean and median of the millions of cases sold. Describe the possible skewness of the data.

**2.** Construct a dotplot of the data. Does the plot support your description of the skewness?

**3.** Which measure of center do you think most accurately reflects the center of the data?

**Calories in Cereal.** For Exercises 4–7, refer to the histogram of calories, and to Table 3.24 (page 132).

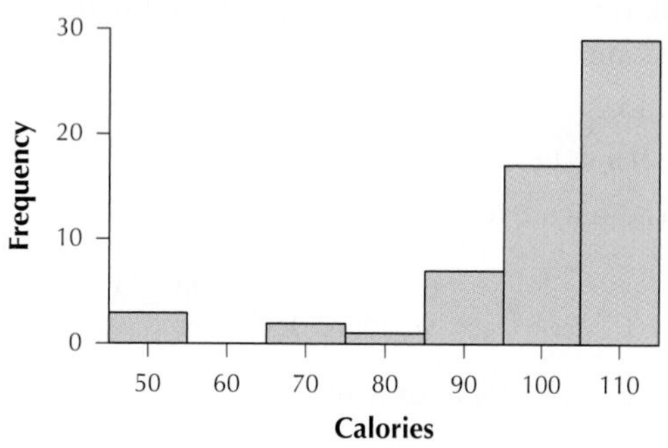


Histogram of calories contained in a sample of breakfast cereals.

**4.** Which is larger, the mean, median, or mode? How do you know?

**5.** If we eliminated from the sample the cereals with less than 60 calories:

**a.** Which measure would not be affected at all? Why?
**b.** Which measure would be affected only a little bit? Why?
**c.** Which measure would be affected the most? Why?

**6.** If we added ten calories to each cereal, how would that affect the mean, median, and mode? Would it affect each of the measures equally?

**7.** If we doubled the calories of each cereal, how would that affect the mean, median, and mode? Would it affect each of the measures equally?

SECTION 3.2

**Women's Volleyball Team Heights.** Refer to Table 3.9 (page 102) for Exercises 8–10.

**8.** Suppose a new player joins the NCU team. She is 7 feet tall (84 inches) and replaces the 72-inch-tall player.

**a.** Would you expect the standard deviation to go up or down, and why?
**b.** Now find the standard deviation for the team including the new player. Was your intuition correct?

**9.** Add 4 inches to the height of each player on the WMU team.

**a.** Recalculate the range and standard deviation.
**b.** Formulate a rule for the behavior of these measures of variability when a constant (like 4) is added to each member of the data set.
**c.** By hand, draw a dotplot of the heights of the original WMU team. Now draw a dotplot of the heights of the team after you added 4 inches to each player.
  **i.** Does the spread of the data set appear to have increased, decreased, or stayed the same?
  **ii.** Does this concur with your answer to part b?

**10.** Starting with the original data, double the height of each player on the NCU team.

**a.** Recalculate the range and standard deviation.
**b.** Formulate a rule for the behavior of these measures of variability when each member of the data set is multiplied by a constant (such as $\alpha$).
**c.** Now compare a dotplot of the original NCU heights with a dotplot of the heights after you doubled each player's height.
  **i.** Does the spread of the data set appear to have increased, decreased, or stayed the same?
  **ii.** Does this concur with your answer to **(b)**?

**Syllable Frequencies.** For Exercises 11 and 12, refer to Table 3.17 (page 117).

**11.** Did you know that the word *the* is the most common syllable in written English? Its frequency is 7310.

**a.** Would you expect the measures of variability to increase or decrease when *the* is included? Why?

**b.** Find the range and standard deviation of syllable frequencies, when the word *the* is included. Has your intuition been borne out?

**12.** Now get rid of *the* again. Also, omit the syllables with the greatest frequency and the smallest frequency, so that you have only 8 syllables in your data set.

**a.** Would you expect the measures of variability to increase or decrease?

**b.** Find the range and standard deviation of the 8 remaining syllable frequencies. Has your intuition been borne out?

## SECTION 3.3

**13. Newborn Baby Weights.** The National Center for Health Statistics published the following data on the weights of newborn babies for 2004. Find the estimated mean and standard deviation of the baby weights.

| Baby weight (kilograms) | Births (1000s) |
|---|---|
| 0–1.99 | 126 |
| 2–2.99 | 935 |
| 3–3.99 | 2698 |
| 4–4.99 | 3440 |

**14. Calculating a Grade Point Average.** At a certain university in Texas, student grade point averages are calculated as follows. For each credit hour, an A is worth 4.0 quality points, an A− is worth 3.7 quality points, a B+ is worth 3.3 quality points, a B is worth 3.0, a B− is worth 2.7, a C+ is worth 2.3, and so on. To find the grade point average, the number of credits for each course is multiplied by the quality points earned for that course; the results are added together; and the sum is divided by the number of credits. This semester, Yolanda's grades are as follows. She got an A in her four-credit honors biology course, an A− in her three-credit calculus course, a B+ in her three-credit English course, a B− in her three-credit anthropology course, and a C+ in her two-credit physical education course. Calculate Yolanda's grade point average for this semester.

**15. AIDS Cases by Age.** The National Center for Health Statistics reported the number of cases of acquired immunodeficiency syndrome (AIDS) by age of patient in 2004.[14] Find the estimated mean and standard deviation of the age of AIDS patients.

| Class: age | Frequency $f_i$ |
|---|---|
| 0–12.99 | 48 |
| 13–14.99 | 60 |
| 15–24.99 | 2,114 |
| 25–34.99 | 9,361 |
| 35–44.99 | 16,778 |
| 45–54.99 | 10,178 |
| 55–64.99 | 3,075 |
| 65–74.99 | 901 |

## SECTION 3.4

**Cholesterol.** Use the cholesterol level data in Table 3.27 (page 141) for Exercises 16–21.

Find the following percentiles of cholesterol levels.

**16.** 10th percentile

**17.** 50th percentile

**18.** 90th percentile

Find the $z$-scores for the following food items.

**19.** French toast

**20.** Scrambled eggs

**21.** Waffles

For Exercises 22 and 23, refer to Figure 3.53.

**TRIPLES**

| | | Frequency |
|---|---|---|
| No. of Triples | 0 | 166 |
| | 1 | 65 |
| | 2 | 36 |
| | 3 | 20 |
| | 4 | 17 |
| | 5 | 7 |
| | 6 | 15 |
| | 7 | 1 |
| | 8 | 2 |
| | 9 | 1 |
| | 11 | 1 |
| | Total | 331 |

Figure 3.53 Frequency distributions of the number of triples in the 2002 American League regular season.

**22.** The mean number of triples hit in 2002 was 1.3 with a standard deviation of 1.9.

**a.** How many of the batters lie between one and two standard deviations above the mean?

**b.** How many of the batters lie between zero and one standard deviation above the mean?

**c.** Which is larger, your answer for part a or part b? Do you think this is typical? Why or why not?

**23.** The mean number of triples hit in 2002 was 1.3 with a standard deviation of 1.9.

**a.** How many of the batters lie between zero and one standard deviation below the mean?

**b.** How many of the batters lie between one and two standard deviations below the mean?

**c.** Comment on your answer in **(b)**.

**d.** Comment on whether you should look for outliers on the low end of the triples distribution.

## SECTION 3.5

Use the following information for Exercises 24–30. The vehicles on Interstate 95 have a mean of 70 miles per hour with a standard deviation of 5 miles per hour.

**24.** If the distribution of vehicle speeds is unknown, what proportion of vehicles travel over 77 miles per hour?

**25.** If the distribution of vehicle speeds is bell-shaped, what proportion of vehicles travel over 77 miles per hour? Obtain an upper and lower bound on the approximation.

**26.** Compare the results from the previous two exercises. Comment on the precision when using the approximation for the Empirical Rule versus Chebyshev's Rule.

**27.** If the distribution of vehicle speeds is unknown, find the 50th percentile, if possible. If not possible, clearly explain why not.

**28.** If the distribution of vehicle speeds is bell-shaped, find the 50th percentile, if possible. If not possible, find an approximation to the 50th percentile.

**29.** If the distribution of vehicle speeds is bell-shaped, find an approximation of the 36th percentile, if possible. If not possible, clearly explain why not.

**30.** If the distribution of vehicle speeds is bell-shaped, find an approximation of the 97th percentile, if possible. If not possible, clearly explain why not.

## SECTION 3.6

Refer to Table 3.27 (page 141) for Exercises 31–33.

**31.** Find the five-number summary of cholesterol levels.

**32.** By hand, construct a boxplot of the cholesterol levels. Report on the symmetry or skewness of the distribution.

**33.** Detect any outliers using the $z$-score method and the robust method.

- **a.** Determine whether there are any outliers in the data set using the $z$-score method.
- **b.** Use the robust detection of outliers method to see if there are any outliers in the data set. Do the two methods identify the same outliers?

Use Table 3.30 for Exercises 34–38. Do you suffer from ragweed pollen? You are not alone. The American Academy of Allergy maintains the ragweed pollen index, which details the severity of the pollen problem for hundreds of communities across the nation. The following table contains the ragweed pollen index on a particular day for ten localities in New York State.

**TABLE 3.30 Ragweed pollen index in New York localities**

| Locality | Ragweed pollen index |
|---|---|
| Albany | 48 |
| Binghamton | 31 |
| Buffalo | 59 |
| Elmira | 43 |
| Manhattan | 25 |
| Rochester | 60 |
| Syracuse | 25 |
| Tupper Lake | 8 |
| Utica | 26 |
| Yonkers | 38 |

**34.** Find the first, second, and third quartiles of the ragweed pollen index.

**35.** Find the interquartile range. Interpret what this value actually means, so that a nonspecialist could understand it.

**36.** Let's draw a boxplot of the ragweed pollen index.

- **a.** What is the five-number summary?
- **b.** By hand, draw a boxplot.
- **c.** Is the data set left-skewed, right-skewed, or symmetric?
- **d.** What should the symmetry or skewness mean in terms of the relative values of the mean and median?
- **e.** Find the mean and standard deviation. Is your prediction in **(d)** supported?

**37.** Detect any outliers using the $z$-score and the robust method.

- **a.** First use the $z$-score method.
- **b.** Now use the robust method of outlier detection.
- **c.** Do the two methods concur or disagree?

**38.** Suppose the ragweed pollen index in Rochester were 600 instead of 60. How would this outlier affect the quartiles and the IQR? What property of these measures is this behavior an example of?

# CHAPTER 3 Quiz

## TRUE OR FALSE

**1.** True or false: If a measure of variability for data set A is larger than for data set B, then data set A is more spread out than data set B.

**2.** True or false: If two data sets have the same mean, median, and mode, then the two data sets are identical.

**3.** True or false: The variance is the square root of the standard deviation.

**4.** True or false: The Empirical Rule applies for any data set.

**5.** True or false: A data value with a positive $z$-score means that the data value is greater than the mean.

## FILL IN THE BLANK

**6.** According to the Empirical Rule, about 95% of the data values will fall within __________ standard deviations of the mean.

**7.** An __________ is an extremely large or extremely small data value relative to the rest of the data set.

**8.** The mean can be viewed as the __________ point of the data.

**9.** The measure of center that is sensitive to the presence of extreme values is the __________.

**10.** If the sample size $n$ is even, then the median is the _______ of the two data values in the middle.

**11.** A __________ is the difference between a data value and the mean of the data set.

**12.** For a population, the variance is the average of the __________ __________ [two words].

**13.** Provide two other terms for the 50th percentile.

**14.** What do we call summary descriptive measures that are not sensitive to the presence of outliers?

## SHORT ANSWER

**15.** Which of the mean, median, and mode may be used for categorical data?

**16.** For any data set, what is the average of the deviations?

**17.** What do we use to estimate the mean for each class in a frequency distribution?

**18.** Consider the following relationships among the three measures of center. Does this indicate evidence that the data distribution is right-skewed, left-skewed, or symmetric unimodal?

- **a.** The mean is much smaller than the median, which is much smaller than the mode.
- **b.** The mode is much smaller than the median, which is much smaller than the mean.
- **c.** The mean, median, and mode are all about equal.

## CALCULATIONS AND INTERPRETATIONS

**Airline Passengers.** Refer to Table 3.31 for Exercises 19–21.

**TABLE 3.31 Passengers arriving at Portland International Airport, January–April 2007, by airline**

| Airline | Passengers |
|---|---|
| Alaska Airlines | 98,008 |
| Delta Air Lines | 31,054 |
| Horizon Air | 117,964 |
| Southwest Airlines | 106,178 |
| United Airlines | 84,059 |

**19.** Calculate the following:

- **a.** Sample mean
- **b.** Sample median

**20.** Calculate the following:

- **a.** Range
- **b.** Sample standard deviation

**21.** Interpret the meaning of the value you calculated for the sample standard deviation.

**22. NFL Points Scored.** The following frequency distribution contains the total points scored during the 2006 regular season by the teams in the National Football League. Find the estimated mean and standard deviation of the number of points scored.

| Points | Teams |
|---|---|
| 140–259.99 | 4 |
| 260–299.99 | 5 |
| 300–339.99 | 10 |
| 340–379.99 | 6 |
| 380–499.99 | 7 |

**23. Deaths Due to Heat.** The following frequency distribution contains the numbers of deaths due to heat, by age group, as reported by the National Weather Service for 2006. Find the estimated mean and standard deviation of age.

| Age | Deaths due to heat |
|---|---|
| 0–39.99 | 22 |
| 40–49.99 | 31 |
| 50–59.99 | 51 |
| 60–69.99 | 47 |
| 70–79.99 | 44 |
| 80–89.99 | 44 |

**24.** The Bureau of Labor Statistics reports that the mean amount spent by all American citizens in 2001 on tobacco products and smoking supplies was $308. Assume the standard deviation is $154. Find the *z*-scores of the persons who spend the following amounts.

- **a.** $308
- **b.** $462
- **c.** $616
- **d.** $154
- **e.** $0

**25.** The Department of Agriculture reports that the mean consumption of carbonated beverages per year per American is greater than 52 gallons. A sample of 30 Americans yielded a sample mean of 60 gallons. Assume the standard deviation is 40 gallons. Find the *z*-scores for the following amounts of carbonated beverage consumption.

- **a.** 120 gallons
- **b.** 20 gallons
- **c.** 100 gallons
- **d.** 0 gallons
- **e.** 60 gallons

**26.** Refer to the information in Exercise 24.

- **a.** If we assume that the distribution is bell-shaped, what is the approximate percentile of a person who spends no money at all on tobacco products?
- **b.** Now, suppose we do not know the distribution.
  - **i.** What percentage of Americans spend between $0 and $616 annually on tobacco products?
  - **ii.** What percentage of Americans spend between $154 and $462? Comment.

**27.** Refer to the information in Exercise 25. Assume the distribution is bell-shaped.

- **a.** Find the 50th percentile.
- **b.** Estimate the proportion of Americans who drink between 29 and 109 gallons per year.
- **c.** Discuss whether we could find the estimate in **(b)** without assuming that the distribution is bell-shaped.
- **d.** Estimate the proportion of Americans who drink more than 109 gallons per year.

Use the information in Table 3.32 for Exercises 28–32.

**TABLE 3.32 2004 SAT scores for the states with the highest participation rate**

| State | SAT I Math |
|---|---|
| New York | 510 |
| Connecticut | 515 |
| Massachusetts | 523 |
| New Jersey | 514 |
| New Hampshire | 521 |
| Maine | 501 |
| Pennsylvania | 502 |
| Delaware | 499 |

**28.** Find the following quartiles for *SAT I Math score:*

**a.** Q1

**b.** Q2

**c.** Q3

**29.** Find the interquartile range of *SAT I Math score.*

**30.** Find the five-number summary for *SAT I Math score.*

**31.** Use robust methods to investigate the presence of outliers.

**32.** Construct a boxplot for *SAT I Math score.*

AP Photo/Manish Swarup

Chapter 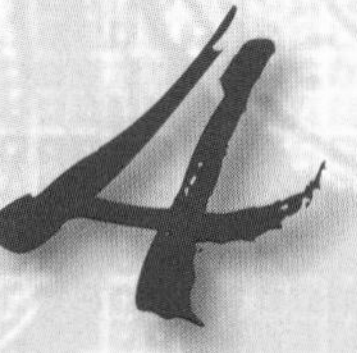

# Describing the Relationship Between Two Variables

CASE STUDY

## Female Literacy and Fertility Worldwide

The United Nations estimates that world population levels will continue to increase, to as high as 10 billion people by 2050. These 10 billion people will strain the finite resources of planet Earth and contribute to environmental degradation around the world. Most of this increase will occur in developing countries, which are least equipped to handle and care for the additional population. It is therefore useful to ask, "What policies might governments adopt to help alleviate overpopulation in developing countries?"

One encouraging trend is that fertility rates tend to decrease as the quality of life improves. As developing countries begin to educate girls as well as boys—that is, as the female literacy rate increases—fertility rates tend to come down. In Section 4.3, we examine details of the relationship between these two quantitative variables.

OVERVIEW



## The Big Picture — Where we are coming from, and where we are headed...

Chapter 3 has shown us methods for summarizing and analyzing data sets using descriptive statistics. However, up to now, we have looked at only one variable at a time. In Chapter 4, we learn how to examine the *relationship between two variables*. In Section 4.1, we are introduced to tables and graphs that can summarize such a relationship. Crosstabulations and clustered bar graphs are used for the relationship between two categorical variables, while scatterplots are used for quantitative variables. In Sections 4.2 and 4.3, we learn ways to quantify the relationship between two quantitative variables using correlation and regression.

This chapter brings to an end our exploration of descriptive statistics. In Chapter 5, we will learn about probability, which will teach us concepts we need in order to reach our eventual goal of statistical inference.

# 4.1 Tables and Graphs for the Relationship Between Two Variables

### Objectives: By the end of this section, I will be able to...

1 Construct and interpret crosstabulations for two categorical variables.

2 Construct and interpret clustered bar graphs for two categorical variables.

3 Construct and interpret scatterplots for two quantitative variables.

So far we have looked at ways to describe only one variable at a time. For example, Table 2.2, in Section 2.1 (page 36), lists the frequencies of the most prestigious career choice of a sample of 20 high school students. Suppose that we now introduce a second categorical variable, *gender*. We would like to examine *the relationship between the two variables: gender* and *career*. For instance, it may be interesting to determine whether more females than males preferred the Doctor career. This section introduces three methods of examining the relationship between two variables: crosstabulations, clustered bar graphs, and scatterplots.

## 1 Crosstabulations

**Crosstabulation** is a tabular method for simultaneously summarizing the data for two categorical (qualitative) variables.

**Steps for Constructing a Crosstabulation**

***Step 1*** Put the categories of one variable at the top of each column, and the categories of the other variable at the beginning of each row.

***Step 2*** For each row and column combination, enter the number of observations that fall in the two categories.

***Step 3*** The bottom of the table gives the column totals, and the right-hand column gives the row totals.

Crosstabulations are also known as **two-way tables** or **contingency tables.** We will introduce crosstabulations using an example. Table 4.1 contains the career survey data from Table 2.1, in Section 2.1, but adds information on each student's gender.

Table 4.1 Prestigious career survey data set

| Student | Career seen as prestigious | Student gender | Student | Career seen as prestigious | Student gender |
|---|---|---|---|---|---|
| 1 | Doctor | F | 11 | Doctor | F |
| 2 | Scientist | M | 12 | Military Officer | M |
| 3 | Military Officer | M | 13 | Scientist | F |
| 4 | Military Officer | F | 14 | Lawyer | F |
| 5 | Doctor | M | 15 | Lawyer | F |
| 6 | Scientist | F | 16 | Military Officer | M |
| 7 | Military Officer | M | 17 | Doctor | F |
| 8 | Athlete | F | 18 | Scientist | M |
| 9 | Doctor | F | 19 | Doctor | F |
| 10 | Scientist | M | 20 | Lawyer | M |

## Example 4.1 Crosstabulation of the prestigious career survey

Construct a crosstabulation of *career* and *gender*.

### Solution

***Step 1*** The crosstabulation for this sample is given in Table 4.2. Note that the categories for the variable *gender* are shown at the top, while the categories for the variable *career* are shown on the left. Each student in the sample is associated with a certain *cell* in the crosstabulation, in the appropriate row and column. For example, a male student who reported Military Officer as the most prestigious career appears as one of the four students in the "Male" column and the "Military Officer" row.

***Step 2*** For each row and column combination, enter the number of observations that fall in the two categories. This is shown in Table 4.2.

***Step 3*** The "Total" column contains the sum of the counts of the cells in each row (category) of the *career* variable and represents the frequency distribution for this variable. Similarly, the "Total" row along the bottom sums the counts of the cells in each column (category) of the *gender* variable, thereby representing the frequency distribution for this variable. Thus, we see that crosstabulations contain the frequency distributions of each of the two variables.

Table 4.2 Crosstabulation of prestigious career survey data set

| | Gender | | |
|---|---|---|---|
| **Career** | **Female** | **Male** | **Total** |
| Doctor | 5 | 1 | 6 |
| Scientist | 2 | 3 | 5 |
| Military Officer | 1 | 4 | 5 |
| Lawyer | 2 | 1 | 3 |
| Athlete | 1 | 0 | 1 |
| Total | 11 | 9 | 20 |

We can use the crosstabulation to look for patterns in the data set. For example, does there appear to be a difference between males and females in their responses to the survey? Examination of the crosstabulation shows that most of the students who responded "Doctor" were females, and most of the students who responded "Military Officer" were males. ■

## 2 Clustered Bar Graphs

**Clustered bar graphs** are useful for comparing two categorical variables and are often used in conjunction with crosstabulations. Each set of bars in a clustered bar graph represents a single category of one variable across all the categories of the other categorical variable (see Figures 4.1a and 4.1b). This allows the analyst to make comparisons easily. One can construct clustered bar graphs using either frequencies or relative frequencies. When the sample sizes for the clustered groups are substantially different, it is better to use a relative frequency clustered bar graph. To construct a clustered bar graph, identify which of the two categorical variables will define the cluster of bars. Then, for each category of the other variable, draw bars for each category of the first variable, the clustering variable.

### Example 4.2 Clustered bar graphs for the emotions felt by females and males on September 11

AP Photo/Carmen Taylor

Recall Example 1.2, in Section 1.1 (page 3). (The original survey question read, "Which one of the following emotions do you feel the most strongly in response to these terrorist attacks: sadness, fear, anger, disbelief, vulnerability?")[1] The results are given in the crosstabulation in Table 4.3. Construct a clustered bar graph of the emotions felt, clustered by gender in order to illustrate any differences between males and females.

Table 4.3 Frequency of survey respondents expressing particular emotions, by gender

| | Emotion | | | | | | |
|---|---|---|---|---|---|---|---|
| **Gender** | **Sadness** | **Fear** | **Anger** | **Disbelief** | **Vulnerability** | **Not sure** | **Total** |
| **Female** | 94 | 21 | 87 | 80 | 28 | 4 | 314 |
| **Male** | 56 | 16 | 141 | 50 | 36 | 5 | 304 |
| **Total** | 150 | 37 | 228 | 130 | 64 | 9 | 618 |

**Solution**

*Gender* is given as the clustering variable. Thus, for each category of the variable *emotion,* we will draw two bars, one representing males and the other representing females. For example, for the first emotion, sadness, we draw one rectangle going up to 56 on the vertical axis, and a separate rectangle going up to 94 on the vertical axis. These two rectangles should touch each other but should not touch any other rectangles. Continue to draw two rectangles for each emotion, one for each of the males' and females' frequencies. The resulting clustered bar graph is shown here as Figure 4.1a. We say that the emotions are *clustered* by gender.

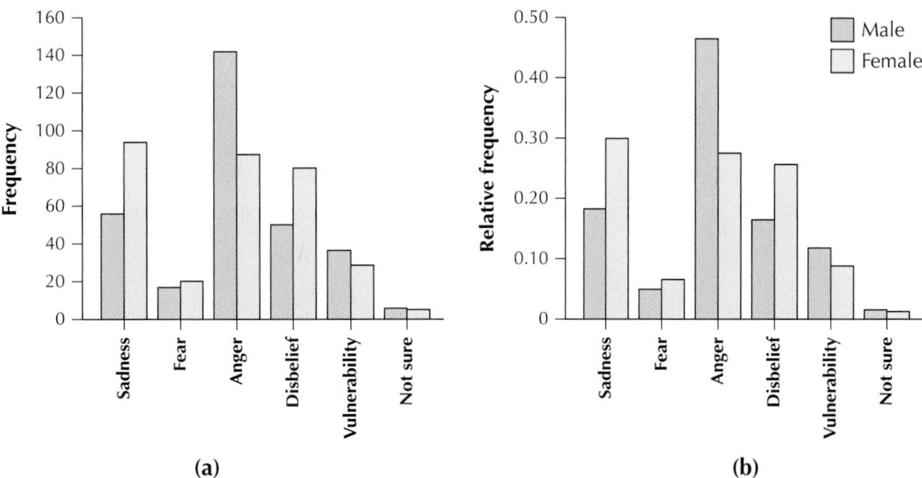


FIGURE 4.1 (a) Clustered bar graph using frequencies; (b) clustered bar graph using relative frequencies.

*Note:* We can use either *percentage* or *proportion* to describe relative frequency. For example, in Table 4.4, we can say either that the *percentage* of females who expressed sadness was 29.9% or that the *proportion* of females who expressed sadness was 0.299.

Now, what if females were underrepresented in this survey, so that there were only 100 females and 304 males? Then, direct comparison of the counts would be misleading. When the sample sizes are substantially different, one should use relative frequency clustered bar graphs. The relative frequencies for the frequencies in Table 4.3 are provided in Table 4.4, and the clustered bar graph is given in Figure 4.1b. Note that we divide the counts by the total for that gender, not by the total for the emotion. ■

Table 4.4 Relative frequencies of emotions, by gender

| | Emotion | | | | | | |
|---|---|---|---|---|---|---|---|
| **Gender** | **Sadness** | **Fear** | **Anger** | **Disbelief** | **Vulnerability** | **Not sure** | **Total** |
| **Females** | 0.299 | 0.067 | 0.277 | 0.255 | 0.089 | 0.013 | 1.000 |
| **Males** | 0.184 | 0.053 | 0.464 | 0.164 | 0.118 | 0.016 | 1.000 |

## 3 Scatterplots

Crosstabulations and clustered bar graphs are useful for summarizing the relationship between pairs of categorical variables. Next we turn to a type of graph used for quantitative variables. A **scatterplot** is used to summarize the relationship between two quantitative variables that have been measured on the same element. An example of a scatterplot is given in Figure 4.2 (see page 169).

*Note:* The predictor variable and response variable are sometimes referred to as the independent variable and dependent variable, respectively. This textbook avoids this terminology, since it may be confused with the definition of independent and dependent events and variables in probability and categorical data analysis.

> A **scatterplot** is a graph of points ($x$, $y$) each of which represents one observation from the data set. One of the variables is measured along the horizontal axis and is called the *x variable*. The other variable is measured along the vertical axis and is called the *y variable*.

Often, the value of the $x$ variable can be used to predict or estimate the value of the $y$ variable. For this reason, the $x$ variable is referred to as the *predictor* variable, and the $y$ variable is called the *response* variable.

## Example 4.3 Lot prices in Glen Ellyn, Illinois

Suppose you are interested in moving to Glen Ellyn, Illinois, and would like to purchase a lot upon which to build a new house. Table 4.5 contains a random sample of eight lots for sale in Glen Ellyn, with their square footage and prices as of March 7, 2007. Identify the predictor variable and the response variable, and construct a scatterplot.

Table 4.5 Lot square footage and sales prices

| Lot location | Square footage | Sales price |
|---|---|---|
| Harding St. | 9,000 | $200,000 |
| Newton Ave. | 13,200 | $423,000 |
| Stacy Ct. | 13,900 | $300,000 |
| Eastern Ave. | 15,000 | $260,000 |
| Second St. | 20,000 | $270,000 |
| Sunnybrook Rd. | 30,000 | $650,000 |
| Ahlstrand Rd. | 40,800 | $680,000 |
| Eastern Ave. | 55,400 | $1,450,000 |

*Source:* **Realtor.com**.

### Solution

It is reasonable to expect that the price of a new lot depends in part on how large the lot is. Thus, we define our predictor variable $x$ to be $x =$ *square footage* and our response variable $y$ to be $y =$ *sales price.*

**Reexpressing the data values in simpler terms.** We can reexpress the data values from Table 4.5 to simplify our calculations and to produce a cleaner plot. Table 4.6 shows the data with square footage expressed in "hundreds of square feet," and the sales price expressed in "thousands of dollars."

Table 4.6 Reexpressed data values

| Lot location | $x$ = square footage (100s of sq. ft.) | $y$ = sales price ($1000s) |
|---|---|---|
| Harding St. | 90 | 200 |
| Newton Ave. | 132 | 423 |
| Stacy Ct. | 139 | 300 |
| Eastern Ave. | 150 | 260 |
| Second St. | 200 | 270 |
| Sunnybrook Rd. | 300 | 650 |
| Ahlstrand Rd. | 408 | 680 |
| Eastern Ave. | 554 | 1450 |

Next we construct the scatterplot using the data from Table 4.6. Draw the horizontal axis so that it can contain all the values of the predictor ($x$) variable, and similarly for the vertical axis. Then, at each data point ($x$, $y$), draw a dot. For example, for the Harding Street lot, move along the $x$ axis to 90, then go up until you reach a spot level with $y = 200$, at which point you draw a dot. Proceed similarly for all eight properties. The result should look similar to the scatterplot in Figure 4.2.

From this scatterplot, we can see that there is a tendency for larger lots to have higher prices. This is not the case for each observation. For example, the Stacy Court property is larger than the Newton Avenue property but has a lower price. Nevertheless, the overall tendency remains. ■

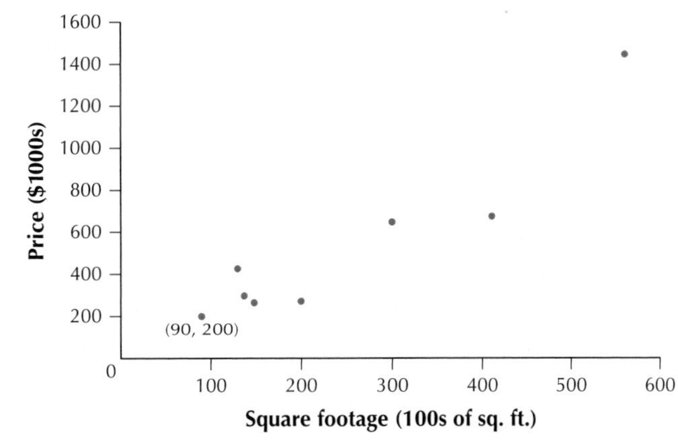


FIGURE 4.2 Scatterplot of sales price versus square footage.

### Scatterplot Terminology

Note the terminology in the caption to Figure 4.2. When describing a scatterplot, always indicate the $y$ variable first and use the term *versus* (*vs.*) or *against* the $x$ variable. This terminology reinforces the notion that the $y$ variable depends on the $x$ variable. ■

The relationship between two quantitative variables can take many different forms. Three of the most common relationships are shown in Figures 4.3a–4.3c.

- **Positive relationship** between $x$ and $y$ (Figure 4.3a): Smaller values of the $x$ variable are associated with smaller values of the $y$ variable; larger values of the $x$ variable are associated with larger values of the $y$ variable. In other words, as the $x$ variable increases in value, the $y$ variable also tends to increase.
- **Negative relationship** between $x$ and $y$ (Figure 4.3b): Smaller values of the $x$ variable are associated with larger values of the $y$ variable; larger values of the $x$ variable are associated with smaller values of the $y$ variable. In other words, as the $x$ variable increases in value, the $y$ variable tends to decrease.
- **No apparent relationship** (Figure 4.3c): The values of the $x$ variable are not especially associated with any particular range of values of the $y$ variable. In other words, as the $x$ variable increases in value, the $y$ variable tends to remain unchanged.

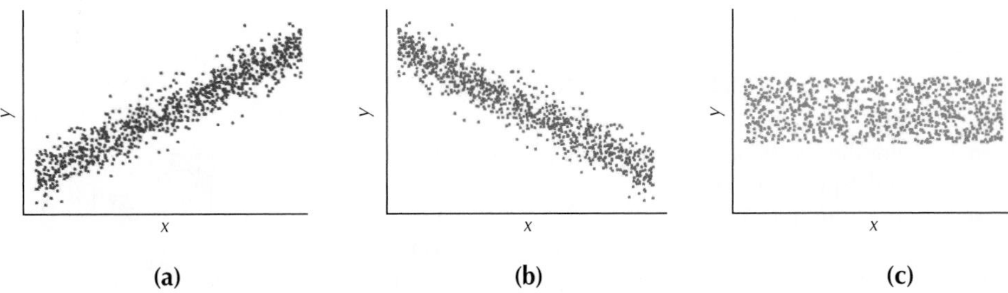


FIGURE 4.3 (a) Scatterplot of a positive relationship; (b) scatterplot of a negative relationship; (c) scatterplot of no apparent relationship.

## Example 4.4 Relationship between lot size and price in Glen Ellyn, Illinois

Using Figure 4.2, investigate the relationship between lot square footage and lot price.

### Solution

The scatterplot in Figure 4.2 most resembles Figure 4.3a, where a positive relationship exists between the variables. Thus, smaller lot sizes tend to be associated with lower prices, and larger lot sizes tend to be associated with higher prices. Put another way, as the lot size increases, the lot price tends to increase as well. ■

## What If Scenario ? A Shift to the Right

*What if* we add 4000 square feet to the size of *each* lot in Example 4.3, while leaving the sales price unchanged. What would happen to the relationship between square footage and sales price?

### Solution

As seen in Figure 4.4, adding 4000 square feet to each lot does not affect the overall relationship between square footage and sales price. The relationship remains positive. This result is true even if we don't know how much square footage is added to each lot, as long as each lot increases by the same amount.

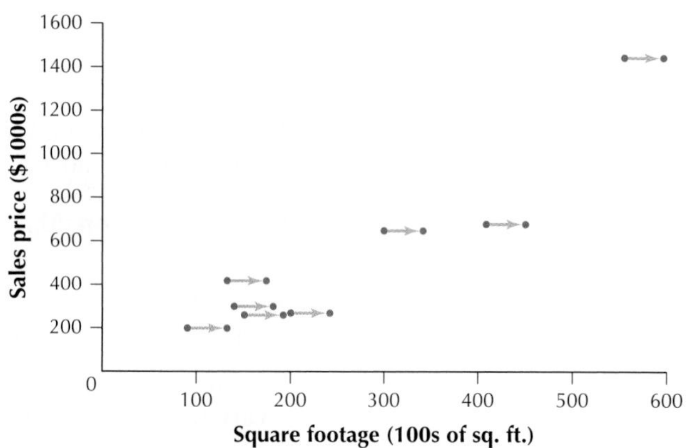


FIGURE 4.4

## STEP-BY-STEP TECHNOLOGY GUIDE: Tables and Graphs for Two Variables

### TI-83/84

**Constructing a Scatterplot for Data in Table 4.6 (page 168)**

***Step 1*** Enter the *x* variable (square footage) into **L1** and the *y* variable (sales price) into **L2**.

***Step 2*** Press **2nd**, then **Y=** for the STAT PLOTS menu.

***Step 3*** Select **1**, and press **ENTER**. Select **ON**, and press **ENTER**.

***Step 4*** Select the scatterplots icon (see Figure 4.5), and press **ENTER**.

***Step 5*** Select **L1** for **Xlist**, and **L2** for **Ylist**.

***Step 6*** Press **ZOOM**, choose **9: ZoomStat**, and press **ENTER**. The scatterplot is shown in Figure 4.6.

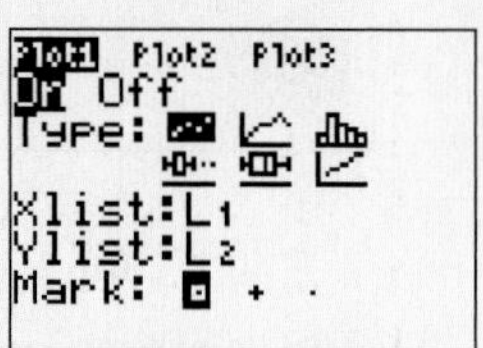


Figure 4.5

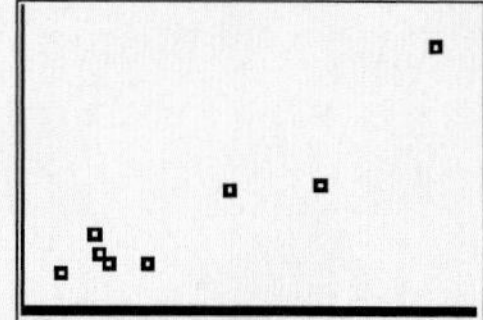

Figure 4.6

## EXCEL

### Clustered Bar Graphs

***Step 1*** Select the crosstabulation.
***Step 2*** Click **Insert > Column**.
***Step 3*** Select **Clustered column**.

### Scatterplots

***Step 1*** Enter your *x* variable and your *y* variable in two neighboring columns, with the *x* variable on the left. Make sure the first entry in each column is the variable name. Select the two columns.
***Step 2*** Click **Insert > Scatter** (in **Chart** section). See Figure 4.7.

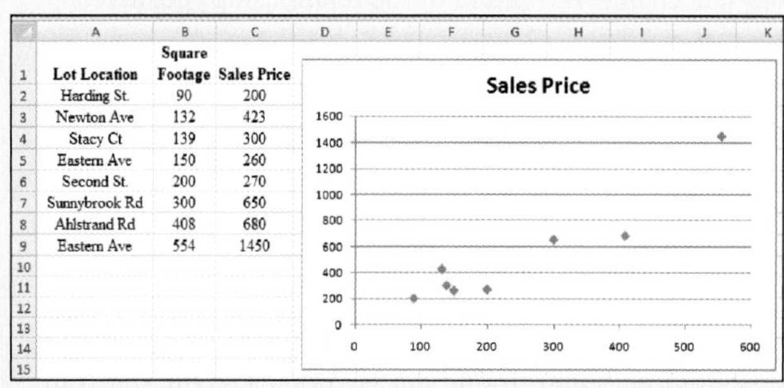

| | A | B | C |
|---|---|---|---|
| 1 | Lot Location | Square Footage | Sales Price |
| 2 | Harding St. | 90 | 200 |
| 3 | Newton Ave | 132 | 423 |
| 4 | Stacy Ct | 139 | 300 |
| 5 | Eastern Ave | 150 | 260 |
| 6 | Second St. | 200 | 270 |
| 7 | Sunnybrook Rd | 300 | 650 |
| 8 | Ahlstrand Rd | 408 | 680 |
| 9 | Eastern Ave | 554 | 1450 |



Figure 4.7

## MINITAB

### Crosstabulation of Career Data

***Step 1*** Enter the data from Table 4.1 (page 165) into two columns, named *career* and *gender*.
***Step 2*** Click **Stat > Tables > Cross-Tabulation**.
***Step 3*** Double click on *career* and *gender*. Select **Counts** under **Display**. Then click **OK**.
***Step 4*** The resulting crosstabulation is shown in Figure 4.8.

```
Rows: Career       Columns: Gender
                  f           m        All

  athlete         1           0          1
  doctor          5           1          6
  lawyer          2           1          3
  military        1           4          5
  scientist       2           3          5
  All            11           9         20
```

Figure 4.8

### Clustered Bar Graphs

If you have the original data set:
***Step 1*** Click **Graph > Bar Chart**.
***Step 2*** Select **Bars Represent: Counts of Unique Values**, and select **Clustered**. Then click **OK**.
***Step 3*** Select your two categorical variables, and click **OK**.

If you have only the crosstabulation and not the original data:
***Step 1*** Click **Graph > Bar Chart**.
***Step 2*** Select **Bars Represent: Values from a Table**, and select **Clustered**. Then click **OK**.
***Step 3*** For **Graph Variables**, choose the variable that contains the frequencies or relative frequencies. For **Categorical Variables for Grouping**, choose your two categorical variables. Then click **OK**.

### Scatterplots

***Step 1*** Enter the data into two columns.
***Step 2*** Click **Graph > Scatterplot**.
***Step 3*** Click on the cell under **Y**, and double-click on your *y* variable; then click on the cell under **X**, and double-click on your *x* variable. Then click **OK**.

## Section 4.1 SUMMARY

Section 4.1 introduces methods for summarizing the relationship between two variables using tables and graphs.

1 Crosstabulation summarizes the relationship between two categorical variables. A crosstabulation is a table that gives the counts for each row-column combination, with totals for the rows and columns.

2 Clustered bar graphs are useful for comparing two categorical variables and are often used in conjunction with crosstabulations.

3 For two numerical variables, scatterplots summarize the relationship by plotting all the $(x, y)$ points.

## Section 4.1 EXERCISES

### CLARIFYING THE CONCEPTS

1. Why would researchers be interested in examining the relationship between two variables?
2. Why can't we use a scatterplot for two categorical variables?
3. Why can't we use crosstabulations for two numerical variables? Is there some way we could recode the variables in order to use crosstabulations?

**4.** If we put the $x$ variable on the $y$ axis, and vice versa, does this change the nature of the relationship (positive, negative, or none)?

## PRACTICING THE TECHNIQUES

Don't Mess with Texas (**dontmesswithtexas.org**) is a Texas statewide antilittering organization. Its 2005 report, *Visible Litter Study 2005,* identified paper, plastic, metals, and glass as the top four categories of litter by composition. The report also identified tobacco, household/personal, food, and beverages as the top four categories of litter by use. Assume a sample of 12 items of litter had the following characteristics. Use the table to answer Exercises 5–8.

| Litter item | Composition | Use |
|---|---|---|
| 1 | Paper | Tobacco |
| 2 | Plastic | Household/personal |
| 3 | Glass | Beverages |
| 4 | Paper | Tobacco |
| 5 | Metal | Household/personal |
| 6 | Plastic | Food |
| 7 | Glass | Beverages |
| 8 | Paper | Household/personal |
| 9 | Metal | Household/personal |
| 10 | Plastic | Beverages |
| 11 | Paper | Tobacco |
| 12 | Plastic | Food |

**5.** Construct a crosstabulation of litter composition by litter use.

**6.** Do you see any patterns in the crosstabulation you constructed in the previous exercise? For example, which cells (combination of composition and use) in the table have the highest frequency? How many cells have zero frequency?

**7.** Construct a clustered bar graph of litter composition and litter use. Cluster by use.

**8.** Use the crosstabulation to construct a frequency distribution of the following variables:

**a.** Litter composition

**b.** Litter use

The following table shows the relative frequencies for the three leading causes of death for Americans aged 16–24 years and 25–34 years.

| Age group | Cause of death | Percentage of all deaths for age group |
|---|---|---|
| 16–24 | Motor vehicle crashes | 0.32 |
| 16–24 | Homicide | 0.16 |
| 16–24 | Suicide | 0.12 |
| 25–34 | Motor vehicle crashes | 0.16 |
| 25–34 | Suicide | 0.12 |
| 25–34 | Homicide | 0.11 |

**9.** Construct a clustered bar graph of the cause of death, clustered by age group.

**10.** Construct a clustered bar graph of age group, clustered by cause of death.

For Exercises 11–13 do the following.

**a.** Construct a scatterplot of the relationship between $x$ and $y$.

**b.** Characterize the relationship as positive, negative, or no apparent relationship.

**c.** Describe the relationship using two different sentences.

**11.**

| $x$ | $y$ |
|---|---|
| −3 | −5 |
| −1 | −15 |
| 1 | −20 |
| 3 | −25 |
| 5 | −30 |

**12.**

| $x$ | $y$ |
|---|---|
| 10 | 100 |
| 20 | 95 |
| 30 | 85 |
| 40 | 85 |
| 50 | 80 |

**13.**

| $x$ | $y$ |
|---|---|
| 0 | 11 |
| 20 | 11 |
| 40 | 16 |
| 60 | 21 |
| 80 | 26 |

**14.** Does it pay to stay in school? Refer to the accompanying table of U.S. Census Bureau data.

**a.** Construct a scatterplot of the relationship between $x$ = the number of years of education and $y$ = unemployment rate.

**b.** Would you characterize the relationship as positive or negative or neither?

**c.** How would you describe the relationship in a sentence?

| $x$ = years of education | $y$ = unemployment rate (%) |
|---|---|
| 5 | 16.8 |
| 7.5 | 17.1 |
| 8 | 15.3 |
| 10 | 20.6 |
| 12 | 11.7 |
| 14 | 8.1 |
| 16 | 3.8 |

## APPLYING THE CONCEPTS

**Majors of College Students.** A sample of 25 college students was selected, and their major, class, and gender were recorded. Use Table 4.7 for Exercises 15–18.

**TABLE 4.7 Major, class, and gender of 25 college students**

| Major | Class | Gender |
|---|---|---|
| English | Sophomore | Male |
| Business | Sophomore | Female |
| Political Science | Senior | Male |
| English | Freshman | Female |
| Psychology | Sophomore | Female |
| History | Junior | Female |
| Philosophy | Senior | Male |
| Communication | Sophomore | Male |
| History | Junior | Female |
| English | Freshman | Female |
| Psychology | Sophomore | Female |
| English | Freshman | Female |
| History | Junior | Male |
| Psychology | Sophomore | Female |
| Economics | Senior | Male |
| Art | Sophomore | Female |
| Psychology | Junior | Male |
| English | Freshman | Female |
| Economics | Senior | Male |
| Business | Sophomore | Male |
| Business | Junior | Female |
| Communication | Sophomore | Male |
| Art | Junior | Female |
| Communication | Freshman | Female |
| Psychology | Junior | Female |

15. Construct a crosstabulation for the variables *major* and *class*.
   a. Do you notice any patterns? How would you characterize the relationship?
   b. Construct the frequency distributions for the individual variables.

16. Construct a crosstabulation for the variables *major* and *gender*.
   a. Do you notice any patterns? How would you characterize the relationship?
   b. Construct the frequency distributions for the individual variables.

17. Construct a clustered bar graph for *major*, clustered by *class*.

18. Construct a clustered bar graph for *major*, clustered by *gender*.

**Shopping Enjoyment and Gender.** Use the information in Table 4.8 for Exercises 19–24. The Pew Internet and American Life Project surveyed 4514 American men and women and asked them, "How much, if at all, do you enjoy shopping?" The results are shown Table 4.8, which, however, is missing some entries.

**TABLE 4.8 Crosstabulation of shopping enjoyment by gender**

| "How much do you enjoy shopping?" | Gender | | |
|---|---|---|---|
| | **Male** | **Female** | **Total** |
| A lot | | 950 | 1338 |
| Some | 582 | | 1255 |
| Only a little | 662 | 497 | |
| Not at all | 497 | | 717 |
| Don't know/refused | | 25 | 45 |
| Total | 2149 | | 4514 |

19. Fill in the missing entries.

20. Convert the table to a relative frequency crosstabulation. Make it so that the "Male" and "Female" proportions in each row add up to 1.0.

21. Did men or women have the higher proportion of respondents who enjoy shopping
   a. a lot?
   b. some?
   c. only a little?
   d. not at all?

22. What proportion of the respondents are female?

23. What proportion are female, of the respondents who like shopping
   a. a lot?
   b. some?
   c. only a little?
   d. not at all?

24. Construct a clustered bar graph of how much the respondents enjoy shopping, clustered by gender.

25. **Teenage Birth Rates.** The National Center for Health Statistics (NCHS) publishes data on state birth rates. The following table contains the overall birth rate and the teenage (ages 15–19) birth rate for ten states. The overall birth rate is defined by the NCHS as "live births per 1000 women," and the teenage birth rate is defined as "live births per 1000 women aged 15–19."
   a. Construct the appropriate scatterplot.
   b. Describe the relationship between overall birth rate and teenage birth rate for these ten states.

| State | $x$ = overall birth rate | $y$ = teenage birth rate |
|---|---|---|
| Alabama | 13.1 | 52.4 |
| Arizona | 16.3 | 60.1 |
| California | 15.2 | 39.5 |
| Florida | 12.5 | 42.4 |
| Georgia | 15.7 | 53.4 |
| New York | 13.0 | 26.9 |
| Ohio | 13.0 | 38.5 |
| Pennsylvania | 11.7 | 30.5 |
| Texas | 17.0 | 62.6 |
| Virginia | 13.9 | 35.2 |

**26. Stock Prices.** Would you expect there to be a relationship between the price ($x$) of a stock and its change ($y$) in price on a particular day? The following table provides stock price and stock price change for August 2, 2004, for a sample of ten stocks listed on the New York Stock Exchange.

**a.** Construct the appropriate scatterplot.

**b.** Describe the relationship between price and change.

| Stock | Price ($) | Change ($) |
|---|---|---|
| Nortel Networks | 3.86 | +0.04 |
| Qwest Communications | 3.41 | −0.56 |
| Tyco International | 32.02 | +0.78 |
| Lucent Technologies | 3.04 | −0.03 |
| Vishay Intertechnology | 13.96 | −1.96 |
| Tenet Healthcare | 10.36 | −0.82 |
| Select Medical Group | 14.74 | +1.62 |
| Cox Communications | 33.19 | +0.03 |
| Verizon Communications | 39.15 | +0.46 |
| General Electric | 33.05 | −0.21 |

**27. NCAA Power Ratings.** The following table shows each team's winning percentage ($x$) and power rating ($y$) for the 2004 NCAA Basketball Tournament, according to **www.teamrankings.com.**

**a.** Construct the appropriate scatterplot.

**b.** Describe the relationship between winning percentage and power rating.

| School | Win % | Rating |
|---|---|---|
| Duke | 0.838 | 96.020 |
| St. Joseph's | 0.938 | 95.493 |
| Connecticut | 0.846 | 95.478 |
| Oklahoma State | 0.882 | 95.320 |
| Pittsburgh | 0.857 | 94.541 |
| Georgia Tech. | 0.737 | 93.091 |
| Stanford | 0.938 | 92.862 |
| Kentucky | 0.844 | 92.692 |
| Gonzaga | 0.903 | 92.609 |
| Miss. State | 0.867 | 91.912 |

**Batting Average and Age.** For Exercises 28 and 29, refer to the accompanying scatterplot of batting average against player age for the 2002 American League season (minimum at-bats 100).

**28.** How would you characterize the relationship?

**29.** In general, would knowing the age of the baseball player help us to estimate the player's batting average very much? Explain why or why not.

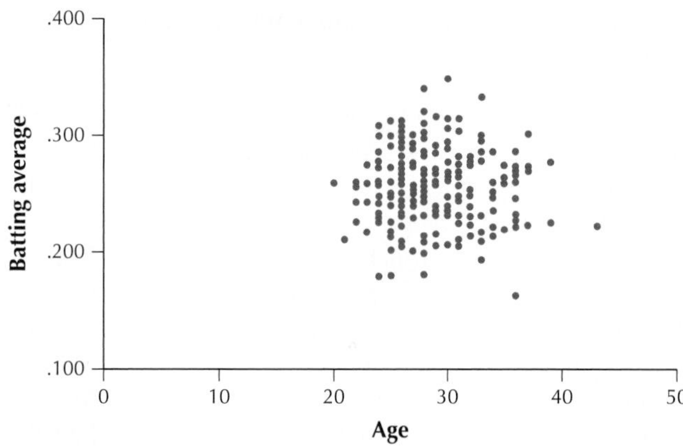


Scatterplot for Exercises 28 and 29.

**Words and Sentences.** Use the information in Table 4.9 for Exercises 30–32. Six advertisements were randomly selected from each of *Fortune* magazine and the *New Yorker* magazine. The number of words and sentences are presented here. Ads 1–6 came from *Fortune*, while ads 7–12 came from *The New Yorker*.

**30.** Construct a scatterplot of words versus sentences for the *Fortune* ads. How would you characterize the relationship?

**31.** Construct a scatterplot of words versus sentences for *The New Yorker* ads. How would you characterize the relationship?

**32.** Construct a scatterplot of words versus sentences for all the ads. How would you characterize the relationship?

**Gaming Survey.** Use the information in Table 4.10 for Exercises 33–36. In June 2003, the Pew Internet and American Life Project released their data on the College Students Gaming Survey. They asked 1162 college students three questions inquiring when the students first began playing video games, computer games, and online games. Table 4.10 is the relative frequency crosstabulation. Each student supplied a response for each type of game.

**TABLE 4.10 Survey: when did college students begin playing?**

| | Elementary school | Jr. high/ high school | College | Do not play |
|---|---|---|---|---|
| Video games | 0.69 | 0.15 | 0.02 | 0.14 |
| Computer games | 0.28 | 0.49 | 0.09 | 0.14 |
| Online games | 0.06 | 0.43 | 0.22 | 0.29 |

**TABLE 4.9 Number of words and sentences in 12 randomly selected ads**

| | Ad | | | | | | | | | | | |
|---|---|---|---|---|---|---|---|---|---|---|---|---|
| | 1 | 2 | 3 | 4 | 5 | 6 | 7 | 8 | 9 | 10 | 11 | 12 |
| Words: $y$ | 215 | 153 | 205 | 80 | 208 | 89 | 125 | 100 | 80 | 34 | 130 | 88 |
| Sentences: $x$ | 16 | 9 | 11 | 13 | 22 | 16 | 10 | 9 | 8 | 6 | 11 | 12 |

**33.** What proportion of students first played computer games in college?

**34.** What percentage of students first played online games in elementary school or junior high/high school?

**35.** Construct a clustered bar graph using the data in Table 4.10

**36.** Explain why the columns do not add up to 100%.

**37. Brain and Body Weight.** A study compared the body weight (in kilograms) and brain weight (in grams) for a sample of mammals, with the results shown in the following table.[2] We are interested in estimating brain weight based on body weight.

- **a.** Construct a scatterplot of the data.
- **b.** Describe the relationship between body weight and brain weight.
- **c.** Is there a particular mammal that is off by itself in the scatterplot?
- **d.** What mammal do you think this might be, given the large brain weight for its body weight?

| $x$ = body weight (kg) | $y$ = brain weight (g) |
|---|---|
| 52.16 | 440 |
| 60 | 81 |
| 27.66 | 115 |
| 85 | 325 |
| 36.33 | 119.5 |
| 100 | 157 |
| 35 | 56 |
| 62 | 1320 |
| 83 | 98.2 |
| 55.5 | 175 |

**38. Forecasting the Weather.** The journal *Nature* collected data on the weather and the weather forecasts for London over 1000 days. On 66 days the forecast was rain, and it did rain. On 156 days the forecast was rain, and it did not rain. On 14 days the forecast was no rain and it did rain. On 764 days the forecast was no rain and it did not rain.

- **a.** Construct the crosstabulation.
- **b.** If the forecast was rain, what would you estimate your chances of being rained on if you forgot your umbrella? How did you figure this out?
- **c.** If the forecast was no rain, what would you estimate your chances were of wasting your energy carrying an umbrella around?

**Scrabble®.** Use the accompanying table for Exercises 39 and 40. Scrabble® is a tile-based word game where players compete for points by placing letter tiles to form words crossword style. There are a certain number of letter tiles in the game, with the rarer letters (for example, Q and Z) worth more points than common letters such as vowels.

**39.** Construct a scatterplot of points versus frequency. Try to characterize the relationship between points and frequency as positive, negative, or not apparent. Comment!

| Letter | Freq. | Points | Letter | Freq. | Points |
|---|---|---|---|---|---|
| A | 9 | 1 | N | 6 | 1 |
| B | 2 | 3 | O | 8 | 1 |
| C | 2 | 3 | P | 2 | 3 |
| D | 4 | 2 | Q | 1 | 10 |
| E | 12 | 1 | R | 6 | 1 |
| F | 2 | 4 | S | 4 | 1 |
| G | 3 | 2 | T | 6 | 1 |
| H | 2 | 4 | U | 4 | 1 |
| I | 9 | 1 | V | 2 | 4 |
| J | 1 | 8 | W | 2 | 4 |
| K | 1 | 5 | X | 1 | 8 |
| L | 4 | 1 | Y | 2 | 4 |
| M | 2 | 3 | Z | 1 | 10 |

**40.** Consider the scatterplot from the previous exercise. Suppose we divided the data set into two parts, in each of which the relationship between points and frequency could be characterized as positive, negative, or not apparent.

- **a.** Where would we make such a division?
- **b.** What type of letters are in each part?
- **c.** For each group, characterize the relationship between points and frequency.

**9/11 Memorial Web site.** Use the following information for Exercises 41–43. In the immediate aftermath of the September 11, 2001, attacks, did you visit a memorial or commemorative Web site? The Pew Internet and American Life Project asked that question of 2458 respondents, along with whether the respondent was a full-time student, a part-time student, or not a student. The results are shown in the following crosstabulation. However, the output is missing some entries.

| | Visited Memorial Web Site | | |
|---|---|---|---|
| **Student status** | **Yes** | **No** | **Total** |
| **Full-time student** | | 171 | |
| **Part-time student** | 53 | | 242 |
| **Not a student** | 371 | | 2000 |
| **Total** | 469 | | 2458 |

Crosstabulation of student status with whether visited memorial Web site.

**41.** Fill in the missing entries.

**42.** Convert the table to a relative frequency crosstabulation so that the "Yes" and "No" proportions in each row sum to 1.0.

**43.** Which group of people had the highest proportion who visited a Web site? The lowest proportion?

**Endangered and Threatened Species.** Use the following table for Exercises 44–47. The table shows the numbers of endangered species and threatened species according to the U.S. Fish and Wildlife Service.

| | Endangered species ($y$) | Threatened species ($x$) |
|---|---|---|
| Mammals | 324 | 33 |
| Birds | 252 | 21 |
| Reptiles | 79 | 39 |
| Amphibians | 21 | 11 |
| Fishes | 87 | 62 |
| Snails, clams, crustaceans | 108 | 23 |
| Insects, arachnids | 52 | 9 |

**44.** Construct a scatterplot of endangered species versus threatened species.

**45.** Characterize the relationship as either positive, negative, or not apparent.

**46.** Write a sentence that describes the number of endangered species as the number of threatened species increases.

**47.** Write a sentence that indicates which values of $y$ that low values of $x$ are associated with, and which values of $y$ that high values of $x$ are associated with.

### CONSTRUCT YOUR OWN DATA SETS

**48.** Think of two variables that have a positive relationship.
- **a.** Make a table of ten observations for these variables, choosing reasonable data values.
- **b.** Construct the scatterplot. Does it verify your positive relationship?

**49.** Think of two variables that have a negative relationship.
- **a.** Make a table of ten observations for these variables, choosing reasonable data values.
- **b.** Construct the scatterplot. Does it verify your negative relationship?

**50.** Think of two variables that have no apparent relationship.
- **a.** Make a table of ten observations for these variables, choosing reasonable data values.
- **b.** Construct the scatterplot. Does it verify that no apparent relationship exists?

Use your knowledge of Minitab or Excel to solve the following problems. Open the data set **Households.**

**51.** The variable *FAM_MPC* contains, for each state and the District of Columbia, the percentage of households with a married-couple family. The variable *FAM_FPC* contains the percentage of households with a female head of household, no husband present.
- **a.** Construct a scatterplot of *FAM_MPC* against *FAM_FPC*.
- **b.** Would you say that there is a positive relationship, a negative relationship, or no apparent relationship between the variables? Comment on why this is so.

**52.** The variable *FAM_TPC* contains, for each state and the District of Columbia, the percentage of total family households (households where the individuals live together as a family). The variable *FAM_MPC* contains the percentage of households with a married-couple family.
- **a.** Construct a scatterplot of *FAM_TPC* against *FAM_MPC*.
- **b.** Would you say that there is a positive relationship, a negative relationship, or no apparent relationship between the variables? Comment on why this is so.

**53.** The variable *FAM_TPC* contains, for each state and the District of Columbia, the percentage of total family households. The variable *NFM_TPC* contains the percentage of total nonfamily households (households where the individuals do not live as a family).
- **a.** Construct a scatterplot of *FAM_TPC* against *NFM_TPC*.
- **b.** Would you say that there is a positive relationship, a negative relationship, or no apparent relationship between the variables? Comment on why this is so.
- **c.** What is unusual about your scatterplot from **(a)**? Why is this so?

## 4.2 Introduction to Correlation

*Objectives:* **By the end of this section, I will be able to…**

1 Calculate and interpret the value of the correlation coefficient.

In Section 4.1 we learned tabular methods (crosstabulations) and graphical methods (clustered bar graphs and scatterplots) for investigating the relationship between two variables. In Sections 4.2 and 4.3 we examine numerical methods for quantifying the relationship between two quantitative variables. Here, in Section 4.2 we discuss the *correlation coefficient,* a numerical measure for describing the linear relationship between two quantitative variables. Then in Section 4.3 we move on to examine the *regression line,* which allows us to make predictions.

Table 4.11 contains the high and low temperatures in degrees Fahrenheit (°F) for ten American cities on a particular winter day. The variables are *high temperature*

Table 4.11 High and low temperatures, in degrees Fahrenheit, of 10 American cities

| City | $x$ = Low temp. | $y$ = High temp. | City | $x$ = Low temp. | $y$ = High temp. |
|---|---|---|---|---|---|
| Boston | 27 | 35 | Memphis | 47 | 63 |
| Chicago | 33 | 44 | Miami | 71 | 79 |
| Cincinnati | 39 | 46 | Minneapolis | 16 | 29 |
| Dallas | 57 | 68 | Philadelphia | 37 | 42 |
| Las Vegas | 45 | 55 | Washington | 39 | 45 |

and *low temperature*. Applying what we have learned in Section 4.1, we construct a scatterplot of the data set, which is presented in Figure 4.9. Whenever you are interested in examining the relationship between two quantitative variables, your best bet is to start with a scatterplot. Here we will regard the low temperatures as the $x$ variable and the high temperatures as the $y$ variable.

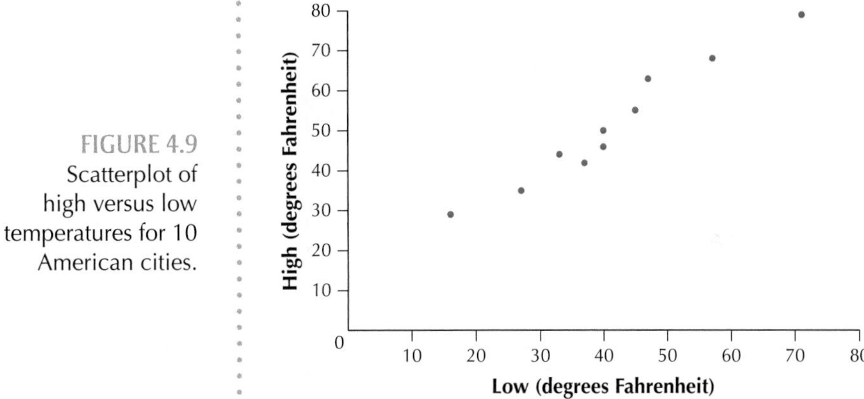


FIGURE 4.9 Scatterplot of high versus low temperatures for 10 American cities.

Getty Images/Nordic Photos

The relationship between the two variables is rather striking visually. There is a strong positive relationship between the high temperature and the low temperature of a city. That is, colder low temperatures are associated with colder high temperatures. Warmer low temperatures are associated with warmer high temperatures. In this section we seek to *quantify this relationship* between two numerical variables. The first measure that we will use to quantify the relationship between two quantitative variables is the **correlation coefficient *r*.**

The correlation coefficient $r$ (sometimes known as the *Pearson product moment correlation coefficient*) measures the strength and direction of the linear relationship between two variables. By *linear,* we mean *straight line*. The correlation coefficient does not measure the strength of a curved relationship between two variables.

**Correlation Coefficient *r***

The **correlation coefficient *r*** measures the strength and direction of the linear relationship between two variables. The correlation coefficient $r$ is

$$r = \frac{\sum(x-\bar{x})(y-\bar{y})}{(n-1)s_x s_y}$$

where $s_x$ is the sample standard deviation of the $x$ data values, and $s_y$ is the sample standard deviation of the $y$ data values.

## Example 4.5 Calculating the correlation coefficient $r$

Find the value of the correlation coefficient $r$ for the temperature data in Table 4.11.

**Solution**

We will outline the steps used in calculating the value of $r$ using the temperature data.

***Step 1*** Calculate the respective sample means, $\bar{x}$ and $\bar{y}$.

$$\bar{x} = \frac{\sum x}{n} = \frac{27 + 33 + 39 + 57 + 45 + 47 + 71 + 16 + 37 + 39}{10} = \frac{411}{10} = 41.1$$

$$\bar{y} = \frac{\sum y}{n} = \frac{35 + 44 + 46 + 68 + 55 + 63 + 79 + 29 + 42 + 45}{10} = \frac{506}{10} = 50.6$$

***Step 2*** Construct a table, as shown here in Table 4.12.

Table 4.12 Calculation table for the correlation coefficient $r$

| | $x$ | $y$ | $(x - \bar{x})$ | $(x - \bar{x})^2$ | $(y - \bar{y})$ | $(y - \bar{y})^2$ | $(x - \bar{x})(y - \bar{y})$ |
|---|---|---|---|---|---|---|---|
| Bos. | 27 | 35 | $27 - 41.1 =$ **−14.1** | $(-14.1)^2 =$ **198.81** | $35 - 50.6 =$ **−15.6** | $(-15.6)^2 =$ **243.36** | $(-14.1)(-15.6) =$ **219.96** |
| Chi. | 33 | 44 | $33 - 41.1 =$ **−8.1** | $(-8.1)^2 =$ **65.61** | $44 - 50.6 =$ **−6.6** | $(-6.6)^2 =$ **43.56** | $(-8.1)(-6.6) =$ **53.46** |
| Cin. | 39 | 46 | $39 - 41.1 =$ **−2.1** | $(-2.1)^2 =$ **4.41** | $46 - 50.6 =$ **−4.6** | $(-4.6)^2 =$ **21.16** | $(-2.1)(-4.6) =$ **9.66** |
| Dal. | 57 | 68 | $57 - 41.1 =$ **15.9** | $(15.9)^2 =$ **252.81** | $68 - 50.6 =$ **17.4** | $(17.4)^2 =$ **302.76** | $(15.9)(17.4) =$ **276.66** |
| Vegas | 45 | 55 | $45 - 41.1 =$ **3.9** | $(3.9)^2 =$ **15.21** | $55 - 50.6 =$ **4.4** | $(4.4)^2 =$ **19.36** | $(3.9)(4.4) =$ **17.16** |
| Mem. | 47 | 63 | $47 - 41.1 =$ **5.9** | $(5.9)^2 =$ **34.81** | $63 - 50.6 =$ **12.4** | $(12.4)^2 =$ **153.76** | $(5.9)(12.4) =$ **73.16** |
| Mia. | 71 | 79 | $71 - 41.1 =$ **29.9** | $(29.9)^2 =$ **894.01** | $79 - 50.6 =$ **28.4** | $(28.4)^2 =$ **806.56** | $(29.9)(28.4) =$ **849.16** |
| Minn. | 16 | 29 | $16 - 41.1 =$ **−25.1** | $(-25.1)^2 =$ **630.01** | $29 - 50.6 =$ **−21.6** | $(-21.6)^2 =$ **466.56** | $(-25.1)(-21.6) =$ **542.16** |
| Phil. | 37 | 42 | $37 - 41.1 =$ **−4.1** | $(-4.1)^2 =$ **16.81** | $42 - 50.6 =$ **−8.6** | $(-8.6)^2 =$ **73.96** | $(-4.1)(-8.6) =$ **35.26** |
| Wash. | 39 | 45 | $39 - 41.1 =$ **−2.1** | $(-2.1)^2 =$ **4.41** | $45 - 50.6 =$ **−5.6** | $(-5.6)^2 =$ **31.36** | $(-2.1)(-5.6) =$ **11.76** |
| | | | | $\sum(x - \bar{x})^2 =$ **2116.9** | | $\sum(y - \bar{y})^2 =$ **2162.4** | $\sum(x - \bar{x})(y - \bar{y}) =$ **2088.4** |

***Step 3*** Calculate the respective sample standard deviations $s_x$ and $s_y$. Using the sums calculated from Table 4.12, we have

$$s_x = \sqrt{\frac{\sum(x - \bar{x})^2}{n - 1}} = \sqrt{\frac{2116.9}{10 - 1}} \approx 15.33659386 \quad \text{and}$$

$$s_y = \sqrt{\frac{\sum(y - \bar{y})^2}{n - 1}} = \sqrt{\frac{2162.4}{10 - 1}} \approx 15.50053763$$

*Note on Rounding:* Whenever you calculate a quantity that will be needed for later calculations, do not round. Round only when you arrive at the final answer. Here, since the quantities $s_x$ and $s_y$ are used to calculate the correlation coefficient $r$, neither of them is rounded until the end of the calculation.

***Step 4*** Put these values all together in the formula for the correlation coefficient $r$:

$$r = \frac{\sum(x - \bar{x})(y - \bar{y})}{(n - 1)s_x s_y} = \frac{2088.4}{(9)(15.33659386)(15.50053763)} \approx 0.9761$$

The correlation coefficient $r$ for the high and low temperatures is 0.9761. ■

*What Does This Formula Mean?*

### The Correlation Coefficient $r$

Let's analyze the definition formula for the correlation coefficient $r$. When would $r$ be positive, and when would it be negative? We see that the formula

$$r = \frac{\sum(x - \bar{x})(y - \bar{y})}{(n - 1)s_x s_y}$$

consists of a ratio. Note that the denominator can never be negative, since it is the product of three non-negative values (standard deviations can never be negative). Therefore, the numerator determines whether $r$ will be positive or negative. So then, when will the numerator be positive and when will it be negative? We know that $x - \bar{x}$ is positive whenever the data value $x$ is greater than $\bar{x}$, and negative when $x$ is less than $\bar{x}$. Similarly for $y - \bar{y}$. The numerator of $r$ is the product $(x - \bar{x}) \cdot (y - \bar{y})$. There are four cases (or regions, illustrated in Figure 4.10) that describe when the product $(x - \bar{x})(y - \bar{y})$ will be positive or negative, shown in Table 4.13. Note that Figure 4.10 is centered at the point $(\bar{x}, \bar{y})$.

Region 2
$(x - \bar{x}) < 0$
$(y - \bar{y}) > 0$
$(x - \bar{x})(y - \bar{y}) < 0$
$r < 0$

Region 1
$(x - \bar{x}) > 0$
$(y - \bar{y}) > 0$
$(x - \bar{x})(y - \bar{y}) > 0$
$r > 0$

point $(\bar{x}, \bar{y})$

line $y = \bar{y}$

Region 3
$(x - \bar{x}) < 0$
$(y - \bar{y}) < 0$
$(x - \bar{x})(y - \bar{y}) > 0$
$r > 0$

Region 4
$(x - \bar{x}) > 0$
$(y - \bar{y}) < 0$
$(x - \bar{x})(y - \bar{y}) < 0$
$r < 0$

line $x = \bar{x}$

FIGURE 4.10 The four regions for determining whether $r$ will tend to be positive or negative.

Table 4.13 Four cases for the product $(x - \bar{x})(y - \bar{y})$

| Case/region | $(x - \bar{x})$ | $(y - \bar{y})$ | $(x - \bar{x})(y - \bar{y})$ | Correlation coefficient $r$ |
|---|---|---|---|---|
| 1 | Positive | Positive | Positive | Positive |
| 2 | Negative | Positive | Negative | Negative |
| 3 | Negative | Negative | Positive | Positive |
| 4 | Positive | Negative | Negative | Negative |

Thus, data values that fall in Regions 1 and 3 will tend to make the value of $r$ positive, while data values that fall in Regions 2 and 4 will tend to make the value of $r$ negative. The summation in the numerator of $r$ acts as a blender, combining the contributions of all the various data values falling in all the various regions.

- If most of the data values fall in Regions 1 and 3, then $r$ will tend to be positive.
- If most of the data values fall in Regions 2 and 4, then $r$ will tend to be negative.
- If the four regions share the data values more or less equally, then $r$ will be near zero. ■

Alamy

Let's explore how our high and low temperature data fit into the above framework. The mean low temperature is $\bar{x} = 41.1$°F, while the mean high temperature is $\bar{y} = 50.6$°F. We find the point $(\bar{x}, \bar{y}) = (41.1, 50.6)$ in our scatterplot of the high and low temperatures, draw the lines $x = \bar{x} = 41.1$ and $y = \bar{y} = 50.6$, and mark out our four regions, as shown in Figure 4.11. Note that all of the data points fall in Regions 1 and 3. Therefore, we expect the value of $r$ for this data set to be positive, which is indeed the case, since we observed $r = 0.9761$ in Example 4.5.

The correlation coefficient $r$ always takes on values between 1 and $-1$, inclusive. For example, $r = 0.90$ indicates a strong **positive correlation,** while $r = -0.85$ indicates a strong **negative correlation.** Here are some standard methods for interpreting the value of $r$.

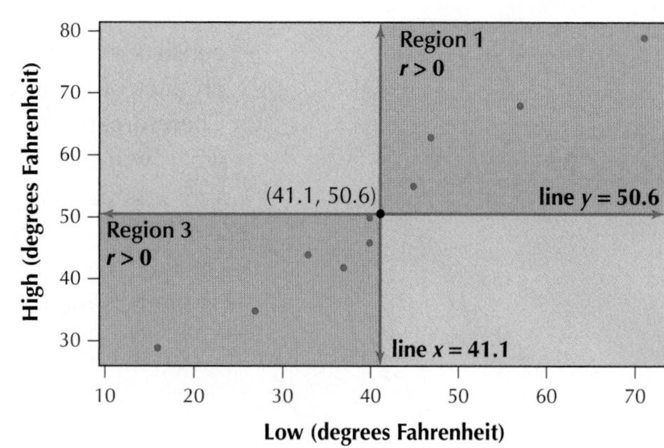


FIGURE 4.11 All of the temperature data points lie in Regions 1 and 3, making $r$ positive.

**Interpreting the Correlation Coefficient $r$**

1. Values of $r$ close to 1 indicate a positive relationship between the two variables.
   - The variables are said to be **positively correlated.**
   - As $x$ increases, $y$ tends to increase as well.
2. Values of $r$ close to $-1$ indicate a negative relationship between the two variables.
   - The variables are said to be **negatively correlated.**
   - As $x$ increases, $y$ tends to decrease.
3. Other values of $r$ indicate the lack of either a positive or negative linear relationship between the two variables.
   - The variables are said to be **uncorrelated.**
   - As $x$ increases, $y$ tends to neither increase nor decrease linearly.

What do we mean by "close to 1"? Although the threshold for determining the presence of correlation depends on the particular data set, we nevertheless offer a *rough* rule of thumb for detecting correlation.

**Guidelines for Interpreting the Correlation Coefficient $r$**

If the correlation coefficient between two variables is

- greater than 0.7, the variables are positively correlated.
- between 0.33 and 0.7, the variables are mildly positively correlated.
- between –0.33 and 0.33, the variables are not correlated.
- between –0.7 and –0.33, the variables are mildly negatively correlated.
- less than –0.7, the variables are negatively correlated.

## Example 4.6 Interpreting the correlation coefficient

Interpret the correlation coefficient found in Example 4.5.

### Solution

In Example 4.5, we found the correlation coefficient for the relationship between high and low temperature to be $r = 0.9761$. This value of $r$ is strongly positive, very close to the maximum value $r = 1$. We would therefore say that high and low temperatures for these 10 American cities are strongly positively correlated. As low temperature increases, high temperatures also tend to increase. ■

### Beyond Black and White

In practice, of course, the presence or absence of a correlation is not black and white. The data analyst needs to exercise his or her judgment. Since the values that $r$ can take range continuously from –1 to 1, the data analyst must be prepared for shades of gray. ■

The *Correlation and Regression* applet allows you to insert your own data values and see how the value of the correlation coefficient changes.

Note: *Correlation does not imply causation*. For example, daily umbrella sales in Miami, Florida, is probably negatively correlated with attendance at the Florida Marlins baseball games. But we would *not* say that umbrella purchases *cause* attendance to fall at the ball games, or vice versa. Instead, both phenomena are presumably caused by a "hidden variable," rainfall.

The following computational formula may be used as an equivalent of the definition formula for the correlation coefficient $r$.

**Equivalent Computational Formula for Calculating the Correlation Coefficient $r$**

$$r = \frac{\sum xy - \left(\sum x \sum y\right)/n}{\sqrt{\left[\sum x^2 - \left(\sum x\right)^2/n\right]\left[\sum y^2 - \left(\sum y\right)^2/n\right]}}$$

## Example 4.7 Using the computational formula to calculate $r$

Use the computational formula to calculate the correlation coefficient $r$ for the relationship between square footage and sales price of the eight home lots for sale in Glen Ellyn from Example 4.3, in Section 4.1.

| $x$ = square footage (100s of sq. ft.) | $y$ = sales price ($1000s) |
|---|---|
| 90 | 200 |
| 132 | 423 |
| 139 | 300 |
| 150 | 260 |
| 200 | 270 |
| 300 | 650 |
| 408 | 680 |
| 554 | 1450 |

### Solution

We are given $n = 8$ and the following summations: $\sum x = 1973$, $\sum y = 4233$, $\sum xy = 1{,}484{,}276$, $\sum x^2 = 670{,}725$, and $\sum y^2 = 3{,}436{,}829$.

Substituting into the computational formula, we have

$$r = \frac{1{,}484{,}276 - (1973)(4233)/8}{\sqrt{(670{,}725 - (1973)^2/8)(3{,}436{,}829 - (4233)^2/8)}} \approx 0.9379$$

Square footage and sales price are positively correlated. As the square footage increases, the sales price tends to increase as well. ■

### A Shift to the Right (Continued)

*What if* we add 4000 square feet to the size of *each* lot in Example 4.7, while leaving the sales price unchanged. What would happen to the correlation coefficient?

**Solution**

As we saw in Section 4.1, such a shift does not affect the relationship between square footage and sales price. The correlation coefficient also remains unchanged. The character of the *relationship* between the variables is unchanged by such a shift.

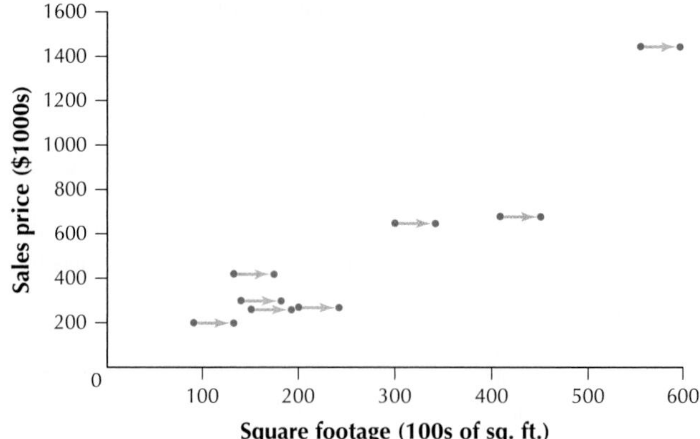


FIGURE 4.4

## Section 4.2 SUMMARY

Section 4.2 introduces the correlation coefficient $r$, a measure of the strength of linear association between two numeric variables. Values of $r$ close to 1 indicate that the variables are positively correlated. Values of $r$ close to –1 indicate that the variables are negatively correlated. Values of $r$ close to 0 indicate that the variables are not correlated.

## Section 4.2 EXERCISES

**CLARIFYING THE CONCEPTS**

1. When investigating the relationship between two quantitative variables, what graph should you use first?
2. In your own words, explain what the correlation coefficient measures. What is the symbol that we use for the correlation coefficient?
3. What is the range of values the correlation coefficient can take?
4. What do the following values of $r$ indicate about the relationship between two variables? What can we say about the variables?
   a. A value of $r$ close to 1
   b. A value of $r$ close to $-1$
   c. A value of $r$ close to 0

**PRACTICING THE TECHNIQUES**

For the scatterplots in **i–v**, identify which plot represents the data set with the following correlation coefficients:

5. Near 1
6. Near 0.5
7. Near zero
8. Near $-0.5$
9. Near $-1$

**i.**

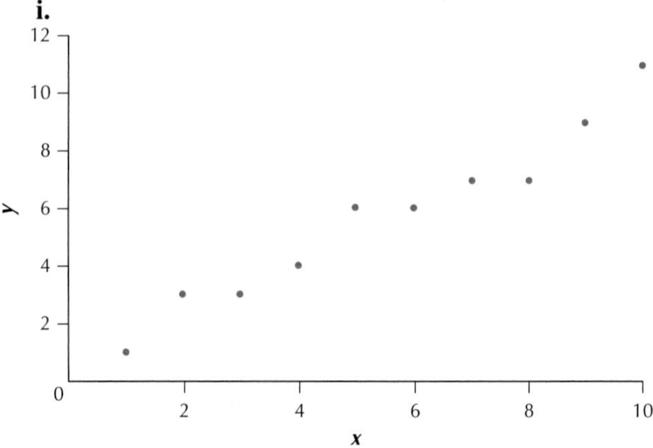


**ii.**

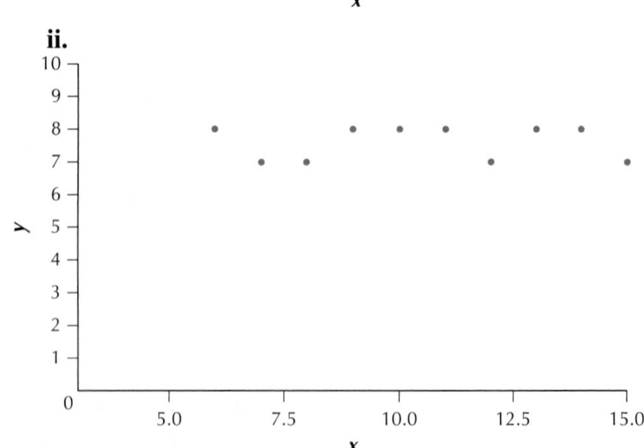

iii.

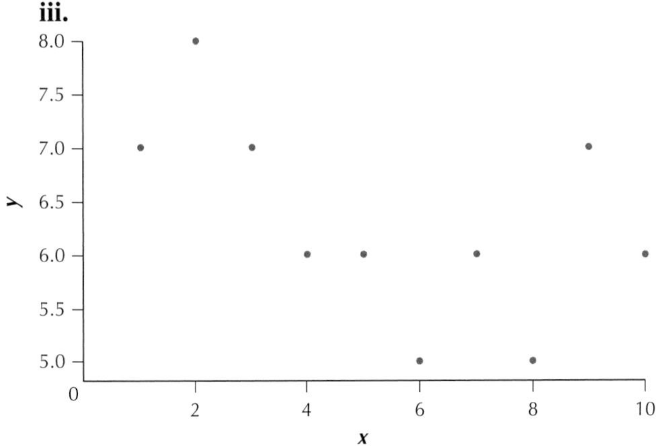

iv.

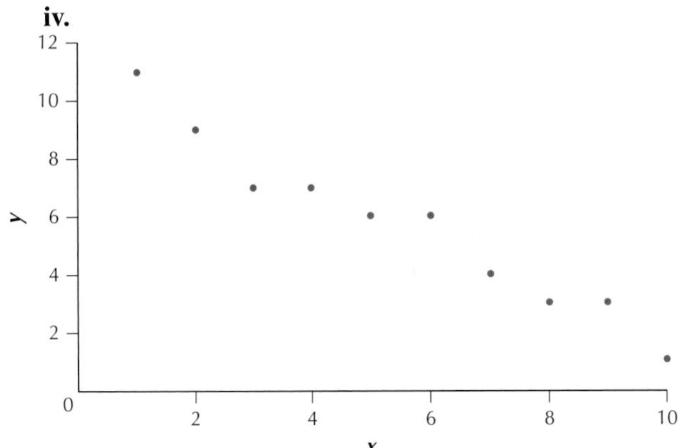

v.

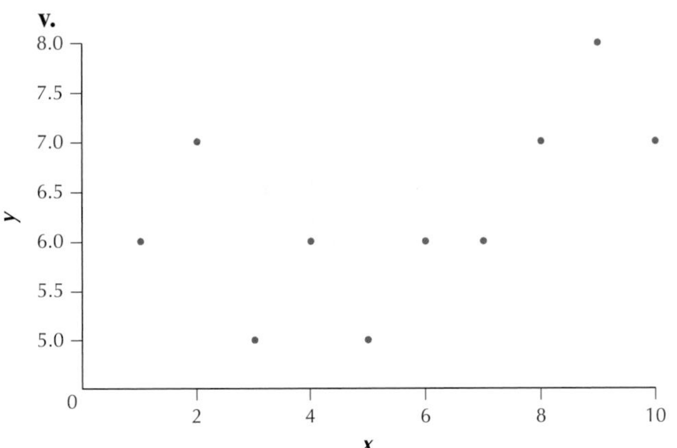

The values for $x$ and $y$ in each scatterplot above are integer-valued. For each scatterplot, (a) reconstruct the original data set, and (b) calculate the correlation coefficient for the data.

**10.** The data in scatterplot i
**11.** The data in scatterplot ii
**12.** The data in scatterplot iii
**13.** The data in scatterplot iv
**14.** The data in scatterplot v

Refer to Table 4.14 for Exercises 15–18.

**TABLE 4.14 Top five NASCAR winners in the modern era**

| Driver | Super speedway wins | Short tracks wins |
|---|---|---|
| Darrell Waltrip | 18 | 47 |
| Dale Earnhardt | 29 | 27 |
| Jeff Gordon | 15 | 15 |
| Cale Yarborough | 15 | 29 |
| Richard Petty | 19 | 23 |

**15.** Construct a scatterplot, with $x$ = short tracks wins and $y$ = super speedways wins.
**16.** Based on the scatterplot, would you say that $x$ and $y$ are positively correlated, negatively correlated, or not correlated?
**17.** Calculate the value of the correlation coefficient, using the following steps.

- **a.** Calculate the respective sample means $\bar{x}$ and $\bar{y}$.
- **b.** Construct a table like Table 4.12, as follows.
  - **i.** For each observation, calculate the deviations $(x - \bar{x})$ and $(y - \bar{y})$.
  - **ii.** For each observation, calculate $(x - \bar{x})^2$, $(y - \bar{y})^2$, and $(x - \bar{x})(y - \bar{y})$.
  - **iii.** Calculate the following sums: $\Sigma(x - \bar{x})^2$, $\Sigma(y - \bar{y})^2$, and $\Sigma(x - \bar{x})(y - \bar{y})$.
- **c.** Calculate the respective sample standard deviations $s_x$ and $s_y$.
- **d.** Put these all together in the formula for the correlation coefficient $r$.
- **e.** Using technology, confirm the value you calculated in **(d)**.

**18.** Interpret the meaning of the correlation coefficient you found in Exercise 17d, using at least two sentences. Does this agree with your judgment from Exercise 16?

Refer to Table 4.15 for Exercises 19–21.

**TABLE 4.15 Mean SAT scores for the five states with the best participation rate**

| State | SAT I Verbal | SAT I Math |
|---|---|---|
| New York | 497 | 510 |
| Connecticut | 515 | 515 |
| Massachusetts | 518 | 523 |
| New Jersey | 501 | 514 |
| New Hampshire | 522 | 521 |

**19.** Construct a scatterplot of the data, with $x$ = SAT I Verbal and $y$ = SAT I Math. Describe the apparent relationship, if any, between the variables. Based on the scatterplot, would you say that $x$ and $y$ are positively correlated, negatively correlated, or not correlated?

**20.** Calculate the value of the correlation coefficient, using the following steps.

**a.** Calculate the respective sample means $\bar{x}$ and $\bar{y}$.

**b.** Construct a table like Table 4.12, as follows.

**i.** For each observation, calculate the deviations $(x - \bar{x})$ and $(y - \bar{y})$.

**ii.** For each observation, calculate $(x - \bar{x})^2$, $(y - \bar{y})^2$, and $(x - \bar{x})(y - \bar{y})$.

**iii.** Calculate the following sums: $\sum(x - \bar{x})^2$, $\sum(y - \bar{y})^2$, and $\sum(x - \bar{x})(y - \bar{y})$.

**c.** Calculate the respective sample standard deviations $s_x$ and $s_y$.

**d.** Put these all together in the formula for the correlation coefficient $r$.

**e.** Using technology, confirm the value you calculated in **(d)**.

**21.** Interpret the meaning of the correlation coefficient you found in Exercise 20d, using at least two sentences. Does this agree with your judgment from Exercise 19?

## APPLYING THE CONCEPTS

**Brain and Body Weight.** A study compared the body weight (in kilograms) and brain weight (in grams) for a sample of mammals, with the results shown in the following table. We are interested in estimating brain weight ($y$) based on body weight ($x$).

| Body weight (kg) | Brain weight (g) |
|---|---|
| 52.16 | 440 |
| 60 | 81 |
| 27.66 | 115 |
| 85 | 325 |
| 36.33 | 119.5 |
| 100 | 157 |
| 35 | 56 |
| 62 | 1320 |
| 83 | 98.2 |
| 55.5 | 175 |

**22.** Construct a scatterplot of the data. Describe the apparent relationship, if any, between the variables. Based on the scatterplot, would you say that $x$ and $y$ are positively correlated, negatively correlated, or not correlated? (*Hint:* The scatterplot was constructed in Exercise 37 [page 175].)

**23.** Calculate the value of the correlation coefficient $r$.

**24.** Interpret the meaning of the correlation coefficient you found, using at least two sentences. Does this agree with your judgment from Exercise 22?

**Calories and Sugar Content.** Use Figure 4.12 for Exercises 25–27.

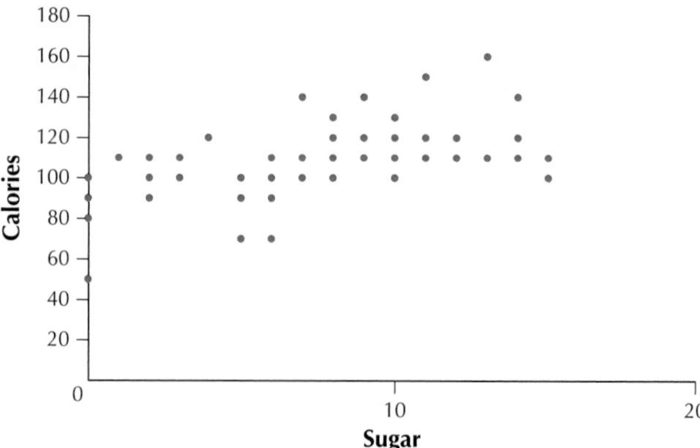


Figure 4.12 Scatterplot of calories versus sugar content for breakfast cereals.

**25.** Would you expect the correlation coefficient to be positive, negative, or near zero?

**26.** Would you estimate that the value of $r$ is nearer 1, nearer 0, or somewhere in the middle?

**27.** Interpret the estimated value of $r$ for this data set. Use two different sentences.

**Batting Average and Age.** For Exercises 28 and 29, refer to the following scatterplot of batting average against player age for the 2002 American League season (minimum at-bats 100).

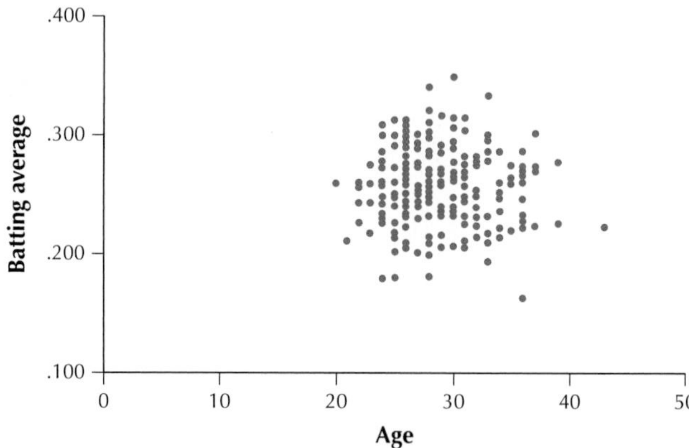


**28.** Is the correlation coefficient positive, negative, or near zero?

**29.** Interpret the estimated value of $r$ for this data set. Use two different sentences.

**Words and Sentences.** Use the information in Table 4.9 (page 174) for Exercises 30–32. The table contains the number of words and the number of sentences in 12 randomly selected advertisements in two magazines.

**30.** Just by looking at the raw data in the table (without considering the scatterplot), can you tell what type of relationship exists between the variables?

31. Now consider the scatterplot of the same data. What type of relationship do you think exists between the variables? Will the value of the correlation coefficient be positive or negative?

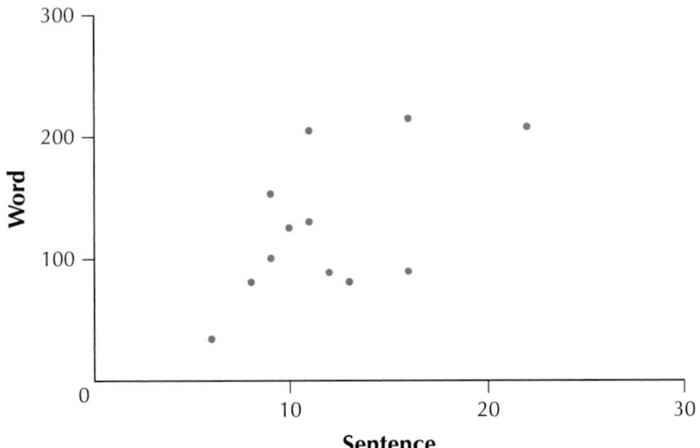


32. Interpret the meaning of the estimated value of $r$ for this data set. Use two different sentences.

33. **Age of Californians.** The data set **State Age** contains the percentages of townspeople (a) under 18, (b) between 18 and 65, and (c) over 65 for all towns in California. Each of these three age variables was compared with the total population of each town:

- The correlation coefficient of the under-18 group versus the total population is $-0.185$.
- The correlation coefficient of the 18–65 group versus the total population is 0.174.
- The correlation coefficient of the over-65 group versus the total population is 0.031.

  **a.** Which group is most positively correlated with population?
  **b.** Are any of the groups even mildly positively or negatively correlated with total population?

**Country and Hip-Hop CDs.** Use the information in Table 4.16 for Exercises 34 and 35. The table contains the number of country music CDs and the number of hip-hop CDs owned by 6 randomly selected students.

**TABLE 4.16 Number of country and hip-hop CDs owned by six students**

| | Student | | | | | |
|---|---|---|---|---|---|---|
| | 1 | 2 | 3 | 4 | 5 | 6 |
| **Hip-hop CDs owned ($y$)** | 10 | 12 | 1 | 3 | 6 | 1 |
| **Country CDs owned ($x$)** | 1 | 3 | 11 | 8 | 5 | 27 |

34. Investigate the correlation.
  **a.** Construct a scatterplot of the variables. Make sure the $y$ variable goes on the $y$ axis.
  **b.** What type of relationship do these variables have: positive, negative, or no apparent linear relationship?
  **c.** Will the correlation coefficient be positive, negative, or near zero?

35. Calculate and interpret the correlation coefficient.
  **a.** Compute the value of the correlation coefficient using either the definition formula (page 178) or the computational formula (page 181). Check your answer using a computer or calculator.
  **b.** Does this value for $r$ concur with your earlier judgment?
  **c.** Interpret the meaning of this value of the correlation coefficient. Use two different sentences.

**Babies in Brisbane.** Use the following information for Exercises 36 and 37. On a busy day at Brisbane Hospital in Australia, the maternity ward recorded the birth weights (*WT_GRAMS*) of 44 newborns along with the time in minutes after midnight (*AFTMID*) that they were born.

36. Would you expect these two variables to be correlated? Why or why not? What would be your estimate of $r$?

37. The value of the correlation coefficient $r$ is 0.08.
  **a.** Does this agree with your expectation?
  **b.** Interpret this value of $r$ using two different sentences.

## CONSTRUCT YOUR OWN DATA SETS

38. Describe two variables from real life that would have a value of $r$ close to 1. Explain why they are positively correlated.

39. Create a sample of five observations from each of your variables in the previous exercise, and put them into a table similar to Table 4.11 (page 177). Next, construct a scatterplot of the variables. Finally, draw a single straight line through the data points in the plot in a manner that you think best approximates the relationship between the variables.

40. Describe two variables that would have a value of $r$ close to –1. Explain why they are negatively correlated.

41. Describe two variables that would have a value of $r$ close to 0. Explain why they are uncorrelated.

Use the *Correlation and Regression* applet for Exercises 42–45.

42. Create a set of $n = 10$ points such that the correlation coefficient $r$ takes approximately the following values. Note that you can drag points up or down to adjust your value of $r$.
  **a.** $r = 0.90$
  **b.** $r = -0.90$
  **c.** $r = 0.00$

43. Describe the relationship between the variables for each of the sets of points in the previous exercise.

44. Select "Show mean X and mean Y lines." Create a set of $n = 4$ points such that the correlation coefficient $r$ takes approximately the following values. Note that you can drag points up or down to adjust your value of $r$.
  **a.** $r = 0.70$
  **b.** $r = -0.70$
  **c.** $r = 0.00$

45. Consider Figure 4.10 (page 179), which shows the four regions. For each set of points in the previous exercise, describe the regions in which these points lie. Explain how this makes sense in terms of each value of $r$.

# 4.3 Introduction to Regression

## *Objectives:* By the end of this section, I will be able to...

1. Calculate the value and understand the meaning of the slope and the $y$ intercept of the regression line.
2. Predict values of $y$ for given values of $x$.

## 1 The Regression Line

In Section 4.2 we learned about the *correlation coefficient,* a numerical measure for describing the linear relationship between two variables. Here in Section 4.3 we will learn how to approximate the relationship between two numerical variables using the regression line and the regression equation.

Table 4.11 High and low temperatures in degrees Fahrenheit of 10 American cities

| City | $x$ = low temp. | $y$ = high temp. | City | $x$ = low temp. | $y$ = high temp. |
|---|---|---|---|---|---|
| Boston | 27 | 35 | Memphis | 47 | 63 |
| Chicago | 33 | 44 | Miami | 71 | 79 |
| Cincinnati | 39 | 46 | Minneapolis | 16 | 29 |
| Dallas | 57 | 68 | Philadelphia | 37 | 42 |
| Las Vegas | 45 | 55 | Washington | 39 | 45 |

Consider again Figure 4.9 (page 177), the scatterplot of the high and low temperatures for 10 American cities, from Table 4.11. The data points generally seem to follow a roughly linear path. We may in fact draw a straight line from the lower left to the upper right to approximate this relatively linear path. Such a straight line, called a **regression line,** is shown in Figure 4.13.

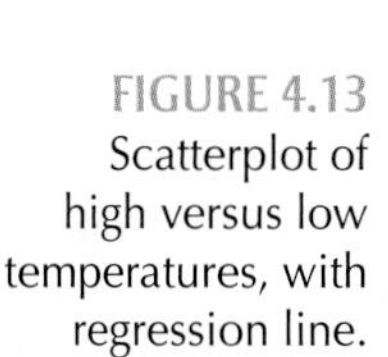

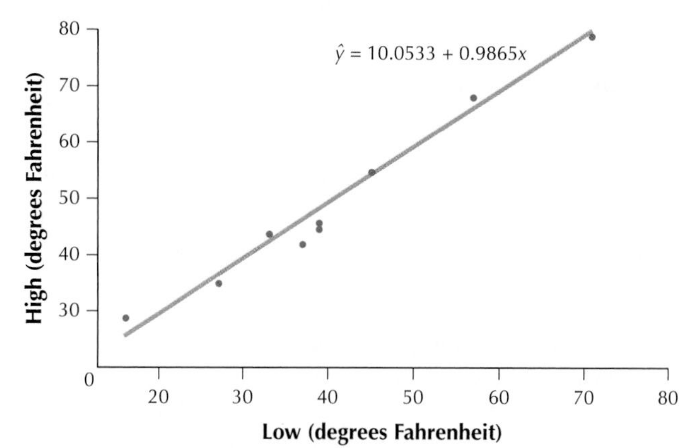


FIGURE 4.13 Scatterplot of high versus low temperatures, with regression line.

As you may recall from high school algebra, the equation of a straight line may be written as $y = mx + b$. In statistics, we write the **equation of the regression line** equivalently as $\hat{y} = b_0 + b_1x$.

**Equation of the Regression Line**

The **equation of the regression line** that approximates the relationship between $x$ and $y$ is

$$\hat{y} = b_0 + b_1 x$$

where the *regression coefficients* are the **slope,** $b_1$, and the ***y* intercept,** $b_0$. The "hat" over the $y$ (pronounced "y-hat") indicates that this is an estimate of $y$ and not necessarily an actual value of $y$. The equations of these coefficients are

$$b_1 = \frac{\sum(x - \bar{x})(y - \bar{y})}{\sum(x - \bar{x})^2} \qquad b_0 = \bar{y} - (b_1 \cdot \bar{x})$$

Note that all of the quantities needed to calculate $b_0$ and $b_1$ have already been computed in the formula for $r$. In particular, the numerators for $b_1$ and $r$ are exactly the same.

There are an infinite number of different straight lines that could approximate the relationship between high and low temperatures. Why did we choose this one? Because this is the so-called *least-squares* regression line, which is the most widely used linear approximation for bivariate relationships. We will learn more about least squares in Chapter 13.

## Example 4.8 Calculating the regression coefficients $b_0$ and $b_1$

Find the value of the regression coefficients $b_0$ and $b_1$ for the temperature data in Table 4.11.

### Solution

We will outline the steps used in calculating the value of $b_1$ using the temperature data.

***Step 1*** Calculate the respective sample means $\bar{x}$ and $\bar{y}$. We have already done this in Example 4.5: $\bar{x} = 41.1$ and $\bar{y} = 50.6$.

***Step 2*** Construct a table like Table 4.12, except that the $(y - \bar{y})^2$ column is not needed. Write down the values of $\sum(x - \bar{x})^2$ and $\sum(x - \bar{x})(y - \bar{y})$. We have already done this in Example 4.5, and we have

$$\sum(x - \bar{x})^2 = 2116.9 \text{ and } \sum(x - \bar{x})(y - \bar{y}) = 2088.4$$

***Step 3*** Use the statistics from Steps 1 and 2 to calculate $b_1$:

$$b_1 = \frac{\sum(x - \bar{x})(y - \bar{y})}{\sum(x - \bar{x})^2} = \frac{2088.4}{2116.9} = 0.9865369172$$

***Step 4*** Use the statistics from Steps 1–3 to calculate $b_0$:

$$b_0 = \bar{y} - (b_1 \cdot \bar{x}) = 50.6 - (0.9865369172)(41.1) \approx 10.0533$$

Thus, the equation of the regression line for the temperature data is

$$\hat{y} = 10.0533 + 0.9865x$$

Since $y$ and $x$ represent high and low temperatures, respectively, this equation is read as follows: "The estimated high temperature for an American city is 10.0533 degrees Fahrenheit plus 0.9865 times the low temperature for that city." ■

## Example 4.9 Correlation and regression using technology

Use technology to find the correlation coefficient $r$ and the regression equation for the temperature data in Example 4.5.

### Solution

The instructions for using technology for correlation and regression are provided in the Step-by-Step Technology Guide at the end of this section (page 193). The TI-83/84 scatterplot is shown in Figure 4.14, and the TI-83/84 regression results are shown in Figure 4.15. (Note that the TI-83/84 indicates the $y$ intercept $b_0$ as $a$, and the slope $b_1$ as $b$.) Figure 4.16 shows partial Excel results, with the $y$ intercept ("Intercept") and the slope ("Low") highlighted. Figure 4.17 shows the Minitab results, with the $y$ intercept ("Constant") and the slope ("Low") highlighted. ■

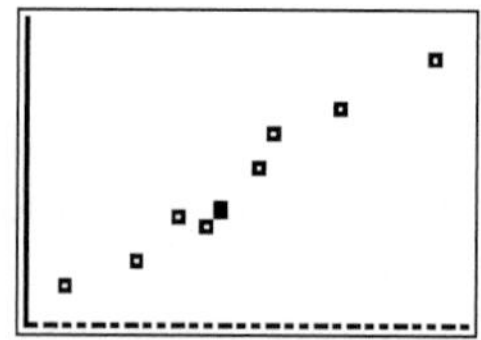

FIGURE 4.14 TI-83/84 scatterplot.

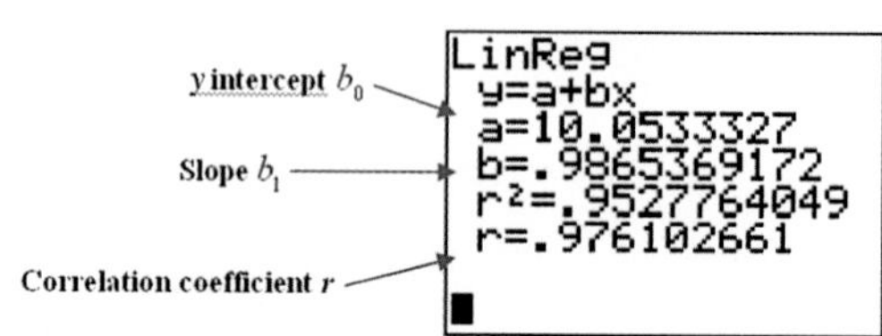


FIGURE 4.15 TI-83/84 regression results.

| | Coefficients | Standard Error | t Stat | P-value |
|---|---|---|---|---|
| Intercept | 10.05333 | 3.385572 | 2.969464 | 0.017886 |
| Low | 0.986537 | 0.077652 | 12.7046 | 1.39E-06 |

FIGURE 4.16 Excel results.

**Correlations: Low, High**

```
Pearson correlation of Low and High = 0.976
```

**Regression Analysis: High versus Low**

```
The regression equation is High = 10.1 + 0.987 Low

Predictor     Coef   SE Coef      T      P
Constant    10.053     3.386   2.97  0.018
Low        0.98654   0.07765  12.70  0.000
```

FIGURE 4.17 Minitab results.

The following computational formula is equivalent to the definition formula for the slope $b_1$.

**Equivalent Computational Formula for Calculating the Slope $b_1$**

$$b_1 = \frac{\sum xy - \left(\sum x \sum y\right)/n}{\sum x^2 - \left(\sum x\right)^2/n}$$

## Example 4.10 Using the computational formula to calculate the slope $b_1$

Use the computational formula to calculate the slope $b_1$ for the relationship between square footage and sales price of the eight home lots for sale in Glen Ellyn from Example 4.3 in Section 4.1. Then find the $y$ intercept $b_0$ and the regression equation.

| $x$ = square footage (100s of sq. ft.) | $y$ = sales price ($1000s) |
|---|---|
| 90 | 200 |
| 132 | 423 |
| 139 | 300 |
| 150 | 260 |
| 200 | 270 |
| 300 | 650 |
| 408 | 680 |
| 554 | 1450 |

**Solution**

We are given $n = 8$ and the following summations: $\sum x = 1973$, $\sum y = 4233$, $\sum xy =$ 1,484,276, and $\sum x^2 = 670{,}725$.

Substituting into the computational formula, we have

$$b_1 = \frac{1{,}484{,}276 - (1973)(4233)/8}{670{,}725 - 1973^2/8} = \frac{440{,}312.375}{184{,}133.875} \approx 2.39126$$

To find $b_0$ we first calculate

$$\bar{y} = \frac{\sum y}{n} = \frac{4233}{8} = 529.125 \text{ and } \bar{x} = \frac{\sum x}{n} = \frac{1973}{8} = 246.625$$

Then,

$$b_0 = \bar{y} - (b_1 \cdot \bar{x}) = 529.125 - (2.39126)(246.625) = -60.62$$

This gives us the following regression equation:

$$\hat{y} = b_0 + b_1 x = -60.62 + 2.39126x$$

## What If Scenario — The Sensitivity of the Regression Line to Extreme Values

*What if* the sales price of the largest lot for sale (55,400 square feet) was not $1,450,000 but $145,000. What would happen to the slope and the $y$ intercept of the regression line?

**Solution**

The correlation coefficient and the regression line are both sensitive to extreme values. As shown in Figure 4.18, the change to a much lower price for the largest lot acts as a weight pulling down on the regression line. The slope decreases from $b_1 = 2.39126$ to $b_1 = 0.2128$.

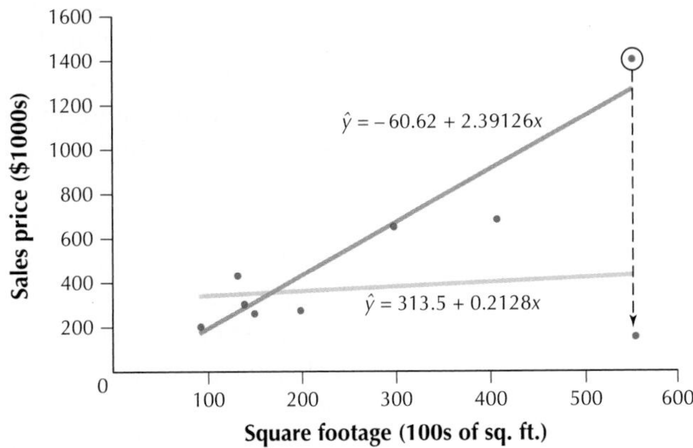


FIGURE 4.18

Consequently, the $y$ intercept increases from $b_0 = -60.62$ to $b_0 = 313.5$, giving us the new regression equation:

$$\hat{y} = 313.5 + 0.2128x$$

The *Correlation and Regression* applet allows you to insert your own data values and see how the regression line changes.

## 2 Using the Regression Equation to Make Predictions

We can use the regression equation to make estimates or predictions. For any particular value of $x$, the predicted value for $y$ lies on the regression line.

### Example 4.11 Using the regression equation to make an estimate

Suppose we are considering moving to a city that has a low temperature of 47 degrees Fahrenheit (°F) on this particular winter's day. What would the estimated high temperature be for this city?

**Solution**

To generate an estimate of the high temperature, we plug the value of 47°F for the variable *low* into the regression equation from Example 4.8:

$$\begin{aligned}\hat{y} &= 10.0533 + 0.9865(low)\\ &= 10.0533 + 0.9865(47)\\ &= 56.4188\end{aligned}$$

We would say: "The estimated high temperature for an American city with a low of 47°F, is 56.4188°F." ■

**Actual Data versus Predicted (Estimated) Data**

Now, we do have one city in our data table (Table 4.11) whose low temperature is 47°F: Memphis, Tennessee. Its high temperature, however, is 63°F, not 56.4188°F as we predicted. Note the conceptual difference between the Memphis high temperature of $y = 63$°F and our prediction of $\hat{y} = 56.4188$°F. The high temperature $y$ in Memphis is a fact; real, empirical, observed data. On the other hand, our prediction $\hat{y}$ is nothing more than an estimate. The actual data point for Memphis in Figure 4.19 is the circle in the scatterplot at (47, 63).

The "Predicted high temp." arrow in Figure 4.19 points to a location on the regression line directly beneath Memphis where the regression equation predicted the high temperature to be. Note that our prediction's position in the graph is at (47, 56.4188). Our prediction was too low by $y - \hat{y} = 63 - 56.4188 = 6.5812$°F. This is the vertical distance from the Memphis data point to the regression line. This distance is the *prediction error,* a topic we will discuss much more deeply in Chapter 13. Of course, we need not restrict our predictions to values of $x$ (*low*) that are in our data set (though see the warning on extrapolation below). Suppose we were interested in estimating the high temperature for a city whose low temperature is 25. Then the estimated high temperature for a city in which $low = 25$°F is

$$\hat{y} = 10.0533 + 0.9865(25) = 34.7158°\text{F}$$ ■

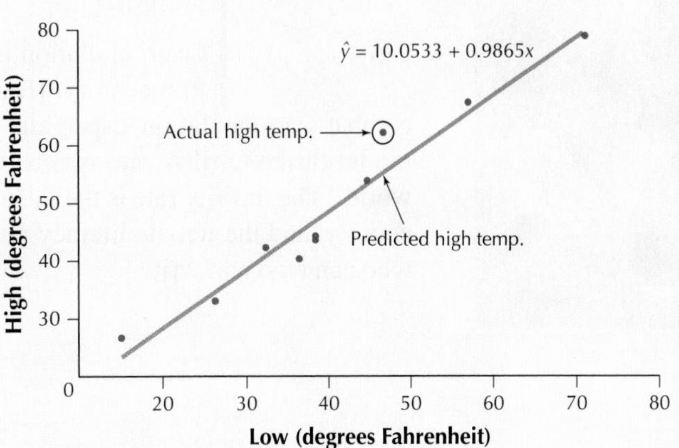


FIGURE 4.19 Actual versus predicted high temperature.

## Extrapolation

The $y$ intercept is the estimated value for $y$ when $x$ equals zero. However, in many regression problems, a value of zero for the $x$ variable would not make sense. For example, suppose we were trying to predict freshman GPA ($y$) based on SAT Verbal scores. However, there is no zero score for the SAT Verbal exam, so the $y$ intercept would not be meaningful. With our temperature data set, however, a value of zero for the low temperature does make sense. Therefore, we would be tempted to predict 10.78°F as the high temperature for a city with a low of zero degrees. However, $low = 0$°F is not within the range of the data set. Making predictions based on $x$-values that are beyond the range of the $x$-values in our data set is called *extrapolation*. It may be dangerous and should be avoided. We discuss extrapolation further in Chapter 13.

### Interpreting the Slope

What is the meaning of the slope coefficient $b_1$? In data analysis, we interpret the slope of the regression line as the *estimated change in y per unit increase in x*. In our temperature example, the units are degrees Fahrenheit, and the value of $b_1$ is 0.9865. Therefore, $b_1 = 0.9865$ means the following: "For each increase of 1°F in low temperature, the estimated high temperature increases by 0.9865°F." Also note that the slope is positive, which concurs with our assessment of the overall relationship between high and low temperatures. Further, is it a coincidence that both the slope and the correlation coefficient are positive? Not at all.

**Relationship Between Slope and Correlation Coefficient**

The slope $b_1$ of the regression line and the correlation coefficient $r$ always have the same sign.

- $b_1$ is positive if and only if r is positive.
- $b_1$ is negative if and only if $r$ is negative.

Hence, when we found in Section 4.2 that the correlation coefficient between high and low temperatures was positive, we could have immediately concluded that the slope of the regression line was also positive. ■

David Greedy/Getty Images

## CASE STUDY Female Literacy and Fertility Worldwide

Overpopulation is a serious concern for planet Earth in the twenty-first century. However, there are some encouraging signs of how to combat overpopulation, especially in developing countries. Figure 4.20 shows a scatterplot of the fertility rate versus the female literacy rate in 204 countries around the world.[3] The fertility rate is the mean number of children born to a typical woman in the country, and the female literacy rate is the percentage of women at least 15 years old who can read and write.

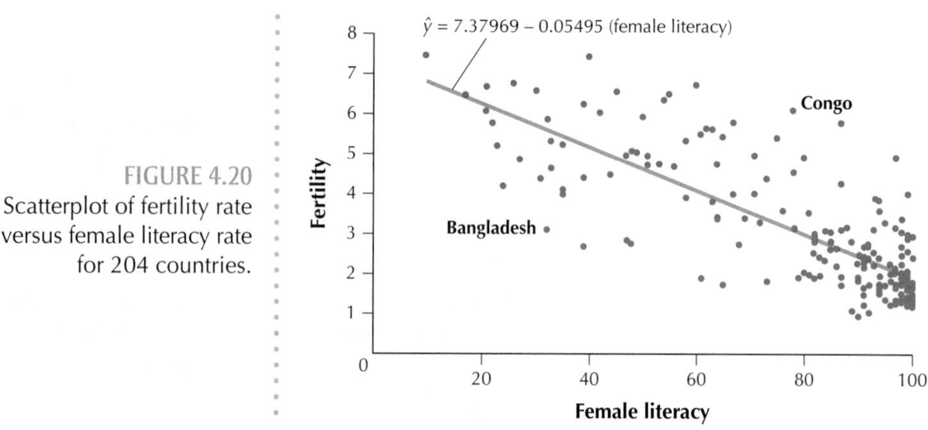


FIGURE 4.20 Scatterplot of fertility rate versus female literacy rate for 204 countries.

Based on the scatterplot in Figure 4.20, we can state that there is a *negative* relationship between female literacy and fertility. That is, as female literacy increases, fertility tends to decrease. Since the relationship is negative, the correlation coefficient $r$ must therefore be negative, $-1 < r < 0$. Excel provides the correlation coefficient in Figure 4.21, $r = -0.7962$.

As expected, the correlation coefficient is negative, so that female literacy and fertility are negatively correlated. An increase in female literacy is associated with a decrease in fertility. The $y$ intercept and slope of the regression line are shown in the Excel output in Figure 4.22, giving us the regression line

$$\hat{y} = 7.37969 - 0.05495(\text{female literacy})$$

The estimated fertility rate is 7.37969 children minus 0.05495 times the female literacy rate. We interpret the slope $b_1$ as follows: an increase of 1% in the female literacy rate is associated with a decrease of 0.05495 children in the fertility rate. Thus, for example, an increase of 20% in the female literacy rate is associated with a decrease in the estimated number of births per woman of $20(0.05495) = 1.099$ children. The $y$ intercept is interpreted as follows: when the female literacy rate is zero percent, the estimated fertility rate equals $b_0 = 7.37969$ children.

| | Female Literacy | Fertility |
|---|---|---|
| Female Literacy | 1 | |
| Fertility | -0.7962 | 1 |

FIGURE 4.21

| | Coefficients | |
|---|---|---|
| Intercept | 7.37969 | y intercept |
| Female Literacy | -0.05495 | slope |

FIGURE 4.22

For a country with a 50% literacy rate, the estimated number of births per woman is

$$\hat{y} = 7.37969 - 0.05495(50) \approx 4.6$$

Note the two countries singled out in Figure 4.20. The Republic of Congo has a relatively high female literacy rate, so one would expect the fertility rate to be low. But the fertility rate in the Republic of Congo is 6.45 births per woman, which is higher than expected. Bangladesh has a relatively low female literacy rate, so one would expect the fertility rate to be high. But the fertility rate in Bangladesh is 3.11 births per woman, which is lower than expected. Note that the data points for both Congo and Bangladesh lie far from the regression line. In general, unusual observations in a linear regression problem are points that lie at an unusually large vertical distance from the regression line. We will learn more about such points in Chapter 13.

## STEP-BY-STEP TECHNOLOGY GUIDE: Correlation and Regression

We illustrate using Example 4.5, the temperature data (page 178).

### TI-83/84

***Step 1*** Turn diagnostics on as follows. Press **2nd 0**. Scroll down and select **DiagnosticOn** (Figure 4.23). Press **ENTER** twice to turn diagnostics on.
***Step 2*** **Enter** the **X (Low Temp)** data in **L1**, and the **Y (High Temp)** data in **L2**.
***Step 3*** Press **STAT** and highlight **CALC**.
***Step 4*** Press **8** (not **4**) to choose **LinReg(a+bx)**, as shown in Figure 4.24.
***Step 5*** On the home screen, **LinReg(a+bx)** appears. Press **ENTER**.

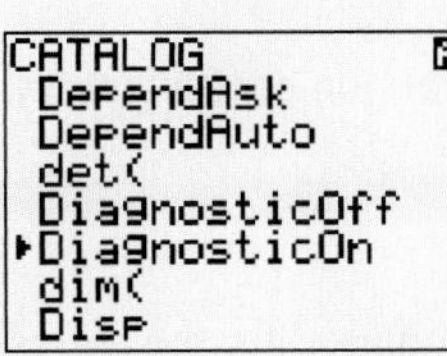


Figure 4.23

Figure 4.24

### EXCEL

***Step 1*** Enter the $x$ variable in column **A** and the $y$ variable in column **B**.
***Step 2*** Click on **Data > Data Analysis > Regression** and click **OK**.
***Step 3*** For **Input Y Range**, select cells **B1–B10**. For **Input X Range**, select cells **A1–A10**. Click **OK**.

### MINITAB

#### Correlation Coefficient *r*

***Step 1*** Select **Stat > Basic Statistics > Correlation…**
***Step 2*** Select the variables you want to analyze and click **OK**.

#### Regression

***Step 1*** Enter the $x$ variable variable in **C1** and the $y$ variable in **C2**.
***Step 2*** Click on **Stat > Regression > Regression**.
***Step 3*** Select the $y$ variable for the **Response Variable** and the $x$ variable for the **Predictor Variable**. Click **OK**.

## Section 4.3 SUMMARY

1 Section 4.3 introduces regression, where the linear relationship between two numerical variables is approximated using a straight line, called the regression line. The equation of the regression line is written as $\hat{y} = b_0 + b_1 x$ where the regression coefficients are the $y$ intercept, $b_0$, and the slope, $b_1$.

2 The regression equation can be used to make predictions about values of $y$ for particular values of $x$.

## Section 4.3 EXERCISES

### CLARIFYING THE CONCEPTS

1. What is the objective of regression analysis?
2. What is the regression equation?
3. Describe how we use the regression equation to make predictions.
4. Explain the difference between $y$ and $\hat{y}$.

5. Describe what is meant by extrapolation.
6. What is the relationship between the slope of the regression line and the correlation coefficient?

## PRACTICING THE TECHNIQUES

Refer to the following table for Exercises 7–9.

| Driver | $y$ = super speedway wins | $x$ = short tracks wins |
|---|---|---|
| Darrell Waltrip | 18 | 47 |
| Dale Earnhardt | 29 | 27 |
| Jeff Gordon | 15 | 15 |
| Cale Yarborough | 15 | 29 |
| Richard Petty | 19 | 23 |

7. Calculate the values for the regression coefficients $b_0$ and $b_1$, using the following steps.
   **a.** Compute the slope $b_1$.
   **b.** Calculate the $y$ intercept $b_0$.
   **c.** Write down the regression equation for the regression of $y$ = Super Speedways wins versus $x$ = Short Tracks wins. Express this equation in words that a nonspecialist would understand.

*Note:* When we say, "Perform a regression of Variable 1 *versus* Variable 2," Variable 1 is always the *y* variable, and Variable 2 is always the *x* variable. Similarly, if we say, "Regress Variable 1 *on* Variable 2, Variable 1 is the *y* variable and Variable 2 is the *x* variable.

8. Clearly interpret the meaning of the following:
   **a.** Slope $b_1$
   **b.** $y$ intercept $b_0$
9. If appropriate, estimate the number of Super Speedways wins for drivers with the following numbers of Short Tracks wins. If not appropriate, clearly explain why not.
   **a.** 30
   **b.** 50
   **c.** 20

Refer to the following table for Exercises 10–12.

| State | $x$ = Mean SAT I Verbal | $y$ = Mean SAT I Math |
|---|---|---|
| New York | 497 | 510 |
| Connecticut | 515 | 515 |
| Massachusetts | 518 | 523 |
| New Jersey | 501 | 514 |
| New Hampshire | 522 | 521 |

10. Calculate the values for the regression coefficients $b_0$ and $b_1$, using the following steps.
   **a.** Compute the slope $b_1$.
   **b.** Calculate the $y$ intercept $b_0$.
   **c.** Write down the regression equation for the regression of $y$ = SAT I Math versus $x$ = SAT I Verbal scores. Express this equation in words that a nonspecialist would understand.
11. Clearly interpret the meaning of the following:
   **a.** Slope $b_1$
   **b.** $y$ intercept $b_0$
   **c.** Comment on the usefulness of the literal interpretation of the $y$ intercept in this case.
12. If appropriate, estimate the SAT I Math scores for states with the following SAT I Verbal scores. If not appropriate, clearly explain why not.
   **a.** 500
   **b.** 510
   **c.** 490

## APPLYING THE CONCEPTS

**Brain and Body Weight.** A study compared the body weight (in kilograms) and brain weight (in grams) for a sample of mammals. We are interested in estimating brain weight ($y$) based on body weight ($x$).

| Body weight (kg) | Brain weight (g) |
|---|---|
| 52.16 | 440 |
| 60 | 81 |
| 27.66 | 115 |
| 85 | 325 |
| 36.33 | 119.5 |
| 100 | 157 |
| 35 | 56 |
| 62 | 1320 |
| 83 | 98.2 |
| 55.5 | 175 |

13. Calculate the values for the regression coefficients $b_0$ and $b_1$.
   **a.** Write down the regression equation for the regression of brain weight versus body weight.
   **b.** Express this equation in words that a nonspecialist would understand.
14. Clearly interpret the meaning of the following:
   **a.** Slope $b_1$
   **b.** $y$ intercept $b_0$
   **c.** Comment on the usefulness of the literal interpretation of the $y$ intercept in this case.
15. If appropriate, estimate the brain weight for mammals with the following body weights. If not appropriate, clearly explain why not.
   **a.** 50 kg
   **b.** 100 kg
   **c.** 200 kg

**Scrabble.** Use the information in the following table and in Figure 4.25 for Exercises 16–18. Scrabble® (**www.hasbro.com/scrabble/home.cfm**) is one of the most popular games in the world. We are interested in approximating the relationship between the frequency ($x$) of the letter tiles in the game and their point value ($y$). Figure 4.25 is a scatterplot of the point value versus the letter frequency, along with the regression line, as plotted by Minitab.

| Letter | Freq. | Points | Letter | Freq. | Points |
|---|---|---|---|---|---|
| A | 9 | 1 | N | 6 | 1 |
| B | 2 | 3 | O | 8 | 1 |
| C | 2 | 3 | P | 2 | 3 |
| D | 4 | 2 | Q | 1 | 10 |
| E | 12 | 1 | R | 6 | 1 |
| F | 2 | 4 | S | 4 | 1 |
| G | 3 | 2 | T | 6 | 1 |
| H | 2 | 4 | U | 4 | 1 |
| I | 9 | 1 | V | 2 | 4 |
| J | 1 | 8 | W | 2 | 4 |
| K | 1 | 5 | X | 1 | 8 |
| L | 4 | 1 | Y | 2 | 4 |
| M | 2 | 3 | Z | 1 | 10 |

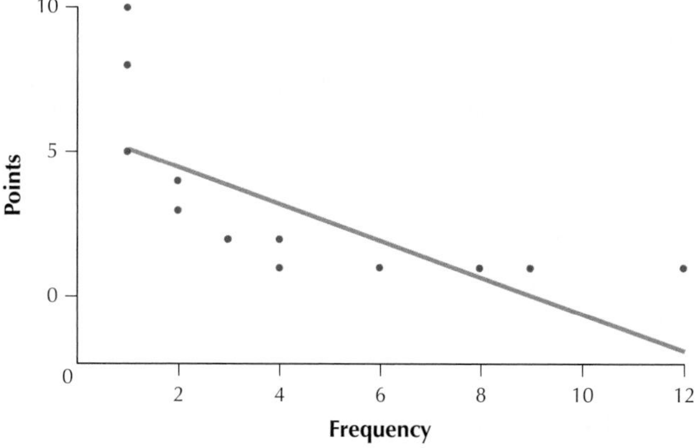


Figure 4.25 Scatterplot of Scrabble® point value versus letter frequency, with regression line.

**16.** Does the relationship between frequency and points seem to be positive or negative?

**17.** Does the relationship between frequency and points seem to be a straight-line relationship or a curved relationship?

**18.** Do you think that we should use a regression line to approximate the relationship between frequency and points? If not, explain clearly what you think is wrong.

**DC Households.** Use the following information for Exercises 19–21. The data set **Households,** located on your CD and companion Web site, contains information on the number and type of households in the fifty states and the District of Columbia. For each state, there are seven variables. Two of these variables are the percentage of households headed by women ($y = HHLD\_WOMEN$) and the total number of households in the state ($x = TOT\_HHLD$). Minitab provides the following regression equation:

**Regression Analysis**

```
The regression equation is
HHLD_Women = 10.5 + 2.82E-07 TOT_HHLD
```

*Note:* The notation 2.82E-07 refers to the scientific notation method of writing numbers. Often, software and calculators will present you with this type of notation, so you need to know how to read it. The number 2.82E-07 represents 2.82 times $10^{-7}$, or 0.000000282.

**19.** In this exercise, we explore the regression coefficients and the regression equation.

**a.** What is the value of the $y$ intercept?

**b.** Interpret the meaning of this value for the $y$ intercept. Does it make sense?

**c.** Would the estimate in **(b)** be considered extrapolation? Why or why not?

**d.** What is the value of the slope coefficient?

**e.** Interpret the meaning of the slope coefficient as the total number of households in the state increases.

**f.** Write the regression equation. Now state in words what the regression equation means.

**g.** Is the correlation coefficient positive or negative? How do you know?

**20.** Estimate the increase or decrease in the percentage of households headed by women, using a sentence, for the following situations.

**a.** Suppose State A has one million more households than State B.

**b.** Suppose State C has five million fewer households than State D.

**21.** The number of households per state ranges from about 170,000 to about ten million.

**a.** Estimate the percentage of households headed by women for a state with seven million households, if appropriate.

**b.** Estimate the percentage of households headed by women for a state with 100,000 households, if appropriate.

**Chapter 3 Case Study (Continued).** Use the following information for Exercises 22–25. Shown here is the regression equation for the linear relationship between the randomly selected Darts portfolio and the Dow Jones Industrial Average (DJIA), from the Chapter 3 Case Study.

**Regression Analysis**

```
The regression equation is
Darts = -2.49 + 1.032 DJIA
```

**22.** In this exercise, we examine the $y$ intercept.

**a.** Which variable is the $x$ variable and which is the $y$ variable? Work by analogy with the previous exercises.

**b.** What is the value of the $y$ intercept?

**c.** Interpret the meaning of this value for the $y$ intercept. Does it make sense?

**d.** Would the estimate in **(c)** be considered extrapolation? Why or why not?

**e.** Would this value for the $y$ intercept be used as evidence that the Darts are outperforming the market? Clearly explain why or why not, without reference to the discussion in Section 3.3.

**23.** In this exercise, we look at the slope and the regression equation.

**a.** What is the value of the slope coefficient?

**b.** Interpret the meaning of the slope coefficient, as the DJIA increases.

**c.** Write the regression equation. Now state in words what the regression equation means.
**d.** Is the correlation coefficient positive or negative? How do we know?

**24.** Estimate the increase or decrease in the net price of the Darts portfolio, using a sentence, for the following situations.
**a.** Suppose that for Contest A the DJIA increased by 10% more than for Contest B.
**b.** Suppose that for Contest C the DJIA decreased by 5% more than for Contest D.

**25.** The net change in the DJIA ranged from –13.1% to 22.5%.
**a.** Estimate the net price change for the Darts portfolio when the DJIA is up by 22%, if appropriate.
**b.** Estimate the net price change for the Darts portfolio when the DJIA is down by 10%, if appropriate.
**c.** Estimate the net price change for the Darts portfolio when the DJIA is down by 22%, if appropriate.

WHAT IF? **26.** Consider again the temperature data in Example 4.8. *What if* there was a typo, and all the low temperatures in the data set needed to be adjusted downward by the same amount? Explain how this change would affect the following, and why. Increase, decrease, or no change?
**a.** $\bar{x}$
**b.** $\bar{y}$
**c.** $y$ intercept $b_0$
**d.** Slope $b_1$
**e.** Correlation coefficient $r$

WHAT IF? **27. Challenger Exercise.** Consider again the temperature data in Example 4.8. *What if* we add a new city with a low temperature of 10ºF and a high temperature of at least 50ºF? Explain how this would affect the slope of the regression line.

**28.** Describe two variables from real life whose regression line would have a positive slope $b_1$.
**a.** Explain why the $y$ variable depends on the $x$ variable.
**b.** Explain why the slope is positive.

**29.** Create a sample of five observations from each of your variables from Exercise 28, and put them into a table similar to Table 4.11 in Section 4.2.
**a.** Construct a scatterplot of the variables.
**b.** Draw a single straight line through the data points in the plot in a manner that you think best approximates the relationship between the variables.
**c.** Using your regression line from **(b)**, estimate the slope $b_1$ and the $y$ intercept $b_0$.
**d.** Write your results from **(c)** in the form of a regression equation.

**30.** Describe two variables whose regression line would have a negative slope $b_1$. Explain why the slope is negative.

**31.** Describe two variables whose regression line would have a slope $b_1$ whose value was close to zero. Explain why the slope is close to zero.

APPLET Use the *Correlation and Regression* applet for Exercises 32 and 33.

**32.** Create a set of $n = 10$ points such that the slope of the regression line has the following characteristic. (Note that you can drag points up or down to adjust your regression line.)
**a.** The slope is positive.
**b.** The slope is negative.
**c.** The slope is neither positive nor negative.

**33.** Describe the relationship between the variables for each of the sets of points in the previous exercise.

## *Try This in Class!* In-class activities to enhance your understanding of statistics

### CORRELATION AND REGRESSION

For each student in the class, obtain the following information: the distance (in miles) his or her (off-campus) home is from the statistics classroom and the time (in minutes) it takes to commute that distance. If the class is large, take a random sample of the students in the class using APPLET either (a) technology, (b) the *Random Sample* applet, or (c) the method shown in the Chapter 1 Try This in Class! (page 29).

1. Discuss which variable is the predictor ($x$) variable, and which variable is the response ($y$) variable.
2. Construct a scatterplot of time versus distance.
3. Is there a positive relationship? Negative relationship? No apparent relationship?
4. Estimate the value of the correlation coefficient. Will it be positive or negative?
5. Calculate the correlation coefficient.
6. Interpret the correlation coefficient.
7. Next, consider the regression of time on distance. Will the slope be positive or negative?
8. Using the scatterplot, visually estimate the slope and $y$ intercept of the regression line.
9. Calculate the actual values of the slope and $y$ intercept.
10. Carefully interpret the meaning of the value for the slope, using words that someone who has not taken statistics would understand.
11. Interpret the meaning of the $y$ intercept.
12. For the person who lives farthest from campus, use the regression equation to estimate the time it takes him or her to get to class.
13. Calculate the prediction error for this student (the difference between the actual commuting time and the estimated commuting time).
14. Can you estimate commuting time for a new student who lives farther away than any other student in the class? If not, clearly explain why not.

# CHAPTER 4 FORMULAS AND VOCABULARY

## SECTION 4.1

- **Crosstabulation (two-way table, contingency table)** (page 164). A tabular summary of the relationship between two categorical variables. To construct a crosstabulation, put the categories of the first variable at the top of each column, and the categories of the second variable at the beginning of each row. And then, for each combination of categories in row and column, fill in the count of the observations that take those values. At the bottom of the table are the column totals, and along the right-hand side of the table are the row totals.
- **Clustered bar graph** (page 166). Graphical equivalent of a crosstabulation. To construct, identify which of the two categorical variables will define the cluster of bars. Then, for each category of the other variable, draw bars for each category of the first variable, the clustering variable.
- **Scatterplot** (page 167). Summarizes the relationship between two quantitative variables. One of the variables is measured along the horizontal axis and is called the *x variable*. The other variable is measured along the vertical axis and is called the *y variable*.

## SECTION 4.2

- **Correlation coefficient** ***r*** (page 177). Measures the strength of the linear relationship between two numerical variables. Definition formula:

$$r = \frac{\sum(x - \bar{x})(y - \bar{y})}{(n - 1)s_x s_y}$$

where $s_x$ and $s_y$ represent the sample standard deviation of the $x$ data values and the $y$ data values, respectively.

Computational formula:

$$r = \frac{\sum xy - \left(\sum x \sum y\right)/n}{\sqrt{\left[\sum x^2 - \left(\sum x\right)^2/n\right]\left[\sum y^2 - \left(\sum y\right)^2/n\right]}}$$

- **Positive and negative correlation** (page 179). Values of $r$ close to 1 indicate that the two variables are positively correlated. Values of $r$ close to –1 indicate that the two variables are negatively correlated.

## SECTION 4.3

- **Regression equation (regression line)** (page 187). $\hat{y} = b_0 + b_1x$, where $\hat{y}$ is the estimated value of $y$, $b_0$ is the $y$ intercept, and $b_1$ is the slope of the regression line.
- **Slope of the regression line** (page 187).

Definition formula:

$$b_1 = \frac{\sum(x - \bar{x})(y - \bar{y})}{\sum(x - \bar{x})^2}$$

Computational formula:

$$b_1 = \frac{\sum xy - \left(\sum x \sum y\right)/n}{\sum x^2 - \left(\sum x\right)^2/n}$$

- ***y*** **Intercept of the regression line** (page 187). $b_0 = \bar{y} - (b_1 \cdot \bar{x})$

# CHAPTER 4 REVIEW EXERCISES

## SECTION 4.1

**Happiness in Marriage.** The General Social Survey tracks trends in American society through annual surveys. Use the following contingency table for Exercises 1–5.

| | Happiness of Marriage | | | |
|---|---|---|---|---|
| **Respondents' gender** | **Very happy** | **Pretty happy** | **Not too happy** | **Total** |
| **Male** | 242 | 115 | 9 | 366 |
| **Female** | 257 | 149 | 17 | 423 |
| **Total** | 499 | 264 | 26 | 789 |

**1.** What proportion of the males responded that they were very happy in their marriage?
**2.** What proportion of the females responded that they were very happy in their marriage?
**3.** What proportion of the males responded that they were not too happy in their marriage?
**4.** What proportion of the females responded that they were not too happy in their marriage?
**5.** Construct a clustered bar graph of the data.

**Olympic Gold.** Table 4.17 presents data on the men's gold medal performance in the Olympics, from 1900 to 1984.

**TABLE 4.17 Gold medal performance (in inches)**

| Year | High jump | Discus | Long jump |
|---|---|---|---|
| 1900 | 74.8 | 1418.9 | 282.875 |
| 1904 | 71 | 1546.5 | 289 |
| 1908 | 75 | 1610 | 294.5 |
| 1912 | 76 | 1780 | 299.25 |
| 1920 | 76.25 | 1759.25 | 281.5 |
| 1924 | 78 | 1817.125 | 293.125 |
| 1928 | 76.375 | 1863 | 304.75 |
| 1932 | 77.625 | 1948.875 | 300.75 |
| 1936 | 79.9375 | 1987.375 | 317.3125 |
| 1948 | 78 | 2078 | 308 |
| 1952 | 80.32 | 2166.85 | 298 |
| 1956 | 83.25 | 2218.5 | 308.25 |
| 1960 | 85 | 2330 | 319.75 |
| 1964 | 85.75 | 2401.5 | 317.75 |
| 1968 | 88.25 | 2550.5 | 350.5 |
| 1972 | 87.75 | 2535 | 324.5 |
| 1976 | 88.5 | 2657.4 | 328.5 |
| 1980 | 92.75 | 2624 | 336.25 |
| 1984 | 92.5 | 2622 | 336.25 |

6. Construct the following scatterplots:
   a. High jump versus discus
   b. High jump versus long jump
   c. Discus versus long jump
7. For each of your scatterplots,
   a. characterize the relationship as positive, negative, or not apparently linear.
   b. write a sentence that describes the behavior of the $y$ variable as the $x$ variable increases.
   c. Write a sentence that indicates which values of $y$ that low values of $x$ are associated with, and which values of $y$ that high values of $x$ are associated with.

## SECTION 4.2

**Chapter 3 Case Study (Continued).** Use the following scatterplot of the Pros versus Darts from the Chapter 3 Case Study data for Exercises 8 and 9. Recall that the professional financial experts compared the performance of the stocks they chose with the stocks chosen at random by darts thrown at the financial pages of the *Wall Street Journal*.

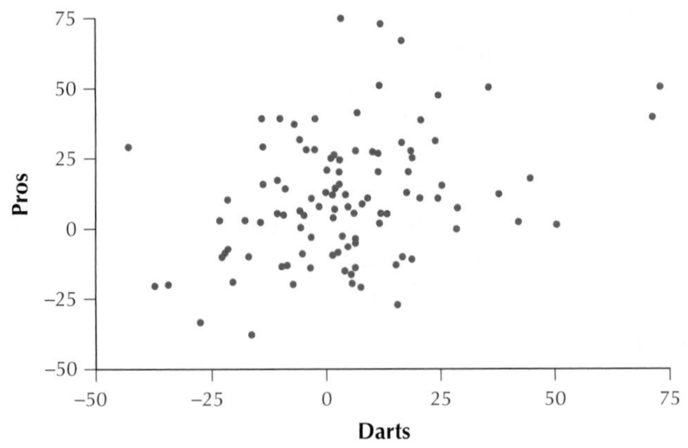


8. Would you characterize the relationship as positive, negative, or not apparently linear?
9. Based on your answer to the previous question, would the correlation coefficient be positive, negative, or near zero?

**Income and Hours Worked.** Use the following information for Exercises 11 and 12. Table 4.18 contains the weekly gross income ($y$) and the number of hours worked ($x$) for a random selection of 5 fast-food workers.

**TABLE 4.18 Gross income (in dollars) and hours worked for five fast-food workers**

| | Fast-Food Worker | | | | |
|---|---|---|---|---|---|
| | 1 | 2 | 3 | 4 | 5 |
| **Gross income: $y$** | 80 | 120 | 160 | 200 | 240 |
| **Hours worked: $x$** | 10 | 15 | 20 | 25 | 30 |

10. Construct a scatterplot of the variables.
11. What type of relationship do these variables have: positive, negative, or no apparent linear relationship?
12. Calculating and interpreting the correlation coefficient.
   a. Before calculating the value of the correlation coefficient, try to guess what it might be, based on your scatterplot.
   b. Compute the value of the correlation coefficient, using a table similar to Table 4.12. Check your answer using a computer or calculator.
   c. Does this value for $r$ agree with your earlier estimate?
   d. Interpret in two different ways the meaning of this value of the correlation coefficient.

## SECTION 4.3

**Income and Hours Worked.** Use the fast-food worker data from Table 4.18 for Exercises 13–16.

13. In your scatterplot from Exercise 10, draw a single straight line that you think best approximates the relationship between the variables.
   a. Estimate the gross income for a fast-food worker with 27 hours worked.
   b. Estimate the gross income for a fast-food worker with 24 hours worked.
   c. Comment on the usefulness of using this $x$ variable to predict this $y$ variable.
14. Using your scatterplot only, estimate the value of the slope coefficient $b_1$. If you have trouble, think about its interpretation as the estimated change in $y$ for every unit increase in $x$. Put this interpretation into the words of this exercise to see if you can figure out what $b_1$ is.
15. Using your scatterplot only, estimate the value of the $y$ intercept $b_0$. If you have trouble, think about its interpretation as the estimated value for $y$ when $x$ equals zero. Put this interpretation into the words of this exercise, to see if you can figure out what $b_0$ is.
16. Now calculate the equation of the regression line. Are the actual values of $b_0$ and $b_1$ close to your estimates above?

**TRUE OR FALSE**

**1.** True or false: A scatterplot is used to summarize the relationship between two quantitative variables.

**2.** True or false: Scatterplots are constructed with the *y* variable on the horizontal axis and the *x* variable on the vertical axis.

**3.** True or false: The *y* intercept measures the strength of the linear relationship between two numerical variables.

**FILL IN THE BLANK**

**4.** Other terms for a crosstabulation are __________ and __________.

**5.** The "hat" over the *y* in $\hat{y}$ indicates that it is an __________ of *y*.

**6.** The *y* intercept is the estimated value for *y* when *x* equals __________.

**7.** We interpret the slope of the regression line as the estimated change in *y* per __________ increase in *x*.

**SHORT ANSWER**

**8.** Which type of graph is often used in conjunction with crosstabulations?

**9.** Making predictions based on *x*-values that are beyond the range of the *x*-values in our data set is called what?

**10.** Values of *r* close to –1 indicate what type of relationship between the two variables?

**CALCULATIONS AND INTERPRETATIONS**

**Religious Preference and Excitement in Life.** Use the information in Table 4.19 for Exercises 11–15. The General Social Survey asked 967 respondents, "Do you find life exciting, routine, or dull?" along with other questions, including religious preference. Table 4.19 shows the interrelationship of these variables. Unfortunately, however, not all cell counts are provided.

**TABLE 4.19 Crosstabulation of religious preference with excitement in life**

| Religious preference | Is Life Exciting or Dull? | | | |
|---|---|---|---|---|
| | **Exciting** | **Routine** | **Dull** | **Total** |
| Protestant | 264 | | 33 | 623 |
| Catholic | 107 | 128 | 6 | |
| Jewish | 12 | 8 | | 20 |
| None | 38 | 29 | 2 | 69 |
| Other | 9 | 5 | | 14 |
| Total | | | 41 | 967 |

**11.** Fill in the missing entries in the crosstabulation.

**12.** Convert the table to a relative frequency crosstabulation so that the "Exciting," "Routine," and "Dull" proportions in each row add up to 1.0.

**13.** Which religious preference (including None and Other) has the largest proportion of those who find life dull?

**14.** Which religious preference (including None and Other) has the largest proportion of those who find life routine?

**15.** Which religious preference (including None and Other) has the largest proportion of those who find life exciting?

**Violent Crime.** Use the following information for Exercises 16–18. The Federal Bureau of Investigation publishes crime statistics, including those in the following table, which shows the percentage of violent crime committed per month nationwide for the years 2002 and 2004.[4]

| Month | 2002 | 2004 |
|---|---|---|
| January | 7.9 | 7.8 |
| February | 6.8 | 7.0 |
| March | 7.9 | 8.3 |
| April | 8.1 | 8.2 |
| May | 8.7 | 9.0 |
| June | 8.8 | 8.6 |
| July | 9.3 | 9.2 |
| August | 9.3 | 9.0 |
| September | 9.2 | 8.5 |
| October | 8.6 | 8.6 |
| November | 7.7 | 7.8 |
| December | 7.7 | 7.9 |

**16.** Construct a scatterplot of 2004 monthly crime versus 2002 monthly crime.

**17.** Based on your scatterplot, would you characterize the linear relationship, if any, as positive or negative?

**18.** How would you verbalize the relationship in a sentence?

**Equities and Investment Price Changes.** Use the information in Table 4.20 for Exercises 19–21. The table contains the net price change for equities (*y*) and the net price change for guaranteed income investments (*x*) for a random selection of 6 investors.

**TABLE 4.20 Net price change (in percent) for equities and guaranteed income investments for six investors**

| | Investor | | | | | |
|---|---|---|---|---|---|---|
| | **1** | **2** | **3** | **4** | **5** | **6** |
| **Equities: *y*** | 10 | −10 | −5 | 5 | 8 | −8 |
| **Guaranteed income: *x*** | 3 | 3 | 4 | 4 | 5 | 5 |

**19.** Investigating the correlation.

**a.** Construct a scatterplot of the variables.

**b.** What type of relationship do these variables have: positive, negative, or no apparent linear relationship?

**c.** Will the correlation coefficient be positive, negative, or near zero? Will the slope of the regression line be positive, negative, or near zero?

**20.** Calculating and interpreting the correlation coefficient.

**a.** Compute the value of the correlation coefficient, using a table similar to Table 4.12. Check your answer using a computer or calculator.

**b.** Does this value for $r$ concur with your earlier estimate?

**c.** Interpret in two different ways the meaning of this value of the correlation coefficient for this data set.

**21.** In your scatterplot, draw a single straight line that you think best approximates the relationship between the variables.

**a.** Estimate the equities net price change for an investor who has guaranteed income investment net price change of 3.5. (*Hint:* This estimate lies on your regression line.)

**b.** Estimate the equities net price change for an investor who has guaranteed income investment net price change of 4.5. Comment.

**c.** Comment on the usefulness of using this $x$ variable to predict this $y$ variable.

**Calories and Fat Content.** Use the following information for Exercises 22–24. Minitab calculated the regression equation, shown here, for the linear relationship between the number of calories ($y$) and the fat content in grams ($x$) in a sample of breakfast cereals.

```
Regression Analysis
The regression equation is
Calories = 97.1 + 9.65 Fat
```

**22.** The regression coefficients and the regression equation.

**a.** What is the value of the $y$ intercept?

**b.** Interpret the meaning of this value for the $y$ intercept. Does it make sense?

**c.** In view of the fact that there are many cereals in the data set with zero fat content, would the value for the $y$ intercept in **(b)** be considered extrapolation? Why or why not?

**d.** Interpret the meaning of the slope coefficient $b_1 = 9.65$ as fat content increases.

**e.** Write the regression equation. Now state in words what the regression equation means.

**f.** Is the correlation coefficient positive or negative? How do you know?

**23.** Estimate the increase or decrease in calories, using a sentence, for the following situations.

**a.** Suppose Cereal A has two more grams of fat than Cereal B.

**b.** Suppose Cereal C has three fewer grams of fat than Cereal D.

**24.** The maximum fat content in the data set is five grams.

**a.** Estimate the number of calories in a cereal with four grams of fat, if appropriate.

**b.** Estimate the number of calories in a cereal with ten grams of fat, if appropriate.

Chapter 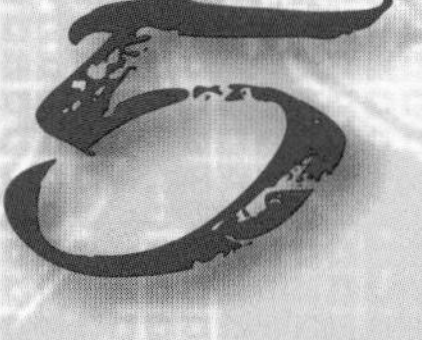

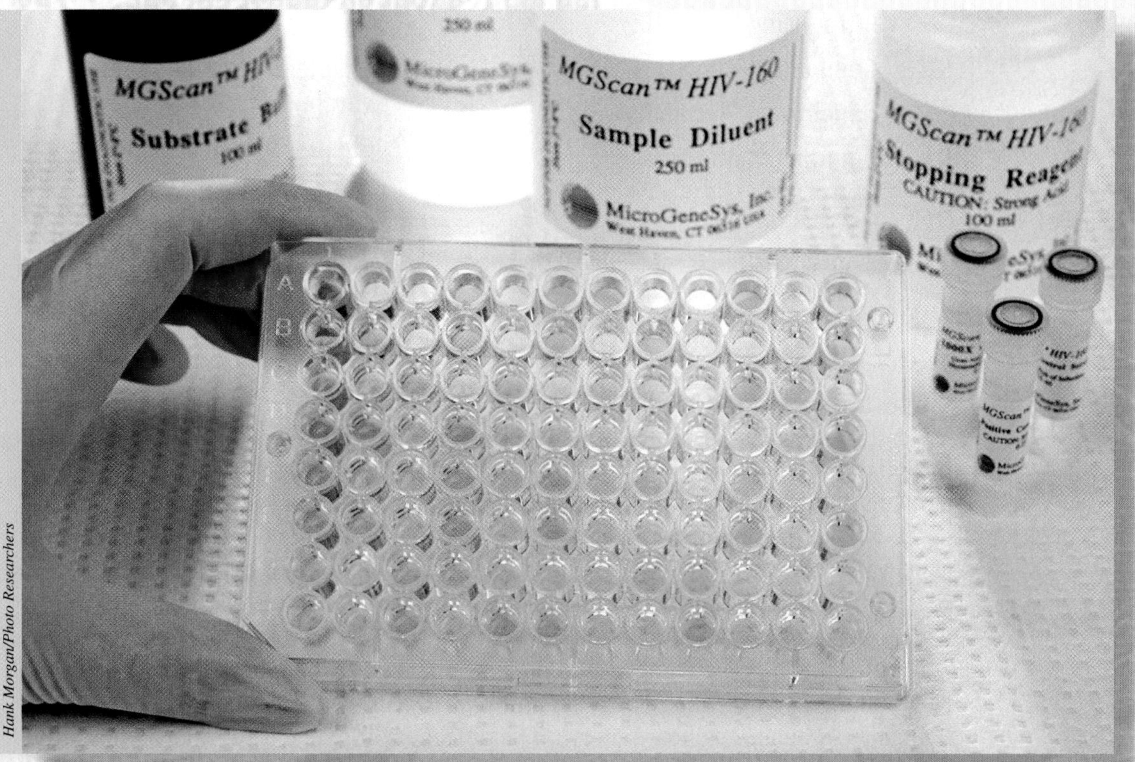

Hank Morgan/Photo Researchers

# Probability

## CASE STUDY — The ELISA Test for the Presence of HIV

If someone suspects that he or she is at increased risk of HIV infection, then he or she might be interested in going for an HIV ELISA test. The ELISA test is used to screen blood for the presence of HIV. Sometimes called an HIV enzyme immunoassay (EIA), an HIV ELISA is the most basic test for finding out if an individual is carrying a particular pathogen, such as HIV.

Like most diagnostic procedures, the ELISA test is not foolproof. In this chapter's Case Study we study the types of errors the ELISA test can make and what this means for those who carry the HIV virus and for those who do not. For example, did you know that if your ELISA test comes back positive, then the chances are eight out of ten that you do *not* carry the virus?

OVERVIEW



## The Big Picture

### Where we are coming from, and where we are headed...

The first part of this textbook dealt with descriptive statistics, which help us summarize the characteristics of data sets. In later chapters, we will learn to use the information contained in a sample to make generalizations about the characteristics of a population. However, whenever you proceed from a part of the whole (the sample) to the whole (the population), there is *uncertainty* involved in drawing conclusions.

In this chapter, we learn how to discuss the likelihood of events using the language of uncertainty: **probability**. The chapter explains the tools of probability, which enable data analysts to quantify the level of uncertainty in their statistical inference.

In Section 5.1, we learn how to analyze and solve probability problems like the following: You are striding down the midway of your local town fair, when a particular game of chance catches your eye. The object of this game is to roll a 6 on a single roll of a single fair die. If you do so, you win \$5. It costs \$1 to play the game. What is the likelihood of winning?

To solve this and other problems, we introduce the building blocks of probability, learn how to assign probabilities, and learn how to simulate probabilities using technology. In Section 5.2, we learn about combining events using the concepts of union, intersection, and complement. In Section 5.3, we examine conditional probability, independent events, and sampling with and without replacement. Finally, in Section 5.4 we learn about the counting methods used to find the number of combinations and permutations of a set of objects.

Later, in Chapter 6, we will acquire further tools for putting probability to work for us, such as how to work with probability distributions. The two most important probability distributions, the binomial and the normal, will be our companions for the remainder of the text. ■

# 5.1 Introducing Probability

**Objectives:** By the end of this section, I will be able to...

1. Understand the meaning of an experiment, an outcome, an event, and a sample space.
2. Describe the classical method of assigning probability.
3. Explain the Law of Large Numbers and the relative frequency method of assigning probability.

## 1 Building Blocks of Probability: Experiments, Outcomes, Events, and Sample Spaces

Our daily lives are filled with *uncertainty,* seemingly governed by *chance.* We try to cope with uncertainty by estimating the *chances* that a particular event will occur. We are daily called on to make intelligent decisions about probabilities.

Consider the following scenarios, and think about how the italicized words all refer to uncertainty.

- What is the *chance* that there will be a speed trap on this stretch of I-95 on a particular day?
- What is the *likelihood* that this lottery ticket will make me rich?
- What is the *probability* that this throw of the dice will come up a seven?

Sometimes, the amount of uncertainty in our daily lives is so great that there appears to be no order to the world whatsoever. However, if you look closely, there are *patterns in randomness,* predictable structures in the seeming chaos. In this chapter, we learn to become better decision makers by organizing and structuring our thought process. We learn the language and rules of probability so that we can better quantify the myriad uncertainties of everyday life.

**A Different Perspective**

As you read this chapter, notice that the perspective differs from that in previous chapters. Earlier, we were looking at a data set and trying to describe it graphically and numerically. Now, instead of trying to describe a data set, we are faced with an experimental situation, and our task is to calculate probabilities associated with various outcomes in the experiment. ■

The **probability** of an outcome is defined as the long-term proportion of times the outcome occurs.

We discuss what we mean by "long-term proportion" later in this section. But first we acquaint ourselves with the building blocks of probability, starting with the concept of an experiment. In probability, an **experiment** is any activity for which the outcome is uncertain. Consider the stock market, for example. Suppose you own 100 shares of Consolidated Widgets and are interested in what the share price will be at the end of trading tomorrow. Will the share price increase or decrease? The actual result is uncertain, so this is an example of an experiment. Each of the possible results of the experiment is called an **outcome**. Another example of an experiment is when you toss a coin. In the coin toss experiment, the result may be heads or it may be tails. The collection of all possible outcomes is called the **sample space**. The sample space for the coin toss experiment is {heads, tails} or {H, T} (where the braces are used to enclose a set of outcomes). Following are some common experiments, together with their sample spaces.

| Experiment | Sample space |
|---|---|
| Roll a single six-sided die | {1, 2, 3, 4, 5, 6} |
| Toss two coins | {HH, HT, TH, TT} |
| Play a video game | {win, lose} |

We use the building blocks of probability to investigate the likelihood of an outcome or **event**.

**Building Blocks of Probability**

An **experiment** is any activity for which the outcome is uncertain.

An **outcome** is the result of a single performance of an experiment.

The collection of all possible outcomes is called the **sample space**. We denote the sample space *S*.

An **event** is a collection of outcomes from the sample space. To find the probability of an event, add up the probabilities of all the outcomes in the event.

When we talk about the probability of some outcome, we are referring to a number that indicates how likely the particular outcome is. The notation $P(A)$ stands for "the probability that outcome $A$ occurred." Say we define outcome $W$ to be $W$ = "you win the video game." Then "the probability that you win the video game" can be denoted as $P(W)$. Probabilities abide by the following rules.

**Rules of Probability**

1. The probability $P(E)$ for any event $E$ is always between 0 and 1, inclusive. That is, $0 \leq P(E) \leq 1$.
2. **Law of Total Probability:** For any experiment, the sum of all the outcome probabilities in the sample space must equal 1.

From the definition, the probability of an event is a proportion, so the probability cannot be negative because proportions cannot be negative and it cannot be greater than 1 (100%) because an event cannot occur more than 100% of the time. A **probability model** is a table or listing of all the possible outcomes of an experiment, together with the probability of each outcome. A probability model must follow the Rules of Probability.

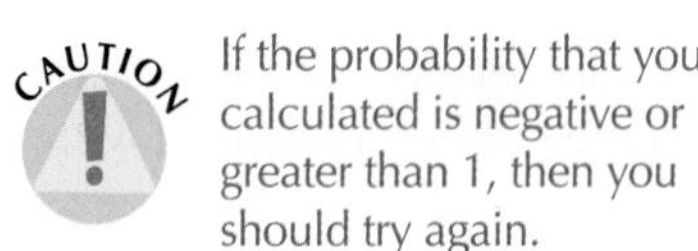

CAUTION
If the probability that you calculated is negative or greater than 1, then you should try again.


Throughout the remainder of this book, you will often be asked to calculate the probability of various events. Following are the meanings of some probabilities.

| Probability value | Meaning |
|---|---|
| Near 0 | Outcome or event is very unlikely. |
| Equal to 0 | Outcome or event cannot occur. |
| Near 1 | Outcome or event is nearly certain to occur. |
| Equal to 1 | Outcome or event is certain to occur. It's "a sure thing." |
| Low | Outcome or event is unusual. |
| High | Outcome or event is not unusual. |

Higher probability values are associated with higher likelihood of occurrence. An outcome with probability 0.5 will happen about half of the time. An outcome with probability 0.95 is very likely. We say that an outcome or event is *unusual* if its probability is below a certain threshold, say, 0.05. (What determines this threshold depends on the specific experiment; the 0.05 is not set in stone.) When we perform an experiment, it is a "sure thing" that one of the outcomes in the sample space will occur. For example, when you toss a coin, you know that it will be either heads or tails. Put into probability terms, the sum of the probabilities of all the individual outcomes must equal 1, the **Law of Total Probability**.

The following table shows some typical events for the experiments in the probability value table.

| Experiment | Sample space | Typical events |
|---|---|---|
| Roll a single die | {1, 2, 3, 4, 5, 6} | E: roll an even number = {2, 4, 6}<br>L: roll a 4 or larger = {4, 5, 6} |
| Toss two coins | {HH, HT, TH, TT} | H: exactly one head = {HT, TH}<br>T: at most one tail = {HH, HT, TH} |
| Play a video game | {win, lose} | W: win = {win}<br>L: lose = {lose} |

## 2 Classical Method of Assigning Probability

Many people have a certain degree of intuition when it comes to assigning probabilities. For example, when asked what the chances are of rolling a 6 on a single toss of a fair die, many people would quite correctly answer 1/6. However, intuition can often let us down. For example, when asked what the chances are of observing two heads when you toss a fair coin twice, many people would incorrectly respond 1/3 ("Well, it's either both heads or both tails or one of each." The correct answer is in fact 1/4.) In this section, we learn how to quantify our methods of assigning probabilities so that we don't have to depend on intuition alone.

There are three methods for assigning probabilities:

- Classical method
- Relative frequency method
- Subjective method

We first take a close look at the classical method. Later in this section we will examine the relative frequency method and the subjective method.

Many experiments are structured so that each experimental outcome is equally likely. **Equally likely outcomes** are outcomes that have the same probability of occurring. For example, if you toss a fair coin, the probability of observing either of the outcomes heads or tails is the same. The **classical method of assigning probabilities** is used when an experiment has equally likely outcomes.

**Classical Method of Assigning Probabilities**

Let $N(E)$ and $N(S)$ denote the number of outcomes in event $E$ and the sample space $S$, respectively. If the experiment has equally likely outcomes, then the probability of event $E$ is then

$$P(E) = \frac{\text{number of outcomes in } E}{\text{number of outcomes in sample space}} = \frac{N(E)}{N(S)}$$

### Example 5.1 Probability of drawing an ace

Find the probability of drawing an ace when drawing a single card at random from a deck of cards.

**Solution**

The sample space for the experiment where a subject chooses a single card at random from a deck of cards is given in Figure 5.1. If the card is chosen truly at random, then it is

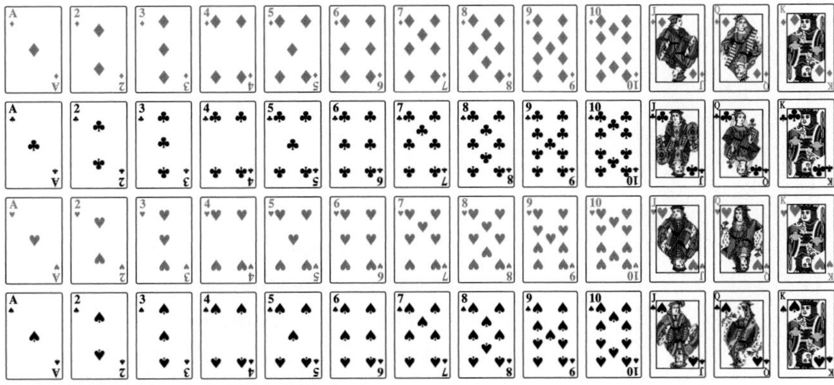

FIGURE 5.1 Sample space for drawing a card at random from a deck of cards.

reasonable to assume that each card has the same chance of being drawn. Since each card is equally likely to be drawn, we can use the classical method to assign probabilities.

There are 52 outcomes in this sample space, so $N(S) = 52$. Let $E$ be the event that an ace is drawn. Event $E$ consists of the four aces {A♥, A♦, A♣, A♠}, so $N(E) = 4$. Therefore, the probability of drawing an ace is

$$P(E) = \frac{N(E)}{N(S)} = \frac{4}{52} = \frac{1}{13}$$

## Tree Diagrams

Often we need to count the number of outcomes of an experiment to solve a probability problem. A **tree diagram** is a device we can use to count the outcomes of an experiment like tossing a coin twice.

Think of this experiment as a two-stage process:

- Stage 1: Toss the coin the first time.
- Stage 2: Toss the coin the second time.

A tree diagram is a graphical display that helps us to visualize a multistage experiment. Figure 5.2 shows the tree diagram for the experiment of tossing a fair coin twice. Note the branches for Stage 1: the first time the coin is tossed, it can come up heads or tails. At Stage 2, the tree diagram again has branches for either heads or tails. The tree diagram helps us to construct the sample space for a multistage experiment. In this case, the sample space for the experiment of tossing a coin twice is {HH, HT, TH, TT}. There are $N(S) = 4$ outcomes in the sample space.

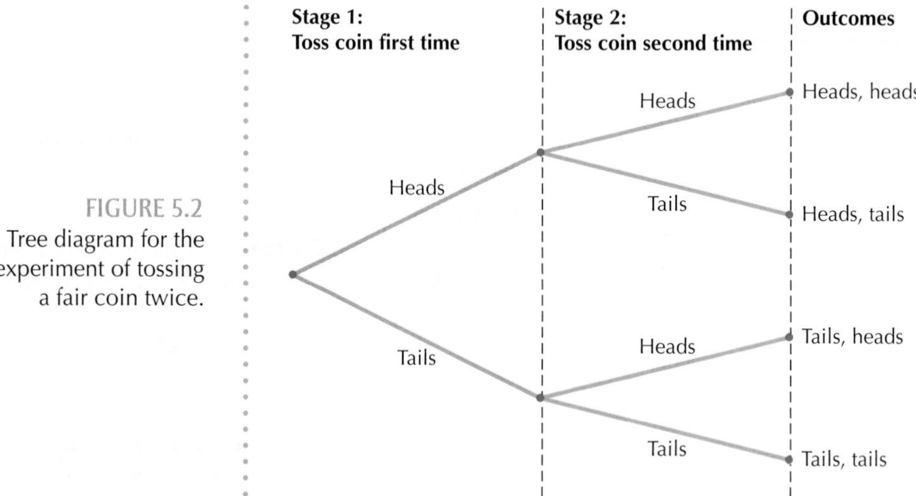


FIGURE 5.2 Tree diagram for the experiment of tossing a fair coin twice.

Note that there are two possible outcomes at Stage 1 of this two-stage experiment and two possible outcomes when flipping the coin at Stage 2. To determine how many outcomes there are in the entire experiment, the *counting rule* is simply to multiply the number of possible outcomes at each stage. In this two-stage experiment, 2 times 2 equals 4 possible outcomes, which is the number of outcomes we see in the sample space.

## Example 5.2 Tossing a coin twice

Find the probability of obtaining one heads and one tails when a fair coin is tossed twice.

**Solution**

It is reasonable to assume that the $N(S) = 4$ outcomes in the sample space {HH, HT, TH, TT} are equally likely. The coin doesn't remember what occurred at Stage 1, so the probabilities at Stage 2 are precisely the same as at Stage 1. Also, recall from the Law of Total Probability that the sum of the probabilities of all the outcomes in the sample space must equal 1. Thus, each of the four outcomes must have probability 1/4. Let $E$ be the event that one heads and one tails is obtained. Then $E = \{\text{HT, TH}\}$, so $N(E) = 2$. Thus,

$$P(E) = \frac{\text{number of outcomes in } E}{\text{number of outcomes in sample space}} = \frac{N(E)}{N(S)} = \frac{2}{4} = \frac{1}{2}$$

## Example 5.3 Probabilities when tossing two dice

Punchstock/CutandDeal

Imagine that you are playing Monopoly with your dormitory roommate, and the loser has to do the laundry for both of you for the rest of the semester. You have a hotel on Boardwalk, and if your roommate lands on it, you will surely win. Right now your roommate's piece is on Short Line: if he or she rolls a 4, you get your laundry done gratis for the remainder of the semester! Put into statistical terms, the experiment is to toss two fair dice and observe the sum of the two dice. Find the probability of rolling a sum of 4 when tossing two fair dice.

**Solution**

It is reasonable to assume that each of these $N(S) = 36$ outcomes in the sample space (Figure 5.3) is equally likely. If you wish, the experiment of tossing two dice can be viewed as a two-stage experiment, where we add the result from the first die to the result from the second die. If a 5 appears on the first (say, dark green) die, and a 3 appears on the second (light green) die, the overall outcome is (5,3), with the resulting sum equal to 8. Note that the outcome (5,3) is not the same as the outcome (3,5), where the dark green die comes up 3 and the light green die comes up 5.

*Note:* Did you know? People have been tossing dice for a long time. Archaeologists have dug up dice from Roman ruins looking just the same as ours. These three dice were uncovered from the ruins of Pompeii buried by the eruption of Mount Vesuvius in the first century A.D.

Réunion des Musées Nationaux/Art Resource, NY

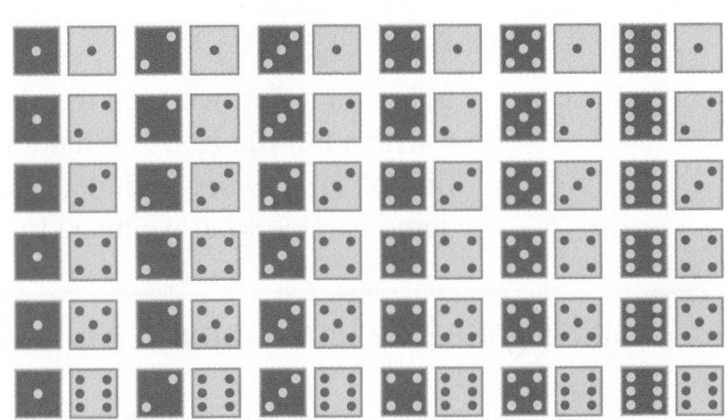

FIGURE 5.3 Sample space for tossing two fair dice.

Let $E$ denote the event that your roommate rolls a sum equal to 4. Then the outcomes that belong in this event are $E$: {(3,1) (2,2) (1,3)}, so $N(E) = 3$. Since the outcomes are equally likely, we can use the classical method for finding probabilities of events.

$$P(E) = \frac{\text{number of outcomes in } E}{\text{number of outcomes in sample space}} = \frac{N(E)}{N(S)} = \frac{3}{36} = \frac{1}{12}$$

The probability that your roommate will land on Boardwalk on this throw of the dice is 1/12.

## Example 5.4 Inappropriate use of classical method: teenagers who own a personal computer

A recent study[1] showed that 59% of teenagers owned a computer (either a desktop or a laptop). Suppose we choose one teenager at random. Define the following events:

$C$: The randomly chosen teenager owns a computer.
$D$: The randomly chosen teenager does not own a computer.

Determine whether the classical method can be used to assign probability to events $C$ and $D$.

**Solution**

Because a majority of teenagers own a computer, if we choose a teenager at random, we are more likely to select a teenager who owns a computer than to select one who does not. Therefore, the events $C$ and $D$ are not equally likely. It would be inappropriate to use the classical method of assigning probabilities for this experiment because the classical method can be used only when all the outcomes of an experiment are equally likely. ■

## Example 5.5 Fair die toss outcomes are equally likely

Recall the town fair example from The Big Picture at the beginning of this chapter (page 202). In the game, you win if you roll a 6 on a single roll of a single fair die. Find the probability of winning the game.

**Solution**

The sample space for a single die toss consists of *six* outcomes, $\{1, 2, 3, 4, 5, 6\}$. When the six outcomes are equally likely, we say that the die is *fair*. If the outcomes are not equally likely, then the die is *loaded* or defective. If we assume the die is fair, then, since the sum of the probabilities of the $n = 6$ outcomes must equal 1, the probability of any particular outcome must equal 1/6, using the classical method. We write

$$\text{probability of winning} = P(W) = 1/6$$ ■

## 3 Law of Large Numbers and Relative Frequency Method of Assigning Probability

In Example 5.5, we need the classical method to find that the probability of rolling a 6 with a fair die is 1/6. What does this probability mean? Remember that the definition of *probability* included the phrase "long-term proportion." The next example demonstrates what we mean by "long-term."

## Example 5.6 Simulating the long-term proportion of 6's in a fair die roll

Suppose we would like to investigate the proportion of 6's we observe if we roll a fair die 100 times. We can use technology, such as the TI-83/84 used here, to help us simulate rolling a fair die a large number of times. A **simulation** uses methods such as rolling dice or computer generation of random numbers to generate results from an experiment. The actual die rolls from our simulation are shown here, in order, with the 6's in boldface.

144**6**2432134334 33**6**3551535521311155**6**3**6**21**6**5544**6**54114**6**
4222**6**3255**6**1131**6**54**66**5552553424**6**4551**6**311135423 33**6**253

Thus, the first die roll was a 1, so the proportion of 6's was 0/1. The second and third die rolls were 4's, so the proportion of 6's after 3 rolls was 0/3. On the fourth roll a

6 appeared, so the proportion of 6's after the fourth roll was 1/4. Figure 5.4 provides a graph of the proportion of 6's in this simulation as the number of die rolls increased. Note that as the number of die rolls increases, the proportion of 6's tends to get closer to the horizontal line, $0.1667 \approx 1/6$.

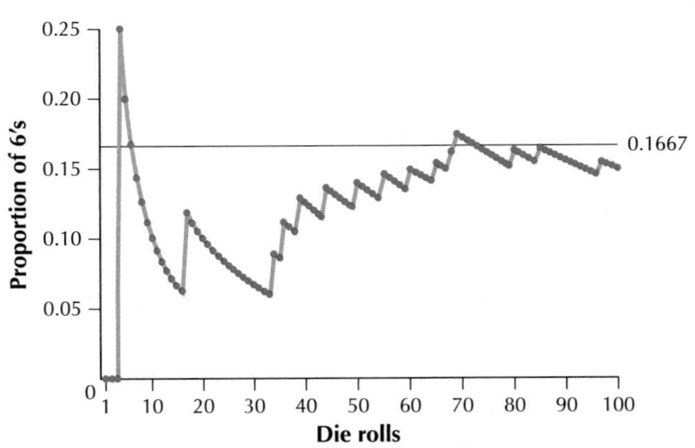


FIGURE 5.4 Proportion of 6's, 100 die rolls.

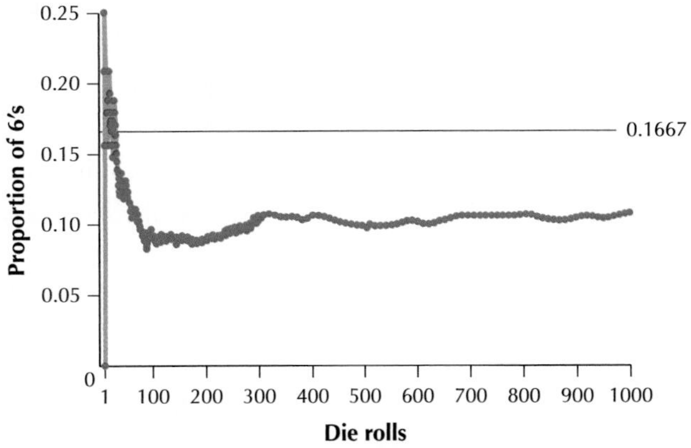

FIGURE 5.5 Proportion of 6's, 1000 die rolls.

The simulation was rerun, this time with 1000 die rolls. The resulting graph of the proportion of 6's is provided in Figure 5.5. Note that as the number of die rolls increases, the proportion of 6's approaches the line $0.1667 \approx 1/6$, and the fit is tighter with 1000 die rolls than with 100. This is what we mean by "long-term proportion."

This example leads directly to the following law.

**Law of Large Numbers**

As the number of times that an experiment is repeated increases, the relative frequency (proportion) of a particular outcome tends to approach the *probability* of the outcome. For quantitative data, as the number of times that an experiment is repeated increases, the mean of the outcomes tends to approach the population mean. For categorical (qualitative) data, as the number of times that an experiment is repeated increases, the proportion of times a particular outcome occurs tends to approach the population proportion.

The *Law of Large Numbers for Proportions* applet allows you to simulate coin tossing and observe the proportion of heads as the number of tosses increases.

## Relative Frequency Method

*Note:* Tree diagrams can be used for the relative frequency method as well as the classical method of assigning probability.

If we can't use the classical method for assigning probabilities, then the **Law of Large Numbers** gives us a hint about how we can estimate the probability of an event. It often happens that previous information is available about the relative frequency of an event. Relative frequency information can be used to estimate the probability of the event.

**Relative Frequency Method of Assigning Probabilities**

The probability of event $E$ is approximately equal to the relative frequency of event $E$. That is,

$$P(E) \approx \text{relative frequency of } E = \frac{\text{frequency of } E}{\text{number of trials of experiment}}$$

The relative frequency method is also known as the **empirical method**.

## Example 5.7 Teen bloggers

A recent study found that 35% of all online teen girls are bloggers, compared to 20% of online teen boys.[1] Suppose that the 35% came from a random sample of 100 teen girls who use the Internet, 35 of whom are bloggers. If we choose one teen girl at random, find the probability that she is a blogger.

**Solution**

Define the event.

*B*: The online girl is a blogger.

We use the relative frequency method to find the probability of event *B*:

$$P(B) \approx \text{relative frequency of } B = \frac{\text{frequency of } B}{\text{number of trials in experiment}} = \frac{35}{100} = 0.35$$

We can also use the relative frequency method to build a probability model with data that have been summarized in a table.

## Example 5.8 Probability models based on tabular data

Table 5.1 contains the employment type for a sample of 1000 employed citizens of Fairfax County, Virginia.[2] Use the data to generate the relative frequencies and use the frequencies to estimate the probabilities for each employment type.

Table 5.1 Employment types

| Employment type | Count |
|---|---|
| Private company | 597 |
| Federal government | 141 |
| Self-employed | 97 |
| Private nonprofit | 92 |
| Local government | 59 |
| State government | 12 |
| Other | 2 |

**Solution**

We calculate the relative frequencies of each employment group by dividing the count (frequency) for each group by the sample size 1000. For example, the relative frequency for "Private Company" is $\frac{597}{1000} = 0.597$. The relative frequency is then used to estimate the probability of selecting citizens who work at private companies in Fairfax County, Virginia. Filling in the remaining calculations produces the *probability model* in Table 5.2. Note that the table follows the Rules of Probability in that (a) each outcome has probability between 0 and 1 and (b) the sum of the probabilities of all the outcomes equals 1.0.

Table 5.2 Probability model

| Employment type | Probability |
|---|---|
| Private company | 0.597 |
| Federal government | 0.141 |
| Self-employed | 0.097 |
| Private nonprofit | 0.092 |
| Local government | 0.059 |
| State government | 0.012 |
| Other | 0.002 |

## Example 5.9 Random draws using a probability model

Suppose we consider the probabilities in Table 5.2 as population values. Use technology to simulate random draws using the probability model in Table 5.2.

**Solution**

Using the Step-by-Step Technology Guide below, we draw samples of sizes 10, 100, 1000, and 10,000 from the probability model in Table 5.2. The results are shown in Table 5.3.

Table 5.3 Relative frequencies from random draws of different sizes

| **Employment type** | **Rel freq** $n = 10$ | **Rel freq** $n = 100$ | **Rel freq** $n = 1000$ | **Rel freq** $n = 10{,}000$ |
|---|---|---|---|---|
| Private company | 0.60 | 0.62 | 0.566 | 0.596 |
| Federal government | 0.20 | 0.15 | 0.15 | 0.143 |
| Self-employed | 0.10 | 0.11 | 0.109 | 0.991 |
| Private nonprofit | 0.10 | 0.07 | 0.106 | 0.914 |
| Local government | 0.00 | 0.04 | 0.055 | 0.056 |
| State government | 0.00 | 0.01 | 0.012 | 0.012 |
| Other | 0.00 | 0.00 | 0.002 | 0.002 |

Note that each relative frequency tends to approach its respective probability as the sample sizes grow larger. ■

### Subjective Method

There are cases where the outcomes are not equally likely (so classical method does not apply) and there has been no previous research (so the relative frequency approach does not apply). For example, what is the probability that the Dow Jones Industrial Average will decrease today? In cases like this, there is no absolutely correct probability. Reasonable people can disagree reasonably over these probabilities. The idea is to consider all available information, tempered by our experience and intuition, and then assign a probability value that expresses our estimate of the likelihood that the outcome will occur. For example, we might say, "The Chairman of the Federal Reserve warned against inflation in a major speech yesterday, so we expect that the probability that the Dow Jones Industrial Average will go down today is about 90%." Finally, it should be noted that the subjective method should be used when the event is not (even theoretically) repeatable.

> **Subjective probability** refers to the assignment of a probability value to an outcome based on personal judgment.

## STEP-BY-STEP TECHNOLOGY GUIDE: Probability Simulations Using Technology

### TI-83/84

**Simulating 100 Die Rolls**

***Step 1*** Set the random number seed as follows. (The random number seed is a number that the calculator uses to generate random numbers.) Enter any number on the home screen. Press **STO→**, then **MATH**, highlight **PRB**, select **1: rand**, and press **ENTER**. On the home screen press **ENTER**.

***Step 2*** Press **MATH**, highlight **PRB**, select **5: randInt(**, and press **ENTER**.

*(Continued)*

***Step 3*** Enter **1**, comma, **6**, comma, **100**, close parenthesis (Figure 5.6).
***Step 4*** Store the data in list **L1** as follows. Press **STO→**, then **2nd**, then **1**, then press **ENTER**.
***Step 5*** To examine the die rolls, press **STAT**, select **1: EDIT**, and press **ENTER** (Figure 5.7).

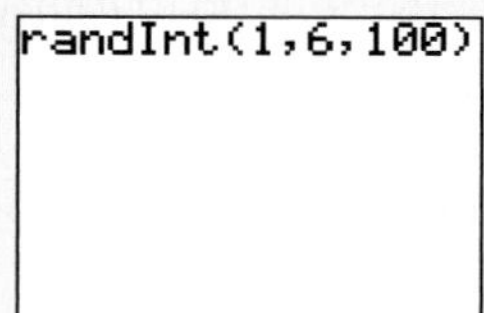

Figure 5.6

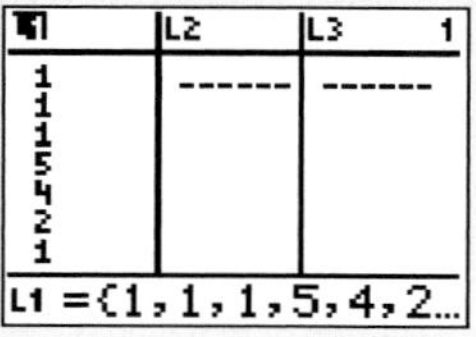

Figure 5.7

## Simulating Coin Flips

You can simulate coin flips instead of die rolls by coding "heads" as 1 and "tails" as 0. Use the instructions for simulating 100 die rolls with the following changes: Enter **0**, comma, **1**, comma, **100**, close parenthesis, so that the home screen shows **randInt(0, 1, 100)**.

# EXCEL

## Simulating 100 Die Rolls

***Step 1*** Select cell **A1**. Click the **Insert Function** icon $f_x$.
***Step 2*** For **Search for a Function**, type **randbetween** and click **OK**.
***Step 3*** For **Bottom**, enter **1**. For **Top**, enter **6** (Figure 5.8). Click **OK**. Cell **A1** now contains a simulated random die roll.
***Step 4*** Select cell **A1**, copy it, and paste the contents into cells **A2** through **A100**.

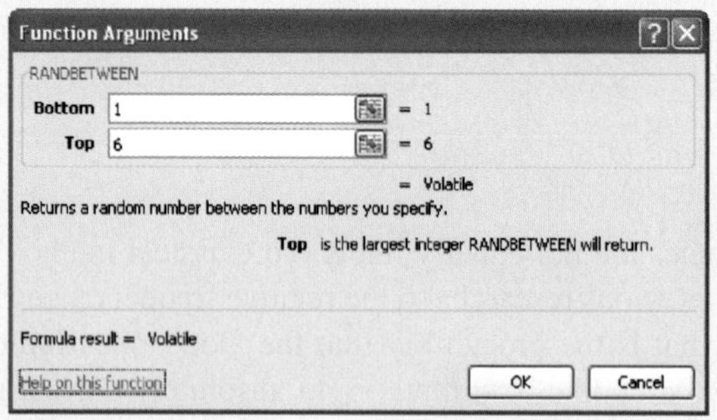

Figure 5.8

## Simulating the Sum of Two Dice

***Step 1*** Generate 100 die rolls in column A and another 100 die rolls in column B.
***Step 2*** Select cell **C1**. Enter **=(A1+B1)**, and press **ENTER**.
***Step 3*** Select cell **C1**, copy it, and paste the contents into cells **C2** through **C100**. Column C then represents 100 randomly generated sums of two dice.

## Simulating Random Draws from a Probability Table

We illustrate using Example 5.9 (page 211). Excel and Minitab both require that the categories in the probability model be coded as numeric. We therefore code "Private company" as 1, "Federal government" as 2, and so on.
***Step 1*** Type the model categories (for example, "Employment type") in column A, their numeric codes in column B, and the respective probabilities in column C.
***Step 2*** Click **Data > Data Analysis > Random Number Generation**, then **OK**.
***Step 3*** For **Number of Variables**, enter **1**.
***Step 4*** For **Number of Random Numbers**, enter the desired sample size.
***Step 5*** For **Distribution**, select **Discrete**.
***Step 6*** For **Value & Prob. Input Range**, click and drag to select the coded categories and their probabilities, for example, **B1:C7**.

Repeat Steps 1–6 for increasing sample sizes.

## Simulating Coin Flips Using Technology

You can simulate coin flips instead of die rolls by coding "heads" as 1 and "tails" as 0. Use the die roll instructions with the following changes: For **Bottom**, enter **0**. For **Top**, enter **1**.

# MINITAB

## Simulating 100 Die Rolls

***Step 1*** Click on **Calc > Random Data > Integer**.
***Step 2*** For **Generate ___ rows of data**, enter **100**.
***Step 3*** For **Store in column(s)**, select **C1**.
***Step 4*** For **Minimum value**, enter **1**. For **Maximum value**, enter **6**.
***Step 5*** Click **OK**.

## Simulating the Sum of Two Dice

***Step 1*** Generate 100 die rolls in **C1** and another 100 die rolls in **C2**.
***Step 2*** Click **Calc > Calculator**. For **Store result in variable**, enter **C3**. For **Expression**, enter **C1 + C2**. Click **OK**. Column **C3** then represents 100 randomly generated sums of two dice.

## Simulating Random Draws from a Probability Table

***Step 1*** Type the model categories in **C1**, their numeric codes in **C2**, and the respective probabilities in **C3** (Figure 5.9).

| ↓ | C1-T | C2 | C3 | C4 | C5 | C6 | C7 |
|---|---|---|---|---|---|---|---|
| | Type | Type-N | Prob | 10 | 100 | 1000 | 10000 |
| 1 | Private Company | 1 | 0.597 | 1 | 3 | 1 | 1 |
| 2 | Federal Government | 2 | 0.141 | 1 | 5 | 1 | 4 |
| 3 | Self-Employed | 3 | 0.097 | 1 | 1 | 1 | 2 |
| 4 | Private Nonprofit | 4 | 0.092 | 1 | 1 | 4 | 4 |
| 5 | Local Government | 5 | 0.059 | 1 | 1 | 1 | 1 |
| 6 | State Government | 6 | 0.012 | 3 | 1 | 2 | 3 |
| 7 | Other | 7 | 0.002 | 4 | 1 | 1 | 2 |
| 8 | | | | 2 | 1 | 1 | 1 |
| 9 | | | | 2 | 1 | 3 | 1 |
| 10 | | | | 1 | 4 | 1 | 6 |
| 11 | | | | | 1 | 1 | 5 |
| 12 | | | | | 1 | 3 | 1 |
| 13 | | | | | 4 | 1 | 3 |

Figure 5.9

*Step 2* Click on **Calc > Random Data > Discrete**.
*Step 3* For **Generate ___ rows of data**, enter the desired sample size.
*Step 4* For **Store in column(s)**, select the next available column, such as **C4**.
*Step 5* For **Values in**, enter the column with the numerically coded categories, such as **C2**.
*Step 6* For **Probabilities in**, enter the column with the probabilities, such as **C3**.
*Step 7* Click **OK**.

Repeat Steps 1–7 for increasing sample sizes, as shown in Figure 5.9.

### Simulating Coin Flips

You can simulate coin flips instead of die rolls by coding "heads" as 1 and "tails" as 0. Use the die roll instructions with the following changes: For **Minimum value**, enter **0**. For **Maximum value**, enter **1**.

## Section 5.1 SUMMARY

1 Section 5.1 introduces the building blocks of probability, including the concepts of probability, outcome, experiment, and sample space. Probabilities always take values between 0 and 1, where 0 means that the outcome cannot occur and 1 means that the outcome is certain.

2 The classical method of assigning probability is used if all outcomes are equally likely. The classical method states that the probability of an event *A* equals the number of outcomes in *A* divided by the number of outcomes in the sample space.

3 The Law of Large Numbers states that, as an experiment is repeated many times, the relative frequency (proportion) of a particular outcome tends to approach the probability of the outcome. The relative frequency method of assigning probability uses prior knowledge about the relative frequency of an outcome. The subjective method of assigning probability is used when the other methods are not applicable.

## Section 5.1 EXERCISES

### CLARIFYING THE CONCEPTS

1. Describe in your own words how chance and uncertainty affect you in your life. List some synonyms that we use in everyday life for the word *probability*.
2. Why do you think we use numerical values for probability rather than only qualitative terms such as "likely" or "impossible"?
3. Give three examples from your own life of experiments, as the term is used in this chapter.
   a. For each experiment, what are some of the outcomes?
   b. Write out the sample space of one of these experiments.
   c. Describe how the Law of Total Probability applies to the sample.
4. List the three methods for assigning probabilities.
5. What assumption do we need to make to use the classical method?
6. When can we use the relative frequency method?
7. If we can't use either the classical method or the relative frequency method, explain how we go about using the subjective method.
8. The experiment is to toss 10 fair coins 25 times each. Which methods can we use to assign probabilities?
9. How would you find the probability that a randomly chosen student at your college likes hip-hop music? What method would you use?
10. Describe the meaning of the following probabilities.
    a. Near 0
    b. 0
    c. Near 1
    d. 1

### PRACTICING THE TECHNIQUES

Determine whether each table in Exercises 11–15 is a probability model. If not, clearly explain why it is not a probability model.

11. Customers at a clothing store at the mall

| Gender | Probability |
|---|---|
| Females | 1.5 |
| Males | 0.2 |

12. Singers in the church choir

| Voice | Probability |
|---|---|
| Soprano | 0.25 |
| Alto | 0.25 |
| Tenor | −0.25 |
| Bass | 0.50 |

13. Voters at a town meeting

| Party | Probability |
|---|---|
| Democrat | 0.3 |
| Republican | 0.25 |
| Independent | 0.25 |
| Green | 0.1 |
| Libertarian | 0.1 |
| Other | 0.1 |

14. Students taking undergraduate introductory statistics

| Class | Probability |
|---|---|
| Freshmen | 0.15 |
| Sophomores | 0.25 |
| Juniors | 0.40 |
| Seniors | 0.20 |

15. Reasons why Hurricane Katrina survivors did not evacuate

| Reason | Probability |
|---|---|
| I did not have a car or a way to leave | .36 |
| I thought the storm and its aftermath would not be as bad as it was | .29 |
| I just didn't want to leave | .10 |
| I had to care for someone who was physically unable to leave | .07 |
| I waited too long | .07 |
| I was physically unable to leave | .05 |
| I worried that my possessions would be stolen or damaged if I left | .04 |
| I didn't want to leave my pet | .01 |

For Exercises 16–21, the experiment is to roll a fair die once. Find the probabilities.

16. Observing a 3
17. Observing an even number
18. Observing a number greater than 3
19. Observing a number less than 3
20. Observing a 3 or a 5
21. Observing a 3 and a 5

For Exercises 22–25, the experiment is to draw a card at random from a shuffled deck of 52 cards. Find the probabilities.

22. Drawing a king
23. Drawing a heart
24. Drawing the king of hearts
25. Drawing a black card

For Exercises 26–30, consider the experiment of tossing a fair coin three times.

26. Construct the tree diagram for the experiment.
27. What is the sample space for the experiment?
28. How does the tree diagram help to construct the sample space?
29. How do we find each outcome using the tree diagram?
30. What is the probability of observing three heads? What method of assigning probability are you using?

For Exercises 31 and 32, refer to Exercises 16–21.

31. For each of Exercises 16–21, was the probability you found for an event or an outcome?
32. Explain in your own words why the probability of observing a 3 cannot be less than the probability of observing a 3 or a 5.
33. The experiment is to roll two fair dice.
    - **a.** What is the most likely event?
    - **b.** What is the probability of this event?

## APPLYING THE CONCEPTS

**34. Picnic Lunch.** Picnickers at the Fourth of July Fair have the following preferences for grilled lunch: cheeseburger 50%, hot dog 25%, veggieburger 25%. Consider the experiment of two picnickers chosen at random choosing their preferred lunch.
- **a.** Construct the tree diagram for the experiment.
- **b.** What is the sample space?

**35. Rainy Days.** Students at the local middle school have been keeping track of the number of days it has rained. Of the past 100 days, it rained on 33 days.
- **a.** What is the probability that it rains on a randomly chosen day?
- **b.** What is the probability that it doesn't rain on a randomly chosen day?
- **c.** Which method of assigning probability did you use?

**36. Illegal Parking.** Suppose you find that 10 out of 50 cars parked in handicapped zones did so illegally.
- **a.** You would like to find the probability that a car parked in the handicapped zone is parked illegally. Explain whether or not you can use the classical method of assigning probability and why.
- **b.** Find the probability that a car parked in a handicapped zone is illegally parked. Which method did you use?

**37. Cutting Classes.** Over her college career, a student has cut 25 of the total 1000 class periods in all her courses.
- **a.** Find the probability that the student cuts a particular class.
- **b.** Which method did you use?
- **c.** Does it make sense to use the classical method? Why or why not?

**38. Basketball.** Your college's basketball team is playing a game next week.
- **a.** What is the probability that the team will win the game?
- **b.** What reasoning went into your probability assessment?
- **c.** Which method did you use?

**39. Die Rolls.** A die is rolled twice. Define the following events for each roll: low = $\{1, 2\}$, medium = $\{3, 4\}$, high = $\{5, 6\}$.
- **a.** Construct a tree diagram for this experiment.
- **b.** Construct the sample space for this experiment.
- **c.** What is the probability of observing two low die results? What method of assigning probability are you using?

**40.** Refer to Exercise 39. Find the following probabilities:

**a.** 2 high die results
**b.** 1 medium die result
**c.** 0 low die results
**d.** At least 1 high die result
**e.** At most 1 medium die result

**41. Brisbane Babies.** The table shows the births of babies at a Brisbane, Australia, hospital on a particular day.

| | | | | | | | | |
|---|---|---|---|---|---|---|---|---|
| Girl | Girl | Boy | Boy | Boy | Girl | Girl | Boy | Boy |
| Boy | Boy | Boy | Girl | Girl | Boy | Girl | Girl | Boy |
| Boy | Boy | Boy | Girl | Girl | Girl | Girl | Boy | Boy |
| Boy | Girl | Boy | Girl | Boy | Boy | Boy | Boy | Boy |
| Girl | Boy | Boy | Boy | Boy | Girl | Girl | Girl | |

**a.** Construct a relative frequency distribution of the numbers of girls and boys born.
**b.** Use the relative frequencies to construct a probability model.
**c.** Confirm that your probability model follows the Rules of Probability.

**Mozart Effect.** Use this information for Exercises 42 and 43. The Mozart effect refers to the hypothesis that listening to a piano sonata by the great Austrian master improves cognitive performance. Assume that we have an experiment where two children have each listened to Mozart's Sonata for Two Pianos in D Major (K 448). The children's cognitive skills are then tested. Assume that for each child, there are three possible outcomes: cognitive performance increased, cognitive performance decreased, and cognitive performance remained unchanged.

**42. a.** Construct a tree diagram for the experiment.
**b.** Construct the sample space for the experiment.
**c.** For how many outcomes do both children improve their cognitive performance?
**d.** Explain why it is not appropriate to use the classical method to assign probabilities for the Mozart experiment.

**43. Challenger Exercise.** Using your own subjective probabilities, find the probability that both children in the Mozart experiment will improve their cognitive performance.

**Letters in English.** Use Table 5.4 for Exercises 44 and 45 (it is the frequency distribution of letters in the English alphabet that we used in the Chapter 2 Case Study, The Caesar Cipher).

Imagine yourself on the television game show *Wheel of Fortune,* where contestants guess the letters contained in a hidden phase. You would like to ask for the letters that have the greatest chance of occurring, so you want to know the various probabilities of the letters in the English alphabet.

The experiment is to choose one letter at random from a sample of 1000 letters. The sample space is the English alphabet consisting of the usual 26 letters. Since the total sample size is 1000, you can find the relative frequencies of the letters in English simply by dividing each frequency by the total sample size (remember this from Chapter 2?).

**TABLE 5.4 Frequency distribution of English letters**

| A | B | C | D | E | F | G |
|---|---|---|---|---|---|---|
| 73 | 9 | 30 | 44 | 130 | 28 | 16 |
| **H** | **I** | **J** | **K** | **L** | **M** | **N** |
| 35 | 74 | 2 | 3 | 35 | 25 | 78 |
| **O** | **P** | **Q** | **R** | **S** | **T** | **U** |
| 74 | 27 | 3 | 77 | 63 | 93 | 27 |
| **V** | **W** | **X** | **Y** | **Z** | | |
| 13 | 16 | 5 | 19 | 1 | | |

**44.** Using the relative frequency method, construct the probability model for the letters in the English alphabet. Verify that your probability model satisfies the Rules of Probability.

**45.** Answer the following questions about the English letters probability model.

**a.** What is the most likely letter to be drawn? The least likely letter? Do these answers make sense?
**b.** Choosing one letter at random, which is more likely, a consonant or a vowel?
**c.** Of the 26 letters in the alphabet, what proportion are vowels?
**d.** Of the 8 letters with the highest frequencies, what proportion are vowels?
**e.** Compare your answers to questions **(c)** and **(d)**, and explain why *Wheel of Fortune* contestants are not allowed to guess vowels (they have to buy them).
**f.** If you were a contestant on *Wheel of Fortune* and had no money to buy a vowel, what would be your first five letter choices?

**46. Draw an Ace.** If you draw the ace of spades from a deck of cards, you win \$100.

**a.** What is the probability of winning this game?
**b.** What would be a fair price for playing this game? *Hint:* A fair price might be determined by balancing out the winnings and the price in the long run.

**47. A Bazaar Game.** Lenny has gone to the church bazaar with his family. In one of the games at the bazaar, if Lenny rolls two dice and gets a sum of at least 9 he wins \$5; otherwise he wins nothing.

**a.** Find the probability of winning \$5.
**b.** Find the probability of winning nothing.
**c.** What would you suggest would be a fair (break-even) price for playing this game?

**Teenagers' Music and Lifestyle.** Every year, *USA Weekend* conducts a survey of the nation's teenagers, asking them various questions about lifestyle and music. In 2002, nearly 60,000 teenagers responded to the survey, conducted in part through a Web site. Use this information for Exercises 48 and 49.

**48.** One *USA Weekend* survey question asked teenagers, "Do you listen to music while you are . . . ?" and listed several options. The most common responses are shown in the table. Respondents could choose more than one response. Explain why you cannot construct a probability model with these percentages.

| | |
|---|---|
| Doing chores | 79% |
| On the computer | 73% |
| Doing homework | 72% |
| Eating meals at home | 33% |
| In the classroom | 18% |

**49.** Another survey question asked by *USA Weekend* was "If you had to choose just one type of music to listen to exclusively, which would it be?" The results are shown in the table.

| | |
|---|---|
| Hip-hop/rap | 27% |
| Pop | 23% |
| Rock/punk | 17% |
| Alternative | 7% |
| Christian/gospel | 6% |
| R&B | 6% |
| Country | 5% |
| Techno/house | 4% |
| Jazz | 1% |
| Other | 4% |

**a.** Construct a probability model.
**b.** Confirm that your probability model follows the Rules of Probability.
**c.** Is it unusual for a respondent to prefer jazz?
**d.** Use technology to draw random samples of sizes 10, 100, 1000, and 10,000 from your probability model.
**e.** What can you conclude about the relative frequencies as the sample size increases?

**50. Fairfax County Income.** The following table contains a probability model for the distribution of income in Fairfax County, Virginia.[2]

| Annual income | Probability |
|---|---|
| Under $25,000 | 0.083 |
| $25,000 to $49,999 | 0.166 |
| $50,000 to $74,999 | 0.169 |
| $75,000 to $99,999 | 0.160 |
| $100,000 to $149,999 | 0.200 |
| $150,000 or more | 0.222 |

**a.** Use technology to draw random samples of sizes 10, 100, 1000, and 10,000 from this probability model.
**b.** What can you conclude about the relative frequencies as the sample size increases?

APPLET Use the *Law of Large Numbers for Proportions* applet for Exercises 51 and 52.

**51.** Set the probability of heads to 0.5 and the number of tosses to 40. Click **Toss**.
**a.** Record the proportion of heads observed.
**b.** Without pressing **Reset**, continue to click **Toss** until the total number of tosses is 120. Again record the proportion of heads.
**c.** Without pressing **Reset**, continue to click **Toss** until the total number of tosses is 240. Again record the proportion of heads.
**d.** Without pressing **Reset**, continue to click **Toss** until the total number of tosses is 480. Again record the proportion of heads.

**52.** The proportions you recorded in Exercise 51 are relative frequencies of heads. What can you conclude about the relative frequencies as the sample size increases?

# 5.2 Combining Events

*Objectives:* **By the end of this section, I will be able to...**

1 Understand how to combine events using complement, union, and intersection.

2 Apply the Addition Rule to events in general and to mutually exclusive events in particular.

## 1 Combining Events Using Complement, Union, and Intersection

In Example 5.3, if your roommate rolled a 4, then your roommate was to do your laundry for the rest of the semester. Your roommate is keenly interested in *not* rolling a 4. If $A$ is an event, then the collection of outcomes not in event $A$ is called the

**complement of $A$**, denoted $A^C$. The term *complement* comes from the word "to complete," meaning that any event and its complement together make up the complete sample space.

## Example 5.10 Finding the probability of the complement of an event

If $A$ is the event "observing a sum of 4 when the two fair dice are rolled," then your roommate is interested in the probability of $A^C$, the event that a 4 is not rolled. Find the probability that your roommate does not roll a 4.

**Solution**

Which outcomes belong to $A^C$? By the definition, $A^C$ is all the outcomes in the sample space that do not belong in $A$. There are the following outcomes in $A$: $\{(3,1)(2,2)(1,3)\}$.

Figure 5.10 shows all the outcomes except the outcomes from $A$ in the two-dice sample space. There are 33 outcomes in $A^C$ and 36 outcomes in the sample space. The classical probability method then gives the probability of not rolling a 4 to be

$$P(A^C) = \frac{N(A^C)}{N(S)} = \frac{33}{36} = \frac{11}{12}$$

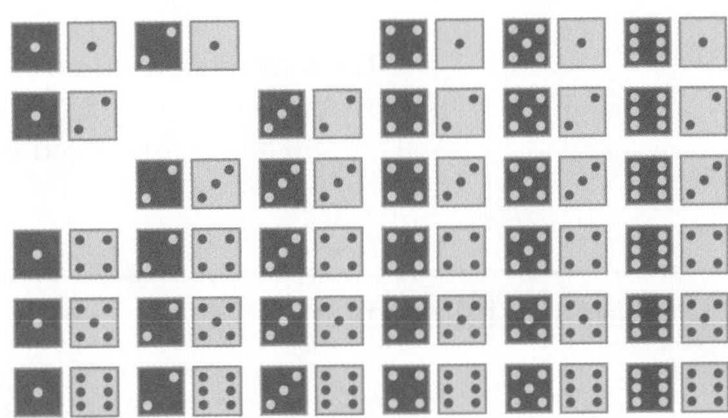

FIGURE 5.10 Outcomes in $A^C$.

So your roommate may be breathing a little easier. The probability is high that, on this roll at least, your roommate will not land on Boardwalk. (But you never know until you roll' dem bones. . . .)

For event $A$ in Example 5.10, note that

$$P(A) + P(A^C) = \frac{1}{12} + \frac{11}{12} = 1$$

Is this a coincidence, or does the sum of the probabilities of an event and its complement always add to 1? Recall the Law of Total Probability (Section 5.1), which states that the sum of all the outcome probabilities in the sample space must be equal to 1. Since any event $A$ and its complement $A^C$ together make up the entire sample space, then it always happens that $P(A) + P(A^C) = 1$.

**Probabilities for Complements**

For any event $A$ and its complement $A^C$, $P(A) + P(A^C) = 1$. Applying a touch of algebra gives the following:

- $P(A) = 1 - P(A^C)$
- $P(A^C) = 1 - P(A)$

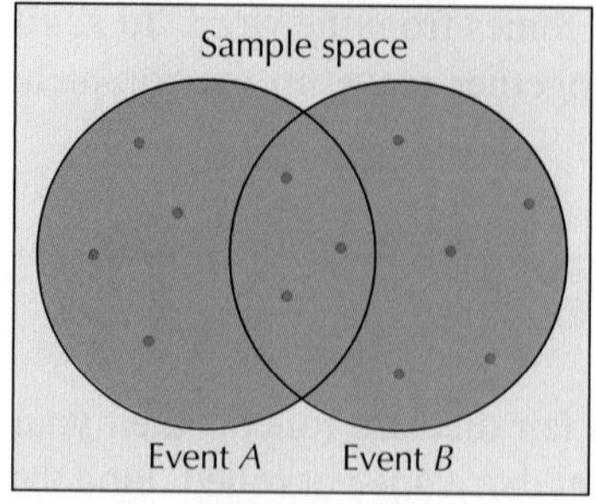


FIGURE 5.11 Union of events A, B.

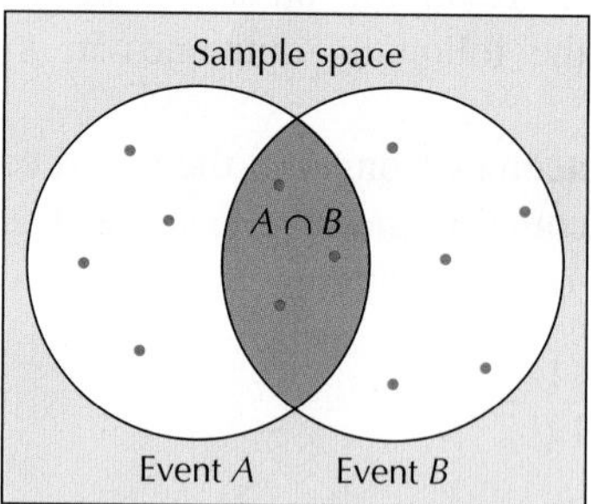


FIGURE 5.12 Intersection of Events A, B.

Sometimes we need to find the probability of a combination of events. For example, consider the casino game of craps where you roll two dice. One way of winning is by rolling the sum 7 or 11. We can find the probability of the following two events: the sum is 7 or the sum is 11. First, we need some tools for finding the probability of a combination of events.

The **union** of two events $A$ and $B$ is defined as the event containing all the outcomes that belong to $A$ or $B$ or both. The union of events $A$ and $B$ is denoted $A \cup B$. The English word associated with union is "or." So if you are asked to find the probability of "$A$ or $B$," you should find the probability of $A \cup B$. Figure 5.11 shows the union of two events, with the red dots indicating the outcomes. Note from Figure 5.11 that the union of the events $A$ and $B$ refers to all outcomes in $A$ or $B$ or both.

The **intersection** of two events $A$ and $B$ is the event containing the outcomes that belong to both $A$ and $B$. Figure 5.12 shows that the intersection of the two events is the part where $A$ and $B$ overlap. The intersection of events $A$ and $B$ is denoted as $A \cap B$ and is associated with the English word "and." Both union and intersection are commutative. That is, $A \cup B = B \cup A$ and $A \cap B = B \cap A$.

**Union and Intersection of Events**

The **union** of two events $A$ and $B$ is the event representing all the outcomes that belong to $A$ or $B$ or both. The union of $A$ and $B$ is denoted as $A \cup B$ and is associated with "or."

The **intersection** of two events $A$ and $B$ is the event representing all the outcomes that belong to both $A$ and $B$. The intersection of $A$ and $B$ is denoted as $A \cap B$ and is associated with "and."

## Example 5.11 Union and intersection

Let our experiment be to draw a single card at random from a deck of cards. Define the following events:

$A$: The card drawn is an ace.

$H$: The card drawn is a heart.

**a.** Find $A \cup H$.
**b.** Find $A \cap H$.

### Solution

**a.** The union of $A$ and $H$ is the event containing all the outcomes that are either aces or hearts or both (the ace of hearts). That is, the event $A \cup H$ consists of the set of outcomes (the cards) shown in Figure 5.13.

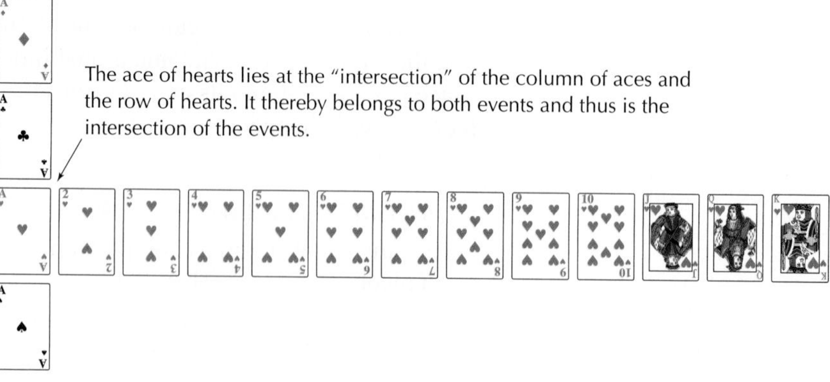

The ace of hearts lies at the "intersection" of the column of aces and the row of hearts. It thereby belongs to both events and thus is the intersection of the events.


FIGURE 5.13

**b.** The intersection of $A$ and $H$ is the event containing the outcomes that are common to both $A$ and $H$. There is only one such outcome: the ace of hearts (see Figure 5.13).

## 2 Addition Rule

We are often interested in finding the probability that either one event *or* another event may occur. The formula for finding these kinds of probabilities is called the **Addition Rule.**

**Addition Rule**

$$P(A \text{ or } B) = P(A \cup B) = P(A) + P(B) - P(A \cap B)$$

*What Does the Addition Rule Mean?*

We can use Figure 5.11 to understand the Addition Rule. We are trying to find the probability of all the outcomes in $A$ or $B$ or both. The formula says to first count the probabilities of the outcomes in $A$ and then add the probabilities of the outcomes in $B$. But what about the overlap between $A$ and $B$, outcomes that belong to both events? To avoid counting the outcomes in the overlap (intersection) twice, we have to subtract the probability of the intersection, $P(A \cap B)$. ■

### Example 5.12 Addition Rule applied to a deck of cards

Suppose you pay \$1 to play the following game. You choose one card at random from a deck of 52 cards, and you will win \$3 if the card is either an ace *or* a heart. Find the probability of winning this game.

**Solution**

Using the same events defined in Example 5.11, we find $P(A \text{ or } H) = P(A \cup H)$. By the Addition Rule, we know that

$$P(A \cup H) = P(A) + P(H) - P(A \cap H)$$

There are 4 aces in a deck of 52 cards, so by the classical method (equally likely outcomes), $P(A) = 4/52$. There are 13 hearts in a deck of 52 cards, so $P(H) = 13/52$. From Example 5.11, we know that $A \cap H$ represents the ace of hearts. Since each card is equally likely to be drawn, then $P(\text{ace of hearts}) = P(A \cap H) = 1/52$. Thus,

$$\begin{aligned} P(A \cup H) &= P(A) + P(H) - P(A \cap H) \\ &= \frac{4}{52} + \frac{13}{52} - \frac{1}{52} = \frac{16}{52} = \frac{4}{13} \end{aligned}$$

The intersection of two events may be represented by the intersection of a row and a column in a two-way table. Recall from Section 4.1 (pages 164–165) that a *two-way table* (also known as a *crosstabulation* or a *contingency table*) is a tabular summary of the relationship between two categorical variables.

## Example 5.13 Addition Rule applied to a two-way table

A study of online dating behavior found that users of a particular online dating service self-reported their physical appearance according to the counts given in Table 5.5.[3]

Table 5.5 Gender and self-reported physical appearance

| | Physical Appearance | | | | |
|---|---|---|---|---|---|
| **Gender** | **Very attractive** | **Attractive** | **Average** | **Prefer not to answer** | **Total** |
| Female | 3113 | 16,181 | 6,093 | 3478 | 28,865 |
| Male | 1415 | 12,454 | 7,274 | 2809 | 23,952 |
| Total | 4528 | 28,635 | 13,367 | 6287 | 52,817 |

Using this information, find the probability that a randomly chosen online dater has the following characteristics.

**a.** Is female
**b.** Self-reported as attractive
**c.** Is a female who self-reported as attractive
**d.** Is a female *or* self-reported as attractive

### Solution

**a.** There are a total of $N(S) = 52{,}817$ online daters in the entire data set. Of these, 28,865 are female, denoted as event $F$. Therefore,

$$P(F) = P(\text{Female}) = \frac{N(\text{Female})}{N(S)} = \frac{N(F)}{N(S)} = \frac{28{,}865}{52{,}817} \approx 0.5465$$

**b.** There are 28,635 people who self-reported their physical appearance as attractive, denoted as event $A$. Therefore,

$$P(A) = P(\text{Self-reported attractive}) = \frac{N(\text{Self-reported attractive})}{N(S)} = \frac{N(A)}{N(S)}$$

$$= \frac{28{,}635}{52{,}817} \approx 0.5422$$

**c.** The online daters who are both female and self-reported as attractive are shown in the highlighted cell in Table 5.5. This cell is located at the intersection of the row of females and the column of people who self-reported as attractive. Therefore, this cell reports the frequency of people who belong to both events. Thus,

$$P(F \text{ and } A) = P(F \cap A) = P(\text{Female and self-reported attractive}) = \frac{N(F \cap A)}{N(S)}$$

$$= \frac{16{,}181}{52{,}817} \approx 0.3064$$

**d.** Here we seek $P(F \text{ or } A) = P(F \cup A)$. By the Addition Rule,

$$P(F \cup A) = P(F) + P(A) - P(F \cap A) = 0.5465 + 0.5422 - 0.3064 = 0.7823$$

### Mutually Exclusive Events

When drawing a card at random from a deck of 52 cards, the events "a heart is drawn" and "a diamond is drawn" have no outcomes in common. That is, no card is both a heart and a diamond. We say that these two events are **mutually exclusive**.

Two events are said to be **mutually exclusive** or *disjoint* if they have no outcomes in common.

Note that any event and its complement are always mutually exclusive. Other examples of mutually exclusive events are given in Table 5.6.

Table 5.6 Examples of mutually exclusive events

| Experiment | Mutually exclusive events |
|---|---|
| Toss fair coin | Observe heads; observe tails |
| Draw a single card from a deck of 52 cards | Card is red; card is a spade |
| Select a student at random | Student is female; student is male |
| Choose a digit at random | Digit is even; digit is odd |

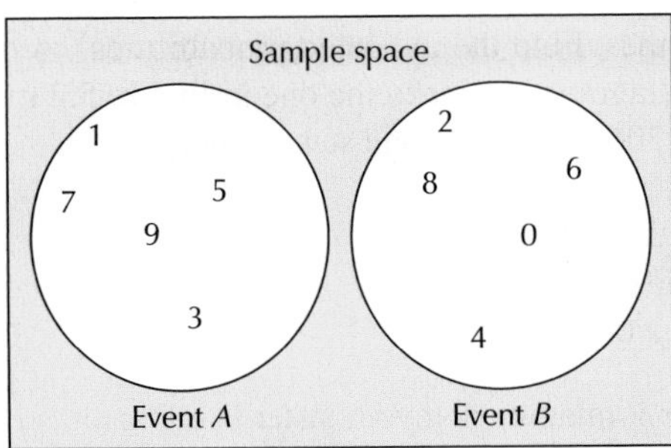


FIGURE 5.14 Mutually exclusive events.

Figure 5.14 shows how mutually exclusive events are represented graphically. It shows the events

$$A = \{1, 3, 5, 7, 9\} \qquad \text{and} \qquad B = \{0, 2, 4, 6, 8\}$$

Note that there is no overlap between the two events. When two events are mutually exclusive, they share no outcomes, and therefore the intersection of mutually exclusive events is empty. Since the intersection $(A \cap B)$ is empty, then for mutually exclusive events, $P(A \cap B) = 0$. Therefore, we can formulate a special case of the **Addition Rule for Mutually Exclusive Events *A* and *B***:

$$P(A \cup B) = P(A) + P(B) - P(A \cap B) = P(A) + P(B) - 0 = P(A) + P(B)$$

**Addition Rule for Mutually Exclusive Events**

If $A$ and $B$ are mutually exclusive events, $P(A \cup B) = P(A) + P(B)$.

## Example 5.14 Addition Rule for Mutually Exclusive Events

Using Table 5.5 from Example 5.13, find the probability that a randomly chosen online dater self-reported as either attractive or very attractive.

### Solution

From Table 5.5, there are 28,635 online daters who self-reported as attractive and 4528 who self-reported as very attractive, yielding the following probabilities:

$$P(A) = P(\text{Self-reported attractive}) = \frac{N(A)}{N(S)} = \frac{28{,}635}{52{,}817} \approx 0.5422$$

$$P(V) = P(\text{Self-reported very attractive}) = \frac{N(V)}{N(S)} = \frac{4528}{52{,}817} \approx 0.08573$$

Since no online daters self-reported as both attractive and very attractive, the two groups are mutually exclusive. Thus, by the Addition Rule for Mutually Exclusive Events,

$$P(A \cup V) = P(A) + P(V) = 0.5422 + 0.08573 = 0.62793$$

## Section 5.2 SUMMARY

1 Combinations of events may be formed using the concepts of complement, union, and intersection.

2 The Addition Rule provides the probability of Event *A* or Event *B* to be the sum of their two probabilities minus the probability of their intersection. Mutually exclusive events have no outcomes in common.

## Section 5.2 EXERCISES

### CLARIFYING THE CONCEPTS

**1.** Describe in your own words what it means for two events to be mutually exclusive.

**2.** Describe the intersection of two mutually exclusive events.

**3.** Is it true that the union of two events always contains at least as many outcomes as the intersection of two events? Use Figures 5.11 and 5.12 to help you visualize this problem.

**4.** If we choose a student at random from your college or university, is it more likely that we choose a male or a male football player? Why?

**5.** What is your personal estimate of the probability that it will rain on any given day? How about the probability that it won't rain? Why do these numbers have to add up to 1 (or 100%)?

### PRACTICING THE TECHNIQUES

For Exercises 6–11, consider the experiment of rolling a fair die twice. Find the indicated probabilities.

**6.** One of the dice is a 4

**7.** Neither die is a 4

**8.** Sum of the two dice equals 3

**9.** Sum of the two dice equals 3 and one of the dice is a 4

**10.** Sum of the two dice equals 3 or one of the dice is a 4

**11.** Sum of the two dice equals 3 or neither of the dice is a 4

For Exercises 12–17, consider the experiment of drawing a card at random from a shuffled deck of 52 cards. Find the indicated probabilities.

**12.** Drawing a king and a heart

**13.** Drawing a king or a heart

**14.** Drawing a card that is neither a king nor a heart

**15.** Drawing a heart or a spade

**16.** Drawing a heart and a spade

**17.** Drawing a card that is not the king of hearts

For Exercises 18–23, consider the experiment of drawing a card at random from a shuffled deck of 52 cards. Find the indicated probabilities.

**18.** Drawing a face card (king, queen, or jack)

**19.** Drawing a card that is not red

**20.** Drawing a card that is not a face card

**21.** Drawing a face card that is not a diamond

**22.** Drawing a face card or a diamond

**23.** Drawing a face card and a diamond

For Exercises 24–27, consider the experiment of tossing a fair coin three times. Find the indicated probabilities. (*Hint*: Use a tree diagram similar to the one in Figure 5.2 in Section 5.1 [page 206] but going one step farther.)

**24.** Observing 3 heads

**25.** Not observing 3 heads

**26.** Observing 2 tails

**27.** Not observing 2 tails

For Exercises 28–33, imagine that your sister is going to have triplets. Assume that the probability of a baby boy or a baby girl is equally likely. (In fact, it is not quite.)

**28.** Construct the sample space.

**29.** Find the probability of 1 girl and 2 boys.

**30.** Find the probability of 1 boy and 2 girls.

**31.** Find the probability of 2 of one gender and 1 of the other gender.

**32.** Find the probability of 1 girl or 1 boy.

**33.** Find the probability of getting 3 girls.

### APPLYING THE CONCEPTS

**34. Game of Craps.** You win the casino game of craps if you roll a 7 or 11. Find the probability of rolling a sum of 7 or 11 when two dice are rolled.

**35. Acing College Courses.** Jessica has gotten A's in 10 of her 30 college courses.

- **a.** Find the probability that Jessica gets an A in a randomly chosen college course.
- **b.** Find the probability that Jessica gets a grade other than an A in a randomly chosen college course.

**Best Sellers.** Use Table 5.7 for Exercises 36 and 37. The table contains the Top 10 Best Sellers List from *USA Today* for the week of March 3, 2007. In the information listed is the rank, title, first author, whether the work is fiction ($F$) or nonfiction ($NF$), and whether the first author is female or male.

**36.** For a book chosen from the Top 10 List, find the following probabilities if we select a book at random

- **a.** $P(F)$
- **b.** $P(F^C)$
- **c.** $P(F \cup F^C)$

**TABLE 5.7 Top 10 best sellers list from *USA Today* (March 3, 2007)**

| Rank | Title | Author | Type | Gender |
|---|---|---|---|---|
| 1 | The Secret | Rhonda Byrne | NF | Female |
| 2 | Innocent in Death | Nora Roberts | F | Female |
| 3 | Step on a Crack | James Patterson | F | Male |
| 4 | Bridge to Terabithia | Katherine Paterson | F | Female |
| 5 | Sisters | Danielle Steel | F | Female |
| 6 | The Measure of a Man: A Spiritual Autobiography | Sidney Poitier | NF | Male |
| 7 | The Memory Keeper's Daughter | Kim Edwards | F | Female |
| 8 | The Audacity of Hope | Barack Obama | NF | Male |
| 9 | You: On a Diet | Michael Roizen | NF | Male |
| 10 | The Best Life Diet | Bob Greene | NF | Male |

**37.** For a book chosen from the Top 10 List, find the following probabilities if we select a book at random.

- **a.** Author is Katherine Paterson and type is fiction
- **b.** Author is Katherine Paterson or type is fiction

**38. Traffic Lights.** Let $A$ be the event that you encounter a green light at your next traffic light.

- **a.** What outcomes make up $A^C$?
- **b.** What is the probability of $A$? Which method did you use?
- **c.** What is the probability of $A^C$?

**39. Monopoly.** You are playing Monopoly with your roommate (the loser does laundry for both for the rest of the semester). Your piece is sitting on Water Works, and you are about to roll your two fair dice. Your opponent has hotels on Park Place and Boardwalk, and if you land on either one, you will lose. Park Place lies 9 spaces away and Boardwalk lies 11 spaces away.

- **a.** What is the probability that you will land on Park Place and get stuck with laundry duty?
- **b.** What is the probability that you will land on Boardwalk and still get stuck with laundry duty?
- **c.** What is the probability that you will lose the game on this roll?

**40. Monopoly.** As well as the conditions in Exercise 39, you have hotels on the green properties (Pacific, North Carolina, and Pennsylvania Avenues), and if your roommate lands on one of them, your roommate will be sorting your laundry starting tonight. Your roommate's piece is on B & O Railroad.

- **a.** What is the probability that your roommate will land on Pacific Avenue, 6 spaces away?
- **b.** What is the probability that your roommate will land on North Carolina Avenue, 7 spaces away?
- **c.** What is the probability that your roommate will land on Pennsylvania Avenue, 9 spaces away?
- **d.** What is the probability that your roommate will lose the game on this roll?
- **e.** For this roll only, who has the greater chance of sorting your permanent press from your towels for the rest of the semester?

**41. High School Students.** In a local high school of 500 students, there are 200 females, 100 sophomores, and 50 female sophomores.

- **a.** If we choose a student at random, what is the probability that we choose a female or a sophomore?
- **b.** Find the probability that a randomly chosen student is a male or a sophomore.
- **c.** Find the probability that a randomly chosen student is a female or is not a sophomore.

**42. Halloween Candy.** At Halloween, the accent is on taste—taste of candy, that is. In a sample of 100 children, 70 like chocolate bars, 60 like peanut butter cups, and 50 like both.

- **a.** If we choose one child at random, find the probability that the child likes either chocolate bars or peanut butter cups.
- **b.** In **(a)**, suppose you forgot to subtract the probability of the intersection. How would you know that your answer is wrong?

**43. Pick a Card.** If we draw a single card at random from a deck of 52 playing cards, find the probability that the card is

- **a.** a heart or a diamond.
- **b.** a red card or a jack.
- **c.** a club or a face card (king, queen, jack).
- **d.** a heart and a diamond.
- **e.** not a spade.

**Don't Mess with Texas.** Don't Mess with Texas (**dontmesswithtexas.org**) is a Texas statewide antilittering organization. Its 2005 report, *Visible Litter Study 2005,* identified paper, plastic, metals, and glass as the top four categories of litter by composition. The report also identified tobacco, household / personal, food, and beverages, as the top four categories of litter by use. Assume that a sample of 12 items of litter had the following characteristics. Use Table 5.8 for Exercises 44–46.

**TABLE 5.8 Litter composition and use**

| Litter item | Composition | Use |
|---|---|---|
| 1 | Paper | Tobacco |
| 2 | Plastic | Household/personal |
| 3 | Glass | Beverages |
| 4 | Paper | Tobacco |
| 5 | Metal | Household/personal |
| 6 | Plastic | Food |
| 7 | Glass | Beverages |
| 8 | Paper | Household/personal |
| 9 | Metal | Household/personal |
| 10 | Plastic | Beverages |
| 11 | Paper | Tobacco |
| 12 | Plastic | Food |

44. A litter item is chosen at random.
   - **a.** Find the probability that the composition of the item is paper.
   - **b.** Find the probability that the composition of the item is not paper. Calculate this probability in two different ways.

45. A litter item is chosen at random.
   - **a.** Find the probability that the use of the item is tobacco.
   - **b.** Find the probability that the use of the item is not tobacco. Calculate this probability in two different ways.

46. A litter item is chosen at random.
   - **a.** Find the probability that the composition of the item is paper *and* its use is tobacco.
   - **b.** Find the probability that the composition of the item is paper *or* its use is tobacco.

47. **A New Sonata.** Music researchers have discovered a new sonata from the classical period of the late eighteenth century. The probability that the sonata was written by the following composers is as follows: Mozart 35%, Haydn 50%, Beethoven 15%. It is known that the sonata composers are mutually exclusive, meaning that no sonata was written by more than one of these composers. Find the following probabilities.
   - **a.** The sonata was written by either Mozart or Haydn.
   - **b.** The sonata was written by either Mozart or Beethoven.
   - **c.** The sonata was written by either Haydn or Beethoven.
   - **d.** The sonata was written by both Mozart and Haydn.

48. **Cheeseburger or Chicken?** The college football team (60 players) stopped off at a hamburger joint on the way home from a big road win. Thirty players ordered a bacon cheeseburger. Twenty players ordered a chicken deluxe. No player ordered both.
   - **a.** Find the probability that a randomly chosen player ordered either a bacon cheeseburger or a chicken deluxe.
   - **b.** Suppose we didn't know that no player ordered both. Would we have been able to answer **(a)**? Why or why not?

49. **Global Terrorism.** The U.S. State Department, in its 2003 report "Patterns of Global Terrorism" (**http://www.state.gov**), reported that 20 of the 60 anti-American terrorist incidents that occurred in 2003 took place in the Middle East.
   - **a.** What is the probability that a randomly selected anti-American terrorist incident in 2003 occurred in the Middle East?
   - **b.** Which method of assigning probability did you use to answer **(a)**?
   - **c.** Find the probability that a randomly selected anti-American terrorist incident in 2003 did not occur in the Middle East.

50. **Classic Lotto.** The Connecticut Lottery Corporation runs a game called Connecticut Classic Lotto. You pick six different numbers from 1 to 44 and pay $1 to play.

   If your picks match three of the six numbers chosen (probability 0.02381), you win $2.

   If you match four out of six (probability 0.001495), you win $50.

   If you match five out of six (probability 0.0000323), you win $2000.

   If you match all six numbers (probability 0.0000001417), you win the jackpot. The jackpot on May 20, 2008, was $1,800,000.
   - **a.** Find the probability that fewer than three of the six numbers match.
   - **b.** Find the probability that fewer than four of the six numbers match.
   - **c.** Find the probability that three or four of the six numbers match.

51. **Baseball Strategy.** Earnshaw Cook analyzed 20,000 baseball games and used the relative frequency method to assign the following probabilities.[4] The probability of scoring a run with a runner on first and no outs is 0.43. The probability of scoring a run with a runner on second and one out is 0.45. Actually, considering all the bunting and sacrifice fly strategies used by baseball managers to get from the first situation to the second, it seems hardly worth it, increasing the probability of scoring by only 0.02. Using these probabilities, answer the following questions.
   - **a.** If there is a runner on first with no outs, what is the probability of not scoring?
   - **b.** If there is a runner of second with one out, what is the probability of not scoring?

52. **First Beer.** Nearly half of all young people aged 10–19 have never drunk beer, which is good because for most of them it is illegal. However, according to the United States Bureau of Justice Statistics, of the young people who drank beer before age 20, 25% were from 6 to 9 years old, 16% were 10 or 11 years old, 27% were 12 or 13 years old, 24% were 14 or 15 years old, 7% were 16 or 17 years old, and 1% were 18 or 19 years old when they had their first beer. Find

the following probabilities for the people who drank their first beer before age 20.

**a.** A randomly selected person drank his or her first beer either at age 16–17 years old or at age 18–19 years old.

**b.** A randomly selected person drank his or her first beer at age 16–17 years old and at age 18–19 years old.

**c.** What can we say about the two events $A$ = first beer at 16–17 years old and $B$ = first beer at 18–19 years old?

**53. Online Dating Data.** Refer to Table 5.9 (Table 5.5 from Example 5.13). Find the probability that a randomly selected online dater has the characteristics that follow the table.

**a.** Is male

**b.** Prefers not to describe physical appearance

**c.** Is male and prefers not to describe physical appearance

**d.** Is male or prefers not to describe physical appearance

**Causes of Death.** Refer to Table 5.10 for Exercises 54–56.

**54.** Find the following probabilities.

**a.** The cause of death was accidents.

**b.** The cause of death was not accidents.

**55.** Find the following probabilities.

**a.** The cause of death was accidents and Parkinson's disease.

**b.** The cause of death was accidents or Parkinson's disease.

**56.** Are the causes of death mutually exclusive?

**TABLE 5.10 Causes of death**

| Cause of death | Deaths |
|---|---|
| Diseases of heart | 654,092 |
| Malignant neoplasms | 550,270 |
| All other causes | 435,769 |
| Cerebrovascular diseases | 150,147 |
| Chronic lower respiratory diseases | 123,884 |
| Accidents | 108,694 |
| Diabetes mellitus | 72,815 |
| Alzheimer's disease | 65,829 |
| Influenza and pneumonia | 61,472 |
| Nephritis | 42,762 |
| Septicemia | 33,464 |
| Intentional self-harm | 31,647 |
| Chronic liver disease | 26,549 |
| Hypertension | 22,953 |
| Parkinson's disease | 18,018 |
| Total | 2,398,365 |

*Source:* Centers for Disease Control, 2004.

**TABLE 5.9 Online dating data**

| | Physical Appearance | | | | |
|---|---|---|---|---|---|
| **Gender** | **Very attractive** | **Attractive** | **Average** | **Prefer not to answer** | **Total** |
| Female | 3113 | 16,181 | 6,093 | 3478 | 28,865 |
| Male | 1415 | 12,454 | 7,274 | 2809 | 23,952 |
| Total | 4528 | 28,635 | 13,367 | 6287 | 52,817 |

## 5.3 Conditional Probability

*Objectives:* **By the end of this section, I will be able to…**

1. Calculate conditional probabilities.
2. Explain independent and dependent events.
3. Solve problems using the Multiplication Rule.
4. Recognize the difference between sampling with replacement and sampling without replacement.

### 1 Introduction to Conditional Probability

As we progress through this book, you will notice a recurring theme: *the more information available, the better*. Very often, when we are investigating the probability of a certain event $A$, we learn that another event $B$ has occurred. If events $A$ and $B$ are related, then the occurrence of event $B$ often influences the probability that event $A$ will occur.

## Example 5.15 Having more information often affects the probability of an event

In Section 5.1, we found that the probability of rolling a sum of 4 when tossing two dice is $3/36 \approx 0.0833$. But what if we were told that at least one of the dice shows a 1. How does this extra information affect the probability of rolling a 4?

### Solution

Figure 5.15 shows the 11 outcomes from the two-dice sample space in which at least one die shows a 1. The extra information reduces the number of possible outcomes in the sample space from 36 to 11. We see that two of these outcomes have a sum equal to 4. Thus, the probability of observing a sum of 4, *given that* at least one of the dice shows a 1, is $2/11 \approx 0.1818$. ■

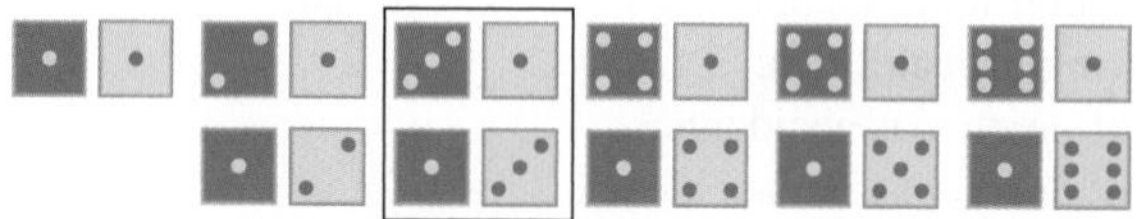

FIGURE 5.15 Using the extra knowledge changes the probability.

The extra information about a related event changed the probability of the event of interest. This type of probability is an example of what is called **conditional probability**.

> For two related events $A$ and $B$, the probability of $B$ given $A$ is called a **conditional probability** and denoted $P(B|A)$.

Thus, if we let $A$ represent the event that one of the dice shows a 1, and let $B$ represent the event that the sum of the two dice equals 4, then

$$P(B) = \frac{3}{36} \approx 0.0833 \qquad \text{but} \qquad P(B \mid A) = \frac{2}{11} \approx 0.1818$$

Figure 5.16 can help us visualize how conditional probability works. The idea is that, once event $A$ has occurred, the only chance for event $B$ to occur is in the overlap, the intersection $A \cap B$. Therefore, the conditional probability that $B$ will occur, given that event $A$ has already taken place, is found by taking the ratio $P(A \cap B)/P(A)$.

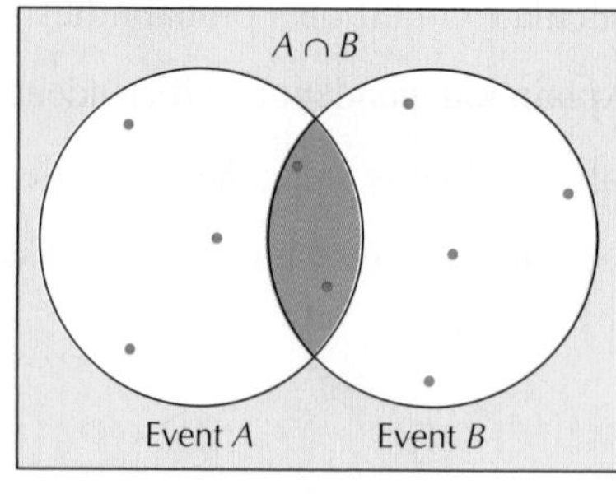


(a)

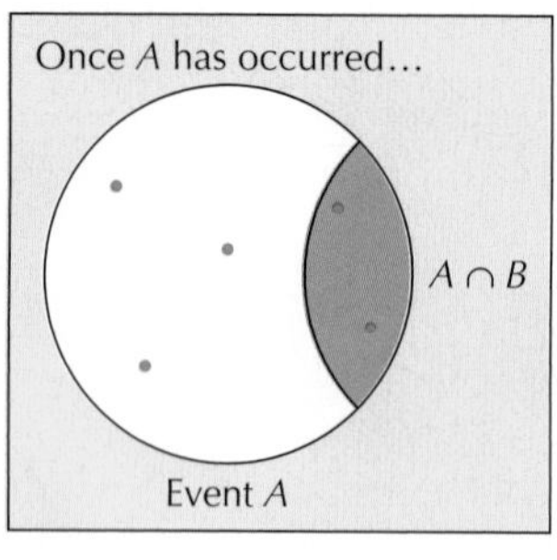


(b)

FIGURE 5.16 How conditional probability works.

**Calculating Conditional Probability**

The conditional probability that $B$ will occur, given that event $A$ has already taken place, equals

$$P(B|A) = \frac{P(A \cap B)}{P(A)} = \frac{N(A \cap B)}{N(A)}$$

## Example 5.16 Probability of responding to a direct mail marketing campaign

Punchstock/Photodisc

Market research represents a fast-growing field in need of analysts with strong statistical expertise. Table 5.11 is adapted from a study on direct mail marketing. It contains the numbers of customers who either responded or did not respond to a direct mail marketing campaign, along with whether they had a credit card on file with the company. The two events are

$R$: Responded to direct mail marketing campaign.
$C$: Has a credit card on file.

Table 5.11 Credit card status and marketing response

| | Credit Card on File? | | |
|---|---|---|---|
| **Response** | **No** | **Yes** | **Totals** |
| Did not respond | 161 | 79 | 240 |
| Did respond | 17 | 31 | 48 |
| Totals | 178 | 110 | 288 |

*Source:* Daniel Larose, *Data Mining Methods and Models* (Wiley Interscience, 2006).

**a.** Find the probability that a randomly chosen customer responded to the marketing campaign.
**b.** Find the conditional probability that a randomly selected customer responded, given that the customer has a credit card on file.

### Solution

**a.** $P(R) = \frac{N(R)}{N(S)}$. There are $N(R) = 48$ customers who did respond, and there are $N(S) = 288$ customers in this experiment. Thus,

$$P(R) = \frac{N(R)}{N(S)} = \frac{48}{288} \approx 0.1667$$

**b.** We will use $P(R|C) = N(R \cap C)/N(C)$ because in this example it is easier to work directly with the numbers of outcomes rather than the probabilities. Now, $R \cap C$ represent customers who did respond and had a credit card on file. From Table 5.11, there are $N(R \cap C) = 31$ such customers. Also, there are $N(C) = 110$ customers total who had a credit card on file. Therefore,

$$P(R|C) = \frac{N(R \cap C)}{N(C)} = \frac{31}{110} \approx 0.2818$$

That is, the probability that a randomly chosen customer responded to the direct mail marketing campaign, given that the customer had a credit card on file, is 0.2818. ■

What Do These Numbers Mean?

### Conditional Probability

Conditional probabilities can often be interpreted as percentages of some *subset* of a population. For example, the conditional probability that a customer responded, given that the customer has a credit card on file, may be interpreted as the percentage of customers with credit cards who responded. ■

Students sometimes confuse the meanings of $P(B \cap A)$ and $P(B \mid A)$. For $P(B \mid A)$, we *assume that the event A has occurred* and now need to find the probability of event $B$, given event $A$. On the other hand, for $P(B \cap A)$, we do not assume that event $A$ has occurred and instead need to determine the probability that both events occurred.

## 2 Independent Events

Since having a credit card on file increased the probability of a customer responding from 0.1667 to 0.2818, we can therefore say that the probability of responding *depends* in part on whether the customer has a credit card on file. In other words, the events $R$ and $C$ are **dependent events**.

On the other hand, if the probability of responding had been unaffected by whether the customer had a credit card on file, then we would have said that $R$ and $C$ were **independent events**. That is, $R$ and $C$ would have been independent events had $P(R \mid C)$ equaled $P(R)$. In general, if the occurrence of an event does not affect the probability of a second event, then the two events are independent.

Events $A$ and $B$ are **independent** if

$$P(A|B) = P(A) \qquad \text{or if} \qquad P(B|A) = P(B)$$

Otherwise the events are said to be **dependent**.

**Strategy for Determining Whether Two Events Are Independent**

1. Find $P(B)$.
2. Find $P(B \mid A)$.
3. Compare the two probabilities. If they are equal, then $A$ and $B$ are independent events. Otherwise, $A$ and $B$ are dependent events.

### Example 5.17 Are successive coin tosses independent?

Let our experiment be tossing a fair coin twice. Determine whether the following events are independent.

$A$: Heads is observed on the first toss.
$B$: Heads is observed on the second toss.

#### What Results Might We Expect?

Intuitively, we may consider that the coin has no memory, and that therefore, the result of previous tosses would not affect the probability of future outcomes. Thus, we would expect that successive coin tosses are independent. ■

### Solution

Figure 5.17 shows the tree diagram for this experiment.

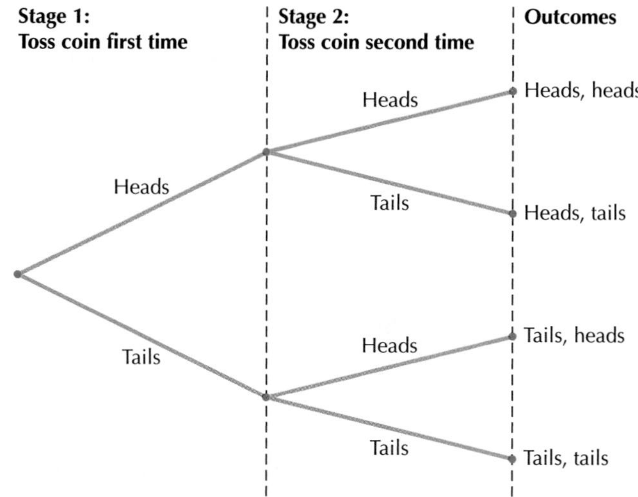


FIGURE 5.17 Tree diagram for the experiment of tossing a fair coin twice.

***Step 1*** To find $P(B)$, the probability of observing heads on the second toss, we concentrate on the second flip of the experiment, where the coin is tossed for the second time. At Stage 2, there are four branches, two of which are labeled "Heads." Therefore, since each outcome is equally likely, the probability of observing heads on the second toss is $P(B) = 2/4 = 1/2$.

***Step 2*** Next we find $P(B \mid A)$. Again consider Stage 2, but this time restrict your attention to the two "upper branches," which correspond to having observed heads on the first toss. One upper branch is labeled "Heads." Therefore, since each outcome is equally likely, $P(B \mid A) = 1/2$.

***Step 3*** Finally, we have $P(B) = P(B \mid A)$, and we conclude that events $A$ and $B$ are *independent events*. That is, *successive coin tosses are independent*. (Exercises 35–39 [page 238] show that successive die rolls are independent.)

## Example 5.18 Gambler's Fallacy

Suppose we have tossed a fair coin ten times and have observed "heads" come up every time. Find the probability of tails on the next toss.

### Solution

Since we have observed an unusual number of heads, we might think that the probability of tails on the next toss is increased. However, the short answer is "Not so." We learned in Example 5.17 that successive tosses of a coin are independent. Thus, what happened on the first ten tosses has no effect on the next toss. Probability theory tells us that in the long run, the proportion of heads and tails will eventually even out, if the coin is fair. Therefore, the probability of tails on the next toss is 0.5. This is an example of the Gambler's Fallacy.

## 3 Multiplication Rule

Just as the Addition Rule is used to find probabilities of unions of events, the **Multiplication Rule** is used to find probabilities of intersections of events. Recall the formula for the conditional probability of event $B$ given event $A$:

$$P(B \mid A) = \frac{P(A \cap B)}{P(A)}$$

We solve for $P(A \cap B)$ by multiplying each side by $P(A)$:

$$P(A \cap B) = P(A)\, P(B \mid A)$$

Similarly, consider the conditional probability of event $A$ given event $B$:

$$P(A \mid B) = \frac{P(A \cap B)}{P(B)}$$

Solving for $P(A \cap B)$ gives us a second equation for $P(A \cap B)$:

$$P(A \cap B) = P(B)\, P(A \mid B)$$

The two equations for $P(A \cap B)$ lead directly to the Multiplication Rule.

**Multiplication Rule**

$P(A \cap B) = P(A)\, P(B \mid A)$ or equivalently $P(A \cap B) = P(B)\, P(A \mid B)$

## Example 5.19 Probability of heads on both tosses of a fair coin

Again suppose that our experiment consists of tossing a fair coin twice. What is the probability of observing heads on both tosses?

**Solution**

Define the following events:

$A$: Heads is observed on the first toss.
$B$: Heads is observed on the second toss.

Since the coin is fair, $P(A) = 1/2$, and from Example 5.17, $P(B \mid A) = 1/2$. Thus, using the Multiplication Rule,

$$P(A \cap B) = P(A)\, P(B \mid A) = (1/2)(1/2) = 1/4$$

This value concurs, of course, with the result in Example 5.17 for the probability of observing two heads. ■

When events $A$ and $B$ are independent, $P(A \mid B) = P(A)$ or $P(B \mid A) = P(B)$. Using these identities, we can formulate a special case of the Multiplication Rule. Using $P(A \mid B) = P(A)$, we can write the Multiplication Rule as

$$P(A \cap B) = P(B)\, P(A \mid B) = P(B)\, P(A) = P(A)\, P(B)$$

Equivalently, the Multiplication Rule also states that $P(A \cap B) = P(A)\, P(B \mid A)$, but if $A$ and $B$ are independent, $P(B \mid A) = P(B)$, so again $P(A \cap B) = P(A)\, P(B)$.

**Multiplication Rule for Two Independent Events**

If $A$ and $B$ are any two independent events, $P(A \cap B) = P(A)\, P(B)$.

## Example 5.20 Multiplication Rule for Independent Events

Metropolitan Washington, D.C., has the highest proportion of female top-level executives in the United States: 27%.[5] We take a random sample of two top-level executives and find the probability that both are female.

### Solution

Define the following events:

*A*: First top-level executive is female.
*B*: Second top-level executive is female.

Because we are choosing the executives at random, it makes sense to assume that the events are independent. That is, if the executives are chosen at random, then we may expect that the outcome of choosing the first executive will not affect the outcome of choosing the second executive.

We are interested in finding $P(A \cap B)$. Using the Multiplication Rule for Independent Events,

$$P(A \cap B) = P(A)\,P(B) = (0.27)(0.27) = 0.0729$$ ■

Note that the Multiplication Rule for Independent Events provides us with an alternative method for determining whether two events are indeed independent.

**Alternative Method for Determining Independence**

If $P(A)\,P(B) = P(A \cap B)$, then events $A$ and $B$ are **independent**.
If $P(A)\,P(B) \neq P(A \cap B)$, then events $A$ and $B$ are **dependent**.

## Example 5.21 Determining independence using the alternative method

In 2004, 50% of professional and managerial positions were held by women.[6] Suppose that 30% of such positions are held by women who worked on their days off and that the proportion of all (male and female) such employees who worked on their days off is 53.5%. Determine whether the following events are independent or dependent.

*F*: Professional or managerial position is held by a woman.
*W*: Professional or managerial employee works on days off.

### Solution

We have $P(F) = 0.5$ and $P(W) = 0.535$. The product is then

$$P(F)\,P(W) = (0.5)\,(0.535) = 0.2675$$

We are given $P(F \cap W) = 0.30$. Since $0.30 \neq 0.2675$, we conclude that events $F$ and $W$ are dependent. ■

## Example 5.22 Conditional probability for mutually exclusive events

Suppose two events $A$ and $B$ are mutually exclusive, with $P(A) > 0$ and $P(B) > 0$.

**a.** Find $P(B \mid A)$.

**b.** Are events $A$ and $B$ independent or dependent?

**Solution**

**a.** Since $A$ and $B$ are mutually exclusive, $P(A \cap B) = 0$. Then

$$P(B \mid A) = \frac{P(A \cap B)}{P(A)} = 0$$

That is, *if event A has occurred, then event B cannot occur*. This is a natural consequence of events $A$ and $B$ being mutually exclusive.

**b.**

*What Results Might We Expect?*

Two events are independent if the occurrence of one does not affect the probability that the other will occur. However, as we saw in **(a)**, if event $A$ occurs, then the probability that event $B$ will occur is 0. Thus, we would expect events $A$ and $B$ to be dependent. ■

We are given that $P(A) > 0$ and $P(B) > 0$. Hence the product $P(A)\,P(B)$ is also greater than 0. However, from **(a)**, $P(A \cap B) = 0$. Thus, $P(A)\,P(B) \neq P(A \cap B)$, and from the alternative method for determining independence, we conclude that events $A$ and $B$ are dependent. ■

We can extend the Multiplication Rule to cover $n$ independent events.

**Multiplication Rule for $n$ Independent Events**
If $A, B, C, \ldots$ are independent events, then $P(A \cap B \cap C \cap \ldots) = P(A)\,P(B)\,P(C)\ldots$

## Example 5.23 Multiplication Rule for $n$ Independent Events

According to the National Health Interview Survey, 24% of Americans aged 18–44 smoke tobacco.

**a.** In a random sample of $n = 3$ Americans aged 18–44, find the probability that all 3 smoke.

**b.** In a random sample of $n = 10$ Americans aged 18–44, find the probability that all 10 smoke.

**Solution**

Since we are taking random samples, it is reasonable to assume that the events are independent. Let $S_i$ denote the event that the $i$th American aged 18–44 smokes.

**a.** $P(S_1) = P(S_2) = P(S_3) = 0.24$. Then, using the Multiplication Rule for Independent Events,

$$P(S_1 \cap S_2 \cap S_3) = P(S_1) \cdot P(S_2) \cdot P(S_3) = (0.24)(0.24)(0.24) = (0.24)^3 = 0.013824$$

**b.** $P(S_1) = P(S_2) = \cdots = P(S_{10}) = 0.24$. Then, using the Multiplication Rule for Independent Events,

$$P(S_1 \cap S_2 \cap \ldots \cap S_{10}) = P(S_1) \cdot P(S_2) \cdot \ldots \cdot P(S_{10}) = (0.24)^{10} \approx 0.0000006$$ ■

## Example 5.24 Solving an "at least" problem

Using information in Example 5.23, find the probability that, in a random sample of three Americans aged 18–44, at least one of them smokes.

**Solution**

The phrase "at least" means that one or more of the three Americans smoke. Using the complement, the probability for this event may be written

$$P(\text{At least one of the three Americans smokes})$$
$$= P(\text{One or more of the three Americans smoke})$$
$$= 1 - P(\text{None of the three Americans smokes})$$

The probability of not smoking for the first American is

$$P(N_1) = 1 - P(S_1) = 1 - 0.24 = 0.76$$

and similarly for each American in the sample. Thus,

$$P(\text{None of the three Americans smokes}) = P(N_1) \cdot P(N_2) \cdot P(N_3) = (0.76)^3 = 0.438976$$

Hence, the probability that at least one of the three Americans smokes is

$$1 - P(\text{None of the three Americans smokes}) = 1 - 0.438976 = 0.561024$$

What If Scenario ? Give the Calculator a Rest

Suppose that the percentage of Americans aged 18–44 who smoke tobacco this year has decreased to less than 24%, though we are not sure how much less. Determine whether the following quantities will increase or decrease from the values calculated in Examples 5.23 and 5.24.

**a.** In a random sample of $n = 3$ Americans aged 18–44, the probability that all 3 smoke

**b.** In a random sample of $n = 3$ Americans aged 18–44, the probability that none of them smokes

**Solution**

**a.** Let $P(S_1^*) < 0.24$ represent the revised probability that an American aged 18–44 smokes. Then

$$P(S_1^* \cap S_2^* \cap S_3^*) = P(S_1^*) \cdot P(S_2^*) \cdot P(S_3^*)$$
$$< P(S_1) \cdot P(S_2) \cdot P(S_3) = P(S_1 \cap S_2 \cap S_3)$$

Thus, the probability that all three will smoke will decrease.

**b.** If $P(S_1^*) < 0.24$, then $P(N_1^*) = 1 - P(S_1^*) > 1 - 0.24 = P(N_1)$; that is, the probability that an American aged 18–44 doesn't smoke has increased. Thus,

$$P(\text{None of the three Americans smokes}) = P(N_1^*) \cdot P(N_2^*) \cdot P(N_3^*) > (0.76)^3 = 0.438976.$$

Therefore, the probability that none of the three Americans aged 18–44 smokes will increase.

## 4 Sampling With and Without Replacement

Recall from Chapter 1 that there are various methods for **sampling**, or selecting elements from a population. Choosing the right sampling method is important because an inappropriate sampling method can lead to a faulty conclusion. In this section, we uncover the association between independent events and a certain kind of sampling called **sampling with replacement**.

In **sampling with replacement**, the randomly selected unit is returned to the population after being selected. When sampling with replacement, it is possible for the same unit to be sampled more than once.

In **sampling without replacement**, the randomly selected unit is not returned to the population after being selected. When sampling without replacement, it is not possible for the same unit to be sampled more than once.

## Example 5.25 Sampling with replacement: What are the chances of drawing two aces in a row?

We draw a card at random from a shuffled deck, observe the card, and return it to the deck. The deck is then reshuffled, and we are to draw another card at random. What is the probability that both cards we select will be aces?

### Solution

Define the following events:

$A$: Observe an ace on the first draw.
$B$: Observe an ace on the second draw.

We want to find $P(A \cap B)$, the probability of observing an ace on the first draw *and* an ace on the second draw. From the Multiplication Rule, $P(A \cap B) = P(A)\,P(B \mid A)$. To find

$$P(A) = \text{probability of observing an ace on the first draw}$$

recall that there are 4 aces in the deck of 52 cards. It is reasonable to assume that all cards are equally likely to be selected, so using the classical method, $P(A) = 4/52$. Similarly, $P(B) = 4/52$.

Next we need to find $P(B \mid A)$, the probability of observing an ace on the second draw, given that we observe an ace on the first draw. Since *the deck of 52 cards has not changed* (except for shuffling), there are still 52 cards, 4 of which are aces. So regardless of which card we drew first (even an ace), the probability of observing an ace on the second draw is again 4/52. Therefore, $P(B \mid A) = 4/52$. Thus, the probability that both cards we select will be aces is $(4/52)(4/52) \approx 0.0059$.

Note that $P(B \mid A) = P(B) = 4/52$. Thus, by the alternative method for determining independence, $A$ and $B$ are independent events when sampling with replacement. ■

We can generalize as follows.

When sampling with replacement, successive draws can be considered **independent**.

## Example 5.26 Sampling without replacement: Now what are the chances of drawing two aces in a row?

Suppose we alter the experiment in Example 5.25 as follows: We draw a card at random from a shuffled deck, hold onto the card (do not replace it) while the deck is reshuffled, and then select another card at random. What is the probability that both cards we select will be aces?

Punchstock/Charles Sturge

### Solution

Define events $A$ and $B$ as in Example 5.25. Again we use the Multiplication Rule to find $P(A \cap B)$, the probability of observing an ace on the first draw *and* an ace on the second draw. $P(A)$ is still the same as in Example 5.24: $P(A) = 4/52$. The difference in this experiment comes when finding $P(B \mid A)$, the probability of observing an ace on the second draw given an ace on the first draw. Once we select the first ace, we do not replace it in the deck. Therefore, when the deck is reshuffled, it has only 51 cards left, only 3 of which are aces. The classical method then gives the probability of observing an ace on the second draw:

$$P(B \mid A) = \frac{\text{Number of aces in the deck}}{\text{Number of cards in the deck}} = \frac{3}{51}$$

Thus, the probability that both cards we select will be aces is

$$P(A \cap B) = P(A)P(B \mid A) = \frac{4}{52} \cdot \frac{3}{51} = \frac{12}{2652} \approx 0.0045$$

This probability is somewhat less than the probability that both cards will be aces when sampling with replacement. Note that here we found that $P(B \mid A)$ was not equal to $P(B)$. Thus, by the alternative method for determining independence, $A$ and $B$ are not independent events; they are dependent events. ■

We can generalize as follows.

When sampling without replacement, successive draws should be considered **dependent**.

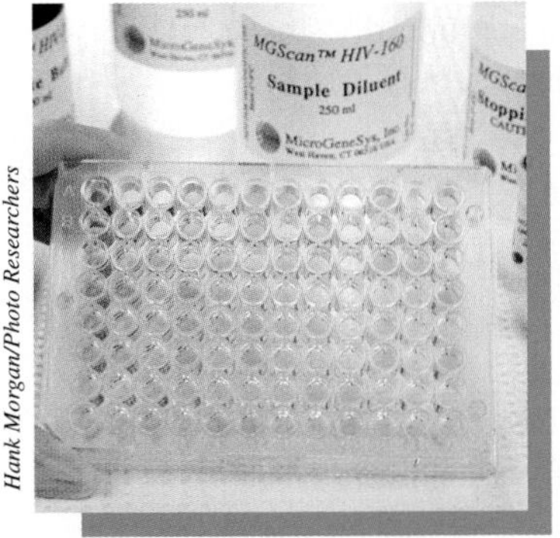

Hank Morgan/Photo Researchers

## CASE STUDY The ELISA Test for the Presence of HIV

The ELISA test is used to screen blood for the presence of HIV. Like most diagnostic procedures, the test is not foolproof.

- When a blood sample contains HIV, the ELISA test will give a positive result 99.6% of the time. That is, the *false negative rate,* the percentage of tests returning a negative result when the HIV virus is actually present, is $1 - 0.996 = 0.004$.
- When the blood does not contain HIV, the ELISA test will give a negative result 98% of the time. That is, the *false positive rate,* the percentage of tests returning a positive result when the HIV virus is not actually present, is $1 - 0.98 = 0.02$.

A positive result means that the test says that the person has the HIV infection. A negative result means that the test says that the person does not have the HIV virus. The *prevalence rate* for HIV in the general population is 0.5%. That is, 5 of 1000 persons in the general population have HIV.

Suppose we have samples of blood from 100,000 randomly chosen people.

### Problem 1

How many people in the sample of 100,000 have HIV? How many do not?

**Solution.** The prevalence rate of 0.5% means that 0.005 (100,000) = 500 people in the sample have HIV. The remainder—99,500—do not.

### Problem 2

A positive result is given 99.6% of the time for blood containing HIV. For the 500 people with HIV, how many positive results will the ELISA test return? How many of the 500 people with HIV will receive a negative result?

**Solution.** The ELISA test will return a positive result for 0.996 (500) = 498 of the 500 people. Thus, two people who actually have HIV will receive a test result indicating that they do not have the virus.

### Problem 3

A negative result is given 98% of the time for blood without HIV. For the 99,500 people without HIV, how many negative results will the ELISA test return? Positive results?

**Solution.** The ELISA test will return a negative result for 0.98 (99,500) = 97,510 of the 99,500 people without HIV. The remaining 2%, or 1990 people will receive positive ELISA test results, even though they do not have the virus.

We can use the counts we found to fill in the following table.

| | In Reality | | |
|---|---|---|---|
| **ELISA test results** | **Person has HIV** | **Person does not have HIV** | **Total** |
| Positive | 498 | 1,990 | 2,488 |
| Negative | 2 | 97,510 | 97,512 |
| Total | 500 | 99,500 | 100,000 |

We will use the information in the ELISA test contingency table to solve Problems 4 and 5. If a person is chosen at random from the sample of 100,000, define the following events:

$A$: Person has HIV.

$A^C$: Person does not have HIV.

Pos: ELISA test returned positive results.

Neg: ELISA test returned negative results.

### Problem 4

What is the probability that a randomly chosen person actually does have HIV, given that the ELISA results are negative? In other words, find $P(A \mid \text{Neg})$.

### Solution

$$P(A \mid \text{Neg}) = \frac{N(A \cap \text{Neg})}{N(\text{Neg})} = \frac{2}{97{,}512} \approx 0.0000205$$

### Problem 5

What is the probability that a randomly chosen person actually does not have HIV, given that the ELISA test results are positive? In other words, find $P(A^C \mid \text{Pos})$.

### Solution

$$P(A^C \mid \text{Pos}) = \frac{N(A^C \cap \text{Pos})}{N(\text{Pos})} = \frac{1990}{2488} \approx 0.7998 \approx 0.80$$

**Which Error Is More Dangerous?**

In Problems 4 and 5, we examined the probabilities of the two ways that the ELISA test can be wrong. Which error do you think is more dangerous? $P(A \mid \text{Neg})$ represents the probability that HIV is present, even though the ELISA test says otherwise. $P(A^C \mid \text{Pos})$ represents the probability that HIV is not present, even though the ELISA test says it is present. The designers of the ELISA test worked hard to reduce the false negative rate $P(A \mid \text{Neg})$ to as low a level as possible. They rightly considered that it is the more dangerous type of error because of the epidemic nature of the illness. A person who receives a false negative ELISA result could spread the infection further. Therefore, the designers tried to keep this probability as low as they could.

There is price to be paid, however, which is the high false positive rate, $P(A^C \mid \text{Pos})$, a very high 80%. Thus, if a random person receives a positive ELISA test result, the probability that the person does not have HIV is 80%. When the ELISA test comes back positive, a second batch of tests that have a more reasonable false positive rate is usually administered. ■

## Section 5.3 SUMMARY

1 Section 5.3 discusses conditional probability $P(B \mid A)$, the probability of an event $B$ given that an event $A$ has occurred.

2 We can compare $P(B \mid A)$ to $P(B)$ to determine whether the events $A$ and $B$ are independent. Events are independent if the occurrence of one event does not affect the probability that the other event will occur.

3 The Multiplication Rule for Independent Events is the product of the individual probabilities.

4 Sampling with replacement is associated with independence, while sampling without replacement means that the events are not independent.

## Section 5.3 EXERCISES

### CLARIFYING THE CONCEPTS

**1.** Suppose you are the coach of a football team, and your star quarterback is injured.
- **a.** Does the injury affect the chances that your team will win the big game this weekend?
- **b.** How would you describe this situation in the terminology presented in this section?

**2.** Write a sentence or two about a situation in your life similar to Exercise 1, where the probability of some event was affected by whether or not some other event occurred.

**3.** Explain clearly the difference between $P(A \cap B)$ and $P(A \mid B)$.

**4.** Give an example from your own experience of two events that are independent. Describe how they are independent.

**5.** Picture yourself explaining to your friends about the Gambler's Fallacy. How would you explain the Gambler's Fallacy in your own words?

**6.** Explain why two events $A$ and $B$ cannot have the following characteristics: $P(A) = 0.25$, $P(B) = 0.25$ and $P(A \cap B) = 0.30$. *Hint*: Figure 5.16b might help.

**7.** Explain why each of the following events is either dependent or independent.
- **a.** Drawing a ball from a box, replacing it, and then drawing a second ball
- **b.** Drawing a ball from a box, not replacing it, and then drawing a second ball

**8.** Explain why the following events are either dependent or independent, and provide support for your assertion.
- **a.** Tossing a coin and drawing a card from a deck of playing cards
- **b.** Drawing a card from a deck, not replacing it, and drawing another card

### PRACTICING THE TECHNIQUES

For Exercises 9–12, let $A$ and $B$ be two independent events, with $P(A) = 0.6$ and $P(B) = 0.4$. Find the indicated probabilities.

**9.** $P(A \cap B)$
**10.** $P(A \mid B)$
**11.** $P(B \mid A)$
**12.** $P(A \cup B)$

For Exercises 13–16, let $A$ and $B$ be two independent events, with $P(A) = 0.5$ and $P(B) = 0.2$. Find the indicated probabilities.

**13.** $P(A \cap B)$
**14.** $P(A \mid B)$
**15.** $P(B \mid A)$
**16.** $P(A \cup B)$

**17.** Suppose that $A$ and $B$ are two events with $P(A) = 0.3$ and $P(A \cap B) = 0.05$. Find $P(B \mid A)$.

**18.** Suppose that $A$ and $B$ are two events, with $P(A) = 0.9$ and $P(B \mid A) = 0.6$. Find $P(A \cap B)$.

For Exercises 19–22, use the Multiplication Rule for $n$ Independent Events to find the probabilities.

**19.** Observing tails on each of three successive tosses of a fair coin

**20.** Observing tails on each of four successive tosses of a fair coin

**21.** Observing tails on each of five successive tosses of a fair coin

**22.** Observing tails on each of six successive tosses of a fair coin

For Exercises 23–28, let $A$ and $B$ be independent events such that $P(A) = 0.4$ and $P(B) = 0.5$. Find the indicated probabilities.

**23.** $P(A \text{ and } B)$

**24.** $P(A \mid B)$

**25.** $P(B \mid A)$

**26.** $P(A \text{ or } B)$

**27.** $P(A \text{ and } B)^C$

**28.** $P(A \text{ or } B)^C$

For Exercises 29–32, let $C$ and $D$ be events such that $P(C) = 0.7$, $P(D) = 0.3$, and $P(C \text{ and } D) = 0.21$.

**29.** Find $P(C \mid D)$.

**30.** Find $P(D \mid C)$.

**31.** Are events $C$ and $D$ independent? How can you tell?

**32.** Are events $C$ and $D$ mutually exclusive? How can you tell?

For Exercises 33 and 34, let $E$ and $F$ be events such that $P(E) = 0.5$ and $P(F) = 0.6$.

**33.** What further information do we need to know to determine whether events $E$ and $F$ are independent?

**34.** What further information do we need to know to determine whether events $E$ and $F$ are mutually exclusive?

For Exercises 35–38, a single fair die is rolled twice in succession. Find the indicated probabilities.

**35.** Observe a 1 on the second roll

**36.** Observe a 1 on the second roll, given that you observe a 1 on the first roll

**37.** Observe an even number on the second roll

**38.** Observe an even number on the second roll, given that you observe an even number on the first roll

**39.** Based on the probabilities in Exercises 35–38, what can you say about the dependence or independence of successive rolls of a single fair die?

## APPLYING THE CONCEPTS

**40. Teen Birth Rate.** The Federal Interagency Forum on Child and Family Statistics (**www.childstats.gov**) reported that the teenage birth rate in 2000 was 0.0453.

- **a.** Find the probability that two randomly selected births are to teenagers.
- **b.** Find the probability that five randomly selected births are to teenagers.
- **c.** Find the probability that at least one of four randomly selected births is to a teenager.

**41. Fair Die Experiment.** The experiment is to roll a fair die once.

- **a.** What is the probability of rolling a 6 with a fair die?
- **b.** What if we are told that the number rolled is an even number? How does this knowledge affect the probability of rolling a 6?
- **c.** We saw that the probability of rolling a six increased when we were told that an even number was observed. Does this always happen? Can you think of an example where the probability of an event decreases given the occurrence of another event?

**Gender and Pet Preference.** Use Table 5.12 for Exercises 42–46. Do you think your gender affects what type of pet you own?

**TABLE 5.12 Pet preference**

| Gender | Cats | Dogs | Other pets | Total |
|---|---|---|---|---|
| Female | 100 | 50 | 30 | 180 |
| Male | 50 | 50 | 20 | 120 |
| Total | 150 | 100 | 50 | 300 |

**42.** Find the probabilities that a randomly chosen person has the following characteristics.

- **a.** Is female, $P(F)$
- **b.** Is male, $P(M)$
- **c.** Owns a cat, $P(C)$
- **d.** Owns some other kind of pet, $P(O)$

**43.** Find the probability that a randomly chosen person has the following characteristics.

- **a.** Is female and owns a cat, $P(F \cap C)$
- **b.** Is female and owns some other kind of pet, $P(F \cap O)$
- **c.** Is male and owns a cat, $P(M \cap C)$
- **d.** Is male and owns some other kind of pet, $P(M \cap O)$

**44.** Find the following conditional probabilities for a randomly chosen person.

- **a.** Owns a cat, given that the person is female, $P(C \mid F)$
- **b.** Owns a cat, given that the person is male, $P(C \mid M)$
- **c.** Owns some other kind of pet, given that the person is female, $P(O \mid F)$
- **d.** Owns some other kind of pet, given that the person is male, $P(O \mid M)$

**45.** Are gender and pet preference independent? Why or why not?

**46.** If you were a cat-food manufacturer, would you advertise more in men's magazines or women's magazines? Why? Cite your evidence.

**47. Balls in a Box.** A box contains four blue balls and three red balls. If we select two balls at random, what is the probability that both balls will be blue if

- **a.** we sample with replacement.
- **b.** we sample without replacement.

**48. Acceptance Sampling.** You are in charge of purchasing for a large computer retailer. Your wholesaler delivers computers to you in batches of 100. You either accept or reject an entire batch based on a random sample of two computers: if both computers you sample are defective, then you reject the entire batch. Suppose that (unknown to you, of course) there are 10 defective computers in the batch of 100 computers.

- **a.** Should you conduct your sampling with or without replacement? Why?
- **b.** What is the probability that the first computer you select is defective?
- **c.** What is the probability that the second computer you select is defective, given that the first was defective, if you sample without replacement?
- **d.** What is the probability that you will accept the batch?
- **e.** What is the probability that you will reject the batch?
- **f.** Usually you accept each batch of computers from this wholesaler. Do you think that is a wise move, considering that 10% of their product is defective? How could you make your test stricter so that there is a smaller chance of accepting a batch with 10% defectives?

**49. 9/11 Information.** Within two weeks of the terrorist attacks of September 11, 2001, the Pew Internet and American Life project conducted a survey, asking, among other things, from which medium the respondent received most of his/her information regarding these events. The results indicated that 1.9% of Americans received most of their information about September 11 from the Internet. Assuming independence, find the following probabilities.

- **a.** 2 randomly selected people both received most of their information from the Internet
- **b.** 2 randomly selected people both received most of their information from sources other than the Internet
- **c.** 5 randomly selected people received most of their information from sources other than the Internet
- **d.** At least 1 of 3 randomly selected persons received most information from the Internet

**50. Treasury Bonds.** One of the most important tasks for economists is to make forecasts for the performance (up or down) of investments such as 30-year Treasury bonds. *The Journal of Investing* (Vol. 6, No. 2, page 8, 1997) reports that, in a sample of 30 six-month surveys, the consensus estimate of performance for the 30-year Treasury bond has been wrong 20 out of the 30 times!

- **a.** Find the probability that two randomly selected consensus estimates were correct.
- **b.** Find the probability that three randomly selected consensus estimates were wrong.
- **c.** If we choose two consensus estimates and if we sample with replacement, find the probability that the second consensus estimate was right, given that the first consensus estimate was right. Are the successive draws independent? Why or why not?
- **d.** Repeat **(c)**, this time sampling without replacement. Are the successive draws independent? Why or why not?

**51. Female Students.** At Northern Connecticut University, 55% of the students are female. Of these, 10% are business majors. Find the following probabilities, assuming sampling with replacement.

- **a.** A randomly chosen student is a female business major.
- **b.** Two randomly selected students are females.
- **c.** The second of two randomly selected students is a female, given that the first is a female.
- **d.** The second of two randomly selected students is a female, given that the first is a male.
- **e.** Two randomly selected students are female business majors.
- **f.** If possible, find the probability that two randomly selected students are business majors. If not possible, explicitly explain why not.

**52. Adjustable Rate Mortgages.** Half of the 20 mortgages provided by a certain mortgage lending company last week are adjustable rate mortgages (ARMs). Suppose we sample three mortgages without replacement. Find the following probabilities.

- **a.** The first mortgage is an ARM.
- **b.** The second mortgage is an ARM, given that the first mortgage is an ARM.
- **c.** The third mortgage is an ARM, given that the first two mortgages are ARMs.

**53. Sampling Songs.** Juan is sampling music at his local CD store. Suppose Juan is sampling one particular CD, and on this CD are ten songs, of which Juan would like six. Juan will sample only two songs. If he likes both songs, he will purchase the CD. Assuming that the songs are sampled without replacement, find the following probabilities.

- **a.** Juan will like the first song he hears.
- **b.** Juan will not like the first song he hears.
- **c.** Juan will like both the first and the second songs he hears.
- **d.** Juan will like the second song, given that he liked the first song.
- **e.** Juan will purchase the CD.

**9/11 and Pearl Harbor.** What were the feelings of Americans in the days immediately following the events of September 11, 2001? In an NBC News Terrorism Poll, the following question was asked: "Would you say that Tuesday's attacks are more serious than, equal to, or not as serious as the Japanese attack on Pearl Harbor?" This poll was conducted on September 12, 2001. Use the following crosstabulation of the poll results for Exercises 54–57.

| | SEX | | |
|---|---|---|---|
| | Male | Female | Total |
| More serious than Pearl Harbor | 200 | 212 | 412 |
| Equal to Pearl Harbor | 70 | 84 | 154 |
| Not as serious as Pearl Harbor | 23 | 6 | 29 |
| Not sure | 11 | 12 | 23 |
| | 304 | 314 | 618 |

**54.** Find the probabilities that a randomly chosen person has the following characteristics.

**a.** Is female, $P(F)$

**b.** Is male, $P(M)$

**c.** Believes September 11 is more serious than Pearl Harbor, $P(\text{More})$

**55.** Find the probability that a randomly chosen person has the following characteristics.

**a.** Is female and believes September 11 is more serious than Pearl Harbor, $P(F \cap \text{More})$

**b.** Is male and believes September 11 is more serious than Pearl Harbor, $P(M \cap \text{More})$

**56.** Find the following conditional probabilities for a randomly chosen person.

**a.** Given that the person is female, believes September 11 is more serious than Pearl Harbor, $P(\text{More} \mid F)$

**b.** Given that the person is male, believes September 11 is more serious than Pearl Harbor, $P(\text{More} \mid M)$

**57.** Are gender and the belief whether September 11 was more or less serious than Pearl Harbor independent? Why or why not?

# 5.4 Counting Methods

**Objectives:** By the end of this section, I will be able to...

1 Apply the Multiplication Rule for Counting to solve certain counting problems.

2 Use permutations and combinations to solve certain counting problems.

3 Compute probabilities using combinations.

Counting methods allow us to solve a range of problems, including how to compute certain probabilities.

## 1 Multiplication Rule for Counting

Let us begin with an example illustrating a general rule of counting.

### Example 5.27 Design your own T-shirt

Bruce Laurance/The Image Bank/Getty Images

A store at the local mall allows customers to design their own T-shirts. The store offers the following options to its customers:

- **Sleeve type:** Long-sleeve or short-sleeve
- **Color:** White, black, or red
- **Image:** Stock picture or uploaded photo

Figure 5.18 is a tree diagram that shows all the different T-shirts that can be designed.

**Solution**

There are two choices for type of sleeve. For each sleeve type, there are three choices for color. For each color, there are two choices of image: stock picture or uploaded photo. All together, customers have a choice from among

$$2 \cdot 3 \cdot 2 = 12$$

different T-shirt options. ■

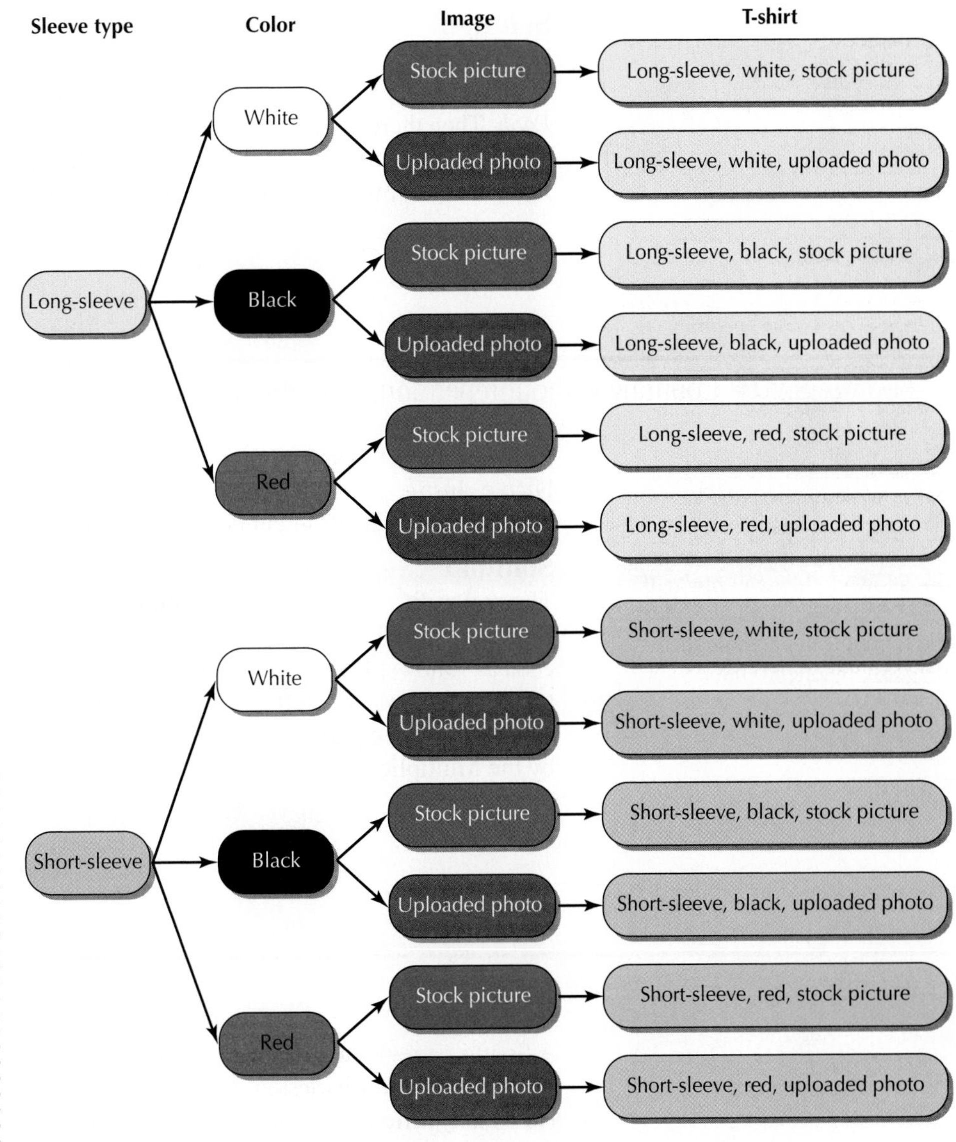


FIGURE 5.18 Tree diagram for the different T-shirt options.

We can generalize the result as the **Multiplication Rule for Counting**.

> **Multiplication Rule for Counting**
> Suppose an activity consists of a series of events in which there are *a* possible outcomes for the first event, *b* possible outcomes for the second event, *c* possible outcomes for the third event, and so on. Then the total number of different possible outcomes for the series of events is:
>
> $$a \cdot b \cdot c \cdot \ldots$$

## Example 5.28 Counting with repetition: Famous initials

Some Americans in history are uniquely identified by their initials. For example, "JFK" stands for John Fitzgerald Kennedy, and "FDR" stands for Franklin Delano Roosevelt. How many different possible sets of initials are there for people with a first, middle, and last name?

**Solution**

Let us consider the three initials as an activity consisting of three events. Note that a particular letter may be repeated, as in "AAM" for A. A. Milne, author of *Winnie the Pooh.* Then there are $a = 26$ ways to choose the first initial, $b = 26$ ways to choose the second initial, and $c = 26$ ways to choose the third initial. Thus, by the Multiplication Rule for Counting, the total number of different sets of initials is

$$26 \cdot 26 \cdot 26 = 17{,}576$$

## Example 5.29 Counting without repetition: Intramural singles tennis

*Note:* To summarize the key difference between Examples 5.28 and 5.29, if repetitions are allowed, then $a = b = c$. If repetitions are not allowed, then the numbers being multiplied decrease by one from left to right.

A local college has an intramural singles tennis league with five players, Ryan, Megan, Nicole, Justin, and Kyle. The college presents a trophy to the top three players in the league. How many different possible sets of three trophy winners are there?

**Solution**

The major difference between Examples 5.28 and 5.29 is that in Example 5.29 there can be no repetition. Ryan cannot finish in first place *and* second place. So we proceed as follows. Five possible players could finish in first place, so $a = 5$. Now there are only four players left, one of whom will finish in second place, so $b = 4$. That leaves only three players, one of whom will finish in third place, giving $c = 3$. Thus, by the Multiplication Rule for Counting, the number of different possible sets of trophy winners is

$$5 \cdot 4 \cdot 3 = 60$$

## Example 5.30 Traveling salesman problem

A Southeast regional salesman has eight destinations that he must travel to this month: Atlanta, Raleigh, Charleston, Nashville, Jacksonville, Richmond, Mobile, and Jackson. How many different possible routes could he take?

**Solution**

The salesman has $a = 8$ different choices for where to go first. Once the first destination has been chosen, there are only $b = 7$ choices for where to go second. And once the first two destinations have been chosen, there are only $c = 6$ choices for where to go third. And so on. Thus, by the Multiplication Law for Counting, the number of different possible routes for the salesman is

$$a \cdot b \cdot c \cdot d \cdot e \cdot f \cdot g \cdot h = 8 \cdot 7 \cdot 6 \cdot 5 \cdot 4 \cdot 3 \cdot 2 \cdot 1 = 40{,}320$$

The calculation in Example 5.30 leads us to introduce the **factorial symbol,** which is used for the counting rules we will learn in the remainder of this section.

For any integer $n \geq 0$, the **factorial symbol** $\boldsymbol{n!}$ is defined as follows:

- $0! = 1$
- $1! = 1$
- $n! = n(n-1)(n-2) \cdots 3 \cdot 2 \cdot 1$

For example:

- $2! = 2 \cdot 1 = 2$
- $3! = 3 \cdot 2 \cdot 1 = 6$
- $4! = 4 \cdot 3 \cdot 2 \cdot 1 = 24$
- $5! = 5 \cdot 4 \cdot 3 \cdot 2 \cdot 1 = 120$
- $6! = 6 \cdot 5 \cdot 4 \cdot 3 \cdot 2 \cdot 1 = 720$
- $7! = 7 \cdot 6 \cdot 5 \cdot 4 \cdot 3 \cdot 2 \cdot 1 = 5040$
- $8! = 8 \cdot 7 \cdot 6 \cdot 5 \cdot 4 \cdot 3 \cdot 2 \cdot 1 = 40{,}320$, as in Example 5.30.

## 2 Permutations and Combinations

### Example 5.31 Traveling to some but not all of the cities

Example 5.30 calculated the number of possible routes for traveling to $n = 8$ cities. However, suppose, we are interested in traveling to *some but not all* of the cities? For example, suppose that the salesman is traveling to three of the eight cities. Find the number of possible routes.

**Solution**

There are eight choices for the first city, seven choices for the second city, and six choices for the third city. Since the salesman is traveling to three cities only, the number of possible routes is thus

$$8 \cdot 7 \cdot 6 = 336$$

This result may be rewritten using factorial notation, as follows:

$$8 \cdot 7 \cdot 6 = \frac{8 \cdot 7 \cdot 6 \cdot (5 \cdot 4 \cdot 3 \cdot 2 \cdot 1)}{(5 \cdot 4 \cdot 3 \cdot 2 \cdot 1)} = \frac{8!}{5!} = \frac{8!}{(8-3)!}$$

Example 5.31 leads us to the following definition.

**Permutations**

A **permutation** is an arrangement of items, such that

- $r$ items are chosen at a time from $n$ distinct items.
- repetition of items is not allowed.
- the order of the items is important.

The number of permutations of $n$ items chosen $r$ at a time is denoted as ${}_nP_r$ and given by the formula:

$${}_nP_r = \frac{n!}{(n-r)!}$$

In Example 5.31, we are looking for the number of permutations of 8 cities taken 3 at a time. We have $n = 8$, $r = 3$,

$$_nP_r = {_8P_3} = \frac{n!}{(n-r)!} = \frac{8!}{(8-3)!} = \frac{8!}{5!} = 8 \cdot 7 \cdot 6 = 336$$

## Example 5.32 Counting permutations: Secret Santas

"Secret Santa" refers to a method whereby each member of group anonymously buys a holiday gift for another member of the group. Each person is secretly assigned to buy a gift for another randomly chosen person in the group. Suppose Jessica, Laverne, Samantha, and Luisa share a dorm suite and would like to do Secret Santa this holiday season.

**a.** Verify that in this instance one woman purchasing a gift for another woman represents a permutation.

**b.** Calculate how many possible different permutations of gift buying there are for the four women.

### Solution

**a.**
- There are $n = 4$ women, and $r = 2$ two people are associated with each gift, the giver and the receiver.
- Each person can buy only one gift, so repetition is not allowed.
- Finally, there is a difference between Jessica buying for Laverne and Laverne buying for Jessica. Thus, order is important, and thus, buying a gift represents a permutation.

**b.** The number of permutations is calculated as follows:

$$_nP_r = {_4P_2} = \frac{4!}{(4-2)!} = \frac{4 \cdot 3 \cdot 2!}{2!} = 12$$

■

## Example 5.33 Calculating numbers of permutations

Find the following numbers of permutations.

**a.** $_5P_2$ **b.** $_6P_2$ **c.** $_6P_6$

### Solution

**a.** $_5P_2 = \frac{5!}{(5-2)!} = \frac{5 \cdot 4 \cdot 3!}{3!} = 20$

**b.** $_6P_2 = \frac{6!}{(6-2)!} = \frac{6 \cdot 5 \cdot 4!}{4!} = 30$

**c.** $_6P_6 = \frac{6!}{(6-6)!} = \frac{6 \cdot 5 \cdot 4 \cdot 3 \cdot 2 \cdot 1}{0!} = 720$

■

In a permutation, order is important. For example, in Example 5.32, there was a difference between Jessica buying a gift for Laverne and Laverne buying one for Jessica. However, what if we consider shaking hands instead? Then Jessica shaking hands with Laverne is considered the same as Laverne shaking hands with Jessica. Hence, sometimes order is not important. What is important here is the **combination** of Jessica and Laverne.

**Combinations**

A **combination** is an arrangement of items in which

- $r$ items are chosen from $n$ distinct items.
- repetition of items is not allowed.
- the order of the items is not important.

The number of combinations of $r$ items chosen from $n$ different items is denoted as

$$_nC_r$$

## Example 5.34 How many combinations in the intramural tennis league?

We return to the intramural singles tennis league at the local college. There are five players, Ryan, Megan, Nicole, Justin, and Kyle. Each player must play each other once.

**a.** Confirm that a match between two players represents a combination.

**b.** How many matches will be held?

### Solution

**a.** Let {Ryan, Megan} denote a tennis match between Ryan and Megan. *Note:*

- There are $r = 2$ players chosen from $n = 5$ players.
- Each player plays each other player once, so repetition is not allowed.
- There is no difference between {Ryan, Megan} and {Megan, Ryan}, so order is not important.

Thus, a tennis match between two players represents a combination.

**b.** The list of all matches is as follows.

| | | |
|---|---|---|
| {Ryan, Megan} | {Megan, Nicole} | {Nicole, Justin} |
| {Ryan, Nicole} | {Megan, Justin} | {Nicole, Kyle} |
| {Ryan, Justin} | {Megan, Kyle} | {Justin, Kyle} |
| {Ryan, Kyle} | | |

Thus there are $_5C_2 = 10$ possible matches of $r = 2$ players chosen from $n = 5$ players. ■

We saw in Example 5.33 that $_5P_2 = 20$ and in Example 5.34 that $_5C_2 = 10$. Permutations and combinations differ only in that ordering is ignored for combinations. To calculate the number of combinations $_nC_r$, we simply do not count however many rearrangements there are of the same items. For example, in Example 5.34, there are $r! = 2! = 2$ rearrangements of the same players, such as {Ryan, Megan} and {Megan, Ryan}. Thus,

$$_5C_2 = \frac{_5P_2}{2!} = \frac{20}{2} = 10$$

In general, the **number of combinations** can be computed as the number of permutations divided by factorial of the number of items chosen.

**Formula for the Number of Combinations**

The number of combinations of $r$ items chosen from $n$ different items is given by

$$_nC_r = \frac{n!}{r!(n-r)!}$$

For instance, in Example 5.34, the formula for the number of combinations is

$$_5C_2 = \frac{5!}{2!(5-2)!} = \frac{5!}{2!3!} = \frac{5 \cdot 4 \cdot 3!}{2 \cdot 1 \cdot 3!} = \frac{20}{2} = 10$$

Thus the relation: $_5C_2 = {_5P_2}/2!$ is verified.

## Example 5.35 Calculating numbers of combinations

*Note:* Following are some special combinations you may find useful. For any integer $n$:

$_nC_n = 1$
$_nC_0 = 1$
$_nC_1 = n$
$_nC_{n-1} = n$

Find the following numbers of combinations.

**a.** $_6C_2$ **b.** $_6C_3$ **c.** $_6C_4$

### Solution

**a.** $_6C_2 = \frac{6!}{2!(6-2)!} = \frac{6 \cdot 5 \cdot 4!}{2 \cdot 1 \cdot 4!} = \frac{30}{2} = 15$

**b.** $_6C_3 = \frac{6!}{3!(6-3)!} = \frac{6 \cdot 5 \cdot 4 \cdot 3!}{3 \cdot 2 \cdot 1 \cdot 3!} = \frac{120}{6} = 20$

**c.** $_6C_4 = \frac{6!}{4!(6-4)!} = \frac{6!}{(6-4)!4!} = \frac{6 \cdot 5 \cdot 4!}{2 \cdot 1 \cdot 4!} = \frac{30}{2} = 15$

Note that in **(c)** we used the commutative property of multiplication ($a \cdot b = b \cdot a$) and found that $_6C_4 = {_6C_2} = 15$. In general, $_nC_r = {_nC_{n-r}}$ for this reason. ■

## Example 5.36 Calculating the number of permutations and combinations using technology

Use the TI-83/84 and Excel to calculate the following.

**a.** $_9P_6$ **b.** $_{10}C_7$

### Solution

We use the instructions provided in the Step-by-Step Technology Guide at the end of this section (page 250).

**a.** From Figures 5.19 and 5.20 we find that $_9P_6 = 60{,}480$.

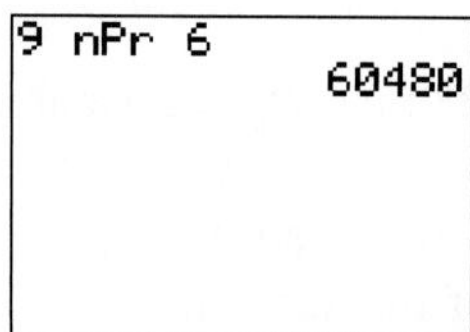

FIGURE 5.19 TI-83/84 permutation results.

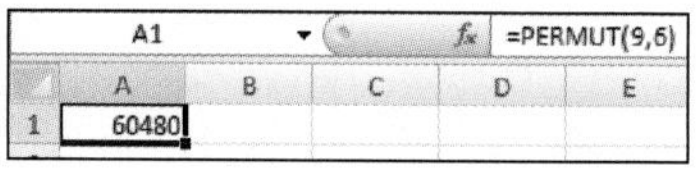

| A1 | =PERMUT(9,6) | | | |
|---|---|---|---|---|
| A | B | C | D | E |
| 60480 | | | | |

FIGURE 5.20 Excel permutation results.

**b.** From Figures 5.21 and 5.22 we find that $_{10}C_7 = 120$. ■

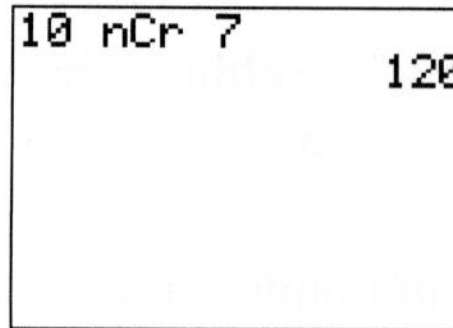

FIGURE 5.21 TI-83/84 combination results.

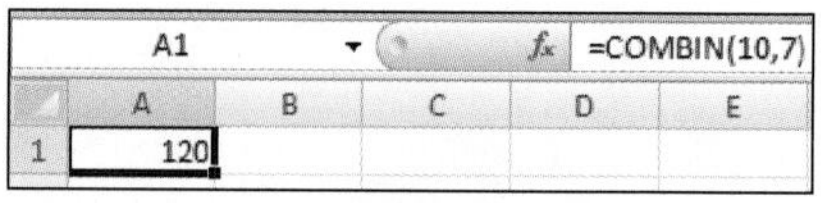

| A1 | =COMBIN(10,7) | | | |
|---|---|---|---|---|
| A | B | C | D | E |
| 120 | | | | |

FIGURE 5.22 Excel combination results.

Sometimes we wish to find the number of permutations of items where some of the items are not distinct.

## Example 5.37 Permutations with nondistinct items

How many distinct strings of letters can we make by using all the letters in the word STATISTICS?

### Solution

Each string will be ten letters long and include 3 S's, 3 T's, 2 I's, 1 A, and 1 C. The ten positions shown here need to be filled.

| ___ | ___ | ___ | ___ | ___ | ___ | ___ | ___ | ___ | ___ |
|---|---|---|---|---|---|---|---|---|---|
| **1** | **2** | **3** | **4** | **5** | **6** | **7** | **8** | **9** | **10** |

The string-forming process is as follows:

***Step 1*** Choose the positions for the three S's.

***Step 2*** Choose the positions for the three T's.

***Step 3*** Choose the positions for the two I's.

***Step 4*** Choose the position for the one A.

***Step 5*** Choose the position for the one C.

There are ${}_{10}C_3$ ways to place the three S's in Step 1. Once Step 1 is done, there are seven slots left, leaving ${}_7C_3$ positions for the three T's. Once Step 2 is done, there are four slots left, so there are ${}_4C_2$ ways to place the two I's. Once Step 3 is done, there are only 2 slots left, so there are only ${}_2C_1$ ways to position the A. Finally, there is only ${}_1C_1$ way to place the C.

Putting Steps 1–5 together, we calculate the number of distinct letter strings as

$$
\begin{aligned}
{}_{10}C_3 \cdot {}_7C_3 \cdot {}_4C_2 \cdot {}_2C_1 \cdot {}_1C_1 &= \frac{10!}{3!\cancel{7!}} \cdot \frac{\cancel{7!}}{3!\cancel{4!}} \cdot \frac{\cancel{4!}}{2!\cancel{2!}} \cdot \frac{\cancel{2!}}{1!\cancel{1!}} \cdot \frac{\cancel{1!}}{1!0!} \\
&= \frac{10!}{3!3!2!1!1!} = \frac{3{,}628{,}800}{72} \\
&= 50{,}400
\end{aligned}
$$

There are 50,400 distinct strings of letters that can be made using the letters in the word STATISTICS. ■

This example can be generalized in the following result.

> **Permutations of Nondistinct Items**
>
> The number of permutations of $n$ items of which $n_1$ are of the first kind, $n_2$ are of the second kind, . . ., and $n_k$ are of the $k$th kind is calculated as
>
> $$\frac{n!}{n_1! \cdot n_2! \cdot \ldots \cdot n_k!}$$
>
> where $n = n_1 + n_2 + \ldots + n_k$.

## Example 5.38 Number of permutations of nondistinct items

Brandon brings a healthy snack to school each day, consisting of 5 carrot sticks, 4 celery sticks, and 2 cherry tomatoes. If Brandon eats one item at a time, in how many different ways can he eat his snack?

**Solution**

We are seeking the number of permutations of $n = 11$ items, of which $n_1 = 5$ are carrot sticks, $n_2 = 4$ are celery sticks, and $n_3 = 2$ are cherry tomatoes. Using the formula for the number of permutations of nondistinct items,

$$\frac{n!}{n_1! \cdot n_2! \cdot n_3!} = \frac{11!}{5! \cdot 4! \cdot 2!} = \frac{39{,}916{,}800}{120 \cdot 24 \cdot 2} = 6930$$

There are 6930 distinct ways in which Brandon can eat his snack. ■

**Acceptance sampling** refers to the process of (1) selecting a random sample from a batch of items, (2) evaluating the sample for defectives, and (3) either accepting or rejecting the entire batch based on the evaluation of the sample.

### Example 5.39 Acceptance sampling uses combinations

Suppose we have a batch of 20 cell phones, of which, unknown to us, 3 are defective and 17 are non-defective. We will take a random sample of size 2 and evaluate both items once.

**a.** Are the arrangements in acceptance sampling permutations or combinations?

**b.** Find the number of ways that both sampled cell phones are defective.

**Solution**

**a.** Both permutations and combinations require the following:

- $r$ items are chosen from $n$ distinct items. Here we are selecting $r = 2$ phones from a batch of $n = 20$.
- Repetition of the items is not allowed. Each item is evaluated only once.

The difference between permutations and combinations is that, for permutations order is important while for combinations order is not important. In acceptance sampling, the order of the items is not important. Thus, acceptance sampling uses combinations.

**b.** The number of ways of choosing 2 of the 3 defectives is

$${}_3C_2 = \frac{3!}{2!(3-2)!} = \frac{3 \cdot 2!}{2! \cdot 1!} = 3$$

Selecting 2 defectives means that we are choosing 0 of the 17 non-defectives. The number of ways this can happen is

$${}_{17}C_0 = \frac{17!}{0!(17-0)!} = \frac{17!}{1 \cdot 17!} = 1$$

By the Multiplication Rule for Counting, the number of ways that both sampled cell phones are defective is

$${}_3C_2 \cdot {}_{17}C_0 = 3 \cdot 1 = 3$$

■

## 3 Computing Probabilities Using Combinations

The counting methods we have learned in this section may be used to compute probabilities. We assume that each possible outcome in a random sample is *equally likely,* and thus we use the classical method for assigning the probability of an event $E$:

$$P(E) = \frac{\text{number of outcomes in } E}{\text{number of outcomes in sample space}} = \frac{N(E)}{N(S)}$$

## Example 5.40 Probability using combinations: acceptance sampling

Continuing with Example 5.39, if both cell phones in the sample of size 2 are defective, we will reject the batch and cancel our contract with the supplier.

**a.** What is the number of ways that both cell phones will be defective?
**b.** What is the number of outcomes in this sample space?
**c.** What is the probability that both cell phones will be defective?

**Solution**

**a.** From Example 5.39, the number of ways that both cell phones will be defective is

$$_3C_2 \cdot {}_{17}C_0 = 3 \cdot 1 = 3$$

**b.** The number of outcomes in the sample space is given by the number of ways of selecting 2 cell phones out of a batch of 20, that is,

$$N(S) = {}_{20}C_2 = \frac{20!}{2!(20-2)!} = \frac{20 \cdot 19 \cdot 18!}{2! \cdot 18!} = \frac{380}{2} = 190$$

**c.** Therefore, the probability that both cell phones will be defective is given by

$$P(\text{Both defective}) = \frac{\text{number of ways both defective}}{\text{number of outcomes in sample space}} = \frac{3}{190} \approx 0.01579$$

## Example 5.41 Florida Lotto

You can win the jackpot in the Florida Lotto by correctly choosing all 6 winning numbers out of the numbers 1–53.

**a.** What is the number of ways of winning the jackpot by choosing all 6 winning numbers?
**b.** What is the number of outcomes in this sample space?
**c.** If you buy a single ticket for \$1, what is your probability of winning the jackpot?
**d.** If you mortgage your house and buy 500,000 tickets, what is your probability of winning the jackpot (assuming that all the tickets are different)?

**Solution**

**a.** The number of ways of winning the jackpot by correctly choosing all 6 of the winning numbers and none of the losing numbers is

$$N(\text{Jackpot}) = {}_6C_6 \cdot {}_{47}C_0 = 1 \cdot 1 = 1$$

**b.** The size of the sample space is

$$N(S) = {}_{53}C_6 = \frac{53!}{6!(53-6)!} = \frac{53 \cdot 52 \cdot 51 \cdot 50 \cdot 49 \cdot 48 \cdot 47!}{6! \cdot 47!}$$

$$= \frac{16{,}529{,}385{,}600}{720} = 22{,}957{,}480$$

**c.** Therefore, if you buy a single ticket for \$1, your probability of winning the jackpot is given by

$$P(\text{Jackpot}) = \frac{1}{22{,}957{,}480} \approx 0.00000004356$$

**d.** If you buy 500,000 tickets and they are all unique, then your probability of winning becomes

$$P(\text{Jackpot}) = \frac{500{,}000}{22{,}957{,}480} \approx 0.02178$$

This is because the unique tickets are mutually exclusive, and the Addition Rule for Mutually Exclusive Events allows us to add the probabilities of the 500,000 tickets. After mortgaging your $500,000 house and buying lottery tickets with the proceeds, there is a better than 97% probability that you will *not* win the lottery. ■

## STEP-BY-STEP TECHNOLOGY GUIDE: Permutations and Combinations

### TI-83/84

#### Factorials $n!$

***Step 1*** On the home screen, enter the value of $n$.
***Step 2*** Press **MATH**, highlight **PRB**, and select **4: !** (Figure 5.23).
***Step 3*** Press **ENTER**.

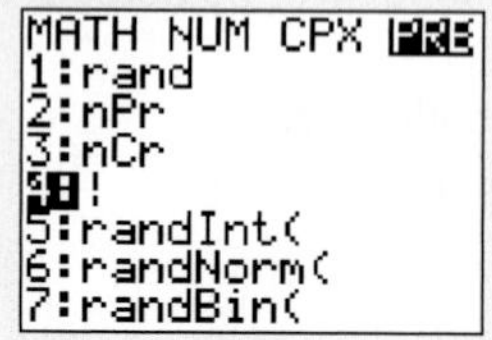


Figure 5.23

#### Permutations $_nP_r$ and Combinations $_nC_r$

***Step 1*** On the home screen, enter the value of $n$.
***Step 2*** **a.** For permutations, press **MATH**, highlight **PRB**, and select **2:nPr**.
**b.** For combinations, press **MATH**, highlight **PRB**, and select **3:nCr**.
***Step 3*** On the home screen, enter the value of $r$.
***Step 4*** Press **ENTER** (see Figure 5.19 and Figure 5.21 in Example 5.36 [page 246]).

### EXCEL

#### Factorials $n!$

Calculate **9!**

***Step 1*** Select an empty cell, and type = **FACT(9)**.
***Step 2*** Press **ENTER**.

#### Permutations $_nP_r$

We illustrate Example 5.36a (page 246): $_9P_6$.

***Step 1*** Select an empty cell and type = **PERMUT(9,6)**.
***Step 2*** Press **ENTER**. See Figure 5.20 in Example 5.36 for the result.

#### Combinations $_nC_r$

We illustrate Example 5.36b (page 246): $_{10}C_7$.

***Step 1*** Select an empty cell and type =**COMBIN(10,7)**.
***Step 2*** Press **ENTER**. See Figure 5.22 in Example 5.36 for the result.

## Section 5.4 SUMMARY

1 The Multiplication Rule for Counting provides the total number of different possible outcomes for a series of events.

2 A permutation $_nP_r$ is an arrangement in which

- $r$ items are chosen from $n$ distinct items.
- repetition of items is not allowed.
- the order of the items is important.

In a permutation, order is important. In a combination, order does not matter. A combination $_nC_r$ is an arrangement in which

- $r$ items are chosen from $n$ distinct items.
- repetition of items is not allowed.
- the order of the items is not important.

3 Combinations may be used to calculate certain probabilities. For such problems, use the following steps.

*Step 1* Confirm that the desired probability involves a combination.
*Step 2* Find $N(E)$, the number of outcomes in event $E$.
*Step 3* Find $N(S)$, the number of outcomes in the sample space.
*Step 4* Assuming that each possible combination is equally likely, find the probability of event $E$ as follows:

$$P(E) = \frac{N(E)}{N(S)}$$

## Section 5.4 EXERCISES

### CLARIFYING THE CONCEPTS

**1.** What type of diagram is helpful in itemizing the possible outcomes of a series of events?

**2.** Explain in words how 5! is calculated.

**3.** What is the difference between a permutation and a combination?

**4.** Does ${}_8P_9$ make sense? Explain why or why not.

**5.** Describe in your own words what is meant by acceptance sampling.

**6.** The counting methods that we have learned in this section may be used to compute probabilities.

- **a.** For assigning probability, which method is used, classical, relative frequency, or subjective?
- **b.** Referring to part a, what assumption must be made to apply the method?

### PRACTICING THE TECHNIQUES

For Exercises 7–12, find the value of each factorial.

**7.** 6!

**8.** 9!

**9.** 0!

**10.** 11!

**11.** 1!

**12.** 15!

For Exercises 13–20, find the value of each permutation ${}_nP_r$.

**13.** ${}_7P_3$

**14.** ${}_7P_4$

**15.** ${}_8P_5$

**16.** ${}_8P_3$

**17.** ${}_{100}P_1$

**18.** ${}_{100}P_0$

**19.** ${}_{100}P_{100}$

**20.** ${}_{100}P_{99}$

For Exercises 21–28, find the value of each combination ${}_nC_r$. Then answer Exercises 29 and 30.

**21.** ${}_7C_3$

**22.** ${}_7C_4$

**23.** ${}_{11}C_8$

**24.** ${}_{11}C_9$

**25.** ${}_{11}C_{10}$

**26.** ${}_{11}C_{11}$

**27.** ${}_{100}C_0$

**28.** ${}_{100}C_1$

**29.** Explain why the answers to Exercises 21 and 22 are equal. Use the commutative property of multiplication (for example, $2 \cdot 7 = 7 \cdot 2$) in your answer.

**30.** Use the idea behind your answer to Exercise 29 to find a combination that is equal to ${}_{11}C_8$. Verify your answer.

**31.** List all the permutations of the following people taken three at a time: Amy, Bob, Chris, Danielle. What is ${}_4P_3$?

**32.** List all the combinations of the following people taken three at a time: Amy, Bob, Chris, Danielle. What is ${}_4C_3$?

**33.** Explain in your own words why ${}_4P_3$ is larger than ${}_4C_3$.

**34.** What quantity do we divide ${}_4P_3$ by to get ${}_4C_3$? Express this quantity as a factorial. (*Hint:* For example, if the quantity were 120, we would express it as 5!).

**35.** In general, what do we divide ${}_nP_r$ by to get ${}_nC_r$?

### APPLYING THE CONCEPTS

**36. Fast Food.** A fast-food restaurant has three types of sandwiches: chicken sandwich, fish sandwich, and beef burger. The restaurant has two types of side dishes: French fries and salad.

- **a.** Draw a tree diagram to find all the different meals a customer can order at this restaurant.
- **b.** How many different meals can a customer order at this restaurant?

**37. What to Eat?** A sit-down restaurant has two types of appetizers: garden salad and buffalo wings. It has three entrees: spaghetti, steak, and chicken. And it offers three kinds of desserts: ice cream, cake, and pie.

- **a.** Draw a tree diagram to find all the different meals a customer can order at this restaurant.
- **b.** How many different meals can a customer order at this restaurant?

**38. Greek Alphabet.** The ancient Greek alphabet had 24 letters. How many different possible initials are there for people with a first and last name?

**39. Baseball Lineup.** A baseball manager has to submit to the umpire the ordering of his lineup of nine players. How many ways can the manager arrange the nine players?

**40. Facebook Friends.** A student has 10 friends on her FaceBook page. How many ways can she arrange her 10 friends top to bottom?

**41. Document Delivery.** A document delivery person must deliver documents to five different destinations within a particular city. How many different routes are possible?

**42. Country Music Mix.** Madison has just burnt 15 songs onto a country music mix CD. She chooses "random" on the CD player when she listens to the CD. How many different orderings of the 15 songs are there?

**43. Traveler Fellow.** A corporate sales executive must travel to the following countries this quarter: China, Russia, Germany, Brazil, India, and Nigeria. How many different routes are possible?

**44. Sales Traveler.** A corporate sales executive has the choice of traveling to four of the following six countries this quarter: China, Russia, Germany, Brazil, India, and Nigeria. How many different routes are possible?

**45. Playing Catch.** Five children are playing catch with a ball. How many different ways can one child throw a ball to another child once?

**46. Chimp Grooming.** Six chimpanzees are grooming each other at the city zoo. In how many different ways can one chimp groom another?

**47. Shake Hands.** In an ice-breaker exercise, each of 25 students is asked to shake hands with each of the other students. How many handshakes will there be in all?

**48. Statistics Competition.** Three students from the Honors Statistics class of 15 students will be chosen to represent the school at the state statistics competition. How many different possible groupings of three students are there?

**49.** How many random samples of size 1 can be chosen from a population of size 20?

**50.** How many random samples of size 20 can be chosen from a population of size 20?

**51.** How many random samples of size 10 can be chosen from a population of size 20?

**52.** How many distinct strings of letters can be made using all the letters in the word MATHEMATICS?

**53.** How many distinct strings of letters can be made using all the letters in the word BUSINESS?

**54. CD Burning.** You would like to use your new CD burner to make two new music-mix CDs. Your box of disks contains 14 CDs and 7 DVDs. If you choose 2 at random, find the number of ways that both disks will be CDs.

**55. Grocery Produce Manager.** A grocery produce manager inspects the corn supplied by a nearby farm. In a bushel of 30 ears of corn, the manager inspects 3 ears of corn. If 2 or more of the ears of corn are defective, the entire bushel is rejected. In this bushel, 5 ears of corn are defective.

- **a.** Explain whether the grocery produce manager is using a permutation or a combination.
- **b.** Find the number of ways that 2 or more ears of corn will be found defective by the produce manager.
- **c.** Find the probability that the bushel will be rejected.

**56. Acceptance Sampling.** A shipment of 25 personal digital assistants (PDAs) contains 3 that are defective. A quality control specialist inspects 2 of the 25 PDAs. If both are defective, then the shipment is rejected.

- **a.** Explain whether a permutation or a combination is being used.
- **b.** Find the number of ways that both PDAs will be defective.
- **c.** Find the probability of rejecting the shipment.

**57. Downloading Music.** You have just downloaded a new album of 12 songs from an online music retailer and made a playlist of the 12 songs. It turns out that you like 7 of the songs. Setting the playlist to random mode, find the probability that, of the first five songs you hear,

- **a.** you like all of them.
- **b.** you like none of them.
- **c.** you like two of them.
- **d.** you like three of them.

## *Try This in Class!* In-class activities to enhance your understanding of statistics

**1.** Have you ever taken a multiple-choice test where you couldn't decide between the alternatives and wished you could flip a coin or roll a die to decide? Suppose you are facing a four-question multiple-choice exam, with three alternatives (A, B, or C) for each question, and you have no clue as to what the right answers are. In this activity you will calculate the probability of getting one or more answers right.

- **a.** Why does it make sense to use the classical method?
- **b.** Draw a tree diagram for this four-question test—the experiment.
- **c.** What is the probability that you will answer a particular question correctly?
- **d.** What is the probability that you will answer two questions correctly?
- **e.** What is the probability that you will answer three questions correctly?
- **f.** What is the probability that you will answer all four questions correctly?
- **g.** Suppose that you need to answer at least three out of the four questions correctly to pass. What is the probability that you will pass?

**2.** We use simulation to estimate the probabilities we calculated in Activity 1. We use a single fair die to determine our responses.

- **a.** How should we "code" the die results so that there is an even chance of choosing A, B, or C?
- **b.** Your answer to each exam question will be either right or wrong. Why not use a coin toss rather than a die roll for this simulation?
- **c.** We need to code the probability of getting a correct answer. For each question, A, B, or C is correct. We want to simulate the probability of guessing correctly, which is 1/3. How do you code the die roll so that the probability of guessing correctly is 1/3?
- **d.** Perform 20 trials of the experiment. Rolling the die one time is considered to be answering one test question. So rolling the die four times is considered to be one "trial" of the experiment (taking the test). For each trial, keep track of the number of "right" and "wrong" guesses. Answer questions **(e)**–**(h)** using your simulation results.
- **e.** Estimate the probability that you will answer exactly one particular question correctly.
- **f.** Estimate the probability that you will answer exactly two questions correctly.
- **g.** Estimate the probability that you will answer exactly three questions correctly.
- **h.** Estimate the probability that you will answer all four questions correctly.
- **i.** Estimate the probability that you will pass.

# CHAPTER 5 FORMULAS AND VOCABULARY

## SECTION 5.1

- **Classical method of assigning probabilities** (p. 205). Let $N(E)$ and $N(S)$ denote the number of outcomes in event $E$ and the sample space $S$, respectively. The probability of event $E$ is then

$$P(E) = \frac{\text{number of outcomes in } E}{\text{number of outcomes in sample space}} = \frac{N(E)}{N(S)}$$

- **Event** (p. 203). Collection of outcomes from the sample space of an experiment.
- **Experiment** (p. 203). Any activity for which the outcome is uncertain.
- **Law of Large Numbers** (p. 209). As the number of times that an experiment is repeated increases, the relative frequency (proportion) of a particular outcome tends to approach the probability of the outcome.
- **Law of Total Probability** (p. 204). For any experiment, the sum of all the outcome probabilities in the sample space must equal 1.
- **Outcome** (p. 203). Result of a single trial of an experiment.
- **Probability** (p. 203). Long-term proportion of times the outcome or event occurs.
- **Probability model** (p. 204). Table or listing of all the possible outcomes of an experiment and the probability of each outcome.
- **Probability tree diagram** (p. 206). Tree diagram that specifies the probabilities for each branch.
- **Relative frequency method of assigning probabilities** (p. 209). Probability of event $E$ is approximately equal to the relative frequency of event $E$:

$$P(E) \approx \frac{\text{frequency of } E}{\text{number of trials of experiment}}$$

Also known as the **empirical method**.

- **Sample space** (p. 203). Collection of all possible outcomes in an experiment.
- **Simulation** (p. 208). Uses methods such as dice or computerized random numbers to generate results from an experiment.
- **Subjective probability** (p. 211). Assignment of a probability value to an outcome based on personal judgment.

## SECTION 5.2

- **Addition Rule** (p. 219). $P(A \text{ or } B) = P(A \cup B) = P(A) + P(B) - P(A \cap B)$.
- **Addition Rule for Mutually Exclusive Events** (p. 221). If $A$ and $B$ are mutually exclusive, then $P(A \cup B) = P(A) + P(B)$.
- **Complement of an event $A$** (p. 217). Collection of outcomes in the sample space that are *not* in the event $A$. It is denoted as $A^C$.
- **Intersection of two events $A$ and $B$** (p. 218). Event representing all the outcomes that belong to both $A$ and $B$. It is denoted as $A \cap B$ or as "$A$ and $B$."
- **Mutually exclusive** (p. 220). Events with no outcomes in common.
- **Probabilities for complements** (p. 217). For any event $A$ and its complement $A^C$, $P(A) + P(A^C) = 1$, $P(A) = 1 - P(A^C)$, and $P(A^C) = 1 - P(A)$.
- **Union of two events $A$ and $B$** (p. 218). Event representing all the outcomes that belong to $A$ or $B$ or both. It is denoted as $A \cup B$ or as "$A$ or $B$."

## SECTION 5.3

- **Conditional probability** (p. 226). For two related events $A$ and $B$, the probability of $B$ given $A$, denoted $P(B|A)$, and calculated as

$$P(B|A) = \frac{P(A \cap B)}{P(A)} = \frac{N(A \cap B)}{N(A)}$$

- **Independent events** (p. 228). Events $A$ and $B$ are *independent* if $P(A|B) = P(A)$ or if $P(B|A) = P(B)$. Otherwise the events are said to be *dependent*.
- **Multiplication Rule** (p. 229). $P(A \cap B) = P(B)\,P(A \mid B)$, or equivalently, $P(A \cap B) = P(A)\,P(B \mid A)$.
- **Multiplication Rule for Independent Events** (p. 230). If Events $A$ and $B$ are independent, then $P(A \cap B) = P(A)\,P(B)$.
- **Multiplication Rule for $n$ Independent Events** (p. 232). If $A, B, C, \ldots$ are independent events, then $P(A \cap B \cap C \cap \ldots) = P(A)\,P(B)\,P(C) \ldots$.
- **Sampling with replacement** (p. 234). The randomly selected unit is returned to the population after being selected. It is possible for the same unit to be sampled more than once.
- **Sampling without replacement** (p. 234). The randomly selected unit is not returned to the population after being selected. It is not possible for the same unit to be sampled more than once.

## SECTION 5.4

- **Acceptance sampling** (p. 248). Process of (1) selecting a random sample from a batch of items, (2) evaluating the sample for defectives, and (3) either accepting or rejecting the entire batch based on the evaluation of the sample.
- **Combination** (p. 244). Arrangement of items in which (1) $r$ items are chosen from $n$ distinct items, (2) repetition of items is not allowed, and (3) the order of the items is not important. The number of combinations of $r$ items chosen from $n$ different items is denoted ${}_nC_r$ and given by the formula

$${}_nC_r = \frac{n!}{r!(n-r)!}$$

- **Factorial symbol $n!$** (p. 242). $0! = 1$; $1! = 1$; $n! = n(n-1)(n-2)\ldots 3 \cdot 2 \cdot 1$
- **Multiplication Rule for Counting** (p. 241). Suppose an activity consists of a series of events in which there are $a$ possible outcomes for the first event, $b$ possible outcomes

for the second event, $c$ possible outcomes for the third event, and so on. Then the total number of different possible outcomes for the series of events is: $a \cdot b \cdot c \cdot \ldots$.

- **Permutation** (p. 243). An arrangement of items in which (1) $r$ items are chosen at a time from $n$ distinct items, (2) repetition of items is not allowed, and (3) the order of the items is important. The number of permutations of $n$ items chosen $r$ at a time is denoted as ${}_nP_r$ and given by the formula

$$_nP_r = \frac{n!}{(n-r)!}$$

- **Permutations of nondistinct items** (p. 247). The number of permutations of $n$ items of which $n_1$ are of the first kind, $n_2$ are of the second kind, . . . , and $n_k$ are of the $k$th kind, is calculated as

$$\frac{n!}{n_1! \cdot n_2! \cdot \ldots \cdot n_k!}$$

where $n = n_1 + n_2 + \ldots + n_k$.

## CHAPTER 5 REVIEW EXERCISES

### SECTION 5.1

For Exercises 1–5, consider the experiment of tossing a fair coin three times and find the probabilities.

1. 2 heads
2. At least 2 heads
3. 4 heads
4. 2 tails
5. At most 1 tail
6. **A New Sonnet.** Literature researchers have unearthed a sonnet that they know to be by either William Shakespeare or Christopher Marlowe. The probability that the sonnet is by Marlowe is 25%.
   a. What is the probability that the sonnet is by Shakespeare?
   b. What method of assigning probability do you think was used here? Why was this method used, and not the others?

### SECTION 5.2

7. **Farmworkers' Educational Level.** The U.S. Department of Agriculture reports on the demographics of hired farmworkers.[7] An excerpt of the results is provided in the table, showing the percentage of noncitizen and citizen farmworkers who attained various educational levels. The educational levels are mutually exclusive. Find the following probabilities.
   a. The probability that a noncitizen farmworker is a high school graduate or has some college.
   b. The probability that a citizen farmworker is a high school graduate or has some college.
   c. The probability that a noncitizen has less than a ninth-grade education and has some college.

| | Noncitizens | Citizens |
|---|---|---|
| Less than 9th grade | 238,008 | 61,776 |
| 9th–12th grade (no diploma) | 57,904 | 152,880 |
| High school graduate | 59,784 | 222,144 |
| Some college | 20,304 | 187,200 |

**Ecstasy and Methamphetamine Use.** The following table contains the numbers of 12th-grade students who said that they had used Ecstasy (MDMA) or methamphetamine in the past 30 days for data collected in 2006 and 2007, as reported by the National Institute on Drug Abuse (www.drugabuse.gov). Use the table to answer Exercises 8–10.

| Drug | 2006 | 2007 |
|---|---|---|
| Ecstasy | 185 | 232 |
| Methamphetamine | 128 | 87 |
| Sample size | 14,200 | 14,500 |

8. Find the indicated probabilities.
   a. A randomly selected 12th-grader in 2006 used Ecstasy in the past 30 days.
   b. A randomly selected 12th-grader in 2006 used methamphetamine in the past 30 days.
   c. A randomly selected 12th-grader in 2007 used Ecstasy in the past 30 days.
   d. A randomly selected 12th-grader in 2007 used methamphetamine in the past 30 days.
9. Assume that no student in the samples repeated twelfth grade. Answer the following questions.
   a. Are these two events mutually exclusive: $A$ = 12th-grader in 2006 used Ecstasy and $B$ = 12th-grader in 2007 used Ecstasy? Explain why or why not.
   b. Find the probability that a randomly selected 12th-grader in 2006 used Ecstasy or a randomly selected 12th-grader in 2007 used Ecstasy.
   c. Find the probability that a randomly selected 12th-grader in 2006 used methamphetamine or a randomly selected 12th-grader in 2007 used methamphetamine.
10. Answer the following questions.
   a. Are these two events mutually exclusive: $C$ = used Ecstasy in the past 30 days and $D$ = used methamphetamine in the past 30 days? Explain why or why not.
   b. Discuss why we cannot find the probability that a randomly chosen 2006 12th-grader used both Ecstasy and methamphetamine in the past 30 days. What extra information do we need?

## SECTION 5.3

**11. Drug Research Studies.** The *Annals of Internal Medicine* reported that 39 of the 40 research studies sponsored by a drug company had outcomes favoring a certain drug. Find the following probabilities.

- **a.** Three randomly selected research studies all favor this drug.
- **b.** None of the three randomly selected research studies favors this drug.
- **c.** At least one of three randomly selected research studies favors this drug.

**12. Gun Control.** A Harris poll conducted in June 2004 of 2408 American adults found 52% who supported tighter gun control. Find the following probabilities.

- **a.** Two randomly selected respondents both supported tighter gun control.
- **b.** At least two of three randomly selected respondents supported tighter gun control.
- **c.** At most one of three randomly selected respondents did not support tighter gun control.

**13. Politics of Gay Marriage.** The Pew Research Center for the People and the Press (**http://people-press.org**) reported in 2004 that 322 out of 1149 people surveyed would not vote for a political candidate who disagreed with their views on gay marriage. Find the probabilities.

- **a.** Three out of three randomly selected respondents would vote for a political candidate who disagreed with his or her views on gay marriage.
- **b.** At least one of three randomly selected respondents would not vote for a political candidate who disagreed with his or her views on gay marriage.

**14. Drug Research Studies.** Use the information in Exercise 11. Suppose we sample two research studies without replacement. Find the probability that the second study does not favor this drug given that the first study does not favor this drug.

**Gender and Pet Preference.** Do you think your gender affects what type of pet you own? For Exercises 15–18, use the following table, showing preferences for various pets by owner gender.

| Gender of owner | Cats | Dogs | Other pets | Total |
|---|---|---|---|---|
| Female | 100 | 50 | 30 | 180 |
| Male | 50 | 50 | 20 | 120 |
| Total | 150 | 100 | 50 | 300 |

**15.** Find the probabilities that a randomly chosen person has the following characteristics.

- **a.** Owns a cat, $P(C)$
- **b.** Owns a dog, $P(D)$

**16.** Find the probability that a randomly chosen person has the following characteristics.

- **a.** Is female and owns a dog, $P(F \cap D)$
- **b.** Is male and owns a dog, $P(M \cap D)$

**17.** Find the following conditional probabilities for a randomly chosen person.

- **a.** Owns a dog, given that the person is female, $P(D \mid F)$
- **b.** Owns a dog, given that the person is male, $P(D \mid M)$

**18.** If you were a dogfood manufacturer, would you advertise more on the men's TV channel or the women's TV channel? Why? Cite your evidence.

## SECTION 5.4

**19.** How many distinguishable strings of letters can be made using all the letters in the word MISSISSIPPI?

**20. Statistics Quiz.** On a statistics quiz, there are 5 true/false questions, 4 fill-in-the-blank questions, and 3 short-answer questions. How many different ways are there of taking this quiz?

**21. Inspection Time.** A U.S. Army drill instructor will perform inspection on 2 soldiers in a squad of 18 soldiers. If both soldiers fail the inspection because their rifles are not clean, the entire squad will have to run a five-mile course in full gear. Three of the 18 soldiers have rifles that are not clean.

- **a.** Explain whether the drill instructor is using a permutation or a combination.
- **b.** Find the number of ways that both soldiers will fail the inspection.
- **c.** Find the probability that the entire squad will have to run a five-mile course in full gear.

# CHAPTER 5 Quiz

### TRUE OR FALSE

**1.** True or false: An outcome is a collection of a series of events from the sample space of an experiment.

**2.** True or false: It is fine for an event to consist of only one outcome.

**3.** True or false: For any event $A$ (even events like $A$: the moon is made of green cheese) the probability of $A$ plus the probability of $A^C$ always add up to 1.

### FILL IN THE BLANK

**4.** The minimum value that a probability can take is ______ and the maximum value is _________.

**5.** The union of two events is associated with the English word _________, and the intersection of two events is associated with the English word _________.

**6.** Someone has told you that there is a 50-50 chance of rain tomorrow. This means that the probability of rain tomorrow equals __________.

7. A __________ is an arrangement in which $r$ items are chosen from $n$ distinct items, repetition of items is not allowed, and the order of the items is important.

## SHORT ANSWER

8. For any experiment, what is the sum of all the outcome probabilities in the sample space?
9. For which type of sampling are consecutive draws independent?
10. For two events $A$ and $B$, what do we call the event containing only those outcomes that belong to both $A$ and $B$?
11. Which rule do we use for calculating the number of different possible outcomes of a series of events?

## CALCULATIONS AND INTERPRETATIONS

For Exercises 12–15, the experiment is to roll a fair die twice. Find the probabilities.

12. Pair of 1's
13. Pair of 2's
14. One of the dice shows 5
15. Sum of the two dice equals 7

For Exercises 16–20, consider the experiment of rolling a fair die twice. Find the probabilities.

16. Sum of the two dice equals 5
17. Sum of the two dice does not equal 5
18. One of the dice shows 2
19. Sum of the two dice equals 5 and one of the dice shows 2
20. Sum of the two dice equals 5 or one of the dice shows 2
21. Suppose that $A$ and $B$ are any two events, with $P(B) = 0.75$ and $P(A \cap B) = 0.15$. Find $P(A | B)$.
22. Suppose that $A$ and $B$ are any two events, with $P(B) = 0.85$ and $P(A | B) = 0.25$. Find $P(A \cap B)$.
23. **Pick a Card.** Consider the experiment of drawing a single card from a deck of 52 cards. Find the probability of observing the following events.
    - **a.** Heart
    - **b.** Face card (king, queen, or jack)
    - **c.** Seven
    - **d.** Red card
    - **e.** Seven of hearts
    - **f.** Red queen
24. **Direct Mail Marketing.** The following table was taken from a study on modeling response to direct mail marketing.

| | **Credit Card on File?** | | |
|---|---|---|---|
| **Response** | **No** | **Yes** | **Total** |
| Did not respond | 16,102 | 7,935 | 24,037 |
| Did respond | 1,666 | 3,096 | 4,762 |
| Total | 17,768 | 11,031 | 28,799 |

*Source:* Daniel Larose, *Data Mining Methods and Models* (Wiley Interscience, 2006).

The events are $A$ = "responded to direct mail marketing campaign" and $B$ = "has a credit card on file." Find the probability that a randomly selected customer has the following characteristics.

- **a.** Has a credit card on file
- **b.** Responded to the direct mail marketing campaign
- **c.** Has a credit card on file and responded to the direct mail marketing campaign
- **d.** Has a credit card on file or responded to the direct mail marketing campaign

**Happiness in Marriage.** The General Social Survey tracks trends in American society through annual surveys. The married respondents were asked to characterize their feelings about being married. The results, crosstabulated with gender, are shown in the following figure. Use this information for Questions 25–28.

**RESPONDENTS SEX * HAPPINESS OF MARRIAGE Crosstabulation**

Count

| | | HAPPINESS OF MARRIAGE | | | |
|---|---|---|---|---|---|
| | | VERY HAPPY | PRETTY HAPPY | NOT TOO HAPPY | Total |
| RESPONDENTS SEX | MALE | 242 | 115 | 9 | 366 |
| | FEMALE | 257 | 149 | 17 | 423 |
| Total | | 499 | 264 | 26 | 789 |

25. Find the probabilities that a randomly chosen person has the following characteristics.
    - **a.** Is female, $P(F)$
    - **b.** Is male, $P(M)$
    - **c.** Is not too happily married, $P(\text{Not})$
26. Find the probabilities that a randomly chosen person has the following characteristics.
    - **a.** Is female and not too happily married, $P(F \cap \text{Not})$
    - **b.** Is male and not too happily married, $P(M \cap \text{Not})$
27. Find the following conditional probabilities for a randomly chosen person.
    - **a.** Is not too happily married, given that the person is female, $P(\text{Not} | F)$.
    - **b.** Is not too happily married, given that the person is male, $P(\text{Not} | M)$.
28. Are gender and being not too happily married independent? Why or why not?
29. **Football Teams.** The four teams in the AFC South division of the National Football League are Indianapolis Colts, Jacksonville Jaguars, Tennessee Titans, and Houston Texans. Suppose the top three teams in the division this year will make the playoffs. How many different sets of teams making the playoffs are there?
30. **State Lottery.** In a state lottery, balls numbered 1 to 20 are placed in an urn. To win, you must choose numbers that match the three balls chosen in the order that they're chosen.
    - **a.** Explain whether a permutation or a combination is being used.
    - **b.** How many possible outcomes are there?
    - **c.** Find the probability of winning this lottery if your ticket contains a single ordering of three numbers?

Chapter 6

David Madison/Digital Vision/Getty Images

# Random Variables and the Normal Distribution

CASE STUDY

## Be Careful What You Assume: Major League Baseball Salaries

Many fans of Major League Baseball (MLB) are concerned about the continuing increase in player salaries, making a trip to the ballpark an expensive family outing. In this Case Study, we consider a scenario where the MLB owners wish to "cap" a team's top salary, with the owners suggesting the 75th percentile of all salaries. In our scenario, the players' union states that they will agree to the salary cap only if management adds a bonus to the salaries of the lowest-paid players, as measured by the 10th percentile. We find out what transpires in Section 6.5.

OVERVIEW



## The Big Picture

### Where we are coming from, and where we are headed...

In Chapter 5, we were introduced to probability, which allows us to quantify the uncertainty we will face when performing statistical inference in the later chapters. Probability provides us with a framework when we are called upon to make decisions, such as:

- Should we invest in this mutual fund?
- What is a typical profit for a small capital investor?
- What would be an unusual number of runs scored in a baseball game?

However, to answer these questions we need a new set of tools in our probability toolbox: *random variables* and *probability distributions*. With these new tools, we can increase the efficiency of our decision making. Sections 6.1 and 6.2 introduce us to discrete random variables, including the binomial random variable. In Section 6.3 we learn about continuous random variables, including the normal random variable. Section 6.4 covers the special case of the standard normal distribution, and Section 6.5 provides some applications of the normal distribution to everyday problems.

Chapter 6 lies within the scope of probability, because we are still faced with an experimental situation, and our task is to calculate probabilities associated with various events in the experiment. In Chapter 7, we will shift our perspective toward sampling and learn about sampling distributions. Chapter 7 is pivotal, because all the statistical inference we will learn in the remainder of the course depends on the behavior of statistics, which is quantified using sampling distributions.

## 6.1 Discrete Random Variables

**Objectives:** By the end of this section, I will be able to...

1. Identify random variables.
2. Explain what a discrete probability distribution is and construct probability distribution tables and graphs.
3. Calculate the mean, variance, and standard deviation of a discrete random variable.

## 1 Random Variables

Life is constantly calling upon us to make decisions in the face of uncertainty. As a way of helping us to make these decisions, we can sometimes list all the possible outcomes of the situation and the probabilities associated with each outcome. In this section, we will learn about random variables and how they can help us quantify the uncertainty.

### Example 6.1 Decision making: should Kristin invest?

In her position as junior manager Kristin, fresh out of college, has managed to save up $2000. She wants to invest in the stock market, in particular Cape Ann Biotech, a nano-biotechnology company. Based on the company's recent performance, she is optimistic about its prospects, despite the *uncertainty* involved in the investment. Based on her research and professional estimates about the direction of the overall economy, she estimates the probabilities of the five outcomes shown in Table 6.1 for making or losing money over the next 12 months. Note that the two loss outcomes are expressed as negative gains, and that the probabilities add up to 1.0.

**a.** What method is Kristin using to assign probabilities?

**b.** Use Table 6.1 to find the probability that Kristin will gain $1000.

**c.** Will Kristin make money or lose money?

Table 6.1 Kristin's probability estimates for five investment outcomes for a period of 12 months (initial investment $2000)

| Scenario | Financial gain | Kristin's estimated probability |
|---|---|---|
| Cape Ann Biotech does very well. | Gain $1000 | 0.15 |
| Cape Ann Biotech does fairly well. | Gain $500 | 0.30 |
| Cape Ann Biotech treads water. | Gain $0 | 0.25 |
| Cape Ann Biotech does not do very well. | Lose $200 (Gain −$200) | 0.20 |
| Cape Ann Biotech folds. | Lose $2000 (Gain −$2000) | 0.10 |

**Solution**

**a.** Kristin is using the subjective method to assign probabilities. The outcomes are not equally likely, and she is not using relative frequencies to assign the probabilities. Instead, she is assigning probabilities based on her own subjective estimates of the company's prospects and the direction of the economy.

**b.** Kristin's estimated probability that her financial gain will be $1000 is 0.15.

**c.** We do not know yet whether Kristin will make money or lose money, because the outcome is uncertain, which fits the definition of an *experiment*.

We can represent Kristin's financial gain with a variable. Recall from Chapter 1 that a *variable* is a characteristic of an element and can assume different values. Examples of variables include *age* and *gender*. Sometimes we use letters of the alphabet, such as $X$ or $Z$, to represent variables. Kristin's financial gain is a variable that takes on different values. However, for this variable, there is an element of *chance* that influences which value it takes. The outcome of Kristin's investment is uncertain and contains elements of randomness. To help her decide how to quantify this uncertainty, we need to define a new kind of variable, called a **random variable**.

**Random Variable**

A **random variable** is a variable whose values are determined by chance.

**Random Variables Must Be Random!**

The role of chance in the definition of a random variable is crucial. For example, a variable that is based on the results of an experiment is a random variable because an experiment by definition has outcomes that are unknown. For example, is your age a random variable? If we are just talking about you and no one else, and we know your age, then there is no chance involved. In that case, your age is not a random variable. On the other hand, what if we select students *at random* by picking names from a hat? Then the age of the person drawn is a random variable because its value depends at least partly on chance (on which name is drawn at random). ■

## Example 6.2 Notation for random variables

Comstock/Jupiter Images

Suppose our experiment is to toss a single fair die, and we are interested in the number rolled. We define our random variable $X$ to be the outcome of a single die roll.

**a.** Why is the variable $X$ a random variable?
**b.** What are the possible values that the random variable $X$ can take?
**c.** What is the notation used for rolling a 5?
**d.** Use random variable notation to express the probability of rolling a 5.

### Solution

**a.** We don't know the value of $X$ before we toss the die, which introduces an element of chance into the experiment, thereby making $X$ a random variable.
**b.** The possible values for $X$ are 1, 2, 3, 4, 5, and 6.
**c.** When a 5 is rolled, then $X$ equals the outcome 5, and we write $X = 5$.
**d.** Recall from Section 5.1 that the probability of rolling a 5 for a fair die is 1/6. In random variable notation, we denote this as $P(X = 5) = 1/6$. ■

There are two main types of random variables: **discrete random variables** and **continuous random variables**. The difference between the two types relates to the possible values that each type of random variable can assume.

**Discrete and Continuous Random Variables**

- A **discrete random variable** can take either a finite or a countable number of values. Since these values may be written as a list of numbers, each value can be graphed as a separate point on a number line, with space between each point.
- A **continuous random variable** can take infinitely many values. Because there are infinitely many values, the values of a continuous random variable form an interval on the number line.

Examples of discrete random variables include the number of children a randomly selected person has and the number of times a randomly chosen student has been pulled over for speeding on the interstate. Continuous random variables often need to be measured, not counted. For example, the temperature in Atlanta, Georgia, at noon today may be reported as 77 degrees, but this value represents actual temperatures that may lie anywhere between 76.5 degrees and 77.5 degrees.

## Example 6.3 Discrete and continuous random variables

For the following random variables, (i) determine whether they are discrete or continuous, and (ii) indicate the possible values they can take.

**a.** The number of automobiles owned by a family
**b.** The width of your desk in this classroom
**c.** The number of games played in the next World Series
**d.** The weight of model year 2007 SUVs

### Solution

**a.** Since the possible number of automobiles owned by a family is finite and may be written as a list of numbers, it represents a discrete random variable. The possible values are {0, 1, 2, 3, 4, . . .}.

**b.** Width is something that must be measured, not counted. Width can take infinitely many different possible values, with these values forming an interval on the number line. Thus, the width of your desk is a continuous random variable. The possible values might be 1 ft $\leq W \leq$ 10 ft.

**c.** The number of games played in the next World Series can be counted and thus represents a discrete random variable. The possible values are finite and may be written as a list of numbers: {4, 5, 6, 7}.

**d.** The weight of model year 2007 SUVs must be measured, not counted, and so represents a continuous random variable. Weight can take infinitely many different possible values, with these values forming an interval on the number line: 2500 lb $\leq Y \leq$ 7000 lb. ■

We will return to continuous random variables in Section 6.3. Sections 6.1 and 6.2 concentrate on discrete random variables. Let's start with an example aimed at helping you move from the language of probability (experiments and outcomes) to the language of random variables.

## Example 6.4 The number of heads in a two-coin-toss experiment is a discrete random variable

Recall the experiment of tossing two fair coins from Section 5.1. The sample space is

{HH, HT, TH, TT}

Now, suppose we are interested in the number of heads observed on the two tosses.

**a.** Define our random variable $X$.
**b.** What are the possible values of $X$?
**c.** Is $X$ a discrete or continuous random variable, and why?
**d.** For each possible value of $X$, use random variable notation to express the probability of that value.

### Solution

**a.** We define our random variable $X$ as follows:

$X$ = the number of heads observed on the two tosses

**b.** When tossing two fair coins, we may observe either 0, 1, or 2 heads. Thus, the possible values of $X$ are 0, 1, and 2.

**c.** $X$ is a discrete random variable because it takes a finite number of possible values that may be written as a list.

**d.** In Section 5.1 we learned that the probability of observing two heads is 1/4. Therefore, the probability that $X = 2$ is 1/4, denoted $P(X = 2) = 1/4$. The probability of observing one head is 1/2. Therefore, the probability that $X = 1$ is 1/2, written as $P(X = 1) = 1/2$. The probability of observing zero heads is 1/4. Therefore, the probability that $X = 0$ is 1/4, written as $P(X = 0) = 1/4$. ■

## 2 Discrete Probability Distributions

In this section, we consider discrete probability distributions, which inform us of the probabilities associated with the various values that the discrete random variable can take.

> A **probability distribution of a discrete random variable** provides all the possible values that the random variable can assume, together with the probability associated with each value. The probability distribution can take the form of a table, graph, or formula. Probability distributions describe populations, not samples.

When constructing the tabular form of a **probability distribution of a discrete random variable**, create a table with two rows:

- The top row will contain all the possible values of $X$.
- The bottom row will contain the probability associated with each value of $X$.

### Example 6.5 Probability distribution table

Construct the probability distribution table of the number of heads observed when tossing a fair coin twice.

**Solution**

The probability distribution table given in Table 6.2 uses probabilities we found in Example 6.4. ■

Table 6.2 Probability distribution of number of heads on two fair coin tosses

| **$X$ = number of heads observed** | 0 | 1 | 2 |
|---|---|---|---|
| **$P(X)$ = probability of observing that many heads** | 1/4 | 1/2 | 1/4 |

Note that the probabilities in the bottom row of Table 6.2 add up to 1. This is no coincidence. According to the Law of Total Probability from Chapter 5, the sum of the probabilities of all the outcomes in the sample space must equal 1. Put into the language of random variables, this means that the sum of the probabilities of all the possible values of a discrete random variable $X$ must equal 1. This requirement is written as $\sum P(X) = 1$. Also, since each value in the bottom row is a probability, each value must be between 0 and 1, inclusive, that is, $0 \leq P(X) \leq 1$. There are no such restrictions on the values that $X$ can take except that they be countable.

> **Requirements for the Probability Distribution of a Discrete Random Variable**
>
> - The sum of the probabilities of all the possible values of a discrete random variable must equal 1. That is, $\sum P(X) = 1$.
> - The probability of each value of $X$ must be between 0 and 1, inclusive. That is, $0 \leq P(X) \leq 1$.

The probabilities assigned to the various values of $X$ are found using either the classical method, the relative frequency method, or the subjective method. The probabilities in Table 6.2 were assigned using the classical method, since we assumed that tossing a fair coin would result in equally likely outcomes. In the next example, we turn to the subjective method to assign probabilities.

## Example 6.6 Probability distribution for Kristin's financial gain

Recall Example 6.1, where Kristin assigned probabilities to various levels of financial gain associated with an investment of \$2000 in Cape Ann Biotech. Table 6.1 is repeated here.

Table 6.1 Kristin's probability estimates for five investment outcomes for a period of 12 months (initial investment \$2000)

| Scenario | Financial gain | Kristin's estimated probability |
|---|---|---|
| Cape Ann Biotech does very well. | Gain \$1000 | 0.15 |
| Cape Ann Biotech does fairly well. | Gain \$500 | 0.30 |
| Cape Ann Biotech treads water. | Gain \$0 | 0.25 |
| Cape Ann Biotech does not do very well. | Lose \$200 (Gain −\$200) | 0.20 |
| Cape Ann Biotech folds. | Lose \$2000 (Gain −\$2000) | 0.10 |

Once we define our random variable, Table 6.1 can be converted into a probability distribution table. Since Kristin is interested in what happens to her money, we define our random variable $X$ to be

$$X = \text{financial gain (the financial gain associated with the five possible scenarios)}$$

**a.** Why is the variable $X$ = financial gain a random variable?
**b.** What are the possible values that $X$ can take?
**c.** Use random variable notation to express the probabilities associated with each possible outcome of $X$.
**d.** Construct the probability distribution of $X$ = financial gain.
**e.** Find the probability that Kristin will make a profit on her investment.
**f.** Find the probability that Kristin will take a loss on her investment.

### Solution

**a.** $X$ = financial gain is a random variable because we do not know, before the investment is made, the value the variable will take.

**b.** The possible values of $X$ are

$$X = \{-\$2000, -\$200, \$0, \$500, \$1000\}$$

**c.** Table 6.1 gives the probabilities of each of these outcomes:

$$P(X = -2000) = P(\text{gain } -\$2000) = 0.10$$
$$P(X = -200) = P(\text{gain } -\$200) = 0.20$$
$$P(X = 0) = P(\text{gain } \$0) = 0.25$$
$$P(X = 500) = P(\text{gain } \$500) = 0.30$$
$$P(X = 1000) = P(\text{gain } \$1000) = 0.15$$

**d.** For consistency, we retain the term "gain" when a loss is involved, inserting a negative value for the gain, which means the same as a loss of that value. Using this information, we may now proceed to construct the probability distribution as shown in Table 6.3. By convention, the values of $X$ are usually shown in increasing order.

Table 6.3 Probability distribution of Kristin's financial gain

| **$X$ = financial gain in dollars** | −2000 | −200 | 0 | 500 | 1000 |
|---|---|---|---|---|---|
| **$P(X)$** | 0.10 | 0.20 | 0.25 | 0.30 | 0.15 |

*Note:* In discrete probability distributions, the outcomes of the random variable are always mutually exclusive. For example, it is not possible for the result of Kristin's investment to be both gain \$1000 and gain \$500. This is why we may always use the Addition Rule for Mutually Exclusive Events to find the probability of two or more outcomes for a discrete random variable.

**e.** The probability distribution table makes it straightforward to combine probabilities. For example, the probability that Kristin will make a profit on the investment is

$$P(\text{gain } \$1000) + P(\text{gain } \$500) = P(X = 1000) + P(X = 500) = 0.15 + 0.30 = 0.45$$

So there is a 45% chance that Kristin will profit from this investment. Since the outcomes $X = 1000$ and $X = 500$ are mutually exclusive, we may add the probabilities of these outcomes by the Addition Rule for Mutually Exclusive Events: $P(A \text{ or } B) = P(A) + P(B)$.

**f.** The probability that she will take a loss is

$$P(X = -200) + P(X = -2000) = 0.20 + 0.10 = 0.30$$

## Discrete Probability Distribution as a Graph

We can construct graphs of probability distributions that show all the information contained in probability distribution tables. The graphical representation allows us to identify patterns more quickly.

The horizontal axis in the probability distribution graph in Figure 6.1 is the usual $x$ axis. The range is all the possible values that the random variable $X$ can take. Here the random variable $X$ takes on values $-2000$, $-200$, 0, 500, and 1000, from Table 6.3. The horizontal axis gives the same information as the top row of the probability distribution table. The vertical axis represents probability. A vertical bar is drawn at each value of $X$; the height of the bar represents the probability of that value of $X$. The bar of probability at $X = 0$ goes up to 0.25 and represents the probability that Kristin will neither gain nor lose money from her investment. The heights of these bars give the same information as the bottom row in the probability distribution table.

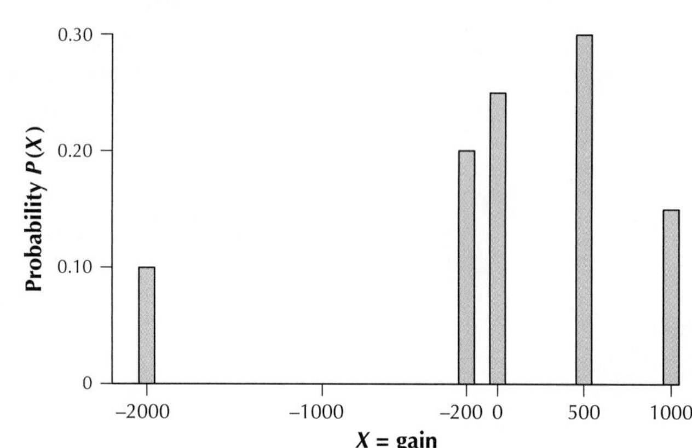


FIGURE 6.1 Graph of probability distribution for Kristin's financial gain.

Given a graph of a probability distribution, you should know how to construct the probability distribution table, and vice versa.

# 3 Mean and Variability of a Discrete Random Variable

Suppose you are about to perform a new experiment, such as finding the number of siblings that a randomly chosen classmate has. You have assigned probabilities to the outcomes, but the experiment has not been performed yet. Wouldn't it be nice to know the mean number of siblings? If you had a particular outcome in mind, wouldn't it be nice to know whether this outcome was unusual? Questions like these are addressed in this section.

The **mean $\mu$ of a discrete random variable** represents the mean result when the experiment is repeated an indefinitely large number of times.

Note that this meaning of *mean* differs from that of the sample mean we learned in Chapter 3. The sample mean $\bar{x}$ is simply the arithmetic mean of a finite set of sample values.

## Example 6.7 The meaning of the mean of a discrete probability distribution

The U.S. Department of Health and Human Services reports that there were 250,000 babies born to teenagers aged 15–18 in 2004. Of these 250,000 births, 7% were to 15-year-olds, 17% were to 16-year-olds, 29% were to 17-year-olds, and 47% were to 18-year-olds. The following table contains the probability distribution of the random variable $X$ = age.

| $X$ = age | $P(X)$ |
|---|---|
| 15 | 0.07 |
| 16 | 0.17 |
| 17 | 0.29 |
| 18 | 0.47 |

**a.** Graph the probability distribution of the random variable $X$ = age.
**b.** Estimate the mean of $X$ using the balance point idea from page 88.
**c.** If one of the teenagers represented in the table is chosen at random, what is the most likely age of that teenager when her baby was born?

### Solution

**a.** The probability distribution graph of $X$ = age is given in Figure 6.2.

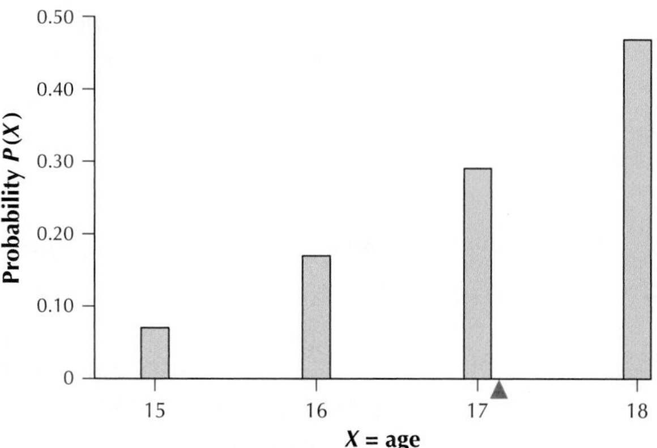


FIGURE 6.2

**b.** Recall that one interpretation of the mean is the *balance point* of the distribution. The balance point for the probability distribution graph appears to be somewhere near 17 years old. Hence, we would estimate the mean age to be about 17 years old, as indicated by the red fulcrum in Figure 6.2. We will calculate the actual value for the mean in Example 6.8.

**c.** Since $P(X = 18)$ has a larger value than any other probability in the table, and the bar for $X = 18$ is the tallest bar in the probability distribution graph, 18 is the most likely age. ■

**Finding the Mean of a Discrete Random Variable**

The mean $\mu$ of a discrete random variable is found as follows:

1. Multiply each possible value of $X$ by its probability.
2. Add the resulting products.

This procedure is denoted as

$$\mu = \sum[X \cdot P(X)]$$

## Example 6.8 Calculating the mean of a discrete random variable

Find the mean age of the mother for the babies born to teenagers aged 15−18 in 2004, from Example 6.7.

### Solution

First we need to multiply each possible outcome (value of $X$) by its probability $P(X)$. We multiply the value $X = 15$ by its probability $P(X) = 0.07$, the value $X = 16$ by its probability $P(X) = 0.17$, and so on. Then we add these four products to find the mean:

$$\mu = 15(0.07) + 16(0.17) + 17(0.29) + 18(0.47) = 17.16$$

The mean age of the mother for the babies born to teenagers aged 15–18 is 17.16 years old, which is near 17 years old, as we estimated in Example 6.7 using the balance point method. ■

**What Does This Number Mean?**

What does it mean to say that $\mu = 17.16$ is the mean of the random variable $X$? First of all, the mean of the random variable $X$ is definitely not the same as the mean of a sample of teenage mothers. The latter is a sample mean. For example, suppose that, for a certain hospital, the teenage mothers' ages for the last 5 such births were 16, 18, 18, 17, 18. The mean of this sample of 5 births is $\bar{x} = 17.4$. However, if we were to consider an *infinite number* of births to mothers aged 15–18, then the mean of this large (!) sample would indeed converge to 17.16. So the mean $\mu$ of a discrete random variable is interpreted as the mean of the results from the *population* of all possible repetitions of the experiment. That is why we denote the mean of a random variable as $\mu$, since it denotes a population mean (population of all possible results). Fortunately, since we cannot take a sample of infinite size, we have the method shown in Example 6.8 for calculating the mean. ■

**Why Does This Formula Work?**

The formula for the mean of a discrete random variable works because it is a special case of the weighted mean (page 119). Of the population of 250,000 babies, 7%, or 17,500, were born to 15-year-olds. Thus, $w_1 = 17{,}500$. Similarly, we can find, $w_2 = (0.17)(250{,}000) = 42{,}500$, $w_3 = (0.29)(250{,}000) = 72{,}500$, and $w_4 = (0.47)(250{,}000) = 117{,}500$. Thus, the population weighted mean is

$$\mu = \frac{\sum w_i x_i}{\sum w_i} = \frac{(17{,}500)(15) + (42{,}500)(16) + (72{,}500)(17) + (117{,}500)(18)}{250{,}000}$$

Dividing through and rearranging terms give us

$$\mu = (15)(0.07) + (16)(0.17) + (17)(0.29) + (18)(0.47) = \sum[X \cdot P(X)]$$ ■

The mean $\mu$ of a random variable is also called the **expected value** or the **expectation of the random variable $X$**. It does not necessarily follow that the expected value of $X$ is the most likely value of $X$ (although it does follow if the distribution is symmetric with a single mode). However, the expected value of $X$ (that is, the mean $\mu$) is often a good indication of the center of the distribution of the random variable.

The **expected value, or expectation, of a random variable $X$** is the mean $\mu$ of $X$. It is denoted as $E(X)$. This definition holds for both discrete and continuous random variables.

## Example 6.9 Expected value of Kristin's financial gain from her investment

Calculate the expected value for Kristin's financial gain from her investment described in Example 6.1.

**Solution**

The following table contains the probability distribution we found earlier for the discrete random variable $X$ = financial gain for Kristin's investment.

| $X$ = financial gain in dollars | −2000 | −200 | 0 | 500 | 1000 |
|---|---|---|---|---|---|
| $P(X)$ | 0.10 | 0.20 | 0.25 | 0.30 | 0.15 |

We calculate the expected value of Kristin's financial gain as follows. We multiply each value of $X$ by its associated probability (entry in the top row by the entry below it), and we add up all these products.

$$\mu = (-2000)(0.10) + (-200)(0.20) + 0(0.25) + 500(0.30) + 1000(0.15) = 60$$

Since the unit for $X$ is dollars, the mean is \$60, which is the expected financial gain for Kristin's investment. Note that, according to Table 6.1, in no particular 12-month period can Kristin's earnings exactly equal the expected gain of \$60. The \$60 is the long-run mean. ■

### Variability of a Discrete Random Variable

In Chapter 3 we discussed methods for detecting the presence of outliers in data sets. Similarly, in this section, we use the **variance** or **standard deviation of a random variable** to help us determine whether a particular value of that random variable is unusual. Say you are playing a game that requires you to toss two dice and you toss a pair of ones. Is this an unusual result? We can answer this question using the concepts of variation and standard deviation. Recall in Chapter 3 we examined measures of center and measures of spread. Just as a random variable has a mean ($\mu$), which is a measure of center, so a random variable also has a standard deviation ($\sigma$) and variance ($\sigma^2$), which are measures of spread. The variance of a discrete random variable is given by

$$\sigma^2 = \sum[(X - \mu)^2 \cdot P(X)]$$

Notice that this formula includes $\mu$ as one of its terms, so that you must first find the mean of a discrete random variable before you find the variance (or standard deviation). This definition formula can be tedious in that it requires you to find each of the deviations $(X - \mu)$. There is a computational formula for the variance of a discrete random variable that is equivalent to the definition formula but sometimes is computationally simpler.

**Formulas for the Variance and Standard Deviation of a Discrete Random Variable**

| Definition Formulas | Computational Formulas |
|---|---|
| $\sigma^2 = \sum[(X - \mu)^2 \cdot P(X)]$ | $\sigma^2 = \sum[X^2 \cdot P(X)] - \mu^2$ |
| $\sigma = \sqrt{\sum[(X - \mu)^2 \cdot P(X)]}$ | $\sigma = \sqrt{\sum[X^2 \cdot P(X)] - \mu^2}$ |

Recall from Chapter 3 that the standard deviation is simply the square root of the variance. The definition and computational formulas for the standard deviation of a discrete random variable are the square roots of the formulas for the variance.

## Example 6.10 Finding the variance and standard deviation of a discrete random variable

Find the variance and standard deviation of Kristin's financial gain, using (a) the definition formula, and (b) the computational formula. Assume we have already calculated the mean $\mu = \$60$.

**Solution**

**a.** Refer to Table 6.4. The first two columns correspond to the probability distribution of Kristin's financial gain. The next two columns are intermediate steps for the calculation in the rightmost column, $(X - \mu)^2 \cdot P(X)$. Summing the values in the rightmost column provides the variance. Taking the square root of the variance gives us the standard deviation.

Table 6.4 Using the definition formula

| $X$ | $P(X)$ | $X - \mu$ | $(X - \mu)^2$ | $(X - \mu)^2 \cdot P(X)$ |
|---|---|---|---|---|
| −2000 | 0.10 | −2060 | 4,243,600 | 424,360 |
| −200 | 0.20 | −260 | 67,600 | 13,520 |
| 0 | 0.25 | −60 | 3,600 | 900 |
| 500 | 0.30 | 440 | 193,600 | 58,080 |
| 1000 | 0.15 | 940 | 883,600 | 132,540 |

$$\sigma^2 = \sum[(X - \mu)^2 \cdot P(X)] = 629{,}400$$

$$\sigma = \sqrt{\sigma^2} = 793.3473388 \approx \$793.35$$

**b.** Refer to Table 6.5. The rightmost column contains the values $X^2 \cdot P(X)$. Summing the values in the rightmost column provides $\sum[X^2 \cdot P(X)]$. To find the variance, we must subtract $\mu^2$. Taking the square root of the variance gives us the standard deviation.

Table 6.5 Using the computational formula

| $X$ | $P(X)$ | $X^2$ | $X^2 \cdot P(X)$ |
|---|---|---|---|
| −2000 | 0.10 | 4,000,000 | 400,000 |
| −200 | 0.20 | 40,000 | 8,000 |
| 0 | 0.25 | 0 | 0 |
| 500 | 0.30 | 250,000 | 75,000 |
| 1000 | 0.15 | 1,000,000 | 150,000 |

$$\sum[X^2 \cdot P(X)] = 633{,}000$$

$$\sigma^2 = \sum[X^2 \cdot P(X)] - \mu^2 = 629{,}400$$

$$\sigma = \sqrt{\sigma^2} = 793.3473388 \approx \$793.35$$

The standard deviation of $X$ = Kristin's financial gain is approximately \$793. ■

*What Does This Value for the Standard Deviation Mean?*

The standard deviation of a random variable is a measure of spread of the distribution of $X$. We have $\mu + 1\sigma \approx \$60 + \$793 = \$853$, and $\mu - 1\sigma \approx \$60 - \$793 = -\$733$, as shown in Figure 6.3. Note that most of the values of $X$ lie between $\mu - 1\sigma$ and $\mu + 1\sigma$. Like the mean, the standard deviation $\sigma$ of a random variable $X$ is interpreted as the standard deviation of the results from the population of all possible repetitions of the experiment. ■

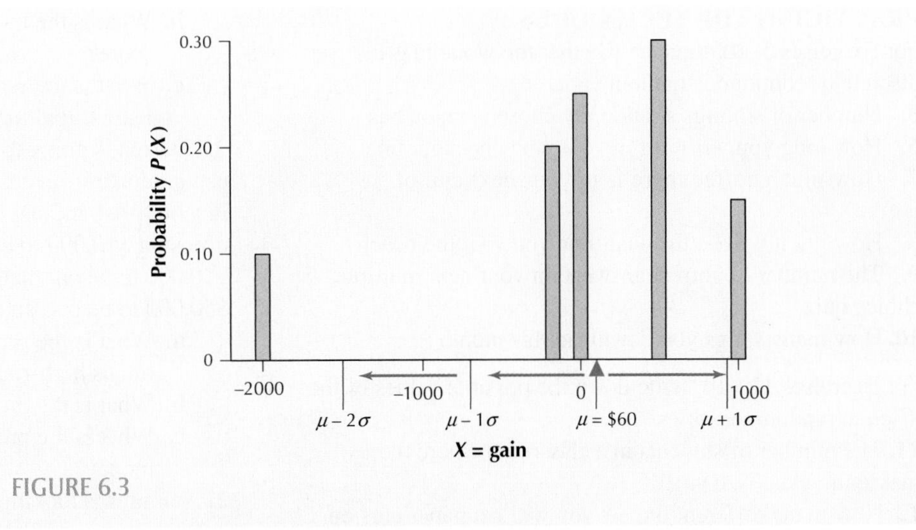

FIGURE 6.3

## Example 6.11 An unusual result?

Using the information from Example 6.10, determine whether $X = -\$2000$ is an unusual result.

### Solution

Recall from Section 3.4 (page 131) that a data value farther than two standard deviations from the mean may be considered moderately unusual. How many standard deviations below the mean does $X = -\$2000$ lie? We have the mean $\mu = \$60$ and the standard deviation $\sigma = \$793.35$. The deviation for $X = -\$2000$ is

$$|X - \mu| = |-2000 - 60| = 2060$$

Expressed in terms of standard deviations, we see that the value $X = -\$2000$ lies $2060/793.35 \approx 2.6$ standard deviations below the mean. Thus, it would be considered moderately unusual for Kristin to lose \$2000 on her investment. ■

## Section 6.1 SUMMARY

1 Section 6.1 introduces the idea of random variables, a crucial concept that we will use to assess the behavior of variable processes for the remainder of the text. Random variables are variables whose value is determined at least partly by chance. Discrete random variables take values that are either finite or countable and may be put in a list. Continuous random variables take an infinite number of possible values, represented by an interval on the number line.

2 Discrete random variables can be described using a probability distribution, which specifies the probability of observing each value of the random variable. Such a distribution can take the form of a table, graph, or formula. Probability distributions describe populations, not samples.

3 We can find the mean $\mu$, standard deviation $\sigma$, and variance $\sigma^2$ of a discrete random variable using formulas.

## Section 6.1 EXERCISES

### CLARIFYING THE CONCEPTS

**1.** Explain in your own words what a random variable is. Give an example of a random variable from your own life experience.

**2.** Is your height a random variable? Under what circumstances would your height be considered a random variable? Under what circumstance would your height not be considered a random variable?

**3.** What is the difference between a discrete random variable and a continuous random variable?

**4.** What is the difference between a discrete random variable and a discrete probability distribution?

## PRACTICING THE TECHNIQUES

For Exercises 5–10, indicate whether the variable is a discrete or continuous random variable.

5. Number of siblings a randomly chosen person has
6. How long you will wait in your next checkout line
7. How much coffee there is in your next cup of coffee
8. How hot it will be the next time you visit the beach
9. The number of correct answers on your next multiple-choice quiz
10. How many songs you download this month

For Exercises 11–14, write down the possible values of the discrete random variables.

11. The number of students in a classroom where the maximum class size is 15
12. How many different fingers you will get paper cuts on next week
13. The number of games that the California Angels will win the next time they are in the World Series (maximum = 4)
14. The number of Donald Duck's three nephews, Huey, Duey, and Luey, who will get into trouble in their next cartoon adventure

For Exercises 15–18, determine whether the distribution represents a valid probability distribution. If it does not, explain why.

15.

| $X$ | −10 | 0 | 10 |
|---|---|---|---|
| $P(X)$ | 1/5 | 1/2 | 1/5 |

16.

| $X$ | 15 | 16 | 17 | 20 |
|---|---|---|---|---|
| $P(X)$ | 0.98 | 0.005 | 0.005 | 0.01 |

17.

| $X$ | 1 | 2 | 3 | 4 | 5 |
|---|---|---|---|---|---|
| $P(X)$ | −0.5 | 0.5 | 0.7 | 0.1 | 0.2 |

18.

| $X$ | −100,000 | 50,000 | 100,000 |
|---|---|---|---|
| $P(X)$ | 0.5 | 0.1 | 1.1 |

For Exercises 19–22, use the given information to construct a probability distribution table and a probability distribution graph. Then answer the questions.

19. Shirelle enjoys listening to rap CDs. The probabilities that she will listen to 0, 1, 2, 3, or 4 CDs tonight are 6%, 24%, 38%, 22%, and 10%, respectively.
   - **a.** What is the probability that Shirelle will listen to at most 1 CD tonight?
   - **b.** What is the probability that Shirelle will not listen to any CDs tonight?
   - **c.** What is the probability that Shirelle will listen to 5 CDs tonight?

20. Josefina is the star athlete on her college soccer team. She especially loves to score goals, and she does this on a regular basis. The probability that she will score 0, 1, 2, or 3 goals tonight are 0.25, 0.35, 0.25, and 0.15.
   - **a.** What is the probability that Josefina will score fewer than 3 goals tonight?
   - **b.** What is the most likely number of goals Josefina will score?
   - **c.** What is the probability that Josefina will score at least 1 goal tonight?

21. Darren is going to make it big on Wall Street, if only he can graduate from college first. Darren has invested money in a high-risk mutual fund and has figured his probability of losing $10,000 to be one-third, his probability of gaining $10,000 to be one-half, and his probability of gaining $50,000 to be one-sixth.
   - **a.** What is the probability that Darren will make money on his high-risk investment?
   - **b.** What is the most likely outcome of this investment?
   - **c.** What is the probability that Darren will not make money?

22. Jennifer is looking for a roommate. She likes pets, but not too many. She would prefer to move in with a roommate who had either one or two pets. Of the 10 possible roommates who answered Jennifer's ad, 5 have no pets, 3 have one pet, 1 has two pets, and 1 has more than two pets.
   - **a.** What is the most likely number of pets among the 10 possible roommates?
   - **b.** What is the probability that, if 1 of the 10 possible roommates is chosen at random, he or she will have the number of pets that Jennifer prefers?
   - **c.** What is the probability that a possible roommate has more than one pet, if the roommate is chosen at random?

For Exercises 23–26, find the mean, variance, and standard deviation of the given random variables. Then draw a probability distribution graph for each variable. Plot the mean of the random variable on each graph. Does this value for the mean make sense as the point where the distribution balances?

23. Kari is trying to estimate the number of guests who will come to her party next Friday. She has drawn up the following probability distribution of $X$ = number of guests.

| $X$ | 2 | 3 | 4 |
|---|---|---|---|
| $P(X)$ | 1/4 | 1/2 | 1/4 |

24. Paige has kept track of the number of students in her classrooms over the years. She has drawn up the following probability distribution of $X$ = number of students.

| $X$ | 22 | 23 | 24 | 25 |
|---|---|---|---|---|
| $P(X)$ | 0.3 | 0.2 | 0.1 | 0.4 |

25. Paul is trying to estimate the number of times he will have to go watch a baseball game before the home team finally wins. Using subjective probability, he has drawn up the following probability distribution of $X$ = number of games.

| $X$ | 1 | 2 | 3 | 4 | 5 |
|---|---|---|---|---|---|
| $P(X)$ | 0.1 | 0.2 | 0.3 | 0.3 | 0.1 |

26. Shona is an English major who would like to estimate the number of term papers she will have to write next semester. Using subjective probability, she has drawn up the following probability distribution for $X$ = the number of term papers.

| $X$ | 1 | 2 | 3 | 4 | 5 | 6 |
|---|---|---|---|---|---|---|
| $P(X)$ | 0.05 | 0.15 | 0.2 | 0.25 | 0.25 | 0.1 |

## APPLYING THE CONCEPTS

27. **Stanley Cup Finals.** Figure 6.4 is the probability distribution graph of the number of games ($X$) in the Stanley Cup Finals for the years 1956–2007. Estimate the mean of $X$ using the balance point method.

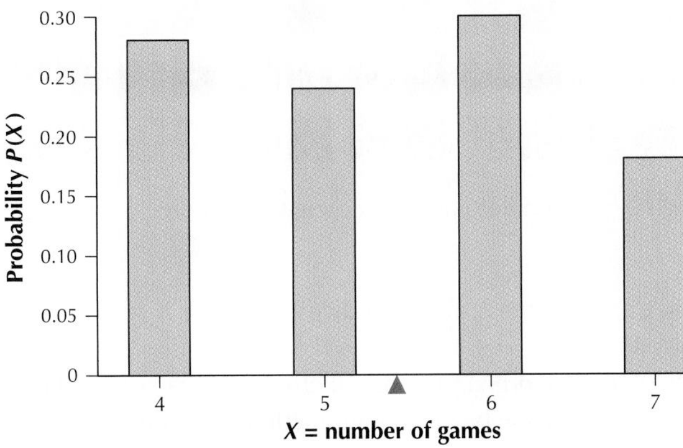


Figure 6.4

28. **Smokers.** The National Survey on Drug Use and Health (2005) reported that 5 million young people aged 12–18 had tried tobacco products in the previous month. The table contains the proportions of these 5 million who had done so, at each age level. Let $X$ = age of the person who had tried tobacco products in the previous month.

| $X$ = age | $P(X)$ |
|---|---|
| 12 | 0.01 |
| 13 | 0.04 |
| 14 | 0.07 |
| 15 | 0.13 |
| 16 | 0.18 |
| 17 | 0.23 |
| 18 | 0.34 |

**a.** Construct the graph of the probability distribution.
**b.** Estimate the mean of the random variable $X$ = age.
**c.** Calculate the expected age.
**d.** Clearly interpret the meaning of the expected age.

29. **Stanley Cup Finals.** The table contains the number of games ($X$) in the Stanley Cup Finals for the years 1956–2007, along with the relative frequency ($P(X)$) for that number of games.

| $X$ = games | $P(X)$ |
|---|---|
| 4 | 0.28 |
| 5 | 0.24 |
| 6 | 0.30 |
| 7 | 0.18 |

**a.** Find the mean of the random variable $X$ = number of games in the Stanley Cup Finals.
**b.** How close is $\mu$ to your estimate from Exercise 27?

30. Refer to Exercise 28.
**a.** Find the standard deviation of $X$ = age.
**b.** Determine whether a 12-year-old who had tried tobacco products in the previous month would be considered unusual.

31. **Number of Courses Taught.** The table provides the probability distribution for $X$ = number of courses taught by faculty at all degree-granting institutions of higher learning in the United States in the fall 2003 semester.[1]

| $X$ = courses taught | $P(X)$ |
|---|---|
| 1 | 0.233 |
| 2 | 0.334 |
| 3 | 0.243 |
| 4 | 0.122 |
| 5 | 0.068 |

**a.** Construct the graph of the probability distribution.
**b.** If a faculty member is chosen at random, what is the most likely number of courses taught? The least likely?
**c.** Find the expected number of courses taught.
**d.** Clearly interpret the meaning of the expected number of courses taught.

32. **California Rooms.** The table provides the probability distribution for $X$ = number of rooms in California housing units.[2]

| $X$ = rooms | $P(X)$ |
|---|---|
| 1 | 0.050 |
| 2 | 0.094 |
| 3 | 0.138 |
| 4 | 0.157 |
| 5 | 0.189 |
| 6 | 0.165 |
| 7 | 0.103 |
| 8 | 0.059 |
| 9 | 0.045 |

**a.** Construct the graph of the probability distribution.
**b.** If a housing unit is chosen at random, what is the most likely number of rooms? The least likely?
**c.** Find the expected number of rooms in a California housing unit.
**d.** Interpret the meaning of the expected number of rooms.

33. **Number of Courses Taught.** For the data in Exercise 31:
**a.** Find the standard deviation of the number of courses taught.
**b.** Determine whether teaching 5 courses would be considered unusual.

**34. California Rooms.** For the data in Exercise 32:

**a.** Find the standard deviation of the number of rooms.

**b.** Determine whether a house with 9 rooms would be considered unusual.

**35. Florida Vehicle Ownership.** The table provides the probability distribution for $X$ = number of vehicles owned by residents of Florida.[3]

| $X$ = vehicles | $P(X)$ |
|---|---|
| 0 | 0.081 |
| 1 | 0.414 |
| 2 | 0.382 |
| 3 | 0.123 |

**a.** Construct the graph of the probability distribution.

**b.** If a resident is chosen at random, what is the most likely number of vehicles? The least likely?

**c.** Find the mean number of vehicles.

**d.** Clearly interpret the meaning of the mean number of vehicles.

**36. Youngsters and Beer.** Nearly half of all youngsters (aged 10–19) have never drunk beer, which is good because for most of them it is illegal. However, the U.S. Bureau of Justice Statistics keeps track of, among other things, the age at which young persons imbibe their first brew. For the young people who drank beer before age 20, the table shows the proportions who had their first beer in each age group. Suppose you are a member of a task force investigating methods for curbing youth alcohol abuse.

| Age group | Proportion |
|---|---|
| 6–9 | 0.25 |
| 10–11 | 0.16 |
| 12–13 | 0.27 |
| 14–15 | 0.24 |
| 16–17 | 0.07 |
| 18–19 | 0.01 |

**a.** State why this table does not represent a true probability distribution.

**b.** Define the values of the random variable $X$ = age at first beer as the midpoints of the classes given in the table. For example, for the first age group, use $X = 7.5$.

**c.** Find the age group with the highest probability.

**d.** Find the mean age. Use the midpoints of the age group as your values of $X$.

**e.** Suppose that your task force wanted to capture the attention of the news media. Which measure would you report?

**37. Florida Vehicle Ownership.** For the data in Exercise 35:

**a.** Find the standard deviation of the number of vehicles, and

**b.** Determine whether owning no vehicles would be considered unusual.

**The Two-Dice Experiment.** Use the following information and Figure 6.5 for Exercises 38–45. Your experiment is to toss a pair of fair dice and find

$$X = \text{sum of the two dice}$$

Recall the sample space for the two-dice experiment, reproduced in Figure 6.5.

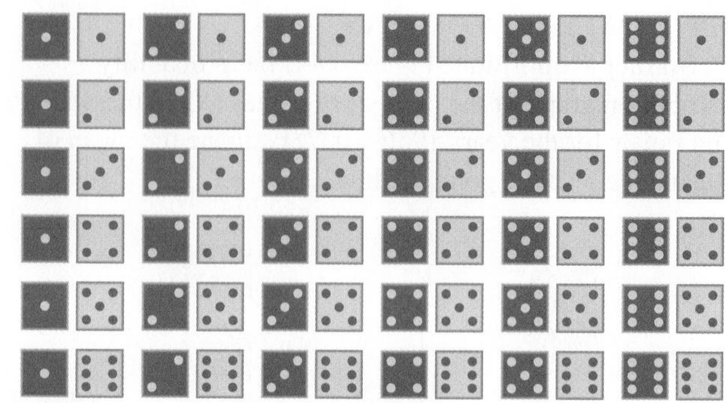

Figure 6.5 Sample space for tossing two fair dice.

**38.** Construct the probability distribution for $X$ = sum of the two dice.

**39.** Graph the probability distribution for $X$ = sum of the two dice. Use the balance point method to estimate the mean $\mu$ of $X$.

**40.** Calculate the mean $\mu$ of $X$ = sum of the two dice. How close is your estimate from the previous exercise to the mean $\mu$?

**41.** Calculate the standard deviation of $X$ = sum of the two dice.

**42.** In your graph of the probability distribution of $X$, label the mean $\mu$ and indicate the size of the standard deviation $\sigma$, as in Figure 6.3 (page 269).

**43.** Determine whether snake eyes ($X = 2$) is an unusual result. By symmetry, apply your finding to another value of $X$.

**44.** Note that the mean of $X$ also happens to be the value of $X$ that occurs most often.

**a.** Does it always happen that the mean of a discrete random variable is the same as the most likely value of that variable? If not, give a counterexample.

**b.** Specify the conditions when it is true that the mean of a discrete random variable is the same as the most likely value of that variable.

WHAT IF? **45.** *What if* we add the same unknown amount $k$ to each value of $X$? Describe what would happen to the following, and why.

**a.** The mean of $X$

**b.** The standard deviation of $X$

# 6.2 Binomial Probability Distribution

## Objectives: By the end of this section, I will be able to...

1 Explain what constitutes a binomial experiment.

2 Compute probabilities using the binomial probability formula.

3 Find probabilities using the binomial tables.

4 Calculate and interpret the mean, variance, and standard deviation of the binomial random variable.

## 1 Binomial Experiment

There are many different types of discrete probability distributions. Perhaps the most important is the *binomial* distribution, which we will learn about in this section. Life is full of situations where there are *only two possible outcomes* to a process.

- A baby is about to be born. Will it be a boy or a girl?
- A basketball player is about to take a foul shot. Will she make it or miss?
- A friend of yours is also taking statistics. Will he pass or fail?

Many experiments having more than two outcomes can often be defined so that there are only two outcomes. For example, the answer to a multiple-choice question that has five answer choices may be recorded as either correct or incorrect. Because situations where there are only two possible outcomes are so widespread, methods have been developed to make it more convenient to analyze them. These methods begin with the definition of a **binomial experiment**.

**Binomial Experiment**

A probability experiment that satisfies the following four requirements is said to be a **binomial experiment:**

1. Each trial of the experiment has only *two possible outcomes* (or is defined in such a way that the number of outcomes is reduced to two). One outcome is considered a *success* and the other a *failure*.
2. There is a *fixed number of trials,* known in advance of the experiment.
3. The experimental outcomes are *independent* of each other.
4. The *probability* of observing a success remains the same from trial to trial.

Let's take a moment to discuss what these requirements really mean. First, a *success* denotes simply the outcome we are interested in. It does not mean that the outcome is desirable. For example, a researcher may be interested in the proportion of college students who drop out. In this case, dropping out of college would be considered a success in the context of a binomial experiment.

Next, to qualify as a binomial experiment, there must be *a fixed number of trials,* known in advance. For example, tossing a coin ten times is a binomial experiment because we know the number of trials in advance. In contrast, going fishing and continuing to fish until you catch a rainbow trout is not a binomial experiment. Certainly, each fish you catch is either a rainbow trout or is not a rainbow trout, so there are two possible outcomes. However, since you don't know how many fish you will catch before your rainbow trout shows up, there is not a fixed number of trials known in advance.

Binomial experiments also require that each trial be *independent*. What experiment does not meet this condition? Suppose a market researcher at a shopping mall is asking consumers if they use Fib detergent. She asks a pair of men, the first of whom is clearly the employer of the second. Since the response given by the employer is likely to affect the response given by the employee, the outcomes are not independent.

Finally, for a binomial experiment, the *probability of observing a success must remain the same from trial to trial*. In Chapter 5, we learned that sampling without replacement violates this requirement because the probability of a success is not the same from trial to trial. However, when the sample is quite small compared to the size of the population, the change in probability from one trial to another is so small that we can consider the probabilities to be the same.

The outcomes of a binomial experiment, together with their probabilities, generate a special discrete probability distribution called the **binomial probability distribution**. For binomial probability distributions, there are always only two outcomes, and each outcome has a probability associated with it. The *binomial random variable*, denoted by $X$, represents the number of successes observed in the $n$ trials. Note that $0 \leq X \leq n$.

## Example 6.12 Recognizing binomial experiments

Determine whether each of the following experiments fulfills the conditions for a binomial experiment. If the experiment is binomial, identify the random variable $X$, the number of trials, the probability of success, and the probability of failure. If the experiment is not binomial, explain why not.

**a.** We flip a fair coin three times and observe the number of heads.

**b.** A basketball player with 75% free-throw accuracy is going to the line to shoot "one-and-one," meaning that she will get the chance to shoot a second free throw if she succeeds on the first free throw.

**c.** The National Burglar and Fire Alarm Association reports that 34% of burglars get in through the front door. A random sample of 36 burglaries is taken, and the number of entries through the front door is noted.

### Solution

**a.** This is a binomial experiment because it fulfills the requirements:

**i.** There are only two possible outcomes on each trial, with heads defined as success and tails as failure.

**ii.** We know in advance that we are tossing the coin three times.

**iii.** The coin doesn't remember its result from toss to toss, and so the trials are independent.

**iv.** The coin is fair on each toss, and so the probability of observing heads is the same on each toss.

The binomial random variable $X$ is the number of heads observed on the three trials; since the coin is fair, the probability of success is 0.5 and the probability of failure is 0.5. The possible values for $X$ are 0, 1, 2, or 3.

**b.** This is not a binomial experiment, because there is not a fixed number of trials. If she gets the first basket, then $n = 2$, but if she misses the first basket, then $n = 1$.

**c.** This is a binomial experiment because it fulfills the requirements:

**i.** There are only two possible outcomes on each trial: entering through the front door or not entering through the front door.

**ii.** We know in advance that the size of the random sample is 36 burglaries.

**iii.** Since the sample is random, the trials are independent.

iv. Since the sample is quite small compared to the size of the population, the probability of entering through the front door remains the same from burglary to burglary.

The binomial random variable $X$ is the number of front-door-entry burglaries noted for the 36 break-ins; the probability of success is 0.34 and the probability of failure is $1 - 0.34 = 0.66$.

Table 6.6 gives some notation regarding binomial experiments and the binomial distribution. Using this notation in the experiment in Example 6.12a, we have

$$S = \text{the outcome is heads, and } F = \text{the outcome is tails}$$

Table 6.6 Notation for binomial experiments and the binomial distribution

| Symbol | Meaning |
|---|---|
| $S$ | The outcome denoted as a success |
| $F$ | The outcome denoted as a failure |
| $P(S) = p$ | The probability of observing a success |
| $P(F) = 1 - p$ | The probability of observing a failure |
| $n$ | The number of trials |

Since the coin is fair, the probability of observing a success (that is, the probability of observing heads on a single toss) is 0.5. Thus,

$$P(S) = p = 0.5, \text{ and } P(F) = 1 - p = 1 - 0.5 = 0.5$$

## 2 Binomial Probability Distribution Formula

Before we examine the binomial probability distribution formula, let us recall from Section 5.4 (page 245) the formula for the **number of combinations**.

*Note:* In Section 5.4, we used ${}_nC_r$ to indicate the number of combinations. Now that we have learned about random variables, which can be denoted $X$, we use ${}_nC_X$ to represent the number of combinations.

**Formula for the Number of Combinations**

The **number of combinations** of $X$ items chosen from $n$ different items is given by

$${}_nC_X = \frac{n!}{X!\,(n - X)!}$$

where $n!$ represents $n$ **factorial**, which equals $n(n - 1)(n - 2) \cdots (2)(1)$; and $0!$ is defined to be 1.

### Example 6.13 How many team combinations in the intramural volleyball league?

Jeffrey is in charge of drawing up a schedule for his college's intramural volleyball league. This year five teams have been fielded, and they must play each other once. How many games will be held?

**Solution**

The number of combinations of $n = 5$ volleyball teams taken $x = 2$ at a time is

*Note:* You may find the following special combinations useful. For any integer $n$:

$${}_nC_n = 1 \quad {}_nC_0 = 1$$

$${}_nC_1 = n \quad {}_nC_{n-1} = n$$

$${}_5C_2 = \frac{5!}{2!(5 - 2)!} = \frac{5 \cdot 4 \cdot 3 \cdot 2 \cdot 1}{(2 \cdot 1)(3 \cdot 2 \cdot 1)} = \frac{120}{(2)(6)} = 10$$

Ten games will be held.

We are often interested in finding probabilities associated with a binomial experiment. For example, out of five randomly selected cars on the highway, what is the probability that four out of five drivers are obeying the speed limit if we assume that there is a 50% chance that a driver is obeying the speed limit? Pretty small, you think? Example 6.14 shows how to solve a problem like this.

## Example 6.14 Online dating

Lori Lee Miller/Alamy

A recent study reported that about 40% of online dating-survey respondents are "hoping to start a long-term relationship" (LTR).[4] Consider the experiment of choosing three online daters at random, and let

$$X = \text{the number of "LTRers"}$$

so that a success is defined as choosing someone hoping to start a long-term relationship.

**a.** Construct a tree diagram for this experiment.

**b.** Suppose that we are interested in finding the probability that exactly two of the three online daters would be LTRers, $P(X = 2)$. In the tree diagram, highlight in blue the outcomes where exactly two of the three online daters are LTRers. Find the probability for each outcome, and use these to find $P(X = 2)$.

**c.** Suppose that we are interested in finding $P(X = 1)$. In the tree diagram, highlight in red the different outcomes where exactly one of the three online daters is an LTRer. Find the probability for each outcome, and use these to find $P(X = 1)$.

### Solution

**a.** Figure 6.6 shows the tree diagram for this experiment.

**b.** As we can see from Figure 6.6, there are $({}_nC_X) = ({}_3C_2) = 3$ different ways that exactly two of the three online daters could be LTRers (highlighted in blue).

| 1st Trial | 2nd Trial | 3rd Trial | Outcome | Number of successes, $X$ | Probability of outcome |
|---|---|---|---|---|---|
| S | S | S | S, S, S | 3 | $(0.4) \cdot (0.4) \cdot (0.4) = 0.064$ |
| | | F | S, S, F | 2 | $(0.4) \cdot (0.4) \cdot (0.6) = 0.096$ |
| | F | S | S, F, S | 2 | $(0.4) \cdot (0.6) \cdot (0.4) = 0.096$ |
| | | F | S, F, F | 1 | $(0.4) \cdot (0.6) \cdot (0.6) = 0.144$ |
| S | S | S | F, S, S | 2 | $(0.6) \cdot (0.4) \cdot (0.4) = 0.096$ |
| | | F | F, S, F | 1 | $(0.6) \cdot (0.4) \cdot (0.6) = 0.144$ |
| | F | S | F, F, S | 1 | $(0.6) \cdot (0.6) \cdot (0.4) = 0.144$ |
| | | F | F, F, F | 0 | $(0.6) \cdot (0.6) \cdot (0.6) = 0.216$ |

FIGURE 6.6 Tree diagram and binomial probabilities for Example 6.14.

For each of these three outcomes, the probability that $X = 2$ is $(0.4)^2(0.6) = 0.096$. In particular, the outcome $S, S, F$ (second row in Figure 6.6) has probability

$P(S) \cdot P(S) \cdot P(F) = (0.4)(0.4)(0.6) = 0.096$. Similarly, the outcome $S$, $F$, $S$ has probability $P(S) \cdot P(F) \cdot P(S) = (0.4)(0.6)(0.4) = 0.096$, and the outcome $F$, $S$, $S$ has probability $P(F) \cdot P(S) \cdot P(S) = (0.6)(0.4)(0.4) = 0.096$. Note that, since $P(S) = p$, this probability is $(p)^2\,(1 - p)$, with $p$ having exponent $X = 2$, and $(1 - p)$ having exponent $n - X = 3 - 2 = 1$. Thus,

$$\begin{aligned} P(X = 2) &= ({}_3C_2)(0.4)^2(0.6) \\ &= 3(0.096) \\ &= 0.288. \end{aligned}$$

**c.** Similarly, suppose that we are interested in whether exactly one ($X = 1$) of the three online daters is an LTRer. Then, Figure 6.6 shows us, highlighted in red, that there are $({}_nC_X) = ({}_3C_1) = 3$ different ways this could happen. Each of these outcomes has probability $(p)(1 - p)^2 = (0.4)(0.6)^2 = 0.144$, where $p$ has exponent $X = 1$, and $(1 - p)$ has exponent $n - X = 3 - 1 = 2$. Thus,

$$\begin{aligned} P(X = 1) &= ({}_3C_1)(0.4)(0.6)^2 \\ &= 3(0.144) \\ &= 0.432. \end{aligned}$$

We can generalize these procedures and use the **binomial probability distribution formula** to find probabilities for the number of successes for any binomial experiment.

**The Binomial Probability Distribution Formula**

The probability of observing exactly $X$ successes in $n$ trials of a binomial experiment is

$$P(X) = ({}_nC_X)\,p^X\,(1 - p)^{n-X}$$

The binomial probability distribution formula is often called simply the binomial formula.

## Example 6.15 Can Joshua ace the multiple-choice stats quiz by guessing?

Suppose that Joshua is about to take a four-question multiple choice statistics quiz. Josh did not study for the quiz, so he will have to take random guesses on each of the four questions. Each question has five possible alternatives, only one of which is correct.

**a.** What is the probability that Joshua will ace the quiz by answering all the questions correctly?

**b.** What is the probability that Joshua will pass the quiz by answering at least three questions correctly?

### Solution

When dealing with binomial probability problems like this one, first find the values for $n$, $p$, $1 - p$, and $X$. Then plug these values into the binomial formula. There are four questions on the quiz, so the number of trials is $n = 4$. Next we know that $p = 1/5$, since there are five choices and Joshua has a 1 in 5 chance of being correct if he chooses randomly. Thus,

$$p = \text{probability of success} = 1/5 = 0.2$$

Four of the five possible alternatives are incorrect. So,

$$(1 - p) = \text{probability of failure} = 4/5 = 0.8$$

**a.** To find the probability of correctly guessing the right answer on all four questions, Joshua is interested in observing $X = 4$ successes. Using the binomial formula, we obtain

$$P(X = 4) = ({}_4C_4)(0.2)^4\,(1-0.2)^{4-4} = (1)\,(0.0016)(1) = 0.0016$$

So Joshua's chance of acing this quiz by making random guesses is very small, less than one-fifth of 1%. (Note that in this formula we used the shortcut ${}_nC_n = 1$ and the fact that any real number raised to the zero power equals 1.)

**b.** Answering at least three questions correctly does not change the values $n = 4$, $p = 0.2$, and $1 - p = 0.8$. To answer at least three questions correctly, Joshua must answer either $X = 3$ *or* $X = 4$ questions correctly. Since these events are mutually exclusive, we find the required probability by using the Addition Rule for Mutually Exclusive Events,

$$P(X \geq 3) = P(X = 3) + P(X = 4)$$

We already found $P(X = 4) = 0.0016$ in **(a)**. Now we find

$$P(X = 3) = ({}_4C_3)\,(0.2)^3\,(1-0.2)^{4-3} = (4)(0.008)(0.8) = 0.0256$$

Therefore, the probability that Joshua will pass this quiz by random guessing is $0.0016 + 0.0256 = 0.0272$. Since he has less than a 3% chance of even passing this quiz, we would tell Joshua that perhaps random guessing isn't the best strategy for stats quizzes. ■

## 3 Binomial Distribution Tables

As you can imagine, calculations involving binomial probabilities can sometimes get tedious. For example, to find the probability of observing at least 60 heads on 100 tosses of a fair coin, we would have to use the binomial formula for $X = 60$, $X = 61$, $X = 62$, and so on, right up to $X = 100$. For this type of problem, you can use Table B, Binomial Distribution, in the Appendix (pages T-3–T-8). If you are trying to answer a question involving unusual values of $n$, such as 103, or unusual values of $p$, such as 0.47, then you can use technology instead.

### Example 6.16 Using the binomial table

Use the binomial table and the binomial distribution from Example 6.15 to find the following probabilities:

**a.** Joshua will answer all four questions correctly

**b.** Joshua will answer at least three of the four questions correctly

**Solution**

**a.** To find the probability of answering all four questions correctly, we first need to find $n$ and $p$. There are four questions, so $n = 4$. The probability of success is 0.2 (1 out of 5 choices), so $p = 0.2$. In Figure 6.7:

- Look under the $n$ column until you find $n = 4$. That is the portion of the table you will use.
- Then go across the top of the table until you get to $p = 0.20$. That gives you your column.

- We are interested in finding the probability of observing $X = 4$, where $X$ is the number of successes. So go down the column until you see 4 under the $X$ column on the left (and in the subgroup with $n = 4$).
- The number in the $p$ column is 0.0016, which is the same answer we calculated in Example 6.15.

| | | $p$ | | | | |
|---|---|---|---|---|---|---|
| $n$ | $X$ | **0.10** | **0.15** | **0.20** | **0.25** | **0.30** |
| 2 | 0 | 0.8100 | 0.7225 | 0.6400 | 0.5625 | 0.4900 |
| | 1 | 0.1800 | 0.2550 | 0.3200 | 0.3750 | 0.4200 |
| | 2 | 0.0100 | 0.0225 | 0.0400 | 0.0625 | 0.0900 |
| 3 | 0 | 0.7290 | 0.6141 | 0.5120 | 0.4219 | 0.3430 |
| | 1 | 0.2430 | 0.3251 | 0.3840 | 0.4219 | 0.4410 |
| | 2 | 0.0270 | 0.0574 | 0.0960 | 0.1406 | 0.1890 |
| | 3 | 0.0010 | 0.0034 | 0.0080 | 0.0156 | 0.0270 |
| 4 | 0 | 0.6561 | 0.5220 | 0.4096 | 0.3164 | 0.2401 |
| | 1 | 0.2916 | 0.3685 | 0.4096 | 0.4219 | 0.4116 |
| | 2 | 0.0486 | 0.0975 | 0.1536 | 0.2109 | 0.2646 |
| | 3 | 0.0036 | 0.0115 | 0.0256 | 0.0469 | 0.0756 |
| | 4 | 0.0001 | 0.0005 | 0.0016 | 0.0039 | 0.0081 |

FIGURE 6.7 Excerpt from the binomial tables.

**b.** In this case, "at least three" means three or four. So find the probabilities for $X = 3$ and $X = 4$ and add them up. We already found that the probability that $X = 4$ is 0.0016. The probability that $X = 3$ is 0.0256 in Figure 6.7. Adding, we get the probability $0.0016 + 0.0256 = 0.0272$ that at least three questions were answered correctly, just as we found in Example 6.15. ■

Next, a word about *cumulative probability*. Recall that Joshua will pass if he answers at least three of the four questions correctly. What is the probability that Joshua will not pass? In other words, what is the probability that Joshua answers at most two of the four questions correctly? This is what is known as a cumulative probability, since it is asking for the accumulated probability for $X = 0$, $X = 1$, and $X = 2$. Statistical software and the TI-83/84 graphing calculator each have a function that will find cumulative binomial probabilities for you.

## Example 6.17 Using technology to find binomial probabilities

Recall that we have $n = 4$ questions, each with 5 possible alternatives, so that $p$ = probability of success on any given question is 0.2 (1/5). Use the TI-83 to find the following probabilities:

**a.** $P(X = 4)$, the probability that Joshua will get all four correct

**b.** $P(X \leq 2)$, the (cumulative) probability that Joshua will flunk the quiz

**Solution**

We use the instructions in the Step-by-Step Technology Guide at the end of this section (page 282).

**a.** Figure 6.8 shows that we use the function **binompdf** with $n = 4$, $p = 0.2$, and $X = 4$. Figure 6.9 shows the result, $P(X = 4) = 0.0016$.

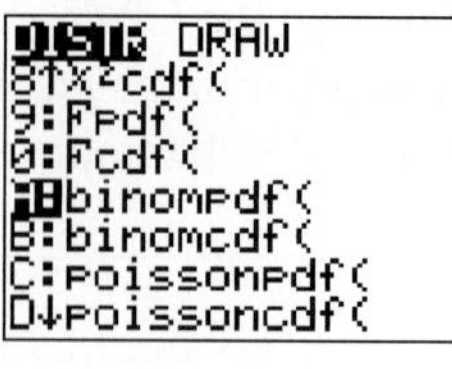


FIGURE 6.8

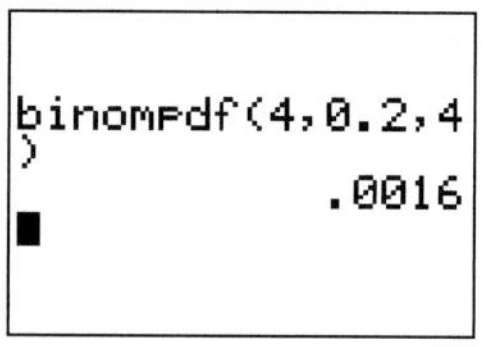

binompdf(4,0.2,4)
.0016

FIGURE 6.9

**b.** Figure 6.10 shows that we use the function **binomcdf** with $n = 4$, $p = 0.2$, and $X = 2$. Figure 6.11 shows the result, $P(X \leq 2) = 0.9728$.

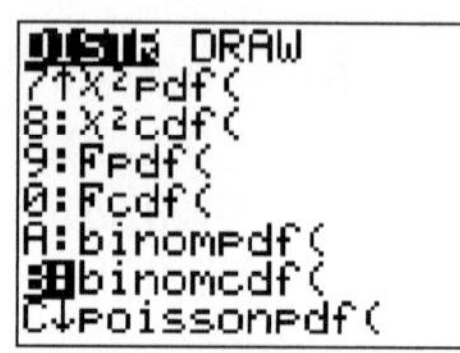


FIGURE 6.10

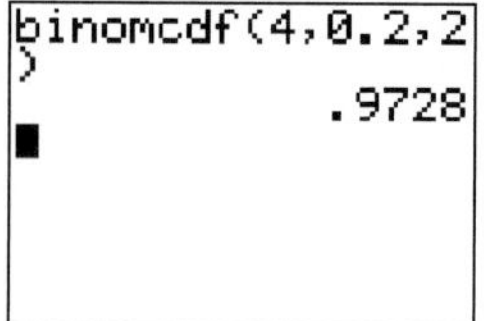


FIGURE 6.11

## 4 Binomial Mean, Variance, and Standard Deviation

In Section 6.1, we examined the mean, variance, and standard deviation of a discrete random variable. Since the binomial random variable $X$ is discrete, it also has a **mean, variance,** and **standard deviation,** shown here.

**Mean, Variance, and Standard Deviation of a Binomial Random Variable $X$**

- Mean (or expected value): $\mu = n \cdot p$
- Variance: $\sigma^2 = n \cdot p \cdot (1 - p)$
- Standard deviation: $\sigma = \sqrt{n \cdot p \cdot (1 - p)}$

## Example 6.18 Mean, variance, and standard deviation of left-handed students

Suppose we know that the population proportion $p$ of left-handed students is 0.10.

**a.** In a sample of 200 students, how many would we expect to be left-handed?
**b.** Would 40 left-handed students out of 200 be considered unusual?

### Solution

The binomial random variable here is $X$ = the number of left-handed students.

**a.** Here, $n = 200$, and $p = 0.10$. So the expected number of left-handed students in a sample of 200 is

$$E(X) = \mu = n \cdot p = (200)(0.10) = 20$$

**b.** To determine whether 40 lefties is unusual:

1. Find the standard deviation of $X$ = the number of lefties in a sample of 200 students.
2. Calculate how many standard deviations 40 lies from the mean $\mu = 20$.

The standard deviation of $X$ is

$$\sigma = \sqrt{n \cdot p \cdot (1 - p)} = \sqrt{(200)(0.1)(1 - 0.1)} = \sqrt{18} \approx 4.2426$$

How many standard deviations does 40 lie above the mean of 20? Simply take the difference between 40 and 20, and divide the result by the standard deviation 4.2426:

$$\frac{X - \mu}{\sigma} = \frac{40 - 20}{4.2426} \approx 4.714$$

Finding 40 lefties in a sample of 200 is unusual because this value lies 4.7 standard deviations above the mean. ■

**What Does This Value for the Standard Deviation Mean?**

The standard deviation of a random variable $X$ is a measure of spread of the binomial random variable $X$. Suppose the binomial experiment in Example 6.18 is performed an infinitely large number of times. Then the standard deviation of this infinitely large collection of values of $X$ will be $\sigma \approx 4.2426$. ■

## Example 6.19 The ballot that changed history

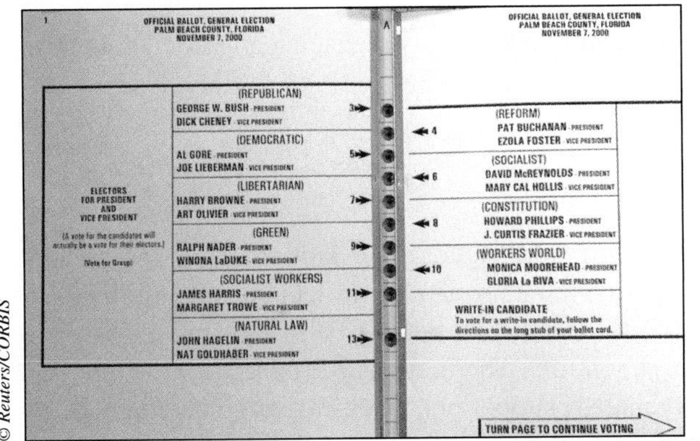




Recall from Section 1.1 that many Palm Beach County voters were confused by the presidential election ballot of 2000. We will examine evidence that many confused Palm Beach County voters chose Reform Party candidate Pat Buchanan when they had intended to vote for Democrat Al Gore. Here $X$ = the number of votes for Buchanan. The scatterplot in Figure 6.12 shows, county-by-county for all Florida's counties, the number of votes for Buchanan versus the total number of votes. There is a clear linear (straight-line) relationship except for one glaring outlier (an observation that lies outside the straight-line pattern), which is Palm Beach County.

The scatterplot shows that Buchanan received many more votes in Palm Beach County than elsewhere. Now, we investigate the likelihood of observing such results. Of the 5,961,531 ballots cast statewide in Florida's 2000 presidential election, Buchanan received 17,317. Thus, statewide, the proportion of Buchanan voters is $p = 17{,}317/5{,}961{,}531 \approx 0.0029$. For this analysis, we will consider all of the votes in Florida statewide to be the population, so that Buchanan's population proportion of success is $p = 0.0029$. We will consider the votes in Palm Beach County to be a sample from this population. There were 432,286 votes cast in this county, of which 3407 votes were for Buchanan.

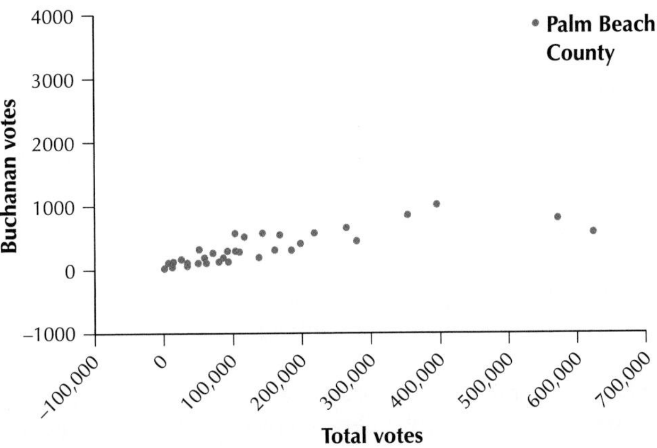


FIGURE 6.12 Palm Beach County shows an unlikely attraction to Buchanan.

**Solution**

**Step 1** Find the expected number of votes cast for Buchanan in Palm Beach County. In our sample, we have $n = 432{,}286$ and $p = 0.0029$, so the mean number of votes is $n \cdot p = 432{,}286(0.0029) \approx 1254$. Buchanan may have expected, based on his statewide proportion, to receive 1254 votes in Palm Beach County.

**Step 2** Find the standard deviation of the number of votes cast for Buchanan in Palm Beach County:

$$\sqrt{n \cdot p(1-p)} = \sqrt{(432{,}286)(0.0029)(0.9971)} \approx 35.3553$$

So, typically, the number of votes for Buchanan in a county the size of Palm Beach County would deviate from the mean by about 35 votes.

**Step 3** Compute the number of standard deviations above the mean Buchanan's 3407 votes lie. We have a mean of 1254 votes and a standard deviation of 35.3553 votes. That gives us

$$\frac{X - \mu}{\sigma} = \frac{3407 - 1254}{35.3553} \approx 60.9$$

In other words, Buchanan's votes in Palm Beach County register at an astonishing 60.9 standard deviations above the mean. ■

The probability of observing a statistic 60.9 standard deviations above the mean is infinitesimally small. For comparison, the probability of winning the multistate Powerball lottery is 0.000000008. Buchanan's results in Palm Beach County are at least a million times less likely than this. The inescapable conclusion is that $p$, the population proportion of Buchanan supporters, has been augmented by another group of voters who were evidently confused by the controversial ballot. The confusion of these voters affected the state results, changing Florida from Gore's column to Bush's column. ■

## STEP-BY-STEP TECHNOLOGY GUIDE: Finding Binomial Probabilities

For Example 6.17 (page 279).

### TI-83/84

***Step 1*** Press **2nd** > **DISTR** (the **VARS** key).
***Step 2*** Do one of **(a)** or **(b)**:
**a.** For individual binomial probabilities, highlight **binompdf(** and press **ENTER**.
**b.** For cumulative binomial probabilities, highlight **binomcdf(** and press **ENTER**.
***Step 3*** Enter the values for $n$, $p$, and ***K***, separated by commas.
***Step 4*** Press **ENTER**.

### EXCEL

***Step 1*** Select cell **A1**. Click the **Insert Function** icon $f_x$.
***Step 2*** In the **Search for a function** area, type **BINOMDIST**, and click **OK**.
***Step 3*** For **Number_s**, enter the number of successes, ***K***. For **Trials**, enter the sample size, **n**. For **Probability_s**, enter the probability of success, **p**.
***Step 4*** Do one of **(a)** or **(b)**:
**a.** For individual binomial probabilities, next to **Cumulative**, enter **false**.
**b.** For cumulative binomial probabilities, next to **Cumulative**, enter **true**.
***Step 5*** Click **OK**. See Figures 6.13 and 6.14 for illustrations using Example 6.17.

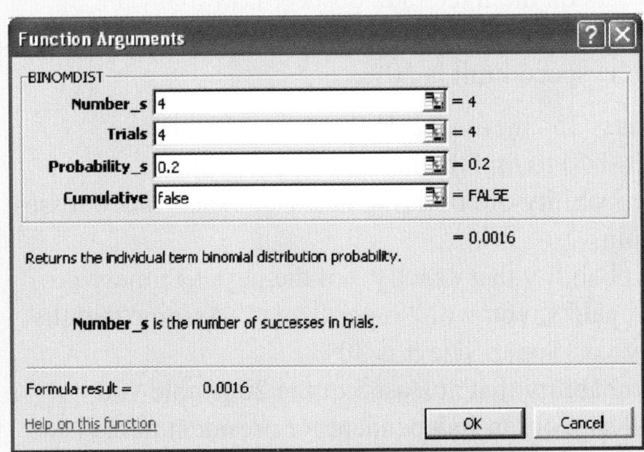


Figure 6.13 Example 6.17a using Excel.

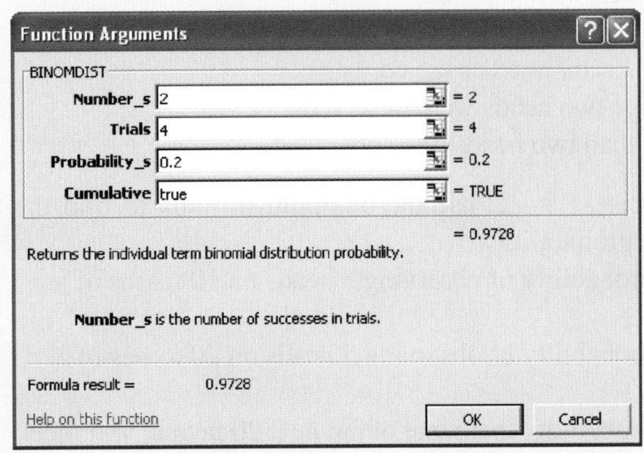


Figure 6.14 Example 6.17b using Excel.

## MINITAB

***Step 1*** Click **Calc > Probability Distributions > Binomial**.
***Step 2*** Do one of **(a)** or **(b)**:
**a.** For individual binomial probabilities, select **Probability** and enter the number of trials **n** and probability of success **p**.
**b.** For cumulative binomial probabilities, select **Cumulative Probability** and enter the number of trials **n** and probability of success **p**.
***Step 3*** Select **Input Constant**, enter ***K*** and click **OK**.

## Section 6.2 SUMMARY

1 The most important discrete distribution is the binomial distribution, where there are two possible outcomes, each with probability of success $p$, and $n$ independent trials.

2 The probability of observing a particular number of successes can be calculated using the binomial probability distribution formula.

3 Binomial probabilities can also be found using the binomial tables or using technology.

4 There are formulas for finding the mean, variance, and standard deviation of a binomial random variable.

## Section 6.2 EXERCISES

### CLARIFYING THE CONCEPTS

For Exercises 1–6, indicate whether the experiment is binomial. If it is not binomial, explain why not.

**1.** Asking 10 of your friends to come to your party. (*Hint:* Remember the independence assumption.)

**2.** Selecting a student in the class at random until you come across a left-handed student.

**3.** Recording the ethnicity of the next 20 babies born at City Hospital.

**4.** Answering the next 10 multiple choice questions correctly or incorrectly on a test.

**5.** Recording the age of 10 randomly selected students in your class.

**6.** Selecting 4 cards at random from a deck of 52 cards, without replacement, and counting the number of aces.

### PRACTICING THE TECHNIQUES

For Exercises 7–12, determine whether or not the experiment is binomial. If the experiment is binomial, identify the random variable $X$, the number of trials $n$, the probability of success $p$, and the probability of failure $1 - p$. If the experiment is not binomial, say why not.

**7.** Toss a fair die 3 times, and note the total number of spots.

**8.** Toss a fair die 3 times, and note the number of sixes.

**9.** Toss a fair die 3 times, and note how often an "even" result is observed.

**10.** Toss a fair die 3 times, and note whether a six was observed.

**11.** Bob has paid to play 2 games at a carnival. The probability that he wins a particular game is 0.25.

**12.** Bob is playing a game at a carnival where he gets to play until he loses. The probability that he wins a particular game is 0.25.

For Exercises 13–18, the experiment is to toss a fair coin three times. Use the binomial formula to find the indicated probabilities.

**13.** No heads were observed.

**14.** One head was observed.
**15.** Two heads were observed.
**16.** Three heads were observed.
**17.** At most two heads were observed.
**18.** More than two heads were observed.

For Exercises 19–22, use the binomial formula to find the indicated probabilities.
**19.** The probability of observing 5 heads on 10 tosses of a fair coin.
**20.** The probability of observing 5 heads on 20 tosses of a fair coin.
**21.** The probability that none of the next 20 people you meet will be left-handed. Assume that the proportion of lefties is 10%.
**22.** The probability that at least 3 of the next 4 dentists you survey will recommend sugarless gum for their patients who chew gum. Assume that a given doctor will make such a recommendation 95% of the time.

For each of the binomial experiments in Exercises 23–26, find the mean, variance, and standard deviation. Interpret the mean for each.
**23.** Toss a fair coin 10 times. Observe the number of heads.
**24.** Toss a fair coin 20 times. Observe the number of heads.
**25.** You are interested in how many of the next 20 people you meet will be left-handed. Assume that the proportion of lefties is 10%.
**26.** You ask 4 dentists whether they recommend sugarless gum for their patients who chew gum. Assume that a given dentist will make such a recommendation 95% of the time.

For Exercises 27–30, use the binomial formula to find the following probabilities.
**27.** The probability that you will roll doubles on at least 2 of your next 3 tosses of a pair of fair dice. (*Hint:* $P(\text{doubles}) = 1/6$.)
**28.** The probability that you will roll doubles on at most 2 of your next 3 tosses of a pair of fair dice. (*Hint:* $P(\text{doubles}) = 1/6$.)
**29.** The probability that at most 2 of your 17 classmates are from Canada. Assume that 4.5% of students are from Canada.
**30.** The probability that 4 of the next 5 cars on the interstate are obeying the speed limit, if the probability that a car obeys the speed limit is 50%.

For each of the binomial experiments in Exercises 31–38, find the mean, variance, and standard deviation. Interpret the mean for each.
**31.** Observe how many of your 17 classmates are from Canada. Assume that 4.5% of students are from Canada.
**32.** Toss a pair of fair dice 3 times. Observe the number of doubles that you roll. (*Hint:* $P(\text{doubles}) = 1/6$.)
**33.** Of 5 cars on the interstate, observe how many are obeying the speed limit. Assume that the probability that a car obeys the speed limit is 50%.
**34.** Of 10 cars on the interstate, observe how many are obeying the speed limit. Assume that the probability that a car obeys the speed limit is 50%.

For Exercises 35–38, use Table B, in the Appendix (pages T-3–T-8) to find the indicated probabilities.
**35.** The probability of observing at least 5 heads on 10 tosses of a fair coin.
**36.** The probability that exactly 3 of the next 10 times you visit McDonald's, you will "Super-Size it." Assume that the probability you Super-Size it is 30%.
**37.** The probability that at least 5 of the 20 people you survey will support an Independent for president in the next election. Assume that 15% of Americans would support an Independent for president.
**38.** The probability that you will pass a quiz and get at least 4 correct answers by randomly guessing on each question of a 6-question multiple-choice test with 5 choices for each question.

For each of the binomial experiments in Exercises 39–42, find the mean, variance, and standard deviation. Interpret the mean for each.
**39.** Toss a fair coin 40 times. Observe the number of heads.
**40.** On 10 visits to McDonald's, observe the number of times you "Super-Size it." Assume that the probability you Super-Size it is 30%.
**41.** You survey 20 people and are interested in the number of people who will support an Independent for president in the next election. Assume that 15% of Americans would support an Independent for president.
**42.** On a 6-question multiple-choice test with 5 choices for each question, observe how many correct answers you get by randomly guessing.

## APPLYING THE CONCEPTS

**43. Statistics Students.** Suppose that the proportion of statistics students who are sophomores is 25%. We take a random sample of 12 statistics students.
- **a.** Find the probability that the sample contains exactly 3 sophomores.
- **b.** Find the probability that the sample contains at most 3 sophomores.
- **c.** What is the most likely number of sophomores in the sample?

**44. Vowels.** In the written English language, 37.8% of letters in a randomly chosen text are vowels.
- **a.** If you choose 15 letters at random, what is the most likely number of vowels?
- **b.** Find the probability that the sample contains at most 5 vowels.
- **c.** Find the probability that the sample contains exactly 5 vowels.

**45. Statistics Students.** Refer to the experiment in Exercise 43.
- **a.** Find the mean, variance, and standard deviation of the number of sophomores. Interpret the mean.
- **b.** Suppose that the sample contains no sophomores at all. Is this unusual? Explain how you would determine this.

**46. Vowels.** Refer to the experiment in Exercise 44.

**a.** Find the variance and standard deviation of the number of vowels.

**b.** Suppose that a sample of 15 letters contains 1 vowel. Is this unusual? Explain how you would determine this.

**47. Women in Management.** We take a random sample of size $n = 20$ from a population of students in management courses. Assume that the population proportion of women in management courses is 0.40.

**a.** Find the probability that the sample contains exactly 10 women.

**b.** Find the probability that the sample contains at least 10 women.

**c.** What is the most likely number of women the sample contains?

**48. Shaquille O'Neal.** In the 2003–4 National Basketball Association regular season, Shaquille O'Neal led the league with a 58.4% field goal percentage (proportion of shots that are baskets). Suppose that we take a sample of 50 of O'Neal's shots.

**a.** Find the probability that the sample contains exactly 25 baskets.

**b.** What is the most likely number of baskets that O'Neal will make in 50 shots?

**c.** Find and interpret the mean number of baskets.

**49. Women in Management.** Refer to the experiment in Exercise 47.

**a.** Find the mean, variance, and standard deviation of the number of women taking management courses. Interpret the mean.

**b.** Suppose your sample contains 12 women. Is this unusual? Explain how you determine this.

**50. Shaquille O'Neal.** Refer to the experiment in Exercise 48.

**a.** Find the variance and standard deviation of the number of baskets.

**b.** Suppose that O'Neal makes only half the baskets. Is he "cold"? Is this unusual? Explain how you would determine this.

**51. What Women Want.** In their 2001 study *What Women Want: Five Secrets to Better Ratings*, the Arbitron Company reported that "Music I Like" is the biggest factor in deciding which radio station to tune to, and that women choose this reason 87% of the time.

**a.** Find the probability that, in a random sample of 10 women, 8 reported that "Music I Like" is the biggest factor in deciding which radio station to tune to.

**b.** What is the most likely number of women in a random sample of 10 who would report that "Music I Like" is the biggest factor in deciding which radio station to tune to?

**c.** Find and interpret the mean.

**52. Spinal Cord Injuries.** The National Spinal Cord Injury Statistical Center reports that 46.9% of all spinal cord injuries are caused by vehicle crashes. Suppose that we choose 20 spinal cord injuries at random.

**a.** What is the most likely number of such injuries to result from vehicle crashes?

**b.** What is the expected number of such injuries to result from vehicle crashes?

**53. What Women Want.** Refer to the experiment in Exercise 51.

**a.** Find the variance and standard deviation of the number of women who would report that "Music I Like" is the biggest factor in deciding which radio station to tune to.

**b.** Suppose that a sample of 10 contains 2 women who report "Music I Like" as the biggest factor in deciding which radio station to tune to. Is this unusual? Explain how you would determine this.

**54. Spinal Cord Injuries.** Refer to the experiment in Exercise 52.

**a.** Find the variance and standard deviation of the number of spinal cord injuries that are the result of vehicle crashes.

**b.** If we choose 20 spinal cord injuries at random, and observe that 10 are from vehicle crashes, is this unusual? Clearly explain why or why not.

**55. Internet Access.** The Pew Internet and American Life Project reported in 2003 that 61% of women had access to the Internet. Consider a random sample of 100 women.

**a.** Find the probability that the sample contains exactly 60 women who had access to the Internet.

**b.** Find the probability that the sample contains 60, 61, or 62 women who had access to the Internet.

**c.** What is the most likely number of women in the sample who had access to the Internet?

**d.** Find and interpret the mean number of women with access to the Internet.

**56. AIDS and Drug Use.** The Centers for Disease Control and Prevention reported that, in 2002, 11% of white males living with AIDS contracted it through injection drug use. A random sample of 120 white males living with AIDS is examined.

**a.** Find the probability that the sample contains exactly 10 men who contracted AIDS through injection drug use.

**b.** Find the probability that the sample contains at most 3 men who contracted AIDS through injection drug use.

**c.** What is the most likely number of white males living with AIDS who contracted it through injection drug use?

**d.** Find and interpret the mean.

**57. Internet Access.** Refer to the experiment in Exercise 55.

**a.** Find the variance and standard deviation of the number of women with access to the Internet.

**b.** Suppose that the sample contains only 49 women who had access to the Internet. Is this unusual? Explain how you would determine this.

**58. AIDS and Drug Use.** Refer to the experiment in Exercise 56.

a. Find the variance and standard deviation.

b. Suppose that the sample contains 20 white males who contracted AIDS through injection drug use. Is this unusual? Explain how you would determine this.

**59. Women and Depression.** According to the National Institute for Mental Health, nearly twice as many American women (12%) as men (6.6%) are affected by a depressive disorder each year. A random sample of 100 women is examined.

a. Find the probability that the sample contains at most 10 women who are affected by a depressive disorder.

b. What is the most likely number of women who are affected by a depressive disorder?

c. Find and interpret the mean number of women who are affected by a depressive disorder.

**60. Small Business Jobs.** According the U.S. Small Business Administration, small businesses provide 75% of the net new jobs added to the economy. Consider a random sample of 1000 new jobs.

a. Find the probability that the sample contains 800 new jobs provided by small businesses.

b. What is the most likely number of new jobs to be provided by small businesses?

c. Find and interpret the mean.

**61. Women and Depression.** Refer to the experiment in Exercise 59.

a. Find the variance and standard deviation of the number of women affected by a depressive disorder.

b. Suppose that the sample contained 10 women who were affected by a depressive disorder. Is this unusual? Explain how you would determine this.

**62. Small Business Jobs.** Refer to the experiment in Exercise 60.

a. Find the variance and standard deviation.

b. Suppose that the sample of 1000 contains 180 new jobs provided by small businesses. Is this unusual? Explain how you would determine this.

**63. Downloading Music.** The Pew Internet and American Life Project reported in 2004 that "17% of current music downloaders say they are using paid services." Consider a random sample of 400 music downloaders.

a. Find the probability that the sample contains 60 music downloaders who are using paid services.

b. What is the most likely number of music downloaders who are using paid services?

c. Find and interpret the mean.

**64. Violent Crime at School.** The National Center for Education Statistics reports that, in 2000, the percentage of 18-year-olds who were victims of violent crime at school was 3.4%. Consider a random sample of 1000 18-year-olds.

a. Find the probability that at most 40 are victims of violent crime at school.

b. Find and interpret the mean.

**65. Downloading Music.** Refer to the experiment in Exercise 63.

a. Find the variance and standard deviation.

b. Suppose that the sample contains 90 music downloaders who are using paid services. Is this unusual? Explain how you would determine this.

**66. Violent Crime at School.** Refer to the experiment in Exercise 64.

a. Find the variance and standard deviation.

b. Suppose that the sample contains 35 students who are victims of violent crime at school. Is this unusual? Explain how you would determine this.

## 6.3 Continuous Random Variables and the Normal Probability Distribution

*Objectives:* **By the end of this section, I will be able to...**

1 Identify a continuous probability distribution.

2 Explain the properties of the normal probability distribution.

### 1 Continuous Probability Distributions

Continuous random variables assume infinitely many possible values, with no gap between the values. A continuous random variable can take any numerical value in an interval on the number line. For example, the grade point average (GPA) of a randomly chosen classmate of yours is a continuous random variable because it can take any value between 0.0 and 4.0, that is, any value in the interval [0.0, 4.0].

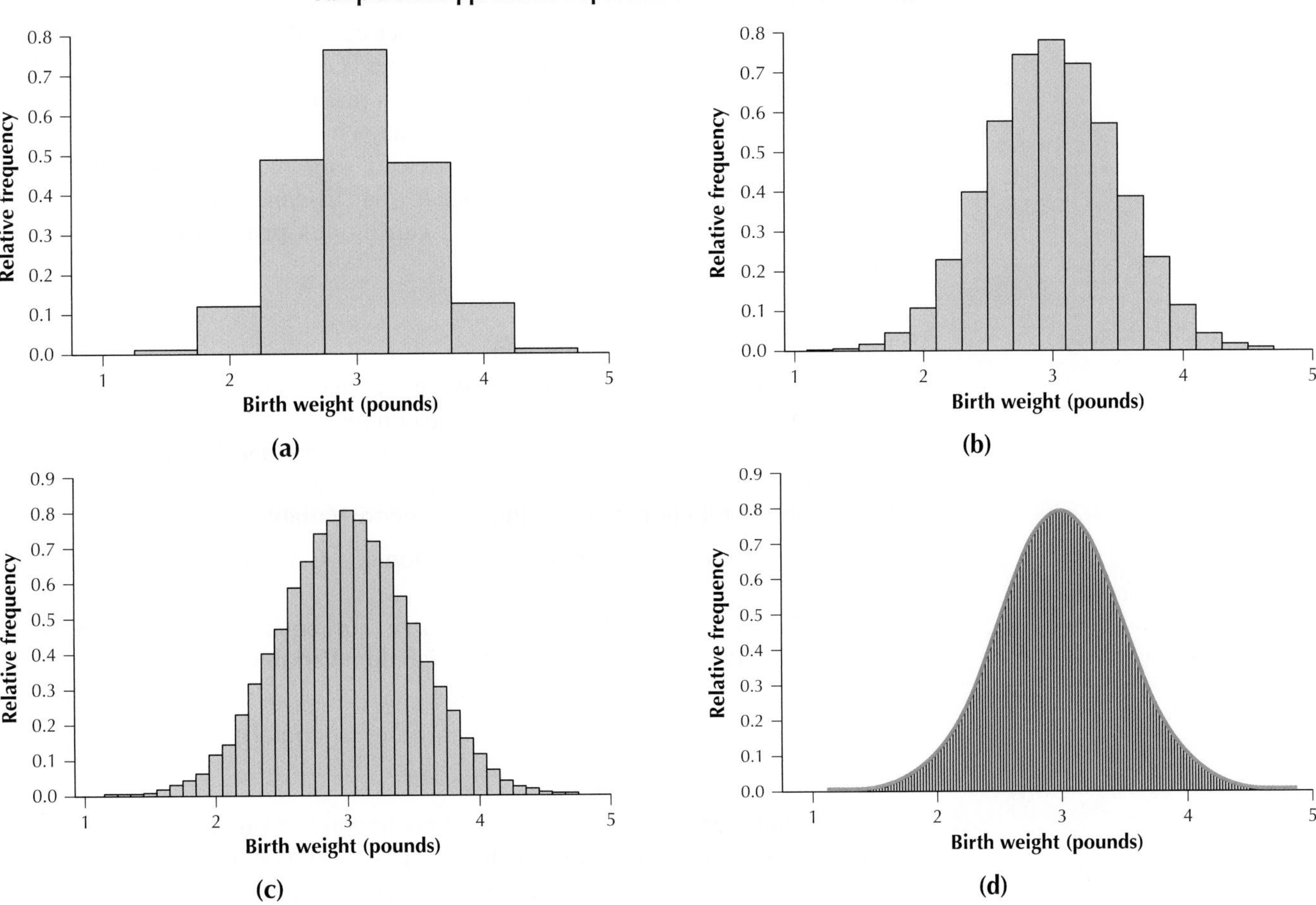


FIGURE 6.15 (a) Relatively small sample ($n = 100$) with large class widths (0.5 lb). (b) Large sample ($n = 200$) with smaller class widths (0.2 lb). (c) Very large sample ($n = 400$) with very small class widths (0.1 lb). (d) Eventually, theoretical histogram of entire population becomes smooth curve with class widths arbitrarily small.

Since a continuous random variable represents a "smooth" process, that is, a process that takes on all values in an interval, we need a graph that accurately illustrates this process. Figures 6.15a–d show four histograms of the birth weights of high-risk newborn babies in progressively larger samples taken from a nationwide population. The first histogram (a) provides a rough picture with very wide class widths. Then, as the sample size increases (b), the class widths are decreased, which presents a somewhat smoother picture. Next, as the sample size becomes very large (c), the class widths become much smaller, providing a good approximation of the shape of the population distribution. Notice that, as the class widths become smaller and smaller, the histograms get closer and closer to the smooth curve (d) that represents the shape of the distribution of the birth weights of high-risk newborns. For the remainder of Section 6.3, we will work with curves that are smoothed histograms, representing the theoretical histogram of the entire population.

For a given continuous random variable $X$, we are not interested in whether $X$ equals any particular value. Rather, we are interested in whether $X$ is

- greater than a particular value, or
- less than a particular value, or
- between two particular values.

That is, we are interested in whether $X$ is located in an *interval*.

We are not interested in the probability that $X$ equals some particular value, because this probability *always equals zero.* Sounds crazy, no? Well, consider the following example. How much soda does a "12-ounce can" of soda actually contain? Are you sure it's 12 ounces and not 11.99999999 ounces? Or could it contain 12.00000001 ounces? In fact, the can could contain any amount that is close to 12 ounces (assuming that quality control is doing its job). In fact, any given weight of soda in the can is so unlikely that its probability is essentially zero. Therefore, the probability that you will get exactly 12.00000000 ounces of soda in your 12-ounce can is zero.

The graph in Figure 6.15d is called a **continuous probability distribution**, defined as follows.

**Continuous Probability Distribution**

A **continuous probability distribution** is a graph that indicates on the horizontal axis the range of values that the continuous random variable $X$ can take, and above which is drawn a curve, called the **density curve**. A continuous probability distribution must follow the Requirements for the Probability Distribution of a Continuous Random Variable.

**Requirements for the Probability Distribution of a Continuous Random Variable**

1. The total area under the density curve must equal 1 (this is the **Law of Total Probability for Continuous Random Variables**).
2. The vertical height of the density curve can never be negative. That is, the density curve never goes below the horizontal axis.

*Note:* It is important to remember that the height of the density curve above a point does *not* represent the probability of that point (all such probabilities equal zero, remember?).

For example, grade point averages are continuous and defined on an interval ranging from 0.0 to 4.0. The various values of $X$ are found on the horizontal number line (see Figure 6.16). Therefore, we label the horizontal line as $X$. The curve in Figure 6.16 is the density curve for the grade point averages.

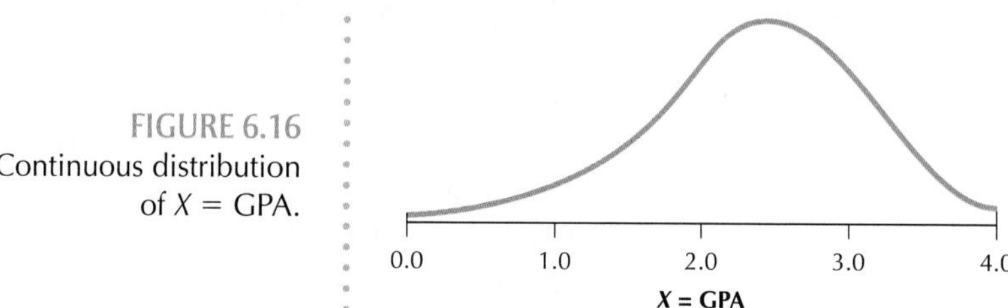


FIGURE 6.16
Continuous distribution of $X$ = GPA.

Recall that the heights of the rectangles in a histogram represent the relative frequency of the data values in the distribution. In the smoothed histogram world of continuous distributions, the mode of the distribution is the value of $X$ directly below the highest point on the curve. So, how do we represent probability for continuous random variables?

**Probability for Continuous Distributions** is represented by area under the curve above an interval.

## Example 6.20 Finding the area to the right of an $X$-value

Suppose that students at a certain university will be named to the Dean's List if they maintain a grade point average of at least 3.0. Figure 6.17 shows the graph of student GPAs. Graph the area that represents the probability that $X$ is between 3.0 and 4.0.

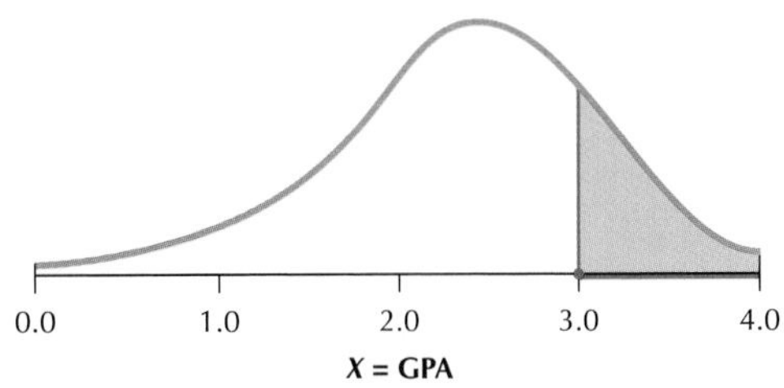


FIGURE 6.17 The area to the right of $X = 3.0$ represents the probability that the student makes the Dean's List.

**Solution**

**Step 1** The cutoff score for making the Dean's List is 3.0, so draw a vertical line starting from the horizontal axis at the point $X = 3.0$ and extending up to the density curve.

**Step 2** Next, to be on the Dean's List, the student must have a GPA of *at least* 3.0, which means 3.0 or more. So, the values of $X$ to the right of (and including) 3.0 should be highlighted, up to $X = 4.0$. This forms an *interval* on the number line from 3.0 to 4.0.

**Step 3** Finally, shade in the area under the density curve and above the interval from $X = 3.0$ to $X = 4.0$ that you highlighted in Step 2. This area represents the probability that $X$ is between 3.0 and 4.0. In other words, this area represents the probability that a randomly chosen GPA is at least 3.0. ■

Probabilities for a random variable are given by *areas* under the curve and above an interval. Later we will discover how to compute the exact value of this probability. For now, it is important to understand the two essential types of information in the graph of a continuous random variable:

- The values of $X$ are always found on the horizontal axis.
- The probabilities associated with intervals of values of $X$ are represented by *areas under the curve*. Probabilities are not found on the horizontal axis. To find a probability, locate the interval of values, and then determine the area under the curve for this interval. ■

Another interesting feature of continuous distributions is, to use our example, the fact that the probability that $X$ is at least 3.0 is the same as the probability that $X$ is greater than 3.0. That is, $P(X \geq 3.0) = P(X > 3.0)$. This is because $P(X \geq 3.0) = P(X > 3.0) + P(X = 3.0)$, and, as we have seen, $P(X = 3.0) = 0$ for a true continuous random variable. Similarly, we have $P(X \leq 3.0) = P(X < 3.0)$, if $X$ is continuous.

"But wait a minute," you say. "My friend got all B's her first semester in college, so that her GPA was equal to 3.0! Doesn't that mean that $P(X = 3.0) > 0$?" Well, there's no arguing with your friend's grades, and it does show that the probability that someone's GPA is 3.0 is not zero. A density curve is a way of *modeling* GPA as a continuous random variable, even though GPAs are not quite truly continuous, as, for example, time is continuous. We are using an approximation to reality because it is convenient for analysis.

## 2 Introduction to the Normal Probability Distribution

Much of the behavior of the natural world can be described using a very special continuous probability distribution: the **normal probability distribution**. The normal distribution is often referred to as the bell-shaped curve, which we worked with in Chapter 3. The normal distribution has been found to accurately model such disparate phenomena as

- the amount of rainfall in Imperial Valley, California,
- the heights and weights of high-risk infants in New York City,
- the emotional intelligence test scores of college students in Texas, and
- the errors in manufacturing machine bolts in a Pennsylvania factory.

The normal probability distribution is considered to be the most important probability distribution in the world. In Sections 6.4 and 6.5, we will become intimately familiar with the properties and techniques associated with the normal distribution. Remember that, like all probability distributions, what we are dealing with here is a *population* of data values. When a population is said to be *normally distributed,* the data values follow a normal probability distribution, with a specific population mean $\mu$ and a specific population standard deviation $\sigma$. Together, $\mu$ and $\sigma$ constitute the *parameters* of the normal distribution, that is, the characteristics of the population of the normally distributed data.

The mean $\mu$ determines the center of the distribution on the number line, and the standard deviation $\sigma$ determines the spread or shape of the distribution curve (see Figure 6.18). The mean $\mu$ can be any real number, positive, negative, or zero; the standard deviation $\sigma$ can never be negative.

Figure 6.18 illustrates the *inflection points* of the normal curve. Inflection points are points on the graph where the curve changes from "cup upward" to "cup downward."

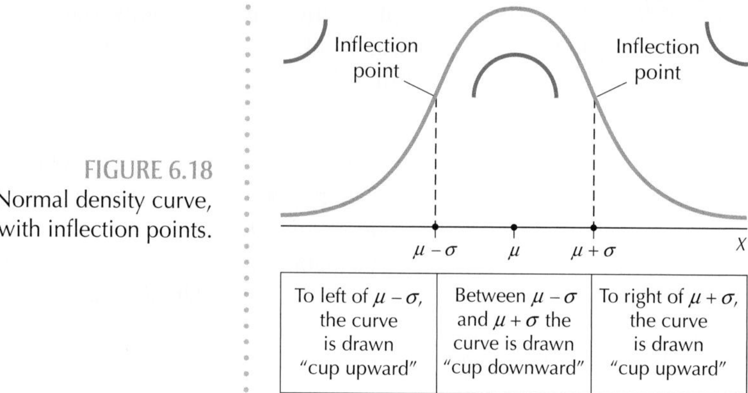


FIGURE 6.18
Normal density curve, with inflection points.

From Figure 6.19 we can see that normal distribution curve is symmetric about $\mu$. If you slice the curve neatly in half at the mean $\mu$, the result will be two pieces that are perfect mirror images of each other, as in Figure 6.19.

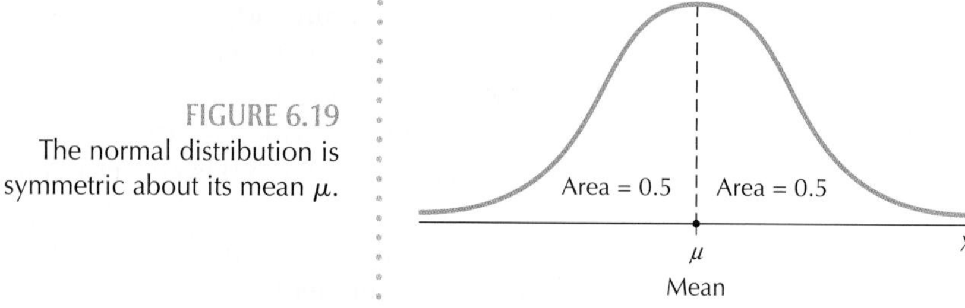


FIGURE 6.19
The normal distribution is symmetric about its mean $\mu$.

**Properties of the Normal Density Curve (Normal Curve)**

1. It is symmetric about the mean $\mu$.
2. The highest point occurs at $X = \mu$, because symmetry implies that the mean equals the median, which equals the mode of the distribution.
3. It has inflection points at $\mu - \sigma$ and $\mu + \sigma$.
4. The total area under the curve equals 1.
5. Symmetry also implies that the area under the curve to the left of $\mu$ and the area under the curve to the right of $\mu$ are both equal to 0.5 (Figure 6.19).
6. The normal distribution is defined for values of $X$ extending indefinitely in both the positive and negative directions. As $X$ moves farther from the mean, the density curve approaches but never quite touches the horizontal axis.

Figure 6.20 shows two normal density curves, with different means but the same standard deviation. Note that the two curves have precisely the same spread or shape, because each distribution has the same standard deviation, $\sigma = 2$. However, because the mean of the curve on the right is $\mu = 6$ while the mean of the curve on the left in $\mu = 2$, the curve on the right is shifted four units to the right.

Since $\sigma$ is a measure of spread, the larger the value of $\sigma$, the more spread out the distribution of $X$ will be. This is illustrated in Figure 6.21. The normal distribution with the smaller standard deviation ($\sigma = 1$) has a curve with a higher peak in the center and thinner "tails" than the distribution with a larger standard deviation ($\sigma = 2$). When the standard deviation is small, a greater proportion of the data values will be located near the mean, which also indicates that a smaller proportion of data values will be located in the tails (far from the mean).

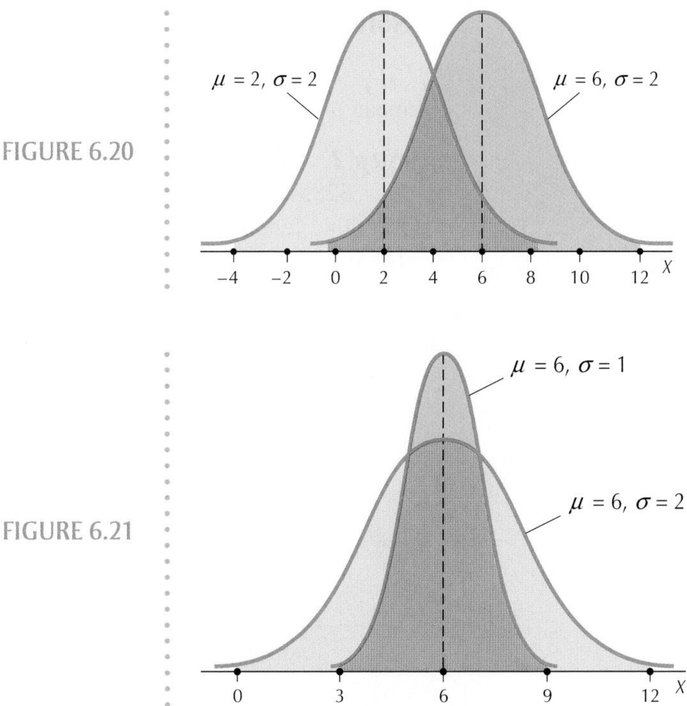


FIGURE 6.20

FIGURE 6.21

**Steps in Drawing a Graph to Help You Solve Normal Probability Problems**

1. Draw a "generic" bell-shaped curve, with a horizontal number line under it that is labeled as the random variable $X$. Insert the mean $\mu$ in the center of the number line.
2. Mark on the number line the value of $X$ indicated in the problem. Shade in the desired area under the normal curve to the right or left of this value.
3. Proceed to find the desired area or probability.

Example 6.21 illustrates these steps.

## Example 6.21 Visits by prenatal nurses lead to healthier teen births

A statistical study found that, when nurses made home visits to pregnant teenagers to provide support services, discourage smoking, and otherwise provide care, the mean birth weight of the babies was higher for this treatment group (3285 grams) than for a control group of teenagers who were not visited (2922 grams), when the visits began before midgestation.[5] Suppose the birth weights for the babies whose mothers were visited by the nurses (treatment group) follow a normal distribution. Then our random variable is

$$X = \text{birth weight of babies in the treatment group}$$

The mean is $\mu = 3285$ grams. Assume that the standard deviation is $\sigma = 500$ grams. We are interested in the probability that a randomly chosen baby from the treatment group weighs less than 3285 grams.

**a.** Draw a graph of the appropriate normal curve, with the desired area shaded.

**b.** Find the probability that a randomly chosen baby from the treatment group weighs less than 3285 grams.

### Solution

***Step 1*** We draw a bell-shaped curve, label the horizontal line as $X$, and insert the mean $\mu = 3285$ in the center of the number line, as shown in Figure 6.22a.

***Step 2*** We are asked for the probability that $X$ is less than 3285 grams. Note that $X = 3285$ also happens to be the mean $\mu$. Since we need the probability of a baby weighing *less than* 3285 grams, we are looking for an area *to the left of* 3285 grams. The desired area is shaded in Figure 6.22b.

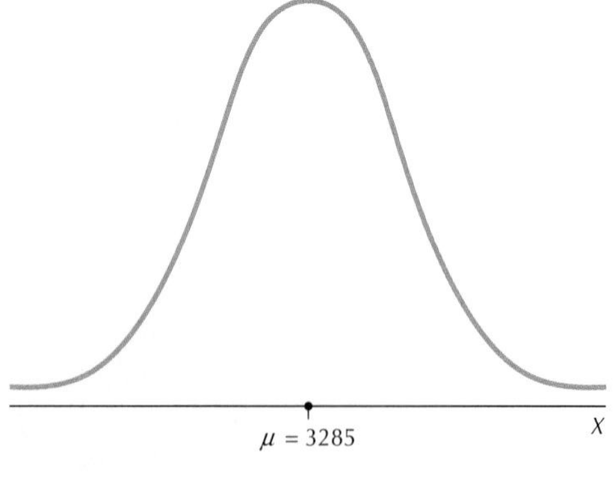


FIGURE 6.22a

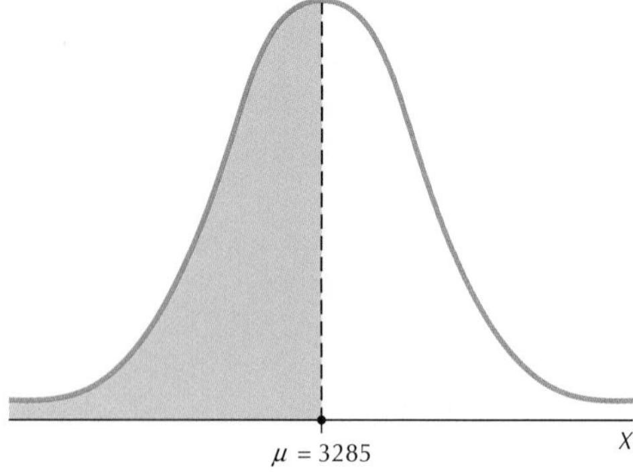


FIGURE 6.22b

**Step 3** Since the curve is a normal curve, the mean $\mu = 3285$ divides the area under the curve in half, because the normal distribution is symmetric about its mean. Thus, the shaded region has area 0.5, and the probability that a randomly chosen baby from the treatment group will weigh less than 3285 grams equals 0.5. ■

In Chapter 3, we learned that according to the Empirical Rule the area under the normal curve has the following properties (see Figure 6.23).

1. About 68% of the area under the curve lies within one standard deviation of the mean.
2. About 95% of the area under the curve lies within two standard deviations of the mean.
3. About 99.7% of the area under the curve lies within three standard deviations of the mean.

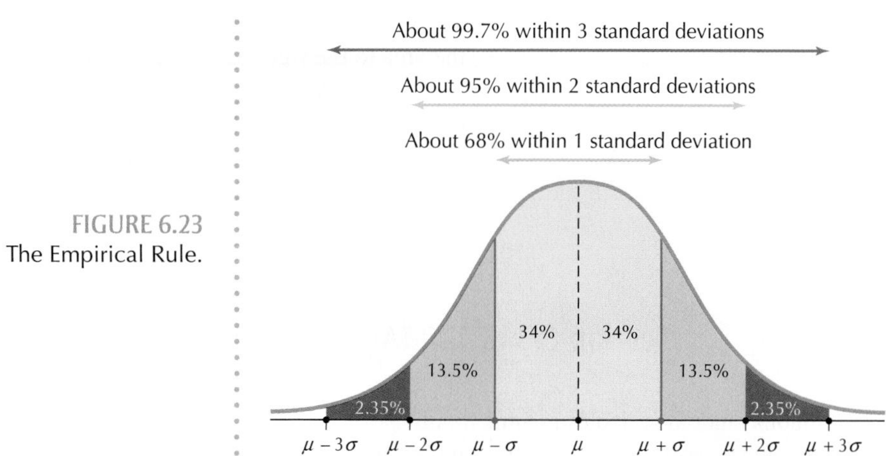


FIGURE 6.23
The Empirical Rule.

We will verify the Empirical Rule in Section 6.4.

## Example 6.22 Grade inflation

Many educators are concerned about grade inflation. One study shows that one low-SAT-score high school (with mean combined SAT score = 750) had higher mean grade point average (mean GPA = 3.6) than a high-SAT-score school (with mean combined SAT score = 1050 and mean GPA = 2.6).[6] Define the following random variable:

$$X = \text{GPA at the high-SAT-score school}$$

Assume that $X$ is normally distributed with mean $\mu = 2.6$ and standard deviation $\sigma = 0.46$.

**a.** What is the probability that a randomly chosen GPA at the high-SAT-score school will be between 3.06 and 3.52?

**b.** Find the probability that a randomly chosen GPA at the high-SAT-score school will be greater than 3.52.

### Solution

**a.** Figure 6.24 shows the distribution of GPA at the high-SAT-score school. The area under the curve between 3.06 and 3.52 represents the area between $\mu + \sigma$ and $\mu + 2\sigma$. Courtesy of the Empirical Rule, Figure 6.23 tells us that the area between $\mu + \sigma$ and $\mu + 2\sigma$ is about 13.5% of the area under the curve.

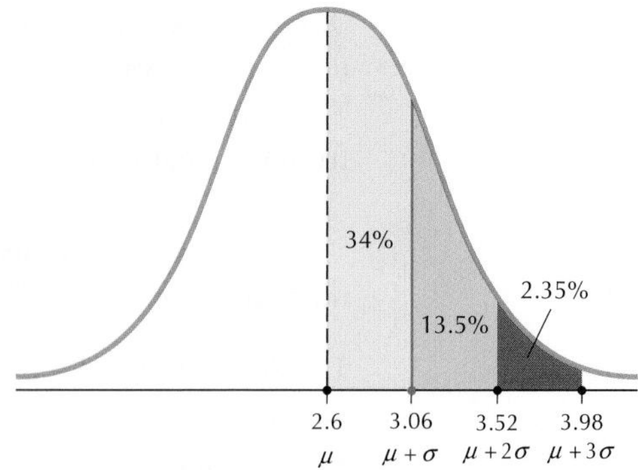


FIGURE 6.24

Therefore, the probability that a randomly chosen GPA at the high-SAT-score school will be between 3.06 and 3.52 is about 0.135.

**b.** The area under the normal density curve to the right of the mean equals 0.5. So the area to the right of $\mu = 2.6$ equals 0.5, or 50% of the area under the curve. To find the area to the right of $X = 3.52$, we need to subtract the yellow area (34%) and the light green area (13.5%) from 50%: 50% – 34% – 13.5% = 2.5%. Therefore, the probability that a randomly chosen GPA at the high-SAT-score school will be greater than 3.52 is about 0.025. ■

## Section 6.3 SUMMARY

1 Continuous random variables assume infinitely many possible values, with no gap between the values. Probability for continuous random variables consists of area above an interval on the number line and under the distribution curve.

2 The normal distribution is the most important continuous probability distribution. It is symmetric about its mean $\mu$ and has standard deviation $\sigma$. One should always sketch a picture of a normal probability problem to help solve it.

## Section 6.3 EXERCISES

### CLARIFYING THE CONCEPTS

**1.** For a continuous random variable $X$, why are we not interested in whether $X$ equals some particular value?

**2.** In the graph of a probability distribution, what is represented on the number line?

**3.** How is probability represented in the graph of a continuous probability distribution?

**4.** What are the possible values for the mean of a normal distribution? For the standard deviation?

### PRACTICING THE TECHNIQUES

Use the normal distribution from Example 6.21 for Exercises 5–12. Birth weights are normally distributed with a mean weight of $\mu = 3285$ grams and a standard deviation of $\sigma = 500$ grams.

**5.** What is the probability of a birth weight equal to 3285 grams?

**6.** What is the probability of a birth weight more than 3285 grams?

**7.** What is the probability of a birth weight of at least 3285 grams?

**8.** Is the area to the right of $X = 4285$ grams greater than or less than 0.5? How do you know this?

**9.** Is the area to the left of $X = 4285$ grams greater than or less than 0.5? How do you know this?

**10.** What is the probability of a birth weight between 2785 and 3785 grams?

**11.** What is the probability of a birth weight between 1785 and 4785 grams?

**12.** What is the probability of a birth weight between 785 and 5785 grams?

**13.** The two normal distributions in the accompanying figure have the same standard deviation of 5 but different means. Which normal distribution has mean 10 and which has mean

25? Explain how you know this.

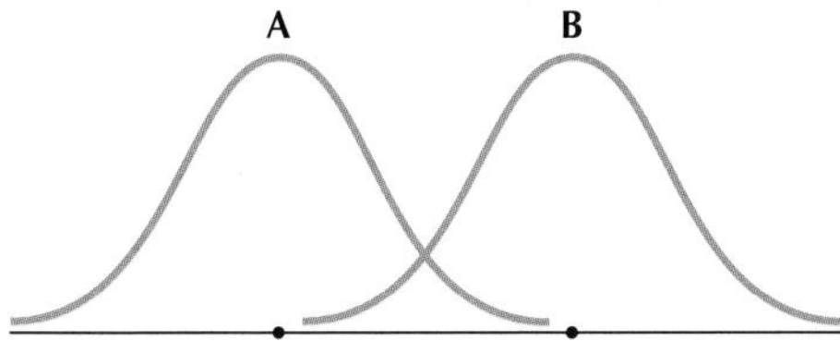


**14.** The two normal distributions in the figure below have the same mean of 100 but different standard deviations. Which normal distribution has standard deviation 3 and which has standard deviation 6? Explain how you know this.

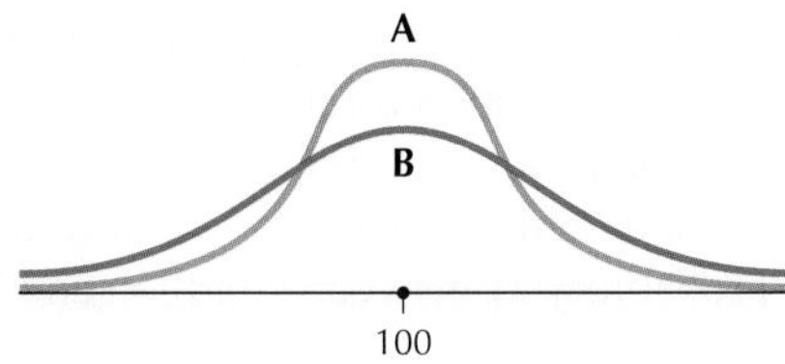


For Exercises 15–22, use the graph of the normal distribution to determine the mean and standard deviation.

**15.** 

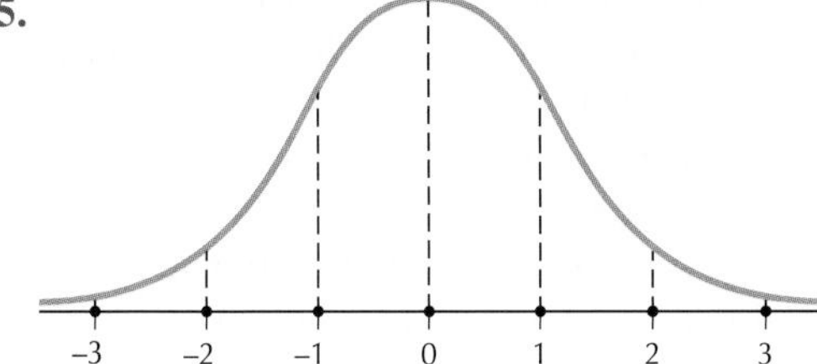


**16.**

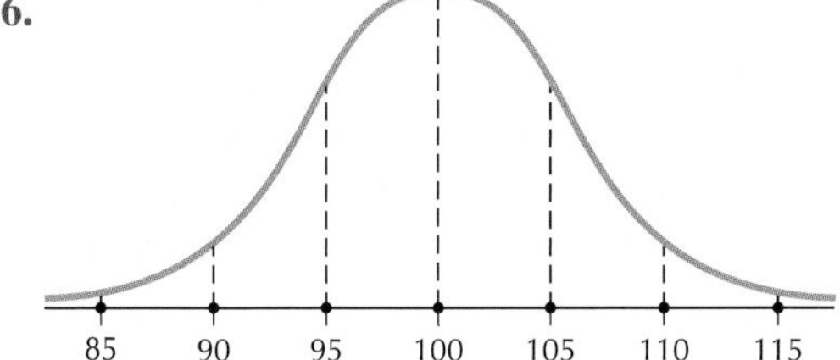


**17.**

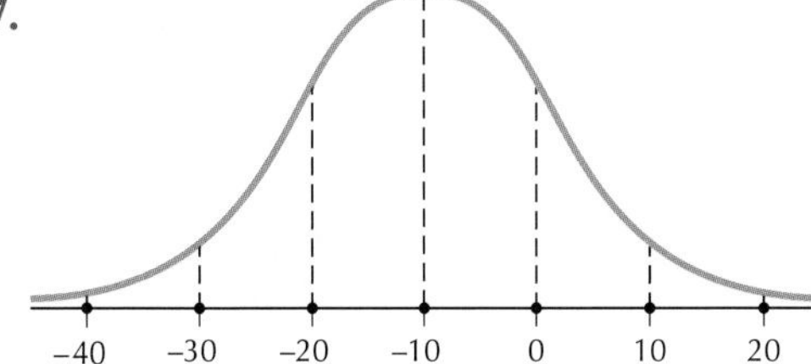


**18.**

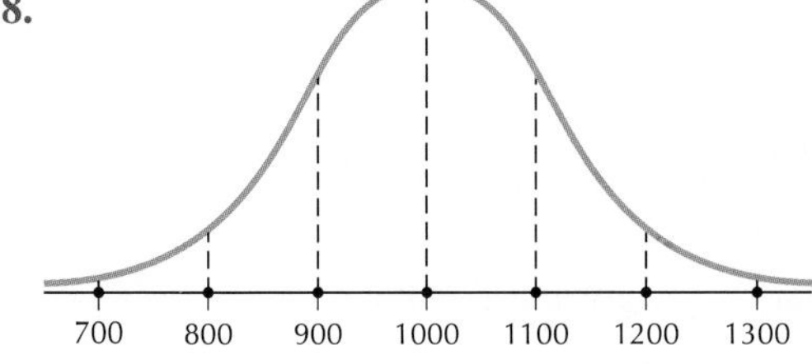


**19.**

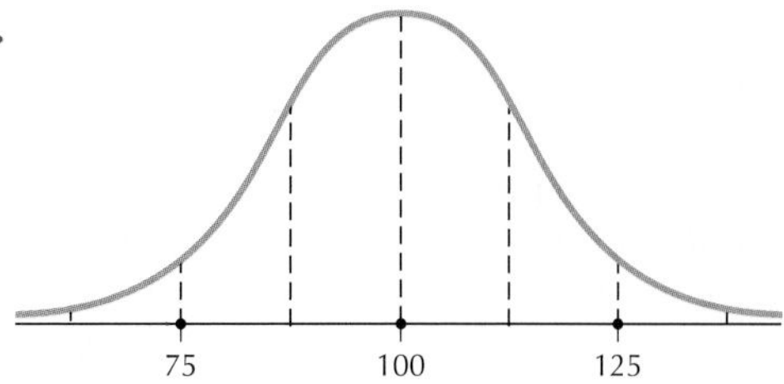


**20.**

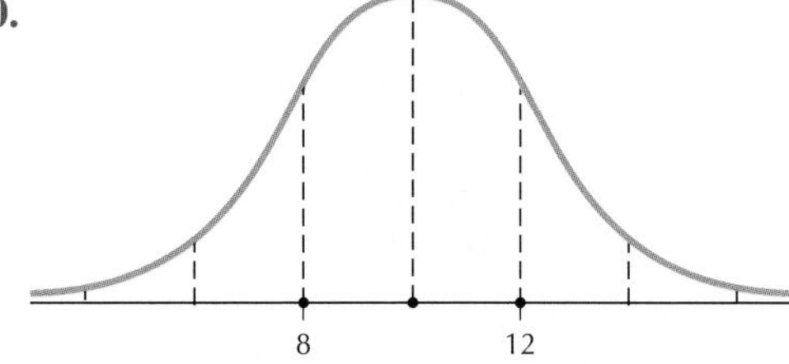


**21.**

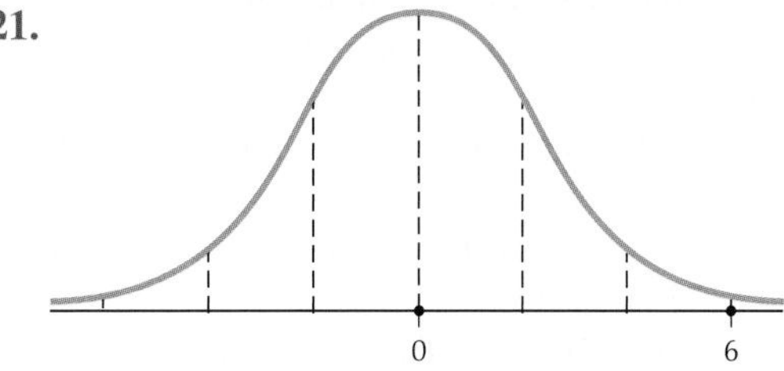


**22.**

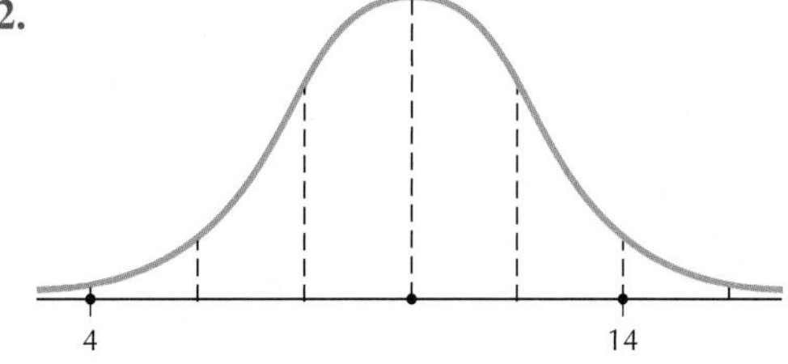


## APPLYING THE CONCEPTS

For Exercises 23–32, sketch the distribution, showing $\mu$, $\mu + \sigma$, $\mu + 2\sigma$, $\mu + 3\sigma$, $\mu - \sigma$, $\mu - 2\sigma$, and $\mu - 3\sigma$. Then answer the questions.

**23. Windy Frisco.** The average wind speed in San Francisco in July is 13.6 miles per hour (mph) in July, according to the U.S. National Oceanic and Atmospheric Administration. Suppose that the distribution of the wind speed in July in San Francisco is normal with mean $\mu = 13.6$ mph and standard deviation $\sigma = 4$ mph.

**a.** Shade the region that represents wind speeds between 9.6 and 17.6 mph.

**b.** What is the proportion of wind speeds between 9.6 and 17.6? (*Hint:* See Figure 6.23 [page 293].)

**24. Magic McGrady.** In the 2003–2004 National Basketball Association regular season, Tracy McGrady of the Orlando Magic led the league in points per game with 28. Suppose that McGrady's points per game follow a normal distribution with mean $\mu = 28$ and standard deviation $\sigma = 8$.

**a.** Shade the region that represents more than 36 points scored in a game.

**b.** What is the probability that McGrady scored more than 36 points in a game?

**25. Viewers of *60 Minutes*.** Nielsen Media Research reported that, for the week ending July 11, 2004, 12 million viewers watched the television show *60 Minutes*. Suppose that the distribution of viewers of *60 Minutes* is normal with mean $\mu = 12$ million and standard deviation $\sigma = 4$ million.

**a.** Shade the region that represents fewer than 4 million viewers.

**b.** What is the probability that fewer than 4 million viewers will watch *60 Minutes*?

**26. Math Scores.** The National Center for Education Statistics reports that in 2005 the mean score on the standardized mathematics test for eighth-graders increased by 7 points from 2000. The mean score in 2005 was $\mu = 273$; assume a standard deviation of $\sigma = 7$.

**a.** Shade the region that represents scores between 266 and 273.

**b.** What is the probability that a student scored between 266 and 273 on the test?

**27. Los Angeles Temperature.** The U.S. Department of Commerce reports that the mean temperature in downtown Los Angeles is 66.2 degrees Fahrenheit (ºF). Assume that the distribution of temperature in Los Angeles is normal with mean $\mu = 66.2$ºF and standard deviation $\sigma = 8$ºF.

**a.** Shade the region that represents temperatures between 50.2ºF and 82.2ºF.

**b.** What is the probability that the temperature in Los Angeles is between 50.2ºF and 82.2ºF?

**28. Hospital Patient Length of Stays.** A study of Pennsylvania hospitals showed that the mean patient length of stay in 2001 was 4.87 days with a standard deviation of 0.97 days.[7] Assume that the distribution of patient length of stays is normal.

**a.** Shade the region that represents patient length of stay of less than 3.9 days.

**b.** Find the probability that a randomly selected patient has a length of stay of less than 3.9 days.

**29. Engineers' Salaries.** According to the U.S. Census Bureau, students in their first year after graduating with bachelor's degrees in engineering earned an average salary of \$3189 per month. Assume that this distribution is normal with mean $\mu = 3189$ and standard deviation $\sigma = 600$. What proportion of salaries lie between \$3789 and \$4389?

**30. Tobacco-Related Deaths.** The World Health Organization states that tobacco is the second leading cause of death in the world. Every year, an average of 5 million people die of tobacco-related causes. Assume that the distribution is normal with mean $\mu = 5$ (in millions) and standard deviation $\sigma = 1$ (in millions). What is the probability of between 4 million and 7 million deaths?

**31. Median Household Income.** The Census Bureau reports that the median household income was \$48,201 in 2006. Assume that the distribution of income is normal with mean $\mu = \$48{,}201$ and standard deviation $\sigma = \$16{,}000$. Find the probability that a randomly selected household has an income of greater than \$80,201.

**32. Stock Shares Traded.** The average number of shares traded on the New York Stock Exchange in June 2004 was 2.4 billion per day. Assume that the distribution of shares traded is normal with mean $\mu = 2.4$ (in billions of shares) and standard deviation $\sigma = 0.6$ (in billions of shares). What is the probability that between 1.8 billion and 2.4 billion shares were traded?

**33. Birth Weights.** Can you think of a problem with assuming that the birth weights from Example 6.21 follow a normal distribution? Is there a minimum value of $X$ for a normal distribution? Is there a minimum birth weight?

## 6.4 Standard Normal Distribution

*Objectives:* **By the end of this section, I will be able to...**

1 Find areas under the standard normal curve, given a $Z$-value.

2 Find the standard normal $Z$-value, given an area.

## Finding Areas Under the Standard Normal Curve for a Given Z-Value

*Note:* Understanding the techniques explained in this section will allow you to analyze a whole world of data sets, even those that are not themselves normally distributed (see the Central Limit Theorem in the next chapter). Beyond this chapter, these techniques help you to find and understand *p*-values in Chapters 9–13.

There are billions of data sets in the world that are normally distributed, from test scores to student heights. But there is one very special normal distribution called the **standard normal distribution**. To solve questions involving normally distributed test scores, for example, we *transform* the original normal distribution to the standard normal distribution, using a process called *standardizing*. More on this in Section 6.5.

The mean and standard deviation of the standard normal distribution make it unique.

**The Standard Normal (*Z*) Distribution**

The **standard normal distribution** is a normal distribution with

- mean $\mu = 0$ and
- standard deviation $\sigma = 1$.

Because of its importance, the standard normal random variable is always denoted as a capital *Z*. The graph of the standard normal random variable *Z* is given in Figure 6.25. The standard normal curve is symmetric about its mean $\mu = 0$. Since the entire area under the curve equals 1, the area to the left of 0 and the area to the right of 0 both equal 0.5.

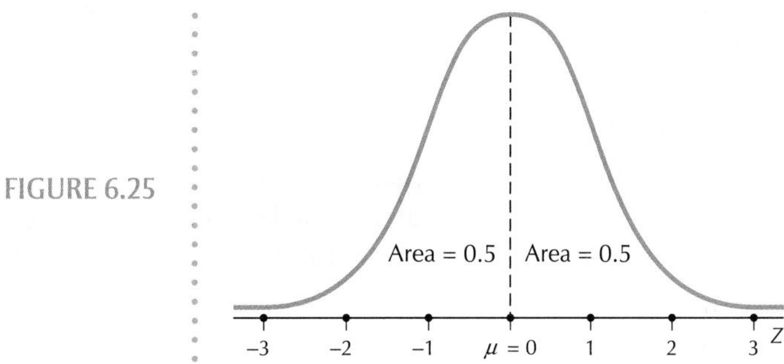

FIGURE 6.25

We will discuss two methods for finding probabilities associated with *Z*, using (a) the table for finding standard normal probabilities, called the ***Z* table,** and (b) technology. For the *Z* table, see Table C in the Appendix (pages T-9–T-10). *The Z table provides areas under the standard normal curve* <u>*to the left*</u> *of a specified value of Z, denoted as* $Z_1$ (see Figure 6.26).

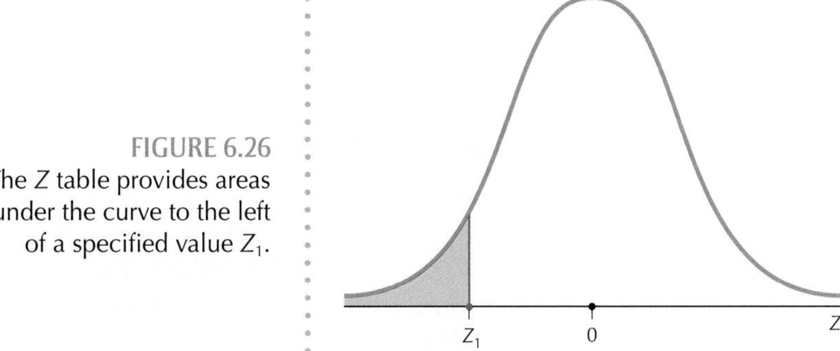

FIGURE 6.26
The *Z* table provides areas under the curve to the left of a specified value $Z_1$.

*Note:* Although your $Z$ table contains only values between $Z = -3.49$ and $Z = 3.49$, there is no upper or lower limit to the values that $Z$ may take. The curve essentially goes on forever in both the positive and the negative directions, always getting closer and closer to the horizontal axis but never quite touching it (there's a great plot for a love story in there somewhere).

Let's get acquainted with the $Z$ table (see excerpt in Figure 6.28). Along the left side and across the top of the $Z$ table are possible values of $Z$. These numbers, which in the table run from –3.49 to 3.49, are the values of $Z$ found on the number line when you draw a graph. Down the left are the ones and tenths digits of the $Z$-value, and across the top is the hundredths digit. The body of the $Z$ table contains areas (probabilities). These numbers, which run from 0.0002 to 0.9998, are areas under the standard normal curve that represent probabilities *to the left* of the specified value of $Z$. Table 6.7 shows the steps for finding areas under the standard normal curve, that is, for finding probabilities for specified values of $Z$.

Table 6.7 Steps for finding areas under the standard normal curve

**Case 1**

**Find the area to the left of $Z_1$.**

***Step 1*** Draw the standard normal curve. Label the $Z$-value $Z_1$.

***Step 2*** Shade in the area to the left of $Z_1$.

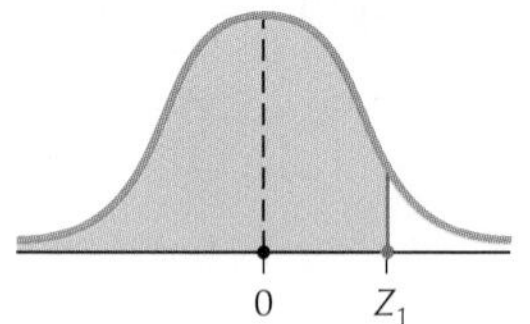


***Step 3*** Use the $Z$ table to find the area to the left of $Z_1$.

**Case 2**

**Find the area to the right of $Z_1$.**

***Step 1*** Draw the standard normal curve. Label the $Z$-value $Z_1$.

***Step 2*** Shade in the area to the right of $Z_1$.

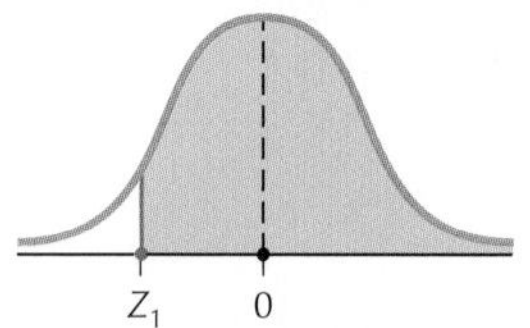


***Step 3*** Use the $Z$ table to find the area to the left of $Z_1$. The area to the right of $Z_1$ is then equal to 1 – (area to the left of $Z_1$).

**Case 3**

**Find the area between $Z_1$ and $Z_2$.**

***Step 1*** Draw the standard normal curve. Label the $Z$-values $Z_1$ and $Z_2$.

***Step 2*** Shade in the area between $Z_1$ and $Z_2$.

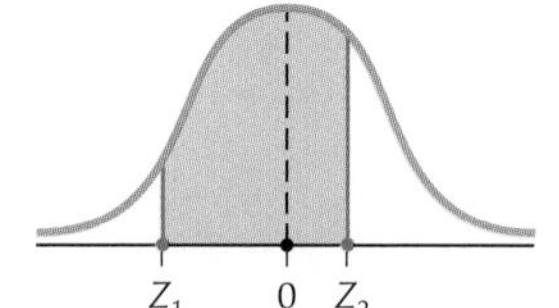


***Step 3*** Use the $Z$ table to find the area to the left of $Z_1$ and the area to the left of $Z_2$. The area between $Z_1$ and $Z_2$ is then equal to (area to the left of $Z_2$) – (area to the left of $Z_1$).

## Example 6.23 Case 1: Find the area to the left of a value of $Z$

Find the area to the left of $Z = 0.57$.

### Solution

***Step 1*** First draw the standard normal curve and label $Z = 0.57$.

***Step 2*** Shade the area to the left of 0.57, as shown in Figure 6.27.

***Step 3*** Now, $Z = 0.57 = 0.5 + 0.07$. In the $Z$ table, excerpted here as Figure 6.28, go down the left-hand column to 0.5 and select that row. Then go across the top row to .07 and select that column. The quantity at the intersection of this row and column represents the area to the left of $Z = 0.57$. That is, the area to the left of $Z = 0.57$ is 0.7157. ■

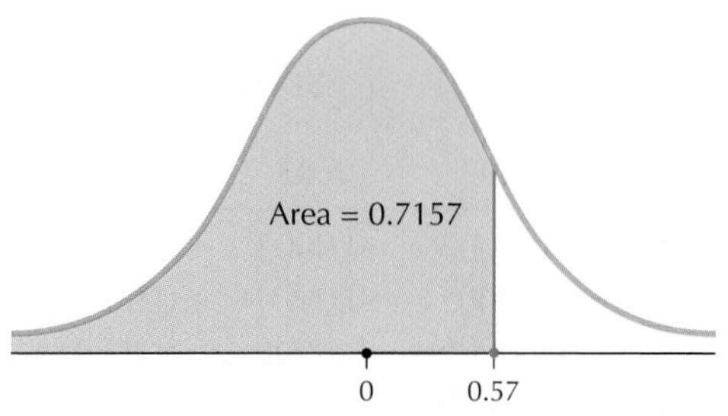


FIGURE 6.27

**Standard Normal Distribution**

| Z | 0.00 | 0.01 | 0.02 | 0.03 | 0.04 | 0.05 | 0.06 | 0.07 | 0.08 | 0.09 |
|---|---|---|---|---|---|---|---|---|---|---|
| **0.0** | 0.5000 | 0.5040 | 0.5080 | 0.5120 | 0.5160 | 0.5199 | 0.5239 | 0.5279 | 0.5319 | 0.5359 |
| **0.1** | 0.5398 | 0.5438 | 0.5478 | 0.5517 | 0.5557 | 0.5596 | 0.5636 | 0.5675 | 0.5714 | 0.5753 |
| **0.2** | 0.5793 | 0.5832 | 0.5871 | 0.5910 | 0.5948 | 0.5987 | 0.6026 | 0.6064 | 0.6103 | 0.6141 |
| **0.3** | 0.6179 | 0.6217 | 0.6255 | 0.6293 | 0.6331 | 0.6368 | 0.6406 | 0.6443 | 0.6480 | 0.6517 |
| **0.4** | 0.6554 | 0.6591 | 0.6628 | 0.6664 | 0.6700 | 0.6736 | 0.6772 | 0.6808 | 0.6844 | 0.6879 |
| **0.5** | 0.6915 | 0.6950 | 0.6985 | 0.7019 | 0.7054 | 0.7088 | 0.7123 | 0.7157 | 0.7190 | 0.7224 |
| **0.6** | 0.7257 | 0.7291 | 0.7324 | 0.7357 | 0.7389 | 0.7422 | 0.7454 | 0.7486 | 0.7517 | 0.7549 |

FIGURE 6.28

## Example 6.24 Case 2: Find the area to the right of a value of *Z*

Find the area to the right of $Z = -1.25$.

**Solution**

***Step 1*** First draw the standard normal curve and label $Z = -1.25$.

***Step 2*** Shade the area to the right of $-1.25$, as shown in Figure 6.29.

***Step 3*** In the *Z* table, excerpted here as Figure 6.30, go down the left-hand column to $-1.2$ and select that row. Then go across the top row to .05 and select that column. The area to the left of $Z = -1.25$ is therefore 0.1056. From Case 2 in Table 6.7, the area to the right of $-1.25$ is then

$$1 - (\text{area to the left of } -1.25) = 1 - 0.1056 = 0.8944$$

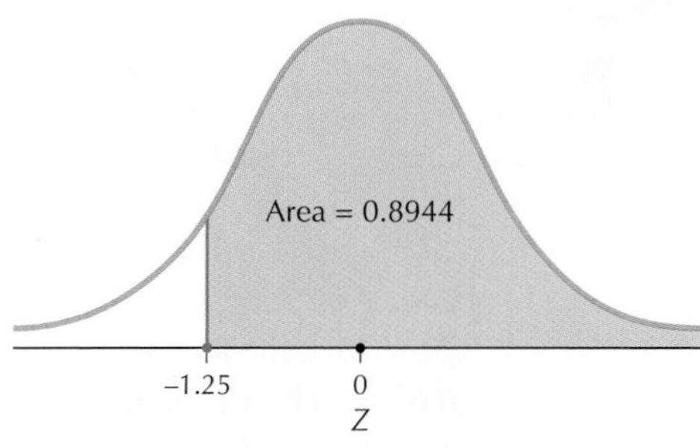


FIGURE 6.29

**Standard Normal Distribution**

| Z | 0.00 | 0.01 | 0.02 | 0.03 | 0.04 | 0.05 | 0.06 | 0.07 | 0.08 | 0.09 |
|---|---|---|---|---|---|---|---|---|---|---|
| **–3.4** | 0.0003 | 0.0003 | 0.0003 | 0.0003 | 0.0003 | 0.0003 | 0.0003 | 0.0003 | 0.0003 | 0.0002 |
| **–3.3** | 0.0005 | 0.0005 | 0.0005 | 0.0004 | 0.0004 | 0.0004 | 0.0004 | 0.0004 | 0.0004 | 0.0003 |
| **–3.2** | 0.0007 | 0.0007 | 0.0006 | 0.0006 | 0.0006 | 0.0006 | 0.0006 | 0.0005 | 0.0005 | 0.0005 |
| **–3.1** | 0.0010 | 0.0009 | 0.0009 | 0.0009 | 0.0008 | 0.0008 | 0.0008 | 0.0008 | 0.0007 | 0.0007 |
| **–3.0** | 0.0013 | 0.0013 | 0.0013 | 0.0012 | 0.0012 | 0.0011 | 0.0011 | 0.0011 | 0.0010 | 0.0010 |
| **–1.4** | 0.0808 | 0.0793 | 0.0778 | 0.0764 | 0.0749 | 0.0735 | 0.0721 | 0.0708 | 0.0694 | 0.0681 |
| **–1.3** | 0.0968 | 0.0951 | 0.0934 | 0.0918 | 0.0901 | 0.0885 | 0.0869 | 0.0853 | 0.0838 | 0.0823 |
| **–1.2** | 0.1151 | 0.1131 | 0.1112 | 0.1093 | 0.1075 | 0.1056 | 0.1038 | 0.1020 | 0.1003 | 0.0985 |
| **–1.1** | 0.1357 | 0.1335 | 0.1314 | 0.1292 | 0.1271 | 0.1251 | 0.1230 | 0.1210 | 0.1190 | 0.1170 |
| **–1.0** | 0.1587 | 0.1562 | 0.1539 | 0.1515 | 0.1492 | 0.1469 | 0.1446 | 0.1423 | 0.1401 | 0.1379 |

FIGURE 6.30

Remember that, although values of $Z$ can be negative, *probabilities (or areas) can never be negative.*

## Example 6.25 Case 3: Find the area between two $Z$-values (checking the accuracy of the Empirical Rule)

Recall that the Empirical Rule (page 293) states that about 68% of the area under the curve lies within one standard deviation of the mean, that is, between $\mu - \sigma$ and $\mu + \sigma$. Check this result for the standard normal distribution by using the $Z$ table.

### Solution

For the standard normal random variable $Z$, $\mu = 0$ and $\sigma = 1$, so that $\mu - \sigma = 0 - 1$ and $\mu + \sigma = 0 + 1 = 1$. Thus, using Case 3, we have $Z_1 = -1$ and $Z_2 = 1$.

***Step 1*** Draw the standard normal curve. Label the $Z$-values $Z_1 = -1$ and $Z_2 = 1$.

***Step 2*** Shade the area between $-1$ and 1, as shown in Figure 6.31a.

***Step 3*** Find the area to the left of $Z_1 = -1$ and the area to the left of $Z_2 = 1$. The $Z$ table gives these areas as follows: area to the left of $Z_1 = -1$ is 0.1587, and area to the left of $Z_2 = 1$ is 0.8413. We subtract the smaller area from the larger to give us the area between $-1$ and 1, as shown in Figures 6.31a–6.31c. Thus we have:

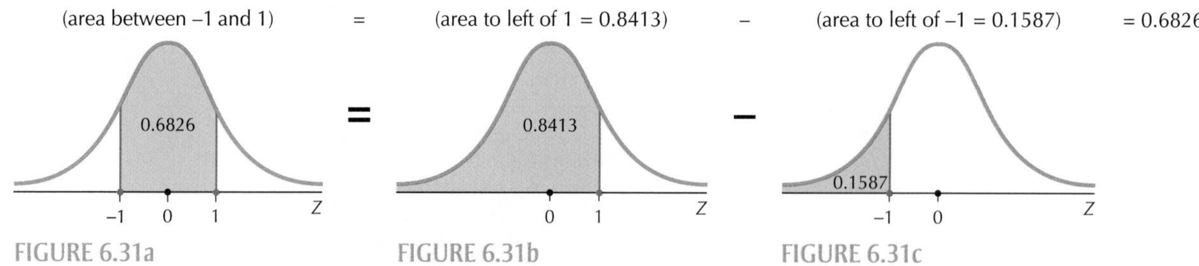


FIGURE 6.31a FIGURE 6.31b FIGURE 6.31c

Thus, the area under the $Z$ curve within one standard deviation of the mean equals 0.6826. The Empirical Rule does very well for an approximation, missing the actual area by only 0.0026. Checking the accuracy of the Empirical Rule for other values of $Z$ is left as an exercise. ■

## Example 6.26 Checking the symmetry of the standard normal curve

Check the symmetry of the standard normal curve by finding the following two areas and confirming that they are equal.

**a.** Area to the left of $Z = -1$

**b.** Area to the right of $Z = 1$

### Solution

**a.** In Example 6.25, we found the area to the left of $Z = -1$ to be 0.1587.

**b.** This represents a Case 2 problem from Table 6.7.

***Step 1*** Figure 6.32 shows the standard normal curve with $Z = 1$ indicated.

***Step 2*** The area to the right of $Z = 1$ is shaded.

***Step 3*** In Example 6.25, we found the area to the left of $Z = 1$ to be 0.8413. The area to the right of 1 is then $1 - 0.8413 = 0.1587$.

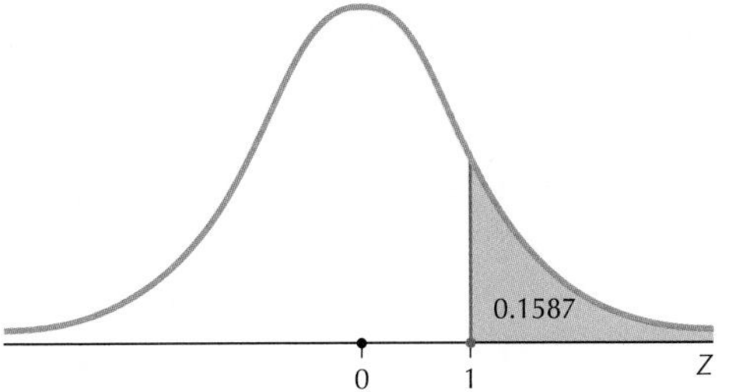


FIGURE 6.32

Thus,

$$(\text{area to the left of } Z = -1) = (\text{area to the right of } Z = 1) = 0.1587$$

Thus, the two tail areas are equal. Since this result holds for any given value of Z, the standard normal curve is symmetric. ■

## Example 6.27 Using technology to find the area under a standard normal curve

In Example 6.23, we found the area under the standard normal curve to the left of $Z = 0.57$ to be 0.7157. Confirm this result using technology.

### Solution

We follow the instructions in the Step-by-Step Technology Guide at the end of Section 6.5 (pages 322–323). Figures 6.33a–c show the results from TI-83/84, Excel, and Minitab, respectively.

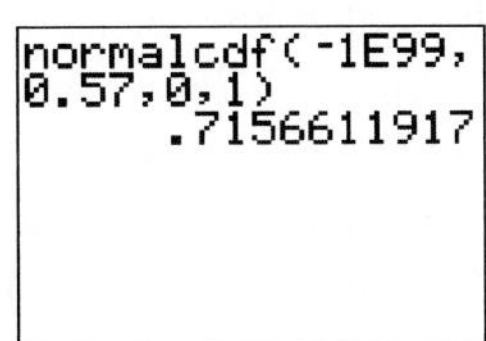


FIGURE 6.33a

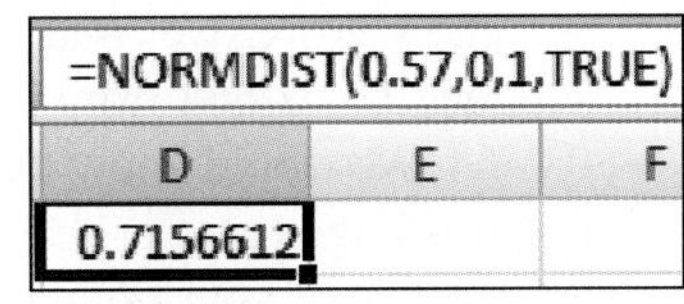


FIGURE 6.33b

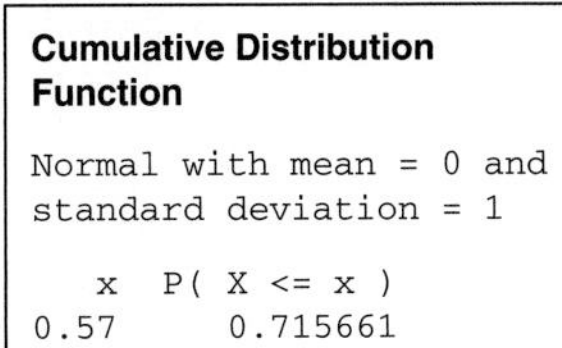


FIGURE 6.33c

The word "cumulative" in the Minitab output means "less than or equal to." Each of these results provides the area under the standard normal curve for values of Z that are less than or equal to 0.57. Each technology rounds to a different number of decimal places. ■

## Example 6.28 Estimating areas

For each of the graphs in Figure 6.34,

**i.** estimate the shaded area, and

**ii.** calculate the area.

**a.**

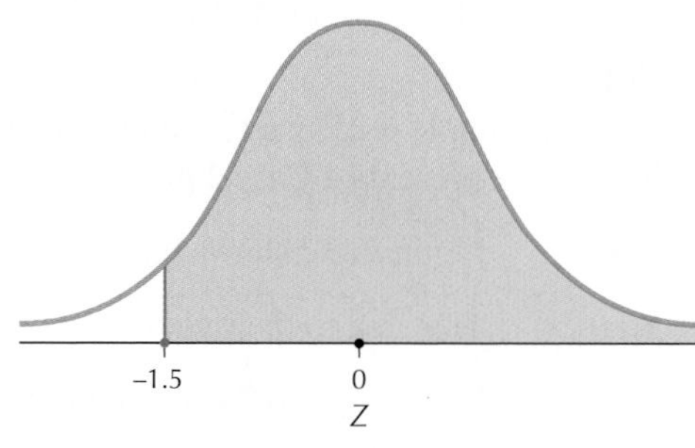


FIGURE 6.34a

**b.**

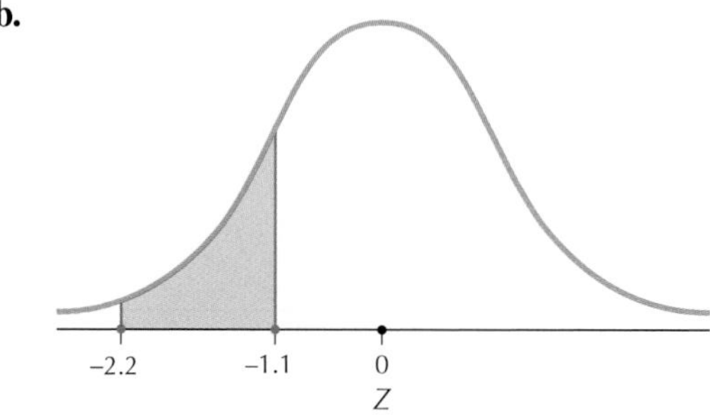


FIGURE 6.34b

### Solution

**a.** **i.** If the area to the right of $Z = 0$ is 0.5, then the shaded area must be greater than 0.5.

**ii.** This is a Case 2 problem from Table 6.7. Looking up $Z = -1.5$ in the $Z$ table, we find the area to the left of $Z = -1.5$ to be 0.0668. Therefore, the area to the right of $Z = -1.5$ must be $1 - 0.0668 = 0.9332$, which is indeed larger than 0.5.

**b.** **i.** This area must be less than 0.5, since the area to the left of $Z = 0$ is 0.5, and the shaded area is smaller than this.

**ii.** This is a Case 3 problem from Table 6.7. We look up $Z = -1.1$ in the $Z$ table and find 0.1357. We look up $Z = -2.2$ in the $Z$ table and find 0.0139. Subtracting the smaller area from the larger gives us the shaded area $0.1357 - 0.0139 = 0.1218$. ■

### Checking That Your Answer Makes Sense!

As you are finding probabilities for values of $Z$, you should always be checking to see that your answer makes sense. In Example 6.28a, what if we had forgotten to subtract the 0.0668 from 1? That would have given us an incorrect desired area of 0.0668. We would know that this answer is incorrect because the shaded area in the graph is large (greater than 0.5) while 0.0668 is small (less than 0.5). See how the artistic side of your brain can help out the mathematical side? ■

The *Normal Density Curve* applet allows you to find areas associated with various values of $Z$.

Note that the areas we have been finding in this section may also be expressed as probabilities. For continuous distributions probabilities are represented by areas under the curve above an interval. Specifically, for the standard normal distribution, probability is represented under the standard normal curve. For instance, in Example 6.23, we found that the area under the standard normal curve to the left of $Z = 0.57$ is 0.7157. This may be reexpressed as follows:

"The probability that $Z$ is less than 0.57 is 0.7157"

or

$$P(Z < 0.57) = 0.7157$$

## Example 6.29 Expressing areas under the standard normal curve as probabilities

Reexpress the following areas as probabilities.

**a.** In Example 6.24, we found the area under the standard normal curve to the right of $Z = -1.25$ to be 0.8944.

**b.** In Example 6.25, we found the area under the standard normal curve between $Z = -1$ and $Z = 1$ to be 0.6826.

**Solution**

**a.** The probability that $Z$ is greater than $-1.25$ is 0.8944. That is, $P(Z > -1.25) = 0.8944$.

**b.** The probability that $Z$ is between $-1$ and 1 is 0.6826. That is, $P(-1 < Z < 1) = 0.6826$. ■

## 2 Finding Standard Normal Values for a Given Area

In the previous examples, you were asked to locate a $Z$-value on the outside of the $Z$ table (that is, using the first column and the top row) and look up the area on the inside of the table. What if we turn this around, so that we know what the area is but need to find the $Z$-value? We will call these problems "backward" problems because we need to use the $Z$ table in reverse. Let's check out an example.

## Example 6.30 Find the value of $Z$ with area 0.90 to the left

*Note:* The value of $Z$ with area 0.90 to the left of it is $Z = 1.28$. Since the area to the left is 0.90, 90% of the $Z$-values are less than $Z = 1.28$. Recall that the *rth percentile* is the value in the data set such that $r$ percent of the observations fall at or below that value. Thus, $Z = 1.28$ represents the 90th percentile of the $Z$ distribution, since it is greater than 90% of $Z$-values.

Find the standard normal $Z$-value that has area 0.90 to the left of it.

**Solution**

***Step 1*** Draw the standard normal curve. Label the $Z$-value $Z_1$.

***Step 2*** Shade the area to the left of $Z_1$. Remember that *we are given an area and are looking for a value of* $Z$. Label the area to the left of $Z_1$ with the given area (0.90), as shown in Figure 6.35.

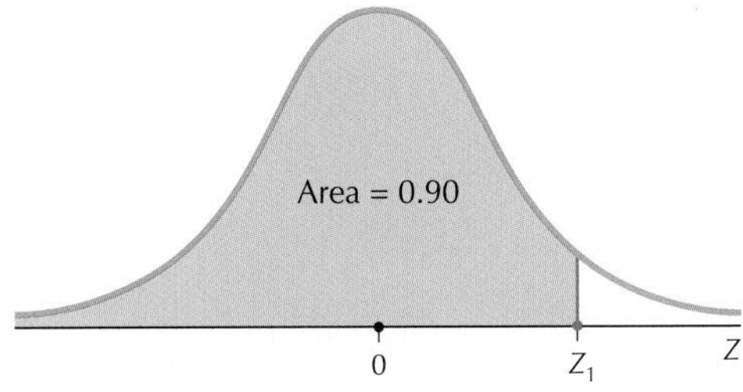


FIGURE 6.35

**Step 3** Look for 0.90 on the inside of the $Z$ table (that is, in the body of the table), since the values *inside the table* represent areas. Because there is no 0.90 inside the table, by convention we take the area that is closest to 0.90, which is 0.8997. Next comes the trick of the backward problems, and the reason for that name. Move from 0.8997 to the left until you reach 1.2 in the first column, and then move up from 0.8997 until you get to .08 (see Figure 6.36). Putting these values together, we get $Z = 1.2 + 0.08 = 1.28$. ■

Standard Normal Distribution

| Z | 0.00 | 0.01 | 0.02 | 0.03 | 0.04 | 0.05 | 0.06 | 0.07 | 0.08 | 0.09 |
|---|---|---|---|---|---|---|---|---|---|---|
| **0.0** | 0.5000 | 0.5040 | 0.5080 | 0.5120 | 0.5160 | 0.5199 | 0.5239 | 0.5279 | 0.5319 | 0.5359 |
| **0.1** | 0.5398 | 0.5438 | 0.5478 | 0.5517 | 0.5557 | 0.5596 | 0.5636 | 0.5675 | 0.5714 | 0.5753 |
| **0.2** | 0.5793 | 0.5832 | 0.5871 | 0.5910 | 0.5948 | 0.5987 | 0.6026 | 0.6064 | 0.6103 | 0.6141 |
| **0.3** | 0.6179 | 0.6217 | 0.6255 | 0.6293 | 0.6331 | 0.6368 | 0.6406 | 0.6443 | 0.6480 | 0.6517 |
| **0.4** | 0.6554 | 0.6591 | 0.6628 | 0.6664 | 0.6700 | 0.6736 | 0.6772 | 0.6808 | 0.6844 | 0.6879 |
| **0.5** | 0.6915 | 0.6950 | 0.6985 | 0.7019 | 0.7054 | 0.7088 | 0.7123 | 0.7157 | 0.7190 | 0.7224 |
| **0.6** | 0.7257 | 0.7291 | 0.7324 | 0.7357 | 0.7389 | 0.7422 | 0.7454 | 0.7486 | 0.7517 | 0.7549 |
| **0.7** | 0.7580 | 0.7611 | 0.7642 | 0.7673 | 0.7704 | 0.7734 | 0.7764 | 0.7794 | 0.7823 | 0.7852 |
| **0.8** | 0.7881 | 0.7910 | 0.7939 | 0.7967 | 0.7995 | 0.8023 | 0.8051 | 0.8078 | 0.8106 | 0.8133 |
| **0.9** | 0.8159 | 0.8186 | 0.8212 | 0.8238 | 0.8264 | 0.8289 | 0.8315 | 0.8340 | 0.8365 | 0.8389 |
| **1.0** | 0.8413 | 0.8438 | 0.8461 | 0.8485 | 0.8508 | 0.8531 | 0.8554 | 0.8577 | 0.8599 | 0.8621 |
| **1.1** | 0.8643 | 0.8665 | 0.8686 | 0.8708 | 0.8729 | 0.8749 | 0.8770 | 0.8790 | 0.8810 | 0.8830 |
| **1.2** | 0.8849 | 0.8869 | 0.8888 | 0.8907 | 0.8925 | 0.8944 | 0.8962 | 0.8980 | 0.8997 | 0.9015 |
| **1.3** | 0.9032 | 0.9049 | 0.9066 | 0.9082 | 0.9099 | 0.9115 | 0.9131 | 0.9147 | 0.9162 | 0.9177 |

FIGURE 6.36

## Example 6.31 Find the value of $Z$ with area 0.03 to the right

Find the standard normal $Z$-value that has area 0.03 to the right of it.

### Solution

**Step 1** Draw the standard normal curve. Label the $Z$-value $Z_1$. Shade the area to the right of it with the given area, as shown in Figure 6.37.

**Step 2** Since the $Z$ table contains areas to the left of values of $Z$, we must find the area to the left of the specific value $Z_1$, as follows:

$$\text{area to left of } Z_1 = 1 - \text{area to right of } Z_1$$

So, the area to the left of $Z_1$ is $1 - 0.03 = 0.97$.

**Step 3** Look up 0.97 on the inside of the $Z$ table. The closest area is 0.9699. Move from 0.9699 to the left until you reach 1.8, and then move up from 0.9699 until you get to 0.08 (see Figure 6.38). Putting these values together, we get $Z = 1.8 + 0.08 = 1.88$. In other words, the $Z$-value with area 0.03 to its right is $Z = 1.88$. ■

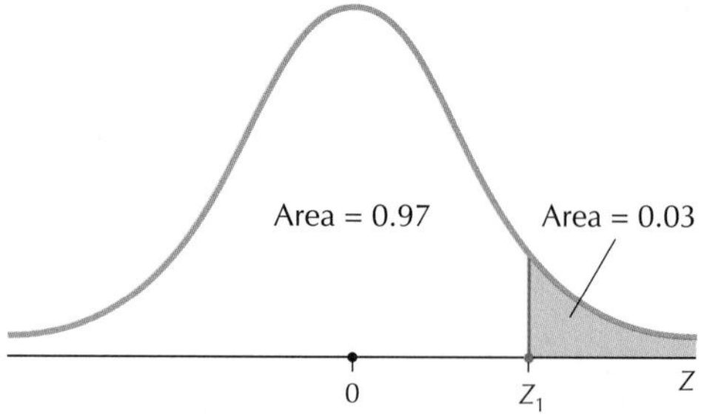


FIGURE 6.37

**Standard Normal Distribution**

| Z | 0.00 | 0.01 | 0.02 | 0.03 | 0.04 | 0.05 | 0.06 | 0.07 | 0.08 | 0.09 |
|---|---|---|---|---|---|---|---|---|---|---|
| 0.0 | 0.5000 | 0.5040 | 0.5080 | 0.5120 | 0.5160 | 0.5199 | 0.5239 | 0.5279 | 0.5319 | 0.5359 |
| 0.1 | 0.5398 | 0.5438 | 0.5478 | 0.5517 | 0.5557 | 0.5596 | 0.5636 | 0.5675 | 0.5714 | 0.5753 |
| 0.2 | 0.5793 | 0.5832 | 0.5871 | 0.5910 | 0.5948 | 0.5987 | 0.6026 | 0.6064 | 0.6103 | 0.6141 |
| 1.6 | 0.9452 | 0.9463 | 0.9474 | 0.9484 | 0.9495 | 0.9505 | 0.9515 | 0.9525 | 0.9535 | 0.9545 |
| 1.7 | 0.9554 | 0.9564 | 0.9573 | 0.9582 | 0.9591 | 0.9599 | 0.9608 | 0.9616 | 0.9625 | 0.9633 |
| 1.8 | 0.9641 | 0.9649 | 0.9656 | 0.9664 | 0.9671 | 0.9678 | 0.9686 | 0.9693 | 0.9699 | 0.9706 |
| 1.9 | 0.9713 | 0.9719 | 0.9726 | 0.9732 | 0.9738 | 0.9744 | 0.9750 | 0.9756 | 0.9761 | 0.9767 |

FIGURE 6.38

When we learn statistical inference in later chapters, we will need to identify which *Z*-values divide the middle 90%, 95%, or 99% of the area under the standard normal curve from the tail area.

## Example 6.32 Find the values of *Z* that mark the boundaries of the middle 95% of the area

Find the two values of *Z* that mark the boundaries of the middle 95% of the area under the standard normal curve.

### Solution

***Step 1*** Draw the standard normal curve, showing the desired middle area (95%) with boundaries labeled as $Z_1$ and $Z_2$, as shown in Figure 6.39. By symmetry, there is area $= (1 - 0.95)/2 = 0.025$ in each tail.

***Step 2*** Look up 0.025 on the inside of the *Z* table. Find $Z_1$ by moving to the left and up from 0.025 in the *Z* table, giving us $Z_1 = -1.96$, as shown in Figure 6.40.

***Step 3*** Since the area in the right tail is 0.025 as well, the area to the left of $Z_2$ is $1 - 0.025 = 0.975$. Looking up 0.975 in the *Z* table gives us $Z_2 = 1.96$.

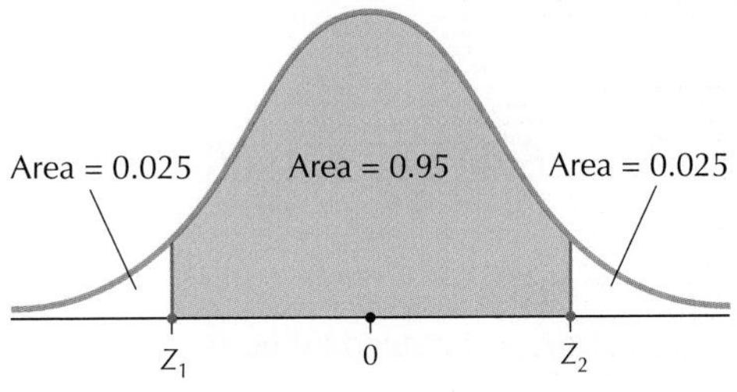


FIGURE 6.39

Thus, the two *Z*-values that mark the boundaries of the middle 95% of the area under the standard normal curve are –1.96 and 1.96.

*Note:* Is it a coincidence that the two values of *Z* that determine the middle 95% of the area under the standard normal curve are 1.96 and −1.96? Not at all. Since the standard normal curve is symmetric about the mean 0, the values −1.96 and 1.96 that form the boundaries of the middle 95% must be equidistant from zero. Of course, −1.96 and 1.96 are called additive inverses.

**Standard Normal Distribution**

| Z | 0.00 | 0.01 | 0.02 | 0.03 | 0.04 | 0.05 | 0.06 | 0.07 | 0.08 | 0.09 |
|---|---|---|---|---|---|---|---|---|---|---|
| **–3.4** | 0.0003 | 0.0003 | 0.0003 | 0.0003 | 0.0003 | 0.0003 | 0.0003 | 0.0003 | 0.0003 | 0.0002 |
| **–3.3** | 0.0005 | 0.0005 | 0.0005 | 0.0004 | 0.0004 | 0.0004 | 0.0004 | 0.0004 | 0.0004 | 0.0003 |
| **–3.2** | 0.0007 | 0.0007 | 0.0006 | 0.0006 | 0.0006 | 0.0006 | 0.0006 | 0.0005 | 0.0005 | 0.0005 |
| **–2.0** | 0.0228 | 0.0222 | 0.0217 | 0.0212 | 0.0207 | 0.0202 | 0.0197 | 0.0192 | 0.0188 | 0.0183 |
| **–1.9** | 0.0287 | 0.0281 | 0.0274 | 0.0268 | 0.0262 | 0.0256 | 0.0250 | 0.0244 | 0.0239 | 0.0233 |
| **–1.8** | 0.0359 | 0.0351 | 0.0344 | 0.0336 | 0.0329 | 0.0322 | 0.0314 | 0.0307 | 0.0301 | 0.0294 |
| **–1.7** | 0.0446 | 0.0436 | 0.0427 | 0.0418 | 0.0409 | 0.0401 | 0.0392 | 0.0384 | 0.0375 | 0.0367 |

FIGURE 6.40

## Example 6.33 Using technology to find values of Z, given an area

In Example 6.30, we found that the value of $Z$ with area 0.90 to its left is $Z = 1.28$. Confirm this result using technology.

**Solution**

We follow the instructions in the Step-by-Step Technology Guide at the end of Section 6.5 (pages 322–323). Figures 6.41a–c show the results from TI-83/84, Excel, and Minitab, respectively.

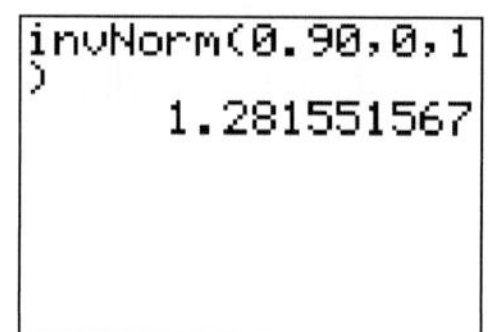


FIGURE 6.41a

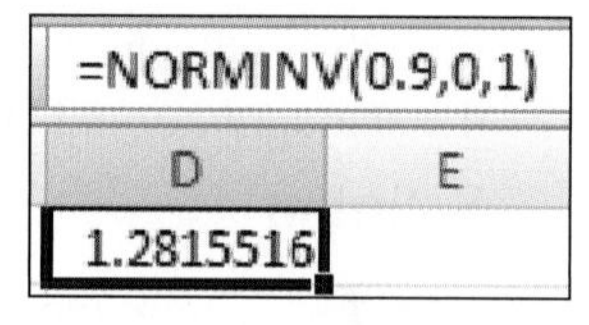


FIGURE 6.41b

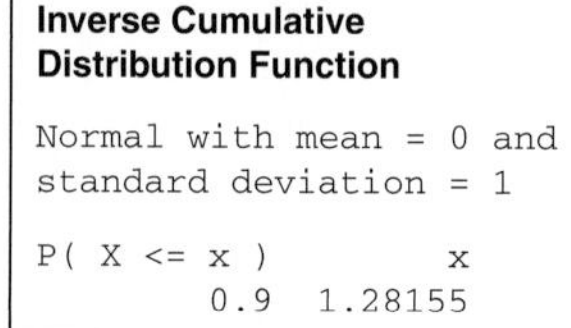


FIGURE 6.41c

## Section 6.4 SUMMARY

1 The standard normal distribution has mean $\mu = 0$ and standard deviation $\sigma = 1$. This distribution is often called the $Z$ distribution. The $Z$ table and technology can be used to find areas under the standard normal curve. In the $Z$ table, the numbers on the outside are values of $Z$, and the numbers inside are areas to the left of values of $Z$.

2 The $Z$ table and technology can also be used to find a value of $Z$, given a probability or an area under the curve.

## Section 6.4 EXERCISES

**CLARIFYING THE CONCEPTS**

**1.** What is the value for the mean of the standard normal distribution?

**2.** What is the value for the standard deviation of the standard normal distribution?

**3.** True or false: The area under the $Z$ curve to the right of $Z = 0$ is 0.5.

**4.** True or false: $P(Z = 0) = 0$.

**PRACTICING THE TECHNIQUES**

For Exercises 5–16, use the graph of the standard normal distribution to

- **a.** estimate the shaded area,
- **b.** find the shaded area using the $Z$ table or technology, and
- **c.** check your estimate in **(a)** against your answer in **(b)**.

5.

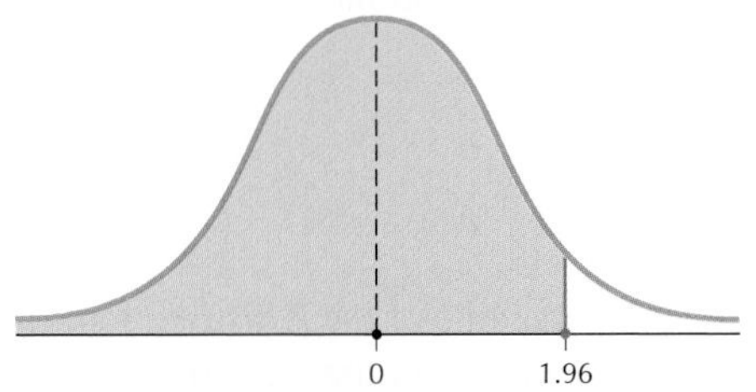


6.

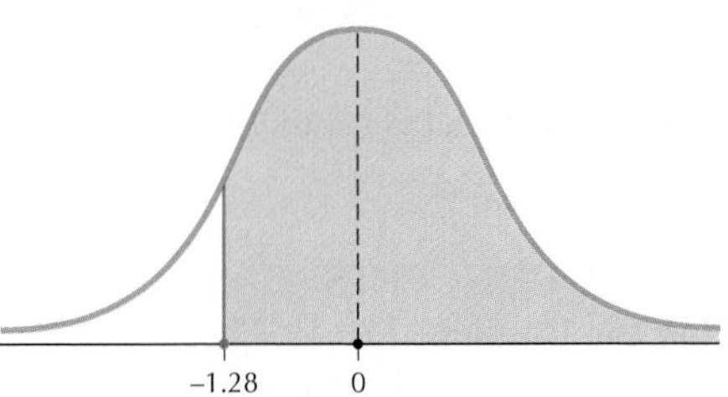


7.

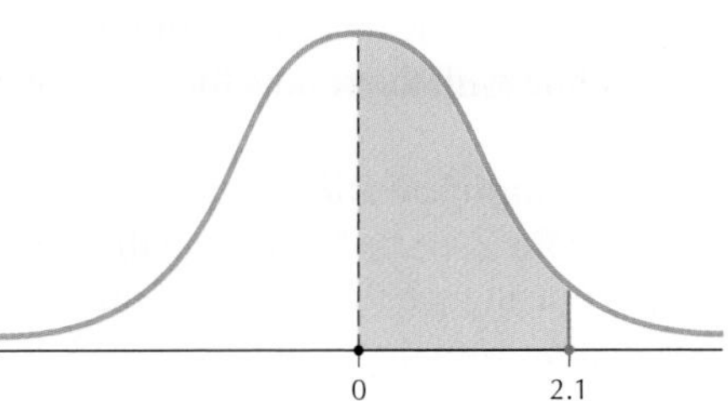


8.

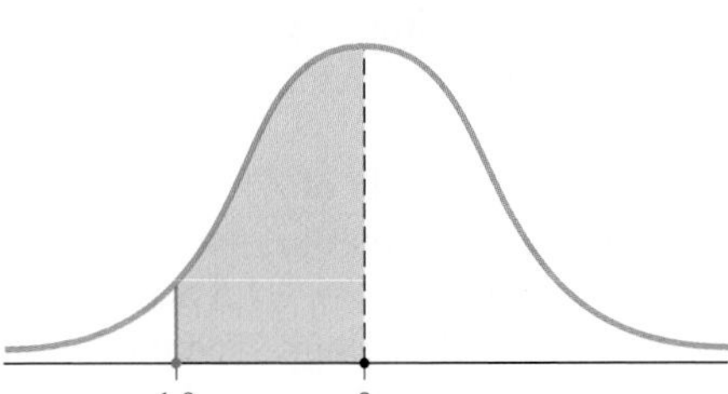


9.

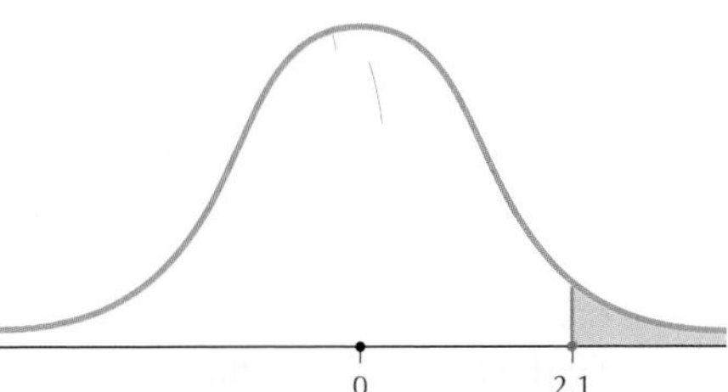


10.

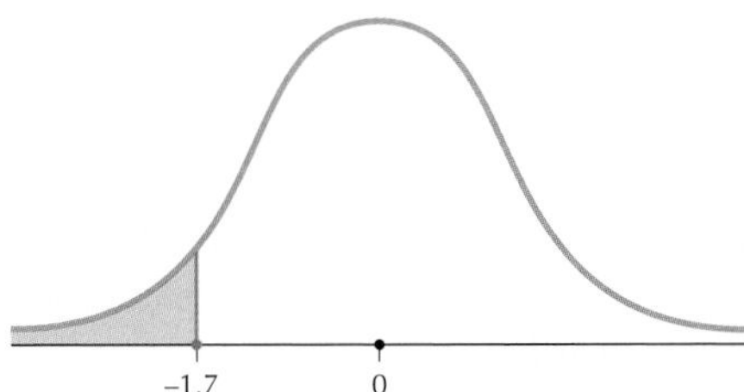


11.

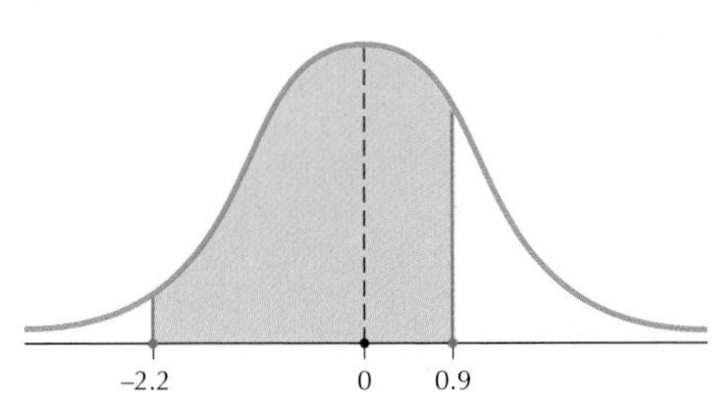


12.

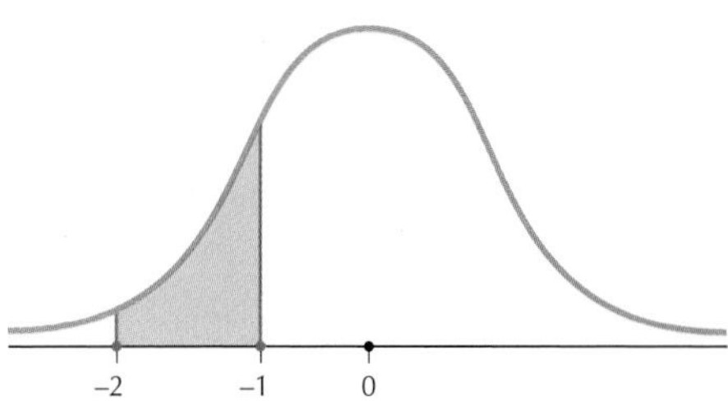


13.

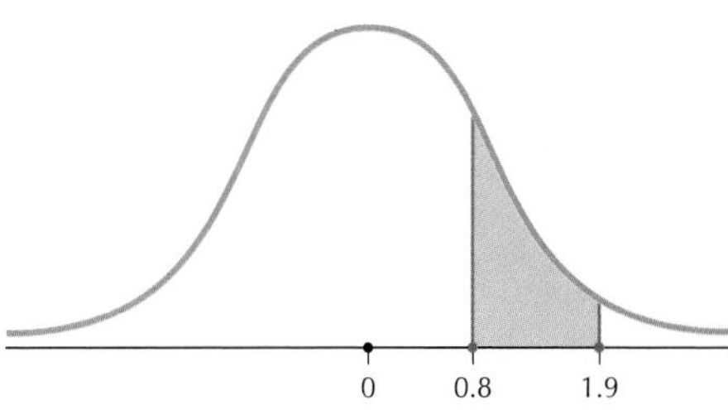


14.

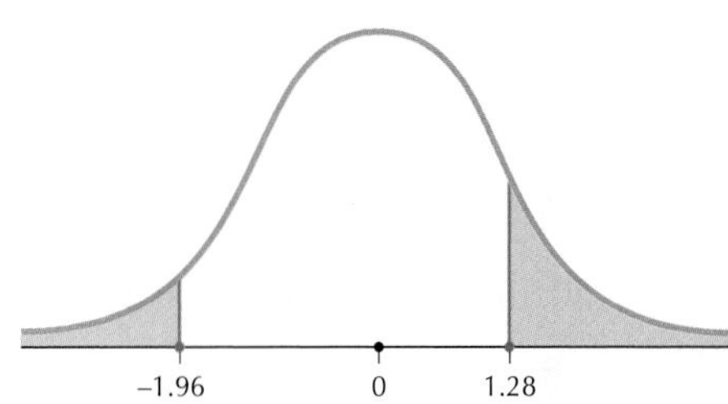


15.

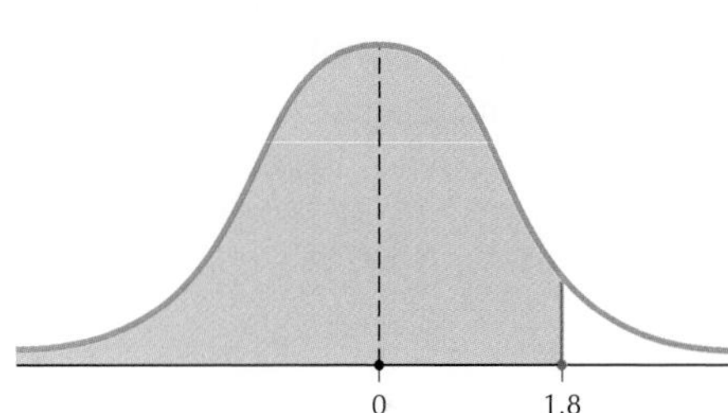


16.

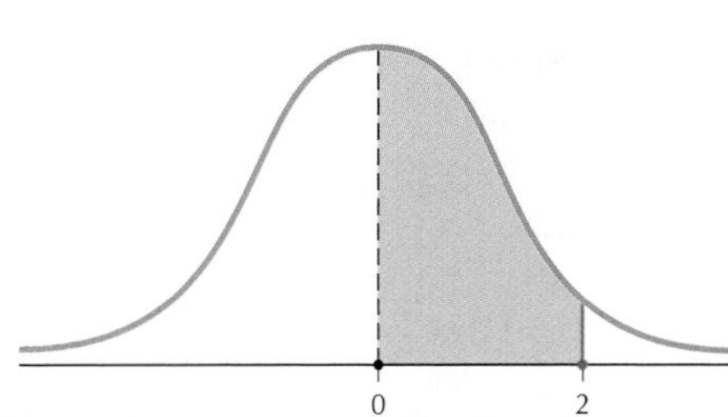


For Exercises 17–38,

**a.** draw the graph,
**b.** estimate the area,
**c.** find the area using the $Z$ table or technology,
**d.** check your estimate against your final answer.

Find the area under the standard normal curve that lies to the left of the following.

**17.** $Z = 1$
**18.** $Z = 2$
**19.** $Z = 3$
**20.** $Z = 0.5$
**21.** $Z = -2.7$

**22.** $Z = -0.9$
**23.** $Z = -0.2$
**24.** $Z = -1.2$

Find the area under the standard normal curve that lies to the right of the following.
**25.** $Z = 1.27$
**26.** $Z = 2.12$
**27.** $Z = -3.01$
**28.** $Z = -0.69$

Find the area under the standard normal curve that lies between the following.
**29.** $Z = 0$ and $Z = 1$
**30.** $Z = 1$ and $Z = 2$
**31.** $Z = 2$ and $Z = 3$
**32.** $Z = 1.28$ and $Z = 1.96$
**33.** $Z = -1$ and $Z = 0$
**34.** $Z = -2$ and $Z = -1$
**35.** $Z = -3$ and $Z = -2$
**36.** $Z = -1.96$ and $Z = -1.28$
**37.** $Z = -1.28$ and $Z = 1.28$
**38.** $Z = -2.01$ and $Z = 2.37$

For Exercises 39–50, find the indicated probability for the standard normal *Z*.

**a.** Draw the graph.
**b.** Estimate the area.
**c.** Find the area using the *Z* table or technology.
**d.** Check your estimate against your final answer.

**39.** $P(Z = 0)$
**40.** $P(Z < 0)$
**41.** $P(Z < 10)$
**42.** $P(Z > 1.29)$
**43.** $P(Z < -2.17)$
**44.** $P(Z < 0.57)$
**45.** $P(-1.96 < Z < 1.96)$
**46.** $P(-2.07 < Z < 0.46)$
**47.** $P(-3.05 < Z < -0.94)$
**48.** $P(1.54 < Z < 2.20)$
**49.** $P(-100 < Z < 0)$
**50.** $\text{P}(-1.72 < Z < -1.57)$

For Exercises 51–54, find the *Z*-value with the following areas under the standard normal curve to its left. Draw the graph first and estimate the *Z*-value. Then find the value and check your estimate against your final answer.
**51.** 0.3336
**52.** 0.4602
**53.** 0.3264
**54.** 0.4247

For Exercises 55–58, find the *Z*-value with the following areas under the standard normal curve to its right. Draw the graph first and estimate the *Z*-value. Then find the value and check your estimate against your final answer.
**55.** 0.8078
**56.** 0.3085
**57.** 0.9788
**58.** 0.5120
**59.** Find the 50th percentile of the *Z* distribution.
**60.** Find the 75th percentile of the *Z* distribution.
**61.** Find the value of *Z* that is larger than 99.5% of all values of *Z*.
**62.** Find the value of *Z* that is smaller than 99.5% of all values of *Z*.

## APPLYING THE CONCEPTS

**63. Checking the Empirical Rule.** Check the accuracy of the Empirical Rule for $Z = 2$. That is, find the area between $Z = -2$ and $Z = 2$ using the techniques of this section. Then compare your finding with the results for $Z = 2$ using the Empirical Rule.

**64. Checking the Empirical Rule.** Check the accuracy of the Empirical Rule for $Z = 3$. That is, find the area between $Z = -3$ and $Z = 3$ using the techniques of this section. Then compare your finding with the results for $Z = 3$ using the Empirical Rule.

**65. Without Tables or Technology.** Find the following areas *without* using the *Z* table or technology. The area to the left of $Z = -1.5$ is 0.0668.

**a.** Find the area to the right of $Z = 1.5$.
**b.** Find the area to the right of $Z = -1.5$.
**c.** Find the area between $Z = -1.5$ and $Z = 1.5$.

**66. Without Tables or Technology.** Find the following areas *without* using the *Z* table or technology. The area to the right of $Z = 2.7$ is 0.0035.

**a.** Find the area to the left of $Z = 2.7$.
**b.** Find the area to the left of $Z = -2.7$.
**c.** Find the area between $Z = -2.7$ and $Z = 2.7$.

**67. Values of Z That Mark the Middle 99%.** Find the two values of *Z* that contain the middle 99% of the area under the standard normal curve.

**68. Values of Z That Mark the Middle 90%.** Find the two values of *Z* that contain the middle 90% of the area under the standard normal curve.

APPLET Use the *Normal Density Curve* applet for Exercises 69 and 70.

**69.** Find the quartiles of the standard normal distribution. That is, find the 25th, 50th, and 75th percentiles of the standard normal distribution.

**70.** Use the applet to find the answers to the following exercises from this section.

**a.** Exercise 6
**b.** Exercise 7
**c.** Exercise 17
**d.** Exercise 18
**e.** Exercise 28
**f.** Exercise 29

# 6.5 Applications of the Normal Distribution

**Objectives:** By the end of this section, I will be able to...

1 Compute probabilities for a given value of any normal random variable.

2 Find the appropriate value of any normal random variable, given an area or probability.

3 Calculate binomial probabilities using the normal approximation to the binomial distribution.

## 1 Finding Probabilities for Any Normal Distribution

The data in problems that we face in the real world do not usually follow the standard normal distribution. Instead, a problem may be stated in terms of some normal random variable $X$ that has a mean other than 0 or a standard deviation other than 1. In cases like these, $X$ needs to be standardized to $Z$ so that we can use the Section 6.4 techniques.

### Standardizing $X$ to $Z$

To *standardize* things means to make them all the same, uniform, or equivalent. For example, college applicants take standardized tests so that the admissions officers can compare students according to a uniform assessment tool. To standardize a normal random variable $X$, we *transform* that normal random variable $X$ into the standard normal random variable $Z$ that we worked so extensively with in Section 6.4. In this section, we will focus on real-world applications of the normal distribution.

Suppose that $X$ is a normal random variable with population mean $\mu$ and population standard deviation $\sigma$. We standardize $X$ by subtracting the mean $\mu$ and dividing by the standard deviation $\sigma$. The result of this transformation is the familiar **standard normal random variable Z**.

**Standardizing a Normal Random Variable**

Any normal random variable $X$ can be transformed into the standard normal random variable $Z$ by *standardizing* using the formula

$$Z = \frac{X - \mu}{\sigma}$$

The key here is the following: *for a given area of interest for a normal random variable X, the corresponding area after the transformation to Z is exactly the same.* For any normal random variable $X$, the area between $a$ and $b$ is exactly the same as the area between $Z_a = (a - \mu)/\sigma$ and $Z_b = (b - \mu)/\sigma$ (see Figure 6.42).

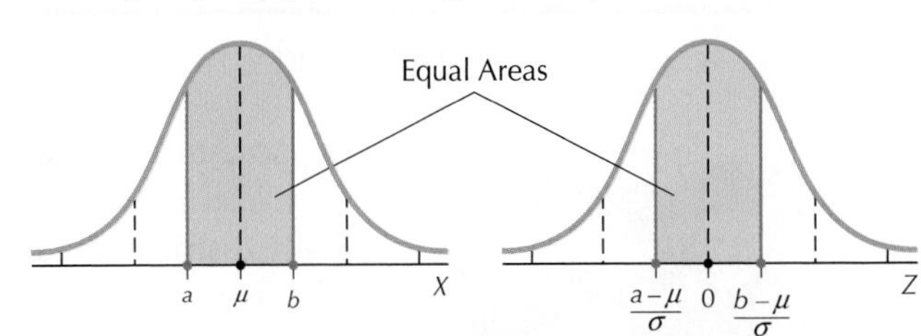


FIGURE 6.42
Corresponding areas are equal.

## Example 6.34 April in Georgia

The state of Georgia reports that the average temperature statewide for the month of April from 1949 to 2006 was $\mu = 61.5°F$. Assume that the standard deviation is $\sigma = 8°F$

Carol Christensen/Jupiter Images

and that temperature in Georgia in April is normally distributed. Draw the normal curve for temperatures between 45.5°F and 77.5°F, and the corresponding $Z$ curve. Find the probability that the temperature is between 45.5°F and 77.5°F in April in Georgia.

**Solution**

Here we have $a = 45.5$ and $b = 77.5$, giving us

$$Z_a = \frac{a - \mu}{\sigma} = \frac{45.5 - 61.5}{8} = -2 \quad \text{and} \quad Z_b = \frac{b - \mu}{\sigma} = \frac{77.5 - 61.5}{8} = 2$$

In Figure 6.43, the area between 45.5°F and 77.5°F is the same as between $Z = -2$ and $Z = 2$. In other words,

$$P(45.5 < X < 77.5) = P(-2 < Z < 2)$$

This is a Case 3 problem from Table 6.7. The $Z$ table tells us that the area to the left of $Z_1 = -2$ is 0.0228, and the area to the left of $Z_2 = 2$ is 0.9772. The area between $-2$ and 2 is then equal to $0.9772 - 0.0228 = 0.9544$. The probability that temperature is between 45.5°F and 77.5°F in April in Georgia is 0.9544. ■

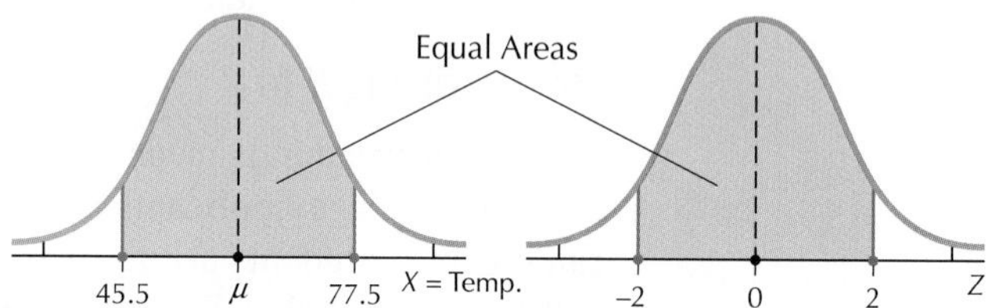


FIGURE 6.43

**Finding Probabilities for Any Normal Distribution**

***Step 1*** Determine the random variable $X$, the mean $\mu$, and the standard deviation $\sigma$. Draw the normal curve for $X$, and shade the desired area.

***Step 2*** Standardize by using the formula $Z = (X - \mu)/\sigma$ to find the values of $Z$ corresponding to the $X$-values.

***Step 3*** Draw the standard normal curve and shade the area corresponding to the shaded area in the graph of $X$.

***Step 4*** Find the area under the standard normal curve using either (a) the $Z$ table or (b) technology. This area is equal to the area under the normal curve for $X$ drawn in Step 1.

## Example 6.35 Ordering caps and gowns for graduation

Punchstock/Digital Vision

Suppose that you are in charge of ordering caps and gowns for senior graduation in May. You know that the heights of the population of students at your college are normally distributed with a mean of 68 inches (5 feet 8 inches) and a standard deviation of 3 inches. The gown manufacturer wants to know how many students will need to special order their gowns because they are very tall. Find the proportion of students who are above 74 inches tall.

**Solution**

***Step 1*** **Determine $X$, $\mu$, and $\sigma$, and draw the normal curve for $X$.** You know that the heights of students at your college are normally distributed with a mean of

68 inches and a standard deviation of 3 inches. Thus, the normal random variable $X$ = heights of students with $\mu = 68$ inches and $\sigma = 3$ inches. In the center of the number line, mark the mean $\mu$. Also mark on the number line the value of $X$ that the problem is asking about. Figure 6.44 shows the graph of $X$ (the heights of students) with the mean of 68 inches and the height of 74 inches marked.

Since you need to know the proportion of students taller than (above) 74 inches, shade the area under the curve and above 74 inches. We can express this proportion as a probability, the probability that a randomly chosen student will be taller than 74 inches, or $P(X > 74)$. *Just by looking at Figure 6.44 you should be able to get a rough idea of what the proportion of students taller than 74 inches will be.* Certainly this proportion will be less than 50%, and probably pretty small. If you get an answer like "60%" for your proportion, you would surely know that it is wrong.

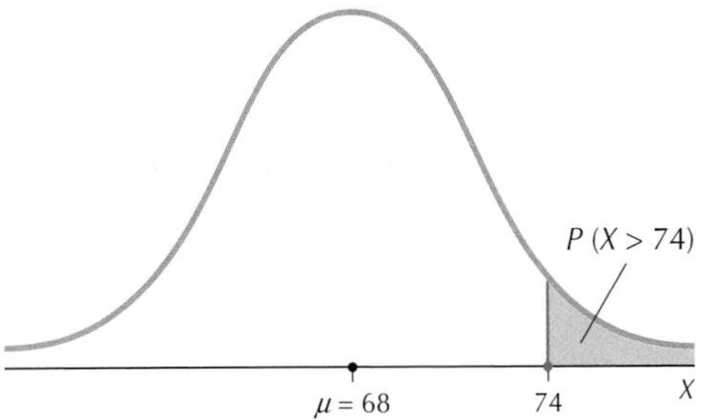


FIGURE 6.44 Graph of proportion of college students taller than 74 inches.

**Step 2** **Standardize.** Now standardize the random variable $X$ to the standard normal $Z$:

$$Z = \frac{X - \mu}{\sigma} = \frac{X - 68}{3}$$

Find the $Z$-value corresponding to the height of 74 inches:

$$\frac{74 - \mu}{\sigma} = \frac{74 - 68}{3} = 2$$

So the $Z$-value associated with 74 inches is 2, which indicates that the height of 74 inches is 2 standard deviations above the mean of 68 inches.

**Step 3** **Draw the standard normal curve.** Heights above 74 inches are more than 2 standard deviations above the mean, so shade the area to the right of 2 in Figure 6.45. Now find the area to the right of $Z = 2$ using the methods of Section 6.4.

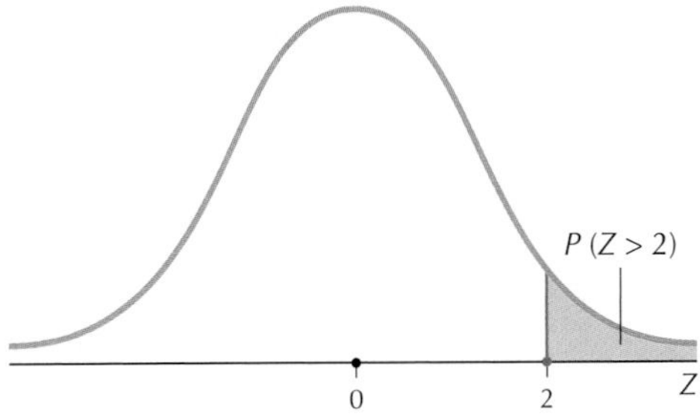


FIGURE 6.45 Graph of $P(Z > 2)$.

***Step 4*** **Find the area under the standard normal curve.** Figure 6.45 represents a Case 2 problem from Table 6.7. The $Z$ table tells us that the area to the left of $Z = 2.00$ is 0.9772. Thus, the area to the right is

$$P(Z > 2) = 1 - 0.9772 = 0.0228$$

The proportion of students taller than 74 inches is 0.0228, or 2.28%. Note that this value for $P(X > 74)$ agrees with our earlier intuition that the proportion was surely less than 50% and most likely very small. ■

## Example 6.36 Finding probability that $X$ lies between two given values

Continuing the cap-and-gown problem, what percentage of students are between 60 and 70 inches tall?

### Solution

***Step 1*** **Determine $X$, $\mu$, and $\sigma$, and draw the normal curve for $X$.** We have already seen that $X$ = heights of students, $\mu = 68$ inches, and $\sigma = 3$ inches. Once again, draw a graph of the distribution of heights $X$, with the mean 68 inches in the middle, the height 60 inches to the left of the mean, and the height 70 inches to the right of the mean, as in Figure 6.46.

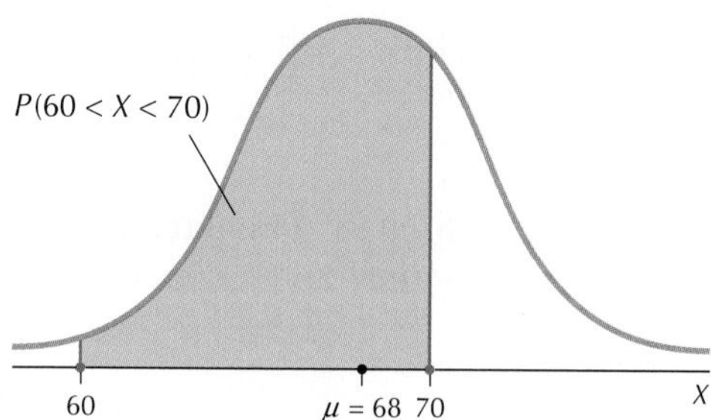


FIGURE 6.46 Graph of percentage of students between 60 and 70 inches tall.

***Step 2*** **Standardize.** This is a "between" example, where two values of $X$ are given, and we are asked to find the area between them. In this case, just standardize both of these values of $X$ to get a $Z$-value for each:

$$Z = \frac{60 - \mu}{\sigma} = \frac{60 - 68}{3} \approx -2.67 \quad \text{and} \quad Z = \frac{70 - \mu}{\sigma} = \frac{70 - 68}{3} \approx 0.67$$

***Step 3*** **Draw the standard normal curve.** Draw a graph of $Z$, shading the area between $Z = -2.67$ and $Z = 0.67$, as shown in Figure 6.47. Again, the key is that the area between $Z = -2.67$ and $Z = 0.67$ is exactly the same as the area between $X = 60$ inches and $X = 70$ inches.

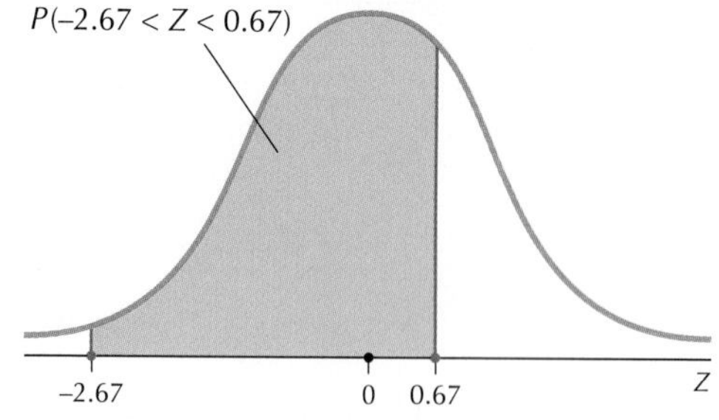


FIGURE 6.47 Graph of percentage of $Z$-values between $-2.67$ and $0.67$.

**Step 4** **Find area under the standard normal curve.** Figure 6.47 looks like a Case 3 problem from Table 6.7. Find the area to the left of 0.67, which is 0.7486, and the area to the left of −2.67, which is 0.0038. Subtracting the smaller from the larger gives us

$$P(-2.67 < Z < 0.67) = 0.7486 - 0.0038 = 0.7448$$

Thus, the percentage of students who are between 60 and 70 inches tall is 74.48%. ■

**Check Your Answer!** According to the Empirical Rule, almost all Z-values lie between –3 and 3, so it is unlikely that a randomly selected value of Z lies outside this range. You should remember this when you are doing your calculations. If you are standardizing a normal random variable X and get a very large Z-value (such as, say, 50), you should recheck your calculations because the probability that Z takes such a large value is very small.

## 2 Finding Normal Data Values for Specified Probabilities

Sometimes we are given a probability (or proportion or area), and we are asked to find the associated value of X. Questions like these are similar to the "backward" problems of Section 6.4, so called because we must use the Z table backward. Since the formula for standardizing X gives the value for Z, we need to use our algebra skills to find the equation for X: Start with the standard normal formula $Z = (X - \mu)/\sigma$. Multiply both sides by $\sigma$ to get $Z\sigma = X - \mu$. Then add $\mu$ to both sides, giving us $X = Z\sigma + \mu$.

**Finding Normal Data Values for Specified Probabilities**

***Step 1*** **Determine X, $\mu$, and $\sigma$, and draw the normal curve for X.** Shade the desired area. Mark the position of $X_1$, the unknown value of X.

***Step 2*** **Find the Z-value corresponding to the desired area.** Look up the area you identified in Step 1 on the *inside* of the Z table. If you do not find the exact value of your area, by convention choose the area that is closest.

***Step 3*** **Transform this value of Z into a value of X, which is the solution.** Use the formula $X_1 = Z\sigma + \mu$.

### Example 6.37 How tall do you have to be before you have to special-order your cap and gown?

Suppose that we wanted only the tallest 1% of our students to have to special-order gowns. What is the height at which tall students will have to special-order their gowns?

**Solution**

Notice that we are not asked to find a probability (or proportion or area). Instead, we are given a percentage (1%) and asked to find the value of X (the height) that is associated with this 1%.

**Step 1** **Determine X, $\mu$, and $\sigma$, and draw the normal curve for X.** We already know that X = heights of students, $\mu$ = 68 inches, and $\sigma$ = 3 inches. The value of X we are interested in refers to very tall students, so that $X_1$ will be at the far right of the distribution of X. Only 1% of students will be taller than this height, so the area to the right of $X_1$ is 1%, as shown in Figure 6.48.

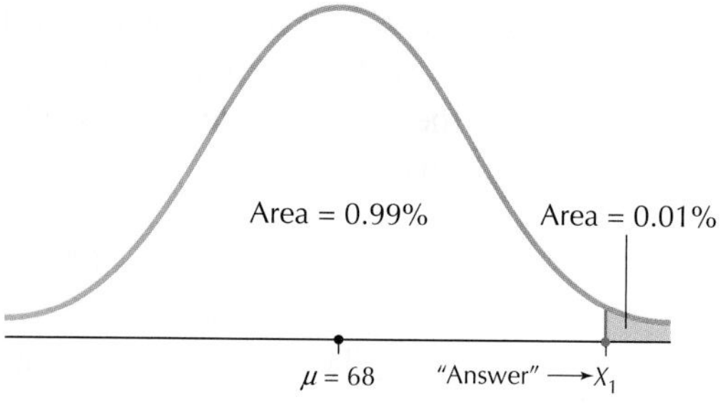


FIGURE 6.48

**Step 2** **Find the Z-value corresponding to the desired area.** The area to the right of $X_1$ equals 0.01, so that the area to the left of $X_1$ equals $1 - 0.01 = 0.99$. Looking up 0.99 on the inside of the Z table gives us $Z = 2.33$.

**Step 3** **Transform using the formula $X_1 = Z\sigma + \mu$.** We calculate

$$X_1 = Z\sigma + \mu = (2.33)(3) + 68 = 74.99$$

If we want only the tallest 1% of our students to have to special-order their gowns, the height at which tall students will have to special-order their gowns is 74.99 inches. ■

## Example 6.38 Finding the X-values that mark the boundaries of the middle 95% of X-values

**Edmunds.com** reported that the average amount that people were paying for a 2007 Toyota Camry XLE was \$23,400. Let $X$ = price, and assume that price follows a normal distribution with $\mu = \$23{,}400$ and $\sigma = \$1000$. Find the prices that separate the middle 95% of 2007 Toyota Camry XLE prices from the bottom 2.5% and the top 2.5%.

### Solution

**Step 1** **Determine X, μ, and σ, and draw the normal curve for X.** Let $X$ = price, $\mu = \$23{,}400$, and $\sigma = \$1000$. The middle 95% of prices are delimited by $X_1$ and $X_2$, as shown in Figure 6.49.

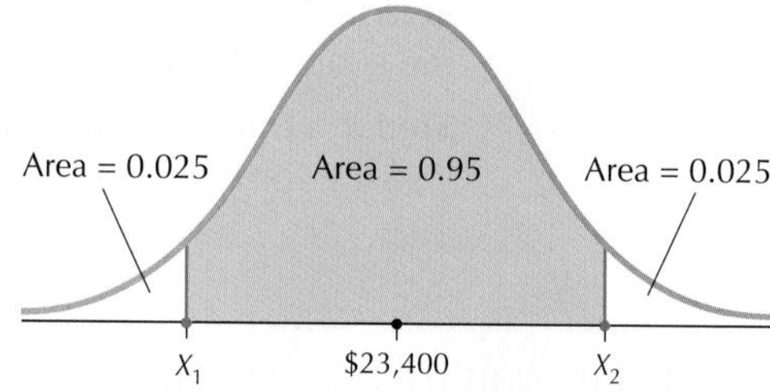


FIGURE 6.49

**Step 2** **Find the Z-values corresponding to the desired area.** The area to the left of $X_1$ equals 0.025, and the area to the left of $X_2$ equals 0.975. Looking up area 0.025 on the inside of the Z table gives us $Z_1 = -1.96$. Looking up area 0.975 on the inside of the Z table gives us $Z_2 = 1.96$.

**Step 3** **Transform using the formula $X_1 = Z\sigma + \mu$.** We calculate

$$X_1 = Z_1\sigma + \mu = (-1.96)(1000) + 23{,}400 = 21{,}440$$

$$X_2 = Z_2\sigma + \mu = (1.96)(1000) + 23{,}400 = 25{,}360$$

The prices that separate the middle 95% of 2007 Toyota Camry XLE prices from the bottom 2.5% of prices and the top 2.5% of prices are \$21,440 and \$25,360. ■

## How Change in Spread Affects Camry Prices

In Example 6.38, *what if* we ask the same question again, but this time the standard deviation $\sigma$ of 2007 Toyota Camry XLE prices is not \$1000 but some value less than \$1000. How and why would this affect the following?

**a.** The values $Z_1$ and $Z_2$ found in Step 2

**b.** The value $X_1$ separating the middle 95% of prices from the bottom 2.5%

**c.** The value $X_2$ separating the middle 95% of prices from the top 2.5%

### Solution

Figure 6.50 illustrates the distribution of 2007 Toyota Camry XLE prices, where everything is the same as in Figure 6.49 except that the standard deviation of the prices is smaller by an unknown amount. Thus, the spread of the distribution is smaller.

FIGURE 6.50

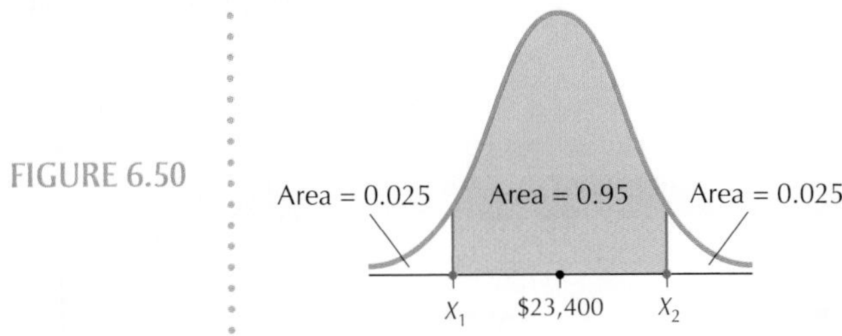


**a.** Since we are still asking for the middle 95% of prices, the $Z$-values remain the same, $-1.96$ and $1.96$.

**b.** Reexpress the formula $X_1 = Z_1\sigma + \mu$ as $X_1 = \$23{,}400 - 1.96 \cdot \sigma$. If $\sigma$ is smaller than \$1000, then the quantity $1.96 \cdot \sigma$, which represents the difference between the mean price and $X_1$, will also be smaller.

Since $X_1$ is less than the mean $\mu = \$23{,}400$, the smaller difference between the mean price and $X_1$ leads us to conclude that $X_1$ will be *larger* than in Example 6.38. For example, if the new standard deviation is $\sigma = \$500$, then $X_1 = \$23{,}400 - 1.96 \cdot 500 = \$22{,}420$, which is larger than the \$21,440 in Example 6.38.

**c.** Similarly, a smaller $\sigma$ means a smaller quantity $1.96 \cdot \sigma$, which means that $X_2 = \$23{,}400 + 1.96 \cdot \sigma$ will be closer to the mean $\mu = \$23{,}400$. Since $X_2$ is larger than the mean, the new value for $X_2$ will be *smaller* than in Example 6.38.

The *Normal Density Curve* applet allows you to find areas associated with various values of any normal random variable.

## Example 6.39 Normal probabilities and percentiles using technology

Applying the information on Toyota Camry prices from Example 6.38, use the TI-83/84, Excel, or Minitab to find the following.

**a.** The proportion of 2007 Camry XLEs costing between \$22,000 and \$24,000, $P(22{,}000 \le X \le 24{,}000)$

**b.** The 99th percentile of Camry XLE prices, that is, find the value of $X$, namely, $X_1$, such that $P(X \le X_1) = 0.99$

### Solution

The instructions for finding these quantities are given in the Step-by-Step Technology Guide at the end of this section (pages 322–323).

### TI-83/84

**a.** Figure 6.51 shows that $P(22{,}000 \le X \le 24{,}000) = 0.6449902243 \approx 0.6450$.

**b.** Figure 6.52 shows that the value for $X_1$ such that $P(X \le X_1) = 0.99$ is given by $X_1 = \$25{,}726.34788 \approx \$25{,}726.35$.

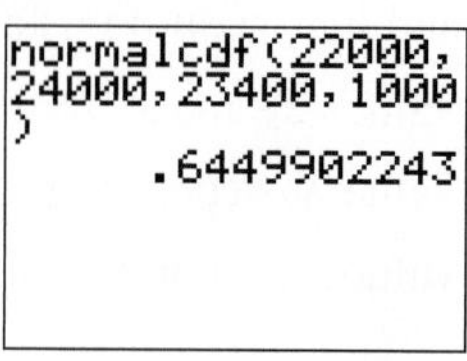


FIGURE 6.51

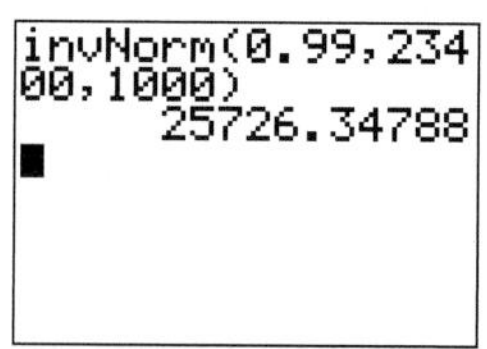


FIGURE 6.52

### Excel

**a.** Excel provides the cumulative probabilities $P(X \le 22{,}000)$ in Figure 6.53 and $P(X \le 24{,}000)$ in Figure 6.54. To find $P(22{,}000 \le X \le 24{,}000)$, we subtract $P(X \le 22{,}000)$ from $P(X \le 24{,}000)$:

$$P(22{,}000 \le X \le 24{,}000) = 0.725746882 - 0.080756659 = 0.644990223$$

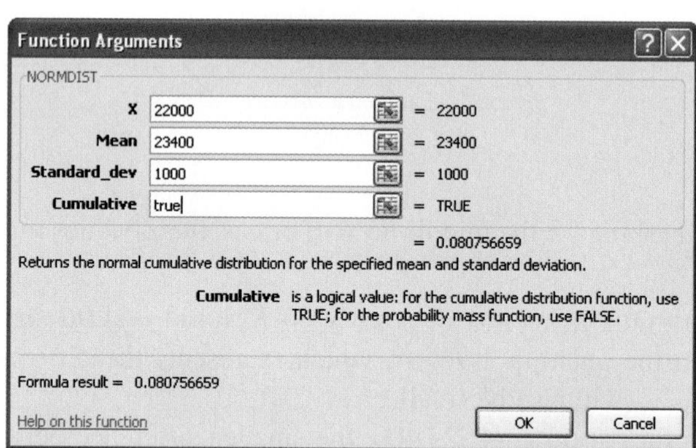


FIGURE 6.53

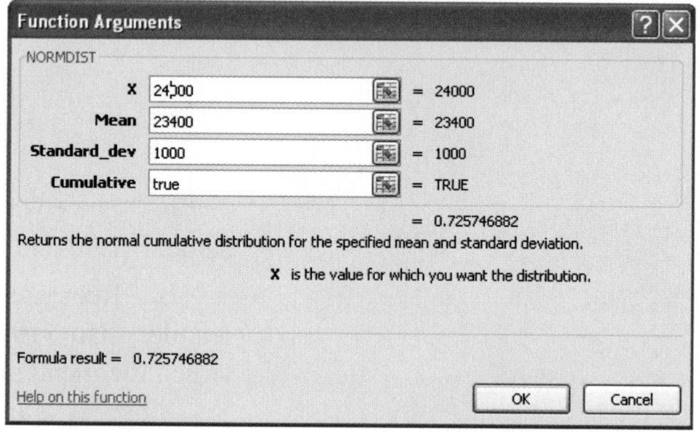


FIGURE 6.54

**b.** Excel provides the result shown in Figure 6.55, $X_1 = \$25{,}726.34787 \approx \$25{,}726.35$.

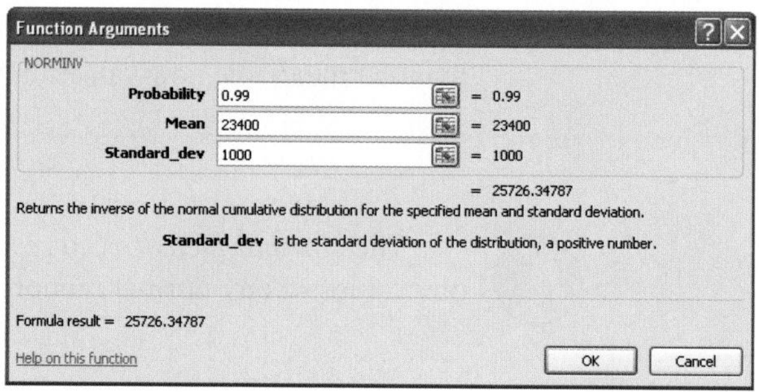


FIGURE 6.55

### Minitab

**a.** Like Excel, Minitab asks you to take the difference of two cumulative probabilities, $P(X \le 22{,}000)$ in Figure 6.56 and $P(X \le 24{,}000)$ in Figure 6.57:

$$P(22{,}000 \le X \le 24{,}000) = 0.725747 - 0.0807567 = 0.6449903 \approx 0.6450$$

**Cumulative Distribution Function**

```
Normal with mean = 23400 and standard deviation = 1000

    x  P( X <= x )
22000    0.0807567
```

FIGURE 6.56

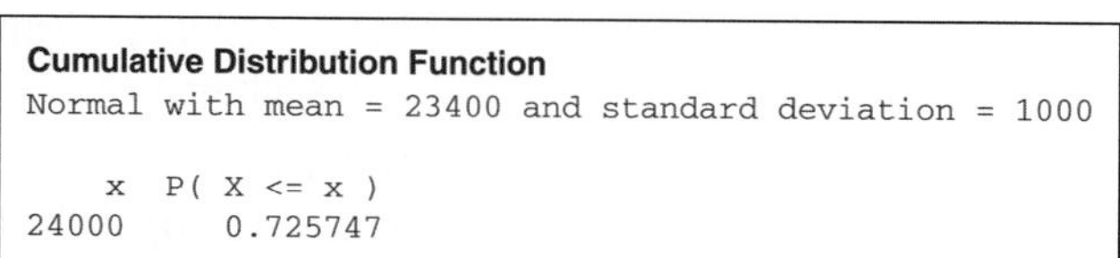

**Cumulative Distribution Function**

```
Normal with mean = 23400 and standard deviation = 1000

    x  P( X <= x )
24000     0.725747
```

FIGURE 6.57

**b.** The results are given in Figure 6.58; $X_1 = \$25{,}726.30$.

**Inverse Cumulative Distribution Function**

```
Normal with mean = 23400 and standard deviation = 1000

P( X <= x )        x
      0.99  25726.3
```

FIGURE 6.58

## 3 The Normal Approximation to the Binomial Probability Distribution

Recall from Section 6.2 that a binomial experiment satisfies the following four requirements: (1) Each trial must have two possible outcomes. (2) There is a fixed number of trials, $n$. (3) The experimental outcomes are independent of each other. (4) The probability of observing a success is the same from trial to trial.

The binomial random variable $X$ represents the number of successes in $n$ trials and thus depends on the sample size $n$ and the probability of success $p$. For a given probability of success $p$, if the sample size $n$ gets large enough, the binomial distribution begins to resemble the normal distribution. Figure 6.59 shows the binomial probability distribution for Example 6.15 (page 277), where Joshua tried to guess ($p = 0.2$) his way to success on a quiz of $n = 4$ questions. The distribution of $X$ = the number of correct responses in Figure 6.59 is clearly not normal.

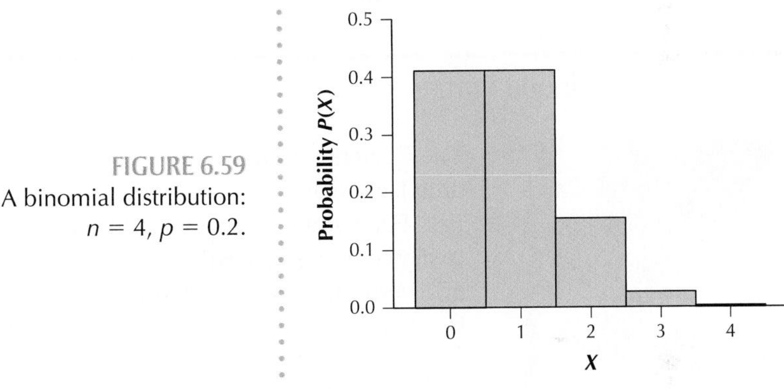


FIGURE 6.59
A binomial distribution: $n = 4, p = 0.2$.

If we increase the sample size to $n = 34$ while retaining the same $p = 0.2$ (Figure 6.60), the distribution of $X$ becomes somewhat more bell-shaped but is still a bit right-skewed. Finally, if we further increase the sample size to $n = 64$ (Figure 6.61), the binomial distribution of $X$, which is discrete, looks like it can be nicely approximated by the normal distribution, which is continuous. We generalize this behavior as follows.

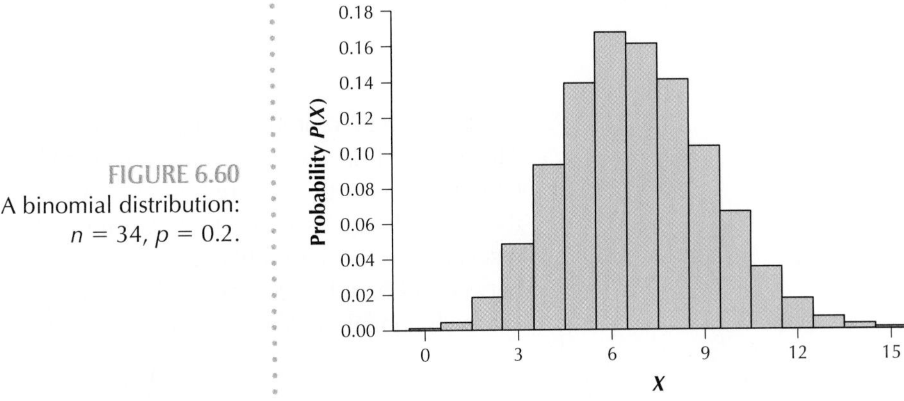


FIGURE 6.60
A binomial distribution: $n = 34, p = 0.2$.

FIGURE 6.61
A binomial distribution: $n = 64, p = 0.2$.

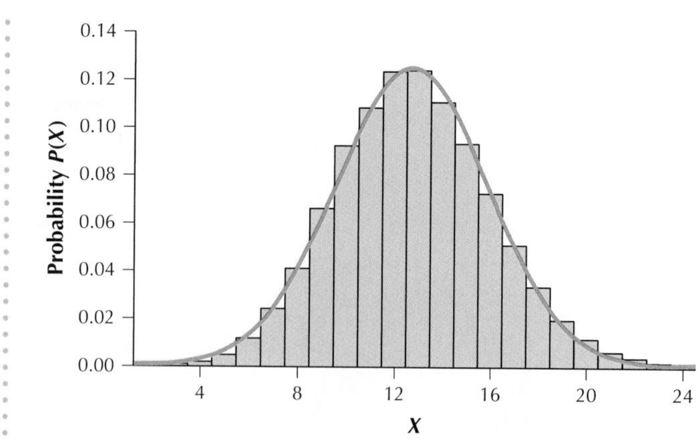


**The Normal Approximation to the Binomial Probability Distribution**

For the binomial random variable $X$ with probability of success $p$ and number of trials $n$: if $n \cdot p \geq 5$ and $n \cdot (1 - p) \geq 5$, the binomial distribution may be approximated by a normal distribution with mean $\mu_X = np$ and standard deviation $\sigma_X = \sqrt{np(1-p)}$.

The *Normal Approximation to the Binomial* applet allows you to choose your own values of $n$ and $p$ and see how changes in these values affect the normal approximation to the binomial distribution.

## Example 6.40 Childhood immunizations

Punchstock/Blend

The Centers for Disease Control and Prevention reported that 20% of preschool children lack required immunizations, thereby putting themselves and their classmates at risk. For a group of $n = 64$ children with $p = 0.2$, the binomial probability distribution is shown in Figure 6.61.

**a.** Verify that this distribution can be approximated by a normal distribution.

**b.** Find the mean and standard deviation of this normal distribution.

### Solution

**a.** The normal approximation is valid if $n \cdot p \geq 5$ and $n \cdot (1 - p) \geq 5$. Substituting $n = 64$ and $p = 0.2$, we get

$$n \cdot p = (64)(0.2) = 12.8 \geq 5 \quad \text{and} \quad n \cdot (1 - p) = (64)(0.8) = 51.2 \geq 5$$

Thus, the normal approximation is valid.

**b.** The mean and standard deviation of the normal distribution are

$$\mu_X = np = (64)(0.2) = 12.8$$

$$\sigma_X = \sqrt{np(1-p)} = \sqrt{(64)(0.2)(0.8)} = 3.2$$

Figure 6.62 reproduces Figure 6.61, with the rectangle for $X = 12$ highlighted. The height of the rectangle represents the binomial probability that exactly 12 of the 64 children lack the required immunizations, that is, $P(X = 12)$. Since the width of the rectangle equals 12.5 – 11.5 = 1, it follows that the area of the rectangle also represents the binomial probability that $X = 12$. Now, the area under the normal curve between 11.5 and 12.5 is approximately equal to this rectangle, which is $P(X = 12)$ for the binomial random variable $X$, with $n = 64$ and $p = 0.2$. That is

$$P(X_{\text{binomial}} = 12) \approx P(11.5 \leq Y_{\text{normal}} \leq 12.5)$$

where $Y_{\text{normal}}$ is the normal random variable from Example 6.40b, with mean $\mu_X = 12.8$ and standard deviation $\sigma_X = 3.2$.

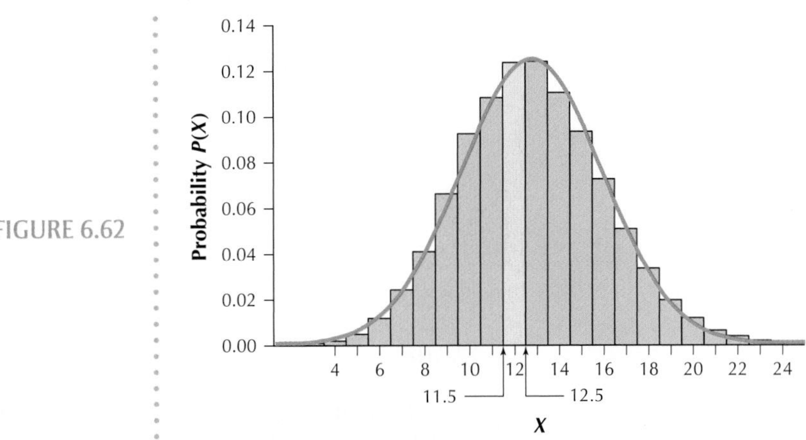


FIGURE 6.62

The 0.5 that we add and subtract from 12 when approximating the binomial distribution with the normal distribution is called the **continuity correction**, since it is an adjustment for approximating a discrete probability with a continuous one. When using the normal approximation to the binomial, the analyst must determine which binomial rectangles are included and apply the continuity correction accordingly. This is shown in Table 6.8, which provides a listing of several types of binomial probabilities and their normal probability approximations.

Table 6.8 Binomial probabilities and approximate normal probabilities

| Exact binomial probability | Approximate normal probability |
|---|---|
| $P(X_{\text{binomial}} = a)$ | $P(a - 0.5 \leq Y_{\text{normal}} \leq a + 0.5)$ |
| $P(X_{\text{binomial}} \leq a)$ | $P(Y_{\text{normal}} \leq a + 0.5)$ |
| $P(X_{\text{binomial}} \geq a)$ | $P(Y_{\text{normal}} \geq a - 0.5)$ |
| $P(X_{\text{binomial}} < a)$ | $P(Y_{\text{normal}} < a - 0.5)$ |
| $P(X_{\text{binomial}} > a)$ | $P(Y_{\text{normal}} > a + 0.5)$ |
| $P(a \leq X_{\text{binomial}} \leq b)$ | $P(a - 0.5 < Y_{\text{normal}} \leq b + 0.5)$ |
| $P(a < X_{\text{binomial}} < b)$ | $P(a + 0.5 < Y_{\text{normal}} \leq b - 0.5)$ |

## Example 6.41 The normal approximation to the binomial probability distribution

For a group of $n = 64$ preschool children with probability of lack of immunization $p = 0.2$, perform the following approximations.

**a.** Approximate the probability that there are at most 12 children without immunization.

**b.** Approximate the probability that there are more than 12 children without immunization.

### Solution

Once again we have a binomial experiment with $n = 64$ and $p = 0.2$.

**a.** "At most" 12 children means 12 or fewer children. That is, $X = 12$ and $X = 11$ and $X = 10$, and so on; that is, $P(X_{\text{binomial}} \leq 12)$. In this case, we see that $X = 12$ is included in the probability we seek, as shown in Figure 6.63. From Table 6.8, we see that $P(X_{\text{binomial}} \leq 12)$ is of the form $P(X_{\text{binomial}} \leq a)$. Thus, our continuity correction takes the form $P(Y_{\text{normal}} \leq a + 0.5)$, where we add 0.5 to 12, so that

$$P(X_{\text{binomial}} \leq 12) \approx P(Y_{\text{normal}} \leq 12.5)$$

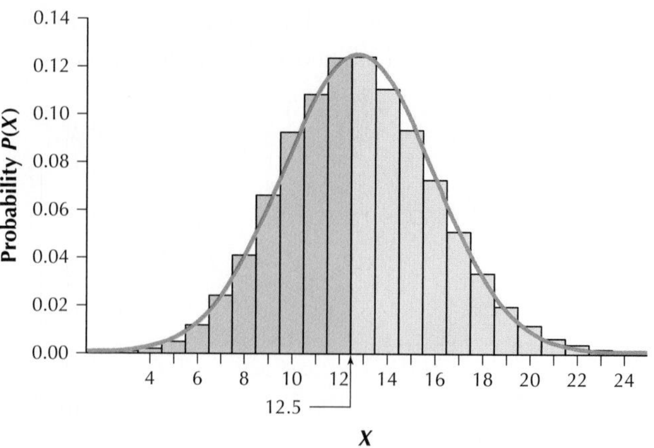


FIGURE 6.63

Recall that $\mu_X = 12.8$ and $\sigma_X = 3.2$. We use the TI-83/84, as shown in Figures 6.64 and 6.65, and find that the probability that at most 12 children lack immunization is $0.4626221269 \approx 0.4626$.

FIGURE 6.64

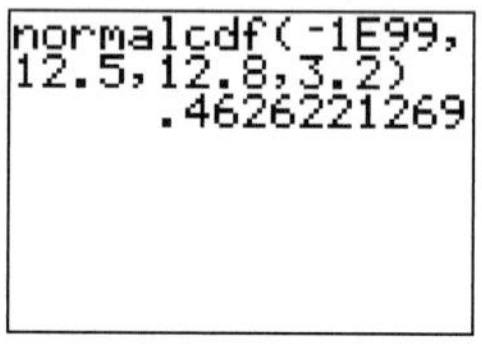

```
normalcdf(-1E99,
12.5,12.8,3.2)
          .4626221269
```

FIGURE 6.65

**b.** "More than" 12 children means $X = 13$ and $X = 14$, and so on. In other words, $X = 12$ is not included, as shown in Figure 6.66. That is, we want $P(X_{\text{binomial}} > 12)$. From Table 6.8, we see that $P(X_{\text{binomial}} > 12)$ is of the form $P(X_{\text{binomial}} > a)$. Thus, our continuity correction takes the form $P(Y_{\text{normal}} > a + 0.5)$, where we add 0.5 to 12, so that

$$P(X_{\text{binomial}} > 12) \approx P(Y_{\text{normal}} > 12.5)$$

Since the green area in Figure 6.66 is the complement of the green area in Figure 6.63, we can find the answer like this:

$$P(X_{\text{binomial}} > 12) \approx P(Y_{\text{normal}} > 12.5) = 1 - P(Y_{\text{normal}} \leq 12.5) = 1 - 0.4626 = 0.5374$$

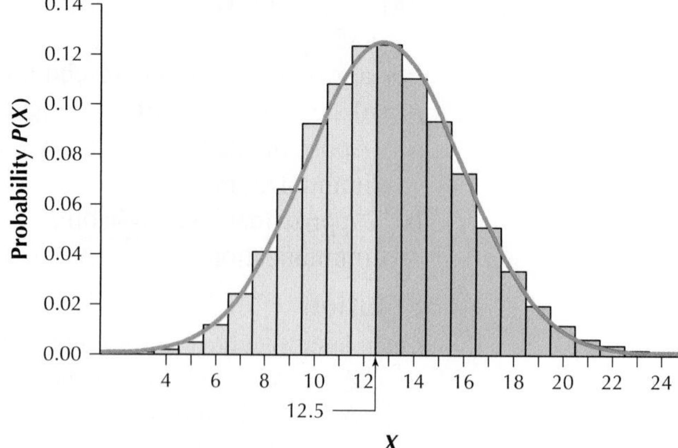


FIGURE 6.66

The probability that more than 12 preschool children will not have the required immunizations is 0.5374. ■

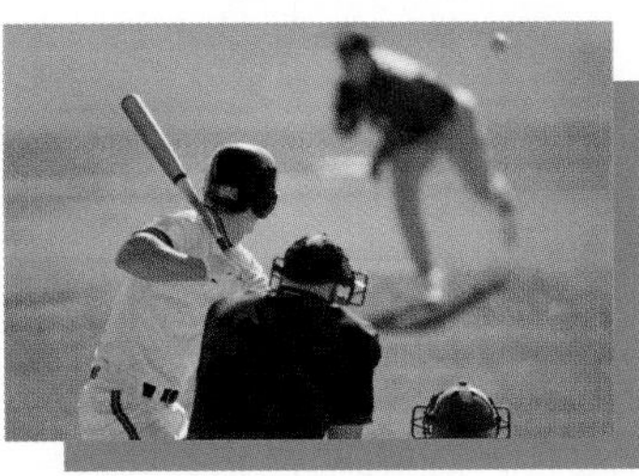
David Madison/Digital Vision/ Getty Images

## CASE STUDY | Be Careful What You Assume: Major League Baseball Salaries

The following are the summary characteristics for the population of all baseball salaries, as of opening day, 2004:

$$\text{population mean salary } \mu = \$2{,}482{,}534$$

$$\text{population standard deviation of salary } \sigma = \$3{,}520{,}375$$

Assume that the baseball salaries are normally distributed. Suppose MLB players and owners are huddled around their statistics texts, negotiating the new MLB contract.

**Problem 1.** In order to decrease operating costs, MLB owners have proposed to "cap" a team's top salary at the 75th percentile of all current salaries. Find the 75th percentile of MLB player salaries on the assumption that the salaries are normally distributed.

### Solution

Because of the equivalence of corresponding areas under the $X$ curve and the $Z$ curve, we look up 0.75 in the inside of the $Z$ table and work backward. It turns out that the closest $Z$ is $Z = 0.67$. Reexpressing in terms of salaries, we get

$$X_1 = Z\sigma + \mu = (0.67)(3{,}520{,}375) + 2{,}482{,}534 \approx \$4{,}841{,}185$$

The owners feel that approximately \$4.841 million per year is plenty of compensation to play baseball.

**Problem 2.** The players' union states that they will agree to the cap in Problem 1 only if management adds a bonus to the salaries of the lowest-paid players, as measured by the 10th percentile. Find the 10th percentile of MLB player salaries on the assumption that the salaries are normally distributed.

### Solution

To find the 10th percentile, we look up 0.10 in the inside of the $Z$ table and work backward. Here, we get $Z = -1.28$. Reexpressing in terms of salaries, we get

$$X_1 = Z\sigma + \mu = (-1.28)(3{,}520{,}375) + 2{,}482{,}534 \approx -\$2{,}023{,}546$$

What went wrong? How could a salary be negative? Figure 6.67 provides a histogram (in green) of the distribution of actual player salaries, overlaid with the assumed normal distribution (in orange) of player salaries. Both are centered at \$2,482,534 (red triangle), and both have standard deviation \$3,520,375, but that is about all the two distributions have in common. Clearly, we cannot assume that the salaries are normally distributed.

The salaries are in fact quite remarkably right-skewed and decidedly non-normal. Our finding that the 10th percentile salary is negative is a direct result of the inappropriate assumption of normality. *Be careful what you assume*. We provide methods for assessing the normality of a data set in Chapter 7.

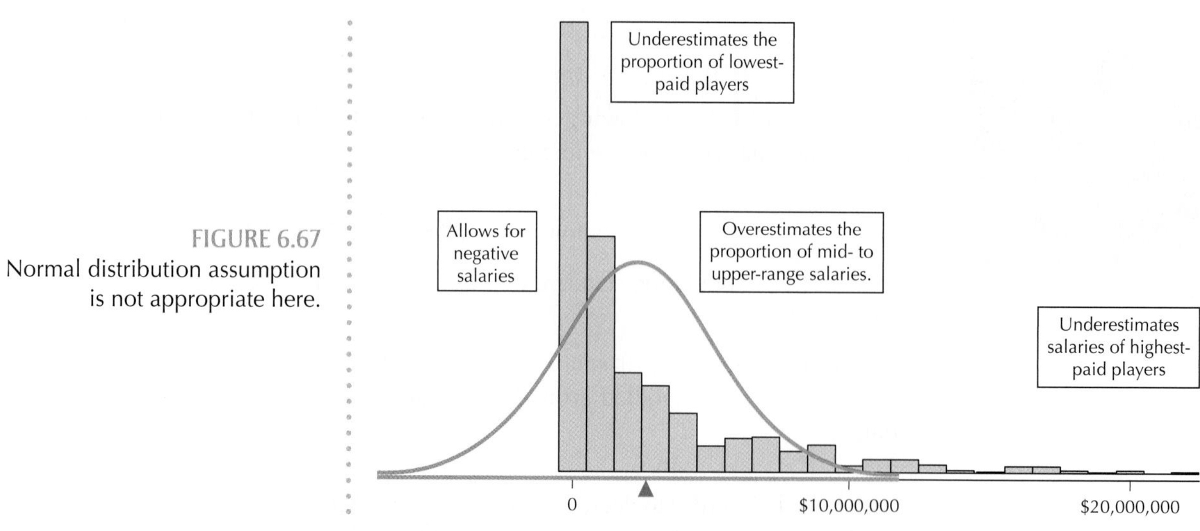
Underestimates the proportion of lowest-paid players
Allows for negative salaries
Overestimates the proportion of mid- to upper-range salaries.
Underestimates salaries of highest-paid players
0
$10,000,000
$20,000,000

FIGURE 6.67
Normal distribution assumption is not appropriate here.

# STEP-BY-STEP TECHNOLOGY GUIDE: Finding Areas, Probabilities, and Percentiles for Any Normal Distribution

## TI-83/84

### Finding Areas or Probabilities for Any Normal Distribution

***Step 1*** Press **2nd**, then **DISTR** (the **VARS** key).
***Step 2*** Press **2** to choose **normalcdf(**.
***Step 3*** On the home screen, enter the smaller value of *X*, comma, the larger value of *X*, comma, the mean of *X*, comma, the standard deviation of *X*, then close parenthesis. See Figure 6.51 (page 316).
***Step 4*** Press **ENTER**.

*Note:* When finding the area to the right of a value of *X*, use **1E99** as the larger value. When finding the area to the left of a value of *X*, use **-1E99** as the smaller value. Also, the shortcut for using the standard normal distribution is to specify only the lower and higher values of *X*. If you enter only two values, the calculator assumes you want the standard normal distribution.

### Finding Percentiles for Any Normal Distribution

***Step 1*** Press **2nd**, then **DISTR** (the **VARS** key).
***Step 2*** Press **3** to choose **invNorm(**.
***Step 3*** On the home screen, enter the probability value or area, then the mean of *X*, then the standard deviation of *X*, then close parenthesis. See Figure 6.52 (see page 316).
***Step 4*** Press **ENTER**.

*Note:* A shortcut for finding standard normal percentiles is to enter only the value of *X* for this function, in which case the calculator assumes you want the standard normal distribution.

## EXCEL

### Finding Areas or Probabilities for Any Normal Distribution

***Step 1*** Select cell **A1** and click the **Insert Function** icon $F_x$.
***Step 2*** In the **Search for a function**, type **NORMDIST**, click **GO**, then **OK**.
***Step 3*** For **X**, enter the *X*-value that you want to find the probability for. For **Mean,** enter the value of $\mu$. For **Standard_dev**, enter the value of $\sigma$. For **Cumulative**, always enter **true**. Click **OK**. See Figure 6.53 (page 316).
***Step 4*** Excel provides the cumulative probability, $P(X \leq X_1)$ (see Example 6.39). If you need to find $P(X > X_1)$, subtract the result from 1. If you need to find $P(X_1 \leq X \leq X_2)$, find the two cumulative probabilities, and subtract the lesser from the greater, as in Example 6.39 .

### Finding Percentiles for Any Normal Distribution

***Step 1*** Select cell **A1** and click the **Insert Function** icon $f_x$ .
***Step 2*** In the **Search for a function**, type **NORMINV**, click **GO**, then **OK**.
***Step 3*** For **Probability**, enter the desired percentile in decimal form (for example, 0.99). For **Mean**, enter the value of $\mu$. For **Standard_dev**, enter the value of $\sigma$. Click **OK**. See Figure 6.55 (page 316).

## MINITAB

### Finding Areas or Probabilities for Any Normal Distribution

***Step 1*** Click **Calc > Probability Distributions > Normal**.
***Step 2*** Select **Cumulative Probability**, enter the mean $\mu$ and standard deviation $\sigma$.
***Step 3*** Select **Input Constant**, enter the $X$-value that you want to find the probability for.
***Step 4*** Minitab provides the cumulative probability, $P(X \le X_1)$ (see Example 6.39 on page 315). If you need to find $P(X > X_1)$, subtract the result from 1. If you need to find $P(X_1 \le X \le X_2)$, find the two cumulative probabilities, and subtract the lesser from the greater, as in Example 6.39.

### Finding Percentiles for Any Normal Distribution

***Step 1*** Click **Calc > Probability Distributions > Normal**.
***Step 2*** Select **Inverse Cumulative Probability**, and enter the mean $\mu$ and standard deviation $\sigma$.
***Step 3*** Select **Input Constant**. For the constant, enter the desired percentile in decimal form (for example, 0.99).

## Section 6.5 SUMMARY

Section 6.5 showed how to solve normal probability problems for any conceivable normal random variable by first standardizing and then using the methods of Section 6.4. Methods for finding probabilities for a given value of the normal random variable $X$ were discussed.

For any normal probability distribution, values of $X$ can be found for given probabilities using the "backward" methods of Section 6.4 and the formula $X_1 = Z\sigma + \mu$.

The normal distribution can be used to approximate binomial probabilities when $np \ge 5$ and $np(1 - p) \ge 5$.

## Section 6.5 EXERCISES

### CLARIFYING THE CONCEPTS

**1.** What does the word *standardize* mean? Explain how we use standardization in solving normal probability problems.
**2.** When finding a data value for a specified probability, explain why we can't just report the $Z$-value but must transform back to the original normal distribution.
**3.** Think of a data distribution that is clearly non-normal. What would happen if you assumed that the distribution was normal and reported probabilities based on that assumption?
**4.** How do we determine whether it is appropriate to use the normal approximation to the binomial probability distribution?

### PRACTICING THE TECHNIQUES

For Exercises 5–16, assume that the random variable $X$ is normally distributed with mean $\mu = 70$ and standard deviation $\sigma = 10$. Draw a graph of the normal curve with the desired probability and value of $X$ indicated. Find the indicated probabilities by standardizing $X$ to $Z$.

**5.** $P(X > 70)$
**6.** $P(X > 80)$
**7.** $P(X < 80)$
**8.** $P(X > 95)$
**9.** $P(X \ge 95)$
**10.** $P(X \ge 60)$
**11.** $P(X \ge 55)$
**12.** $P(60 < X < 100)$
**13.** $P(60 \le X \le 100)$
**14.** $P(90 \le X \le 100)$
**15.** $P(90 \le X \le 91)$
**16.** $P(60 \le X \le 70)$

For Exercises 17–22, assume that the random variable $X$ is normally distributed with mean $\mu = 70$ and standard deviation $\sigma = 10$. Draw a graph of the normal curve with the desired probability and value of $X$ indicated. Find the indicated values of $X$ using the formula $X_1 = Z\sigma + \mu$.

**17.** The value of $X$ larger than 95% of all $X$-values (that is, the 95th percentile)
**18.** The 10th percentile
**19.** The 90th percentile (note that the 10th and 90th percentiles are symmetric values of $X$ that contain the central 80% of the area under the curve between them)
**20.** The two symmetric values of $X$ that contain the central 90% of $X$ between them
**21.** The two symmetric values of $X$ that contain the central 95% of $X$ between them
**22.** The two symmetric values of $X$ that contain the central 99% of $X$ between them

For Exercises 23–28, determine whether it is appropriate to use the normal approximation to the binomial probability distribution.

**23.** $X$ is a binomial random variable with $n = 10$ and $p = 0.5$
**24.** $X$ is a binomial random variable with $n = 8$ and $p = 0.5$

**25.** $X$ is a binomial random variable with $n = 10$ and $p = 0.4$
**26.** $X$ is a binomial random variable with $n = 13$ and $p = 0.4$
**27.** $X$ is a binomial random variable with $n = 45$ and $p = 0.1$
**28.** $X$ is a binomial random variable with $n = 50$ and $p = 0.1$

For Exercises 29–36, let $X$ be a binomial random variable with $n = 40$ and $p = 0.5$. Use the normal approximation to find the following probabilities.
**29.** $P(X = 20)$
**30.** $P(X \geq 20)$
**31.** $P(X > 20)$
**32.** $P(X \leq 20)$
**33.** $P(X < 20)$
**34.** $P(18 \leq X \leq 22)$
**35.** $P(18 < X < 22)$
**36.** $P(18 \leq X < 22)$

For Exercises 37–44, let $X$ be a binomial random variable with $n = 120$ and $p = 0.1$. Use the normal approximation to find the following probabilities.
**37.** $P(X = 10)$
**38.** $P(X \geq 10)$
**39.** $P(X > 10)$
**40.** $P(X \leq 8)$
**41.** $P(X < 8)$
**42.** $P(9 \leq X \leq 11)$
**43.** $P(9 < X < 11)$
**44.** $P(9 < X \leq 11)$

## APPLYING THE CONCEPTS

**Finding Probabilities and Data Values for Any Normal Distribution.** The concepts exercises for the normal approximation to the binomial distribution begin at Exercise 62.

**45. Hungry Babies.** Six-week-old babies consume a mean of $\mu = 15$ ounces of milk per day, with a standard deviation $\sigma$ of 2 ounces. Assume that the distribution is normal.

- **a.** Find the probability that a randomly chosen baby consumes between 17 and 19 ounces of milk per day.
- **b.** Find the probability that a randomly selected baby consumes less than 11 ounces of milk per day.

**46. Trading Volume.** The Associated Press reports that the mean trading volume for equity and index options contracts was 3.6 million in July 2007. Assume that the distribution is normal with mean $\mu = 3.6$ (in millions) and standard deviation $\sigma = 0.5$ (in millions).

- **a.** Find the probability that a randomly selected trading volume exceeds 4.1 million.
- **b.** Find the trading volume greater than 90% of all trading volumes. (*Hint:* Use the formula $X_1 = Z\sigma + \mu$.)

**47. Windy Frisco.** The mean wind speed in San Francisco is 13.6 mph in July, according to the U.S. National Oceanic and Atmospheric Administration. Suppose that the distribution of the wind speed in July in San Francisco is normal with mean $\mu = 13.6$ mph and standard deviation $\sigma = 6$ mph.

- **a.** Find the probability that a randomly selected wind speed in San Francisco in July is 7.2 mph or less.
- **b.** What proportion of days in July have wind speed greater than 20 mph?
- **c.** What percentage of days in July have wind speed between 15 and 20 mph?
- **d.** Tours to Alcatraz Island are canceled if the day is too windy, specifically if the wind speed is higher than 99% of all other wind speeds in July. Find the cutoff wind speed.
- **e.** Suppose that a particular day in July has no wind at all. Should this be considered unusual? On what do you base your answer?

**48. Magic McGrady.** In the 2003–2004 National Basketball Association regular season, Tracy McGrady of the Orlando Magic led the league in points per game with 28. Suppose that McGrady's points per game follow a normal distribution with mean $\mu = 28$ and standard deviation $\sigma = 8$.

- **a.** Find the probability that McGrady will score more than 30 points in a randomly chosen game.
- **b.** In what percentage of games does McGrady score fewer than 10 points?
- **c.** In what proportion of games does McGrady score between 10 and 20 points?
- **d.** Everybody gets the cold hand now and then. Find McGrady's points in a game in which he scores fewer points than in 95% of his games.
- **e.** In one particular game, McGrady scored 40 points. Is this unusual? On what do you base your answer?

**49. Viewers of *60 Minutes*.** Nielsen Media Research reported that, for the week ending July 11, 2004, 12 million viewers watched the television show *60 Minutes*. Suppose that the distribution of viewers of *60 Minutes* is normal with mean $\mu = 12$ million and standard deviation $\sigma = 4$ million.

- **a.** Find the probability that fewer than 10 million people will watch *60 Minutes*.
- **b.** What percentage of broadcasts are watched by more than 18 million viewers?
- **c.** What proportion of broadcasts are watched by between 10 million and 11 million viewers?
- **d.** Find the number of viewers that represents the 75th percentile.
- **e.** On one particular night, 24 million people watched *60 Minutes*. Is this unusual? On what do you base your answer?

**50. Math Scores.** The National Center for Education Statistics reports that mean scores on the standardized math test for 12th-graders in 2000 declined significantly from previous years. The mean score in 2000 was $\mu = 147$; assume $\sigma = 10$.

- **a.** Find the probability that the test score of a randomly selected 12th-grader was greater than 150.
- **b.** What percentage of test scores were below 125?
- **c.** What proportion of test scores were between 155 and 160?
- **d.** Suppose students who scored at the 5th percentile or lower could not graduate. Find the 5th percentile test score.

e. Suppose that you know someone who scored 118 on the test. Is this unusual? On what do you base your answer?

**51. Long-Lived Tortoise.** The Philadelphia Zoo reports that the mean life span of a Galapagos tortoise is 100 years. Assume that their life span follows a normal distribution with mean $\mu = 100$ years and standard deviation $\sigma = 15$ years.

**a.** Find the probability that a randomly chosen tortoise will live longer than 70 years.
**b.** What percentage of tortoises live longer than 125 years?
**c.** What proportion of tortoises live for between 115 and 145 years?
**d.** Suppose that a particular zoo claims to have a tortoise older than 99% of all such tortoises. Find the age of this tortoise.
**e.** A particular tortoise is 150 years old. Is this unusual? On what do you base your answer?

**52. Hospital Patient Length of Stays.** A study of Pennsylvania hospitals showed that the mean patient length of stay in 2001 was 4.87 days with a standard deviation of 0.97 days. Assume that the distribution of patient length of stays is normal.

**a.** Find the probability that a randomly selected patient has a length of stay of greater than 7 days.
**b.** What percentage of patient lengths of stay are less than 4 days?
**c.** What proportion of patient lengths of stay are between 3 and 5 days?
**d.** Find the 50th percentile of patient lengths of stay. What is the relationship between the mean and the median for normal distributions?
**e.** A particular patient had a length of stay of 8 days. Is this unusual? On what do you base your answer?

**53. Tobacco-Related Deaths.** The World Health Organization states that tobacco is the second leading cause of death in the world. Every year, a mean of 5 million people die of tobacco-related causes. Assume that the distribution is normal with $\mu = 5$ (in millions) and $\sigma = 2$ (in millions).

**a.** Find the probability that more than 4 million people will die of tobacco-related causes in a particular year.
**b.** Find the probability that fewer than 1 million people will die from tobacco-related causes in a particular year.
**c.** What is the probability that more than 3 million people will die from tobacco-related causes in a particular year?
**d.** Find the 75th percentile of the distribution of tobacco-related deaths.
**e.** In one particular year, 8 million people died from tobacco-related causes. Is this unusual? On what do you base your answer?

**54. Grade Inflation.** Many educators are concerned about grade inflation. One study showed that one low SAT-score high school (with mean combined SAT score = 750) had higher mean grade point average (mean GPA = 3.6) than a high SAT-score school (with mean combined SAT score = 1050 and mean GPA = 2.6).[7] Assume that GPA at the high-SAT-score school is normally distributed with mean $\mu = 2.6$ and standard deviation $\sigma = 0.5$.

**a.** What is the probability that a randomly selected student at the high-SAT-score school has a GPA of 3.6 or higher?
**b.** What percentage of students at the school have a GPA of 3.0 or higher?
**c.** What proportion of students have GPAs between 2.0 and 2.5?
**d.** Suppose that students with the lowest 5% of GPAs get put on academic probation. Find the cutoff grade.
**e.** One particular student had a GPA of 4.0. Is this unusual? Explain your answer.

**55. Patriots Passing.** ESPN reports that the 2003 New England Patriots football team had a mean of 11.4 yards passing per play. Assume that the distribution of yards passing per play is normal with $\mu = 11.4$ yards and $\sigma = 6$ yards.

**a.** What is the probability that a randomly chosen passing play loses yardage?
**b.** What percentage of passing plays gain more than 20 yards?
**c.** What proportion of passing plays gain between 5 and 10 yards?
**d.** Between which two yardage amounts do the central 95% of passing plays fall?
**e.** One particular passing play went for 35 yards. Is this unusual? Explain your answer.

**56. Stock Shares Traded.** The mean number of shares traded on the New York Stock Exchange in June 2004 was 2.4 billion per day. Assume that the distribution of shares traded is normal with $\mu = 2.4$ and $\sigma = 0.6$ (both in billions of shares).

**a.** Find the probability that the number of shares traded on a randomly selected day falls below 0.6 billion.
**b.** What percentage of days finds a volume of shares traded of more than 2.5 billion shares?
**c.** What proportion of days finds the volume of shares traded between 1 billion and 2 billion?
**d.** A slow trading day has fewer shares traded than 99% of all other days. Find the number of shares traded that represents this amount.
**e.** On one particular day, 3 billion shares were traded. Is this unusual? Explain your answer.

**57. Sociology Salaries.** Sociology graduates, upon entering the workforce, earn a mean salary of $30,000 with a standard deviation of $4000.

**a.** Jessica is an honors sociology student. Upon graduation she would like to take a job that starts at $40,000. What is the probability that a randomly chosen salary exceeds $40,000?
**b.** Jeremy squeezed through his sociology major with a C– average. He would take any job he could lay hands on, as long as it's not tossing fries, which he's

been doing for four years. What is the probability that a randomly chosen salary is less than $22,000?

**c.** Jolene would like to earn between $27,000 and $38,000 when she graduates. What is the probability that a randomly chosen salary lies within this range?

**d.** Josefina didn't pull those all-nighters for nothing. She would like to make between $32,000 and $39,000. What is the probability that a randomly chosen salary lies within this range?

**e.** Jamal looks forward to spending that big salary after graduation. Anything over $27,000 would do just fine. What is the probability that a randomly chosen salary exceeds $27,000?

**f.** Jillian struggled through college, having to care for her children while holding down a full-time job. She hopes that her salary is at least the 10th percentile. Find the 10th percentile of sociology salaries.

**g.** Jung is not sure how much he will make upon graduation. He would be happy if his salary was in the middle 60% of salaries. Find the two salaries that are the cutoff values for the middle 60% of salaries.

**58. Emotional Maturity.** An emotional maturity test was administered at a local university hospital. The scores are normally distributed with a mean of 100 and a standard deviation of 10.

**a.** Find the probability that a randomly chosen subject scored better than average.

**b.** Find the probability that a randomly chosen subject scored lower than 72.

**c.** Find the probability that a randomly chosen subject scored between 95 and 101.

**d.** Find the scores that mark off the middle 95% of scores.

**59. Calories per Gram.** Figure 6.68 shows the number of calories per gram for 961 food items. Assume that the population mean calories per gram is 2.25 with a standard deviation of 2.

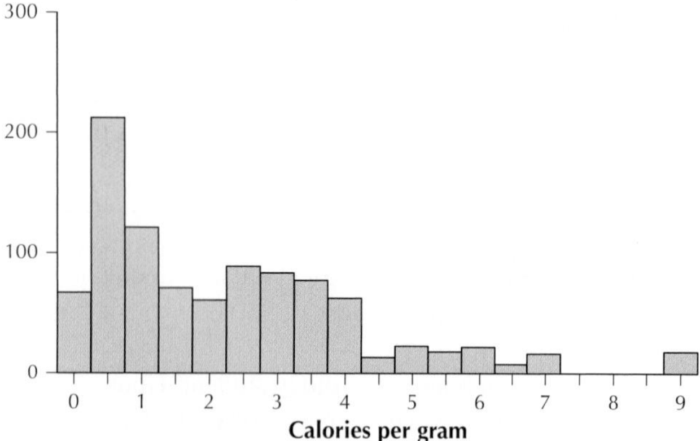


Figure 6.68

**a.** Assuming that the data follow a normal distribution, what is the 5th percentile of calories per gram?

**b.** Comment on whether your answer from **(a)** makes any sense.

**c.** The actual 5th percentile for this data set is 0.2 calories per gram. Looking at the histogram, does this make more sense than your answer from **(a)**?

**d.** Why is your answer in **(a)** wrong?

APPLET Use the *Normal Density Curve* applet for Exercises 60 and 61.

**60.** Use the applet to find the answers to the following exercises from this section.

**a.** Exercise 57a

**b.** Exercise 57b

**c.** Exercise 57c

**d.** Exercise 57d

**61.** Use the applet to find the answers to the following exercises from this section.

**a.** Exercise 58b

**b.** Exercise 58c

**c.** Exercise 58d

**Normal Approximation to the Binomial Distribution**

**62. Gas Tax.** A *New York Times*/CBS News Poll conducted in April 2007 reported that 64% of Americans would favor an increased federal tax on gasoline if it would reduce dependence on foreign oil. For a sample of 200 Americans, approximate the following probabilities.

**a.** Exactly 128 would favor such a tax.

**b.** At least 128 would favor such a tax.

**c.** More than 128 would favor such a tax.

**d.** Between 120 and 130 would favor such a tax.

**63. Dress Casual.** A survey conducted in 2007 by the Society for Human Resource Management found that the number of businesses allowing employees to "dress casually" every day has dropped from 53% in 2002 to 38% in 2007. Suppose we have a sample of 50 businesses from 2002 and a sample of 50 businesses from 2007. Approximate the following probabilities.

**a.** At least 25 businesses allowed casual dress every day in 2002.

**b.** At least 25 businesses allowed casual dress every day in 2007.

**c.** Fewer than 15 businesses allowed casual dress every day in 2002.

**d.** Fewer than 15 businesses allowed casual dress every day in 2007.

**64.** A survey found that 19% of respondents in New Orleans rated the overall response by government and volunteer agencies to major hurricanes in the past three years as good or excellent, while 57% of those living in other areas did so.[8] Suppose that we have a sample of 100 people living in New Orleans and 100 people living in other areas. Approximate the following probabilities.

**a.** At least 30 of the respondents living in New Orleans rated the response as good or excellent.

**b.** At least 30 of the respondents living in other areas rated the response as good or excellent.

**c.** Fewer than 20 of the respondents living in New Orleans rated the response as good or excellent.

**d.** Fewer than 20 of the respondents living in other areas rated the response as good or excellent.

**65.** A survey found that only 9% of Americans were "very confident" that the U.S. government is prepared to handle a major outbreak of an infectious disease.[9] Suppose that we have a sample of 100 Americans. Approximate the following probabilities.

**a.** Exactly 9 Americans are very confident.

**b.** At least 9 Americans are very confident.

**c.** More than 9 Americans are very confident.

**d.** At most 9 Americans are very confident.

**e.** Fewer than 9 Americans are very confident.

**66.** A September 2006 *New York Times*/CBS News Poll found that 69% of New York City residents were personally very concerned about a terrorist attack occurring in the area in which they live, compared with 22% of Americans as a whole. Suppose we have a sample of 100 New York City residents and 100 non–New Yorkers. Approximate the following probabilities.

**a.** More than 50 New Yorkers are so concerned.

**b.** More than 50 non–New Yorkers are so concerned.

**c.** Fewer than 25 New Yorkers are so concerned.

**d.** Fewer than 25 non–New Yorkers are so concerned.

APPLET Use the *Normal Approximation to the Binomial Distribution* applet for Exercises 67 and 68.

**67.** Select $n$ (trials) = 10 and $p$ (probability) = 0.2. The rectangles represent the binomial probabilities and the area under the curve represents the normal probabilities.

**a.** For $n = 10$ and $p = 0.2$, is there a tight fit between the rectangles and the curve?

**b.** What does this mean for whether the normal approximation should be used for a binomial distribution with $n = 10$ and $p = 0.2$?

**c.** Verify whether the conditions are met for applying the normal approximation.

**68.** Select $n$ (trials) = 63 and $p$ (probability) = 0.2. The rectangles represent the binomial probabilities and the area under the curve represents the normal probabilities.

**a.** For $n = 63$ and $p = 0.2$, is there a tight fit between the rectangles and the curve?

**b.** What does this mean for whether the normal approximation should be used for a binomial distribution with $n = 63$ and $p = 0.2$?

**c.** Verify whether the conditions are met for applying the normal approximation.

## Try This in Class! In-class activities to enhance your understanding of statistics

### PROBABILITY DISTRIBUTIONS AND A DECK OF CARDS

Bring in a deck of cards and perform the following experiment.

1. Take out the 12 face cards (kings, queens, and jacks).
2. Construct the probability distribution of the face value of the remaining cards (ace $\rightarrow X = 1$, deuce $\rightarrow X = 2$, and so on).
3. Find the population mean $\mu$ and standard deviation $\sigma$ of this probability distribution.
4. Designate one student as shuffler, another student as selector. Have the selector draw one card at random from the abridged deck, observe it, record the value of $X$, and replace it.
5. The shuffler then shuffles the cards. The selector then selects another card at random, until 20 cards have been sampled with replacement.
6. Using the relative frequencies from Step 5, construct a relative frequency distribution of your observed face values. Is it very different from the population distribution from Step 2?
7. Calculate the mean and standard deviation of your relative frequency distribution. Are they very different from the population values calculated in Step 3? How would you define "very different"?

### A BINOMIAL PROBABILITY ACTIVITY

There is a famous problem in binomial probability. Which parents would wind up with the greater proportion of boy babies: the parents who simply stopped having babies after a certain number of babies or the parents who had babies until they had a boy? It might seem that the answer is that the parents who stopped after the first boy would have more boys, but the answer is that both sets of parents could expect to have the same proportion of boys. You will use simulation to try to confirm this. Assume that the probability that a randomly chosen newborn is a boy is 50%. Then, clearly, the parents who simply stopped having babies after a certain number of babies could expect to have the same proportion (long-run probability!) of boys and girls. Now, take a single die and define the results of your rolls as follows: an odd-number result will represent a boy, and an even-number result will represent a girl.

1. Have your group roll the die. If you get a girl the first time, record it and roll again. Continue rolling and recording until you get a boy. That ends that particular trial.
2. Repeat until you have a large number (say, 30) of trials.
3. Then find the proportions of boys and girls in your 30 trials (representing 30 reproducing families). Your results shouldn't be too far away from a 50-50 split between boys and girls.

## CHAPTER 6 FORMULAS AND VOCABULARY

### SECTION 6.1

• **Continuous random variable** (p. 260). Takes infinitely many values, with no gap between the values. Because there are infinitely many values, the values of a continuous random variable form an interval on the number line.

• **Discrete random variable** (p. 260). Can take either a finite or a countable number of values. Since these values may be written as a list of numbers, each value can be graphed as a separate point on a number line, with space between each point.

• **Expected value, or expectation, of a random variable *X*** (p. 266). The mean $\mu$ of $X$, denoted $E(X)$. This definition holds for both discrete and continuous random variables.

• **Mean $\mu$ of a discrete random variable** (p. 265). Multiply each possible outcome by its probability, and add up the resulting products. That is, $\mu = \Sigma[X \cdot P(X)]$.

• **Probability distribution of a discrete random variable** (p. 262). A listing of all the values that the random variable can assume, together with the probability associated with each value. The probability distribution can take the form of a table, graph, or formula.

• **Random variable** (p. 259). A variable whose values are determined by chance.

• **Requirements for the probability distribution of a discrete random variable** (p. 262). $\Sigma P(X) = 1$ and $0 \le P(X) \le 1$.

• **Variance and standard deviation of a discrete random variable** (pp. 267–268).

**Definition formulas:**

$$\sigma^2 = \sum[(X - \mu)^2 \cdot P(X)]$$

$$\sigma = \sqrt{\sum[(X - \mu)^2 \cdot P(X)]}$$

**Computational formulas:**

$$\sigma^2 = \sum[X^2 \cdot P(X)] - \mu^2$$

$$\sigma = \sqrt{\sum[X^2 \cdot P(X)] - \mu^2}$$

### SECTION 6.2

• **Binomial experiment** (p. 273). An experiment that satisfies each of the following four criteria.

1. Each trial of the experiment has only two possible outcomes (or is defined in such a way that the number of outcomes is reduced to two). One outcome is considered a *success* and the other a *failure*.
2. There is a fixed number of trials, known in advance of the experiment.
3. The experimental outcomes are independent of each other.
4. The probability of observing a success remains the same from trial to trial.

• **Binomial probability distribution formula** (p. 277). The probability of observing exactly $X$ successes in $n$ trials of a binomial experiment is $P(X) = ({}_nC_X)\, p^X (1 - p)^{n-X}$.

• **Factorial (!)** (p. 275). $n! = n(n - 1)(n - 2) \ldots (2)(1)$.

• **Mean, variance, and standard deviation of a binomial random variable *X*** (p. 280). For a binomial experiment with $n$ trials and probability of success $p$:

| | |
|---|---|
| Mean (or expected value): | $\mu = n \cdot p$ |
| Variance: | $\sigma^2 = n \cdot p \cdot (1 - p)$ |
| Standard deviation: | $\sigma = \sqrt{n \cdot p \cdot (1 - p)}$ |

• **Number of combinations** (p. 275). In general, the number of combinations of $n$ objects taken $X$ at a time is denoted by ${}_nC_X$ and is given by ${}_nC_X = \dfrac{n!}{X!(n - X)!}$

### SECTION 6.3

• **Density curve** (p. 288). A graph of the distribution of a continuous random variable. It must follow the Rules of Probability for Continuous Random Variables (p. 288).

• **Law of Total Probability for Continuous Random Variables** (p. 288). For continuous distributions, the total area (which represents probability) under the curve equals 1.

• **Probability for continuous distributions** (p. 288). For continuous distributions, probability is represented by area under the curve above an interval.

• **Properties of the normal density curve (normal curve)** (p. 291). The normal density curve is symmetric about the mean $\mu$, has a highest point at $X = \mu$, and has inflection points at $\mu - \sigma$ and $\mu + \sigma$. The total area under the density curve equals 1. The area under the curve to the left of $\mu$ and the area under the curve to the right of $\mu$ are both equal to 0.5. The normal distribution is defined for values of $X$ that extend indefinitely in both the positive and negative directions. As $X$ moves farther from the mean, the density curve approaches but never quite touches the horizontal axis.

### SECTION 6.4

• **Standard normal (*Z*) distribution** (p. 297). A normal distribution with mean $\mu = 0$ and standard deviation $\sigma = 1$.

• ***Z* table for the standard normal random variable *Z*** (p. 297). Provides probabilities representing areas to the left of a given value of $Z$.

### SECTION 6.5

• **Finding an *X*-value, given a *Z*-value** (p. 313). $X_1 = Z\sigma + \mu$.

• **Normal approximation to the binomial probability distribution** (p. 318). If $n \cdot p \ge 5$ and $n \cdot (1 - p) \ge 5$, the binomial distribution may be approximated by a normal distribution with mean $\mu_X = np$ and standard deviation $\sigma_X = \sqrt{np\,(1 - p)}$.

• **Standardizing a normal random variable** (p. 309). Any random variable can be transformed into the standard normal random variable $Z$ by standardizing using the formula

$$Z = \frac{X - \mu}{\sigma}$$

## CHAPTER 6 REVIEW EXERCISES

### SECTION 6.1

For Exercises 1 and 2, use the given information to construct a probability distribution table and a probability distribution graph. Then answer the questions.

**1. Take me to the Ball Game.** Andrew loves to go to the ballpark to watch Major League Baseball games. If this year is like most years, there is only a 5% chance that he will not catch a single game, 20% that he will see one game, 35% that he will see two games, 25% that he will see three games, and 15% that he will see four games.

- **a.** What is the probability that Andrew will see at least one game this year?
- **b.** What is the most likely number of games that Andrew will see?
- **c.** What is the probability that Andrew will see at most two games this year?

**2. Early Lunch.** Chad has gotten to lunch early and is waiting for his friends to catch up. He figures that the probability that one friend shows up is 25%; two friends, 35%; three friends, 20%; and more than three friends, 5%.

- **a.** What is the probability that no friends show up?
- **b.** What is the probability that more than one friend shows up?
- **c.** What is the least likely outcome?

**3. Investment Profits.** Chang is looking forward to a profitable career as a day trader on Wall Street, after he has completed college. His present investments have the following probability distribution, with $X$ = profit.

| $X$ | −10,000 | 5000 | 10,000 |
|---|---|---|---|
| $P(X)$ | 0.3 | 0.5 | 0.2 |

- **a.** Find the mean, variance, and standard deviation of $X$.
- **b.** Draw a probability distribution graph.
- **c.** Plot the mean of the random variable on the graph from **(b)**. Does this value for the mean make sense as the point where the distribution balances?

**4. Connecticut Lotto.** The Connecticut Lottery Corporation runs a game called Connecticut Classic Lotto. The player picks six different numbers from 1 to 44 and pays $1 to play.

- If your picks match 3 of the 6 numbers chosen (probability 0.02381), you win $2.
- If you match 4 out of 6 (probability 0.001495), you win $50.
- If you match 5 out of 6 (probability 0.0000323), you win $2000.
- If you match all 6 numbers (probability 0.0000001417), you win the jackpot. The jackpot on July 23, 2004, was $2,600,000.
  - **a.** Construct the probability distribution of your winnings. Make sure to include the probability of not winning anything (in effect, losing $1), which equals 1 minus the four probabilities specified above.
  - **b.** Find the expected winnings. Compare this with the price to play.

**5. NCAA Basketball Championships.** The following table lists schools that have won more than one NCAA national championship in men's basketball, as of 2008.

| School | No. of NCAA men's basketball championships |
|---|---|
| UCLA | 11 |
| Kentucky | 7 |
| Indiana | 5 |
| North Carolina | 4 |
| Duke | 3 |
| Kansas | 3 |
| Michigan State | 2 |
| Oklahoma State | 2 |
| San Francisco | 2 |
| Cincinnati | 2 |
| Connecticut | 2 |
| Louisville | 2 |
| North Carolina State | 2 |

- **a.** Convert the table to a probability distribution by defining $X$ to be the number of championships for those schools that have won more than one. Use the relative frequency method to assign probabilities.
- **b.** Construct the graph of the probability distribution.
- **c.** What is the most common number of national championships for those schools that have won more than one?
- **d.** Find the mean number of national championships.

**6. Ages of College Students.** The National Center for Education Statistics published the following table, which provides the number of students attending degree-granting institutions of higher learning in fall 2004 by age group.

| Age group | Number of students (in 1000s) |
|---|---|
| 14–17 | 198 |
| 18–19 | 3671 |
| 20–21 | 3508 |
| 22–24 | 3138 |
| 25–29 | 2280 |
| 30–34 | 1319 |
| 35 or older | 3157 |

- **a.** State two reasons why this table does not represent a proper probability distribution.
- **b.** Define the values of the random variable $X$ = age as the midpoints of the classes given in the table. For example, for the first age group, use $X = 15.5$. For the last age group, use $X = 40$.
- **c.** Assign the probabilities for each value of $X$ using the relative frequency method.
- **d.** Construct the graph of the probability distribution.
- **e.** Find the mean student age.
- **f.** Clearly interpret the meaning of the mean student age.

**7. Ages of College Students.** For the data in the previous exercise:

**a.** find the standard deviation of student age, and
**b.** determine whether a 40-year-old college student would be considered unusual.

## SECTION 6.2

**8. Gestational Diabetes.** Gestational diabetes occurs in 8% of pregnancies, according to the American Diabetes Association. A random sample of 20 pregnancies is taken.

**a.** Find the probability that none of the pregnancies results in gestational diabetes.
**b.** Find the probability that at least 1 of the pregnancies results in gestational diabetes.
**c.** Find the probability that at most 2 of the pregnancies result in gestational diabetes.

**9. Female Students at DuPage.** According to the National Center for Education Statistics, 57% of the students at the College of DuPage in Glen Ellyn, Illinois, are female. A random sample of 10 students from the College of DuPage is taken.

**a.** Find the probability that all 10 of the students are female.
**b.** Find the probability that at most 9 of the students are female.
**c.** Find the probability that at least 8 of the students are female.

**10. Watching the Price of Gas.** The Pew Research Center reports that, for the week of May 28, 2008, 35% of Americans said that the price of gasoline was the news story that they followed more closely than any other news item. A random sample of 15 Americans is taken.

**a.** Find the probability that 5 Americans said that the price of gas was the news story they followed most closely.
**b.** Find the probability that 4 Americans said that the price of gas was the news story they followed most closely.
**c.** Find the probability that between 3 and 5 Americans said that the price of gas was the news story they followed most closely.

For Exercises 11–13, find the mean, variance, and standard deviation. Interpret the mean for each.

**11. Gestational Diabetes.** Gestational diabetes occurs in 8% of pregnancies, according to the American Diabetes Association. A random sample of 20 pregnancies is taken and the number of those with gestational diabetes is recorded.

**12. Female Students at DuPage.** According to the National Center for Education Statistics, 57% of the students at the College of DuPage in Glen Ellyn, Illinois, are female. A random sample of 10 students from the College of DuPage is taken and the number of females is recorded.

**13. Watching the Price of Gas.** The Pew Research Center reports that, for the week of May 28, 2008, 35% of Americans said that the price of gasoline was the news story that they followed more closely than any other news item. A random sample of 15 Americans is taken and the number who follow the price of gasoline most closely is recorded.

**14. Asian American Californians.** The *Statistical Abstract of the United States* reported that, in 2000, 11% of California residents were of Asian ethnicity. Suppose a random sample of 400 California residents is taken.

**a.** Find the probability that the sample contains 10 residents of Asian ethnicity.
**b.** Find the probability that the sample contains at most 10 residents of Asian ethnicity.

**15. Eighth-Grade Alcohol Consumption.** According to a study, 17% of eighth-graders report having consumed alcohol during the past month.[10] Suppose a random sample of 100 eighth-graders is taken.

**a.** Find the probability that the sample contains 17 eighth-graders who had consumed alcohol during the past month.
**b.** Find the probability that the sample contains 18 or 19 eighth-graders who had consumed alcohol during the past month.

**16.** Refer to Exercise 14.

**a.** Find the mean, variance, and standard deviation. Interpret the mean.
**b.** Suppose that the sample contains 50 residents of Asian ethnicity. Is this unusual? Explain how you determine this.

**17.** Refer to Exercise 15.

**a.** Find the mean, variance, and standard deviation. Interpret the mean.
**b.** Suppose that the sample contains 27 eighth-graders who had consumed alcohol during the past month. Is this unusual? Explain how you determine this.

**18. Children and Secondhand Smoke.** The U.S. Environmental Protection Agency reported that 11% of children under age 6 are exposed to secondhand smoke.[11] A random sample is taken this year of 200 children under age 6.

**a.** Find the probability that the sample contains 20 children under age 6 who were exposed to secondhand smoke.
**b.** Find the probability that the sample contains 21 or 22 children under age 6 who where exposed to secondhand smoke.

**19. Baptists in America.** The American Religious Identification Survey found that 16.3% of Americans identified themselves as Baptists.[12] A random sample of 100 American church members is taken this year.

**a.** Find the probability that the sample contains 16 Baptists.
**b.** Find the probability that the sample contains at least 16 Baptists.

**20.** Refer to Exercise 18.

**a.** Find the mean, variance, and standard deviation. Interpret the mean.
**b.** Suppose that the sample contains 30 children under age 6 who were exposed to secondhand smoke. Is this unusual? Explain how you determine this.

**21.** Refer to Exercise 19.

**a.** Find the mean, variance, and standard deviation. Interpret the mean.

   **b.** Suppose that the sample contains 25 Baptists. Is this unusual? Explain how you determine this.

**22. Exercise in High School.** What did you do in high school for vigorous physical activity? The National Center for Health Statistics reported in 1998 that 54% of high school females engaged in aerobics or dancing. A random sample of 10 high school females is taken this year.

   **a.** Find the probability that the sample contains 6 females who engaged in aerobics or dancing.
   **b.** Find the probability that the sample contains at most 3 females who engaged in aerobics or dancing.
   **c.** What is the most likely number of females in the sample who engaged in aerobics or dancing?

**23. Motor Vehicle Accidents.** According to the National Highway Traffic Safety Administration, motor vehicle accidents are the number one cause of death of young people aged 16–24, with 32% of deaths. A random sample of 100 deaths of people aged 16–24 is taken.

   **a.** Find the probability that the sample contains 32 deaths due to motor vehicle accidents.
   **b.** Find the probability that the sample contains at most 32 deaths due to motor vehicle accidents.

**24.** Refer to Exercise 22.

   **a.** Find the mean, variance, and standard deviation. Interpret the mean.
   **b.** Suppose that the sample contains 2 females who engaged in aerobics or dancing. Is this unusual? Explain how you determine this.

**25.** Refer to Exercise 23.

   **a.** Find the mean, variance, and standard deviation. Interpret the mean.
   **b.** Suppose that the sample contains 25 deaths caused by motor vehicle accidents. Is this unusual? Explain how you determine this.

## SECTION 6.3

**Systolic Blood Pressure.** Use the following information for Exercises 26–33. A study found that the mean systolic blood pressure was 106 mm Hg.[13] Assume that systolic blood pressure follows a normal distribution with mean $\mu = 106$ and standard deviation $\sigma = 8$.

**26.** What is the probability that a randomly chosen systolic blood pressure is equal to 106 mm Hg?

**27.** What is the probability that a randomly selected systolic blood pressure is more than 106 mm Hg?

**28.** What is the probability that a randomly chosen systolic blood pressure is at least 106 mm Hg?

**29.** Is the area to the right of $X = 110$ mm Hg greater than or less than 0.5? How do you know this?

**30.** Is the area to the left of $X = 110$ mm Hg greater than or less than 0.5? How do you know this?

**31.** What is the probability that a randomly selected systolic blood pressure is between 98 and 114 mm Hg?

**32.** What is the probability that a randomly chosen systolic blood pressure is between 90 and 122 mm Hg?

**33.** What is the probability that a randomly selected systolic blood pressure is between 82 and 130 mm Hg?

**34. Temperature in Los Angeles.** The U.S. Department of Commerce reports that the mean temperature in downtown Los Angeles is 66.2°F. Assume that the distribution of temperature in Los Angeles is normal with mean $\mu = 66.2$°F and standard deviation $\sigma = 8$°F.

   **a.** Draw a graph of the normal curve and shade the region that represents temperature between 50.2°F and 82.2°F.
   **b.** What is the probability that a randomly selected temperature in Los Angeles is between 50.2°F and 82.2°F?

## SECTION 6.4

For Exercises 35–44, **(a)** draw the graph, **(b)** estimate the area, **(c)** find the indicated area using the $Z$ table or technology, and **(d)** check your estimate against your final answer.

Find the area under the standard normal curve that lies to the left of the following.

**35.** $Z = 2.1$
**36.** $Z = 2.9$
**37.** $Z = -2.2$
**38.** $Z = -2.9$

Find the area under the standard normal curve that lies between the following.

**39.** $Z = -1$ and $Z = 1$
**40.** $Z = -2$ and $Z = 2$
**41.** $Z = -3$ and $Z = 3$
**42.** $Z = -1.28$ and $Z = 1.28$
**43.** $Z = -1.04$ and $Z = 1.51$

**44.** Find the following areas *without* using the $Z$ table or technology. The area to the right of $Z = 0.5$ is 0.3085.

   **a.** The area to the left of $Z = 0.5$
   **b.** The area to the left of $Z = -0.5$
   **c.** The area between $Z = -0.5$ and $Z = 0.5$

## SECTION 6.5

**45. Playing Computer Games.** In a 2004 poll of its readers conducted by the magazine *Computer Gaming World,* the respondents reported that they played computer games for a mean of 15 hours per week. Denote the mean to be $\mu =$ 15 hours, and assume that the distribution is normal and $\sigma = 5$ hours.

   **a.** Draw a normal curve with the mean labeled.
   **b.** We are interested in the probability that a reader of *Computer Gaming World* plays computer games for more than 20 hours per week. Sketch the appropriate area under the curve.
   **c.** Suppose that we found out that the area under the curve to the right of $X = 20$ equals 0.1587. What does this mean in terms of the question asked in **(b)**?
   **d.** What is the probability that a randomly selected reader plays computer games for fewer than 20 hours per week?

**e.** What is the probability that a randomly chosen reader plays computer games for fewer than 15 hours per week? More than 15 hours?

**46. Babysitter Rates.** A survey by Kid's Money (www.kidsmoney.org) reported that the mean hourly rate for an 18-year-old babysitter for a single child was \$5. Denote the mean to be $\mu = \$5$, and assume that the distribution is normal and $\sigma = \$1$.

**a.** Draw a normal curve with the mean labeled.

**b.** We are interested in the probability that a randomly selected 18-year-old babysitter earns less than \$3 per hour. Sketch the appropriate area under the curve.

**c.** Suppose that we found out that the area under the curve to the left of $X = 3$ equals 0.0228. What does this mean in terms of the question asked in **(b)**?

**d.** What is the probability that a randomly chosen 18-year-old babysitter earns more than \$3 per hour?

**e.** What is the probability that a randomly selected 18-year-old babysitter earns less than \$5 per hour? More than \$5 per hour?

**47. Generation Worse Off?** A *New York Times*/CBS News Poll conducted in 2007 found that 48% of young Americans (aged 17–29) think that their generation will be worse off than the previous generation. Suppose we take a random sample of 100 young Americans. Approximate the following probabilities.

**a.** Exactly 50 think that their generation will be worse off.

**b.** At least 50 think that their generation will be worse off.

**c.** More than 50 think that their generation will be worse off.

**d.** At most 50 think that their generation will be worse off.

**e.** Fewer than 50 think that their generation will be worse off.

**f.** Between 48 and 52 think that their generation will be worse off.

**48. Playing Computer Games.** Use the information in Exercise 45 to answer the following.

**a.** Find the probability that a randomly chosen reader of *Computer Gaming World* plays computer games for more than 20 hours per week.

**b.** What percentage of readers do not play computer games at all?

**c.** What proportion of readers play between 10 and 25 hours per week?

**d.** Suppose carpal tunnel syndrome sets in for people playing computer games in the highest 1%. How many hours a week does this represent?

**e.** Suppose a roommate plays computer games for 25 hours per week. Is this unusual? On what do you base your answer?

**49. Babysitter Rates.** Use the information in Exercise 46 to answer the following.

**a.** Find the probability that a randomly selected 18-year-old babysitter earns less than \$3 per hour.

**b.** What percentage of 18-year-old babysitters earn more than \$7 per hour?

**c.** What proportion of 18-year-old babysitters earn between \$5 and \$6 per hour?

**d.** Suppose that your neighbors' child is particularly difficult, so that 18-year-old babysitters charge the top 1% hourly rate. What rate is this?

**e.** Suppose that you know an 18-year-old who babysits for \$6 per hour. Is this unusual? On what do you base your answer?

**50. South Dakota Speeds.** The National Motorists Association reports that, in South Dakota, the mean speed on interstate highways is 68.3 mph. Denote the mean to be $\mu = 68.3$ mph, and assume that the distribution is normal and $\sigma = 4$ mph.

**a.** Find the probability that a randomly chosen vehicle is traveling faster than the 65 mph speed limit.

**b.** What percentage of vehicles travel slower than 60 mph?

**c.** What proportion of vehicles travel at speeds between 65 and 68.3 mph?

**d.** The National Motorists Association asserts that speeding tickets should be issued only for drivers whose speeds exceed the 85th percentile. If the police in South Dakota followed this rule, then at what speed would they start handing out speeding tickets?

**e.** Suppose that someone from South Dakota never drives faster than 55 mph on the interstate. Is this unusual? On what do you base your answer?

**51. Income for White Americans.** The U.S. Census Bureau reported that the mean income for white Americans living alone in 2002 was \$32,257. Denote the mean to be $\mu = \$32{,}257$, and assume that the distribution is normal and $\sigma = \$8000$.

**a.** Find the probability that a randomly selected white American living alone earned more than \$40,000.

**b.** What percentage of white Americans living alone earned less than \$30,000?

**c.** What proportion of white Americans living alone earned between \$30,000 and \$35,000?

**d.** Suppose the lowest 2% of white Americans living alone are eligible for an earned income tax credit. Find the cutoff income for this tax credit.

**e.** Suppose you know a white American living alone who makes \$55,000. Is this unusual? On what do you base your answer?

**52. Reinstate the Draft.** A *New York Times*/CBS News Poll conducted in 2007 found that 97% of young Americans (aged 17–29) oppose reinstating the military draft. Suppose we take a random sample of 400 young Americans. Approximate the following probabilities.

**a.** Exactly 388 oppose reinstating the military draft.

**b.** All 400 oppose reinstating the military draft.

**c.** More than 388 oppose reinstating the military draft.

**d.** At most 388 oppose reinstating the military draft.

**e.** Fewer than 388 oppose reinstating the military draft.

**f.** Between 385 and 390 oppose reinstating the military draft.

**53. Challenger Exercise.** Refer to Exercise 51. In the same report, the U.S. Census Bureau also stated that the median income for white Americans living alone was \$22,133.

**a.** How does this affect our assumption that the distribution of incomes is normal?
**b.** How does this affect our results in **(a)–(e)** of Exercise 51?

**54. Drunk-Driving Deaths.** In 2002 in the United States, a mean of 48 people *per day* were killed in vehicle accidents involving a drunk-driver. Assume that the distribution of drunk-driving accident deaths per day is normal, $\mu = 48$, and $\sigma = 12$.
**a.** Find the probability that at most 12 people will be killed in drunk-driving accidents today.
**b.** Find the probability that between 50 and 80 people will be killed in drunk-driving accidents today.
**c.** Suppose that Mothers against Drunk Driving holds special memorial services for the innocent people killed on days when the number of drunk-driving accident deaths reaches the 99.5th percentile. What is the cutoff number of deaths for the services to be held?
**d.** Suppose on one particular day in the United States, 60 people are killed in drunk-driving accidents. Is this unusual? On what do you base your answer?

**55. Hispanic Favorite: *Copa America*.** According to Nielsen Media Research, the highest-rated television program among Hispanic Americans is *Copa America,* with a rating of 23.4 among Hispanic viewers for the week ending July 18, 2004. Assume that the ratings for this program are normally distributed with mean $\mu = 23.4$ and standard deviation $\sigma = 5$.
**a.** Find the probability that the ratings for *Copa America* will be lower than 18.4.
**b.** Find the probability that the ratings for *Copa America* will be between 15 and 20.
**c.** Suppose that the show's producer must report to the UNI network executives any time the ratings fall below the 10th percentile. What is the ratings value that will trigger this report?
**d.** Suppose that the ratings for *Copa America* in one particular week were 8.4. Is this unusual? On what do you base your answer? Will the producer have to file a report?

**56. Fathers of Children Born to Teenagers.** The Web portal **www.catholiceducation.org** cites a 1990 study which states that fathers of children born to teenagers 15 years old and younger are on average 6.7 years older than the mothers of the children. Assume that the age difference distribution is normal with mean $\mu = 6.7$ years and standard deviation $\sigma = 3$ years.
**a.** Find the probability that a randomly selected father of a child born to a teenager under 16 is the same age as the mother or younger.
**b.** What percentage of fathers of children born to teenagers under 16 are less than two years older than the mothers?
**c.** What proportion of fathers of children born to teenagers under 16 are between 10 and 15 years older than the mothers?
**d.** Suppose that the highest 10% of age differences comes under scrutiny by child advocate organizations. How much older than the mother does the father of the child have to be to come under scrutiny? Find the cutoff value.
**e.** Suppose that you know of a teenager under 16 who had a child with a man who was 10 years older than her. Is this unusual? On what do you base your answer?

**57. Do Women Shop More?** The British Office of National Statistics reports that females spend more than twice as much time shopping than males. The weekly mean for females is 5.8 hours, and for males, 2.8 hours. We are interested in the time females spend shopping, which we assume to be normally distributed, with mean $\mu = 5.8$ hours and standard deviation $\sigma = 1.8$.
**a.** Find the probability that a randomly chosen British female spends more than 6 hours per week shopping.
**b.** What percentage of British females spend between 3 and 4 hours shopping per week?
**c.** Suppose that marketing executives are trying to reach the women in Britain who spend the most time shopping, as measured by the 98th percentile. How much time does this represent per week?
**d.** Suppose one randomly selected British female spends only half an hour per week shopping. Is this unusual? On what do you base your answer?

**58. Junior High Study Time.** The Carnegie Council on Adolescent Development reports that American junior high school students spend a mean of 3.2 hours per week studying. Assume that the distribution of time spent studying by American junior high school students is normally distributed with mean $\mu = 3.2$ hours and standard deviation $\sigma = 1$ hour.
**a.** Find the probability that a randomly chosen American junior high school student spent more than 6 hours studying this week.
**b.** What percentage of American junior high school students spent between 2.2 and 4.2 hours studying this week?
**c.** Suppose that a parent-teacher organization plans intervention strategies for students who spend the least amount of time studying, as measured by the 10th percentile. How much time per week does this represent?
**d.** Suppose you know of an American junior high school student who spends no time studying. Is this unusual? On what do you base your answer?

# CHAPTER 6 Quiz

## TRUE OR FALSE

1. True or false: The following is a continuous and not a discrete random variable: How much coffee there is in your next cup of coffee.
2. True or false: The following is an example of a binomial experiment: Rolling a pair of dice 3 times and observing the sum of the two dice.
3. True or false: Our distributions for continuous random variables are for samples and not for populations.

## FILL IN THE BLANK

4. The probability that a randomly chosen value of a normally distributed random variable will be greater than the mean is __________.
5. The probability that a randomly chosen value of a normally distributed random variable will be equal to the mean is __________.
6. The standard deviation of a normal random variable can never take a value that is less than __________.

## SHORT ANSWER

7. Is the following a discrete or continuous random variable: The number of goals your college soccer team will score in its next game.
8. Recording the gender of the next 20 babies born at City Hospital is an example of what kind of experiment?
9. What are the values for the mean and standard deviation of the standard normal distribution?

## CALCULATIONS AND INTERPRETATIONS

10. **Church Bazaar.** Lenny has gone down to the church bazaar with his family. There is a game there where if you roll two dice and get a sum of at least 9 you will win $5; otherwise, you don't win anything.
    - **a.** Construct the probability distribution for the amount you win playing this game.
    - **b.** What are the expected winnings?
    - **c.** What would be a fair (break-even) price for the church to ask you to pay to play this game?

11. **CEOs Driving Luxury Cars.** According to CareerBuilder.com, 19% of company CEOs drive luxury cars. Suppose a random sample is taken of 100 company CEOs.
    - **a.** Find the probability that the sample contains 20 CEOs who drive luxury cars.
    - **b.** What is the most likely number of CEOs who drive luxury cars?
    - **c.** Find the mean, variance, and standard deviation. Interpret the mean.
    - **d.** Suppose the sample contains 40 CEOs who drive luxury cars. Is this unusual? Explain how you determine this.

12. **Gambling Losses.** Treatment providers for problem gamblers report that men who approached them for intervention had lost a mean of $2849 in the preceding four weeks, according to a 2002 report.[14] Assume that the distribution of gambling losses is normally distributed with mean $\mu = \$2849$ and standard deviation $\sigma = \$900$.
    - **a.** Find the probability that a randomly selected male had lost more than $4000.
    - **b.** What percentage of males lost between $3000 and $4000?
    - **c.** Suppose that a gambling support group is trying to identify those who lose the most, as measured by the 95th percentile. How much money in gambling losses does this represent?
    - **d.** Suppose you know of a male problem gambler who lost $1000 in four weeks and then approached a treatment provider. Is this amount unusual? On what do you base your answer?

13. **Cats and Dogs.** The American Veterinary Medical Association reported that there were 61.1 million cats in America and 53.6 million dogs. It also reported that 31.4% of American households had a cat while 34.3% of American households had a dog. A random sample of 1000 households is taken.
    - **a.** Find the probability that the sample contains 320 households with a cat.
    - **b.** Find the probability that the sample contains 320 households with a dog.
    - **c.** Find the mean, variance, and standard deviation of the number of cats. Interpret the mean. Now do the same for dogs.
    - **d.** Suppose that the sample contains 290 households with a cat. Is this unusual? Explain how you determine this.

14. **Attendance at the Magic Kingdom.** The Magic Kingdom at Walt Disney World in Orlando, Florida, has a mean of 38,000 customers per day. Assume that the distribution of attendance is normal, $\mu = 38{,}000$, and $\sigma = 10{,}000$.
    - **a.** Find the probability that more than 40,000 people visit the Magic Kingdom in one day.
    - **b.** Find the probability that between 35,000 and 41,000 people visit the Magic Kingdom in one day.
    - **c.** Suppose that special overflow parking lots are opened for the days with the highest 10% attendance levels. What is the cutoff attendance number that triggers the opening of the overflow parking areas?
    - **d.** Suppose that on one rainy day, only 15,000 people visit the Magic Kingdom. Is this unusual? On what do you base your answer?

© Image 100/CORBIS

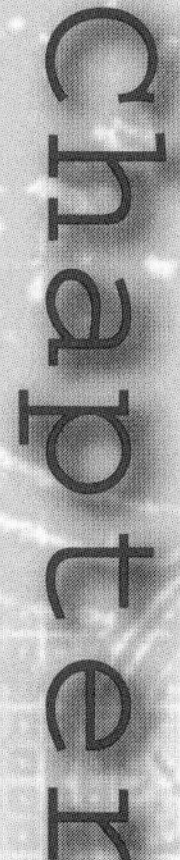
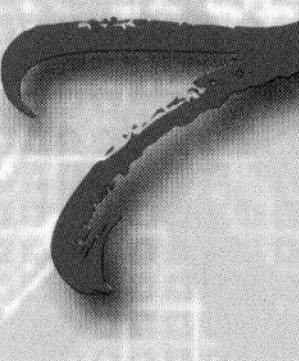

Chapter 7

# Sampling Distributions

CASE STUDY

## Trial of the Pyx: How Much Gold Is in Your Gold Coins?

The kings of bygone England had a problem: How much gold should they put into their gold coins? After all, the very commerce of the kingdom depended on the purity of the currency. How did the lords of the realm ensure that the coins floating around the kingdom contained reliable amounts of gold?

From the year 1282, the Trial of the Pyx has been held annually in London to ensure that newly minted coins adhere to the standards of the realm. It is the responsibility of the presiding judge to ensure that the trial proceeds lawfully and to inform Her Majesty's Treasury of the verdict. Six members of the Company of Goldsmiths compose the jury, who are given two months to test the coins. It works like this: A ceremonial boxwood chest, called the Pyx, is brought forth, and a sample of 100 of the coins cast that year at the mint is put into it. The Pyx is then weighed. In times past, each gold coin, called a guinea, had an expected weight of 128 grams, so the total weight of the guineas in the Pyx was expected to be 12,800 grams.

If the weight of the coins in the Pyx was much less than 12,800 grams, the jury concluded that the Master of the Mint was cheating the crown by pocketing the excess gold, and he was severely punished. On the other hand, if the coins in the Pyx weighed much more than 12,800 grams, that wasn't good either, since it cut down on the profits produced by the kings' coin-minting monopoly.

By how much could the Master of the Mint debase the coinage before getting caught? We shall see in the chapter's Case Study, The Trial of the Pyx, which unfolds in Section 7.2.

OVERVIEW



## The Big Picture

### Where we are coming from, and where we are headed...

We have reached a major crossroads in our journey. Chapters 1–4 introduced us to the language of data analysis and methods for summarizing a data set using graphical or numerical methods. Next, to help us deal with the uncertainty implicit in generalizing from a sample to a population, we acquired the tools of probability in Chapters 5 and 6. We examined some of the rules of probability and learned how random variables can help us describe patterns in randomness.

Here, in Chapter 7, Sampling Distributions, we approach a significant turning point. In the chapters on probability, we assumed that we knew the population mean and standard deviation. However, in the real world, such knowledge is rare. Uncovering the true values of population characteristics represents the very heart of data analysis: the notion of statistical inference. Statistical inference refers to learning about population characteristics by studying the same characteristics in a sample.

In this chapter, we discover that seemingly random statistics, such as the sample mean, have *predictable behaviors*. Taking repeated samples from the same population, we find that the sample means do not vary randomly but rather form a distinctive pattern or distribution. This special type of distribution is called the **sampling distribution** of the sample mean. We explore sampling distributions and uncover perhaps the most important result in statistical inference, the Central Limit Theorem.

Thus, the key to all the statistical inference in Chapters 8–13 is the set of patterns implicit in the sampling distributions that we learn about here in Chapter 7. When trying to infer the true value of a population parameter, data analysts often start with a point estimate. However, in Chapter 8, Confidence Intervals, we will learn that point estimates, though widespread, have serious drawbacks. We need something better, confidence interval estimates, our first application of statistical inference. ■

## 7.1 Introduction to Sampling Distributions

*Objectives:* By the end of this section, I will be able to...

1. Compute point estimates and sampling error.
2. Explain the sampling distribution of the sample mean $\bar{x}$.
3. Describe the sampling distribution of the sample mean $\bar{x}$ when the population is normal.
4. Use normal probability plots to assess normality.
5. Find probabilities and percentiles for the sample mean when the population is normal.

## 1 Point Estimates and Sampling Error

Recall from Chapter 1 that **statistical inference** consists of methods for estimating and drawing conclusions about population characteristics based on the information contained in a subset (sample) of that population. Sample characteristics, like the sample mean $\bar{x}$, are called **statistics**. Population characteristics, like the population mean $\mu$, are called **parameters**. Table 7.1 summarizes the relationship between three of the most common statistics and their associated parameters. Using known statistics to estimate unknown parameters and reporting a single number as the estimate is called **point estimation**, and the value of the statistic is called the **point estimate**.

Table 7.1 Point estimation: Use statistics to estimate unknown population parameters

| | **Sample statistic** | **. . . estimates . . .** | **Population parameter** |
|---|---|---|---|
| Mean | $\bar{x}$ | . . . estimates . . . | $\mu$ |
| Standard deviation | $s$ | . . . estimates . . . | $\sigma$ |
| Proportion | $\hat{p}$ | . . . estimates . . . | $p$ |

### Example 7.1 Using a sample mean to estimate a population mean

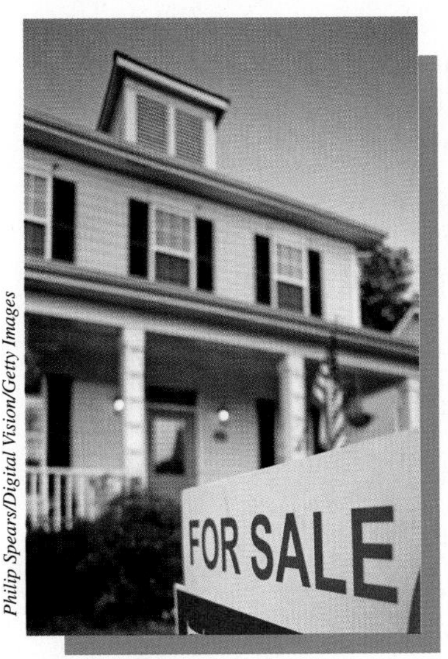

Philip Spears/Digital Vision/Getty Images

In Example 3.2 (page 89) we calculated the mean of a sample of 6 home sales prices for Broward County, Florida (Table 7.2). The mean price was

$$\bar{x} = \frac{\sum x}{n} = \frac{3{,}849{,}000}{6} = \$641{,}500$$

Find a point estimate for the mean price $\mu$ of all home sales prices for Broward County, Florida.

Table 7.2 Home sales prices, Broward County, Florida

| **Location** | **Price** |
|---|---|
| Coral Springs | \$415,000 |
| Weston | \$449,000 |
| Coconut Creek | \$472,000 |
| Davie | \$649,000 |
| Hallandale | \$889,000 |
| Fort Lauderdale | \$975,000 |

**Solution**

We use the sample mean $\bar{x} = \$641{,}500$ as the point estimate for $\mu$. ■

It is not likely that the population mean home sales price $\mu$ for all houses in Broward County precisely equals the sample mean $\bar{x} = \$641{,}500$ because the sample is a subset of the population. Therefore, when using point estimates, we nearly always "miss" our target by a certain amount. This type of error is called **sampling error**. We define sampling error as *distance*, so it is always positive. Table 7.3 presents the sampling error for most common characteristics.

> **Sampling error** is the distance between the point estimate and its target parameter.

Table 7.3 Sampling error for common characteristics

| Characteristic | Sample statistic | Population parameter | Sampling error |
|---|---|---|---|
| Mean | $\bar{x}$ | $\mu$ | $\lvert\bar{x} - \mu\rvert$ |
| Standard deviation | $s$ | $\sigma$ | $\lvert s - \sigma\rvert$ |
| Proportion | $\hat{p}$ | $p$ | $\lvert\hat{p} - p\rvert$ |

## Example 7.2 Sampling error for $\bar{x}$

Suppose that someone did an exhaustive search of the multiple listing service for all of Broward County and found the population mean home sales price $\mu$ to be \$650,000. Find the sampling error for $\bar{x}$ for the sample in Table 7.2.

**Solution**

From Example 7.1, we have $\bar{x}$ = \$641,500, and from Table 7.3 we have the sampling error for $\bar{x}$ as $\lvert\bar{x} - \mu\rvert$. The sampling error is therefore

$$\lvert\bar{x} - \mu\rvert = \lvert\$641{,}500 - \$650{,}000\rvert = \lvert-\$8{,}500\rvert = \$8{,}500.$$

Unfortunately, the true value of the sampling error usually is unknown because we rarely know the actual value of the population parameter. Since a sample is a subset (usually a tiny subset) of the population, generalizing from the sample to the population carries risks. Point estimates are not always useful because we cannot confidently tell how accurate the estimates are. That is, the sampling error may be small or large. In Chapter 8, we will learn another way to calculate estimates. But first we need to learn more about how statistics behave.

The *Law of Large Numbers for Means* applet allows you to simulate dice rolls and observe the behavior of the mean of the rolls as the number of rolls increases. As the sample size $n$ increases, the value of $\bar{x}$ tends to settle down at $\mu$. Thus, as $n$ increases, the sampling error $\lvert\bar{x} - \mu\rvert$ tends to decrease.

The *Law of Large Numbers for Proportions* applet allows you to simulate coin tosses, and observe the proportion of heads as the number of tosses increases. As the sample size $n$ increases, the value of $\hat{p}$ tends to settle down at $p$. Thus, as $n$ increases, the sampling error $\lvert\hat{p} - p\rvert$ tends to decrease.

## 2 Sampling Distribution of the Sample Mean $\bar{x}$

We begin by taking some samples from a population and seeing what we can learn.

## Example 7.3 Commuting times for student government members

We are interested in how long it takes the five members of the student government to commute to school. The times (in minutes) are given in Table 7.4. Since these five people are *all* the members of the student government, we can consider them to constitute a *population*.

Table 7.4 Commuting times for the five members of the student government

| Amber | Brandon | Chantal | Dave | Emma |
|---|---|---|---|---|
| 10 | 20 | 5 | 30 | 15 |

**a.** Calculate the population mean commuting time $\mu$.

**b.** Take a sample of the following student government members: Amber, Brandon, and Chantal. Find the sample mean commuting time $\bar{x}$ and the sampling error $|\bar{x} - \mu|$.

**Solution**

**a.** The mean commuting time of this population is

$$\mu = \frac{\sum x}{N} = \frac{10 + 20 + 5 + 30 + 15}{5} = 16 \text{ minutes}$$

**b.** For Amber, Brandon, and Chantal, the sample mean commuting time is

$$\bar{x}_1 = \frac{\sum x}{n} = \frac{10 + 20 + 5}{3} \approx 11.67 \text{ minutes}$$

The sampling error of this sample mean is $|\bar{x}_1 - \mu| = |11.67 - 16| = 4.33$ minutes. ■

Your Statistical Sense

**Sampling Variability**

Before we take a second sample, consider this question: When we take another sample, is the sample mean commuting time from the new sample likely to be exactly the same as the $\bar{x}_1 = 11.67$ minutes from the first sample? Probably not, since different samples provide different values for the sample statistics. This is called *sampling variability.* For the second sample, Amber, Brandon, and Dave, the sample mean commuting time is $\bar{x}_2 = (10 + 20 + 30)/3 = 20$ minutes. This time the sampling error is $|\bar{x}_2 - \mu| = |20 - 16| = 4$ minutes. ■

Table 7.5 shows all possible samples of size 3 from the five students, along with the respective sample means and sampling errors. Note that, while the value for the sample mean varies, the population mean $\mu$ is constant: $\mu = 16$. A population parameter like $\mu$ is a constant (a fixed number). However, with each sample we take, we obtain a different estimate of $\mu$. This is because the sample mean $\bar{x}$ is a random variable and exhibits sampling variability.

Table 7.5 All possible samples of size 3 from population of student government members

| Sample | Amber Brandon Chantal | Amber Brandon Dave | Amber Brandon Emma | Amber Chantal Dave | Amber Chantal Emma | Amber Dave Emma | Brandon Chantal Dave | Brandon Chantal Emma | Brandon Dave Emma | Chantal Dave Emma |
|---|---|---|---|---|---|---|---|---|---|---|
| | 10 | 10 | 10 | 10 | 10 | 10 | 20 | 20 | 20 | 5 |
| Data | 20 | 20 | 20 | 5 | 5 | 30 | 5 | 5 | 30 | 30 |
| | 5 | 30 | 15 | 30 | 15 | 15 | 30 | 15 | 15 | 15 |
| $\bar{x}$ | 11.67 | 20 | 15 | 15 | 10 | 18.33 | 18.33 | 13.33 | 21.67 | 16.67 |
| $\mu$ | 16 | 16 | 16 | 16 | 16 | 16 | 16 | 16 | 16 | 16 |
| $\lvert\bar{x} - \mu\rvert$ | 4.33 | 4 | 1 | 1 | 6 | 2.33 | 2.33 | 2.67 | 5.67 | 0.67 |

Fortunately, there are patterns in how the sample mean varies. These patterns depend on the sample size, on the shape of the population distribution, and on the underlying value of the population parameters, $\mu$ and $\sigma$. The highlighted row in Table 7.5 represents the sample means for all possible samples of size $n = 3$. This is called the **sampling distribution for the sample mean $\bar{x}$** for this tiny population and sample size, obtained by examining every possible sample of size $n = 3$.

The **sampling distribution of the sample mean $\bar{x}$** for a given sample size $n$ consists of the collection of the sample means of all possible samples of size $n$ from the population.

Sampling distributions can tell us about the expected location and variability of a statistic. We are especially interested in the center and the spread of the sampling distribution. Like any distribution, the sampling distribution for the sample mean has a balance point and therefore a mean. Figure 7.1 provides a dotplot of the sample means in Table 7.5, along with the mean of these sample means, indicated at the balance point $\mu = 16$. Figure 7.1 represents the sampling distribution of the sample mean for this example.

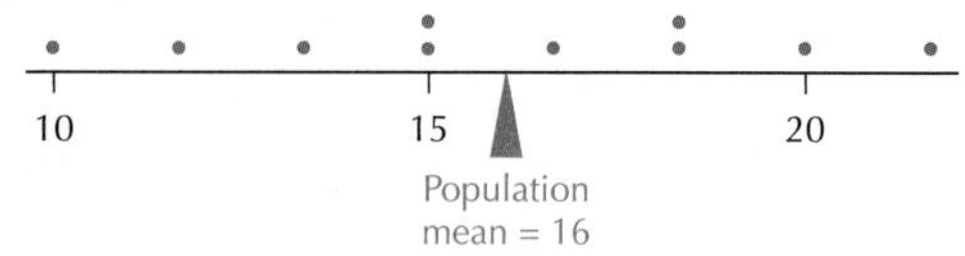

FIGURE 7.1
Sampling distribution of the sample mean for Example 7.3.



The mean value of the ten sample means is

$$\frac{11.67 + 20 + 15 + 15 + 10 + 18.33 + 18.33 + 13.33 + 21.67 + 16.67}{10} = 16$$

Note that this value is exactly equal to the population mean $\mu = 16$. In fact, the sampling distribution of the sample mean $\bar{x}$ has a mean value of $\mu$. That is, the sampling distribution of $\bar{x}$ is centered at $\mu$. We generalize this result as follows.

*Note:* It is convenient to number a set of important *facts*, as we build toward the Central Limit Theorem for Means and the Central Limit Theorem for Proportions.

**Fact 1: Mean of the Sampling Distribution of the Sample Mean $\bar{x}$**
The mean of the sampling distribution of the sample mean $\bar{x}$ is the value of the population mean $\mu$. It can be denoted as $\mu_{\bar{x}} = \mu$ and read as "the mean of the sampling distribution of $\bar{x}$ is $\mu$."

Next we would like to uncover information regarding the spread of the sampling distribution of $\bar{x}$. Now the population standard deviation of the original commute times in Table 7.4 is

$$\sigma = \sqrt{\frac{\sum(x-\mu)^2}{N}}$$
$$= \sqrt{\frac{\sum\left[(10-16)^2 + (20-16)^2 + (5-16)^2 + (30-16)^2 + (15-16)^2\right]}{5}}$$
$$\approx 8.6023$$

And the population standard deviation of the sampling distribution of $\bar{x}$ is

$$\sigma_{\bar{x}} = \sqrt{\frac{\sum(\bar{x}-\mu)^2}{10}} = \sqrt{\frac{\sum\left[(11.67-16)^2 + (20-16)^2 + \ldots + (16.67-16)^2\right]}{10}}$$
$$\approx 3.5119$$

Note that the standard deviation of the sample means is smaller than the original standard deviation. This is a good thing, since smaller variability is better.

*Note:* In this example, the precise relationship between the two standard deviations is

$$\sigma_{\bar{x}} = \sqrt{\frac{N-n}{N-1}} \cdot \frac{\sigma}{\sqrt{n}}$$

where $N$ is the population size and $n$ is the sample size. This gives

$$\sigma_{\bar{x}} = \sqrt{\frac{5-3}{5-1}} \cdot \frac{8.6023}{\sqrt{3}} \approx 3.5119$$

However, the coefficient

$$\sqrt{\frac{N-n}{N-1}}$$

called the *finite population correction factor,* is required only for special cases (like this textbook example) where the population is not much larger than the sample. This coefficient does not apply when sampling with replacement, and its value tends to zero as the sample size approaches the population size. However, for most real-world problems, and for the remainder of this book, we dispense with this coefficient and assume that the population size is very large compared to the sample size.

**Fact 2: Standard Deviation of the Sampling Distribution of the Sample Mean $\bar{x}$**

The standard deviation of the sampling distribution of the sample mean $\bar{x}$ is $\sigma_{\bar{x}} = \sigma/\sqrt{n}$, where $\sigma$ is the population standard deviation and $n$ is the sample size.

Note the $\sqrt{n}$ in the denominator of the formula. Because of this factor, the larger the sample size, the tighter the resulting sampling distribution. Larger sample sizes result in more precise estimation.

## Example 7.4 Finding the mean and standard deviation of the sampling distribution of $\bar{x}$

According to CanEquity Mortgage company, the mean age of mortgage applicants in the City of Toronto is 37 years old. Assume that the standard deviation is 6 years. Find the mean and standard deviation for the sampling distribution of $\bar{x}$ for the following sample sizes: (a) 4, (b) 9, (c) 25, (d) 49, (e) 100, (f) 225.

### Solution

We have $\mu_{\bar{x}} = \mu = 37$. Note that this value for $\mu_{\bar{x}}$ does not depend on the sample size, so the value is true for any sample size. We also have $\sigma = 6$.

**a.** $n = 4$. Then $\sigma_{\bar{x}} = \dfrac{\sigma}{\sqrt{n}} = \dfrac{6}{\sqrt{4}} = 3$.

**b.** $n = 9$. Then $\sigma_{\bar{x}} = \dfrac{\sigma}{\sqrt{n}} = \dfrac{6}{\sqrt{9}} = 2$.

**c.** $n = 25$. Then $\sigma_{\bar{x}} = \dfrac{\sigma}{\sqrt{n}} = \dfrac{6}{\sqrt{25}} = 1.2$.

**d.** $n = 49$. Then $\sigma_{\bar{x}} = \dfrac{\sigma}{\sqrt{n}} = \dfrac{6}{\sqrt{49}} \approx 0.86$.

**e.** $n = 100$. Then $\sigma_{\bar{x}} = \dfrac{\sigma}{\sqrt{n}} = \dfrac{6}{\sqrt{100}} = 0.6$.

**f.** $n = 225$. Then $\sigma_{\bar{x}} = \dfrac{\sigma}{\sqrt{n}} = \dfrac{6}{\sqrt{225}} = 0.4$. ■

*What Does This Number Mean?*

Consider $\sigma_{\bar{x}} = 2$ for $n = 9$. This is a measure of the variability of the sampling distribution of $\bar{x}$ for this sample size. That is, if we take samples of size 9, our sampling error in estimating the population mean age $\mu$ of all mortgage applicants in Toronto will be within 2 years most of the time. ■

Figure 7.2 shows the relationship between $n$ and $\sigma_{\bar{x}}$ for the six samples in Example 7.4. Note that, as the sample size begins to get larger, there is rapid decrease in $\sigma_{\bar{x}}$. This is good, since we want the variability to be small. Less variability means more precise analysis. However, the decrease in $\sigma_{\bar{x}}$ slows gradually, until, somewhere around $n = 70$, further increases in sample size lead to very small decreases in $\sigma_{\bar{x}}$. That is, increasing the sample size past a certain point yields *diminishing returns*. The data analyst needs to think carefully about balancing the need for precision against the cost of a larger sample.

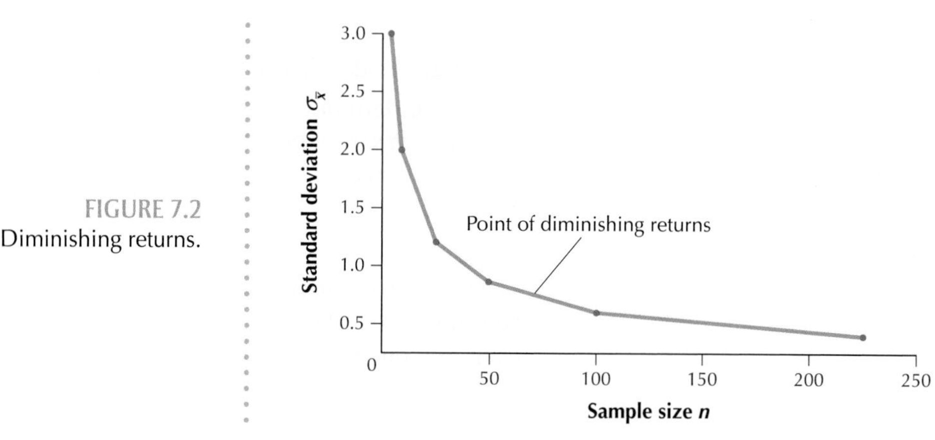


FIGURE 7.2
Diminishing returns.

## 3 Sampling Distribution of $\bar{x}$ for a Normal Population

To find out what form the sampling distribution of $\bar{x}$ takes when the population is normal, we consider the following example in which we examine a small, normally distributed population and find the sample means for all possible samples of a certain size.

### Example 7.5 Sampling distribution of $\bar{x}$ for a normal population

In Example 6.34 (pages 309–310), the average statewide temperature in Georgia in the month of April was normally distributed with a mean of $\mu = 61.5°F$ and a standard deviation of $\sigma = 8°F$. Using Minitab, 1000 samples of size $n = 2$ were generated from this normal distribution, and the sample means $\bar{x}$ were calculated for each sample. Construct a histogram and observe the shape of the distribution.

**Solution**

Figure 7.3 shows the histogram of the means from the 1000 samples of size $n = 2$. As you may have expected, the histogram looks quite normal, even with this tiny sample size of $n = 2$. ■

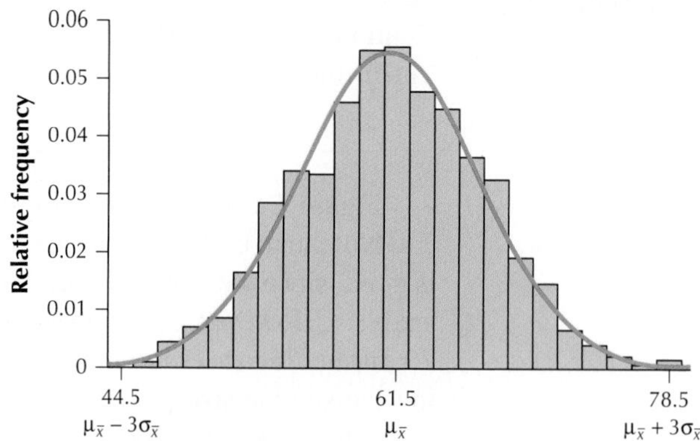


FIGURE 7.3 Sampling distribution of sample means of size $n = 2$ from a normal population.

Actually, this outcome is true generally and may be stated formally as follows.

**Fact 3: Sampling Distribution of the Sample Mean for a Normal Population**
The sampling distribution of the sample mean for a normal population is itself normal, regardless of sample size.

In Example 7.5, we chose the "worst-case scenario," the smallest possible sample size, $n = 2$. Still, the sampling distribution of $\bar{x}$ from a normal population was normal, even with this tiny sample size. We can combine Fact 3 with Facts 1 and 2, and produce the following completely specified normal distribution for the sampling distribution of the sample mean for a normal population.

*Note:* Let the notation

normal$(\mu, \sigma/\sqrt{n})$

denote a normal distribution with mean of $\mu$ and standard deviation of $\sigma/\sqrt{n}$.

**Fact 4: Sampling Distribution of the Sample Mean for a Normal Population**
For a normal population, the sampling distribution of the sample mean $\bar{x}$ is distributed as normal $(\mu, \sigma/\sqrt{n})$, where $\mu$ is the population mean and $\sigma$ is the population standard deviation.

Once we know that the sample mean is normally distributed, we can standardize and produce Z, just as we would for any normal random variable.

**Fact 5: Standardizing a Normal Sampling Distribution for Means**
When the sampling distribution of $\bar{x}$ is normal, we may standardize to produce the standard normal random variable $Z$ as follows:

$$Z = \frac{\bar{x} - \mu_{\bar{x}}}{\sigma_{\bar{x}}} = \frac{\bar{x} - \mu}{\sigma/\sqrt{n}}$$

where $\mu$ is the population mean, $\sigma$ is the population standard deviation, and $n$ is the sample size.

## 4 Assessing Normality Using Normal Probability Plots

Recall from Section 2.2 that histograms for the same data set can take different shapes, depending on the class width and number of classes. We therefore introduce a type of graph called the **normal probability plot**, which is a more reliable tool for assessing normality. A normal probability plot is a scatterplot of the estimated cumulative normal probabilities (expressed as percents) against the corresponding data values in the data set.

Figure 7.4 shows the normal probability plot for a sample of normally distributed data. The points are arrayed nicely along the straight line, and all the points lie within the curved bounds. Figure 7.5 shows the normal probability plot for a sample of right-skewed data. The points do not line up in a straight line, and many points lie outside the curved bounds, indicating that the data set is not normal.

**Analyzing Normal Probability Plots**
If the points in the normal probability plot either cluster around a straight line or nearly all fall within the curved bounds, then it is likely that the data set is normal. Systematic deviations off the straight line are evidence against the claim that the data set is normal.

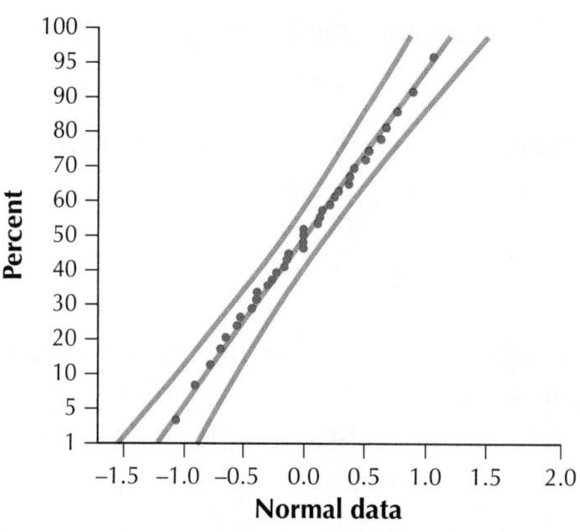


FIGURE 7.4 Normal probability plot of normal data.

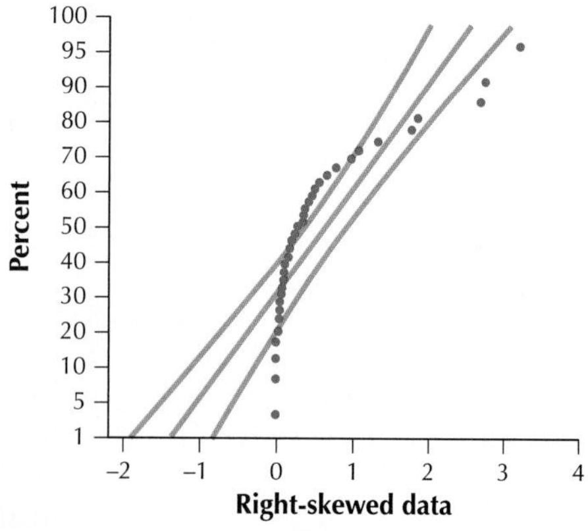


FIGURE 7.5 Normal probability plot of right-skewed data.

We illustrate how to construct normal probability plots using technology in the Step-by-Step Technology Guide at the end of this section (page 348).

## 5 Finding Probabilities and Percentiles Using a Sampling Distribution

Since we know that the sampling distribution of the sample mean $\bar{x}$ is normal when the population is normally distributed, we can use the techniques of Chapter 6 to answer questions about the means of samples taken from normal populations. First, we recall how to solve a normal probability problem.

### Example 7.6 Sampling distribution of sample mean statistics quiz score

Suppose that statistics quiz scores for a certain instructor are normally distributed with mean 70 and standard deviation 10.

**a.** Find the probability that a randomly chosen student's score will be above 80.

**b.** Suppose that, over the years, the instructor has had many sections of size $n = 25$. What kind of distribution is the sampling distribution of the sample mean?

**Solution**

**a.** This is a normal probability problem, which we learned how to do in Chapter 6. Since the student's score is randomly chosen, it is a random variable. Therefore, the normal random variable $X$ is the stats quiz scores. Standardizing, we have the corresponding $Z$-values:

$$\frac{X - 70}{10} = Z \qquad \text{and} \qquad \frac{80 - 70}{10} = 1$$

So the probability that a randomly chosen student's score will be above 80 is

$$P(X > 80) = P\left(\frac{X - 70}{10} > \frac{80 - 70}{10}\right)$$

$$= P(Z > 1) = 1 - 0.8413 = 0.1587$$

using Case 2 from Table 6.7 in Section 6.4 (page 298). Therefore, there is a 15.87% probability that a randomly chosen student will have a quiz score above 80 (Figure 7.6a).

**b.** We are given $\mu = 70$ and $\sigma = 10$. So by Fact 1, $\mu_{\bar{x}} = \mu = 70$. And by Fact 2, $\sigma_{\bar{x}} = \dfrac{\sigma}{\sqrt{n}} = \dfrac{10}{\sqrt{25}} = 2$.

Next, we are given that the population of quiz scores is normal. Therefore, by Fact 3, the sampling distribution of $\bar{x}$ is also normal. Putting this information together, we find that the sampling distribution of $\bar{x}$ is distributed as normal(70, 2), which we could have found out directly from Fact 4 (see Figure 7.6b).

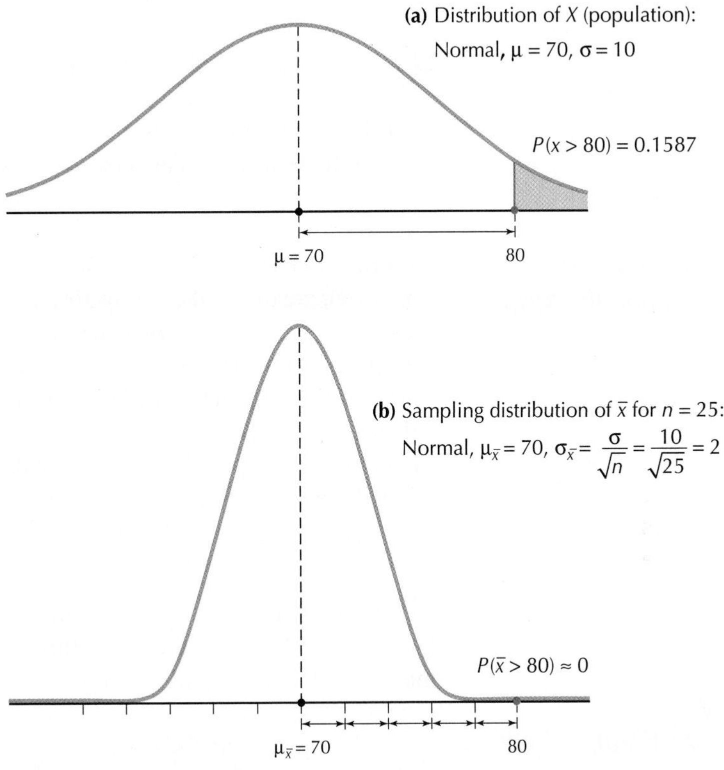


FIGURE 7.6 Distribution of $X$ and sampling distribution of $\bar{x}$ for Examples 7.6 and 7.7.

## Example 7.7 Finding probabilities using the sample mean

Using the information in Example 7.6, find the probability that a sample of 25 quiz scores will have a mean score greater than 80.

### Solution

We combine our information from Example 7.6b with our knowledge of how to solve normal probability problems. Note that, once we know that the sample mean is normally distributed, we can standardize and produce $Z$, just as we would for any normal random variable. Just be sure to use $\sigma_{\bar{x}} = 2$, the standard deviation of the sampling distribution of $\bar{x}$, and not $\sigma = 10$, the standard deviation for the population. $\sigma_{\bar{x}}$ is always smaller.

Applying Fact 5,

$$Z = \frac{\bar{x} - \mu_{\bar{x}}}{\sigma_{\bar{x}}} = \frac{\bar{x} - \mu}{\sigma/\sqrt{n}} = \frac{\bar{x} - 70}{10/\sqrt{25}} = \frac{\bar{x} - 70}{2}$$

We are looking for the probability that the sample mean $\bar{x}$ is greater than 80, that is, $P(\bar{x} > 80)$. So we need to standardize this score of 80 as well. Make sure to use $\sigma_{\bar{x}} = 2$

rather than $\sigma = 10$.

$$Z = \frac{80 - \mu_{\bar{x}}}{\sigma_{\bar{x}}} = \frac{80 - \mu}{\sigma/\sqrt{n}} = \frac{80 - 70}{10/\sqrt{25}} = \frac{80 - 70}{2} = 5$$

Hence,

$$P(\bar{x} > 80) = P\left(\frac{\bar{x} - 70}{10/\sqrt{25}} > \frac{80 - 70}{10/\sqrt{25}}\right) = P\left(\frac{\bar{x} - 70}{2} > \frac{80 - 70}{2}\right) = P(Z > 5) \approx 0$$

as shown in Figure 7.6b. Since $Z$ is standard normal, nearly all observations lie between $-3$ and 3. Thus, the $Z$ table does not go up to 5 since the probabilities are so close to zero. The TI-83 provides the more precise probability of $P(Z > 5) = 0.000000287$, or about 3 in 10 million. This instructor just does not give easy quizzes! ■

*What Does This Probability Mean?*

There is essentially no chance that the mean on one of the quizzes will be greater than 80. Compare this to the nearly 16% chance that a *particular* student's score would be above 80. Figure 7.6 shows the graphs of the distributions of quiz scores and class means. Both distributions are centered at $\mu_{\bar{x}} = \mu = 70$, but the standard deviations differ. The arrow in Figure 7.6a represents the standard deviation of $X$, $\sigma = 10$, and it shows that $x = 80$ is only one standard deviation above the mean $\mu = 70$. The arrows in Figure 7.6b represent the standard deviation of the sampling distribution of $\bar{x}$, $\sigma_{\bar{x}} = 2$, and they illustrate that $\bar{x} = 80$ lies five standard deviations above the mean $\mu_{\bar{x}} = 70$. Thus, class means are less variable than individual student scores. ■

In Chapter 6, we found the percentiles of normally distributed random variables (the so-called "backward problems"). Since the sampling distribution of $\bar{x}$ is normal, we are able to find the percentiles of the $\bar{x}$'s as well.

*What Results Might We Expect?*

### Estimating the Percentiles of the Sampling Distribution of $\bar{x}$

Consider Figure 7.7, which shows the sampling distribution of $\bar{x}$ for the data from Examples 7.6 and 7.7. The sample mean that is greater than 95% of all sample means is indicated by the green dot on the right. *Before we do any calculations,* estimate the value of the 95th percentile of the sample means. Clearly, it lies between 70 and 75, somewhat closer to 75. So our estimate of the 95th percentile is somewhere near 73. ■

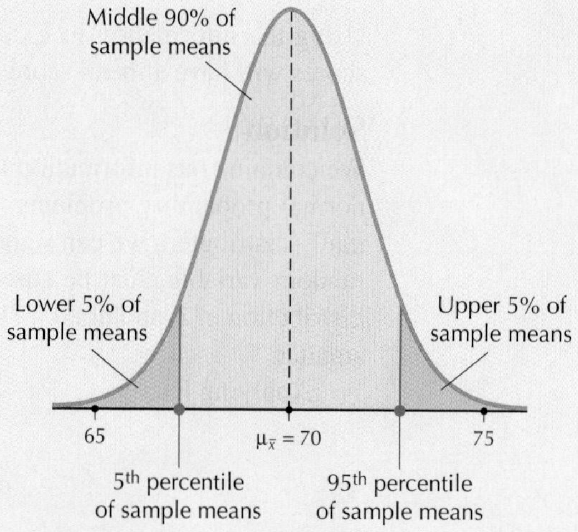


FIGURE 7.7 Sampling distribution of $\bar{x}$ with 5th and 95th percentiles.

## Example 7.8 Finding a value of $\bar{x}$, given a probability or area

Using the information in Example 7.7, find the 95th percentile of the class mean quiz scores.

### Solution

Since we want the 95th percentile, we consider the right-hand tail area in Figure 7.7, which has area 0.05, leaving area 0.95 to the left of the 95th percentile. We then use the $Z$ table *backward*. We seek 0.95 on the *inside* of the $Z$ table. Since 0.95 is not in the $Z$ table, we take the closest value. Since the two closest values, 0.9495 and 0.9505, are equally close, we split the difference. Working backward from 0.9495, we find $Z = 1.64$, and for 0.9505 we find $Z = 1.65$. Splitting the difference, we get $Z = 1.645$. This value of $Z = 1.645$ is the 95th percentile of the standard normal distribution.

Since we are looking for a sample mean quiz score, 1.645 cannot be the answer. We need to "unstandardize" by transforming this value of $Z$ using a formula similar to the formula $x = Z \cdot \sigma + \mu$ that we used in Section 6.5:

$$\bar{x} = Z \cdot \sigma_{\bar{x}} + \mu = 1.645(2) + 70 = 73.29$$

Thus, the 95th percentile of the sample means for the statistics quizzes is 73.29. This is close to our earlier estimate of 73. ■

## Example 7.9 Finding probabilities and percentiles using sample means

Use the information in Example 7.6.

**a.** Find the 5th percentile of the class mean quiz scores.

**b.** If a sample mean is chosen at random, what is the probability that it will lie between 66.71 and 73.29?

**c.** What two symmetric values for the sample mean quiz score contain 90% of all sample means between them?

### Solution

**a.** Since the sampling distribution is normal, it is also symmetric. Thus, the 95th percentile and the 5th percentile are the same distance away from the mean. Since the 95th percentile is $(73.29 - 70) = 3.29$ above the mean, the 5th percentile must be 3.29 below the mean, or $(70 - 3.29) = 66.71$.

**b.** We seek $P(66.71 < \bar{x} < 73.29)$.

#### What Results Might We Expect?

From **(a)**, we know that 66.71 represents the 5th percentile of the sampling distribution of $\bar{x}$ and from Example 7.8 we know that 73.29 is the 95th percentile. Since 90% of the sample means lie between the 5th percentile and the 95th percentile, $P(66.71 < \bar{x} < 73.29)$ must be 0.90, as shown in Figure 7.8. ■

Proceeding with the calculations, we have, as expected,

$$P(66.71 < \bar{x} < 73.29) = P\left(\frac{66.71 - 70}{2} < \frac{\bar{x} - 70}{2} < \frac{73.29 - 70}{2}\right)$$
$$= P(-1.645 < Z < 1.645) = 0.95 - 0.05 = 0.90$$

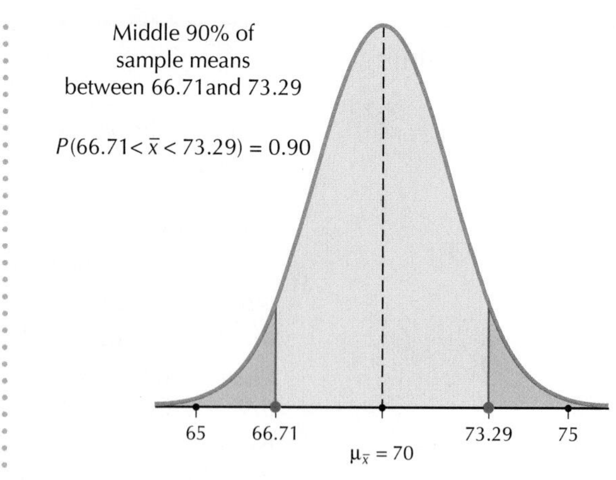


FIGURE 7.8
Middle 90% of the sample means.

**c.** This is just another way of asking for the 5th and 95th percentiles. The answer is 66.71 and 73.29.

In Section 7.2, we tackle the more challenging problem of finding the sampling distribution of the sample mean for non-normal populations.

## STEP-BY-STEP TECHNOLOGY GUIDE: Constructing Normal Probability Plots

### TI-83/84

Assume that the data set is in list L1.

***Step 1*** Access STAT PLOTS by pressing **2nd Y**.

***Step 2*** Select **1:Plot1**. Press **ENTER**.

***Step 3*** Move the cursor over **On** and press **ENTER**.

***Step 4*** Select the normal probability plot type by moving the cursor to the lower-right plot among the choices for Type. Press **ENTER**.

***Step 5*** For **Data List**, enter **L1**.

***Step 6*** For **Data Axis**, choose **X**.

***Step 7*** Press **ZOOM**, then **9: ZoomStat**.

### MINITAB

Assume that the data set is in column **C1**.

***Step 1*** From the menu, select **Graph**, then click **Probability Plot**.

***Step 2*** Select **Single** and click **OK**.

***Step 3*** In the Probability Plot dialog box, select **C1**, and click **OK**. The normal probability plot for the data set in **C1** is then generated.

## Section 7.1 SUMMARY

1 We can use sample statistics as point estimates of the unknown population parameters. For each statistic, sampling error is the distance between the point estimate and its target parameter.

2 The sampling distribution of the sample mean $\bar{x}$ for a given sample size $n$ consists of the collection of the sample means of all possible samples of size $n$ from the population.

The mean of the sampling distribution of the sample mean $\bar{x}$ is the value of the population mean $\mu$ (Fact 1). The standard deviation of the sampling distribution of the sample mean $\bar{x}$ is $\sigma_{\bar{x}} = \sigma/\sqrt{n}$, where $\sigma$ is the population standard deviation (Fact 2).

3 The sampling distribution of the sample mean for a normal population is itself normal, regardless of sample size (Fact 3).

For a normal population, the sampling distribution of the sample mean $\bar{x}$ is distributed as normal $(\mu, \sigma/\sqrt{n})$, where $\mu$ is the population mean and $\sigma$ is the population standard deviation (Fact 4).

4 Normal probability plots are used to assess the normality of a data set.

5 We can use Fact 4 to find probabilities and percentiles using sample means.

## Section 7.1 EXERCISES

### CLARIFYING THE CONCEPTS

**1.** Explain in your own words what sampling error means. In the real world, when would we know the actual value of sampling error?

**2.** Explain what a sampling distribution is. Why are sampling distributions so important?

**3.** Is it always good to keep getting larger and larger samples, even when they are very expensive? Explain. Is there a point at which the trade-off becomes less beneficial? Explain.

**4.** For a normal population, what can we say about the sampling distribution of the sample mean?

**5.** What is the relationship between the population standard deviation and the standard deviation of the sampling distribution for $\bar{x}$?

**6.** Explain what we use a normal probability plot for. What should we look for in a normal probability plot?

### PRACTICING THE TECHNIQUES

For Exercises 7–12, find the sampling error.

**7.** $\mu = 100, \bar{x} = 95$

**8.** $\mu = 0, \bar{x} = 1.5$

**9.** $\sigma = 20, s = 15$

**10.** $\sigma = 1, s = 0.9$

**11.** $p = 0.25, \hat{p} = 0.30$

**12.** $p = 0.75, \hat{p} = 0.70$

For Exercises 13–18, find $\mu_{\bar{x}}$ and $\sigma_{\bar{x}}$, the mean and standard deviation of the sampling distribution of $\bar{x}$.

**13.** $\mu = 100, \sigma = 20, n = 25$

**14.** $\mu = 100, \sigma = 20, n = 100$

**15.** $\mu = 0, \sigma = 10, n = 9$

**16.** $\mu = 0, \sigma = 10, n = 25$

**17.** $\mu = -10, \sigma = 5, n = 100$

**18.** $\mu = -10, \sigma = 5, n = 400$

For Exercises 19–26, assume that the random variable $X$ is normally distributed, with mean $\mu = 10$ and standard deviation $\sigma = 4$. That is, $X$ is normal (10,4). Let $n = 16$.

**19.** Describe the sampling distribution of $\bar{x}$ for $n = 16$.

**20.** Find the probability that $\bar{x}$ exceeds 11.

**21.** Without using your calculator, find the probability that $\bar{x}$ is less than 9.

**22.** Without using your calculator, find the probability that $\bar{x}$ lies between 9 and 11.

**23.** Find the value of $\bar{x}$ that is greater than 97.5% of all values of $\bar{x}$.

**24.** Find the value of $\bar{x}$ that is less than 97.5% of all values of $\bar{x}$.

**25.** What proportion of $\bar{x}$ lies between the values you found in Exercises 23 and 24?

**26.** What proportion of $\bar{x}$ lie outside the values you found in Exercises 23 and 24?

### APPLYING THE CONCEPTS

**Vehicle Weights.** Use the following information for Exercises 27–32. The data set **Crash**, located on your data disk, contains the results of crash tests performed in 1992 for 352 different vehicles. The variable "weight" contains the weight in pounds of the vehicles. A random sample of 46 of these vehicles was selected and their weights noted. The following table contains descriptive statistics for the weights of the 46 vehicles in the sample and population parameter values for the weights of all 352 vehicles in the population of vehicles in the data set **Crash**.

| | **Sample statistic** | **Population parameter** |
|---|---|---|
| **Mean** | $\bar{x} = 3021$ | $\mu = 2930.34$ |
| **Median** | 3040 | 2855 |
| **Minimum** | 1650 | 1590 |
| **Standard deviation** | $s = 607$ | $\sigma = 627.13$ |

**27. a.** What is the point estimate for the population mean weight of all crash test vehicles?

**b.** If we took another sample of vehicles, would you expect to get the same sample mean weight? Explain why or why not.

**c.** Is it likely that the population mean weight of all crash test vehicles equals the sample mean weight from **(b)**? Why or why not?

**28. a.** What is the point estimate for the minimum weight of all crash test vehicles?

**b.** Do you think that this estimate (using the sample of 46 vehicles) tends to overestimate or underestimate the population minimum weight of all 352 crash test vehicles? Clearly explain why you think so.

**c.** If we took another sample of vehicles, would you expect to get the same minimum weight? Explain why or why not.

**29. a.** What is the population mean weight for all the vehicles?

**b.** Find the sampling error for the mean weight.

**30. a.** What is the minimum weight for the population?
**b.** What is the sampling error for the minimum weight?

**31. a.** What is the population standard deviation?
**b.** What is the value of the sample standard deviation?
**c.** What is the sampling error for the standard deviation?

**32. a.** What is the median of the population?
**b.** What is the value of the sample median?
**c.** What is the sampling error for the median?

**33. Lab Rat Reaction Time.** A laboratory rat's mean reaction time to a stimulus is 1.7 seconds, with a standard deviation of 0.2 seconds. For each of the given sample sizes, find $\mu_{\bar{x}}$ and $\sigma_{\bar{x}}$.
**a.** Sample size $n = 9$ rats
**b.** Sample size $n = 16$ rats
**c.** Sample size $n = 25$ rats
**d.** Sample size $n = 36$ rats

**34. Student Heights.** The heights of a population of students have a mean of 68 inches (5 feet 8 inches) and a standard deviation of 3 inches. For each of the following sample sizes, find $\mu_{\bar{x}}$ and $\sigma_{\bar{x}}$.
**a.** Sample size $n = 10$ students
**b.** Sample size $n = 100$ students
**c.** Sample size $n = 1000$ students

**35.** For the lab rat reaction times in Exercise 33:
**a.** Describe what happens to $\mu_{\bar{x}}$ as $n$ increases.
**b.** Describe the pattern you see in $\sigma_{\bar{x}}$ as $n$ increases.

**36.** For the student heights in Exercise 34:
**a.** Describe what happens to $\mu_{\bar{x}}$ as $n$ increases.
**b.** Describe the pattern you see in $\sigma_{\bar{x}}$ as $n$ increases.

**37. Initial Public Offerings.** A prominent monitor of initial public offerings (IPOs) reports that the mean amount of stock offered was \$100 million with a standard deviation of \$40 million. Suppose that IPO amounts are normally distributed and that we take a sample of 4 IPOs. Find the following probabilities.
**a.** The sample mean IPO amount will be greater than \$125 million.
**b.** The sample mean IPO amount will be between \$120 million and \$140 million.
**c.** The sample mean IPO amount will be less than \$95 million.

**38. Teacher Salaries.** Suppose the salaries of teachers in your city are normally distributed with a mean of \$50,000 and a standard deviation of \$5000. Suppose we take samples of size 25 teachers.
**a.** Find the probability that the sample mean salary will exceed \$52,000.
**b.** Find the probability that the sample mean salary will be less than \$47,000.
**c.** Find the probability that the sample mean salary will be between \$52,000 and \$53,000.

**39. Initial Public Offerings.** For the initial public offerings in Exercise 37:
**a.** Find the 50th percentile of sample mean IPO amounts. Comment.
**b.** Find the 95th percentile of sample mean IPO amounts.
**c.** Find the 5th percentile of sample mean IPO amounts.
**d.** Describe the area that lies between the values you found in **(a)** and **(b)**.

**40. Teacher Salaries.** For the teacher salaries in Exercise 38:
**a.** Find the median sample mean salary. How does it compare to the population mean?
**b.** Find the 95th percentile of sample mean salaries.
**c.** Find the 5th percentile of sample mean salaries.
**d.** Between which two values do the middle 90% of sample mean salaries lie?

**41. Carbon Dioxide Concentration.** The Environmental Protection Agency reports that the mean concentration of the greenhouse gas carbon dioxide is 365 parts per million (ppm). Suppose the population is normally distributed, assume that the standard deviation is 100 ppm, and suppose we take atmospheric samples from 25 cities.
**a.** Find the probability that the sample mean carbon dioxide concentration will be above 385 ppm.
**b.** Find the probability that the sample mean carbon dioxide concentration will be between 305 and 425 ppm.
**c.** Find the probability that the sample mean carbon dioxide concentration will be below 345.

**42. Short-Term Memory.** In a famous research paper in the psychology literature, George Miller found that the amount of information humans could process in short-term memory was 7 bits (pieces of information), plus or minus 2 bits.[1] Assume that the mean number of bits is 7 and the standard deviation is 2, and that the distribution is normal. Suppose we take a sample of 100 people and test their short-term memory skills.
**a.** Find the probability that the sample mean number of bits retained in short-term memory is greater than 7.5.
**b.** Find the probability that the sample mean number of bits retained in short-term memory is between 6.8 and 7.2.
**c.** Find the probability that the sample mean number of bits retained in short-term memory is less than 6.75.

**43. Carbon Dioxide Concentration.** For the carbon dioxide concentrations in Exercise 41, do the following.
**a.** Find the sample mean carbon dioxide concentration higher than 97.5% of all such sample means.
**b.** Find the sample mean carbon dioxide concentration lower than 97.5% of all such sample means.
**c.** Describe the area that lies between the values you found in **(a)** and **(b)**.

**44. Short-Term Memory.** For the numbers of bits in Exercise 42:
**a.** Find the 97.5th percentile of the sample mean number of bits retained in short-term memory.
**b.** Find the 2.5th percentile of the sample mean number of bits retained in short-term memory.
**c.** Between which two values do the middle 95% of sample mean number of bits retained in short-term memory lie?

**2006 Olympic Figure Skating Scores.** Use this information for Exercises 45–48. The table shows the Ladies' Short

Program scores for the top five finishers in the 2006 Winter Olympic Games in Torino, Italy. *Consider these scores to be a population.*

| Skater | Nation | Score |
|---|---|---|
| Sasha Cohen | United States | 66.73 |
| Irina Slutskaya | Russia | 66.70 |
| Shizuka Arakawa | Japan | 66.02 |
| Fumie Suguri | Japan | 61.75 |
| Kimmie Meissner | United States | 59.40 |

**45.** **a.** How many samples of size $n = 2$ can we generate from this population of size 5?

**b.** Compute the population mean $\mu$. Do we usually know the value of the population mean in a typical real-world problem? Why or why not?

**c.** Calculate the population standard deviation $\sigma$. Do we usually know the value of the population standard deviation in a typical real-world problem? Why or why not?

**46.** Take every possible sample of size $n = 2$ skaters from this tiny population.

**a.** Find the mean score $\bar{x}$ of each sample.

**b.** Construct a dotplot of the sample mean scores.

**47.** Refer to the dotplot in your answer to Exercise 46.

**a.** Where would the balance point be located in the dotplot? Indicate it on the plot.

**b.** Recall that the balance point is an estimate of the mean. What is your estimate of $\mu_{\bar{x}}$, using this balance point?

**48.** **a.** Find the mean of all the sample mean scores in Exercise 46.

**b.** Is your estimate for Exercise 47b close to your answer to Exercise 48a?

**c.** Does your answer to Exercise 48a agree with value of the population mean that you found in Exercise 45b?

**49. A Fair Die.** Consider a fair six-sided die. Suppose we take samples of size 16 and are interested in the population mean of the die rolls.

**a.** Find $\mu_{\bar{x}}$.

**b.** Find $\sigma_{\bar{x}}$. (*Hint:* First find the standard deviation of a fair die roll using a frequency distribution.)

**Stock Portfolios.** Use this information for Exercises 50–54. A stockbroker is examining her track record. Over the last 12 quarters, the mean net gain in stock price for all her clients' portfolios was $4, with a standard deviation of $6. She investigates and finds that the distribution is normal.

**50.** What is the probability that a randomly selected client will have a net loss in stock price (a gain less than 0)?

**51.** Suppose the stockbroker takes random samples of size $n = 9$ from all her clients.

**a.** Find $\mu_{\bar{x}}$.

**b.** Find $\sigma_{\bar{x}}$.

**c.** What can you say about the sampling distribution of the sample mean? How do you know this?

**52.** Refer to Exercise 51.

**a.** What is the probability that a sample of nine clients have a sample mean net loss in stock price?

**b.** Why is the probability so much lower for the sample mean than for a particular client?

**c.** Which figure would look better on the broker's resume?

WHAT IF? **53.** Suppose the population standard deviation was less than $6. Explain how this would affect the following, if at all.

**a.** $\mu$

**b.** $n$

**c.** Probability that a randomly selected client will have a net loss in stock price

**d.** $\mu_{\bar{x}}$

**e.** $\sigma_{\bar{x}}$

WHAT IF? **54.** Suppose the population standard deviation was less than $6. Explain how this would affect the following, if at all.

**a.** Sampling distribution of the sample mean

**b.** Probability that a sample of nine clients have a sample mean net loss in stock price

**SAT Math Scores.** Use this information for Exercises 55–59. The College Board (**www.collegeboard.com**) reports that the nationwide mean math SAT score was 515 in 2007. Assume that the standard deviation is 116 and that the scores are normally distributed.

**55.** What is the probability that a randomly selected SAT math score will be less than 500?

**56.** As a researcher, you are looking at samples of SAT math scores of size 16.

**a.** Find $\mu_{\bar{x}}$.

**b.** Find $\sigma_{\bar{x}}$.

**c.** What can you say about the sampling distribution of the sample mean? How do you know this?

**57.** Refer to Exercise 56.

**a.** What is the probability that a sample of 16 students will have a sample mean math SAT score below 500?

**b.** Why is the probability so much lower for the sample mean than for a particular student?

**58.** Refer to Exercise 57. Suppose the population standard deviation was greater than 116. Explain how this would affect the following, if at all.

**a.** Probability that a randomly selected SAT math score will be less than 500

**b.** $\mu_{\bar{x}}$

**c.** $\sigma_{\bar{x}}$

**d.** Distribution of the sample mean

**59.** Suppose the population standard deviation was greater than $116. Explain how this would affect the following, if at all.

**a.** Probability that a sample of 16 SAT math scores will have a mean less than 500

**b.** 99.5th percentile of the sample mean SAT math scores

**c.** 0.5th percentile of the sample mean SAT math scores

**Law of Large Numbers for Means Applet.** Use the *Law of Large Numbers for Means* applet for Exercises 60 and 61.

**60.** Set the number of **Rolls** to 10, and select **Show Mean.** Click **Roll dice.**

**a.** Observe the behavior of the sample mean $\bar{x}$ compared to the population mean $\mu = 3.5$.

**b.** Estimate the sampling error $|\bar{x} - \mu|$.

**c.** Without pressing **Reset**, click **Roll dice** until the total number of dice equals $n = 20, 40, 60, 80$, and 100.

**d.** In each case in **(c)**, estimate the sampling error $|\bar{x} - \mu|$.

**61.** What can you conclude about the sampling error $|\bar{x} - \mu|$ as the sample size $n$ increases?

**Law of Large Numbers for Proportions Applet.** Use the *Law of Large Numbers for Proportions* applet for Exercises 62 and 63.

**62.** Set the probability of heads to $p = 0.5$ and the number of tosses to $n = 40$. Click **Toss.**

**a.** Record the sample proportion $\hat{p}$ of heads observed and the sampling error $|\hat{p} - p|$.

**b.** Without pressing **Reset**, continue to click **Toss** until the total number of tosses is $n = 120$. Again record the sample proportion $\hat{p}$ of heads observed and the sampling error $|\hat{p} - p|$.

**c.** Without pressing **Reset**, continue to click **Toss** until the total number of tosses is $n = 180$. Again record the sample proportion $\hat{p}$ of heads observed and the sampling error $|\hat{p} - p|$.

**d.** Without pressing **Reset**, continue to click **Toss** until the total number of tosses is $n = 240$. Again record the sample proportion $\hat{p}$ of heads observed and the sampling error $|\hat{p} - p|$.

**63.** What can you conclude about the sampling error $|\hat{p} - p|$ as the sample size $n$ increases?

## 7.2 Central Limit Theorem for Means

*Objectives:* **By the end of this section, I will be able to...**

1 Describe the sampling distribution of $\bar{x}$ for skewed and symmetric populations as the sample size increases.

2 Apply the Central Limit Theorem for Means to solve probability questions about the sample mean.

### 1 Sampling Distribution of $\bar{x}$ for Skewed and Symmetric Populations

In Section 7.1, we discovered that the sampling distribution for the sample mean for a normal population is also normal. What if the population is not normal? In this section, we use simulation studies to learn how the sampling distribution of the sample mean $\bar{x}$ for non-normal populations becomes approximately normal as the sample size increases. We start with small sample sizes and work our way up, keeping our eye on the sampling distribution of the sample mean $\bar{x}$.

### Example 7.10 Simulation study: Sample means from strongly skewed population

The data set **Nutrition** on your CD and the companion Web site contains nutrition information on a population of 961 foods.

**a.** Construct a histogram of the potassium content of these 961 foods, and describe the shape of the population distribution.

**b.** Using Minitab, take 500 random samples of sizes $n = 10$, 20, and 30 from the population. Assess the normality of the resulting sampling distributions of $\bar{x}$ using histograms and normal probability plots.

**Solution**

**a.** A histogram of the potassium content of these foods is shown in Figure 7.9, revealing a strongly right-skewed, non-normal data set.

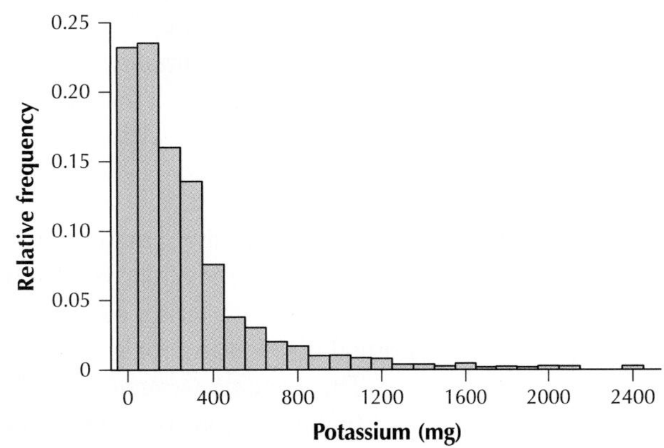


FIGURE 7.9 Potassium content is strongly right-skewed, not normal.

**b.** Using Minitab, we take 500 random samples of size $n = 10$ from the population. We find the means of the 500 samples shown in the graphs in Figure 7.10.

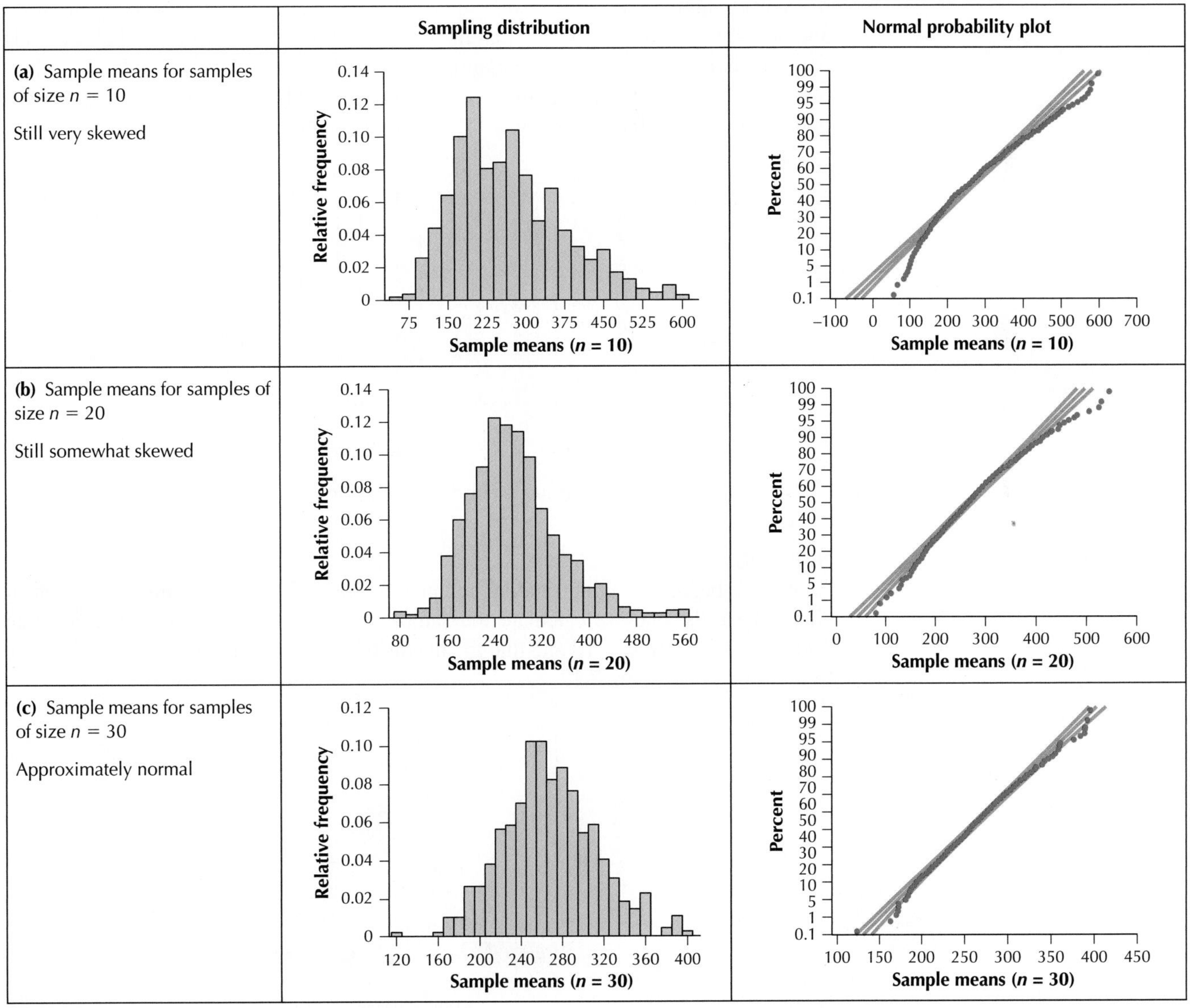

| | Sampling distribution | Normal probability plot |
|---|---|---|
| **(a)** Sample means for samples of size $n = 10$<br>Still very skewed | Relative frequency; Sample means ($n$ = 10) | Percent; Sample means ($n$ = 10) |
| **(b)** Sample means for samples of size $n = 20$<br>Still somewhat skewed | Relative frequency; Sample means ($n$ = 20) | Percent; Sample means ($n$ = 20) |
| **(c)** Sample means for samples of size $n = 30$<br>Approximately normal | Relative frequency; Sample means ($n$ = 30) | Percent; Sample means ($n$ = 30) |

FIGURE 7.10 Sampling distribution of $\bar{x}$ and normal probability plots for $n$ = 10, 20, and 30.

- $n = 10$: The sampling distribution of $\bar{x}$ is skewed (Figure 7.10a).
- $n = 20$: The sampling distribution of $\bar{x}$ is still somewhat skewed (Figure 7.10b).
- $n = 30$: Despite a few outliers, the sampling distribution of $\bar{x}$ is approximately normal (Figure 7.10c).

For a skewed population, we have seen that the sampling distribution of the sample mean becomes approximately normal as the sample size reaches 30. For a less skewed population, we can expect that the sampling distribution of $\bar{x}$ approximates a normal distribution for smaller sample sizes.

This is demonstrated in the following simulation of a symmetric population. Recall that a distribution is **symmetric** if an axis of symmetry splits the image in half so that one side is the mirror image of the other.

## Example 7.11 Simulation study: Sample means from a symmetric distribution

Figure 7.11 shows the population distribution of values from a single roll of a fair die. Since each number is equally likely to be rolled, the distribution is perfectly uniform and therefore symmetric. Recall that the population mean, or the mean of all possible die rolls, is $\mu = 3.5$, and the population standard deviation of all possible die rolls is $\sigma = 1.7078$. Take random samples of sizes $n = 10$ and $n = 20$ from this symmetric population. Assess the normality of the resulting sampling distribution of $\bar{x}$.

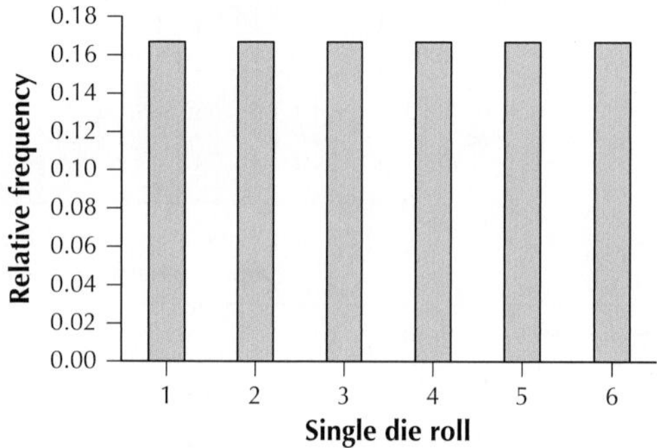


FIGURE 7.11 Distribution of a single fair die roll is symmetric.

### Solution

Again, we begin by taking random samples of size $n = 10$ from this population of single die rolls, and we calculate the sample mean $\bar{x}$. Using Minitab, we generate 100 samples of size $n = 10$ and find the means of each of these 100 samples, shown in the histogram in Figure 7.12.

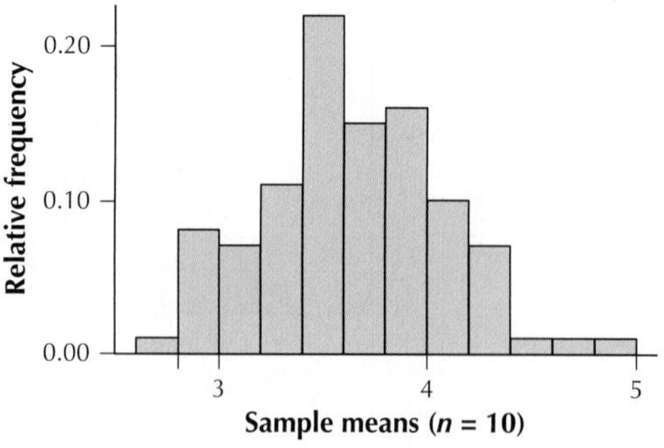


FIGURE 7.12 Sample means of size $n = 10$: already approaching normality.

We see that the sample means are already approaching normality, even for a small sample size like $n = 10$. Clearly the sample means are already "settling down." Note also that the variability in the histogram has been reduced. For example, there are no sample means of 1 or 6. This is analogous to the situation where you would not be surprised if an individual student got 100 on an exam but you would be very surprised if the mean on the exam was 100.

Figure 7.13 shows the sample means for samples of size $n = 20$. In the normal probability plot in Figure 7.14 only a few points toward the right tail are off the line—nothing systematic. We can therefore conclude that, already at $n = 20$, the sampling distribution of the sample means is approximately normal. (It should be noted that the discrete, non-continuous nature of the population seems to be the main obstacle to an even better normal approximation for the means, which could be expected with a uniform continuous population.)

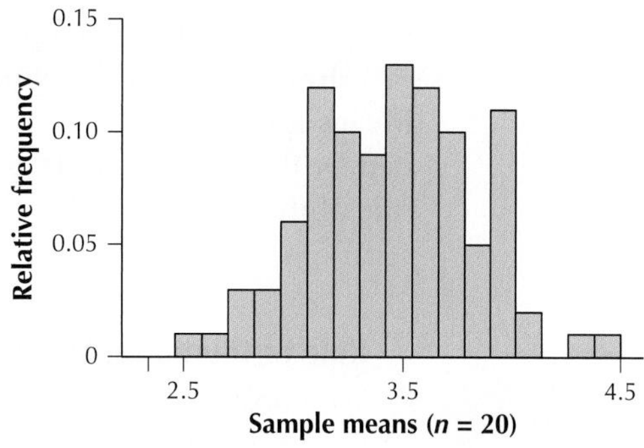


FIGURE 7.13 Sample means of size $n = 20$: approximately normal.

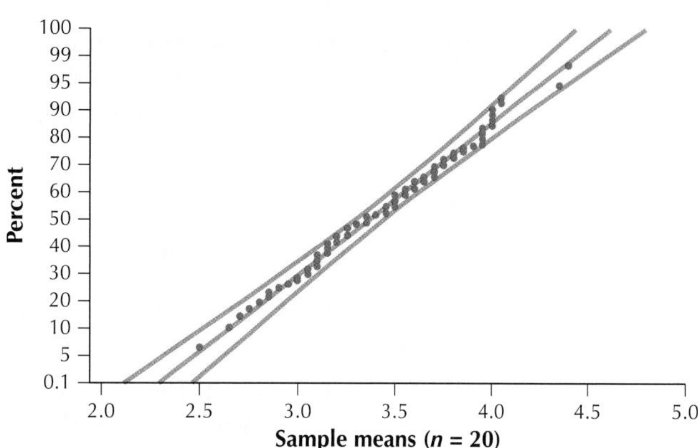


FIGURE 7.14 Normal probability plot for $n = 20$: acceptable normality.

## 2 Using Central Limit Theorem for Means to Solve Probability Problems About the Sample Mean

We can summarize our investigation into the behavior of the sample means for non-normal populations as follows: *Regardless of the population, the sampling distribution of the sample mean becomes approximately normal as the sample size gets larger*. We can then combine this statement with Facts 3 and 4 to form the **Central Limit Theorem for Means**.

**Central Limit Theorem for Means**

Given a population with mean $\mu$ and standard deviation $\sigma$, the sampling distribution of the sample mean $\bar{x}$ becomes approximately normal $(\mu, \sigma/\sqrt{n})$ as the sample size gets larger, regardless of the shape of the population.

The Central Limit Theorem (CLT) is one of the most important results in statistics. Worldwide, much statistical inference is based on the CLT. It actually makes fairly intuitive sense, doesn't it? If we find the mean of a sample of data values, in many cases the extreme values will tend to be blended away. However, remember that the mean is very sensitive to outliers. In a small sample, there may not be enough non-extreme values to balance the influence of the outliers. This is what was happening early in the potassium simulation (for example, Figure 7.10a). However, as the sample sizes increase, the influence of extreme values diminishes and the resulting sample means start to migrate toward the center.

How large does the sample size have to be before the Central Limit Theorem for Means takes effect? As we have seen, it depends on the degree of symmetry exhibited by the population. In general, however, we shall abide by the following rough rule of thumb.

**Rule of Thumb for When Central Limit Theorem for Means Takes Effect**

We consider $n \geq 30$ as large enough to apply the Central Limit Theorem for any population.

The *Central Limit Theorem* applet allows you to experiment with various sample sizes and see how the Central Limit Theorem for Means behaves in action.

Combining Fact 4 and the Central Limit Theorem for Means, we can identify three cases for the sampling distribution of $\bar{x}$.

**Three Cases for the Sampling Distribution of the Sample Mean $\bar{x}$**

- **Case 1: The population is normal.** Then the sampling distribution of $\bar{x}$ is *normal* (Fact 3).
- **Case 2: The population is either non-normal or of unknown distribution *and* the sample size is at least 30**. Then the sampling distribution of $\bar{x}$ is *approximately normal* (Central Limit Theorem for Means).
- **Case 3: The population is either non-normal or of unknown distribution *and* the sample size is less than 30**. Then we have *insufficient information* to conclude that the sampling distribution of the sample mean $\bar{x}$ is either normal or approximately normal.

Of course, in the real world, no one will tell you which of the three cases applies. You need to investigate the assumptions of each of the cases to determine for yourself which case applies.

## Example 7.12 Sometimes the solution to a problem lies beyond our available means

The U.S. Small Business Administration (SBA) provides information on the number of small businesses for each metropolitan area in the United States.[2] Figure 7.15 shows a histogram of our population for this example, the number of small businesses in each of the 328 cities nationwide. (For example, Austin, Texas, has 22,305 small businesses, while Pensacola, Florida, has 6020.) The mean is $\mu = 12{,}485$ and the standard deviation is $\sigma = 21{,}973$. Find the probability that a random sample of size $n = 10$ cities will have a mean number of small businesses greater than 17,000.

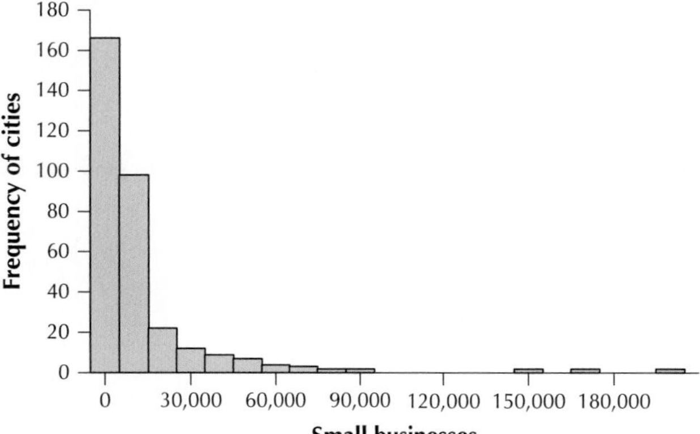


FIGURE 7.15

### Solution

First we try to apply Case 1. Clearly, the population is not normal, so Case 1 does not apply. We cannot apply Case 2, since the sample size $n = 10$ is too small. We must default to Case 3. The population is skewed and the sample size is small. Therefore, we have insufficient information to conclude that the sampling distribution of the sample mean $\bar{x}$ is either normal or approximately normal.

Using the methods in this textbook, we cannot find the probability that a random sample of size $n = 10$ cities will have a mean number of small businesses greater than 17,000. ■

## Example 7.13 Application of the Central Limit Theorem for the Mean

Suppose we have the same data set as in Example 7.12, but this time we increase our sample size to 36. Now, try again to find the probability that a random sample of size $n = 36$ cities will have a mean number of small businesses greater than 17,000.

### Solution

We try to apply Case 2. Since the sample size $n = 36$ is large enough, the Central Limit Theorem applies. The sampling distribution of the sample mean $\bar{x}$ is approximately normal. Now that we know that the sampling distribution of $\bar{x}$ is approximately normal, we need to find $\mu_{\bar{x}}$ and $\sigma_{\bar{x}}$. Facts 1 and 2 tell us that

$$\mu_{\bar{x}} = \mu = 12{,}485 \qquad \text{and} \qquad \sigma_{\bar{x}} = \frac{\sigma}{\sqrt{n}} = \frac{21{,}973}{\sqrt{36}} \approx 3662.1667$$

Therefore, as the CLT indicates, the sampling distribution of $\bar{x}$ is approximately normal($\mu_{\bar{x}} = 12{,}485$, $\sigma_{\bar{x}} = 3662.1667$). We are then left to solve a normal probability problem, just as we did in Chapter 6. Figure 7.16 shows the sampling distribution of $\bar{x}$ and the probability we are interested in, $P(\bar{x} > 17{,}000)$. Using Fact 5, we standardize:

$$Z = \frac{17{,}000 - \mu_{\bar{x}}}{\sigma_{\bar{x}}} = \frac{17{,}000 - 12{,}485}{3662.1667} \approx 1.2329 \approx 1.23$$

Thus, $P(\bar{x} > 17{,}000) \approx P(Z > 1.23)$, as shown in Figure 7.17.

We therefore look up $Z = 1.23$ in the $Z$ table and subtract this table area (0.8907) from 1 to get the desired tail area:

$$P(Z > 1.23) = 1 - 0.8907 = 0.1093$$

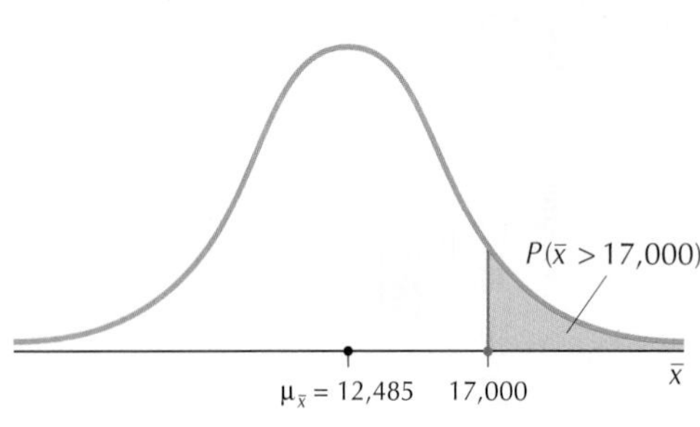


FIGURE 7.16

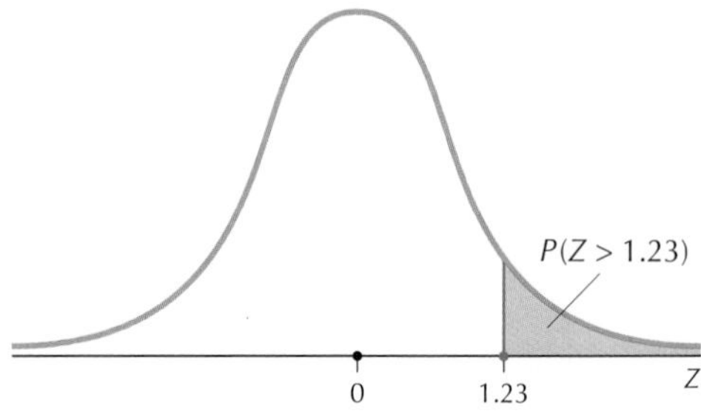


FIGURE 7.17

There is a 10.93% probability that a random sample of 36 cities will have a mean number of small businesses greater than 17,000. ■

## What If Scenario — Give the Calculator a Rest

In this scenario, we explore how increasing the sample size affects the sampling distribution of $\bar{x}$. Put down your calculator and think through the problem in the scenario. We continue our analysis of the small business data in Examples 7.12 and 7.13, where $\mu = 12{,}485$ and $\sigma = 21{,}973$. In Example 7.13, we took samples of size $n = 36$. *What if* we increase our sample size $n$ to an *unspecified* number larger than 36.

**a.** Describe how this change will affect $\sigma_{\bar{x}}$.

**b.** How will this change affect the probability that the sample mean number of small businesses exceeds 17,000?

**c.** How will this change affect the 99th percentile of the sample mean number of small businesses?

**d.** How will this change affect the 1st percentile?

### Solution

**a.** Note the presence of $n$ in the denominator of the formula $\sigma_{\bar{x}} = \frac{\sigma}{\sqrt{n}}$. This has the effect, as $n$ increases, of decreasing the value of $\sigma_{\bar{x}}$. So, if $n$ is larger than 36, then $\sigma_{\bar{x}}$ will become smaller than the 3662.1667 we found in Example 7.13.

**b.** Note the presence of $\sigma_{\bar{x}}$ in the denominator of $Z = \frac{\bar{x} - \mu_{\bar{x}}}{\sigma_{\bar{x}}}$. If $\sigma_{\bar{x}}$ is decreased (from **(a)**), the quantity $Z = (17{,}000 - \mu_{\bar{x}})/\sigma_{\bar{x}}$ will increase, producing a bigger value of $Z$ than the 1.23 we saw in Example 7.13. Call this bigger value $z_1$. Then the area to the right of $z_1$ will necessarily be smaller than the area to the right of $Z = 1.23$, as shown in Figure 7.18. Thus, increasing the number of cities in our sample will decrease the probability that the sample mean number of small businesses exceeds 17,000.

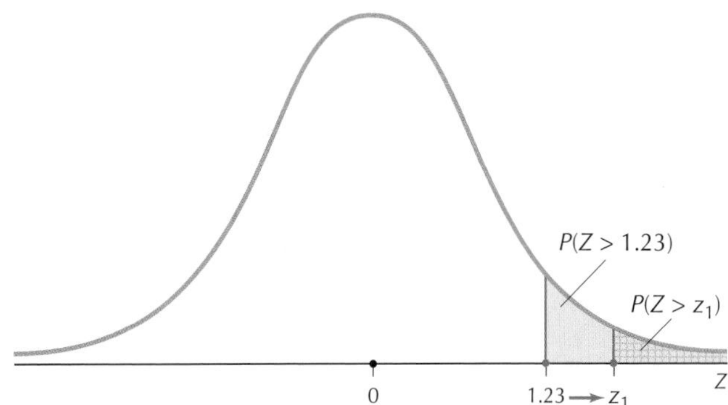


FIGURE 7.18
As $Z$ increases, the area to its right decreases.

**c.** To find the 99th percentile, we use the formula from Example 7.8, $\bar{x} = Z \cdot \sigma_{\bar{x}} + \mu$, where the Z-value is chosen so that it is larger than 99% of standard normal values. From **(a)** we know that increasing the sample size will decrease $\sigma_{\bar{x}}$. Therefore, since $Z$ and $\mu$ are constants in this formula, it follows that, if $\sigma_{\bar{x}}$ is smaller, then the quantity $\bar{x} = Z \cdot \sigma_{\bar{x}} + \mu$ will also be smaller. Therefore, the 99th percentile of the sample means will be smaller.

**d.** The sampling distribution of $\bar{x}$, being normal, is thereby also symmetric. Thus, if the 99th percentile is decreasing (getting close to the mean), then the 1st percentile must be increasing (also getting close to the mean). This is confirmed by substituting the smaller $\sigma_{\bar{x}}$ into the formula $\bar{x} = Z \cdot \sigma_{\bar{x}} + \mu$. Therefore, the 1st percentile will be larger.

The key for all these behaviors is to remember that *increasing the sample size will decrease the variability of the sampling distribution.* Roughly speaking, the sample means will tend to gravitate toward the center ($\mu$). In this way, the extremely large values (for example, the 99th percentile) become smaller, and the extremely small values (for example, the 1st percentile) become larger.

CASE STUDY

## Trial of the Pyx: How Much Gold Is in Your Gold Coins?

Medieval English kings devised a procedure to ensure that the coins of the realm contained the proper amount of gold. A sample of 100 of the gold coins that were cast each year was placed in a ceremonial box called the Pyx. At the chosen time, the Company of Goldsmiths jury weighed the gold coins. The weight of the entire sample of coins was supposed to be 12,800 grams. If the weight was much less than 12,800 grams, the jury concluded that the Master of the Mint was cheating the crown by pocketing the excess gold, and he was severely punished.

### Problem 1

How did the jury determine what was "much less than 12,800 grams"?

### Solution

If the weight of the coins was within 32 grams of the expected 12,800 grams, the jury accepted the year's gold coins as pure. Thus, the *sum* of the weights of the gold coins had to lie between 12,768 grams and 12,832 grams. However, the methods we have learned in this chapter have dealt with the sampling distribution of the sample *mean,* not the sum. Since the sample mean $\bar{x} = \frac{\sum X}{n}$, we divide the total weight by $n = 100$. So the probability that the Master of the Mint passed the Trial of the Pyx is

$$P(12{,}768 < \textstyle\sum X < 12{,}832)$$

which, divided through by $n = 100$, becomes

$$P\left(127.68 < \frac{\sum X}{n} < 128.32\right) \qquad \text{or} \qquad P(127.68 < \bar{x} < 128.32)$$

That is, the throne would have accepted the guineas as pure if the mean weight of the guineas lay between 127.68 and 128.32 grams, as shown in Figure 7.19.

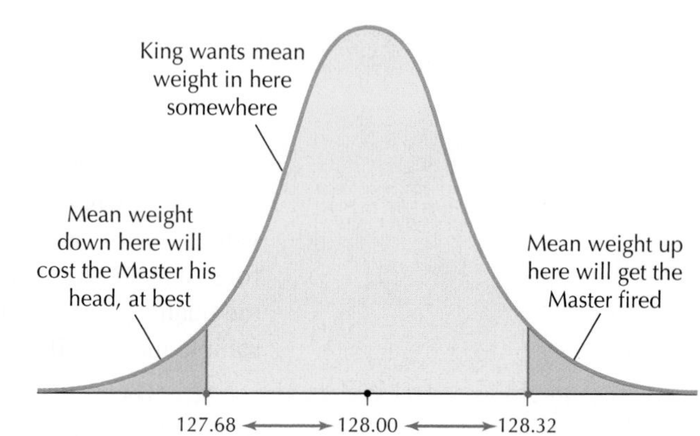


FIGURE 7.19 Sampling distribution of mean weight of 100 gold coins if population mean weight is 128 grams.

## Problem 2

Can we estimate what the jury used for a standard deviation?

## Solution

Let's assume that "much less than" indicated a measurement that is 2 or more standard deviations below average. For the sampling distribution of $\bar{x}$, then, this would indicate a range of $0.32 = 2\,\sigma_{\bar{x}}$ between 127.68 and the mean 128, as indicated by the arrows in Figure 7.19. Therefore, $\sigma_{\bar{x}} = 0.16$. And therefore, by the Empirical Rule, for instance, approximately 95% of the sample mean observations for the Trial of the Pyx would have been between 127.68 and 128.32. Since $\sigma_{\bar{x}} = \sigma/\sqrt{n}$, it follows that $\sigma = \sqrt{100} \cdot 0.16 = 1.6$ grams.

## Problem 3

What were the chances that the Master of the Mint would have been caught and punished if he were in fact cheating the throne?

## Solution

What if the Master of the Mint set the mean amount of gold in the population of all coins to be $\mu = 127.9$ grams instead of the required 128, shortchanging the crown by a tenth of a gram of gold per coin? The jury would never have noticed this, would they?

Let's calculate the probability that the Master of the Mint would have passed the Trial of the Pyx if the mean amount of gold in the coins had been only 127.9 grams. We've seen that the Master of the Mint would have passed the Trial of the Pyx if $127.68 < \bar{x} < 128.32$. Now, because 100 is a large sample size, the Central Limit Theorem tells us that the sampling distribution of $\bar{x}$ is approximately normal, with $\mu_{\bar{x}} = \mu = 127.9$ and $\sigma_{\bar{x}} = \frac{\sigma}{\sqrt{n}} = \frac{1.6}{\sqrt{100}} = 0.16$.

Standardizing using Fact 5,

$$Z = \frac{127.68 - \mu_{\bar{x}}}{\sigma_{\bar{x}}} = \frac{127.68 - 127.9}{0.16} \approx -1.38 \qquad \text{and}$$

$$Z = \frac{128.32 - \mu_{\bar{x}}}{\sigma_{\bar{x}}} = \frac{128.32 - 127.9}{0.16} \approx 2.63$$

Solving using Table 6.7 in Section 6.4 (page 298)

$$P(-1.38 < Z < 2.63) = 0.9957 - 0.0838 = 0.9119$$

That is, the chances of the crown accepting the coins as pure, even if the Master of the Mint had been shortchanging by a tenth of a gram per coin, were over 91% (Figure 7.20).

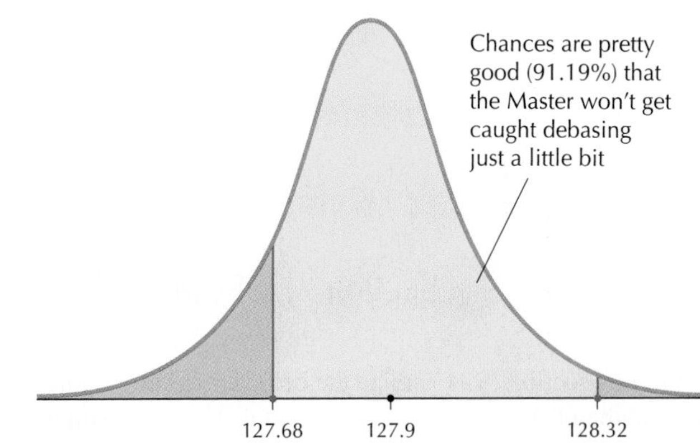
Chances are pretty good (91.19%) that the Master won't get caught debasing just a little bit
127.68
127.9
128.32

FIGURE 7.20 Sampling distribution if population mean gold weight is reduced to 127.9 grams.

*Note:* Sir William Sharington, 1493–1553, Master of the Mint during the turbulent Tudor era in England. He debased the currency, issued worthless coinage, and diverted the real gold to fund Thomas Seymour's conspiracy to topple the government and seize young King Edward VI. Sharington was arrested in 1548 or 1549, but he later received pardon and became Sheriff of Wiltshire for a short time before he died.

## Problem 4

Would the Master of the Mint have been satisfied with this small amount of debasement? Would he have quit while he was ahead?

## Solution

No way! The following year the Master of the Mint decided to debase the currency even further, setting the mean amount of gold in the guineas to be $\mu = 127.3$ grams per coin.

We need to find the probability of the Master passing the Trial of the Pyx if the mean amount of gold in a coin was 127.3 grams instead of the required 128 grams per coin. We use the same calculations, with $\mu_{\bar{x}} = 127.3$ grams. Standardizing:

$$Z = \frac{127.68 - \mu_{\bar{x}}}{\sigma_{\bar{x}}} = \frac{127.68 - 127.3}{0.16} \approx 2.38 \qquad \text{and}$$

$$Z = \frac{128.32 - \mu_{\bar{x}}}{\sigma_{\bar{x}}} = \frac{128.32 - 127.3}{0.16} \approx 6.38$$

Then $P(2.38 < Z < 6.38) \approx 1 - 0.9913 = 0.0087$.

In other words, the Master of the Mint actually would have stood very little chance—less than 1% probability—of passing the Trial of the Pyx if he cheated by this much (Figure 7.21).

England is a great country for retaining fine old traditions. Today England's Company of Goldsmiths still operates the London Assay Office where the purity of the kingdom's coin is tested at the annual Trial of the Pyx.

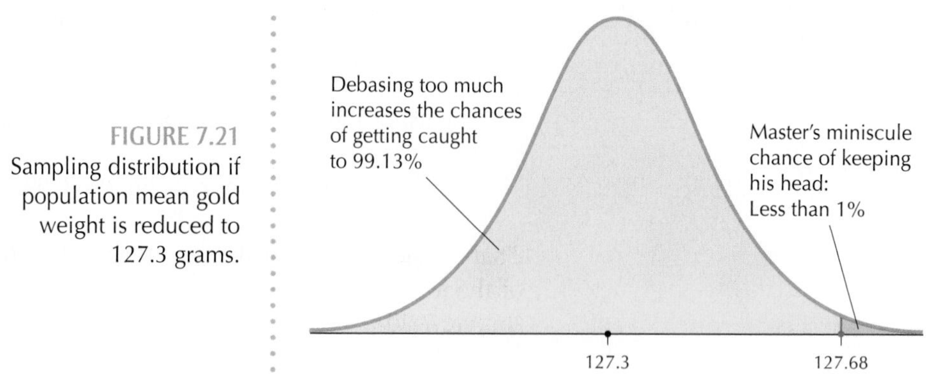


FIGURE 7.21 Sampling distribution if population mean gold weight is reduced to 127.3 grams.

## Section 7.2 SUMMARY

1 In this section, we examine the behavior of the sample mean when the population is not normal. The approximate normality of the sampling distribution of the sample mean kicks in much quicker when the original population is symmetric rather than skewed.

2 The Central Limit Theorem is one of the most important results in statistics and is stated as follows:

Given a population with mean $\mu$ and standard deviation $\sigma$, the sampling distribution of the sample mean $\bar{x}$ becomes approximately normal($\mu$, $\sigma/\sqrt{n}$) as the sample size gets larger, regardless of the shape of the population. This approximation applies for smaller sample sizes when the original distribution is more symmetric.

## Section 7.2 EXERCISES

### CLARIFYING THE CONCEPTS

**1.** For which types of population distribution does the sampling distribution of the sample mean become approximately normal quicker: symmetric or skewed distributions? Explain in your own words why this happens.

**2.** Compare the $x$ axes (the scale) of the histograms in Figure 7.10. What do you notice as the sample size increases? How does the scale change?

**3.** Refer to Exercise 2. What does this change indicate about the variability in the sampling distribution?

**4.** True or false: The Central Limit Theorem states that the sampling distribution of $\bar{x}$ is approximately normal, regardless of sample size.

### PRACTICING THE TECHNIQUES

In Exercises 5–12, samples are taken. Determine which of the three cases for the sampling distribution of the sample mean applies.

**5.** SAT scores are normally distributed and the sample size is small.

**6.** SAT scores are not normally distributed and the sample size is large.
**7.** Accountant incomes are normally distributed and the sample size is small.
**8.** Accountant incomes are not normally distributed and the sample size is small.
**9.** Systolic blood pressure readings are not normally distributed and the sample size is large.
**10.** Systolic blood pressure readings are normally distributed and the sample size is small.
**11.** Body mass indices of college students are not normally distributed and the sample size is small.
**12.** Body mass indices of college students are not normally distributed and the sample size is large.

In Exercises 13–20, samples are taken. Provide (a) $\mu_{\bar{x}}$ and (b) $\sigma_{\bar{x}}$, and (c) determine whether the sampling distribution of $\bar{x}$ is normal, approximately normal, or unknown.
**13.** SAT scores are normally distributed, with $\mu = 516$ and $\sigma = 116$. A sample of size 9 is taken.
**14.** SAT scores are not normally distributed, with $\mu = 516$ and $\sigma = 116$. A sample of size 36 is taken.
**15.** Accountant incomes are normally distributed, with $\mu = \$60{,}000$ and $\sigma = \$10{,}000$. A sample of size 16 is taken.
**16.** Accountant incomes are not normally distributed, with $\mu = \$60{,}000$ and $\sigma = \$10{,}000$. A sample of size 16 is taken.
**17.** Systolic blood pressure readings are not normally distributed, with $\mu = 80$ and $\sigma = 8$. A sample of size 64 is taken.
**18.** Systolic blood pressure readings are normally distributed, with $\mu = 80$ and $\sigma = 8$. A sample of size 25 is taken.
**19.** The gas mileage for 2007 Toyota Prius hybrid automobiles is not normally distributed, with $\mu = 50$ miles per gallon and $\sigma = 6$. A sample of size 16 is taken.
**20.** The gas mileage for 2007 Toyota Prius hybrid automobiles is not normally distributed, with $\mu = 50$ miles per gallon and $\sigma = 6$. A sample of size 64 is taken.

For Exercises 21–28, if possible find the indicated probability. If not possible, explain why not.
**21.** The exchange rate for U.S.$100 in euros for 2007 is not normally distributed, with $\mu = 74$ euros and $\sigma = 5$. A sample of size 9 is taken. Find $P(\bar{x} > 75)$.
**22.** The exchange rate for U.S.$100 in euros for 2007 is not normally distributed, with $\mu = 74$ euros and $\sigma = 5$. A sample of size 36 is taken. Find $P(\bar{x} > 75)$.
**23.** 2007 prices for boned trout are normally distributed, with $\mu = \$3.10$ per pound and $\sigma = \$0.30$. A sample of size 16 is taken. Find $P(\bar{x} < \$3)$.
**24.** 2007 prices for boned trout are not normally distributed, with $\mu = \$3.10$ per pound and $\sigma = \$0.30$. A sample of size 16 is taken. Find $P(\bar{x} < \$3)$.
**25.** Systolic blood pressure readings are not normally distributed, with $\mu = 80$ and $\sigma = 8$. A sample of size 64 is taken. Find $P(78 < \bar{x} < 82)$.
**26.** Systolic blood pressure readings are normally distributed, with $\mu = 80$ and $\sigma = 8$. A sample of size 25 is taken. Find $P(78 < \bar{x} < 82)$.
**27.** The pollen count distribution for Los Angeles in September is not normally distributed, with $\mu = 8.0$ and $\sigma = 1.0$. A sample of size 16 is taken. Find $P(\bar{x} > 9.0)$.
**28.** The pollen count distribution for Los Angeles in September is not normally distributed, with $\mu = 8.0$ and $\sigma = 1.0$. A sample of size 64 is taken. Find $P(\bar{x} > 9.0)$.

For Exercises 29–38, find the indicated value of $\bar{x}$. If it is not possible, explain why not.
**29.** SAT scores are normally distributed, with $\mu = 516$ and $\sigma = 116$. A sample of size 9 is taken. Find the sample mean SAT score larger than 95% of all sample means.
**30.** SAT scores are not normally distributed, with $\mu = 516$ and $\sigma = 116$. A sample of size 36 is taken. Find the sample mean SAT score larger than 95% of all sample means.
**31.** 2007 prices for boned trout are normally distributed, with $\mu = \$3.10$ per pound and $\sigma = \$0.30$. A sample of size 16 is taken. Find the sample mean price that is smaller than 90% of all sample means.
**32.** 2007 prices for boned trout are not normally distributed, with $\mu = \$3.10$ per pound and $\sigma = \$0.30$. A sample of size 16 is taken. Find the sample mean price that is smaller than 90% of all sample means.
**33.** Accountant incomes are normally distributed, with $\mu = \$60{,}000$ and $\sigma = \$10{,}000$. A sample of size 16 is taken. Find the sample mean accountant income that is smaller than 90% of all sample means.
**34.** Accountant incomes are not normally distributed, with $\mu = \$60{,}000$ and $\sigma = \$10{,}000$. A sample of size 16 is taken. Find the sample mean accountant income that is smaller than 90% of all sample means.
**35.** Systolic blood pressure readings are not normally distributed, with $\mu = 80$ and $\sigma = 8$. A sample of size 64 is taken. Find the 95th percentile of sample means.
**36.** Systolic blood pressure readings are normally distributed, with $\mu = 80$ and $\sigma = 8$. A sample of size 25 is taken. Find the 95th percentile of sample means.
**37.** The pollen count distribution for Los Angeles in September is not normally distributed, with $\mu = 8.0$ and $\sigma = 1.0$. A sample of size 16 is taken. Find the 75th percentile of sample means.
**38.** The pollen count distribution for Los Angeles in September is not normally distributed, with $\mu = 8.0$ and $\sigma = 1.0$. A sample of size 64 is taken. Find the 75th percentile of sample means.

## APPLYING THE CONCEPTS

**39. Cholesterol Levels.** The Centers for Disease Control reports that the mean serum cholesterol level in Americans in 2005 was 202. Assume that the standard deviation is 45.

There is no information about the distribution. We take a sample of 36 Americans.

**a.** Does Case 1 apply? If so, find $P(\bar{x} > 212)$. If not, explain why not.

**b.** Does Case 2 apply? If so, find $P(\bar{x} > 212)$. If not, explain why not.

**40. Stock Prices.** A stockbroker was examining her track record. The mean net gain in stock price for all her clients' portfolios was \$4, with a standard deviation of \$6. She has no information about the distribution of net gains in stock price, and she takes a sample of 16 stocks.

**a.** Does Case 1 apply? If so, find the probability that the sample of 16 stocks will have a mean net loss in stock price (again less than 0). If not, explain why not.

**b.** Does Case 2 apply? If so, find the probability that the sample of 16 stocks will have a mean net loss in stock price. If not, explain why not.

**41. Shaquille O'Neal.** In the 2004 National Basketball Association playoffs, Los Angeles Lakers star Shaquille O'Neal averaged 21.5 points per game. Suppose that the standard deviation is 5 points and that points per game for Shaquille are normally distributed. Suppose we take a sample of 4 of Shaquille's playoff games and are interested in whether the sample mean points per game is better than 25 points.

**a.** Does Case 1 apply? If so, find $P(\bar{x} > 25)$. If not, explain why not.

**b.** Does Case 2 apply? If so, find $P(\bar{x} > 25)$. If not, explain why not.

**42. Tennessee Temperatures.** According to the National Oceanic and Atmospheric Administration, the mean temperature for Nashville, Tennessee, in the month of January between 1872 and 2006 was 38.6°F. Assume that the standard deviation is 10 degrees and the distribution is normal. Suppose we take a sample of 25 days.

**a.** Does Case 1 apply? If so, find $P(\bar{x} > 40)$. If not, explain why not.

**b.** Does Case 2 apply? If so, find $P(\bar{x} > 40)$. If not, explain why not.

**43. Sociology Professors' Salaries.** The American Sociological Association reports that the mean salary of all new sociology professors in 2004 was \$45,722. Suppose that the standard deviation is \$6000 and that the distribution of salaries is right-skewed. Suppose we take a sample of 16 new sociology professors and are interested in whether the sample mean salary exceeds \$48,000.

**a.** Does Case 1 apply? If so, find $P(\bar{x} > 48{,}000)$. If not, explain why not.

**b.** Does Case 2 apply? If so, find $P(\bar{x} > 48{,}000)$. If not, explain why not.

**44. Computers per School.** The National Center for Educational Statistics (**http://nces.ed.gov**) reported that the mean number of instructional computers per public school nationwide was 124. Assume that the standard deviation is 49 computers and that there is no information about the shape of the distribution. Suppose we take a sample of size 75 public schools, and suppose we are interested in the probability that the sample mean number of instructional computers is less than 100.

**a.** Does Case 1 apply? If so, proceed to find the $P(\bar{x} < 100)$. If not, explain why not.

**b.** Does Case 2 apply? If so, proceed to find the $P(\bar{x} < 100)$. If not, explain why not.

**c.** What is the probability that a sample of size 75 public schools will have more than 100 instructional computers? (*Hint:* Use your answers to **(a)** and **(b)** and take the easy way to a solution.)

**45. Sociology Professors' Salaries.** Refer to Exercise 43. Suppose we take a sample of 36 new sociology professors and are interested in whether the sample mean salary exceeds \$48,000.

**a.** Does Case 1 apply? If so, find $P(\bar{x} > 48{,}000)$. If not, explain why not.

**b.** Does Case 2 apply? If so, find $P(\bar{x} > 48{,}000)$. If not, explain why not.

**Strikeouts.** Use this information for Exercises 46 and 47. The population mean number of strikeouts per player in the 2002 American League baseball season for the 331 batters is $\mu = 42.695 \approx 42.7$, with standard deviation $\sigma = 38.614 \approx 38.6$. Figure 7.22 shows the normal probability plot of the strikeout data.

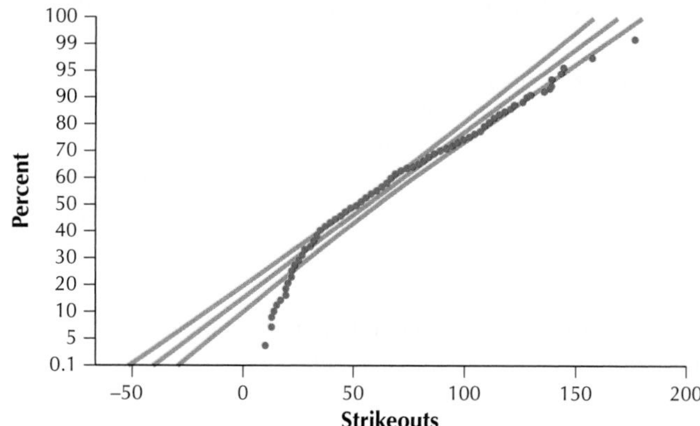


Figure 7.22

**46.** If possible, find the probability that a random sample of size $n = 10$ batters will have a mean of more than 50 strikeouts. If it is not possible, explain why not.

**47.** If possible, find the probability that a random sample of size $n = 36$ batters will have a mean of more than 50 strikeouts. If it is not possible, explain why not.

Use the *Central Limit Theorem* applet for Exercises 48 and 49.

**48.** Describe the shape of the sampling distribution of $\bar{x}$ for the following sample sizes.

**a.** 2

**b.** 5

**c.** 30

**49.** At what sample size would you say the sampling distribution of $\bar{x}$ becomes approximately normal?

**Coaching and SAT Scores.** Use this information for Exercises 50–57. The College Board reports that the mean increase in SAT math scores of students who attend review courses (coaching) is 18 points. Assume that the standard deviation of the change in score is 12 points and that changes in score are not normally distributed. We are interested in the probability that the sample mean score increase is negative, indicating a loss of points after coaching.

**50.** Suppose we take a sample of size 20.

**a.** Does Case 1 apply? If so, proceed to find $P(\bar{x} < 0)$, the probability that the sample mean score increase is negative, indicating a loss of points after coaching. If not, explain why not.

**b.** Does Case 2 apply? If so, proceed to find $P(\bar{x} < 0)$. If not, explain why not.

**51.** Suppose we take a sample of size 40, and we are interested in $P(\bar{x} < 0)$, indicating a loss of points after coaching.

**a.** Does Case 1 apply? If so, proceed to find $P(\bar{x} < 0)$. If not, explain why not.

**b.** Does Case 2 apply? If so, proceed to find $P(\bar{x} < 0)$. If not, explain why not.

**52.** Suppose we take a sample of size 40. What is the probability that the sample mean of a sample of size 40 students will gain points after coaching? (*Hint:* Use your answers to Exercise 51 and take the easy way to a solution.)

**53.** Suppose we take a sample of size 80, and we are interested in $P(\bar{x} < 0)$.

**a.** *Before doing any calculations,* predict whether this probability will be greater or smaller than if we had used $n = 40$. Why?

**b.** Verify your intuition by finding $P(\bar{x} < 0)$ for $n = 80$.

**54.** *What if* the sample size used was some unspecified value greater than 80. Describe how and why this change would have affected the following, if at all. Would the quantities increase, decrease, remain unchanged? Or is there insufficient information to tell what would happen? Explain your answers.

**a.** $\mu_{\bar{x}}$

**b.** $\sigma_{\bar{x}}$

**c.** $Z = \dfrac{\bar{x} - \mu_{\bar{x}}}{\sigma_{\bar{x}}}$

**d.** $P(\bar{x} < 0)$

**55.** Refer to Exercise 54. Describe how and why this change would have affected the following, if at all. Would the means increase, decrease, remain unchanged? Or is there insufficient information to tell what would happen? Explain your answers.

**a.** Sample mean increase in SAT math score larger than 97.5% of all sample means

**b.** Sample mean increase in SAT math score smaller than 97.5% of all sample means

**c.** Two sample means that, together, span the middle 95% of all sample means

**56.** Suppose we take a sample of 100 students.

**a.** Find $P(\bar{x} > 0)$, the probability that the sample of 100 students had an increase in mean SAT score.

**b.** Find the 95th percentile of SAT score sample mean increases after coaching.

**c.** Find the 5th percentile of SAT score sample mean increases after coaching.

**d.** Between which two symmetric values lie 90% of all sample mean SAT score increases?

**e.** By hand, draw a normal curve with your answer to **(d)** indicated.

**57.** *What if* the sample size is decreased from 100 to some value between 30 and 100.

**a.** Would $\sigma_{\bar{x}} = \sigma/\sqrt{n}$ increase or decrease? Why?

**b.** How would the effect in **(a)** affect $P(\bar{x} > 0)$? Would it be larger or smaller? Why?

Refer to the Case Study on pages 359–362 for Exercises 58–64.

**58.** What were the chances that the Master of the Mint would have been accused of cheating if he had in fact been completely honest?

In the Case Study, we interpreted the phrase "much less than" to mean a measurement that is 2 or more standard deviations below average. For Exercises 59–62, suppose we interpret the phrase "much less than" to mean a measurement that is 3 or more standard deviations below average.

**59.** Answer the following questions. Use the methods shown in the Case Study.

**a.** What would be the new value of $\sigma_{\bar{x}}$?

**b.** What would be the standard deviation $\sigma$ for each gold piece?

**c.** By the Empirical Rule, about what proportion of the mean weights would you now expect to lie between 127.68 and 128.32?

**60.** Assume just "a little debasement" from 128 grams per coin to 127.9 grams per coin.

**a.** Find $P(127.68 < \bar{x} < 128.32)$.

**b.** What was the probability that the Master of the Mint would get away with the debasement? That he would be caught?

c. Compare the probabilities in **(b)** with those in the Case Study in the text. Which values favor the Master of the Mint?

**61.** Assume that greed attacked the Master of the Mint and he went for the big debasement from 128 grams per coin to 127.3 grams per coin.

a. Find $P(127.68 < \bar{x} < 128.32)$.

b. What was the probability that the Master of the Mint would get away with the debasement? That he would be caught?

c. Compare the probabilities in **(b)** with those in the Case Study in the text. Which values favor the Master of the Mint?

**62.** The Master of the Mint wanted to choose a mean value so that his chances of failing the Trial of the Pyx were 25%. Which mean value is this?

**63.** Suppose we interpret the phrase "much less than" to mean a measurement that is 2 or more standard deviations below average. Again, the Master of the Mint wanted to choose a mean value so that his chances of failing the Trial of the Pyx were 25%. Which mean value is this?

**64.** Compare the two values you derived in Exercises 63 and 64.

a. Which value would make the Master of the Mint less likely to be caught?

b. Which value would divert more gold to him?

## 7.3 Central Limit Theorem for Proportions

*Objectives:* **By the end of this section, I will be able to...**

1. Explain the sampling distribution of the sample proportion $\hat{p}$.
2. Describe the sampling distribution of the sample proportion $\hat{p}$ for extreme and moderate values of $p$.
3. Apply the Central Limit Theorem for Proportions to solve probability questions about the sample proportion.

### 1 Sampling Distribution of the Sample Proportion $\hat{p}$

The sample mean is not the only statistic that can have a sampling distribution. Every sample statistic has a sampling distribution. One of the most important is the sampling distribution of the **sample proportion $\hat{p}$**.

Suppose each individual in a population either has or does not have a particular characteristic. If we take a sample of size $n$ from this population, the **sample proportion $\hat{p}$** (read "p-hat") is

$$\hat{p} = \frac{x}{n}$$

where $x$ represents the number of individuals in the sample that have the particular characteristic. We use $\hat{p}$ to estimate the unknown value of the population proportion $p$. In Section 6.2, we were introduced to $\hat{p}$ as the sample proportion of successes in a binomial experiment.

Just as there is a Central Limit Theorem for Means, there is a Central Limit Theorem for Proportions. In Section 7.2, we learned how the Central Limit Theorem for Means allows us to consider the sampling distribution of the sample mean as approximately normal for sufficiently large sample sizes. Here in Section 7.3, we develop a Central Limit Theorem for Proportions, where the **sampling distribution of the sample proportion** also becomes approximately normal if the right conditions are satisfied.

> The **sampling distribution of the sample proportion $\hat{p}$** for a given sample size $n$ consists of the collection of the sample proportions of all possible samples of size $n$ from the population.
>
> In general, the **sampling distribution of any particular statistic** for a given sample size $n$ consists of the collection of the values of that sample statistic across all possible samples of size $n$.

## Example 7.14 Genders of the student government members

Photofusion Picture Library/Alamy

Suppose we are interested in the proportion of female members of the student government. Since 3 of the 5 members of the student government are female, the population proportion of females is $p = 3/5 = 0.6$.

| Amber | Brandon | Chantal | Dave | Emma |
|---|---|---|---|---|
| Female | Male | Female | Male | Female |

Suppose we take all possible samples of size 3. For example, the first sample of Amber, Brandon, and Chantal contains 2 females and 1 male, so the sample proportion of females is $\hat{p} = 2/3$. The second sample of Amber, Brandon, and Dave contains 1 female and 2 males, so $\hat{p} = 1/3$.

**a.** Make a table of the sample proportion $\hat{p}$ and the sampling error $|\hat{p} - p|$ for all possible samples of size $n = 3$.

**b.** Construct a dot plot of the values of $\hat{p}$ for all possible samples of size $n = 3$.

### Solution

Table 7.6 shows the results from all possible samples of size 3. The highlighted row contains the values of the sampling distribution of the statistic $\hat{p}$. These values are plotted in Figure 7.23.

Table 7.6 All possible samples of size 3 from population of student government members

| Sample | Amber Brandon Chantal | Amber Brandon Dave | Amber Brandon Emma | Amber Chantal Dave | Amber Chantal Emma | Amber Dave Emma | Brandon Chantal Dave | Brandon Chantal Emma | Brandon Dave Emma | Chantal Dave Emma |
|---|---|---|---|---|---|---|---|---|---|---|
| Data | F M F | F M M | F M F | F F M | F F F | F M F | M F M | M F F | M M F | F M F |
| $\hat{p}$ | 2/3 | 1/3 | 2/3 | 2/3 | 3/3 | 2/3 | 1/3 | 2/3 | 1/3 | 2/3 |
| $p$ | 3/5 | 3/5 | 3/5 | 3/5 | 3/5 | 3/5 | 3/5 | 3/5 | 3/5 | 3/5 |
| $\|\hat{p} - p\|$ | 1/15 | 4/15 | 1/15 | 1/15 | 6/15 | 1/15 | 4/15 | 1/15 | 4/15 | 1/15 |

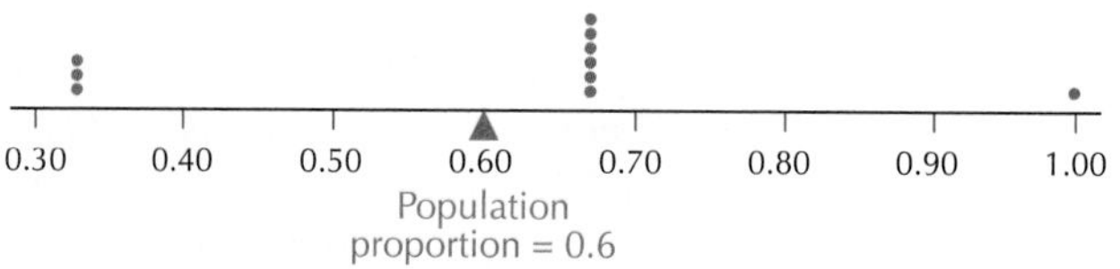


FIGURE 7.23 Sampling distribution of the sample proportion for Example 7.14.

Just as the population parameter $\mu$ was a constant, so also is the target parameter $p$, the population proportion of females in the student government. However, with each sample we take, we obtain a different estimate of the population proportion $p$ because the sample proportion $\hat{p}$ is a random variable. The sample proportion is a random variable because we do not know its value before the random sample of student government members is chosen.

## Example 7.15 Mean of sample proportions

Calculate the mean of the ten sample proportions $\hat{p}$ from Table 7.6.

**Solution**

The mean of these ten sample proportions is

$$\frac{\left(\frac{2}{3}+\frac{1}{3}+\frac{2}{3}+\frac{2}{3}+\frac{3}{3}+\frac{2}{3}+\frac{1}{3}+\frac{2}{3}+\frac{1}{3}+\frac{2}{3}\right)}{10}=0.6$$

It is no coincidence that the mean of all the sample proportions in the sampling distribution of $\hat{p}$ exactly equals the population proportion of females for the original population, $p = 3/5 = 0.6$. We generalize this result in Fact 6. ■

**Fact 6: Mean of the Sampling Distribution of the Sample Proportion $\hat{p}$**

The mean of the sampling distribution of the sample proportion $\hat{p}$ is the value of the population proportion $p$. This may be denoted as $\mu_{\hat{p}} = p$ and read as "the mean of the sampling distribution of $\hat{p}$ is $p$."

Fact 6 provides a measure of center for the sampling distribution of the sample proportion $\hat{p}$, and Fact 7 provides a measure of spread. We will use these and other results as we build toward the Central Limit Theorem for Proportions.

*Note:* Just like for $\sigma_{\bar{x}}$ (see page 340), the *finite population correction factor*

$$\sqrt{\frac{N-n}{N-1}}$$

is used when the population is not much larger than the population. (See Exercise 59 on page 381.)

**Fact 7: Standard Deviation of the Sampling Distribution of the Sample Proportion p**

The standard deviation of the sampling distribution of the sample proportion $\hat{p}$ is $\sigma_{\hat{p}} = \sqrt{\frac{p \cdot (1-p)}{n}}$, where $p$ is the population proportion and $n$ is the sample size.

Developing Your Statistical Sense

**Why Learn This Stuff?**

The mean and standard deviation of a sampling distribution are crucial because they help us understand what value of the statistic can be considered unusual, and they give us an idea of where the bulk of the values of the statistic lie. These considerations are important for building toward *statistical inference*. For example, in *hypothesis testing* (Chapter 9), a hypothesized value of a parameter is rejected if the value of a certain statistic based on that value is too unusual. ■

## Example 7.16 Color blindness in men

The National Institutes of Health reported that color blindness linked to the X chromosome afflicts 8% of men. Suppose we take a random sample of 100 men and let $p$ denote the proportion of men in the population who have color blindness linked to the X chromosome. Find $\mu_{\hat{p}}$ and $\sigma_{\hat{p}}$.

**Solution**

First, we note that this is a binomial experiment with $p = 0.08$ and $n = 100$. Fact 6 tells us that $\mu_{\hat{p}} = p$, that is, the sampling distribution of the sample proportion $\hat{p}$ has a mean of $p = 0.08$. Fact 7 states that

$$\sigma_{\hat{p}} = \sqrt{\frac{p \cdot (1 - p)}{n}} = \sqrt{\frac{0.08 \cdot (1 - 0.08)}{100}} = \sqrt{0.000736} \approx 0.02713$$ ■

What Do These Numbers Mean?

Imagine that we repeatedly draw random samples of 100 men and observe the proportion of men $\hat{p}$ in each sample who have color blindness linked to the X chromosome. Each sample provides us with a value for $\hat{p}$. Eventually, the values of $\hat{p}$, when graphed, form the sampling distribution shown in Figure 7.24.

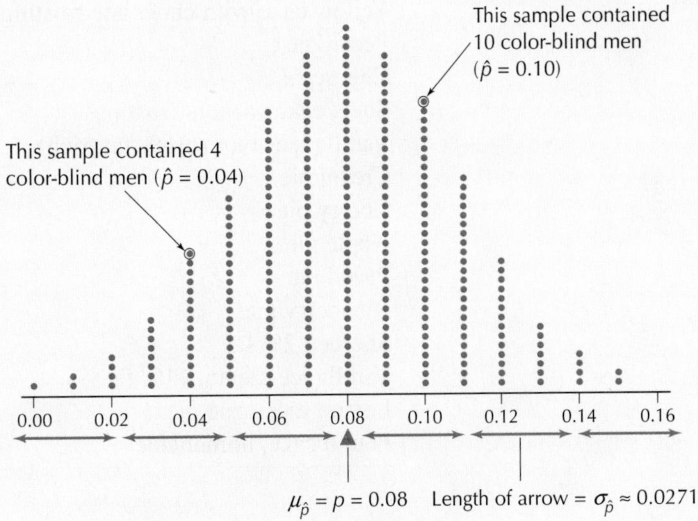


FIGURE 7.24 Sampling distribution of sample proportion $\hat{p}$.

Note that $\mu_{\hat{p}} = p = 0.08$ is located at the balance point of this distribution, which we should expect since the mean proportion of these samples is $\mu_{\hat{p}} = p = 0.08$. Each arrow represents one standard deviation of this distribution, a measure of spread given by $\sigma_{\hat{p}} = 0.02713$. Note that nearly all the sample proportions lie within three standard deviations of the mean. ■

## 2 Sampling Distribution of $\hat{p}$ for Extreme and Moderate Values of $p$

Next, we investigate whether we can develop a Central Limit Theorem for Proportions using a simulation study as we did for the means.

### Example 7.17 Simulation study: Proportion of high-calorie foods

We return to the **Nutrition** data set, this time examining the calorie content in the 961 foods. We define a high-calorie food item as one that contains at least 2000 calories. There are $X = 20$ high-calorie foods in the data set, shown in Table 7.7.

Defining "success" as a high-calorie food item, we find that the population proportion of successes is $p = \frac{20}{961} \approx 0.02081$. The very small proportion was chosen deliberately as an extreme case.

Table 7.7 Food items with 2000 calories or more

| Food item | Serving size | Calories |
|---|---|---|
| Carrot cake with cream cheese frosting | 1 cake | 6175 |
| Fruitcake, dark, homemade | 1 cake | 5185 |
| White cake with white frosting, commercial | 1 cake | 4170 |
| Sheet cake with white frosting, homemade | 1 cake | 4020 |
| Yellow cake with chocolate frosting, commercial | 1 cake | 3895 |
| Devil's food cake with chocolate frosting, mix | 1 cake | 3755 |
| Yellow cake with chocolate frosting, mix | 1 cake | 3735 |
| Pecan pie | 1 pie | 3450 |
| Cheesecake | 1 cake | 3350 |
| Sheet cake without frosting | 1 cake | 2830 |
| Vanilla ice cream, 16% fat (rich) | ½ gallon | 2805 |
| Cream pie | 1 pie | 2710 |
| Cherry pie | 1 pie | 2465 |
| Apple pie | 1 pie | 2420 |
| Peach pie | 1 pie | 2410 |
| Blueberry pie | 1 pie | 2285 |
| Sherbet, 2% fat | ½ gallon | 2160 |
| Vanilla ice cream, 11% fat | ½ gallon | 2155 |
| Lemon meringue pie | 1 pie | 2140 |
| Pound cake, homemade | 1 loaf | 2025 |

Start by taking random samples of size $n = 10$ from the population. Generate 100 samples of size $n = 10$ and find the proportion $\hat{p}$ of each sample. Then gradually increase the sample size, repeating the process for sample size $n = 50$, 100, 200, and 250. Assess the normality of each sampling distribution of $\hat{p}$.

**Solution**

- $n = 10$: The sampling distribution of $\hat{p}$ for these 100 samples is quite right-skewed. In fact, the skewness is so severe that we increase the sample size to $n = 50$ (Figure 7.25a).

- $n = 50$: The sampling distribution of $\hat{p}$ for these 100 samples is still skewed (Figure 7.25b).
- $n = 100$: The sampling distribution of $\hat{p}$ for these 100 samples is still somewhat skewed (Figure 7.25c).
- $n = 200$: The sampling distribution of $\hat{p}$ for these 100 samples is finally approaching approximate normality (Figure 7.25d).
- $n = 250$: Generating 100 samples of size $n = 250$, we seem to have finally achieved approximate normality (Figure 7.25e).

(a) Sample proportions for sample size $n = 10$ extremely skewed

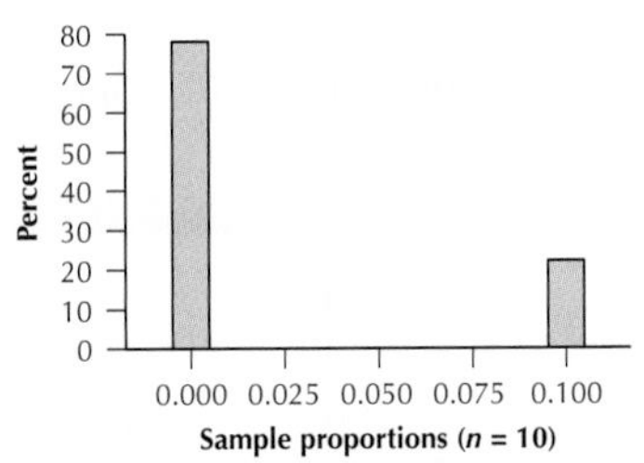


(b) Sample proportions for sample size $n = 50$ still skewed

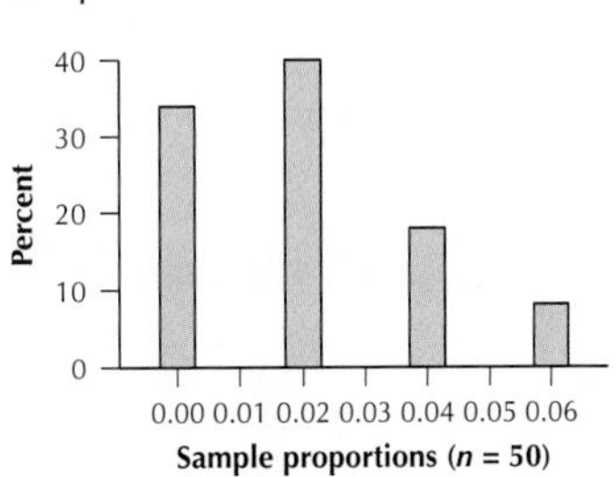


(c) Sample proportions for sample size $n = 100$ still somewhat skewed

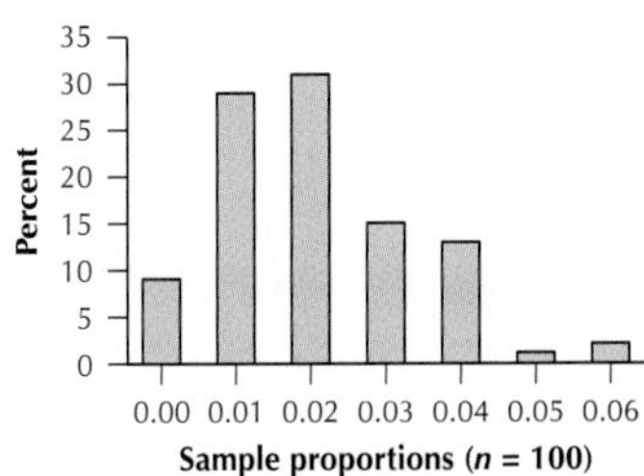


(d) Sample proportions for sample size $n = 200$ approaching approximate normality

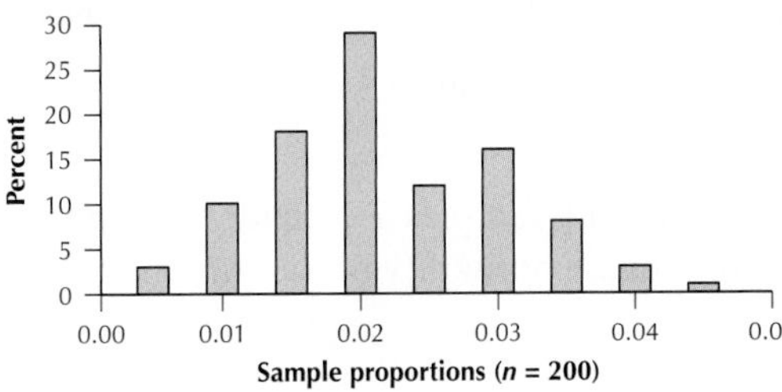


(e) Approximate normality achieved at $n = 250$

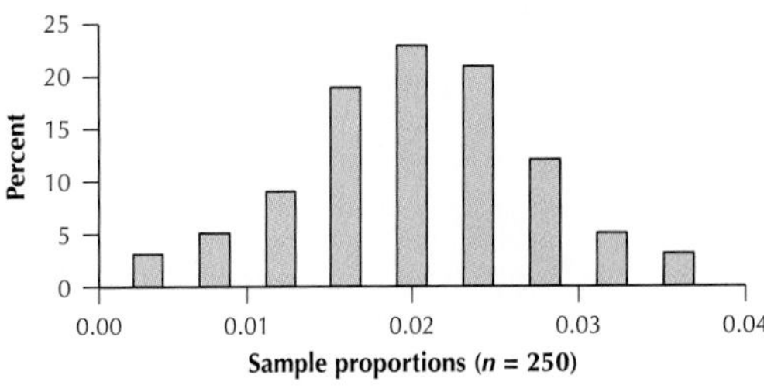


FIGURE 7.25 Sampling distribution of sample proportion for $n = 10, 50, 100, 200$, and 250.

Approximate normality for the sampling distribution of the sample proportion $\hat{p}$ appears to finally be achieved when the sample size has attained the rather dizzying size of $n = 250$. This is more than eight times larger than the sample size ($n = 30$) needed for the sampling distribution for sample means. Clearly, we need a new rule of thumb for proportions. Recall that the population proportion of high-calorie foods was very low, $p = 0.02081$, which is an extreme case. We'll have a better idea of what this new rule of thumb should be after we look at a "better behaved" example. ■

## Example 7.18 Simulation study: Proportion of foods containing vitamin C

Many different foods have some vitamin C (ascorbic acid). Of the 961 food items listed in the **Nutrition** data set, 457 have some vitamin C. Defining "success" as a food item containing vitamin C, we find that the population proportion of successes is

$$\hat{p} = \frac{457}{961} \approx 0.4755$$

This value of $\hat{p}$ near 0.5 is in fact nearly a best-case scenario for the sampling distribution of the sample proportion, as we shall see. Generate 100 random samples of size $n = 10$ and size $n = 12$ and find the sample proportion $\hat{p}$ for each sample.

**Solution**

- $n = 10$: The sampling distribution of the sample proportion $\hat{p}$ for these 100 samples is already almost normal (Figure 7.26a).
- $n = 12$: Even with a modest increase in sample size, the sampling distribution of $\hat{p}$ achieves approximate normality (Figure 7.26b).

Thus, for this moderate case of $p \approx 0.4755$, the sampling distribution of $\hat{p}$ achieved approximate normality much quicker than for $p \approx 0.02081$. The sample size $n = 250$ required for the extreme case in Example 7.17 was about 21 times the moderate case $n = 12$. ■

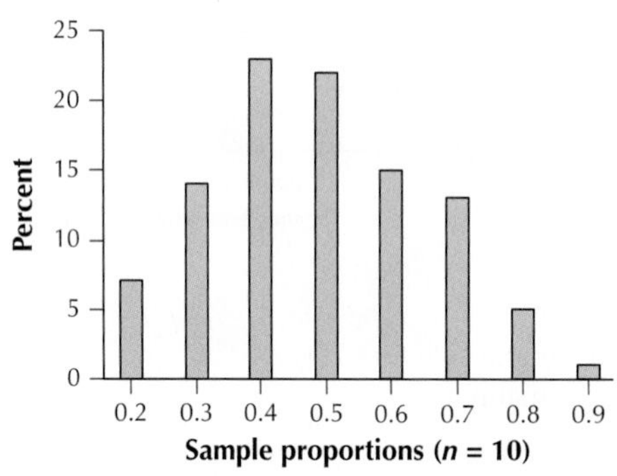


(b) For $n = 12$, approximately normal

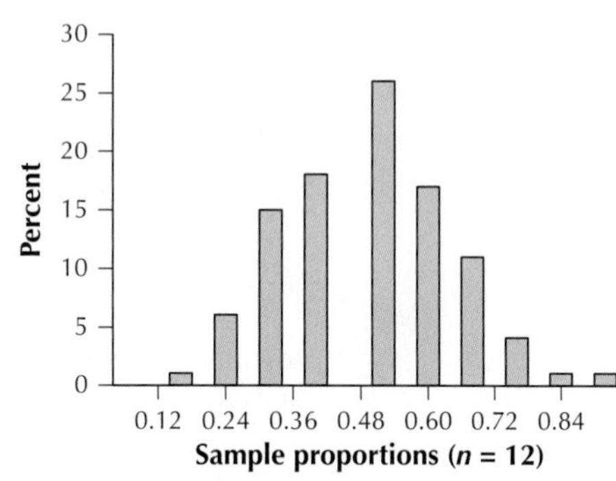


FIGURE 7.26 Sampling distribution of sample proportion for $n = 10$ and 12.

The simulation studies in Examples 7.17 and 7.18 lead us to state the following conditions for the approximate normality of the sampling distribution of $\hat{p}$.

**Fact 8: Conditions for Approximate Normality for the Sampling Distribution of the Sample Proportion $\hat{p}$**

The sampling distribution of the sample proportion $\hat{p}$ may be considered approximately normal only if both the following conditions hold:

$$(1)\ np \geq 5 \qquad \text{and} \qquad (2)\ n(1-p) \geq 5$$

The *minimum sample size* required to produce approximate normality in the sampling distribution of $\hat{p}$ is the *larger* of either

$$n_1 = \frac{5}{p} \qquad \text{or} \qquad n_2 = \frac{5}{(1-p)}$$

(rounded up to the next integer).

To find out where these two conditions for approximate normality come from, see *Discovering the Minimum Sample Size for Normality* at the end of this section (page 377).

## 3 Central Limit Theorem for Proportions

Based on the simulation studies in Examples 7.17 and 7.18, we are finally ready to express the Central Limit Theorem for Proportions, using information from Facts 6, 7, and 8.

**Central Limit Theorem for Proportions**
The sampling distribution of the sample proportion $\hat{p}$ follows an approximately normal distribution with mean $\mu_{\hat{p}} = p$ and standard deviation $\sigma_{\hat{p}} = \sqrt{\frac{p \cdot (1-p)}{n}}$ when both the following conditions are satisfied: (1) $np \geq 5$ *and* (2) $n(1-p) \geq 5$.

## Example 7.19 Color blindness in men

In Example 7.16, we learned that color blindness linked to the X chromosome afflicts 8% of men. Describe the sampling distribution of $\hat{p}$, the proportion of men who have color blindness linked to the X chromosome, for samples of size (a) 50 and (b) 100.

### Solution

We need to check both conditions to find whether the sampling distribution of $\hat{p}$ is approximately normal.

**a.** We are given that $p = 0.08$ and $n = 50$.

$$np = 50 \cdot 0.08 = 4 \qquad \text{and} \qquad n(1-p) = 50 \cdot (0.92) = 46$$

Since 4 is not $\geq 5$, then the first condition is not satisfied. The Central Limit Theorem for Proportions cannot be used. We cannot conclude that the sampling distribution of $\hat{p}$ is approximately normal.

**b.** Here $p = 0.08$ and $n = 100$.

$$np = 100 \cdot 0.08 = 8 \qquad \text{and} \qquad n(1-p) = 100 \cdot (0.92) = 92$$

Since both 8 and 92 are $\geq 5$, both conditions are satisfied. The Central Limit Theorem for Proportions takes effect, and we can conclude that the sampling distribution of $\hat{p}$ is approximately normal. From Example 7.16 we have $\mu_{\hat{p}} = 0.08$ and $\sigma_{\hat{p}} = 0.02713$. Thus, the sampling distribution of $\hat{p}$ is approximately normal with $\mu_{\hat{p}} = 0.08$ and $\sigma_{\hat{p}} = 0.02713$. ■

## Example 7.20 Unemployment rate in Texas, March 2007

The Texas Workforce Commission reported that the state unemployment rate in March 2007 was 4.3%. Let $p = 0.043$ represent the population proportion of unemployed workers in Texas.

**a.** Find the minimum size of the samples that produces a sampling distribution of $\hat{p}$ that is approximately normal.

**b.** Describe the sampling distribution of $\hat{p}$ if we use this minimum sample size.

### Solution

**a.**

***What Results Might We Expect?***

The proportion of unemployed workers in Texas is only 0.043, compared to 0.957 for the proportion of employed workers. Since these proportions are relatively imbalanced, we would expect the sample size required to be fairly large. ■

Using Fact 8, the minimum sample size required is the *larger* of either

$$n_1 = \frac{5}{p} \qquad \text{or} \qquad n_2 = \frac{5}{1-p}$$

Here

$$n_1 = \frac{5}{p} = \frac{5}{0.043} \approx 116.3 \qquad \text{and} \qquad n_2 = \frac{5}{1-p} = \frac{5}{0.957} \approx 5.2$$

The larger of $n_1$ and $n_2$ is $n_1 = 116.3$. However, it is unclear what "0.3" of a worker means. So we round up to the next integer: $n = 117$. Therefore, the minimum sample size required to produce a sampling distribution of $\hat{p}$ that is approximately normal is $n = 117$ Texas workers. We confirm that this satisfies our conditions:

$$np = (117)(0.043) = 5.031 \geq 5, \quad \text{and} \quad n(1-p) = (117)(0.957) = 111.969 \geq 5$$

This sample size of $n = 117$ confirms our earlier expectation that this required sample size would be fairly large compared to the means case, where the required sample size is $n = 30$.

**b.** We have $\mu_{\hat{p}} = 0.043$ and

$$\sigma_{\hat{p}} = \sqrt{\frac{p(1-p)}{n}} = \sqrt{\frac{0.043(0.957)}{117}} \approx \sqrt{0.00035172} \approx 0.01875$$

Since the conditions are met, we invoke the Central Limit Theorem for Proportions. The sampling distribution of the sample proportion $\hat{p}$ is an approximately normal$\left(p, \sqrt{\frac{p(1-p)}{n}}\right)$ distribution. That is, normal$\left(0.043, \sqrt{\frac{0.043(0.957)}{117}}\right)$, or normal($\mu_{\hat{p}} = 0.043$, $\sigma_{\hat{p}} = 0.01875$). ■

*Developing Your Statistical Sense*

### The Normality Is Approximate, Not Exact

Figure 7.27 shows a histogram of the sample proportion for a simulation of 200 random samples of size $n = 117$ from a population with $p = 0.043$. Note that the sampling distribution of $\hat{p}$ is *only approximately normal, not exactly normal*. The actual normal(0.043, 0.01875) distribution (Figure 7.28) contains a certain amount of negative values, which are not appropriate for modeling proportions.

The lesson here is that we need to be careful when working with sampling distributions for "extreme" values of $\hat{p}$, especially when finding extreme percentiles, because the normality is only *approximate*. ■

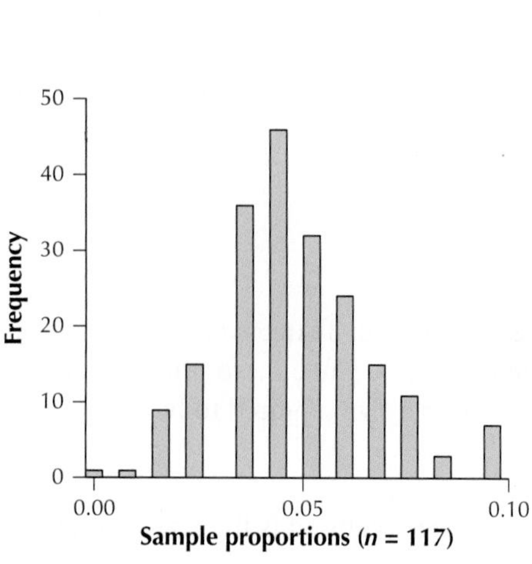


FIGURE 7.27 Sampling distribution of $\hat{p}$ is approximately normal.

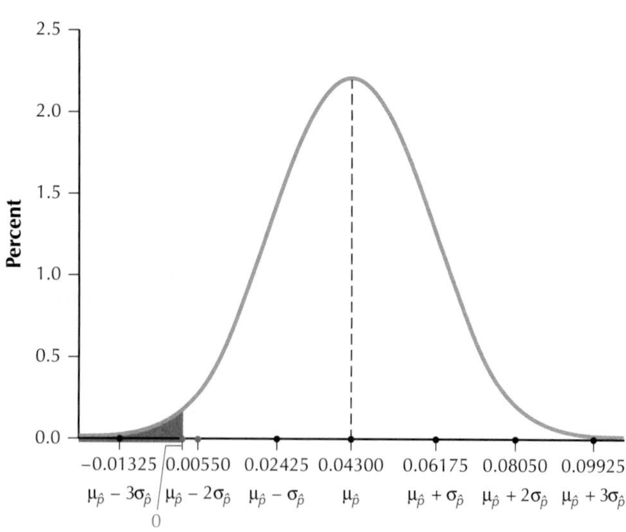


FIGURE 7.28 Actual normal(0.043, 0.01875) distribution contains negative values.

In those cases where we determine that the sampling distribution of $\hat{p}$ is approximately normal, we can then proceed to determine probabilities or find percentiles using the normal distribution methods we have learned in Chapter 6. Fact 9 is similar to Fact 5.

**Fact 9: Standardizing a Normal Sampling Distribution for Proportions**
When the sampling distribution of $\hat{p}$ is normal (or approximately normal), we can standardize to produce the standard normal $Z$:

$$Z = \frac{\hat{p} - \mu_{\hat{p}}}{\sigma_{\hat{p}}} = \frac{\hat{p} - p}{\sqrt{\frac{p(1-p)}{n}}}$$

where $p$ is the population proportion of successes and $n$ is the sample size.

## Example 7.21 Using the sampling distribution of $\hat{p}$ to find a probability

Using the information in Example 7.20, find the probability that a sample of Texas workers will have a proportion unemployed greater than 9% for samples of size (a) 30 respondents and (b) 117 respondents.

### Solution

**a.** We found in Example 7.20a that this sample size of $n = 30$ does not meet the minimum sample size required for the sampling distribution of $\hat{p}$ to be approximately normal, so we cannot conclude that the sampling distribution of $\hat{p}$ is approximately normal. Thus, we cannot solve this problem.

**b.** From Example 7.20b, the sampling distribution of $\hat{p}$ is approximately normal with mean $\mu_{\hat{p}} = 0.043$ and standard deviation $\sigma_{\hat{p}} = 0.01875$. We are then faced with a normal probability problem similar to those in Section 6.5 (page 309). Figure 7.29 shows the sampling distribution of $\hat{p}$ and the probability we are interested in, $P(\hat{p} > 0.09)$. Using Fact 9, we standardize as follows:

$$Z = \frac{0.09 - \mu_{\hat{p}}}{\sigma_{\hat{p}}} = \frac{0.09 - 0.043}{0.01875} \approx 2.51$$

Thus, $P(\hat{p} > 0.09) = P(Z > 2.51)$, as shown in Figure 7.30.

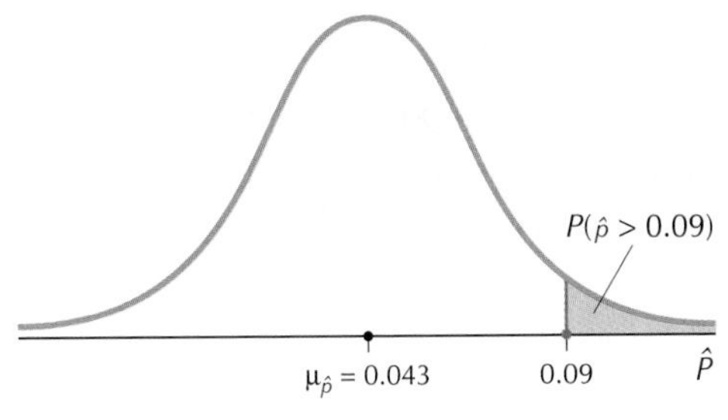

FIGURE 7.29

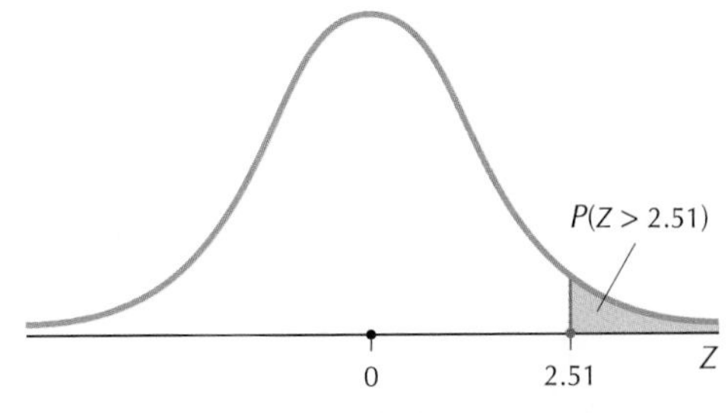

FIGURE 7.30

Following Table 6.7 (page 298), we look up $Z = 2.51$ in the $Z$ table and subtract this table area (0.9940) from 1 to get the desired tail area. That is,

$$P(Z > 2.51) = 1 - 0.9940 = 0.0060$$

So the probability that the sample proportion of unemployed Texas workers will exceed 0.09 is 0.60%. ■

## Example 7.22 Using the sampling distribution of $\hat{p}$ to find a percentile

Using the information from Example 7.20, find the 99th percentile of the sampling distribution of $\hat{p}$ for $n = 117$.

**Solution**

The 99th percentile shown in Figure 7.31 separates the top 1% of sample proportions from the lower 99%. Thus, the area to the left of the 99th percentile is 0.99. We look up $Z = 0.99$ on the inside of the $Z$ table, and the closest value we can find is 0.9901. The $Z$-value associated with 0.9901 is 2.33.

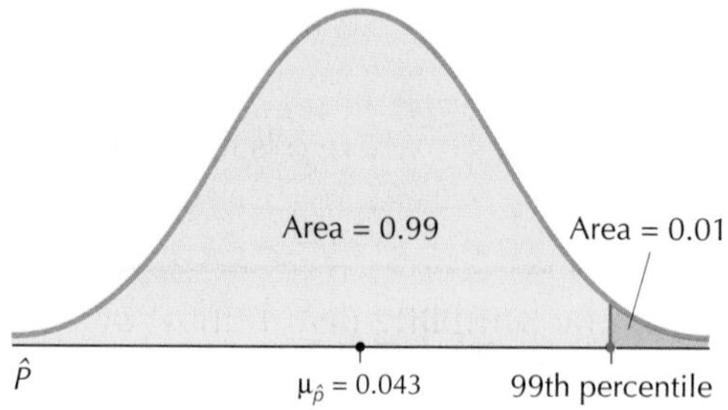


FIGURE 7.31

We need to transform this $Z$-value back to the scale of sample proportions. Use

$$\hat{p} = Z \cdot \sigma_{\hat{p}} + \mu_{\hat{p}} = (2.33)(0.01875) + 0.043 \approx 0.0867$$

The 99th percentile of the sampling distribution of $\hat{p}$ is 0.0867.

## Example 7.23 Pitfalls of using an approximation

Using the information from Example 7.20, find the 1st percentile of the sampling distribution of $\hat{p}$ for $n = 117$.

*Note:* What can we do to estimate the 1st percentile? One way is to use simulation. Generate samples of size $n = 117$ from the population of the original survey respondents, record the sample proportion from each, and simply choose the 1st percentile. Proceeding in this manner, our estimate of the 1st percentile is 0.0128.

**Solution**

By symmetry, the 1st percentile will be the same distance below the mean that the 99th percentile is above the mean. The 99th percentile, 0.0867, lies $(0.0867 - 0.043) = 0.0437$ above the mean. Therefore, the 1st percentile lies 0.0437 below the mean:

$$\hat{p} = (0.043 - 0.0437) = -0.0007$$

However, this value of $-0.0007$ is negative and cannot represent a sample proportion of any kind. This is because the normality of the sampling distribution of $\hat{p}$ is only approximate and not exact.

## What If Scenario: Give the Calculator a Rest

Put your calculator down and think through the following problem, in which we explore how increasing the sample size in Example 7.20 affects the sampling distribution of the sample proportion $\hat{p}$.

**Scenario**

*What if* we increase our sample size $n$ to some *unspecified* higher level.

**a.** How will this change affect the standard deviation of the sampling proportion, $\sigma_{\hat{p}}$?

**b.** How will this change affect the probability that $p$ will exceed 0.09 (Example 7.21)?

**c.** How will this change affect the 99th percentile (Example 7.22)?

**d.** How will this change affect the 1st percentile (Example 7.23)?

## Solution

**a.** Note the presence of $n$ in the denominator of the formula:

$$\sigma_{\hat{p}} = \sqrt{\frac{p \cdot (1-p)}{n}}$$

As $n$ increases, the value of $\sigma_{\hat{p}}$ decreases. So if $n$ is larger than 117, then $\sigma_{\hat{p}}$ will get smaller than the 0.01875 we found in Example 7.20.

**b.** If $\sigma_{\hat{p}}$ is decreased, the quantity $Z = \left(\frac{0.09 - 0.043}{\sigma_{\hat{p}}}\right)$ will increase, producing a bigger value of $Z$ than the 2.51 we saw in Example 7.21. Thus, increasing the sample size will decrease the probability of exceeding this new, larger $Z$-value. That is, for a smaller $\sigma_{\hat{p}}$,

$$P(\hat{p} > 0.09) = P\left(\frac{\hat{p} - 0.043}{\sigma_{\hat{p}}} > \frac{0.09 - 0.043}{\sigma_{\hat{p}}}\right) = P(Z > \text{some value of } Z \text{ bigger than } 2.51)$$

So the probability must be smaller.

**c.** If $\sigma_{\hat{p}}$ is smaller, then the quantity $\hat{p} = Z \cdot \sigma_{\hat{p}} + \mu_{\hat{p}}$ used to find the 99th percentile will also be smaller. (The $Z$-value used in this formula has area 0.01 to the right of it and is therefore constant.) Therefore, the 99th percentile will be smaller.

**d.** If $\sigma_{\hat{p}}$ is smaller, then the quantity $\hat{p} = Z \cdot \sigma_{\hat{p}} + \mu_{\hat{p}}$ used to find the 1st percentile will be larger. Alternatively, since the sampling distribution of $\hat{p}$ is normal and therefore symmetric, the decrease in the 99th percentile must be reflected by an increase in the 1st percentile.

---

### Discovering the Minimum Sample Size for Normality

How are we going to construct a general rule of thumb if the sample size required for approximate normality for certain population proportions is about 20 times higher than the sample size for others? The key is to notice that when $p$ was small (0.02081 for Example 7.17), the required sample size (say, $n^*$) was large, and when $p$ was moderate (0.4755 for Example 7.18), the required sample size $n^*$ was small. Thus, the relationship between $n$ and $p$ is an *inverse* relationship. That is, the relationship between $n^*$ and $p$ can be expressed in the form

$$n^* = \frac{k}{p} \quad \text{for some constant } k \text{ (and for } 0 < p \leq 0.5)$$

Since we are looking for the minimum value of $n^*$ for normality, we are thus looking for some value of $n^*$ that is equal to $\frac{k}{p}$ or greater. That is, we are seeking $n^*$ such that

$$n^* \geq \frac{k}{p}$$

For simplicity, reexpress this as $n^*p \geq k$. Table 7.8 shows the values of $n^*p$ for the distributions we considered in Examples 7.17 and 7.18.

Table 7.8 Searching for the threshold value for $n^*p \geq k$

| High Calories Example | | | |
|---|---|---|---|
| $n^*$ | $p$ | $n^*p$ | **Approximately normal?** |
| 10 | 0.02081 | 0.2081 | No |
| 50 | 0.02081 | 1.0405 | No |
| 100 | 0.02081 | 2.081 | No |
| 200 | 0.02081 | 4.162 | Almost |
| 250 | 0.02081 | 5.2025 | Yes |
| **Vitamin C Example** | | | |
| $n^*$ | $p$ | $n^*p$ | **Approximately normal?** |
| 10 | 0.4755 | 4.755 | Almost |
| 12 | 0.4755 | 5.706 | Yes |

We see that, as $n^*p$ approaches the value of

$$n^*p = k = 5$$

the sampling distribution of $\hat{p}$ becomes approximately normal. We therefore conclude that $k = 5$ is the threshold value for our rule of thumb. (Note that, had $p$ exceeded 0.5, then it would be $(1 - p)$ that would have the inverse relationship with $n^*$. This is the reason for having to check two quantities in Fact 8.)

## Section 7.3 SUMMARY

1 The sampling distribution of the sample proportion $\hat{p}$ for a given sample size $n$ consists of the collection of the sample proportions of all possible samples of size $n$ from the population.

2 The approximate normality of the sampling distribution of the sample proportion kicks in much quicker when the population proportion is moderate rather than extreme.

3 According to the Central Limit Theorem for Proportions, the sampling distribution of the sample proportion $\hat{p}$ follows an approximately normal distribution with mean $\mu_{\hat{p}} = p$ and standard deviation $\sigma_{\hat{p}} = \sqrt{p \cdot (1 - p)/n}$ when both the following conditions are satisfied: (1) $np \geq 5$ and (2) $n(1 - p) \geq 5$. We can use Fact 9 to find probabilities and percentiles for sample proportions.

## Section 7.3 EXERCISES

### CLARIFYING THE CONCEPTS

1. For the rule of thumb governing sample size given in Fact 8, explain why we can't just use what we did in Section 7.2, $n$ at least 30?

2. Explain why we have to round up to the next integer in Fact 8.

3. Explain in your own words why larger sample sizes are needed to achieve normality for the sampling distributions of $p$ when the value of $p$ is extreme.

4. Explain in your own words the meaning of the Central Limit Theorem for Proportions.

5. Compare the $x$ axis (scale) for Figures 7.25a–e (page 371).
   **a.** What is happening to the scale as the sample size is increasing?
   **b.** Why is this happening? Use $\sigma_{\hat{p}}$ in your response.
6. Explain why we have to check both $np \geq 5$ and $n(1 - p) \geq 5$ in Fact 8.

## PRACTICING THE TECHNIQUES

In Exercises 7–12, samples are taken. Find (a) $\mu_{\hat{p}}$ and (b) $\sigma_{\hat{p}}$, and (c) determine whether the sampling distribution of $\hat{p}$ is approximately normal or unknown.

7. $p = 0.5, n = 100$
8. $p = 0.5, n = 5$
9. $p = 0.01, n = 100$
10. $p = 0.01, n = 500$
11. $p = 0.9, n = 40$
12. $p = 0.9, n = 50$

In Exercises 13–18, find the minimum sample size that produces a sampling distribution of $\hat{p}$ that is approximately normal.

13. $p = 0.5$
14. $p = 0.25$
15. $p = 0.1$
16. $p = 0.05$
17. $p = 0.01$
18. $p = 0.001$

For each of Exercises 19–28, sketch the sampling distribution of $\hat{p}$. Make sure to indicate the locations of $\mu_{\hat{p}}$, $\mu_{\hat{p}} \pm \sigma_{\hat{p}}$, $\mu_{\hat{p}} \pm 2\sigma_{\hat{p}}$, and $\mu_{\hat{p}} \pm 3\sigma_{\hat{p}}$. Use Figure 7.29 as a guide.

19. $p = 0.5, n = 100$
20. $p = 0.5, n = 400$
21. $p = 0.02, n = 400$
22. $p = 0.02, n = 625$
23. $p = 0.9, n = 64$
24. $p = 0.9, n = 144$
25. $p = 0.1, n = 64$
26. $p = 0.1, n = 144$
27. $p = 0.98, n = 400$
28. $p = 0.98, n = 625$

For Exercises 29–34, if possible find the indicated probability. If it is not possible, explain why not.

29. $p = 0.5, n = 100, P(\hat{p} > 0.55)$
30. $p = 0.5, n = 5, P(\hat{p} > 0.55)$
31. $p = 0.01, n = 100, P(\hat{p} > 0.011)$
32. $p = 0.01, n = 500, P(\hat{p} > 0.011)$
33. $p = 0.9, n = 40, P(0.88 < \hat{p} < 0.91)$
34. $p = 0.9, n = 50, P(0.88 < \hat{p} < 0.91)$

For Exercises 35–40, find the indicated value of $\hat{p}$. If it is not possible, explain why not.

35. $p = 0.5, n = 100$, value of $\hat{p}$ larger than 90% of all values of $\hat{p}$
36. $p = 0.5, n = 400$, value of $\hat{p}$ larger than 90% of all values of $\hat{p}$
37. $p = 0.9, n = 64$, 95th percentile of values of $\hat{p}$
38. $p = 0.9, n = 144$, 95th percentile of values of $\hat{p}$
39. $p = 0.1, n = 64$, 10th percentile of values of $\hat{p}$
40. $p = 0.1, n = 144$, 10th percentile of values of $\hat{p}$

## APPLYING THE CONCEPTS

41. **The Green Party.** At a certain university, 25% of students support the Green Party. Suppose we take samples of size 100 students and are interested in the population proportion of students who support the Green Party.
    **a.** What is the mean of the sampling distribution of $\hat{p}$?
    **b.** What is the standard deviation of the sampling distribution of $\hat{p}$?
    **c.** Describe the sampling distribution of $\hat{p}$.
42. **Women's Radio Preferences.** In their 2001 study *What Women Want: Five Secrets to Better Ratings,* the Arbitron company reported that "Music I Like" is the biggest factor in deciding which radio station to tune to, chosen by 87% of women.
    **a.** Find the minimum sample size $n^*$ that produces a sampling distribution of $\hat{p}$ that is approximately normal.
    **b.** Confirm that this sample size satisfies the conditions in Fact 8.
    **c.** Describe the sampling distribution of $\hat{p}$ if we use this minimum sample size. Which fact allows us to say this?
    **d.** Find $\mu_{\hat{p}}$ and $\sigma_{\hat{p}}$ for $n = 50$.
    **e.** Find the probability that, in a sample of 50 women, more than 45 chose "Music I Like" as the reason they decide which radio station to tune to.
43. **The Green Party.** Refer to Exercise 41.
    **a.** Find the probability that the sample mean proportion supporting the Green Party exceeds 0.30.
    **b.** Find the 95th percentile of sample proportions.
44. **Shaquille O'Neal.** In the 2003–2004 National Basketball Association regular season, Shaquille O'Neal led the league with a 58.4% field goal percentage.
    **a.** Find the minimum sample size that produces a sampling distribution of $\hat{p}$ that is approximately normal.
    **b.** Find $\mu_{\hat{p}}$ and $\sigma_{\hat{p}}$ for $n = 50$.
    **c.** Find the probability that, in a sample of 200 shots, Shaquille would score more than 120 baskets.
45. **Women's Access to the Internet.** The Pew Internet and American Life Project reported in 2003 that 61% of

women had access to the Internet. Consider a sample of 100 women.

**a.** Confirm that the Central Limit Theorem for Proportions applies.

**b.** Find $\mu_{\hat{p}}$ and $\sigma_{\hat{p}}$.

**c.** Find the probability that fewer than 60 of these 100 women had access to the Internet.

**46. High-Calorie Foods.** Refer to Example 7.17.

**a.** Find the values of $\sigma_{\hat{p}}$ for the data in the graphs in Figure 7.25.

**b.** Plot the values against the sample size.

**c.** Describe the relationship between $\sigma_{\hat{p}}$ and sample size.

**Women in Management Courses.** Use this information for Exercises 47–50. Suppose we are interested in the population proportion of women in management courses and we take samples of size $n = 20$ from a population of management courses. Assume that the population proportion of women in management courses is 0.40.

**47.** Follow steps **(a)**–**(e)**.

**a.** Find the minimum sample size $n^*$ that produces a sampling distribution of $\hat{p}$ that is approximately normal.

**b.** Confirm that this sample size satisfies the conditions in Fact 8.

**c.** Describe the sampling distribution of $\hat{p}$ if we use this minimum sample size. What allows us to say this?

**d.** Find $\mu_{\hat{p}}$ and $\sigma_{\hat{p}}$ for $n = 20$.

**e.** Find, if possible, the probability that a sample of 20 people in management courses will have a proportion of women that is smaller than 38%. If it is not possible, clearly explain why not, and suggest what we should do.

**48.** Refer to Exercise 47.

**a.** Explain in your own words what we mean by the sampling distribution of the sample proportion of women in management courses.

**b.** Where will the sampling distribution of the sample proportion be centered?

**c.** Provide the value of a measure of spread of the sampling distribution.

**d.** Suppose a particular management course has 65% women. Is this unusual? How do we determine whether this is unusual?

**e.** Suppose a particular management course has 75% women. Is this unusual? How do we determine whether this is unusual?

**49.** Refer to Exercise 48.

**a.** Suppose we find the measure of spread in Exercise 48 to be too large and we would like a more precise measure of spread. How can we reduce the size of the measure of spread?

**b.** Provide a value of $n$ that would result in a smaller standard deviation of this statistic.

**50.** Suppose a particular management course has 65% women.

**a.** Is this proportion to be considered unusual with our new value for the standard deviation for the statistic?

**b.** How does this proportion compare with the first time you checked whether it was unusual?

**c.** What has changed?

**People with No One to Turn To.** Use this information for Exercises 51–54. The Pew Internet and American Life Project conducts surveys on Americans' use of the Internet in everyday life. In the weeks following the attacks of September 11, 2001, they asked respondents the following question: "When you need help, would you say that you can turn for support to many people, just a few people, or hardly any people at all?" The results are shown in the table. Consider the 4395 respondents to form a population.

| Response | Frequency |
|---|---|
| Many people | 2058 |
| Just a few people | 1806 |
| Hardly any people | 485 |
| No one/None | 46 |
| Total | 4395 |

Note that the response "No one/None" was not part of the question, but some people volunteered the response anyway. We are interested in the proportion of respondents who say that they have no one to turn to when they need help.

**51.** Construct a relative frequency distribution from the information provided in the table of survey results.

**52.** Consider the table of survey results.

**a.** What is the value of $p$, the population proportion of people who have no one to turn to?

**b.** What effect do you think $p$ being so small will have on the sample size that produces a sampling distribution of $\hat{p}$ that is approximately normal?

**c.** Find the minimum sample size $n^*$ required to produce a sampling distribution of $\hat{p}$ that is approximately normal.

**d.** Confirm that this sample size satisfies the conditions in Fact 8.

**e.** Comment on the size of this minimum required sample size.

**f.** Describe the sampling distribution of $\hat{p}$ if we use this minimum sample size.

**53.** Consider the table of survey results. Use the sample size $n^*$ you calculated in Exercise 52c for **(a)** and **(b)**.

**a.** Find $\mu_{\hat{p}}$ and $\sigma_{\hat{p}}$ for those who responded "No one/ None."

**b.** Would you say that this value of $\sigma_{\hat{p}}$ is smaller or larger than most of the values of $\sigma_{\hat{p}}$ that we have dealt with so far. Why?

c. If possible, find the probability that a sample of 200 respondents will have a proportion of people with no one to turn to that is greater than 2%. If it is not possible, clearly explain why not, and suggest what we should do.

**54.** Use the minimum sample size $n^*$ you calculated in Exercise 52c to follow steps **(a)–(e)**.

**a.** If possible, find the probability that a sample of $n^*$ respondents will have a proportion of people with no one to turn to that is greater than 2%. If it is not possible, clearly explain why not.

**b.** Find the 97.5th percentile of the sampling distribution of $\hat{p}$ for $n = n^*$.

**c.** Find the 2.5th percentile of the sampling distribution of $\hat{p}$ for $n = n^*$.

**d.** Between what two symmetric values of $\hat{p}$ lies 95% of the sample proportions from the sampling distribution of $\hat{p}$?

**e.** Based on your response to Exercise 53, what would you say about the likelihood of observing $\hat{p} = 0.03$?

For the What If Scenarios in Exercises 55–58, describe how and why the change would have affected the indicated quantities, if at all. Would they increase, decrease, remain unchanged? Or is there insufficient information to tell what would happen? Explain your answers.

**55.** *What if* the population proportion of people with no one to turn to was some unspecified value larger than 0.01.

**a.** $p$

**b.** $(1 - p)$

**c.** $\mu_{\hat{p}}$

**d.** $\sigma_{\hat{p}}$

**e.** Minimum sample size required for approximate normality in the sampling distribution for $\hat{p}$

**56.** *What if* the population proportion of people with no one to turn to was some unspecified value larger than 0.01.

**a.** $P(p > 0.02)$

**b.** Variability of the sampling distribution of $\hat{p}$

**c.** Location of the sampling distribution of $\hat{p}$

**d.** 97.5th percentile of sample proportions

**e.** 2.5th percentile of sample proportions

**57.** *What if* the sample size used was some unspecified value greater than $n^*$.

**a.** $p$

**b.** $(1 - p)$

**c.** $\mu_{\hat{p}}$

**d.** $\sigma_{\hat{p}}$

**e.** Minimum sample size required for approximate normality in the sampling distribution for $\hat{p}$

**58.** *What if* the sample size used was some unspecified value greater than $n^*$.

**a.** $P(p > 0.02)$

**b.** Variability of the sampling distribution of $\hat{p}$

**c.** Location of the sampling distribution of $\hat{p}$

**d.** 97.5th percentile of sample proportions

**e.** 2.5th percentile of sample proportions

**59. Challenger Exercise**

**a.** Find the standard deviation of the set of $\hat{p}$'s in Table 7.6.

**b.** Use the finite population correction factor to show that your answer in **(a)** equals $\sqrt{\dfrac{N-n}{N-1}} \cdot \sigma_{\bar{x}}$.

## *Try This in Class!* In-class activities to enhance your understanding of statistics

### SAMPLING DISTRIBUTIONS

Perform the following activity as a class. Use the *Random Sample* applet to generate the random samples requested.

**1.** Each student in the class should write down an estimate of the length of the following line segment. Use your eyes only! No rulers.

---

**2.** Consider your class to be a population. Find the mean $\mu$ and the standard deviation $\sigma$ of all student estimates.

**3.** Choose one student at random from the class. Ask for this student's estimate.

**4.** Take a random sample of two students from the class, and find the mean $\bar{x}$ of their estimates. What is the mean $\mu_{\bar{x}}$ and standard deviation $\sigma_{\bar{x}}$ of the sampling distribution of $\bar{x}$? Given these values for $\mu_{\bar{x}}$ and $\sigma_{\bar{x}}$, which values for the length of the line segment would be considered unusual? (*Hint:* Which would be outliers?)

**5.** Take a random sample of 10 students from the class, and find the mean $\bar{x}$ of their estimates. What are $\mu_{\bar{x}}$ and $\sigma_{\bar{x}}$ for $n = 10$? Now, which values for the length of the line segment would be considered unusual?

**6.** Take a random sample of 20 students and find their estimates. What are $\mu_{\bar{x}}$ and $\sigma_{\bar{x}}$? Now, which values for the length of the line segment would be considered unusual?

**7.** As the sample size increases, what is happening to $\sigma_{\bar{x}}$, and what is happening to the estimates that are considered unusual? How can you state this phenomenon in terms of sample size and precision of the estimate?

## CHAPTER 7 FORMULAS AND VOCABULARY

### SECTION 7.1

- **Mean of the sampling distribution of the sample mean $\bar{x}$** (p. 340). Value of the population mean $\mu$, denoted as $\mu_{\bar{x}} = \mu$, and read as "the mean of the sampling distribution of $\bar{x}$ is $\mu$."
- **Normal probability plot** (p. 343). Scatterplot of the data values against corresponding values of the standard normal distribution. If the points cluster around a straight line, then the data set is normal. If there are systematic deviations off the straight line, then that is evidence against the claim that the data set is normal.
- **Parameter** (p. 337). A characteristic of a population.
- **Point estimation** (p. 337). Using a single known value of a statistic to estimate the associated population parameter. The value of the statistic is called the **point estimate**.
- **Sampling distribution of the sample mean $\bar{x}$** (p. 339). For a given sample size $n$, the collection of the sample means of all possible samples of size $n$ from the population.
- **Sampling error** (p. 337). Distance between the point estimate and its target parameter.
- **Standard deviation of the sampling distribution of the sample mean $\bar{x}$** (p. 341). Value of $\sigma_{\bar{x}} = \sigma/\sqrt{n}$, where $\sigma$ is the population standard deviation.
- **Standardizing a normal sampling distribution for means** (p. 343). When the sampling distribution of $\bar{x}$ is normal, we can standardize to produce the standard normal $Z$:

$$Z = \frac{\bar{x} - \mu_{\bar{x}}}{\sigma_{\bar{x}}} = \frac{\bar{x} - \mu}{\sigma/\sqrt{n}}$$

where $\mu$ is the population mean, $\sigma$ is the population standard deviation, and $n$ is the sample size.
- **Statistic** (p. 337). A characteristic of a sample.

### SECTION 7.2

- **Central Limit Theorem for Means** (p. 356). Given a population with mean $\mu$ and standard deviation $\sigma$, the sampling distribution of the sample mean $\bar{x}$ becomes approximately normal($\mu$, $\sigma/\sqrt{n}$) as the sample size gets larger, regardless of the shape of the population.
  - Consider $n \geq 30$ as large enough to apply the Central Limit Theorem for any population.

### SECTION 7.3

- **Central Limit Theorem for Proportions** (p. 373). The sampling distribution of the sample proportion $\hat{p}$ follows an approximately normal distribution with mean $\mu_{\hat{p}} = p$ and standard deviation $\sigma_{\hat{p}} = \sqrt{p \cdot (1-p)/n}$ when both the following conditions are satisfied: (1) $np \geq 5$ and (2) $n(1-p) \geq 5$.
- **Mean of the sampling distribution of the sample proportion $\hat{p}$** (p. 368). The value of the population proportion $p$, denoted as $\mu_{\hat{p}} = p$ and read as "the mean of the sampling distribution of $\hat{p}$ is $p$."
- **Sampling distribution for any statistic** (p. 367). For a given sample size $n$, consists of the collection of that sample statistic across all possible samples of size $n$.
- **Sampling distribution of the sample proportion $\hat{p}$** (p. 367). For a given sample size $n$ consists of the collection of the sample proportion of all possible samples of size $n$ from the population.
- **Standard deviation of the sampling distribution of the sample proportion $p$** (p. 368). The value of $\sigma_{\hat{p}} = \sqrt{p \cdot (1-p)/n}$, where $p$ is the population proportion and $n$ is the sample size.
- **Standardizing a normal sampling distribution for proportions** (p. 375). When the sampling distribution of $\hat{p}$ is normal, we can standardize to produce the standard normal $Z$:

$$Z = \frac{p - \mu_{\hat{p}}}{\sigma_{\hat{p}}} = \frac{\hat{p} - p}{\sqrt{\frac{p(1-p)}{n}}}$$

where $p$ is the population proportion of successes and $n$ is the sample size.

## CHAPTER 7 REVIEW EXERCISES

### SECTION 7.1

For Exercises 1–3, find the sampling error.

1. $\mu = 95, \bar{x} = 95$
2. $\sigma = 5, s = 7$
3. $p = 0.1, \hat{p} = 0$

For Exercises 4–7, find $\mu_{\bar{x}}$ and $\sigma_{\bar{x}}$, the mean and standard deviation of the sampling distribution of $\bar{x}$.

4. $\mu = 2, \sigma = 0.5, n = 25$
5. $\mu = 2, \sigma = 0.5, n = 36$
6. $\mu = 50, \sigma = 40, n = 4$
7. $\mu = 50, \sigma = 40, n = 16$

For Exercises 8–11, assume that $X$ is normal(10, 4) and $n = 25$.

8. Find the sampling distribution of $\bar{x}$ for $n = 25$.

9. Find the probability that $\bar{x}$ exceeds 11.

10. Without using your calculator, find the probability that $\bar{x}$ is less than 9.

11. Without using your calculator, find the probability that $\bar{x}$ lies between 9 and 11.

**12. Calories in Breakfast Cereals.** Refer to the sample of 12 breakfast cereals in the table.

| Cereal | Manufacturer | Calories |
|---|---|---|
| Apple Jacks | Kellogg's | 110 |
| Basic 4 | General Mills | 130 |
| Bran Chex | Ralston Purina | 90 |
| Bran Flakes | Post | 90 |
| Cap'n Crunch | Quaker Oats | 120 |
| Cheerios | General Mills | 110 |
| Cinnamon Toast Crunch | General Mills | 120 |
| Cocoa Puffs | General Mills | 110 |
| Corn Chex | Ralston Purina | 110 |
| Corn Flakes | Kellogg's | 100 |
| Corn Pops | Kellogg's | 110 |
| Count Chocula | General Mills | 110 |

**a.** Find a point estimate of the population standard deviation of calories in all breakfast cereals. What is the notation for this estimate? What is the notation for the parameter it is estimating?

**b.** Find a point estimate for the population proportion of all cereals manufactured by General Mills. What is the notation for this estimate? What is the notation for the parameter it is estimating?

**c.** Find a point estimate for the population mode cereal manufacturer.

**Vehicle Weights.** Descriptive statistics for the weights of a sample of 46 crash test vehicles taken from a population of 352 vehicles are as follows (in pounds): mean = 3020.83, range = 2562, standard deviation = 607.23. Use this information for Exercises 13–15.

13. **a.** What is the point estimate for the population range of weights for all crash test vehicles?

**b.** Is this estimate in **(a)** a statistic or a parameter? Is the population range a statistic or a parameter?

**c.** Do you think that this estimate in **(a)** tends to overestimate or underestimate the true range of weights for all crash test vehicles? Clearly explain why you think so.

14. **a.** What is the point estimate for the population standard deviation of weights for all crash test vehicles?

**b.** What is the notation for this estimate in **(a)**?

**c.** What is the notation for the parameter it is trying to estimate?

15. The population of 352 crash test vehicles has a range of 4029 pounds.

**a.** By how much did the sample range of weights miss the population range of weights?

**b.** What is the terminology for the quantity in **(a)**?

**c.** Did the sample range underestimate or overestimate the population range?

**d.** Why do you think this happened?

**e.** Does your answer to **(b)** concur with your earlier prediction?

## SECTION 7.2

In Exercises 16 and 17, determine which of the three cases for the sampling distribution of the sample mean applies.

16. Hummingbird wing beats are not normally distributed, and the sample size is small.

17. Hummingbird wing beats are not normally distributed, and the sample size is large.

In Exercises 18 and 19, provide (a) $\mu_{\bar{x}}$ and (b) $\sigma_{\bar{x}}$, and (c) specify whether the sampling distribution of $\bar{x}$ is normal, approximately normal, or unknown.

18. Hummingbird wing beats are not normally distributed, with $\mu = 50$ beats per second and $\sigma = 10$ beats per second. A sample of size 9 is taken.

19. Hummingbird wing beats are not normally distributed, with $\mu = 50$ beats per second and $\sigma = 10$ beats per second. A sample of size 100 is taken.

If possible, for Exercises 20 and 21 find the indicated probability. If it is not possible, explain why not.

20. Hummingbird wing beats are not normally distributed, with $\mu = 50$ beats per second and $\sigma = 10$ beats per second. A sample of size 9 is taken. Find $P(\bar{x} > 53)$.

21. Hummingbird wing beats are not normally distributed, with $\mu = 50$ beats per second and $\sigma = 10$ beats per second. A sample of size 100 is taken. Find $P(\bar{x} > 53)$.

For Exercises 22 and 23, find the indicated value of $\bar{x}$. If it is not possible, explain why not.

22. Hummingbird wing beats are not normally distributed, with $\mu = 50$ beats per second and $\sigma = 10$ beats per second. A sample of size 9 is taken. Find the 10th percentile of sample means.

23. Hummingbird wing beats are not normally distributed, with $\mu = 50$ beats per second and $\sigma = 10$ beats per second. A sample of size 100 is taken. Find the 10th percentile of sample means.

**24. Black Women's Radio Usage.** The Arbitron Corporation tracks media usage. In their study *Black Radio Today 2004 Edition,* they found that black women between the ages of 18 and 24 spend a mean of 21.5 hours per week listening to the radio. Assume that the distribution of time spent listening to the radio is right-skewed, with a standard deviation of 10 hours. Suppose we take a sample of 100 black women.

**a.** Find the 95th percentile of the mean amount of time spent listening.

**b.** Find the 5th percentile of the mean amount of time spent listening.

**c.** Between which two symmetric mean scores lie 90% of all times spent listening to the radio per week for black women?

**d.** By hand, sketch a plot of how this would look.

**25. Cocaine and Heart Attacks.** The American Medical Association reported: "During the first hour after using cocaine, the user's risk of heart attack increases nearly 24 times. The average age of people in the study who suffered heart attacks soon after using cocaine was only 44. That's about 17 years younger than the average heart attack patient. Of the 38 cocaine users who had heart attacks, 29 had no prior symptoms of heart disease."[4] Assume that the standard deviation of the age of people who suffered heart attacks soon after using cocaine was 10 years and we take a sample of size 38.

- **a.** Find the 97.5th percentile of the mean age at heart attack after using cocaine.
- **b.** Find the 2.5th percentile of the mean age at heart attack after using cocaine.
- **c.** Between which two sample mean ages that are symmetric about the population mean lie 95% of mean ages of all people who suffered heart attacks soon after using cocaine?
- **d.** By hand, sketch a plot of how this would look.

## SECTION 7.3

In Exercises 26 and 27, provide (a) $\mu_{\hat{p}}$ and (b) $\sigma_{\hat{p}}$, and (c) determine whether the sampling distribution of $\hat{p}$ is approximately normal or unknown.

**26.** $p = 0.1, n = 40$

**27.** $p = 0.1, n = 50$

In Exercises 28 and 29, find the minimum sample size that produces a sampling distribution of $\hat{p}$ that is approximately normal.

**28.** $p = 0.75$

**29.** $p = 0.9$

For each of Exercises 30 and 31, sketch the sampling distribution of $p$. Make sure to indicate the locations of $\mu_{\hat{p}}$, $\mu_{\hat{p}} \pm \sigma_{\hat{p}}$, $\mu_{\hat{p}} \pm 2\sigma_{\hat{p}}$, and $\mu_{\hat{p}} \pm 3\sigma_{\hat{p}}$.

**30.** $p = 0.1, n = 64$

**31.** $p = 0.1, n = 144$

For Exercises 32 and 33, if possible find the indicated probability. If it is not possible, explain why not.

**32.** $p = 0.1, n = 40, P(\hat{p} < 0.12)$

**33.** $p = 0.1, n = 50, P(\hat{p} < 0.12)$

For Exercises 34 and 35, find the indicated value of $\hat{p}$. If it is not possible, explain why not.

**34.** $p = 0.02, n = 400$, the value of $\hat{p}$ smaller than 75% of all $p$ values

**35.** $p = 0.02, n = 625$, the value of $\hat{p}$ smaller than 75% of all $p$ values

**36. Violent Crime at School.** The National Center for Education Statistics reports that the percentage of 18-year-olds who are victims of violent crime at school is 3.4%.

- **a.** Find the minimum sample size $n^*$ that produces a sampling distribution of $\hat{p}$ that is approximately normal.
- **b.** Confirm that this sample size satisfies the conditions in Fact 8.
- **c.** Describe the sampling distribution of $\hat{p}$ if we use this minimum sample size. What allows us to say this?
- **d.** Consider a sample of 1000 18-year-olds in school. Find $\mu_{\hat{p}}$ and $\sigma_{\hat{p}}$.
- **e.** Find the probability that, in a sample of 1000 18-year-olds in school, more than 50 of them have been victims of violent crime at school.

**37. Living with AIDS.** The Centers for Disease Control reported that, in 2004, 11% of white males living with AIDS contracted it through injection drug use. A sample of 120 white males living with AIDS is examined.

- **a.** Confirm that the Central Limit Theorem for Proportions applies.
- **b.** Find $\mu_{\hat{p}}$ and $\sigma_{\hat{p}}$.
- **c.** Find the probability that less than 10 of the 120 males contracted AIDS through injection drug use.

**38. Small Business Jobs.** According the the U.S. Small Business Administration, small businesses provide 75% of the net new jobs added to the economy. Consider a sample of 1000 new jobs.

- **a.** Confirm that the Central Limit Theorem for Proportions applies.
- **b.** Find $\mu_{\hat{p}}$ and $\sigma_{\hat{p}}$.
- **c.** Find the probability that more than 750 of the 1000 new jobs were provided by a small business.

# CHAPTER 7 Quiz

## TRUE OR FALSE

**1.** True or false: For a normal population, the sampling distribution of the sample mean is always normal.

**2.** True or false: Since the Central Limit Theorem takes effect at $n = 30$, it doesn't make sense to get larger samples.

**3.** True or false: Compared to a symmetric distribution, a skewed distribution would generate means that behave more nicely, so the approximation to normality takes effect for smaller samples.

## FILL IN THE BLANK

**4.** The distance between the point estimate and its target parameter is called the __________ __________ [two words].

**5.** The rule of thumb for when to apply the Central Limit Theorem for Means is when the sample size is at least __________.

**6.** If the population is either non-normal or of unknown distribution and the sample size is large, then the sampling distribution of $\bar{x}$ is __________ __________ (two words).

## SHORT ANSWER

7. What type of plot is used to determine whether a data set is normal?

8. If the population is either non-normal or of unknown distribution and the sample size is small, then do we know the sampling distribution of $\bar{x}$?

9. The sampling distribution of the sample proportion $\hat{p}$ may be considered approximately normal only if *both* the following conditions hold: (1) __________ and (2) __________.

## CALCULATIONS AND INTERPRETATIONS

**Soybean Crop.** Protein content in a particular farmer's soybean crop is normally distributed, with a mean of 40 grams and a standard deviation of 20 grams. Suppose we take samples of size 100 soy plants. Use this information for Exercises 10 and 11.

**10. a.** Find the probability that the sample mean protein content will be less than 38 grams.

**b.** Find the probability that the sample mean protein content will be between 36.08 and 43.92 grams.

**c.** Find the probability that the sample mean protein content will be greater than 42.5 grams.

**11.** Refer to Exercise 10.

**a.** Find the sample mean protein content higher than 99.5% of all such sample means.

**b.** Find the sample mean protein content lower than 99.5% of all such sample means.

**c.** Between which two values does the middle 99% of sample mean protein content lie?

**Student Heights.** Use this information for Exercises 12 and 13. The heights of the population of students at a college are normally distributed with a mean of 68 inches (5 feet 8 inches) and a standard deviation of 3 inches. Suppose we take samples of 100 students.

**12. a.** Find the probability that the sample mean height will exceed 68.6 inches.

**b.** Find the probability that the sample mean height will be less than 67.4 inches.

**c.** Find the probability that the sample mean height will be between 67.4 and 68.6 inches.

**13. a.** Find the 99.5th percentile of sample mean heights.

**b.** Find the 0.5th percentile of sample mean heights.

**c.** Between which two values do the middle 99% of sample mean heights lie?

In Exercises 14 and 15, determine which of the three cases for the sampling distribution of the sample mean applies.

**14.** Scores on a psychological test are not normally distributed and the sample size is small.

**15.** Scores on a psychological test are normally distributed and the sample size is small.

In Exercises 16 and 17, find (a) $\mu_{\bar{x}}$ and (b) $\sigma_{\bar{x}}$, and (c) determine whether the sampling distribution of $\bar{x}$ is normal, approximately normal, or unknown.

**16.** Scores on a psychological test are not normally distributed, with $\mu = 100$ and $\sigma = 15$. A sample of size 25 is taken.

**17.** Scores on a psychological test are normally distributed, with $\mu = 100$ and $\sigma = 15$. A sample of size 25 is taken.

For Exercises 18 and 19, if possible find the indicated probability. If it is not possible, explain why not.

**18.** Scores on a psychological test are not normally distributed, with $\mu = 100$ and $\sigma = 15$. A sample of size 25 is taken. Find $P(94 < \bar{x} < 103)$.

**19.** Scores on a psychological test are normally distributed, with $\mu = 100$ and $\sigma = 15$. A sample of size 25 is taken. Find $P(94 < \bar{x} < 103)$.

For Exercises 20 and 21, find the indicated value of $\bar{x}$. If it is not possible, explain why not.

**20.** Scores on a psychological test are not normally distributed, with $\mu = 100$ and $\sigma = 15$. A sample of size 25 is taken. Find the 50th percentile of sample means.

**21.** Scores on a psychological test are normally distributed, with $\mu = 100$ and $\sigma = 15$. A sample of size 25 is taken. Find the 50th percentile of sample means.

**Computers per School.** Use this information for Exercises 22 and 23. The National Center for Educational Statistics (**http://nces.ed.gov**) reported that, in 2001, the mean number of instructional computers per public school nationwide was 124. Assume that the standard deviation is 49 computers and that there is no information about the shape of the distribution.

**22.** We are interested in finding the probability that a randomly selected sample of 15 schools has a mean number of instructional computers smaller than 100.

**a.** Does Case 1 apply? If so, find $P(\bar{x} < 100)$. If not, explain why not.

**b.** Does Case 2 apply? If so, find $P(\bar{x} < 100)$. If not, explain why not.

**23.** We are interested in finding the probability that a randomly selected sample of 36 schools has a mean number of instructional computers smaller than 100.

**a.** Does Case 1 apply? If so, find $P(\bar{x} < 100)$. If not, explain why not.

**b.** Does Case 2 apply? If so, find $P(\bar{x} < 100)$. If not, explain why not.

In Exercises 24 and 25, find (a) $\mu_{\hat{p}}$ and (b) $\sigma_{\hat{p}}$, and (c) determine whether the sampling distribution of $\hat{p}$ is approximately normal or unknown.

**24.** $p = 0.99$, $n = 100$

**25.** $p = 0.99$, $n = 500$

In Exercises 26 and 27, find the minimum sample size that produces a sampling distribution of $\hat{p}$ that is approximately normal.

**26.** $p = 0.95$

**27.** $p = 0.99$

For Exercises 28 and 29, sketch the sampling distribution of $p$. Make sure to indicate the locations of $\mu_{\hat{p}}$, $\mu_{\hat{p}} \pm \sigma_{\hat{p}}$, $\mu_{\hat{p}} \pm 2\sigma_{\hat{p}}$, and $\mu_{\hat{p}} \pm 3\sigma_{\hat{p}}$.

**28.** $p = 0.98$, $n = 400$

**29.** $p = 0.98$, $n = 625$

For Exercises 30 and 31, if possible find the indicated probability. If it is not possible, explain why not.

**30.** $p = 0.99$, $n = 100$, $P(0.985 < \hat{p} < 0.994)$

**31.** $p = 0.99$, $n = 500$, $P(0.985 < \hat{p} < 0.994)$

For Exercises 32 and 33, find the indicated value of $\hat{p}$. If it is not possible, explain why not.

**32.** $p = 0.98$, $n = 400$, the 50th percentile of values of $\hat{p}$

**33.** $p = 0.98$, $n = 625$, the 50th percentile of values of $\hat{p}$

**34. Depressive Disorders.** According to the National Institute for Mental Health, nearly twice as many American women (12%) as men (6.6%) are affected by a depressive disorder each year. A sample of 100 women is examined.

- **a.** Confirm that the Central Limit Theorem for Proportions applies.
- **b.** Find $\mu_{\hat{p}}$ and $\sigma_{\hat{p}}$.
- **c.** Find the probability that more than 10 of the 100 women are affected by a depressive disorder.

**35. Music Downloaders.** The Pew Internet and American Life Project reported in 2004 that 17% of current music downloaders say they are using paid services. Consider a sample of 400 music downloaders.

- **a.** Confirm that the Central Limit Theorem for Proportions applies.
- **b.** Find $\mu_{\hat{p}}$ and $\sigma_{\hat{p}}$.
- **c.** Find the probability that more than 175 of the 400 music downloaders are using paid services.

Chapter 8

© Scott Wintrow/Getty Images

# Confidence Intervals

## CASE STUDY | Analyzing Stock Performance on the New York Stock Exchange

Corporate compensation packages for executives often include stocks and stock options to attract and retain the best employees. Also, stocks form the backbone of many 401K, 403B, and similar retirement accounts. Therefore, more Americans than ever before have a stake in the performance of stock markets, especially the New York Stock Exchange (NYSE). Financial analysts scrutinize stock prices, stock price changes, share volume, and a host of other variables to gain the competitive edge in the marketplace and to turn a profit for themselves and for their clients.

In this chapter, we shall examine the performance of a set of stocks on the NYSE on a certain date, and we'll apply our new analytical tools to help us understand and quantify the variability in the markets.

## OVERVIEW



## The Big Picture

### Where we are coming from, and where we are headed...

We stand on the threshold of the two most important statistical inference methods: confidence intervals and hypothesis testing. In a way, everything that we have studied thus far has been a preparation for this moment.

The descriptive statistical methods we learned in Chapters 1–4 enable us to explore new data sets. The probability methods we studied in Chapters 5 and 6 provide us with the concepts and tools about uncertainty needed when generalizing from a sample to a population. In Chapter 7, we learned that point estimation cannot determine how close a point estimate is to its target parameter. There has to be a better way—and there is: **confidence intervals**.

In Chapter 8, we learn about confidence interval estimation, using our knowledge of sampling distributions from Chapter 7. By studying the patterns implicit in the sampling distribution of a statistic (such as the sample mean or sample proportion), we can infer with a certain degree of confidence that the associated population parameter lies within a certain interval.

Have you ever heard the phrase "plus or minus 3 percentage points" in a news report of some poll results? The famous "3 percentage points" represents the *interval* within which the pollster has a certain degree of *confidence* in the results. Thus, the poll result "plus or minus 3 percentage points" represents a confidence interval. In this chapter, we build confidence intervals for a population mean, for a population proportion, and for the population variance and standard deviation. We also discuss the relationship between sample size and the precision of the interval estimate.

Confidence intervals are only our first topic in inferential statistics. Every chapter from here to the end of the book will uncover a new and different aspect of statistical inference. In Chapter 9, Hypothesis Testing, we'll learn about an inferential method even more prevalent than confidence intervals.

## 8.1 *Z* Interval for the Mean

### Objectives: By the end of this section, I will be able to...

1. Explain how the formula is developed for the *Z* interval for the population mean $\mu$.
2. Interpret the meaning of a confidence interval.

3 Calculate and interpret a $Z$ interval for the population mean $\mu$, when $\sigma$ is known, for two different cases.

4 Explain the meaning of the margin of error.

## 1 Introduction to Confidence Intervals: Developing the Formula

Recall that when we take a random sample and calculate the sample mean $\bar{x}$ to estimate the population mean $\mu$, we are finding a point estimate $\bar{x}$ that probably is not equal to the population mean $\mu$. For example, in Chapter 7, we showed that the mean of a sample of students' commuting times did not equal the population mean. Moreover, Table 7.6 (Section 7.1, page 367) showed that the means of different samples differ as well, which led us to consider the sampling distribution of the sample mean $\bar{x}$. We know from the Central Limit Theorem that the sampling distribution of the mean behaves like the normal distribution for large enough sample sizes.

Like any continuous distribution, probability for the normal distribution is defined as area under the curve above an *interval*. In Chapter 6, we learned that for any continuous random variable, the probability that the random variable *equaled* any particular value was precisely *zero*. This is because there is no area above a point on the number line, and for continuous distributions, probability equals area. So the sad truth is that, for the point estimate $\bar{x}$, $P(\bar{x} = \mu) = 0$. Thus, we have no *measure of confidence* associated with the estimate, telling us how confident we are that it is correct.

Although we cannot put a measure of confidence on $\bar{x}$ as a point estimate for $\mu$, *we can use $\bar{x}$ to find an interval that is likely to contain $\mu$.* In Chapter 7, we learned that the mean of the sampling distribution of $\bar{x}$ is $\mu$, and the standard deviation of this sampling distribution of $\bar{x}$ is $\sigma_{\bar{x}} = \sigma/\sqrt{n}$. Since the sampling distribution of $\bar{x}$ is approximately normal for large samples, using $\mu$ and $\sigma/\sqrt{n}$, we can standardize $\bar{x}$ to obtain the standard normal distribution $Z$ (Fact 5, page 343):

$$Z = \frac{\bar{x} - \mu}{\sigma/\sqrt{n}}$$

Further, an exercise in Chapter 6 testing the accuracy of the Empirical Rule showed that 95.44% of the values of a normal random variable lie within 2 standard deviations of the mean, as in Figure 8.1a. Since the standard normal distribution has mean 0 and standard deviation 1, then we may say that 95.44% of the standard normal values lie within the following interval: $-2 < Z < 2$ (Figure 8.1b).

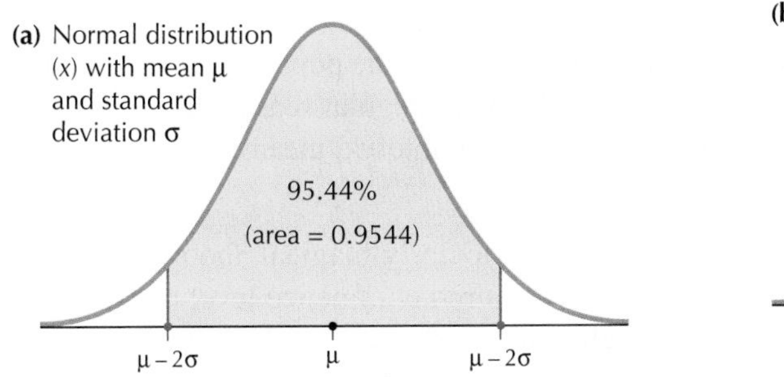


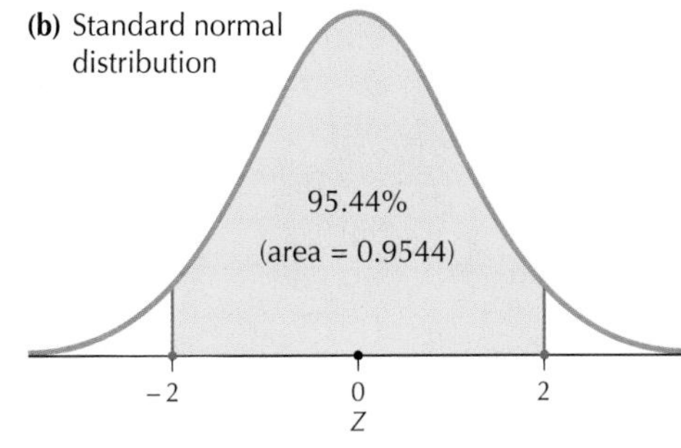


FIGURE 8.1 95.44% of normal random variable values lie within two standard deviations of the mean.

Plugging the formula for $Z$ back into this inequality gives

$$-2 < \frac{\bar{x} - \mu}{\sigma/\sqrt{n}} < 2$$

We then use algebra to isolate $\mu$ as the middle term:

$$\bar{x} - 2\,(\sigma/\sqrt{n}) < \mu < \bar{x} + 2\,(\sigma/\sqrt{n})$$

Thus, we can be 95.44% confident that the unknown population mean $\mu$ lies between the *lower bound,* $\bar{x} - 2(\sigma/\sqrt{n})$, and the *upper bound,* $\bar{x} + 2(\sigma/\sqrt{n})$. The values on the number line in the interval between the lower bound and upper bound make up the *confidence interval*. The 95.44% is called the **confidence level**.

A **confidence interval estimate** of a parameter consists of an interval of numbers generated by a point estimate, together with an associated **confidence level** specifying the probability that the interval contains the parameter.

## Example 8.1 95.44% confidence interval estimate of mean runs per game

Use the given random sample of runs scored in 36 Major League Baseball games in 2006 to find the 95.44% confidence interval estimate for the population mean number of runs $\mu$ per MLB game in 2006. Assume that the population standard deviation $\sigma = 3$.

**Sample of 2006 MLB runs per game**

| | | | | | | | | | | | | | | | | | |
|---|---|---|---|---|---|---|---|---|---|---|---|---|---|---|---|---|---|
| 5 | 4 | 7 | 6 | 6 | 8 | 3 | 4 | 6 | 4 | 2 | 4 | 4 | 4 | 1 | 0 | 3 | 4 |
| 6 | 3 | 5 | 5 | 5 | 10 | 6 | 7 | 8 | 3 | 1 | 3 | 2 | 4 | 7 | 5 | 3 | 4 |

### Solution

We would like to estimate $\mu$ = the population mean number of runs per game for all MLB teams in 2006. We are given that $\sigma = 3$ and we know that $n = 36$ because a sample of 36 games was taken. We calculate:

$$\bar{x} = \frac{\sum x}{n} = \frac{162}{36} = 4.5$$

*Note:* In Example 8.1, we are given $\sigma = 3$. In Section 8.1, we need to assume that we know the value of the population standard deviation $\sigma$. In Section 8.2, we cover the case where $\sigma$ is not known.

Then, since $n = 36$ is large ($\geq 30$), we can construct our 95.44% confidence interval estimate as follows:

$$\text{lower bound} = \bar{x} - 2(\sigma/\sqrt{n}) = 4.5 - 2(3/\sqrt{36}) = 4.5 - 1 = 3.5$$

$$\text{upper bound} = \bar{x} + 2(\sigma/\sqrt{n}) = 4.5 + 2(3/\sqrt{36}) = 4.5 + 1 = 5.5$$

We are 95.44% confident that $\mu$, the population mean number of runs scored per game for all teams in the 2006 MLB regular season, lies between 3.5 and 5.5 runs. Later we find out what this confidence interval means. ■

There is nothing particularly magical about the choice of 95.44%. We can construct intervals of any desired confidence level between 0% and 100% (though only intervals with high confidence are useful). Let us now generalize this procedure to produce confidence interval estimates of any desired confidence level.

Suppose $\alpha$ (alpha) is some small constant, usually $0 < \alpha \leq 0.10$. (Later we shall learn just what $\alpha$ means.) Then let's define $Z_{\alpha/2}$ to be the value of the standard normal ($Z$) distribution that has area $\alpha/2$ to the right of it (Figure 8.2). For example, if we let $\alpha = 0.05$, then $\alpha/2 = 0.025$ and the value of the $Z$ distribution with area 0.025 to the right of it is $Z_{0.025} = 1.96$.

The $Z$ value with area 0.025 to the right of it is the 97.5th percentile of the standard normal distribution, which we learned how to find in Chapter 7. For the case when $\alpha = 0.0456$, $\alpha/2 = 0.0228$, and we have $Z_{\alpha/2} = Z_{0.0228} = 2$, as in Example 8.1.

FIGURE 8.2
$Z_{\alpha/2}$ is the value of $Z$ that has area $\alpha/2$ to the right of it.

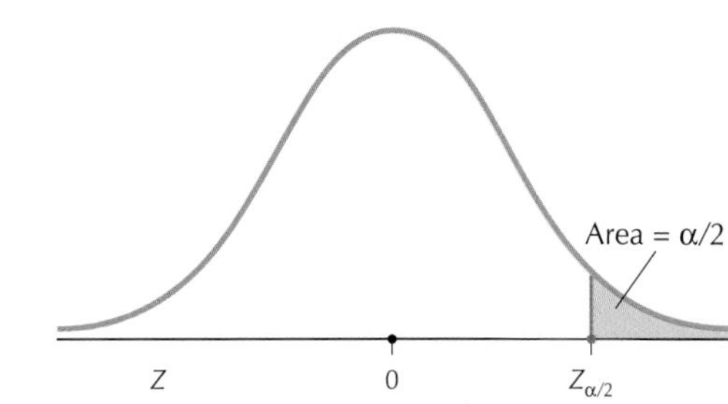


Since the standard normal distribution $Z$ is symmetric, we know that the area to the left of $-Z_{\alpha/2}$ is also $\alpha/2$. For example, the area to the left of –1.96 is also 0.025. For any value of $\alpha$, the area between $-Z_{\alpha/2}$ and $Z_{\alpha/2}$ is $1 - \alpha$, as shown in Figure 8.3.

FIGURE 8.3

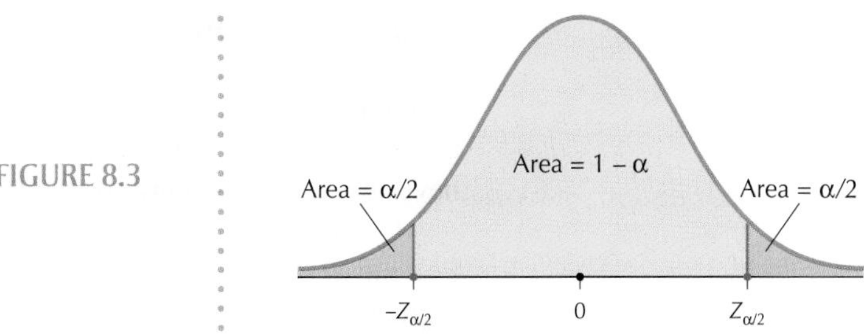


Therefore, since areas represent probabilities, we can write

$$P(-Z_{\alpha/2} < Z < Z_{\alpha/2}) = 1 - \alpha$$

We can follow the same logic that we used earlier to develop the general form for confidence intervals for the mean. Since this confidence interval for $\mu$ is based on the standard normal $Z$ distribution, it is called the **$Z$ interval for the population mean $\mu$.**

> ***Z* Interval for the Population Mean $\mu$**
>
> The $Z$ interval for $\mu$ may be constructed only when either of the following two conditions are met:
>
> Case 1: The population is normally distributed, and the value of $\sigma$ is known.
> Case 2: The sample size is large, and the value of $\sigma$ is known.
>
> When a random sample of size $n$ is taken from a population, a $100(1 - \alpha)\%$ confidence interval for $\mu$ is given by
>
> $$\text{lower bound} = \bar{x} - Z_{\alpha/2}(\sigma/\sqrt{n})$$
> $$\text{upper bound} = \bar{x} + Z_{\alpha/2}(\sigma/\sqrt{n})$$
>
> where $1 - \alpha$ is the confidence level. The $Z$ interval can also be written as
>
> $$\bar{x} \pm Z_{\alpha/2}(\sigma/\sqrt{n})$$
>
> and is denoted
>
> (lower bound, upper bound)

*Note:* The $\pm$ notation is shorthand for writing two numbers. For example, $\bar{x} \pm 2$ refers to the pair of numbers $\bar{x} - 2$ and $\bar{x} + 2$.

Table 8.1 provides a listing of $Z_{\alpha/2}$ values for the most common confidence levels.

Table 8.1 $Z_{\alpha/2}$ values for common confidence levels

| **Confidence level $(1 - \alpha)100\%$** | $\alpha$ | $\alpha/2$ | $Z_{\alpha/2}$ |
|---|---|---|---|
| 100(1 – 0.10)% = 90% | 0.10 | 0.05 | 1.645 |
| 100(1 – 0.05)% = 95% | 0.05 | 0.025 | 1.96 |
| 100(1 – 0.01)% = 99% | 0.01 | 0.005 | 2.576 |

The *Normal Density Curve* applet may be used to find $Z_{\alpha/2}$ critical values for confidence levels not listed in Table 8.1.

## 2 What Do Confidence Intervals Mean?

Recall that we found the 95.44% confidence interval for the population mean $\mu$ in the baseball example was

$$\text{lower bound} = 3.5, \text{ upper bound} = 5.5$$

meaning that we are 95.44% confident that the population mean number of runs scored per game for all teams in the 2006 MLB regular season lies between 3.5 and 5.5 runs. This means that we now know what the value is of the population mean runs per game, right? Actually, not. If this were a real-world problem (that we couldn't just look up), then we would never find out what the population mean is, no matter how much statistical analysis was done. For real-world problems, all we can say is that we are pretty sure the population mean is captured by this interval, but we will never be absolutely certain.

However, since this is a textbook example, we do know the value of the population mean $\mu$, the mean number of runs per game for all teams in all the 2006 MLB regular season. It is 4.86 runs per game, as reported by **mlb.com**. Did our 95.44% confidence interval contain the population mean $\mu$? Yes, our confidence interval does contain 4.86. But it did not have to happen that way. If we had had an unusual sample of games (with lots of corked bats and steroids) with a higher than usual sample mean $\bar{x}$, then our confidence interval may not have captured the population mean $\mu$.

### What Is Random Here?

It is important to note that *it is the interval that is random,* not the parameter. The interval is shaped by the sample statistics like $\bar{x}$, which are random variables, so the interval is random. The parameter $\mu$, though unknown, is nevertheless constant. For each different sample we take, we may expect to get different values for our sample statistics. Since the sample statistics differ, then the intervals differ as well. ■

## Example 8.2 The interval is random; the parameter is constant

Randomly select 20 different samples of size $n = 36$ from the 2006 MLB season, calculate the 95% confidence interval for each sample, and determine which confidence intervals actually contain the population mean $\mu = 4.86$ runs per game. Assume that $\sigma = 3$.

**Solution**

Table 8.2 shows the sample mean $\bar{x}$ and the lower and upper bounds of the 95% confidence interval for the population mean $\mu$ for each randomly selected sample. Note that the intervals are random while the parameter $\mu = 4.86$ is constant. As it turned out, 19 of the 20 samples (95%) produced 95% confidence intervals that contained $\mu = 4.86$. Did it have to happen this way? No. We were just lucky. For example, the confidence interval for Sample 5 nearly missed capturing the population mean $\mu$. A slightly different sample, perhaps including a shutout game, would have meant that the confidence interval for this sample would not have captured the population mean $\mu$. ■

Table 8.2 The confidence interval (CI) is random. The parameter is constant

| Sample | Sample mean $\bar{x}$ | Lower bound $x - 1.96(\sigma/\sqrt{n}) = \bar{x} - 1.96(3/\sqrt{36})$ | Upper bound $x + 1.96(\sigma/\sqrt{n}) = \bar{x} + 1.96(3/\sqrt{36})$ | $\mu$ in CI? | $\mu = 4.86$ (axis: 3 4 5 6 7) |
|---|---|---|---|---|---|
| 1 | 4.8 | 3.82 | 5.78 | Yes | |
| 2 | 5.2 | 4.22 | 6.18 | Yes | |
| 3 | 4.2 | 3.22 | 5.18 | Yes | |
| **4** | **6.1** | **5.12** | **7.08** | **No** | |
| 5 | 3.9 | 2.92 | 4.88 | Yes | |
| 6 | 5.0 | 4.02 | 5.98 | Yes | |
| 7 | 4.9 | 3.92 | 5.88 | Yes | |
| 8 | 5.2 | 4.22 | 6.18 | Yes | |
| 9 | 5.7 | 4.72 | 6.68 | Yes | |
| 10 | 5.1 | 4.12 | 6.08 | Yes | |
| 11 | 4.1 | 3.12 | 5.08 | Yes | |
| 12 | 4.5 | 3.52 | 5.48 | Yes | |
| 13 | 5.1 | 4.12 | 6.08 | Yes | |
| 14 | 4.6 | 3.62 | 5.58 | Yes | |
| 15 | 5.3 | 4.32 | 6.28 | Yes | |
| 16 | 4.9 | 3.92 | 5.88 | Yes | |
| 17 | 5.6 | 4.62 | 6.58 | Yes | |
| 18 | 5.1 | 4.12 | 6.08 | Yes | |
| 19 | 5.7 | 4.72 | 6.68 | Yes | |
| 20 | 5.0 | 4.02 | 5.98 | Yes | |

### What Does This Confidence Interval Mean?

What does the 95% mean in the phrase *95% confidence interval?* If we take sample after sample for a very long time, then in the long run, the proportion of intervals that will contain the parameter $\mu$ will equal 95%.

It is always a good idea to interpret the confidence intervals you construct. You can use the following generic interpretation: "We are 95% (or 90% or 99% and so on) confident that the population mean ________ lies between ____ and ____." ■

The *Confidence Interval* applet allows you to see for yourself how individual samples generate intervals that either do or do not contain the population mean.

## 3 Z Interval for the Population Mean $\mu$: Two Cases

The $Z$ interval for $\mu$ can be used only under certain conditions. There are two cases:

- Case 1: The population is normally distributed, and the value of $\sigma$ is known.
- Case 2: The sample size is large ($n \geq 30$), and the value of $\sigma$ is known.

If $\sigma$ is unknown, then we may be able to use the confidence interval we learn about in Section 8.2.

## Example 8.3 Z Interval for the mean

AP Photo/Jeff T. Green

The Washington State Department of Ecology reported that the mean lead contamination in trout in the Spokane River is 1 part per million (ppm), with a standard deviation of 0.5 ppm.[1] Suppose a sample of $n = 100$ trout has a mean lead contamination of $\bar{x} = 1$ ppm. Assume that $\sigma = 0.5$ ppm.

**a.** Determine whether Case 1 or Case 2 applies.

**b.** Construct a 95% confidence interval for $\mu$, the population mean lead contamination in all trout in the Spokane River.

**c.** Interpret the confidence interval.

### Solution

**a.** We are not given any information about the distribution of the population, so we don't know if the population is normally distributed. Therefore Case 1 does not apply. Case 2 applies since the sample size $n = 100$ is greater than 30 and the value of $\sigma = 0.5$ is known. We can proceed to construct the confidence interval.

**b.** The formula for the confidence interval is given by

$$\text{lower bound} = \bar{x} - Z_{\alpha/2}(\sigma/\sqrt{n})$$

$$\text{upper bound} = \bar{x} + Z_{\alpha/2}(\sigma/\sqrt{n})$$

We are given $n = 100$, $\bar{x} = 1$, and $\sigma = 0.5$. For a confidence level of 95%, Table 8.1 provides the value of $Z_{\alpha/2} = Z_{0.025} = 1.96$. Plugging in the formula:

$$\text{lower bound} = 1 - 1.96(0.5/\sqrt{100}) = 1 - 1.96(0.05) = 1 - 0.098 = 0.902$$

$$\text{upper bound} = 1 + 1.96(0.5/\sqrt{100}) = 1 + 1.96(0.05) = 1 + 0.098 = 1.098$$

**c.** We are 95% confident that $\mu$, the population mean lead contamination for all trout on the Spokane River, lies between 0.902 ppm and 1.098 ppm. ■

## Example 8.4 Watching that sodium: We cannot use the Z interval if the conditions are not met

Research has shown that the amount of sodium consumed in food has been associated with hypertension (high blood pressure). Table 8.3 provides a list of 28 breakfast cereals, along with their sodium content, in milligrams per serving. Determine whether Case 1 or Case 2 applies. If appropriate, construct a 99% confidence interval for $\mu$, the population mean sodium content per serving for all breakfast cereals. Assume that the population standard deviation $\sigma$ equals 90 mg.

### Solution

Since $n = 28$, Case 2 does not apply, since we do not have $n \geq 30$. We thus turn to Case 1 and evaluate the normality of the data set. On the right-hand side of the normal probability plot in Figure 8.4 (sodium content greater than 100 mg), the points all lie within the bounds. However, on the left side a couple of points lie outside the bounds, for sodium content of 0 mg. We should probably look further into this. A histogram is called for, and it is presented in Figure 8.5. This histogram has a "spike" on the left, representing the five breakfast cereals that have zero sodium content. While very healthy for your diet, these five cereals seem to be undermining our assumption of normality. Clearly, the distribution of sodium in breakfast cereals is *not* normally distributed and Case 1 does not apply. It would not be appropriate to construct the Z interval in this case. ■

Table 8.3 Sodium content of 28 breakfast cereals

| Cereal | Sodium (mg) | Cereal | Sodium (mg) |
|---|---|---|---|
| Apple Jacks | 125 | Life | 150 |
| Cap'n Crunch | 220 | Lucky Charms | 180 |
| Cheerios | 290 | Maypo | 0 |
| Cinnamon Toast Crunch | 210 | Mueslix Crispy | 150 |
| Cocoa Puffs | 180 | Puffed Rice | 0 |
| Corn Flakes | 290 | Quaker Oatmeal | 0 |
| Count Chocula | 180 | Raisin Bran | 210 |
| Cream of Wheat | 80 | Rice Chex | 240 |
| Froot Loops | 125 | Rice Krispies | 290 |
| Frosted Flakes | 200 | Shredded Wheat | 0 |
| Frosted Mini-Wheats | 0 | Special K | 230 |
| Fruity Pebbles | 135 | Total Whole Grain | 200 |
| Grape Nuts Flakes | 140 | Trix | 140 |
| Kix | 260 | Wheaties | 200 |

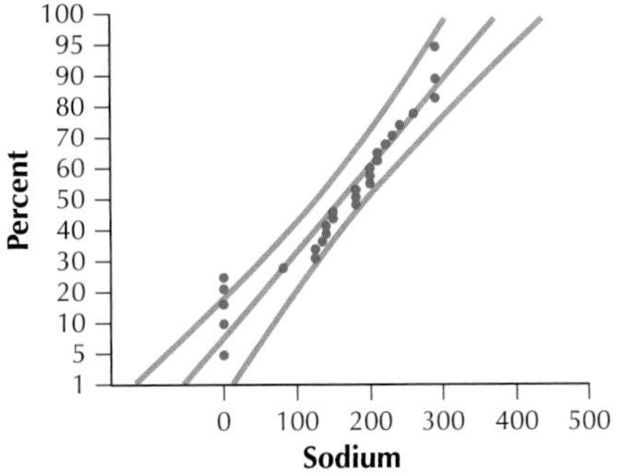


FIGURE 8.4 Normal probability plot of sodium content.

FIGURE 8.5 Histogram of sodium content.

Developing Your Statistical Sense

## The Importance of Assumptions

The explosive growth in statistical software and scientific calculators has led to an explosion in the amount of statistical analysis. Unfortunately, however, not all of the analysis is valid. The availability of statistical inference at the touch of a button has led to neglect of the assumptions underlying the procedures, that is, the conditions required to perform the analysis. If these conditions are not met, then the statistical inference lacks foundation, and any conclusions drawn are questionable. ■

## Example 8.5 Watching that sodium (continued)

We continue with the breakfast cereals data set in Example 8.4.

**a.** Identify the five cereals in Table 8.3 that contained zero sodium, and omit them from the calculations.

**b.** Assess the normality of the reduced data set.

**c.** If appropriate, construct a 99% confidence interval for the population mean sodium content. The sample mean sodium is 192.39 mg and the sample size is $n = 23$. Assume that $\sigma$ equals 50 mg of sodium.

**d.** Discuss the meaning of this confidence interval.

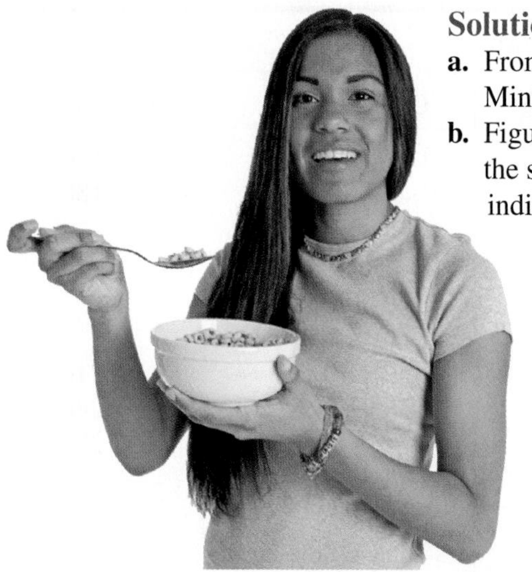

©Dynamic Graphics/Jupiterimages

**Solution**

**a.** From Table 8.3, the five cereals containing zero sodium are Frosted Mini-Wheats, Maypo, Puffed Rice, Quaker Oatmeal, and Shredded Wheat.

**b.** Figure 8.6 contains the normal probability plot for the data set after omitting the sodium-free cereals. Though not perfect, all points lie within the bounds, indicating a better fit to normality. Thus, Case 1 applies and we proceed to construct the 99% confidence interval.

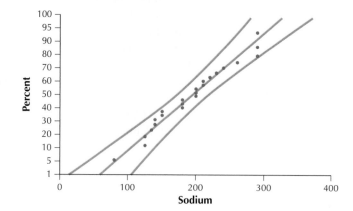

FIGURE 8.6 Normal probability plot of sodium content with five cereals omitted.

**c.** For a 99% confidence level, we find $Z_{\alpha/2} = 2.576$ from Table 8.1. Thus,

$$\begin{aligned}\text{lower bound} &= \bar{x} - Z_{\alpha/2}(\sigma/\sqrt{n}) \\ &= 192.39 - 2.576(50/\sqrt{23}) \\ &\approx 192.39 - 26.86 \\ &= 165.53\end{aligned}$$

$$\begin{aligned}\text{upper bound} &= \bar{x} + Z_{\alpha/2}(\sigma/\sqrt{n}) \\ &= 192.39 + 2.576(50/\sqrt{23}) \\ &\approx 192.39 + 26.86 \\ &= 219.25\end{aligned}$$

**d.** Note, however, that we can no longer claim to be constructing a confidence interval for the mean sodium content of *all* breakfast cereals, since we are ignoring an entire important class of breakfast cereals: those with no sodium at all. Therefore, our confidence interval will be for the population mean sodium content of *all breakfast cereals that contain sodium.* We are 99% confident that $\mu$, the population mean sodium content per serving of all breakfast cereals that contain sodium, lies between 165.53 grams and 219.25 grams. ■

## Example 8.6 $Z$ intervals for $\mu$ using technology: TI-83/84, Minitab, and the WHFStat Macros for Excel

The U.S. Small Business Administration (SBA) provides information on the number of small businesses for each metropolitan area in the United States. Table 8.4 contains a random sample of 30 moderately large cities and the number of small businesses in each city. Use the TI-83/84, Minitab, and the WHFStat Macros for Excel to construct a 95% $Z$ confidence interval for the population mean number of small businesses in cities nationwide. Assume that the standard deviation is $\sigma = 4300$ for the number of small businesses in moderately large cities.

**Solution**

We shall use the instructions provided in the Step-by-Step Technology Guide at the end of this section (pages 401–402). Since the sample size $n = 30$ is large ($\geq 30$), Case 2 applies, and it is not necessary to check for normality (Case 1).

Table 8.4 Small businesses in a sample of 30 cities

| City | Small businesses | City | Small businesses | City | Small businesses |
|---|---|---|---|---|---|
| Orlando, FL | 32,751 | Cincinnati, OH | 25,618 | Nashville, TN | 21,736 |
| Kansas City, MO | 32,750 | Salt Lake City, UT | 25,107 | New Orleans, LA | 21,565 |
| San Jose, CA | 30,921 | Las Vegas, NV | 24,867 | Oklahoma City, OK | 21,102 |
| West Palm Beach, FL | 30,226 | Monmouth, NJ | 24,255 | Hartford, CT | 20,677 |
| Charlotte, NC | 28,739 | Columbus, OH | 23,786 | Jacksonville, FL | 20,168 |
| Indianapolis, IN | 27,397 | Raleigh, NC | 23,566 | Grand Rapids, MI | 18,636 |
| Sacramento, CA | 27,189 | Providence, RI | 23,205 | Buffalo, NY | 18,285 |
| Milwaukee, WI | 26,456 | Norfolk, VA | 22,844 | Richmond, VA | 18,015 |
| Fort Worth, TX | 25,735 | Greensboro, NC | 22,359 | Louisville, KY | 17,754 |
| Middlesex, NJ | 25,726 | Austin, TX | 22,305 | Greenville, SC | 16,791 |

The results for the TI-83/84 in Figure 8.7 show that the 95% $Z$ confidence interval for the population mean number of small businesses per city is approximately

$$\text{lower bound} = 22{,}479, \text{ upper bound} = 25{,}556$$

It also shows the sample mean $\bar{x} = 24{,}017.7$, the sample standard deviation $s = 4322.473886$, and the sample size $n = 30$.

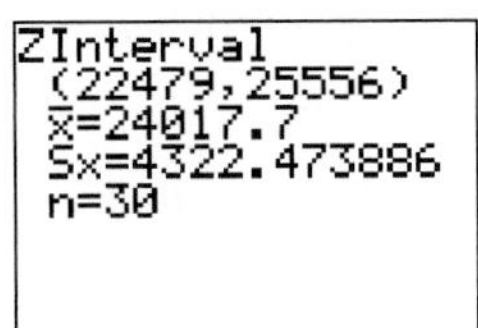


FIGURE 8.7 TI-83/84 results.

The Minitab results are provided in Figure 8.8. The "assumed standard deviation" is indicated to be $\sigma = 4300$. Then the sample size $n = 30$, the sample mean $\bar{x} = 24{,}018$ (rounded), and the sample standard deviation $s = 4322$ (rounded) are displayed. "SE Mean" refers to the standard error of the mean and is not used here. Finally, the 95% confidence interval is given as (Lower bound = 22,479, Upper bound = 25,556).

**One-Sample Z: Small Businesses**

```
The assumed standard deviation = 4300

Variable          N   Mean  StDev  SE Mean        95% CI
Small Businesses  30  24018   4322      785  (22479, 25556)
```

FIGURE 8.8 Minitab results.

The results from the WHFStat Macros for Excel are shown in Figure 8.9.

SUMMARY STATISTICS

| Sample Mean | Sample Size | Standard Deviation |
|---|---|---|
| 24017.7 | 30 | 4300 |

ONE SAMPLE Z CONFIDENCE INTERVAL

| Confidence Level | | Critical Z Value |
|---|---|---|
| 95 % | | 1.96 |
| **Confidence Interval** | | |
| 24017.7 | +/- | 1538.735 |
| 22478.96 | to | 25556.43 |

FIGURE 8.9 WHFStat Macros results.

The sample mean $\bar{x} = 24{,}017.7$, the sample size $n = 30$, and the population standard deviation $\sigma = 4300$ are displayed. The confidence level 95% is shown, along with the critical Z value, $Z_{\alpha/2} = 1.96$. The confidence interval is then shown:

$$\text{lower bound} = 22{,}478.96, \text{ upper bound} = 25{,}556.43$$

You will note in Example 8.6 that the TI-83/84 and Minitab report the confidence interval in the following form:

(lower bound, upper bound)

This form is widespread in the statistical world. For example, in Example 8.5, we found the 99% confidence interval for the population mean sodium content per serving of all breakfast cereals that contain sodium to be

$$\text{lower bound} = 165.53, \text{ upper bound} = 219.25$$

This 99% confidence interval can also be expressed as (165.53, 219.25).

## 4 Margin of Error

Suppose we wish to estimate the population mean $\mu$ to within a certain measure of accuracy. For example, when we estimate the mean age of when children learn to read, we might estimate it to within 1 year. This estimate of accuracy (1 year, in this case) is called the **margin of error**. In Example 8.5, we found that we are 99% confident that $\mu$, the population mean sodium content for breakfast cereals that contain sodium lies within 26.86 grams of our sample mean 192.39:

$$\begin{aligned}\text{lower bound} &= \bar{x} - Z_{\alpha/2}(\sigma/\sqrt{n}) \\ &= 192.39 - 2.576(50/\sqrt{23}) \\ &\approx 192.39 - 26.86 \\ &= 165.53\end{aligned}$$

$$\begin{aligned}\text{upper bound} &= \bar{x} + Z_{\alpha/2}(\sigma/\sqrt{n}) \\ &= 192.39 + 2.576(50/\sqrt{23}) \\ &\approx 192.39 + 26.86 \\ &= 219.25\end{aligned}$$

In these calculations, note that we are adding (+) and subtracting (−) the same quantity $Z_{\alpha/2}(\sigma/\sqrt{n}) = 26.86$ to and from $\bar{x}$. The quantity $Z_{\alpha/2}(\sigma/\sqrt{n}) = 26.86$ is called the *margin of error* ***E***, and the *Z* interval

$$(\bar{x} - Z_{\alpha/2}(\sigma/\sqrt{n}), \bar{x} + Z_{\alpha/2}(\sigma/\sqrt{n}))$$

can be written

$$\bar{x} \pm Z_{\alpha/2}(\sigma/\sqrt{n})$$

which consists of the point estimate $\bar{x}$ plus and minus the margin of error $Z_{\alpha/2}(\sigma/\sqrt{n})$.

Note the form that the confidence interval takes:

CAUTION: Remember that the "±" notation *always* represents a pair of numbers.

point estimate ± margin of error

Of course, we would like our confidence interval estimates to be as precise as possible. Therefore, we would like the margin of error to be as small as possible, which would in turn result in a shorter confidence interval. Shorter, tighter confidence intervals are better, since the likely maximum difference between the sample mean and the population mean is reduced.

> The **margin of error *E*** is a measure of the precision of the confidence interval estimate.
>
> For the *Z* interval, the margin of error takes the form
>
> $$E = Z_{\alpha/2}(\sigma/\sqrt{n})$$

## Example 8.7 Confidence interval as point estimate ± margin of error

In Example 8.3, the $Z$ interval for the population mean lead contamination (in ppm) for all trout on the Spokane River is

$$\text{lower bound} = 1 - 1.96(0.5/\sqrt{100}) = 1 - 1.96(0.05) = 1 - 0.098 = 0.902$$
$$\text{upper bound} = 1 + 1.96(0.5/\sqrt{100}) = 1 + 1.96(0.05) = 1 + 0.098 = 1.098$$

**a.** Find the margin of error $E$.
**b.** Express the confidence interval in the form "point estimate ± margin of error."

### Solution

**a.** We find the margin of error as follows:

$$E = Z_{\alpha/2}(\sigma/\sqrt{n}) = 1.96(0.5/\sqrt{100}) = 1.96(0.05) = 0.098$$

**b.** The point estimate is $\bar{x} = 1$. Thus, the 95% confidence interval for the population mean lead contamination (in ppm) for all trout on the Spokane River takes the following form:

$$\begin{aligned} &\text{point estimate} \pm \text{margin of error} \\ &= \bar{x} \pm Z_{\alpha/2}(\sigma/\sqrt{n}) \\ &= 1 \pm 0.098 \end{aligned}$$

*Note:* When it comes to the margin of error $E$, smaller is better!

**Interpreting the Margin of Error**
Interpret the margin of error $E$ for a $(1 - \alpha)100\%$ confidence interval for $\mu$ as follows:
"We can estimate $\mu$ to within $E$ units with $(1 - \alpha)100\%$ confidence."

## Example 8.8 Interpreting the margin of error

Find and interpret the margin of error $E$ for the confidence interval for the mean sodium content of the 23 breakfast cereals containing sodium in Example 8.5.

### Solution

The 99% confidence interval from Example 8.5 took the form

$$\begin{aligned} &\text{point estimate} \pm \text{margin of error} \\ &= 192.39 \pm 26.86 \\ &= (192.39 - 26.86 = 165.53,\ 192.39 + 26.86 = 219.25) \\ &= (165.53, 219.25) \end{aligned}$$

The point estimate of the population mean sodium content is $\bar{x} = 192.39$. The margin of error is

$$E = Z_{\alpha/2}(\sigma/\sqrt{n}) = 2.576(50/\sqrt{23}) \approx 26.86$$

We interpret $E$ by saying that we can estimate the population mean sodium content for breakfast cereals containing sodium *to within 26.86 mg with 99% confidence.*

If a smaller margin of error is better, how do we make it smaller? Let's look at the margin of error for the $Z$ interval:

$$E = Z_{\alpha/2}(\sigma/\sqrt{n})$$

Since the population standard deviation $\sigma$ is fixed, only $Z_{\alpha/2}$ and $n$ can vary. To make the margin of error smaller, we would need to make $Z_{\alpha/2}$ smaller, which is achieved by *decreasing the confidence level.* Next, notice that $n$ is in the denominator of the formula. Therefore to make the margin of error smaller, we would need to make $n$ larger, thereby *increasing the sample size.* Thus, there are two strategies for decreasing the margin of error:

1. Decrease the confidence level.
2. Increase the sample size.

## Example 8.9 Decreasing the margin of error by decreasing the confidence level

For the confidence interval for the population mean sodium content in Example 8.5, suppose we reduce the confidence level from 99% to 90% and leave everything else unchanged. Find the new margin of error. Describe how the margin of error has changed.

### Solution

For confidence level 90%, $Z_{\alpha/2} = 1.645$, giving the following margin of error:

$$E = Z_{\alpha/2}(\sigma/\sqrt{n}) = 1.645\,(50/\sqrt{23}) \approx 17.15$$

Decreasing the confidence level from 99% to 90% decreases the margin of error from 26.86 mg to 17.15 mg. ■

**There's No Free Lunch**

The margin of error in Example 8.9 is smaller than the one in Example 8.8. A smaller margin of error is good because it gives a more precise estimate of the unknown population mean. However, this smaller margin of error is due entirely to the decrease in the confidence level, which in turn decreased the value of $Z_{\alpha/2}$. And a decrease in the confidence level is not good. In statistical data analysis, there is rarely a free lunch. The trade-off here is that, while the margin of error went down, *so did the confidence level,* from 99% to 90%. For a given sample, the only difference in the formula between a 99% and a 90% confidence interval is the value of $Z_{\alpha/2}$, which is larger for larger confidence levels. The larger value of $Z_{\alpha/2}$ for the higher confidence level means that the margin of error $E = Z_{\alpha/2}(\sigma/\sqrt{n})$ is larger, which gives a wider confidence interval. Confidence intervals that are too wide are ineffectual and often useless. For example, we can be 99.9999% confident that the population mean age of college students in Florida lies between 15 and 75 years old. But, so what? The interval is too wide to be of practical use. More useful would be a 95% confidence interval that the population mean age of college students in Florida lies between 20 and 27. ■

This leads us to Strategy 2 for reducing the margin of error: Increase the sample size. *The only way to have both high confidence and a tight interval is to boost the sample size.*

## Example 8.10 Decreasing the margin of error by increasing the sample size

For the confidence interval for the population mean sodium content in Example 8.5, suppose the results were based on a sample of size $n = 100$ rather than $n = 23$. Leaving everything else unchanged, find the new margin of error, and describe how the margin of error has changed.

**Solution**

For $n = 100$, the margin of error is

$$E = Z_{\alpha/2}(\sigma/\sqrt{n}) = 2.576(50/\sqrt{100}) = 12.88$$

Increasing the sample size from $n = 23$ to $n = 100$ has decreased the margin of error from 26.86 mg to 12.88 mg. ■

"More data" is a familiar refrain in statistical analysis circles. Of course, increasing the sample size often raises pocketbook issues, since large samples can get very expensive ("We would like a large sample estimate of the amount of damage sustained by Corvettes hitting a wall at 90 mph"). Sometimes obtaining large samples is simply impossible. Suppose an astronomer has developed a new technique for predicting corona effects during solar eclipses; she will have to wait a while (say, a few hundred years) to build up a large sample. So, take samples as large as realistically possible to keep the width of the confidence interval down.

## STEP-BY-STEP TECHNOLOGY GUIDE: Z Confidence Intervals

We illustrate how to construct the confidence interval for Example 8.6 (page 396).

### TI-83/84

**If you have the data values:**

1. Enter the data into list **L1** (Figure 8.10).
2. Press **STAT**, highlight **TESTS**.
3. Press 7 (for **ZInterval**).
4. For input (**Inpt**), highlight **Data** and press **ENTER** (Figure 8.11).

**a.** For $\sigma$, enter the assumed value of **4300**.
**b.** For **List**, press **2nd** then **L1**.
**c.** For **Freq**, enter **1**.
**d.** For **C-Level** (confidence level), enter the appropriate confidence level (e.g., **0.95**), and press **ENTER**.
**e.** Highlight **Calculate** and press **ENTER**. The results are shown in Figure 8.7 in Example 8.6.

**If you have the summary statistics:**

1. Press **STAT**, highlight **TESTS**.
2. Press 7 (for **ZInterval**).
3. For input (**Inpt**), highlight **Stats** and press **ENTER** (Figure 8.12).

**a.** For $\sigma$, enter the assumed value of **4300**.
**b.** For $\bar{x}$, enter the sample mean 24017.7.
**c.** For $n$, enter the sample size **30**.
**d.** For **C-Level** (confidence level), enter the appropriate confidence level (e.g., **0.95**), and press **ENTER**.
**e.** Highlight **Calculate** and press **ENTER**. The results are shown in Figure 8.7 in Example 8.6.

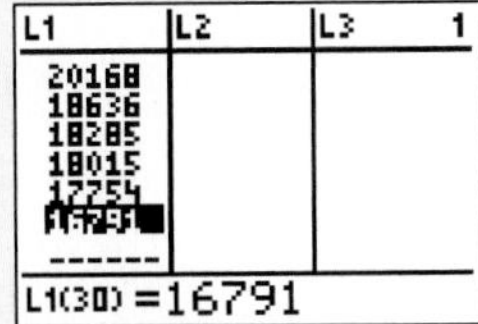


Figure 8.10

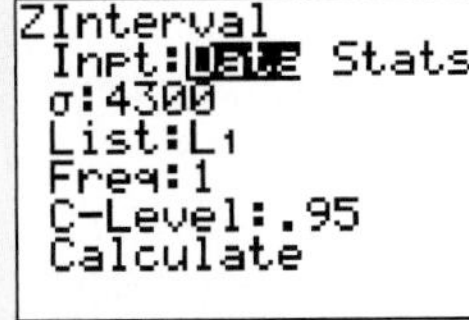


Figure 8.11

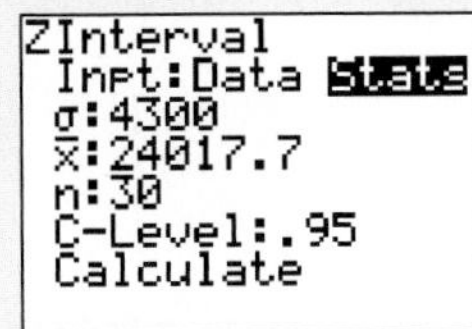


Figure 8.12

### EXCEL

**If you have the data values:**

1. Enter the data into column A.
2. Load the **WHFStat Macros**.
3. Select **Add-Ins > Macros > Estimating a Mean > Z Confidence Interval**.
4. Click **Select Dataset Range**, highlight A1–A30, and click **OK**.
5. Input **4300** for the **Population Standard Deviation**, select the **95%** confidence level, and click **OK**.

The results are displayed in Figure 8.9 in Example 8.6.

*(Continued)*

**If you have the summary statistics:**

*1.* Load the **WHFStat Macros.**

*2.* Select **Add-Ins > Macros > Estimating a Mean > Z Confidence Interval.**

*3.* Click **Input Summary Statistics,** enter 24017.7 for the **Sample Mean**, enter **30** for the **Sample Size**, and click **OK.**

*4.* Input **4300** for the **Population Standard Deviation**, select the **95%** confidence level, and click **OK.**

The results are displayed in Figure 8.9 in Example 8.6.

## MINITAB

**If you have the data values:**

*1.* Enter the data into column C1.

*2.* Click **Stat > Basic Statistics > 1-Sample Z.**

*3.* Click **Samples in Columns** and select **C1.**

*4.* Click **Options**, enter **95** as the **Confidence Level**, and click **OK.**

*5.* Enter **4300** for **Sigma** and click **OK.**

The results are displayed in Figure 8.8 in Example 8.6.

**If you have the summary statistics:**

*1.* Click **Stat > Basic Statistics > 1-Sample Z.**

*2.* Click **Summarized Data.**

*3.* Enter the **Sample Size 30** and the **Sample Mean 24017.7.**

*4.* Enter **4300** for the **Standard Deviation.**

*5.* Click **Options**, enter **95** as the **Confidence Level**, click **OK**, and click **OK** again.

The results are displayed in Figure 8.8 in Example 8.6.

## Section 8.1 SUMMARY

1 A confidence interval estimate of a parameter consists of an interval of numbers generated by a point estimate, together with an associated confidence level specifying the probability that the interval contains the parameter.

2 The meaning of a $100(1 - \alpha)\%$ confidence interval is as follows: If we take sample after sample for a very long time, then in the long run, the proportion of intervals that will contain the parameter $\mu$ will equal $100(1 - \alpha)\%$.

3 The $100(1 - \alpha)\%$ confidence interval for $\mu$ is given by the interval

$$\text{lower bound} = \bar{x} - Z_{\alpha/2}(\sigma/\sqrt{n})$$
$$\text{upper bound} = \bar{x} + Z_{\alpha/2}(\sigma/\sqrt{n})$$

where $1 - \alpha$ is the confidence level. The conditions for applying this confidence interval are as follows:

- Case 1: The original population is normal and $\sigma$ is known.
- Case 2: The sample size is large ($n \geq 30$) and $\sigma$ is known.

If $\sigma$ is not known, then the $Z$ interval cannot be used.

4 The margin of error $E$ is a measure of the precision of the confidence interval estimate. For the $Z$ interval, the margin of error takes the form

$$E = Z_{\alpha/2}(\sigma/\sqrt{n})$$

We interpret the margin of error $E$ for a $(1 - \alpha)100\%$ confidence interval for $\mu$ as follows: "We can estimate $\mu$ to within $E$ with $(1 - \alpha)100\%$ confidence." In general, our confidence intervals take the form

point estimate ± margin of error

## Section 8.1 EXERCISES

### CLARIFYING THE CONCEPTS

**1.** What is the difference between *confidence interval* and *confidence level*?

**2.** Describe $Z_{\alpha/2}$ in terms of percentiles.

**3.** Assume that the confidence level increases.

**a.** What happens to the value of $Z_{\alpha/2}$?

**b.** Explain why this happens. Draw a sketch to help you.

**4.** Assume that the confidence level increases.

**a.** What happens to the value of $\alpha$? $\alpha/2$?

**b.** Explain why this happens. Draw a sketch to help you.

**5.** Explain why tighter (narrower) confidence intervals are better. Explain what is wrong with confidence intervals that are too wide.

**6.** Explain the two strategies for decreasing the margin of error.

**7.** Suppose your supervisor says to you, "Our competition is using 99% confidence intervals. Why on earth are we only using 95% confidence intervals? I want all our intervals from now on to be 99%!" How would you explain to your supervisor that increasing the confidence level has advantages and disadvantages?

**8.** Suppose your supervisor wants to (a) increase the confidence level from 95% to 99% and (b) keep the width of the confidence interval small. What is the only way to accomplish this?

## PRACTICING THE TECHNIQUES

**9.** For the following situations, random samples are drawn. Indicate whether or not we can use the $Z$ interval.

- **a.** The sample size is large ($n \geq 30$) and $\sigma$ is unknown.
- **b.** The original population is normal and $\sigma$ is known.
- **c.** The sample size is large ($n \geq 30$) and $\sigma$ is known.

**10.** For the following situations, random samples are drawn. Indicate whether or not we can use the $Z$ interval:

- **a.** The sample size is small, the original population is normal, and $\sigma$ is known.
- **b.** The sample size is large ($n \geq 30$) the original population is not normal, and $\sigma$ is known.
- **c.** The original population is not normal, and $\sigma$ is not known.

For Exercises 11–18, find the value of $Z_{\alpha/2}$.

**11.** Confidence level = 99%
**12.** $\alpha = 0.05$
**13.** $\alpha/2 = 0.005$
**14.** Confidence level = 95%
**15.** $\alpha = 0.10$
**16.** $\alpha/2 = 0.025$
**17.** Confidence level = 90%
**18.** $\alpha = 0.01$

For Exercises 19–24, answer the following questions.

- **a.** Which case applies?
- **b.** Calculate $\sigma/\sqrt{n}$.
- **c.** Find $Z_{\alpha/2}$ for a confidence interval for $\mu$ with 95% confidence.
- **d.** Compute $E$, the margin of error for a confidence interval $\mu$ with 95% confidence. Interpret this value for the margin of error.
- **e.** Construct and interpret a 95% confidence interval for $\mu$.

**19.** A random sample of $n = 64$ with sample mean $\bar{x} = 10$ is drawn from a population in which $\sigma = 4$.

**20.** A random sample of $n = 49$ with sample mean $\bar{x} = 20$ is drawn from a population in which $\sigma = 7$.

**21.** A random sample of $n = 16$ with sample mean $\bar{x} = 35$ is drawn from a normal population in which $\sigma = 2$.

**22.** A random sample of $n = 25$ with sample mean $\bar{x} = 50$ is drawn from a normal population in which $\sigma = 5$.

**23.** A random sample of $n = 81$ with sample mean $\bar{x} = 100$ is drawn from a population in which $\sigma = 18$.

**24.** A random sample of $n = 9$ with sample mean $\bar{x} = 15$ is drawn from a normal population in which $\sigma = 6$.

**25.** A random sample of $n = 25$ is drawn from a normal population in which $\sigma = 2$. The sample mean is $\bar{x} = 10$. For **(a)**–**(c)**, construct and interpret confidence intervals for $\mu$ with the indicated confidence levels. Then answer the questions in **(d)** and **(e)**.

- **a.** 90%
- **b.** 95%
- **c.** 99%
- **d.** What can you conclude about the width of the interval as the confidence level increases?
- **e.** Which case did you apply?

**26.** A random sample of $n = 100$ is drawn from a population in which $\sigma = 5$. The sample mean is $\bar{x} = 50$. For parts **(a)**–**(c)**, construct and interpret confidence intervals for $\mu$ with the indicated confidence levels. Then answer the questions in **(d)** and **(e)**.

- **a.** 90%
- **b.** 95%
- **c.** 99%
- **d.** What can you conclude about the width of the interval as the confidence level increases?
- **e.** Which case did you apply?

**27.** Random samples are drawn from a normal population in which $\sigma = 10$. The sample mean is $\bar{x} = 90$. For **(a)**–**(c)**, construct and interpret 95% confidence intervals for $\mu$ for the following samples sizes. Then answer the questions in **(d)** and **(e)**.

- **a.** $n = 25$
- **b.** $n = 100$
- **c.** $n = 400$
- **d.** What can you conclude about the width of the interval, as the sample size increases?
- **e.** Which case did you apply?

**28.** Random samples are drawn from a normal population in which $\sigma = 20$. The sample mean is $\bar{x} = 200$. For parts **(a)**–**(c)**, construct and interpret 95% confidence intervals for $\mu$ for the following sample sizes. Then answer the questions in **(d)** and **(e)**.

- **a.** $n = 16$
- **b.** $n = 81$
- **c.** $n = 225$
- **d.** What can you conclude about the width of the interval as the sample size increases?
- **e.** Which case did you apply?

## APPLYING THE CONCEPTS

**29. Old Soda Machine.** A random sample of 36 cups of soda dispensed from the old soda machine in the lobby had a mean amount of soda of 7 ounces. Assume that the population standard deviation is 1 ounce.

- **a.** Find the point estimate of the amount of soda dispensed.
- **b.** Calculate $\sigma/\sqrt{n}$.
- **c.** Find $Z_{\alpha/2}$ for a confidence interval with 95% confidence.

**d.** Compute and interpret the margin of error for a confidence interval with 95% confidence.

**e.** Construct and interpret a 95% confidence interval for $\mu$, the population mean amount of soda dispensed from the old soda machine in the lobby.

**30. Consumption of Carbonated Beverages.** The U.S. Department of Agriculture reports that the mean American consumption of carbonated beverages per year is greater than 52 gallons. A random sample of 30 Americans yielded a sample mean of 69 gallons. Assume that the population standard deviation is 20 gallons.

**a.** Find the point estimate of the amount of carbonated beverages consumed by all Americans per year.

**b.** Calculate $\sigma/\sqrt{n}$.

**c.** Find $Z_{\alpha/2}$ for a confidence interval with 95% confidence.

**d.** Compute and interpret the margin of error for a confidence interval with 95% confidence.

**e.** Construct and interpret a 95% confidence interval for $\mu$, the population mean amount of carbonated beverages consumed by all Americans per year.

**31. Catholic Church Membership.** Ten years ago, the Lily Endowment funded a study that found that the mean number of members in a Roman Catholic Church parish was 2723. This year, a random sample of 50 Roman Catholic Church parishes had a mean size of 2600 members. Assume that the population standard deviation is 1000 members.

**a.** Find the point estimate of the number of members in a Roman Catholic Church parish.

**b.** Calculate $\sigma/\sqrt{n}$.

**c.** Find $Z_{\alpha/2}$ for a confidence interval with 99% confidence.

**d.** Compute and interpret the margin of error for a confidence interval with 99% confidence.

**e.** Construct and interpret a 99% confidence interval for $\mu$, the population mean number of members in a Roman Catholic Church parish.

**32. Stock Shares Traded.** The *Statistical Abstract of the United States* reports that the mean daily number of shares traded on the New York Stock Exchange (NYSE) in 2002 was 1441 million. Assume that the population standard deviation equals 500 million shares. Suppose that, in a random sample of 36 days from the present year, the mean daily number of shares traded equals 1.5 billion.

**a.** Find the point estimate of the number of shares traded daily on the NYSE.

**b.** Calculate $\sigma/\sqrt{n}$.

**c.** Find $Z_{\alpha/2}$ for a confidence interval with 95% confidence.

**d.** Compute and interpret the margin of error for a confidence interval with 95% confidence.

**e.** Construct and interpret a 95% confidence interval for $\mu$, the population mean number of shares traded daily on the NYSE.

**33. Engaging with Science.** A psychological study found that the mean length of time that boys remained engaged with a science exhibit at a museum was 107 seconds with a standard deviation of 117 seconds.[2] Assume that the 117 seconds represents the population standard deviation. The sample size is not given, and no evidence is given that the lengths of time are normally distributed.

**a.** Suppose the sample size is 10. Discuss whether 95% confidence interval should be constructed for the population mean length of time that boys remained engaged with the science exhibit at the museum.

**b.** Suppose the sample size is 36. Calculate $\sigma/\sqrt{n}$.

**c.** Find $Z_{\alpha/2}$ for a confidence interval with 95% confidence for $n = 36$.

**d.** Compute and interpret the margin of error for a confidence interval with 95% confidence for $n = 36$.

**e.** Construct and interpret a 95% confidence interval for the population mean length of time that boys remained engaged with the science exhibit for $n = 36$.

**34. Short-Term Memory.** In a famous research paper in the psychology literature, the researcher found that the amount of information humans could process in short-term memory was 7 bits (pieces of information), plus or minus 2 bits.[3] Let us assume that Miller's title (The Magical Number Seven, Plus or Minus Two) refers to a confidence interval.

**a.** What is the point estimate for the amount of information all humans can process in short-term memory.

**b.** What is the margin of error?

**c.** The most common confidence level in the psychological literature is 95%. Which value for $Z_{\alpha/2}$ is associated with 95% confidence?

**d.** What is the value of $\sigma/\sqrt{n}$?

**e.** Describe the confidence interval given by Miller in his title.

**35. Commuting Distances.** A university is trying to attract more commuting students from the local community. As part of the research into the modes of transportation students use to commute to the university, a survey was conducted asking how far commuting students commuted from home to school each day. A random sample of 30 students provided the distances (in miles) shown in the table. Assume that the standard deviation is $\sigma = 3$ miles.

| | | | | | | | | | | | | | | |
|---|---|---|---|---|---|---|---|---|---|---|---|---|---|---|
| 14 | 10 | 14 | 12 | 12 | 11 | 5 | 6 | 9 | 14 | 9 | 9 | 4 | 7 | 15 |
| 9 | 7 | 7 | 12 | 10 | 15 | 10 | 6 | 11 | 9 | 11 | 10 | 11 | 7 | 12 |

**a.** Compute and interpret the margin of error for a confidence interval with 95% confidence.

**b.** Construct and interpret a 95% confidence interval for the population mean commuting distance.

**36. Working Women's Incomes.** A random sample of 15 working women yielded a sample mean income of \$50,000. Assume that the standard deviation of all incomes of working women is $5000 and that the distribution of these incomes is normal.

a. Compute and interpret the margin of error for a confidence interval with 99% confidence.

b. Provide and interpret a 99% confidence interval for the population mean income of the population of all working women.

**37. Herbicide Concentration.** Human exposure to herbicides is a potential hazard in America's grain-producing areas. One study found that, in a random sample of 112 homes in Iowa, the median concentration of the herbicide dicamba was 179.5 ng/g (nanograms per gram).[4] Suppose that the sample mean was 180 ng/g, and assume that the population standard deviation is 20 ng/g.

a. Compute and interpret the margin of error for a confidence interval with 95% confidence.

b. Construct and interpret a 95% confidence interval for the population mean concentration of the herbicide dicamba in Iowa homes.

**38. Asthma and Quality of Life.** A study examined the relationship between perceived neighborhood problems, quality of life, and mental health among adults with asthma.[5] Among those reporting the most serious neighborhood problems, the 95% confidence interval for the population mean quality of life score was (2.7152, 9.1048).

a. What is $Z_{\alpha/2}$?

b. Find $\bar{x}$.

c. Compute and interpret the margin of error.

d. Suppose $n = 36$. Find the value for $\sigma$.

**39. Mozart Effect.** A random sample of 45 children showed a mean increase of 8.5 IQ points after listening to a Mozart piano sonata for about 10 minutes. The distribution of such increases is unknown, but the standard deviation is assumed to be 3 IQ points.

a. Compute and interpret the margin of error for a confidence interval with 90% confidence.

b. Construct and interpret a 90% confidence interval for the mean increase in IQ points for all children after listening to the sonata for about 10 minutes.

**40. Hispanic Tobacco Consumption.** The Bureau of Labor Statistics reported that the mean amount spent by all American citizens in 2001 on tobacco products and smoking supplies was $308; the mean for American Hispanics was $177. Assume that $\sigma$, the standard deviation for American Hispanics, equals $150. Assume that the data on American Hispanics represents a sample of size 36.

a. Find the point estimate of the amount spent by all American Hispanics on tobacco products.

b. Compute the margin of error for a confidence interval with 90% confidence.

c. Construct and interpret a 90% confidence interval for the population mean amount spent by all American Hispanics on tobacco products.

**41. TV Viewing and Physical Activity.** A study examined the relationship between television viewing and physical activity.[6] The study found that, for each additional hour of television viewing, subjects walked between 12 and 276 fewer steps that day, as measured by a pedometer. This interval was reported with a 95% confidence level suppose $n = 100$.

a. What is $Z_{\alpha/2}$?

b. Find $\bar{x}$.

c. Compute and interpret the margin of error.

d. Find the value for $\sigma$.

**Lead Contamination in Trout.** Use this information for Exercises 42–44. In Example 8.3, a 95% confidence interval was constructed for the population mean lead contamination in Spokane River trout, based on a sample of 100 trout.

**42.** *What if* the sample size is increased to some unspecified higher number. Describe what effect, if any, the increase has on the following:

a. $\sigma$

b. $\sigma/\sqrt{n}$

c. Width of the confidence interval

d. Margin of error

e. Confidence level

**43.** *What if* the confidence level is decreased from 95% to some unspecified lower value. Explain whether and how this decrease would affect the following:

a. $\sigma$

b. $\sigma/\sqrt{n}$

c. Width of the confidence interval

d. $\bar{x}$

e. Margin of error

**44.** *What if* the sample size is reduced from 100 to some unspecified amount (>30), and no other changes are made. Explain whether and how this reduction would affect the following:

a. $\sigma/\sqrt{n}$

b. Width of the confidence interval

c. Margin of error

**Small Businesses.** Use this information for Exercises 45 and 46. The United States Small Business Administration publishes data on the number of small businesses in each of 327 metropolitan areas. This data is in the data file **Small Businesses.**

**45.** Follow steps **(a)**–**(e)**.

a. Find the sample mean number of small firms per metropolitan area.

b. Generate a histogram of the number of small firms per metropolitan area.

c. Generate a normal probability plot of the number of small firms in each metropolitan area. What is your conclusion regarding the normality of the distribution of the number of firms?

d. Construct and interpret a 95% confidence interval for the population number of small firms per metropolitan area. Assume that the standard deviation is 25,000 firms.

e. On the histogram, indicate the location of the confidence interval.

**46.** Follow steps **(a)–(e)**.

**a.** Compute a new variable, the payroll per firm by dividing the "Payroll 1000's" column by the "Firms" column. Find the sample mean payroll per firm. Note that payroll is expressed in \$1000's.

**b.** Generate a histogram of payroll per firm.

**c.** Generate a normal probability plot of the payroll per firm. What is your conclusion regarding the normality of the distribution?

**d.** Construct and interpret a 95% confidence interval for the population payroll per firm. Assume that the standard deviation is \$177,000.

**e.** On the histogram, indicate the location of the confidence interval.

APPLET Use the *Confidence Interval* applet for Exercises 47 and 48.

**47.** Set the confidence level to 90%. Click "Sample 50" to produce 50 simple random samples (SRSs) and display the resulting 90% confidence intervals for $\mu$.

**a.** What is the percent hit, that is, the proportion of the confidence intervals that actually contain the true value of $\mu$?

**b.** Keep clicking "Sample 50" until 1000 confidence intervals are generated. What is the percent hit?

**c.** It is not likely (though it is possible) that the percent hit in **(b)** exactly equals 90%. Explain why the percent hit is not equal to 90% when we asked for a confidence level of 90%.

**48.** Set the confidence level to 99%. Click "Sample 50" to produce 50 99% confidence intervals for $\mu$.

**a.** Compare the widths of these intervals with the width of the 90% confidence intervals. Which set of intervals is wider?

**b.** Explain why the intervals you identified in **(a)** are wider, using the concept of *margin of error*.

APPLET Use the *Normal Density Curve* applet for Exercises 49 and 50.

**49.** Use the applet to find $Z_{\alpha/2}$ critical values for unusual confidence levels. Select **2-Tail**, and click and drag the flags so that the central area and not the tail area is highlighted. Verify that the $Z_{\alpha/2}$ critical value for 95% confidence is 1.96.

**50.** Use the applet to find $Z_{\alpha/2}$ critical values for the following confidence levels.

**a.** 80%

**b.** 85%

**c.** 98%

## 8.2 *t* Interval for the Mean

*Objectives:* **By the end of this section, I will be able to...**

1 Describe the characteristics of the *t* distribution.

2 Calculate and interpret a *t* interval for the mean, for either of two cases.

3 Compute and interpret the standard error.

### 1 Introducing the *t* Distribution

In Section 8.1 we estimated the population mean assuming that the population standard deviation $\sigma$ was known. This assumption may be valid for certain fields such as quality control. However, in many real-world problems, we do not know the value of $\sigma$. Therefore, we cannot use a $Z$ interval to estimate the mean. In this section, we use the sample standard deviation $s$ to estimate $\sigma$ to construct a confidence interval that is likely to contain the population mean.

Suppose we do not know the value of the population standard deviation $\sigma$. What should a data analyst do? Estimate it! Recall that the sample standard deviation $s$ may be used as an estimate of the population standard deviation $\sigma$. Fact 5 from Chapter 7 showed us that we could standardize $\bar{x}$ to derive the standard normal random variable:

$$Z = \frac{\bar{x} - \mu}{\sigma/\sqrt{n}}$$

Unfortunately, however, replacing the unknown $\sigma$ in this equation with the known $s$, we can no longer obtain the standard normal $Z$ because $s$ is itself a random variable. Instead, $\frac{\bar{x} - \mu}{s/\sqrt{n}}$ follows an entirely new and different distribution, called the ***t* distribution**.

> **t Distribution**
>
> For a normal population, the distribution of
>
> $$t = \frac{\bar{x} - \mu}{s/\sqrt{n}}$$
>
> follows a $t$ distribution, with $n - 1$ degrees of freedom, where $\bar{x}$ is the sample mean, $\mu$ is the unknown population mean, $s$ is the sample standard deviation, and $n$ is the sample size.

### Degrees of Freedom

Notice that the definition of the $t$ distribution includes a new concept called *degrees of freedom*. Degrees of freedom is a measure that determines how the $t$ distribution changes as the sample size changes. The idea of degrees of freedom is that, in a sum of $n$ numbers, you need to know only the first $n - 1$ of these numbers to find the $n$th number because you already know the sum. For example, suppose you know that the sum of $n = 3$ numbers is 10 and are told that the first two numbers are 5 and 1. Then you can deduce that the last number is $10 - (5 + 1) = 4$. The first two numbers have the freedom to take on any values, but the third number must take a particular value. Thus, there are only $n - 1$ independent pieces of information. The concept is similar for the $t$ distribution. Since we use the sample standard deviation $s$ to estimate the unknown $\sigma$ and since $s$ is known, only $n - 1$ independent pieces of information are needed to find the value of $t$. Thus, we say that $t = \frac{\bar{x} - \mu}{s/\sqrt{n}}$ follows a $t$ distribution with $n - 1$ degrees of freedom. ■

Figure 8.13 displays a comparison of some $t$ curves with the $Z$ curve. Note first of all that, although there is one and only one $Z$ distribution (or curve), there are many different $t$ curves. In fact, there is a different $t$ curve for every different degrees of freedom (df), that is, for every different sample size. The degrees of freedom, $\text{df} = n - 1$, determines the shape of the $t$ distribution, just as the mean and variance uniquely determine the shape of the normal distribution. All $t$ curves have several characteristics in common.

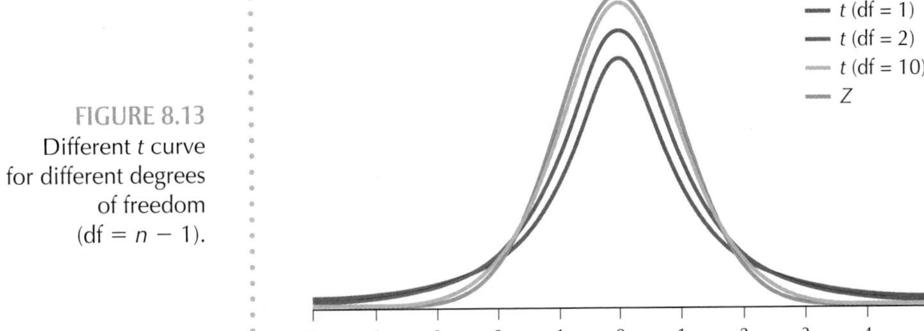


FIGURE 8.13 Different $t$ curve for different degrees of freedom ($\text{df} = n - 1$).

**Characteristics of the $t$ Distribution**

- Centered at zero. The mean of $t$ is zero, just as with $Z$.
- Symmetric about its mean zero, just as with $Z$.
- As df decreases, the $t$ curve gets flatter, and the area under the $t$ curve decreases in the center and increases in the tails. That is, the $t$ curve has heavier tails than the $Z$ curve.
- As df increases toward infinity, the $t$ curve approaches the $Z$ curve, and the area under the $t$ curve increases in the center and decreases in the tails.

Similar to the definition of $Z_{\alpha/2}$ in Section 8.1, we can define $t_{\alpha/2}$ to be the value of the $t$ distribution with area $\alpha/2$ to the right of it, as seen in Figure 8.14. Table 8.1 in Section 8.1 provides the $Z_{\alpha/2}$ values for certain common confidence levels. Unfortunately, because there is a different $t$ curve for each sample size, there are many possible $t_{\alpha/2}$ values. You will need to use the $t$ table in the Appendix (page T-11) to find the value of $t_{\alpha/2}$. Here is how to find $t_{\alpha/2}$ in the $t$ table.

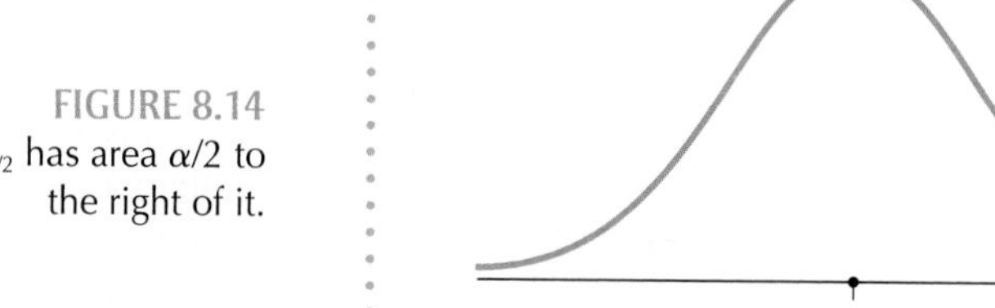


FIGURE 8.14
$t_{\alpha/2}$ has area $\alpha/2$ to the right of it.

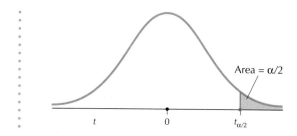

**Procedure for Finding $t_{\alpha/2}$**

***Step 1*** Go across the row marked "Confidence level" in the $t$ table (Table D in the Appendix, page T-11) until you find the column with the desired confidence level at the top. The $t_{\alpha/2}$ value is in this column somewhere.

***Step 2*** Go down the column until you see the correct number of degrees of freedom on the left. The number in that row and column is the desired value of $t_{\alpha/2}$.

## Example 8.11 Finding $t_{\alpha/2}$

Find the value of $t_{\alpha/2}$ that will produce a 95% confidence interval for $\mu$ if the sample size is $n = 20$.

**Solution**

***Step 1*** We go across the row labeled "Confidence level" in the $t$ table (Figure 8.15) until we see the 95% confidence level. Our $t_{\alpha/2}$ is somewhere in this column.

***Step 2*** The degrees of freedom are df $= n - 1 = 20 - 1 = 19$. We go down the column until we see 19 on the left. The number in that row is our $t_{\alpha/2}$, 2.093. ■

**t-Distribution**

| | Confidence level 80% | 90% | 95% | 98% | 99% |
|---|---|---|---|---|---|
| | **Area in one tail** | | | | |
| | 0.10 | 0.05 | 0.025 | 0.01 | 0.005 |
| | **Area in two tails** | | | | |
| | 0.20 | 0.10 | 0.05 | 0.02 | 0.01 |
| **df 1** | 3.078 | 6.314 | 12.706 | 31.821 | 63.657 |
| **2** | 1.886 | 2.920 | 4.303 | 6.965 | 9.925 |
| **3** | 1.638 | 2.353 | 3.182 | 4.541 | 5.841 |
| **14** | 1.345 | 1.761 | 2.145 | 2.624 | 2.977 |
| **15** | 1.341 | 1.753 | 2.131 | 2.602 | 2.947 |
| **16** | 1.337 | 1.746 | 2.120 | 2.583 | 2.921 |
| **17** | 1.333 | 1.740 | 2.110 | 2.567 | 2.898 |
| **18** | 1.330 | 1.734 | 2.101 | 2.552 | 2.878 |
| **19** | 1.328 | 1.729 | 2.093 | 2.539 | 2.861 |
| **20** | 1.325 | 1.725 | 2.086 | 2.528 | 2.845 |
| **21** | 1.323 | 1.721 | 2.080 | 2.518 | 2.831 |

FIGURE 8.15 *t* table (excerpt).

## 2 *t* Interval for the Mean: Two Cases

The *t* distribution provides the following confidence interval for the unknown population mean $\mu$, called the ***t* interval**.

**t Interval for $\mu$**

A random sample of size $n$ is taken from a population with unknown mean $\mu$. A $100(1 - \alpha)\%$ confidence interval for $\mu$ is given by the interval

$$\text{lower bound} = \bar{x} - t_{\alpha/2}(s/\sqrt{n}),\ \text{upper bound} = \bar{x} + t_{\alpha/2}(s/\sqrt{n})$$

where $\bar{x}$ is the sample mean, $t_{\alpha/2}$ is associated with the confidence level and $n - 1$ degrees of freedom, and $s$ is the sample standard deviation. The *t* interval may also be written as

$$\bar{x} \pm t_{\alpha/2}(s/\sqrt{n})$$

and is denoted

(lower bound, upper bound)

The *t* interval applies whenever *either* of the following conditions is met:

- Case 1: The population is normal.
- Case 2: The sample size is large ($n \geq 30$).

*Note:* Suppose that $\sigma$ is unknown, *and* the population is either non-normal or of unknown distribution, *and* the sample size is not large. Then we should not use the *t* interval. Rather, we need to turn to nonparametric methods, for example the *sign interval* or the *Wilcoxon interval*.

## Example 8.12 Fourth-grade feet

Suppose a children's shoe manufacturer is interested in estimating the population mean length of fourth graders' feet. A random sample of 20 fourth graders' feet yielded the following foot lengths, in centimeters.[7]

| | | | | | | | | | |
|---|---|---|---|---|---|---|---|---|---|
| 22.4 | 23.4 | 22.5 | 23.2 | 23.1 | 23.7 | 24.1 | 21.0 | 21.6 | 20.9 |
| 25.5 | 22.8 | 24.1 | 25.0 | 24.0 | 21.7 | 22.0 | 22.7 | 24.7 | 23.5 |

Construct a 95% confidence interval for $\mu$, the population mean length of all fourth graders' feet.

### Solution

We do not know the population standard deviation $\sigma$, so we cannot use the $Z$ interval. We can construct a $t$ interval whenever either the population is normal (Case 1) or the sample size is large (Case 2). The sample size here is 20, which is not large ($n \geq 30$), so Case 2 does not apply. The only remaining choice is to check for normality. Figure 8.16 shows the normal probability plot of the foot lengths. The points generally line up along the line, so the assumption of normality is validated for this data set. By Case 1, we can then proceed to construct the $t$ interval for $\mu$.

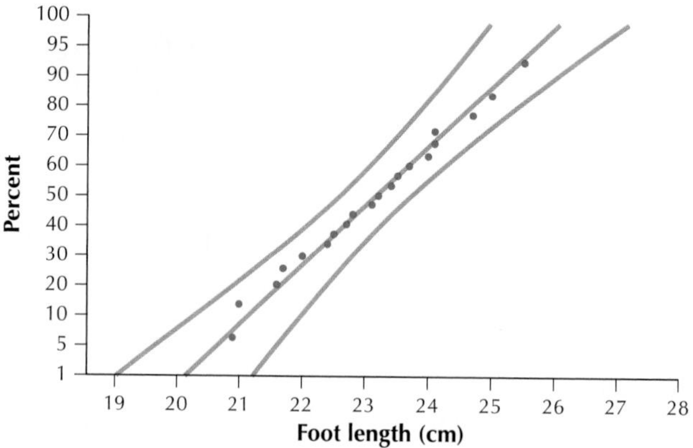


FIGURE 8.16 Fourth-grade foot lengths are normally distributed.

The TI-83/84 provides the summary statistics shown here, giving $n = 20$, $\bar{x} = 23.095$, and $s \approx 1.280$.

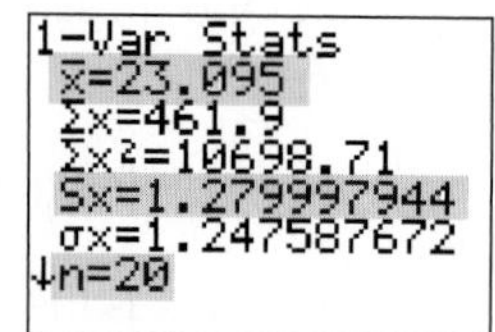


All that is left is to find $t_{\alpha/2}$. In Example 8.11, we found the value of $t_{\alpha/2}$ for confidence level = 95% and $n = 20$ to be $t_{\alpha/2} = 2.093$. The 95% confidence interval then becomes

$$\begin{aligned} \text{lower bound} &= \bar{x} - t_{\alpha/2}(s/\sqrt{n}) \\ &= 23.095 - 2.093(1.280/\sqrt{20}) \approx 23.095 - 0.599 = 22.496 \end{aligned}$$

$$\begin{aligned} \text{upper bound} &= \bar{x} + t_{\alpha/2}(s/\sqrt{n}) \\ &= 23.095 + 2.093(1.280/\sqrt{20}) \approx 23.095 + 0.599 = 23.694 \end{aligned}$$

We are 95% confident that the population mean length of fourth graders' feet lies between 22.496 and 23.694 cm. ■

Your Statistical Sense

### *t* Intervals May Offer More Peace of Mind than *Z* Intervals

In Example 8.12, if we had assumed that the population standard deviation $\sigma$ was known ($\sigma = 1.280$), then the 95% $Z$ interval for the population mean length of fourth-grade feet would have been

$$\text{lower bound} = \bar{x} - Z_{\alpha/2}(\sigma/\sqrt{n}) = 23.095 - 1.96\left(1.280/\sqrt{20}\right) \approx 23.095 - 0.561 = 22.534$$

$$\text{upper bound} = \bar{x} + Z_{\alpha/2}(\sigma/\sqrt{n}) = 23.095 + 1.96\left(1.280/\sqrt{20}\right) \approx 23.095 + 0.561 = 23.656$$

Note that this $Z$ interval (22.534, 23.656) is only slightly more precise than the $t$ interval (22.496, 23.694). However, the $Z$ interval depends on prior knowledge of the value of $\sigma$. If the value of $\sigma$ is inaccurate, then the $Z$ interval will be misleading and overly optimistic. With even moderate sample sizes, reporting the $t$ interval rather than the $Z$ interval may offer peace of mind to the data analyst. ■

If the degrees of freedom needed to find $t_{\alpha/2}$ do not appear in the df column of the $t$ table, a conservative solution is to take the next row with smaller df. Alternatively, we can use interpolation. Both methods are illustrated in Example 8.13.

## Example 8.13 Water use in Israel: degrees of freedom not in the *t* table

© Ed Kashi/CORBIS

Water scarcity is a continuing serious problem for the countries in the Middle East, including Israel. The Jordan River (shown here) feeds the Sea of Galilee, which provides Israel with 40% of its potable water and 80% of the water used for agricultural purposes. The mean daily water consumption in Israel is reported to be 72 cubic meters per person.[8] Suppose a random sample of 49 Israelis showed a sample mean daily water consumption of 72 cubic meters, with a sample standard deviation of 28 cubic meters. Construct a 99% confidence interval for the population mean daily water consumption for all Israelis. Use (a) the conservative method and (b) interpolation.

### Solution

Since the population standard deviation is unknown and the sample size is large, we proceed to construct the $t$ interval for $\mu$. Three of the four numbers we need to complete the $t$ interval are sample size $n = 49$, sample mean $\bar{x} = 72$, and sample standard deviation $s = 28$. Now we must find $t_{\alpha/2}$. The confidence level is 99% and the degrees of freedom are $n - 1 = 49 - 1 = 48$. *Unfortunately, the value of 48 for the df does not appear in the df column.*

**a.** The next row with df smaller than 48 would be df = 40. Thus, the "conservative" $t_{\alpha/2}$ is 2.704. We then proceed to construct the 99% confidence interval:

$$\bar{x} \pm t_{\alpha/2}(s/\sqrt{n}) = 72 \pm 2.704(28/\sqrt{49}) = 72 \pm 10.816 = (61.184, 82.816)$$

**b.** Alternatively, you could interpolate, as follows. Since df = 48 is 8/20 of the distance between 40 and 60, we can estimate $t_{\alpha/2}$ by taking 8/20 of the distance between the associated $t$ values for df = 40 and df = 60 and subtracting that result from the $t$ value for df = 40:

$$\frac{8}{20}\left[\left(t_{\alpha/2} \text{ for df} = 60\right) - \left(t_{\alpha/2} \text{ for df} = 40\right)\right] = \frac{8}{20}(2.704 - 2.660) = 0.0176$$

Thus, $t_{\alpha/2}$ for df = 49 would be $2.704 - 0.0176 = 2.6864$, using interpolation. The 99% confidence interval using interpolation is thus

$$\bar{x} \pm t_{\alpha/2}(s/\sqrt{n}) = 72 \pm 2.6864(28/\sqrt{49}) = 72 \pm 10.7456 = (61.2544, 82.7456)$$

Note that the confidence interval using the conservative method is somewhat wider, reflecting the conservative choice of $t_{\alpha/2}$. ■

## 3 Margin of Error for the *t* Interval

Recall that the margin of error for the $Z$ interval equals $Z_{\alpha/2} \cdot (\sigma/\sqrt{n})$. For the $t$ interval, since $\sigma$ is unknown, the margin of error is given as follows.[9]

**Margin of Error for the *t* Interval**

$$E = t_{\alpha/2} \cdot \left(\frac{s}{\sqrt{n}}\right)$$

The margin of error $E$ for a $(1 - \alpha)100\%$ $t$ interval for $\mu$ can be interpreted as follows: "We can estimate $\mu$ to within $E$ units with $(1 - \alpha)100\%$ confidence."

### Example 8.14 Margin of error for the fourth-grader foot lengths

Use the statistics observed in Example 8.12.

**a.** Find the margin of error for the 95% confidence interval for mean foot lengths.
**b.** Interpret the margin of error.

**Solution**

**a.** From Example 8.12, $n = 20$ and $s = 1.280$. Also, for a confidence level of 95%, $t_{\alpha/2} = 2.093$. Therefore, the margin of error of fourth-grade foot length is

$$E = t_{\alpha/2} \cdot \left(\frac{s}{\sqrt{n}}\right) = (2.093) \cdot \frac{1.280}{\sqrt{20}} \approx 0.599.$$

**b.** We can estimate the population mean of fourth-grade foot lengths to within 0.599 centimeters with 95% confidence. ■

*What Does the Margin of Error Mean?*

The margin of error $E = 0.599$ provides an indication of the accuracy of the confidence interval estimate for confidence level = 95%. That is, if we repeatedly take many samples of size 20 fourth graders, our sample mean $\bar{x}$ will be within $E = 0.599$ centimeters of the unknown population mean $\mu$ in 95% of those samples. ■

### Example 8.15 *t* Intervals for $\mu$ using technology

We continue Example 8.6 from Section 8.1, where we are interested in estimating the population mean number of small businesses for moderately large cities. The U.S. Small Business Administration provides information on the number of small businesses for each metropolitan area in the United States. In Example 8.6, we considered a sample of 30 randomly selected moderately large cities and counted the number of small businesses in each city (see Table 8.4, page 397). We found that the sample mean $\bar{x} = 24{,}017.7$ and the sample standard deviation $s = 4322.473886$. However, this time we are not assuming that we know the value of the population standard deviation, $\sigma$. Use the TI-83/84, Minitab, and the WHFStat Macros for Excel to construct a 95% $t$ confidence interval for the population mean number of small businesses in moderately sized cities nationwide.

**Solution**

We use the instructions provided in the Step-by-Step Technology Guide at the end of this section (pages 415–416). Since the sample size $n = 30$ is large ($\geq 30$), Case 2 applies, and it is not necessary to check for normality (Case 1).

The results for the TI-83/84 in Figure 8.17 display the 95% $t$ confidence interval for the population mean number of small businesses per city to be

$$(\text{lower level} = 22{,}404, \text{upper level} = 25{,}632)$$

It also shows the sample mean $\bar{x} = 24{,}017.7$, the sample standard deviation $s = 4322.473886$, and the sample size $n = 30$.

```
TInterval
 (22404,25632)
 x̄=24017.7
 Sx=4322.473886
 n=30
```

FIGURE 8.17 TI-83/84 results.

The Minitab results are shown in Figure 8.18, providing the sample size $n = 30$, the sample mean $\bar{x} = 24{,}017.7$, the sample standard deviation $s = 4322.5$, the standard error (SE mean) $s_{\bar{x}} = \dfrac{s}{\sqrt{n}} = \dfrac{4322.5}{\sqrt{30}} = 789.2$, and the 95% $t$ confidence interval (22,403.7, 25,631.7).

**One-Sample T: Small Business**

| Variable | N | Mean | StDev | SE Mean | 95% CI |
|---|---|---|---|---|---|
| Small Business | 30 | 24017.7 | 4322.5 | 789.2 | (22403.7, 25631.7) |

FIGURE 8.18 Minitab results.

The results from the WHFStat Macros for Excel are shown in Figure 8.19. Displayed are the sample mean $\bar{x} = 24{,}017.7$, the sample size $n = 30$, the degrees of freedom $\text{df} = n - 1 = 29$, the sample standard deviation $s = 4322.474$, and the standard error $s_{\bar{x}} = \dfrac{s}{\sqrt{n}} = \dfrac{4322.474}{\sqrt{30}} \approx 789.1722$.

| SUMMARY STATISTICS | | | |
|---|---|---|---|
| **Sample Mean** | **Sample Size** | | **Sample Standard Deviation** |
| 24017.7 | 30 | | 4322.474 |
| | **Deg of Freedom** | | **Standard Error (SE)** |
| | 29 | | 789.1722 |
| **ONE SAMPLE T CONFIDENCE INTERVAL** | | | |
| **Confidence Level** | | | **Critical T Value** |
| 95 % | | | 2.045231 |
| **Confidence Interval** | | | |
| 24017.7 | +/- | 1614.039 | |
| 22403.66 | to | 25631.74 | |

FIGURE 8.19 WHFStat macros.

The confidence level 95% is shown, along with the critical $t$ value, $t_{\alpha/2} = 2.045231$. The confidence interval is then shown in the form

$$\text{point estimate} \pm \text{margin of error}$$
$$= 24{,}017.7 \pm 1614.039$$

so the margin of error is

$$E = t_{\alpha/2}\,(s/\sqrt{n}) = 1614.039$$

The confidence interval is also shown as "22,403.66 to 25,631.74." ■

© Scott Wintrow/Getty Images

CASE STUDY

## Analyzing Stock Performance on the New York Stock Exchange

Table 8.5 shows the 49 most active stocks on the New York Stock Exchange (NYSE) on August 15, 2003. We are interested in finding a 95% confidence interval for the population mean change in stock prices. The sample mean is −0.193. For now, we assume that the population standard deviation $\sigma$ is 1.6.

Table 8.5 49 most active stocks on the New York Stock Exchange, August 15, 2003

| Company name | Price change | Company name | Price change | Company name | Price change |
|---|---|---|---|---|---|
| SUN | 0.02 | DELIAS | 0.12 | CRYSTALLEX | 0.59 |
| CISCO | 0.21 | NEWELL | −4.79 | COMCAST | −0.34 |
| INTEL | 0.40 | EMC | 0.59 | TAIWAN | 0.52 |
| MICROSOFT | 0.18 | ADC | 0.03 | SIEBEL | 0.23 |
| SPDR | 0.23 | NORTEL | 0.09 | REDBACK | 0.02 |
| ORACLE | 0.11 | FLEXTRONICS | −0.52 | HEWLETT | 0.15 |
| DYNEGY | −0.45 | SIRIUS | 0.01 | TEXAS | 0.17 |
| STORAGENET | 0.21 | WAVE | 1.41 | ATANDT | 0.03 |
| LUCENT | 0.00 | NEXTEL | 0.11 | CHINA.COM | −0.42 |
| BROADCOM | −0.44 | EXXON | 0.26 | CORNING | 0.14 |
| GENERAL | 0.47 | NETWORK | −1.25 | MOTOROLA | −0.06 |
| JDS | 0.06 | CORVIS | −0.07 | CALPINE | 0.12 |
| DELL | 0.55 | CITIGROUP | −0.43 | AMGEN | −0.21 |
| PFIZER | −0.03 | FORD | 0.41 | EL | −0.02 |
| APPLIED | 0.24 | ANADARKO | 1.10 | TYCO | −0.10 |
| CARDINAL | −9.71 | RF | 0.02 | DIAMONDS | 0.31 |
| AOL | 0.26 | | | | |

*Q.* Are the conditions satisfied for constructing a *Z* interval for the population mean stock price change?

A. Since $n = 49 \geq 30$, then Case 2 applies and we may proceed.

*Q.* Construct and interpret a 95% *Z* confidence interval for the population mean stock price change.

A.
$$\bar{x} \pm Z_{\alpha/2}(\sigma/\sqrt{n}) = -0.193 \pm 1.96(1.6/\sqrt{49})$$
$$= -0.193 \pm 0.448 = (-0.641, 0.255)$$

This confidence interval can be interpreted as meaning that we are 95% confident that the population mean stock price change for all stocks lies between −\$0.641 and \$0.255.

*Q.* Could there be a problem with the interpretation of this particular confidence interval?

A. Yes. The *sample was not randomly selected* from among all NYSE stocks, so it therefore does not necessarily reflect the behavior of the broader market. The sample was taken from the most actively traded stocks, which often exhibit more volatile behavior than usual, since the high volume may be related to some positive or negative industry news. We would therefore have to *interpret the interval as applying only to high-volume stocks.*

*Q.* How comfortable are you with the assumption that $\sigma = 1.6$? How do you think this value was arrived at? How would you go about verifying this assumption?

A. Making unsupported assumptions should always ring alarm bells in the back of your head. Perhaps this value of $\sigma = 1.6$ is well known in the finance literature. On the other hand, perhaps the value is simply a rough guess. We should at least try to verify this assumption by finding the sample standard deviation $s$ of the stock price changes. The sample standard deviation $s = \$1.606$. Because $s = 1.606$ is relatively close to 1.6, it appears that our assumption that $\sigma$ equaled 1.6 is reasonable. However, remember that assumptions should be checked whenever possible.

*Q.* Recall that the sample mean price change was $-0.193$ for the $n = 49$ stocks. Find the value of $t_{\alpha/2}$ for a 99% confidence interval, and construct a 99% confidence interval ($t$ interval) for the population mean stock price change of high-volume stocks.

A. The degrees of freedom $n - 1 = 48$ are not listed in the $t$ table. Therefore, we use the conservative df = 40, providing $t_{\alpha/2} = 2.704$. The 99% $t$ confidence interval is then given by

$$\bar{x} \pm t_{\alpha/2}\,(s/\sqrt{n}) = -0.193 \pm 2.704(1.606/\sqrt{49}) \approx -0.193 \pm 0.620 = (-0.813, 0.427)$$

WHAT IF? *Q. What if* the confidence level was decreased from 99% to some unspecified lower level. Explain whether and how this decrease would affect $s/\sqrt{n}$, the margin of error $E$, and the width of the confidence interval.

A. A lower confidence level will result in a higher value for $t_{\alpha/2}$. This has no effect on $s/\sqrt{n}$, but it increases both the margin of error and the width of the confidence interval. Lower confidence $\Leftrightarrow$ wider interval.

WHAT IF? *Q. What if* the sample size was decreased from 49 to some unspecified number (still greater than 30). Explain whether and how this decrease would affect $s/\sqrt{n}$, the margin of error, and the width of the confidence interval.

A. A smaller $n$ will result in a larger $s/\sqrt{n}$. This results in a larger margin of error and a wider confidence interval. Smaller sample size $\Leftrightarrow$ wider interval.

We continue our exploration of this data set further in the exercises throughout this chapter.

## STEP-BY-STEP TECHNOLOGY GUIDE: *t* Confidence Intervals

We illustrate how to construct the *t* confidence interval for Example 8.15 (page 412).

### TI-83/84

**If you have the data values:**

*1.* Enter the data into list **L1**.
*2.* Press **STAT**, highlight **TESTS**.
*3.* Press **8** (for **TInterval,** see Figure 8.20).
*4.* For input (**Inpt**), highlight **Data** and press **ENTER** (Figure 8.21).

**a.** For **List**, press **2nd** then **L1**.
**b.** For **Freq**, enter **1**.
**c.** For **C-Level** (confidence level), enter the appropriate confidence level (for example, **0.95**), and press **ENTER**.
**d.** Highlight **Calculate** and press **ENTER**. The results are shown in Figure 8.17 in Example 8.15.

**If you have the summary statistics:**

*1.* Press **STAT**, highlight **TESTS**.
*2.* Press 8 (for **TInterval,** see Figure 8.20).
*3.* For input (**Inpt**), highlight **Stats** and press **ENTER** (Figure 8.22).

**a.** For $\bar{x}$, enter the sample mean 24017.7.
**b.** For **Sx**, enter the sample standard deviation 4322.473886.
**c.** For *n*, enter the sample size **30**.
**d.** For **C-Level** (confidence level), enter the appropriate confidence level (for example, **0.95**), and press **ENTER**.
**e.** Highlight **Calculate** and press **ENTER**. The results are shown in Figure 8.17 in Example 8.15.

*(Continued)*

Figure 8.20

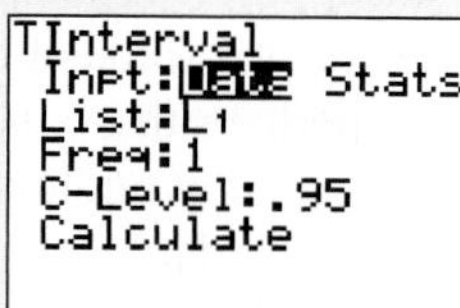


Figure 8.21

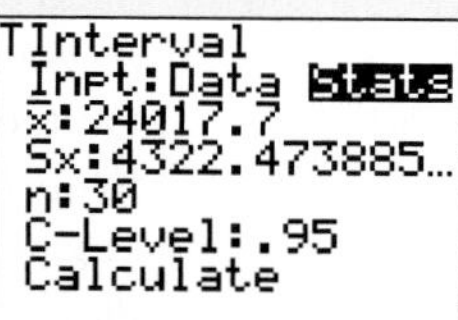


Figure 8.22

## EXCEL

**If you have the data values:**

*1.* Enter the data into column A.
*2.* Load the **WHFStat Macros.**
*3.* Select **Add-Ins > Macros > Estimating a Mean > t Confidence Interval.**
*4.* Click **Select Dataset Range**, highlight A1–A30, and click **OK.**
*5.* Select the **95%** confidence level, and click **OK.**
The results are shown in Figure 8.19 in Example 8.15.

**If you have the summary statistics:**

*1.* Load the **WHFStat Macros.**
*2.* Select **Add-Ins > Macros > Estimating a Mean > Z Confidence Interval.**
*3.* Click **Input Summary Statistics,** enter 24017.7 for the **Sample Mean**, enter **30** for the **Sample Size**, enter **4322.473886** for the **Sample Standard Deviation**, and click **OK.**
*4.* Select the **95%** confidence level and click **OK.**
The results are shown in Figure 8.19 in Example 8.15.

## MINITAB

**If you have the data values:**

*1.* Enter the data into column C1.
*2.* Click **Stat > Basic Statistics > 1-Sample t.**
*3.* Click **Samples in Columns** and select **C1.**
*4.* Click **Options**, enter **95** as the **Confidence Level**, click **OK**, and click **OK** again.
The results are shown in Figure 8.18 in Example 8.15.

**If you have the summary statistics:**

*1.* Click **Stat > Basic Statistics > 1-Sample t.**
*2.* Click **Summarized Data.**
*3.* Enter the **Sample Size 30,** the **Sample Mean 24017.7**, and **4322.473886** for the **Standard Deviation.**
*4.* Click **Options**, enter **95** as the **Confidence Level**, click **OK**, and click **OK** again.

The results are shown in Figure 8.18 in Example 8.15.

## Section 8.2 SUMMARY

1 For a normal population, the distribution of

$$t = \frac{\bar{x} - \mu}{s/\sqrt{n}}$$

follows a $t$ distribution, with $n - 1$ degrees of freedom, where $\bar{x}$ is the sample mean, $\mu$ is the unknown population mean, $s$ is the sample standard deviation, and $n$ is the sample size. The $t$ distribution is symmetric about its mean 0, just like the $Z$ distribution. However, the $t$ distribution is flatter.

2 A $100(1 - \alpha)\%$ confidence interval for $\mu$ is given by the interval

$$\bar{x} \pm t_{\alpha/2}\,(s/\sqrt{n})$$

where $\bar{x}$ is the sample mean, $t_{\alpha/2}$ is associated with the confidence level and $n - 1$ degrees of freedom, $s$ is the sample standard deviation, and $n$ is the sample size. We can construct a $t$ interval whenever *either* of the following conditions is met: Case 1: The population is normal, or Case 2: The sample size is large ($n \geq 30$).

3 The margin of error for the $t$ interval is given by

$$E = t_{\alpha/2} \cdot (s/\sqrt{n})$$

## Section 8.2 EXERCISES

### CLARIFYING THE CONCEPTS

**1.** Why do we need the $t$ interval? Why can't we always use $Z$ intervals?

**2.** Suppose that $\sigma$ is known. Can we still use a $t$ interval?

**3.** As the sample size gets larger and larger, what happens to the $t$ curve?

**4.** What information do we need to be able to find $t_{\alpha/2}$?

## PRACTICING THE TECHNIQUES

5. For the following scenarios, we are taking a random sample from a normal population with $\sigma$ unknown. Find $t_{\alpha/2}$.
   **a.** Confidence level 90%, sample size 10
   **b.** Confidence level 95%, sample size 10
   **c.** Confidence level 99%, sample size 10

6. For the following scenarios we are taking a random sample from a normal population with $\sigma$ unknown. Find $t_{\alpha/2}$.
   **a.** Confidence level 95%, sample size 10
   **b.** Confidence level 95%, sample size 15
   **c.** Confidence level 95%, sample size 20

7. Refer to Exercise 5.
   **a.** Describe what happens to the value of $t_{\alpha/2}$, as the confidence level increases, for a given sample size.
   **b.** Draw a sketch of the $t$ curve for sample size $n = 10$, and explain why the value of $t_{\alpha/2}$ changes as it does.

8. Refer to Exercise 6.
   **a.** Describe what happens to the value of $t_{\alpha/2}$, as the sample size increases, for a given confidence level.
   **b.** Draw a sketch of the $t$ curve for a confidence level of 95%, and explain why the value of $t_{\alpha/2}$ changes as it does.

For each of Exercise 9–12, use two different methods to find $t_{\alpha/2}$, one conservative and one by interpolating.

9. Confidence level 95%, sample size 55
10. Confidence level 99%, sample size 117
11. Confidence level 90%, sample size 46
12. Confidence level 95%, sample size 46

For Exercises 13–18, we are taking a random sample from a normal population with $\sigma$ unknown. Find the measures in **(a)**–**(c)**. Then sketch the confidence interval on a number line.

   **a.** $t_{\alpha/2}$
   **b.** Margin of error $E$
   **c.** Confidence interval for $\mu$ with the indicated confidence level

13. Confidence level 95%, sample size 25, sample mean 10, sample standard deviation 5
14. Confidence level 90%, sample size 9, sample mean 22, sample standard deviation 3
15. Confidence level 95%, $n = 4, \bar{x} = 50, s = 6$
16. Confidence level 99%, $n = 16, \bar{x} = 0, s = 8$
17. Confidence level 90%, $n = 9, \bar{x} = -20, s = 6$
18. Confidence level 95%, $n = 25, \bar{x} = 0, s = 15$

For Exercises 19–24, we are taking a random sample from a population with $\sigma$ unknown. However, do not assume that the population is normally distributed. Find the measures in **(a)**–**(c)**. Then sketch the confidence interval on a number line.

   **a.** $t_{\alpha/2}$
   **b.** Margin of error $E$
   **c.** Confidence interval for $\mu$ with the indicated confidence level

19. Confidence level 95%, sample size 100, sample mean 100, sample standard deviation 10.
20. Confidence level 90%, sample size 64, sample mean 250, sample standard deviation 20.
21. Confidence level 99%, $n = 64, \bar{x} = 35, s = 8$
22. Confidence level 95%, $n = 400, \bar{x} = 42, s = 10$
23. Confidence level 90%, $n = 81, \bar{x} = -20, s = 6$
24. Confidence level 95%, $n = 225, \bar{x} = 0, s = 15$

For each of Exercises 25–32, we are taking a random sample from a population with $\sigma$ unknown. Determine whether Case 1 or Case 2 applies. If appropriate, construct the indicated confidence interval. If it is not appropriate, explain why not.

25. Confidence level 95%, sample size 25, sample mean 100, sample standard deviation 10
26. Confidence level 90%, sample size 16, sample mean 250, sample standard deviation 20
27. Confidence level 95%, sample size 225, sample mean 10, sample standard deviation 5, normal population
28. Confidence level 90%, sample size 81, sample mean 22, sample standard deviation 3
29. Confidence level 99%, $n = 16, \bar{x} = 35, s = 8$
30. Confidence level 95%, $n = 25, \bar{x} = 42, s = 10$, normal population
31. Confidence level 95% $n = 36, \bar{x} = 50, s = 6$
32. Confidence level 99% $n = 64, \bar{x} = 0, s = 8$

## APPLYING THE CONCEPTS

33. **Parking Meters.** A tried-and-true revenue stream for large cities has been the funds collected from parking meters. A random sample of 75 parking meters yielded a mean of \$120 per meter with a standard deviation of \$30.
   **a.** Find $t_{\alpha/2}$ for a confidence interval with 95% confidence.
   **b.** Compute and interpret the margin of error $E$ for a confidence interval with 95% confidence.
   **c.** Construct and interpret a 95% confidence interval for the population mean revenue collected from all parking meters.

34. **Teachers Graded.** A 2007 study reported in *Science* magazine stated that fifth-grade teachers scored a mean of 3.4 (out of 7) points for "providing evaluative feedback to students on their work."[10] Assume that the sample size was 36 and the sample standard deviation was 1.5.
   **a.** Find $t_{\alpha/2}$ for a confidence interval with 90% confidence.
   **b.** Compute and interpret the margin of error $E$ for a confidence interval with 90% confidence.
   **c.** Construct and interpret a 90% confidence interval for the population mean points scored by fifth-grade teachers for providing evaluative feedback.

35. **Hybrid Car Gas Mileage.** The accompanying table shows the city gas mileage for 6 hybrid cars, as reported by the Environmental Protection Agency and **www.hybridcars.com** in 2007.

| Vehicle | Mileage (mpg) |
|---|---|
| Honda Accord | 30 |
| Ford Escape (2 wd) | 36 |
| Toyota Highlander | 33 |
| Saturn VUE Green Line | 27 |
| Lexus RX 400h | 31 |
| Lexus GS 450h | 25 |

**a.** Use technology to construct a normal probability plot of the gas mileages. Confirm that the distribution appears to be normal.
**b.** Find $t_{\alpha/2}$ for a confidence interval with 90% confidence.
**c.** Compute and interpret the margin of error $E$ for a confidence interval with 90% confidence.
**d.** Construct and interpret a 90% confidence interval ($t$ interval) for the population mean mileage.

**36. Hybrid Car Gas Mileage II.** The accompanying table contains the complete listing of 12 hybrid vehicle gas mileages shown on **www.hybridcars.com** in 2007.

| Vehicle | Mileage (mpg) |
|---|---|
| Honda Insight | 61 |
| Toyota Prius | 60 |
| Honda Civic | 50 |
| Toyota Camry | 43 |
| Honda Accord | 30 |
| Ford Escape (2wd) | 36 |
| Ford Escape | 33 |
| Mercury Mariner | 33 |
| Toyota Highlander | 33 |
| Saturn VUE Green Line | 27 |
| Lexus RX 400h | 31 |
| Lexus GS 450h | 25 |

**a.** Use technology to construct a normal probability plot of the gas mileages.
**b.** Is there evidence that the distribution is not normal?
**c.** Can you proceed to construct a $t$ interval? Why or why not?

**37. Calories in Breakfast Cereals.** What is the mean number of calories in a bowl of breakfast cereal? A random sample of 6 well-known breakfast cereals yielded the following calorie data.

| Cereal | Calories |
|---|---|
| Apple Jacks | 110 |
| Cocoa Puffs | 110 |
| Mueslix | 160 |
| Cheerios | 110 |
| Corn Flakes | 100 |
| Shredded Wheat | 80 |

**a.** Use technology to construct a normal probability plot of the number of calories.
**b.** Is there evidence that the distribution is not normal?
**c.** Can we proceed to construct a $t$ interval? Why or why not?

**38. Cigarette Consumption.** Health officials are interested in estimating the population mean number of cigarettes smoked per capita in order to evaluate the efficacy of their antismoking campaign. A random sample of 8 U.S. counties yielded the following numbers of cigarettes smoked per capita: 2206, 2391, 2540, 2116, 2010, 2791, 2392, 2692.

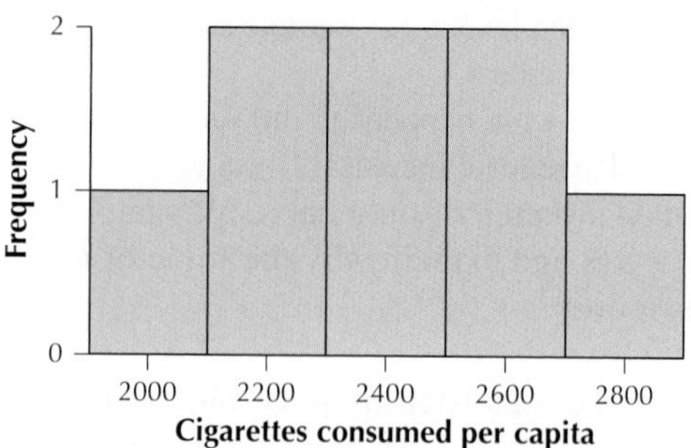


Figure 8.23 Histogram of cigarettes smoked.

**a.** Evaluate the normality assumption using the histogram in Figure 8.23. Is it appropriate to construct a $t$ interval using this data set? Why or why not? What is it about the histogram that tells you one way or the other?
**b.** Find the point estimate of $\mu$, the population mean number of cigarettes smoked per capita.
**c.** Compute the sample standard deviation $s$.
**d.** Find $t_{\alpha/2}$ for a confidence interval with 90% confidence.
**e.** Compute and interpret the margin of error $E$ for a confidence interval with 90% confidence. What is the meaning of this number?
**f.** Construct and interpret a 90% confidence interval for the population mean number of cigarettes smoked per capita.

**39. Commuting Distances.** A university is trying to attract more commuting students from the local community. As part of the research into the modes of transportation students use to commute to the university, a survey was conducted asking how far commuting students commuted from home to school each day. A random sample of 30 students provided the distances (in miles) shown.

| | | | | | | | | | | | | | | |
|---|---|---|---|---|---|---|---|---|---|---|---|---|---|---|
| 14 | 10 | 14 | 12 | 12 | 11 | 5 | 6 | 9 | 14 | 9 | 9 | 4 | 7 | 15 |
| 9 | 7 | 7 | 12 | 10 | 15 | 10 | 6 | 11 | 9 | 11 | 10 | 11 | 7 | 12 |

**a.** Find $t_{\alpha/2}$ for a confidence interval with 95% confidence.
**b.** Compute and interpret the margin of error for a confidence interval with 95% confidence.
**c.** Construct and interpret a 95% $t$ confidence interval for the population mean commuting distance.

**40. Van Gogh's Palette.** A random sample of 30 paintings by Vincent van Gogh yielded a sample mean number of colors in his palette of 15 with a sample standard deviation of 3. Assume that the number of palette colors is normally distributed.

**a.** Calculate and interpret a 90% confidence interval for the mean number of colors in van Gogh's palette for all his paintings.

**b.** Calculate and interpret a 95% confidence interval for the mean number of colors in van Gogh's palette for all his paintings

**c.** Calculate and interpret a 99% confidence interval for the mean number of colors in van Gogh's palette for all his paintings.

**d.** Write a sentence explaining what happens to the width of the confidence interval as the confidence level increases.

**e.** Is it possible to find a 100% confidence interval for the mean number of colors in van Gogh's palette for all his paintings? Why or why not?

**41. HIV-Related Deaths.** A random sample of 10 large American cities found a sample mean number of HIV-related deaths per week of 42 and a sample standard deviation of 12. Assume that the numbers of HIV-related deaths are normally distributed with a population standard deviation of 12.

**a.** Construct and interpret a 99% confidence interval for the mean weekly number of HIV-related deaths in all large American cities.

**b.** Suppose your research supervisor now tells you that the interval you reported in the previous exercise is too wide. What are two possible solutions to the problem? Discuss the benefits and drawbacks of each.

**42. Baby Weights.** A random sample of 20 babies born in Brisbane (Australia) Hospital had the histogram of the babies' weights shown in Figure 8.24. The sample mean is 3227 grams, and the sample standard deviation is 560 grams.

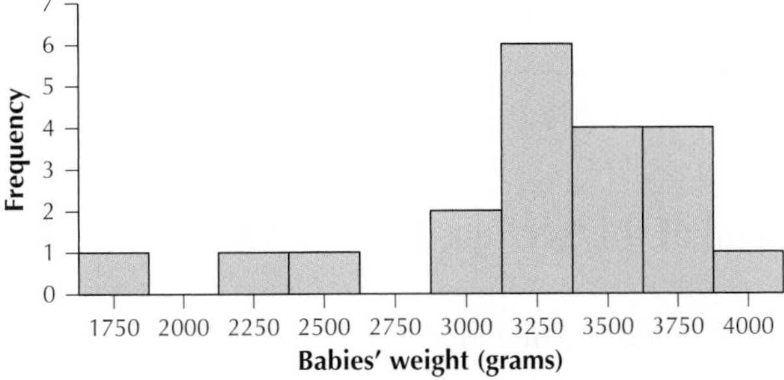


Figure 8.24 Histogram of the baby weights.

**a.** Discuss the normality of the data. Do the data appear acceptably normal?

**b.** Is it appropriate to apply the $t$ interval or not? Explain why or why not.

**c.** Is there something you could do to make the data set more normal?

**d.** How would the changes described in **(c)** affect the meaning and interpretation of the confidence interval?

**e.** Would it be possible to find a valid value for the population standard deviation? Discuss how you might go about finding this value or something close to it.

**43. Most Active Stocks.** The ten most traded stocks on the New York Stock Exchange on October 3, 2007, are shown in Table 8.6, together with their closing prices and net change, in dollars. Use only the net change data for this analysis.

**TABLE 8.6 New York Stock Exchange, October 3, 2007**

| Stock | Closing price | Net change |
|---|---|---|
| Micron Technology | \$10.74 | −1.05 |
| Ford Motor Company | \$ 8.43 | −0.14 |
| Citigroup | \$47.89 | 0.03 |
| Advanced Micro Devices | \$13.23 | 0.03 |
| EMC Corporation | \$21.13 | −0.24 |
| Commerce Bancorp | \$38.84 | −0.63 |
| General Electric | \$41.55 | −0.57 |
| Avaya | \$16.95 | −0.07 |
| Sprint Nextel Corporation | \$18.76 | −0.24 |
| iShares:Taiwan | \$17.18 | −0.18 |

Source: *USA Today.* **http://markets.usatoday.com/custom/usatoday=com/html=mktscreener.asp**

**a.** Check the normality assumption using a normal probability plot or histogram.

**b.** What are the sample size, sample mean, and sample standard deviation?

**c.** What is $t_{\alpha/2}$?

**d.** Estimate the mean net change for the population of high-volume NYSE stocks using a 95% confidence interval.

**e.** What is the margin of error? Explain what this number means.

**44. Assistant Professor Salaries.** As a background to a wage discrimination lawsuit, officials compiled the salaries of twelve randomly selected assistant professors nationwide. Their salaries were as follows, in thousands of dollars:

| | | | | | |
|---|---|---|---|---|---|
| 46.10 | 44.50 | 43.80 | 41.50 | 59.40 | 49.40 |
| 44.86 | 45.50 | 43.80 | 43.90 | 43.00 | 46.70 |

**a.** Check the normality assumption using a normal probability plot or histogram.

**b.** Do you find an outlier? If you do, how should you deal with it?

**c.** Assuming that the distribution of salaries is normal, provide and interpret a 90% confidence interval for the mean salary of all American assistant professors. Omit the outlier from the sample on the assumption that it is a typo.

**45. Minority Loan Refusal Rates.** Banks are sometimes accused of racial discrimination based on differential mortgage lending practices between white and minority applicants. A random sample of ten banks revealed the following minority refusal rates (in percent):

| | | | | |
|---|---|---|---|---|
| 25.9 | 23.2 | 33.2 | 30.4 | 42.7 |
| 39.5 | 38.4 | 30.2 | 35.7 | 34.3 |

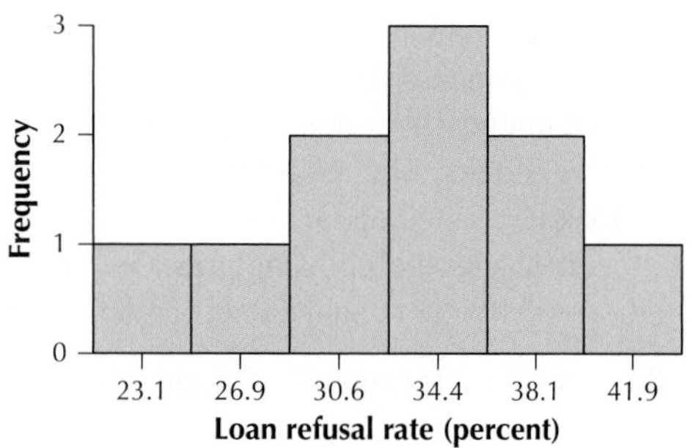


Figure 8.25 Histogram of minority loan refusal rates.

**a.** Evaluate the normality assumption using the histogram in Figure 8.25.

**b.** Calculate and intepret the 95% confidence interval for the population mean rate of minority mortgage lending refusals for all banks.

**Case Study: Analyzing Stock Performance on the New York Stock Exchange**

Suppose we are interested in finding a 99% confidence interval for the population mean number of shares traded for the companies in Table 8.5 (page 414). The sample mean of the 49 stocks in Table 8.5 is 23,188,877 shares, with a sample standard deviation of 16,430,060 shares. Use this information for Exercises 46–48.

**46. a.** Based on the available information, can you proceed with finding an interval for the population mean number of shares traded? Explain why or why not.

**b.** Find the point estimate of the population mean number of shares traded.

**c.** Find $t_{\alpha/2}$ for a confidence interval with 95% confidence. Use the conservative method.

**d.** Compute the margin of error $E$ for a confidence interval with 95% confidence. Explain the meaning of this number.

**e.** Construct and interpret a 95% confidence interval ($t$ interval) for the population mean number of shares traded.

WHAT IF? **47.** *What if* the confidence level is decreased from 95% to some unspecified lower level. Explain whether and how this decrease would affect (a) the margin of error $E$ and (b) the width of the confidence interval.

WHAT IF? **48.** *What if* the sample size is increased from 49 to some unspecified number. Explain whether and how this increase would affect (a) the margin of error $E$ and (b) the width of the confidence interval.

WHAT IF? **49.** Consider the confidence interval we found for the fourth-graders' foot lengths in Example 8.12. *What if* we increased the sample size to some unspecified value but everything else stayed the same. Describe what, if anything, would happen to each of the following measures and why.

**a.** $t_{\alpha/2}$

**b.** Margin of error $E$

**c.** Width of the confidence interval

WHAT IF? **50.** Consider the 95% confidence interval we found in Example 8.12 for the mean length of fourth-graders' feet. Suppose that we increased the confidence level to some unspecified value but everything else stayed the same. Describe what, if anything, would happen to each of the following, and why.

**a.** $t_{\alpha/2}$

**b.** Margin of error $E$

**c.** Width of the confidence interval.

## 8.3 Z Interval for a Population Proportion

*Objectives:* **By the end of this section, I will be able to…**

1. Calculate the point estimate $\hat{p}$ of the population proportion $p$.
2. Construct and interpret a $Z$ interval for the population proportion $p$.
3. Compute and interpret the margin of error for the $Z$ interval for $p$.

### 1 Point Estimate $\hat{p}$ of the Population Proportion $p$

So far we have dealt with interval estimates of the population mean $\mu$ only. However, we may also be interested in an interval estimate for the population proportion of successes, $p$. Recall from Section 7.1 that the sample proportion of successes

$$\hat{p} = \frac{x}{n} = \frac{\text{number of successes}}{\text{sample size}}$$

is a point estimate of the population proportion $p$.

## Example 8.16 Community College Survey of Student Engagement

Collaborative learning in college helps students prepare for life in the business world, where employees are required to work together in teams. The Community College Survey of Student Engagement reports on the proportion of students who have worked with classmates outside class to prepare a group assignment during the current academic year.[11] Suppose that a random sample of 300 students is polled, and 174 students respond that they did indeed work on a group project this year. Calculate the point estimate $\hat{p}$ of the population proportion $p$.

**Solution**

We have $n = 300$ students and $x = 174$. Thus,

$$\hat{p} = \frac{x}{n} = \frac{174}{300} = 0.58$$

The point estimate of the population proportion $p$ of community college students who have worked with classmates outside class to prepare a group assignment during the current academic year is $\hat{p} = 0.58$. ■

Of course, different samples of community college students may turn up different sample proportions $\hat{p}$. These are point estimates, and thus they carry no measure of confidence in their accuracy. The point estimates are probably close to the true values, but it's possible that they are not. They may be far from the true values. Only by using confidence intervals can we make probability statements about the accuracy of the estimates.

## 2 *Z* Interval for the Population Proportion *p*

Recall the Central Limit Theorem for Proportions in Section 7.3.

**Central Limit Theorem for Proportions**
The sampling distribution of the sample proportion $\hat{p}$ follows an approximately normal distribution with mean $\mu_{\hat{p}} = p$ and standard deviation $\sigma_{\hat{p}} = \sqrt{\frac{p \cdot (1-p)}{n}}$ when *both* the following conditions are satisfied: (1) $np \geq 5$ and (2) $n(1-p) \geq 5$.

We can use the Central Limit Theorem for Proportions to construct confidence intervals for the population proportion $p$. Because the confidence interval for $p$ is based on the standard normal $Z$ distribution, it is called the ***Z* interval for the population proportion *p***. Because $p$ is unknown, the conditions and the formula for $\sigma_{\hat{p}}$ substitute $\hat{p}$ for $p$.

***Z* Interval for *p***
The $Z$ interval for $p$ may be performed only if *both* the following conditions apply: $n\hat{p} \geq 5$ and $n(1-\hat{p}) \geq 5$. When a random sample of size $n$ is taken from a binomial population with unknown population proportion $p$, the $100(1-\alpha)\%$ confidence interval for $p$ is given by

$$\text{lower bound} = \hat{p} - Z_{\alpha/2}\sqrt{\frac{\hat{p}(1-\hat{p})}{n}}$$

$$\text{upper bound} = \hat{p} + Z_{\alpha/2}\sqrt{\frac{\hat{p}(1-\hat{p})}{n}}$$

Alternatively,

$$\hat{p} \pm Z_{\alpha/2}\sqrt{\frac{\hat{p}(1-\hat{p})}{n}}$$

where $\hat{p}$ is the sample proportion of successes, $n$ is the sample size, and $Z_{\alpha/2}$ depends on the confidence level.

## Example 8.17 $Z$ interval for the population proportion $p$

Using the survey data from Example 8.16, (a) verify that the conditions for constructing the $Z$ interval for $p$ have been met, and (b) construct a 95% confidence interval for the population proportion of community college students who have worked with classmates outside class to prepare a group assignment during the current academic year.

### Solution

**a.** We have $n = 300$ students and $x = 174$. We check the conditions for the confidence interval:

$$n\hat{p} = (300) \cdot (0.58) = 174 \geq 5 \quad \text{and} \quad n(1-\hat{p}) = (300) \cdot (0.42) = 126 \geq 5.$$

The conditions for constructing the $Z$ interval for $p$ have been met.

**b.** From Table 8.7, the confidence level of 95% gives $Z_{\alpha/2} = 1.96$. Thus, the confidence interval is

$$\text{lower bound} = \hat{p} - Z_{\alpha/2}\sqrt{\frac{\hat{p}(1-\hat{p})}{n}} = 0.58 - 1.96\sqrt{\frac{0.58(0.42)}{300}}$$

$$= 0.58 - 1.96(0.0284956137) \approx 0.58 - 0.05585 = 0.52415$$

$$\text{upper bound} = \hat{p} + Z_{\alpha/2}\sqrt{\frac{\hat{p}(1-\hat{p})}{n}} = 0.58 + 1.96\sqrt{\frac{0.58(0.42)}{300}}$$

$$= 0.58 + 1.96(0.0284956137) \approx 0.58 + 0.05585 = 0.63585$$

We are 95% confident that the population proportion of community college students who have worked with classmates outside class to prepare a group assignment during the current academic year lies between 0.52415 and 0.63585. ■

Table 8.7 $Z_{\alpha/2}$ values for common confidence levels

| Confidence level | $\alpha$ | $\alpha/2$ | $Z_{\alpha/2}$ |
|---|---|---|---|
| 90% | 0.10 | 0.05 | 1.645 |
| 95% | 0.05 | 0.025 | 1.96 |
| 99% | 0.01 | 0.005 | 2.576 |

*What Does This Confidence Interval Mean?*

Suppose we repeatedly take random samples of size 300 from a population of community college students, and observe the sample proportion $\hat{p}$ who have worked with classmates outside class to prepare a group assignment during the academic year. Then, suppose that we use these sample proportions to construct, for each sample, a 95% confidence interval for the population proportion $p$. In the long run, we may expect 95% of these confidence intervals to contain the population proportion $p$. In Figure 8.26, the first interval is our 95% confidence interval for $p$ from Example 8.17, obtained from the sample proportion $\hat{p} = 0.58$: (0.52415, 0.63585). A second sample yielded $\hat{p} = 0.60$, giving us a 95% confidence interval of (0.54456, 0.65544), and a third sample yielded $\hat{p} = 0.56$, giving us a 95% confidence interval of (0.50383, 0.61617). Figure 8.26 shows confidence intervals generated by seven further samples as well. For any particular confidence interval, we do not know whether or not the interval contains the population proportion $p$. The true value of the population proportion $p$ is unknown, but a possible value of $p$ is shown in Figure 8.26 in order to underscore the fact that this value of $p$ is not random and does not change from sample to sample. ■

**Possible value of $p$**
**(true value unknown)**

(------------•------------)
0.52415 $\hat{p} = 0.58$ 0.63585

(------------•------------)
0.54456 $\hat{p} = 0.60$ 0.65544

(------------•------------)
0.50383 $\hat{p} = 0.56$ 0.61617

(------------•------------)
0.51398 $\hat{p} = 0.57$ 0.62602

(------------•------------)
0.51737 $\hat{p} = 0.573$ 0.62930

(------------•------------)
0.55481 $\hat{p} = 0.61$ 0.66519

(------------•------------)
0.54798 $\hat{p} = 0.6033$ 0.65869

(------------•------------)
0.51059 $\hat{p} = 0.5667$ 0.62274

(------------•------------)
0.56507 $\hat{p} = 0.62$ 0.67493

(------------•------------)
0.52076 $\hat{p} = 0.5766$ 0.63258

FIGURE 8.26 The confidence interval random. The population proportion is constant.

## Example 8.18 *Z* intervals for *p* using technology

A 2005 poll by the Center for Social Research at Stony Brook University asked, "Should high school athletes who test positive for steroids or other performance-enhancing drugs be banned from high school athletic teams, or not?" Of the 830 randomly selected respondents, 631 responded, "Yes, they should be banned." Use technology to find a 95% confidence interval for the population proportion of all Americans who think such athletes should be banned.

### Solution

We use the instructions provided in the Step-by-Step Technology Guide at the end of this section (page 425). The results for the TI-83/84 in Figure 8.27 display the 95% confidence interval for the population proportion of Americans who think such athletes should be banned to be

$$(\text{lower bound} = 0.7312, \text{upper bound} = 0.78929)$$

It also shows the sample proportion $\hat{p} = 0.7602409639$ and the sample size $n = 830$.

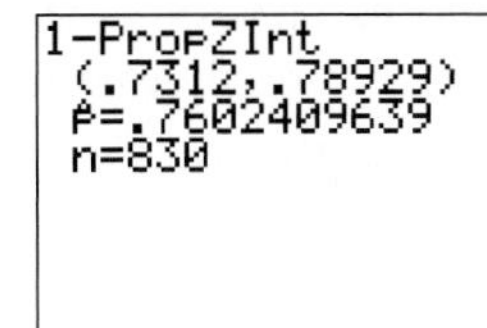


FIGURE 8.27 TI-83/84 results.

The results for Minitab are shown in Figure 8.28. At this point, we consider only the statistics in blue. The remaining material will be explained in Chapter 9. Minitab provides the sample number of successes $X = 631$, the sample size $n = 830$, the sample proportion $\hat{p} = 0.7602409639$ (rounded to 0.760241), and the 95% confidence interval for $p$ (0.731196, 0.789286). ■

**Test and CI for One Proportion**

```
Test of p = 0.5 vs p not = 0.5

Sample     X    N  Sample p          95% CI              Z-Value  P-Value
1        631  830  0.760241  (0.731196, 0.789286)         14.99    0.000
```

FIGURE 8.28 Minitab results.

## 3 Margin of Error for the $Z$ Interval for the Population Proportion $p$

For the $Z$ interval for the population proportion $p$, the margin of error is given as follows.

**Margin of Error for the $Z$ Interval for $p$**

$$E = Z_{\alpha/2} \cdot \sqrt{\frac{\hat{p}(1-\hat{p})}{n}}$$

The margin of error $E$ for a $(1-\alpha)100\%$ $Z$ interval for $p$ can be interpreted as follows:

"We can estimate $p$ to within $E$ with $(1-\alpha)100\%$ confidence."

Note that, just like the confidence interval for $\mu$, the $Z$ interval for $p$ takes the form

$$\begin{aligned}&\text{point estimate} \pm \text{margin of error}\\&= \hat{p} \pm Z_{\alpha/2}\sqrt{\frac{\hat{p}(1-\hat{p})}{n}}\\&= \hat{p} \pm E\end{aligned}$$

### Example 8.19 Polls and the famous "plus or minus 3 percentage points"

There is hardly a day that goes by without some new poll coming out. Especially during election campaigns, polls influence the choice of candidates and the direction of their policies. In October 2004, the Gallup organization polled 1012 American adults, asking them, "Do you think there should or should not be a law that would ban the possession of handguns, except by the police and other authorized persons?" Of the 1012 randomly chosen respondents, 638 said that there should NOT be such a law.

**a.** Check that the conditions for the $Z$ interval for $p$ have been met.

**b.** Find and interpret the margin of error $E$.

**c.** Construct and interpret a 95% confidence interval for the population proportion of all American adults who think there should not be such a law.

**Solution**

The sample size is $n = 1012$. The observed proportion is $\hat{p} = \frac{638}{1012} \approx 0.63$, so $(1-\hat{p}) = 0.37$.

**a.** We next check the conditions for the confidence interval:

$$n\hat{p} = (1012)\cdot(0.63) = 637.56 \geq 5 \quad \text{and} \quad n(1-\hat{p}) = (1012)\cdot(0.37) = 374.44 \geq 5$$

**b.** The confidence level of 95% implies that our $Z_{\alpha/2}$ equals 1.96 (from Table 8.7). Thus, the margin of error equals

$$E = Z_{\alpha/2} \cdot \sqrt{\frac{\hat{p}(1-\hat{p})}{n}} = 1.96 \cdot \sqrt{\frac{0.63(0.37)}{1012}} \approx 0.02975 \approx 0.03$$

**c.** The 95% confidence interval is

$$\begin{aligned}\text{point estimate} &\pm \text{margin of error}\\ &= \hat{p} \pm Z_{\alpha/2}\sqrt{\frac{\hat{p}(1-\hat{p})}{n}}\\ &= \hat{p} \pm E\\ &\approx 0.63 \pm 0.03\\ &= (\text{lower bound} = 0.60,\ \text{upper bound} = 0.66)\end{aligned}$$

Thus, we are 95% confident that the population proportion of all American adults who think that there should not be such a law lies between 60% and 66%. ■

### Famous "Plus or Minus 3 Points"

Note that this confidence interval was obtained by adding and subtracting 3% from the 63% point estimate. That is, the poll has a margin of error of $E = 3$ percentage points = 0.03. This is the famous "plus or minus 3 percentage points" used in many news reports. However, newscasters rarely announce the confidence level of the poll. National pollsters almost always use 95% as their confidence level and usually try to select the sample size necessary to create a margin of error of about 3%. We learn how they do this in Section 8.5. ■

## STEP-BY-STEP TECHNOLOGY GUIDE: *Z* Confidence Intervals for *p*

We illustrate how to construct the *Z* confidence interval for *p* from Example 8.18 (page 423).

### TI-83/84

1. Press **STAT** and highlight **TESTS**.
2. Scroll down to **A** (for **1-PropZInt,** see Figure 8.29), and press **ENTER**.
3. For **x**, enter the number of success, **631**.
4. For **n**, enter the sample size **830**.
5. For **C-Level** (confidence level), enter the appropriate confidence level (e.g., **0.95**), and press **ENTER** (Figure 8.30).
6. Highlight **Calculate** and press **ENTER**. The results are shown in Figure 8.27 in Example 8.18.

Figure 8.29

```
1-PropZInt
 x:631
 n:830
 C-Level:.95
 Calculate
```

Figure 8.30

### MINITAB

1. Click **Stat > Basic Statistics > 1-Proportion**.
2. Click **Summarized Data**.
3. Enter the **Number of Trials** (n) **830** and the **Number of Events** (X) **631**.
4. Click on **Options**, enter **95** as the **Confidence Level**, select **Use test and interval based on normal distribution**, and click **OK**. Then click **OK** again.

The results are shown in Figure 8.28 in Example 8.18.

## Section 8.3 SUMMARY

**1** The sample proportion of successes

$$\hat{p} = \frac{x}{n} = \frac{\text{number of successes}}{\text{sample size}}$$

is a point estimate of the population proportion $p$.

**2** The $100(1-\alpha)\%$ confidence interval for the population proportion $p$, is given by

$$\hat{p} \pm Z_{\alpha/2}\sqrt{\frac{\hat{p}(1-\hat{p})}{n}}$$

where $\hat{p}$ is the sample proportion of successes, $n$ is the sample size, and $Z_{\alpha/2}$ depends on the confidence level. The $Z$ interval for $p$ may be constructed only if *both* the following conditions apply: $n\hat{p} \geq 5$ and $n(1-\hat{p}) \geq 5$.

3 Note that the confidence interval for $p$ takes on the form

point estimate $\pm$ margin of error

where $\hat{p}$ is the point estimate of $p$ and $E = Z_{\alpha/2}\sqrt{\hat{p}(1-\hat{p})/n}$ is the margin of error.

## Section 8.3 EXERCISES

### CLARIFYING THE CONCEPTS

**1.** Suppose the population proportion of successes $p$ is known. Does it make sense to construct a confidence interval for $p$?

**2.** A news broadcast mentions that the sample size of a poll is about 1000 and that the margin of error is plus or minus three percentage points. How do we know that the pollsters are using a 95% confidence level?

### PRACTICING THE TECHNIQUES

For Exercises 3–10, do the following.

- **a.** Verify that the conditions are met.
- **b.** Find $Z_{\alpha/2}$
- **c.** Calculate the margin of error $E$.

**3.** Confidence level 95%, sample size 25, sample proportion 0.5

**4.** Confidence level 90%, sample size 81, sample proportion 0.1

**5.** Confidence level 99%, sample size 100, sample proportion 0.95

**6.** Confidence level 99%, sample size 400, sample proportion 0.02

**7.** Confidence level 95% $n = 64, \hat{p} = 0.4$

**8.** Confidence level 99% $n = 144, \hat{p} = 0.6$

**9.** Confidence level 90% $n = 49, \hat{p} = 0.3$

**10.** Confidence level 95% $n = 225, \hat{p} = 0.03$

For Exercises 11–18, use the information from Exercises 3–10 to help you do the following.

- **a.** Construct a confidence interval for $p$ with the indicated confidence level.
- **b.** Sketch the confidence interval on a number line.

**11.** Confidence level 95%, sample size 25, sample proportion 0.5

**12.** Confidence level 90%, sample size 81, sample proportion 0.1

**13.** Confidence level 99%, sample size 100, sample proportion 0.95

**14.** Confidence level 99%, sample size 400, sample proportion 0.02

**15.** Confidence level 95%, $n = 64, \hat{p} = 0.4$

**16.** Confidence level 99%, $n = 144, \hat{p} = 0.6$

**17.** Confidence level 90%, $n = 49, \hat{p} = 0.3$

**18.** Confidence level 95%, $n = 225, \hat{p} = 0.03$

For Exercises 19–26, do the following.

- **a.** Find $Z_{\alpha/2}$
- **b.** Determine whether the conditions are met.
- **c.** If the conditions are met, calculate the margin of error, $E = Z_{\alpha/2} \cdot \sqrt{\hat{p}(1-\hat{p})/n}$.
- **d.** If the conditions are met, construct a confidence interval for $p$ with the indicated confidence level, and sketch the confidence interval on a number line. If the conditions are not met, state why not.

**19.** Confidence level 95%, sample size 121, sample proportion 0.1

**20.** Confidence level 99%, sample size 121, sample proportion 0.04

**21.** Confidence level 90%, sample size 16, sample proportion 0.5

**22.** Confidence level 99%, sample size 16, sample proportion 0.2

**23.** Confidence level 95%, $n = 25, \hat{p} = 0.1$

**24.** Confidence level 99%, $n = 25, \hat{p} = 0.2$

**25.** Confidence level 90%, $n = 36, \hat{p} = 0.15$

**26.** Confidence level 95%, $n = 36, \hat{p} = 0.12$

**27.** For the following samples, find the margin of error $E$ for a 95% confidence interval for $p$.

- **a.** 5 successes in 10 trials
- **b.** 50 successes in 100 trials
- **c.** 500 successes in 1000 trials
- **d.** 5000 successes in 10,000 trials

**28.** For the following samples, find the margin of error $E$ for a 95% confidence interval for $p$.

- **a.** 10 successes in 100 trials
- **b.** 20 successes in 100 trials
- **c.** 30 successes in 100 trials
- **d.** 40 successes in 100 trials
- **e.** 50 successes in 100 trials

**29.** Refer to Exercise 27.

- **a.** Write a sentence describing what happens to the margin of error as the sample size increases while $\hat{p}$ remains constant.
- **b.** What effect will the behavior you observed in (a) have on the width of the confidence interval?

**30.** Refer to Exercise 28.

- **a.** Write a sentence describing what happens to the margin of error as the sample proportion approaches 0.5 while the sample size remains constant.
- **b.** What effect will the behavior you observed in (a) have on the width of the confidence interval?

### APPLYING THE CONCEPTS

For Exercises 31–36, do the following.

**a.** Find $Z_{\alpha/2}$

**b.** Determine whether the conditions are met for constructing a confidence interval for $p$.

**c.** If the conditions are met, calculate the margin of error $E$ and explain what this number means.

**d.** If the conditions are met, construct and interpret a confidence interval for $p$ with the indicated confidence level, and sketch the confidence interval on the number line. If the conditions are not met, state why not.

**31. Global Assignments.** On June 8, 2004, Cendant Mobility (now Cartus) reported the results of a survey it conducted asking randomly selected workers if they would be more likely, less likely, or just as likely to accept a global assignment rather than a domestic assignment in view of recent world events. Of the 548 subjects polled, 367 said that they would be more likely to accept a global assignment. Use a 99% confidence level.

**32. Rather Be Fishing?** A study found that Minnesota, at 38%, leads the nation in the proportion of people who go fishing.[12] Assume that the study sample size was 12 and use a 95% confidence level.

**33. Race-Conscious College Admissions.** A Marist College poll found that 64% of minorities oppose race-conscious college admissions.[13] Assume that the sample size was 225 and that we are interested in a 95% confidence interval.

**34. Spring Break and Drinking.** A study released by the American Medical Association in 2006 found that 83% of college female respondents agreed that heavier drinking occurs on spring break trips than is typically found on campus. Assume that the sample size was 25 and use a 90% confidence level.

**35. NASCAR Fans and Pickup Trucks.** In April 2004, *American Demographics* magazine reported that 40% of a sample of NASCAR racing attendees said they owned a pickup truck. Suppose the sample size was 1000. Construct a 95% confidence interval for the population proportion of NASCAR racing attendees who own a pickup truck.

**36. Information on 9/11.** Within two weeks of the terrorist attacks of September 11, 2001, the Pew Internet and American Life project conducted a survey asking, among other things, which medium the respondent got most of his/her information from regarding these events. Of the 3733 randomly selected respondents, 71 responded "the Internet." We are interested in a 99% confidence interval for the population proportion of all Americans who have gotten most of their information about the September 11 events from the Internet.

**37. Terrorism in the Middle East.** The U.S. State Department, in its 2003 report "Patterns of Global Terrorism" (**www.state.gov**), reported that 20 of the 60 anti-American terrorist incidents that occurred in 2003 took place in the Middle East (Figure 8.31).

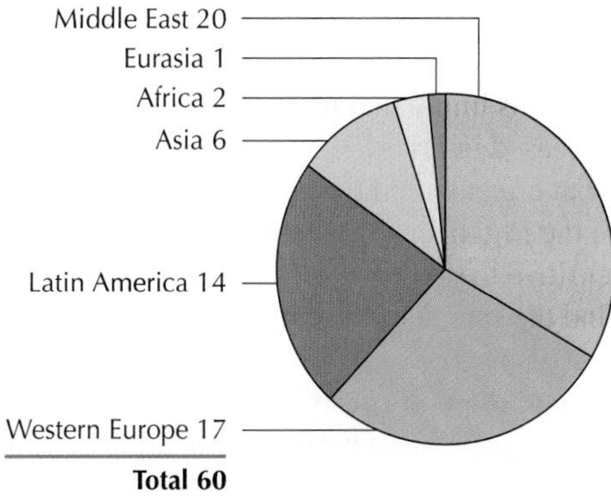


Figure 8.31 Total 2003 anti-United States attacks by region.

**a.** Is it appropriate to use the $Z$ interval to estimate the population proportion of all anti-American terrorist incidents that have taken place in the Middle East since 2003?

**b.** Find the margin of error for confidence level 90%. What does this number mean?

**c.** Construct and interpret a 90% confidence interval for the population proportion of such terrorist incidents.

**38. Politics of Gay Marriage.** The Pew Research Center for the People and the Press (**http://people-press.org**) reported in 2004 that 322 out of 1149 randomly selected people surveyed would not vote for a political candidate who disagreed with their views on gay marriage.

**a.** Is it appropriate to apply the $Z$ interval for the population proportion?

**b.** Find the margin of error. What does this number mean?

**c.** Construct and interpret a 95% confidence interval for the population proportion of all people who would not vote for a political candidate who disagreed with their views on gay marriage.

**39. Ecstasy Use in Twelfth Grade.** According to the National Institute on Drug Abuse (**www.drugabuse.gov**), in 2005 there was a decline in the proportion of twelfth-graders who were using MDMA (Ecstasy).[14] Among whites, 3.9% reported use of the drug. Assume that the sample size is 200.

**a.** Is it appropriate to use the $Z$ interval to estimate the population proportion?

**b.** Find the margin of error. What does this number mean?

**c.** Construct and interpret a 99% confidence interval for the population proportion of all white twelfth-graders who were using Ecstasy.

### Case Study: Analyzing Stock Performance on the New York Stock Exchange

For Exercises 40 and 41, let us continue to examine the Case Study data in Table 8.5 in Section 8.2 (page 414). Now we are interested in the population proportion of stocks that posted price gains.

**40.** We noted in Section 8.2 that the mean change in stock price was negative.

**a.** Does this imply that most of the stocks in the data set decreased in price?

**b.** What effect would the two strongly negative outliers on the negative side have on the mean, which is sensitive to outliers?

**c.** Find the sample proportion of stocks which posted price increases.

**41.** We are interested in constructing a 95% confidence interval for the population proportion of all high-volume stocks to post an increase in price.

**a.** If appropriate, find the margin of error for confidence level 95%. What does this number mean?

**b.** Construct, if appropriate, a 95% confidence interval for the population proportion $p$.

**42. Objective News Source?** A random sample of 1113 American adults found 240 who cited CNN as the media outlet that provides the most objective news.

**a.** If appropriate, find the margin of error for confidence level 99%. What does this number mean?

**b.** If appropriate, construct a 99% confidence interval for the population proportion of all American adults who cite CNN as the media outlet that provides the most objective news.

**43. Most Churchgoers Are Not Drinkers.** According to Gallup.com, 60% of Americans who attend church weekly do not drink alcohol. The poll reported that the sample size was 1006 respondents. Assuming 95% confidence, what is the margin of error of this poll?

**44. Doubling the sample size.** Refer to Exercise 43. *What if* both the sample size and the number of successes were doubled, while nothing else changed. Explain precisely what would happen to the following statistics and why.

**a.** Margin of error

**b.** $Z_{\alpha/2}$

**c.** Width of the confidence interval

**45.** Refer to Exercise 43. Try to think about the relationship between the various statistics rather than calculating the statistics. *What if* the sample size is increased from 1006 and everything else stays the same.

**a.** What will be the effect on the margin of error?

**b.** What will be the effect on the reported sample proportion?

**c.** What will be the effect on the confidence level?

**46. Mozart Effect.** Harvard University's Project Zero (**pzweb.harvard.edu**) found that listening to certain kinds of music, including Mozart, improved spatial-temporal reasoning abilities in children. Suppose that, in a sample of 100 randomly chosen fifth-graders, 65 performed better on a spatial-temporal achievement test after listening to a Mozart sonata.

**a.** If appropriate, find a 95% confidence interval for the population proportion of all fifth-graders who performed better after listening to a Mozart sonata.

**b.** *What if* we increase the confidence level to 99% while changing nothing else. Explain what would happen to the following statistics and why.

**i.** $Z_{\alpha/2}$

**ii.** Margin of error

**iii.** Width of the confidence interval

**47. Taxes Too High?** In a Gallup poll of 1005 randomly selected Americans in April 2006, 482 responded "Yes" to the question "Do you consider the amount of federal income tax you have to pay as too high?"

**a.** If appropriate, find a 90% confidence interval for the population proportion of all Americans who consider that the amount of federal income tax they pay is too high.

**b.** If appropriate, find a 99% confidence interval for the population proportion of all Americans who consider that the amount of federal income tax they pay is too high.

**The Famous ± 3 Percentage Points.** Use the information from Example 8.19 for Exercises 48–50.

**48.** *What if* the sample size is higher than 1012, but otherwise everything else is the same as the example. How would this affect the following?

**a.** Margin of error

**b.** $Z_{\alpha/2}$

**c.** Width of the confidence interval

**49.** *What if* the confidence level is lower than 95%, but otherwise everything else is the same as the example. How would this affect the following?

**a.** Margin of error

**b.** $Z_{\alpha/2}$

**c.** Width of the confidence interval

**50.** *What if* the sample proportion $p$ is higher than 0.63, but otherwise everything else is the same as the example. How would this affect the following?

**a.** Margin of error

**b.** $Z_{\alpha/2}$

**c.** Width of the confidence interval

**Drug Companies and Research Studies.** Use this information for Exercises 51–54. The *Annals of Internal Medicine* reported that 39 of the 40 research studies with acknowledged sponsorship by a drug company had outcomes favoring the drug under investigation.[15]

**51.** If appropriate, construct and interpret a 90% confidence interval for the population proportion of all studies sponsored by drug companies that have outcomes favoring the drug. If not appropriate, clearly state why not.

**52.** Suppose that 34 of the 40 research studies sponsored by a drug company favored the drug.

**a.** If appropriate, find a 90% confidence interval for the population proportion of all studies sponsored by drug companies which have outcomes favoring the drug.

**b.** If appropriate, find a 95% confidence interval for the population proportion of all studies sponsored by drug companies which have outcomes favoring the drug.

**c.** If appropriate, find a 99% confidence interval for the population proportion of all studies sponsored by drug companies which have outcomes favoring the drug.

**d.** Using the confidence intervals from **(a)**–**(c)**, construct a general statement about the change in confidence intervals for the population proportion as the confidence level is increased.

**53.** The article in the *Annals of Internal Medicine* found that 89 of the 112 studies *without* acknowledged drug company support had outcomes favoring the drug.

**a.** If appropriate, construct a 95% confidence interval for the population proportion of all studies without acknowledged drug company support which have outcomes favoring the drug. If not appropriate, clearly state why not.

**b.** What if we decrease the confidence level to 90%, while changing nothing else? Explain precisely what would happen to the following statistics and why.

**i.** $Z_{\alpha/2}$

**ii.** Margin of error

**iii.** Width of the confidence interval

**54.** The article in the *Annals of Internal Medicine* found that 89 of the 112 studies *without* acknowledged drug company support had outcomes favoring the drug of interest. If appropriate, construct and interpret a 99% confidence interval for the population proportion of all studies without acknowledged drug company support which had outcomes favoring the drug of interest.

WHAT IF? **55. State of the Economy.** The Roper Center for Public Opinion Research at the University of Connecticut reported the results of a 2003 poll that asked the question "Would you describe the state of the nation's economy these days as excellent, good, not so good, or poor?" The percentage of randomly selected respondents who replied either "not so good" or "poor" was 50%. The sample was of size 1004. To answer the following questions, try to think about the relationship between the various statistics rather than calculating the statistics. *What if* the sample size is decreased from 1004. What will be the effect on the following?

**a.** Margin of error

**b.** Reported sample proportion

**c.** Confidence level

**56. Construct Casinos in Rhode Island?** The proportion of all citizens in Rhode Island who favor amending the state constitution to allow the construction of casinos is unknown. However, a poll of $n = 578$ randomly selected people by Brown University in September 2006 found 208 respondents in favor of such a change.

**a.** If appropriate, find the margin of error for confidence level 95%. What does this number mean?

**b.** Find a 95% confidence interval for the population proportion of Rhode Island citizens who favor amending the state constitution to allow the construction of casinos.

## 8.4 Confidence Intervals for the Population Variance and Standard Deviation

**Objectives:** By the end of this section, I will be able to…

1. Describe the properties of the $\chi^2$ (chi-square) distribution.
2. Find critical values for the $\chi^2$ (chi-square) distribution.
3. Construct confidence intervals for the population variance and standard deviation.

We have seen how confidence intervals can be used to estimate the unknown value of a population mean or a population proportion. However, the variability of a population is also important. As we have learned, less variability is almost always better. For example, a tool manufacturer relies on a quality control technician (who has a strong background in statistics) to make sure that the tools the company is making do not vary appreciably from the required specifications. Otherwise, the tools may be too large or too small. Data analysts therefore construct confidence intervals to estimate the unknown value of the population parameters that measure variability: the population variance $\sigma^2$ and the population standard deviation $\sigma$.

The $\chi^2$ **(chi-square) distribution** is used to construct these confidence intervals. So we first need to become acquainted with the $\chi^2$ distribution.

## 1 Properties of the $\chi^2$ (Chi-Square) Distribution

The $\chi^2$ (pronounced *ky-square*, to rhyme with "my square") distribution was discovered in 1875 by the German physicist Friedrich Helmert and further developed in 1900 by the English statistician Karl Pearson.

The $\chi^2$ random variable is continuous. Just as we did with the normal and *t* distributions, we can find probabilities associated with values of $\chi^2$, and vice versa. Like any continuous distribution, probability is represented by area below the curve above an interval. We examine the properties of the $\chi^2$ distribution and then learn how to use the $\chi^2$ table to find the critical values of the $\chi^2$ distribution.

**Properties of the $\chi^2$ Distribution**

- Just as for any continuous random variable, the total area under the $\chi^2$ curve equals 1.
- The value of the $\chi^2$ random variable is never negative, so the $\chi^2$ curve starts at 0. However, it extends indefinitely to the right, with no upper bound.
- Because of the characteristics just described, the $\chi^2$ curve is right-skewed.
- There is a different curve for every different degrees of freedom, $n - 1$. As the number of degrees of freedom increases, the $\chi^2$ curve begins to look more symmetric (Figure 8.32).

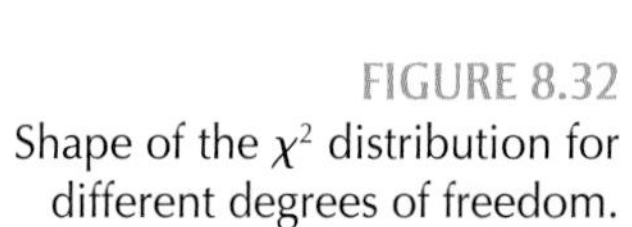

FIGURE 8.32
Shape of the $\chi^2$ distribution for different degrees of freedom.

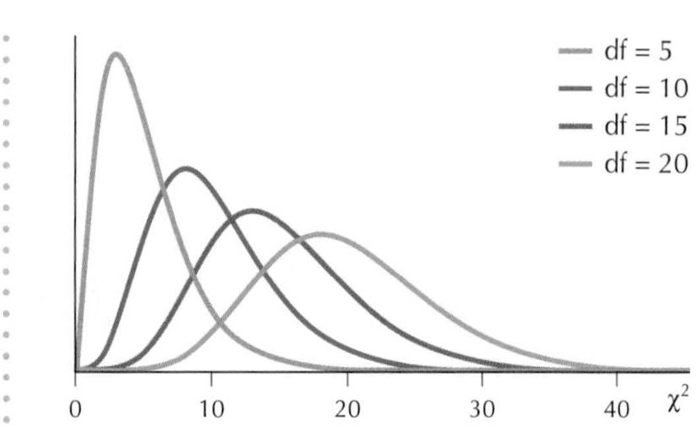


## 2 Finding Critical Values for the $\chi^2$ (Chi-Square) Distribution

To construct the confidence intervals in this section, we shall need to find the critical values of a $\chi^2$ distribution for the given confidence level $100(1 - \alpha)\%$. To find these $\chi^2$ critical values, we can use either the $\chi^2$ table (Table E in the Appendix on page T-12) or technology. The $\chi^2$ table is somewhat similar to the *t* table (Table D in the Appendix); both tables show the degrees of freedom in the left column. The area to the right of the $\chi^2$ critical value is given across the top of the table.

Since the $\chi^2$ distribution is not symmetric, we cannot construct the confidence interval for $\sigma^2$ using the "Point estimate $\pm$ Margin of error" method. Rather, the lower bound and upper bound for the confidence interval are determined using different critical values. Thus we must find two different $\chi^2$ critical values for each confidence interval. That is, for a confidence interval for $\sigma^2$ with confidence level $100(1 - \alpha)\%$, we need to find the following two $\chi^2$ critical values:

- $\chi^2_{1-\alpha/2}$, which represents the value of the $\chi^2$ distribution with area $1 - \alpha/2$ to the right of it.
- $\chi^2_{\alpha/2}$, which represents the value of the $\chi^2$ distribution with area $\alpha/2$ to the right of it (Figure 8.33).

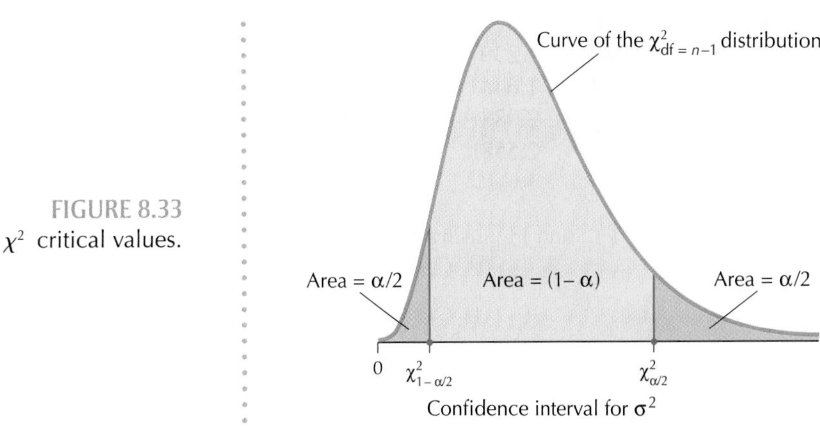


FIGURE 8.33
$\chi^2$ critical values.

For instance, for a 95% confidence interval $(1 - \alpha) = 0.95$, $\alpha/2 = 0.025$ and $1 - \alpha/2 = 0.975$. Thus, $\chi^2_{0.975}$ represents the value of the $\chi^2$ distribution with area $1 - \alpha/2 = 0.975$ to the right of the $\chi^2$ critical value. This first critical value $\chi^2_{0.975}$ is used to determine the upper bound of the confidence interval (see page 433). The second critical value $\chi^2_{0.025}$ represents the value of the $\chi^2$ distribution with area $\alpha/2 = 0.025$ to the right of the $\chi^2$ critical value. This value $\chi^2_{0.025}$ is used to determine the lower bound of the confidence interval.

## Example 8.20 Finding the $\chi^2$ critical values

Find $\chi^2$ critical values for a 90% confidence interval, where we have a sample size of size $n = 10$.

**Solution**

For a 90% confidence interval,

$$(1 - \alpha) = 0.90 \qquad \frac{\alpha}{2} = \frac{0.10}{2} = 0.05 \qquad 1 - \frac{\alpha}{2} = 1 - 0.05 = 0.95$$

*Note:* If the appropriate degrees of freedom are not given in the $\chi^2$ table, the conservative solution is to take the next row with the smaller df.

So we are seeking (1) $\chi^2_{0.95}$, the critical value with area $1 - \alpha/2 = 0.95$ to the right of it, and (2) $\chi^2_{0.05}$, the critical value with area $\alpha/2 = 0.05$ to the right of it.

Since $n = 10$, the degrees of freedom is df $= n - 1 = 10 - 1 = 9$. To find $\chi^2_{0.95}$ for df $= 9$, go across the top of the $\chi^2$ table (Table E in the Appendix) until you see 0.95 (Figure 8.34). $\chi^2_{0.95}$ is somewhere in that column. Now go down that column until you see your number of degrees of freedom df $= 9$. Thus, for df $= 9$, $\chi^2_{0.95} = 3.325$. For a $\chi^2$ distribution with 9 degrees of freedom, there is area $= 0.95$ to the right of 3.325. Similarly, $\chi^2_{0.05}$ is found in the column labeled "0.05" and the row corresponding to df $= 9$. We find that $\chi^2_{0.05} = 16.919$, as shown in Figure 8.35. ■

**Chi-Square ($\chi^2$) Distribution**
**Area to the Right of Critical Value**

| Degrees of Freedom | 0.995 | 0.99 | 0.975 | 0.95 | 0.90 | 0.10 | 0.05 | 0.025 | 0.01 | 0.005 |
|---|---|---|---|---|---|---|---|---|---|---|
| 1 | — | — | 0.001 | 0.004 | 0.016 | 2.706 | 3.841 | 5.024 | 6.635 | 7.879 |
| 2 | 0.010 | 0.020 | 0.051 | 0.103 | 0.211 | 4.605 | 5.991 | 7.378 | 9.210 | 10.597 |
| 3 | 0.072 | 0.115 | 0.216 | 0.352 | 0.584 | 6.251 | 7.815 | 9.348 | 11.345 | 12.838 |
| 4 | 0.207 | 0.297 | 0.484 | 0.711 | 1.064 | 7.779 | 9.488 | 11.143 | 13.277 | 14.860 |
| 5 | 0.412 | 0.554 | 0.831 | 1.145 | 1.610 | 9.236 | 11.071 | 12.833 | 15.086 | 16.750 |
| 6 | 0.676 | 0.872 | 1.237 | 1.635 | 2.204 | 10.645 | 12.592 | 14.449 | 16.812 | 18.548 |
| 7 | 0.989 | 1.239 | 1.690 | 2.167 | 2.833 | 12.017 | 14.067 | 16.013 | 18.475 | 20.278 |
| 8 | 1.344 | 1.646 | 2.180 | 2.733 | 3.490 | 13.362 | 15.507 | 17.535 | 20.090 | 21.955 |
| 9 | 1.735 | 2.088 | 2.700 | 3.325 | 4.168 | 14.684 | 16.919 | 19.023 | 21.666 | 23.589 |
| 10 | 2.156 | 2.558 | 3.247 | 3.940 | 4.865 | 15.987 | 18.307 | 20.483 | 23.209 | 25.188 |

FIGURE 8.34 Finding $\chi^2_{0.95}$ and $\chi^2_{0.05}$ using the $\chi^2$ table.

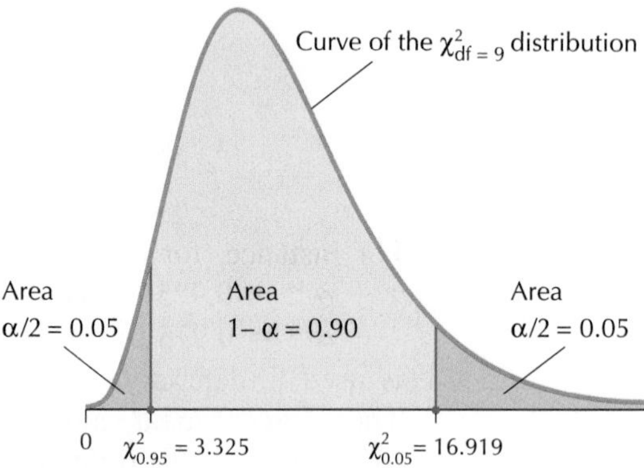


FIGURE 8.35

## 3 Constructing Confidence Intervals for the Population Variance and Standard Deviation

We derive the formula for a $100(1-\alpha)\%$ confidence interval for the population variance $\sigma^2$. Suppose we take a random sample of size $n$ from a normal population with mean $\mu$ and standard deviation $\sigma$. Then the statistic

$$\chi^2 = \frac{(n-1)s^2}{\sigma^2}$$

follows a $\chi^2$ distribution with $n - 1$ degrees of freedom, where $s^2$ represents the sample variance. From Figure 8.33, we see that $100(1-\alpha)\%$ of the values of $\chi^2$ lie between $\chi^2_{1-\alpha/2}$ and $\chi^2_{\alpha/2}$. These values are described as

$$\chi^2_{1-\alpha/2} < \frac{(n-1)s^2}{\sigma^2} < \chi^2_{\alpha/2}$$

Rearranging this inequality so that $\sigma^2$ is in the numerator gives us the formula for the $100(1-\alpha)\%$ confidence interval for $\sigma^2$:

$$\frac{(n-1)s^2}{\chi^2_{\alpha/2}} < \sigma^2 < \frac{(n-1)s^2}{\chi^2_{1-\alpha/2}}$$

Thus the lower bound of the confidence interval for $\sigma^2$ is $\frac{(n-1)s^2}{\chi^2_{\alpha/2}}$, and the upper bound is $\frac{(n-1)s^2}{\chi^2_{1-\alpha/2}}$.

**Confidence Interval for the Population Variance $\sigma^2$**

Suppose we take a sample of size $n$ from a normal population with mean $\mu$ and standard deviation $\sigma$. Then a $100(1-\alpha)\%$ confidence interval for the population variance $\sigma^2$ is given by

$$\text{lower bound} = \frac{(n-1)s^2}{\chi^2_{\alpha/2}}, \quad \text{upper bound} = \frac{(n-1)s^2}{\chi^2_{1-\alpha/2}}$$

where $s^2$ represents the sample variance and $\chi^2_{1-\alpha/2}$ and $\chi^2_{\alpha/2}$ are the critical values for a $\chi^2$ distribution with $n-1$ degrees of freedom.

**Confidence Interval for the Population Standard Deviation $\sigma$**

A $100(1-\alpha)\%$ confidence interval for the population standard deviation $\sigma$ is then given by

$$\text{lower bound} = \sqrt{\frac{(n-1)s^2}{\chi^2_{\alpha/2}}}, \quad \text{upper bound} = \sqrt{\frac{(n-1)s^2}{\chi^2_{1-\alpha/2}}}$$

## Example 8.21 Constructing confidence intervals for the population variance $\sigma^2$ and population standard deviation $\sigma$

© Mario Tama/Getty Images

The accompanying table shows the city gas mileage for 6 hybrid cars, as reported by the Environmental Protection Agency and **www.hybridcars.com** in 2007.

| Vehicle | Mileage (mpg) |
|---|---|
| Honda Accord | 30 |
| Ford Escape (2 wd) | 36 |
| Toyota Highlander | 33 |
| Saturn VUE Green Line | 27 |
| Lexus RX 400h | 31 |
| Lexus GS 450h | 25 |

**a.** Confirm that the distribution of gas mileage is normal, using a normal probability plot.

**b.** Find the critical values $\chi^2_{1-\alpha/2}$ and $\chi^2_{\alpha/2}$ for a confidence interval with a 95% confidence level.

**c.** Construct and interpret a 95% confidence interval for the population variance of hybrid gas mileage.

**d.** Construct and interpret a 95% confidence interval for the population standard deviation of hybrid gas mileage.

### Solution

**a.** The normal probability plot in Figure 8.36 indicates that the data are normally distributed.

**b.** There are $n = 6$ hybrid cars in our sample, so the degrees of freedom equal $n - 1 = 5$. For a 95% confidence interval,

$$(1-\alpha) = 0.95 \qquad \alpha/2 = 0.025 \qquad 1-\alpha/2 = 0.975$$

From the $\chi^2$ table (Table E in the Appendix), therefore,

$$\chi^2_{1-\alpha/2} = \chi^2_{0.975} = 0.831 \qquad \chi^2_{\alpha/2} = \chi^2_{0.025} = 12.833$$

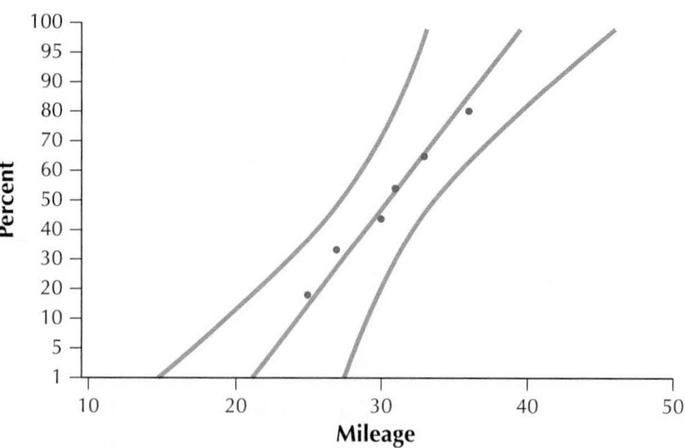


FIGURE 8.36 Normal probability plot of mileage.

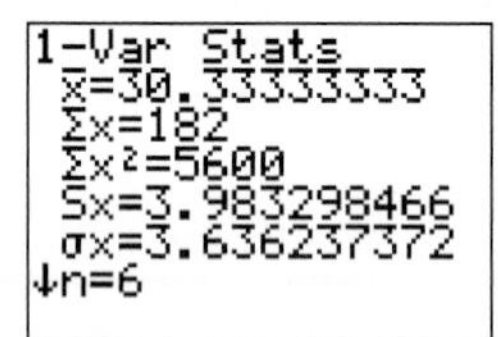


FIGURE 8.37 TI-83/84 results.

**c.** Figure 8.37 shows the descriptive statistics for the hybrid car gas mileages, as obtained by the TI-83/84. The sample standard deviation is $s = 3.983298466$. Thus, our 95% confidence interval for $\sigma^2$ is given by

$$\text{lower bound} = \frac{(n-1)s^2}{\chi^2_{\alpha/2}} = \frac{(5)3.983298466^2}{12.833} \approx 6.181978754 \approx 6.18$$

$$\text{upper bound} = \frac{(n-1)s^2}{\chi^2_{1-\alpha/2}} = \frac{(5)3.983298466^2}{0.831} \approx 95.46730848 \approx 95.47$$

We are 95% confident that the population variance $\sigma^2$ lies between 6.18 and 95.47 miles per gallon squared, that is, $(\text{mpg})^2$. (Recall that the variance is measured in *units* squared.) Since it is unclear what miles per gallon squared means, we prefer to construct a confidence interval for the population standard deviation $\sigma$.

**d.** Using the results from **(c)**.

$$\text{lower bound} = \sqrt{\frac{(n-1)s^2}{\chi^2_{\alpha/2}}} = \sqrt{6.181978754} \approx 2.486358533 \approx 2.49$$

$$\text{upper bound} = \sqrt{\frac{(n-1)s^2}{\chi^2_{1-\alpha/2}}} = \sqrt{95.46730848} \approx 9.770737356 \approx 9.77$$

We are 95% confident that the population standard deviation $\sigma$ lies between 2.49 and 9.77 miles per gallon. ■

## Example 8.22 Using technology to find $\chi^2_{1-\alpha/2}$, $\chi^2_{\alpha/2}$, and the confidence interval for $\sigma$

**a.** Using the data from Example 8.21, find the critical values $\chi^2_{1-\alpha/2}$ and $\chi^2_{\alpha/2}$ using Excel and Minitab.

**b.** Find a 95% confidence interval for $\sigma$ using Minitab.

### Solution

We use the instructions provided in the Step-by-Step Technology Guide at the end of this section.

**a.** Figure 8.38a shows the Excel results for $\chi^2_{1-\alpha/2} = \chi^2_{0.975} = 0.831212$, and Figure 8.38b shows the Excel results for $\chi^2_{\alpha/2} = \chi^2_{0.025} = 12.8325$. Figure 8.39a shows the Minitab results for $\chi^2_{1-\alpha/2} = \chi^2_{0.975} = 0.831212$, and Figure 8.39b shows the Minitab results for $\chi^2_{\alpha/2} = \chi^2_{0.025} = 12.8325$.

| | A | B | C | D | E |
|---|---|---|---|---|---|
| 1 | 0.831212 | | | | |
| 2 | 12.8325 | | | | |

A1: =CHIINV(0.975,5)

(a)

| | A | B | C | D | E |
|---|---|---|---|---|---|
| 1 | 0.831212 | | | | |
| 2 | 12.8325 | | | | |

A2: =CHIINV(0.025, 5)

(b)

FIGURE 8.38 Excel results.

```
Inverse Cumulative Distribution Function
Chi-Square with 5 DF

P( X <= x )         x
      0.025  0.831212
```

(a)

```
Inverse Cumulative Distribution Function
Chi-Square with 5 DF

P( X <= x )         x
      0.975   12.8325
```

(b)

FIGURE 8.39 Minitab results.

**b.** Figure 8.40 shows an excerpt from a Minitab printout showing the 95% confidence interval for $\sigma$ and the 95% confidence interval for $\sigma^2$. ■

```
                              CI for         CI for
Variable   Method            StDev          Variance
Mileage    Standard    (2.49, 9.77)    (6.2, 95.4)
```

FIGURE 8.40 Minitab results showing the confidence intervals (excerpt).

## STEP-BY-STEP TECHNOLOGY GUIDE: $\chi^2$ Distribution

### EXCEL

#### Finding the Critical Values $\chi^2_{1-\alpha/2}$ and $\chi^2_{\alpha/2}$

*1.* Select cell **A1**. Click the **Insert Function** icon $f_x$.
*2.* For **Search for a Function**, type **chiinv**, click **GO**, then click **OK**.
*3.* To find $\chi^2_{1-\alpha/2}$: For **Probability**, enter $1 - \alpha/2$ (such as **0.975** for a 95% confidence interval), and for **Deg_freedom** enter the degrees of freedom. Excel displays the value of $\chi^2_{1-\alpha/2}$ in the cell.
*4.* To find $\chi^2_{\alpha/2}$: Repeat Steps 1 – 2. For **Probability**, enter $\alpha/2$ (such as **0.025** for a 95% confidence interval), and for **Deg_freedom** enter the degrees of freedom. Excel displays the value of $\chi^2_{\alpha/2}$ in the cell.

### MINITAB

#### Finding the Critical Values $\chi^2_{1-\alpha/2}$ and $\chi^2_{\alpha/2}$

*1.* Click **Calc > Probability Distributions > Chi-Square**.
*2.* Select **Inverse cumulative probability**, and enter the **Degrees of freedom**.
*3.* To find $\chi^2_{1-\alpha/2}$: For **Input constant**, enter the area to the *left* of the desired critical value. For $\chi^2_{1-\alpha/2}$, this will be $\alpha/2$ (such as **0.025**). Click **OK**.
*4.* To find $\chi^2_{\alpha/2}$: Repeat Steps 1 – 2. For **Input constant**, enter the area to the *left* of the desired critical value. For $\chi^2_{\alpha/2}$, this will be $1 - \alpha/2$ (such as **0.975**). Click **OK**.
*5.* Minitab displays the values of $\chi^2_{1-\alpha/2}$ and $\chi^2_{\alpha/2}$ in the session window.

#### Finding a 100(1 − $\alpha$)% Confidence Interval for $\sigma$

*1.* Enter the data into column **C1**.
*2.* Select **Stat > Basic Statistics > 1 Variance ...**
*3.* For **Samples in columns**, select **C1**.
*4.* Click **Options**, choose the confidence level, and click **OK**. The confidence interval for $\sigma$ is reported in the output, as shown in Figure 8.40.

## Section 8.4 SUMMARY

*1* The $\chi^2$ continuous random variable takes values that are never negative, so the $\chi^2$ distribution curve starts at 0 and extends indefinitely to the right. Thus, the $\chi^2$ curve is right-skewed and not symmetric. There is a different curve for every different degrees of freedom, $n - 1$.

*2* To find $\chi^2$ critical values, we can use either the $\chi^2$ table or technology.

3 If the population is normally distributed, we use the $\chi^2$ distribution to construct a $100(1 - \alpha)\%$ confidence interval for the population variance $\sigma^2$, which is given by

$$\text{lower bound} = \frac{(n-1)s^2}{\chi^2_{\alpha/2}}, \quad \text{upper bound} = \frac{(n-1)s^2}{\chi^2_{1-\alpha/2}}$$

where $s^2$ represents the sample variance and $\chi^2_{1-\alpha/2}$ and $\chi^2_{\alpha/2}$ are the critical values for a $\chi^2$ distribution with $n - 1$ degrees of freedom. A $100(1 - \alpha)\%$ confidence interval for the population standard deviation $\sigma$ is then given by

$$\text{lower bound} = \sqrt{\frac{(n-1)s^2}{\chi^2_{\alpha/2}}}$$

$$\text{upper bound} = \sqrt{\frac{(n-1)s^2}{\chi^2_{1-\alpha/2}}}$$

## Section 8.4 EXERCISES

### CLARIFYING THE CONCEPTS

1. To construct a confidence interval for $\sigma^2$ or $\sigma$, what must be true about the population?
2. Explain the difference between $\sigma^2$ and $s^2$.
3. Explain why we need to find two different critical values to construct the confidence intervals in this section. Why can't we just use the "point estimate $\pm$ margin of error" method we used earlier in this chapter?

Determine whether each proposition in Exercises 4–7 is true or false. If it is false, restate the proposition correctly.

4. The $\chi^2$ curve is symmetric.
5. The value of the $\chi^2$ random variable is never negative.
6. The $\chi^2$ curve is right-skewed.
7. The total area under the $\chi^2$ curve equals 1.
8. Provide an example from the real world where it would be important to estimate the variability of a data set.

### PRACTICING THE TECHNIQUES

For Exercises 9–14, find the critical values $\chi^2_{1-\alpha/2}$ and $\chi^2_{\alpha/2}$ for the given confidence level and sample size.

9. Confidence level 90%, $n = 25$
10. Confidence level 95%, $n = 25$
11. Confidence level 99%, $n = 25$
12. Confidence level 95%, $n = 10$
13. Confidence level 95%, $n = 15$
14. Confidence level 95%, $n = 20$
15. Consider the critical values you calculated in Exercises 9–11. Describe what happens to the critical values for a given sample size as the confidence level increases.
16. Consider the critical values you calculated in Exercises 12–14. Describe what happens to the critical values for a given confidence level as the sample size increases.

In Exercises 17–22, a random sample is drawn from a normal population. The sample of size $n = 25$ has a sample variance of $s^2 = 10$. Construct the specified confidence interval.

17. 90% confidence interval for the population variance $\sigma^2$
18. 95% confidence interval for the population variance $\sigma^2$
19. 99% confidence interval for the population variance $\sigma^2$
20. 90% confidence interval for the population standard deviation $\sigma$
21. 95% confidence interval for the population standard deviation $\sigma$
22. 99% confidence interval for the population standard deviation $\sigma$
23. Consider the confidence intervals you constructed in Exercises 17–19. Describe what happens to the lower bound and upper bound of a confidence interval for $\sigma^2$ as the confidence level increases but the sample size stays the same.
24. Consider the confidence intervals you constructed in Exercises 20–22. Describe what happens to the lower bound and upper bound of a confidence interval for $\sigma$ as the confidence level increases but the sample size stays the same.

In Exercises 25–30, a random sample is drawn from a normal population. The sample variance is $s^2 = 10$. Construct the specified confidence interval.

25. 95% confidence interval for the population variance $\sigma^2$ for a sample of size $n = 10$
26. 95% confidence interval for the population variance $\sigma^2$ for a sample of size $n = 15$
27. 95% confidence interval for the population variance $\sigma^2$ for a sample of size $n = 20$
28. 95% confidence interval for the population standard deviation $\sigma$ for a sample of size $n = 10$
29. 95% confidence interval for the population standard deviation $\sigma$ for a sample of size $n = 15$
30. 95% confidence interval for the population standard deviation $\sigma$ for a sample of size $n = 20$
31. Consider the confidence intervals you constructed in Exercises 25–27. Describe what happens to the lower bound and upper bound of a confidence interval for $\sigma^2$ as the sample size increases but the confidence level stays the same.
32. Consider the confidence intervals you constructed in Exercises 28–30. Describe what happens to the lower bound and upper bound of a confidence interval for $\sigma$ as the sample size increases but the confidence level stays the same.

### APPLYING THE CONCEPTS

33. **Prisoner Deaths.** The table contains the numbers of prisoners who died in state custody in 2005 for a random sample of 5 states. The normal probability plot in Figure 8.41 indicates acceptable normality.

| State | Prisoner deaths |
|---|---|
| New York | 171 |
| Pennsylvania | 149 |
| Michigan | 140 |
| Ohio | 121 |
| Georgia | 122 |

Source: U.S. Bureau of Justice Statistics.

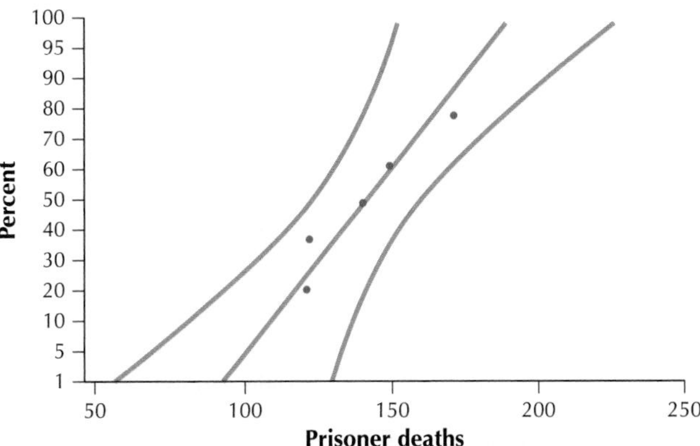


Figure 8.41 Normal probability plot.

**a.** Find the critical values $\chi^2_{1-\alpha/2}$ and $\chi^2_{\alpha/2}$ for a 95% confidence interval for $\sigma^2$.

**b.** Construct and interpret a 95% confidence interval for the population variance $\sigma^2$ of the number of prisoners who died in state custody.

**c.** Construct and interpret a 95% confidence interval for the population standard deviation $\sigma$ of the number of prisoners who died in state custody.

**34. Most Active Stocks.** Table 8.6 shows the ten most traded stocks on the New York Stock Exchange on October 3, 2007, together with their closing prices and net change in price, in dollars. Use only the net change data for this analysis. Assume that the net change data are normally distributed.

**TABLE 8.6 New York Stock Exchange, October 3, 2007**

| Stock | Closing price | Net change |
|---|---|---|
| Micron Technology | \$10.74 | −1.05 |
| Ford Motor Company | \$ 8.43 | −0.14 |
| Citigroup | \$47.89 | 0.03 |
| Advanced Micro Devices | \$13.23 | 0.03 |
| EMC Corporation | \$21.13 | −0.24 |
| Commerce Bancorp | \$38.84 | −0.63 |
| General Electric Company | \$41.55 | −0.57 |
| Avaya | \$16.95 | −0.07 |
| Sprint Nextel Corporation | \$18.76 | −0.24 |
| iShares:Taiwan | \$17.18 | −0.18 |

Source: *USA Today*. **http://markets.usatoday.com/custom/usatoday=com/html=mktscreener.asp**

**a.** Find the critical values $\chi^2_{1-\alpha/2}$ and $\chi^2_{\alpha/2}$ for a 95% confidence interval for $\sigma^2$.

**b.** Construct and interpret a 95% confidence interval for the population variance $\sigma^2$ of net price changes

**35. Prisoner Deaths.** Refer to Exercise 33.

**a.** What are the units you used to interpret your confidence interval in **(b)**?

**b.** Do you think that those units would be easily understood by most people?

**c.** What are the units you used to interpret your confidence interval in **(c)**?

**d.** Do you think that these units are more easily understood by most people?

**36. Most Active Stocks.** Refer to Exercise 34.

**a.** What are the units you used to interpret your confidence interval in **(b)**?

**b.** Do you think that those units would be easily understood by most people?

**c.** What would the units be for a confidence interval for the population standard deviation $\sigma$?

**d.** Proceed to construct and interpret a 95% confidence interval for $\sigma$.

**37. Fourth-Grade Feet.** Suppose a children's shoe manufacturer is interested in estimating the variability of fourth graders' feet. A random sample of 20 fourth graders' feet yielded the following foot lengths, in centimeters.[16] The normality of the data was verified in Example 8.12 in Section 8.2.

| | | | | | | | | | |
|---|---|---|---|---|---|---|---|---|---|
| 22.4 | 23.4 | 22.5 | 23.2 | 23.1 | 23.7 | 24.1 | 21.0 | 21.6 | 20.9 |
| 25.5 | 22.8 | 24.1 | 25.0 | 24.0 | 21.7 | 22.0 | 22.7 | 24.7 | 23.5 |

**a.** Find the critical values $\chi^2_{1-\alpha/2}$ and $\chi^2_{\alpha/2}$ for a 95% confidence interval for $\sigma^2$.

**b.** Construct and interpret a 95% confidence interval for $\sigma^2$.

**c.** Construct and interpret a 95% confidence interval for $\sigma$.

**d.** Is the interpretation in **(b)** or **(c)** easier to understand?

**38. Sodium in Breakfast Cereal.** In Table 8.8 is a list of 23 breakfast cereals, along with their sodium content in milligrams per serving. The normality of the data was confirmed in Example 8.5 in Section 8.1.

**a.** Find the critical values $\chi^2_{1-\alpha/2}$ and $\chi^2_{\alpha/2}$ for a 95% confidence interval for $\sigma^2$.

**b.** Construct and interpret a 95% confidence interval for $\sigma^2$.

**c.** Construct and interpret a 95% confidence interval for $\sigma$.

**d.** Is the interpretation in **(b)** or **(c)** easier to understand? Explain.

**TABLE 8.8 Sodium content of 23 breakfast cereals**

| Cereals | Sodium (mg) |
|---|---|
| Apple Jacks | 125 |
| Cap'n Crunch | 220 |
| Cheerios | 290 |
| Cinnamon Toast Crunch | 210 |
| Cocoa Puffs | 180 |
| Corn Flakes | 290 |
| Count Chocula | 180 |
| Cream of Wheat | 80 |
| Froot Loops | 125 |
| Frosted Flakes | 200 |
| Fruity Pebbles | 135 |
| Grape Nuts Flakes | 140 |
| Kix | 260 |
| Life | 150 |
| Lucky Charms | 180 |
| Mueslix Crispy | 150 |
| Raisin Bran | 210 |
| Rice Chex | 240 |
| Rice Krispies | 290 |
| Special K | 230 |
| Total Whole Grain | 200 |
| Trix | 140 |
| Wheaties | 200 |

**39. Biomass Power Plants.** Power plants around the country are retooling in order to consume biomass instead of or in addition to coal. Table 8.9 contains a random sample of 10 such power plants and the amount of biomass they consumed in 2006 in trillions of BTUs (British thermal units). The normal probability plot in Figure 8.42 indicates acceptable normality.

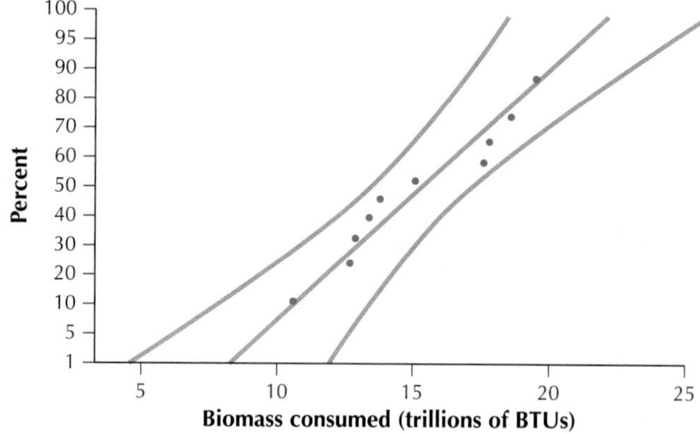


Figure 8.42 Normal probability plot.

a. Find the critical values for a 99% confidence interval for $\sigma$.

b. Construct and interpret a 99% confidence interval for $\sigma$.

**TABLE 8.9 Ten power plants that consume biomass instead of fuel**

| Power plant | Location | Biomass consumed (trillions of BTUs) |
|---|---|---|
| Georgia Pacific Naheola Mill | Choctaw, AL | 13.4 |
| Jefferson Smurfit Fernandina Beach | Nassau, FL | 12.9 |
| International Paper Augusta Mill | Richmond, GA | 17.8 |
| Gaylord Container Bogalusa | Washington, LA | 15.1 |
| Escanaba Paper Company | Delta, MI | 19.5 |
| Weyerhaeuser Plymouth NC | Martin, NC | 18.6 |
| International Paper Georgetown Mill | Georgetown, SC | 13.8 |
| Bowater Newsprint Calhoun Operation | McMinn, TN | 10.6 |
| Covington Facility | Covington, VA | 12.7 |
| Mosinee Paper | Marathon, WI | 17.6 |

Sources: Energy Information Administration, Form EIA-860, "Annual Electric Generator Report" and Form EIA-906, "Power Plant Report."

## 8.5 Sample Size Considerations

*Objectives:* By the end of this section, I will be able to...

1. Calculate the sample size needed to estimate the population mean when $\sigma$ is known, given the confidence level and margin of error.
2. Compute the sample size needed to estimate the population proportion of successes, given the confidence level and margin of error, without prior information about the value of $\hat{p}$.

3 Compute the sample size needed to estimate the population proportion of successes, given the confidence level and margin of error, in cases with prior information about the value of $\hat{p}$.

In Section 8.1 we discussed that, for a fixed sample size, you cannot simultaneously maximize the confidence level and minimize the width of the confidence interval. All other things being equal, a 99% confidence interval is wider than a 95% confidence interval, which is wider than a 90% confidence interval. For a fixed sample size, the higher the confidence level, the wider the confidence interval. To reduce the width of the confidence interval, just increase the sample size. Sometimes larger samples are not possible because the cost is prohibitive. However, when samples are plentiful and cheap, *arbitrarily precise confidence intervals with arbitrarily high confidence are possible simply by taking sufficiently large samples.*

Therefore, the question arises: *How large a sample size do I need* to get a tight confidence interval with high confidence level? We will discuss the following cases:

- Estimating the population mean $\mu$
- Estimating the population proportion with no prior information about $p$
- Estimating the population proportion when we have prior information about $p$

## 1 Sample Size for Estimating the Population Mean

Recall that the general form of the $Z$ interval for the mean is

$$\text{point estimate} \pm \text{margin of error}$$

where the point estimate is the sample mean $\bar{x}$ and the margin of error $Z_{\alpha/2}(\sigma/\sqrt{n})$ measures the precision of the estimate. A smaller margin of error is associated with an estimate with higher precision. A larger margin of error has less precision and results in a wider confidence interval.

### Example 8.23 Estimating the mean salary of business majors

Suppose we would like to estimate to within \$1000 the mean salary $\mu$ of all graduates who were business majors in college. Estimating "to within \$1000" means that our margin of error is $E = \$1000$. Suppose we would like to be 95% confident that the resulting confidence interval contained the population mean salary of business majors. Once we specify the desired margin of error and confidence level, we can then ask the question: How many business majors would we have to sample to estimate the mean salary to within \$1000 with 95% confidence?

**Solution**

Recall that the margin of error for 95% confidence is given by

$$E = 1.96\ (\sigma/\sqrt{n}),$$

where 1.96 is the $Z_{\alpha/2}$ value associated with 95% confidence. Since the desired margin of error is 1000,

$$E = 1000 = 1.96\ (\sigma/\sqrt{n})$$

To find the value of $n$ that is the sample size that will produce an estimate within \$1000 with 95% confidence, we solve for $n$:

$$1000 = 1.96\ (\sigma/\sqrt{n})$$

Multiply both sides by $\sqrt{n}$:

$$1000\sqrt{n} = 1.96\ \sigma$$

Divide both sides by 1000:

$$\sqrt{n} = \left(\frac{1.96\sigma}{1000}\right)$$

Square both sides to get the formula for $n$:

$$n = \left(\frac{1.96\sigma}{1000}\right)^2$$

Thus, if we know the value of $\sigma$, then we can calculate the value of $n$ that will produce an estimate within \$1000 with 95% confidence. ■

We generalize the result from Example 8.23 as follows.

**Sample Size for Estimating the Population Mean**

The sample size for a $Z$ interval that estimates the population mean $\mu$ to within a margin of error $E$ with confidence $100(1 - \alpha)\%$ is given by

$$n = \left(\frac{(Z_{\alpha/2})\sigma}{E}\right)^2$$

where $Z_{\alpha/2}$ is the value associated with the desired confidence level (Table 8.1), $E$ is the desired margin of error, and $\sigma$ is the population standard deviation. By convention, whenever this formula yields a sample size with a decimal, *always round up to the next whole number.*

## Example 8.24 Business majors' salary example revisited

Suppose that we wanted higher precision in Example 8.23 and preferred to make our estimate to within \$100 of the population mean rather than within \$1000 of the population mean. What effect would this have on the required sample size, if we wanted to keep the same confidence level of 95%? Suppose we know that $\sigma$ = \$5000.

**Solution**

Since we want a smaller margin of error, we know that we either decrease the confidence level or increase the sample size. Since we want the same 95% confidence level, we must increase the sample size. For $E$ = \$1000,

$$n = \left(\frac{(Z_{\alpha/2})\sigma}{E}\right)^2 = \left(\frac{(1.96)(5000)}{1000}\right)^2 = 96.04$$

Thus, we need a sample size of $n = 97$ for a margin of error at 95% confidence. To obtain a margin of error of only \$100, we need

$$n = \left(\frac{(Z_{\alpha/2})\sigma}{E}\right)^2 = \left(\frac{(1.96)(5000)}{100}\right)^2 = 9604$$

That is, if we want a confidence interval that estimates the mean salary of business majors to within \$100 with 95% confidence, we must take a sample of business majors of size 9604. ■

**Balance Desired Accuracy with Cost**

To obtain an arbitrarily accurate estimate, the sample size necessary to achieve it increases very quickly. Notice that when we increase our precision from "within \$1000" to "within \$100," that is, by a factor of 10, the sample size increases from 96.04 to 9604, a factor of $10^2 = 100$. Thus, when deciding on the precision of your estimate, you must consider the time and expense required to achieve a desired accuracy. ■

Increasing the confidence level inevitably increases the required sample size, other things being equal. For example, suppose we are interested in estimating the population mean accounting error to within \$1, and we know that the standard deviation of all accounting errors is \$15. Figure 8.43 shows the sample size requirements as the confidence level increases from 90% to 99.95%. Note that the sample size requirements climb rather slowly until we pass 95% confidence, when the rate of increase increases. This is the point of diminishing returns and is one reason why 95% is often a more popular confidence level choice than 99%.

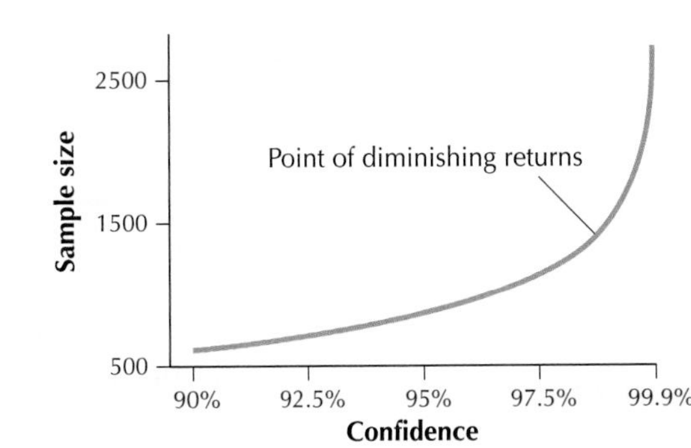


FIGURE 8.43 Sample size requirement increases as confidence level increases.

## 2 Sample Size for Estimating the Population Proportion with No Prior Information About *p*

Next we consider the question: How large a sample size do I need to estimate the population proportion $p$ to within margin of error $E$ with $100(1 - \alpha)\%$ confidence? Section 8.3 showed us that the margin of error of the confidence interval for proportions equals

$$E = Z_{\alpha/2} \cdot \sqrt{\frac{\hat{p}(1-\hat{p})}{n}}.$$

To derive the formula for sample size, solve for the required sample size $n$:

$$E = Z_{\alpha/2}\sqrt{\frac{\hat{p}(1-\hat{p})}{n}}$$

$$\sqrt{n}\,E = Z_{\alpha/2}\sqrt{\hat{p}(1-\hat{p})}$$

$$\sqrt{n} = \sqrt{\hat{p}(1-\hat{p})}\left(\frac{Z_{\alpha/2}}{E}\right)$$

$$n = \hat{p}(1-\hat{p})\left(\frac{Z_{\alpha/2}}{E}\right)^2 \qquad \text{(Equation 8.1)}$$

Equation 8.1 has $\hat{p}$ in it, which implies that we need to have information about $\hat{p}$ before we can find the required sample size $n$. This is a problem. In Section 8.3, we had already taken a sample, so we knew the value of $\hat{p}$. However, in this section, we have no such sample. We are in fact still trying to figure out how large a sample we need to take from which we will eventually compute $\hat{p}$.

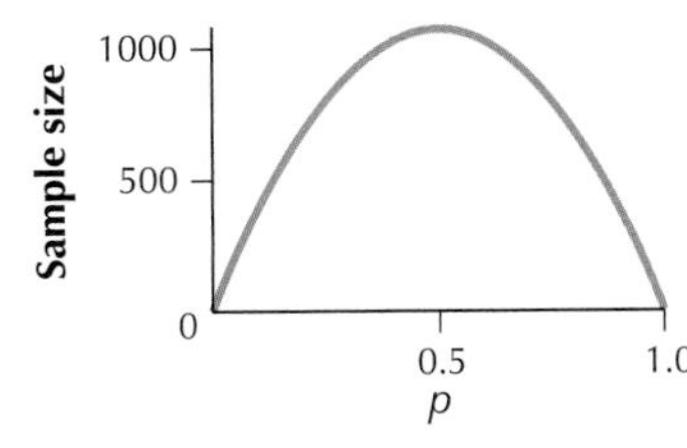


FIGURE 8.44 Sample size required for the range of values of *p*.

Figure 8.44 plots the sample size requirements for a 95% confidence interval for $p$, with a desired margin of error of 0.03, for values of $\hat{p}$ ranging from 0.01 to 0.99, representing all sample proportions from 1% to 99%. Note that the plot is symmetric, and therefore the largest required sample size occurs at the midpoint $\hat{p} = 0.5$. This is why $\hat{p} = 0.5$ is used as the most conservative value for $\hat{p}$ in the absence of further prior information. The value $\hat{p} = 0.5$ always gives the largest required sample size for a given confidence level and margin of error. Thus, we use $\hat{p} = 0.5$ to calculate $n$.

**Sample Size for Estimating a Population Proportion with No Prior Information About *p***

The sample size needed to estimate the population proportion $p$ to within a margin of error $E$ with confidence $100(1-\alpha)\%$ is given by

$$n = \left(\frac{0.5 \cdot Z_{\alpha/2}}{E}\right)^2$$

where $Z_{\alpha/2}$ is the value associated with the desired confidence level, $E$ is the desired margin of error, and 0.5 is a constant representing the most conservative estimate.

## Example 8.25 Required sample size for polls

Suppose the Dimes-Newspeak organization would like to take a poll on the proportion of Americans who will vote Republican in the next Presidential election. How large a sample size does the Dimes-Newspeak organization need to estimate the proportion to within plus or minus three percentage points ($E = 0.03$) with 95% confidence?

### Solution

The 95% confidence implies that the value for $Z_{\alpha/2}$ is 1.96. Since there is no information available about the value of the population proportion of all Americans who will vote Republican in the next election, we use 0.5 as our "worst case scenario" value of $p$:

$$n = \left(\frac{0.5 \cdot Z_{\alpha/2}}{E}\right)^2 = \left(\frac{(0.5)(1.96)}{0.03}\right)^2 \approx 1067.11$$

So if the pollsters would like to estimate the population proportion of all American voters who will vote Republican in the upcoming election to within 3% with 95% confidence, they will need a sample of 1068 voters (don't forget to round up!).

Most pollsters require a sample size of about 1000, although some use only about 500. What happens to the precision of their estimate if they use a smaller sample size? It certainly goes down, which explains why you sometimes hear that polling results are accurate only to within plus or minus (±) 4 (or 5) percentage points. Figure 8.45 plots the required sample size for a 95% confidence interval with no known prior information about $p$. Note that decreasing the desired margin of error will lead to an increase in sample size.

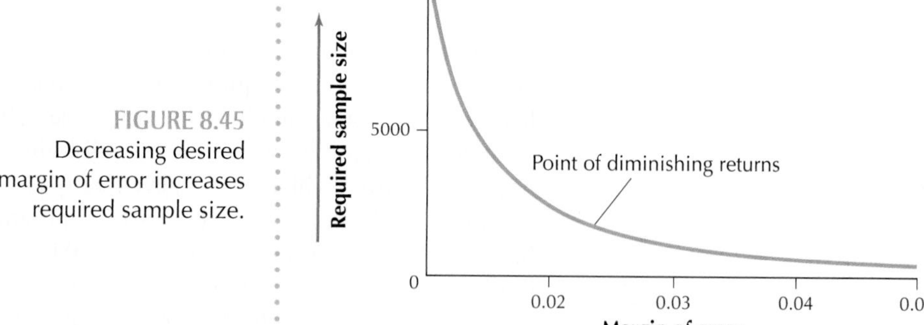


FIGURE 8.45 Decreasing desired margin of error increases required sample size.

Decreases in margin of error come rather cheaply until the margin of error reaches 0.02 or so. At that point, the required sample size begins to rise steeply: we are reaching the *point of diminishing returns*. This is why so many polling organizations have found that a margin of error of about 0.03 (the famous ± 3 percentage points) and therefore a sample size of about 1068 maintain a good balance of fairly high precision with relatively low cost.

## 3 Sample Size Requirement When Prior Information About *p* Is Available

Suppose we do have some prior information about the value of $p$ in the form of a sample estimate $\hat{p}$ from a previous sample. How large a sample size would we need to estimate the population proportion $p$ to within 3 percentage points (0.03) with 95% confidence (just as before)?

*What Results Might We Expect?*

Do you think we will require a larger or smaller sample than we did when we had no information about $\hat{p}$?

In nearly all cases, having more relevant information leads to better results. For example, suppose that a previous poll showed that $\hat{p} = 0.45$ of the respondents will vote Republican in the next presidential election. We can use Equation 8.1 and plug in our estimate $\hat{p} = 0.45$:

$$n = \hat{p}(1 - \hat{p})\left(\frac{Z_{\alpha/2}}{E}\right)^2 = 0.45(0.55)\left(\frac{(1.96)}{0.03}\right)^2 \approx 1056.44$$

The required sample size is 1057, which is less than the 1068 we found in Example 8.25. The reduction is small because the value of $\hat{p}$ is so close to the worst-case scenario value of $\hat{p} = 0.5$. ■

**Sample Size for Estimating a Population Proportion When Prior Information About *p* Is Available**

The sample size needed to estimate the population proportion $p$ to within a margin of error $E$ with confidence $100(1 - \alpha)\%$ is given by

$$n = \hat{p}(1 - \hat{p})\left(\frac{Z_{\alpha/2}}{E}\right)^2$$

where $Z_{\alpha/2}$ is the value associated with the desired confidence level, $E$ is the desired margin of error, and $\hat{p}$ is the sample proportion of successes available from some earlier sample.

## Section 8.5 SUMMARY

1 To use a $Z$ interval to estimate the population mean $\mu$ to within a margin of error $E$ with confidence $100(1 - \alpha)\%$, the required sample size is given by

$$n = \left(\frac{(Z_{\alpha/2})\sigma}{E}\right)^2$$

where $Z_{\alpha/2}$ is associated with the desired confidence level (Table 8.1), $E$ is the desired margin of error, and $\sigma$ is the population standard deviation.

2 Suppose we would like to estimate the population proportion $p$ to within a margin of error $E$ with confidence $100(1 - \alpha)\%$. If there is no prior information about $\hat{p}$, then the required sample size needed is given by

$$n = \left(\frac{0.5 \cdot Z_{\alpha/2}}{E}\right)^2$$

3 If there is prior information about $\hat{p}$, then the required sample size needed is given by

$$n = \hat{p}(1 - \hat{p})\left(\frac{Z_{\alpha/2}}{E}\right)^2$$

where $\hat{p}$ is the sample proportion of successes available from some earlier sample.

## Section 8.5 EXERCISES

### CLARIFYING THE CONCEPTS

1. Explain why we cannot simultaneously maximize the confidence level and minimize the width of the confidence interval if the sample size is fixed.
2. What happens to the required sample size for estimating the population mean as the confidence level is increased? Decreased?
3. What happens to the required sample size for estimating the population mean as the margin of error is increased? Decreased?
4. Suppose data set A has larger variance than data set B. We would like to estimate the mean for each data set to within the same margin of error with the same confidence level.
   a. Which data set will require a larger sample size?
   b. Clearly explain why, in a way a non-statistician would understand.

### PRACTICING THE TECHNIQUES

Suppose we are estimating $\mu$. For Exercises 5–7, find the required sample size.

5. $\sigma = 10$, confidence level 90%, margin of error 32
6. $\sigma = 10$, confidence level 90%, margin of error 16
7. $\sigma = 10$, confidence level 90%, margin of error 8
8. Using Exercises 5–7, write a sentence summarizing what happens to the required sample size when the margin of error is halved and $\sigma$ and the confidence level stay fixed.

Suppose we are estimating $\mu$. For Exercises 9–11, find the required sample size.

9. $\sigma = 10$, confidence level 90%, margin of error 8
10. $\sigma = 10$, confidence level 95%, margin of error 8
11. $\sigma = 10$, confidence level 99%, margin of error 8
12. Using Exercises 9–11, write a sentence summarizing what happens to the required sample size when the confidence level increases and the margin of error and $\sigma$ stay fixed.

Suppose we are estimating $\mu$. For Exercises 13–15, find the required sample size.

13. $\sigma = 10$, confidence level 90%, margin of error 16
14. $\sigma = 20$, confidence level 90%, margin of error 16
15. $\sigma = 40$, confidence level 90%, margin of error 16
16. Using Exercises 13–15, write a sentence summarizing what happens to the required sample size for a different data set with double the standard deviation and the margin of error and confidence level stay fixed.

Suppose we are estimating $p$, and suppose there is no prior information about $p$. For Exercises 17–20, find the required sample size.

17. Confidence level 95%, margin of error 0.03
18. Confidence level 95%, margin of error 0.015
19. Confidence level 95%, margin of error 0.0075
20. Confidence level 95%, margin of error 0.00375
21. Refer to Exercises 17–20.
    a. Write a sentence summarizing what happens to the required sample size when the margin of error is halved and the confidence level stays fixed.
    b. When would you say that the point of diminishing returns is achieved?

Suppose we are estimating $p$, and suppose we have prior information about $p$ in the form of an earlier estimate, $\hat{p}$. For Exercises 22–25, find the required sample size.

22. Confidence level 95%, margin of error 0.03, $\hat{p} = 0.2$
23. Confidence level 95%, margin of error 0.03, $\hat{p} = 0.1$
24. Confidence level 95%, margin of error 0.03, $\hat{p} = 0.01$
25. Confidence level 95%, margin of error 0.03, $\hat{p} = 0.001$
26. Using Exercises 22–25, write a sentence summarizing what happens to the required sample size when the previous sample proportion approaches very small values and the margin of error and the confidence level stay fixed.

Suppose we are estimating $p$, and suppose there is no prior information about $p$. For Exercises 27–29, find the required sample size.

27. Confidence level 90%, margin of error 0.03
28. Confidence level 95%, margin of error 0.03
29. Confidence level 99%, margin of error 0.03
30. Using Exercises 27–29, write a sentence summarizing what happens to the required sample size when the confidence level increases and the margin of error stays fixed.

Suppose we are estimating $p$. Suppose we have prior information about $p$ in the form of an earlier estimate, $\hat{p}$. For Exercises 31–33, find the required sample size.

31. Confidence level 95%, margin of error 0.03, $\hat{p} = 0.3$
32. Confidence level 95%, margin of error 0.03, $\hat{p} = 0.5$
33. Confidence level 95%, margin of error 0.03, $\hat{p} = 0.7$
34. Using Exercises 31–33, write a sentence summarizing what happens to the required sample size when the previous sample proportion approaches and then surpasses 0.5, and the margin of error and the confidence level stay fixed. Use the concept of symmetry in your explanation.

### APPLYING THE TECHNIQUES

35. **State of the Economy.** The Roper Center for Public Opinion Research at the University of Connecticut reported the results of a June–July 2003 poll that asked the following question: "Would you describe the state of the nation's economy these days as excellent, good, not so good, or poor?" The poll, also cited in Exercise 56 of Section 8.3, had a sample size of 1004 randomly selected respondents and an assumed confidence level of 95%. How large a sample size do we need to estimate the population proportion of those who replied either "not so good" or "poor" to within 3% with 99% confidence?
36. **Does Heavy Debt Lead to Ulcers?** An AP–AOL poll reported on June 9, 2008, that 27% of respondents carrying

heavy mortgage or credit card debt also said that they had stomach ulcers.[17] How large a sample size is needed to estimate the population proportion of respondents carrying heavy debt who also have stomach ulcers to within 1% with 99% confidence?

**37. Consumption of Carbonated Beverages.** The U.S. Department of Agriculture reports that the mean American consumption of carbonated beverages per year is greater than 52 gallons. Assume that the population standard deviation is 20 gallons.

- **a.** How large a sample size is needed to estimate $\mu$ to within 25 gallons with 95% confidence?
- **b.** How large a sample size is needed to estimate $\mu$ to within 5 gallons with 95% confidence?

**38. Engaging with Science.** A psychological study found that the mean length of time that boys remained engaged with a science exhibit at a museum was 107 seconds with a standard deviation of 117 seconds.[18] Assume that the 117 seconds represents the population standard deviation.

- **a.** How large a sample size is required to estimate $\mu$ to within 30 seconds with 95% confidence?
- **b.** How large a sample size is required to estimate $\mu$ to within 3 seconds with 95% confidence?

**39. Male Binge Drinking.** The American Medical Association is interested in commissioning a study on the population proportion of males who binge drink during spring break.

- **a.** How large a sample size is required to estimate $p$ to within 0.05 with 90% confidence?
- **b.** How large a sample size is required to estimate $p$ to within 0.01 with 90% confidence?

**40. Global Assignment.** On June 8, 2004, Cendant Mobility (now Cartus) reported the results of a survey it conducted asking workers if they would be more likely, less likely, or just as likely to accept a global assignment in view of recent world events. Of the 548 randomly chosen subjects polled, 367 said that they would be more likely to accept a global assignment.

- **a.** How large a sample size would have been required if Cendant wanted to estimate $p$ to within 0.03 with 95% confidence?
- **b.** How large a sample size would have been required if Cendant wanted to estimate $p$ to within 0.01 with 95% confidence?

**41. Most Churchgoers Are Not Drinkers.** According to Gallup.com, 60% of Americans who attend church weekly do not drink alcohol. The poll reported in the Section 8.3 exercises had a sample size of 1006 respondents and an assumed confidence level of 95%. How large a sample size do we need to estimate the population proportion to within 3% with 99% confidence?

**42. Egyptian Hieroglyphics.** An archaeologist would like to estimate the mean number of hieroglyphs in a given archaeological site. Assume that the standard deviation is 50 hieroglyphs.

- **a.** How many sites must she examine before she has an estimate that is accurate to within 2 hieroglyphs with 99% confidence?
- **b.** The sample size in **(a)** would be too expensive. Give two helpful suggestions to lower the required sample size.
- **c.** Follow through on your suggestions in **(b)** and verify how they would lower the required sample size.

**43. Clinical Psychology.** A clinical psychologist would like to estimate the population mean number of episodes her patients have suffered through in the past year. Assume that the standard deviation is 3 episodes.

- **a.** How many patients will she have to examine if she wants her estimate to be within 5 episodes with 99% confidence?
- **b.** Suppose the sample size in **(a)** would be too large. Give two helpful suggestions to lower the required sample size.
- **c.** Follow through on your suggestions in **(b)** and verify how they would lower the required sample size.

**44. Too Many Erasures.** The Iowa Test of Basic Skills is a comprehensive national test given annually to many third-graders and fifth-graders. At one school, investigators found that 3 to 5 times as many erasures were made to the tests, compared to other schools.[19] They also found that 89% of the erasures resulted in a correct answer, compared to less than 70% for other schools.

- **a.** Has there been tampering? Say you are an investigator interested in estimating the national proportion of Iowa Test answers that, after erasure, were found to result in a correct answer. How large a sample size do you need to estimate this proportion to within 1% with 99% accuracy?
- **b.** Suppose the sample size in **(a)** would be too large. Give two helpful suggestions to lower the required sample size.
- **c.** Follow through on your suggestions in **(b)** and show how they would lower the required sample size.

**45. Hispanic Tobacco Expenditures.** The Bureau of Labor Statistics (BLS) reported that the mean amount spent by all American citizens in 2001 on tobacco products and smoking supplies was \$308, while the mean for American Hispanics was \$177. Assume that the population standard deviation of the amount spent by American Hispanics, $\sigma = \$150$.

- **a.** How large a sample size would have been required if BLS had wanted to estimate the population mean amount spent by American Hispanics to within \$50 with 95% confidence?
- **b.** How large a sample size would have been required if BLS had wanted to estimate the population mean amount spent by American Hispanics to within \$10 with 95% confidence?
- **c.** How large a sample size would have been required if BLS had wanted to estimate the population mean amount spent by all American citizens to within \$50 with 95% confidence?
- **d.** How large a sample size would have been required if BLS had wanted to estimate the population mean

amount spent by all American citizens to within \$10 with 95% confidence?

**e.** Based on your results in **(a)–(d)**, what role does the value of the mean play in calculating the required sample size?

**46. Short-Term Memory.** In a famous research paper in the psychology literature, George Miller found that the amount of information humans could process in short-term memory was seven bits (pieces of information), plus or minus two bits.[20] Let us assume that Miller's title ("The Magical Number Seven, Plus or Minus Two" ) refers to a confidence interval and that $\sigma = 10$ bits.

**a.** The 95% confidence level is the most common in the psychological literature. How large a sample size did Miller use to find the confidence interval in the title, assuming that he used 95% confidence?

**b.** Suppose he had wanted the title to read "The Magical Number Seven, Plus or Minus One"? How large a sample size would he have needed?

**47. Stock Shares Traded.** The *Statistical Abstract of the United States* reports that the mean daily number of shares traded on the New York Stock Exchange (NYSE) in 2002 was 1441 million. Assume that the population standard deviation equals 500 million shares.

**a.** How large a sample size (trading days) is needed to estimate the population mean number of shares traded per day to within 100 million shares?

**b.** How large a sample size (trading days) is needed to estimate the population mean number of shares traded per day to within 10 million shares? How many years does this number of days translate into?

**Case Study: Analyzing Stock Performance on the New York Stock Exchange**

In Exercises 48 and 49, we continue our analysis of the NYSE data set in Table 8.5 (page 414). The 95% confidence interval for the population mean stock price change is $(-0.641, 0.255)$, with a margin of error of 0.448 (measured in dollars).

**48.** Suppose the financial analyst would like to lower the margin of error of the interval to \$0.25 (25 cents), with 95% confidence. How large a sample size is needed? Recall that we assumed that $\sigma = 1.6$.

**49.** Suppose the financial analyst wants to impress his boss and report a 95% confidence interval with a margin of error of a single penny. How large a sample size is needed? Comment.

## *Try This in Class!* In-class activities to enhance your understanding of statistics

CONFIDENCE INTERVALS FOR THE POPULATION MEAN:
**Estimating the Mean Shoe Size for All Female and Male Students at Your School**

**1.** Make a guess at what the population mean shoe size might be for all female students at your school. Do the same for the males.

**2.** Collect data on the shoe sizes of the female students in your class. Find the mean of this sample.

**3.** What would you say is the probability that the value for the sample mean from Activity 2 is exactly equal to the mean foot size of all female students at your school?

**4.** Use technology to determine if the female shoe sizes are normally distributed. If not, and if the number of females is not at least 30, discuss whether it is appropriate to construct a *t*-confidence interval for the population mean shoe size of all females at your school, and go to Activity 7.

**5.** Construct a 95% confidence interval for the population mean shoe size for all female students at your school.

**6.** Does your guess for the population mean shoe size from Activity 1 fall within the confidence interval from Activity 6?

**7.** Repeat Activities 2–6 for the males in your class.

## CHAPTER 8 FORMULAS AND VOCABULARY

### SECTION 8.1

- **Confidence interval** (p. 390). A *confidence interval estimate* of a parameter consists of an interval of numbers generated by a point estimate, together with an associated *confidence level* specifying the probability that the interval contains the parameter. Most confidence intervals take the form

$$\text{point estimate} \pm \text{margin of error}$$

- **Margin of error $E$ for the $Z$ interval for $\mu$** (p. 398). A measure of the *precision* of the confidence interval estimate. For the $Z$ interval, the margin of error takes the form

$$E = Z_{\alpha/2}(\sigma/\sqrt{n})$$

- **Z interval for $\mu$** (p. 391). The $100(1-\alpha)\%$ confidence interval for the population mean $\mu$ is given by

$$\text{lower bound } \bar{x} - Z_{\alpha/2}(\sigma/\sqrt{n}),$$
$$\text{upper bound } \bar{x} + Z_{\alpha/2}(\sigma/\sqrt{n})$$

where $1-\alpha$ is the confidence level and $\sigma$ represents the population standard deviation. The $Z$ interval may be used whenever either of the following conditions is met: Case 1: The original population is normal, and $\sigma$ is known. Case 2: The sample size is large ($n \geq 30$), and $\sigma$ is known.

## SECTION 8.2

- **Margin of error $E$ for the $t$ interval for $\mu$** (p. 412). For the $t$ interval, the margin of error takes the form

$$E = t_{\alpha/2}(s/\sqrt{n})$$

- ***t* Distribution** (p. 407). For a normal population, the distribution of

$$t = \frac{\bar{x} - \mu}{s/\sqrt{n}}$$

follows a $t$ distribution, with $n-1$ degrees of freedom, where $\bar{x}$ is the sample mean, $\mu$ is the unknown population mean, $s$ is the sample standard deviation, and $n$ is the sample size. The $t$ distribution is centered at its mean, zero, just like the $Z$ distribution.
- ***t* interval for $\mu$** (p. 409). A $100(1-\alpha)\%$ confidence interval for $\mu$ is given by the interval

$$\text{lower bound } \bar{x} - t_{\alpha/2}(s/\sqrt{n}),$$
$$\text{upper bound } \bar{x} + t_{\alpha/2}(s/\sqrt{n})$$

where $\bar{x}$ is the sample mean, $t_{\alpha/2}$ is associated with the confidence level and $n-1$ degrees of freedom, $s$ is the sample standard deviation, and $n$ is the sample size. The $t$ interval may be used whenever *either* of the following conditions is met: Case 1: The population is normal. Case 2: The sample size is large ($n \geq 30$).

## SECTION 8.3

- **Margin of error $E$ for the $Z$ interval for $p$** (p. 424). For the $Z$ interval for $p$, the margin of error takes the form

$$E = Z_{\alpha/2}\sqrt{\frac{\hat{p}(1-\hat{p})}{n}}$$

- **Z interval for $p$** (p. 421). The $100(1-\alpha)\%$ confidence interval for the population proportion $p$ is given by

$$\text{lower bound } \hat{p} - Z_{\alpha/2}\sqrt{\frac{\hat{p}(1-\hat{p})}{n}},$$
$$\text{upper bound } \hat{p} + Z_{\alpha/2}\sqrt{\frac{\hat{p}(1-\hat{p})}{n}}$$

where $\hat{p}$ is the sample proportion of successes, $n$ is the sample size, and $Z_{\alpha/2}$ is associated with the confidence level. The $Z$ interval for $p$ may be performed only if *both* of the following conditions apply: $n\hat{p} \geq 5$ and $n(1-\hat{p}) \geq 5$.

## SECTION 8.4

**$\chi^2$ (chi square) distribution properties** (p. 430). (1) Total area under the $\chi^2$ curve equals 1. (2) $\chi^2$ is never negative, so the $\chi^2$ curve starts at 0 and extends indefinitely to the right. (3) The $\chi^2$ curve is right-skewed. (4) There is a different curve for all different degrees of freedom, $n-1$.
- **$\chi^2$ (chi square) critical values** (p. 431). (1) $\chi^2_{1-\alpha/2}$ represents the value of the $\chi^2$ distribution with area $1-\alpha/2$ to the right of it, (2) $\chi^2_{\alpha/2}$ represents the value of the $\chi^2$ distribution with area $\alpha/2$ to the right of it.
- **Confidence interval for the population variance $\sigma^2$** (p. 433). Suppose we take a sample of size $n$ from a normal population with mean $\mu$ and standard deviation $\sigma$. Then a $100(1-\alpha)\%$ confidence interval for the population variance $\sigma^2$ is given by

$$\text{lower bound} = \frac{(n-1)s^2}{\chi^2_{\alpha/2}}, \qquad \text{upper bound} = \frac{(n-1)s^2}{\chi^2_{1-\alpha/2}}$$

where $s^2$ represents the sample variance and $\chi^2_{1-\alpha/2}$ and $\chi^2_{\alpha/2}$ are the critical values for a $\chi^2$ distribution with $n-1$ degrees of freedom.
- **Confidence interval for the population standard deviation $\sigma$** (p. 433). A $100(1-\alpha)\%$ confidence interval for the population standard deviation $\sigma$ is given by

$$\text{lower bound} = \sqrt{\frac{(n-1)s^2}{\chi^2_{\alpha/2}}}, \qquad \text{upper bound} = \sqrt{\frac{(n-1)s^2}{\chi^2_{1-\alpha/2}}}$$

## SECTION 8.5

- **Sample size for estimating the population mean** (p. 440). The sample size needed to estimate the population mean $\mu$ to within a margin of error $E$ with confidence $100(1-\alpha)\%$ is given by

$$n = \left(\frac{(Z_{\alpha/2})\sigma}{E}\right)^2$$

where $Z_{\alpha/2}$ is the $Z$ critical value associated with the desired confidence level (Table 8.1), $E$ is the desired margin of error, and $\sigma$ is the population standard deviation.
- **Sample size for estimating a population proportion with no prior information about $p$** (p. 442). The sample size needed to estimate the population proportion $p$ to within a margin of error $E$ with confidence $100(1-\alpha)\%$ is given by

$$n = \left(\frac{(0.5)(Z_{\alpha/2})}{E}\right)^2$$

where $Z_{\alpha/2}$ is the $Z$ critical value associated with the desired confidence level and $E$ is the desired margin of error.
- **Sample size for estimating a population proportion when prior information about $p$ is available** (p. 443). The sample size needed to estimate the population proportion $p$ to within a margin of error $E$ with confidence $100(1-\alpha)\%$ is given by

$$n = \hat{p}(1-\hat{p})\left(\frac{Z_{\alpha/2}}{E}\right)^2$$

where $Z_{\alpha/2}$ is the $Z$ critical value associated with the desired confidence level, $E$ is the desired margin of error, and $\hat{p}$ is the sample proportion of successes available from some earlier sample.

## CHAPTER 8 REVIEW EXERCISES

### SECTION 8.1

For Exercises 1–3, answer the following questions.

**a.** Which case applies? If neither case applies, state that neither case applies and do not do **(b)**–**(e)**.

**b.** Calculate $\sigma/\sqrt{n}$.

**c.** Find $Z_{\alpha/2}$ for a confidence interval for $\mu$ with 95% confidence.

**d.** Compute $E$, the margin of error for a confidence interval $\mu$ with 95% confidence. Interpret this value for the margin of error.

**e.** Construct and interpret a 95% confidence interval for $\mu$.

**1.** A sample of $n = 25$ with sample mean $\bar{x} = 50$ is drawn from a normal population in which $\sigma = 10$.

**2.** A sample of $n = 25$ with sample mean $\bar{x} = 50$ is drawn from a population in which $\sigma = 10$.

**3.** A sample of $n = 100$ with sample mean $\bar{x} = 50$ is drawn from a population in which $\sigma = 10$.

**The Mozart Effect.** Use this information for Exercises 4–6. A random sample of 45 children showed a mean increase of 7 IQ points after listening to a Mozart piano sonata for about 10 minutes. The distribution of such increases is unknown, but the standard deviation is assumed to be 2 IQ points.

**4.** Follow steps **(a)**–**(e)**.

**a.** Find the point estimate of the increase in IQ points for all children after listening to Mozart.

**b.** Calculate $\sigma/\sqrt{n}$.

**c.** Find $Z_{\alpha/2}$ for a confidence interval with 90% confidence.

**d.** Compute and interpret the margin of error for a confidence interval with 90% confidence.

**e.** Construct and interpret a 90% confidence interval for the mean increase in IQ points for all children after listening to a Mozart piano sonata for about 10 minutes.

**5.** Follow steps **(a)**–**(c)**.

**a.** Find a 95% confidence interval for the mean increase in IQ points for all children after listening to a Mozart piano sonata for about 10 minutes.

**b.** Find a 99% confidence interval for the mean increase in IQ points for all children after listening to a Mozart piano sonata for about 10 minutes.

**c.** Using your confidence intervals from **(a)** and **(b)**, write a general statement about what happens to the confidence interval as you increase the confidence level.

**6.** Suppose your research partner wanted to increase the confidence level to 99% without increasing the sample size. What would you tell your partner about the effect this would have on your confidence interval?

**Working Women's Salaries.** Use the following information for Exercises 7–9. A random sample of 25 working women in California yielded a sample mean income of \$45,000. Assume that the population standard deviation is \$6000 and that the distribution of such incomes is normal.

**7.** Follow steps **(a)**–**(e)**.

**a.** Find the point estimate of the income of all working women.

**b.** Find $\sigma/\sqrt{n}$.

**c.** Find $Z_{\alpha/2}$ for a confidence interval with 99% confidence.

**d.** Compute and interpret the margin of error for a confidence interval with 99% confidence.

**e.** Construct and interpret a 99% confidence interval for the mean income of the population of all working women.

WHAT IF? **8.** *What if* we changed the confidence level from 99% to 95%. Explain how this change would affect the following measures, if at all.

**a.** $\bar{x}$

**b.** $\sigma/\sqrt{n}$

**c.** $Z_{\alpha/2}$

**d.** Margin of error

**9.** Is it possible to construct a 100% confidence interval for the mean income of the population of all working women in California? Clearly explain why or why not.

### SECTION 8.2

For Exercises 10–12, determine whether Case 1 or Case 2 applies, if either. If appropriate, construct the indicated confidence interval. If it is not appropriate, explain why not.

**10.** Confidence level 90%, $n = 25$, $\bar{x} = 22$, $s = 5$, non-normal population

**11.** Confidence level 90%, $n = 25$, $\bar{x} = 22$, $s = 5$, normal population

**12.** Confidence level 90%, $n = 100$, $\bar{x} = 22$, $s = 5$, non-normal population

**Cigarette Consumption.** Health officials are interested in estimating the population mean number of cigarettes smoked per capita in order to evaluate the efficacy of the antismoking campaign. A random sample of eight U.S. counties yielded the following numbers of cigarettes smoked per capita: 2206, 2391, 2540, 2116, 2010, 2791, 2392, 2692. Use this information for Exercises 13 and 14.

**13.** Follow steps **(a)**–**(c)**.

**a.** Construct a 95% confidence interval for the population mean per capita number of cigarettes smoked in all U.S. counties.

**b.** Construct a 99% confidence interval for the population mean per capita number of cigarettes smoked in all U.S. counties.

**c.** Compare the intervals from **(a)** and **(b)**. Name one strength and one weakness of each interval.

**14.** Discuss whether it is possible to find a 100% confidence interval for the population mean number of cigarettes smoked. Why or why not?

**Parking Meters.** Use this information for Exercises 15 and 16. A random sample of 75 parking meters yielded a mean of $120 per meter with a standard deviation of $30.

**15.** Follow steps **(a)**–**(c)**.

**a.** Find $t_{\alpha/2}$ for a confidence interval with 90% confidence.

**b.** Compute and interpret the margin of error $E$ for a confidence interval with 90% confidence.

**c.** Construct and interpret a 90% $t$ confidence interval for the population mean amount of money collected by all parking meters.

**16.** Suppose the sample size 75 was an error, and the real sample size was some lower value, left unspecified but greater than 30. Otherwise everything else was the same. Describe what would happen to each of the following, if anything, and why.

**a.** Margin of error

**b.** Width of the confidence interval

**17. Van Gogh's Palette.** A random sample of 30 paintings by Vincent van Gogh yielded a sample mean number of colors used in his palette of 15 with a sample standard deviation of 3. Assume that the number of palette colors are normally distributed.

**a.** Why should we ever settle for 90% confidence when we could easily have a 99% confidence level? What are the costs and benefits of each?

**b.** Find the margin of error for a 90% confidence interval for the population mean number of van Gogh's palette colors. Do the same for a 99% confidence interval. Write a sentence that describes what happens to the margin of error as the confidence level increases.

**18. Women's Heart Rates.** Suppose that a sample of $n =$ 15 women had a mean heart rate of 73.93 and a standard deviation of 8.42. Assume that the distribution of women's heart rates is normal.

**a.** Construct and interpret a 90% confidence interval for the population mean heart rate of all women.

**b.** Construct and interpret a 99% confidence interval for the population mean heart rate of all women.

**c.** Which of the intervals from **(a)** and **(b)** has the smaller margin of error? Which has the higher confidence level? Write a sentence describing the trade-off between confidence level and margin of error.

WHAT IF? **Calories in Breakfast Cereal.** Use this information for Exercises 19 and 20. In Section 8.2, we estimated the population mean number of calories in a bowl of breakfast cereal using the six cereals in the accompanying table.

| Cereal | Calories |
|---|---|
| Apple Jacks | 110 |
| Cocoa Puffs | 110 |
| Mueslix | 160 |
| Cheerios | 110 |
| Corn Flakes | 100 |
| Shredded Wheat | 80 |

*What if* a seventh cereal is added to the data set and the number of calories contained in this seventh cereal is exactly equal to the mean number of calories for the other six cereals.

**19.** Try to answer the following questions by *thinking* about the data set rather than by recalculating the answers. (*Hint:* A dotplot might be helpful.) How would the following measures be affected by this seventh cereal?

**a.** Sample mean

**b.** Sample standard deviation

**c.** Degrees of freedom for the $t$ interval

**d.** $t_{\alpha/2}$

**20.** Follow steps **(a)**–**(c)**.

**a.** If we construct a 95% confidence interval based on the seven cereals, will the resulting confidence interval be more or less precise than the one based on six cereals? Why?

**b.** Generalize **(a)** to the case of adding not one but many new observations, at or quite near the mean.

**c.** Generalize the phenomenon we just witnessed in this exercise in a sentence relating the spread of the data set with the width of the resulting confidence interval.

## SECTION 8.3

For Exercises 21–24, follow steps **(a)**–**(d)**.

**a.** Find $Z_{\alpha/2}$.

**b.** Determine whether the conditions are met.

**c.** Calculate and interpret the margin of error, $E = Z_{\alpha/2}\sqrt{\hat{p}(1-\hat{p})/n}$

**d.** If the conditions are met, construct a confidence interval for $p$ with the indicated confidence level, and sketch the confidence interval on the number line. If the conditions are not met, then state why not.

**21.** Confidence level 95%, $n = 36, \hat{p} = 0.1$

**22.** Confidence level 95%, sample size 100, sample proportion 0.1

**23.** Confidence level 95%, sample size 100, sample proportion 0.99

**24.** Confidence level 95%, $n = 500, \hat{p} = 0.99$

**25. Terrorism in Western Europe.** The United States State Department, in its 2003 report "Patterns of Global Terrorism" (www.state.gov), reported that 17 of the 60 anti-American terrorist incidents occurring in 2003 took place in Western Europe.

**a.** Is it appropriate to apply the $Z$ interval for the population proportion?

b. Find the margin of error. What does this number mean?
c. Construct and interpret a 95% confidence interval for the population proportion of all anti-American terrorist incidents taking place in Western Europe since 2003.

**26. Ecstasy and Emergency Room Visits.** According to the National Institute on Drug Abuse (**www.drugabuse.gov**), in 2001, 77% of the emergency room patients who mentioned MDMA (Ecstasy) as a factor in their admission were age 25 and under. Assume that the sample size is 200.
a. Check the conditions for performing the confidence interval for $p$.
b. Calculate and interpret the margin of error.
c. Construct and interpret a 95% confidence interval for the population proportion of all emergency room patients mentioning MDMA (Ecstasy) as a factor in their admission who are age 25 and under.

**27. Too Many Erasures?** The Iowa Test of Basic Skills is a comprehensive national test given annually to many third-graders and fifth-graders. At one school, investigators found that 3 to 5 times as many erasures were made to the tests, compared to other schools. They also found that 89% of the erasures resulted in a correct answer, compared to less than 70% for other schools. Assume that the sample size is 1000. Let's investigate whether there is evidence of tampering.
a. Check the conditions for performing the confidence interval for $p$ the population proportion of erasures that resulted in a correct answer at the school.
b. Calculate and interpret the margin of error.
c. Construct a 99% confidence level for $p$.
d. Compare this confidence interval to the proportion for other such schools. Do you think there is evidence of tampering?

## SECTION 8.4

For Exercises 28–33, a random sample is drawn from a normal population. The sample of size $n = 36$ has a sample variance of $s^2 = 100$. Construct the specified confidence interval.

**28.** 90% confidence interval for the population variance $\sigma^2$
**29.** 95% confidence interval for the population variance $\sigma^2$
**30.** 99% confidence interval for the population variance $\sigma^2$
**31.** 90% confidence interval for the population standard deviation $\sigma$
**32.** 95% confidence interval for the population standard deviation $\sigma$
**33.** 99% confidence interval for the population standard deviation $\sigma$

**34. Union Membership.** The table contains the total union membership for seven randomly selected states in 2006. The normal probability plot in Figure 8.46 indicates acceptable normality. Construct and interpret a 95% confidence interval for $\sigma$.

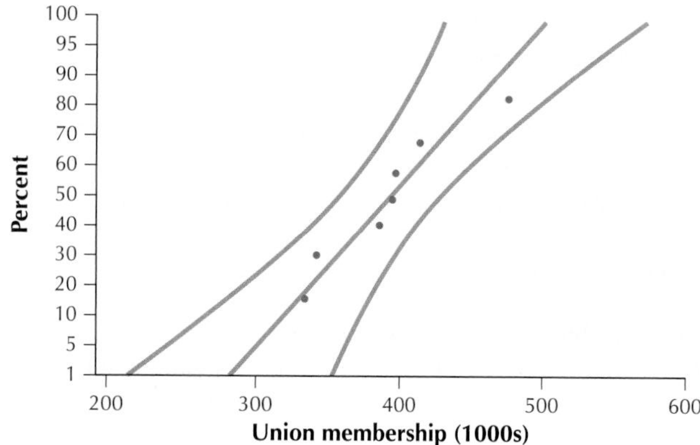


Figure 8.46 Normal probability plot.

| State | Union membership (thousands) |
|---|---|
| Florida | 397 |
| Indiana | 334 |
| Maryland | 342 |
| Massachusetts | 414 |
| Minnesota | 395 |
| Texas | 476 |
| Wisconsin | 386 |

Source: U.S. Bureau of Labor Statistics.

## SECTION 8.5

Suppose we are estimating $\mu$. For Exercises 35–37, find the required sample size.

**35.** $\sigma = 5$, confidence level 95%, margin of error 10
**36.** $\sigma = 3$, confidence level 95%, margin of error 10
**37.** $\sigma = 1$, confidence level 95%, margin of error 10

Suppose we are estimating $p$ and there is no prior information about $p$. For Exercises 38–40, find the required sample size.

**38.** Confidence level 90%, margin of error 0.05
**39.** Confidence level 90%, margin of error 0.03
**40.** Confidence level 90%, margin of error 0.01

Suppose we are estimating $p$ and we have prior information about $p$ in the form of an earlier estimate, $\hat{p}$. For Exercises 41–43, find the required sample size.

**41.** Confidence level 99%, margin of error 0.03, $\hat{p} = 0.9$
**42.** Confidence level 95%, margin of error 0.03, $\hat{p} = 0.99$
**43.** Confidence level 95%, margin of error 0.03, $\hat{p} = 0.999$

**44. Clinical Psychology.** A clinical psychologist would like to estimate the population mean number of episodes her patients have suffered through in the past year. Assume that the standard deviation is 3 episodes.
a. How many patients will she have to examine if she wants her estimate to be within 4 episodes with 90% confidence?
b. How many patients will she have to examine if she wants her estimate to be within 4 episodes with 95% confidence?

**c.** How many patients will she have to examine if she wants her estimate to be within 4 episodes with 99% confidence?

**d.** Using your answers to **(a)**–**(c)**, describe what happens to the required sample size as the desired confidence level increases.

**45. Literacy Levels.** The growth of HMOs has brought about an increasing reliance on printed reading material to inform patients about medical matters. However, much of the reading material is written at a higher level than the literacy level of the patients. Suppose a consumer's group would like to estimate the population proportion of all printed reading material that is beyond patients' reading level. Also, they have information from a prior study that 20% of all the reading material is written at a higher level than the literacy level of the patients.

**a.** To estimate this proportion to within 3% with 90% confidence, how large a sample size do they need?

**b.** To estimate this proportion to within 3% with 95% confidence, how large a sample size do they need?

**c.** To estimate this proportion to within 3% with 99% confidence, how large a sample size do they need?

# CHAPTER 8 *Quiz*

## TRUE OR FALSE

**1.** True or false: In Table 8.2 (page 393), since the confidence level is 95%, then 95% of the intervals must contain $\mu$. Explain your answer.

**2.** True or false: The $t$ curve is symmetric about 0, just like the $Z$ curve was. Therefore we can use all our symmetry techniques with the $t$ curve as well.

**3.** True or false: For any confidence interval, the value of the sample proportion will always be considered a plausible value for $p$.

## FILL IN THE BLANK

**4.** In a real-world problem, the probability is __________ that a point estimate exactly equals the parameter it is trying to estimate.

**5.** Suppose we cut a margin of error in half. The sample size requirement then becomes __________ times larger.

**6.** Our estimate of $\mu$ is __________ precise using the $t$ curve rather than the $Z$ curve.

## SHORT ANSWER

**7.** $\alpha$ is used to find the value of $Z_{\alpha/2}$. Is $\alpha$ a probability or a value of $x$ or a value of $Z$?

**8.** What are the conditions for constructing a $t$ interval?

**9.** If $p$ is unknown, state two methods to proceed with finding the required sample size. Which method will always deliver the larger sample size?

## CALCULATIONS AND INTERPRETATIONS

**10. College Education Costs.** A random sample of 49 colleges yielded a mean cost of college education of $30,500 per year. Assume that the population standard deviation is $3000.

**a.** Find the point estimate of the mean cost of college education.

**b.** Calculate $\sigma/\sqrt{n}$.

**c.** Find $Z_{\alpha/2}$ for a confidence interval with 90% confidence.

**d.** Compute and interpret the margin of error for a confidence interval with 90% confidence.

**e.** Construct and interpret a 90% confidence interval for the population mean cost of college education.

**11. Crash Test Data.** The National Highway Traffic Safety Administration collects data on crash tests for new motor vehicles. They reported that the mean femur load (force applied to the femur) in a frontal crash for the passenger in a 2005 Ford Equinox SUV was 1003 pounds. Assume that the population standard deviation was 210 pounds and the sample size was 49.

**a.** Calculate $\sigma/\sqrt{n}$.

**b.** Find $Z_{\alpha/2}$ for a confidence interval with 90% confidence.

**c.** Compute and interpret the margin of error for a confidence interval with 90% confidence.

**d.** Construct and interpret a 90% confidence interval for the population mean femur load in a frontal crash for the passenger in a 2005 Ford Equinox SUV.

**12. Wolves Returning to Isle Royale?** A random sample of 81 day-counts of wolves from Isle Royale in the middle of Lake Superior resulted in a mean of 57 wolves and a standard deviation of 10 wolves. Assume that the data are normally distributed.

**a.** Calculate and interpret a 90% confidence interval for the population mean day-count of wolves in Isle Royale.

**b.** Since 90% is the minimum confidence level in common usage, what is the only way we can decrease the margin of error?

**13. The Fog Index.** The Fog Index is a scale that measures the readability of a particular English passage. For example, a Fog Index of 11.5 indicates that the material is about eleventh-grade level. A random sample of 64 passages by a well-known horror author yielded a mean Fog Index of 9.5 with a standard deviation of 0.8.

**a.** Construct and interpret a 95% confidence interval for the mean Fog Index for all the books by this horror author.

**b.** If we now construct a 99% confidence interval for the mean Fog Index for all of this horror author's material, would such an interval be more (have smaller width) or less precise (have greater width) than the 95% confidence interval?

**c.** Discuss the strengths and weakness of the confidence intervals in **(a)** and **(b)**, one against the other. Which would you prefer to use?

**d.** Would it be appropriate to use a *Z* interval for this data set? Explain why or why not.

**14. Women's Heart Rates.** A random sample of the heart rates of 15 women had the normal probability plot shown here. The sample mean was 75.6 beats per minute, with a standard deviation of 6.65 beats per minute.

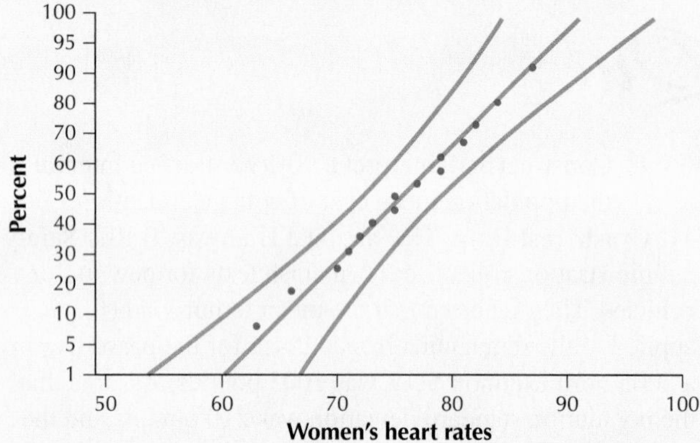


**a.** Discuss the evidence for or against the normality assumption. Should we use the *t* interval? Why or why not? What specific characteristics of the plot tell you this?

**b.** Assume that the plot does not contradict the normality assumption, and construct and interpret a 90% confidence interval for the population mean heart rate of all women.

**15. 9/11 and Religious Attendance.** The Pew Research Center reported that, in a survey of 3733 randomly selected respondents, 991 had attended a religious service in response to the attacks on the World Trade Center and the Pentagon.

**a.** If appropriate, find the margin of error for confidence level 95%. What does this number mean?

**b.** Construct, if appropriate, a 95% confidence interval for the population proportion of Americans who attended a religious service in response to the attacks on the World Trade Center and the Pentagon.

**16. Independence for Quebec?** A January 2006 poll by the newspaper *La Presse* reported that 340 of 1000 randomly chosen Quebec adults surveyed would vote "Yes" in a referendum for independence from Canada.

**a.** If appropriate, find the margin of error for confidence level 99%. What does this number mean?

**b.** If appropriate, find a 99% confidence interval for the population proportion of all Quebec residents who favor independence for the Province of Quebec.

**17.** Recall from Section 3.2 that Amanda and Bethany have contracted with a local document dispatch service to deliver documents in midtown Manhattan. Their clients would like to be able to count on a consistent delivery time schedule. Not only do the documents have to arrive in a timely fashion, but the delivery times should not vary a lot. One particularly important client kept careful track of the delivery times (in minutes) for documents delivered by Amanda and Bethany, shown in the accompanying table.

| | | | | | | | | | | |
|---|---|---|---|---|---|---|---|---|---|---|
| **Amanda** | 15 | 17 | 12 | 15 | 10 | 16 | 14 | 20 | 16 | 15 |
| **Bethany** | 5 | 22 | 15 | 25 | 4 | 18 | 10 | 29 | 17 | 5 |

**a.** Use technology to verify that both Amanda's delivery times and Bethany's delivery times are normally distributed.

**b.** Construct and interpret a 95% confidence interval for the population standard deviation of Amanda's delivery times.

**c.** Construct and interpret a 95% confidence interval for the population standard deviation of Bethany's delivery times.

**d. Challenger Exercise.** Discuss how the confidence intervals in **(b)** and **(c)** show that there is strong statistical evidence that Bethany's delivery times are more highly variable than Amanda's.

**18. Rather Be Fishing?** A study found that Minnesota, at 38%, leads the nation in the proportion of people who go fishing.[21]

**a.** How large a sample size is required to estimate *p* to within 0.05 with 95% confidence?

**b.** How large a sample size is required to estimate *p* to within 0.05 with 99% confidence?

**19. Genetic Analysis.** A biotechnician would like to estimate the mean number of genes that she will have to analyze in her research. Assume that the standard deviation is 120 genes.

**a.** How many genes does she have to sample if she would like the estimate to be accurate to within 10 genes with 95% confidence?

**b.** How many genes does she have to sample if she would like the estimate to be accurate to within 5 genes with 95% confidence?

**20. Quality of Education in America.** The National Assessment of Educational Progress (NAEP) administers exams to a nationwide sampling of students to assess the quality of education in America. Suppose NAEP would like to estimate the population proportion of American schoolchildren who would answer a given question correctly.

**a.** They would like the estimate to be within 20% of the population proportion with 90% confidence. How large a sample size do they need?

**b.** What do you think of the precision of the estimate that will result from the sample size mentioned in the previous exercise?

**c.** Find a sample size which would give a margin of error of 0.03 with 90% confidence.

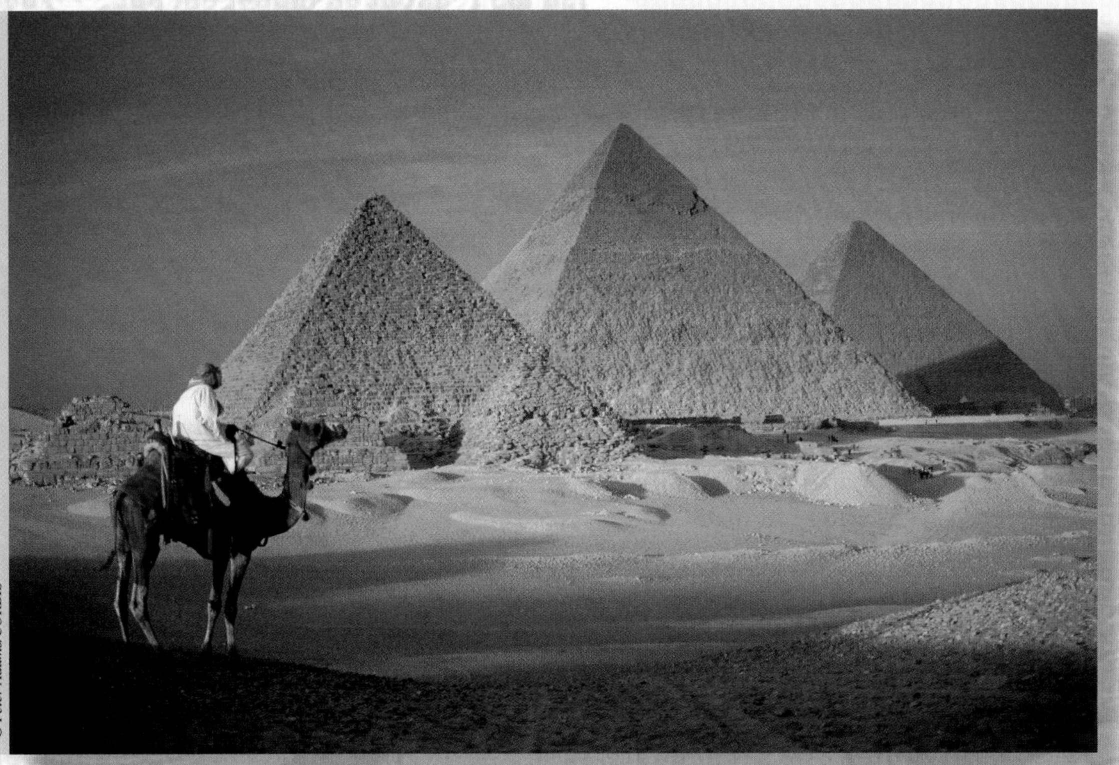



Chapter 9

# Hypothesis Testing

CASE STUDY

## The Golden Ratio

What do Euclid's *Elements,* the Parthenon of ancient Greece, the *Mona Lisa,* and the beadwork of the Shoshone tribe of Native Americans have in common? An appreciation for the *golden ratio*. Suppose we have two quantities $A$ and $B$, with $A > B > 0$. Then, $A/B$ is called the golden ratio if

$$\frac{A + B}{A} = \frac{A}{B}$$

that is, if the ratio of the sum of the quantities to the larger quantity equals the ratio of the larger to the smaller.

The golden ratio permeates ancient, medieval, Renaissance, and modern art and architecture. For example, the Egyptians constructed their great pyramids using the golden ratio. (Specifically, in Figure 9.1, if $A = \overline{XY}$ is the height from the top vertex to the base, and $B = \overline{YZ}$ is the distance from the center of the base to the edge, then $(A + B)/A = A/B$.) Some mathematicians have said that the golden ratio may be intrinsically pleasing to the human species. Support for this conjecture would

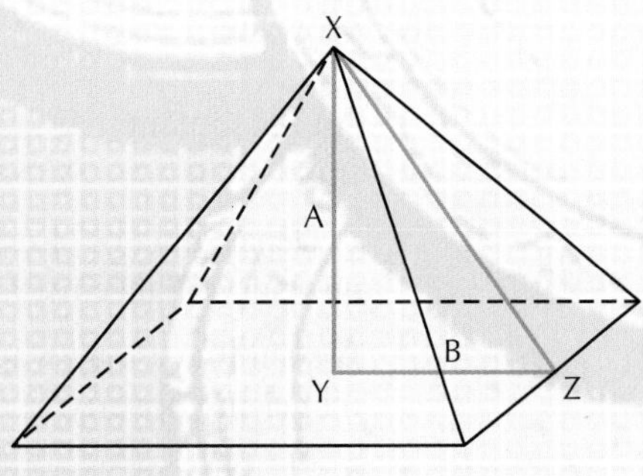


FIGURE 9.1

be especially strong if evidence was found for the use of the golden ratio in non-Western artistic traditions. In the Case Study on pages 506–509, we examine whether the decorative beaded rectangles sewn by the Shoshone tribe of Native Americans follow the golden ratio.

## OVERVIEW



## The Big Picture

### Where we are coming from, and where we are headed...

Hypothesis testing is the most widely used method for statistical inference in the world. Hypothesis testing forms the bedrock of the scientific method, and it touches nearly every field of scientific endeavor, from biology and genetics, through physics and chemistry, to sociology and psychology. It also forms the basis for decision science, and thus for all business-oriented decision-making methods. In this chapter we will discover what hypothesis testing is all about.

In Chapter 8, we learned about confidence interval estimation. But confidence intervals represent only the first of a large family of topics in inferential statistics. In fact, each of Chapters 8–13 covers a fresh and distinct facet of statistical inference. In Chapters 7 and 8, we learned that, since the value of the target population parameter is unknown, we need to estimate it using point and interval estimates provided by sample data. However, because the target population parameter is unknown, people may disagree about its value. In this chapter, we introduce the statistical method for resolving conflicting claims about the value of a population parameter, hypothesis testing.

Researchers are interested in investigating many different types of questions, such as the following:

- An accountant may wish to examine whether evidence exists for corporate tax fraud.
- A Department of Homeland Security executive may want to test whether a new surveillance method will uncover terrorist activity.
- A sociologist may want to examine whether the mayor's economic policy is increasing poverty in the city.

Questions such as these can be tackled using statistical **hypothesis testing,** which is a process for rendering a decision about the unknown value of a population parameter. We will learn how to make decisions about the value of a population mean, and we will examine different types of errors that can be made. We will learn how to conduct hypothesis tests about the value of a population proportion. Finally, we will perform hypothesis tests about the population standard deviation.

In Chapter 10, we will begin discussion of further topics in statistical inference with a look at two-sample problems. We will continue to apply both confidence interval estimation and hypothesis-testing methodology throughout the remaining chapters.

## 9.1 Introduction to Hypothesis Testing

**Objectives:** By the end of this section, I will be able to…

1. Construct the null hypothesis and the alternative hypothesis from the statement of the problem.
2. Explain the role of chance variation in establishing reasonable doubt.
3. State the two types of errors made in hypothesis tests: the Type I error, made with probability $\alpha$, and the Type II error, made with probability $\beta$.

### 1 Constructing the Hypotheses

Recall that inferential statistics refers to methods for learning about the unknown value of a population parameter by studying the information in a sample. The basic idea of hypothesis testing is the following:

1. We need to make a *decision* about the value of a population parameter (such as $\mu$ or $p$).
2. Unfortunately, the true value of that parameter is *unknown*.
3. Therefore, there may be different *hypotheses* about the true value of this parameter.

Statistical hypothesis testing is a way of formalizing the decision-making process so that a decision can be rendered about the value of the parameter. **Hypothesis testing** is a procedure where claims about the value of a population parameter may be investigated using the sample evidence. We craft *two competing statements (hypotheses)* about the value of the population parameter (either $\mu$, $p$, or $\sigma$). The two hypotheses cover all possible values of the parameter, and they cannot both be correct. We must find evidence to conclude that one of the hypotheses is likely to be true.

**The Hypotheses**

- The status quo hypothesis represents what has been tentatively assumed about the value of the parameter and is called the **null hypothesis,** denoted as $H_0$.
- The **alternative hypothesis,** or **research hypothesis,** denoted as $H_a$, represents an alternative claim about the value of the parameter.

Hypothesis testing is like conducting a criminal trial. In a trial in the United States, the defendant is innocent until proven guilty, and the jury must evaluate the truth of two competing hypotheses:

$H_0$: defendant is not guilty versus $H_a$: defendant is guilty

The not-guilty hypothesis is considered the **null hypothesis $H_0$** because the jurors must assume it is true until proven otherwise. The **alternative hypothesis $H_a$**, that the defendant is guilty, *must be demonstrated* to be true, beyond a reasonable doubt. How does a court of law determine whether the defendant is convicted or acquitted? This judgment is based upon the *evidence,* the hard facts heard in court. Similarly, in hypothesis testing, the researcher draws a conclusion based on the evidence provided by the sample data.

In Sections 9.1–9.4, we will examine hypotheses for the unknown mean $\mu$. The null hypothesis will be a claim about certain values for $\mu$, and the alternative hypothesis will be a claim about other values for $\mu$. We will begin by considering hypotheses about the population mean $\mu$ for some fixed-value $\mu_0$ (that is, $\mu_0$ is just a specified value for the unknown mean $\mu$). The hypotheses have one of the three possible forms shown in Table 9.1. The right-tailed test and the left-tailed test are called one-tailed tests. In Section 9.3 we will find out why we use this terminology.

Table 9.1 The three possible forms for the hypotheses for a test for $\mu$

| **Form** | **Null and alternative hypotheses** |
|---|---|
| Right-tailed test | $H_0$: $\mu \le \mu_0$ versus $H_a$: $\mu > \mu_0$ |
| Left-tailed test | $H_0$: $\mu \ge \mu_0$ versus $H_a$: $\mu < \mu_0$ |
| Two-tailed test | $H_0$: $\mu = \mu_0$ versus $H_a$: $\mu \ne \mu_0$ |

## Example 9.1 Invasive species in the Great Lakes

Edward Kinsman/Photo Researchers

Invasive species are a growing problem in the Great Lakes. Transoceanic ships from around the world discharge their ballast tanks into the Great Lakes. These tanks carry invasive species non-native to North America. The round goby is an invasive species that carries the parasite *Neochasmus umbellus.* A 2007 study reported that the mean intensity (mean number of parasites per infected host) was 7.2 parasites.[1] Scientists worry that the population mean number of parasites in the round goby has *increased* since 2007 and want to test that hypothesis. Scientists would prefer that the population mean number of parasites were some number at or below 7.2 parasites per host. Construct the null and alternative hypotheses.

### Solution

We can write this situation as two hypotheses about the mean number of parasites. Note that the term *increased* means *greater than.* Thus, we write a null hypothesis and an alternative hypothesis for a right-tailed test:

$$H_0: \mu \le 7.2 \quad \text{versus} \quad H_a: \mu > 7.2$$

The null hypothesis $H_0$ states that the population mean $\mu$ is *at most* 7.2 parasites per fish. The alternative hypothesis $H_a$ states that the population mean number of parasites is *more than* 7.2. Here, $\mu_0 = 7.2$, which is the possible value of $\mu$ specified in the example. Figure 9.2 shows that $H_0$ and $H_a$ cover all possible values of $\mu$. ■

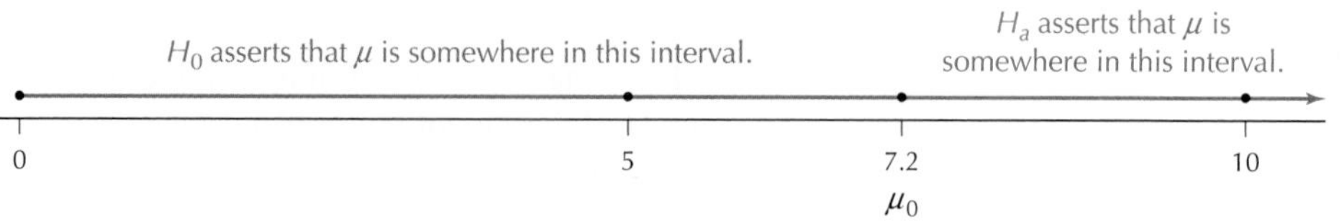
$H_0$ asserts that $\mu$ is somewhere in this interval.
$H_a$ asserts that $\mu$ is somewhere in this interval.
0
5
7.2
$\mu_0$
10

FIGURE 9.2 The hypotheses divide all possible values of $\mu$ between them.

Recall that each hypothesis is a claim that the statement is true. We must find evidence that is consistent with one of the two claims $H_0$ or $H_a$. The first task in hypothesis testing is to form hypotheses. To convert a word problem into two hypotheses, look for certain key words that can be expressed mathematically. Table 9.2 shows how to convert words typically found in word problems into symbols.

Table 9.2 Key English words, with mathematical symbols and synonyms

| English words | Symbols | Synonyms |
|---|---|---|
| Equal | $=$ | Is; is the same as |
| Not equal | $\neq$ | Is different from; has changed from; differs from |
| Greater than | $>$ | Is more than; is larger than; exceeds |
| Less than | $<$ | Is below; is smaller than |
| At least | $\geq$ | Is this much or more; is greater than or equal to |
| At most | $\leq$ | Is this much or less; is less than or equal to |

Once you have identified the key words, use the associated mathematical symbol to write two hypotheses. The following strategies can be used to write the hypotheses.

**Strategy for Constructing the Hypotheses About $\mu$**

***Step 1*** Search the word problem for certain key English words and select the associated symbol from Table 9.2.

***Step 2*** Determine the form of the hypotheses listed in Table 9.1 that uses this symbol.

***Step 3*** Find the value of $\mu_0$ (the number that answers the question: "greater than what?" or "less than what?") and write your hypotheses in the appropriate forms.

## Example 9.2 Has mean gas mileage continued to fall?

According to the U.S. Department of Transportation, the mean gas mileage of U.S. passenger cars (exclusive of SUVs, trucks, and vans) decreased from 21.2 miles per gallon (mpg) in 2001 to 20.8 mpg in 2004. Has this downward trend continued until today? Economists are interested in testing whether the population mean gas mileage for U.S. passenger cars has decreased from the mean of 20.8 mpg in 2004. Write a null hypothesis and an alternative hypothesis that describe this situation.

### Solution

Let's use our strategy to construct the hypotheses needed to test this claim.

***Step 1*** **Search the word problem for certain key English words and select the appropriate symbol.** The problem uses the words "decreased from," which means "less than." Thus, we will write a hypothesis that contains the $<$ symbol.

***Step 2*** **Determine the form of the hypotheses.** Since each symbol appears only once in Table 9.1, we see that the symbol $<$ means that we use a left-tailed test:

$$H_0: \mu \geq \mu_0 \quad \text{versus} \quad H_a: \mu < \mu_0$$

**Step 3** **Find the value for $\mu_0$ and write your hypotheses.** The alternative hypothesis $H_a$ states that the mean gas mileage $\mu$ is less than some value $\mu_0$. Less than what? The 20.8 mpg from 2004. Write the two hypotheses with $\mu_0 = 20.8$.

$$H_0: \mu \geq 20.8 \quad \text{versus} \quad H_a: \mu < 20.8$$

Do not blindly apply this strategy without thinking about what you are doing. Rather, use the strategy to help formulate your own hypotheses. *There is no substitute for thinking through the problem!*

## 2 Significant Difference or Chance Variation?

In hypothesis testing for $\mu$, you find the mean of the sample data, and compare it with the population mean that you would expect if the null hypothesis were true. If the difference between what you observe and what you expected is extreme enough, this difference may represent evidence against the null hypothesis. The question is, what is extreme? How do we know whether this difference is **statistically significant** or simply due to chance variation?

**Statistical Significance**

A result is said to be **statistically significant** if it is unlikely to have occurred due to chance.

### Example 9.3 Does the new medication have side effects?

Suppose that you are a researcher for a pharmaceutical research company. You are investigating the side effects of a new cholesterol-lowering medication and would like to determine whether the medication will decrease the population mean systolic blood pressure level from the current mean of 110. If so, then a warning will have to be given not to prescribe the new medication to patients whose blood pressure is already low. Construct the appropriate hypotheses.

**Solution**

We apply our strategy for constructing hypotheses about $\mu$.

**Step 1** **Find the key word.** Note that we want to determine whether the medication will *decrease* blood pressure. The key word, "decrease," is a synonym for "less than," <.

**Step 2** **Determine the form of the hypotheses.** The symbol < indicates that we should use the hypotheses for the left-tailed test in Table 9.1.

$$H_0: \mu \geq \mu_0 \quad \text{versus} \quad H_a: \mu < \mu_0$$

**Step 3** **Find the value for $\mu_0$ and write your hypotheses.** Asking "less than what?" (or "decreased from what?"), we see that $\mu_0 = 110$, so the hypotheses are

$$H_0: \mu \geq 110 \quad \text{versus} \quad H_a: \mu < 110$$

To determine which of the hypotheses in Example 9.3 is correct, we take a sample of randomly selected patients who are taking the medication. We record their systolic blood pressure levels and calculate the sample mean $\bar{x}$ and sample standard deviation $s$. Most likely, the mean of this sample of patients' systolic blood pressure levels will not be exactly equal to 110. Recall that the sample means vary about the population mean. So, even if the null hypothesis is true, the mean of the sample $\bar{x}$ will not, in most cases, be exactly equal to $\mu = 110$.

Now, suppose that the sample mean blood pressure $\bar{x}$ is less than the hypothesized population mean of 110. *Is the difference due simply to chance variation, or is it evidence of a side effect of the cholesterol medication?* Let's consider some possible values for $\bar{x}$:

- $\bar{x} = 109$: The difference between $\bar{x}$ and $\mu = 110$ is only $|109 - 110| = 1$. Depending on the variability present in the sample, the researcher would probably conclude that this small difference is due to chance variation. Thus, the researcher would not reject the null hypothesis, because the value of $\bar{x}$ is not evidence that the null hypothesis $\mu \geq 110$ is not true. The result is not statistically significant.
- $\bar{x} = 90$: The difference between $\bar{x}$ and $\mu = 110$ is $|90 - 110| = 20$. Depending on the variability present in the sample, the researcher would probably conclude that this difference is so large that it is unlikely that it is due to chance variation alone. Thus, the researcher would reject the null hypothesis (that $\mu \geq 110$) in favor of the alternative hypothesis (that $\mu < 110$). The result is statistically significant.

To summarize: *in a hypothesis test, we compare the sample mean with the value $\mu_0$ of the population mean used in the $H_0$ hypothesis. If the difference is large, then $H_0$ is rejected. If the difference is not large, then $H_0$ is not rejected.* The question is, "Where do you draw the line?" Just how large a difference is large enough? This is exactly the issue we will face throughout the remainder of Chapter 9 (actually, throughout the rest of the book).

## 3 Type I and Type II Errors

Next, we take a closer look at some of the thorny issues involved in performing a hypothesis test. Let's return to the example of a criminal trial. The jury will convict the defendant if they find evidence compelling enough to reject the null hypothesis of "not guilty" *beyond a reasonable doubt.* However, jurors are only human; sometimes their decisions are correct and sometimes they are not. Thus, the jury's verdict will be one of the following outcomes:

Tetra/Getty Images

1. An innocent defendant is wrongfully convicted.
2. A guilty defendant is convicted.
3. A guilty defendant is wrongfully acquitted.
4. An innocent defendant is acquitted.

Recall that we can write the two hypotheses for a criminal trial as

$$H_0\text{: defendant is not guilty} \quad \text{versus} \quad H_a\text{: defendant is guilty}$$

Table 9.3 shows the four possible jury verdicts in terms of these hypotheses. The possible verdicts are in the left-hand column, and the two hypotheses appear in the headings of the middle and right-hand columns.

Table 9.3 Four possible outcomes of a criminal trial

| **Jury's decision** | **Reality** $H_0$ true: Defendant did not commit the crime | $H_0$ false: Defendant did commit the crime |
|---|---|---|
| Reject $H_0$: Find defendant guilty | Type I error | Correct decision |
| Do not reject $H_0$: Find defendant not guilty | Correct decision | Type II error |

Let's look at the two possible decisions the jury can make. It can find the defendant guilty. In this case, the jury is rejecting what is stated in the null hypothesis $H_0$, so we say that the jury "rejects $H_0$." On the other hand, the jury can find the defendant not guilty. In this case, the jury does not reject the null hypothesis $H_0$. There are two ways for the jury to render the *correct decision*. (Note that in Table 9.3 a hypothesis and decision with the same color lead to correct decisions.)

**Two Ways of Making the Correct Decision**

- To not reject $H_0$ when $H_0$ is true. Example: To find the defendant not guilty when in reality he did not commit the crime.
- To reject $H_0$ when $H_0$ is false. Example: To find the defendant guilty when in reality he did commit the crime.

Unfortunately, there are also two ways for the jury to render an incorrect decision. Judges, juries, and researchers are just people, and all can make mistakes. In statistics, the two incorrect decisions are called Type I and Type II errors.

**Two Types of Errors**

- **Type I error**: To reject $H_0$ when $H_0$ is true. Example: To find the defendant guilty when in reality he did not commit the crime.
- **Type II error**: To not reject $H_0$ when $H_0$ is false. Example: To find the defendant not guilty when in reality he did commit the crime.

Note that in Table 9.3 a decision and hypothesis with different colors lead to either a Type I or a Type II error. In the courtroom, errors can be made because of prejudice, legal ineptitude, or some legal technicality. In statistics, errors in decisions can be made because of the *pervasive presence of chance and sampling variability*.

### A Decision Is Not Proof

*It is important to understand that the decision to reject or not reject $H_0$ does not prove anything.* The decision represents whether or not there is sufficient evidence against the null hypothesis. This is our best judgment given the data available. You cannot claim to have *proven* anything about the value of a population parameter unless you elicit information from the entire population, which in most cases is not possible. ■

How can we make decisions about population parameters using the limited information available in a sample? The answer is that we base our decisions on *probability.* When the difference between the sample mean $\bar{x}$ and the hypothesized population mean $\mu_0$ is large, then the null hypothesis is *probably* not correct. When the difference is small, then the data are *probably* consistent with the null hypothesis. But we don't know for sure.

The probability of a Type I error is denoted as **α (alpha).** We set the value of $\alpha$ to be some small constant, such as 0.01, 0.05, or 0.10, so that there is only a small probability of rejecting a true null hypothesis. To say that $\alpha = 0.05$ means that, if this hypothesis test were repeated over and over again, the long-term probability of rejecting a true null hypothesis would be 5%. The **level of significance** of a hypothesis test is another name for $\alpha$, the probability of rejecting $H_0$ when $H_0$ is true. A smaller $\alpha$ makes it harder to wrongfully reject $H_0$ just by chance. If the consequences of making a Type I error are serious, then the level of significance should be small, such as $\alpha = 0.01$. If the consequences of making a Type I error are not so serious, then one may choose a larger value for the level of significance, such as $\alpha = 0.05$ or $\alpha = 0.10$.

The probability of a Type II error is denoted as **β (beta).** This is the probability of not rejecting $H_0$ when $H_0$ is false, such as acquitting someone who is really guilty. Making $\alpha$ smaller inevitably makes $\beta$ larger (for a fixed sample size). Of course, our goal is to simultaneously minimize both $\alpha$ and $\beta$. Unfortunately, the only way to do this is to increase the sample size.

*There are only two possible hypothesis-testing conclusions:*

- Reject $H_0$, or
- Do not reject $H_0$.

*Note:* When we reject $H_0$, we say that the results are statistically significant. If we do not reject $H_0$, the results are not statistically significant.

The material in this chapter is challenging but well worth the effort. You are learning critical thinking through statistics. What you are learning is a bedrock component of the scientific method, which is the backbone of scientific research all over the planet.

## Section 9.1 SUMMARY

1 Statistical hypothesis testing is a way of formalizing the decision-making process so that a decision can be rendered about the unknown value of the parameter. The status quo hypothesis that represents what has been tentatively assumed about the value of the parameter is called the null hypothesis and is denoted as $H_0$. The alternative hypothesis, or research hypothesis, denoted as $H_a$, represents an alternative conjecture about the value of the parameter.

2 In a hypothesis test, we compare the sample mean $\bar{x}$ with the value $\mu_0$ of the population mean used in the $H_0$ hypothesis. If the difference is large, then $H_0$ is rejected. If the difference is not large, then $H_0$ is not rejected.

3 When performing a hypothesis test, there are two ways of making a correct decision: to not reject $H_0$ when $H_0$ is true and to reject $H_0$ when $H_0$ is false. Also, there are two types of error: a Type I error is to reject $H_0$ when $H_0$ is true, and a Type II error is to not reject $H_0$ when $H_0$ is false. The probability of a Type I error is denoted as $\alpha$ (alpha). The probability of a Type II error is denoted as $\beta$ (beta).

## Section 9.1 EXERCISES

### CLARIFYING THE CONCEPTS

1. Explain in your own words why we need hypothesis testing.
2. What are some characteristics of the null hypothesis?
3. What are some characteristics of the alternative hypothesis?
4. Explain what is meant by $\mu_0$.
5. In the hypothesis test for the population mean, how many forms of the hypotheses are there? Write out these forms.

**6.** Say we are interested in testing whether the population mean is less than 100, and the sample we take yields a mean of 90. Is this sufficient evidence that the population mean is less than 100? Explain clearly why or why not.
**7.** In a criminal trial, what are the two possible decision errors? What do statisticians call these errors?
**8.** Do hypothesis tests *prove* anything? Why or why not?

## PRACTICING THE TECHNIQUES

For Exercises 9–16, provide the null and alternative hypotheses.
**9.** Test whether $\mu$ is greater than 10.
**10.** Test whether $\mu$ is less than 100.
**11.** Test whether $\mu$ is different from 0.
**12.** Test whether $\mu$ is at least 40.
**13.** Test whether $\mu$ equals 4.0.
**14.** Test whether $\mu$ is at most 25.
**15.** Test whether $\mu$ has changed from 36.
**16.** Test whether $\mu$ exceeds $-4$.

For Exercises 17–21, do the following.
- **a.** Provide the null and alternative hypotheses.
- **b.** Determine if a correct decision has been made. If an error has been made, indicate which type of error.

**17. Child Abuse.** The U.S. Administration for Children and Families reported that the national rate for child abuse referrals was 43.9 per 1000 children in 2005. A hypothesis test was carried out that tested whether the population mean referral rate had increased this year from the 2005 level. The null hypothesis was not rejected. Suppose that, in actuality, the population mean child abuse referral rate for this year is 45 per 1000 children.
**18. California Warming.** *Science Daily* reported in 2007 that the mean temperature in California has increased from 1950 to 2000 by 2 degrees Fahrenheit (°F).[2] Suppose that we perform a hypothesis test today to determine whether the population mean temperature increase exceeds two degrees, and suppose that the actual population mean temperature increase was three degrees.
**19. Travel Costs.** In 2007, a motorists' guide reported that travel costs had increased over the previous two years. Suppose that this report was based on a hypothesis test and that in actuality the population mean travel costs were lower.
**20. Eating Trends.** According to the NPD Group, higher gasoline prices are causing consumers to go out to eat less and eat at home more.[3] Suppose that this report found that the mean number of meals prepared and eaten at home is less than 700 per year, and that in actuality the population mean number of such meals is 600.
**21. Hybrid Vehicles.** A 2006 study by **Edmunds.com** showed that owners of hybrid vehicles can recoup their increased initial cost through reduced fuel consumption in at most three years. Suppose that this report was based on a hypothesis test and that in actuality the population mean number of years it takes to recoup their initial cost is two years.

## APPLYING THE CONCEPTS

For Exercises 22–27, do the following.
- **a.** Provide the null and alternative hypotheses.
- **b.** Describe the two ways a correct decision could be made.
- **c.** Describe what a Type I error would mean in the context of the problem.
- **d.** Describe what a Type II error would mean in the context of the problem.

**22. Shares Traded on the Stock Market.** The *Statistical Abstract of the United States* reports that the mean daily number of shares traded on the New York Stock Exchange in 2005 was 1.602 billion. Based on a sample of this year's trading results, a financial analyst would like to test whether the mean number of shares traded will be larger than the 2005 level.
**23. Traffic Light Cameras.** The Ministry of Transportation in the province of Ontario reported in 2003 that the installation of cameras that take pictures at traffic lights has decreased the mean number of fatal and injury collisions to 339.1 per year. A hypothesis test was performed this year to determine whether the population mean number of such collisions has changed since 2003.
**24. Price of Milk.** The Bureau of Labor Statistics reports that the mean price for a gallon of milk in 2005 was \$3.24. Suppose that we conduct a hypothesis test to investigate if the population mean price of milk this year has increased.
**25. Americans' Height.** Americans used to be on average the tallest people in the world. That is no longer the case, according to a study by Dr. Richard Steckel, professor of economics and anthropology at The Ohio State University. The Norwegians and Dutch are now the tallest, at 178 centimeters, followed by the Swedes at 177, and then the Americans, with a mean height of 175 centimeters (approximately 5 feet 9 inches). According to Dr. Steckel, "The average height of Americans has been pretty much stagnant for 25 years."[4] Suppose that we conduct a hypothesis test to investigate whether the population mean height of Americans this year has changed from 175 centimeters.
**26. Credit Score in Florida.** According to **CreditReport.com**, the mean credit score in Florida in 2006 was 673. Suppose that a hypothesis test was conducted to determine if the mean credit score in Florida has decreased since that time.
**27. Salary of College Grads.** According to the U.S. Census Bureau, the mean salary of college graduates in 2002 was \$52,200. Suppose that a hypothesis test was carried out to determine whether the population mean salary of college graduates has increased.

# 9.2 Z Test for the Population Mean $\mu$: p-Value Method

## Objectives: By the end of this section, I will be able to...

1. Explain the essential idea about hypothesis testing for the mean.
2. State what a $p$-value is and how to interpret it.
3. Perform and interpret the $Z$ test for the population mean using the $p$-value method.
4. Assess the strength of evidence against the null hypothesis.

## 1 The Essential Idea About Hypothesis Testing for the Mean

In Example 9.3, we compared the sample mean blood pressure level directly with the hypothesized population mean blood pressure level of 110. We indicated that a large difference between the two means could be taken as evidence of inconsistency between the data and the null hypothesis. However, in hypothesis testing, we do not base our decision solely on the size of the difference between the sample mean $\bar{x}$ and the hypothesized population mean $\mu_0$. We pointed out that the decision also depended on the variability present in the sample.

For example, suppose that you were conducting a hypothesis test about whether the mean price of a new-release music CD exceeded \$10:

$$H_0: \mu \leq 10 \quad \text{versus} \quad H_a: \mu > 10$$

A random sample of 36 new-release CDs yielded a mean price of \$15, as illustrated in Figure 9.3. Most new-release CDs cost about the same, so the variability in their prices is relatively small. Thus, the \$5 difference between the sample mean price of \$15 and the hypothesized mean price of \$10 can be considered large. The \$5 difference is *large compared with the variability in the sample.*

FIGURE 9.3

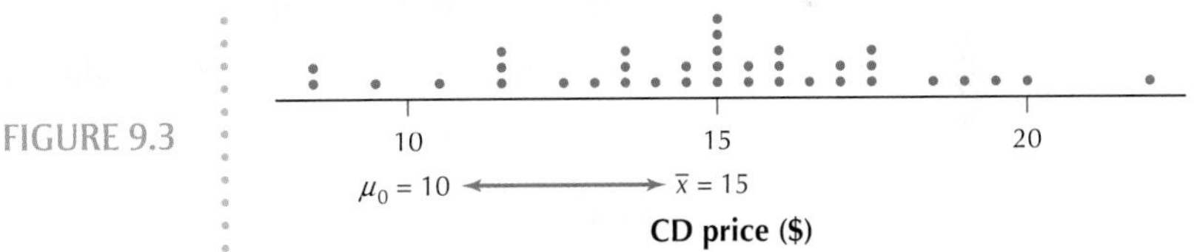


However, now suppose that you were conducting a hypothesis test about whether the mean cost of a certain surgical procedure was less than \$6000:

$$H_0: \mu \geq 6000 \quad \text{versus} \quad H_a: \mu < 6000$$

A sample of 36 surgical procedures yielded a sample mean cost of \$5995, as illustrated in Figure 9.4. However, the procedures varied in price by thousands of dollars. Hence, the \$5 difference between the sample mean cost of \$5995 and the hypothesized mean cost of \$6000 is *small compared with the variability in the sample.*

FIGURE 9.4

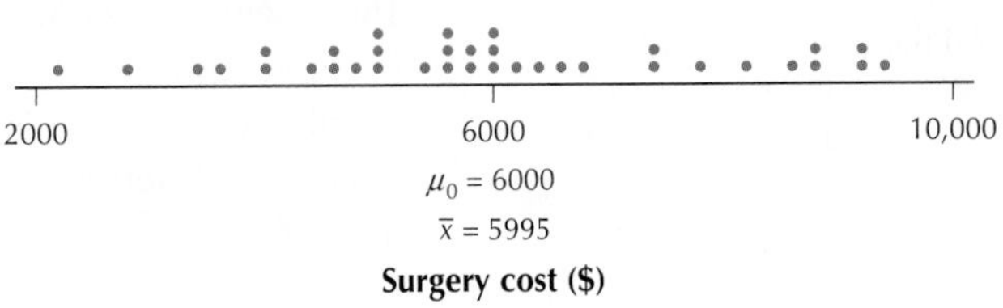

2000 6000 10,000

$\mu_0 = 6000$

$\bar{x} = 5995$

Surgery cost (\$)

To account for variability, we use the sampling distribution of the sample mean $\bar{x}$. You recall from Chapters 7 and 8 that the sampling distribution of the sample mean $\bar{x}$, for large samples, is approximately normal with mean $\mu_{\bar{x}} = \mu$ and standard deviation $\sigma_{\bar{x}} = \sigma/\sqrt{n}$. Now, if we assume $H_0$ is true, then the sampling distribution of the sample mean will have a mean of $\mu_{\bar{x}} = \mu_0$, the value of $\mu$ given in $H_0$. Figure 9.5 shows the sampling distribution of $\bar{x}$ under the assumption that the population mean equals $\mu_0$. Notice that the distribution of $\bar{x}$ is centered at $\mu_0$, the value of $\mu$ that is claimed in $H_0$.

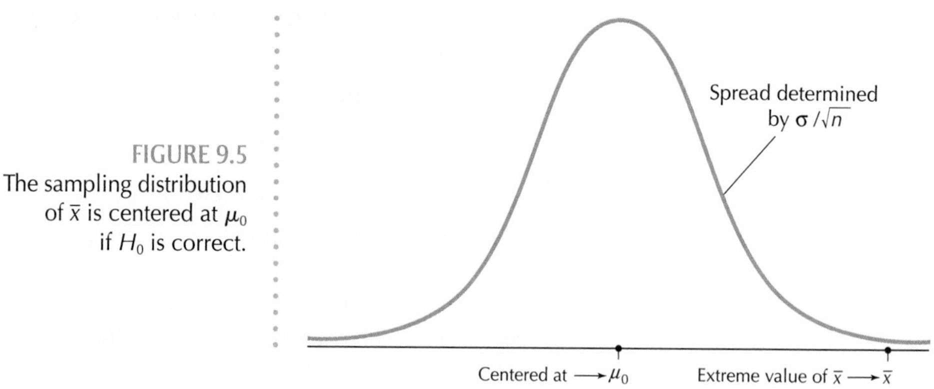


FIGURE 9.5
The sampling distribution of $\bar{x}$ is centered at $\mu_0$ if $H_0$ is correct.

Note, however, that we do not actually observe this distribution of $\bar{x}$. We observe only a single value of $\bar{x}$, the sample mean of the observed data set. We need to determine whether this value of $\bar{x}$ is *unusual or extreme,* according to this distribution. By *unusual or extreme,* we mean a value of $\bar{x}$ located in one of the tails of the distribution, far from the hypothesized mean $\mu_0$.

Suppose that our observed value $\bar{x}$ is indeed unusual or extreme. Then this indicates a conflict between what we hypothesize to be true about $\mu$ (that is, that $\mu = \mu_0$) and the actual sample data. In this case, we are faced with choosing between one of the following two scenarios:

1. $H_0$ is correct, the value of $\mu_0$ is accurate, and our observation of this unusual value of $\bar{x}$ is an amazingly unlikely event.
2. $H_0$ is not correct, and the true value of $\mu$ must be closer to $\bar{x}$.

*Developing Your Statistical Sense*

### The Data Prevail!

Since we are reluctant to base our scientific methodology on "amazingly unlikely events," we would therefore conclude that $H_0$ is not correct. Recall that the null hypothesis is simply a conjecture. On the other hand, the sample mean $\bar{x}$ represents data that are directly observable and not conjectural. The modern scientific method states that, in the face of a conflict between a conjecture and observed data, the data prevail and we need to rethink our null hypothesis. This leads us to the following statement of the **essential idea about hypothesis testing for the mean.** ■

*Note:* All the remaining parts of Sections 9.2–9.4, all the steps and all the calculations, are really explanations of how to implement this essential idea of hypothesis testing for the mean.

### The Essential Idea About Hypothesis Testing for the Mean

When the observed value of $\bar{x}$ is unusual or extreme in the sampling distribution of $\bar{x}$ that is based on the assumption that $H_0$ is correct, we should reject $H_0$. Otherwise, there is insufficient evidence against $H_0$, and we should not reject $H_0$.

## Example 9.4 An extreme value

Let us illustrate this key concept by revisiting our CD cost example. The hypotheses are

$$H_0: \mu \leq 10 \quad \text{versus} \quad H_a: \mu > 10$$

Specifically, the null hypothesis asserts that the mean cost of CDs is at most \$10. Suppose the population standard deviation is \$9. Show that the observed value of $\bar{x} = \$15$ is unusual if we assume that $H_0$ is correct.

### Solution

*Assuming that $H_0$ is correct,* samples of size $n = 36$ are approximately normally distributed, because $n$ is at least 30. The mean of the sampling distribution is \$10, with standard deviation $\sigma_{\bar{x}} = \sigma/\sqrt{n} = 9/\sqrt{36} = \$1.50$.

Figure 9.6 shows the sampling distribution of $\bar{x}$ if $H_0$ is correct. Note that the observed value of $\bar{x} = \$15$ is deep in the right-hand tail of the distribution, indicating that this is an unusual value, assuming $H_0$ is correct. Thus, the observed value $\bar{x} = \$15$ is not consistent with $H_0$: $\mu \leq 10$. Since we must base our conclusion on the evidence we uncovered, we reject the null hypothesis that $\mu \leq 10$. ■

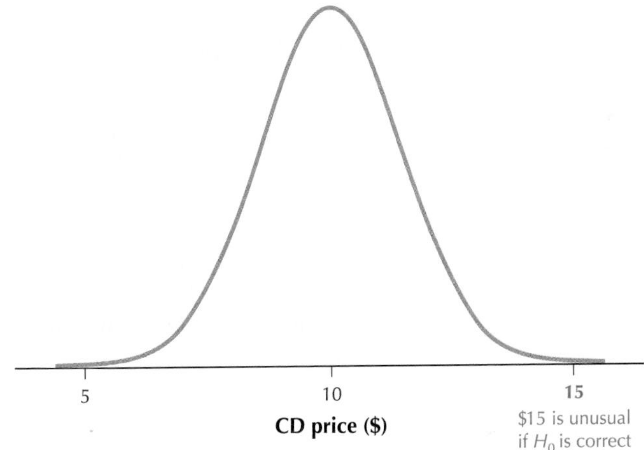


FIGURE 9.6 Sampling distribution of sample mean CD prices if $H_0$ is correct.

Of course, not all examples are so straightforward. Therefore, we need to find a way to determine in general whether or not we have an extreme value of $\bar{x}$. In Chapter 6, we were able to answer any normal distribution problem by standardizing. Here we again turn to standardizing to provide a general method to help us determine which sample means $\bar{x}$ are extreme. Recall that, if $H_0$ is correct, then, for large samples, the sampling distribution of $\bar{x}$ is approximately normal with mean $\mu_0$ and standard deviation $\sigma_{\bar{x}} = \sigma/\sqrt{n}$. Also recall from Chapter 6 that, if we standardize this *Normal* $(\mu_0, \sigma/\sqrt{n})$ distribution, we get the standard normal random variable $Z$:

$$Z = \frac{\bar{x} - \mu_0}{\sigma/\sqrt{n}}$$

However, as you may have guessed, this particular form of $Z$ is very special, and we give it a special name, $Z_{\text{data}}$.

*Note:* The test statistic is called $Z_{\text{data}}$ because its value is derived largely from the sample data and the population data. The sample mean $\bar{x}$ and the sample size $n$ are characteristics of the sample data, and the population standard deviation $\sigma$ is a characteristic of the population data.

**The Test Statistic $Z_{\text{data}}$**

The test statistic used for the $Z$ test for the mean is

$$Z_{\text{data}} = \frac{\bar{x} - \mu_0}{\sigma_{\bar{x}}} = \frac{\bar{x} - \mu_0}{\sigma/\sqrt{n}}$$

$Z_{\text{data}}$ represents how far the sample mean $\bar{x}$ lies above or below the hypothesized mean $\mu_0$, expressed in terms of the number of standard deviations of the sampling distribution of $\bar{x}$, $\sigma_{\bar{x}} = \sigma/\sqrt{n}$.

This $Z_{\text{data}}$ is an example of a **test statistic,** a statistic generated from a data set for the purposes of testing a statistical hypothesis. We will meet several other test statistics throughout the remainder of the text. This type of hypothesis test is called the *Z test* because the test statistic $Z_{\text{data}}$ comes from the standard normal $Z$ distribution.

## Example 9.5 Finding $Z_{\text{data}}$ for the price of CDs

Calculate $Z_{\text{data}}$ for the data provided in Example 9.4.

**Solution**

We have the hypothesized mean $\mu_0 = \$10$, $\bar{x} = \$15$, and $\sigma_{\bar{x}} = \sigma/\sqrt{n} = 9/\sqrt{36} = \$1.50$. Thus,

$$Z_{\text{data}} = \frac{\bar{x} - \mu_0}{\sigma_{\bar{x}}} = \frac{\bar{x} - \mu_0}{\sigma/\sqrt{n}} = \frac{15 - 10}{9/\sqrt{36}} = \frac{5}{1.5} \approx 3.33$$

Extreme values of $\bar{x}$, that is, values of $\bar{x}$ that are far from the hypothesized $\mu_0$, will translate into extreme values of $Z_{\text{data}}$. In other words, when $\bar{x}$ is far from $\mu_0$ (as in Figure 9.6), $Z_{\text{data}}$ will be far from 0, as shown in Figure 9.7. We interpret $Z_{\text{data}} = 3.33$ to mean that the sample mean $\bar{x}$ lies 3.33 standard deviations above the mean $\mu_0 = \$10$ in the sampling distribution of $\bar{x}$. ■

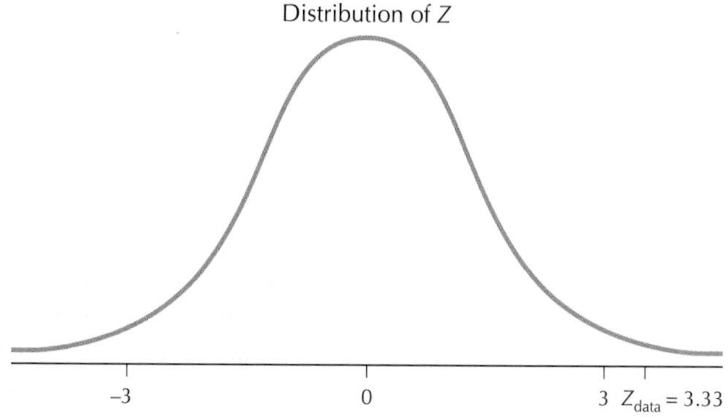


FIGURE 9.7

## 2 p-Values

The two main methods for carrying out a hypothesis test are

- the $p$-value method, which we will learn in this section, and
- the critical-value method, which we will learn in Section 9.3.

The two methods are in fact equivalent for the same level of $\alpha$. That is, they will give you the same conclusion. The ***p*-value** is a measure of how well (or how poorly) the data fit the null hypothesis. For this reason, the $p$-value method is more widespread in the real world.

**p-Value**

The ***p*-value** is the probability of observing a sample statistic (such as $\bar{x}$ or $Z_{data}$) at least as extreme as the statistic actually observed if we assume that the null hypothesis is true. Roughly speaking, the *p*-value represents the probability of observing the sample statistic if the null hypothesis is true. Since the term "*p*-value" means "probability value," its value must always lie between 0 and 1.

There are three possible cases for the *p*-value, summarized in Table 9.4. Note that the method for calculating the *p*-value depends on the form of the hypothesis test.

Table 9.4 Finding the *p*-value depends on the form of the hypothesis test

| Type of hypothesis test | *p*-Value is tail area associated with $Z_{data}$ |
|---|---|
| **Right-tailed test**<br>$H_0: \mu \le \mu_0$ versus $H_a: \mu > \mu_0$<br>*p*-value $= P(Z > Z_{data})$<br>Area to right of $Z_{data}$ | 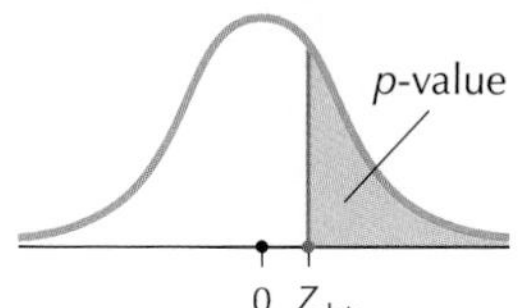<br> |
| **Left-tailed test**<br>$H_0: \mu \ge \mu_0$ versus $H_a: \mu < \mu_0$<br>*p*-value $= P(Z < Z_{data})$<br>Area to left of $Z_{data}$ | 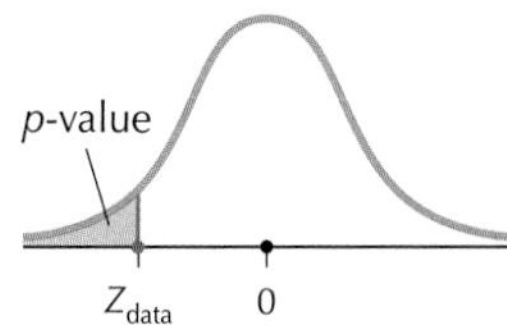<br> |
| **Two-tailed test**<br>$H_0: \mu = \mu_0$ versus $H_a: \mu \ne \mu_0$<br>*p*-value $= P(Z > \lvert Z_{data}\rvert) + P(Z < -\lvert Z_{data}\rvert)$<br>$= 2 \cdot P(Z > \lvert Z_{data}\rvert)$<br>Sum of the two tail areas. | 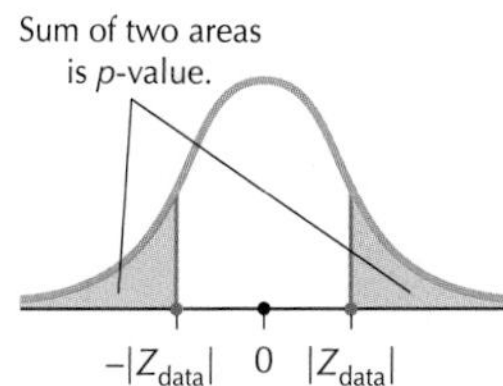<br> |

## Example 9.6 *p*-value for the price of CDs

Find the *p*-value for Example 9.5.

**Solution**

Since we have a right-tailed test,

$$H_0: \mu \le 10 \quad \text{versus} \quad H_a: \mu > 10$$

and $Z_{data} = 3.33$, we have

$$p\text{-value} = P(Z > Z_{data}) = P(Z > 3.33)$$

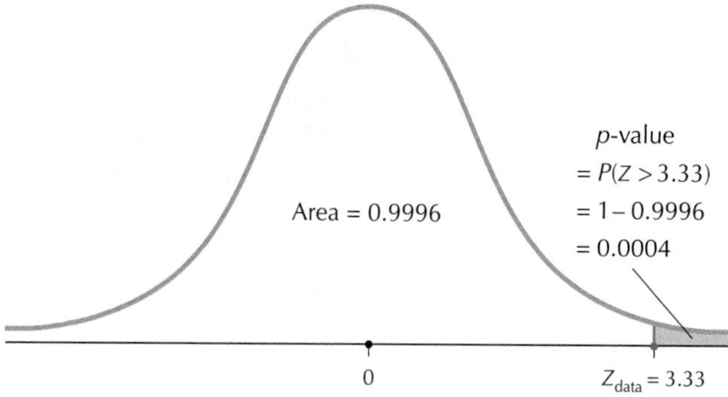


FIGURE 9.8

as shown in Figure 9.8. The $Z$ table tells us that the area to the left of $Z_{\text{data}} = 3.33$ is 0.9996. Thus, using Case 2 from Table 6.7 in Section 6.4 (page 298), we find the area to the right of $Z_{\text{data}} = 3.33$ to be

$$p\text{-value} = 1 - 0.9996 = 0.0004$$

Alternatively, we can use the TI-83/84 to help us calculate the $p$-value. Press **2nd DISTR** then **2: normalcdf(-1E99, 3.33)** (Figure 9.9). The TI-83/84 provides the area to the left of $Z = 3.33$ as 0.9995657137 (Figure 9.10). Thus,

$$\begin{aligned} p\text{-value} &= P(Z > Z_{\text{data}}) = P(Z > 3.33) \\ &= 1 - 0.9995657137 = 0.0004342863 \approx 0.0004 \end{aligned}$$

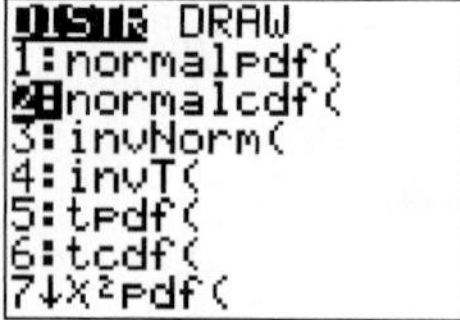


FIGURE 9.9

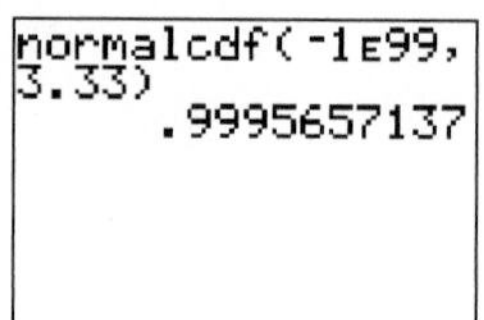


FIGURE 9.10

Unusual and extreme values of $\bar{x}$, and therefore of $Z_{\text{data}}$, will have a small $p$-value, while values of $\bar{x}$ and $Z_{\text{data}}$ nearer to the center of the distribution will have a large $p$-value.

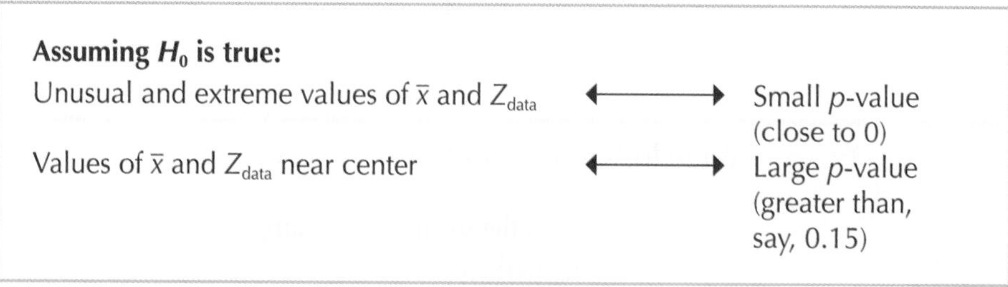

**Assuming $H_0$ is true:**

| | | |
|---|---|---|
| Unusual and extreme values of $\bar{x}$ and $Z_{\text{data}}$ | ⟷ | Small $p$-value (close to 0) |
| Values of $\bar{x}$ and $Z_{\text{data}}$ near center | ⟷ | Large $p$-value (greater than, say, 0.15) |

Keep in mind that *the p-value is a probability.* In fact, the term "$p$-value" actually means "probability value." Since it is a probability, the $p$-value must always be between 0 and 1. A small $p$-value indicates a conflict between your sample data and the null hypothesis. Therefore, since the data are real and the hypothesis is a conjecture, this *small p-value will lead us to reject $H_0$.*

However, *how small is small?* We have already met another fellow in this chapter, a type of probability called $\alpha$, the probability of a Type I error. (Recall that a Type I error is rejecting $H_0$ when $H_0$ is true.) The researcher chooses $\alpha$ to be some small constant like 0.01, 0.05, or 0.10 and then compares the $p$-value with $\alpha$. A $p$-value is small if it is smaller than $\alpha$. This leads us to the **rejection rule** that tells us the conditions under which we may reject the null hypothesis. In fact, this rejection rule can be applied to any type of $p$-value hypothesis test we perform.

> The rejection rule for performing a hypothesis test using the $p$-value method is
> **Reject $H_0$ when the $p$-value is less than $\alpha$.**

Note that this rejection rule calls for a new role for $\alpha$.

> **Level of Significance, $\alpha$**
> The value of $\alpha$ represents the boundary between results that are statistically significant (where we reject $H_0$) and results that are not statistically significant (where we do not reject $H_0$). Thus, $\alpha$ is called the level of significance of the hypothesis test.

## Example 9.7 Compare $p$-value with $\alpha$ to draw conclusion

Suppose we let our level of significance be $\alpha = 0.05$ for the hypothesis test in Example 9.6. What is our conclusion?

### Solution

#### What Results Might We Expect?

The value of the test statistic $Z_{\text{data}} = 3.33$ was shown to lie deep in the right-hand tail of Figure 9.8. This is evidence that our observed sample mean $\bar{x} = \$15$ is unusual or extreme if we assume that $H_0$ is correct. Thus, by the "essential idea about hypothesis testing for the mean," we expect our conclusion to be "Reject $H_0$." ■

Our hypothesis test is

$$H_0\colon \mu \leq 10 \quad \text{versus} \quad H_a\colon \mu > 10$$

where $\mu$ represents the population mean cost of a CD. Since $\alpha = 0.05$, we will reject $H_0$ if $p$-value $< 0.05$. Now, in Example 9.6, we found $p$-value $= 0.0004$. Since 0.0004 is less than our level of significance $\alpha = 0.05$, our conclusion is therefore to reject $H_0$, as expected. The difference between $\bar{x} = \$15$ and $\mu_0 = \$10$ is statistically significant. ■

#### What Does This Conclusion Mean?

#### Interpreting Your Conclusion for Nonspecialists

Recall that a data analyst needs to interpret the results so that nonspecialists can understand them. You can use the following generic interpretation for the two possible conclusions. Just remember that generic interpretations are no substitute for thinking clearly about the problem and the implications of the conclusion. ■

**Interpreting the Conclusion**

- If you reject $H_0$, the interpretation is *"There is evidence that [whatever $H_a$ says]."*
- If you do not reject $H_0$, the interpretation is *"There is insufficient evidence that [whatever $H_a$ says]".*

## Example 9.8 Interpreting the conclusion

Interpret the conclusion for the hypothesis test in Example 9.7.

**Solution**

Our hypothesis test is

$$H_0: \mu \leq 10 \quad \text{versus} \quad H_a: \mu > 10$$

where $\mu$ represents the population mean cost of a CD. Since the conclusion for the hypothesis test in Example 9.7 was to reject $H_0$, the generic interpretation is: "There is evidence that [whatever $H_a$ says]." Here, $H_a$ states that $\mu > 10$; that is, the population mean cost of a CD is greater than \$10. Hence, we interpret our conclusion as follows: "There is evidence that the population mean cost of a CD is greater than \$10."

## 3 Performing the *Z* test for the Population Mean Using the *p*-Value Method

We summarize the methods used in Examples 9.4–9.8 in the following step-by-step method for performing the *Z* test for $\mu$ using the *p*-value method.

**Z Test for the Population Mean $\mu$: p-Value Method**

When a random sample of size *n* is taken from a population where the population standard deviation $\sigma$ is known, you can use the *Z* test if either of the following conditions is satisfied:

- Case 1: the population is normal, or
- Case 2: the sample size is large ($n \geq 30$).

***Step 1*** **State the hypotheses and the rejection rule.** Use one of the forms from Table 9.1 to write $H_0$ and $H_a$. Define $\mu$ (that is, "$\mu$ represents the population mean ____"). The rejection rule is "Reject $H_0$ if the *p*-value is less than $\alpha$."

***Step 2*** **Find $Z_{data}$.** Either use technology to find the value of the test statistic $Z_{data}$, or calculate the value of $Z_{data}$ as follows:

$$Z_{data} = \frac{\bar{x} - \mu_0}{\sigma_{\bar{x}}} = \frac{\bar{x} - \mu_0}{\sigma/\sqrt{n}}$$

***Step 3*** **Find the *p*-value.** Either use technology to find the *p*-value, or calculate it using the form in Table 9.4 that corresponds to your hypotheses.

***Step 4*** **State the conclusion and interpretation.** If the *p*-value is less than $\alpha$, then reject $H_0$. Otherwise, do not reject $H_0$. Interpret your conclusion so that a nonspecialist can understand.

## Example 9.9 The $Z$ test for the mean using the $p$-value method: left-tailed test

The technology Web site **www.cnet.com** publishes user reviews of computers, software, and other electronic gadgetry. The mean user rating, on a scale of 1–10, for the Dell XPS 410 desktop computer as of September 10, 2007, was 7.2. Assume that the population standard deviation of user ratings is known to be $\sigma = 0.9$. A random sample taken this year of $n = 81$ user ratings for the Dell XPS 410 showed a mean of $\bar{x} = 7.05$. Using level of significance $\alpha = 0.05$, test whether the population mean user rating for this computer has fallen since 2007.

### Solution

The sample size $n = 81$ is large, and the population standard deviation is known. We may therefore perform the $Z$ test for the mean.

***Step 1*** **State the hypotheses and the rejection rule.** The key words here are "has fallen," which means "is less than." The only form of the test with the < sign in it is the left-tailed test (from Table 9.1):

$$H_0: \mu \geq \mu_0 \quad \text{versus} \quad H_a: \mu < \mu_0$$

We find the value of $\mu_0$ by asking, "Is less than what?" The answer is, "Less than the mean user rating from 2007, which was 7.2." Thus, $\mu_0 = 7.2$, and our hypotheses are

$$H_0: \mu \geq 7.2 \quad \text{versus} \quad H_a: \mu < 7.2$$

where $\mu$ refers to the population mean user rating for the Dell XPS 410 computer. We will reject $H_0$ if the $p$-value is less than $\alpha = 0.05$.

***Step 2*** **Find $Z_{data}$.** We have $\bar{x} = 7.05$, $\mu_0 = 7.2$, $n = 81$, and $\sigma = 0.9$. Thus, our test statistic is

$$Z_{data} = \frac{\bar{x} - \mu_0}{\sigma_{\bar{x}}} = \frac{\bar{x} - \mu_0}{\sigma/\sqrt{n}} = \frac{7.05 - 7.2}{0.9/\sqrt{81}} = \frac{-0.15}{0.1} = -1.5$$

***Step 3*** **Find the $p$-value.** Our hypotheses represent a left-tailed test from Table 9.4. Thus,

$$p\text{-value} = P(Z < Z_{data}) = P(Z < -1.5)$$

This is a Case 1 problem from Table 6.7 (page 298). The $Z$ table provides us with the area to the left of $Z = -1.5$ (Figure 9.11):

$$P(Z < -1.5) = 0.0668$$

Thus, the $p$-value is 0.0668.

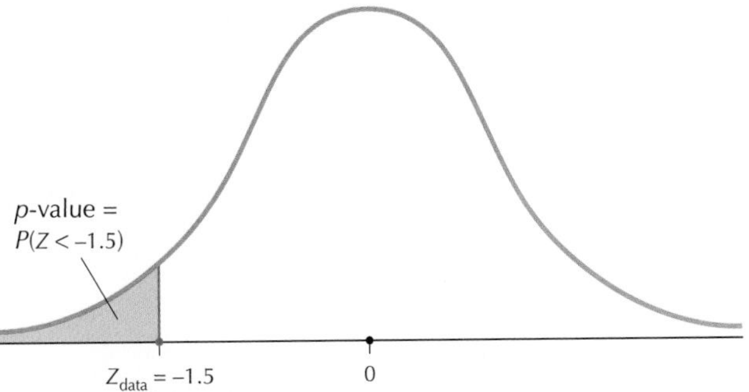


FIGURE 9.11

**Step 4** **State the conclusion and interpretation.** In Step 1, we said that we would reject $H_0$ if the $p$-value was less than our level of significance $\alpha = 0.05$. From Step 3, we have $p$-value $= 0.0668$. Since 0.0668 is *not* $< 0.05$, we therefore do *not* reject $H_0$. The difference between $\bar{x} = 7.05$ and $\mu_0 = 7.2$ is not statistically significant. The generic interpretation is "There is insufficient evidence that [whatever $H_a$ says]." Here, $H_a$ states that $\mu < 7.2$; that is, the population mean user rating for the Dell XPS 410 computer is less than 7.2. Hence, we interpret our conclusion as follows:

"There is insufficient evidence that the population mean user rating for a Dell XPS 410 computer is less than 7.2."

## Example 9.10 The $p$-value method using technology: right-tailed test

Here are the birth weights, in grams (1000 grams = 1 kilogram ≈ 2.2 pounds), of a random sample of 44 babies from Brisbane, Australia.

| | | | | | | | | | | | |
|---|---|---|---|---|---|---|---|---|---|---|---|
| 3837 | 3334 | 3554 | 3838 | 3625 | 2208 | 1745 | 2846 | 3166 | 3520 | 3380 | 3294 |
| 2576 | 3208 | 3521 | 3746 | 3523 | 2902 | 2635 | 3920 | 3690 | 3430 | 3480 | 3116 |
| 3428 | 3783 | 3345 | 3034 | 2184 | 3300 | 2383 | 3428 | 4162 | 3630 | 3406 | 3402 |
| 3500 | 3736 | 3370 | 2121 | 3150 | 3866 | 3542 | 3280 | | | | |

Health care officials are interested in whether there is evidence that recent prenatal nutrition campaigns are working. Formerly, the mean birth weight of babies was 3200 grams. Assume that the population standard deviation $\sigma = 528$ grams. The sample mean birth weight is $\bar{x} = 3276$ grams. Is there evidence that the population mean birth weight of Brisbane babies now exceeds 3200 grams? Use technology to perform the appropriate hypothesis test, with level of significance $\alpha = 0.10$.

### What Results Might We Expect?

Note from Figure 9.12 that the sample mean birth weight $\bar{x} = 3276$ grams is close to the hypothesized mean birth weight of $\mu_0 = 3200$ grams. This value of $\bar{x}$ is not extreme and thus does not seem to offer strong evidence that the hypothesized mean birth weight is wrong. Therefore, we might expect to *not reject* the hypothesis that $\mu_0 = 3200$ grams.

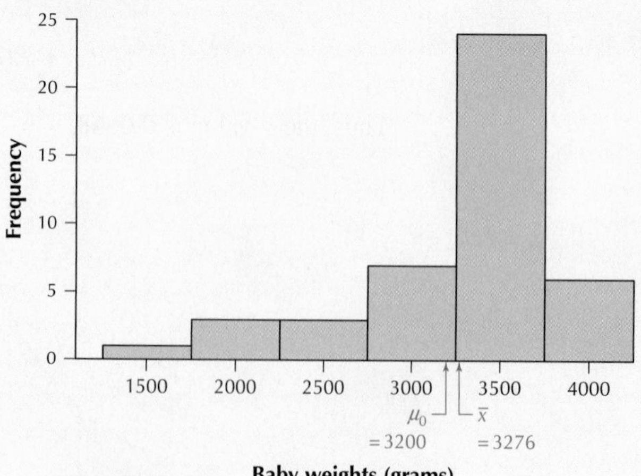


FIGURE 9.12 Sample mean is close to hypothesized mean.

## Solution

Since the sample size $n = 44$ is large and $\sigma = 528$ is known, we may proceed with the hypothesis test, using Case 2 ($n \geq 30$).

**Step 1** **State the hypotheses and the rejection rule.** The key word "exceeds" means that we have a right-tailed test:

$$H_0: \mu \leq 3200 \quad \text{versus} \quad H_a: \mu > 3200$$

where $\mu$ refers to the population mean birth weight of Brisbane babies. We will reject $H_0$ if the $p$-value is less than $\alpha = 0.10$.

**Step 2** **Find $Z_{data}$.** We will use the instructions provided in the Step-by-Step Technology Guide at the end of this section (page 478). Figure 9.13 shows the TI-83/84 results from the $Z$ test for $\mu$:

$$Z_{data} = \frac{\bar{x} - \mu_0}{\sigma/\sqrt{n}} = \frac{3276 - 3200}{528/\sqrt{44}} = 0.9547859245 \approx 0.9548$$

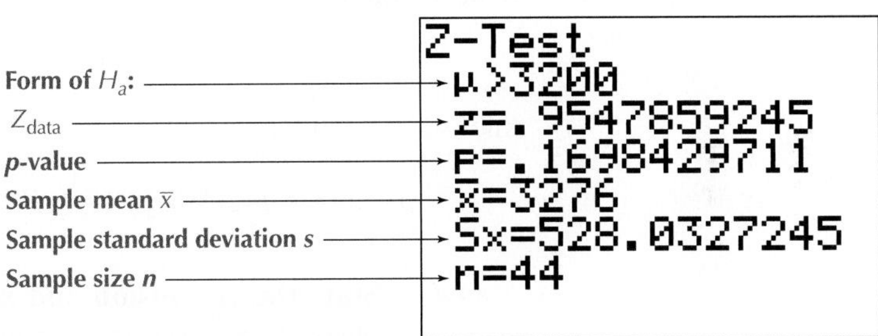


FIGURE 9.13 TI-83/84 results.

Figure 9.14 shows the Minitab results, where

- **Test of mu = 3200 vs. > 3200** refers to the hypotheses being tested, $H_0 : \mu \leq 3200$ versus $H_a : \mu > 3200$.
- **The assumed standard deviation = 528** refers to our assumption that $\sigma = 528$.
- **SE Mean** refers to the standard deviation of the sampling distribution of $\bar{x}$, that is, $\sigma/\sqrt{n}$. You can see that $528/\sqrt{44} \approx 79.60$.
- **95.0% Lower Bound** refers to a one-way confidence interval and will not be covered here.
- ***Z*** refers to our test statistic:

$$Z_{data} = \frac{\bar{x} - \mu_0}{\sigma/\sqrt{n}} = (3276 - 3200)/(528/\sqrt{44}) \approx 0.95$$

- ***P*** represents our $p$-value of 0.170.

```
One-Sample Z: Baby Weights

Test of mu = 3200 vs > 3200
The assumed standard deviation = 528
                                                     95%
                                                   Lower
Variable        N      Mean   StDev  SE Mean       Bound      Z      P
Baby Weights   44   3276.00  528.03    79.60     3145.07   0.95  0.170
```

FIGURE 9.14 Minitab results.

Figure 9.15 shows output from WHFStat Macros for Excel.

| SUMMARY STATISTICS | | |
|---|---|---|
| Sample Mean | Sample Size | Standard Deviation |
| 3276 | 44 | 528 |

| ONE SAMPLE Z TEST - CONFIDENCE INTERVAL | | | |
|---|---|---|---|
| Confidence Level | | Z Value | Critical Z Value |
| 90 % | | 0.954786 | 1.645 |
| Confidence Interval | | | |
| 3276 +/- | 130.9403 | | |
| 3145.06 to | 3406.94 | | |
| Population Mean (Null Hypothesis Value) | | | |
| 3200 | | | |
| Alternative Hypothesis | | P-Value | |
| > 3200 | | 0.169843 | |
| | | | 1-Sided |
| < 3200 | | 0.830157 | |
| Not = 3200 | | 0.339686 | 2-Sided |

FIGURE 9.15 WHFStat Macros for Excel results.

**Step 3** **Find the *p*-value.** We have a right-tailed test from Step 1, so that from Table 9.4 our *p*-value is

$$p\text{-value} = P(Z > Z_{\text{data}}) = P(Z > 0.9548)$$

And from Figure 9.13 we have

$$p\text{-value} = P(Z > Z_{\text{data}}) = P(Z > 0.9548) = 0.1698429711 \approx 0.1698$$

**Step 4** **State the conclusion and interpretation.** This *p*-value of 0.1698 represents the probability of observing a sample mean birth weight at least as extreme as the sample mean birth weight of 3276 grams that we actually observed if we assume the null hypothesis is true. In Step 1 we said that we would reject $H_0$ if the *p*-value is less than our level of significance $\alpha = 0.10$. Since 0.1698 is *not* less than 0.10, we do *not* reject $H_0$. This is as we expected from the fact that $\bar{x} = 3276$ was not far away from the hypothesized mean birth weight of $\mu_0 = 3200$ grams. We interpret our conclusion as follows: "There is insufficient evidence that the population mean birth weight is greater than 3200 grams." ■

The *P-Value* applet allows you to experiment with various hypotheses, means, standard deviations, and sample sizes in order to see how changes in these values affect the *p*-value.

Next, we will learn how to find the *p*-value manually. Notice that each of the formulas from Table 9.4 for finding the *p*-value uses the standard normal methods for finding probability shown in Table 6.7 (page 298).

## Example 9.11 Hemoglobin levels in adult males undergoing cardiac surgery: two-tailed test

When the level of hemoglobin in the blood is too low, the person is anemic. Unusually high levels of hemoglobin are undesirable as well and can be associated with dehydration. A study by the Harvard Medical School Cardiogenomics Group recorded the hemoglobin levels, in grams per deciliter (g/dL) of blood, in patients undergoing cardiac surgery.[5] A random sample of 20 male cardiac patients yielded a sample mean hemoglobin level of 12.35 g/dl. Assume that the population standard deviation is 2.8 g/dl. Test whether the population mean hemoglobin level for males undergoing cardiac surgery *differs from* 13.8 g/dl using the *p*-value method at level of significance $\alpha = 0.05$.

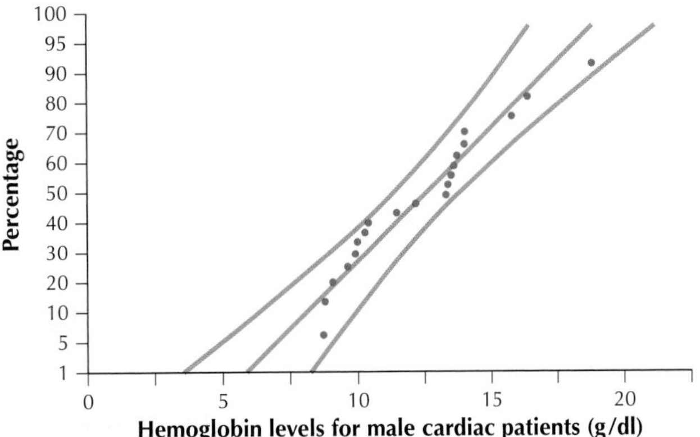


FIGURE 9.16 Normal probability plot.

## Solution

The normal probability plot shown in Figure 9.16 indicates an acceptable level of normality, allowing us to proceed with the hypothesis test.

**Step 1** **State the hypotheses and the rejection rule.** Since *either* too much or too little hemoglobin is bad, we have a two-tailed test here: $H_0 : \mu = 13.8$ versus $H_a : \mu \neq 13.8$, where $\mu$ is the population mean hemoglobin level in male cardiac patients. We will reject $H_0$ if the *p*-value is less than $\alpha = 0.05$.

**Step 2** **Find $Z_{data}$.** We have a sample of $n = 20$ patients, for which $\bar{x} = 12.35$. We know that $\sigma = 2.8$. Plugging into the formula for $Z_{data}$, we get

$$Z_{data} = \frac{\bar{x} - \mu_0}{\sigma_{\bar{x}}} = \frac{\bar{x} - \mu_0}{\sigma/\sqrt{n}} = \frac{12.35 - 13.8}{2.8/\sqrt{20}} \approx -2.32$$

When calculating the *p*-value manually, round $Z_{data}$ to two decimal places.

**Step 3** **Find the *p*-value.** Because we have a two-tailed test, Table 9.4 tells us that

$$\begin{aligned} p\text{-value} &= P(Z > |Z_{data}|) + P(Z < -|Z_{data}|) \\ &= 2 \cdot P(Z > |Z_{data}|) \\ &= 2 \cdot P(Z > |-2.32|) \end{aligned}$$

Using Case 2 from Table 6.7 (page 298),

$$\begin{aligned} P(Z > |Z_{data}|) &= P(Z > 2.32) \\ &= 1 - 0.9898 = 0.0102 \end{aligned}$$

By symmetry, we also have

$$P(Z < -2.32) = 0.0102$$

Thus, our *p*-value, which is the sum of these two tail areas as shown in Figure 9.17, equals

$$p\text{-value} = 2 \cdot P(Z > |-2.32|) = 2(0.0102) = 0.0204$$

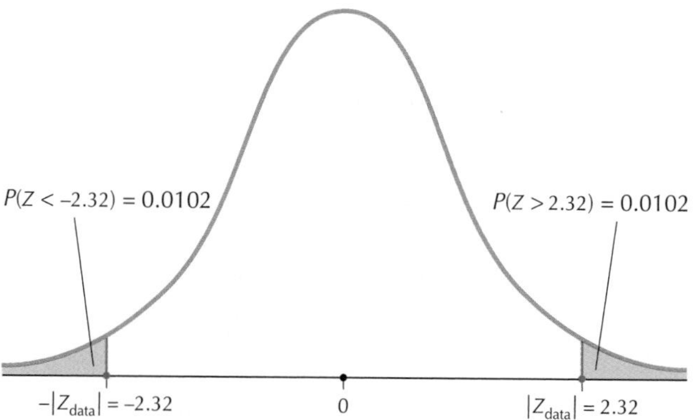


FIGURE 9.17 Finding the $p$-value for a two-tailed test.

*Step 4* **State the conclusion and interpretation.** The $p$-value 0.0204 *is* smaller than $\alpha = 0.05$. Therefore, our conclusion is to reject $H_0$. The difference between $\bar{x} = 12.35$ g/dl and $\mu_0 = 13.8$ g/dl is statistically significant. Using the generic interpretation, we can say: "There is evidence that the population mean hemoglobin level in male cardiac patients is not equal to 13.8 g/dl." ■

## 4 Assessing the Strength of Evidence Against the Null Hypothesis

The hypothesis-testing methods we have shown so far deliver a simple "yes-or-no" conclusion: either "Reject $H_0$" or "Do not reject $H_0$." There is no indication of how strong the evidence is for rejecting the null hypothesis. Was the decision close? Was it a no-brainer? On the other hand, the $p$-value itself implicitly represents the *strength of evidence against the null hypothesis*. **There is extra information here, which we should not ignore.**

For instance, we can directly compare the results of hypothesis tests. Suppose that we have two hypothesis tests that both result in not rejecting the null hypothesis, with level of significance $\alpha = 0.05$. However, Test A has a $p$-value of 0.06, while Test B has a $p$-value of 0.57. Clearly, Test A came very close to rejecting the null hypothesis and shows a fair amount of evidence against the null hypothesis, while Test B shows no evidence at all against the null hypothesis. A simple statement of the "yes-or-no" conclusion misses the clear distinction between these two hypothesis tests.

The $p$-value provides us with the smallest level of significance at which the null hypothesis would be rejected, that is, the smallest value of $\alpha$ at which the results would be considered significant. Of course, we are free to determine whether the results are significant using whatever $\alpha$ level we wish. For example, Test A above would have rejected $H_0$ for any $\alpha$ level higher than 0.06. Some data analysts in fact do not think in terms of rejecting or not rejecting the null hypothesis. Rather, they think completely in terms of *assessing the strength of evidence against the null hypothesis*.

For many (though not all) data domains, Table 9.5 provides a thumbnail impression of the strength of evidence against the null hypothesis for various $p$-values. Remember, however, that for certain domains (such as the physical sciences), alternative interpretations are appropriate. Nevertheless, you may use Table 9.5 for all exercises that ask for an assessment of the strength of evidence against the null hypothesis.

Table 9.5 Strength of evidence against the null hypothesis for various levels of $p$-value

| **$p$-Value** | **Strength of evidence against $H_0$** |
|---|---|
| $p$-value $\leq 0.001$ | Extremely strong evidence |
| $0.001 < p$-value $\leq 0.01$ | Very strong evidence |
| $0.01 < p$-value $\leq 0.05$ | Solid evidence |
| $0.05 < p$-value $\leq 0.10$ | Mild evidence |
| $0.10 < p$-value $\leq 0.15$ | Slight evidence |
| $0.15 < p$-value | No evidence |

## Example 9.12 Assessing the strength of evidence against $H_0$

Assess the strength of evidence against $H_0$ shown by the $p$-values in (a) Example 9.9, (b) Example 9.10, and (c) Example 9.11.

**Solution**

**a.** In Example 9.9 we tested $H_0: \mu \geq 7.2$ versus $H_a: \mu < 7.2$ where $\mu$ refers to the population mean user rating for the Dell XPS 410 computer. Our $p$-value of 0.0668 implies that there is *mild* evidence against the null hypothesis that the population mean user rating for the Dell XPS 410 computer is 7.2 or higher.

**b.** In Example 9.10 we tested $H_0: \mu \leq 3200$ versus $H_a: \mu > 3200$ where $\mu$ refers to the population mean birth weight of Brisbane babies (in grams). Our $p$-value of 0.1698 implies that there is *no* evidence against the null hypothesis that the population mean birth weight of Brisbane babies is at most 3200 grams.

**c.** In Example 9.11 we tested $H_0: \mu = 13.8$ versus $H_a: \mu \neq 13.8$ where $\mu$ is the population mean hemoglobin level in male cardiac patients. Our $p$-value of 0.0204 implies that there is *solid* evidence against the null hypothesis that the population mean hemoglobin level in male cardiac patients equals 13.8 g/dl. ■

### The Role of the Level of Significance $\alpha$

Suppose that in Example 9.11, our level of significance $\alpha$ was 0.01 rather than 0.05. Would this have changed anything? Certainly. Since our $p$-value of 0.0204 is *not* less than the new $\alpha = 0.01$, we would *not* reject $H_0$. Think about that for a moment. *The data haven't changed at all, but our conclusion is reversed simply by changing $\alpha$.* What is a data analyst to make of a situation like this? There are two alternatives.

1. Since we don't want the choice of $\alpha$ to dictate our conclusion, then perhaps we should turn to a direct assessment of the strength of evidence against the null hypothesis, as provided in Table 9.5. In this case, the $p$-value of about 0.0204 would offer solid evidence against the null hypothesis, *regardless of the value of $\alpha$.*
2. Obtain more data, perhaps through a call for further research. This option is especially relevant for this example, where the sample size $n = 20$ is rather small. ■

## STEP-BY-STEP GUIDE TO TECHNOLOGY: $Z$ test for $\mu$

We will use the weight data from Example 9.10 (page 472).

### TI-83/84

#### If you have the data values:

***Step 1*** Enter the data into list **L1**.
***Step 2*** Press **STAT**, highlight **TESTS**, and press **ENTER**.
***Step 3*** Press **1** (for **Z-Test**; see Figure 9.18).
***Step 4*** For input (**Inpt**), highlight **Data** and press **ENTER** (Figure 9.19).

**a.** For $\boldsymbol{\mu_0}$, enter the value of $\mu_0$, **3200**.
**b.** For $\boldsymbol{\sigma}$, enter the value of $\sigma$, **528**.
**c.** For **List**, press **2**[nd], then **L1**.
**d.** For **Freq**, enter **1**.
**e.** For $\boldsymbol{\mu}$, select the form of $H_a$. Here we have a right-tailed test, so highlight **>** $\boldsymbol{\mu_0}$ and press **ENTER**.
**f.** Highlight **Calculate** and press **ENTER**. The results are shown in Figure 9.13 in Example 9.10.

#### If you have the summary statistics:

***Step 1*** Press **STAT**, highlight **TESTS**, and press **ENTER**.
***Step 2*** Press **1** (for **Z-Test**; see Figure 9.19).
***Step 3*** For input (**Inpt**), highlight **Stats** and press **ENTER** (Figure 9.20).

**a.** For $\boldsymbol{\mu_0}$, enter the value of $\mu_0$, **3200**.
**b.** For $\boldsymbol{\sigma}$, enter the value of $\sigma$, **528**.
**c.** For $\bar{x}$, enter the sample mean **3276**.
**d.** For **n**, enter the sample size **44**.
**e.** For $\boldsymbol{\mu}$, select the form of $H_a$. Here we have a right-tailed test, so highlight **>** $\boldsymbol{\mu_0}$ and press **ENTER**.
**f.** Highlight **Calculate** and press **ENTER**. The results are shown in Figure 9.13 in Example 9.10.

Figure 9.18

Figure 9.19

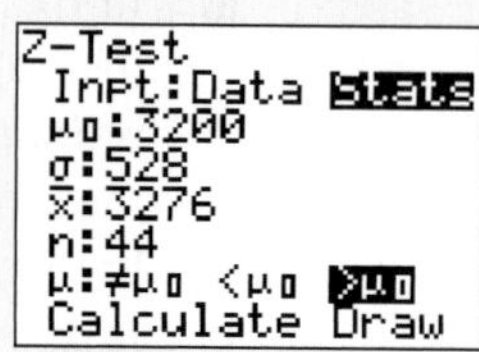


Figure 9.20

### EXCEL

#### WHFStat Macros

***Step 1*** Enter the data into column A. (If you have only the summary statistics, go to Step 2.)
***Step 2*** Load the **WHFStat Macros**.
***Step 3*** Select **Add-Ins > Macros > Testing a Mean > Z Test – Confidence Interval – One Sample**.
***Step 4*** Select cells A1 to A44 as the **Dataset Range**. (Alternatively, you may enter the summary statistics.)
***Step 5*** Select your **Confidence level**, which should be $1 - \alpha$. Here, because $\alpha = 0.10$, we select **90%**.
***Step 6*** Enter the **Population Standard Deviation**, $\sigma = 528$.
***Step 7*** Enter the **Null Hypothesis Value**, $\mu_0 = 3200$, and click **OK**. The results are shown in Figure 9.15 in Example 9.10.

### MINITAB

#### If you have the data values:

***Step 1*** Enter the data into column C1.
***Step 2*** Click **Stat > Basic Statistics > 1-Sample Z**.
***Step 3*** Click **Samples in Columns** and select **C1**.
***Step 4*** Enter **528** as **Standard Deviation**.
***Step 5*** For **Test Mean**, enter 3200.
***Step 6*** Click **Options**.

**a.** Choose your **Confidence Level** as $100(1 - \alpha)$. Our level of significance $\alpha$ here is 0.10, so the confidence level is 90.0.
**b.** Select **Greater Than** to symbolize the right-tailed test.

***Step 7*** Click **OK** and click **OK** again. The results are shown in Figure 9.14 in Example 9.10.

#### If you have the summary statistics:

***Step 1*** Click **Stat > Basic Statistics > 1-Sample Z**.
***Step 2*** Click **Summarized Data**.
***Step 3*** Enter the **Sample Size 44** and the **Sample Mean 3276**.
***Step 4*** Click **Options**.

**a.** Choose your **Confidence Level** as $100(1 - \alpha)$. Our level of significance $\alpha$ here is 0.10, so the confidence level is 90.0.
**b.** Select **Greater Than** to symbolize the right-tailed test.

***Step 5*** Click **OK** and click **OK** again. The results are shown in Figure 9.14 in Example 9.10.

## Section 9.2 SUMMARY

1 The essential idea about hypothesis testing for the mean may be described as follows. When the observed value of $\bar{x}$ is unusual or extreme in the sampling distribution of $\bar{x}$ that is based on the assumption that $H_0$ is correct, we should reject $H_0$. Otherwise, there is insufficient evidence against $H_0$, and we should not reject $H_0$. All of the remaining steps and

calculations are simply explanations of how to implement this essential idea about hypothesis testing for the mean.

2 The $p$-value can be thought of as the probability of observing a sample statistic at least as extreme as the statistic in your sample if we assume that the null hypothesis is true. The rejection rule for performing a hypothesis test using the $p$-value method is to reject $H_0$ when the $p$-value is less than the level of significance $\alpha$.

3 Use the $Z$ test only when the population standard deviation $\sigma$ is known and when either (Case 1) the population is normal or (Case 2) the sample size is large ($n \geq 30$).

4 The $p$-value can be used to assess the strength of evidence against the null hypothesis.

## Section 9.2 EXERCISES

### CLARIFYING THE CONCEPTS

**1.** Explain in your own words how a researcher determines if a particular value of the sample mean is extreme.
**2.** What do we mean by the term *test statistic*? On what does the value of the test statistic depend?
**3.** How do we tell when a $p$-value is small? That is, to what do we compare the $p$-value to determine if it is small?
**4.** In terms of area under the standard normal curve, what does the $p$-value represent for a right-tailed test? For a left-tailed test? For a two-tailed test?
**5.** Refer to Example 9.9. What would the conclusion have been if our level of significance $\alpha$ were set to be 0.10?
**6.** Does $\alpha$ have anything to do with the sample data?

### PRACTICING THE TECHNIQUES

For Exercises 7–12, find the value of $Z_{\text{data}}$.

**7.** $\bar{x} = 10, \sigma = 2, n = 36, \mu_0 = 12$
**8.** $\bar{x} = 12, \sigma = 2, n = 36, \mu_0 = 12$
**9.** $\bar{x} = 14, \sigma = 2, n = 36, \mu_0 = 12$
**10.** $\bar{x} = 95, \sigma = 5, n = 64, \mu_0 = 100$
**11.** $\bar{x} = 97.5, \sigma = 5, n = 64, \mu_0 = 100$
**12.** $\bar{x} = 99, \sigma = 5, n = 64, \mu_0 = 100$

For Exercises 13–18, suppose that we are testing the hypotheses $H_0: \mu \geq \mu_0$ versus $H_a: \mu < \mu_0$ using level of significance $\alpha = 0.05$, and that the conditions for performing the $Z$ test are satisfied. For each exercise, provide

**a.** the conclusion and
**b.** the interpretation of the conclusion.

**13.** $p$-value = 0.01
**14.** $p$-value = 0.06
**15.** $p$-value = 1.0
**16.** $p$-value = 0.5
**17.** $p$-value = 0.0
**18.** $p$-value = 0.05

For Exercises 19–23, suppose that we are testing the hypotheses $H_0: \mu \leq \mu_0$ versus $H_a: \mu > \mu_0$, and that the conditions for performing the $Z$ test are satisfied. For each exercise,

**a.** draw a sketch of the $Z$ distribution, showing $Z_{\text{data}}$ as a point on the number line and the $p$-value as an area under the curve, and
**b.** calculate the $p$-value.

**19.** $Z_{\text{data}} = 1.0$ **20.** $Z_{\text{data}} = 1.5$
**21.** $Z_{\text{data}} = 2.0$ **22.** $Z_{\text{data}} = 2.5$
**23.** $Z_{\text{data}} = 3.0$
**24.** What pattern do you observe in the $p$-value for right-tailed tests as the value of $Z_{\text{data}}$ increases (gets more extreme)?

For Exercises 25–28:

**a.** State the hypotheses and the rejection rule.
**b.** Find $Z_{\text{data}}$, and round it to two decimal places.
**c.** Find the $p$-value, using Table 9.4. Draw the standard normal curve, with $Z_{\text{data}}$ and the $p$-value indicated on it.
**d.** Formulate and interpret your conclusion.

**25.** We are interested in testing at level of significance $\alpha = 0.05$ whether the population mean is greater than 100. A random sample of size 64 is taken, with a mean of 102. Assume $\sigma = 8$.
**26.** We would like to test at level of significance $\alpha = 0.10$ whether the population mean is less than 3.0. A random sample of size 36 is taken, with a mean of 2.9. Assume $\sigma = 0.5$.
**27.** We want to test at level of significance $\alpha = 0.01$ whether the population mean differs from 20. A random sample of size 49 is taken, with a mean of 27. Assume $\sigma = 5$.
**28.** We are interested in testing at level of significance $\alpha = 0.05$ whether the population mean is greater than 0. A random sample of size 16 is taken, with a mean of 2. Assume $\sigma = 4$ and that the population is normally distributed.

For the hypothesis tests in Exercises 29–36, assess the strength of evidence against the null hypothesis using Table 9.5.

**29.** Exercise 13 **30.** Exercise 14
**31.** Exercise 15 **32.** Exercise 16
**33.** Exercise 17 **34.** Exercise 18
**35.** Exercise 19 **36.** Exercise 20

### APPLYING THE CONCEPTS

For Exercises 37–44, do the following.

**a.** State the hypotheses and the rejection rule.
**b.** Find $Z_{\text{data}}$.
**c.** Find the $p$-value.
**d.** State the conclusion and the interpretation.

**37. Stock Market.** The *Statistical Abstract of the United States* reports that the mean daily number of shares traded on the New York Stock Exchange in 2005 was 1.6 billion. Let this value represent the hypothesized population mean, and assume that the population standard deviation equals 0.5 billion shares. Suppose that, in a random sample of 36 days from the present year, the mean daily number of shares traded equals 1.5 billion. We are interested in testing, whether the population mean daily number of shares traded has increased since 2005, using level of significance $\alpha = 0.05$.

**38. Child Abuse.** The U.S. Administration for Children and Families reported that the national rate for child abuse referrals was 43.9 per 1000 children in 2005. Suppose that a random sample of 1000 children taken this year shows 47 child abuse referrals. Assume $\sigma = 5$. Test whether the population mean referral rate has increased this year from the 2005 level, using level of significance $\alpha = 0.10$.

**39. California Warming.** A 2007 report found that the mean temperature in California has increased from 1950 to 2000 by 2 degrees Fahrenheit (°F). Suppose that a random sample of 36 California locations showed a mean increase of 4°F over 1950 levels. Assume $\sigma = 0.5$. Test whether the population mean temperature increase in California is greater than 2°F, at level of significance $\alpha = 0.05$.

**40. Eating Trends.** According to an NPD Group report the mean number of meals prepared and eaten at home is less than 700 per year. Suppose that a random sample of 100 households showed a sample mean number of meals prepared and eaten at home of 650. Assume $\sigma = 25$. Test whether the population mean number of such meals is less than 700, using level of significance $\alpha = 0.10$.

**41. DDT in Breast Milk.** Researchers compared the amount of DDT in the breast milk of 12 Hispanic women in the Yakima Valley of Washington State with the amount of DDT in breast milk in the general U.S. population.[6] They measured the mean DDT level in the general population to be 47.2 parts per billion (ppb) and the mean DDT level in the 12 Hispanic women to be 219.7 ppb. Assume $\sigma = 36$ and a normally distributed population. Test whether the population mean DDT level in the breast milk of Hispanic women in the Yakima Valley is greater than that of the general population, using level of significance $\alpha = 0.01$.

**42. Tree Rings.** Do trees grow more quickly when they are young? The International Tree Ring Data Base collected data on a particular 440-year-old Douglas fir tree.[7] The mean annual ring growth in the tree's first 80 years of life was 1.4261 millimeters (mm). A random sample of size 100 taken from the tree's later years showed a sample mean growth of 0.56 mm per year. Assume $\sigma = 0.5$ mm and a normally distributed population. Test whether the population mean annual ring growth in the tree's later years is less than 1.4261 mm, using level of significance $\alpha = 0.05$.

**43. Hybrid Vehicles.** A 2006 study by **Edmunds.com** examined the time it takes for owners of hybrid vehicles to recoup their additional initial cost through reduced fuel consumption. Suppose that a random sample of 9 hybrid cars showed a sample mean time of 2.1 years. Assume that the population is normal with $\sigma = 0.2$. Test using level of significance $\alpha = 0.01$ whether the population mean time it takes owners of hybrid cars to recoup their initial cost is less than three years.

**44. Americans' Height.** Americans used to be on average the tallest people in the world. That is no longer the case, according to a study by Dr. Richard Steckel, professor of economics and anthropology at The Ohio State University. The Norwegians and Dutch are now the tallest, at 178 centimeters, followed by the Swedes at 177, and then the Americans, with a mean height of 175 centimeters (approximately 5 feet 9 inches). According to Dr. Steckel, "The average height of Americans has been pretty much stagnant for 25 years."[8] Suppose a random sample of 100 Americans taken this year shows a mean height of 174 centimeters, and we assume $\sigma = 10$ centimeters. Test using level of significance $\alpha = 0.01$ whether the population mean height of Americans this year has changed from 175 centimeters.

**Sodium in Breakfast Cereal.** Use the following information for Exercises 45–48. A random sample of 23 breakfast cereals containing sodium had a mean sodium content per serving of 192.39 grams. Assume that the population standard deviation equals 50 grams. We are interested in whether the population mean sodium content per serving is less than 210 grams.

**45. a.** The normal probability plot of the sodium content is shown in Figure 9.21. Should we proceed to apply the $Z$ test? Why or why not?

**b.** Test whether the population mean sodium content per serving is less than 210 grams, using level of significance $\alpha = 0.01$.

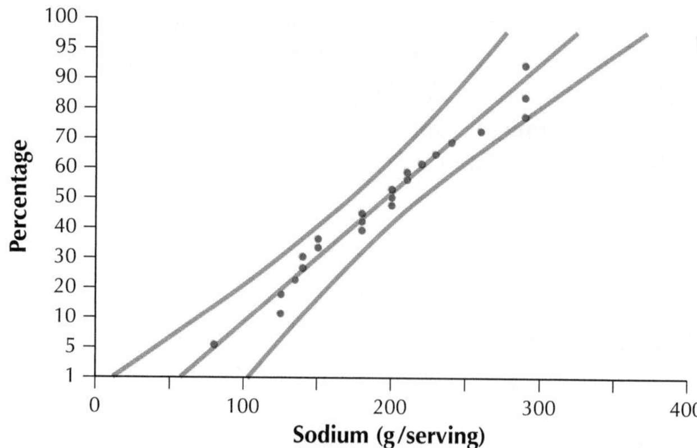


Figure 9.21

WHAT IF? **46.** *What if* the population standard deviation of 50 grams had been a typo, and the actual population standard deviation was smaller. How would this have affected the following?

**a.** The standard deviation of the sampling distribution

**b.** $Z_{data}$

**c.** $p$-value

**d.** The conclusion

**47.** *What if* our level of significance $\alpha$ equaled 0.05 instead of 0.01.

**a.** Perform the appropriate hypothesis test using the manual $p$-value method, but this time using level of significance $\alpha = 0.05$.

**b.** Note that your conclusion differs from that obtained using level of significance $\alpha = 0.01$. Have the data changed? Why did your conclusion change?

**c.** Suggest two alternatives for addressing the contradiction between Exercise 45b and Exercise 47a.

**48.** Assess the strength of the evidence against the null hypothesis.

**49. Women's Heart Rates.** A random sample of 15 women produced the normal probability plot in Figure 9.22 for their heart rates. The sample mean was 75.6 beats per minute. Suppose the population standard deviation is known to be 9.

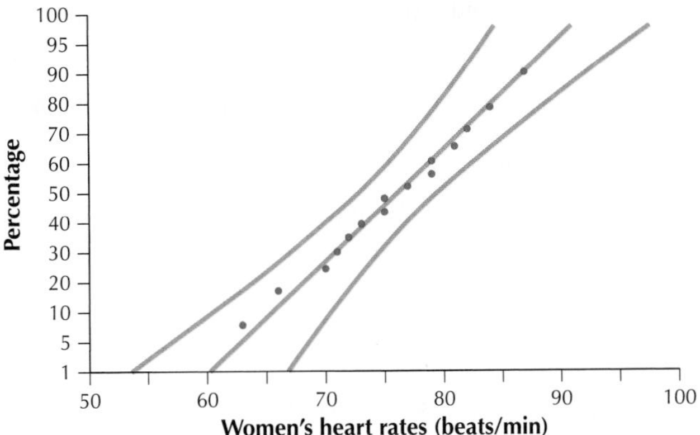


Figure 9.22

**a.** Discuss the evidence for or against the normality assumption. Should we use the $Z$ test? Why or why not?

**b.** Assume that the plot does not contradict the normality assumption; test whether the population mean heart rate for all women *is less than* 78, using level of significance $\alpha = 0.05$.

**c.** Test whether the population mean heart rate for all women *differs from* 78, using $\alpha = 0.05$.

**50. Challenger Exercise.** Refer to the previous exercise.

**a.** Compare your conclusions from **(b)** and **(c)**. Note that the conclusions differ but the meanings of the hypotheses tested also differ. Combine the two conclusions into a single sentence. Do you find this sentence difficult to explain?

**b.** Explain in your own words the difference between the hypotheses in **(b)** and **(c)**. Also, explain how there could be evidence that the population mean heart rate is *less than 78 but not different from 78.*

**c.** Assess the strength of the evidence against the null hypothesis for the hypothesis tests in **(b)** and **(c)**.

**Mean Family Size.** Use the following information for Exercises 51–54. According to the *Statistical Abstract of the United States,* the mean family size in 2002 was 3.15 persons, reflecting a slow decrease since 1980, when the mean family size was 3.29 persons. Has this trend continued to the present day? Suppose a random sample of 225 families taken this year yields a sample mean size of 3.05 persons, and suppose we assume that the population standard deviation of family sizes is 1 person.

**51.** We are interested in testing whether the population mean family size in America has decreased since 2002, using the $p$-value method and level of significance $\alpha = 0.05$. (Try using the *P-Value* applet to help you solve this problem.)

**a.** State the hypotheses and the rejection rule.

**b.** Find $Z_{data}$.

**c.** Find the $p$-value.

**d.** State the conclusion and interpretation.

**52.** Refer to the previous exercise. Suppose that we used level of significance $\alpha = 0.10$ rather than $\alpha = 0.05$. Redo **(a)–(d)**. Comment on any apparent contradictions. Suggest two ways to resolve this dilemma.

**53.** Refer to Exercise 51, with our original hypothesis test (with $\alpha = 0.05$).

**a.** What is the smallest $p$-value for which you will reject $H_0$?

**b.** What do you think of the assumption of 1 person for the value of $\sigma$? How would this have been obtained, if at all possible?

**c.** Which type of error is it possible that we are making, a Type I error or a Type II error? Which type of error are we certain we are not making?

**d.** Suppose a newspaper headline referring to the study was "Mean Family Size Decreasing." Is the headline supported or not supported by the data and the hypothesis test?

**54.** Refer to the previous exercise and our original hypothesis test (with level of significance $\alpha = 0.05$). Try to answer the following questions by thinking about the relationship between the statistics rather than by redoing all the calculations. *What if* the 3.05 persons had been a typo, and the actual sample mean was 3.00 persons. How would this have affected the following?

**a.** The standard deviation of the sampling distribution

**b.** $Z_{data}$

**c.** The $p$-value

**d.** The conclusion

**e.** Write a sentence describing what happens to the elements in **(a)–(d)** when the sample mean becomes more extreme.

**Online Shopping Privacy Protection.** Use the *P-Value* applet and the following information for Exercises 55–58. A 2007 Carnegie Mellon University study reports that online shoppers were willing to pay an average of an extra 60 cents on a $15 purchase in order to have better online privacy protection.[9] Assume that the population standard deviation is 20 cents. Suppose that a study conducted this year found that the mean amount online shoppers were willing to pay for privacy was $\bar{x} = 65$ cents. Does this represent evidence that the mean amount online shoppers are willing to pay for privacy has increased from 60 cents? As it turns out, the answer depends on the sample size. If this $\bar{x} = 65$ cents comes from a large sample, then the evidence against $H_0$ will be stronger than if it comes from a small sample. Use the *P-Value* applet to see this, using level of significance $\alpha = 0.05$.

**55.** Enter $H_0 : \mu \leq 60$, $H_a : \mu > 60$, $\sigma = 20$, and $\bar{x} = 65$.

**a.** Enter $n = 30$. Click **Show P** to get the $p$-value. What is the $p$-value? Sketch the normal curve, showing the $p$-value.

**b.** Would we reject the null hypothesis based on these values of $\bar{x}$ and $n$?

**56.** Repeat Exercise 55 for $n = 40$. Compare the $p$-value with that from Exercise 55.

**57.** Repeat Exercise 55 for $n = 50$. Compare the $p$-value with those from Exercises 55 and 56.

**58.** Note that $\bar{x}$ has not changed in Exercises 55–57; only the sample size has changed. Write a sentence to describe the pattern you see in the $p$-values as the sample size increases.

## 9.3 Z Test for the Population Mean $\mu$: Critical-Value Method

**Objectives:** By the end of this section, I will be able to...

1. Explain the meaning of the critical region and the critical value $Z_{crit}$.
2. Perform and interpret the $Z$ test for the population mean using the critical-value method.
3. Describe the relationship between the $p$-value method and the critical-value method.
4. Use confidence intervals for $\mu$ to perform two-tailed hypothesis tests about $\mu$.
5. Calculate the probability of a Type II error, and compute the power of a hypothesis test.

### 1 Critical Regions and the Critical Value $Z_{crit}$

In Section 9.2, we learned how to perform a $Z$ test using the $p$-value method. The $p$-value method compares one probability (the $p$-value) with another probability (level of significance $\alpha$). In Section 9.3, we learn another way to perform a $Z$ test, this time using the critical-value method. In the critical-value method, we compare one $Z$-value ($Z_{data}$) with another $Z$-value ($Z_{crit}$).

Extreme values of $\bar{x}$ are associated with extreme values of $Z_{data}$, and extreme values of $Z_{data}$ will lead us to reject the null hypothesis. The question is: *How do we measure what extreme is?* In Section 9.2, we found the probability that we would observe such an extreme $Z_{data}$. Here, we compare $Z_{data}$ directly with a *threshold value*, or **critical value** of $Z$, called $Z_{crit}$. Figure 9.23 shows that $Z_{crit}$ separates the values of $Z_{data}$ for which we reject $H_0$ (the **critical region**) from the values of $Z_{data}$ for which we will not reject $H_0$ (the **noncritical region**).

- The **critical region** consists of the range of values of the test statistic $Z_{data}$ for which we reject the null hypothesis.
- The **noncritical region** consists of the range of values of the test statistic $Z_{data}$ for which we do not reject the null hypothesis.
- The value of $Z$ that separates the critical region from the noncritical region is called the **critical value, $Z_{crit}$**.

Example 9.13 shows how the critical value $Z_{crit}$ is determined.

## Example 9.13 Finding the critical value $Z_{crit}$

Do you have a debit card? How often do you use it? ATM network operator Star System of San Diego reported in 2006 that active users of debit cards used them an average of 11 times per month.[10] Suppose that we are interested in testing whether debit card usage has increased since 2006. For level of significance $\alpha = 0.01$, find the critical value $Z_{crit}$ for this hypothesis test.

### Solution

The key word here is "increased," which leads (through our strategy) to the following hypotheses:

$$H_0: \mu \leq 11 \quad \text{versus} \quad H_a: \mu > 11$$

where $\mu$ is the population mean number of times consumers use debit cards per month. Because our level of significance $\alpha$ is chosen to be 0.01, the critical region will consist of the values of $Z_{data}$ that lie in the uppermost 1% of the $Z$ distribution (see Figure 9.23). These are the extreme values of $Z_{data}$. The extreme values of $\bar{x}$ that are associated with the values of $Z$ in the critical region will thus lead us to reject $H_0$. The critical value $Z_{crit}$ is that value of $Z$ which separates the uppermost 1% of the values of $Z_{data}$ from the remaining 99%.

We find the actual value of $Z_{crit}$ using either technology or the standard normal table (see Table C in the Appendix, pages T-9–T-10). We are looking for the value of the $Z$ distribution that has 0.01 area to the right of it and thus has an area of 0.99 to the left of it. We look up the area 0.99 on the *inside* of the $Z$ table and take the closest area we can find, which is 0.9901. Then working the $Z$ table backward, we find that $Z_{crit}$ equals 2.33, as shown in Figure 9.24. Any values of $Z_{data}$ greater than $Z_{crit} = 2.33$ will lead us to reject the null hypothesis. Any values of $Z_{data}$ less than $Z_{crit} = 2.33$ will lead us to not reject the null hypothesis. ■

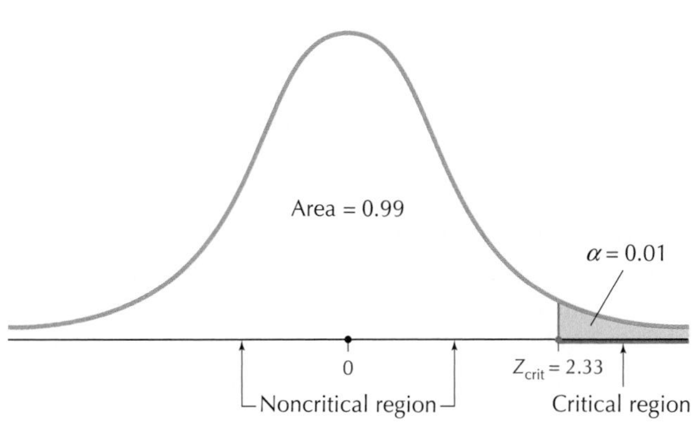


FIGURE 9.23 $Z_{crit}$ separates the critical region from the noncritical region.

| Z | 0.00 | 0.01 | 0.02 | 0.03 | 0.04 |
|---|---|---|---|---|---|
| **0.0** | 0.5000 | 0.5040 | 0.5080 | 0.5120 | 0.5160 |
| **0.1** | 0.5398 | 0.5438 | 0.5478 | 0.5517 | 0.5557 |
| **0.2** | 0.5793 | 0.5832 | 0.5871 | 0.5910 | 0.5948 |
| **0.3** | 0.6179 | 0.6217 | 0.6255 | 0.6293 | 0.6331 |
| **0.4** | 0.6554 | 0.6591 | 0.6628 | 0.6664 | 0.6700 |
| **2.0** | 0.9772 | 0.9778 | 0.9783 | 0.9788 | 0.9793 |
| **2.1** | 0.9821 | 0.9826 | 0.9830 | 0.9834 | 0.9838 |
| **2.2** | 0.9861 | 0.9864 | 0.9868 | 0.9871 | 0.9875 |
| **2.3** | 0.9893 | 0.9896 | 0.9898 | 0.9901 | 0.9904 |
| **2.4** | 0.9918 | 0.9920 | 0.9922 | 0.9925 | 0.9927 |

FIGURE 9.24 Finding $Z_{crit}$.

*Developing Your Statistical Sense*

**We Haven't Seen the Sample Data Yet**

Notice that so far we have not actually looked at any sample data for the debit card example. Since we were not given the sample mean, sample size, or population standard deviation, we cannot find $Z_{data}$ and thus cannot form a conclusion. The critical value $Z_{crit}$ does not depend at all on the sample data. $Z_{crit}$ depends only on the value of our level of significance $\alpha$ and the form of the hypothesis test. ■

Table 9.6 Table of critical values $Z_{\text{crit}}$ for common values of the level of significance $\alpha$

| | Form of Hypothesis Test | | |
|---|---|---|---|
| Level of significance $\alpha$ | **Right-tailed** $H_0: \mu \le \mu_0$ $H_a: \mu > \mu_0$ | **Left-tailed** $H_0: \mu \ge \mu_0$ $H_a: \mu < \mu_0$ | **Two-tailed** $H_0: \mu = \mu_0$ $H_a: \mu \ne \mu_0$ |
| 0.10 | $Z_{\text{crit}} = 1.28$ | $Z_{\text{crit}} = -1.28$ | $Z_{\text{crit}} = 1.645$ |
| 0.05 | $Z_{\text{crit}} = 1.645$ | $Z_{\text{crit}} = -1.645$ | $Z_{\text{crit}} = 1.96$ |
| 0.01 | $Z_{\text{crit}} = 2.33$ | $Z_{\text{crit}} = -2.33$ | $Z_{\text{crit}} = 2.58$ |

Table 9.6 shows values of $Z_{\text{crit}}$ for the most commonly used values of the level of significance $\alpha$ : 0.01, 0.05, and 0.10. In Example 9.13, the critical region consists of the values of the test statistic $Z_{\text{data}}$ that are *greater than* $Z_{\text{crit}}$ because the alternative hypothesis states that the population mean is *greater than* $\mu_0$. This is why this form of the test is called a right-tailed test—because the critical region is in the right (upper) tail.

Similarly, when you perform a left-tailed test, $H_0$: $\mu \ge \mu_0$ versus $H_a$: $\mu < \mu_0$, the critical region will consist of the values of $Z_{\text{data}}$ that are *less than* $Z_{\text{crit}}$ because the alternative hypothesis states that the population mean is *less than* $\mu_0$. This is why this form of the hypothesis test is called a left-tailed test—because the critical region is in the left (lower) tail (Figure 9.25).

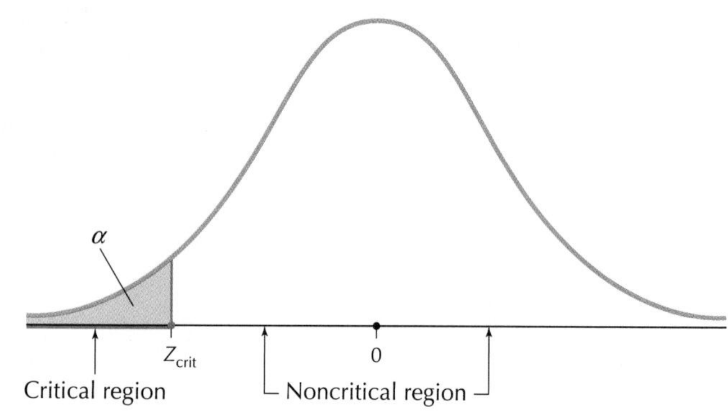


FIGURE 9.25 Critical region for a left-tailed test.

## Example 9.14 Finding the critical region for a left-tailed test

According to a report in the *New England Journal of Medicine,* a regimen of estrogen plus progestin therapy can help reduce cholesterol levels in postmenopausal women with high cholesterol.[11] The study examined whether the population mean cholesterol level could be lowered by this regimen to below 305 milligrams per deciliter (mg/dl). Find the critical region for this hypothesis test, using level of significance $\alpha = 0.05$.

### Solution

Our hypotheses are $H_0$: $\mu \ge 305$ versus $H_a$: $\mu < 305$ where $\mu$ is the population mean cholesterol level in milligrams per deciliter. Since this is a left-tailed test, and $\alpha = 0.05$, we obtain from Table 9.6 that the critical value $Z_{\text{crit}}$ is $-1.645$. Because the alternative hypothesis states that the population mean is *less than* $\mu_0$, we will reject the null hypothesis for values of the test statistic that are less than $-1.645$. ■

Finally, for a two-tailed hypothesis test, $H_0$: $\mu = \mu_0$ versus $H_a$: $\mu \ne \mu_0$, the alternative hypothesis states that the population mean is *not equal to* $\mu_0$. This can happen in either of two ways: (1) the population mean may be greater than $\mu_0$, or (2) the population mean may be less than $\mu_0$. So the critical region must contain values of $Z_{\text{data}}$ that are either extremely large (in the right tail) or extremely large negatively (in the left tail). The critical region will therefore consist of the values of $Z_{data}$ that are *greater than* $Z_{crit}$ and the values of $Z_{data}$ that are *less than* $-Z_{crit}$. This is why this form of the hypothesis test is called a two-tailed test—because the critical region is in both of the tails. Figure 9.26 shows what the critical region looks like for a two-tailed test.

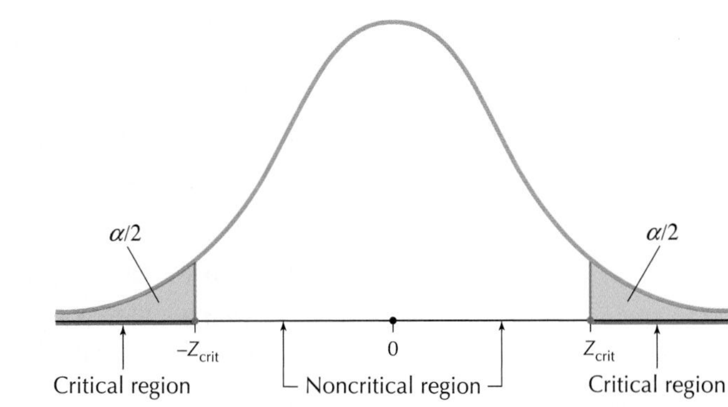


FIGURE 9.26 Critical region for a two-tailed test.

## Example 9.15 Finding the critical region for a two-tailed test

Recall Example 9.11 in Section 9.2, where either too much hemoglobin or too little hemoglobin was not desirable. We were interested in testing whether the population mean hemoglobin level differed from 13.8 g/dl, using level of significance $\alpha = 0.05$. Find the critical region for this test.

**Solution**

Recall from Example 9.11 that the hypotheses are

$$H_0: \mu = 13.8 \quad \text{versus} \quad H_a: \mu \neq 13.8$$

where $\mu$ is the population mean hemoglobin level. Then, from Table 9.6, your critical value $Z_{crit}$ is 1.96. Because it is a two-tailed hypothesis test, we reject $H_0$ for values of $Z_{data}$ that are either larger than 1.96 or smaller than $-1.96$.

Table 9.7 summarizes the rejection rules for each form of the hypothesis test. Notice that you are essentially comparing the test statistic $Z_{data}$ with the critical value $Z_{crit}$ and basing your decision on their relative values.

Table 9.7 Rejection rules for $Z$ test for the mean

| Form of test | | Rejection rules: "Reject $H_0$ if..." |
|---|---|---|
| Right-tailed | $H_0: \mu \leq \mu_0$ vs. $H_a: \mu > \mu_0$ | $Z_{data} > Z_{crit}$ |
| Left-tailed | $H_0: \mu \geq \mu_0$ vs. $H_a: \mu < \mu_0$ | $Z_{data} < Z_{crit}$ |
| Two-tailed | $H_0: \mu = \mu_0$ vs. $H_a: \mu \neq \mu_0$ | $Z_{data} > Z_{crit}$ or $Z_{data} < -Z_{crit}$ |

## 2 Performing the $Z$ Test for the Population Mean Using the Critical-Value Method

Now that we know how to find the critical value and the critical region, we are ready to carry out an actual hypothesis test. *It is important to understand that the critical-value method for the Z test is exactly equivalent to the p-value method for the Z test.*

Therefore, the conditions under which we may use the $Z$ test critical-value method are the same as the conditions for using the $Z$ test $p$-value method. If the population standard deviation $\sigma$ is not known, then the $Z$ test should not be applied.

**$Z$ Test for the Population Mean $\mu$: Critical-Value Method**

When a random sample of size $n$ is taken from a population where the population standard deviation $\sigma$ is known, you can use the $Z$ test if either of the following conditions is satisfied:

- Case 1: the population is normal, or
- Case 2: the sample size is large ($n \geq 30$).

***Step 1*** **State the hypotheses.** Use one of the forms from Table 9.1. Define $\mu$.

***Step 2*** **Find $Z_{crit}$ and state the rejection rule.** Use Tables 9.6 and 9.7.

***Step 3*** **Find $Z_{data}$.** Either use technology to find the value of the test statistic $Z_{data}$ or calculate the value of $Z_{data}$ as follows:

$$Z_{data} = \frac{\bar{x} - \mu_0}{\sigma_{\bar{x}}} = \frac{\bar{x} - \mu_0}{\sigma/\sqrt{n}}$$

***Step 4*** **State the conclusion and the interpretation.** If $Z_{data}$ falls in the critical region, then reject $H_0$. Otherwise, do not reject $H_0$. Interpret your conclusion so that a nonspecialist can understand.

## Example 9.16 Critical-value method for the $Z$ test for $\mu$

Let's use the critical-value method to test, using level of significance $\alpha = 0.01$, whether people use debit cards on average more than 11 times per month (Example 9.13). Suppose a random sample of 36 people used debit cards last month an average of $\bar{x} = 11.5$ times. Assume that the population standard deviation equals 3.

**Solution**

***Step 1*** **State the hypotheses.** In Example 9.13 our hypotheses were

$$H_0: \mu \leq 11 \quad \text{versus} \quad H_a: \mu > 11$$

where $\mu$ is the population mean number of times people use their debit cards per month.

***Step 2*** **Find $Z_{crit}$ and state the rejection rule.** We have a right-tailed test and level of significance $\alpha = 0.01$, which, from Table 9.6, tell us that $Z_{crit} = 2.33$. Using Table 9.7, because we have a right-tailed test, the rejection rule will be "Reject $H_0$ if $Z_{data}$ is greater then $Z_{crit}$"; that is, "Reject $H_0$ if $Z_{data}$ is greater than 2.33" (see Figure 9.27).

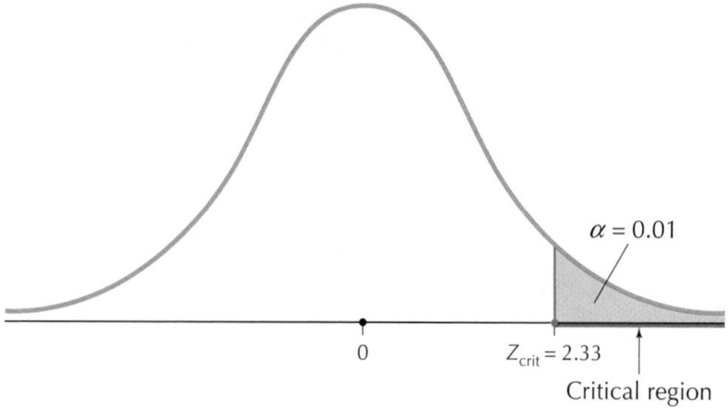


FIGURE 9.27

**Step 3** **Find $Z_{data}$.** We have a sample of $n = 36$ people who used their debit cards an average of $\bar{x} = 11.5$ times, and we assume $\sigma = 3$. Also, $\mu_0$ is the hypothesized value of $\mu$, in this case 11. Using these values, we get

$$Z_{data} = \frac{\bar{x} - \mu_0}{\sigma_{\bar{x}}} = \frac{\bar{x} - \mu_0}{\sigma/\sqrt{n}} = \frac{11.5 - 11.0}{3/\sqrt{36}} = 1.0$$

**Step 4** **State the conclusion and interpretation.** Our rejection rule states that we will reject $H_0$ if $Z_{data}$ is greater than 2.33. Since $Z_{data}$ equals 1.0, which is *not* greater than 2.33, the conclusion is to *not* reject $H_0$ (Figure 9.28). Even though the sample mean of 11.5 exceeds 11.0, it does not do so by a wide enough margin to dispel the reasonable doubt that the difference between this sample mean and $\mu$ may have been due to chance. We interpret our conclusion as follows: "There is insufficient evidence that the population mean monthly debit card use is greater than 11 times per month." ■

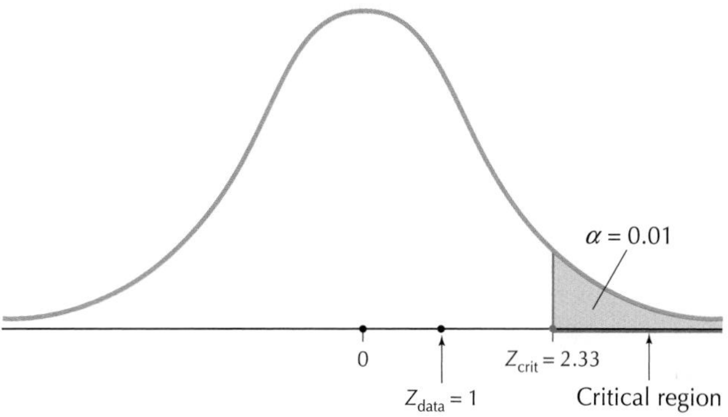


FIGURE 9.28

## 3 The Relationship Between the *p*-Value Method and the Critical-Value Method

Figure 9.29 shows the relationships between the *p*-value method and the critical-value method. The top half represents values of *Z* and the critical-value method that we studied in this section. The bottom half represents probabilities and the *p*-value method that we studied in Section 9.2. The left half represents statistics associated with the observed sample data. The right half represents critical-value thresholds for significance that these statistics are compared against.

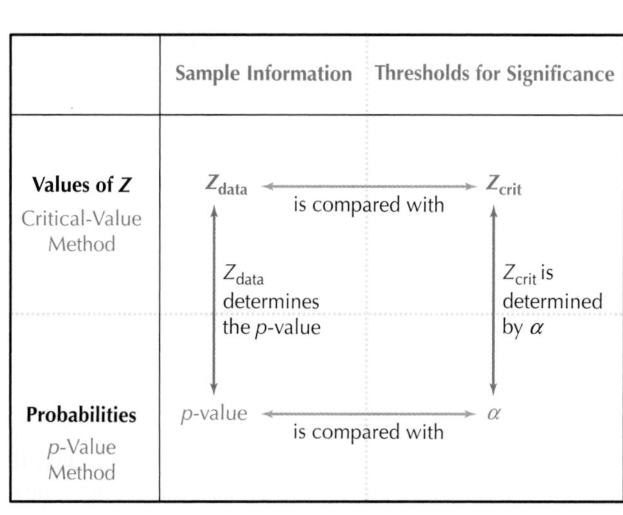


FIGURE 9.29
Critical-value method and *p*-value method are equivalent.

Since $Z_{\text{data}}$ helps us to determine the $p$-value, these two values are related. Similarly, since the level of significance $\alpha$ helps to determine the value of $Z_{\text{crit}}$, these two values are related. Moreover, just as we compare $Z_{\text{data}}$ with the threshold $Z_{\text{crit}}$, we compare the $p$-value statistic with the $\alpha$ threshold to determine significance. Thus, the two methods for carrying out hypothesis tests are equivalent and, in fact, are quite thoroughly interwoven.

Figures 9.30a and 9.30b illustrate this equivalence for a right-tailed test. The rejection rule for the $p$-value method is to reject $H_0$ when the $p$-value is less than $\alpha$. The rejection rule for the critical-value method is to reject $H_0$ when $Z_{\text{data}} > Z_{\text{crit}}$. Note in Figures 9.30a and 9.30b how the $p$-value is determined by $Z_{\text{data}}$, and $\alpha$ is determined by $Z_{\text{crit}}$. In Figure 9.30a, when $Z_{\text{data}} < Z_{\text{crit}}$, it must also happen that the $p$-value is larger than $\alpha$. In both cases we do not reject $H_0$. However, in Figure 9.30b when $Z_{\text{data}} > Z_{\text{crit}}$, it also follows that the $p$-value is smaller than $\alpha$. In both cases we reject $H_0$. Thus, the $p$-value method and the critical-value method are equivalent.

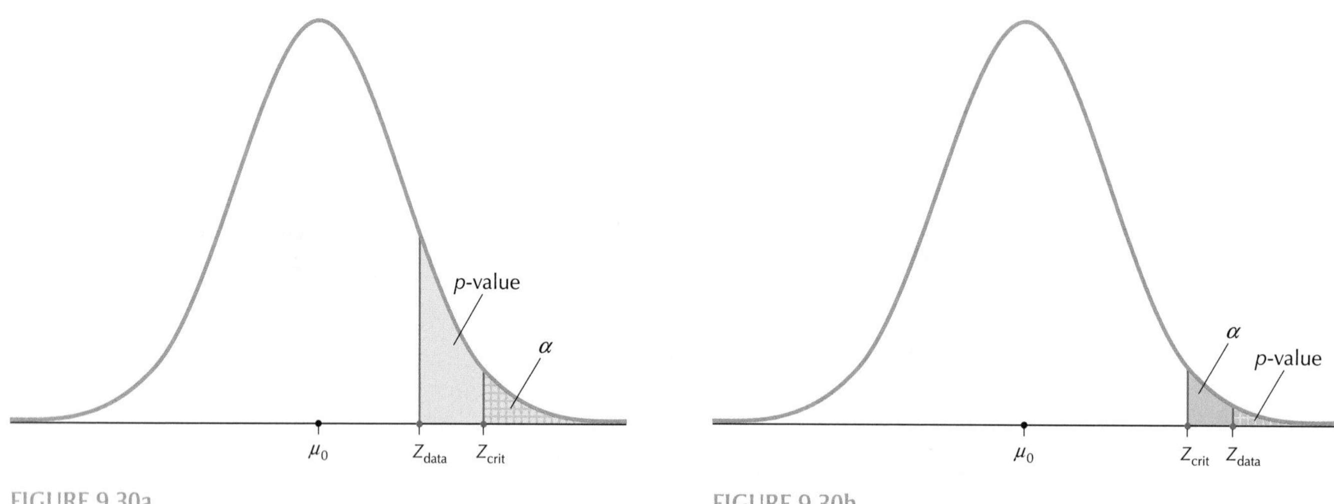


FIGURE 9.30a

FIGURE 9.30b

## 4 Using Confidence Intervals for $\mu$ to Perform Two-Tailed Hypothesis Tests About $\mu$

Since our $Z$ interval for $\mu$ from Chapter 8 and our $Z$ test here in Chapter 9 are both founded on the behavior of the $Z$ statistic, namely

$$Z = \frac{\bar{x} - \mu_0}{\sigma_{\bar{x}}} = \frac{\bar{x} - \mu_0}{\sigma/\sqrt{n}}$$

these two inference methods are equivalent. Here we show how to use a single confidence interval to test as many possible values for $\mu_0$ as one wants. If a certain value for $\mu_0$ is rejected in a two-tailed $Z$ test with level of significance $\alpha$, then $\mu_0$ will lie *outside* the $100(1 - \alpha)\%$ $Z$ confidence interval for $\mu$. Thus, $\mu_0$ is not a plausible value for $\mu$ (see Figure 9.31).

FIGURE 9.31 Reject $H_0$ for values of $\mu_0$ that lie outside confidence interval.

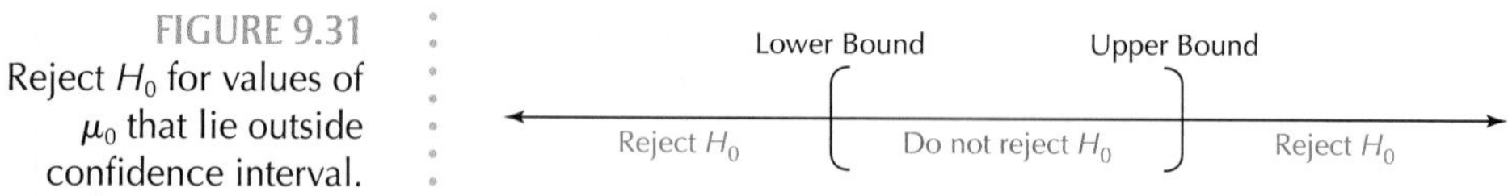


Alternatively, if a certain value for $\mu_0$ is not rejected in a two-tailed $Z$ test with level of significance $\alpha$, then $\mu_0$ will lie *inside* the $100(1 - \alpha)\%$ $Z$ confidence interval

Table 9.8 Confidence levels for equivalent $\alpha$ levels of significance

| **Confidence level for $100(1-\alpha)\%$ confidence interval** | **Level of significance $\alpha$ for two-tailed test $H_0$: $\mu = \mu_0$ versus $H_a$: $\mu \neq \mu_0$** |
|---|---|
| 90% | 0.10 |
| 95% | 0.05 |
| 99% | 0.01 |

for $\mu$ and is a plausible value for $\mu$. Table 9.8 shows the confidence levels and associated $\alpha$ levels of significance that will produce the equivalent inference.

## Example 9.17 Breakfast cereal sodium content, revisited

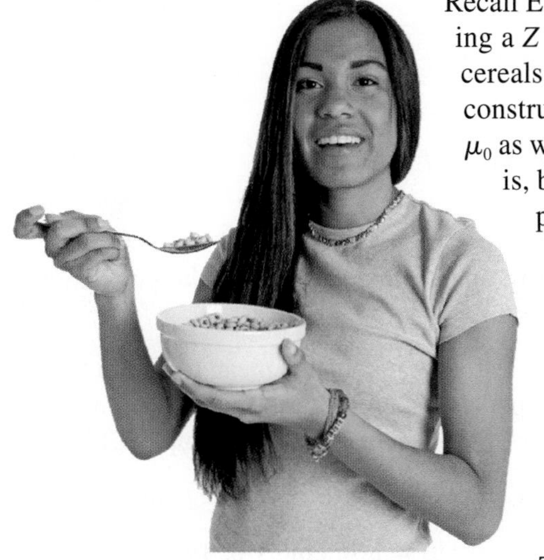

©Dynamic Graphics/Jupiterimages

Recall Example 8.5 from Section 8.1 (page 395), where we were 99% confident using a $Z$ interval that the population mean sodium content per serving for breakfast cereals that contain sodium lies between 165.53 and 219.25 grams. Once we have constructed the 99% confidence interval, we may test as many possible values for $\mu_0$ as we like. If any prospective values of $\mu_0$ lie inside the confidence interval, that is, between 165.53 and 219.25, we will not reject $H_0$ for this value of $\mu_0$. If any prospective values of $\mu_0$ lie outside the confidence interval, that is, either to the left of 165.53 or to the right of 219.25, we will reject $H_0$, as shown in Figure 9.32.

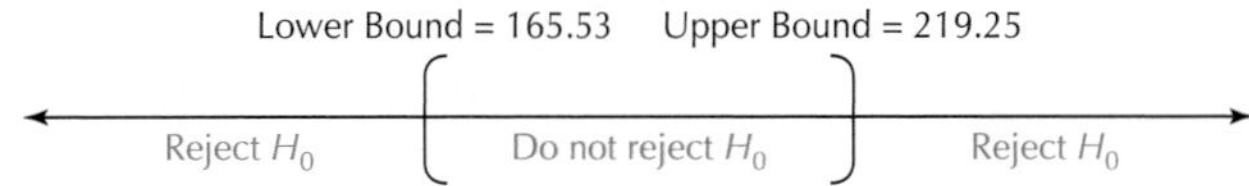


FIGURE 9.32 Reject $H_0$ for values of $\mu_0$ that lie outside (165.53, 219.25).

Test using level of significance $\alpha = 0.01$ whether the population mean sodium content per serving differs from these values: **(a)** 165, **(b)** 166, **(c)** 220.

### Solution

We set up the three two-tailed hypothesis tests as follows:

**a.** $H_0: \mu = 165$ vs. $H_a: \mu \neq 165$
**b.** $H_0: \mu = 166$ vs. $H_a: \mu \neq 166$
**c.** $H_0: \mu = 220$ vs. $H_a: \mu \neq 220$

To perform each hypothesis test, simply observe where each value of $\mu_0$ falls on the number line shown in Figure 9.32. For example, in the first hypothesis test, the hypothesized value $\mu_0 = 165$ lies outside the interval (165.53, 219.25). Thus, we reject $H_0$. The three hypothesis tests are summarized here. ■

| **Value of $\mu_0$** | **Form of hypothesis test, with $\alpha = 0.01$** | **Where $\mu_0$ lies in relation to 99% confidence interval** | **Conclusion of hypothesis test** |
|---|---|---|---|
| **a.** 165 | $H_0: \mu = 165$ vs. $H_a: \mu \neq 165$ | Outside | Reject $H_0$ |
| **b.** 166 | $H_0: \mu = 166$ vs. $H_a: \mu \neq 166$ | Inside | Do not reject $H_0$ |
| **c.** 220 | $H_0: \mu = 220$ vs. $H_a: \mu \neq 220$ | Outside | Reject $H_0$ |

## 5 Probability of a Type II Error and the Power of a Hypothesis Test

In Section 9.1 we defined a Type II error as follows:

Type II error: not rejecting $H_0$ when $H_0$ is false

In this section we learn how to calculate the probability of making a Type II error, called $\beta$ (beta), and to use the value of $\beta$ to compute the power of a hypothesis test.

### Calculating $\beta$, the Probability of a Type II Error

Use the following steps to calculate $\beta$, the probability of a Type II error.

***Step 1*** Recall that $Z_{crit}$ divides the critical region from the noncritical region. Let $\bar{x}_{crit}$ be the value of the sample mean $\bar{x}$ associated with $Z_{crit}$. The following table shows how to calculate $\bar{x}_{crit}$ for the three forms of the hypothesis test.

| Form of test | | Value of $\bar{x}_{crit}$ |
|---|---|---|
| **Right-tailed** | $H_0: \mu \le \mu_0$ vs. $H_a: \mu > \mu_0$ | $\bar{x}_{crit} = \mu_0 + Z_{crit} \cdot \frac{\sigma}{\sqrt{n}}$ |
| **Left-tailed** | $H_0: \mu \ge \mu_0$ vs. $H_a: \mu < \mu_0$ | $\bar{x}_{crit} = \mu_0 - Z_{crit} \cdot \frac{\sigma}{\sqrt{n}}$ |
| **Two-tailed** | $H_0: \mu = \mu_0$ vs. $H_a: \mu \ne \mu_0$ | $\bar{x}_{crit,lower} = \mu_0 - Z_{crit} \cdot \frac{\sigma}{\sqrt{n}}$ <br> $\bar{x}_{crit,upper} = \mu_0 + Z_{crit} \cdot \frac{\sigma}{\sqrt{n}}$ |

Here, $\mu_0$ is the hypothesized value of the population mean, $\sigma$ is the population standard deviation, and $n$ is the sample size.

***Step 2*** Let $\mu_a$ represent a particular value for the population mean $\mu$ chosen from the values indicated in the alternative hypothesis $H_a$. Draw a normal curve centered at $\mu_a$, with the value or values of $\bar{x}_{crit}$ from Step 1 indicated (see Example 9.18).

***Step 3*** Calculate $\beta$ for the particular $\mu_a$ chosen using the following table.

| Form of test | | $\beta$ = probability of Type II error |
|---|---|---|
| **Right-tailed** | $H_0: \mu \le \mu_0$ vs. $H_a: \mu > \mu_0$ | The area under the normal curve drawn in Step 2 to the left of $\bar{x}_{crit}$. |
| **Left-tailed** | $H_0: \mu \ge \mu_0$ vs. $H_a: \mu < \mu_0$ | The area under the normal curve drawn in Step 2 to the right of $\bar{x}_{crit}$. |
| **Two-tailed** | $H_0: \mu = \mu_0$ vs. $H_a: \mu \ne \mu_0$ | The area under the normal curve drawn in Step 2 between $\bar{x}_{crit,lower}$ and $\bar{x}_{crit,upper}$. |

Let us illustrate the steps for calculating $\beta$, the probability of a Type II error, using an example.

### Example 9.18 Calculating $\beta$, the probability of a Type II error

In Example 9.16, we tested whether people use debit cards on average more than 11 times per month. The hypotheses are

$$H_0: \mu \le 11 \quad \text{versus} \quad H_a: \mu > 11$$

where $\mu$ represents the population mean debit card usage per month. From Example 9.16 we have $n = 36$, $\bar{x} = 11.5$, $Z_{crit} = 2.33$, and $\sigma = 3$.

**a.** State what a Type II error would be in this case.

**b.** Let $\mu_a = 13$. That is, suppose the population mean debit card usage is actually 13 times per month. Calculate $\beta$, the probability of making a Type II error when $\mu_a = 13$.

**Solution**

**a.** We make a Type II error when we do not reject $H_0$ when $H_0$ is false. In this case, a Type II error would be to conclude that the population mean debit card usage was at most 11 times per month when in actuality it was more than 11 times per month.

**b.** We follow the steps for calculating $\beta$.

***Step 1*** We have a right-tailed test, so that

$$\bar{x}_{\text{crit}} = \mu_0 + Z_{\text{crit}} \cdot \frac{\sigma}{\sqrt{n}} = 11 + 2.33 \cdot \frac{3}{\sqrt{36}} = 12.165$$

***Step 2*** Figure 9.33 shows the normal curve centered at $\mu_a = 13$, with $\bar{x}_{\text{crit}} = 12.165$ labeled.

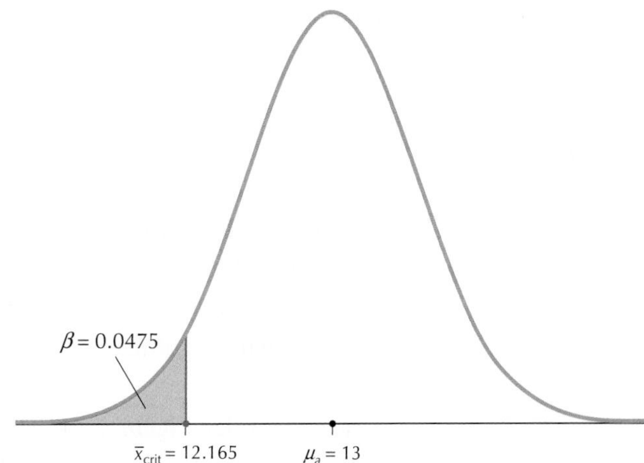


FIGURE 9.33

***Step 3*** The right-tailed test tells us that $\beta$ equals the area under the normal curve drawn in Step 2 to the left of $\bar{x}_{\text{crit}} = 12.165$. This is the shaded area in Figure 9.33. Since area represents probability, we have

$$\beta = P(\bar{x} < 12.165) \text{ when } \mu_a = 13$$

Standardizing with $\mu_a = 13$, $\sigma = 3$, and $n = 36$:

$$\begin{aligned}\beta &= P(\bar{x} < 12.165)\\ &= P\left(Z < \frac{12.165 - 13}{3/\sqrt{36}}\right)\\ &= P(Z < -1.67) = 0.0475\end{aligned}$$

Thus, $\beta = 0.0475$. This represents the probability of making a Type II error, that is, of not rejecting the hypothesis that the population mean debit card usage is at most 11 times per month when in actuality it is 13 times per month. ■

## Power of a Hypothesis Test

It is a correct decision to reject the null hypothesis when the null hypothesis is false. The probability of making this type of correct decision is called the **power of the test**.

**Power of a Hypothesis Test**

The **power of a hypothesis test** is the probability of rejecting the null hypothesis when the null hypothesis is false. Power is calculated as

$$\text{power} = 1 - \beta$$

## Example 9.19 Power of a hypothesis test

Calculate the power, for a particular alternative value of the mean, of the hypothesis test in Example 9.18.

### Solution

The probability of a Type II error was found in Example 9.18 to be $\beta = 0.0475$. Thus, the power of the hypothesis test is

$$\text{power} = 1 - \beta = 1 - 0.0475 = 0.9525$$

The probability of correctly rejecting the null hypothesis is 0.9525.

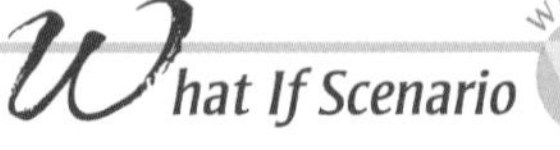


## Type II Error and Power of the Test

Suppose that we have the same hypothesis test from Example 9.18 and the same value $\bar{x}_{\text{crit}} = 12.165$. Now, *what if* we decrease $\mu_a$ such that it is less than 13 but still larger than 12.165. Describe what will happen to the following, and why.

**a.** The probability of a Type II error, $\beta$

**b.** The power of the test, $1 - \beta$

### Solution

**a.** Consider Figure 9.34. The distribution of sample means remains centered at $\mu_a$, so that a smaller $\mu_a$ will "slide" the normal curve toward the value of $\bar{x}_{\text{crit}} = 12.165$. This results in a larger area to the left of 12.165, as you can see by comparing Figure 9.34 with Figure 9.33. Therefore, a smaller $\mu_a$ leads to an *increase* in the probability of a Type II error, $\beta$.

**b.** As $\beta$ increases, $1 - \beta$ decreases. Therefore, a smaller $\mu_a$ leads to a decrease in the power of the test.

FIGURE 9.34

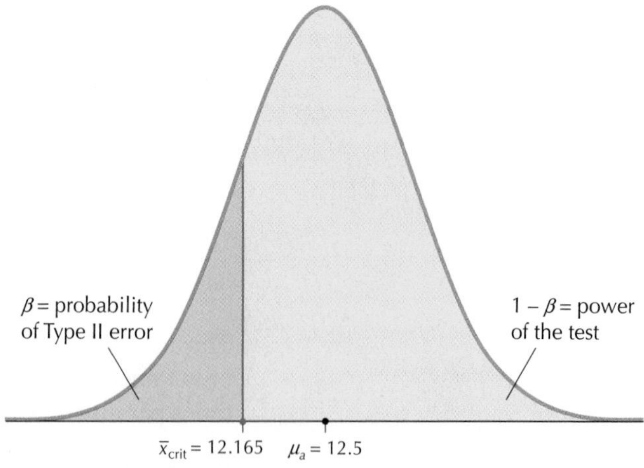


A **power curve** plots the values for the power of the test versus the values of $\mu_a$.

## Example 9.20 Power curve

**a.** Calculate the power of the hypothesis test from Example 9.18 for the following values of $\mu_a$: 11.0, 11.5, 12.0, 12.165, 12.5, 13.5.

**b.** Construct the power curve by graphing the values for the power of the test on the vertical axis against the values of $\mu_a$ on the horizontal axis.

### Solution

**a.** We have $\bar{x}_{\text{crit}} = 12.165$, $\sigma = 3$, and $n = 36$. The calculations are provided in the following table.

| $\mu_a$ | Probability of Type II error: $\beta$ | Power of the test: $1 - \beta$ |
|---|---|---|
| 11.0 | $P\left(Z < \frac{12.165 - 11}{3/\sqrt{36}}\right) = P(Z < 2.33) = 0.9901$ | $1 - 0.9901 = 0.0099$ |
| 11.5 | $P\left(Z < \frac{12.165 - 11.5}{3/\sqrt{36}}\right) = P(Z < 1.33) = 0.9082$ | $1 - 0.9082 = 0.0918$ |
| 12.0 | $P\left(Z < \frac{12.165 - 12}{3/\sqrt{36}}\right) = P(Z < 0.33) = 0.6293$ | $1 - 0.6293 = 0.3707$ |
| 12.165 | $P\left(Z < \frac{12.165 - 12.165}{3/\sqrt{36}}\right) = P(Z < 0.00) = 0.5$ | $1 - 0.5 = 0.5$ |
| 12.5 | $P\left(Z < \frac{12.165 - 12.5}{3/\sqrt{36}}\right) = P(Z < -0.67) = 0.2514$ | $1 - 0.2514 = 0.7486$ |
| 13.5 | $P\left(Z < \frac{12.165 - 13.5}{3/\sqrt{36}}\right) = P(Z < -2.67) = 0.0038$ | $1 - 0.0038 = 0.9962$ |

**b.** Figure 9.35 represents a power curve, since it plots the values for the power of the test on the vertical axis against the values of $\mu_a$ on the horizontal axis. Note that, as $\mu_a$ moves farther away from the hypothesized mean $\mu_0 = 11$, the power of the test increases. This is because it is more likely that the null hypothesis will be correctly rejected as the actual value of the mean $\mu_a$ gets farther away from the hypothesized value $\mu_0$. ■

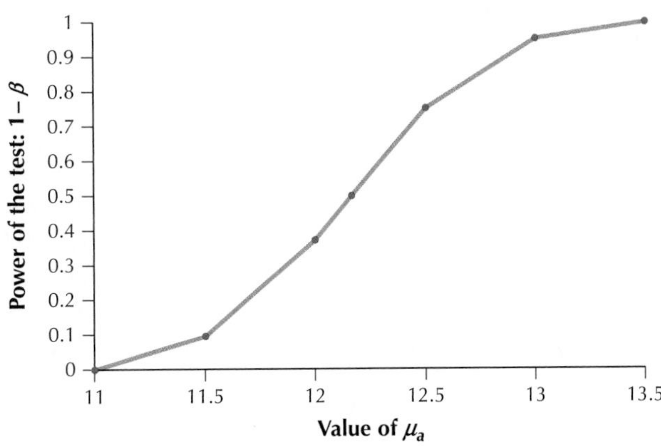


FIGURE 9.35

## Section 9.3 SUMMARY

1 The critical region consists of the range of values of the test statistic $Z_{data}$ for which we reject the null hypothesis. The value of $Z$ that separates the critical region from the noncritical region is called the critical value $Z_{crit}$.

2 The critical-value method of performing the $Z$ test for the mean is equivalent to the $p$-value method in that the same conclusion would be drawn using either method. In the critical-value method, we compare one $Z$-value ($Z_{data}$) with another $Z$-value ($Z_{crit}$).

3 The $p$-value method and the critical-value method are equivalent. In fact, $Z_{data}$, the $p$-value, the level of significance $\alpha$, and $Z_{crit}$ are all interrelated.

4 We can use a single $100(1 - \alpha)\%$ confidence interval for $\mu$ to help us perform any number of two-tailed hypothesis tests about $\mu$.

5 We can calculate $\beta$, the probability of a Type II error, for a given value of $\mu_a$. The power of the hypothesis test is the probability of rejecting the null hypothesis when the null hypothesis is false. Power is calculated as power $= 1 - \beta$.

## Section 9.3 EXERCISES

### CLARIFYING THE CONCEPTS

For Exercises 1–3, indicate whether the quantity represents sample information or a threshold for significance.

1. $Z_{crit}$
2. $p$-value
3. $\alpha$

For Exercises 4–6, indicate whether or not the quantity represents a probability.

4. $Z_{data}$
5. $p$-value
6. $\alpha$

For each of the following pairs, indicate whether they are directly related. Describe how they are related. (What do they have in common, if anything? Can we find one given the other?)

7. $Z_{data}$ and $Z_{crit}$
8. $p$-value and $\alpha$
9. $Z_{data}$ and $p$-value
10. $Z_{data}$ and $\alpha$
11. $p$-value and $Z_{crit}$
12. $Z_{crit}$ and $\alpha$

### PRACTICING THE TECHNIQUES

For Exercises 13–20, assume that the conditions for performing the $Z$ test are met. Do the following.

**a.** Find the value of $Z_{crit}$.
**b.** Find the critical-value rejection rule.
**c.** Draw a standard normal curve and indicate the critical region.

13. $H_0: \mu \geq \mu_0$ vs. $H_a: \mu < \mu_0$, $\alpha = 0.10$
14. $H_0: \mu \leq \mu_0$ vs. $H_a: \mu > \mu_0$, $\alpha = 0.05$
15. $H_0: \mu \geq \mu_0$ vs. $H_a: \mu < \mu_0$, $\alpha = 0.01$
16. $H_0: \mu \leq \mu_0$ vs. $H_a: \mu > \mu_0$, $\alpha = 0.10$
17. $H_0: \mu \geq \mu_0$ vs. $H_a: \mu < \mu_0$, $\alpha = 0.05$
18. $H_0: \mu = \mu_0$ vs. $H_a: \mu \neq \mu_0$, $\alpha = 0.05$
19. $H_0: \mu = \mu_0$ vs. $H_a: \mu \neq \mu_0$, $\alpha = 0.01$
20. $H_0: \mu = \mu_0$ vs. $H_a: \mu \neq \mu_0$, $\alpha = 0.10$

For each of the following hypothesis tests in Exercises 21–26, assume that the conditions for performing the $Z$ test are met. Do the following.

**a.** Find the value of $Z_{crit}$.
**b.** Find the critical-value rejection rule.
**c.** Draw a standard normal curve and indicate the critical region.
**d.** State the conclusion.
**e.** State the interpretation.

21. $H_0: \mu \leq \mu_0$ vs. $H_a: \mu > \mu_0$, $\alpha = 0.05$, $Z_{data} = 2.0$
22. $H_0: \mu \leq \mu_0$ vs. $H_a: \mu > \mu_0$, $\alpha = 0.05$, $Z_{data} = 1.5$
23. $H_0: \mu \leq \mu_0$ vs. $H_a: \mu > \mu_0$, $\alpha = 0.05$, $Z_{data} = 1.6$
24. $H_0: \mu \geq \mu_0$ vs. $H_a: \mu < \mu_0$, $\alpha = 0.10$, $Z_{data} = -1.0$
25. $H_0: \mu \geq \mu_0$ vs. $H_a: \mu < \mu_0$, $\alpha = 0.10$, $Z_{data} = -1.3$
26. $H_0: \mu = \mu_0$ vs. $H_a: \mu \neq \mu_0$, $\alpha = 0.01$, $Z_{data} = 2.0$

For Exercises 27–30, do the following.

**a.** State the hypotheses.
**b.** Find $Z_{crit}$ and the rejection rule.
**c.** Find $Z_{data}$. Also, draw a standard normal curve showing $Z_{crit}$, the critical region, and $Z_{data}$.
**d.** Formulate and interpret the conclusion.

27. We are interested in testing at level of significance $\alpha = 0.05$ whether the population mean is greater than 100. A random sample of size 64 is taken, with a mean of 102. Assume $\sigma = 8$.

28. We would like to test at level of significance $\alpha = 0.10$ whether the population mean is less than 3.0. A random sample of size 36 is taken, with a mean of 2.9. Assume $\sigma = 0.5$.

29. We want to test at level of significance $\alpha = 0.01$ whether the population mean differs from 20. A random sample of size 49 is taken, with a mean of 27. Assume $\sigma = 5$.

30. We are interested in testing at level of significance $\alpha = 0.05$ whether the population mean differs from 500. A random sample of size 100 is taken, with a mean of 520. Assume $\sigma = 50$.

31. A 95% $Z$ confidence interval for the population mean is (–2.7, 6.9). Using the equivalence between $Z$ intervals and $Z$ tests, test using level of significance $\alpha = 0.05$ whether the population mean differs from each of the following hypothesized values.

**a.** –3 **b.** –2 **c.** 0 **d.** 5 **e.** 7

**32.** A 99% $Z$ confidence interval for the population mean is (45, 55). Using the equivalence between $Z$ intervals and $Z$ tests, test using level of significance $\alpha = 0.01$ whether the population mean differs from each of the following hypothesized values.

**a.** 0 **b.** 44 **c.** 50 **d.** 54 **e.** 56

## APPLYING THE CONCEPTS

For Exercises 33–36, do the following.

- **a.** State the hypotheses.
- **b.** Find $Z_{crit}$ and the critical region. Also, draw a standard normal curve showing the critical region.
- **c.** Find $Z_{data}$. Also, draw a standard normal curve showing $Z_{crit}$, the critical region, and $Z_{data}$.
- **d.** State the conclusion and the interpretation.

**33. Price of Milk.** The Bureau of Labor Statistics reports that the mean price for a gallon of milk in 2005 was \$3.24. A random sample of 100 retail establishments this year provides a mean price of \$3.40. Assume $\sigma = \$0.25$. Perform a hypothesis test using level of significance $\alpha = 0.05$ to investigate whether the population mean price of milk this year has increased from the 2005 value.

**34. Household Size.** The U.S. Census Bureau reported that the mean household size in 2002 was 2.58 persons. A random sample of 900 households this year provides a mean size of 2.5 persons. Assume $\sigma = 0.1$. Conduct a hypothesis test using level of significance $\alpha = 0.10$ to determine whether the population mean household size this year has decreased from the 2002 level.

**35. Salaries of Assistant Professors.** Suppose that the mean salary for assistant professors in science in 2005 was \$50,000. A random sample of 81 assistant professors in science taken this year shows a mean salary of \$51,750. Assume $\sigma = \$1000$. Perform a hypothesis test using level of significance $\alpha = 0.01$ to determine whether the population mean salary of assistant professors in science has increased since 2005.

**36. Americans' Height.** A random sample of 400 Americans yields a mean height of 176 centimeters. Assume $\sigma = 2.5$ centimeters. Conduct a hypothesis test to investigate whether the population mean height of Americans this year has changed from 175 centimeters, using level of significance $\alpha = 0.05$.

**37. Cost of Education.** The College Board reports that the mean annual cost of education at a private four-year college was \$22,218 for the 2006–2007 school year. Suppose that a random sample of 49 private four-year colleges this year gives a mean cost of \$24,000 per year. Assume the population standard deviation is \$3000.

- **a.** Construct a 95% confidence interval for the population mean annual cost.
- **b.** Use the confidence interval to test at level of significance $\alpha = 0.05$ whether the population mean annual cost differs from the following amounts.
  - **i.** \$24,000
  - **ii.** \$23,000
  - **iii.** \$23,200
  - **iv.** \$25,000

**Health Care Premiums.** Use the following information for Exercises 38–40. According to the National Coalition on Health Care, the mean annual premium for an employer health plan covering a family of four cost \$11,500 in 2007. A random sample of 100 families of four taken this year showed a mean annual premium of \$11,750. Assume $\sigma = \$1500$.

**38.** Test whether the population mean annual premium has increased, using level of significance $\alpha = 0.05$.

WHAT IF? **39.** *What if* the sample mean premium equaled some value larger than \$11,750, while everything else stayed the same. Explain how this change would affect the following, if at all.

- **a.** The hypotheses
- **b.** $Z_{crit}$
- **c.** The critical region
- **d.** $Z_{data}$
- **e.** The conclusion

**40.** Test whether the population mean annual premium has increased, using level of significance $\alpha = 0.01$. Comment.

**41. Accountants' Salaries.** According to the *Wall Street Journal,* the mean salary for accountants in Texas in 2007 was \$50,529. A random sample of 16 Texas accountants this year showed a mean salary of \$52,000. We assume that the population standard deviation equals \$4000. The histogram of the salary (in \$1000s) is shown here. If it is appropriate to apply the $Z$ test, then do so, using the critical-value method and level of significance $\alpha = 0.05$. If not, then explain clearly why not.

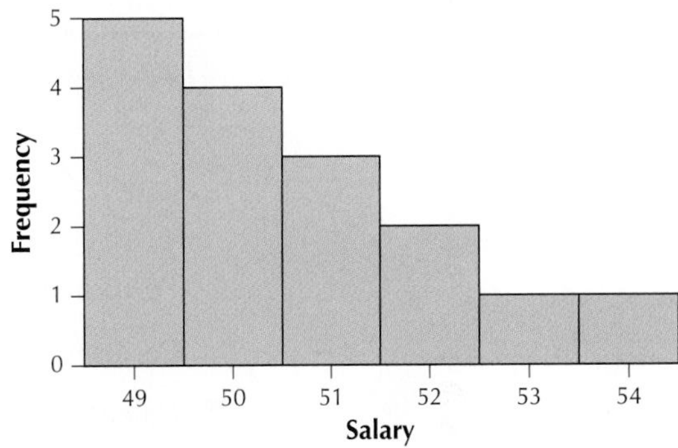


Salaries of 16 accountants.

**42. Honda Civic Gas Mileage.** Cars.com reported in 2007 that the mean city gas mileage for the Honda Civic was 30 mpg. This year, a random sample of 20 Honda Civics had a mean gas mileage of 36 mpg. Assume $\sigma = 5$ mpg. A Minitab histogram of the data is shown here.

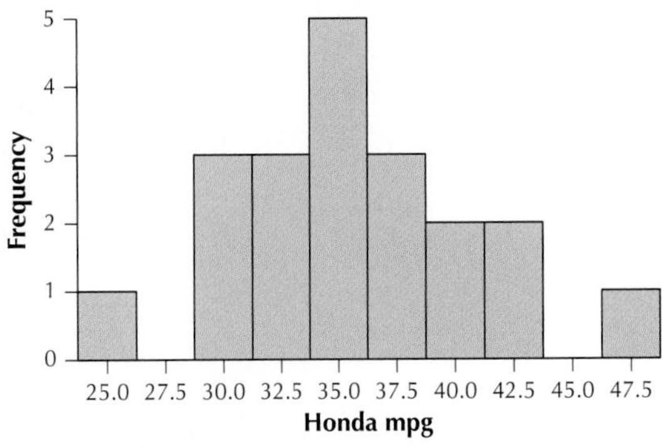


Miles per gallon of 20 imported Hondas.

**a.** Is it appropriate to apply the $Z$ test? Explain clearly why or why not.

**b.** Test at level of significance $\alpha = 0.10$ whether the population mean city gas mileage has increased since 2007.

**c.** What if we now performed the same test on the same data but used $\alpha = 0.05$ instead. Without carrying out the hypothesis test, state whether this would affect our conclusion. Why or why not?

**43. Automobile Operation Cost.** The Bureau of Transportation Statistics reports that in 2002 the mean cost of operating an automobile in the United States, including gas and oil, maintenance and tires, was 5.9 cents per mile. Suppose that a sample taken this year of 100 automobiles shows a mean operating cost of 6.2 cents per mile, and assume that the population standard deviation is 1.5 cents per mile. We are interested in whether the population mean cost has increased since 2002, using the critical-value method and level of significance $\alpha = 0.05$.

**a.** Is it appropriate to apply the $Z$ test? Why or why not?

**b.** We have a sample mean that is greater than the mean in the null hypothesis of 5.9 cents. Isn't this enough by itself to reject the null hypothesis? Explain why or why not.

**c.** What is the standard deviation of $\bar{x}$, $\sigma/\sqrt{n}$?

**d.** How many standard deviations above the mean is the 6.2 cents per mile? Do you think this is extreme?

WHAT IF? **44. Honda Civic Gas Mileage.** Refer to the Exercise 42. Try to answer the following questions by thinking about the relationship between the statistics rather than by redoing all the calculations. Suppose that the 36 mpg is a typo. We are not sure what the actual sample mean is, but it is less than 36 mpg.

**a.** How does this affect $Z_{\text{data}}$?

**b.** How does this affect $Z_{\text{crit}}$?

**c.** How does this affect the conclusion?

**45. Automobile Operation Cost.** Refer to Exercise 43.

**a.** Construct the hypotheses.

**b.** Find the $Z$ critical value and state the rejection rule.

**c.** Calculate the value of the test statistic $Z_{\text{data}}$.

**d.** State the conclusion and the interpretation.

## 9.4 $t$ Test for the Population Mean $\mu$

**Objectives:** By the end of this section, I will be able to…

1. Perform the $t$ test for the mean using the $p$-value method.
2. Perform the $t$ test for the mean using the critical-value method.
3. Calculate the standard error, and use it to explain the meaning of the test statistic.

In many real-world scenarios, the value of the population standard deviation $\sigma$ is unknown. When this occurs, we should use *neither* the $Z$ interval nor the $Z$ test. Recall from Chapter 8 that, when the population standard deviation is unknown, and we have either a normal population or a large sample, we use a $t$ interval for the mean. The situation is similar for hypothesis testing: when the value of $\sigma$ is unknown, we should see whether we may apply the $t$ test instead.

### 1 $t$ Test for $\mu$ Using the $p$-Value Method

Recall that in Section 8.2 we used the $t$ distribution to find a confidence interval for the mean when the population standard deviation was not known.

Let $\bar{x}$ be the sample mean, $\mu$ be the unknown population mean, $s$ be the sample standard deviation, and $n$ be the sample size. The $t$ statistic

$$t = \frac{\bar{x} - \mu}{s/\sqrt{n}}$$

with $n - 1$ degrees of freedom may be used when either (Case 1) the population is normal or (Case 2) the sample size is large. Recall that the shape of the $t$ distribution depends on the degrees of freedom. As $n$ increases, the graph of the $t$ curve gets closer and closer to the standard normal $Z$ (see Figure 9.36).

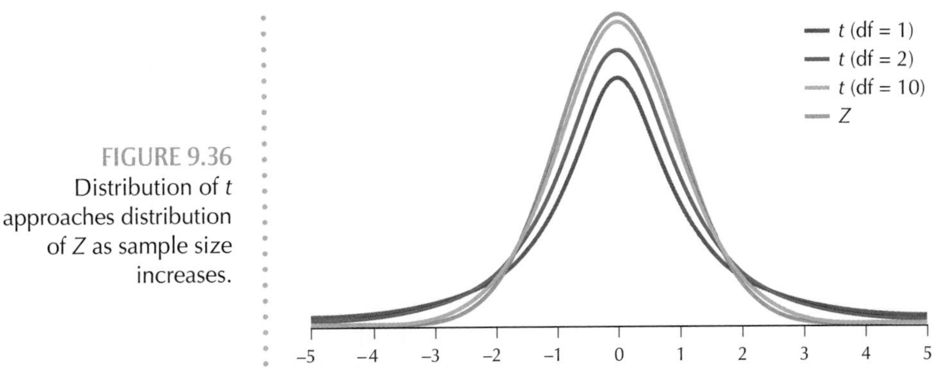


FIGURE 9.36
Distribution of $t$ approaches distribution of $Z$ as sample size increases.

Just as we did for the $Z$ test, let's look at the sampling distribution of $\bar{x}$, but this time, when $\sigma$ is unknown. See Figure 9.37. The sampling distribution of $\bar{x}$ when $\sigma$ is unknown has a $t$ distribution with mean $\mu$. Since $\sigma$ is unknown, the estimate of the standard deviation of the sampling distribution is $s/\sqrt{n}$. Just as for the $Z$ tests, if we assume that $H_0$ is true, then this sampling distribution will have a mean of $\mu_0$, the value of $\mu$ given in $H_0$.

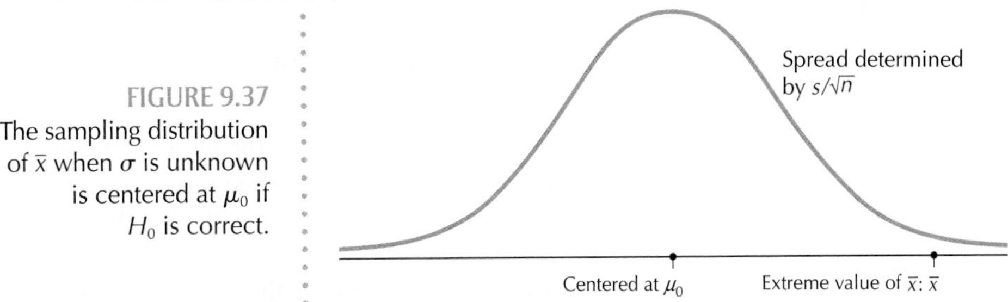


FIGURE 9.37
The sampling distribution of $\bar{x}$ when $\sigma$ is unknown is centered at $\mu_0$ if $H_0$ is correct.

Recall from Section 8.2 that the $t$ distribution may be used only if either of the following conditions is satisfied:

- Case 1: the population is normal, or
- Case 2: the sample size is large ($n \geq 30$).

To test $H_0$, we need to determine whether the observed value of $\bar{x}$ is *unusual or extreme* according to this distribution. By *unusual or extreme*, we mean a value of $\bar{x}$ located in one of the tails of the distribution, far from the hypothesized mean $\mu_0$. We then apply the essential idea about hypothesis testing for the mean:

- If the observed value of $\bar{x}$ is extreme, we reject $H_0$.
- Otherwise, there is insufficient evidence against $H_0$, and we do not reject $H_0$.

When we standardize $\bar{x}$, by subtracting the mean $\mu_0$ and dividing by $s/\sqrt{n}$, the resulting statistic

$$t = \frac{\bar{x} - \mu_0}{s/\sqrt{n}}$$

has a $t$ distribution with $n - 1$ degrees of freedom (if either the population is normal or the sample size is large). We call this $t$ statistic $t_{\text{data}}$ because its value depends largely on the sample data and the population data.

> The test statistic used for the $t$ test for the mean is
>
> $$t_{\text{data}} = \frac{\bar{x} - \mu_0}{s/\sqrt{n}}$$

Extreme values of $\bar{x}$, that is, values of $\bar{x}$ that are significantly far from the hypothesized $\mu$, will translate into extreme values of $t_{\text{data}}$. In other words, just as with $Z_{\text{data}}$, when $\bar{x}$ is far from $\mu_0$, $t_{\text{data}}$ will be far from 0.

**$t$ Test for the Population Mean $\mu$: $p$-Value Method**

When a random sample of size $n$ is taken from a population, you can use the $t$ test if either of the following conditions is satisfied:

- Case 1: the population is normal, or
- Case 2: the sample size is large ($n \geq 30$).

***Step 1*** **State the hypotheses and the rejection rule.** Use one of the forms from Table 9.1. Define $\mu$. The rejection rule is "Reject $H_0$ if the $p$-value is less than $\alpha$."

***Step 2*** **Find $t_{data}$.** Either use technology to find the value of the test statistic $t_{data}$ or calculate the value of $t_{data}$ as follows:

$$t_{data} = \frac{\bar{x} - \mu_0}{s/\sqrt{n}}$$

***Step 3*** **Find the $p$-value.** Either use technology to find the $p$-value or estimate the $p$-value using Table D, $t$-Distribution, in the Appendix (page T-11).

***Step 4*** **State the conclusion and the interpretation.** If the $p$-value is less than $\alpha$, then reject $H_0$. Otherwise, do not reject $H_0$. Interpret your conclusion so that a nonspecialist can understand.

Just as with the $Z$ test, the $p$-value method and the critical-value method are equivalent for the same level of significance $\alpha$, and they will give you the same conclusion. Again, however, the $p$-value method is used more widely than the critical-value method. Therefore, we will begin with the $p$-value method. The definition of $p$-values for $t$ tests is the same as that for the $Z$ test for the mean. Unusual and extreme values of $\bar{x}$, and therefore of $t_{data}$, will have a small $p$-value, while values of $\bar{x}$ and $t_{data}$ nearer to the center of the distribution will have a large $p$-value.

Table 9.9 summarizes the definition of the $p$-value for $t$ tests. Note that we will not be finding these $p$-values manually but will either (a) use a computer or calculator or (b) estimate them using the $t$ table. Once again, based on the $p$-value, we can assess the strength of evidence against the null hypothesis, using Table 9.5 and keeping in mind that, for certain domains, alternative interpretations would be appropriate.

Table 9.9 $p$-Values for $t$ tests

| Form of test | The $p$-value equals... | |
|---|---|---|
| **Right-tailed test**<br>$H_0: \mu \leq \mu_0$<br>$H_a: \mu > \mu_0$ | $P(t > t_{data})$<br>Area to the right of $t_{data}$ | p-value<br>0 $t_{data}$ |
| **Left-tailed test**<br>$H_0: \mu \geq \mu_0$<br>$H_a: \mu < \mu_0$ | $P(t < t_{data})$<br>Area to the left of $t_{data}$ | p-value<br>$t_{data}$ 0 |
| **Two-tailed test**<br>$H_0: \mu = \mu_0$<br>$H_a: \mu \neq \mu_0$ | $P(t > \lvert t_{data}\rvert) + P(t < -\lvert t_{data}\rvert)$<br>$= 2 \cdot P(t > \lvert t_{data}\rvert)$<br>Sum of the two tail areas | Sum of two areas is p-value.<br>$-\lvert t_{data}\rvert$ 0 $\lvert t_{data}\rvert$ |

## Example 9.21 A *t* test for the mean number of home runs

Suppose you have a collection of 14 baseball cards from the American League players who had at least 100 at-bats in the 2007 season (Table 9.10). You are trying to resolve a dispute with a friend about the population mean number of home runs hit per player in the American League. Use Minitab to test whether the population mean number of home runs is less than 16, with level of significance $\alpha = 0.10$.

Table 9.10 Home runs from collection of 14 baseball cards

| Player | Team | Home runs |
|---|---|---|
| Jermaine Dye | Chicago White Sox | 28 |
| Carl Crawford | Tampa Bay Devil Rays | 11 |
| John McDonald | Toronto Blue Jays | 1 |
| Jason Michaels | Cleveland Indians | 7 |
| Melvin Mora | Baltimore Orioles | 14 |
| Jason Varitek | Boston Red Sox | 17 |
| Orlando Cabrera | Los Angeles Angels | 8 |
| Tony Pena | Kansas City Royals | 2 |
| Jason Kubel | Minnesota Twins | 13 |
| Mark Ellis | Oakland Athletics | 19 |
| Jose Vidro | Seattle Mariners | 6 |
| Brad Wilkerson | Texas Rangers | 20 |
| Curtis Granderson | Detroit Tigers | 23 |
| Hideki Matsui | New York Yankees | 25 |

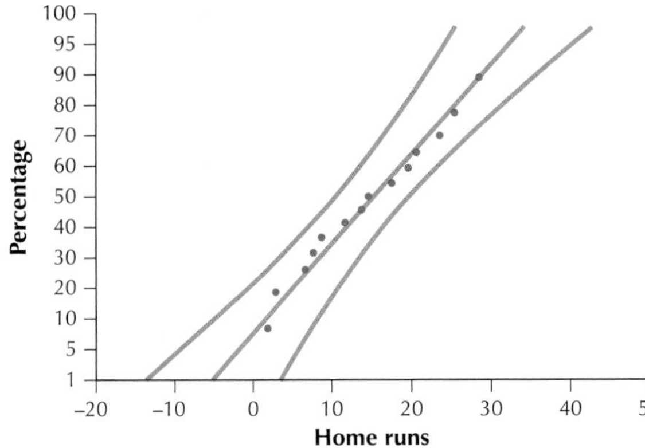


FIGURE 9.38

### Solution

Since we do not know the value of $\sigma$, we cannot perform a $Z$ test. To use a $t$ test, either the population must be normal or the sample size must be large. Since $n = 14$ is not a large sample size, we check whether the distribution of home runs is normal. Figure 9.38 shows the normal probability plot of the number of home runs per player for the baseball card collection. The normality is acceptable, allowing us to proceed with the $t$ test. We use the Minitab instructions provided in the Step-by-Step Technology Guide at the end of the section.

***Step 1*** **State the hypotheses and the rejection rule.** The hypotheses are

$$H_0: \mu \geq 16 \quad \text{versus} \quad H_a: \mu < 16$$

where $\mu$ represents the population mean number of home runs. We will reject $H_0$ if we obtain a $p$-value less than 0.10.

***Step 2*** **Find $t_{\text{data}}$.** In Figure 9.39, Minitab tells us that $n = 14$, $\bar{x} = 13.8571$, $s = 8.5021$, and

$$t_{\text{data}} = \frac{\bar{x} - \mu_0}{s/\sqrt{n}} = \frac{13.8571 - 16}{8.5021/\sqrt{14}} \approx -0.94$$

| Variable | N | Mean | StDev | SE Mean | T | P |
|---|---|---|---|---|---|---|
| Home Runs | 14 | 13.8571 | 8.5021 | 2.2723 | -0.94 | 0.181 |
| | $n$ | $\bar{x}$ | $s$ | $s/\sqrt{n}$ | $t_{\text{data}}$ | $p$-value |

FIGURE 9.39 Minitab results for $t$ test.

**Step 3** **Find the *p*-value.** Also from Figure 9.39, we have the *p*-value, which is

$$P(t < t_{\text{data}}) = P(t < -0.94) = 0.181$$

Note that we could not have found this *p*-value without a computer or calculator. Using the *t* table, the *p*-value may only be estimated.

**Step 4** **State the conclusion and interpretation.** Since the *p*-value 0.181 is not less than $\alpha = 0.10$, we do not reject $H_0$. There is insufficient evidence that the population mean number of home runs per player is less than 16. Even though the sample mean of $\bar{x} = 13.8571$ is less than 16, the difference is not sufficiently great to conclude beyond a reasonable doubt that the population mean $\mu$ is less than 16 home runs per player. ■

## Example 9.22 Performing the *t* test for $\mu$ using technology: nurse-to-patient ratios

The United States is undergoing a nursing shortage. According to the American Association of Colleges of Nursing, the estimated shortage of registered nurses will increase to 340,000 by the year 2020. The table below contains a random sample of ten highly rated cancer care facilities, along with their nursing index (nurse-to-patient ratio), as reported by *U.S. News & World Report* in 2007.[12] Suppose that the population mean nursing index in 2005 was 1.6 nurses per cancer patient, and that we are interested in whether the population mean index has changed since then. Perform the appropriate hypothesis test using level of significance $\alpha = 0.05$.

| Hospital | Index |
|---|---|
| Memorial Sloan Kettering Cancer Center | 1.5 |
| M. D. Anderson Cancer Center | 2.0 |
| Johns Hopkins Hospital | 2.3 |
| Mayo Clinic | 2.8 |
| Dana Farber Cancer Institute | 0.8 |
| Univ. of Washington Medical Center | 2.2 |
| Duke University Medical Center | 1.8 |
| Univ. of Chicago Hospitals | 2.3 |
| UCLA Medical Center | 2.2 |
| UC San Francisco Medical Center | 2.3 |

### Solution

Since the sample size is small, we check normality. The normal probability plot (Figure 9.40) is not perfectly linear, but there are no points outside the bounds, and it is difficult to determine normality for such small sample sizes. We proceed to perform the *t* test using Case 1, with the caveat that the normality assumption could be better supported and that more data would be helpful.

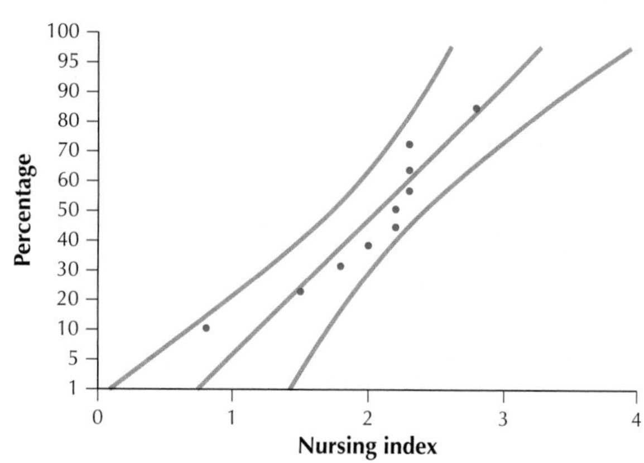

FIGURE 9.40 Normal probability plot of nursing index.

**Step 1** **State the hypotheses and the rejection rule.** The key words "has changed" means that we have a two-tailed test:

$$H_0: \mu = 1.6 \quad \text{versus} \quad H_a: \mu \neq 1.6$$

where $\mu$ represents the population mean nursing index. We will reject $H_0$ if the *p*-value is less than $\alpha = 0.05$.

**Step 2** **Find $t_{data}$.** We use the instructions supplied in the Step-by-Step Technology Guide at the end of this section. Figure 9.41 shows the TI-83/84 results from the $t$ test for $\mu$.

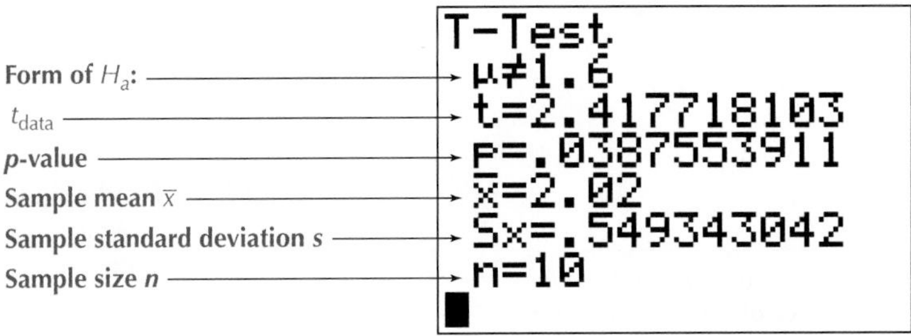


FIGURE 9.41 TI-83/84 results.

Using the statistics from Figure 9.41, we have the test statistic

$$t_{data} = \frac{\bar{x} - \mu_0}{s/\sqrt{n}} = \frac{2.02 - 1.6}{0.549343042/\sqrt{10}} = 2.417718103$$

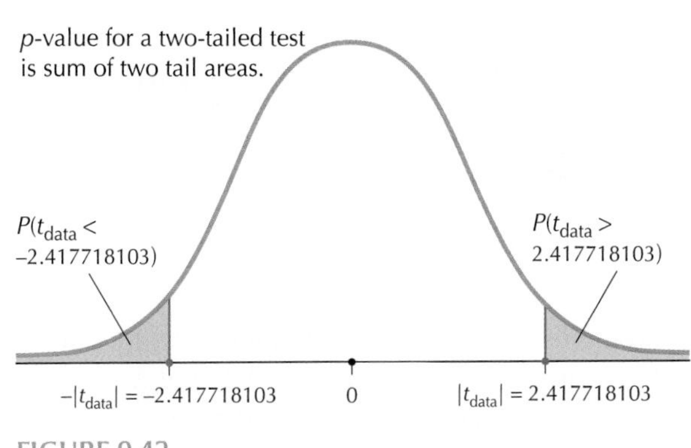


FIGURE 9.42

**Step 3** **Find the $p$-value.** From Figures 9.41 and 9.42, we have

$$\begin{aligned} p\text{-value} &= P(t > |2.417718103|) + P(t < -|2.417718103|) \\ &= 0.0387553911 \end{aligned}$$

**Step 4** **State the conclusion and interpretation.** The $p$-value of 0.0387553911 is less than $\alpha = 0.05$. We therefore reject $H_0$. There is evidence that the population mean nurse-to-patient ratio differs from 1.6. ■

## Example 9.23 The $t$ test for $\mu$ using the estimated $p$-value method

The table below lists the prices in dollars for a sample of 17 mathematics journals in 2005, as reported by the American Mathematical Society.

| Journal | Price | Journal | Price | Journal | Price |
|---|---|---|---|---|---|
| Bull. Amer. Math. | \$402 | Ann. Statist. | \$250 | SIAM J. Math. Anal. | \$717 |
| J. Amer. Math. Soc. | 276 | Statistical Science | 90 | SIAM J. Matrix Anal. | 497 |
| Math. Comp. | 467 | SIAM J. Appl. Math. | 527 | SIAM J. Numer. Anal. | 567 |
| Proc. Amer. Math. | 1022 | SIAM J. Comput. | 536 | SIAM J. Optim. | 438 |
| Ann. Appl. Probab. | 850 | SIAM J. Control Optim. | 633 | SIAM J. Sci. Comput. | 644 |
| Ann. Probab. | 250 | SIAM J. Discrete Math. | 450 | | |

Suppose that the mean cost of all mathematics journals in 2001 was \$400, and we are interested in whether the population mean cost is increasing. Evaluate the normality assumption, and perform the appropriate hypothesis test by *estimating the p-value* and comparing it with level of significance $\alpha = 0.05$.

### Solution

Though not perfect, the normal probability plot indicates an acceptable level of normality of the math journal prices. We may therefore proceed to apply the $t$ test.

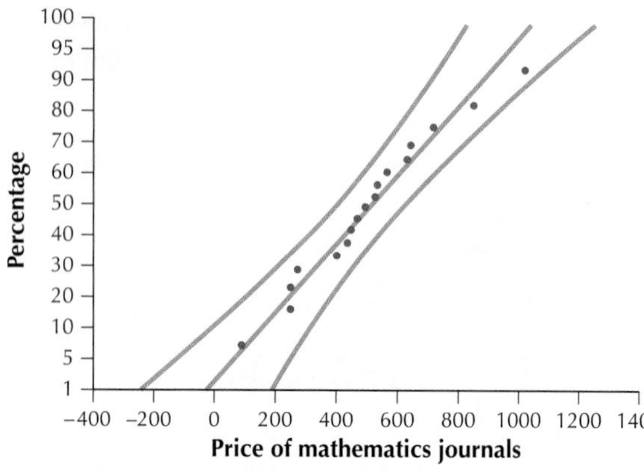


Normal probability plot for price of math journals.

| MINITAB Descriptive Statistics | | | | |
|---|---|---|---|---|
| Variable | N | Mean | SE Mean | StDev |
| Prices | 17 | 506.82 | 55.8 | 230.0 |

**Step 1** **State the hypotheses and the rejection rule.** The hypotheses are

$$H_0: \mu \leq 400 \quad \text{versus} \quad H_a: \mu > 400$$

where $\mu$ represents the population mean math journal price. We will reject the null hypothesis if the $p$-value is less than $\alpha = 0.05$.

**Step 2** **Find $t_{data}$.** Our sample statistics are $\bar{x} = \$506.82$, $s = \$230$, and $n = 17$, so our test statistic is

$$t_{data} = \frac{506.82 - 400}{230/\sqrt{17}} \approx 1.91$$

**Step 3** **Find the $p$-value.** We estimate the $p$-value as follows. We follow the row in the $t$ table for df $= n - 1 = 16$, looking for where $t_{data} = 1.91$ would fit:

| Area in one tail | 0.10 | 0.05 | 0.025 | 0.01 | 0.005 |
|---|---|---|---|---|---|
| Value of $t$ | 1.337 | 1.746 | 2.120 | 2.583 | 23921 |

$t_{data} = 1.91$ would fall here.

We see that our value of $t_{data} = 1.91$ would fit between $t = 1.746$ and $t = 2.120$. Therefore, the tail area associated with $t_{data} = 1.91$ is between area 0.025 and area 0.05. Therefore, since this is a right-tailed test, the estimated $p$-value also lies between 0.025 and 0.05.

**Step 4** **State the conclusion and interpretation.** Since the $p$-value is less than $\alpha = 0.05$, we reject $H_0$. There is evidence that the population mean price of math journals is higher than \$400. ■

## 2 $t$ Test for $\mu$ Using the Critical-Value Method

So far we have learned how to perform a $t$ test using the $p$-value method, which compares one probability (the $p$-value) with another probability ($\alpha$). Now we will learn the critical-value method of performing the $t$ test. We know that extreme values of $\bar{x}$ are associated with extreme values of $t_{data}$ and will thus lead us to reject the null hypothesis. We answer the question "How extreme is extreme?" using the critical-value method by finding a *critical value* of $t$, called $t_{crit}$. This threshold value $t_{crit}$ separates the values of $t_{data}$ for which we reject $H_0$ (the *critical region*) from the values of $t_{data}$ for which we will not reject $H_0$ (the *noncritical region*). Because there is a different $t$ curve for every different sample size, you need to know the following to find the value of $t_{crit}$:

- the form of the hypothesis test (one-tailed or two-tailed)
- your degrees of freedom (df $= n - 1$)
- the level of significance $\alpha$

**$t$ Test for the Population Mean $\mu$: Critical-Value Method**

When a random sample of size $n$ is taken from a population, you can use the $t$ test if either of the following conditions is satisfied:

- Case 1: the population is normal, or
- Case 2: the sample size is large ($n \geq 30$).

***Step 1*** **State the hypotheses.** Use one of the forms from Table 9.11. Define $\mu$.

***Step 2*** **Find $t_{crit}$ and state the rejection rule.** Use Table 9.11.

***Step 3*** **Find $t_{data}$.** Either use technology to find the value of the test statistic $t_{data}$ or calculate the value of $t_{data}$ as follows:

$$t_{data} = \frac{\bar{x} - \mu_0}{s/\sqrt{n}}$$

***Step 4*** **State the conclusion and the interpretation.** If $t_{data}$ falls within the critical region, then reject $H_0$. Otherwise, do not reject $H_0$. Interpret your conclusion so that a nonspecialist can understand.

Table 9.11 contains the rejection rules and critical regions for the $t$ test.

Table 9.11 Critical regions and rejection rules for various forms of the $t$ test for $\mu$

| Form of test | Rejection rule | Critical region |
|---|---|---|
| **Right-tailed test**<br>$H_0: \mu \leq \mu_0$<br>$H_a: \mu > \mu_0$ | Reject $H_0$ if $t_{data} > t_{crit}$ | 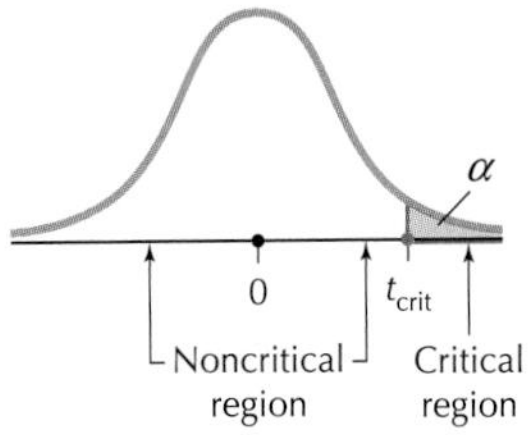 $\alpha$; 0; $t_{crit}$; Noncritical region; Critical region |
| **Left-tailed test**<br>$H_0: \mu \geq \mu_0$<br>$H_a: \mu < \mu_0$ | Reject $H_0$ if $t_{data} < -t_{crit}$ | 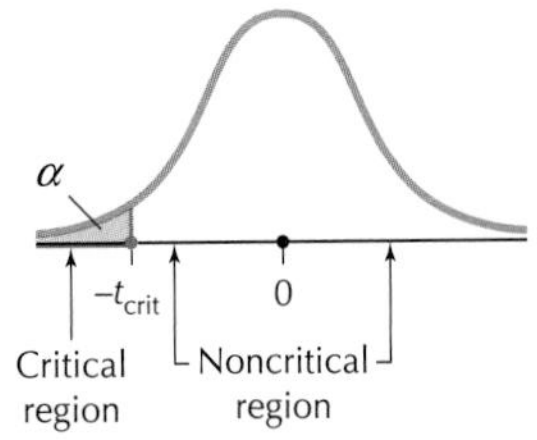 $\alpha$; $-t_{crit}$; 0; Critical region; Noncritical region |
| **Two-tailed test**<br>$H_0: \mu = \mu_0$<br>$H_a: \mu \neq \mu_0$ | Reject $H_0$ if $t_{data} > t_{crit}$ or $t_{data} < -t_{crit}$ | 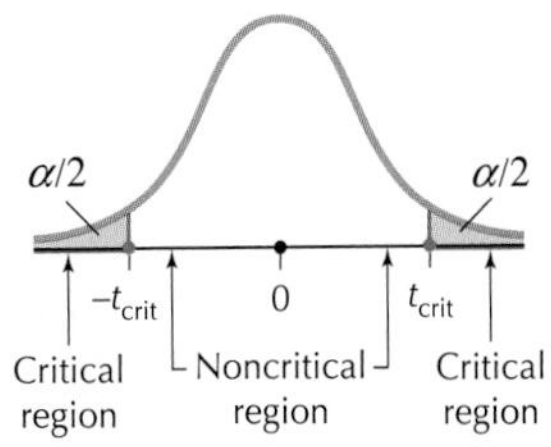 $\alpha/2$; $\alpha/2$; $-t_{crit}$; 0; $t_{crit}$; Critical region; Noncritical region; Critical region |

## Example 9.24 Has the mean age at onset of anorexia nervosa been decreasing?

| Variable | N | Mean | StDev |
|---|---|---|---|
| Patient Age | 20 | 14.251 | 1.512 |

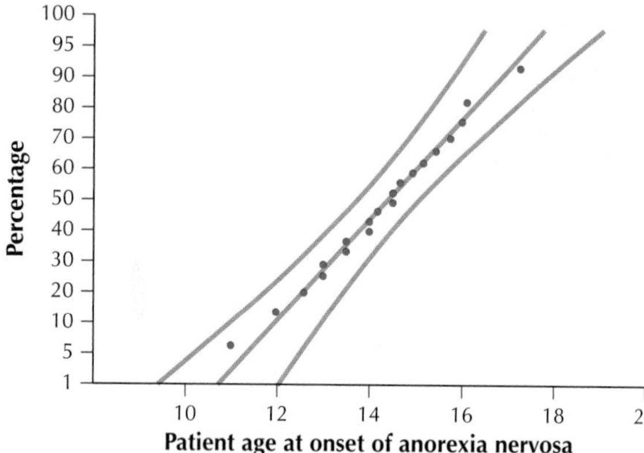


Normal probability plot for age at onset of anorexia nervosa.

We are interested in testing, using level of significance $\alpha = 0.05$, whether the mean age at onset of anorexia nervosa in young women has been decreasing. Assume that the previous mean age at onset was 15 years old. Data were gathered for a study of the onset age for this disease.[13] From these data, a random sample was taken of 20 young women who were admitted under this diagnosis to the Toronto Hospital for Sick Children. The Minitab descriptive statistics shown here indicate a sample mean age of 14.251 years and a sample standard deviation of 1.512 years. If appropriate, perform the $t$ test.

### Solution

Since the sample size $n = 20$ is not large, we need to verify normality. The normal probability plot of the ages at onset, shown here, indicates that the ages in the sample are normally distributed. We may proceed to perform the $t$ test for the mean.

**Step 1** **State the hypotheses.** The key word "decreasing" guides us to state our hypotheses as follows:

$$H_0: \mu \geq 15 \quad \text{versus} \quad H_a: \mu < 15$$

where $\mu$ refers to the population mean age at onset.

**Step 2** **Find $t_{crit}$ and state the rejection rule.** We determine that the hypothesis test is a left-tailed test and the degrees of freedom are df $= 20 - 1 = 19$, and we choose level of significance $\alpha = 0.05$. Referring to the $t$ table with df $= 19$ and $\alpha = 0.05$ using a one-tailed test, we get $t_{crit} = 1.729$. Because we have a left-tailed test, the rejection rule from Table 9.11 is "Reject $H_0$ if $t_{data} < -t_{crit}$"; that is, we will reject $H_0$ if $t_{data} < -1.729$ (Figure 9.43).

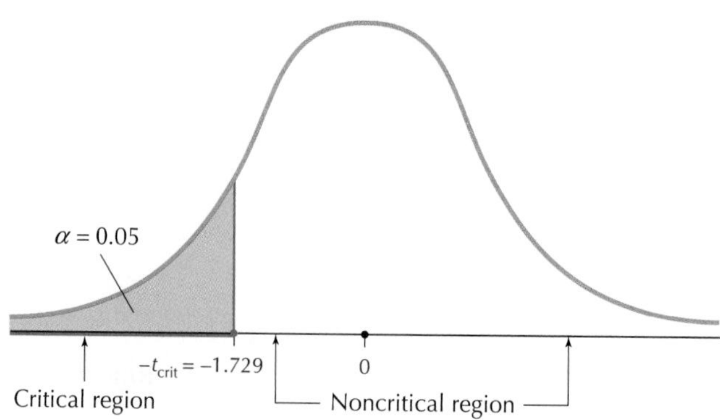


FIGURE 9.43 Critical region for left-tailed $t$ test with $\alpha = 0.05$, df $= 19$.

**Step 3** **Find $t_{data}$.** The sample data show that our sample of $n = 20$ patients has a mean age at onset of $\bar{x} = 14.251$ years with a standard deviation of $s = 1.512$ years. Also, $\mu_0 = 15$, since this is the hypothesized value of $\mu$ stated in $H_0$. Therefore, our test statistic is

$$t_{data} = \frac{\bar{x} - \mu_0}{s/\sqrt{n}} = \frac{14.251 - 15}{1.512/\sqrt{20}} \approx -2.2154$$

***Step 4*** **State the conclusion and interpretation.** The rejection rule from Step 2 says to reject $H_0$ if $t_{data} < -1.729$. From Step 3, we have $t_{data} = -2.2154$. Since $-2.2154$ is less than $-1.729$, our conclusion is to reject $H_0$. If you prefer the graphical approach, consider Figure 9.44, which shows where $t_{data}$ falls in relation to the critical region. Since $t_{data} = -2.2154$ falls within the critical region, our conclusion is to reject $H_0$. There is evidence that the population mean age of onset has decreased from its previous level of 15 years. ■

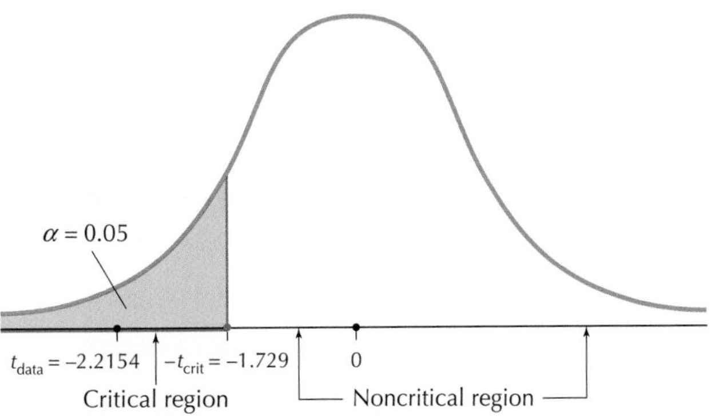


FIGURE 9.44 Our $t_{data} = -2.2154$ falls in the critical region.

## 3 Standard Error of the Sample Mean

In order to interpret the meaning of the test statistic $t_{data}$, we need to learn about the **standard error of the sample mean.** Fact 2 in Section 7.1 (page 340) stated that the standard deviation of the sampling distribution of the sample mean $\bar{x}$ is $\sigma_{\bar{x}} = \sigma/\sqrt{n}$. This quantity was used in Sections 9.2 and 9.3 to calculate the test statistic $Z_{data}$ when $\sigma$ is known. However, here in Section 9.4 we no longer assume that $\sigma$ is known. When $\sigma$ is not known, we use the following quantity to estimate the standard deviation of the sampling distribution of the sample mean.

**Standard Error of the Sample Mean $\bar{x}$**

$$s_{\bar{x}} = \frac{s}{\sqrt{n}}$$

Thus, the test statistic $t_{data}$ is written

$$t_{data} = \frac{\bar{x} - \mu_0}{s_{\bar{x}}} = \frac{\bar{x} - \mu_0}{s/\sqrt{n}}$$

Since the denominator of $t_{data}$ equals the standard error $s_{\bar{x}}$, $t_{data}$ is expressed in terms of standard errors. (The quantity in the denominator is called a *scaling factor.*) This leads us to the following interpretation of the meaning of the test statistic $t_{data}$.

**Interpretation of the Meaning of the Test Statistic $t_{data}$**

When the value of $t_{data}$ is positive, it is interpreted as the number of standard errors above the hypothesized mean $\mu_0$ that the sample mean $\bar{x}$ lies. When the value of $t_{data}$ is negative, it is interpreted as the number of standard errors below the hypothesized mean $\mu_0$ that the sample mean $\bar{x}$ lies.

## Example 9.25 Interpreting the value of $t_{data}$

In Example 9.24, we tested the hypotheses

$$H_0: \mu \geq 15 \quad \text{versus} \quad H_a: \mu < 15$$

where $\mu$ represents the population mean age at onset of anorexia nervosa in young women. The sample data showed that $n = 20$ patients had a mean age at onset of $\bar{x} = 14.251$ years with a standard deviation of $s = 1.512$ years.

**a.** Calculate the standard error of the mean, $s_{\bar{x}} = s/\sqrt{n}$.

**b.** Interpret $t_{data}$ in terms of the number of standard errors.

### Solution

**a.** The standard error is calculated as follows:

$$s_{\bar{x}} = \frac{s}{\sqrt{n}} = \frac{1.512}{\sqrt{20}} \approx 0.3381$$

**b.** From Example 9.24, we had $t_{data} = \dfrac{\bar{x} - \mu_0}{s/\sqrt{n}} = \dfrac{14.251 - 15}{1.512/\sqrt{20}} \approx -2.2154$

Figure 9.45 illustrates the sampling distribution of the sample mean, when $\mu_0 = 15$. The hypothesized mean $\mu_0 = 15$ is shown, along with $\mu_0 - 1s_{\bar{x}} = 15 - 0.3381 = 14.6619$ (one standard error below the mean) and $\mu_0 - 2s_{\bar{x}} = 15 - 2(0.3381) = 14.3238$ (two standard errors below the mean). The sample mean $\bar{x} = 14.251$ lies somewhat more than two standard errors below the mean. Specifically, $t_{data} = -2.2154$ indicates that the sample mean age at onset of anorexia nervosa $\bar{x} = 14.251$ lies 2.2154 standard errors below the hypothesized mean age at onset $\mu_0 = 15$. ■

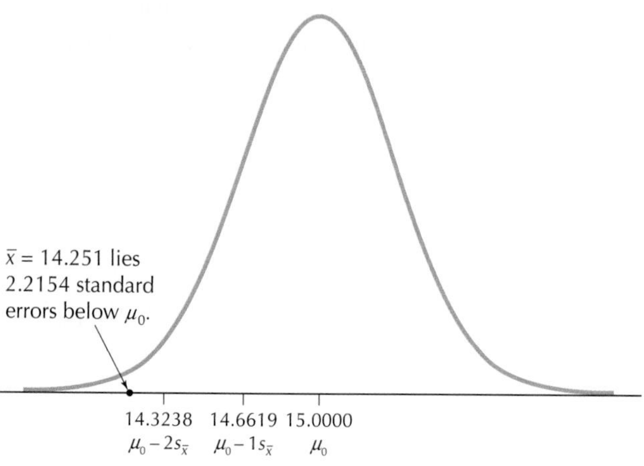


FIGURE 9.45

CASE STUDY

### The Golden Ratio

Euclid's *Elements,* the Parthenon, the *Mona Lisa,* and the beadwork of the Shoshone tribe all have in common an appreciation for the **golden ratio.**

Suppose we have two quantities $A$ and $B$, with $A > B > 0$. Then $A/B$ is called the **golden ratio** if

$$\frac{A+B}{A} = \frac{A}{B}$$

that is, if the ratio of the sum of the quantities to the larger quantity equals the ratio of the larger to the smaller (see Figure 9.46).

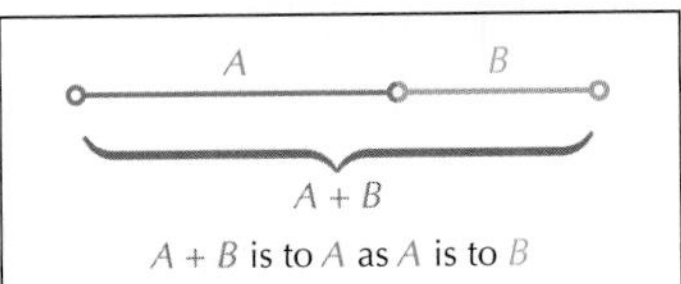


FIGURE 9.46

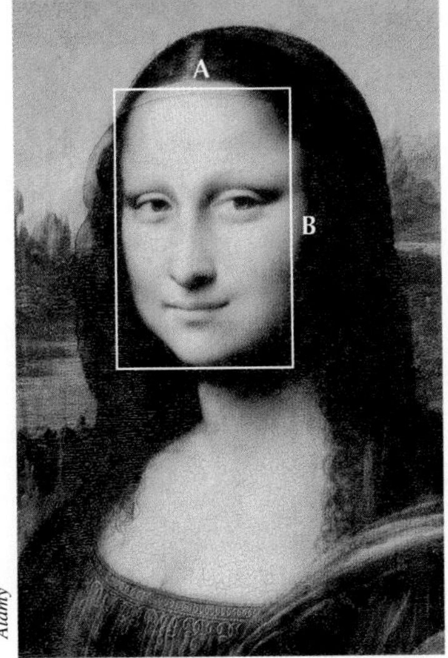


FIGURE 9.47

Euclid wrote about the golden ratio in his *Elements,* calling it the "extreme and mean ratio." The ratio of the width $A$ and height $B$ of the Parthenon, one of the most famous temples in ancient Greece, equals the golden ratio (Figure 9.46). If you enclose the face of Leonardo da Vinci's *Mona Lisa* in a rectangle, the resulting ratio of the long side to the short side follows the golden ratio (Figure 9.47). The golden ratio has a value of approximately 1.6180339887.

Now we will test whether there is evidence for the use of the golden ratio in the artistic traditions of the Shoshone, a Native American tribe from the American West. Perhaps the best-known Shoshone is Sacajawea, whose assistance was crucial to the success of the Lewis and Clark expedition in the early nineteenth century. The Shoshone are superb artisans, with a long tradition of skilled beadwork.

Figure 9.48 shows a fine example of a nineteenth-century Shoshone beaded dress.[14] This dress belonged to Nahtoma, the daughter of Chief Washakie of the Eastern Shoshone. Figure 9.49 shows a detail of the upper part of the dress, including five beaded rectangles. It is intriguing to consider whether Shoshone beaded rectangles such as these follow the golden ratio.

*William R. McIver Collection, American Heritage Center, University of Wyoming*

FIGURE 9.48

FIGURE 9.49

Table 9.12 contains the ratios of lengths to widths of 18 beaded rectangles made by Shoshone artisans.[15] We will perform a hypothesis test to determine whether the population mean ratio of Shoshone beaded rectangles equals the golden ratio of 1.6180339887.

Table 9.12 Ratio of length to width of a sample of Shoshone beaded rectangles

| | | | | |
|---|---|---|---|---|
| 1.44300 | 1.75439 | 1.64204 | 1.66389 | 1.63666 |
| 1.51057 | 1.33511 | 1.52905 | 1.73611 | 1.80832 |
| 1.44928 | 1.48810 | 1.62602 | 1.49254 | |
| 1.65017 | 1.59236 | 1.49701 | 1.65017 | |

Since the population standard deviation for such rectangles is unknown, we must use a $t$ test rather than a $Z$ test. Our sample size $n = 18$ is not large, so we must assess whether the data are normally distributed. Figure 9.50 shows the normal probability plot indicating acceptable support for the normality assumption. We proceed with the $t$ test, using level of significance $\alpha = 0.05$.

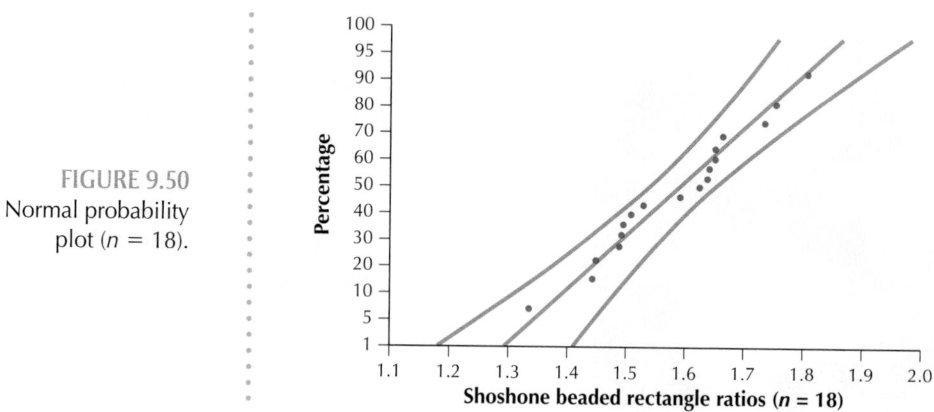


FIGURE 9.50 Normal probability plot ($n = 18$).

## Solution

We use the TI-83/84 to perform this hypothesis test, using the instructions provided at the end of this section.

**Step 1** **State the hypotheses and the rejection rule.** Since we are interested in whether the population mean length-to-width ratio of Shoshone beaded rectangles *equals* the golden ratio of 1.6180339887, we perform a two-tailed test:

$$H_0: \mu = 1.6180339887 \quad \text{versus} \quad H_a: \mu \neq 1.6180339887$$

where $\mu$ represents the population mean length-to-width ratio of Shoshone beaded rectangles. We will reject $H_0$ if the $p$-value is less than 0.05.

```
T-Test
 μ≠1.618033989
 t=-1.183325126
 p=.2529646133
 x̄=1.583599444
 Sx=.1234600663
 n=18
```

FIGURE 9.51

**Step 2** **Find $t_{data}$.** From Figure 9.51 we have

$$t_{data} = \frac{\bar{x} - \mu_0}{s/\sqrt{n}} = \frac{1.583599444 - 1.6180339887}{0.1234600663/\sqrt{18}} = -1.183325126$$

(Note that the TI-83/84 output rounds $\mu_0$ to 1.618033989.)

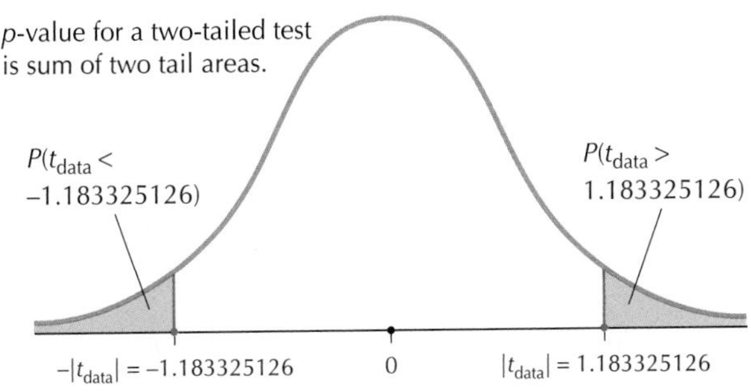


FIGURE 9.52

**Step 3** **Find the *p*-value.** From Figures 9.51 and 9.52, we have

$$p\text{-value} = P(t > |1.183325126|) + P(t < -|1.183325126|)$$
$$= 0.2529646133$$

**Step 4** **State the conclusion and interpretation.** Since *p*-value $= 0.2529646133$ is *not* less than $\alpha = 0.05$, we do *not* reject $H_0$. Thus, there is insufficient evidence that the population mean ratio differs from 1.6180339887. In other words, the data do not reject the claim that Shoshone beaded rectangles follow the same golden ratio exhibited by the Parthenon and the *Mona Lisa.*

## STEP-BY-STEP TECHNOLOGY GUIDE: *t* test for $\mu$

We will use the nurse-to-patient ratio data from Example 9.22 (page 500).

### TI-83/84

#### If you have the data values:

*Step 1* Enter the data into list **L1**.
*Step 2* Press **STAT**, highlight **TESTS**, and press **ENTER**.
*Step 3* Press **2** (for **T-Test**; see Figure 9.53).
*Step 4* For input (**Inpt**), highlight **Data** and press **ENTER** (Figure 9.54).
**a.** For $\mu_0$, enter the value of $\mu_0$, **1.6**.
**b.** For **List**, press **2nd**, then **L1**.
**c.** For **Freq**, enter **1**.
**d.** For $\mu$, select the form of $H_a$. Here we have a two-tailed test, so highlight $\neq \mu_0$ and press **ENTER**.
**e.** Highlight **Calculate** and press **ENTER**. The results are shown in Figure 9.41 in Example 9.22.

#### If you have the summary statistics:

*Step 1* Press **STAT**, highlight **TESTS**, and press **ENTER**.
*Step 2* Press **2** (for **T-Test**; see Figure 9.53).
*Step 3* For input (**Inpt**), highlight **Stats** and press **ENTER** (Figure 9.55).
**a.** For $\mu_0$, enter the value of $\mu_0$, **1.6**.
**b.** For **Sx**, enter the value of *s*, **0.549343042**.
**c.** For $\bar{x}$, enter the sample mean **2.02**.
**d.** For **n**, enter the sample size **10**.
**e.** For $\boldsymbol{\mu}$, select the form of $H_a$. Here we have a two-tailed test, so highlight $\neq \mu_0$ and press **ENTER**.
**f.** Highlight **Calculate** and press **ENTER**. The results are shown in Figure 9.41 in Example 9.22.

Figure 9.53

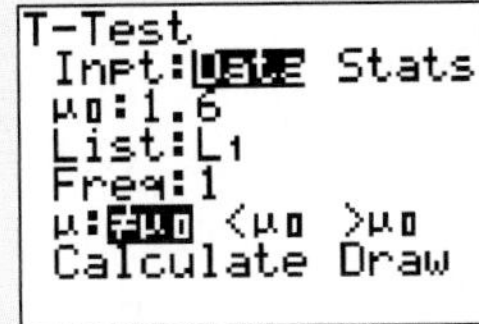


Figure 9.54

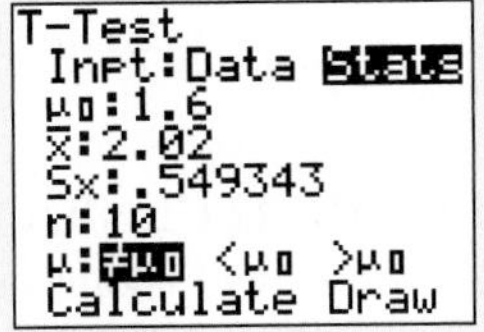


Figure 9.55

### EXCEL

#### WHFStat Macros

*Step 1* Enter the data into column A. (If you have only the summary statistics, go to Step 2.)
*Step 2* Load the **WHFStat Macros**.
*Step 3* Select **Add-Ins > Macros > Testing a Mean > t Test – Confidence Interval – One Sample**.
*Step 4* Select cells A1 to A10 as the **Dataset Range**. (Alternatively, you may enter the summary statistics.)
*Step 5* Select your **Confidence level**, which should be $1 - \alpha$. Here, because $\alpha = 0.05$, we select **95%**.
*Step 6* Enter the **Null Hypothesis Value**, $\mu_0 = 1.6$, and click **OK**.

*(Continued)*

## MINITAB

**If you have the data values:**

*Step 1* Enter the data into column C1.
*Step 2* Click **Stat > Basic Statistics > 1-Sample t**.
*Step 3* Click **Samples in Columns** and select **C1**.
*Step 4* For **Test Mean**, enter 1.6.
*Step 5* Click **Options**.
**a.** Choose your **Confidence Level** as 100(1 − $\alpha$). Our level of significance $\alpha$ here is 0.05, so the confidence level is 95.0.
**b.** Select **not Equal** for the **Alternative**.
*Step 6* Click **OK** and click **OK** again.

**If you have the summary statistics:**

*Step 1* Click **Stat > Basic Statistics > 1-Sample t**.
*Step 2* Click **Summarized Data**,
*Step 3* Enter the **Sample Size 10**, the **Sample Mean 2.02**, and the **Sample Standard Deviation 0.549343042**.
*Step 4* Click **Options**.
**a.** Choose your **Confidence Level** as 100(1 − $\alpha$). Our level of significance $\alpha$ here is 0.05, so the confidence level is 95.0.
**b.** Select **not Equal** for the two-tailed test.
*Step 5* Click **OK** and click **OK** again.

## Section 9.4 SUMMARY

1 The test statistic used for the $t$ test for the mean is

$$t_{\text{data}} = \frac{\bar{x} - \mu_0}{s/\sqrt{n}}$$

with $n - 1$ degrees of freedom. The $t$ test may be used under either of the following conditions: (Case 1) the population is normal, or (Case 2) the sample size is large ($n \geq 30$). We perform hypothesis testing for the mean with the $t$ test, using either the $p$-value or the critical-value method. For the $p$-value method, we reject $H_0$ if the $p$-value is less than $\alpha$.

2 For the critical-value method, we compare the values of $t_{\text{data}}$ and $t_{\text{crit}}$. If $t_{\text{data}}$ falls in the critical region, we reject $H_0$.

3 The value of $t_{\text{data}}$ may be interpreted as the number of standard errors above or below the hypothesized mean $\mu_0$ that the sample mean $\bar{x}$ lies.

## Section 9.4 EXERCISES

### CLARIFYING THE CONCEPTS

1. What assumption is required for performing the $Z$ test that is not required for the $t$ test?
2. What do we use to estimate the unknown population standard deviation $\sigma$?
3. What are the two conditions for using the $t$ test for the mean?
4. What three elements are needed for finding the $t$ critical value?

### PRACTICING THE TECHNIQUES

For each of the following hypothesis tests in Exercises 5–7 and 9–11, find the critical value $t_{\text{crit}}$, and sketch the critical region under the $t$ curve. Assume normality.

5. $H_0$: $\mu \geq 40$ vs. $H_a$: $\mu < 40$, $n = 12$, $\alpha = 0.10$
6. $H_0$: $\mu \geq 40$ vs. $H_a$: $\mu < 40$, $n = 12$, $\alpha = 0.05$
7. $H_0$: $\mu \geq 40$ vs. $H_a$: $\mu < 40$, $n = 12$, $\alpha = 0.01$
8. Describe what happens to the $t$ critical value $t_{\text{crit}}$ for the left-tailed tests in Exercises 5–7 as the level of significance $\alpha$ decreases.
9. $H_0$: $\mu = -10$ vs. $H_a$: $\mu \neq -10$, $n = 18$, $\alpha = 0.10$
10. $H_0$: $\mu = -10$ vs. $H_a$: $\mu \neq -10$, $n = 18$, $\alpha = 0.05$
11. $H_0$: $\mu = -10$ vs. $H_a$: $\mu \neq -10$, $n = 18$, $\alpha = 0.01$
12. Describe what happens to the $t$ critical value $t_{\text{crit}}$ for the two-tailed tests in Exercises 9–11 as the level of significance $\alpha$ decreases.

For Exercises 13–18, find the value of $t_{\text{data}}$.

13. $\bar{x} = 20$, $s = 4$, $n = 36$, $\mu_0 = 22$
14. $\bar{x} = 2$, $s = 1$, $n = 49$, $\mu_0 = 3$
15. $\bar{x} = 10$, $s = 3$, $n = 64$, $\mu_0 = 11$
16. $\bar{x} = 80$, $s = 5$, $n = 64$, $\mu_0 = 82$
17. $\bar{x} = 106$, $s = 10$, $n = 81$, $\mu_0 = 102$
18. $\bar{x} = 99$, $s = 4$, $n = 100$, $\mu_0 = 95$

For Exercises 19–24, estimate the $p$-value. Assume normality.

19. $H_0$: $\mu \leq 100$ vs. $H_a$: $\mu > 100$, $n = 8$, $t_{\text{data}} = 2.5$
20. $H_0$: $\mu \leq 100$ vs. $H_a$: $\mu > 100$, $n = 8$, $t_{\text{data}} = 2.0$
21. $H_0$: $\mu = -10$ vs. $H_a$: $\mu \neq -10$, $n = 18$, $t_{\text{data}} = 2.0$
22. $H_0$: $\mu = -10$ vs. $H_a$: $\mu \neq -10$, $n = 18$, $t_{\text{data}} = 0$
23. $H_0$: $\mu \geq 40$ vs. $H_a$: $\mu < 40$, $n = 12$, $t_{\text{data}} = -2.5$
24. $H_0$: $\mu \geq 40$ vs. $H_a$: $\mu < 40$, $n = 12$, $t_{\text{data}} = -4.0$

For Exercises 25–30, assess the strength of the evidence against the null hypothesis.

25. Exercise 19
26. Exercise 20
27. Exercise 21
28. Exercise 22
29. Exercise 23
30. Exercise 24

For Exercises 31–34, do the following.

- **a.** State the hypotheses and the rejection rule using the $p$-value method.
- **b.** Calculate the test statistic $t_{data}$.
- **c.** Find the $p$-value. (Use technology or estimate the $p$-value.)
- **d.** State the conclusion and the interpretation.

**31.** A random sample of size 9 from a normal population yields $\bar{x} = 1$ and $s = 0.5$. Researchers are interested in finding whether the population mean differs from 0, using level of significance $\alpha = 0.05$.

**32.** A random sample of size 400 from a population with an unknown distribution yields a sample mean of 230 and a sample standard deviation of 5. Researchers are interested in finding whether the population mean is greater than 200, using level of significance $\alpha = 0.05$.

**33.** A random sample of size 100 from a population with an unknown distribution yields $\bar{x} = 27$ and $s = 10$. Researchers are interested in finding whether the population mean is less than 28, using level of significance $\alpha = 0.05$.

**34.** A random sample of size 16 from a normal population yields $\bar{x} = 2.2$ and $s = 0.3$. Researchers are interested in finding whether the population mean differs from 2.0, using level of significance $\alpha = 0.01$.

For Exercises 35–38, do the following.

- **a.** State the hypotheses.
- **b.** Calculate the $t$ critical value $t_{crit}$ and state the rejection rule. Also, sketch the critical region.
- **c.** Find the test statistic $t_{data}$.
- **d.** State the conclusion and the interpretation.

**35.** A random sample of size 25 from a normal population yields $\bar{x} = 104$ and $s = 10$. Researchers are interested in finding whether the population mean exceeds 100, using level of significance $\alpha = 0.01$.

**36.** A random sample of size 100 from a population with an unknown distribution yields a sample mean of $-5$ and a sample standard deviation of 5. Researchers are interested in finding whether the population mean is less than $-4$, using level of significance $\alpha = 0.05$.

**37.** A random sample of size 36 from a population with an unknown distribution yields $\bar{x} = 10$ and $s = 3$. Researchers are interested in finding whether the population mean differs from 9, using level of significance $\alpha = 0.10$.

**38.** A random sample of size 16 from a normal population yields $\bar{x} = 995$ and $s = 15$. Researchers are interested in finding whether the population mean is less than 1000, using level of significance $\alpha = 0.01$.

For Exercises 39–42, do the following.

- **a.** Calculate the standard error of the mean, $s_{\bar{x}}$
- **b.** Interpret the value of $t_{data}$ in terms of the standard error.

**39.** In Example 9.21, we have $n = 14$, $s = 8.5021$, $\mu_0 = 16$, and $t_{data} = -0.94$, where $\mu$ represents the population mean number of home runs.

**40.** In Example 9.22, we have $n = 10$, $s = 0.549343042$, $\mu_0 = 1.6$, and $t_{data} \approx 2.42$, where $\mu$ represents the population mean nursing index.

**41.** In Example 9.23, we have $n = 17$, $s = 230$, $\mu_0 = 400$, and $t_{data} \approx 1.91$, where $\mu$ represents the population mean math journal price.

**42.** In the Case Study on pages 506–509, we have $n = 18$, $s = 0.1234600663$, $\mu_0 = 1.6180339887$, and $t_{data} \approx -1.18$, where $\mu$ represents the population mean length-to-width ratio of Shoshone beaded rectangles.

## APPLYING THE CONCEPTS

**Internet Response Times.** Use the following information for Exercises 43–45. The Web site **www.Internettrafficreport.com** monitors Internet traffic worldwide and reports on the response times of randomly selected servers.

**43.** On July 13, 2004, the Web site reported the following response times to Asia, in milliseconds:

441 271 283 244 789 211 205
148 330 886 238 197 187

We are interested in testing whether the population mean response time is slower than 200 milliseconds, using a $t$ test and level of significance $\alpha = 0.05$.

- **a.** Does Case 2 apply?
- **b.** Here is a boxplot of the data. Does Case 1 apply? (*Hint:* The boxplot is right-skewed and the normal distribution is symmetric.)
- **c.** Can we proceed with the $t$ test? Explain.

**44.** On July 13, 2004, the Web site reported the following response times to North America, in milliseconds:

100 53 67 37 59 75 87
88 78 70 58 71 77 69

We are interested in testing whether the population mean response time is slower than 60 milliseconds, using a $t$ test and $\alpha = 0.05$.

- **a.** Does Case 2 apply?
- **b.** Here is a normal probability plot of the data. Does Case 1 apply?
- **c.** Can we proceed with the $t$ test? Explain.

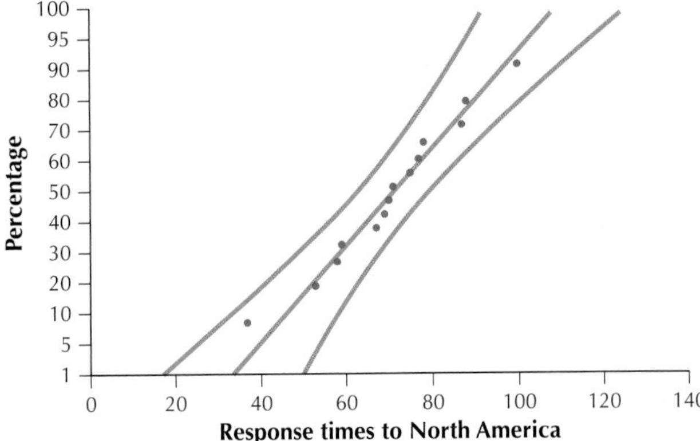

**45.** Refer to the previous exercise.

**a.** Construct the hypotheses.

**b.** Find the $t$ critical value and state the rejection rule.

**c.** Find $t_{data}$.

**d.** State the conclusion and interpretation.

**46. Health Care Costs.** The U.S. Agency for Healthcare Research and Quality (www.ahrq.gov) reports that, in 2002, the mean cost of a stay in the hospital for American women aged 18–44 was \$12,300. A random sample taken this year of 400 hospital stays of women aged 18–44 showed a mean cost of \$15,000, with a standard deviation of \$5000. We are interested in whether the mean cost has increased since then.

**a.** Calculate the standard error of the mean.

**b.** Compute $t_{data}$. Interpet $t_{data}$ using the standard error.

**c.** Find or estimate the $p$-value.

**d.** Assess the strength of the evidence against the null hypothesis.

**Obsession.** Use the following information for Exercises 47–49. Perfume.com listed Obsession by Calvin Klein as their fifth best-selling fragrance for the 2006 holiday season. The mean price per bottle in 1996 was \$46.42, according to the NPD Group. Suppose a random sample of 15 retail establishments yielded a mean retail price for Obsession this year of \$48.92, with a standard deviation of \$2.50. Assume these prices are normally distributed.

**47.** We are interested in whether the population mean price for Obsession perfume has increased.

**a.** Is it appropriate to apply the $t$ test for the mean? Why or why not?

**b.** We have a sample mean of \$48.92, which is greater than the mean of \$46.42 in the null hypothesis. Isn't this enough by itself to reject the null hypothesis? Explain why or why not.

**c.** Find $t_{data}$.

**d.** Has there been a change in the mean price since 1996? Test using the estimated $p$-value method at level of significance $\alpha = 0.01$.

**48.** Answer the following.

**a.** Assess the strength of the evidence against the null hypothesis.

**b.** What value of $\alpha$, if any, would be necessary to alter the conclusion in your hypothesis test?

**c.** What value of $\alpha$, if any, would alter the strength of the evidence against $H_0$?

WHAT IF? **49.** Try to answer the following questions by thinking about the relationship between the statistics rather than by redoing all the calculations. *What if* the sample size was not 15 but rather some number greater than 15. Otherwise, everything else remains unchanged. How does this affect the following?

**a.** $s/\sqrt{n}$

**b.** $t_{data}$

**c.** $t_{crit}$

**d.** The $p$ value

**e.** The conclusion

**f.** The strength of the evidence against the null hypothesis

**g.** Write a sentence summarizing what happens to the elements in **(a)–(f)** when the sample size increases.

**Top Gas Mileage.** Use the following information for Exercises 50–52. The top ten vehicles for city gas mileage in 2007, as reported by the Environmental Protection Agency, are shown in the following table, along with the normal probability plot (Figure 9.56).

| Car | Mileage | Car | Mileage |
|---|---|---|---|
| Toyota Yaris | 39 | Honda Fit | 38 |
| Chevrolet Aveo | 37 | Nissan Versa | 34 |
| Pontiac G5 | 34 | Dodge Caliber | 32 |
| VW Eos | 32 | Ford Escape | 31 |
| Saturn Sky | 30 | BMW 525 | 30 |

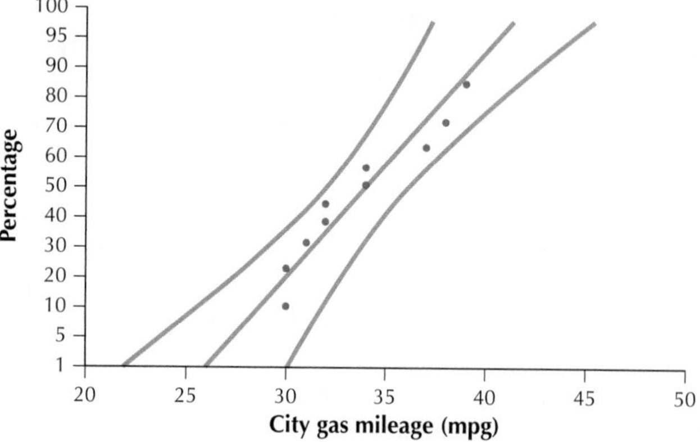


Figure 9.56

**50.** We are interested in testing whether the population mean city mileage of such cars is greater than 30 mpg.

**a.** Is it appropriate to apply the $t$ test for the mean? Why or why not?

**b.** Find the sample mean mileage.

**c.** Calcuate $s/\sqrt{n}$.

**d.** Compute $t_{data}$.

**51.** Answer the following.

**a.** Test, using the estimated $p$-value method at level of significance $\alpha = 0.01$, whether the population mean city mileage exceeds 30 mpg.

**b.** Repeat your test from **(a)**, this time using level of significance $\alpha = 0.001$.

**c.** How do you think we should resolve the apparent contradiction in the preceding two results?

**d.** Assess the strength of the evidence against the null hypothesis. Does this change depend on which level of $\alpha$ you use?

WHAT IF? **52.** *What if* we changed $\mu_0$ to some larger value (though still smaller than $\bar{x}$). Otherwise, everything else remains unchanged. Describe how this change would affect the following, if at all.

**a.** $t_{\text{data}}$
**b.** $t_{\text{crit}}$
**c.** The $p$ value
**d.** The conclusion from Exercise 51(a)
**e.** The conclusion from Exercise 51(b)
**f.** The strength of the evidence against the null hypothesis

**Community College Tuition.** Use the following information for Exercises 53–55. The College Board reported that the mean tuition and fees at community colleges nationwide in 2006 was \$2272. Data were gathered on the total tuition and fees for a random sample of ten community colleges this year. The normal probability plot (Figure 9.57) and Minitab $t$ test output (Figure 9.58) are shown here.

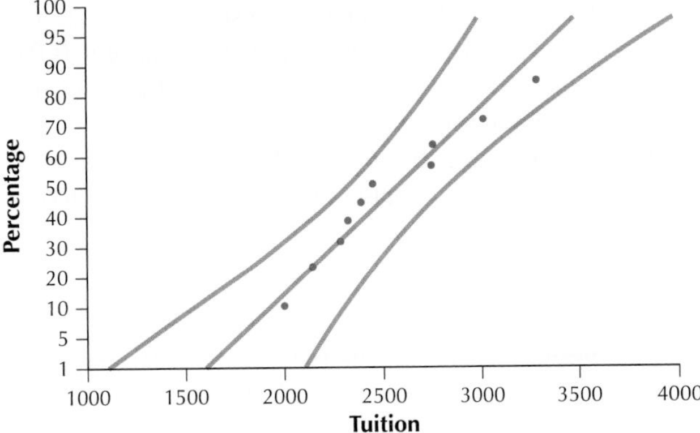


Figure 9.57 Normal probability plot.

```
Test of mu = 2272 vs not = 2272

Variable   N     Mean   StDev  SE Mean            95% CI          T      P
tuition   10  2538.92  404.75   127.99  (2249.38, 2828.46)  2.09  0.067
```

Figure 9.58 Minitab $t$ test output.

**53.** Analysts are interested in whether the population mean tuition and fees this year have increased.

**a.** Is it appropriate to apply the $t$ test for the mean? Why or why not?
**b.** What are the sample size, sample mean cost, and sample standard deviation of the cost?
**c.** Verify the value for the standard error of the mean.
**d.** Find $t_{\text{data}}$ in the Minitab output. Interpet $t_{\text{data}}$ using the standard error.
**e.** It appears that the data analyst who produced the Minitab printout asked for the wrong hypothesis test. How can we tell?

**54.** Refer to your work in the previous exercise.

**a.** Test whether the population mean tuition and fees have increased using the $p$-value method and level of significance $\alpha = 0.05$. How can we use the $p$-value on the Minitab printout to find the $p$-value needed for this right-tailed hypothesis test?
**b.** Compare the conclusion from **(a)** with the conclusion we would have gotten had we not noticed that the data analyst performed the wrong hypothesis test. What are some of the possible consequences of making an error of this sort?
**c.** Based on your experiences in these exercises, write a sentence about the importance of understanding the statistical modeling behind the "point and click" power of statistical software.

**55. Challenger Exercise.** Refer to your work in the previous exercise.

**a.** Note that we have concluded that there is insufficient evidence that the population mean cost has changed, but that there is evidence that the population mean cost has increased. How can the mean cost have increased without changing? Explain what is going on here, in terms of either critical regions or $p$-values.
**b.** Assess the strength of the evidence against the null hypothesis for the test in Exercise 54a.
**c.** Suppose a newspaper story referred to the study under the headline "College Costs Rising." How would you respond? Is the headline supported or not supported by the data and the hypothesis test?

**Marriage Age.** Use the following information Exercises 56–59. The U.S. Census Bureau reported that the median age at first marriage for females in 2003 was 25.3. Let's assume that the mean age at first marriage among females in 2003 was 26 years old. A random sample of 157 females this year yielded a mean age at first marriage of 26.5 with a standard deviation of 4 years. A histogram of the ages was decidedly not normal.

**56.** Answer the following.

**a.** Is it appropriate to apply the $t$ test for the mean? Why or why not?
**b.** Calculate the standard error of the mean.
**c.** Compute $t_{\text{data}}$. Interpret $t_{\text{data}}$ using the standard error.
**d.** Is there evidence that the population mean age at first marriage for females has changed since 2003? Test using the critical-value method at level of significance $\alpha = 0.10$.

**57.** Test using the critical-value method at level of significance $\alpha = 0.10$ whether the population mean age at first marriage for females has increased since 2003.

**58. Challenger Exercise.**

**a.** Compare your conclusions in the previous two exercises. Note that we have concluded that there is insufficient evidence that the population mean marriage age has changed, but that there is evidence that the population mean marriage age has increased. How can the mean marriage age have increased without changing? Explain what is going on here, in terms of either critical regions or $p$-values.
**b.** For which of the two hypothesis tests is the test statistic closer to the critical value?

**59.** Answer the following.

**a.** Find or estimate the $p$-value for the test in Exercise 56.
**b.** Assess the strength of the evidence against the null hypothesis.
**c.** Is the $p$-value closer to 0.05 or 0.10? Why?
**d.** Suppose a newspaper story referred to the study under

the headline "Mean Female Marriage Age Increasing." How would you respond? Is the headline supported or not supported by the data and the hypothesis test?

**60. New York Towns.** Work with the **New York** data set for the following.

**a.** How many observations are in the data set? How many variables?

**b.** Use technology to explore the variable *tot_pop,* which lists the population for each of the towns and cities in New York with at least 1000 people.

**c.** Suppose we are using the data in this data set as a sample of the population of all the towns and cities in the northeastern United States with at least 1000 people. Use technology to test at level of significance $\alpha = 0.05$ whether the population mean population of these towns differs from 50,000.

**61. Texas Towns.** Work with the **Texas** data set for the following.

**a.** How many observations are in the data set? How many variables?

**b.** Use technology to explore the variable *tot_occ,* which lists the total occupied housing units for each county in Texas. Generate numerical summary statistics and graphs for the total occupied housing units. What is the sample mean? The sample standard deviation? Comment on the symmetry or skewness of the data set.

**c.** Suppose we are using the data in this data set as a sample of the total occupied housing units of all the counties in the southwestern United States. Use technology to test at level of significance $\alpha = 0.05$ whether the population mean total occupied housing units for these counties differs from 40,000.

**62. Calories.** Work with the **Nutrition** data set.

**a.** How many observations are in the data set? How many variables?

**b.** Use technology to explore the variable *calories,* which lists the calories in the sample of foods. Generate numerical summary statistics and graphs for the number of calories. What is the sample mean number of calories? The sample standard deviation?

**c.** Use technology to test at level of significance $\alpha = 0.05$ whether the population mean number of calories per food item differs from 300.

**d.** Use technology to test at level of significance $\alpha = 0.05$ whether the population mean number of calories per food item differs from 320.

## 9.5 Z Test for the Population Proportion $p$

*Objectives:* By the end of this section, I will be able to...

1 Explain the essential idea about hypothesis testing for the proportion.

2 Perform and interpret the $Z$ test for the proportion using the $p$-value method.

3 Perform and interpret the $Z$ test for the proportion using the critical-value method.

4 Use confidence intervals for $p$ to perform two-tailed hypothesis tests about $p$.

### 1 The Essential Idea About Hypothesis Testing for the Proportion

Thus far in Chapter 9, we have dealt with testing hypotheses about the population mean $\mu$ only. However, the mean is not the only population characteristic whose value is unknown. As we have seen in Section 8.3, the population proportion of successes $p$ is also unknown for most real-world problems. In that section, we learned how to apply the $Z$ interval to estimate the value of the unknown population proportion $p$. Here in Section 9.5, we will learn how to perform the $Z$ test for the population proportion $p$. We will examine two equivalent methods for performing the $Z$ test for $p$:

1. The $Z$ test for $p$ using the $p$-value method
2. The $Z$ test for $p$ using the critical-value method

What do we use as our point estimate of the unknown population proportion $p$? We use the sample proportion $\hat{p}$. Recall from Section 7.3 that, since $\hat{p}$ is a sample statistic, it has a sampling distribution. We learned that the mean of this distribution is $p$, and the standard deviation of the sampling distribution (the measure of variability) is

$$\sigma_{\hat{p}} = \sqrt{\frac{p\,(1-p)}{n}}$$

We learned that this sampling distribution of $\hat{p}$ is approximately normal whenever both of the following conditions are met: $np \geq 5$ and $n(1-p) \geq 5$. Just as with the $Z$ test for the mean, in the $Z$ test for the proportion the null hypothesis will include a certain hypothesized value for the unknown parameter, which we call $p_0$. For example, the hypotheses for the two-tailed test have the following form:

$$H_0\colon p = p_0 \quad \text{versus} \quad H_a\colon p \neq p_0$$

where $p_0$ represents a particular hypothesized value of the unknown population proportion $p$. For instance, if a researcher is interested in determining whether the population proportion of Americans who support increased funding for higher education differs from 50%, then $p_0 = 0.50$.

If we assume $H_0$ is correct, then the population proportion of successes is $p_0$, and so the sampling distribution of $p$ is centered at $p_0$, as shown in Figure 9.59. The standard deviation of the sampling distribution of $\hat{p}$ takes the form

$$\sigma_{\hat{p}} = \sqrt{\frac{p(1-p)}{n}} = \sqrt{\frac{p_0\,(1-p_0)}{n}}$$

since we claim in $H_0$ that $p = p_0$.

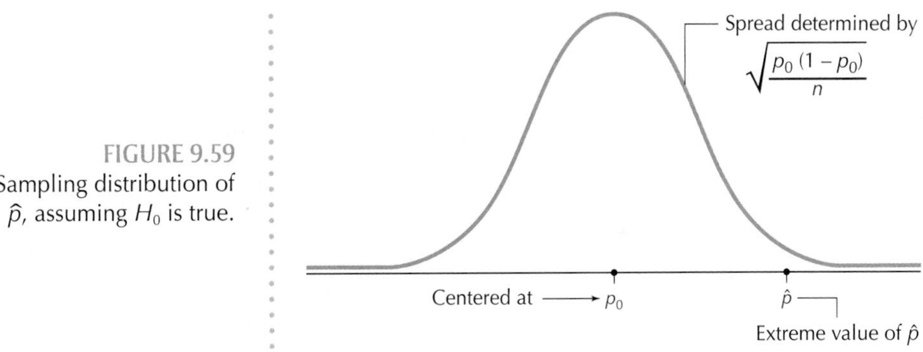


FIGURE 9.59
Sampling distribution of $\hat{p}$, assuming $H_0$ is true.

Finally, this sampling distribution of $\hat{p}$ is approximately normal whenever both of the following normality conditions are met: $np_0 \geq 5$ and $n(1-p_0) \geq 5$ (substituting the hypothesized $p_0$ for $p$ in the earlier form of these conditions). This leads us to the following statement of the essential idea about **hypothesis testing for the proportion.**

**The Essential Idea About Hypothesis Testing for the Proportion**

When the sample proportion $\hat{p}$ is unusual or extreme in the sampling distribution of $\hat{p}$ that is based on the assumption that $H_0$ is correct, we reject $H_0$. Otherwise, there is insufficient evidence against $H_0$, and we should not reject $H_0$.

## 2 Z Test for p Using the p-Value Method

The remainder of Section 9.5 explains the details of implementing hypothesis testing for the proportion. Recall that, in the $Z$ test for the mean, $Z_{\text{data}}$ represents the number of standard deviations ($\sigma_{\bar{x}}$) the observed sample mean $\bar{x}$ lies above or below the hypothesized mean $\mu_0$. In this section, $Z_{data}$ will represent the number of standard deviations ($\sigma_{\hat{p}}$) the sample proportion $\hat{p}$ lies above or below the hypothesized proportion $p_0$.

The test statistic used for the $Z$ test for the proportion is

$$Z_{data} = \frac{\hat{p} - p_0}{\sigma_{\hat{p}}} = \frac{\hat{p} - p_0}{\sqrt{\frac{p_0(1 - p_0)}{n}}}$$

where $\hat{p}$ is the observed sample proportion of successes, $p_0$ is the value of $p$ hypothesized in $H_0$, and $n$ is the sample size. $Z_{data}$ represents the number of standard deviations ($\sigma_{\hat{p}}$) the sample proportion $\hat{p}$ lies above or below the hypothesized proportion $p_0$. Extreme values of $\hat{p}$ will be associated with extreme values of $Z_{data}$.

Just as for hypothesis tests for the mean, there are three possible forms of hypothesis tests for the proportion, as shown in Table 9.13.

Table 9.13 The three possible forms for the hypotheses for a test for $p$

| Form | Null and alternative hypotheses |
|---|---|
| Right-tailed test, one-tailed test | $H_0: p \leq p_0$ versus $H_a: p > p_0$ |
| Left-tailed test, one-tailed test | $H_0: p \geq p_0$ versus $H_a: p < p_0$ |
| Two-tailed test | $H_0: p = p_0$ versus $H_a: p \neq p_0$ |

**$Z$ Test for the Population Proportion $p$: $p$-Value Method**

When a random sample of size $n$ is taken from a population, you can use the $Z$ test for the proportion if both of the normality conditions are satisfied:

$$np_0 \geq 5 \quad \text{and} \quad n(1 - p_0) \geq 5$$

***Step 1*** **State the hypotheses and the rejection rule.** Use one of the forms from Table 9.13. Be sure to explicitly state the meaning of $p$. State the rejection rule as "Reject $H_0$ when the $p$-value is less than $\alpha$."

***Step 2*** **Find $Z_{data}$.** Either use technology to find the value of $Z_{data}$ or calculate $Z_{data}$ as follows:

$$Z_{data} = \frac{\hat{p} - p_0}{\sigma_{\hat{p}}} = \frac{\hat{p} - p_0}{\sqrt{\frac{p_0(1 - p_0)}{n}}}$$

***Step 3*** **Find the $p$-value.** Either use technology to find the $p$-value, or calculate it using the form in Table 9.4 that corresponds to your hypotheses.

***Step 4*** **State the conclusion and the interpretation.** If the $p$-value is less than $\alpha$, then reject $H_0$. Otherwise do not reject $H_0$. Interpret your conclusion so that a nonspecialist can understand.

Note that the $p$-value has precisely the same definition and behavior as in the $Z$ test for the mean. That is, the $p$-value is roughly a measure of how extreme your value of $Z_{data}$ is and takes values between 0 and 1, with small values indicating extreme values of $Z_{data}$.

**Developing Your Statistical Sense**

**The Difference Between the *p*-Value and the Population Proportion *p***

Be careful to distinguish between the *p*-value and the population proportion *p*. The latter represents the population proportion of successes for a binomial experiment and is a population parameter. The *p*-value is the probability of observing a value of $Z_{data}$ at least as extreme as the $Z_{data}$ actually observed. The *p*-value depends on the sample data, but the population proportion *p* does not depend on the sample data. ■

## Example 9.26 Checking the normality conditions

A 2006 study reported that 1% of American Internet users who are married or in a long-term relationship met on a blind date or through a dating service.[16] A survey taken this year of 400 American Internet users who are married or in a long-term relationship found 5 who met on a blind date or through a dating service. If appropriate, test whether the population proportion has increased since 2006.

### Solution

The keyword "increased" gives us the following right-tailed test:

$$H_0: p \leq 0.01 \quad \text{versus} \quad H_a: p > 0.01$$

where $p_0 = 0.01$. The sample size is $n = 400$. Before we perform the hypothesis test, we check the normality conditions. We have

$$np_0 = (400)(0.01) = 4$$

Since 4 is not ≥5, our test fails the first normality condition. (It is not necessary to check the second condition if the first fails, since both must be true in order to perform the test.) We therefore cannot perform this hypothesis test since the sampling distribution of $\hat{p}$ is not approximately normal. ■

## Example 9.27 Are car accidents among young drivers increasing?

Getty Images/Stockbyte Platinum

The National Transportation Safety Board publishes statistics on the number of automobile crashes that people in various age groups have. In 2003, young people aged 18–24 had an accident rate of 12%, meaning that on average 12 out of every 100 young drivers per year had an accident. A more recent study examined 1000 young drivers aged 18–24 and found that 134 had an accident this year. We are interested in whether the population proportion of young drivers having accidents has increased since 2003, using the *p*-value method with level of significance $\alpha = 0.05$.

### Solution

First we check that both of our normality conditions are met. Since we are interested in whether the proportion has increased from 12%, we have $p_0 = 0.12$.

$$np_0 = (1000)\,(0.12) = 120 \geq 5 \quad \text{and} \quad n(1 - p_0) = (1000)(0.88) = 880 \geq 5$$

The normality conditions are met and we may proceed with the hypothesis test.

***Step 1*** **State the hypotheses and the rejection rule.** Our hypotheses are

$$H_0: p \leq 0.12 \quad \text{versus} \quad H_a: p > 0.12$$

where *p* represents the population proportion of young people aged 18–24 who had an accident. We reject the null hypothesis if the *p*-value is less than $\alpha = 0.05$.

**Step 2** **Find $Z_{\text{data}}$.** Our sample proportion is $\hat{p} = 134/1000 = 0.134$. Since $p_0 = 0.12$, the standard deviation of the sampling distribution of $\hat{p}$ is

$$\sigma_{\hat{p}} = \sqrt{\frac{p_0(1-p_0)}{n}} = \sqrt{\frac{(0.12)(0.88)}{1000}} \approx 0.0103$$

Thus, our test statistic is

$$Z_{\text{data}} = \frac{\hat{p}-p_0}{\sigma_{\hat{p}}} = \frac{\hat{p}-p_0}{\sqrt{\frac{p_0(1-p_0)}{n}}} = \frac{0.134-12}{\sqrt{\frac{(0.12)(0.88)}{1000}}} \approx 1.36$$

That is, the sample proportion $\hat{p} = 0.134$ lies approximately 1.36 standard deviations above the hypothesized proportion $p_0 = 0.12$.

**Step 3** **Find the *p*-value.** Since we have a right-tailed test, our *p*-value from Table 9.4 is $P(Z > Z_{\text{data}})$. This is a Case 2 problem from Table 6.7 (page 298), where we find the tail area by subtracting the *Z* table area from 1:

$$P(Z > Z_{\text{data}}) = P(Z > 1.36) = 1 - 0.9131 = 0.0869$$

**Step 4** **State the conclusion and the interpretation.** Since the *p*-value is not less than $\alpha = 0.05$, we do not reject $H_0$. There is insufficient evidence that the population proportion of young people aged 18–24 who had an accident has increased. ■

## What About When the *p*-Value Is Close to $\alpha$?

The sampling distribution of $\hat{p}$ for Example 9.27 is shown in Figure 9.60. Note that the sample proportion $\hat{p} = 0.134$ is not very extreme, but it is not safely snug in the middle either. It is a close call whether to consider this sample proportion extreme, since our *p*-value 0.0869 is not much larger than $\alpha = 0.05$. But who says that $\alpha$ needs to be 0.05? If $\alpha$ were 0.10, then our conclusion would be reversed. Perhaps we should simply report the strength of evidence against the null hypothesis using Table 9.5. In this case, our *p*-value of 0.0869 provides *mild* evidence against the null hypothesis that the population proportion of young people having an accident has not increased. ■

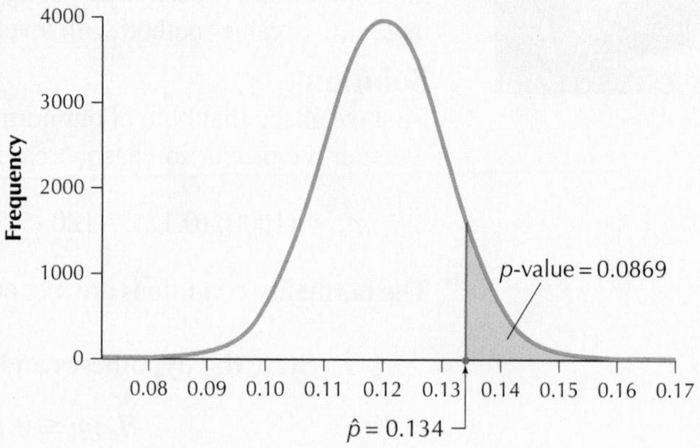


FIGURE 9.60 Sampling distribution of $\hat{p}$, assuming $p_0 = 0.12$: Is $\hat{p} = 0.134$ unusual?

## What If Scenario — What If More Young Drivers Were Having Accidents?

To answer the following question, try to think about the relationships among the various statistics rather than doing all the calculations over again. Consider Example 9.27, where the level of significance $\alpha$ was 0.05. Suppose our sample proportion $\hat{p}$ of young drivers having accidents was somewhat higher than 12%. Otherwise, everything else in the example is the same. Describe how this change would affect the following.

**a.** $\sigma_{\hat{p}}$

**b.** $Z_{\text{data}}$

**c.** The $p$-value

**d.** $\alpha$

**e.** The conclusion

### Solution

**a.** Since $\sigma_{\hat{p}} = \sqrt{p_0(1 - p_0)/n}$ depends only on $p_0$ and $n$, and not on $\hat{p}$, the standard deviation $\sigma_{\hat{p}}$ is *unaffected* by an increase in $\hat{p}$.

**b.** Since $\hat{p}$ is greater than $p_0$, increasing $\hat{p}$ increases the difference $\hat{p} - p_0$, the numerator of $Z_{\text{data}}$. Then, since $\sigma_{\hat{p}}$ (the denominator of $Z_{\text{data}}$) is unchanged, $Z_{\text{data}}$ itself *increases*.

**c.** Since $Z_{\text{data}}$ increases, the $p$-value, which measures (in this case) the area to the right of $Z_{\text{data}}$, *decreases*.

**d.** Since the level of significance $\alpha$ is not affected by sample characteristics, a change in $\hat{p}$ leaves $\alpha$ *unaffected*.

**e.** Previously, our $p$-value was 0.0869, which was not less than $\alpha = 0.05$, so that we did not reject $H_0$. From **(c)**, we know that the $p$-value decreases, but we do not know by how much. Therefore, the answer *cannot be determined*. It is possible that our conclusion may change from nonrejection to rejection, but we are not sure. We would have to specify the new value of $\hat{p}$ and recalculate.

## Example 9.28 Performing the $Z$ test for $p$ using technology

We return to Example 9.26, this time with a larger sample size. This time a survey of 500 American Internet users who are married or in a long-term relationship found 8 who met on a blind date or through a dating service. If appropriate, use technology to test whether the population proportion has increased since 2006. Use the $p$-value method with level of significance $\alpha = 0.05$.

### Solution

We have $p_0 = 0.01$ and $n = 500$. Checking the normality conditions, we have

$$np_0 = (500)(0.01) = 5 \geq 5 \quad \text{and} \quad n(1 - p_0) = (500)(0.99) = 495 \geq 5$$

The normality conditions are met and we may proceed with the hypothesis test.

***Step 1*** **State the hypotheses and the rejection rule.** From Example 9.26, our hypotheses are

$$H_0: p \leq 0.01 \quad \text{versus} \quad H_a: p > 0.01$$

where $p$ represents the population proportion of American Internet users who are married or in a long-term relationship and who met on a blind date or through a dating service. We will reject $H_0$ if $p$-value $< 0.05$.

**Step 2** **Find $Z_{\text{data}}$.** We use the instructions supplied in the Step-by-Step Technology Guide at the end of this section. Figure 9.61 shows the TI-83/84 results from the $Z$ test for $p$, and Figure 9.62 shows the results from Minitab.

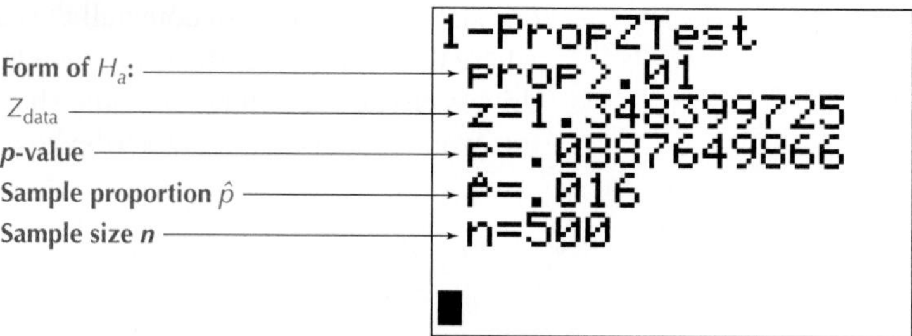


FIGURE 9.61 TI-83/84 results.

Test of p = 0.01 vs p > 0.01

| Sample | X | N | Sample p | 95% Lower Bound | Z-Value | P-Value |
|---|---|---|---|---|---|---|
| 1 | 8 | 500 | 0.016000 | 0.006770 | 1.35 | 0.089 |
| | $X$ | $n$ | $\hat{p}$ | (not used) | $Z_{\text{data}}$ | $p$-value |

FIGURE 9.62 Minitab results.

From Figure 9.61, we have

$$Z_{\text{data}} = \frac{\hat{p} - p_0}{\sigma_{\hat{p}}} = \frac{\hat{p} - p_0}{\sqrt{\dfrac{p_0(1 - p_0)}{n}}} = \frac{0.016 - 0.01}{\sqrt{\dfrac{(0.01)(0.99)}{500}}} \approx 1.348399725$$

**Step 3** **Find the $p$-value.** From Figures 9.61, 9.62, and 9.63, we have

$$p\text{-value} = P(Z > 1.348399725) = 0.0887649866$$

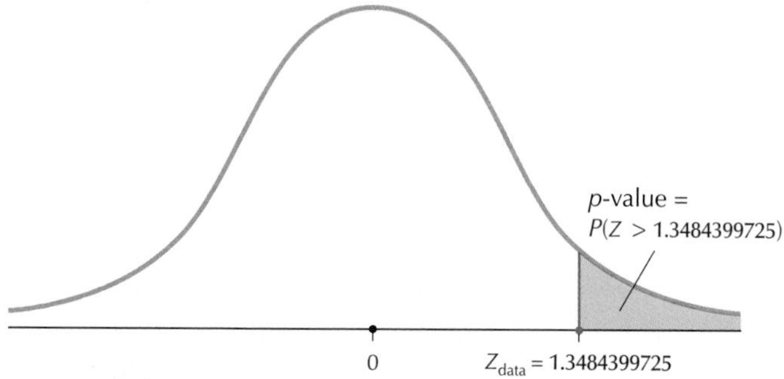


FIGURE 9.63

**Step 4** **State the conclusion and interpretation.** Since $p$-value $\approx 0.08876$ is *not* less than our level of significance $\alpha = 0.05$, we do *not* reject $H_0$. There is insufficient evidence that the population proportion of American Internet users who are married or in a long-term relationship and who met on a blind date or through a dating service has increased since 2006. ■

## 3 Z Test for p Using the Critical-Value Method

Recall that we learned earlier how to apply the hypothesis tests for the mean using the critical-value method. Similarly, we can use the critical-value method to perform the $Z$ test for the proportion. The critical-value method for the $Z$ test for

Table 9.14 Table of critical values $Z_{crit}$ for common values of the level of significance $\alpha$

| | Form of Hypothesis Test | | |
|---|---|---|---|
| **Level of significance $\alpha$** | **Right-tailed** $H_0: p \le p_0$ $H_a: p > p_0$ | **Left-tailed** $H_0: p \ge p_0$ $H_a: p < p_0$ | **Two-tailed** $H_0: p = p_0$ $H_a: p \ne p_0$ |
| 0.10 | $Z_{crit} = 1.28$ | $Z_{crit} = -1.28$ | $Z_{crit} = 1.645$ |
| 0.05 | $Z_{crit} = 1.645$ | $Z_{crit} = -1.645$ | $Z_{crit} = 1.96$ |
| 0.01 | $Z_{crit} = 2.33$ | $Z_{crit} = -2.33$ | $Z_{crit} = 2.58$ |

the proportion is exactly equivalent to the $p$-value method for the $Z$ test for the mean. Therefore, the normality conditions under which you can use the $Z$ test critical-value method are the same as the conditions for using the $Z$ test $p$-value method, namely, $np_0 \ge 5$ and $n(1 - p_0) \ge 5$. In the critical-value method, we compare one $Z$-value ($Z_{data}$) with another $Z$-value ($Z_{crit}$) rather than comparing the $p$-value with $\alpha$. You can use Table 9.14 to find the critical values $Z_{crit}$ (this table is the same as Table 9.6 except that the $\mu$'s have been changed to $p$'s). To find the rejection rules or the critical regions, you can use Table 9.15 (which is the same as Table 9.7 except that the $\mu$'s have been changed to $p$'s).

Table 9.15 Rejection rules for $Z$ test for a proportion

| Form of test | | Rejection rules: "Reject $H_0$ if…" |
|---|---|---|
| Right-tailed | $H_0: p \le p_0$ vs. $H_a: p > p_0$ | $Z_{data} > Z_{crit}$ |
| Left-tailed | $H_0: p \ge p_0$ vs. $H_a: p < p_0$ | $Z_{data} < Z_{crit}$ |
| Two-tailed | $H_0: p = p_0$ vs. $H_a: p \ne p_0$ | $Z_{data} > Z_{crit}$ or $Z_{data} < -Z_{crit}$ |

The $Z$ test for proportions is essentially the same as the $Z$ test for means. They differ only in that $Z_{data}$ is, of course, calculated differently, and the hypotheses and interpretations differ.

**$Z$ Test for the Population Proportion $p$: Critical-Value Method**

When a random sample of size $n$ is taken from a population, you can use the $Z$ test for the proportion if both of the normality conditions are satisfied:

$$np_0 \ge 5 \quad \text{and} \quad n(1 - p_0) \ge 5$$

***Step 1*** **State the hypotheses.** Use one of the forms from Table 9.15. Clearly describe the meaning of $p$.

***Step 2*** **Find $Z_{crit}$ and state the rejection rule.** Use Tables 9.14 and 9.15.

***Step 3*** **Find $Z_{data}$.** Either use technology to find the value of $Z_{data}$ or calculate $Z_{data}$ as follows:

$$Z_{data} = \frac{\hat{p} - p_0}{\sigma_{\hat{p}}} = \frac{\hat{p} - p_0}{\sqrt{\dfrac{p_0(1 - p_0)}{n}}}$$

***Step 4*** **State the conclusion and the interpretation.** If $Z_{data}$ falls in the critical region, then reject $H_0$. Otherwise, do not reject $H_0$. Interpret the conclusion so that a nonspecialist can understand.

## Example 9.29 Proportion of Americans who smoke

The National Center for Health Statistics reports on the proportion of Americans who smoke tobacco. Although the proportion decreased throughout the 1990s, from 2004 to 2006 there was little change in the proportion of Americans who smoke, 21%. A random sample taken this year of 400 Americans found 78 who smoked. Test whether the population proportion of Americans who smoke has changed since 2006, using the critical-value method with level of significance $\alpha = 0.10$.

### Solution

First we check that both of our normality conditions are met. Since we are interested in whether the proportion has changed from 21%, we have $p_0 = 0.21$.

$$np_0 = (400)(0.21) = 84 \geq 5 \quad \text{and} \quad n(1 - p_0) = (400)(0.79) = 316 \geq 5$$

The normality conditions are met and we may proceed with the hypothesis test.

***Step 1*** **State the hypotheses.** Our hypotheses are

$$H_0: p = 0.21 \quad \text{versus} \quad H_a: p \neq 0.21$$

where $p$ represents the population proportion of Americans who smoke tobacco.

***Step 2*** **Find $Z_{\text{crit}}$ and state the rejection rule.** We have a two-tailed test, with $\alpha = 0.10$. This gives us our critical value $Z_{\text{crit}} = 1.645$ from Table 9.14 and the rejection rule from Table 9.15: Reject $H_0$ if $Z_{\text{data}} > 1.645$ or $Z_{\text{data}} < -1.645$.

***Step 3*** **Find $Z_{\text{data}}$.** Our sample proportion is $\hat{p} = 78/400 = 0.195$. Since $p_0 = 0.21$, the standard deviation of the sampling distribution of $\hat{p}$ is

$$\sigma_{\hat{p}} = \sqrt{\frac{p_0(1 - p_0)}{n}} = \sqrt{\frac{(0.21)(0.79)}{400}} \approx 0.020365$$

Thus, our test statistic is

$$Z_{\text{data}} = \frac{\hat{p} - p_0}{\sigma_{\hat{p}}} = \frac{\hat{p} - p_0}{\sqrt{\frac{p_0(1 - p_0)}{n}}} = \frac{0.195 - 0.21}{\sqrt{\frac{(0.21)(0.79)}{400}}} \approx -0.74$$

### *What Results Might We Expect?*

$Z_{\text{data}} = -0.74$ denotes that the sample proportion $\hat{p} = 0.195$ lies 0.74 standard deviations below the hypothesized proportion $p_0 = 0.21$. This is illustrated in Figure 9.64. Clearly, the sample proportion $\hat{p} = 0.195$ is not extreme, and so, by the essential idea about hypothesis testing for the proportion, we would not expect to reject $H_0$. ■

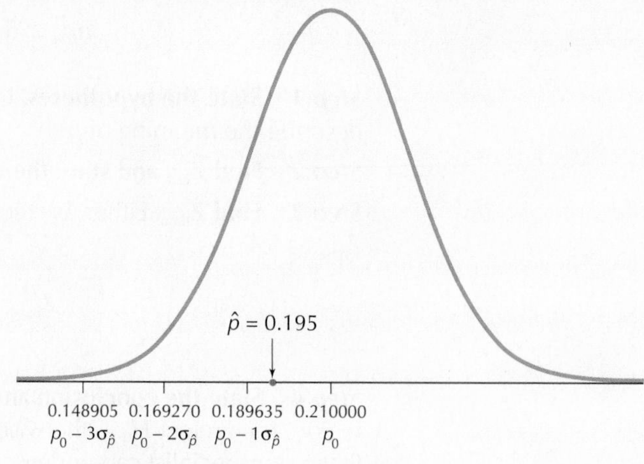


FIGURE 9.64

***Step 4*** **State the conclusion and the interpretation.** Since $Z_{\text{data}} = -0.74$ is not greater than 1.645 and not less than –1.645. Thus, as expected, we do not reject $H_0$. There is insufficient evidence that the population proportion of Americans who smoke tobacco has changed since 2006.

## 4 Using Confidence Intervals for $p$ to Perform Two-Tailed Hypothesis Tests About $p$

Just as for $\mu$, we can use a $100(1 - \alpha)\%$ confidence interval for the population proportion $p$ in order to perform a set of two-tailed hypothesis tests for $p$.

### Example 9.30 Using confidence intervals for $p$ to determine significance

In 2007, the Pew Internet and American Life Project reported that 91% of Americans who have completed a bachelor's degree currently use the Internet. Pew also reports that the margin of error for this survey (confidence level = 95%) was ±3%. The 95% confidence interval for the population proportion of Americans with a bachelor's degree who currently use the Internet is therefore

$$0.91 \pm 0.03 = (0.88, 0.94)$$

Use the confidence interval to test, using level of significance $\alpha = 0.05$, whether the population proportion differs from

**a.** 0.85
**b.** 0.90
**c.** 0.95

**Solution**

There is equivalence between a $100(1 - \alpha)\%$ confidence interval for $p$ and a two-tailed test for $p$ with level of significance $\alpha$. Values of $p_0$ that lie outside the confidence interval lead to rejection of the null hypothesis, while values of $p_0$ within the confidence interval lead to not rejecting the null hypothesis. Figure 9.65 illustrates the 95% confidence interval for $p$.

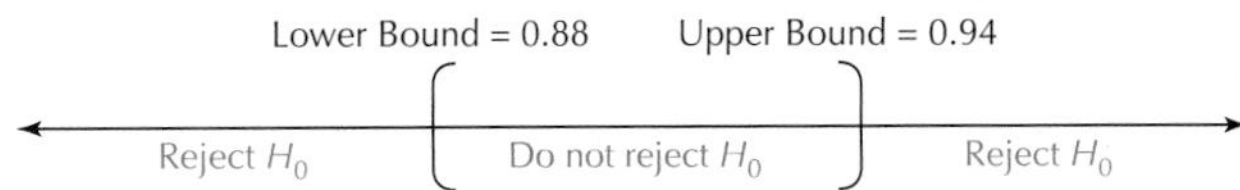


FIGURE 9.65 Reject $H_0$ for values $p_0$ that lie outside the interval (0.88, 0.94).

We would like to perform the following two-tail hypothesis tests:

**a.** $H_0: p = 0.85$ versus $H_a: p \neq 0.85$
**b.** $H_0: p = 0.90$ versus $H_a: p \neq 0.90$
**c.** $H_0: p = 0.95$ versus $H_a: p \neq 0.95$

To perform each hypothesis test, simply observe where each value of $p_0$ falls on the number line. For example, in the first hypothesis test, the hypothesized value $p_0 = 0.85$ lies outside the interval (0.88, 0.94). Thus, we reject $H_0$. The three hypothesis tests are summarized here.

| Value of $p_0$ | Form of hypothesis test, with $\alpha = 0.05$ | Where $p_0$ lies in relation to 95% confidence interval | Conclusion of hypothesis test |
|---|---|---|---|
| **a.** 0.85 | $H_0: p = 0.85$  $H_a: p \neq 0.85$ | Outside | Reject $H_0$ |
| **b.** 0.90 | $H_0: p = 0.90$  $H_a: p \neq 0.90$ | Inside | Do not reject $H_0$ |
| **c.** 0.95 | $H_0: p = 0.95$  $H_a: p \neq 0.95$ | Outside | Reject $H_0$ |

## STEP-BY-STEP TECHNOLOGY GUIDE: *Z* test for *p*

We will use the information from Example 9.28 (page 519).

### TI-83/84

***Step 1*** Press **STAT**, highlight **TESTS**, and press **ENTER**.
***Step 2*** Press 5 (for **1-PropZTest**; see Figure 9.66).
***Step 3*** For $p_0$, enter the value of $p_0$, **0.01**.
***Step 4*** For **x**, enter the number of successes, **8**.
***Step 5*** For **n**, enter the number of trials **500**.
***Step 6*** For **prop**, enter the form of $H_a$. Here we have a right-tailed test, so highlight **>$p_0$** and press **ENTER** (see Figure 9.67).
***Step 7*** Highlight **Calculate** and press **ENTER**. The results are shown in Figure 9.61 in Example 9.28.

Figure 9.66

```
1-PropZTest
 p0:.01
 x:8
 n:500
 prop≠p0 <p0 >p0
 Calculate Draw
```

Figure 9.67

### EXCEL

#### WHFStat Macros

***Step 1*** Enter the data into column A. (If you have only the summary statistics, go to Step 2.)
***Step 2*** Load the **WHFStat Macros**.
***Step 3*** Select **Add-Ins > Macros > Testing a Proportion > One Sample**.
***Step 4*** Enter the **Number of successes 8**.
***Step 5*** Enter the **Sample size 500**.
***Step 6*** Enter the **Testing Proportion, $p_0$= 0.01**.
***Step 7*** Select your **Confidence level**, which should be $1 - \alpha$. Here, because $\alpha = 0.05$, we select **95%**.
***Step 8*** Click **OK**.

### MINITAB

#### If you have the summary statistics:

***Step 1*** Click **Stat > Basic Statistics > 1 Proportion**.
***Step 2*** Click **Summarized Data**.
***Step 3*** Enter the **Number of trials 500** and the **Number of Events 8**.
***Step 4*** Click **Options**.

**a.** Choose your **Confidence Level** as $100(1 - \alpha)$. Our level of significance $\alpha$ here is 0.05, so the confidence level is 95.0.
**b.** Enter **0.01** for the **Test Proportion**.
**c.** Select **Greater than** for the **Alternative**.
**d.** Check **Use test and interval based on normal distribution**.
***Step 5*** Click **OK** and click **OK** again. The results are shown in Figure 9.62 in Example 9.28.

## Section 9.5 SUMMARY

1 The essential idea about hypothesis testing for the proportion may be described as follows. When the sample proportion $\hat{p}$ is unusual or extreme in the sampling distribution of $\hat{p}$ that is based on the assumption that $H_0$ is correct, we should reject $H_0$. Otherwise, there is insufficient evidence against $H_0$, which leads us to the conclusion that we should not reject $H_0$.

2 The test statistic used for the $Z$ test for the proportion is

$$Z_{\text{data}} = \frac{\hat{p} - p_0}{\sigma_{\hat{p}}} = \frac{\hat{p} - p_0}{\sqrt{\dfrac{p_0(1-p_0)}{n}}}$$

where $\hat{p}$ is the observed sample proportion of successes, $p_0$ is the value of $p$ hypothesized in $H_0$, and $n$ is the sample size.

$Z_{\text{data}}$ represents the number of standard deviations ($\sigma_{\hat{p}}$) the sample proportion $\hat{p}$ lies above or below the hypothesized proportion $p_0$. Extreme values of $\hat{p}$ will be associated with extreme values of $Z_{\text{data}}$. The $Z$ test for the proportion may be performed using either the $p$-value method or the critical-value method. For the $p$-value method, we reject $H_0$ if the $p$-value is less than $\alpha$.

3 For the critical-value method, we compare the values of $Z_{\text{data}}$ and $Z_{\text{crit}}$. If $Z_{\text{data}}$ falls in the critical region, we reject $H_0$.

4 We can use a single $100(1 - \alpha)\%$ confidence interval for $p$ to help us perform any number of two-tailed hypothesis tests about $p$.

## Section 9.5 EXERCISES

### CLARIFYING THE CONCEPTS

1. What is the difference between $\hat{p}$ and $p$?
2. What is the standard deviation for the sampling distribution of $\hat{p}$? What does it mean? (*Hint*: Use the concept of variability of the statistic $\hat{p}$.)
3. Explain the essential idea about hypothesis testing for the proportion.
4. Explain what $p_0$ refers to.
5. What possible values can $p_0$ take?
6. What is the difference between $p$ and a $p$-value?

### PRACTICING THE TECHNIQUES

For the following samples, find $\sigma_{\hat{p}}$, the standard deviation of the sampling distribution of $\hat{p}$. Let $p_0 = 0.5$.

7. A sample of size 50 yields 20 successes.
8. A sample of size 50 yields 30 successes.
9. A sample of size 100 yields 70 successes.
10. A sample of size 100 yields 30 successes.
11. A sample of size 1000 yields 900 successes.
12. A sample of size 1000 yields 100 successes.
13. What kind of pattern do we observe in these results?

For Exercises 14–17, find the value of the test statistic $Z_{\text{data}}$ for a right-tailed test with $p_0 = 0.4$.

14. A sample of size 50 yields 20 successes.
15. A sample of size 50 yields 25 successes.
16. A sample of size 50 yields 30 successes.
17. A sample of size 50 yields 35 successes.
18. What kind of pattern do we observe in the value of $Z_{\text{data}}$ for a right-tailed test as the number of successes becomes more extreme?

For Exercises 19–23, find the value of the test statistic $Z_{\text{data}}$ for a two-tailed test with $p_0 = 0.5$.

19. A sample of size 80 yields 20 successes.
20. A sample of size 80 yields 30 successes.
21. A sample of size 80 yields 40 successes.
22. A sample of size 80 yields 50 successes.
23. A sample of size 80 yields 60 successes.
24. What kind of pattern do we observe in the value of $Z_{\text{data}}$ as the sample proportion approaches $p_0$?

For Exercises 25–28, do the following.

- **a.** Check the normality conditions.
- **b.** State the hypotheses and the rejection rule for the $p$-value method, using level of significance $\alpha = 0.05$.
- **c.** Find $Z_{\text{data}}$.
- **d.** Find the $p$-value.
- **e.** Compare the $p$-value with $\alpha = 0.05$. State the conclusion and the interpretation.

25. Test whether the population proportion exceeds 0.4. A random sample of size 100 yields 44 successes.
26. Test whether the population proportion is less than 0.2. A random sample of size 400 yields 75 successes.
27. Test whether the population proportion differs from 0.5. A random sample of size 900 yields 475 successes.
28. Test whether the population proportion exceeds 0.9. A random sample of size 1000 yields 925 successes.

For Exercises 29–31, do the following.

- **a.** Check the normality conditions.
- **b.** State the hypotheses.
- **c.** Find $Z_{\text{crit}}$ and the rejection rule.
- **d.** Calculate $Z_{\text{data}}$.
- **e.** Compare $Z_{\text{crit}}$ with $Z_{\text{data}}$. State the conclusion and the interpretation.

29. Test whether the population proportion is less than 0.5. A random sample of size 225 yields 100 successes. Let $\alpha = 0.05$.
30. Test whether the population proportion differs from 0.3. A random sample of size 100 yields 25 successes. Let $\alpha = 0.01$.
31. Test whether the population proportion exceeds 0.6. A random sample of size 400 yields 260 successes. Let $\alpha = 0.05$.

### APPLYING THE CONCEPTS

32. **Baptists in America.** A study reported in 2001 that 17.2% of Americans identified themselves as Baptists.[17] A survey of 500 randomly selected Americans this year showed that 85 of them were Baptists. If appropriate, test using the $p$-value method at level of significance $\alpha = 0.10$ whether the population proportion of Americans who are Baptists has changed since 2001.
33. **Births to Unmarried Women.** The National Center for Health Statistics reported: "Childbearing by unmarried women increased to record levels for the Nation in 2005."[18] In that year, 36.8% of all births were to unmarried women. Suppose that a random sample taken this year of 1000 births showed 380 to unmarried women. If appropriate, test whether the population proportion has increased since 2005, using level of significance $\alpha = 0.05$.
34. **Twenty-Somethings.** According to the U.S. Census Bureau, 7.1% of Americans living in 2004 were between the ages of 20 and 24. Suppose that a random sample of 400 Americans taken this year yields 35 between the ages of 20 and 24. If appropriate, test whether the population proportion of Americans aged 20–24 is different from the 2004 proportion. Use level of significance $\alpha = 0.01$.
35. **Nonmedical Pain Reliever Use.** The National Survey on Drug Use and Health reported that, from 2002 to 2005, 4.8% of persons aged 12 or older used a prescription pain reliever nonmedically.[19] Suppose that a random sample of 900 persons aged 12 or older taken this year found 54 that had used a prescription pain reliever nonmedically. If appropriate, test whether the population proportion has increased, using level of significance $\alpha = 0.01$.

**36. Ethnic Asians in California.** A research report states that, in 2005, 12.3% of California residents were of Asian ethnicity.[20] Suppose that this year, a random sample of 400 California residents yields 52 of Asian ethnicity. We are interested in whether the population proportion of California residents of Asian ethnicity has risen since 2005.

- **a.** Is it appropriate to perform the $Z$ test for the proportion? Why or why not?
- **b.** What is $\sigma_{\hat{p}}$? What does this number mean?
- **c.** How many standard deviations does $\hat{p}$ lie below $p_0$? Is this extreme? Why or why not?
- **d.** Where in the sampling distribution would the value of $\hat{p}$ lie? Near the tail? Near the center?

**37. Affective Disorders Among Women.** What do you think is the most common nonobstetric (not related to pregnancy) reason for hospitalization among 18- to 44-year-old American women? According to the U.S. Agency for Healthcare Research and Quality (**www.ahrq.gov**), this is the category of *affective* disorders, such as depression. In 2002, 7% of hospitalizations among 18- to 44-year-old American women were for affective disorders. Suppose that a random sample taken this year of 1000 hospitalizations of 18- to 44-year-old women showed 80 admitted for affective disorders. We are interested in whether the population proportion of hospitalizations for affective disorders has changed since 2002. Test using the $p$-value method and level of significance $\alpha = 0.10$.

**38. Ethnic Asians in California.** Refer to Exercise 36. Is there evidence that the population proportion of California residents of Asian ethnicity has risen since 2005? Test using the $p$-value method at level of significance $\alpha = 0.05$.

**39. Hispanic Household Income.** The U.S. Census Bureau reports that, in 2002, 15.3% of Hispanic families had household incomes of at least $75,000. We are interested in whether the population proportion has changed, using the critical-value method and level of significance $\alpha = 0.01$. Suppose that a random sample of 100 Hispanic families taken this year reported 23 with household incomes of at least $75,000.

- **a.** Is it appropriate to perform the $Z$ test for the proportion? Why or why not?
- **b.** Where will the sampling distribution of $\hat{p}$ be centered? What will be the shape of the distribution?
- **c.** What is $\sigma_{\hat{p}}$? What does this number mean?
- **d.** How many standard deviations does $\hat{p}$ lie below $p_0$? Is this extreme? Why or why not?
- **e.** Where in the sampling distribution would the value of $\hat{p}$ lie? Near the tail? Near the center?

**40. Eighth-Grade Alcohol Use.** The National Institute on Alcohol Abuse and Alcoholism reported that 45.6% of eighth-graders had used alcohol.[21] A random sample of 100 eighth-graders this year showed that 41 of them had used alcohol.

- **a.** Is it appropriate to perform the $Z$ test for the proportion? Why or why not?
- **b.** What is $\sigma_{\hat{p}}$? What does this number mean?
- **c.** How many standard deviations does $\hat{p}$ lie below $p_0$? Is this extreme? Why or why not?
- **d.** Where in the sampling distribution would the value of $\hat{p}$ lie? Near the tail? Near the center?
- **e.** Is there evidence that the population proportion of eighth-graders who used alcohol has changed? Test using the $p$-value method at level of significance $\alpha = 0.05$.

**41. Hispanic Household Income.** Refer to Exercise 39.

- **a.** Construct the hypotheses.
- **b.** Find the rejection rule.
- **c.** Calculate the value of the test statistic $Z_{\text{data}}$.
- **d.** Calculate the $p$-value.
- **e.** Assess the strength of evidence against the null hypothesis.
- **f.** Formulate and interpret your conclusion.

**42. Eighth-Grade Alcohol Use.** Refer to Exercise 40.

- **a.** Evaluate the strength of evidence against the null hypothesis.
- **b.** Suppose that we decide to carry out the same $Z$ test—however, this time using the critical-value method. Without actually performing the test, what would the conclusion be and why?
- **c.** Suppose that a newspaper headline reported "Alcohol Use Down Among Eighth-Graders." How would you respond? Does your hypothesis test support this headline?

WHAT IF? **43. Eighth-Grade Alcohol Use.** Refer to Exercise 40. *What if* we cut the sample size down to between 50 and 60 eighth-graders, and we cut the sample number of successes as well, so that the sample proportion $p$ stayed the same. Suppose that everything else stayed the same. Describe what would happen to the following, and why.

- **a.** $\sigma_{\hat{p}}$
- **b.** $Z_{\text{data}}$
- **c.** The $p$-value
- **d.** $\alpha$
- **e.** The conclusion

**Children and Environmental Tobacco Smoke at Home.** Use the following information for Exercises 44–48. The Environmental Protection Agency reported in 2005 that 11% of children aged 6 and under were exposed to environmental tobacco smoke (ETS) at home on a regular basis (at least four times per week).[22] A random sample taken this year of 100 children aged 6 and under showed that 6% of these children had been exposed to ETS at home on a regular basis.

**44.** Answer the following.

- **a.** Is it appropriate to perform the $Z$ test for the proportion? Why or why not?
- **b.** What is $\sigma_{\hat{p}}$? What does this number mean?
- **c.** How many standard deviations does $\hat{p}$ lie below $p_0$? Is this extreme? Why or why not?
- **d.** Where in the sampling distribution would the value of $\hat{p}$ lie? Near the tail? Near the center?

**45.** Test using the critical-value method at level of significance $\alpha = 0.05$ whether the population proportion of children aged 6 and under exposed to ETS at home on a regular basis has decreased since 2005.

46. Refer to Exercise 45.
   - **a.** Which is the only possible error you can be making here, a Type I or a Type II error? What are some consequences of this error?
   - **b.** Suppose that a newspaper headline reported "Second-hand Smoke Prevalence Down." How would you respond? Does your inference support this headline?

47. Refer to your work in Exercise 45.
   - **a.** Test using the $p$-value method at level of significance $\alpha = 0.10$ whether the population proportion of children aged 6 and under exposed to ETS at home on a regular basis has decreased since 2005.
   - **b.** How do you explain the different conclusions you got in the two hypothesis tests above? Is the difference in your conclusions because you used the critical-value method in one test and the $p$-value method in the other? If not, explain how you got different conclusions.
   - **c.** Evaluate the strength of evidence against the null hypothesis.

WHAT IF? 48. Refer to Exercise 45. *What if* the sample proportion $\hat{p}$ was decreased, but everything else stayed the same. Describe what would happen to the following, and why.
   - **a.** $\sigma_{\hat{p}}$
   - **b.** $Z_{\text{data}}$
   - **c.** The $p$-value
   - **d.** $\alpha$
   - **e.** The conclusion

**Car Accidents Among Young Drivers.** For Exercises 49 and 50, refer to Example 9.27.

WHAT IF? 49. Suppose that our sample size and the number of successes are doubled, so that $\hat{p}$ remains the same. Otherwise, everything else is the same as in the original example. Describe how this change would affect the following.
   - **a.** $\sigma_{\hat{p}}$
   - **b.** $Z_{\text{data}}$
   - **c.** The $p$-value
   - **d.** $\alpha$
   - **e.** The conclusion

WHAT IF? 50. Suppose that the hypothesized proportion $p_0$ was no longer 0.12. Instead, $p_0$ takes some value between 0.12 and 0.134. Otherwise, everything else is the same as in the original example. Describe how this change would affect the following.
   - **a.** $\sigma_{\hat{p}}$
   - **b.** $Z_{\text{data}}$
   - **c.** The $p$-value
   - **d.** $\alpha$
   - **e.** The conclusion

## 9.6 Chi-Square Test for the Population Standard Deviation $\sigma$

*Objectives:* **By the end of this section, I will be able to...**

1. Explain the essential idea about hypothesis testing for $\sigma$.
2. Perform and interpret the $\chi^2$ test for $\sigma$ using the $p$-value method.
3. Perform and interpret the $\chi^2$ test for $\sigma$ using the critical-value method.
4. Use confidence intervals for $\sigma$ to perform two-tailed hypothesis tests about $\sigma$.

### 1 The Essential Idea About Hypothesis Testing for the Population Standard Deviation $\sigma$

In Section 8.4 (page 429) we used the $\chi^2$ distribution to help us construct confidence intervals about the population variance and standard deviation. Here, in Section 9.6, we will use the $\chi^2$ distribution to perform hypothesis tests about the population standard deviation $\sigma$. Why might we be interested in doing so? A pharmaceutical company that wishes to ensure the safety of a particular new drug would perform statistical tests to make sure that the drug's effect was consistent and did not vary widely from patient to patient. The biostatisticians employed by the company would therefore perform a hypothesis test to make sure that the population standard deviation $\sigma$ was not too large. Pharmaceutical companies rank as among the largest employers of statisticians in the world, since they are required to demonstrate that their drugs are both safe and effective.

Recall the following from Section 8.4.

**$\chi^2$ (Chi-Square) Distribution**

If we take a random sample of size $n$ from a normal population with mean $\mu$ and standard deviation $\sigma$, then the statistic

$$\chi^2 = \frac{(n-1)s^2}{\sigma^2}$$

follows a **$\chi^2$ distribution** with $n - 1$ degrees of freedom, where $s^2$ represents the sample variance.

Just like our hypothesis tests about the population mean $\mu$ and the population proportion $p$, there are three forms for the hypothesis test about the population standard deviation $\sigma$ (Table 9.16). The notation $\sigma_0$ refers to the value of $\sigma$ to be tested.

Table 9.16 The three possible forms for the hypotheses for a test for $\sigma$

| Form | Null and alternative hypotheses |
|---|---|
| Right-tailed test, one-tailed test | $H_0: \sigma \le \sigma_0$ versus $H_a: \sigma > \sigma_0$ |
| Left-tailed test, one-tailed test | $H_0: \sigma \ge \sigma_0$ versus $H_a: \sigma < \sigma_0$ |
| Two-tailed test | $H_0: \sigma = \sigma_0$ versus $H_a: \sigma \ne \sigma_0$ |

Under the assumption that $H_0$ is true, the $\chi^2$ statistic takes the following form:

$$\chi^2_{\text{data}} = \frac{(n-1)s^2}{\sigma_0^2}$$

For the hypothesis test about $\sigma$, our test statistic is $\chi^2_{\text{data}}$. It is called $\chi^2_{\text{data}}$ because the values of $n - 1$ and $s^2$ come from the observed data. The test statistic $\chi^2_{\text{data}}$ takes a moderate value when the value of $s^2$ is moderate assuming $H_0$ is true, and $\chi^2_{\text{data}}$ takes an extreme value when the value of $s^2$ is extreme assuming $H_0$ is true. This leads us to the following.

**The Essential Idea About Hypothesis Testing for the Standard Deviation**

When the observed value of $\chi^2_{\text{data}}$ is unusual or extreme on the assumption that $H_0$ is true, we should reject $H_0$. Otherwise, there is insufficient evidence against $H_0$, and we should not reject $H_0$.

## 2 $\chi^2$ Test for $\sigma$ Using the $p$-Value Method

The remainder of Section 9.6 explains the details of implementing hypothesis testing for the standard deviation. The $\chi^2$ test for $\sigma$ may be performed using the $p$-value method or the critical-value method.

The $p$-value for the hypothesis test about $\sigma$ may be found using Table 9.17.

Table 9.17 $p$-value for the hypothesis test for $\sigma$

**Right-tailed test**

$H_0: \sigma \le \sigma_0$ versus $H_a: \sigma > \sigma_0$

$p$-value $= P(\chi^2 > \chi^2_{\text{data}})$
Area to right of $\chi^2_{\text{data}}$

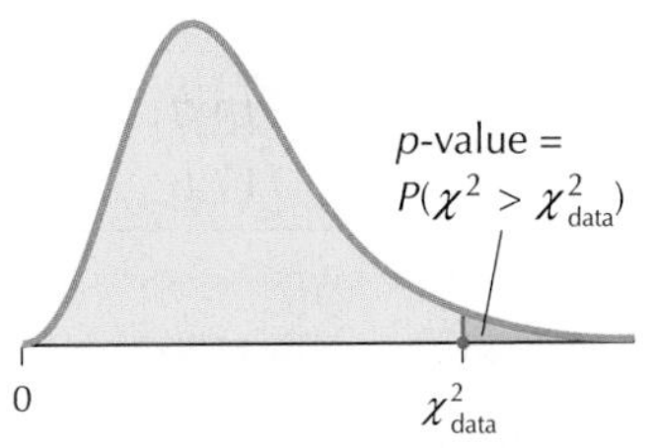


**Left-tailed test**

$H_0: \sigma \ge \sigma_0$ versus $H_a: \sigma < \sigma_0$

$p$-value $= P(\chi^2 < \chi^2_{\text{data}})$

Area to left of $\chi^2_{\text{data}}$

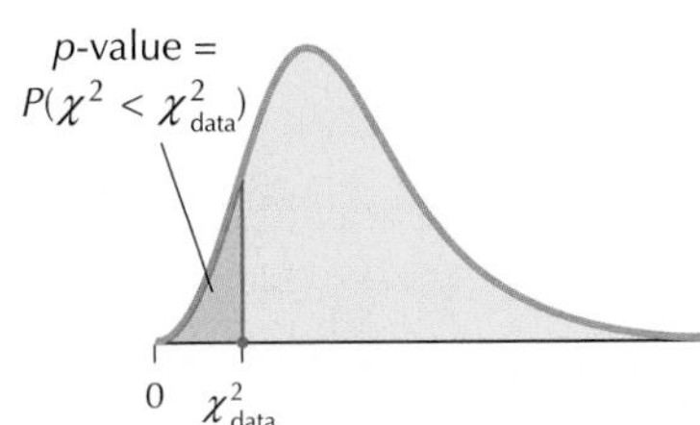


**Two-tailed test**

$H_0: \sigma = \sigma_0$ versus $H_a: \sigma \ne \sigma_0$

If $P(\chi^2 > \chi^2_{\text{data}}) \le 0.5$, then
**a.** $\chi^2_{\text{data}}$ is on the right side of the distribution
**b.** $p$-value $= 2 \cdot P(\chi^2 > \chi^2_{\text{data}})$

If $P(\chi^2 > \chi^2_{\text{data}}) > 0.5$, then
**a.** $\chi^2_{\text{data}}$ is on the left side of the distribution
**b.** $p$-value $= 2 \cdot P(\chi^2 < \chi^2_{\text{data}})$

**$\chi^2$ Test for $\sigma$: $p$-Value Method**

This hypothesis test is valid only if we have a random sample from a normal population.

***Step 1*** **State the hypotheses and the rejection rule.** Use one of the forms in Table 9.17. State the rejection rule as "Reject $H_0$ when the $p$-value is less than $\alpha$." Clearly define $\sigma$.

***Step 2*** **Find $\chi^2_{\text{data}}$.** Either use technology to find the value of the test statistic $\chi^2_{\text{data}}$ or calculate the value of $\chi^2_{\text{data}}$ as follows:

$$\chi^2_{\text{data}} = \frac{(n-1)s^2}{\sigma_0^2}$$

which follows a $\chi^2$ distribution with $n - 1$ degrees of freedom, and where $s^2$ represents the sample variance.

***Step 3*** **Find the $p$-value.** Either use technology to find the $p$-value or estimate the $p$-value using Table E, Chi-square ($\chi^2$) Distribution, in the Appendix (page T-12).

***Step 4*** **State the conclusion and the interpretation.** If the $p$-value is less than $\alpha$, then reject $H_0$. Otherwise, do not reject $H_0$. Interpret your conclusion so that a nonspecialist can understand.

## Example 9.31 $\chi^2$ test for $\sigma$ using the $p$-value method and technology

Power plants around the country are retooling in order to consume biomass instead of or in addition to coal. The following table contains a random sample of 10 such power plants and the amount of biomass they consumed in 2006, in trillions of Btu (British thermal units).[23] Test whether the population standard deviation is greater than 2 trillion Btu using level of significance $\alpha = 0.05$.

Alamy

| Power plant | Location | Biomass consumed (trillions of Btu) |
|---|---|---|
| Georgia Pacific Naheola Mill | Choctaw, Alabama | 13.4 |
| Jefferson Smurfit Fernandina Beach | Nassau, Florida | 12.9 |
| International Paper Augusta Mill | Richmond, Georgia | 17.8 |
| Gaylord Container Bogalusa | Washington, Louisiana | 15.1 |
| Escanaba Paper Company | Delta, Michigan | 19.5 |
| Weyerhaeuser Plymouth NC | Martin, North Carolina | 18.6 |
| International Paper Georgetown Mill | Georgetown, South Carolina | 13.8 |
| Bowater Newsprint Calhoun Operation | McMinn, Tennessee | 10.6 |
| Covington Facility | Covington, Virginia | 12.7 |
| Mosinee Paper | Marathon, Wisconsin | 17.6 |

## Solution

The normal probability plot in Figure 9.68 indicates acceptable normality, allowing us to proceed with the hypothesis test.

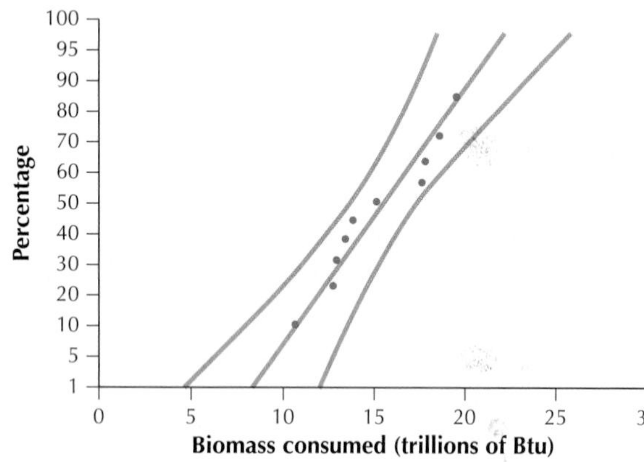


FIGURE 9.68 Normal probability plot.

**Step 1** **State the hypotheses and the rejection rule.** The phrase "greater than" indicates that we have a right-tailed test. The question "Greater than what?" tells us that $\sigma_0 = 2$, giving us

$$H_0: \sigma \leq 2 \quad \text{versus} \quad H_a: \sigma > 2$$

We reject $H_0$ if the $p$-value is less than level of significance $\alpha = 0.05$.

**Step 2** **Find $\chi^2_{\text{data}}$.** The TI-83/84 descriptive statistics in Figure 9.69 tell us that the sample variance is

$$s^2 = 2.990354866^2$$

```
1-Var Stats
 x̄=15.2
 Σx=152
 Σx²=2390.88
 Sx=2.990354866
 σx=2.836899716
↓n=10
```

FIGURE 9.69

Thus,

$$\chi^2_{\text{data}} = \frac{(n-1)s^2}{\sigma_0^2} = \frac{(10-1)\,2.990354866^2}{2^2} \approx 20.12$$

**Step 3** **Find the $p$-value.** For our right-tailed test, Table 9.17 tells us that

$$p\text{-value} = P(\chi^2 > \chi^2_{\text{data}}) = P(\chi^2 > 20.12)$$

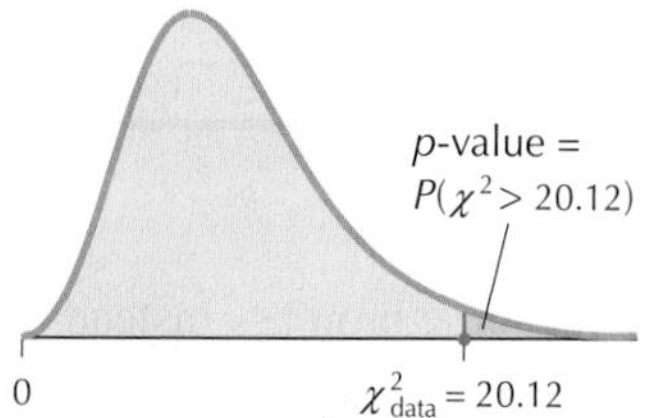


FIGURE 9.70

That is, the $p$-value is the area to the right of $\chi^2_{\text{data}} = 20.12$, as shown in Figure 9.70. To find the $p$-value, we use the instructions provided in the Step-by-Step Technology Guide provided at the end of this section. The TI-83/84 results shown in Figure 9.71a tell us that $p$-value $= P(\chi^2 > 20.12) = 0.0171861114$.

The Excel and Minitab results in Figures 9.71b and 9.71c agree with this $p$-value. (Excel and Minitab do not exactly match the TI-83/84 $p$-value because they round the $p$-values to fewer decimal places.) Instead of providing the $p$-value directly, Minitab gives the area to the left of $\chi^2_{\text{data}}$: $P(X \leq 20.12) = 0.982814$. We therefore need to subtract the given value from 1 to get the $p$-value:

$$p\text{-value} = 1 - 0.982814 = 0.017186$$

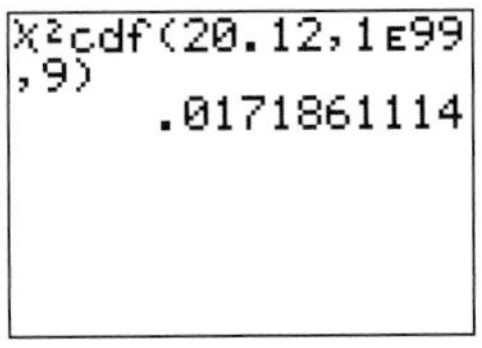

```
X²cdf(20.12,1E99
,9)
      .0171861114
```

FIGURE 9.71a TI-83/84 results.

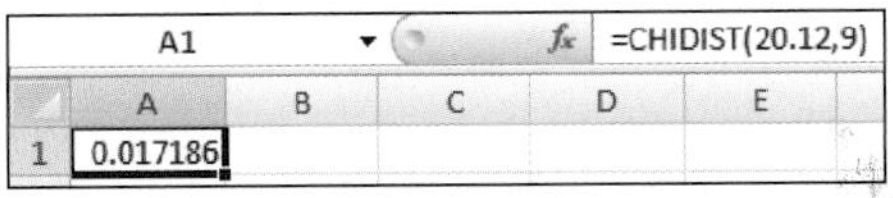

| A1 | =CHIDIST(20.12,9) |
|---|---|
| | A |
| 1 | 0.017186 |

FIGURE 9.71b Excel results.

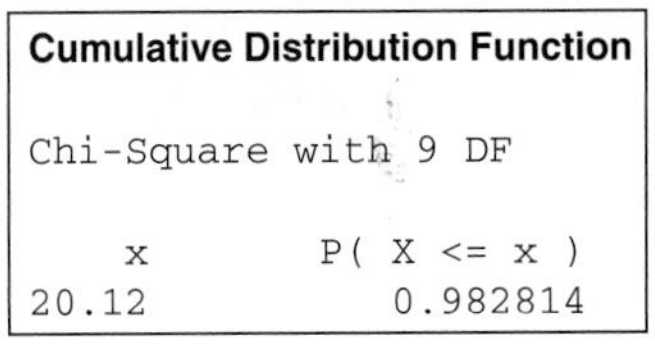

```
Cumulative Distribution Function

Chi-Square with 9 DF

  x        P( X <= x )
20.12         0.982814
```

FIGURE 9.71c Minitab results.

***Step 4*** **State the conclusion and the interpretation.** Since $p$-value $= 0.0171861114$ is less than $\alpha = 0.05$, we reject $H_0$. There is evidence that the population standard deviation is greater than 2 trillion Btu.

## 3 $\chi^2$ Test for $\sigma$ Using the Critical-Value Method

Just like the hypothesis tests for $\mu$ and $p$, we may also use the critical-value method to perform a hypothesis test for the population standard deviation $\sigma$.

**$\chi^2$ Test for $\sigma$: Critical-Value Method**

This hypothesis test is valid only if we have a random sample from a normal population.

***Step 1*** **State the hypotheses.** Use one of the forms in Table 9.17. Clearly define $\sigma$.

***Step 2*** **Find the $\chi^2$ critical value or values and state the rejection rule.** Use Table 9.18.

***Step 3*** **Find $\chi^2_{\text{data}}$.** Either use technology to find the value of the test statistic $\chi^2_{\text{data}}$ or calculate the value of $\chi^2_{\text{data}}$ as follows:

$$\chi^2_{\text{data}} = \frac{(n-1)s^2}{\sigma_0^2}$$

which follows a $\chi^2$ distribution with $n - 1$ degrees of freedom, and where $s^2$ represents the sample variance.

***Step 4*** **State the conclusion and the interpretation.** If $\chi^2_{\text{data}}$ falls in the critical region, then reject $H_0$. Otherwise do not reject $H_0$. Interpret your conclusion so that a nonspecialist can understand.

The $\chi^2$ critical values in the right-tailed, left-tailed, or two-tailed tests use the following notations: $\chi^2_{\alpha}$, $\chi^2_{1-\alpha}$, $\chi^2_{\alpha/2}$, and $\chi^2_{1-\alpha/2}$ (see Table 9.18). In each case, the subscript indicates the area to the right of the $\chi^2$ critical value. Find these values just as you did in Section 8.4, using either technology or Table E, Chi-Square ($\chi^2$) Distribution, in the Appendix (page T-12).

Table 9.18 Critical values and rejection rules for the $\chi^2$ test for $\sigma$

| Right-tailed test | Left-tailed test | Two-tailed test |
|---|---|---|
| $H_0: \sigma \le \sigma_0$ versus $H_a: \sigma > \sigma_0$ | $H_0: \sigma \ge \sigma_0$ versus $H_a: \sigma < \sigma_0$ | $H_0: \sigma = \sigma_0$ versus $H_a: \sigma \ne \sigma_0$ |
| Critical value: $\chi^2_\alpha$<br>Reject $H_0$ if $\chi^2_{\text{data}} > \chi^2_\alpha$ | Critical value: $\chi^2_{1-\alpha}$<br>Reject $H_0$ if $\chi^2_{\text{data}} < \chi^2_{1-\alpha}$ | Critical values: $\chi^2_{\alpha/2}$ and $\chi^2_{1-\alpha/2}$<br>Reject $H_0$ if $\chi^2_{\text{data}} > \chi^2_{\alpha/2}$<br>or if $\chi^2_{\text{data}} < \chi^2_{1-\alpha/2}$ |
| 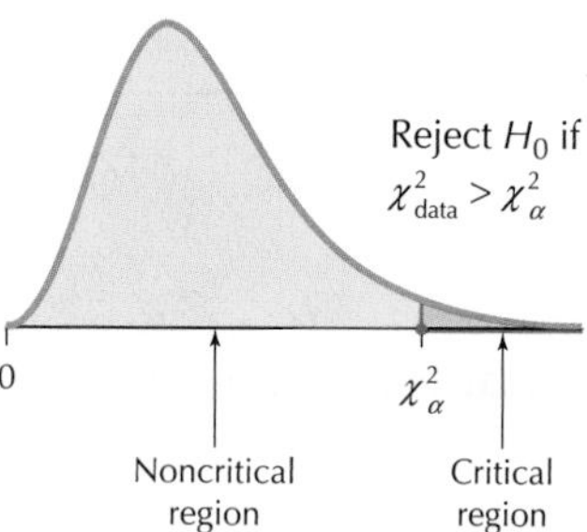 | 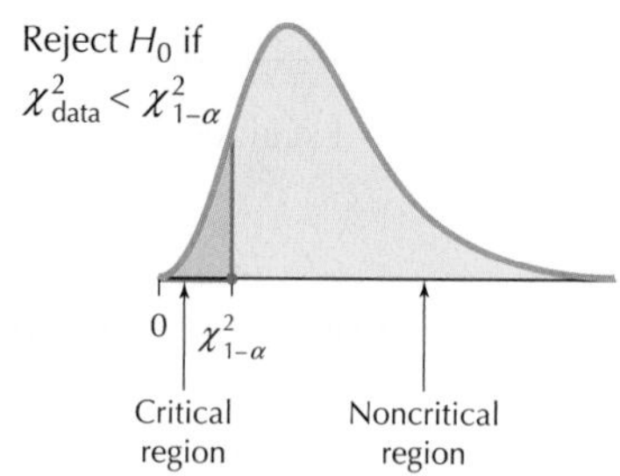 | 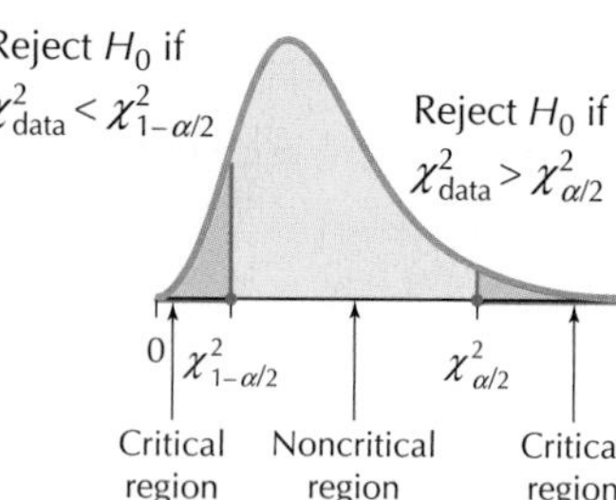 |







## Example 9.32 $\chi^2$ test for $\sigma$ using the critical-value method

The following table contains the numbers of children (in 1000s) living in low-income households without health insurance in 2005 for a random sample of 8 states.[24] Test whether the population standard deviation $\sigma$ of children living in low-income households without health insurance differs from 10,000, using level of significance $\alpha = 0.05$.

| | |
|---|---|
| Alabama | 48 |
| Arkansas | 37 |
| Iowa | 33 |
| Massachusetts | 50 |
| Minnesota | 45 |
| Oregon | 63 |
| South Carolina | 66 |
| Utah | 52 |

### Solution

The normal probability plot indicates acceptable normality.

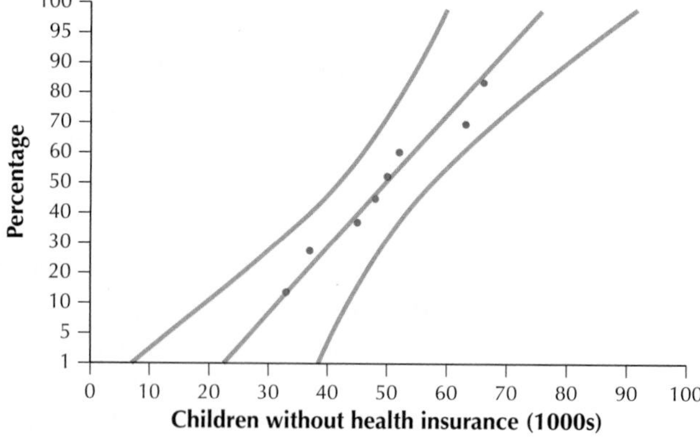



Normal probability plot for children without health insurance.

**Step 1** **State the hypotheses.** The phrase "differs from" indicates that we have a two-tailed test. The value $\sigma_0 = 10$ answers the question "Differs from what?" (Note that $\sigma_0$ is 10, and not 10,000, since the data are expressed in thousands.) Thus, we have our hypotheses:

$$H_0: \sigma = 10 \quad \text{versus} \quad H_a: \sigma \neq 10$$

where $\sigma$ represents the population standard deviation of number of children living in low-income households without health insurance.

**Step 2** **Find the $\chi^2$ critical values and state the rejection rule.** We have $n = 8$, so degrees of freedom $= n - 1 = 7$. Since $\alpha$ is given as 0.05, $\alpha/2 = 0.025$ and $1 - \alpha/2 = 0.975$. Then, from the $\chi^2$ table (Figure 9.72), we have $\chi^2_{\alpha/2} = \chi^2_{0.025} = 16.013$, and $\chi^2_{1-\alpha/2} = \chi^2_{0.975} = 1.690$.

**Chi-Square ($\chi^2$) Distribution**

| Degrees of Freedom | Area to the Right of Critical Value | | | | | | | | |
|---|---|---|---|---|---|---|---|---|---|
| | 0.995 | 0.99 | 0.975 | 0.95 | 0.90 | 0.10 | 0.05 | 0.025 | 0.01 |
| 1 | — | — | 0.001 | 0.004 | 0.016 | 2.706 | 3.841 | 5.024 | 6.635 |
| 2 | 0.010 | 0.020 | 0.051 | 0.103 | 0.211 | 4.605 | 5.991 | 7.378 | 9.210 |
| 3 | 0.072 | 0.115 | 0.216 | 0.352 | 0.584 | 6.251 | 7.815 | 9.348 | 11.345 |
| 4 | 0.207 | 0.297 | 0.484 | 0.711 | 1.064 | 7.779 | 9.488 | 11.143 | 13.277 |
| 5 | 0.412 | 0.554 | 0.831 | 1.145 | 1.610 | 9.236 | 11.071 | 12.833 | 15.086 |
| 6 | 0.676 | 0.872 | 1.237 | 1.635 | 2.204 | 10.645 | 12.592 | 14.449 | 16.812 |
| 7 | 0.989 | 1.239 | 1.690 | 2.167 | 2.833 | 12.017 | 14.067 | 16.013 | 18.475 |
| 8 | 1.344 | 1.646 | 2.180 | 2.733 | 3.490 | 13.362 | 15.507 | 17.535 | 20.090 |

FIGURE 9.72 Finding the $\chi^2$ critical values.

We will reject $H_0$ if $\chi^2_{\text{data}}$ is either greater than $\chi^2_{\alpha/2} = 16.013$ or less than $\chi^2_{1-\alpha/2} = 1.690$.

**Step 3** **Find $\chi^2_{\text{data}}$.** The TI-83/84 descriptive statistics in Figure 9.73 tell us that the sample variance is

$$s^2 = 11.41114743^2$$

Thus,

$$\chi^2_{\text{data}} = \frac{(n-1)s^2}{\sigma_0^2} = \frac{(8-1)11.41114743^2}{10^2} \approx 9.115$$

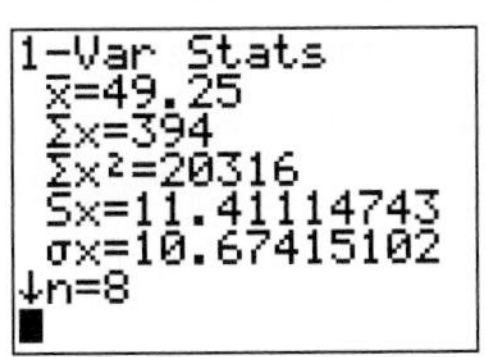


FIGURE 9.73

**Step 4** **State the conclusion and the interpretation.** In Step 2 we said that we would reject $H_0$ if $\chi^2_{\text{data}}$ was either greater than 16.013 or less than 1.690. Since $\chi^2_{\text{data}} = 9.115$ is neither greater than 16.013 nor less than 1.690, we do not reject $H_0$. There is insufficient evidence that the population standard deviation of the numbers of children living in low-income households without health insurance differs from 10,000. ■

## 4 Using Confidence Intervals for $\sigma$ to Perform Two-Tailed Hypothesis Tests for $\sigma$

Recall from Sections 9.3 and 9.5 the equivalence of confidence intervals and two-tailed hypothesis tests. We may apply the same methodology here in hypothesis tests for $\sigma$. Suppose we have a $100(1-\alpha)\%$ confidence interval for $\sigma$, of the form (lower bound, upper bound), and are interested in two-tailed hypothesis tests using $\alpha$ of the form:

$$H_0: \sigma = \sigma_0 \quad \text{versus} \quad H_a: \sigma \neq \sigma_0$$

We will not reject $H_0$ for values of $\sigma_0$ that lie between the lower bound and upper bound of the confidence interval. And we will reject $H_0$ for values of $\sigma_0$ that lie outside this interval.

## Example 9.33 Using confidence intervals for $\sigma$ to determine significance

Here we list the sodium content, in milligrams (mg) per serving, of 23 breakfast cereals. The normality of the data was confirmed in Example 8.4 (page 394).

| | | | |
|---|---|---|---|
| Apple Jacks | 125 | Kix | 260 |
| Cap'n Crunch | 220 | Life | 150 |
| Cheerios | 290 | Lucky Charms | 180 |
| Cinnamon Toast Crunch | 210 | Mueslix Crispy | 150 |
| Cocoa Puffs | 180 | Raisin Bran | 210 |
| Corn Flakes | 290 | Rice Chex | 240 |
| Count Chocula | 180 | Rice Krispies | 290 |
| Cream of Wheat | 80 | Special K | 230 |
| Froot Loops | 125 | Total Whole Grain | 200 |
| Frosted Flakes | 200 | Trix | 140 |
| Fruity Pebbles | 135 | Wheaties | 200 |
| Grape Nuts Flakes | 140 | | |

Construct a 95% confidence interval for the population standard deviation $\sigma$ of sodium content. Test using level of significance $\alpha = 0.05$ whether $\sigma$ differs from the following.

**a.** 80 mg

**b.** 40 mg

### Solution

| 95% Confidence Interval for StDev | |
|---|---|
| 44.53 | 81.50 |

FIGURE 9.74

Figure 9.74 is an excerpt from Minitab output that shows the 95% confidence interval for $\sigma$. (See the Step-by-Step Technology Guide at the end of Section 8.4, page 435.)

**a.** For the hypothesis test $H_0: \sigma = 80$ versus $H_a: \sigma \neq 80$, $\sigma_0 = 80$ lies between the lower bound 44.53 and the upper bound 81.50 of the confidence interval, and we therefore do not reject $H_0$. There is insufficient evidence that the population standard deviation of sodium content differs from 80 mg.

**b.** For the hypothesis test $H_0: \sigma = 40$ versus $H_a: \sigma \neq 40$, $\sigma_0 = 40$ lies outside the confidence interval, and we therefore reject $H_0$. There is evidence that the population standard deviation of sodium content differs from 40 mg. ■

## STEP-BY-STEP TECHNOLOGY GUIDE: Finding $\chi^2$ *p*-value

We will use the information from Example 9.31 (page 529). The steps for finding the $\chi^2$ critical values are given in the Step-by-Step Technology Guide at the end of Section 8.4 (page 435).

### TI-83/84

***Step 1*** Enter the data into List **L1**.

***Step 2*** Press **2nd > DISTR**, then **$\chi^2$cdf(**, and press **ENTER**.

***Step 3*** On the home screen, enter the value of $\chi^2_{\text{data}}$, comma, **1E99**, comma, degrees of freedom, close parenthesis, as shown in Figure 9.71a. (Remember that this "E" is inserted by pressing **2nd**, followed by the comma key.)

***Step 4*** Press **ENTER**. The results for Example 9.31 are shown in Figure 9.71a.

## EXCEL

***Step 1*** Select cell **A1**. Click the **Insert Function** icon $f_x$.
***Step 2*** For **Search for a Function,** type **chidist** and click **OK.**
***Step 3*** For **x**, enter the value of $\chi^2_{\text{data}}$, and for **Deg_freedom,** enter the degrees of freedom. Excel displays the $p$-value in the cell in the dialog box, as shown in Figure 9.71b.

## MINITAB

***Step 1*** Click on **Calc > Probability Distributions > Chi-Square.**
***Step 2*** Select **Cumulative probability**, and enter the **Degrees of freedom.**
***Step 3*** For **Input constant**, enter the value of $\chi^2_{\text{data}}$ and click **OK.**
***Step 4*** Minitab displays the area to the left of $\chi^2_{\text{data}}$ in the session window, as shown in Figure 9.71c. To find the $p$-value, subtract this area from 1.

## Section 9.6 SUMMARY

1 The essential idea about hypothesis testing for the standard deviation $\sigma$ may be described as follows. When the observed value of $\chi^2_{\text{data}}$ is unusual or extreme on the assumption that $H_0$ is true, we should reject $H_0$. Otherwise, there is insufficient evidence against $H_0$, and we should not reject $H_0$.

2 We can perform hypothesis tests to determine whether the population standard deviation $\sigma$ is greater than, less than, or not equal to a certain value $\sigma_0$. Under the assumption that $H_0$ is true, the $\chi^2$ statistic takes the following form:

$$\chi^2_{\text{data}} = \frac{(n-1)s^2}{\sigma_0^2}$$

3 The hypothesis test about $\sigma$ may be performed using the $p$-value method or the critical-value method. Either way, the test is valid only if we have a random sample from a normal population.

4 If we have a $100(1-\alpha)\%$ confidence interval for $\sigma$, of the form (lower bound, upper bound), and are interested in a two-tailed hypothesis test, using $\alpha$, of the form

$$H_0: \sigma = \sigma_0 \quad \text{versus} \quad H_a: \sigma \neq \sigma_0$$

we will not reject $H_0$ for values of $\sigma_0$ that lie between the lower bound and upper bound of the confidence interval. We will reject $H_0$ for values of $\sigma_0$ that lie outside this interval.

## Section 9.6 EXERCISES

### CLARIFYING THE CONCEPTS

**1.** Think of one instance where an analyst would be interested in performing a hypothesis test about the population standard deviation $\sigma$.
**2.** What is the difference between $\sigma$ and $\sigma_0$?
**3.** Does it make sense to test whether $\sigma < 0$? Explain.
**4.** What condition must be fulfilled for us to perform a hypothesis test about $\sigma$?
**5.** Explain how we can use a confidence interval to determine significance.
**6.** In the previous exercise, what must be the relationship between $\alpha$ and the confidence level?

### PRACTICING THE TECHNIQUES

For Exercises 7–10, construct the hypotheses.
**7.** Test whether the population standard deviation is greater than 10.
**8.** Test whether the population standard deviation is less than 5.
**9.** Test whether the population standard deviation differs from 3.
**10.** Test whether the population standard deviation is greater than 100.

For Exercises 11–16, a random sample is drawn from a normal population. Calculate $\chi^2_{\text{data}}$.
**11.** We are testing whether $\sigma > 1$ and have a sample of size $n = 21$ with a sample variance of $s^2 = 3$.
**12.** We are testing whether $\sigma < 5$ and have a sample of size $n = 11$ with a sample variance of $s^2 = 25$.
**13.** We are testing whether $\sigma \neq 3$ and have a sample of size $n = 16$ with a standard deviation of $s = 2.5$.
**14.** We are testing whether $\sigma > 10$ and have a sample of size $n = 14$ with a standard deviation of $s = 12$.

**15.** We are testing whether $\sigma < 20$, using level of significance $\alpha = 0.10$, and have a sample of size $n = 8$ and a sample variance of $s^2 = 350$.

**16.** We are testing whether $\sigma \neq 5$ and have a sample of size $n = 26$ with a standard deviation of $s = 5$.

For Exercises 17–20, a random sample is drawn from a normal population. Do the following.

**a.** Draw a $\chi^2$ distribution and indicate the location of $\chi^2_{\text{data}}$.

**b.** Find the *p*-value and indicate the *p*-value in your distribution in **(a)**.

**c.** Compare the *p*-value with level of significance $\alpha = 0.05$. State the conclusion and interpretation.

**17.** The data in Exercise 11

**18.** The data in Exercise 12

**19.** The data in Exercise 13

**20.** The data in Exercise 14

**21.** The data in Exercise 15

**22.** The data in Exercise 16

For Exercises 23–28, a random sample is drawn from a normal population. The values of $\chi^2_{\text{data}}$ for these exercises are found in Exercises 11–16. Find the critical value or values.

**23.** We are testing whether $\sigma > 1$, using $\alpha = 0.05$, and have a sample of size $n = 21$ and a sample variance of $s^2 = 3$. Find $\chi^2_\alpha$.

**24.** We are testing whether $\sigma < 5$, using $\alpha = 0.05$, and have a sample of size $n = 11$ and a sample variance of $s^2 = 25$. Find $\chi^2_{1-\alpha}$.

**25.** We are testing whether $\sigma \neq 3$, using $\alpha = 0.05$, and have a sample of size $n = 16$ and a standard deviation of $s = 2.5$. Find $\chi^2_{\alpha/2}$ and $\chi^2_{1-\alpha/2}$.

**26.** We are testing whether $\sigma > 10$, using $\alpha = 0.01$, and have a sample of size $n = 14$ and a standard deviation of $s = 12$. Find $\chi^2_\alpha$.

**27.** We are testing whether $\sigma < 20$, using $\alpha = 0.10$, and have a sample of size $n = 8$ and a sample variance of $s^2 = 350$. Find $\chi^2_{1-\alpha}$.

**28.** We are testing whether $\sigma \neq 5$, using $\alpha = 0.05$, and have a sample of size $n = 26$ with a standard deviation of $s = 5$. Find $\chi^2_{\alpha/2}$, and $\chi^2_{1-\alpha/2}$.

For Exercises 29–34, a random sample is drawn from a normal population. The values of $\chi^2_{\text{data}}$ for these exercises were found in Exercises 11–16. The critical values were found in Exercises 23–28. Do the following.

**a.** State the rejection rule.

**b.** Compare $\chi^2_{\text{data}}$ with the critical value or values. State the conclusion and interpretation.

**29.** The data in Exercise 23

**30.** The data in Exercise 24

**31.** The data in Exercise 25

**32.** The data in Exercise 26

**33.** The data in Exercise 27

**34.** The data in Exercise 28

## APPLYING THE CONCEPTS

**35. DDT in Breast Milk.** Researchers compared the amount of DDT in the breast milk of a random sample of 12 Hispanic women in Yakima Valley in Washington State with the amount of DDT in breast milk in the general U.S. population.[25] They measured the standard deviation of the amount of DDT in the general population to be 36.5 parts per billion (ppb). Assume that the population is normally distributed. We are interested in testing whether the population standard deviation of DDT level in the breast milk of Hispanic women in Yakima Valley is greater than that of the general population, using level of significance $\alpha = 0.01$.

**a.** State the hypotheses and the rejection rule.

**b.** The sample variance is $s^2 = 119{,}025$. Calculate $\chi^2_{\text{data}}$.

**c.** Calculate the *p*-value.

**d.** Compare the *p*-value with $\alpha$. State your conclusion.

**e.** Interpret your conclusion.

**36. Tree Rings.** Does the growth of trees vary more when the trees are young? The International Tree Ring Data Base collected data on a particular 440-year-old Douglas fir tree.[26] The standard deviation of the annual ring growth in the tree's first 80 years of life was 0.8 millimeters (mm) per year. Assume that the population is normal. We are interested in testing whether the population standard deviation of annual ring growth in the tree's later years is less than 0.8 mm per year.

**a.** State the hypotheses.

**b.** Find the critical value $\chi^2_{1-\alpha}$ for level of significance $\alpha = 0.05$.

**c.** The sample variance for a random sample of size 100 taken from the tree's later years is $s^2 = 0.3136$. Calculate $\chi^2_{\text{data}}$.

**d.** Compare $\chi^2_{\text{data}}$ with $\chi^2_{1-\alpha}$. State your conclusion.

**e.** Interpret your conclusion.

**37. Union Membership.** The following table contains the total union membership (in 1000s) for 7 randomly selected states in 2006.[27] Assume that the distribution is normal. We are interested in whether the population standard deviation of union membership $\sigma$ differs from 30,000, using level of significance $\alpha = 0.05$.

| | |
|---|---|
| Florida | 397 |
| Indiana | 334 |
| Maryland | 342 |
| Massachusetts | 414 |
| Minnesota | 395 |
| Texas | 476 |
| Wisconsin | 386 |

**a.** State the hypotheses and the rejection rule.

**b.** The sample variance is $s^2 = 2245.67$. Calculate $\chi^2_{\text{data}}$.

**c.** Calculate the *p*-value.

**d.** Compare the *p*-value with $\alpha$. State your conclusion.

**e.** Interpret your conclusion.

**38. Fourth-Grade Feet.** Suppose a children's shoe manufacturer is interested in estimating the variability of fourth-graders' feet. A random sample of 20 fourth-graders' feet yielded the following foot lengths, in centimeters.[28] The normality of the data was verified in Example 8.12 (page 410). We are interested in whether the population standard deviation of foot lengths $\sigma$ is less than 1 centimeter.

| | | | | | | | | | |
|---|---|---|---|---|---|---|---|---|---|
| 22.4 | 23.4 | 22.5 | 23.2 | 23.1 | 23.7 | 24.1 | 21.0 | 21.6 | 20.9 |
| 25.5 | 22.8 | 24.1 | 25.0 | 24.0 | 21.7 | 22.0 | 22.7 | 24.7 | 23.5 |

**a.** State the hypotheses.
**b.** Find the critical value $\chi^2_{1-\alpha}$ for level of significance $\alpha = 0.05$.
**c.** The sample variance is $s^2 = 1.638$. Calculate $\chi^2_{\text{data}}$.
**d.** Compare $\chi^2_{\text{data}}$ with $\chi^2_{1-\alpha}$. State your conclusion.
**e.** Interpret your conclusion.

**39. Does Score Variability Differ by Gender?** Recently, researchers have been examining the evidence for whether there is greater variability in boys' scores than girls' scores on cognitive abilities tests. For example, one study found that boys were overrepresented at both the top and the bottom of nonverbal reasoning tests and quantitative reasoning tests.[29] Suppose that the standard deviation for girls' scores is known to be 50 points for a particular test and that the population of all scores is normal. A random sample of 101 boys has a sample variance of 2600. Test whether the population standard deviation for boys exceeds 50 points, using level of significance $\alpha = 0.05$. Use either the $p$-value method or the critical-value method.

**40. Heart Rate Variability.** A reduction in heart rate variability is associated with elevated levels of stress, since the body continues to pump adrenaline after high-stress situations, even when at rest.[30] Suppose the standard deviation of heartbeats in the general population is 20 beats per minute, and that the population of heart rates is normal. A random sample of 50 individuals leading high-stress lives has a sample variance of 200 beats per minute. Test using level of significance $\alpha = 0.05$ whether the population standard deviation for those leading high-stress lives is lower than that in the general population.

## Try This in Class! In-class activities to enhance your understanding of statistics

### TESTING FOR THE POPULATION MEAN OF THE SUM OF TWO FAIR DICE

Is the population mean $\mu$ of the sum of two fair dice equal to 7? The population standard deviation is $\sigma = 2.4152$. Each student will generate sample data to perform the following hypothesis test with level of significance $\alpha = 0.10$:

$$H_0: \mu = 7 \quad \text{versus} \quad H_a: \mu \neq 7$$

where $\mu$ represents the population mean sum of two fair dice. Assume that the distribution is normal. The instructor should prepare a number line on the board to display the collection of test statistic $Z_{\text{data}}$ values. For $\alpha = 0.10$ and a two-tailed test, the value of $Z_{\text{crit}}$ is 1.645. Thus, the instructor should draw lines separating the critical regions $Z_{\text{data}} < -1.645$ and $Z_{\text{data}} > 1.645$ from the noncritical region $-1.645 < Z_{\text{data}} < 1.645$.

**1.** Each student should toss two dice 10 times, recording the sum each time, and then calculate the mean sum $\bar{x}$ of this sample of size 10.

**2.** Using $n = 10$, $\sigma = 2.4152$, and $\mu_0 = 7$, each student should calculate his or her own value for the test statistic $Z_{\text{data}}$ using his or her own value of $\bar{x}$ generated above.

**3.** Give each student a Post-It Note. Each student should record the following on the note: his or her name and the sample mean $\bar{x}$ and test statistic $Z_{\text{data}}$ for his or her sample.

**4.** Each student should place the Post-It Note on the number line that the instructor prepared earlier.

**5.** Identify any Post-It Notes that indicate evidence against the null hypothesis that the population mean is 7. What is the value of $Z_{\text{data}}$? Of the sample mean $\bar{x}$? Do you agree that these sample means are unusual?

**6.** Approximately what percentage of sample means indicate evidence against $H_0: \mu_0 = 7$? Approximately what percentage of sample means indicate evidence consistent with $H_0: \mu_0 = 7$?

**7.** Overall, what is your personal conclusion about whether or not the population mean of the dice equals 7? What evidence can you point to in support of your conclusion?

## CHAPTER 9 FORMULAS AND VOCABULARY

### SECTION 9.1

- **$\alpha$ (alpha)** (p. 461). The probability of a Type I error; that is, the probability of rejecting $H_0$ when $H_0$ is true.
- **Alternative hypothesis** (p. 455). The research hypothesis, that represents an alternative conjecture about the value of the parameter, denoted as $H_a$.
- **$\beta$ (beta)** (p. 461). The probability of a Type II error; that is, the probability of not rejecting $H_0$ when $H_0$ is false.
- **Null hypothesis** (p. 455). The status quo hypothesis that represents what has been tentatively assumed about the value of the parameter, denoted as $H_0$.

• **Hypothesis testing** (p. 455). A procedure where claims about the value of a population parameter may be investigated using the sample evidence.
• **Statistical significance** (p. 458). A result is said to be statistically significant if it is unlikely to have occurred by chance. When we reject $H_0$, the results are said to be statistically significant.
• **Type I error** (p. 460): To reject $H_0$ when $H_0$ is true.
• **Type II error** (p. 460): To not reject $H_0$ when $H_0$ is false.

## SECTION 9.2

• **Essential idea about hypothesis testing for the mean** (p. 464). When the observed value of $\bar{x}$ is unusual or extreme in the sampling distribution of $\bar{x}$ that is based on the assumption that $H_0$ is correct, we should reject $H_0$. Otherwise, there is insufficient evidence against $H_0$, and we should not reject $H_0$.
• **Level of significance** (p. 469). The value of $\alpha$ that represents the boundary between results that are statistically significant and results that are not statistically significant.
• ***p*-value** (pp. 466–467). The probability of observing a sample statistic at least as extreme as the statistic in your sample if we assume that the null hypothesis is true.
• **Rejection rule for performing a hypothesis test using the *p*-value method** (p. 469). Reject $H_0$ when the *p*-value is less than $\alpha$.
• **Test statistic** (p. 466). A statistic generated from a data set for the purposes of testing a statistical hypothesis.
• $Z_{\text{data}}$ (p. 470). The test statistic used for the *Z* test for the mean is

$$Z_{\text{data}} = \frac{\bar{x} - \mu_0}{\sigma_{\bar{x}}} = \frac{\bar{x} - \mu_0}{\sigma/\sqrt{n}}$$

$Z_{\text{data}}$ represents how far the sample mean $\bar{x}$ lies above or below the hypothesized mean $\mu_0$, expressed in terms of the number of standard deviations of the sampling distribution of $\bar{x}$, $\sigma_{\bar{x}} = \sigma/\sqrt{n}$.

## SECTION 9.3

• **Critical region** (p. 482). The range of values of the test statistic for which the null hypothesis would be rejected.
• **Critical value, $Z_{\text{crit}}$** (p. 482). The value of *Z* that separates the critical region from the noncritical region.
• **Noncritical region** (p. 482). The range of values of the test statistic for which the null hypothesis would not be rejected.
• **Power of a hypothesis test** (p. 492). The probability of rejecting $H_0$ when $H_0$ is false: power $= 1 - \beta$.

## SECTION 9.4

• **Critical value, $t_{\text{crit}}$** (p. 502). The value of *t* that separates the critical region from the noncritical region.
• **Standard error of the sample mean $s_{\bar{x}}$** (p. 505). Formula:

$$s_{\bar{x}} = s/\sqrt{n}$$

Used in the interpretation of $t_{\text{data}}$.

• $t_{\text{data}}$ (p. 498). The test statistic used for the *t* test for the mean is

$$t_{\text{data}} = \frac{\bar{x} - \mu_0}{s_{\bar{x}}} = \frac{\bar{x} - \mu_0}{s/\sqrt{n}}$$

The test statistic $t_{\text{data}}$ represents the number of standard errors the sample mean $\bar{x}$ lies above or below the hypothesized mean $\mu_0$.

## SECTION 9.5

• **Essential idea about hypothesis testing for the proportion** (p. 515). When the sample proportion $\hat{p}$ is unusual or extreme in the sampling distribution of $\hat{p}$ that is based on the assumption that $H_0$ is correct, we reject $H_0$. Otherwise, there is insufficient evidence against $H_0$, and we should not reject $H_0$.
• **$Z_{\text{data}}$ for the hypothesis test for the population proportion** (p. 521). The test statistic used for the *Z* test for the proportion is

$$Z_{\text{data}} = \frac{\hat{p} - p_0}{\sigma_{\hat{p}}} = \frac{\hat{p} - p_0}{\sqrt{\dfrac{p_0(1 - p_0)}{n}}}$$

where $\hat{p}$ is the observed sample proportion of successes, $p_0$ is the value of *p* hypothesized in $H_0$, and *n* is the sample size. $Z_{\text{data}}$ represents the number of standard deviations ($\sigma_{\hat{p}}$) the sample proportion $\hat{p}$ lies above or below the hypothesized proportion $p_0$. Extreme values of $\hat{p}$ will be associated with extreme values of $Z_{\text{data}}$.

## SECTION 9.6

• **$\chi^2$ (chi-square) distribution** (p. 528). If we take a random sample of size *n* from a normal population with mean $\mu$ and standard deviation $\sigma$, then the statistic

$$\chi^2 = \frac{(n - 1)s^2}{\sigma^2}$$

follows a $\chi^2$ distribution with $n - 1$ degrees of freedom, where $s^2$ represents the sample variance.
• $\chi^2_{\text{data}}$ (p. 529). The test statistic used for the $\chi^2$ test for $\sigma$. Under the assumption that $H_0$ is true, the $\chi^2$ statistic takes the following form:

$$\chi^2_{\text{data}} = \frac{(n - 1)s^2}{\sigma_0^2}$$

where degrees of freedom $= n - 1$, $s^2$ represents the sample variance, and $\sigma_0$ represents the hypothesized standard deviation.
• **Essential idea about hypothesis testing for the standard deviation** (p. 528). When the observed value of $\chi^2_{\text{data}}$ is unusual or extreme on the assumption that $H_0$ is true, we should reject $H_0$. Otherwise, there is insufficient evidence against $H_0$, and we should not reject $H_0$.

# CHAPTER 9 REVIEW EXERCISES

## SECTION 9.1

For Exercises 1–3, provide the null and alternative hypotheses.

1. Test whether $\mu$ equals 12 or more.
2. Test whether $\mu$ equals 10 or less.
3. Test whether $\mu$ is below zero.

For Exercises 4–6, do the following.

**a.** Provide the null and alternative hypotheses.
**b.** Describe the two ways a correct decision could be made.
**c.** Describe what a Type I error would mean in the context of the problem.
**d.** Describe what a Type II error would mean in the context of the problem.

4. **Household Size.** The U.S. Census Bureau reported that the mean household size in 2002 was 2.58 persons. We conduct a hypothesis test to determine whether the population mean household size this year has changed.
5. **Speeding-Related Traffic Fatalities.** The National Highway Traffic Safety Administration reports that the mean number of speeding-related traffic fatalities over the Thanksgiving holiday period from 1994 to 2003 was 202.7. We conduct a hypothesis test to examine whether the population mean number of such fatalities has decreased.
6. **Salaries of Assistant Professors.** **Salaries.com** reports that the median salary for assistant professors in science as of 2005 was $49,934. We use this median salary to estimate that the mean salary in 2005 was $50,000. A hypothesis test was conducted to determine if the population mean salary of assistant professors in science has increased.

## SECTION 9.2

For Exercises 7–9, find the value of $Z_{data}$.

7. $\bar{x} = 59, \sigma = 10, n = 100, \mu_0 = 60$
8. $\bar{x} = 59, \sigma = 5, n = 100, \mu_0 = 60$
9. $\bar{x} = 59, \sigma = 1, n = 100, \mu_0 = 60$

For Exercises 10 and 11, perform the following steps.

**a.** State the hypotheses and the rejection rule for the $p$-value method.
**b.** Find $Z_{data}$.
**c.** Find the $p$-value. Draw the standard normal curve, with $Z_{data}$ and the $p$-value indicated on it.
**d.** State the conclusion and the interpretation.

10. We are interested in testing at level of significance $\alpha = 0.05$ whether the population mean differs from 500. A random sample of size 100 is taken, with a mean of 520. Assume $\sigma = 50$.
11. We would like to test at level of significance $\alpha = 0.01$ whether the population mean is less than $-10$. A random sample of size 25 is taken from a normal population. The sample mean is $-12$. Assume $\sigma = 2$.
12. **Health Care Expenditures.** We are interested in whether the population mean per capita annual expenditures on health care have increased since 2007, when the mean was $6096 per person.[31] A random sample taken this year of 100 Americans shows mean annual health care expenditures of $8000. Suppose that prior research has indicated that the population standard deviation of such expenditures is $1600. Perform the appropriate hypothesis test, using the $p$-value method and level of significance $\alpha = 0.01$.

## SECTION 9.3

For each of the following hypothesis tests in Exercises 13–15, do the following.

**a.** Find the value of $Z_{crit}$.
**b.** Find the critical-value rejection rule.
**c.** Draw a standard normal curve and indicate the critical region.
**d.** State the conclusion and interpretation.

13. $H_0 : \mu = \mu_0$ versus $H_0 : \mu \neq \mu_0$, $\alpha = 0.01$, $Z_{data} = -2.5$
14. $H_0 : \mu \leq \mu_0$ versus $H_0 : \mu > \mu_0$, $\alpha = 0.10$, $Z_{data} = 1.5$
15. $H_0 : \mu \leq \mu_0$ versus $H_0 : \mu > \mu_0$, $\alpha = 0.05$, $Z_{data} = -2.5$

For Exercises 16 and 17, do the following.

**a.** State the hypotheses.
**b.** Find the value of $Z_{crit}$ and the rejection rule. Also, draw a standard normal curve, indicating the critical region.
**c.** Calculate $Z_{data}$. Draw a standard normal curve showing $Z_{crit}$, the critical region, and $Z_{data}$.
**d.** State the conclusion and the interpretation.

16. **Credit Scores in Florida.** According to **CreditReport .com**, the mean credit score in Florida in 2006 was 673. A random sample of 144 Florida residents this year shows a mean credit score of 650. Assume $\sigma = 50$. Perform a hypothesis test using level of significance $\alpha = 0.05$ to determine if the population mean credit score in Florida has decreased.
17. **Salary of College Grads.** It pays to stay in school. According to the U.S. Census Bureau, the mean salary of college graduates in 2002 was $52,200, whereas the mean salary of those with "some college" was $36,800. A random sample of 100 college graduates taken this year provides a sample mean salary of $55,000. Assume $\sigma =$ $3000. Perform a hypothesis test to determine whether the population mean salary of college graduates has increased, using level of significance $\alpha = 0.10$.
18. **The Old Coffee Machine.** A random sample of 36 cups of coffee dispensed from the old coffee machine in the lobby had a mean amount of coffee of 7 ounces per cup. Assume that the population standard deviation is 1 ounce.

**a.** Construct a 95% confidence interval for the population mean amount of coffee dispensed by the old coffee machine in the lobby.

**b.** Use the confidence interval to test at level of significance $\alpha = 0.05$ whether the population mean amount of coffee dispensed by the old coffee machine in the lobby differs from the following amounts, in ounces.

**i.** 6.9
**ii.** 7.5
**iii.** 6.7
**iv.** 7

## SECTION 9.4

For Exercises 19–21, find the critical value $t_{crit}$ and sketch the critical region. Assume normality.

**19.** $H_0: \mu \le 100, H_a: \mu > 100, n = 8, \alpha = 0.10$

**20.** $H_0: \mu \le 100, H_a: \mu > 100, n = 8, \alpha = 0.05$

**21.** $H_0: \mu \le 100, H_a: \mu > 100, n = 8, \alpha = 0.01$

**22.** Describe what happens to the $t$ critical value $t_{crit}$ for right-tailed tests as $\alpha$ decreases.

**23.** A random sample of size 16 from a normal population yields a sample mean of 10 and a sample standard deviation of 3. Researchers are interested in finding whether the population mean differs from 9, using level of significance $\alpha = 0.10$. Test using the critical-value method.

**24.** A random sample of size 144 from an unknown population yields a sample mean of 45 and a sample standard deviation of 10. Researchers are interested in finding whether the population mean differs from 45, using level of significance $\alpha = 0.10$. Test using the estimated $p$-value method.

## SECTION 9.5

For Exercises 25 and 26, do the following.

**a.** Check the normality conditions.
**b.** State the hypotheses and the rejection rule for the $p$-value method, using level of significance $\alpha = 0.05$.
**c.** Calculate $Z_{data}$.
**d.** Calculate the $p$-value.
**e.** State the conclusion and the interpretation.

**25.** Test whether the population proportion differs from 0.7. A random sample of size 144 yields 110 successes.

**26.** Test whether the population proportion is less than 0.25. A random sample of size 100 yields 25 successes.

For Exercises 27–29, do the following.

**a.** Check the normality conditions.
**b.** State the hypotheses.
**c.** Find $Z_{crit}$ and the rejection rule.
**d.** Calculate $Z_{data}$.
**e.** State the conclusion and the interpretation.

**27.** Test whether the population proportion exceeds 0.8. A random sample of size 1000 yields 830 successes. Let $\alpha = 0.10$.

**28.** Test whether the population proportion is below 0.2. A random sample of size 900 yields 160 successes. Let $\alpha = 0.05$.

**29.** Test whether the population proportion is not equal to 0.4. A random sample of size 100 yields 55 successes. Let $\alpha = 0.01$.

**30. DSL Internet Service.** The U.S. Department of Commerce reports that, in 2003, 41.6% of Internet users preferred DSL as their method of service delivery.[32] A random sample of 1000 Internet users this year shows 350 who preferred DSL. If appropriate, test whether the population proportion who prefer DSL has decreased, using level of significance $\alpha = 0.05$.

## SECTION 9.6

For Exercises 31 and 32, do the following.

**a.** State the hypotheses and the $p$-value rejection rule for $\alpha = 0.05$.
**b.** Find $\chi^2_{data}$.
**c.** Find the $p$-value. Also, draw a $\chi^2$ distribution and indicate $\chi^2_{data}$ and the $p$-value.
**d.** State the conclusion and the interpretation.

**31.** We are testing whether $\sigma < 35$ and have a random sample of size 8 with a sample variance of 1200.

**32.** We are testing whether $\sigma \ne 50$ and have a random sample of size 26 with a standard deviation of $s = 45$.

For Exercises 33 and 34, do the following.

**a.** State the hypotheses.
**b.** Find the $\chi^2$ critical value or values, and state the rejection rule.
**c.** Find $\chi^2_{data}$. Also, draw a $\chi^2$ distribution and indicate $\chi^2_{data}$ and the $\chi^2$ critical value or values.
**d.** State the conclusion and the interpretation.

**33.** We are testing whether $\sigma > 6$ and have a random sample of size 20 with a standard deviation of $s = 9$. Let $\alpha = 0.05$.

**34.** We are testing whether $\sigma \ne 10$ and have a random sample of size 26 with a sample variance of 90. Let $\alpha = 0.05$.

**35. Prisoner Deaths in State Custody.** The following table contains the numbers of prisoners who died in state custody in 2005 for a random sample of 5 states.[33] Assume normality. Using $\alpha = 0.01$ and the $p$-value method, test whether the population standard deviation of prisoners who died in state custody differs from 50.

| | |
|---|---|
| New York | 171 |
| Pennsylvania | 149 |
| Michigan | 140 |
| Ohio | 121 |
| Georgia | 122 |

# CHAPTER 9 Quiz

## TRUE OR FALSE

**1.** True or false: It is possible that both the null and alternative hypotheses are correct at the same time.

**2.** True or false: The conclusion you draw from performing the critical-value method for the $Z$ test is the same as the conclusion you draw from performing the $p$-value method for the $Z$ test.

**3.** True or false: We do not need the estimated $p$-value method if we have access to a computer or calculator.

## FILL IN THE BLANK

**4.** To reject $H_0$ when $H_0$ is true is a Type __________ error.

**5.** An extreme value of $\bar{x}$ is associated with a __________ $p$-value.

**6.** The rejection rule for performing a hypothesis test using the $p$-value method is to reject $H_0$ when the $p$-value is less than __________.

## SHORT ANSWER

**7.** Under what conditions may we apply the $Z$ test for the population proportion?

**8.** What does a small $p$-value indicate with respect to the null hypothesis? A large $p$-value?

**9.** Does the value of $Z_{\text{data}}$ change when the form of the hypothesis test changes (for example, left-tailed instead of right-tailed)?

## CALCULATIONS AND INTERPRETATIONS

**10. Prison Sentence Length.** The Bureau of Justice Statistics reports that, in 2000, the mean prison sentence length for felons convicted on drug offenses was 47 months. Assume that the standard deviation of all such sentences is 1 year. We are interested in testing using the $p$-value method and level of significance $\alpha = 0.10$ whether the population mean prison sentence length for such offenses has changed since 2000. A random sample taken this year of 100 felons convicted on drug offenses yielded a sample mean sentence length of 54 months.

- **a.** State the hypotheses and the $p$-value rejection rule.
- **b.** Find $Z_{\text{data}}$.
- **c.** Find the $p$-value.
- **d.** State the conclusion and the interpretation.
- **e.** Assess the strength of the evidence against the null hypothesis.

**11. ATM Fees.** Do you hate paying the extra fees imposed by banks when withdrawing funds from an automated teller machine (ATM) not owned by your bank? The Federal Reserve System reports that, in 2002, the mean such fee was \$1.14. A random sample of 36 such transactions yielded a mean of \$1.07 in extra fees. Suppose the population standard deviation of such extra fees is \$0.25. We are interested in testing, using the critical-value method and level of significance $\alpha = 0.05$, whether there has been a reduction in the population mean fee charged on such transactions since 2002.

- **a.** Construct the hypotheses.
- **b.** Find the $Z$ critical value.
- **c.** Find the rejection rule.
- **d.** Calculate the value of the test statistic $Z_{\text{data}}$.
- **e.** Formulate and interpret your conclusion.

**12. ATM Fees.** Refer to Exercise 11. If we performed the same hypothesis test with the same $\alpha = 0.05$, but this time using the $p$-value method, what would be our conclusion? How do you know this?

**13. ATM Fees.** Refer to Exercise 11. Which type of error is it possible that we are making, a Type I error or a Type II error? Which type of error are we certain we are not making?

**14. Alcohol-Related Fatal Car Accidents.** The National Traffic Highway Safety Commission keeps statistics on the "mean years of potential life lost" in alcohol-related fatal automobile accidents. For males in 2001, the mean years of life lost was 32. That is, on average, males involved in fatal drinking-and-driving accidents had their lives cut short by 32 years. We are interested in whether the population mean years of life lost has changed since 2001, using a $t$ test and level of significance $\alpha = 0.10$. A random sample of 36 alcohol-related fatal accidents had a mean years of life lost of 33.8, with a standard deviation of 6 years.

- **a.** Is it appropriate to apply the $Z$ test? Why or why not?
- **b.** Construct the hypotheses.
- **c.** Find the $t$ critical value.
- **d.** Find the rejection rule.
- **e.** Calculate the value of the test statistic $t_{\text{data}}$.
- **f.** Formulate and interpret your conclusion.

**15. Alcohol-Related Fatal Car Accidents.** Refer to the previous exercise.

- **a.** Estimate the $p$-value.
- **b.** Assess the strength of the evidence against the null hypothesis.

**16. Preterm Births.** The U.S. National Center for Health Statistics reports that, in 2005, the percentage of infants delivered at less than 37 weeks of gestation was 12.7%, up from 10.6% in 1990.[34] Has this upward trend continued? We are interested in testing whether the population proportion of preterm births has increased from 12.7%, using the $p$-value method and level of significance $\alpha = 0.05$. A random sample taken this year of 400 births contained 57 preterm births.

- **a.** Is it appropriate to perform the $Z$ test for the proportion? Why or why not?
- **b.** Where will the sampling distribution for $\hat{p}$ be centered? What is the shape of the distribution?
- **c.** What is the standard error? What does this number mean?
- **d.** How many standard errors does $\hat{p}$ lie above $p_0$? Is this extreme? Why or why not?

**e.** Where in the sampling distribution does the value of $\hat{p}$ lie? Near the tail? Near the center?

**17. Preterm Births.** Refer to the previous exercise.

**a.** Construct the hypotheses.

**b.** Find the rejection rule.

**c.** Calculate the value of the test statistic $Z_{\text{data}}$.

**d.** Calculate the *p*-value.

**e.** Assess the strength of the evidence against the null hypothesis.

**f.** Formulate and interpret your conclusion.

**18. Active Stocks.** On October 3, 2007, the ten most traded stocks on the New York Stock Exchange were those shown in the following table, which gives their closing prices and net change in price, in dollars. Use only the net change data for this analysis. Assume normality. Using for level of significance $\alpha = 0.10$ and the critical-value method, test whether the population standard deviation of net price change is less than 25 cents.

| Stock | Closing price | Net change |
|---|---|---|
| Micron Technology, Inc. | \$10.74 | −1.05 |
| Ford Motor Company | \$ 8.43 | −0.14 |
| Citigroup, Inc. | \$47.89 | 0.03 |
| Advanced Micro Devices | \$13.23 | 0.03 |
| EMC Corporation | \$21.13 | −0.24 |
| Commerce Bancorp, Inc. | \$38.84 | −0.63 |
| General Electric | \$41.55 | −0.57 |
| Avaya, Inc. | \$16.95 | −0.07 |
| Sprint Nextel Corporation | \$18.76 | −0.24 |
| iShares:Taiwan | \$17.18 | −0.18 |

Getty Images/Punchstock

Chapter

# Two-Sample Inference

## CASE STUDY | Do Prior Student Evaluations Influence Students' Ratings of Professors?

A study in 1950 reported that instructor reputation affected students' ratings of their instructors.[1] Towler and Dipboye uncovered experimental evidence in support of this phenomenon.[2] They randomly assigned to students one of two summaries of prior student evaluations, one for a "charismatic instructor" and the other for a "punitive instructor." The "charismatic" summary included such phrases as "always lively and stimulating in class" and "always approachable and treated students as individuals." The "punitive" summary included such phrases as "did not show an interest in students' progress" and "consistently seemed to grade students harder." All subjects were then shown the same 20-minute lecture video given by the same instructor. They were asked to rate the instructor using three questions, and a summary rating score was calculated.

Were students' ratings influenced by the prior student evaluations? We examine this question further in the Chapter 10 (page 567) Case Study, Do Prior Student Evaluations Influence Students' Rating of Professors?

OVERVIEW



## The Big Picture

### Where we are coming from, and where we are headed...

Thus far our treatment of statistical inference has been limited to one population and one sample. In Chapter 8 we learned how to use confidence intervals to estimate the mean, proportion, and standard deviation of a single population. In Chapter 9 we learned how to perform hypothesis tests for the mean, proportion, and standard deviation of a single population.

In this chapter we examine differences in the characteristics of two populations. In Section 10.1, we construct confidence intervals and perform hypothesis tests to compare two dependent samples (for example, comparing body fat measurements of the same athletes before and after training). Section 10.2 introduces inference techniques for comparing independent samples (for example, comparing the spending habits of female and male customers of an Internet business). Section 10.3 shows how to construct confidence intervals and hypothesis tests for $p_1 - p_2$, the difference between two population proportions.

Chapter 11 will examine how we conduct hypothesis tests on categorical data sets. Chapter 12 will introduce analysis of variance. Finally, Chapter 13 will investigate regression analysis. The topics in Chapters 10–13 are among the most useful in the text. Consider the analogy of a high-performance automobile, such as a Porsche. After getting your statistical knowledge up to speed in the preceding chapters, you have finally reached fourth gear and can tear down the highway by learning these more powerful and widespread procedures. Pedal to the metal!

## 10.1 Inference for Mean Difference—Dependent Samples

**Objectives:** By the end of this section, I will be able to...

1. Distinguish between independent samples and dependent samples.
2. Construct and interpret confidence intervals for the population mean difference for dependent samples.
3. Perform hypothesis tests for the population mean difference for dependent samples using the $p$-value method and the critical value method.

## 1 Independent Samples and Dependent Samples

Chapter 10 is about two-sample inference. The type of inference we apply depends on whether the data come from **independent samples** or **dependent samples**.

> **Independent Samples and Dependent Samples**
>
> Two samples are **independent** when the subjects selected for the first sample do not determine the subjects in the second sample. Two samples are **dependent** when the subjects in the first sample determine the subjects in the second sample. The data from dependent samples are called **matched-pair** or **paired** samples.

For example, suppose we are interested in comparing the heights of girl-boy fraternal twins. Selecting the girl twin for the first sample automatically results in the boy twin's being selected for the second sample. This is an example of dependent sampling, and the boy-girl pairs are called **matched-pair** samples, or **paired** samples.

### Example 10.1 Dependent or independent sampling?

Indicate whether each of the following experiments uses an independent or dependent sampling method.

**a.** A study wished to compare the differences in price between name-brand merchandise and store-brand merchandise. Name-brand and store-brand items of the same size were purchased from each of the following six categories: paper towels, shampoo, cereal, ice cream, peanut butter, and milk.

**b.** A study wished to compare traditional acupuncture with usual clinical care for a certain type of lower-back pain.[3] The 241 subjects suffering from persistent non-specific lower-back pain were randomly assigned to receive either traditional acupuncture or the usual clinical care. The results were measured at 12 and 24 months.

**Solution**

**a.** For a given store, each name-brand item in the first sample is associated with exactly one store-brand item of that size in the second sample. Therefore, the items in the first sample determine the items in the second sample. This is an example of dependent sampling.

**b.** The subjects were randomly assigned to receive either of the two treatments. Thus, the subjects that received acupuncture did not determine those who received clinical care, and vice versa. This is an example of independent sampling. ■

## 2 Dependent Sample *t* Interval for the Population Mean of the Differences

Table 10.1 shows students' scores on two statistics quizzes. The "After" row (sample 1) contains scores after the students sought help in the Math Center, and the "Before" row (sample 2) shows scores before they had help. The observations are taken from the same students before and after they had help. Thus, sample 1 and sample 2 are dependent, matched-pair data.

Table 10.1 Statistics quiz scores of seven students before and after visiting the Math Center

| Student | Ashley | Brittany | Chris | Dave | Emily | Fran | Greg |
|---|---|---|---|---|---|---|---|
| **After (sample 1)** | 66 | 68 | 74 | 88 | 89 | 91 | 100 |
| **Before (sample 2)** | 50 | 55 | 60 | 70 | 75 | 80 | 88 |

Notice that each student's score improved on the second quiz:

Ashley: 66 − 50 = 16
Brittany: 68 − 55 = 13
Chris: 74 − 60 = 14
Dave: 88 − 70 = 18
Emily: 89 − 75 = 14
Fran: 91 − 80 = 11
Greg: 100 − 88 = 12

The key idea behind dependent sampling is that we consider the set of these seven differences {16, 13, 14, 18, 14, 11, 12} as a sample so that we can perform inference on these differences! In other words, we no longer have two samples. By matching the samples element by element and taking the difference, we have transformed two samples into one that is the sample of differences (Figure 10.1). We have already learned how to perform inference using a single sample, so the remainder of this section uses techniques you have used before.

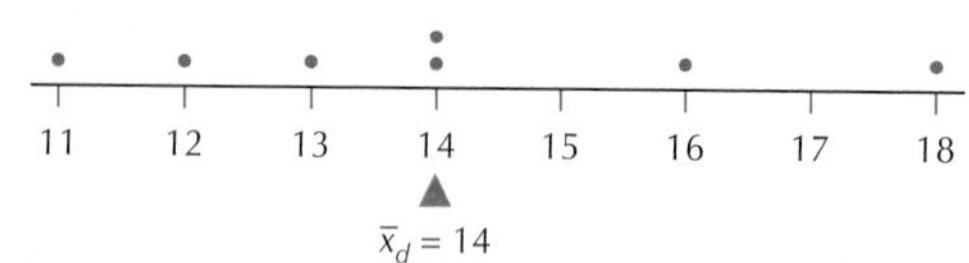


FIGURE 10.1
Taking the differences reduces a two-sample problem to a single sample of differences.

The sample mean of the set of differences is

$$\bar{x}_d = \frac{16 + 13 + 14 + 18 + 14 + 11 + 12}{7} = 14$$

as illustrated in Figure 10.1. The sample of differences can be considered representative of the *population* of these differences, where the population represents all students who took statistics quizzes before and after visiting the Math Center. The sample mean difference $\bar{x}_d = 14$ is a point estimate of the **population mean difference** $\boldsymbol{\mu_d}$, the unknown mean difference in the (after – before) quiz scores for all students who visited the Math Center. Since $\mu_d$ is unknown, we need to construct confidence intervals and perform hypothesis tests to learn about its value.

**Confidence Interval for Population Mean Difference $\boldsymbol{\mu_d}$ (Dependent Samples)**

Suppose we have a set of matched-pair data obtained by taking dependent random samples of two populations and finding the differences to produce a random sample of the difference between the populations. A $100(1 - \alpha)\%$ confidence interval for $\mu_d$, the population mean of the differences, is given by

$$\text{lower bound} = \bar{x}_d - t_{\alpha/2}\left(\frac{s_d}{\sqrt{n}}\right), \qquad \text{upper bound} = \bar{x}_d + t_{\alpha/2}\left(\frac{s_d}{\sqrt{n}}\right)$$

where $\bar{x}_d$ and $s_d$ represent the sample mean and sample standard deviation of the differences, respectively, of the set of $n$ paired differences, $d_1, d_2, d_3, \ldots, d_n$, and where $t_{\alpha/2}$ is based on $n - 1$ degrees of freedom. This $t$ interval applies whenever *either* of the following conditions is met:

- Case 1: the population of differences is normal, or
- Case 2: the sample size of differences is large ($n \geq 30$).

The $100(1 - \alpha)\%$ confidence interval for $\mu_d$ may also be expressed in the form

$$\bar{x}_d \pm t_{\alpha/2}\left(\frac{s_d}{\sqrt{n}}\right)$$

Recall that in Chapter 8 we used the formula $\bar{x} \pm t_{\alpha/2}\,(s/\sqrt{n})$ to calculate the $t$ interval for the population mean $\mu$. Here, to estimate the population mean of the differences $\mu_d$, we use essentially the same formula, substituting $\bar{x}_d$ for $\bar{x}$ and $s_d$ for $s$.

Note that $\mu_d$ always refers to sample 1 minus sample 2, never sample 2 minus sample 1. For example, $\mu_d$ represents the mean difference between the students' "after" scores and the "before" scores on the statistics quizzes in Table 10.1.

## Example 10.2 Confidence interval for $\mu_d$

Construct a 95% confidence interval for the mean of the differences in the statistics quiz scores. Is there evidence that the Math Center tutoring leads to a mean improvement in the quiz scores?

### Solution

The normal probability plot of the differences shows acceptable normality, allowing us to construct the confidence interval.

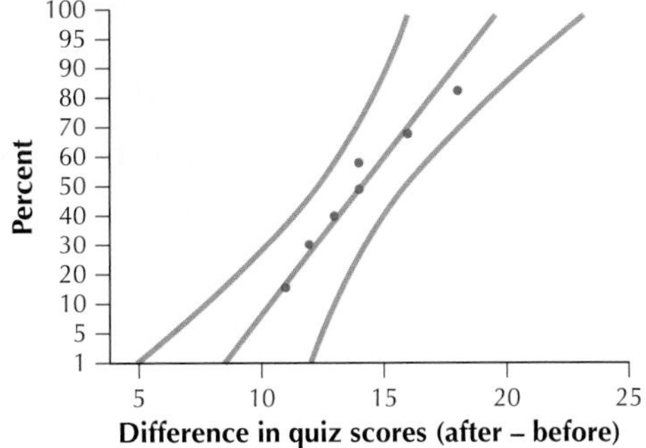


We ignore the original raw data (see Table 10.1) and concentrate only on the set of sample differences: {16, 13, 14, 18, 14, 11, 12}. For the data set of $n = 7$ differences, we find the mean and standard deviation. We found earlier that $\bar{x}_d = 14$. Now we calculate

$$s_d = \sqrt{\frac{\sum(x - \bar{x})^2}{n - 1}}$$

$$= \sqrt{\frac{(16 - 14)^2 + (13 - 14)^2 + (14 - 14)^2 + (18 - 14)^2 + (14 - 14)^2 + (11 - 14)^2 + (12 - 14)^2}{7 - 1}}$$

$$\approx 2.3805$$

For 95% confidence with $n - 1 = 6$ degrees of freedom, $t_{\alpha/2}$ equals 2.447. Using these values,

$$\begin{aligned} \text{lower bound} &= \bar{x}_d - t_{\alpha/2}(s_d/\sqrt{n}) \\ &= 14 - (2.447)(2.3805/\sqrt{7}) \\ &\approx 14 - 2.2017 = 11.7983 \end{aligned} \qquad \begin{aligned} \text{upper bound} &= \bar{x}_d + t_{\alpha/2}(s_d/\sqrt{n}) \\ &= 14 + (2.447)(2.3805/\sqrt{7}) \\ &\approx 14 + 2.2017 = 16.2017 \end{aligned}$$

We are 95% confident that the population mean of the differences between quiz scores before and after visiting the Math Center lies between 11.7983 points and 16.2017 points. If there were no mean change in the quiz scores, the difference would be 0, which is not in this confidence interval. Thus we have evidence that the Math Center tutoring leads to a significant mean improvement in the quiz scores. ■

## 3 Dependent Sample *t* Test for the Population Mean of the Differences

As long as we remember that we are dealing with a sample consisting of the differences of matched pairs, we can go ahead and use the one-sample *t* test as our hypothesis test for the population mean of the differences.

**Paired Sample *t* Test for the Population Mean of the Differences $\mu_d$: *p*-Value Method**

Suppose we have a set of matched-pair data obtained by taking dependent random samples of two populations and finding the differences to produce a random sample of the difference between the populations. We can use the *t* test whenever *either* of the following conditions is met:

- Case 1: the population of differences is normal, or
- Case 2: the sample size of differences is large ($n \geq 30$).

***Step 1*** **State the hypotheses and the rejection rule.** Use one of the hypothesis test forms from Table 10.2. State clearly the meaning of $\mu_d$. The rejection rule is *Reject $H_0$ if the p-value is less than $\alpha$.*

Table 10.2 Forms of the hypothesis test

| Null hypothesis | Alternative hypothesis | Type of test |
|---|---|---|
| $H_0: \mu_d \leq 0$ | $H_a: \mu_d > 0$ | Right-tailed test |
| $H_0: \mu_d \geq 0$ | $H_a: \mu_d < 0$ | Left-tailed test |
| $H_0: \mu_d = 0$ | $H_a: \mu_d \neq 0$ | Two-tailed test |

***Step 2*** **Find $t_{\text{data}}$.**

$$t_{\text{data}} = \frac{\bar{x}_d}{s_d/\sqrt{n}}$$

which follows an approximate *t* distribution with degrees of freedom $n - 1$.

***Step 3*** **Find the *p*-value.** If you have access to technology, use it to find the *p*-value. Otherwise, calculate the *p*-value using one of the test forms in Table 10.3.

***Step 4*** **State the conclusion and interpretation.** Compare the *p*-value with $\alpha$.

Table 10.3 *p*-Value for a paired sample *t* test for $\mu_d$

| Form of test | The *p*-value equals... | Graphic |
|---|---|---|
| Right-tailed test<br>$H_0: \mu_d \leq 0$ versus $H_a: \mu_d > 0$ | $p\text{-value} = P(t > t_{\text{data}})$<br>Area to the right of $t_{\text{data}}$ | |
| Left-tailed test<br>$H_0: \mu_d \geq 0$ versus $H_0: \mu_d < 0$ | $p\text{-value} = P(t < t_{\text{data}})$<br>Area to the left of $t_{\text{data}}$ | |
| Two-tailed test<br>$H_0: \mu_d = 0$ versus $H_0: \mu_d \neq 0$ | $p\text{-value} = 2 \cdot P(t > \lvert t_{\text{data}} \rvert)$ | |

## Example 10.3 Paired-sample $t$ test for $\mu_d$: The $p$-value method

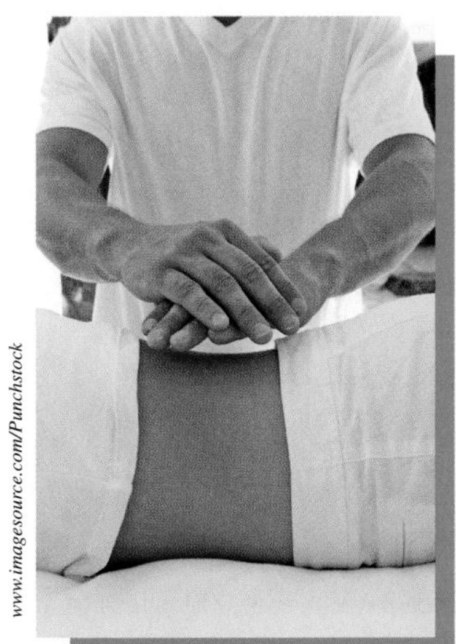


A study was carried out to determine whether Reiki touch therapy was useful in the reduction of chronic pain sufferers, including cancer patients.[4] The pain level reported by a random sample of 13 patients before and after Reiki touch therapy is shown in Table 10.4. Test whether there has been a mean reduction in pain level after the Reiki therapy, using level of significance $\alpha = 0.05$. In other words, test whether the population mean difference $\mu_d$ is less than zero, where $\mu_d$ is defined as the (after – before) difference in pain level.

Table 10.4 Pain level reported by 13 patients before and after Reiki touch therapy

| Patient | 1 | 2 | 3 | 4 | 5 | 6 | 7 | 8 | 9 | 10 | 11 | 12 | 13 |
|---|---|---|---|---|---|---|---|---|---|---|---|---|---|
| After | 3 | 1 | 0 | 0 | 2 | 1 | 2 | 1 | 0 | 4 | 1 | 4 | 8 |
| Before | 6 | 2 | 2 | 3 | 3 | 4 | 2 | 5 | 1 | 6 | 6 | 4 | 8 |
| Difference | –3 | –1 | –2 | –3 | –1 | –3 | 0 | –4 | –1 | –2 | –5 | 0 | 0 |

### Solution

For each patient, we subtract the "before" pain level from the "after" pain level to arrive at a set of $n = 13$ differences, highlighted in Table 10.4. The normal probability plot of the differences indicates acceptable normality, given the small sample size. The Minitab results from the $t$ test are provided here.

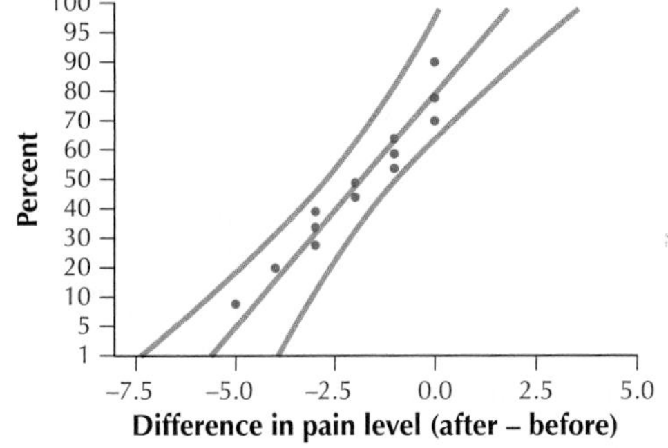


```
Test of mu = 0 vs < 0
                                                    95%
                                                  Upper
Variable   N       Mean    StDev  SE Mean         Bound      T      P
[Diff]    13   -1.92308  1.60528  0.44522      -1.12956  -4.32  0.000
```

***Step 1*** **State the hypotheses and the rejection rule.** We are interested in testing whether there was a mean reduction in pain level, which would mean that the pain level would be lower after the Reiki therapy than before. This implies that the population mean difference in pain level, $\mu_d$ = (after – before), is *less than* 0. Thus, from Table 10.2, the hypotheses are

$$H_0 : \mu_d \geq 0 \qquad H_a : \mu_d < 0$$

where $\mu_d$ represents the population mean difference in pain level. We will reject $H_0$ if the $p$-value $< 0.05$.

***Step 2*** **Find $t_{\text{data}}$.** As provided in the Minitab results,

$$t_{\text{data}} = \frac{\bar{x}_d}{s_d/\sqrt{n}} = \frac{-1.92308}{1.60528/\sqrt{13}} \approx -4.32$$

which follows an approximate $t$ distribution with degrees of freedom $n - 1 = 13 - 1 = 12$.

***Step 3*** **Find the $p$-value.** For a left-tailed test, the $p$-value is the area to the left of $t_{\text{data}}$. This area is essentially 0, as shown in Figure 10.2, and provided by Minitab,

$$P(t < t_{\text{data}}) = P(t < -4.32) \approx 0.000$$

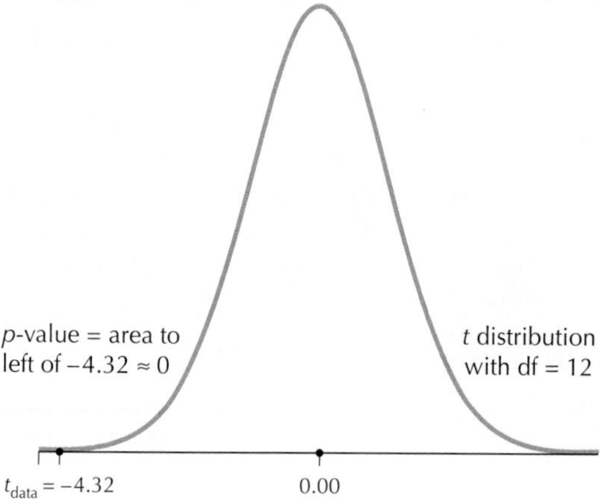


FIGURE 10.2

**Step 4** **State the conclusion and interpretation.** Since $p$-value $\approx 0.000$ is less than $\alpha = 0.05$, we reject $H_0$. There is evidence that $\mu_d < 0$ and hence that the population mean difference in pain level (after − before) has decreased. That is, there is evidence that the Reiki touch therapy has worked to reduce the mean pain level for chronic pain sufferers. ■

The dependent sample $t$ test for the population mean of the differences $\mu_d$ may also be performed using the critical value method.

**Paired Sample $t$ Test for the Population Mean of the Differences $\mu_d$: Critical Value Method**

Suppose we have a set of matched-pair data obtained by taking dependent random samples of two populations and finding the differences to produce a random sample of the difference between the populations. You can use the $t$ test whenever *either* of the following conditions is met:

- Case 1: the population of differences is normal, or
- Case 2: the sample size of differences is large ($n \geq 30$).

***Step 1*** **State the hypotheses.** Use one of the hypothesis test forms in Table 10.2. State clearly the meaning of $\mu_d$.

***Step 2*** **Find $t_{crit}$, and state the rejection rule.** To find $t_{crit}$, use the $t$ table and degrees of freedom $n - 1$. To find the rejection rule, use Table 10.5.

Table 10.5 Rejection rules for the $t$ test for $\mu_d$

| Type of hypothesis test | Rejection rules: Reject $H_0$ if... |
|---|---|
| $H_0: \mu_d \leq 0$ versus $H_a: \mu_d > 0$ | $t_{data} > t_{crit}$ |
| $H_0: \mu_d \geq 0$ versus $H_a: \mu_d < 0$ | $t_{data} < -t_{crit}$ |
| $H_0: \mu_d = 0$ versus $H_a: \mu_d \neq 0$ | $t_{data} > t_{crit}$ or $t_{data} < -t_{crit}$ |

***Step 3*** **Find $t_{data}$.**

$$t_{data} = \frac{\bar{x}_d}{s_d/\sqrt{n}}$$

which follows an approximate $t$ distribution with degrees of freedom $n - 1$.

***Step 4*** **State the conclusion and the interpretation.** Compare $t_{data}$ with $t_{crit}$.

## Example 10.4 Paired $t$ test using the critical value method

Are name-brand groceries more expensive than store-brand groceries? A sample of six randomly selected grocery items yielded the price data shown in Table 10.6. Test at level of significance $\alpha = 0.05$ whether the population mean $\mu_d$ of the differences in price (name brand minus store brand) is greater than zero. Or, more informally, test whether the name-brand items at the grocery store cost more on average than the store-brand items.

Table 10.6 Prices of name-brand and store-brand grocery items

| Item | Paper towels | Shampoo | Cereal | Ice cream | Peanut butter | Milk |
|---|---|---|---|---|---|---|
| **Name brand (sample 1)** | $1.29 | $4.69 | $3.59 | $3.49 | $2.79 | $2.99 |
| **Store brand (sample 2)** | $1.29 | $3.99 | $3.39 | $2.69 | $2.39 | $3.49 |
| **Differences (name brand minus store brand)** | $0.00 | $0.70 | $0.20 | $0.80 | $0.40 | −$0.50 |

### Solution

The normal probability plot of the differences shows acceptable normality, allowing us to proceed with the hypothesis test.

Photodisc/Punchstock

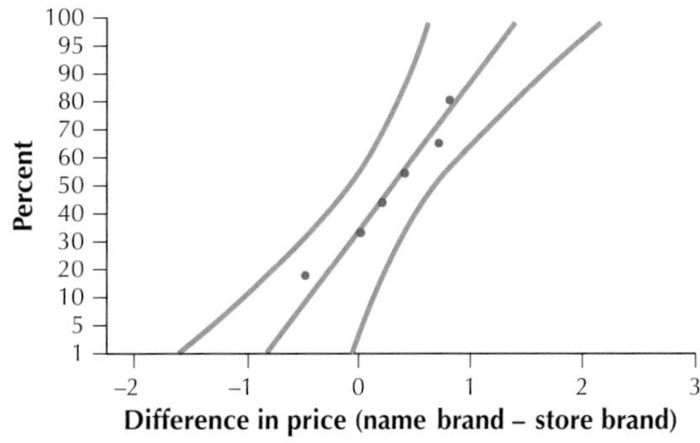


***Step 1*** **State the hypotheses.** "Greater than" implies that $\mu_d > 0$, leading to the hypotheses

$$H_0: \mu_d \leq 0 \qquad H_a: \mu_d > 0$$

where $\mu_d$ represents the population mean difference in price between name-brand and store-brand merchandise.

***Step 2*** **Find the critical value $t_{\text{crit}}$ and state the rejection rule.** Use $n - 1$ degrees of freedom. Here $n = 6$, so df $= n - 1 = 5$. Since we have a right-tailed test with $\alpha = 0.05$, we find our $t$ critical value by choosing the column in the $t$ table with area 0.05 in one tail: $t_{\text{crit}} = 2.015$. The right-tailed test tells us that our rejection rule is to reject $H_0$ when $t_{\text{data}}$ is greater than 2.015.

```
Test of mu = 0 vs > 0
                                                  95%
                                                  Lower
Variable  N       Mean     StDev   SE Mean    Bound      T      P
Diff      6   0.266667  0.480278  0.196073   -0.128   1.36  0.116
```

***Step 3*** **Find $t_{\text{data}}$.** We need to calculate $\bar{x}_d$ and $s_d$.

$$\bar{x}_d = \frac{\sum x}{n} = \frac{0.00 + 0.70 + 0.20 + 0.80 + 0.40 - 0.50}{6} \approx \$0.267$$

$$s_d = \sqrt{\frac{\sum (x - \bar{x})^2}{n - 1}}$$

$$= \sqrt{\frac{(0.00 - 0.267)^2 + (0.70 - 0.267)^2 + (0.20 - 0.267)^2 + (0.80 - 0.267)^2 + (0.40 - 0.267)^2 + (-0.50 - 0.267)^2}{6}}$$

$$\approx \$0.48$$

This gives

$$t_{\text{data}} = \frac{\bar{x}_d}{s_d/\sqrt{n}} = \frac{0.267}{0.48/\sqrt{6}} \approx 1.36$$

***Step 4*** **State conclusion and interpretation.** Since 1.36 is not greater than 2.015 (Figure 10.3), then do not reject $H_0$. There is insufficient evidence that brand-name grocery items cost more on average than store-brand items. It appears that the brand-name milk was on sale; otherwise the conclusion may very well have been different. ■

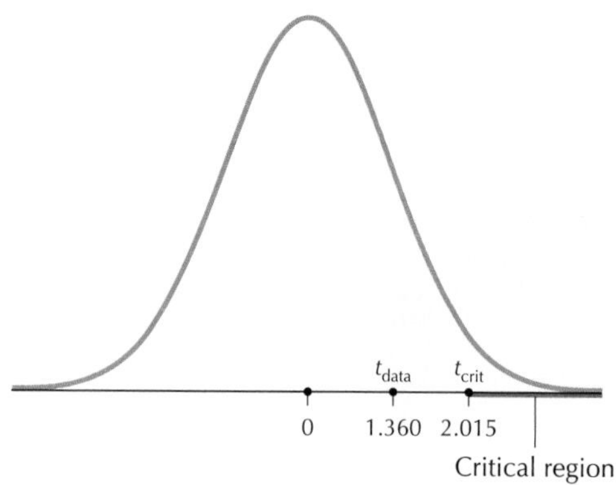


FIGURE 10.3

## STEP-BY-STEP TECHNOLOGY GUIDE: Confidence Intervals and Hypothesis Tests for $\mu_d$

### TI-83/84

**Confidence Interval**

(Example 10.2 is used to illustrate the procedure.)

***Step 1*** Enter samples 1 and 2 in lists L1 and L2.

***Step 2*** Type **(L1 – L2) STO→ L3** and press **ENTER** (Figure 10.4).

***Step 3*** Press **STAT** and highlight **TESTS**.

***Step 4*** Press **8** (for the **TInterval**).

***Step 5*** For input (**Inpt**), highlight **Data** and press **ENTER**. (If given the summary statistics for the differences, choose **STATS**.)

***Step 6*** For **List**, press **2nd** then **L3**. For **Freq**, enter **1**. Enter the **C-Level** (confidence level, such as **0.95** for 95%), and press **ENTER** (Figure 10.5).

***Step 7*** Highlight **Calculate** and press **ENTER**. The results are shown in Figure 10.6.

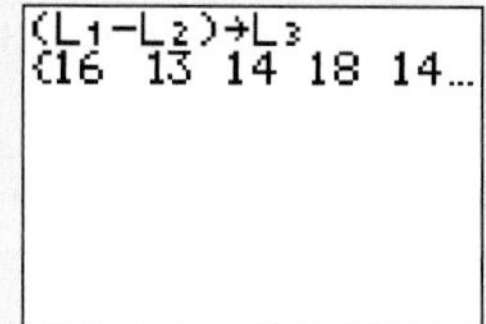


Figure 10.4

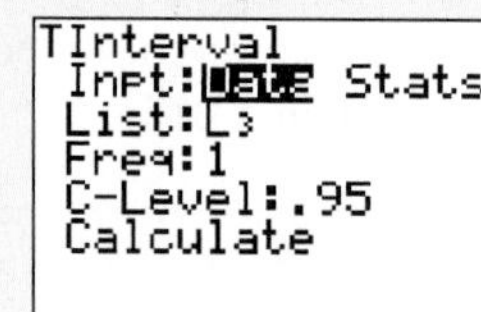


Figure 10.5

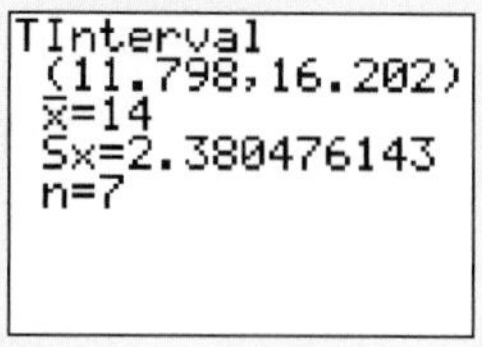

TInterval
(11.798,16.202)
$\bar{x}$=14
Sx=2.380476143
n=7

Figure 10.6

### Hypothesis Test

(Example 10.4 is used to illustrate the procedure.)

***Step 1*** Enter samples 1 and 2 in lists L1 and L2.

***Step 2*** Type **(L1 – L2) STO→ L3** and press **ENTER**.

***Step 3*** Press **STAT** and highlight **TESTS**.

***Step 4*** For the hypothesis test, press **2** (for the **T-Test**). The T-Test menu appears.

***Step 5*** For input (**Inpt**), highlight **Data** and press **ENTER**. (If given the summary statistics for the differences, choose **STATS**.)

***Step 6*** For $\mu_0$, enter the hypothesized value. For **List**, press **2nd** then **L3**. For **Freq**, enter **1**. Choose the form of the hypothesis test, and press **ENTER** (Figure 10.7).

***Step 7*** When the cursor is over **Calculate**, make sure all your entries are correct, and press **ENTER**. The results are shown in Figure 10.8.

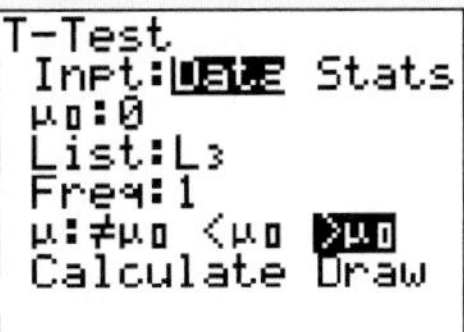


Figure 10.7

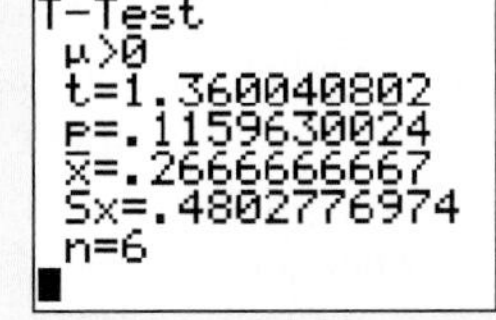


Figure 10.8

## EXCEL

### Hypothesis Test

***Step 1*** Enter samples 1 and 2 in columns A and B.

***Step 2*** Click **Data > Data Analysis > t-Test: Paired Two Sample for Means**, and click **OK**.

***Step 3*** For **Variable 1 Range**, highlight the cells for sample 1 in column A, and for **Variable 2 Range**, highlight the cells for sample 2 in column B.

***Step 4*** Enter the **Hypothesized Mean Difference** (usually 0), and enter a value for **alpha**. Then click **OK**.

## MINITAB

### Confidence Interval and Hypothesis Test

***Step 1*** Enter samples 1 and 2 in columns C1 and C2.

***Step 2*** Click **Stat > Basic Statistics > Paired t**.

***Step 3*** For **First Sample**, enter **C1**, and for **Second Sample**, enter **C2**.

***Step 4*** Click **Options**.

**a.** For the confidence interval, specify the **Confidence Level**, then click **OK** twice.

**b.** For the hypothesis test, specify the form of the alternative hypothesis, then click **OK** twice.

## Section 10.1 SUMMARY

1 Two samples are independent when the subjects selected for the first sample do not determine the subjects in the second sample. Two samples are dependent when the subjects in the first sample determine the subjects in the second sample. The data from dependent samples are called matched-pair or paired samples. The key concept in this section is that we consider the differences of matched pair data as a sample, and perform inference on this sample of differences.

2 A $100(1 - \alpha)\%$ confidence interval for $\mu_d$, the population mean of the differences, is given by $\bar{x}_d \pm t_{\alpha/2}(s_d/\sqrt{n})$, where $\bar{x}_d$ and $s_d$ represent the sample mean and sample standard deviation of the differences, respectively, of the set of $n$ paired differences, $d_1, d_2, d_3, \ldots, d_n$, and where $t_{\alpha/2}$ is based on $n - 1$ degrees of freedom.

3 The paired sample $t$ test for the population mean of the differences $\mu_d$ can be used under either of the following conditions:

- Case 1: the population is normal, or
- Case 2: the sample size is large ($n \geq 30$).

The test may be carried out using either the $p$-value method or the critical value method.

## Section 10.1 EXERCISES

### CLARIFYING THE CONCEPTS

1. When are two samples considered independent?
2. When are two samples considered dependent?
3. What do we call the data obtained from dependent sampling?
4. How do we interpret the meaning of $\mu_d$?

## PRACTICING THE TECHNIQUES

Determine whether the experiments in Exercises 5–8 represent an independent sampling method or a dependent sampling method. Explain your answer.

**5.** The Jacksonville Jaguars are interested in comparing the performance of their first-year players. For each player, a sample is taken of their games from their last year in college and compared to a sample of games taken from their first year in the pros.

**6.** For her senior project, an exercise science major takes a sample of females majoring in exercise science, and a sample of females from her college not majoring in exercise science. She records the body mass index for each subject.

**7.** Before the first lecture, an algebra instructor gives a pre-test to his students to determine the students' algebra readiness. At the end of the course, the instructor gives a post-test to the same students and compares the results with the pre-test.

**8.** The sheriff's department takes a sample of vehicle speeds on a certain stretch of road and compares the results to a sample of vehicle speeds on a certain stretch of a different road. Both roads have the same posted speed limit.

In Exercises 9–12, assume that samples of differences are obtained through dependent sampling and follow a normal distribution.

**a.** Calculate $\bar{x}_d$ and $s_d$.

**b.** Construct a 95% confidence interval for $\mu_d$.

**9.**

| Subject | 1 | 2 | 3 | 4 | 5 |
|---|---|---|---|---|---|
| Sample 1 | 3.0 | 2.5 | 3.5 | 3.0 | 4.0 |
| Sample 2 | 2.5 | 2.5 | 2.0 | 2.0 | 1.5 |

**10.**

| Subject | 1 | 2 | 3 | 4 | 5 | 6 |
|---|---|---|---|---|---|---|
| Sample 1 | 10 | 12 | 9 | 14 | 15 | 8 |
| Sample 2 | 8 | 11 | 10 | 12 | 14 | 9 |

**11.**

| Subject | 1 | 2 | 3 | 4 | 5 | 6 | 7 |
|---|---|---|---|---|---|---|---|
| Sample 1 | 20 | 25 | 15 | 10 | 20 | 30 | 15 |
| Sample 2 | 30 | 30 | 20 | 20 | 25 | 35 | 25 |

**12.**

| Subject | 1 | 2 | 3 | 4 | 5 | 6 | 7 |
|---|---|---|---|---|---|---|---|
| Sample 1 | 1.5 | 1.8 | 2.0 | 2.5 | 3.0 | 3.2 | 4.0 |
| Sample 2 | 1.0 | 1.7 | 2.1 | 2.0 | 2.7 | 2.9 | 3.3 |

**13.** For the data in Exercise 9, test whether $\mu_d > 0$, using the $p$-value method and level of significance $\alpha = 0.05$.

**14.** For the data in Exercise 9, test whether $\mu_d > 0$, using the $p$-value method and level of significance $\alpha = 0.01$.

**15.** For the data in Exercise 10, test whether $\mu_d > 0$, using the $p$-value method and level of significance $\alpha = 0.05$.

**16.** For the data in Exercise 11, test whether $\mu_d < 0$, using the critical value method and level of significance $\alpha = 0.01$.

**17.** For the data in Exercise 11, test whether $\mu_d \neq 0$, using the critical value method and level of significance $\alpha = 0.10$.

**18.** For the data in Exercise 11, test whether $\mu_d < 0$, using the critical value method and level of significance $\alpha = 0.05$.

## APPLYING THE CONCEPTS

**19. New Car Prices.** Kelley's Blue Book (**kbb.com**) publishes data on new and used cars. The table contains the manufacturer's suggested retail price for four vehicles, model years 2006 and 2007. We are interested in the difference in price between the 2006 models and the 2007 models. Assume that the population of price differences is normally distributed.

**a.** Find the mean of the differences, $\bar{x}_d$, and the standard deviation of the differences, $s_d$.

**b.** Construct a 95% confidence interval for $\mu_d$, the population mean difference in price.

| | Subaru Forester | Honda CR-V | Toyota RAV-4 | Nissan Sentra |
|---|---|---|---|---|
| **2006** | \$22,420 | \$20,990 | \$22,980 | \$13,815 |
| **2007** | \$21,820 | \$22,395 | \$23,630 | \$15,375 |

**20. High and Low Temperatures.** The University of Waterloo Weather Station tracks the daily low and high temperatures in Waterloo, Ontario, Canada. Table 10.7 contains a random sample of the daily high and low temperatures for May 1–May 10, 2006, in degrees centigrade. Assume that the temperature differences are normally distributed.

**a.** Find the mean of the differences, $\bar{x}_d$, and the standard deviation of the differences, $s_d$.

**b.** Construct a 99% confidence interval for the population mean difference between high and low temperatures.

**21. New Car Prices.** For the data in Exercise 19, test whether 2007 models are on average more expensive, using level of significance $\alpha = 0.05$.

**22. High and Low Temperatures.** Use the confidence interval from Exercise 20 to test using level of significance

**TABLE 10.7 High and low temperatures**

| May date | 1 | 2 | 3 | 4 | 5 | 6 | 7 | 8 | 9 | 10 |
|---|---|---|---|---|---|---|---|---|---|---|
| High temp | 19.0 | 19.8 | 23.3 | 21.1 | 15.2 | 9.9 | 17.2 | 21.7 | 21.2 | 23.9 |
| Low temp | 7.4 | 3.0 | 3.9 | 7.9 | 4.4 | 0.7 | −1.1 | 2.3 | 6.6 | 5.8 |

$\alpha = 0.01$ whether the population mean difference between high and low temperatures differs from zero.

**23. Falling Home Sales Prices.** A credit crunch gripped the nation in 2007–2008, leading to record numbers of mortgage foreclosures and declines in home sales prices. The following table provides the median home sales prices for four regions of the country in the first quarter (January–March) of 2007 and the first quarter of 2008. Assume that the differences are normally distributed.

**a.** Find the mean of the differences $\bar{x}_d$ and the standard deviation of the differences $s_d$.

**b.** Construct a 90% confidence interval for $\mu_d$, the population mean difference in price.

| | Northeast | Midwest | South | West |
|---|---|---|---|---|
| **Jan.–Mar. 2007** | \$370,300 | \$212,800 | \$222,900 | \$341,500 |
| **Jan.–Mar. 2008** | \$326,600 | \$201,900 | \$204,800 | \$298,900 |

*Source:* U.S. Census Bureau.

**24. Mozart Effect?** A researcher claims that listening to Mozart improves scores on math quizzes. A random sample of five students took math quizzes, first before and then after listening to Mozart. Perform the appropriate hypothesis test for determining whether the results support the researcher's claim, using level of significance $\alpha = 0.01$. Assume normality.

| Student | 1 | 2 | 3 | 4 | 5 |
|---|---|---|---|---|---|
| **Before** | 75 | 50 | 80 | 85 | 95 |
| **After** | 85 | 45 | 85 | 95 | 95 |

**25. Falling Home Sales Prices.** Use the confidence interval you constructed in Exercise 23 and level of significance $\alpha = 0.01$ to test whether the population mean difference $\mu_d$ between the first quarter 2007 median price and the first quarter 2008 median price differs from 0.

**26. Spell-Checking.** A software manufacturer claims that its spell-checker can reduce spelling errors in text documents. A random sample of nine documents were written and the number of spelling errors counted before and after the spell-checker was used. Assume that the differences are normally distributed.

| Document | 1 | 2 | 3 | 4 | 5 | 6 | 7 | 8 | 9 |
|---|---|---|---|---|---|---|---|---|---|
| **Before** | 5 | 3 | 10 | 2 | 0 | 1 | 5 | 0 | 3 |
| **After** | 2 | 1 | 2 | 0 | 0 | 1 | 0 | 0 | 0 |

Is there evidence for the manufacturer's claim? Perform the appropriate hypothesis test, using level of significance $\alpha = 0.10$.

**27. Science Scores Worldwide.** The National Center for Educational Statistics publishes the results from the Trends in International Math and Science Study (TIMSS). Table 10.8 contains the 1995 and 2003 mean science scores for fourth graders from various countries. Assume that the population of score differences is normally distributed.

**a.** Construct a 90% confidence interval for $\mu_d$, the population mean difference in score.

**b.** Using level of significance $\alpha = 0.10$, test whether the 2003 scores are higher than the 1995 scores, on average.

**TABLE 10.8 Fourth-grade science scores**

| Country | 1995 | 2003 |
|---|---|---|
| Singapore | 523 | 565 |
| Japan | 553 | 543 |
| Hong Kong | 508 | 542 |
| England | 528 | 540 |
| United States | 542 | 536 |
| Hungary | 508 | 530 |
| Latvia | 486 | 530 |
| Netherlands | 530 | 525 |
| New Zealand | 505 | 523 |
| Australia | 521 | 521 |
| Scotland | 514 | 502 |
| Slovenia | 464 | 490 |
| Cyprus | 450 | 480 |
| Norway | 504 | 466 |
| Iran | 380 | 414 |

**28. Reading Levels.** An advertisement for a phonetics game claims that it will improve reading scores by at least 10 points. The reading levels of eight randomly selected children were tested without the phonetics game and retested after exposure to the game. Assume that the differences are normally distributed.

| Child | 1 | 2 | 3 | 4 | 5 | 6 | 7 | 8 |
|---|---|---|---|---|---|---|---|---|
| **Pre-test** | 70 | 65 | 85 | 50 | 75 | 80 | 90 | 60 |
| **Post-test** | 75 | 70 | 90 | 70 | 90 | 80 | 95 | 70 |

**a.** Find a 95% confidence interval for the population mean of the differences, $\mu_d$.

**b.** Should we use the confidence interval in **(a)** to test the manufacturer's claim? Why or why not?

**c.** Using level of significance $\alpha = 0.05$, perform the appropriate hypothesis test.

**29. Collisions Before and After.** The Washington Department of Transportation compared collision data on particular sections of roadway before and after a series of road improvements to determine whether road improvements lowered the number of collisions per year.[5] Assume that the differences are normally distributed.

**TABLE 10.9 Collision data**

| Location | Before | After |
|---|---|---|
| Startup | 0.7 | 0.3 |
| Seattle | 77.5 | 43.8 |
| Shoreline | 63.3 | 33.6 |
| Cashmere | 2.1 | 0.0 |
| Othello | 7.4 | 2.0 |
| Royal City | 1.6 | 0.7 |
| Ephrata | 5.7 | 2.9 |
| Orondo | 0.5 | 0.0 |
| Chelan | 5.1 | 0.9 |
| Manson | 1.1 | 0.6 |
| Westport | 3.3 | 2.3 |
| Alderton | 49.9 | 40.3 |
| Chehalis 1 | 2.9 | 0.9 |
| Chehalis 2 | 3.8 | 1.8 |
| Raymond | 2.4 | 3.1 |
| Yakima | 5.5 | 2.3 |
| Ellensburg | 14.9 | 10.7 |
| Snoqualmie | 19.4 | 10.4 |
| Sunnyside | 12.0 | 11.7 |
| Ritzville | 39.0 | 23.7 |
| Milton | 14.5 | 11.2 |
| Skagit | 8.8 | 17.5 |
| Spokane | 114.7 | 77.3 |
| Kent | 25.3 | 13.8 |
| Vancouver | 22.4 | 4.3 |
| Raymond | 0.7 | 0.3 |
| Poulsbo | 5.0 | 12.8 |
| Grant | 1.5 | 0.0 |
| Covington | 2.1 | 2.9 |
| Douglas | 0.0 | 0.0 |

**a.** Find the point estimate of the mean decrease in collisions per year.

**b.** Find a 95% confidence interval for the population mean of the differences, $\mu_d$.

**c.** Using level of significance $\alpha = 0.01$, test whether the improvements have lowered the population mean number of collisions per year.

**30. Press Freedom Worldwide.** The United Nations publishes scores on freedom of the press in various regions around the world: the more press freedom, the higher the score. The scores for 1980 and 2000 for seven major world regions are shown in Table 10.10. Assume that the differences are normally distributed. Test whether press freedom is increasing, using level of significance $\alpha = 0.05$. In other words, test whether the population mean difference $\mu_d$ is less than zero.

**31. NHL Goals Scored.** The entire 2004–2005 National Hockey League season was canceled due to a lockout by the owners looking to introduce a player salary cap. During this time, the league also introduced rule changes designed to increase the number of goals scored per game. Table 10.11 shows the goals scored per game for a random sample of six NHL teams in the season before (2003–2004) and the season after (2005–2006) the lockout. Assume that the differences are normally distributed. Test whether the population mean number of goals scored per game increased, using level of significance $\alpha = 0.01$.

**32. Home Sales.** The number of sales of single-family residences in a random sample of towns in eastern Connecticut is provided in Table 10.12 for the time periods January–September 2006 and January–September 2007. Assume that the differences are normally distributed.

**TABLE 10.12 Home sales in eastern Connecticut**

| Town | 2006 | 2007 |
|---|---|---|
| Andover | 31 | 32 |
| Bolton | 46 | 39 |
| Coventry | 180 | 137 |
| East Hartford | 469 | 405 |
| Ellington | 121 | 121 |
| Hebron | 98 | 74 |
| Manchester | 475 | 479 |
| Somers | 60 | 73 |
| South Windsor | 154 | 161 |
| Stafford | 114 | 89 |
| Suffield | 114 | 121 |
| Tolland | 146 | 141 |
| Vernon | 210 | 213 |
| Windsor | 327 | 288 |
| Windsor Locks | 120 | 128 |

*Source: Manchester* (CT) *Journal-Inquirer,* November 7, 2007.

**TABLE 10.10 United Nations Press Freedom Scores, 1980–2000**

| | High income nations | Latin America | East Asia/ Pacific | Sub-Saharan Africa | Arab states | Eastern Europe/CIS | South Asia |
|---|---|---|---|---|---|---|---|
| **1980** | 2.6 | 2.4 | 1.6 | 1.4 | 1.3 | 1.0 | 2.3 |
| **2000** | 2.8 | 2.4 | 1.9 | 1.9 | 1.5 | 2.0 | 2.2 |

**TABLE 10.11 NHL goals scored**

| | Detroit Red Wings | Tampa Bay Lightning | Phoenix Coyotes | Atlanta Thrashers | Colorado Avalanche | Dallas Stars |
|---|---|---|---|---|---|---|
| **2003–2004 season** | 3.11 | 2.99 | 2.29 | 2.61 | 2.88 | 2.37 |
| **2005–2006 season** | 3.67 | 3.00 | 2.95 | 3.37 | 3.42 | 3.08 |

**a.** Construct a 95% confidence interval for $\mu_d$.

**b.** Using level of significance $\alpha = 0.05$, test whether the population mean number of home sales differs from 2006 to 2007.

WHAT IF? **33. NHL Goals Scored.** Refer to Exercise 31. *What if* we added a certain amount to every entry in the 2005–2006 season results. How would this change affect the conclusion?

WHAT IF? **34. Home Sales.** Refer to Exercise 32. *What if* we added a certain number of home sales to every entry in the table. How would this change affect the conclusion?

## 10.2 Inference for Two Independent Means

**Objectives:** By the end of this section, I will be able to...

1 Describe the sampling distribution of $\bar{x}_1 - \bar{x}_2$.

2 Compute and interpret $t$ intervals for $\mu_1 - \mu_2$.

3 Perform and interpret $t$ tests about $\mu_1 - \mu_2$.

4 Use confidence intervals for $\mu_1 - \mu_2$ to perform two-tailed hypothesis tests about $\mu_1 - \mu_2$.

### 1 Sampling Distribution of $\bar{x}_1 - \bar{x}_2$

Do you think that there may be evidence that the mean body temperature differs between women and men? In Example 3.46 (page 152), we examined a data set that compared the body temperatures for a sample of 65 women and a sample of 65 men.[6] At that time, we compared box plots of these data sets, and found some visual evidence for a difference. However, since the female subjects did not determine the male subjects, and vice versa, the 65 women and 65 men represent independent samples, and we cannot use the dependent sampling methods we learned in Section 10.1. We need to develop new sampling methods to determine beyond a reasonable doubt whether a difference exists between male and female mean body temperatures. The summary statistics are shown in Table 10.13.

Table 10.13 Summary statistics for female versus male body temperatures in °F

| Gender | Sample size | Sample mean body temperature | Sample standard deviation | Population mean body temperature |
|---|---|---|---|---|
| **Females (sample 1)** | $n_1 = 65$ | $\bar{x}_1 = 98.394$ | $s_1 = 0.743$ | $\mu_1 = ?$ |
| **Males (sample 2)** | $n_2 = 65$ | $\bar{x}_2 = 98.105$ | $s_2 = 0.699$ | $\mu_2 = ?$ |

Since we are considering two independent samples, we now have two sample sizes, $n_1$ and $n_2$, two sample means, $\bar{x}_1$ and $\bar{x}_2$, and two sample standard deviations, $s_1$ and $s_2$. These samples are taken from different populations, which have population means $\mu_1$ and $\mu_2$. Since we are interested in the difference in the population means, we consider the quantity

$$\mu_1 - \mu_2$$

Since both $\mu_1$ and $\mu_2$ are population parameters, so also is $\mu_1 - \mu_2$, whose value is unknown.

Your Statistical Sense

**The Difference Difference**

There is a difference in interpretation between $\mu_1 - \mu_2$ and the quantity $\mu_d$ from Section 10.1. Here, $\mu_1 - \mu_2$ refers to the difference in population means, whereas $\mu_d$ represents the population mean of the paired differences. ■

In previous chapters, to estimate $\mu$, we examined the behavior of the statistic $\bar{x}$. For the difference in population means $\mu_1 - \mu_2$, we shall examine the behavior of the statistic $\bar{x}_1 - \bar{x}_2$. Note from Table 10.13 that the value of $\bar{x}_1 - \bar{x}_2$ for these samples is

$$\bar{x}_1 - \bar{x}_2 = 98.394 - 98.105 = 0.289$$

What if we repeat the experiment with two different samples and get $\bar{x}_1 = 98.4$ and $\bar{x}_2 = 97.9$? Then

$$\bar{x}_1 - \bar{x}_2 = 98.4 - 97.9 = 0.5$$

If we repeat the experiment an infinite number of times, then the values of $\bar{x}_1 - \bar{x}_2$ will form a distribution, called the **sampling distribution of $\bar{x}_1 - \bar{x}_2$**. Just as we used the sampling distribution of $\bar{x}$ to estimate $\mu$, we use the sampling distribution of $\bar{x}_1 - \bar{x}_2$ to estimate $\mu_1 - \mu_2$.

It is unlikely that the experimenter will have knowledge of both population standard deviations $\sigma_1$ and $\sigma_2$. Therefore, we use the estimates of $\sigma_1$ and $\sigma_2$ provided by the sample standard deviations $s_1$ and $s_2$. Recall from Chapter 8 that, when the population standard deviation $\sigma$ is unknown, and if either the population was normal or the sample size was large, the quantity

$$t = \frac{\bar{x} - \mu}{s_{\bar{x}}} = \frac{\bar{x} - \mu}{s/\sqrt{n}}$$

has a $t$ distribution with $n - 1$ degrees of freedom. By analogy, we have the following sampling distribution.

**Sampling Distribution of $\bar{x}_1 - \bar{x}_2$**

When random samples are drawn independently from two populations with population means $\mu_1$ and $\mu_2$, and either

- Case 1: the two populations are normally distributed, or
- Case 2: the two sample sizes are large (at least 30), then the quantity

$$t = \frac{(\bar{x}_1 - \bar{x}_2) - (\mu_1 - \mu_2)}{s_{\bar{x}_1 - \bar{x}_2}} = \frac{(\bar{x}_1 - \bar{x}_2) - (\mu_1 - \mu_2)}{\sqrt{\dfrac{s_1^2}{n_1} + \dfrac{s_2^2}{n_2}}}$$

approximately follows a $t$ distribution with degrees of freedom equal to the smaller of $n_1 - 1$ and $n_2 - 1$, where $\bar{x}_1$ and $s_1$ represent the mean and standard deviation of the sample taken from population 1, and $\bar{x}_2$ and $s_2$ represent the mean and standard deviation of the sample taken from population 2.

This $t$ statistic is called *Welch's approximate t,* after the twentieth-century English statistician Bernard Lewis Welch. Although there are other distributions that statisticians use to estimate the difference between two population means, we use this approximation because it is conservative and easy to calculate.

**Standard Error of $\bar{x}_1 - \bar{x}_2$**

The **standard error $s_{\bar{x}_1-\bar{x}_2}$ of the statistic $\bar{x}_1 - \bar{x}_2$** is

$$s_{\bar{x}_1-\bar{x}_2} = \sqrt{\frac{s_1^2}{n_1} + \frac{s_2^2}{n_2}}$$

It measures the size of the typical error in using $\bar{x}_1 - \bar{x}_2$ to estimate $\mu_1 - \mu_2$.

## Example 10.5 Calculating the standard error $s_{\bar{x}_1-\bar{x}_2}$

Calculate the standard error $s_{\bar{x}_1-\bar{x}_2}$ for estimating the difference in population mean body temperatures between women and men, using the data in Table 10.13.

**Solution**

From Table 10.13, $s_1 = 0.743$, $s_2 = 0.699$, $n_1 = 65$, $n_2 = 65$. Thus,

$$s_{\bar{x}_1-\bar{x}_2} = \sqrt{\frac{s_1^2}{n_1} + \frac{s_2^2}{n_2}} = \sqrt{\frac{0.743^2}{65} + \frac{0.699^2}{65}} \approx 0.1265$$

*What Does This Number Mean?*

The size of the typical error in estimating the difference in women's and men's mean body temperatures is about 0.1265 degrees Fahrenheit. ■

## 2 Confidence Intervals for $\mu_1 - \mu_2$

Recall from Chapter 8 that to estimate the unknown population mean $\mu$, we can use a $t$ confidence interval:

$$\bar{x} \pm E = \bar{x} \pm t_{\alpha/2}\,(s_{\bar{x}}) = \bar{x} \pm t_{\alpha/2}\,(s/\sqrt{n})$$

where $E$ is the **margin of error**. By analogy, here the $t$ interval for $\mu_1 - \mu_2$ takes the following form.

**Confidence Interval for $\mu_1 - \mu_2$**

For two independent random samples taken from two populations with population means $\mu_1$ and $\mu_2$, a **100(1 − $\alpha$)% confidence interval for $\mu_1 - \mu_2$** is given by

$$(\bar{x}_1 - \bar{x}_2) \pm t_{\alpha/2}\sqrt{\frac{s_1^2}{n_1} + \frac{s_2^2}{n_2}}$$

where $\bar{x}_1$, $s_1$, and $n_1$ represent the mean, standard deviation, and sample size of the sample taken from population 1 and $\bar{x}_2$, $s_2$, and $n_2$ represent the mean, standard deviation, and sample size of the sample taken from population 2, and $t_{\alpha/2}$ is associated with the confidence level and degrees of freedom the smaller of $n_1 - 1$ and $n_2 - 1$.

The $t$ interval applies whenever *either* of the following conditions is met:

- Case 1: both populations are normally distributed, or
- Case 2: both sample sizes are large.

**Margin of Error $E$**

The **margin of error** for a 100(1 − $\alpha$)% confidence interval for $\mu_1 - \mu_2$ is given by

$$\begin{aligned} E &= t_{\alpha/2} \cdot \text{(standard error)} \\ &= t_{\alpha/2} \cdot (s_{\bar{x}_1-\bar{x}_2}) \\ &= t_{\alpha/2} \cdot \sqrt{\frac{s_1^2}{n_1} + \frac{s_2^2}{n_2}} \end{aligned}$$

Thus, the confidence interval for $\mu_1 - \mu_2$ takes the form $(\bar{x}_1 - \bar{x}_2) \pm E$.

This is a confidence interval for the difference in two population means, which is not the same as in Section 10.1, which was for the population mean of the differences of matched pairs.

## Example 10.6 Confidence interval for difference in women's and men's mean body temperatures

Find a 95% confidence interval for the difference in women's and men's population mean body temperatures, using the data in Table 10.13.

**Solution**

Both sample sizes are large, so the sampling distribution of $\bar{x}_1 - \bar{x}_2$ has a $t$ distribution. We already found the standard error $s_{\bar{x}_1-\bar{x}_2} \approx 0.1265$ in Example 10.5, but we need to find $t_{\alpha/2}$ to use the formula for $E$. The required degrees of freedom is the smaller of $n_1 - 1$ and $n_2 - 1$, which are both equal to $65 - 1 = 64$, so the degrees of freedom for $t_{\alpha/2}$ is also 64. This df = 64 is not listed in the $t$ table, so we choose the next lowest df listed, 60. For 95% confidence, then, $t_{\alpha/2} = 2.00$, and the margin of error is

$$E = t_{\alpha/2} \cdot (s_{\bar{x}_1-\bar{x}_2}) \approx (2.00) \cdot (0.1265) = 0.253$$

The 95% confidence interval is then

$$(\bar{x}_1 - \bar{x}_2) \pm E = (98.394 - 98.105) \pm 0.253 = 0.289 \pm 0.253 = (0.036, 0.542).$$

We are 95% confident that the difference in population means $\mu_1 - \mu_2$ lies between 0.036°F and 0.542°F. ■

Remember that we are estimating the *difference* in population means $\mu_1 - \mu_2$ between women's and men's body temperatures. The margin of error $E \approx 0.253$ means that, when we repeatedly take samples from these two populations, 95% of the time our estimate $\bar{x}_1 - \bar{x}_2$ will be within 0.253°F of the unknown value of $\mu_1 - \mu_2$. ■

## 3 Independent Sample $t$ Test for $\mu_1 - \mu_2$

Researchers are often interested in testing whether the mean of one population is greater than, less than, or different from the mean of another population. Thus, we next learn how to perform hypothesis tests for the difference in population means $\mu_1 - \mu_2$. Usually the most important hypothesized value for $\mu_1 - \mu_2$ is 0. Consider the two-tailed hypothesis test

$$H_0: \mu_1 - \mu_2 = 0 \quad \text{versus} \quad H_a: \mu_1 - \mu_2 \neq 0$$

If $\mu_1 - \mu_2 = 0$, that is, if there is no difference in the population means, then it follows that $\mu_1 = \mu_2$. Thus, the foregoing hypothesis test is equivalent to the hypothesis test

$$H_0: \mu_1 = \mu_2 \quad \text{versus} \quad H_a: \mu_1 \neq \mu_2$$

Hypothesis tests about $\mu_1 - \mu_2$ take one of three forms, shown in Table 10.14.

Table 10.14 Three possible forms for the hypotheses for a test about $\mu_1 - \mu_2$

| Form | Null hypothesis | Alternative hypothesis | Type of test |
|---|---|---|---|
| 1 | $H_0: \mu_1 \leq \mu_2$ | $H_a: \mu_1 > \mu_2$ | Right-tailed test (one-tailed test) |
| 2 | $H_0: \mu_1 \geq \mu_2$ | $H_a: \mu_1 < \mu_2$ | Left-tailed test (one-tailed test) |
| 3 | $H_0: \mu_1 = \mu_2$ | $H_a: \mu_1 \neq \mu_2$ | Two-tailed test |

The hypothesis test is based on the same approximate $t$ distribution as the $t$ interval on page 559, as long as the two samples are independent. The test statistic is

$$t_{\text{data}} = \frac{(\bar{x}_1 - \bar{x}_2) - (\mu_1 - \mu_2)_0}{s_{\bar{x}_1 - \bar{x}_2}} = \frac{(\bar{x}_1 - \bar{x}_2) - (\mu_1 - \mu_2)_0}{\sqrt{\frac{s_1^2}{n_1} + \frac{s_2^2}{n_2}}}$$

where $(\mu_1 - \mu_2)_0$ represents the hypothesized difference between the two population means. In practice, the hypothesized difference is nearly always $(\mu_1 - \mu_2)_0 = 0$. Thus, the test statistic takes the following form.

$$t_{\text{data}} = \frac{(\bar{x}_1 - \bar{x}_2) - 0}{s_{\bar{x}_1 - \bar{x}_2}} = \frac{(\bar{x}_1 - \bar{x}_2)}{s_{\bar{x}_1 - \bar{x}_2}} = \frac{(\bar{x}_1 - \bar{x}_2)}{\sqrt{\frac{s_1^2}{n_1} + \frac{s_2^2}{n_2}}}$$

The test statistic $t_{\text{data}}$ measures the difference between the sample means, expressed in terms of standard errors. Just as in Chapter 10, if $t_{\text{data}}$ is extreme, then it represents evidence against the null hypothesis.

The hypothesis test applies whenever *either* of the following conditions is met:

- Case 1: both populations are normally distributed, or
- Case 2: both sample sizes are large.

The hypothesis test may be carried out using either the $p$-value method or the critical value method.

**Hypothesis Test for the Difference in Two Population Means—*p*-Value Method**

***Step 1*** **State the hypotheses and the rejection rule.** Use one of the forms from Table 10.14. Clearly state the meaning of $\mu_1$ and $\mu_2$. The rejection rule is *Reject $H_0$ if the p-value is less than $\alpha$.*

***Step 2*** **Find $t_{\text{data}}$.**

$$t_{\text{data}} = \frac{(\bar{x}_1 - \bar{x}_2)}{s_{\bar{x}_1 - \bar{x}_2}} = \frac{(\bar{x}_1 - \bar{x}_2)}{\sqrt{\frac{s_1^2}{n_1} + \frac{s_2^2}{n_2}}}$$

which follows an approximate $t$ distribution with degrees of freedom the smaller of $n_1 - 1$ and $n_2 - 1$.

***Step 3*** **Find the *p*-value.** If you have access to technology, use it to find the $p$-value. Otherwise, calculate the $p$-value using one of the forms in Table 10.15.

***Step 4*** **State the conclusion and interpretation.** Compare the $p$-value with $\alpha$.

Table 10.15 $p$-Value

| Form of test | The $p$-value equals... | Graphic |
|---|---|---|
| Form 1<br>Right-tailed test<br>$H_0 : \mu_1 \leq \mu_2$<br>$H_a : \mu_1 > \mu_2$ | $p$-value $= P(t > t_{\text{data}})$<br>Area to the right of $t_{\text{data}}$ | |
| Form 2<br>Left-tailed test<br>$H_0 : \mu_1 \geq \mu_2$<br>$H_a : \mu_1 < \mu_2$ | $p$-value $= P(t > t_{\text{data}})$<br>Area to the left of $t_{\text{data}}$ | |
| Form 3<br>Two-tailed test<br>$H_0 : \mu_1 = \mu_2$<br>$H_a : \mu_1 \neq \mu_2$ | $p$-value $= P(t > \lvert t_{\text{data}}\rvert) + P(t < -\lvert t_{\text{data}}\rvert)$<br>$= 2 \cdot P(t > \lvert t_{\text{data}}\rvert)$ | |

## Example 10.7 Finding the estimated $p$-value from the $t$ table

Recall the independent samples of men's and women's body temperatures from Table 10.13. Test whether women's population mean body temperature differs from that of men using the estimated $p$-value method and level of significance $\alpha = 0.05$.

**Solution**

Both sample sizes are large ($n_1 = n_2 = 65 \geq 30$), so we can perform the hypothesis test.

***Step 1*** **State the hypotheses and the rejection rule.** The key words "differs from" indicate that we will use the two-tailed test from Table 10.14:

$$H_0 : \mu_1 = \mu_2 \quad \text{versus} \quad H_a : \mu_1 \neq \mu_2$$

where $\mu_1$ and $\mu_2$ represent the population mean body temperature for women and men, respectively. We will reject $H_0$ if the $p$-value is less than 0.05.

***Step 2*** **Find $t_{\text{data}}$.** In Example 10.5 we found the standard error $s_{\bar{x}_1 - \bar{x}_2} = 0.1265$. Thus,

$$t_{\text{data}} = \frac{(\bar{x}_1 - \bar{x}_2)}{s_{\bar{x}_1 - \bar{x}_2}} = \frac{(98.394 - 98.105)}{0.1265} = \frac{0.289}{0.1265} \approx 2.28$$

***Step 3*** **Find the $p$-value.** We estimate the $p$-value using Table 10.15:

$$\begin{aligned} p\text{-value} &= P(t > \lvert t_{\text{data}}\rvert) + P(t < -\lvert t_{\text{data}}\rvert) \\ &= P(t > \lvert 2.28\rvert) + P(t < -\lvert 2.28\rvert) \\ &= 2 \cdot P(t > 2.28) \end{aligned}$$

The $p$-value for a two-tailed test is the sum of the areas in the two tails, as shown in Figure 10.9.

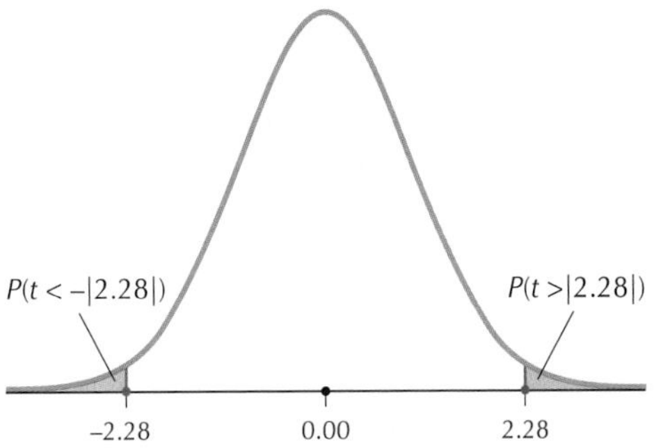


FIGURE 10.9

Using only the $t$ table, we cannot find the exact value for $2 \cdot P(t > 2.28)$. But we can estimate it. The required degrees of freedom is the smaller of $n_1 - 1$ and $n_2 - 1$, which are both equal to $65 - 1 = 64$. Again, this df = 64 is not listed in the $t$ table, so we choose the next lowest df listed, 60. To estimate $2 \cdot P(t > 2.28)$, consider this excerpt from the $t$ table:

| **Two-tailed $\alpha$** | **0.20** | **0.10** | **0.05** | **0.02** | **0.01** |
|---|---|---|---|---|---|
| 60 | 1.296 | 1.671 | 2.000 | 2.390 | 2.660 |

Now, $t_{\text{data}} = 2.28$ lies between 2.000 and 2.390 in the row with df = 60. Note from Table 10.15 that, since this is a two-tailed test, the $p$-value must take into account both tail areas. The area represented by $p$-value $= 2 \cdot P(t > 2.28)$ is therefore between 0.05 and 0.02, as shown by the "two-tailed $\alpha$" at the top of the $t$ table. Since 2.28 is closer to 2.390 than 2.000, the $p$-value is closer to 0.02 than 0.05.

***Step 4*** **State the conclusion and interpretation.** In Step 1 we said that we would reject $H_0$ if the $p$-value were less than 0.05. From Step 3, we know the $p$-value is between 0.02 and 0.05, thus less than 0.05. We therefore reject $H_0$. There is evidence that the population mean body temperatures differs for women and men. ■

## Example 10.8 Finding the $p$-value using technology

Many baseball fans hold that, because of the designated hitter rule, there are more runs scored in the American League than in the National League. Perform an independent samples $t$ test to find out whether that was indeed the case in 2006. Use the TI-83/84, Excel, the $p$-value method, and level of significance $\alpha = 0.05$. Table 10.16 contains the mean runs per game (RPG) for a random sample of six teams from each league.

Table 10.16 Major League Baseball runs scored per game, 2006 regular season

| **American League: Sample 1** | | **National League: Sample 2** | |
|---|---|---|---|
| **Team** | **RPG** | **Team** | **RPG** |
| New York Yankees | 5.74 | Philadelphia Phillies | 5.34 |
| Chicago White Sox | 5.36 | Atlanta Braves | 5.24 |
| Texas Rangers | 5.15 | Colorado Rockies | 5.02 |
| Detroit Tigers | 5.07 | Arizona Diamondbacks | 4.77 |
| Boston Red Sox | 5.06 | Florida Marlins | 4.68 |
| Los Angeles Angels | 4.73 | Houston Astros | 4.54 |

## Solution

Before we can begin, we must determine whether both populations are normally distributed. The normal probability plots for RPG for each league, Figures 10.10 and 10.11, indicate acceptable normality, so that we may perform the hypothesis test.

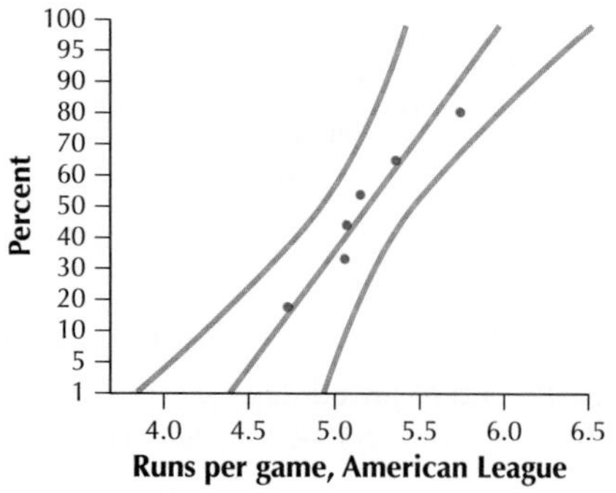


FIGURE 10.10

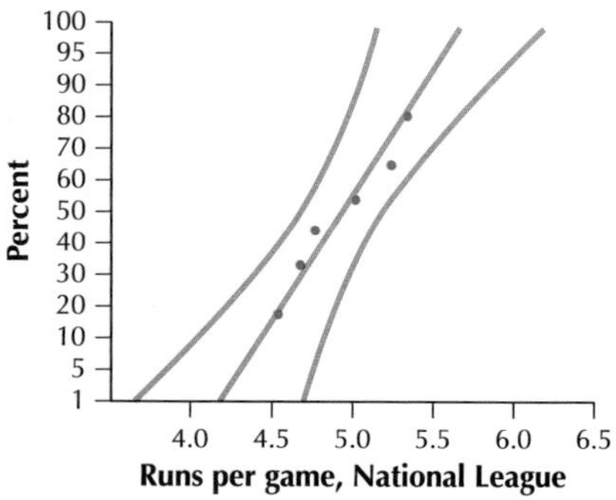


FIGURE 10.11

***Step 1*** **State the hypotheses and the rejection rule.** Since the American League represents sample 1 and we are interested in whether the American League has scored *more* runs than the National League, we have the following hypotheses:

$$H_0 : \mu_1 \leq \mu_2 \quad \text{versus} \quad H_a : \mu_1 > \mu_2$$

where $\mu_1$ and $\mu_2$ represent the population mean runs per game for the American League and National League, respectively. Since we are using the $p$-value method, the rejection rule is to reject $H_0$ if $p$-value $< 0.05$.

*Note:* Our degrees of freedom, the smaller of $n_1 - 1$ and $n_2 - 1$, is $6 - 1 = 5$. However, the TI-83/84 shows df = 9.966314697, and the Excel output rounds this to 10. Why does the technology use different degrees of freedom than we do? Recall that we are using Welch's approximation to the $t$ distribution. The TI-83/84, Excel, Minitab, and other technology calculate the degrees of freedom as follows:[7]

$$\text{df} = \frac{\left(\dfrac{s_1^2}{n_1} + \dfrac{s_2^2}{n_2}\right)^2}{\dfrac{\left(\dfrac{s_1^2}{n_1}\right)^2}{n_1 - 1} + \dfrac{\left(\dfrac{s_2^2}{n_2}\right)^2}{n_2 - 1}}$$

This provides a more accurate determination of the degrees of freedom than our method. However, our method is a conservative estimate that is easier to calculate, and it is recommended for hand calculations.

***Step 2*** **Find $t_{\text{data}}$.** We use the instructions provided in the Step-by-Step Technology Guide at the end of this section. From either Figure 10.12 or Figure 10.13,

$$t_{\text{data}} = \frac{(\bar{x}_1 - \bar{x}_2)}{\sqrt{\dfrac{s_1^2}{n_1} + \dfrac{s_2^2}{n_2}}} \approx \frac{(5.185 - 4.932)}{\sqrt{\dfrac{0.339^2}{6} + \dfrac{0.320^2}{6}}} \approx 1.3301$$

```
2-SampTTest
 μ1 >μ2
 t=1.330123458
 p=.1065564474
 df=9.966314697
 x̄1=5.185
↓x̄2=4.931666667
```

FIGURE 10.12 TI-83/84 output.

| | American | National |
|---|---|---|
| Mean | 5.185 | 4.9317 |
| Variance | 0.11515 | 0.1025 |
| Observations | 6 | 6 |
| Hypothesized Mean Difference | 0 | |
| df | 10 | |
| t Stat | **1.3301** | |
| P(T<=t) one-tail | **0.1065** | |
| t Critical one-tail | 1.8125 | |
| P(T<=t) two-tail | 0.2130 | |
| t Critical two-tail | 2.2281 | |

FIGURE 10.13 Excel output.

***Step 3*** **Find the $p$-value.** From either Figure 10.12 or Figure 10.13,

$$p\text{-value} = P(t > t_{\text{data}}) = P(t > 1.3301) = 0.1065$$

This $p$-value is illustrated in Figure 10.14.

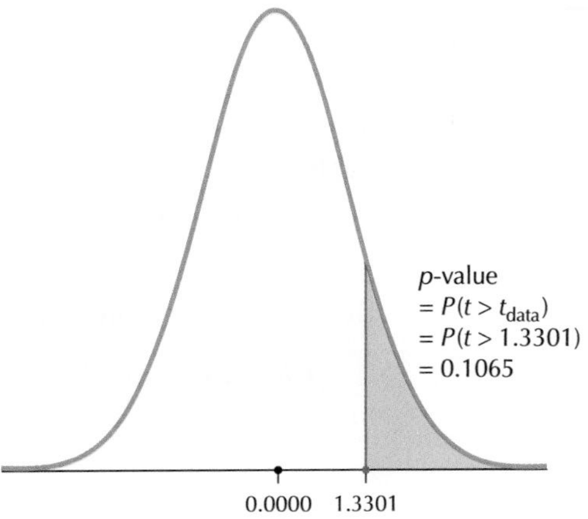


FIGURE 10.14

**Step 4** **State your conclusion and interpretation.** In Step 1, we said we would reject $H_0$ if the $p$-value were less than 0.05. Our $p$-value of 0.1065 is not less than 0.05. Therefore, we do not reject $H_0$. There is insufficient evidence that the population mean runs scored per game is greater in the American League than in the National League. ■

The independent sample $t$ test for the difference in population means $\mu_1 - \mu_2$ may also be performed using the critical value method.

**Hypothesis Test for the Difference in Two Population Means—Critical Value Method**

**Step 1** **State the hypotheses.** Use one of the forms from Table 10.14. Clearly state the meaning of $\mu_1$ and $\mu_2$.

**Step 2** **Find $t_{crit}$ and state the rejection rule.** To find $t_{crit}$, use the $t$ table and degrees of freedom the smaller of $n_1 - 1$ and $n_2 - 1$. To find the rejection rule, use Table 10.17.

Table 10.17 Rejection rules for the $t$ test for $\mu_1 - \mu_2$

| Type of hypothesis test | Rejection rules: Reject $H_0$ if... |
|---|---|
| Right-tailed test<br>$H_0: \mu_1 \leq \mu_2$ versus $H_a: \mu_1 > \mu_2$ | $t_{data} > t_{crit}$ |
| Left-tailed test<br>$H_0: \mu_1 \geq \mu_2$ versus $H_a: \mu_1 < \mu_2$ | $t_{data} < -t_{crit}$ |
| Two-tailed test<br>$H_0: \mu_1 = \mu_2$ versus $H_a: \mu_1 \neq \mu_2$ | $t_{data} > t_{crit}$ or $t_{data} < -t_{crit}$ |

**Step 3** **Find $t_{data}$.**

$$t_{data} = \frac{(\bar{x}_1 - \bar{x}_2)}{s_{\bar{x}_1 - \bar{x}_2}} = \frac{(\bar{x}_1 - \bar{x}_2)}{\sqrt{\frac{s_1^2}{n_1} + \frac{s_2^2}{n_2}}}$$

which follows an approximate $t$ distribution with degrees of freedom the smaller of $n_1 - 1$ and $n_2 - 1$.

**Step 4** **State the conclusion and interpretation.** Compare $t_{data}$ with $t_{crit}$.

## Example 10.9 Critical value method

Recall the independent samples of men's and women's body temperatures from Table 10.13. Test to determine whether women's population mean body temperature differs from that of men, using the critical value method and $\alpha = 0.05$.

**Solution**

Both sample sizes are large, so we can perform the hypothesis test.

***Step 1*** **State the hypotheses.** The hypotheses are the same as for the $p$-value method:

$$H_0: \mu_1 = \mu_2 \quad \text{versus} \quad H_a: \mu_1 \neq \mu_2$$

where $\mu_1$ and $\mu_2$ represent the population mean body temperature for women and men, respectively.

***Step 2*** **Find $t_{crit}$ and state the rejection rule.** Again df = 64, but we use the conservative df = 60 in the $t$ table in the Appendix (page T-11). For $\alpha = 0.05$, this gives $t_{crit} = 2.000$. We have a two-tailed test, so Table 10.17 gives us the following rejection rule:

$$\text{Reject } H_0 \text{ if } t_{data} > 2.000 \text{ or } t_{data} < -2.000$$

***Step 3*** **Find $t_{data}$.**

$$t_{data} = \frac{(\bar{x}_1 - \bar{x}_2)}{s_{\bar{x}_1-\bar{x}_2}} = \frac{(98.394 - 98.105)}{0.1265} = \frac{0.289}{0.1265} \approx 2.28$$

***Step 4*** **State the conclusion and interpretation.** The test statistic $t_{data} = 2.28$ is greater than $t_{crit} = 2.000$. We therefore reject $H_0$. There is evidence that the difference in population mean body temperatures differs for women and men. Note that this result matches the conclusion from Example 10.7. ■

## 4 Using Confidence Intervals for $\mu_1 - \mu_2$ to Perform Two-Tailed Hypothesis Tests About $\mu_1 - \mu_2$

Just as we did in previous chapters, we use the confidence interval to assess significance. Recall that a $100(1 - \alpha)\%$ confidence interval contains all the plausible values for $\mu_1 - \mu_2$ and that any hypothesized values that lie outside this interval are implausible, with confidence $100(1 - \alpha)\%$. Thus, if the $100(1 - \alpha)\%$ confidence interval for $\mu_1 - \mu_2$ contains 0, then it is plausible that $\mu_1 - \mu_2 = 0$ and therefore that $\mu_1 = \mu_2$. But if the interval does not contain 0, then the hypothesis that $\mu_1 - \mu_2 = 0$ is implausible, and we reject this hypothesis. For example, in Example 10.6, we found that for level of significance $\alpha = 0.05$, the 95% confidence interval for the difference in population mean body temperatures $\mu_1 - \mu_2$ was (0.036, 0.542). Since 0 is not contained in this interval, we can conclude that the difference in population mean body temperatures is significant.

**Using Confidence Intervals to Assess Significance**

- If a $100(1 - \alpha)\%$ confidence interval for $\mu_1 - \mu_2$ contains 0, then with $100(1 - \alpha)\%$ confidence we do not reject the hypothesis that there is no difference in the population means.
- If a $100(1 - \alpha)\%$ confidence interval for $\mu_1 - \mu_2$ does not contain 0, then with $100(1 - \alpha)\%$ confidence we reject the hypothesis that there is no difference in the population means.

## Example 10.10 Using a confidence interval to assess significance

**a.** Construct a 95% confidence interval for the difference in runs scored per game in the American League and National League, using the data from Example 10.8.

**b.** Determine with 95% confidence whether the population mean number of runs scored per game in the American League differs significantly from the population mean number of runs scored in the National League.

### Solution

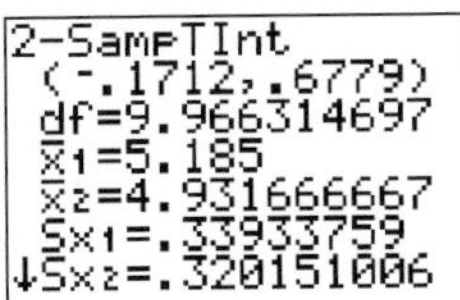


FIGURE 10.15

**a.** Figure 10.15 shows the 95% confidence interval for $\mu_1 - \mu_2$, where $\mu_1$ and $\mu_2$ represent the population mean runs per game in the American and National Leagues, respectively:

$$(-0.1712, 0.6779)$$

**b.** The confidence interval in (**a**) does contain 0. That is, 0 lies between –0.1712 and 0.6779. Therefore, with 95% confidence, we do not reject the hypothesis that there is no difference between population mean runs scored per game in the American and National leagues. ■

Getty Images/Punchstock

## CASE STUDY Do Prior Student Evaluations Influence Students' Ratings of Professors?

Recall that in this case study the students in one sample were shown positive evaluations of an instructor and the students in a second sample were shown negative evaluations of the instructor. Then all subjects were shown the same 20-minute lecture video given by the same instructor. They were then asked to rate the instructor using three questions, and a summary rating score was calculated. Were students' ratings influenced by the prior student evaluations?

We investigate this question by constructing a 95% confidence interval for the difference in population mean ratings $\mu_1 - \mu_2$. Assume that both populations are normally distributed and that the samples are drawn independently.

| Reputation | Subjects | Sample mean rating | Sample standard deviation |
|---|---|---|---|
| **Charismatic (sample 1)** | $n_1 = 25$ | $\bar{x}_1 = 2.613$ | $s_1 = 0.533$ |
| **Punitive (sample 2)** | $n_2 = 24$ | $\bar{x}_2 = 2.236$ | $s_2 = 0.543$ |

The degrees of freedom is the smaller of $n_1 - 1 = 25 - 1 = 24$ and $n_2 - 1 = 24 - 1 = 23$. Thus, df = 23. Then, for 95% confidence, from the $t$ table, $t_{\text{crit}} = 2.069$. Then the 95% confidence interval for $\mu_1 - \mu_2$ is

$$\begin{aligned}
&(\bar{x}_1 - \bar{x}_2) \pm t_{\alpha/2}\sqrt{\frac{s_1^2}{n_1} + \frac{s_2^2}{n_2}} \\
&= (2.613 - 2.236) \pm 2.069\sqrt{\frac{0.533^2}{25} + \frac{0.543^2}{24}} \\
&\approx 0.377 \pm 2.069 \cdot (0.1538) \\
&\approx (0.059, 0.695)
\end{aligned}$$

We are 95% confident that the difference in population mean instructor ratings $\mu_1 - \mu_2$ among the two groups of students lies between 0.059 and 0.695. Since this interval does not contain 0, we can conclude with 95% confidence that the difference in population mean ratings is significant.

# STEP-BY-STEP TECHNOLOGY GUIDE: Two Sample *t* Test and Confidence Interval for $\mu_1 - \mu_2$

## TI-83/84

### Two-Sample *t* Test and Confidence Interval for $\mu_1 - \mu_2$

We use two different examples to illustrate the two different options for performing a two-sample *t* test or confidence interval for $\mu_1 - \mu_2$ using the TI-83/84.

**Data Option.** (Example 10.8 is used to illustrate this method.)

***Step 1*** Enter the American League data into List **L1** and the National League data into List **L2**.

***Step 2*** Press **STAT** and highlight **TESTS**.

***Step 3*** Press **4** (for the **2-Samp TTest**). The **2-Samp TTest** menu appears.

***Step 4*** For input (**INPT**), move the cursor over **Data** and press **ENTER**.

***Step 5*** For **List1** and **List2**, enter **L1** and **L2**.

***Step 6*** For **Freq1** and **Freq2**, enter **1**.

***Step 7*** For $\mu_1$, choose the form of $H_a$. For Example 10.8, choose "$> \mu_2$" and press **ENTER**.

***Step 8*** For **Pooled**, select **No** because we are not assuming the variances are equal, and do not need an estimate of the common variance.

***Step 9*** Press **Calculate**. The results for Example 10.8 are shown in Figure 10.12 on page 564.

**Stats Option.** (Example 10.6 is used to illustrate this method.) Here you enter the summary statistics.

***Step 1*** Press **STAT** and highlight **TESTS**.

***Step 2*** Press **4** (for the **2-Samp TTest**). The **2-Samp TTest** menu appears.

***Step 3*** For input (**Inpt**), move the cursor over **Stats** and press **ENTER**.

***Step 4*** For $\bar{x}_1$, enter **98.394**.

***Step 5*** For $Sx_1$, enter **0.743**.

***Step 6*** For $n_1$, enter **65.**

***Step 7*** For $\bar{x}_2$, enter **98.105**.

***Step 8*** For $Sx_2$, enter **0.699**.

***Step 9*** For $n_2$, enter **65** (Figure 10.17).

***Step 10*** For $\mu_1$, choose the form of $H_a$. For Example 10.6, choose "$\neq \mu_2$" and press **ENTER**.

***Step 11*** For **Pooled**, press **No** (Figure 10.18).

***Step 12*** Press **Calculate**.

**Two-Sample *t* interval for $\mu_1 - \mu_2$**

Follow the same steps as for the two-sample *t* test, except select **0: 2-SampTInt**. Also, to select confidence level (**C-Level**), enter **0.95** for 95%, for example.

Figure 10.16

Figure 10.17

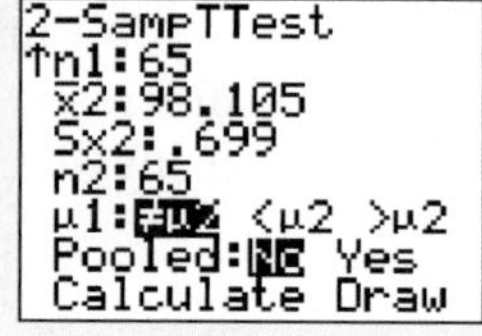


Figure 10.18

## EXCEL

### Two-Sample *t* Test

***Step 1*** Enter Sample 1 and Sample 2 data into columns A and B, respectively.

***Step 2*** Select **Data > Data Analysis > *t*-Test: Two-Sample Assuming Unequal Variances**, and click **OK**.

***Step 3*** For the **Dataset Range**, select the cells in column A for the **Variable 1 range** and the cells in column B for the **Variable 2 range**.

For the hypothesized mean difference, enter **0**, enter your value for **Alpha**, and click **OK**.

## MINITAB

### Two-Sample *t* Test and Confidence Interval

***Step 1*** Enter Sample 1 and Sample 2 data into columns C1 and C2, respectively.

***Step 2*** Click **Stat > Basic Statistics > 2-Sample t.**

***Step 3*** **a.** If you have the data values, select **Samples in different columns**, and select **C1** and **C2** as your two columns.

**b.** If you have the summary statistics, select **summarized data** and enter the **sample size, mean**, and **standard deviation** for each of the **first** and **second** samples.

***Step 4*** Click **Options** and select the form of the **Alternative** hypothesis.

***Step 5*** Click **OK** and click **OK** again.

## Section 10.2 SUMMARY

Section 10.2 examines inferential methods for $\mu_1 - \mu_2$, the difference between the means of two independent populations. The section begins with a discussion of the sampling distribution of $\bar{x}_1 - \bar{x}_2$, which underlies the inference in the remainder of the section.

$100(1 - \alpha)\%$ $t$ confidence intervals for $\mu_1 - \mu_2$ are developed and illustrated.

Two-sample $t$ tests are discussed. These hypothesis tests may be carried out using either the $p$-value method or the critical value method.

The use of confidence intervals for $\mu_1 - \mu_2$ to assess significance is illustrated.

## Section 10.2 EXERCISES

### CLARIFYING THE CONCEPTS

**1.** What are the conditions (cases) that permit us to perform the two-sample $t$ inference we learned about in this section? (Recall that $t$ inference refers to the $t$ test and the $t$ confidence interval).

**2.** Why do we not perform a $Z$ test?

**3.** What is the meaning of the standard error for the difference in two means?

**4.** If a $100(1 - \alpha)\%$ confidence interval for $\mu_1 - \mu_2$ contains 0, then with $100(1 - \alpha)\%$ confidence what is our conclusion regarding the hypothesis that there is no difference in the population means?

### PRACTICING THE TECHNIQUES

The summary statistics in Exercises 5–10 were taken from random samples that were drawn independently. Answer the following questions for each exercise.

**a.** We are interested in constructing a 95% confidence interval for $\mu_1 - \mu_2$. Explain why it is appropriate to do so.

**b.** Provide the point estimate of the difference in population means $\mu_1 - \mu_2$.

**c.** Calculate the standard error $s_{\bar{x}_1 - \bar{x}_2}$.

**d.** Calculate the margin of error for a confidence level of 95%.

**e.** Construct and interpret a 95% confidence interval for $\mu_1 - \mu_2$.

**5.**

| | | | |
|---|---|---|---|
| **Sample 1** | $n_1 = 36$ | $\bar{x}_1 = 10$ | $s_1 = 2$ |
| **Sample 2** | $n_2 = 36$ | $\bar{x}_2 = 8$ | $s_2 = 2$ |

**6.**

| | | | |
|---|---|---|---|
| **Sample 1** | $n_1 = 60$ | $\bar{x}_1 = 100$ | $s_1 = 20$ |
| **Sample 2** | $n_2 = 40$ | $\bar{x}_2 = 90$ | $s_2 = 10$ |

**7.**

| | | | |
|---|---|---|---|
| **Sample 1** | $n_1 = 30$ | $\bar{x}_1 = -10$ | $s_1 = 5$ |
| **Sample 2** | $n_2 = 30$ | $\bar{x}_2 = -5$ | $s_2 = 2$ |

**8.**

| | | | |
|---|---|---|---|
| **Sample 1** | $n_1 = 100$ | $\bar{x}_1 = 50$ | $s_1 = 10$ |
| **Sample 2** | $n_2 = 100$ | $\bar{x}_2 = 75$ | $s_2 = 15$ |

**9.**

| | | | |
|---|---|---|---|
| **Sample 1** | $n_1 = 64$ | $\bar{x}_1 = 0$ | $s_1 = 3$ |
| **Sample 2** | $n_2 = 49$ | $\bar{x}_2 = 1$ | $s_2 = 1$ |

**10.**

| | | | |
|---|---|---|---|
| **Sample 1** | $n_1 = 255$ | $\bar{x}_1 = 103$ | $s_1 = 17$ |
| **Sample 2** | $n_2 = 400$ | $\bar{x}_2 = 95$ | $s_2 = 11$ |

For Exercises 11–16, perform the indicated hypothesis test using the critical value method. Answer the following questions for each exercise.

**a.** State the hypotheses.

**b.** Find the critical value $t_{\text{crit}}$ and the rejection rule for this test.

**c.** Calculate $t_{\text{data}}$. What does this number mean?

**d.** Compare $t_{\text{data}}$ with $t_{\text{crit}}$. State and interpret your conclusion.

**11.** Using the samples from Exercise 5, test at level of significance $\alpha = 0.10$ whether $\mu_1 \neq \mu_2$.

**12.** Using the samples from Exercise 6, test at level of significance $\alpha = 0.05$ whether $\mu_1 < \mu_2$.

**13.** Using the samples from Exercise 7, test at level of significance $\alpha = 0.01$ whether $\mu_1 \neq \mu_2$.

**14.** Using the samples from Exercise 8, test at level of significance $\alpha = 0.10$ whether $\mu_1 < \mu_2$.

**15.** Using the samples from Exercise 9, test at level of significance $\alpha = 0.01$ whether $\mu_1 > \mu_2$.

**16.** Using the samples from Exercise 10, test at level of significance $\alpha = 0.05$ whether $\mu_1 > \mu_2$.

### APPLYING THE CONCEPTS

**17. PC Sales.** A personal computer company launched an advertising campaign in the hopes of boosting sales. A random sample of 16 days before the advertising blitz showed mean sales of 120 computers per day with a standard deviation of 30. A random sample of 15 days after the advertisements appeared showed mean sales of 125 computers per day with a standard deviation of 35. If it is appropriate to perform two-sample $t$ inference, indicate which case applies. If not, explain why not.

**18. Foreclosures.** A random sample of 20 counties in 2007 had a mean number of foreclosures on single-family residences of 50 and a standard deviation of 25. A random sample of 25 counties in 2008 had a mean number of foreclosures of 70 and a standard deviation of 35. Assume that the number of foreclosures per county is normally distributed in both 2007 and 2008. If it is appropriate to perform two-sample $t$ inference, indicate which case applies. If not, explain why not.

**19. Income in California and Los Angeles.** According to the Bureau of Economic Analysis, the mean income for Sacramento County and Los Angeles County, California, in 2004 was \$31,987 and \$33,179, respectively. Suppose random samples from each county are taken, with the following sample statistics.

| | | | |
|---|---|---|---|
| **Sacramento County** | $n_1 = 36$ | $\bar{x}_1 = \$31{,}987$ | $s_1 = \$5000$ |
| **Los Angeles County** | $n_2 = 49$ | $\bar{x}_2 = \$33{,}179$ | $s_2 = \$6000$ |

a. Explain why it is appropriate to apply $t$ inference.
b. Provide the point estimate of the difference in population means $\mu_1 - \mu_2$.
c. Calculate the standard error $s_{\bar{x}_1 - \bar{x}_2}$.
d. Calculate the margin of error for a confidence level of 95%.
e. Construct and interpret a 95% confidence interval for $\mu_1 - \mu_2$.

**20. Math Scores.** The Institute of Educational Sciences published the results of the Trends in International Math and Science Study for 2003. The mean mathematics scores for students from the United States and Hong Kong were 518 and 575, respectively. Suppose independent random samples are drawn from each population, and assume that the populations are normally distributed with the following summary statistics.

| | | | |
|---|---|---|---|
| **USA** | $n_1 = 10$ | $\bar{x}_1 = 518$ | $s_1 = 80$ |
| **Hong Kong** | $n_2 = 12$ | $\bar{x}_2 = 575$ | $s_2 = 70$ |

a. Explain why it is appropriate to apply $t$ inference.
b. Provide the point estimate of the difference in population means $\mu_1 - \mu_2$.
c. Calculate the standard error $s_{\bar{x}_1 - \bar{x}_2}$.
d. Calculate the margin of error for a confidence level of 90%.
e. Construct and interpret a 90% confidence interval for $\mu_1 - \mu_2$.

**21. Income in Sacramento and Los Angeles.** Refer to Exercise 19.

a. Find the critical value $t_{\text{crit}}$ for $\alpha = 0.05$.
b. Calculate $t_{\text{data}}$. What does this number mean?
c. Test at level of significance $\alpha = 0.05$ whether $\mu_1$ differs from $\mu_2$.
d. Explain whether the confidence interval in Exercise 19 could have been used to perform the hypothesis test in (**c**). Why or why not?

**22. Math Scores.** Refer to Exercise 20.

a. Find the critical value $t_{\text{crit}}$ for $\alpha = 0.01$.
b. Calculate $t_{\text{data}}$.
c. Test at level of significance $\alpha = 0.01$ whether $\mu_1$ differs from $\mu_2$.
d. Provide two reasons why the confidence interval in Exercise 20 could not have been used to perform the hypothesis test in (**c**).

**23. Children per Classroom.** According to **www.localschooldirectory.com**, the number of children per teacher in the towns of Cupertino, California, and Santa Rosa, California, are 20.9 and 19.3, respectively. Suppose random samples of classrooms are taken from each county, with the following sample statistics.

| | | | |
|---|---|---|---|
| **Cupertino** | $n_1 = 36$ | $\bar{x}_1 = 20.9$ | $s_1 = 5$ |
| **Santa Rosa** | $n_2 = 64$ | $\bar{x}_2 = 19.3$ | $s_2 = 4$ |

a. Provide the point estimate of the difference in population means $\mu_1 - \mu_2$.
b. Calculate the standard error $s_{\bar{x}_1 - \bar{x}_2}$.
c. Calculate the margin of error for a confidence level of 99%.
d. Construct and interpret a 99% confidence interval for $\mu_1 - \mu_2$.

**24. SAT Verbal Scores.** Suppose that a random sample (sample 1) of 50 males showed a mean verbal SAT score of 517 with a standard deviation of 50 and a random sample (sample 2) of 40 females yielded a mean verbal SAT score of 507 with a standard deviation of 60.

a. Explain why it is appropriate to apply $t$ inference.
b. Provide the point estimate of the difference in population mean SAT scores.
c. Construct and interpret a 95% confidence interval for the difference in population means.

**25. Children per Classroom.** Refer to Exercise 23.

a. Find the critical value $t_{\text{crit}}$ for level of significance $\alpha = 0.01$.
b. Calculate $t_{\text{data}}$. What does this number mean?
c. Test at level of significance $\alpha = 0.01$ whether $\mu_1$ differs from $\mu_2$.

**26. SAT Verbal Scores.** Refer to Exercise 24. Will a test at level of significance $\alpha = 0.05$ of whether the population mean male SAT score differs from the population mean female SAT score be found to be significant or not? Explain your reasoning.

**27. Property Taxes.** Suppose you want to move to either a small town in Ohio or a small town in North Carolina. You did some research on property taxes in each state and chose two random samples shown in the table. The data represent the property taxes in dollars for a residence assessed at \$250,000. Is the difference between mean property taxes significant in the two states? To find out, perform a hypothesis test for the difference in mean property taxes using $\alpha = 0.05$.

| North Carolina | Ohio |
|---|---|
| 206 | 270 |
| 129 | 315 |
| 176 | 177 |
| 120 | 245 |
| 154 | 180 |
| 123 | 292 |
| 164 | 291 |
| 147 | 298 |
| 207 | 270 |
| 138 | 165 |
| 143 | 400 |
| 201 | 268 |
| | 289 |
| | 285 |
| | 225 |

**28. Salaries for Grads.** The National Association of Colleges (NAC) reported in 2003 that the mean starting salary for college graduates majoring in management information systems was $40,915 and for psychology majors was $27,454. Suppose the NAC data is based on surveys of size 144 for each major, with a standard deviation of $10,000 for the management information systems majors and $7000 for the psychology majors.

**a.** Explain why it is appropriate to apply *t* inference.

**b.** Construct and interpret a 95% confidence interval for $\mu_1 - \mu_2$.

**c.** Will a 99% confidence interval for $\mu_1 - \mu_2$ be wider or narrower? Explain your reasoning.

**29. Park Usage.** Suppose that planners for the town of Woodlands, Texas, were interested in assessing usage of their parks. Random samples were taken of the number of daily visitors to Windvale Park and Cranebrook Park, with the statistics as reported here.

| | | | |
|---|---|---|---|
| **Windvale Park** | $n_1 = 36$ | $\bar{x}_1 = 110$ | $s_1 = 60$ |
| **Cranebrook Park** | $n_2 = 30$ | $\bar{x}_2 = 150$ | $s_2 = 75$ |

**a.** Construct and interpret a 95% confidence interval for $\mu_1 - \mu_2$.

**b.** Test at $\alpha = 0.05$ whether $\mu_1$ is less than $\mu_2$. Use the *p*-value method.

**c.** Explain whether the confidence interval in **(a)** could have been used to perform the hypothesis test in **(b)**. Why or why not?

**Coaching for the SAT.** Use this information for Exercises 30–32. The College Board reports that a pre-test and post-test study was done to investigate whether coaching had a significant effect on SAT scores. The improvement from pre-test to post-test was 29 points for the coached sample of students, with a standard deviation of 59 points. For the non-coached students, the pre-test to post-test improvement was 21 points with a standard deviation of 52 points.

**30.** Suppose we consider a sample of 100 students from each group. Perform a test at level of significance $\alpha = 0.05$ for whether the population mean coached SAT pre-test post-test improvement is greater than that for the non-coached students. Use the *p*-value method.

**31.** Refer to Exercise 30.

**a.** Find a point estimate of the difference in population means.

**b.** Find a 99% confidence interval for the difference in population means.

**c.** Perform a hypothesis test of whether the population means differ, at level of significance $\alpha = 0.01$, using the *p*-value method.

WHAT IF? **32.** *What if* the sample sizes for each group were some number greater than $n = 100$.

**a.** How would this affect the width of the confidence interval in Exercise 31b? Is this good? Explain.

**b.** Would this change have any effect on our conclusion in the hypothesis test in Exercise 31c? Explain why or why not.

**33. Nutrition Levels.** A social worker is interested in comparing the nutrition levels of children from the inner city and children from the suburbs. She obtains the following independent random samples of daily calorie intake from children in the inner city and the suburbs. Assume that both samples are taken from normal populations.

Inner city (sample 1): 1125, 1019, 1954, 1546, 1418, 980, 1227

Suburbs (sample 2): 1540, 1967, 1886, 1924, 1863, 1756

**a.** Provide the point estimate of the difference in population means $\mu_1 - \mu_2$.

**b.** Construct and interpret a 90% confidence interval for the difference in population mean daily caloric intake $\mu_1 - \mu_2$.

**c.** Test at level of significance $\alpha = 0.10$ whether the population mean daily caloric intake of inner city children is less than that of children from the suburbs.

**d.** Assess the strength of evidence against the null hypothesis.

**34. Teacher Salaries.** The American Federation of Teachers publishes an annual teacher salary report. In 2003, California led the nation in mean teacher salary with $55,693, with Michigan in second place with $54,020. Suppose the data are from random samples of size 100 each, with a standard deviation of $10,000 in California and $9000 in Michigan.

**a.** Construct and interpret a 95% confidence interval for $\mu_1 - \mu_2$.

**b.** Test at level of significance $\alpha = 0.05$ whether the population mean salary in California is greater than in Michigan.

**c.** Assess the strength of evidence against the null hypothesis.

**35. Working Hours.** Are people working fewer hours? The Bureau of Labor Statistics reported that in 2003 the mean hours worked per week was 40.4, as compared to 41.7 in 1994. Suppose that data are from random samples of size 121 each and that the standard deviation in each sample is 5.5 hours.

**a.** Construct and interpret a 99% confidence interval for $\mu_1 - \mu_2$.

**b.** Test at level of significance $\alpha = 0.01$ whether the population mean number of hours worked in 2003 is less than in 1994.

**c.** Write a sentence summarizing the evidence regarding whether people are working fewer hours.

**36. Nursing Support Services.** A statistical study found that, when nurses made home visits to pregnant teenagers to provide support services, discourage smoking, and otherwise provide care, the sample mean birth weight of the babies was higher for this treatment group (3285 grams) than for the control group (2922 grams), when the visits began before mid-gestation.[8] There were 21 patients in the treatment group and 11 in the control group. Suppose the birth weights for both groups follow a normal distribution. Assume that the standard deviation in each sample is 500 grams.

**a.** Construct and interpret a 95% confidence interval for $\mu_1 - \mu_2$.

**b.** Test at level of significance $\alpha = 0.05$ whether the population birth weight differs between the two groups.

**c.** Assess the strength of evidence against the null hypothesis.

WHAT IF? **37. Working Hours.** Refer to Exercise 35. *What if* the 2003 sample mean is increased by a certain amount so that the 2003 sample mean is still smaller than the 1994 sample mean. Explain how this would affect each of the following.

**a.** $\bar{x}_1 - \bar{x}_2$

**b.** $t_{data}$

**c.** *p*-value

**d.** Conclusion

WHAT IF? **38. Nursing Support Services.** Refer to Exercise 36. *What if* the birth weights of the babies in each group are the same certain amount greater. Explain how this would affect the following.

**a.** $\bar{x}_1 - \bar{x}_2$

**b.** $t_{data}$

**c.** *p*-value

**d.** Conclusion

**39. California and New York.** Use computer software to solve the following problems.

**a.** Open the **Calny** data set. Explore the variable *calif,* which lists the number of residents in the towns and cities in California (minimum of 1000 residents). Generate numerical summary statistics for the number of residents. What is the sample mean number of residents? The sample standard deviation?

**b.** Explore the variable *newyork,* which lists the number of residents in the towns and cities in New York (minimum of 1000 residents). Generate numerical summary statistics for the number of residents. What is the sample mean number of residents? The sample standard deviation?

**c.** A researcher claims that there is no difference in the mean number of residents in the towns and cities in California and New York. Test using independent samples at level of significance $\alpha = 0.05$. Make sure you provide the hypotheses, the rejection rule, the test statistic, the *p*-value, the conclusion, and the interpretation.

**40. Phosphorus and Potassium.** Use computer software to solve the following problems.

**a.** Open the **Nutrition** data set. Explore the variable *phosphor,* which lists the amount of phosphorus (in milligrams) for each food item. Generate numerical summary statistics and graphs for the amount of phosphorus in the food. What is the sample mean amount of phosphorus? The sample standard deviation?

**b.** Explore the variable *potass,* which lists the amount of potassium (in milligrams) for each food item. Generate numerical summary statistics and graphs for the amount of potassium in the food. What is the sample mean amount of potassium? The sample standard deviation?

**c.** A researcher claims that there is no difference in the mean amounts of phosphorus and potassium in food. Test using independent samples at $\alpha = 0.05$. Use the *p*-value method.

**d.** Is the independent sampling method the most appropriate way to test this hypothesis? Why or why not?

**e.** Create a new variable in Excel or Minitab, **phos_pot**, which equals the amount of phosphorus minus the amount of potassium in each food item. Use a paired sample hypothesis test to test at level of significance $\alpha = 0.05$ whether the population mean difference differs from 0.

## 10.3 Inference for Two Independent Proportions

**Objectives:** By the end of this section, I will be able to…

1 Understand the sampling distribution of $\hat{p}_1 - \hat{p}_2$.

2 Compute and interpret confidence intervals for $p_1 - p_2$.

3 Perform and interpret hypothesis tests for $p_1 - p_2$.

So far in this chapter, we have learned how to perform inference about population *means*. In this section, we learn how to perform confidence intervals and hypothesis tests about the difference between two population *proportions*.

## 1 Sampling Distribution of $\hat{p}_1 - \hat{p}_2$

Recall that the sample proportion of success $\hat{p} = x/n$ is the ratio of the number of successes $x$ to the number of trials $n$ in a binomial experiment. Here we consider two independent samples, each of which yields a sample proportion: $\hat{p}_1 = x_1/n_1$ and $\hat{p}_2 = x_2/n_2$. We are interested in performing inference for the difference in population proportions $p_1 - p_2$. In earlier sections, we used the difference in the sample means $\bar{x}_1 - \bar{x}_2$ as a point estimate of the difference in the population means $\mu_1 - \mu_2$. Here we call on the difference in sample proportions $\hat{p}_1 - \hat{p}_2$ as our point estimate of the difference in population proportions $p_1 - p_2$. And just as in earlier sections where we investigated the sampling distributions of $\bar{x}_1 - \bar{x}_2$ to perform inference on $\mu_1 - \mu_2$, here we use the sampling distribution of $\hat{p}_1 - \hat{p}_2$ to help us perform inference about $p_1 - p_2$.

You can imagine how $\hat{p}_1 - \hat{p}_2$ has a sampling distribution by considering the following. Suppose you find the proportion $\hat{p}_1 = 0.30$ of the women in your dormitory who come from out of state; then you find the proportion $\hat{p}_2 = 0.40$ of the men in your dormitory who come from out of state. It is reasonable to assume that the two samples are independent since the subjects in the first sample do not determine the subjects in the second sample. Then you calculate $\hat{p}_1 - \hat{p}_2 = -0.10$. This is one possible value of the statistic $\hat{p}_1 - \hat{p}_2$. In your friend's dormitory, the proportion of women $\hat{p}_1$ and men $\hat{p}_2$ who come from out of state are $\hat{p}_1 = 0.35$ and $\hat{p}_2 = 0.20$. Then $\hat{p}_1 - \hat{p}_2 = 0.15$ is another possible value of the statistic $\hat{p}_1 - \hat{p}_2$. You can imagine doing this for every dormitory on campus and then for every dormitory in your state. The distribution of all possible values of $\hat{p}_1 - \hat{p}_2$ is called the **sampling distribution of $\hat{p}_1 - \hat{p}_2$**.

*Developing Your Statistical Sense*

### Independent Samples Only

The inferential methods of this section are reserved for *independent* samples only. An example of a problem that would not use the methods of this section is the following. In the latest poll, suppose 45% supported the Democrat and 40% supported the Republican. Because each respondent had to choose between the Democratic candidate and the Republican candidate, their respective poll numbers are *not independent*. ■

Just as the mean of the sampling distribution of $\bar{x}_1 - \bar{x}_2$ was $\mu_1 - \mu_2$, here the mean of the sampling distribution of $\hat{p}_1 - \hat{p}_2$ is $p_1 - p_2$. The standard deviation of the sampling distribution of $\hat{p}_1 - \hat{p}_2$ is $\sigma_{\hat{p}_1-\hat{p}_2} = \sqrt{\dfrac{p_1(1-p_1)}{n_1} + \dfrac{p_2(1-p_2)}{n_2}}$.

Let $x_1$ and $x_2$ denote the number of successes in sample 1 and sample 2, respectively. And let $n_1 - x_1$ and $n_2 - x_2$ denote the number of failures in sample 1 and sample 2, respectively. The sampling distribution of $\hat{p}_1 - \hat{p}_2$ is approximately normal when the number of successes and the number of failures in each sample are each at least five—that is, when $x_1 \geq 5$, $(n_1 - x_1) \geq 5$, $x_2 \geq 5$, and $(n_2 - x_2) \geq 5$. Let $q_1 = 1 - p_1$, $q_2 = 1 - p_2$, $\hat{q}_1 = 1 - \hat{p}_1$ and $\hat{q}_2 = 1 - \hat{p}_2$. Since the population proportions are unknown, we cannot directly calculate $\sigma_{\hat{p}_1-\hat{p}_2}$. To estimate $\sigma_{\hat{p}_1-\hat{p}_2}$, we take a sample from each population and substitute the respective sample proportions $\hat{p}_1$ and $\hat{p}_2$ for $p_1$ and $p_2$, providing the **standard error**. Putting it all together, we have the properties of the sampling distribution of $\hat{p}_1 - \hat{p}_2$, as follows.

**Sampling Distribution of $\hat{p}_1 - \hat{p}_2$**

When two random samples are drawn independently from two populations, then the quantity

$$Z = \frac{(\hat{p}_1 - \hat{p}_2) - (p_1 - p_2)}{\sigma_{\hat{p}_1 - \hat{p}_2}} = \frac{(\hat{p}_1 - \hat{p}_2) - (p_1 - p_2)}{\sqrt{\frac{p_1 q_1}{n_1} + \frac{p_2 q_2}{n_2}}}$$

has an approximately standard normal distribution when the following conditions are satisfied:

$$x_1 \geq 5, \qquad (n_1 - x_1) \geq 5, \qquad x_2 \geq 5, \qquad (n_2 - x_2) \geq 5$$

and where $\hat{p}_1$ and $n_1$ represent the sample proportion and sample size of the sample taken from population 1 with population proportion $p_1$; $\hat{p}_2$ and $n_2$ represent the sample proportion and sample size of the sample taken from population 2 with population proportion $p_2$; and $q_1 = 1 - p_1$ and $q_2 = 1 - p_2$.

**Standard Error of $\hat{p}_1 - \hat{p}_2$**

The **standard error $s_{\hat{p}_1 - \hat{p}_2}$ of the statistic $\hat{p}_1 - \hat{p}_2$** is

$$s_{\hat{p}_1 - \hat{p}_2} = \sqrt{\frac{\hat{p}_1 \cdot \hat{q}_1}{n_1} + \frac{\hat{p}_2 \cdot \hat{q}_2}{n_2}}$$

where $\hat{q}_1 = 1 - \hat{p}_1$ and $\hat{q}_2 = 1 - \hat{p}_2$. The standard error $s_{\hat{p}_1 - \hat{p}_2}$ measures the size of the typical error in using $\hat{p}_1 - \hat{p}_2$ to estimate $p_1 - p_2$.

## 2 Independent Sample $Z$-Interval for $p_1 - p_2$

We can use sample statistics to estimate $p_1 - p_2$ using a confidence interval.

**Confidence Interval for $p_1 - p_2$**

For two independent random samples taken from two populations with population proportions $p_1$ and $p_2$, a $100(1 - \alpha)\%$ **confidence interval for $p_1 - p_2$** is given by

$$\hat{p}_1 - \hat{p}_2 \pm Z_{\alpha/2} \sqrt{\frac{\hat{p}_1 \cdot \hat{q}_1}{n_1} + \frac{\hat{p}_2 \cdot \hat{q}_2}{n_2}}$$

where $\hat{p}_1$ and $n_1$ represent the sample proportion and sample size of the sample taken from population 1 with population proportion $p_1$; $\hat{p}_2$ and $n_2$ represent the sample proportion and sample size of the sample taken from population 2 with population proportion $p_2$; the samples are drawn independently; and the following conditions are satisfied: $x_1 \geq 5$, $(n_1 - x_1) \geq 5$, $x_2 \geq 5$, and $(n_2 - x_2) \geq 5$.

**Margin of Error $E$**

The **margin of error** for a $100(1 - \alpha)\%$ confidence interval for $p_1 - p_2$ is given by

$$E = Z_{\alpha/2} \cdot (\text{standard error}) = Z_{\alpha/2} \cdot (s_{\hat{p}_1 - \hat{p}_2}) = Z_{\alpha/2} \cdot \sqrt{\frac{\hat{p}_1 \cdot \hat{q}_1}{n_1} + \frac{\hat{p}_2 \cdot \hat{q}_2}{n_2}}$$

## Example 10.11 Teen privacy in online social networks: Confidence interval

Punchstock/Banana Stock

The Pew Internet and American Life Project (**www.pewinternet.org**) tracks the behavior of Americans on the Internet. In 2007, they published a report that described some of the behaviors of American teenagers in online social networks, such as Facebook.[9] Teenagers who had online profiles were asked: "We'd like to know if your last name is posted to your profile or not." The results are shown in Table 10.18. Assume that the samples are independent.

Table 10.18 Proportions of teenage boys and girls who post their last names in online profiles

| | Boys | Girls |
|---|---|---|
| **Number responding "Yes"** | $x_1 = 195$ | $x_2 = 93$ |
| **Sample size** | $n_1 = 487$ | $n_2 = 487$ |
| **Sample proportion** | $\hat{p}_1 = x_1/n_1 = 195/487 \approx 0.4004$ | $\hat{p}_2 = x_2/n_2 = 93/487 \approx 0.1910$ |

**a.** Find the point estimate $\hat{p}_1 - \hat{p}_2$ for the difference in population proportions $p_1 - p_2$.

**b.** Calculate the standard error

$$s_{\hat{p}_1-\hat{p}_2} = \sqrt{\frac{\hat{p}_1 \cdot \hat{q}_1}{n_1} + \frac{\hat{p}_2 \cdot \hat{q}_2}{n_2}}$$

**c.** For a 95% confidence level, calculate the margin of error.

$$E = Z_{\alpha/2} \cdot \sqrt{\frac{\hat{p}_1 \cdot \hat{q}_1}{n_1} + \frac{\hat{p}_2 \cdot \hat{q}_2}{n_2}}$$

**d.** Construct and interpret a 95% confidence interval for the difference in population proportions of girls and boys whose last name is posted to their online profile.

### Solution

We first check whether the conditions for the $Z$ interval are valid. We have $x_1 = 195 \geq 5$, $(n_1 - x_1) = (487 - 195) = 292 \geq 5$, $x_2 = 93 \geq 5$, and $(n_2 - x_2) = (487 - 93) = 394 \geq 5$. We can therefore proceed.

**a.** The point estimate of the difference in population proportions $p_1 - p_2$ is

$$\hat{p}_1 - \hat{p}_2 \approx 0.4004 - 0.1910 = 0.2094$$

**b.** $\hat{q}_1 = 1 - \hat{p}_1 = 1 - 0.4004 = 0.5996$ $\qquad \hat{q}_2 = 1 - \hat{p}_2 = 1 - 0.1910 = 0.8090$

So the standard error is

$$s_{\hat{p}_1-\hat{p}_2} = \sqrt{\frac{\hat{p}_1 \cdot \hat{q}_1}{n_1} + \frac{\hat{p}_2 \cdot \hat{q}_2}{n_2}} = \sqrt{\frac{(0.4004)(0.5996)}{487} + \frac{(0.1910)(0.8090)}{487}} \approx 0.0285$$

**c.** The $Z_{\alpha/2}$ value for a 95% confidence level is 1.96. Therefore, the margin of error is

$$E = Z_{\alpha/2} \cdot s_{\hat{p}_1-\hat{p}_2} = 1.96(0.0285) \approx 0.0559$$

**d.** The 95% confidence interval is therefore

$$
\begin{aligned}
(\hat{p}_1 - \hat{p}_2) \pm E &= (0.4004 - 0.1910) \pm 0.0559 \\
&= 0.2094 \pm 0.0559 \\
&= (0.1535, 0.2653)
\end{aligned}
$$

We are 95% confident that the difference in population proportions of teenage boys and girls who put their last names on their online profiles lies between 0.1535 and 0.2653. ■

What Do These Numbers Mean?

Consider taking samples over and over again and observing the long-term behavior of the sample statistics. The standard error of $s_{\hat{p}_1 - \hat{p}_2} = 0.0285$ means that the typical error in estimating the unknown quantity $p_1 - p_2$ is 0.0285. The margin of error $E = 0.0559$ means that the point estimate $(\hat{p}_1 - \hat{p}_2)$ will lie within $E = 0.0559$ of the difference in population proportions $p_1 - p_2$ 95% of the time. ■

## 3 Independent Sample $Z$ Test for $p_1 - p_2$

Researchers may also be interested in testing hypotheses regarding the difference in population proportions $p_1 - p_2$. The three possible forms for such a hypothesis test are given in Table 10.19.

Table 10.19 Three possible forms for the hypotheses for a test about $p_1 - p_2$

| Form | Null hypothesis | Alternative hypothesis | Type of test |
|---|---|---|---|
| 1 | $H_0: p_1 \le p_2$ | $H_a: p_1 > p_2$ | Right-tailed test (one-tailed test) |
| 2 | $H_0: p_1 \ge p_2$ | $H_a: p_1 < p_2$ | Left-tailed test (one-tailed test) |
| 3 | $H_0: p_1 = p_2$ | $H_a: p_1 \ne p_2$ | Two-tailed test |

In each form of the hypotheses, the null hypothesis includes the assertion that $p_1 = p_2$. We denote this *common population proportion* as $p$. Since the null hypothesis is assumed true, the test statistic takes the following form.

$$
\begin{aligned}
Z_{\text{data}} &= \frac{(\hat{p}_1 - \hat{p}_2) - (p_1 - p_2)}{\sqrt{\dfrac{p_1(1-p_1)}{n_1} + \dfrac{p_2(1-p_2)}{n_2}}} = \frac{(\hat{p}_1 - \hat{p}_2) - 0}{\sqrt{\dfrac{p_1(1-p_1)}{n_1} + \dfrac{p_2(1-p_2)}{n_2}}} \\
&= \frac{(\hat{p}_1 - \hat{p}_2)}{\sqrt{\dfrac{p(1-p)}{n_1} + \dfrac{p(1-p)}{n_2}}} = \frac{(\hat{p}_1 - \hat{p}_2)}{\sqrt{p(1-p)\left(\dfrac{1}{n_1} + \dfrac{1}{n_2}\right)}}
\end{aligned}
$$

Since the common population proportion $p$ is unknown, we estimate it using the following **pooled estimate of $p$**.

$$
\hat{p}_{\text{pooled}} = \frac{x_1 + x_2}{n_1 + n_2}
$$

Substituting this into the formula for the test statistic gives

$$Z_{\text{data}} = \frac{(\hat{p}_1 - \hat{p}_2)}{\sqrt{\hat{p}_{\text{pooled}} \cdot (1 - \hat{p}_{\text{pooled}})\left(\frac{1}{n_1} + \frac{1}{n_2}\right)}}$$

$Z_{\text{data}}$ measures the distance between the sample proportions. Extreme values of $Z_{\text{data}}$ indicate evidence against the null hypothesis.

## Example 10.12 Finding $\hat{p}_{\text{pooled}}$ and $Z_{\text{data}}$

Use the data from Table 10.18 to find the values of $\hat{p}_{\text{pooled}}$ and $Z_{\text{data}}$.

**Solution**

$$\hat{p}_{\text{pooled}} = \frac{x_1 + x_2}{n_1 + n_2} = \frac{195 + 93}{487 + 487} \approx 0.2957$$

As a check on your calculations, $\hat{p}_{\text{pooled}}$ must always lie between $\hat{p}_1$ and $\hat{p}_2$. Here $\hat{p}_{\text{pooled}} = 0.2957$ lies between $\hat{p}_1 = 0.4004$ and $\hat{p}_2 = 0.1910$.

$$Z_{\text{data}} = \frac{(0.4004 - 0.1910)}{\sqrt{0.2957 \cdot (1 - 0.2957)\left(\frac{1}{487} + \frac{1}{487}\right)}} \approx \frac{0.2094}{0.0292452288} \approx 7.16$$

$Z_{\text{data}} = 7.16$ is extreme, if we recall from Chapter 6 that nearly all values of $Z$ are values between $-3$ and 3. ■

**Hypothesis Test for the Difference in Two Population Proportions: *p*-Value Method**

Suppose we have two independent random samples taken from two populations with population proportions $p_1$ and $p_2$, and the required conditions are met: $x_1 \geq 5$, $(n_1 - x_1) \geq 5$, $x_2 \geq 5$, and $(n_2 - x_2) \geq 5$.

***Step 1*** **State the hypotheses and the rejection rule.** Use one of the forms from Table 10.19. Clearly state the meaning of $p_1$ and $p_2$. The rejection rule is *Reject $H_0$ if the p-value is less than $\alpha$.*

***Step 2*** **Find $Z_{\text{data}}$.**

$$Z_{\text{data}} = \frac{(\hat{p}_1 - \hat{p}_2)}{\sqrt{\hat{p}_{\text{pooled}} \cdot (1 - \hat{p}_{\text{pooled}})\left(\frac{1}{n_1} + \frac{1}{n_2}\right)}}$$

where $\hat{p}_{\text{pooled}} = \frac{x_1 + x_2}{n_1 + n_2}$. $Z_{\text{data}}$ follows an approximately standard normal distribution if the required conditions are satisfied.

***Step 3*** **Find the *p*-value.** If you have access to technology, use it to find the *p*-value. Otherwise, calculate the *p*-value using one of the forms in Table 10.20.

**Step 4.** State the conclusion and interpretation.

Table 10.20 *p*-Value

| Form of test | The *p*-value equals... | Graphic |
|---|---|---|
| Right-tailed test<br>$H_0: p_1 \le p_2$<br>$H_a: p_1 > p_2$ | $p\text{-value} = P(Z > Z_{\text{data}})$<br>Area to right of $Z_{\text{data}}$ |  |
| Left-tailed test<br>$H_0: p_1 \ge p_2$<br>$H_a: p_1 < p_2$ | $p\text{-value} = P(Z < Z_{\text{data}})$<br>Area to left of $Z_{\text{data}}$ |  |
| Two-tailed test<br>$H_0: p_1 = p_2$<br>$H_a: p_1 \ne p_2$ | $p\text{-value} = P(Z > \lvert Z_{\text{data}}\rvert) + P(Z < -\lvert Z_{\text{data}}\rvert)$<br>$= 2 \cdot P(Z > \lvert Z_{\text{data}}\rvert)$ |  |

## Example 10.13 Very happily married: Hypothesis test using the *p*-value method

© Blend Images/Alamy

The General Social Survey tracks trends in American society through annual surveys. Married respondents were asked to characterize their feelings about being married. The results are shown here in a crosstabulation with gender. Test the hypothesis that the proportion of females who report being very happily married is smaller than the proportion of males who report being very happily married. Use the *p*-value method with level of significance $\alpha = 0.05$.

| | Very happy | Pretty happy/ Not too happy | Total |
|---|---|---|---|
| **Female** | 257 | 166 | 423 |
| **Male** | 242 | 126 | 366 |
| **Total** | 499 | 288 | 789 |

### Solution

From the crosstabulation, we assemble the statistics in Table 10.21 for the independent random samples of men and women.

Table 10.21 Sample statistics of very happily married respondents

| | Sample size | Number very happy | Sample proportion very happy |
|---|---|---|---|
| **Females (sample 1)** | $n_1 = 423$ | $x_1 = 257$ | $\hat{p}_1 = \frac{x_1}{n_1} = \frac{257}{423} \approx 0.6076$ |
| **Males (sample 2)** | $n_2 = 366$ | $x_2 = 242$ | $\hat{p}_2 = \frac{x_2}{n_2} = \frac{242}{366} \approx 0.6612$ |

We first check whether the conditions for the $Z$ test are valid: $x_1 = 257 \geq 5$, $(n_1 - x_1) = (423 - 257) = 166 \geq 5$, $x_2 = 242 \geq 5$, and $(n_2 - x_2) = (366 - 242) = 124 \geq 5$. We can therefore proceed.

***Step 1*** **State the hypotheses and the rejection rule.** Since we are interested in whether the proportion of females who report being very happily married *is smaller than* that of males and because the females represent sample 1, the hypotheses are

$$H_0 : p_1 \geq p_2 \qquad H_a : p_1 < p_2$$

where $p_1$ and $p_2$ represent the population proportions of all females and males, respectively, who report being very happily married. We will reject $H_0$ if the $p$-value is less than $\alpha = 0.05$.

***Step 2*** **Find $Z_{\text{data}}$.** First, use the data from Table 10.21 to find the values of $\hat{p}_{\text{pooled}}$.

$$\hat{p}_{\text{pooled}} = \frac{x_1 + x_2}{n_1 + n_2} = \frac{257 + 242}{423 + 366} \approx 0.63245$$

Then

$$Z_{\text{data}} = \frac{(0.6076 - 0.6612)}{\sqrt{0.63245 \cdot (1 - 0.63245)\left(\frac{1}{423} + \frac{1}{366}\right)}} \approx -1.56$$

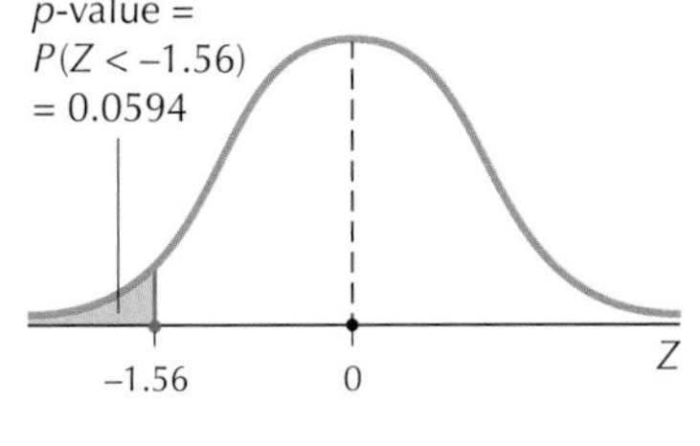


FIGURE 10.19

***Step 3*** **Find the p-value.** Since it is a left-tailed test, the $p$-value is given by Table 10.20 as $P(Z < Z_{\text{data}}) = P(Z < -1.56)$, as shown in Figure 10.19. This amounts to a Case 1 problem from Section 6.4:

$$P(Z < -1.56) = 0.0594$$

***Step 4*** **State and interpret the conclusion.** Since the $p$-value $= 0.0594$ is not less than $\alpha = 0.05$, we do not reject $H_0$. There is insufficient evidence that the proportion of females who report being very happily married is smaller than the proportion of males who do so. ■

*Developing Your Statistical Sense*

### What to Do When the $p$-Value Is Close to $\alpha$

The $p$-value of 0.0594 is close to $\alpha = 0.05$. If we had used level of significance $\alpha = 0.10$ (or $\alpha = 0.06$) instead of $\alpha = 0.05$, then the conclusion would have been to reject $H_0$. In situations like this, when the $p$-value is so close to $\alpha$, many professional data analysts prefer not to report a black-and-white decision. Instead they do one of the following:

1. Simply assess the strength of evidence against the null hypothesis using criteria like those given in Table 9.5 (page 477)
2. Conduct further studies involving larger samples. ■

This type of hypothesis test can also be performed using the critical value method.

**Hypothesis Test for the Difference in Two Population Proportions: Critical Value Method**

Suppose we have two independent random samples taken from two populations with population proportions $p_1$ and $p_2$, and the required conditions are met: $x_1 \geq 5$, $(n_1 - x_1) \geq 5$, $x_2 \geq 5$, and $(n_2 - x_2) \geq 5$.

***Step 1*** **State the hypotheses.** Use one of the forms from Table 10.19. Clearly state the meaning of $p_1$ and $p_2$.

***Step 2*** **Find $Z_{crit}$ and state the rejection rule.** To find $Z_{crit}$, use Table 10.22. To find the rejection rule, use Table 10.23.

Table 10.22 Critical values $Z_{crit}$ for various values of $\alpha$

| Level of significance $\alpha$ | Form of Hypothesis Test: Right-tailed test $H_0: p_1 \leq p_2$, $H_a: p_1 > p_2$ | Left-tailed test $H_0: p_1 \geq p_2$, $H_a: p_1 < p_2$ | Two-tailed test $H_0: p_1 = p_2$, $H_a: p_1 \neq p_2$ |
|---|---|---|---|
| $\alpha = 0.10$ | $Z_{crit} = 1.28$ | $Z_{crit} = -1.28$ | $Z_{crit} = 1.645$ |
| $\alpha = 0.05$ | $Z_{crit} = 1.645$ | $Z_{crit} = -1.645$ | $Z_{crit} = 1.96$ |
| $\alpha = 0.01$ | $Z_{crit} = 2.33$ | $Z_{crit} = -2.33$ | $Z_{crit} = 2.58$ |

Table 10.23 Rejection rules for the $t$ test for $p_1 - p_2$

| Type of hypothesis test | Rejection rules: reject $H_0$ if... |
|---|---|
| Right-tailed test $H_0: p_1 \leq p_2$ versus $H_a: p_1 > p_2$ | $Z_{data} > Z_{crit}$ |
| Left-tailed test $H_0: p_1 \geq p_2$ versus $H_a: p_1 < p_2$ | $Z_{data} < Z_{crit}$ |
| Two-tailed test $H_0: p_1 = p_2$ versus $H_a: p_1 \neq p_2$ | $Z_{data} > Z_{crit}$ or $Z_{data} < -Z_{crit}$ |

***Step 3*** **Find $Z_{data}$.**

$$Z_{data} = \frac{(\hat{p}_1 - \hat{p}_2)}{\sqrt{\hat{p}_{pooled} \cdot (1 - \hat{p}_{pooled})\left(\frac{1}{n_1} + \frac{1}{n_2}\right)}}$$

where

$$\hat{p}_{pooled} = \frac{x_1 + x_2}{n_1 + n_2}$$

$Z_{data}$ follows an approximately standard normal distribution if the required conditions are satisfied.

***Step 4*** **State the conclusion and interpretation.** Compare $Z_{data}$ with $Z_{crit}$.

## Example 10.14 Critical value method using technology

Use the data from Table 10.18 and the TI-83/84 to test whether the population proportion of teenage boys who post their last name in their online profiles is greater than the population proportion of teenage girls who do so. Use the critical value method and level of significance $\alpha = 0.01$.

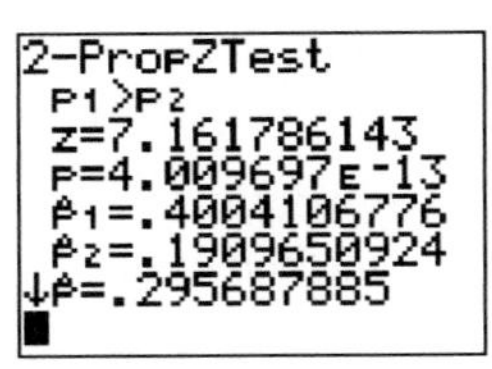


FIGURE 10.20

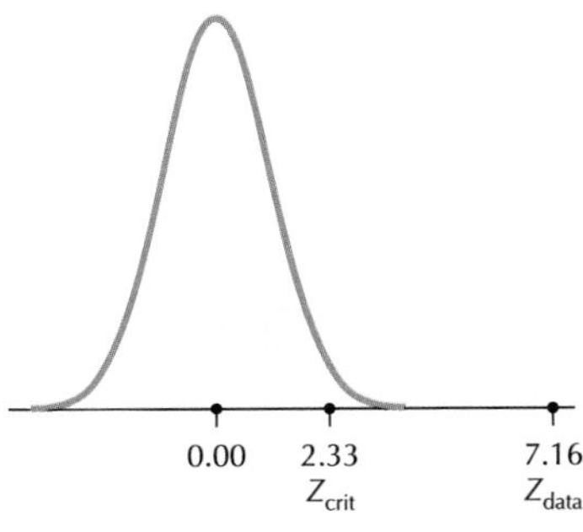


FIGURE 10.21

### Solution

The conditions for performing the hypothesis test are the same as the conditions for the confidence interval, and we checked in Example 10.11 that these conditions were satisfied.

**Step 1** **State the hypotheses.** The key words "greater than" together with the fact that sample 1 represents the boys indicate that we have a right-tailed test:

$$H_0: p_1 \leq p_2 \quad \text{versus} \quad H_a: p_1 > p_2$$

where $p_1$ and $p_2$ represent, respectively, the population proportions of boys and girls who post their last name in their online profiles.

**Step 2** **Find $Z_{crit}$, and state the rejection rule.** Since the test is a right-tailed and $\alpha = 0.01$, Table 10.22 gives us $Z_{crit} = 2.33$. From Table 10.23, we will reject $H_0$ if $Z_{data} > 2.33$.

**Step 3** **Find $Z_{data}$.** From Figure 10.20, we have $Z_{data} \approx 7.16$. This agrees with our calculation from Example 10.12.

**Step 4** **State the conclusion and interpretation.** In Step 2, we said we would reject $H_0$ if $Z_{data} > 2.33$. We have $Z_{data} = 7.16$, which is indeed greater than 2.33, as shown in Figure 10.21. We therefore reject $H_0$. There is evidence that the population proportion of teenage boys who post their last name in their online profiles is greater than the population proportion of teenage girls who do so. ■

## STEP-BY-STEP TECHNOLOGY GUIDE: *Z* Test and *Z* Interval $p_1 - p_2$

(Example 10.14 is used to illustrate the procedure.)

### TI-83/84

#### *Z* Test for $p_1 - p_2$

***Step 1*** Press **STAT** and highlight **TESTS**.
***Step 2*** Select **6** (for the **2-Prop ZTest**).
***Step 3*** For **x1**, enter the number of successes in the first sample, **195**.
***Step 4*** For **n1**, enter the size of the first sample, **487**.
***Step 5*** For **x2**, enter the number of successes in the second sample, **93**.
***Step 6*** For **n2**, enter the size of the second sample, **487**.
***Step 7*** For $p_1$, choose the form of the hypothesis test. For Example 10.14, choose $> p_2$ and press **ENTER** (Figure 10.22).
***Step 8*** Highlight **Calculate** and press **ENTER**. The results are shown in Figure 10.20 in Example 10.14.

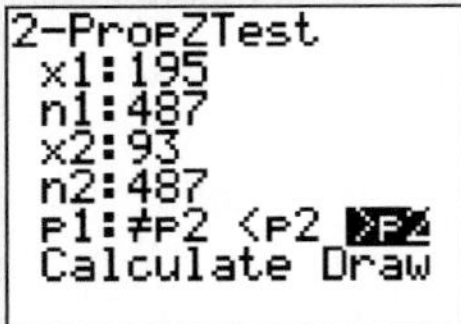


Figure 10.22

#### *Z* Interval for $p_1 - p_2$

Follow the same steps as for the two sample *t* test in Section 10.2, except "Select **B: 2-PropZInt.**" Also, to select confidence level (**C-Level**), enter **0.95** for 95%, for example.

### EXCEL

#### *Z* Test and *Z* Interval for $p_1 - p_2$ Using the WHFStat Macros

***Step 1*** Load the **WHFStat Macros**.
***Step 2*** Select **Add-Ins > Macros > Testing a Proportion > Two Samples**.
***Step 3*** For **Proportion 1**, enter $n_1$ for **Sample Size** and $x_1$ for **Number of Successes**.
***Step 4*** For **Proportion 2**, enter $n_2$ for **Sample Size** and $x_2$ for **Number of Successes**. Select the **Confidence Level** and click **OK**.

### MINITAB

#### *Z* Test and *Z* Interval for $p_1 - p_2$

***Step 1*** Click **Stat > Basic Statistics > 2 Proportions**.
***Step 2*** Select **Summarized Data**.
***Step 3*** For the **First** row, enter $n_1$ for **Trials** and $x_1$ for **Events** $n_1$ for **Trials**.
***Step 4*** For the **Second** row, enter $n_2$ for **Trials** and $x_2$ for **Events**.
***Step 5*** Click **Options** and select the form of the alternative hypothesis and a confidence level. Then click **OK** twice.

## Section 10.3 SUMMARY

1 The section discusses inferential methods for $p_1 - p_2$, the difference between the proportions of two independent populations. The section begins with a discussion of the sampling distribution of $\hat{p}_1 - \hat{p}_2$, which underlies the inference in the remainder of the section.

2 $100(1 - \alpha)\%$ $Z$ confidence intervals for $p_1 - p_2$ are developed and illustrated.

3 Two-sample $Z$ tests are discussed. These hypothesis tests may be carried out using either the $p$-value method or the critical value method.

## Section 10.3 EXERCISES

### CLARIFYING THE CONCEPTS

**1.** $\hat{p}_{\text{pooled}}$ must always lie between which two quantities?

**2.** Provide an example of a situation where the methods from this section should not be applied because the sample proportions are not independent.

**3.** Does it make sense to use $\hat{p}_{\text{pooled}}$ when calculating confidence intervals for $p_1 - p_2$? Why or why not?

**4.** Suppose a 95% confidence interval for $p_1 - p_2$ is (0.1, 0.3). Explain what this means, using the idea of long-term behavior in your answer.

**5.** What does $Z_{\text{data}}$ measure? What do extreme values of $Z_{\text{data}}$ indicate?

**6.** What might we suggest if the $p$-value is very close to the $\alpha$?

### PRACTICING THE TECHNIQUES

The summary statistics in Exercises 7–12 were taken from random samples that were drawn independently. Answer questions a–e for each exercise. Let $n_1$ and $n_2$ denote the size of samples 1 and 2, respectively. Let $x_1$ and $x_2$ denote the number of successes in samples 1 and 2, respectively.

- **a.** We are interested in constructing a 95% confidence interval for $p_1 - p_2$. Is it appropriate to do so? Why or why not? If not appropriate then do not perform **(b)–(e)**.
- **b.** Provide the point estimate of the difference in population means $p_1 - p_2$.
- **c.** Calculate the standard error $s_{\hat{p}_1 - \hat{p}_2}$. What does this number mean?
- **d.** Calculate the margin of error for a confidence level of 95%. What does this number mean?
- **e.** Construct and interpret a 95% confidence interval for $p_1 - p_2$.

**7.**

| | | |
|---|---|---|
| **Sample 1** | $n_1 = 100$ | $x_1 = 80$ |
| **Sample 2** | $n_2 = 40$ | $x_2 = 30$ |

**8.**

| | | |
|---|---|---|
| **Sample 1** | $n_1 = 10$ | $x_1 = 4$ |
| **Sample 2** | $n_2 = 12$ | $x_2 = 5$ |

**9.**

| | | |
|---|---|---|
| **Sample 1** | $n_1 = 200$ | $x_1 = 60$ |
| **Sample 2** | $n_2 = 250$ | $x_2 = 40$ |

**10.**

| | | |
|---|---|---|
| **Sample 1** | $n_1 = 400$ | $x_1 = 250$ |
| **Sample 2** | $n_2 = 400$ | $x_2 = 200$ |

**11.**

| | | |
|---|---|---|
| **Sample 1** | $n_1 = 1000$ | $x_1 = 490$ |
| **Sample 2** | $n_2 = 1000$ | $x_2 = 620$ |

**12.**

| | | |
|---|---|---|
| **Sample 1** | $n_1 = 527$ | $x_1 = 412$ |
| **Sample 2** | $n_2 = 613$ | $x_2 = 498$ |

For Exercises 13–16, perform the indicated hypothesis test using the critical value method. Answer questions a–d for each exercise.

- **a.** State the hypotheses and find the critical value $Z_{\text{crit}}$ and the rejection rule.
- **b.** Calculate $\hat{p}_{\text{pooled}}$.
- **c.** Calculate $Z_{\text{data}}$.
- **d.** Compare $Z_{\text{data}}$ with $Z_{\text{crit}}$. State and interpret your conclusion.

**13.** Using the samples in Exercise 7, test at level of significance $\alpha = 0.10$ whether $p_1 \neq p_2$.

**14.** Using the samples in Exercise 7, test at level of significance $\alpha = 0.10$ whether $p_1 < p_2$.

**15.** Using the samples in Exercise 9, test at level of significance $\alpha = 0.05$ whether $p_1 > p_2$.

**16.** Using the samples in Exercise 9, test at level of significance $\alpha = 0.05$ whether $p_1 \neq p_2$.

For Exercises 17–20, perform the indicated hypothesis test using the $p$-value method. Answer questions a–e for each exercise.

- **a.** State the hypotheses and the rejection rule.
- **b.** Calculate $\hat{p}_{\text{pooled}}$.
- **c.** Calculate $Z_{\text{data}}$.
- **d.** Calculate the $p$-value.
- **e.** Compare the $p$-value with $\alpha$. State and interpret your conclusion.

**17.** Using the samples in Exercise 10, test at level of significance $\alpha = 0.10$ whether $p_1 > p_2$.

**18.** Using the samples in Exercise 11, test at level of significance $\alpha = 0.01$ whether $p_1 < p_2$.

**19.** Using the samples in Exercise 12, test at level of significance $\alpha = 0.05$ whether $p_1 \neq p_2$.

**20.** Using the samples in Exercise 12, test at level of significance $\alpha = 0.05$ whether $p_1 < p_2$.

## APPLYING THE CONCEPTS

**21. Online Photos.** A Pew Internet and American Life Project (**www.pewinternet.org**) 2007 report stated that 74% of teenage boys posted their photo on their online profile, while 83% of teenage girls did so.[10] The sample sizes were both 487. We are interested in finding a 95% confidence interval for the difference in the population proportion of teenage boys who post their photo on their online profile and the population proportion of teenage girls who do so.

- **a.** Explain why it is appropriate to construct the confidence interval.
- **b.** Find a point estimate of the difference in population proportions.
- **c.** Find the standard error $s_{\hat{p}_1-\hat{p}_2}$. What does this number mean?
- **d.** Find the margin of error for a confidence level of 95%. What does this number mean?
- **e.** Construct and interpret a 95% confidence interval for the difference in population proportions.

**22. Medicare Recipients.** The Centers for Medicare and Medicaid Services reported in 2004 that 3,305 of the 50,350 Medicare recipients living in Alaska were age 85 or over, and 73,289 of the 754,642 Medicare recipients living in Arizona were age 85 or over.

- **a.** Find a point estimate of the difference in population proportions.
- **b.** Find the standard error $s_{\hat{p}_1-\hat{p}_2}$. What does this number mean?
- **c.** Compare the point estimate $\hat{p}_1 - \hat{p}_2$ with the standard error. Would you say that the difference is likely to be significant?
- **d.** Find the margin of error for a confidence level of 99%. What does this number mean?
- **e.** Construct and interpret a 99% confidence interval for the difference in population proportions.

**23. Online Photos.** Refer to Exercise 21. Use the confidence interval to test at level of significance $\alpha = 0.05$ whether the population proportion of teenage boys who post their photo on their online profile differs from the population proportion of teenage girls who do so.

- **a.** State the hypotheses. Clearly state the meaning of $p_1$ and $p_2$.
- **b.** Indicate whether 0 falls within the confidence interval.
- **c.** State and interpret your conclusion.

**24. Medicare Recipients.** Refer to Exercise 22. Use the confidence interval to test at level of significance $\alpha = 0.01$ whether the population proportion of Alaska Medicare recipients age 85 or over differs from the population proportion of Arizona Medicare recipients age 85 or over.

- **a.** State the hypotheses. Clearly state the meaning of $p_1$ and $p_2$.
- **b.** Indicate whether 0 falls within the confidence interval.
- **c.** State and interpret your conclusion.

**25. Same-Sex Marriage.** Is there a generation gap in the perception of same-sex marriages? In a University of Pennsylvania/National Annenberg Survey of May 2004, 42% of 18–29-year-olds say that they favor "a law that would allow two men to marry each other or two women to marry each other," while 30% of 30–44-year-olds say that they would favor such a law. Assume that the sample sizes were each 225.

- **a.** Find a point estimate of the difference in population proportions.
- **b.** Find the pooled sample proportion.
- **c.** Perform a hypothesis test at level of significance $\alpha = 0.01$ of whether the population proportions differ, using the $p$-value method. Is there evidence of a generation gap in the perception of same-sex marriages?

**26. Women's Ownership of Businesses.** The U.S. Census Bureau tracks trends in women's ownership of businesses. A random sample of 100 Ohio businesses showed 34 that were woman-owned. A sample of 200 New Jersey businesses showed 64 that were woman-owned.

- **a.** Find a point estimate of the difference in population proportions.
- **b.** Find the pooled sample proportion.
- **c.** Test whether there is a difference in the population proportions of female-owned businesses in Ohio and New Jersey, using level of significance $\alpha = 0.10$.

**27. Fetal Cells and Breast Cancer.** A number of fetal stem cells may cross the placenta from the fetus to the mother during pregnancy and remain in the mother's tissue for decades. A recent study shows that the presence of fetal cells in the mother may offer some protection against the onset of breast cancer.[11] Of the 54 women in the study with breast cancer, 14 had fetal cells. Of the 45 women without breast cancer, 25 had fetal cells.

- **a.** Construct a 95% confidence interval for the difference in the population proportions of women who had fetal cells.
- **b.** Use the confidence interval from **(a)** to test whether there is a difference in the population proportions of women who had fetal cells, using level of significance $\alpha = 0.05$.

**28. Young People Owning Stocks.** The Federal Reserve System reports that the proportion of Americans under 35 years old owning stocks is 17.4%, while the proportion of Americans aged 35–44 owning stocks is 21.6%. Assume that the sample sizes used were 1000 each. Test whether the true proportions differ, at level of significance $\alpha = 0.05$.

**29. Minority Ownership of Businesses.** The U.S. Census Bureau tracks trends in minority ownership of businesses. A random sample of 2000 Michigan businesses showed 106 that were minority-owned. A sample of 1000 Minnesota businesses showed 21 that were minority owned. Test whether the population proportion of minority-owned businesses in Michigan exceeds that of Minnesota, using level of significance $\alpha = 0.05$.

**30. Evidence for Alternative Medical Therapies?** A company called QT, Inc., sells "ionized" bracelets, called Q-Ray bracelets, that it claims help to ease pain through balancing the body's flow of "electromagnetic energy." The Mayo Clinic decided to conduct a statistical experiment to determine whether the claims for the Q-Ray bracelets were justified.[12] At the end of four weeks, of the 305 subjects who wore the "ionized" bracelet, 236 (77.4%) reported improvement in their maximum pain index (where the pain was the worst). Of the 305 subjects who wore the placebo bracelet (a bracelet identical in every respect to the "ionized" bracelet, except that there was no active ingredient—presumably, here, "ionization"), 234 (76.7%) reported improvement in their maximum pain index. Using level of significance $\alpha = 0.05$, test whether the population proportions reporting improvement differ between wearers of the ionized bracelet and wearers of the placebo bracelet.

**Males Listening to the Radio.** Use the following information for Exercises 31–33. The Arbitron corporation tracks trends in radio listening. In their publication *Radio Today*, they reported that 92% of 18–24-year-old males listen to the radio each week, while 87% of males 65 years and older listen to the radio each week. Suppose each sample size was 1000.

**31.** We are interested in constructing a 95% confidence interval for the difference in population proportions.

**a.** Explain why it is appropriate to construct the confidence interval.

**b.** Find a point estimate of the difference in population proportions.

**c.** Find the standard error $s_{\hat{p}_1-\hat{p}_2}$. What does this number mean?

**d.** Find the margin of error for a confidence level of 95%. What does this number mean?

**e.** Construct and interpret a 95% confidence interval for the difference in population proportions.

**32.** Use the confidence interval to test at level of significance $\alpha = 0.05$ whether the population proportion of 18–25-year-old males and the proportion of males 65 and older differ.

**a.** State the hypotheses. Clearly state the meaning of $p_1$ and $p_2$.

**b.** Indicate whether zero falls within the confidence interval.

**c.** State and interpret your conclusion.

WHAT IF? **33.** *What if,* instead of 1000, each sample size was 100. How would this change affect each of the following measures?

**a.** Point estimate of the difference in population proportions

**b.** Standard error $s_{\hat{p}_1-\hat{p}_2}$ for a confidence interval. What does this mean regarding the precision of the inference?

**c.** Width of the 95% confidence interval. How is this related to **(b)**?

**d.** Critical value $Z_{crit}$.

**e.** *p*-value

**f.** Conclusion of the hypothesis test.

## *Try This in Class!* In-class activities to enhance your understanding of statistics

### HYPOTHESIS TEST FOR THE DIFFERENCE IN POPULATION MEAN HEIGHTS OF MEN AND WOMEN

**1.** Suppose we believe that there is no difference between the population mean heights of men and women, and we want to test this hypothesis. State the appropriate hypotheses.

**2.** Suppose we found the difference between sample mean heights of men and women to be 0.1 inches. Do you think this would be sufficient evidence to reject the null hypothesis that both sexes have the same mean height?

**3.** Suppose we found the difference between sample mean heights of men and women to be 1 inch. Do you think this would be sufficient evidence to reject the null hypothesis that both sexes have the same mean height?

**4.** Suppose we found the difference between sample mean heights of men and women to be 5 inches. Do you think this would be sufficient evidence to reject the null hypothesis that both sexes have the same mean height?

APPLET **5.** Generate a random sample of 10 male students from your class, using either (a) technology, (b) the *Random Sample* applet, or (c) the method shown in the Chapter 1 In-Class Activities. Measure the heights of these students.

APPLET **6.** Generate a random sample of size 10 female students from your class, using either (a) technology, (b) the *Random Sample* applet, or (c) the method shown in the Chapter 1 in-class activities. Measure the heights of these students.

**7.** Calculate the mean of each of your samples in Activities 5 and 6. Then calculate the difference in sample means.

**8.** Do you think your difference in sample means is evidence against or in favor of the null hypothesis?

**9.** Each student, using his or her own sample, should perform a hypothesis test to determine whether or not there is no difference between mean heights of men and women. Use level of significance $\alpha = 0.10$. Will everyone in the class necessarily report the same conclusion? Why or why not?

**10.** Use the data from the entire class to perform a hypothesis test to determine whether or not there is no difference between mean heights of men and women. Use level of significance $\alpha = 0.10$. Are you guaranteed to come up with the same conclusion as in Activity 9? Why or why not?

# CHAPTER 10 FORMULAS AND VOCABULARY

## SECTION 10.1

- **100(1 − α)% confidence interval for $\mu_d$** (p. 546):

$$\bar{x}_d \pm t_{\alpha/2}(s_d/\sqrt{n})$$

where $\bar{x}_d$ and $s_d$ represent the sample mean and sample standard deviation of the differences, respectively, of the set of $n$ paired differences, $d_1, d_2, d_3, \ldots, d_n$, and where $t_{\alpha/2}$ is based on $n - 1$ degrees of freedom.

- **Dependent sampling method** (p. 545). A sampling method where the subjects in the first sample determine the subjects for selection in the second sample.
- **Independent sampling method** (p. 545). A sampling method where the subjects selected for the first sample do not determine the subjects for selection in the second sample.
- **Matched-pair data** (p. 545). The data resulting from a dependent sampling method that pairs each observation in the first sample with an observation in the second sample.
- **Test statistic for the paired sample *t* test** (p. 548).

$$t_{\text{data}} = \frac{\bar{x}_d}{s_d/\sqrt{n}}$$

## SECTION 10.2

- **100(1 − α)% confidence interval for $\mu_1 - \mu_2$** (p. 559):

$$(\bar{x}_1 - \bar{x}_2) \pm t_{\alpha/2}\sqrt{\frac{s_1^2}{n_1} + \frac{s_2^2}{n_2}}$$

where $\bar{x}_1$, $s_1$, and $n_1$ represent the mean, standard deviation, and sample size of the sample taken from population 1, and $\bar{x}_2$, $s_2$, and $n_2$ represent the mean, standard deviation, and sample size of the sample taken from population 2, $t_{\alpha/2}$ is associated with the confidence level with degrees of freedom the smaller of $n_1 - 1$ and $n_2 - 1$.

- **Margin of error *E*** (p. 559). For a 100(1 − α)% confidence interval for $\mu_1 - \mu_2$,

$$E = t_{\alpha/2} \cdot \sqrt{\frac{s_1^2}{n_1} + \frac{s_2^2}{n_2}}$$

- **Sampling distribution of $\bar{x}_1 - \bar{x}_2$** (p. 558). When samples are drawn independently from two populations, and *either* (Case 1) the two populations are normally distributed *or* (Case 2) the two sample sizes are large (at least 30), then the quantity

$$t = \frac{(\bar{x}_1 - \bar{x}_2) - (\mu_1 - \mu_2)}{s_{\bar{x}_1 - \bar{x}_2}} = \frac{(\bar{x}_1 - \bar{x}_2) - (\mu_1 - \mu_2)}{\sqrt{\frac{s_1^2}{n_1} + \frac{s_2^2}{n_2}}}$$

approximately follows a *t* distribution, with degrees of freedom equal to the smaller of $n_1 - 1$ and $n_2 - 1$, where $\bar{x}_1$ and $s_1$ represent the mean and standard deviation of the sample taken from population 1, and $\bar{x}_2$ and $s_2$ represent the mean and standard deviation of the sample taken from population 2.

- **Standard error $s_{\bar{x}_1 - \bar{x}_2}$ of $\bar{x}_1 - \bar{x}_2$** (p. 559). Measures the size of the typical error in using $\bar{x}_1 - \bar{x}_2$ to estimate $\mu_1 - \mu_2$:

$$s_{\bar{x}_1 - \bar{x}_2} = \sqrt{\frac{s_1^2}{n_1} + \frac{s_2^2}{n_2}}$$

## SECTION 10.3

- **100(1 − α)% confidence interval for $p_1 - p_2$** (p. 574):

$$\hat{p}_1 - \hat{p}_2 \pm Z_{\alpha/2}\sqrt{\frac{\hat{p}_1 \cdot \hat{q}_1}{n_1} + \frac{\hat{p}_2 \cdot \hat{q}_2}{n_2}}$$

- **Margin of error (*E*)** (p. 574). For a 100(1 − α)% confidence interval for $p_1 - p_2$:

$$E = Z_{\alpha/2} \cdot (\text{standard error}) = Z_{\alpha/2} \cdot (s_{\hat{p}_1 - \hat{p}_2})$$
$$= Z_{\alpha/2} \cdot \sqrt{\frac{\hat{p}_1 \cdot \hat{q}_1}{n_1} + \frac{\hat{p}_2 \cdot \hat{q}_2}{n_2}}$$

- **Pooled estimate of *p*** (p. 576).

$$\hat{p}_{\text{pooled}} = \frac{x_1 + x_2}{n_1 + n_2}$$

- **Sampling distribution of $\hat{p}_1 - \hat{p}_2$** (p. 574). When samples are drawn independently from two populations, then the quantity

$$Z = \frac{(\hat{p}_1 - \hat{p}_2) - (p_1 - p_2)}{\sigma_{\hat{p}_1 - \hat{p}_2}} = \frac{(\hat{p}_1 - \hat{p}_2) - (p_1 - p_2)}{\sqrt{\frac{p_1 q_1}{n_1} + \frac{p_2 q_2}{n_2}}}$$

has an approximately standard normal distribution when the conditions $x_1 \geq 5$, $(n_1 - x_1) \geq 5$, $x_2 \geq 5$ and $(n_2 - x_2) \geq 5$ are satisfied, where $\hat{p}_1$ and $n_1$ represent the sample proportion and sample size of the sample taken from population 1 with population proportion $p_1$; where $\hat{p}_2$ and $n_2$ represent the sample proportion and sample size of the sample taken from population 2 with population proportion $p_2$; and where $q_1 = 1 - p_1$ and $q_2 = 1 - p_2$.

- **Standard error $s_{\hat{p}_1 - \hat{p}_2}$ of the statistic $\hat{p}_1 - \hat{p}_2$** (p. 574):

$$s_{\hat{p}_1 - \hat{p}_2} = \sqrt{\frac{\hat{p}_1 \cdot \hat{q}_1}{n_1} + \frac{\hat{p}_2 \cdot \hat{q}_2}{n_2}}$$

The standard error $s_{\hat{p}_1 - \hat{p}_2}$ measures the size of the typical error in using $\hat{p}_1 - \hat{p}_2$ to estimate $p_1 - p_2$.

- **Test statistic for the independent samples Z test for $p_1 - p_2$** (p. 577):

$$Z_{\text{data}} = \frac{(\hat{p}_1 - \hat{p}_2)}{\sqrt{\hat{p}_{\text{pooled}} \cdot (1 - \hat{p}_{\text{pooled}})\left(\frac{1}{n_1} + \frac{1}{n_2}\right)}}$$

## CHAPTER 10 REVIEW EXERCISES

### SECTION 10.1

**1.** Assume that a sample of differences for the matched pairs in the table follows a normal distribution, and carry out steps a and b.

| Subject | 1 | 2 | 3 | 4 | 5 | 6 | 7 | 8 |
|---|---|---|---|---|---|---|---|---|
| **Sample 1** | 100.7 | 110.2 | 105.3 | 107.1 | 95.6 | 109.9 | 112.3 | 94.7 |
| **Sample 2** | 104.4 | 112.5 | 105.9 | 111.4 | 99.8 | 109.9 | 115.7 | 97.7 |

**a.** Calculate $\bar{x}_d$ and $s_d$.

**b.** Construct a 95% confidence interval for $\mu_d$.

**2.** For the data in Exercise 1, test whether $\mu_d < 0$, using the critical value method and level of significance $\alpha = 0.05$.

**3.** For the data in Exercise 1, test whether $\mu_d < 0$, using the $p$-value method and level of significance $\alpha = 0.05$.

Use this information for Exercises 4–6. The American Mathematical Society publishes information on the prices of scholarly journals. Table 10.24 contains the annual prices for a sample of journals for the years 2004 and 2005. Assume that the differences in price are normally distributed.

**TABLE 10.24 Journal prices**

| Journal | 2004 price | 2005 price |
|---|---|---|
| Bulletin of the American Mathematical Society | \$ 390 | \$ 402 |
| Journal of the American Mathematical Society | \$ 268 | \$ 276 |
| Proceedings of the American Mathematical Society | \$1022 | \$ 992 |
| Transactions of the American Mathematical Society | \$1744 | \$1628 |
| Annals of Mathematics | \$ 260 | \$ 255 |
| Journal of Applied Mathematics | \$ 527 | \$ 498 |
| Journal of Computation | \$ 536 | \$ 506 |
| Journal of Discrete Mathematics | \$ 450 | \$ 414 |
| Journal of Numerical Analysis | \$ 567 | \$ 536 |
| Journal of Scientific Computing | \$ 644 | \$ 506 |

**4.** Calculate the point estimate for the population mean of the differences, $\mu_d$.

**5.** Find a 95% confidence interval for the population mean of the differences, $\mu_d$.

**6.** Test whether the 2005 prices are higher than the 2004 prices, on average, using level of significance $\alpha = 0.05$.

### SECTION 10.2

Refer to the following summary statistics for two independent samples for Exercises 7–11.

| | | | |
|---|---|---|---|
| **Sample 1** | $n_1 = 36$ | $\bar{x}_1 = 14.4$ | $s_1 = 0.01$ |
| **Sample 2** | $n_2 = 81$ | $\bar{x}_2 = 14.3$ | $s_2 = 0.02$ |

**7.** We are interested in constructing a 95% confidence interval for $\mu_1 - \mu_2$. Explain why it is appropriate to do so.

**8.** Provide the point estimate of the difference in population means $\mu_1 - \mu_2$.

**9.** Calculate the standard error $s_{\bar{x}_1-\bar{x}_2}$.

**10.** Calculate the margin of error for a confidence level of 95%.

**11.** Construct and interpret a 95% confidence interval for $\mu_1 - \mu_2$.

**12.** A random sample of 49 young persons with college degrees had a mean salary of \$30,000 with a standard deviation of \$5000. An independent random sample of 36 young persons without college degrees had a mean of \$25,000 and a standard deviation of \$4000.

**a.** Provide the point estimate of the difference in population means $\mu_1 - \mu_2$, where $\mu_1$ and $\mu_2$ refer to the mean salary of all young persons with and without a college degree, respectively.

**b.** Construct and interpret a 90% confidence interval for $\mu_1 - \mu_2$.

**c.** Construct and interpret a 95% confidence interval for $\mu_1 - \mu_2$.

**d.** Test at level of significance $\alpha = 0.10$ whether the population mean salary of college graduates $\mu_1$ is greater than the population mean salary of those without a degree $\mu_2$.

**e.** Test at level of significance $\alpha = 0.05$ whether the population mean salary of college graduates $\mu_1$ differs from the population mean salary of those without a degree $\mu_2$.

### SECTION 10.3

**13.** The Web site www.internettrafficreport.com reports on the current state of Internet data flow around the world. On August 1, 2004, the packet loss from Asian Web sites was 16%, while the packet loss from North American Web sites was 4%.

**a.** Find the minimum sample sizes required to perform $Z$ inference for the difference in population proportions. Use these sizes for **(b)**–**(d)**.

**b.** Find a point estimate of the difference in population proportions.

**c.** Find a 90% confidence interval for the difference in population proportions.

**d.** Perform a hypothesis test of whether the population proportions differ, at level of significance $\alpha = 0.10$, using the $p$-value method.

**14.** Rape continues to be a serious concern in the early twenty-first century. The National Center for Health Statistics reports that, for females under 15 years of age, the percentage for whom their first intercourse was not voluntary is 22.1%. The center also reports that, for females 16 years of age, the percentage for whom their first intercourse was not voluntary is 16.1%. Suppose both statistics are based on samples of size 1000.

**a.** Find a point estimate of the difference in population proportions.

**b.** Find a 90% confidence interval for the difference in population proportions.

**c.** Perform a hypothesis test of whether the population proportions differ, at level of significance $\alpha = 0.10$, using the $p$-value method.

**d.** Test at level of significance $\alpha = 0.05$ whether there is a difference in the population proportion of females in these two groups whose first intercourse was not voluntary.

**15.** The Centers for Disease Control, in their Pregnancy Risk Assessment Monitoring System, reported that 641 of 823 new mothers living in Florida and 658 of 824 new mothers living in North Carolina took their babies in for a checkup within one week of delivery.

**a.** Find a point estimate of the difference in population proportions.

**b.** Find a 99% confidence interval for the difference in population proportions.

**c.** Perform a hypothesis test of whether the population proportions differ, at level of significance $\alpha = 0.01$, using the $p$-value method.

**16.** The *Sourcebook of Criminal Justice Statistics 2002* reported that, for the federal courts, 61% of 3133 suspects were denied bail for violent crimes, while 42.4% of 13,686 suspects were denied bail for property-related crimes.

**a.** Find a point estimate of the difference in population proportions.

**b.** Find a 99% confidence interval for the difference in population proportions.

**c.** Perform a hypothesis test of whether the population proportions differ, at level of significance $\alpha = 0.01$, using the $p$-value method.

**d.** Assess the strength of evidence against the null hypothesis.

# CHAPTER 10 Quiz

## TRUE OR FALSE

**1.** True or false: In a dependent sampling method the subjects in the first sample determine the subjects for selection in the second sample.

**2.** True or false: The pooled estimate of $p$, $\hat{p}_{\text{pooled}} = (x_1 + x_2)/(n_1 + n_2)$, always lies between $\hat{p}_1$ and $\hat{p}_2$.

**3.** True or false: The test statistic $Z_{\text{data}}$ measures the size of the typical error in using $\hat{p}_1 - \hat{p}_2$ to estimate $p_1 - p_2$.

## FILL IN THE BLANK

**4.** The conditions on paired sample data for performing a hypothesis test or constructing a confidence interval on paired sample data are that the population is __________ or the sample size is __________.

**5.** The __________ __________ __________ [three words] for a $100(1 - \alpha)\%$ confidence interval for $p_1 - p_2$ equals the product of $Z_{\alpha/2}$ and the standard error.

**6.** __________ [notation] represents the sample mean of the set of $n$ paired differences.

## SHORT ANSWER

**7.** What is the notation used to indicate the difference in population means for two independent samples?

**8.** What statistic measures the size of the typical error in using $\bar{x}_1 - \bar{x}_2$ to estimate $\mu_1 - \mu_2$?

**9.** If a $100(1 - \alpha)\%$ confidence interval for $\mu_1 - \mu_2$ contains 0, then with $100(1 - \alpha)\%$ confidence what can you conclude about the difference in the population means?

## CALCULATIONS AND INTERPRETATIONS

**10.** Trying to quit smoking? Butt-Enders, a cigarette dependence reduction program, claims to lower the average number of cigarettes smoked for its participants. A sample of 10 participants consumed the following numbers of cigarettes on a randomly chosen day before and after attending Butt-Enders. Assume that the differences are normally distributed.

| **Participant** | 1 | 2 | 3 | 4 | 5 |
|---|---|---|---|---|---|
| **Before** | 40 | 20 | 60 | 30 | 50 |
| **After** | 20 | 0 | 40 | 30 | 20 |

| **Participant** | 6 | 7 | 8 | 9 | 10 |
|---|---|---|---|---|---|
| **Before** | 60 | 20 | 40 | 30 | 20 |
| **After** | 60 | 20 | 20 | 0 | 20 |

**a.** Find a point estimate for the population mean difference in number of cigarettes smoked.

**b.** Find a 90% confidence interval for the population mean difference in number of cigarettes smoked.

**c.** Use your confidence interval to test at level of significance $\alpha = 0.10$ whether the population mean difference in number of cigarettes smoked differs from 0.

**11.** A family is trying to decide where to move. The choice has come down to Suburb A and Suburb B. A random sample of 40 households in Suburb A had a mean income of $50,000 and a standard deviation of $15,000. A random sample of 36 households in Suburb B had a mean income of $65,000 and a standard deviation of $20,000.

**a.** Provide the point estimate of the difference in population means $\mu_1 - \mu_2$.

**b.** Construct and interpret a 90% confidence interval for $\mu_1 - \mu_2$.

**c.** Construct and interpret a 95% confidence interval for $\mu_1 - \mu_2$.

**d.** Test at level of significance $\alpha = 0.10$ whether the population mean income in Suburb A, $\mu_1$, is less than the population mean income in Suburb B, $\mu_2$.

**e.** Test at level of significance $\alpha = 0.05$ whether the population mean income in Suburb A differs from the population mean income in Suburb B.

**f.** Compare your answers to **(c)** and **(e)**. Comment.

Use this information for Exercises 12 and 13. A soft drink company recently performed a major overhaul of one of its bottling machines. Management is eager to determine whether the overhaul has resulted in an increase in productivity for the machine. One hundred "minute segments" are sampled at random from the updated machine (Sample 1) and a machine which was not updated (Sample 2), and the number of bottles processed is noted. The mean and standard deviation of the number of bottles processed by each machine is given in the table.

| | | | |
|---|---|---|---|
| **Updated machine (Sample 1)** | $n_1 = 100$ | $\bar{x}_1 = 200$ | $s_1 = 30$ |
| **Non-updated machine (Sample 2)** | $n_2 = 100$ | $\bar{x}_2 = 190$ | $s_2 = 25$ |

**12. a.** Explain why it is appropriate to apply $t$ inference.

**b.** Provide the point estimate of the difference in population means $\mu_1 - \mu_2$.

**c.** Calculate the standard error $s_{\bar{x}_1 - \bar{x}_2}$.

**d.** Calculate the margin of error for a confidence level of 95%.

**e.** Construct and interpret a 95% confidence interval for $\mu_1 - \mu_2$.

**13.** Refer to the previous exercise.

**a.** Find the critical value $t_{\text{crit}}$ for level of significance $\alpha = 0.05$.

**b.** Calculate $t_{\text{data}}$.

**c.** Test at level of significance $\alpha = 0.05$ whether $\mu_1$ is greater than $\mu_2$.

**d.** Explain whether the confidence interval in Exercise 12e could have been used to perform the hypothesis test in (c). Why or why not?

**14.** The U.S. Census Bureau reported in 2002 that, for people 18–24 years old, the mean annual income for people who never married was $13,539 and for married people was $19,321. Suppose that this information came from a survey of 100 people from each group and that the sample standard deviations were $5000 for the people who never married and $8000 for the married people.

**a.** Provide the point estimate of the difference in population means $\mu_1 - \mu_2$.

**b.** Test at level of significance $\alpha = 0.10$ whether the population mean income for never married people differs from that of married people.

**c.** If we construct a 90% confidence interval for $\mu_1 - \mu_2$, will the interval include 0? Explain why or why not.

**15.** The *2005 National Survey on Drug Use and Health* reported that, in 2004, 38.5% of 18–20 year olds reported having used an illicit drug within the past year, and 37.9% reported use in 2005. Assume $n_1 = n_2 = 1000$.

**a.** Find a point estimate of the difference in population proportions.

**b.** Find the pooled sample proportion.

**c.** Perform a hypothesis test of whether the population proportion of 18–20-year-olds who used an illicit drug decreased from 2004 to 2005, using level of significance $\alpha = 0.05$.

**16.** Are fewer people getting married? The U.S. Census Bureau reported that in 1990, 74.1% of people aged 35–44 were married, while in 2000, 69% of people aged 35–44 were married. Assume that the data are from random samples of size 1000 each. Test whether the population proportion of people aged 35–44 who were married was lower in 2000 than in 1990, using level of significance $\alpha = 0.05$.

Gogo/Punchstock Images

Chapter 11

# Categorical Data Analysis

## CASE STUDY | Online Dating

The Pew Internet and American Life Project reports that about 16 million people, representing 11% of the American Internet-using public, have visited a dating Web site, and 37% of Internet users who are currently seeking partners have gone to a dating Web site.[1] In this chapter, we apply the concepts and methodologies of categorical data analysis to investigate online dating. In Section 11.2, we examine whether women and men report different types of relationships, and whether women and men differ in how they self-report their physical appearance.

OVERVIEW



## The Big Picture — Where we are coming from, and where we are headed...

In Chapters 8–10, we learned how to perform inference for continuous variables. However, not all variables are continuous. In this chapter, we are introduced to the *multinomial* random variable, an extension of the binomial random variable. Multinomial data are not continuous but *categorical* (qualitative). We will learn methods for performing hypothesis tests for this type of data. These methods rely on the $\chi^2$ (chi-square) distribution, which we learned in Chapters 8 and 9.

According to **NetApplications.com**, the market share for the leading Internet browsers in March 2007 was as follows: Microsoft Internet Explorer, 78.5%; Firefox, 15%; Safari, 4.5%; others, 2%. Change is rapid in the online environment. Have these market shares changed since 2007? How would we go about performing a hypothesis test to determine whether market shares have changed significantly? In Section 11.1, we examine this question using a new type of random variable which follows a multinomial distribution, together with a new type of hypothesis test called a $\chi^2$ *goodness of fit test*. In Section 11.2, we learn how to perform a $\chi^2$ *test for independence* between two categorical variables and to test for differences in proportions among $k$ populations. We also learn how to test whether the proportions from several populations are equal or not.

In later chapters we will learn about *analysis of variance* and *regression,* which are perhaps the most widespread data analytic methods we will learn in this text.

# 11.1 $\chi^2$ Goodness of Fit Test

## Objectives: By the end of this section, I will be able to...

1. Explain what a multinomial random variable is and how to calculate expected frequencies.
2. Describe how a $\chi^2$ goodness of fit test works.
3. Perform and interpret the results from the $\chi^2$ goodness of fit test using the critical value method, the exact $p$-value method, and the estimated $p$-value method.

In this chapter, we will learn how to perform hypothesis tests on categorical data. We begin by first considering a new type of random variable that is used to represent categorical data.

### 1 The Multinomial Random Variable

Recall from Chapter 1 that *categorical* (qualitative) variables are not numeric but take values that can be classified into categories. In Chapter 6, we considered binomial random variables, for which there are only two possible outcomes. Now let's consider the following type of random variable, which can have more than two possible values. Suppose we choose a card from a deck of 52 playing cards at random and with replacement, and we define our random variable to be

$$X = \textit{suit of the card}$$

The random variable $X$ has 4 possible outcomes (clubs, spades, hearts, diamonds). We say that $X$ is a **multinomial random variable** because it satisfies the following properties:

**Multinomial Random Variable**

A random variable is *multinomial* if it satisfies each of the following conditions:

- Each independent trial of the experiment has $k$ possible outcomes, $k = 2, 3, 4, \ldots$
- The $i$th outcome (category) occurs with probability $p_i$, where $i = 1, 2, \ldots, k$ (that is, $p_i$ is the population proportion for category $i$)
- $\sum_{i=1}^{k} p_i = 1$ (Law of Total Probability)

Data from a **multinomial random variable** are said to follow a *multinomial distribution*.

*Note:* The binomial distribution may be considered a special case of the multinomial distribution, with $k = 2$.

The random variable $X$ = *suit* of the card has 4 outcomes, and the probability of each outcome is

$$p_{\text{clubs}} = p_{\text{spades}} = p_{\text{hearts}} = p_{\text{diamonds}} = 1/4$$

Successive draws are independent since they are made at random and with replacement. Thus, since

$$\sum_{i=1}^{4} p_i = \frac{1}{4} + \frac{1}{4} + \frac{1}{4} + \frac{1}{4} = 1$$

the random variable $X$ = *suit* is a multinomial random variable.

Next, recall from Chapter 6 that the formula for finding the expected value (mean) of a binomial random variable having $n$ trials and probability of success $p$ is

$$\text{expected value} = n \cdot p$$

For a multinomial random variable, the **expected frequency** of the $i$th category is

$$\text{expected frequency}_i = E_i = n \cdot p_i$$

where $n$ represents the number of trials, and $p_i$ represents the population proportion for the $i$th category.

## Example 11.1 Browser market share

According to **NetApplications.com**, the market share for the leading Internet browsers in March 2007 was as shown in Table 11.1. Let $X$ = *browser* of a randomly selected Internet user.

**a.** Verify that $X$ = *browser* is a valid multinomial random variable.

**b.** Find the expected frequency for each category in a series of 200 trials.

Table 11.1 Distribution of browser market share

| Browser | Relative frequency |
|---|---|
| Microsoft Internet Explorer | 0.785 |
| Firefox | 0.15 |
| Safari | 0.045 |
| Other | 0.02 |

**Solution**

**a.** There are $k = 4$ possible outcomes: Microsoft Internet Explorer, Firefox, Safari, and Other. Assigning probabilities using the relative frequency method, we have the following hypothesized proportions for each browser:

$$p_{\text{MS IE}} = 0.785, \quad p_{\text{Firefox}} = 0.15, \quad p_{\text{Safari}} = 0.045, \quad p_{\text{Other}} = 0.02$$

And

$$\sum_{i=1}^{4} p_i = 0.785 + 0.15 + 0.045 + 0.02 = 1$$

Therefore, $X$ = *browser* is a valid multinomial random variable.

**b.** We have $n = 200$ trials (sample size = 200), so the expected frequencies are as provided in Table 11.2.

Table 11.2 Expected frequencies for browser preference in sample of size 200

| Category | Expected frequency$_i$ = $E_i = n \cdot p_i$ |
|---|---|
| Microsoft Internet Explorer | $E_{\text{MS IE}} = 200 \cdot 0.785 = 157$ |
| Firefox | $E_{\text{Firefox}} = 200 \cdot 0.15 = 30$ |
| Safari | $E_{\text{Safari}} = 200 \cdot 0.045 = 9$ |
| Other | $E_{\text{Other}} = 200 \cdot 0.02 = 4$ |

As a check on the calculations, we have $\sum E_i = n$. In this case,

$$\sum E_i = 157 + 30 + 9 + 4 = 200 = n$$

*What Do These Expected Frequencies Mean?*

Recall that the expected value of a random variable refers to the long-run mean of that random variable after an arbitrarily large number of trials. For example, if we repeatedly took samples of 200 Internet users and asked about browser preference, the mean number of persons who preferred Firefox would approach 30 as the number of trials increased, *if the proportions given in Table 11.1 are correct.* Similarly, since 15% of the entire population of Internet users prefer Firefox, we would *expect* about 15% of any given sample of 200 Internet users to prefer Firefox, since the sample is a subset of the population. This of course begs the question: are the proportions in Table 11.1 still true? That is the type of question we will learn how to address here in Section 11.1. ■

## 2 What Is a $\chi^2$ Goodness of Fit Test?

Market shares often shift, especially in the world of the Internet. Do the 2007 market shares still hold true today? In other words, has the distribution of the multinomial random variable *browser* given in Table 11.1 changed since 2007? To determine this, we introduce a new type of hypothesis test, called a goodness of fit test. Specifically, it is called a **$\chi^2$ goodness of fit test** because we use the $\chi^2$ distribution to perform the hypothesis test.

**$\chi^2$ Goodness of Fit Test**

A **$\chi^2$ goodness of fit test** is a hypothesis test used to determine whether a random variable follows a particular distribution. In a goodness of fit test, the hypotheses are

$H_0$ : The random variable follows a particular distribution.

$H_a$ : The random variable does not follow the distribution specified in $H_0$.

## Example 11.2 Hypotheses for a $\chi^2$ goodness of fit test

Specify the null and alternative hypotheses for a goodness of fit test for the data in Example 11.1.

### Solution

For Example 11.1, the null hypothesis completely specifies each of the probabilities in the relative frequency distribution, as follows:

$$H_0 : p_{\text{MS IE}} = 0.785, p_{\text{Firefox}} = 0.15, p_{\text{Safari}} = 0.045, p_{\text{Other}} = 0.02$$

The alternative hypothesis simply denies the claim made by the null hypothesis:

$$H_a : \text{The random variable does not follow the distribution specified in } H_0.$$

In other words, $H_a$ claims that the browser market shares have changed since 2007. ■

### Fitting the Model to the Data

Now, a goodness of fit test sounds like something you do in a clothing store dressing room. Actually, the analogy to clothes is rather apt. Suppose winter is coming and you are in the market for a new pair of gloves. You find one pair that is especially attractive, but the gloves don't fit your hands. What do you do? You reject the ill-fitting gloves and search for a new pair. In statistics, the gloves represent the models and your hands represent the actual "hard data" observed in the sample.

The null hypothesis $H_0$ represents what is called a *model,* a working theory of how the proportions are distributed. Our working model of how the market shares are distributed is stated in the null hypothesis:

$$\text{Model 1. } H_0 : p_{\text{MS IE}} = 0.785, p_{\text{Firefox}} = 0.15, p_{\text{Safari}} = 0.045, p_{\text{Other}} = 0.02$$

Of course, we could also try other models if we think the market has changed, such as the following:

$$\text{Model 2. } H_0 : p_{\text{MS IE}} = 0.80, p_{\text{Firefox}} = 0.12, p_{\text{Safari}} = 0.05, p_{\text{Other}} = 0.03$$

$$\text{Model 3. } H_0 : p_{\text{MS IE}} = 0.75, p_{\text{Firefox}} = 0.20, p_{\text{Safari}} = 0.04, p_{\text{Other}} = 0.01$$

In hypothesis testing, we "try on" only one model at a time. ■

In statistics, a goodness of fit test determines if the actual "hard data" observed in the sample are consistent with the null hypothesis. Hard data are needed if we are to determine whether or not the market shares have changed. Market researchers would collect data on the actual preferences of a sample of 200 real Internet users. They would come back with a set of *observed frequencies* of Internet users who prefer the various browsers and would compare them with the expected frequencies found in Example 11.1.

**How a Goodness of Fit Test Works**

The goodness of fit test is based on a comparison of the *observed frequencies* (actual data from the field) with the *expected frequencies* when $H_0$ is true. That is, we compare what we actually see with what we would expect to see if $H_0$ were true. If the difference between the observed and expected frequencies is large, we reject $H_0$.

As usual, it comes down to how large a difference is large. The hypothesis test we conduct to answer this question relies on the $\chi^2$ distribution. Before we learn about this hypothesis test, let's review some properties of the $\chi^2$ distribution.

**Properties of the $\chi^2$ Distribution**

1. Just as for any continuous random variable, the total area under the $\chi^2$ curve equals 1.
2. The value of the $\chi^2$ random variable is never negative, so the $\chi^2$ curve starts at 0. However, it extends indefinitely to the right, with no upper bound.
3. Because of the characteristics described in (2), the $\chi^2$ curve is right-skewed.
4. There is a different curve for every different degrees of freedom. As the number of degrees of freedom increases, the $\chi^2$ curve begins to look more symmetric (see Figure 11.1).

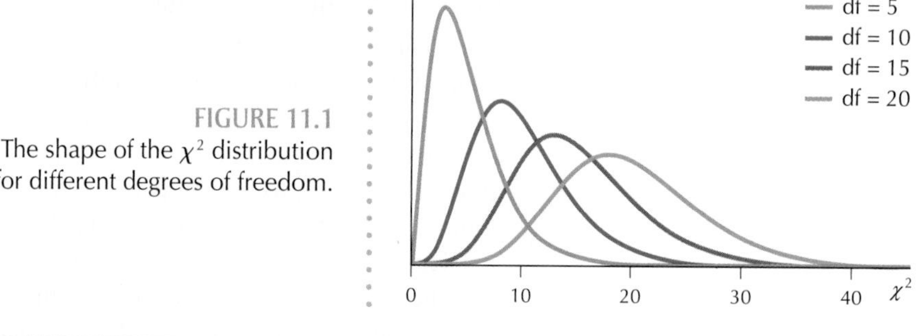


FIGURE 11.1
The shape of the $\chi^2$ distribution for different degrees of freedom.

**Test Statistic for the $\chi^2$ Goodness of Fit Test**

For a multinomial random variable with $k$ categories and $n$ trials, let $O_i$ represent the observed frequency for category $i$, and let $E_i$ represent the expected frequency for category $i$. Then the **test statistic for a goodness of fit test**

$$\chi^2_{\text{data}} = \sum \frac{(O_i - E_i)^2}{E_i}$$

approximately follows a $\chi^2$ (chi-square) distribution with $k - 1$ degrees of freedom, if the following conditions are satisfied:

**a.** None of the expected frequencies is less than 1.

**b.** At most, 20% of the expected frequencies are less than 5.

If the conditions are not satisfied, then it may be possible to combine two or more categories so that the conditions may then be fulfilled.

## Example 11.3 Combining low-frequency categories

Determine if the conditions are met for performing the goodness of fit test for the data in Example 11.1. If necessary, combine the two low-frequency categories to fulfill the conditions.

**Solution**

From Table 11.2, the expected frequencies are

$$E_{\text{MS IE}} = 157 \qquad E_{\text{Firefox}} = 30 \qquad E_{\text{Safari}} = 9 \qquad E_{\text{Other}} = 4$$

Since none of these expected frequencies is less than 1, condition (a) is satisfied. However, one of the four expected frequencies (that is, 25%) is less than 5. This violates condition (b), which states that at most 20% of the expected frequencies be less than 5. We therefore combine the two categories with the smallest expected frequencies, Safari and Other, into a single category, Safari/Other. The relative frequency for this new category is, from Table 11.1, $0.045 + 0.02 = 0.065$. The probabilities may be added because the two categories are mutually exclusive. The expected frequency of the combined category in a sample of size 200 is $E_{\text{Safari/Other}} = 200 \cdot 0.065 = 13$. Now, none of the expected frequencies are less than 5. Therefore, condition (b) is now satisfied, and we may proceed to perform the goodness of fit test. However, we also need to update our hypotheses, as follows:

$H_0 : p_{\text{MS IE}} = 0.785, p_{\text{Firefox}} = 0.15, p_{\text{Safari/Other}} = 0.065$

$H_a$ : The random variable does not follow the distribution specified in $H_0$.

## Example 11.4 Calculating the test statistic $\chi^2_{\text{data}}$

Table 11.3 shows the observed frequencies of browser preference from a survey of 200 Internet users this year. Calculate the test statistic $\chi^2_{\text{data}}$.

Table 11.3 Observed frequencies of browser preference in sample of 200 internet users

| Browser | Observed frequency |
|---|---|
| Microsoft Internet Explorer | 140 |
| Firefox | 40 |
| Safari/Other | 20 |

**Solution**

The observed frequencies $O_i$ are found in Table 11.3 and the expected frequencies are given in Example 11.3. Then

$$\chi^2_{\text{data}} = \sum \frac{(O_i - E_i)^2}{E_i} = \frac{(140 - 157)^2}{157} + \frac{(40 - 30)^2}{30} + \frac{(20 - 13)^2}{13}$$
$$\approx 1.8408 + 3.3333 + 3.7692 = 8.9433$$

Table 11.4 gives the quantities needed to calculate $\chi^2_{\text{data}}$.

Table 11.4 Calculating $\chi^2_{\text{data}}$

| Category | $p_i$ | $O_i$ | $E_i$ | $O_i - E_i$ | $(O_i - E_i)^2$ | $\frac{(O_i - E_i)^2}{E_i}$ |
|---|---|---|---|---|---|---|
| MS IE | 0.785 | 140 | 157 | −17 | 289 | 1.8408 |
| Firefox | 0.15 | 40 | 30 | 10 | 100 | 3.3333 |
| Safari/Other | 0.065 | 20 | 13 | 7 | 49 | 3.7692 |
| Total | 1.0 | 200 | 200 | 0 | | $\chi^2_{\text{data}} = 8.9433$ |

## 3 Performing the $\chi^2$ Goodness of Fit Test

The $\chi^2$ goodness of fit test may be performed using (a) the critical value method, (b) the exact $p$-value method, or (c) the estimated $p$-value method.

**$\chi^2$ Goodness of Fit Test: Critical Value Method**

***Step 1*** **State the hypotheses and check the conditions.**

- The null hypothesis states that the multinomial random variable follows a particular distribution.
- The alternative hypothesis states that the random variable does not follow that distribution.

The following conditions must be met:

**a.** None of the expected frequencies is less than 1.

**b.** At most, 20% of the expected frequencies are less than 5.

The expected frequency for the $i$th category is $E_i = n \cdot p_i$ where $n$ represents the number of trials and $p_i$ represents the population proportion for the $i$th category.

***Step 2*** **Find the $\chi^2$ critical value $\chi^2_{\text{crit}}$ and state the rejection rule.** Reject $H_0$ if $\chi^2_{\text{data}} > \chi^2_{\text{crit}}$.

***Step 3*** **Find the test statistic $\chi^2_{\text{data}}$.**

$$\chi^2_{\text{data}} = \sum \frac{(O_i - E_i)^2}{E_i}$$

where $O_i$ = observed frequency, and $E_i$ = expected frequency.

***Step 4*** **State the conclusion and the interpretation.** Compare $\chi^2_{\text{data}}$ with $\chi^2_{\text{crit}}$.

### Example 11.5 Critical value method for the $\chi^2$ goodness of fit test

Perform the hypothesis test shown in Example 11.3, using the observed frequencies in Table 11.3. Use level of significance $\alpha = 0.05$.

**Solution**

***Step 1*** **State the hypotheses and check the conditions.** The hypotheses are as given in Example 11.3:

$$H_0: p_{\text{MS IE}} = 0.785, p_{\text{Firefox}} = 0.15, p_{\text{Safari/Other}} = 0.065$$

$H_a$: The random variable does not follow the distribution specified in $H_0$.

Example 11.3 shows that the conditions for applying the $\chi^2$ goodness of fit test are satisfied.

***Step 2*** **Find the $\chi^2$ critical value $\chi^2_{crit}$ and state the rejection rule.** We have degrees of freedom $k - 1 = 3 - 1 = 2$ and $\alpha = 0.05$. Turning to the $\chi^2$ table (Table E in the Appendix, page T-12) in the column labeled $\chi^2_{0.05}$ and the row containing df = 2, we find $\chi^2_{crit} = \chi^2_{0.05} = 5.991$ as shown in Figure 11.2. The rejection rule is "Reject $H_0$ if $\chi^2_{data} > 5.991$."

Chi-Square ($\chi^2$) Distribution

| Degrees of freedom | Area to the Right of Critical Value | | | | | | | |
|---|---|---|---|---|---|---|---|---|
| | **0.995** | **0.99** | **0.975** | **0.95** | **0.90** | **0.10** | **0.05** | **0.025** |
| **1** | — | — | 0.001 | 0.004 | 0.016 | 2.706 | 3.841 | 5.024 |
| **2** | 0.010 | 0.020 | 0.051 | 0.103 | 0.211 | 4.605 | 5.991 | 7.378 |
| **3** | 0.072 | 0.115 | 0.216 | 0.352 | 0.584 | 6.251 | 7.815 | 9.348 |

FIGURE 11.2

***Step 3*** **Find the test statistic $\chi^2_{data}$.** From Example 11.4, we have

$$\chi^2_{data} = \sum \frac{(O_i - E_i)^2}{E_i} = 8.9433$$

***Step 4*** **State the conclusion and the interpretation.** Compare $\chi^2_{data}$ with $\chi^2_{crit}$. $\chi^2_{data} = 8.9433$ is greater than $\chi^2_{crit} = 5.991$, as shown in Figure 11.3. Therefore, we reject $H_0$.

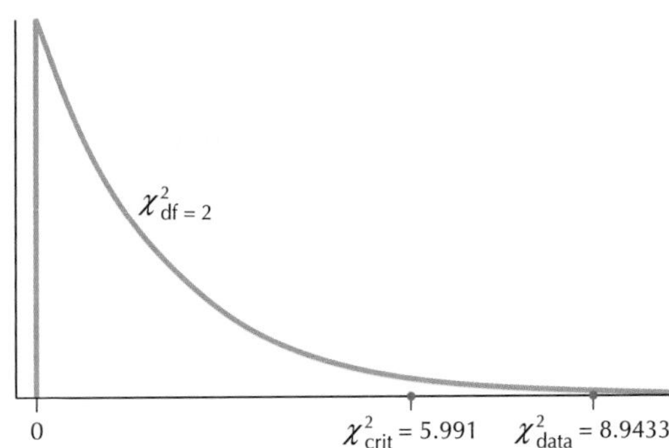


FIGURE 11.3

There is evidence that the random variable *browser* does not follow the distribution specified in $H_0$. In other words, there is evidence that the market shares for Internet browsers have changed. ■

*Developing Your Statistical Sense*

### Be Careful How You Interpret the Conclusion

Note carefully what this conclusion says and what it doesn't say. The $\chi^2$ goodness of fit test shows that there is evidence that the random variable does not follow the distribution specified in $H_0$. In particular, the conclusion does *not* state, for example, that Firefox's proportion is significantly greater. Informally, we can compare the observed frequency of 40 with the expected frequency of 30 for the Firefox browser and note that there appears to be evidence of an increase in market share. But this is only informal and is not part of the hypothesis test. It is a common error in statistical analysis to form conclusions beyond what the hypothesis test is actually testing. ■

The $p$-value for the $\chi^2$ statistic is defined as the area under the $\chi^2$ curve to the right of the test statistic $\chi^2_{\text{data}}$, as shown in Figure 11.4. That is,

$$p\text{-value} = P(\chi^2 > \chi^2_{\text{data}})$$

We can use technology to find the exact $p$-value for a particular value of $\chi^2_{\text{data}}$. Or, alternatively, the $p$-value may be estimated using the $\chi^2$ table.

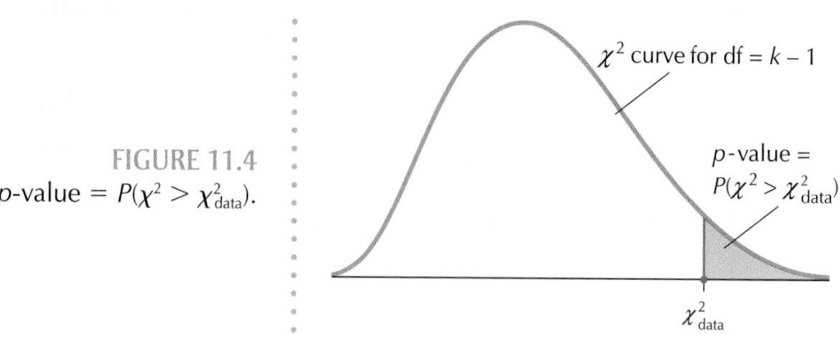


FIGURE 11.4 $p$-value $= P(\chi^2 > \chi^2_{\text{data}})$.

**$\chi^2$ Goodness of Fit Test: $p$-Value Method**

***Step 1*** **State the hypotheses and the rejection rule. Check the conditions.** The null hypothesis states that the multinomial random variable follows a particular distribution. The alternative hypothesis states that the random variable does not follow that distribution. Reject $H_0$ if the $p$-value is less than $\alpha$.

The following conditions must be met:

**a.** None of the expected frequencies is less than 1.

**b.** At most, 20% of the expected frequencies are less than 5.

The expected frequency for the $i$th category is $E_i = n \cdot p_i$ where $n$ represents the number of trials and $p_i$ represents the population proportion for the $i$th category.

***Step 2*** **Find the test statistic $\chi^2_{\text{data}}$.**

$$\chi^2_{\text{data}} = \sum \frac{(O_i - E_i)^2}{E_i}$$

where $O_i$ = observed frequency, and $E_i$ = expected frequency.

***Step 3*** **Find the $p$-value.**

$p$-value $= P(\chi^2 > \chi^2_{\text{data}})$ (see Figure 11.4)

***Step 4*** **State the conclusion and the interpretation.** Compare the $p$-value with $\alpha$.

## Example 11.6 Exact $p$-value method for the $\chi^2$ goodness of fit test using technology

The Pew Internet and American Life Project released the report *Broadband Adoption at Home,* which updated figures on the market share of cable modem, DSL, and wireless broadband from a 2002 survey (Table 11.5). The 2006 survey (Table 11.6) was based on a random sample of 1000 home broadband users. Test whether the population proportions have changed since 2002, using the exact $p$-value method, and level of significance $\alpha = 0.05$.

Table 11.5 2002 broadband adoption survey

| Cable modem | DSL | Wireless/Other |
|---|---|---|
| 67% | 28% | 5% |

Table 11.6 2006 broadband adoption survey

| Cable modem | DSL | Wireless/Other |
|---|---|---|
| 410 | 500 | 90 |

## Solution

**Step 1** **State the hypotheses and the rejection rule. Check the conditions.**

$$H_0 : p_{\text{Cable}} = 0.67, p_{\text{DSL}} = 0.28, p_{\text{Wireless/Other}} = 0.05$$
$$H_a : \text{The random variable does not follow the distribution specified in } H_0.$$

Reject $H_0$ if the $p$-value is less than 0.05.

First we need to find the expected frequencies. We have $n = 1000$, so the expected frequencies are as shown here.

Expected frequencies for broadband access preference in sample of size $n = 1000$

| Category | Expected frequency$_i$ = $E_i = n \cdot p_i$ |
|---|---|
| Cable | $E_{\text{Cable}} = 1000 \cdot 0.67 = 670$ |
| DSL | $E_{\text{DSL}} = 1000 \cdot 0.28 = 280$ |
| Wireless/Other | $E_{\text{Wireless/Other}} = 1000 \cdot 0.05 = 50$ |

### What Results Might We Expect?

Before we do the formal hypothesis test, let's try to figure out what the conclusion might be. Figure 11.5 is a clustered bar graph (see Section 2.1) of the observed and expected frequencies for each of the three categories. If $H_0$ were true, then, for each category, we would expect the green bars (observed frequencies) and yellow bars (expected frequencies) to have similar heights.

Note that the observed frequency for cable is much lower than the expected frequency, while the observed frequency for DSL is much higher than the expected frequency. These both indicate evidence against the null hypothesis. Thus, we might expect to reject $H_0$. ■

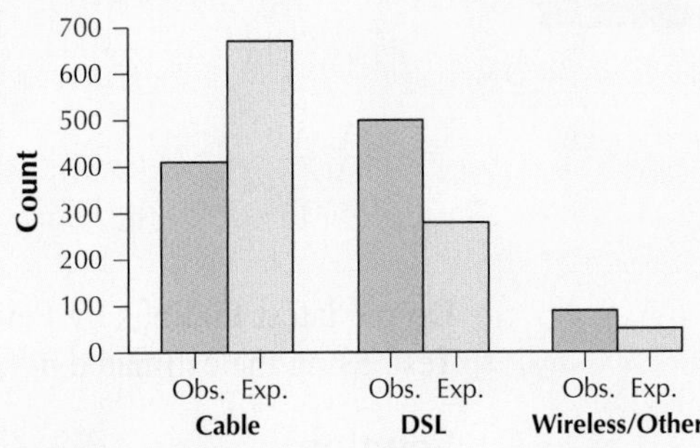


FIGURE 11.5

Next we check whether the requirements for performing the test are met. We verify that (a) none of the expected frequencies is less than 1 and (b) no more than 20% of the expected frequencies are less than 5. We therefore proceed to perform the hypothesis test, using the instructions provided in the Step-by-Step Technology Guide at the end of this section (pages 602–603).

***Step 2*** **Find the test statistic $\chi^2_{\text{data}}$.** The TI-83/84 results in Figure 11.6 tell us that $\chi^2_{\text{data}} = 305.7526652 \approx 305.75$.

***Step 3*** **Find the *p*-value.** Figure 11.6 also tells us that

$$p\text{-value} = P(\chi^2 > 305.75) \approx \text{“4.0425215E–67”} \approx 0$$

Figure 11.7 illustrates why the *p*-value is so small. There is essentially no area to the right of $\chi^2_{\text{data}} = 305.75$ in the $\chi^2_{\text{df}=2}$ distribution.

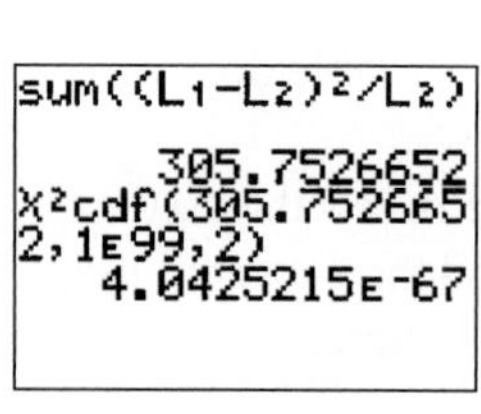


FIGURE 11.6

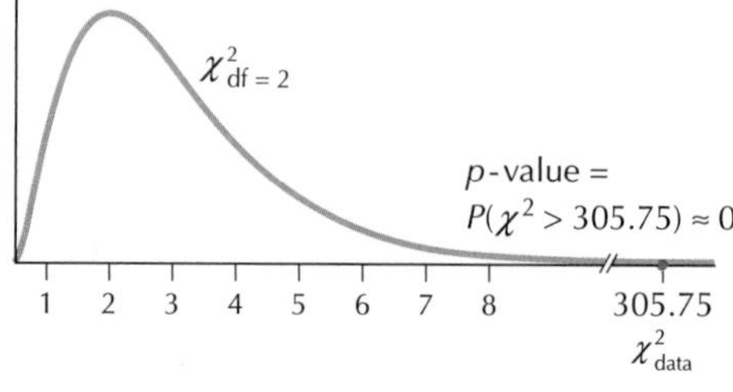


FIGURE 11.7

***Step 4*** **State the conclusion and the interpretation.** Since the *p*-value is less than $\alpha$, we reject $H_0$, which we expected. There is evidence that the proportions of broadband type in 2006 have changed since 2002. As the 2006 Pew report stated, “In a reversal of market share, telephone companies offering digital subscriber line (DSL) services have overtaken cable companies in the broadband world.”

## Example 11.7 Estimated *p*-value method for the $\chi^2$ goodness of fit test

BLOOMimage/Punchstock Images

In June 2003, the Pew Internet and American Life Project released their data on the College Students Gaming Survey. Among the questions they asked college students was “Where do you play video/computer/online games the most?” The responses are summarized in the following table. Suppose previous research had indicated that 40% of college students played video/computer/online games the most at their parents' home, 33% at their friend's home, and 27% in their dorm room.

| Place | Count |
|---|---|
| Parents' home | 360 |
| Friend's home | 314 |
| Dorm room | 267 |

Do the latest findings by Pew indicate that the population proportions have changed? Test, using the estimated *p*-value method, and level of significance $\alpha = 0.10$.

### Solution

We note that we have a multinomial random variable, with three categories and $\text{df} = k - 1 = 3 - 1 = 2$.

**Step 1** **State the hypotheses and the rejection rule. Check the conditions.**

$H_0: p_{\text{Parents}} = 0.40, \quad p_{\text{Friends}} = 0.33, \quad p_{\text{Dorm}} = 0.27$

$H_a$: The random variable does not follow the distribution specified in $H_0$.

We reject $H_0$ if the $p$-value is less than $\alpha = 0.10$. Let's check the conditions. Here we have $n = 360 + 314 + 267 = 941$. Then, assuming $H_0$ is correct, we have the following expected frequencies:

$$E_{\text{Parents}} = n \cdot p_{\text{Parents}} = 941 \cdot 0.40 = 376.40$$
$$E_{\text{Friends}} = n \cdot p_{\text{Friends}} = 941 \cdot 0.33 = 310.53$$
$$E_{\text{Dorm}} = n \cdot p_{\text{Dorm}} = 941 \cdot 0.27 = 254.07$$

The conditions are therefore met and we may proceed with the hypothesis test.

### What Results Might We Expect?

Figure 11.8 represents the observed and expected frequencies for each of the three categories. Note that, for each category, the heights of the bars do not differ very much, indicating that the observed frequencies are similar to the frequencies expected if we assume $H_0$ is correct. Thus, we might expect to *not* reject $H_0$. ■

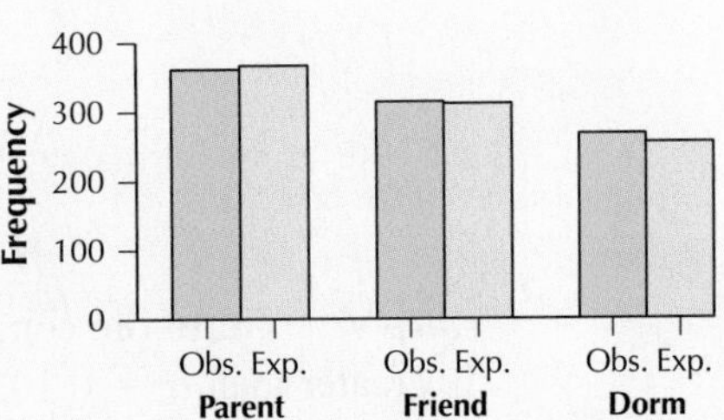


FIGURE 11.8

**Step 2** **Find the test statistic $\chi^2_{\text{data}}$.**

| Category | $p_i$ | $O_i$ | $E_i$ | $O_i - E_i$ | $(O_i - E_i)^2$ | $\frac{(O_i - E_i)^2}{E_i}$ |
|---|---|---|---|---|---|---|
| Parents' | 0.40 | 360 | 376.40 | −16.40 | 268.96 | 0.714559 |
| Friend's | 0.33 | 314 | 310.53 | 3.47 | 12.0409 | 0.038775 |
| Dorm | 0.27 | 267 | 254.07 | 12.93 | 167.1849 | 0.658027 |
| Total | 1.0 | 941 | 941 | 0 | | $\chi^2_{\text{data}} = 1.411361$ |

Thus,

$$\chi^2_{\text{data}} = \sum \frac{(O_i - E_i)^2}{E_i} = 1.411361 \approx 1.41$$

**Step 3** **Find the estimated $p$-value.** Our degrees of freedom is $k - 1 = 3 - 1 = 2$. Turning to our $\chi^2$ table (Figure 11.9), with df $= 2$, we locate where $\chi^2_{\text{data}} = 1.41$ would be. We see that it would be to the left of 4.605. Since the $p$-values increase as $\chi^2_{\text{data}}$ decreases, the estimated $p$-value for $\chi^2_{\text{data}} = 1.41$ is somewhat greater than 0.10. Figure 11.10 shows that the $p$-value is actually much greater than 0.10.

**Chi-Square ($\chi^2$) Distribution**

| Degrees of freedom | Area to the Right of Critical Value | | | | | | | | | |
|---|---|---|---|---|---|---|---|---|---|---|
| | **0.995** | **0.99** | **0.975** | **0.95** | **0.90** | **0.10** | **0.05** | **0.025** | **0.01** | **0.005** |
| 1 | — | — | 0.001 | 0.004 | 0.016 | 2.706 | 3.841 | 5.024 | 6.635 | 7.879 |
| 2 | 0.010 | 0.020 | 0.051 | 0.103 | 0.211 | 4.605 | 5.991 | 7.378 | 9.210 | 10.597 |
| 3 | 0.072 | 0.115 | 0.216 | 0.352 | 0.584 | 6.251 | 7.815 | 9.348 | 11.345 | 12.838 |

$\chi^2_{\text{data}} = 1.41$

FIGURE 11.9

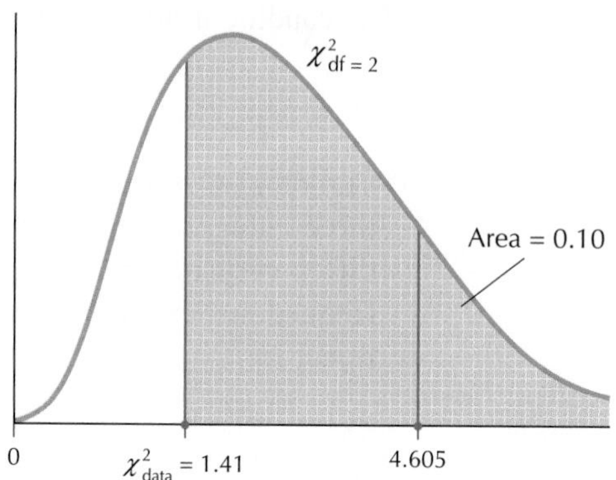


FIGURE 11.10 Estimating the $p$-value for a $\chi^2$-value.

***Step 4*** **State the conclusion and the interpretation.** Since our estimated $p$-value is greater than $\alpha = 0.10$, our conclusion, as expected, is that we do not reject $H_0$. In other words, there is insufficient evidence that the Pew survey data contains significantly different proportions from those found in earlier research.

## STEP-BY-STEP TECHNOLOGY GUIDE: The $\chi^2$ Goodness of Fit Test

We illustrate the use of technology, once the observed and expected frequencies are known, for Example 11.6 (page 598).

### TI-84

***Step 1*** Enter observed frequencies in list **L1** and expected frequencies in list **L2**.

***Step 2*** Press **STAT**, highlight **TESTS**, select **D: $\chi^2$ GOF-Test**, and press **ENTER** (Figure 11.11).

***Step 3*** Highlight **df**, and enter degrees of freedom **2** (Figure 11.12).

***Step 4*** Highlight **Calculate** and press **ENTER**. The results are shown in Figure 11.13, including $\chi^2_{\text{data}}$ and the $p$-value.

Figure 11.11

χ²GOF-Test
Observed:L1
Expected:L2
df:2
Calculate Draw

Figure 11.12

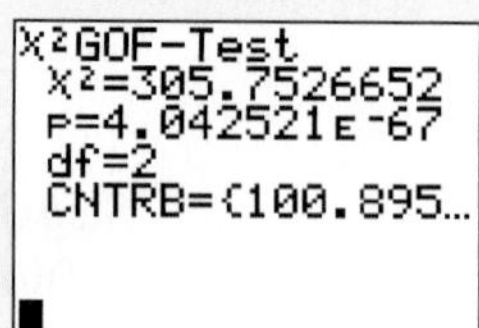


Figure 11.13

## TI-83/84

### To find $\chi^2_{\text{data}}$:

***Step 1*** Enter observed frequencies in list **L1** and expected frequencies in list **L2**. Press **2nd QUIT**.

***Step 2*** Press **2nd LIST**, highlight **MATH**, select **5: sum(**, and press **ENTER** (Figure 11.14).

Figure 11.14

***Step 3*** Type the following: (L1–L2)²/L2) (see Figure 11.6 in Example 11.6) and press **ENTER**.

***Step 4*** The TI-83/84 then displays 305.7526652 as $\chi^2_{\text{data}}$ (see Figure 11.6, page 600).

### To find the *p*-value:

***Step 1*** Select **2nd DISTR**, then **$\chi^2$ cdf(**, and press **ENTER**.

***Step 2*** To get the *p*-value, that is, the area to the right of 305.7526652, enter **305.7526652, comma, 1E99, comma, 2)**, as shown in Figure 11.6, page 600.

## EXCEL

### To find $\chi^2_{\text{data}}$:

***Step 1*** Enter the observed and expected frequencies in rows 1 and 2 (Figure 11.15).

***Step 2*** In cell B3, enter: **=(B1-B2)^2/B2** (Figure 11.15).

B3 | fx =(B1-B2)^2/B2

| | A | B | C | D | E |
|---|---|---|---|---|---|
| 1 | Observed | 410 | 500 | 90 | |
| 2 | Expected | 670 | 280 | 50 | |
| 3 | | 100.8955 | 172.8571 | 32 | 305.7527 |
| 4 | 4.04252E-67 | | | | |

Figure 11.15

***Step 3*** Copy the contents of cell B3 to cells C3 and D3.

***Step 4*** Select an empty cell, enter **=SUM(B3:D3)**, and press **ENTER**. Excel then displays the value $\chi^2_{\text{data}} = 305.7527$ (Figure 11.16).

E3 | fx =SUM(B3:D3)

| | A | B | C | D | E |
|---|---|---|---|---|---|
| 1 | Observed | 410 | 500 | 90 | |
| 2 | Expected | 670 | 280 | 50 | |
| 3 | | 100.8955 | 172.8571 | 32 | 305.7527 |
| 4 | 4.04252E-67 | | | | |

Figure 11.16

### To find the *p*-value:

***Step 1*** Select a cell and enter **=CHITEST(B1:D1,B2:D2)** and press **ENTER**.

***Step 2*** Excel then provides the *p*-value (Figure 11.17).

A4 | fx =CHITEST(B1:D1,B2:D2)

| | A | B | C | D | E | F |
|---|---|---|---|---|---|---|
| 1 | Observed | 410 | 500 | 90 | | |
| 2 | Expected | 670 | 280 | 50 | | |
| 3 | | 100.8955 | 172.8571 | 32 | 305.7527 | |
| 4 | 4.04252E-67 | | | | | |

Figure 11.17

## MINITAB

### To find $\chi^2_{\text{data}}$:

***Step 1*** Enter the observed frequencies (**O**) into **C1** and the hypothesized proportions into **C2**.

***Step 2*** Click **Stat > Tables > Chi-Square Goodness of Fit Test** (one variable).

***Step 3*** For observed counts, enter **C1**.

***Step 4*** Click **Specific Proportions** and enter **C2** in box.

***Step 5*** Click **OK**.

## Section 11.1 SUMMARY

1 A random variable is multinomial if (a) each independent trial has $k$ possible outcomes, $k = 2,3,4, \cdots$; (b) the $i$th outcome (category) occurs with probability $p_i$, where $i = 1,2, \cdots, k$; and (c) $\sum_{i=1}^{k} p_i = 1$ (Law of Total Probability).

2 A goodness of fit test is a hypothesis test used to ascertain whether a random variable follows a particular distribution. In a goodness of fit test, the hypotheses are

$H_0$ : The random variable follows a particular distribution.
$H_a$ : The random variable does not follow the distribution specified in $H_0$.

Compare the observed frequencies (actual data from the field) with the expected frequencies when $H_0$ is true. If the difference between the observed and expected frequencies is large, reject $H_0$.

3 The $\chi^2$ goodness of fit test is performed using (a) the critical value method, (b) the exact $p$-value method, or (c) the estimated $p$-value method.

## Section 11.1 EXERCISES

### CLARIFYING THE CONCEPTS

**1.** What are the conditions required for a random variable to be multinomial?

**2.** Explain in your own words what is meant by a goodness of fit test.

**3.** Explain the meaning of the term "expected frequency." (*Hint:* Use the idea of the long-run mean in your answer.)

**4.** State the hypotheses for a $\chi^2$ goodness of fit test.

### PRACTICING THE TECHNIQUES

For Exercises 5–8, determine whether the random variable is multinomial.

**5.** We take a sample of 10 M&M's at random and with replacement and observe the color of each. Let $X$ = color of the M&M's, where the possible values of $X$ are green, blue, yellow, orange, red, and brown.

**6.** We select 5 students from a group of 25 statistics students at random and without replacement, and we define our random variable to be $X$ = student's class, where the possible values of $X$ are freshman, sophomore, junior, and senior.

**7.** We choose 10 stocks at random and with replacement, and we define our random variable to be $X$ = the exchange that the stock is traded on. The possible values of $X$ are New York Stock Exchange, NASDAQ, London Stock Exchange, and Shenzhen Stock Exchange.

**8.** We pick 10 stocks at random and with replacement, and we define our random variable to be $X$ = the amount that the stock price increased or decreased since the last trading day.

For Exercises 9–12, we have a $\chi^2$ distribution with the given degrees of freedom.

**a.** Estimate the $p$-value for the given $\chi^2_{\text{data}}$ and degrees of freedom.

**b.** Draw the $\chi^2$ curve, showing $\chi^2_{\text{data}}$ on the number line and the area to the right of it equaling the $p$-value.

**9.** $\chi^2_{\text{data}} = 2$, df = 2
**10.** $\chi^2_{\text{data}} = 8$, df = 3
**11.** $\chi^2_{\text{data}} = 16.7$, df = 5
**12.** $\chi^2_{\text{data}} = 9.2$, df = 5

For Exercises 13–16, the alternative hypothesis takes the form

$H_a$ : The random variable does not follow the distribution specified in $H_0$.

**a.** Find the expected frequencies.

**b.** Determine whether the conditions for performing the $\chi^2$ goodness of fit test are met.

**13.** $H_0 : p_1 = 0.50, p_2 = 0.25, p_3 = 0.25; n = 100$
**14.** $H_0 : p_1 = 0.2, p_2 = 0.3, p_3 = 0.4, p_4 = 0.1; n = 10$
**15.** $H_0 : p_1 = 0.9, p_2 = 0.05, p_3 = 0.04, p_4 = 0.01; n = 100$
**16.** $H_0 : p_1 = 0.4, p_2 = 0.35, p_3 = 0.10, p_4 = 0.10, p_5 = 0.05; n = 200$

For Exercises 17–22, calculate the value of $\chi^2_{\text{data}}$.

**17.**

| $O_i$ | $E_i$ |
|---|---|
| 10 | 12 |
| 12 | 12 |
| 14 | 12 |

**18.**

| $O_i$ | $E_i$ |
|---|---|
| 15 | 10 |
| 20 | 25 |
| 25 | 25 |

**19.**

| $O_i$ | $E_i$ |
|---|---|
| 20 | 25 |
| 30 | 25 |
| 40 | 30 |
| 40 | 50 |

**20.**

| $O_i$ | $E_i$ |
|---|---|
| 8 | 6 |
| 10 | 8 |
| 7 | 9 |
| 5 | 7 |

**21.**

| $O_i$ | $E_i$ |
|---|---|
| 1 | 6 |
| 10 | 6 |
| 8 | 6 |
| 0 | 6 |
| 11 | 6 |

**22.**

| $O_i$ | $E_i$ |
|---|---|
| 90 | 100 |
| 100 | 110 |
| 100 | 90 |
| 100 | 80 |
| 110 | 120 |

For Exercises 23–26, do the following.

**a.** Calculate the expected frequencies and verify that the conditions for performing the $\chi^2$ goodness of fit test are met.

**b.** Find $\chi^2_{\text{crit}}$ for the $\chi^2$ distribution with the given degrees of freedom. State the rejection rule.

**c.** Calculate $\chi^2_{\text{data}}$.

**d.** Compare $\chi^2_{\text{data}}$ with $\chi^2_{\text{crit}}$. State the conclusion and the interpretation.

**23.** $H_0: p_1 = 0.4, p_2 = 0.3, p_3 = 0.3; O_1 = 50, O_2 = 25, O_3 = 25$; level of significance $\alpha = 0.05$

**24.** $H_0: p_1 = 1/3, p_2 = 1/3, p_3 = 1/3; O_1 = 40, O_2 = 30, O_3 = 20$; level of significance $\alpha = 0.01$

**25.** $H_0: p_1 = 0.4, p_2 = 0.35, p_3 = 0.10, p_4 = 0.10, p_5 = 0.05; O_1 = 90, O_2 = 75, O_3 = 15, O_4 = 15, O_5 = 5$; level of significance $\alpha = 0.10$

**26.** $H_0: p_1 = 0.3, p_2 = 0.2, p_3 = 0.2, p_4 = 0.2, p_5 = 0.1; O_1 = 63, O_2 = 42, O_3 = 40, O_4 = 38, O_5 = 17$; level of significance $\alpha = 0.05$

For Exercises 27–30, do the following.

**a.** State the rejection rule for the *p*-value method, calculate the expected frequencies, and verify that the conditions for performing the $\chi^2$ goodness of fit test are met.

**b.** Calculate $\chi^2_{\text{data}}$.

**c.** Estimate the *p*-value.

**d.** Compare the *p*-value with level of significance $\alpha$. State the conclusion and the interpretation.

**27.** $H_0: p_1 = 0.50, p_2 = 0.50; O_1 = 40, O_2 = 60; \alpha = 0.05$

**28.** $H_0: p_1 = 0.50, p_2 = 0.25, p_3 = 0.25; O_1 = 52, O_2 = 23, O_3 = 25; \alpha = 0.10$

**29.** $H_0: p_1 = 0.5, p_2 = 0.25, p_3 = 0.15, p_4 = 0.1; O_1 = 90, O_2 = 55, O_3 = 40, O_4 = 15; \alpha = 0.10$

**30.** $H_0: p_1 = 0.4, p_2 = 0.2, p_3 = 0.2, p_4 = 0.1, p_5 = 0.1; O_1 = 90, O_2 = 45, O_3 = 40, O_4 = 15, O_5 = 10; \alpha = 0.05$

## APPLYING THE CONCEPTS

**31. Adult Education.** The National Center for Education Statistics reported on the percentages of adults who enrolled in personal-interest courses in 2005, by the highest education level completed.[2] Of these, 8% had less than a high school diploma, 23% had a high school diploma, 32% had some college, 24% had a bachelor's degree, and 13% had a graduate or professional degree. A survey taken this year of 200 randomly selected adults who enrolled in personal-interest courses showed the following numbers for the highest education level completed. Test whether the distribution of education levels has changed since 2005, using level of significance $\alpha = 0.05$.

| Less than high school | High school diploma | Some college | Bachelor's degree | Graduate or professional degree |
|---|---|---|---|---|
| 12 | 40 | 62 | 54 | 32 |

**32. Mall Restaurants.** The International Council of Shopping Centers publishes monthly sales data for malls in North America. In May 2002, the proportions of meals eaten at food establishments in malls were as follows: fast food, 30%; food court, 46%; and restaurants, 24%. A survey taken this year of 100 randomly selected meals eaten at malls showed that 32 were eaten at fast-food places, 49 were eaten at food courts, and the rest were eaten at restaurants. Test whether the population proportions have changed since 2002, using the critical-value method, and level of significance $\alpha = 0.10$.

**33. Spinal Cord Injuries.** A study in 2002 found that, of the minority patients who suffered spinal cord injury, 30% had a private health insurance provider, 55.6% used Medicare or Medicaid, and 14.4% had other arrangements.[3] Suppose that a study conducted this year of 1000 randomly selected minority patients with spinal cord injuries found that 350 had a private health insurance provider, 500 used Medicare or Medicaid, and 150 had other arrangements. Test whether the proportions have changed since 2002, using level of significance $\alpha = 0.05$.

**34. The College Experience.** A 2007 *New York Times* poll of Americans with at least a four-year college degree asked them how they would rate their overall experience as an undergraduate student. The results were 54% excellent, 39% good, 6% only fair, and 1% poor. A survey held this year of 500 randomly selected Americans with at least a four-year college degree found 275 rated their overall experience as an undergraduate student as excellent, 200 as good, 20 as only fair, and 5 as poor. Test whether the proportions have changed since 2007, using level of significance $\alpha = 0.05$.

**35. Dormitory Life.** The U.S. Census Bureau reported in 2000 that 957,581 males and 1,105,547 females were living in college dormitories. Suppose that a survey taken this year indicates that 1027 males and 1406 females were living in college dormitories. Test whether the population proportions have changed, using level of significance $\alpha = 0.05$.

**36. University Dining.** The university dining service believes there is no difference in student preference among the following four entrees: pizza, cheeseburgers, quiche, and sushi. A sample of 500 students showed that 250 preferred pizza, 215 preferred cheeseburgers, 30 preferred quiche, and 5 preferred sushi. Test at level of significance $\alpha = 0.01$ whether or not there is a difference in student preference among the four entrees.

**37. Car Rentals.** In 2002, the Web site www.consumerwebwatch.org reported that the market shares held by the largest car rental agencies were as follows.

| Hertz | Avis | National | Budget | Alamo | Dollar | All others |
|---|---|---|---|---|---|---|
| 28% | 21.4% | 12.1% | 10.5% | 10.5% | 8.5% | 9% |

This year, a new survey showed that, out of every \$100 spent on car rentals, \$28 went to Hertz, \$22 went to Avis, \$12 went to National, \$10 went to Budget, \$10 went to Alamo, \$8 went to Dollar, and \$10 went to all others. Test whether the population proportions have changed since 2002, using the estimated $p$-value method and level of significance $\alpha = 0.05$.

**38. Census Data.** According to the U.S. Census Bureau, the population of the United States in 2002 was 288,369,000. Of these, 232,647,000 were white, 36,746,000 were black, 2,752,000 were American Indian, 11,559,000 were Asian, 484,000 were Hawaiian or other Pacific Islander, and 4,181,000 were two or more races. A survey of 1000 randomly selected Americans taken this year showed 770 white, 130 black, 10 American Indian, 50 Asian, 1 Hawaiian, and 39 of two or more races. Test whether the population proportions have changed since 2002, using the estimated $p$-value method and level of significance $\alpha = 0.10$.

**39. Weekly Religious Services.** A 2007 *New York Times* poll found that 31% of Americans attend religious services every week, 12% almost every week, 14% once or twice a month, 24% a few times a year, and 19% never. A survey taken this year of 100 randomly selected Americans showed 32 who attend religious services every week, 10 almost every week, 15 once or twice a month, 25 a few times a year, and 18 never. Test whether the population proportions have changed since 2007, using level of significance $\alpha = 0.10$.

**40. Community College Advising.** In 2007, the Community College Survey of Student Engagement found that 50% of students had met with an adviser by the end of their first four weeks at college, while 41% did not do so and 9% did not recall. A survey this year of 1000 randomly selected community college students found that 550 students had met with an adviser by the end of their first four weeks at college, while 370 did not do so and 80 did not recall. Test whether the population proportions have changed since 2007, using the $p$-value method and level of significance $\alpha = 0.05$.

**41. Believing in Angels.** Do you believe in angels? A Gallup poll for May 2004 found that 78% of respondents believed in angels, 12% were not sure or had no opinion, and 10% didn't believe in angels. Suppose that a survey taken this year of 1000 randomly selected people found that 820 people believed in angels, 110 were not sure or had no opinion, and 70 didn't believe in angels. Test whether the population proportions have changed since 2004, using the critical-value method and level of significance $\alpha = 0.05$.

**42. Diversity on Campus.** The University Diversity Officer is interested in determining if the ethnic composition of the university faculty has changed since 1990. In 1990, 67% of the professors were Caucasian, 20% were Asian American, 8% were African American, and 5% were Hispanic. A sample of 100 randomly selected professors this year showed 57 Caucasian, 22 Asian American, 15 African American, and 6 Hispanic. Is there evidence at level of significance $\alpha = 0.05$ that the ethnic composition of the university faculty has changed since 1990? Test using the estimated $p$-value method.

**43. U.S. Defense Budget.** The Department of Defense reported that the 1985 defense budget included 24% for personnel costs, 28% for operations, and the remainder for other budget items. Former Secretary of Defense Donald Rumsfeld reported that, in 2004, the defense budget included \$99 billion for personnel costs, \$117 billion for operations, and \$163.4 billion for other budget items. Test whether the population proportions changed from 1985 to 2004, using the $p$-value method and level of significance $\alpha = 0.05$.

WHAT IF? **44. U.S. Defense Budget.** Refer to the previous exercise. Suppose that the amount spent in 2004 on personnel costs was not \$99 billion but some amount less than \$99 billion. How would that affect the following, and why? Would the following increase, decrease, stay the same, or is there insufficient information to determine?

**a.** $\chi^2_{\text{data}}$
**b.** $p$-value
**c.** Conclusion

## 11.2 $\chi^2$ Tests for Independence and for Homogeneity of Proportions

*Objectives:* **By the end of this section, I will be able to...**

1 Explain what a $\chi^2$ test for the independence of two variables is.

2 Perform and interpret a $\chi^2$ test for the independence of two variables using the critical value method, the exact $p$-value method, and the estimated $p$-value method.

3 Perform and interpret the results of a test for the homogeneity of proportions.

## Introduction to the $\chi^2$ Test for Independence

In Section 11.1, we learned that the $\chi^2$ distribution could help us determine a model's goodness of fit to the data. In Section 11.2, we will learn two more hypothesis tests that use the $\chi^2$ distribution. Recall from Chapter 4 that a *contingency table,* also known as a *crosstabulation* or a *two-way table,* is a tabular summary of the relationship between two categorical variables. The categories of one variable label the rows, and the categories of the other variable label the columns. Each cell in the table contains the number of observations that fit the categories of that row and column. The term *contingency table* derives from the fact that the table covers all possible combinations of the values for the two variables, that is, all possible contingencies. Table 11.7 is a contingency table based on the study *How Young People View Their Lives, Futures, and Politics: A Portrait of "Generation Next."*[4] The researchers asked 1500 randomly selected respondents, "How are things in your life?" Subjects were categorized by age and response. The researchers identified those aged 18–25 in 2007 as representing "Generation Next."

Table 11.7 Contingency table showing relative frequencies of variable categories

| | Age Group | | | |
|---|---|---|---|---|
| **Response** | **Gen Nexter (18–25)** | **26+** | **Total** | **Relative frequency** |
| Very happy | 180 | 330 | **510** | $\frac{510}{1500} = 0.34$ |
| Pretty happy | 378 | 435 | **813** | $\frac{813}{1500} = 0.542$ |
| Not too happy | 42 | 135 | **177** | $\frac{177}{1500} = 0.118$ |
| **Total** | **600** | **900** | **1500** | |
| **Relative frequency** | $\frac{600}{1500} = 0.4$ | $\frac{900}{1500} = 0.6$ | | |

We can use contingency tables like Table 11.7 to determine whether two random variables are independent. Recall that two random variables are *independent* if the value of one variable does not affect the probabilities of the values of the other variable. For example, is a "Gen Nexter" (someone aged 18–25 in 2007) less likely to report that he or she is "very happy" and more likely to report that he or she is "pretty happy" than someone older? If so, then age affects the response, and the variables *age group* and *response* are not independent.

To determine whether two categorical variables are independent, using the data in a contingency table, we use a $\chi^2$ test for independence. Just like our $\chi^2$ goodness of fit test from Section 11.1, the **$\chi^2$ test for independence** is based on a comparison of the observed frequencies with the frequencies that are expected if the null hypothesis is assumed true.

**$\chi^2$ Test for Independence**

To determine whether two categorical variables are independent, using the data from a contingency table, we use a **$\chi^2$ test for independence**. The hypotheses take the form

$H_0$ : Variable $A$ and Variable $B$ are independent.

$H_a$ : Variable $A$ and Variable $B$ are not independent.

We compare the observed frequencies with the frequencies that we expect if we assume that $H_0$ is correct. Large differences lead to the rejection of the null hypothesis.

Here, we are testing whether the variables *age group* and *response* are independent. Thus, the hypotheses are

$$H_0: \textit{Age group} \text{ and } \textit{response} \text{ are independent.}$$
$$H_a: \textit{Age group} \text{ and } \textit{response} \text{ are not independent.}$$

$H_0$ states that a response to the survey question is not affected by age group. $H_a$ says that a response is affected by age group. To calculate the expected frequencies, we begin by recalling the Multiplication Rule for Two Independent Events from Chapter 5:

If $A$ and $B$ are any two independent events, $P(A \cap B) = P(A)\,P(B)$.

To illustrate, let our events be defined as $A$ = 18–25 age group, and $B$ = reported "very happy." Then, on the assumption that these events are independent, we have

$$P(\text{Gen Nexter} \cap \text{very happy}) = P(A \cap B) = P(A)P(B) = \frac{600}{1500} \cdot \frac{510}{1500}$$
$$= 0.4 \cdot 0.34 = 0.136$$

Thus, the probability that a randomly chosen young person is both a Gen Nexter and is very happy is 0.136. Then, to find the expected frequency of this cell (Gen Nexters who are very happy), we multiply this probability 0.136 by the total sample size $n = 1500$, using the result from Section 11.1 that the expected frequency is

$$E = \text{expected frequency} = n \cdot p = 1500 \cdot 0.136 = 204$$

In other words, if the random variables *age group* and *response* are independent, then the expected frequency of Gen Nexters who report being very happy is 204. To recapitulate: the expected frequency for the Gen-Nexters-who-are very-happy cell is calculated as follows:

$$\text{expected frequency}_{\text{Gen Nexter/very happy}} = 1500 \cdot \frac{600}{1500} \cdot \frac{510}{1500} = 204$$

But note that two of the 1500s cancel, providing us with the shortcut

$$\text{expected frequency}_{\text{Gen Nexter/very happy}} = \frac{(600)(510)}{1500} = 204$$

Generalizing, this provides us with the following shortcut method for finding expected frequencies. Table 11.8 shows the expected frequencies calculated using the shortcut method.

**Expected Frequencies for a $\chi^2$ Test for Independence**

The expected frequencies for the cells of a contingency table in a $\chi^2$ test for independence are given by

$$\text{expected frequency} = \frac{\text{(row total)(column total)}}{\text{grand total}}$$

## Example 11.8 Calculating expected frequencies using the shortcut method

Calculate the expected frequencies from Table 11.7 using the shortcut method.

**Solution**

Table 11.8 contains the expected frequencies calculated using the shortcut method. ■

Table 11.8 Expected frequencies using the shortcut method

| | Age Group | | |
|---|---|---|---|
| **Response** | **Gen Nexter (18–25)** | **26+** | **Total** |
| Very happy | $\frac{(510)(600)}{1500} = 204$ | $\frac{(510)(900)}{1500} = 306$ | **510** |
| Pretty happy | $\frac{(813)(600)}{1500} = 325.2$ | $\frac{(813)(900)}{1500} = 487.8$ | **813** |
| Not too happy | $\frac{(177)(600)}{1500} = 70.8$ | $\frac{(177)(900)}{1500} = 106.2$ | **177** |
| **Total** | **600** | **900** | **1500** |

The $\chi^2$ test for independence measures the difference between the observed frequencies and the expected frequencies using the following test statistic.

**Test Statistic for the $\chi^2$ Test for Independence**

Let $O_i$ represent the observed frequency in the $i$th cell, and $E_i$ represent the expected frequency in the $i$th cell. Then the **test statistic for the independence of two categorical variables**

$$\chi^2_{\text{data}} = \sum \frac{(O_i - E_i)^2}{E_i}$$

approximately follows a $\chi^2$ (chi-square) distribution with $(r - 1)(c - 1)$ degrees of freedom, where $r$ is the number of categories in the row variable and $c$ is the number of categories in the column variable, if the following conditions are satisfied:

**a.** None of the expected frequencies is less than 1.

**b.** At most, 20% of the expected frequencies are less than 5.

## Example 11.9 Calculating the test statistic $\chi^2_{\text{data}}$

Find the value of the test statistic $\chi^2_{\text{data}}$ for the data in Example 11.8.

**Solution**

The observed frequencies are found in Table 11.7 and the expected frequencies are found in Table 11.8. Then

$$\chi^2_{\text{data}} = \sum \frac{(O_i - E_i)^2}{E_i} = \frac{(180 - 204)^2}{204} + \frac{(330 - 306)^2}{306} + \frac{(378 - 325.2)^2}{325.2} + \frac{(435 - 487.8)^2}{487.8} + \frac{(42 - 70.8)^2}{70.8} + \frac{(135 - 106.2)^2}{106.2}$$

$$\approx 38.5192$$

■

## 2 Performing the $\chi^2$ Test for Independence

The $\chi^2$ test for independence may be performed using either the critical value method, the exact $p$-value method, or the estimated $p$-value method. We provide examples of each.

**$\chi^2$ Test for Independence: Critical Value Method**

***Step 1*** **State the hypotheses and check the conditions.**

$H_0$ : Variable $A$ and Variable $B$ are independent.

$H_a$ : Variable $A$ and Variable $B$ are not independent.

The following conditions must be met:

**a.** None of the expected frequencies is less than 1.

**b.** At most, 20% of the expected frequencies are less than 5.

The expected frequency for a given cell is

$$\text{expected frequency} = \frac{(\text{row total}) \cdot (\text{column total})}{\text{grand total}}$$

***Step 2*** **Find the critical value $\chi^2_{\text{crit}}$ and state the rejection rule.** Reject $H_0$ if $\chi^2_{\text{data}} > \chi^2_{\text{crit}}$. Use $(r - 1)(c - 1)$ degrees of freedom, where $r$ is the number of categories in the row variable and $c$ is the number of categories in the column variable. Do not include the row or column totals in your counts.

***Step 3*** **Find the test statistic $\chi^2_{\text{data}}$.**

$$\chi^2_{\text{data}} = \sum \frac{(O_i - E_i)^2}{E_i}$$

where $O_i$ = observed frequency and $E_i$ = expected frequency for each cell.

***Step 4*** **State the conclusion and the interpretation.** Compare $\chi^2_{\text{data}}$ with $\chi^2_{\text{crit}}$.

## Example 11.10 Performing the $\chi^2$ test for independence using the critical value method

Using Table 11.7, test whether *age group* is independent of *response*, using level of significance $\alpha = 0.05$.

### Solution

***Step 1*** **State the hypotheses and check the conditions.**

$H_0$ : *Age group* and *response* are independent.

$H_a$ : *Age group* and *response* are not independent.

We note from Table 11.8 that none of the expected frequencies are less than either 1 or 5. Therefore, the conditions are met, and we may proceed with the hypothesis test.

#### *What Results Might We Expect?*

Note from Figure 11.18 that, for the Gen Nexters, the observed (green) frequency of the response "very happy" is *lower* than the expected (yellow) frequency, while for the 26+ group, the observed frequency of the response "very happy" is *higher* than the expected frequency. A similar situation occurs for the "not too happy" response, while the reverse situation occurs for the "pretty happy" response. In other words, the response proportions of very happy, pretty happy, and not too happy seem to *depend* on the age group. Thus, we might expect to reject $H_0$, which states that the response does not depend on the age group. ■

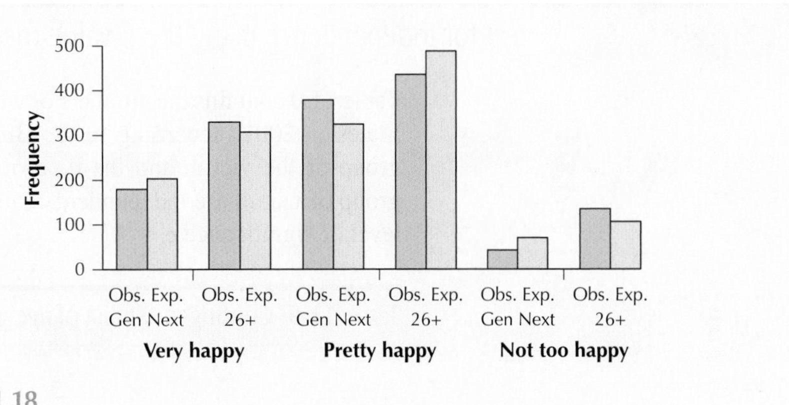


FIGURE 11.18

**Step 2** **Find the critical value $\chi^2_{\text{crit}}$ and state the rejection rule.** The row variable, *response,* has three categories, so $r = 3$. The column variable, *age group*, has two categories, so $c = 2$. Thus,

$$\text{degrees of freedom} = (r - 1)(c - 1) = (3 - 1)(2 - 1) = 2$$

With level of significance $\alpha = 0.05$, this gives us $\chi^2_{\text{crit}} = 5.991$ from the $\chi^2$ table. The rejection rule is therefore

$$\text{Reject } H_0 \text{ if } \chi^2_{\text{data}} > 5.991$$

**Step 3** **Find the test statistic $\chi^2_{\text{data}}$.** From Example 11.9, we have

$$\chi^2_{\text{data}} = \sum \frac{(O_i - E_i)^2}{E_i} \approx 38.5192$$

**Step 4** **State the conclusion and the interpretation.** Our $\chi^2_{\text{data}}$ of 38.5192 is greater than our $\chi^2_{\text{crit}}$ of 5.991, and so we reject $H_0$, which is what we expected. The interpretation is: "There is evidence that *age group* is not independent of *response.*" In other words, there does appear to be evidence that age group affects the response to the question "How are things in your life?" ■

**$\chi^2$ Test for Independence: *p*-Value Method**

***Step 1*** **State the hypotheses and the rejection rule. Check the conditions.**

$H_0$ : Variable *A* and Variable *B* are independent.

$H_a$ : Variable *A* and Variable *B* are not independent.

Reject $H_0$ if the *p*-value is less than $\alpha$.

The following conditions must be met:

**a.** None of the expected frequencies is less than 1.

**b.** At most, 20% of the expected frequencies are less than 5.

The expected frequency for a given cell is

$$\text{expected frequency} = \frac{(\text{row total})\ (\text{column total})}{\text{grand total}}$$

***Step 2*** **Find the test statistic $\chi^2_{\text{data}}$.**

$$\chi^2_{\text{data}} = \sum \frac{(O_i - E_i)^2}{E_i}$$

where $O_i$ = observed frequency and $E_i$ = expected frequency for each cell.

***Step 3*** **Find the *p*-value.**

$$p\text{-value} = P(\chi^2 > \chi^2_{\text{data}})$$

***Step 4*** **State the conclusion and the interpretation.** Compare the *p*-value with $\alpha$.

## Example 11.11 $\chi^2$ test for independence using the p-value method and technology

Table 11.9 contains the numbers of work-related homicides that took place in the United States in 2002, according to the Bureau of Labor Statistics, categorized by the age group of the victim and the type of homicide. Test whether homicide type and age group of victim are independent, using the TI-83/84, Minitab, the $p$-value method, and level of significance $\alpha = 0.01$.

Table 11.9 Contingency table of age group of victim versus type of homicide

| | Age Group of Victim | | | |
|---|---|---|---|---|
| **Type of homicide** | **Under 25** | **25 to 44** | **Over 44** | **Total** |
| Shooting | 31 | 258 | 180 | 469 |
| Stabbing | 5 | 21 | 37 | 63 |
| Total | 36 | 279 | 217 | 532 |

### What Results Might We Expect?

Figure 11.19 is a clustered bar graph of the data in Table 11.9. Note that there is no information about the expected frequencies in this graph since we have not calculated them yet. Consider the relative heights of the yellow bars and the green bars for each group. For the shootings group, the yellow bar is longer, whereas for the stabbings group, the green bar is longer. This means that workplace homicides involving shooting are more common for the 25–44 age group than for the over 44 age group, while the reverse is true for workplace homicides involving stabbing.

We see some evidence that the homicide type depends in part on the age group, and that the two variables may not be independent. We thus might expect to reject $H_0$. ■

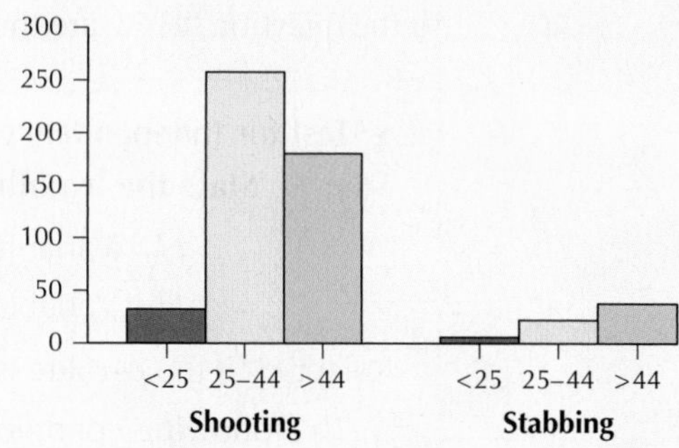


FIGURE 11.19

### Solution

**Step 1** **State the hypotheses and the rejection rule. Check the conditions.**

$H_0$ : Age group of victim and homicide type are independent.

$H_a$ : Age group of victim and homicide type are not independent.

Reject $H_0$ if the $p$-value is less than 0.01.

Note that Minitab provides the expected counts (frequencies) below the observed counts. We can then verify that none of the expected frequencies is less than 1. We do have 1 of the 6 expected frequencies (4.26) with a value less than 5. But this represents 1/6 ≈ 0.1667, which is less than 20%, as required. The $\chi^2$ hypothesis test is therefore valid.

**Step 2** **Find the test statistic $\chi^2_{\text{data}}$.** We use the instructions found in the Step-by-Step Technology Guide at the end of this section. The TI-83/84 results in Figure 11.20 tell us that $\chi^2_{\text{data}} = 10.76001797$. The Minitab results in Figure 11.21 round this to "Chi-Sq" = $\chi^2_{\text{data}} = 10.760$.

```
X²-Test
 X²=10.76001797
 P=.0046077805
 df=2
```

FIGURE 11.20 TI-83/84 $\chi^2$ results.

```
Expected counts are printed below observed counts
Chi-Square contributions are printed below expected counts
                     Age 25
        Age < 25      - 44  Age > 44  Total
    1         31       258       180    469
           31.74    245.96    191.30
           0.017     0.589     0.668

    2          5        21        37     63
            4.26     33.04     25.70
           0.127     4.387     4.971

Total         36       279       217    532

Chi-Sq = 10.760, DF = 2, P-Value = 0.005
1 cells with expected counts less than 5.
```

FIGURE 11.21 Minitab $\chi^2$ results.

**Step 3** **Find the $p$-value.** From the TI-83/84 results in Figure 11.20, we have

$$p\text{-value} = P(\chi^2 > \chi^2_{\text{data}}) = 0.0046077805$$

The Minitab results in Figure 11.21 round this to $p$-value = 0.005.

**Step 4** **State the conclusion and the interpretation.** Since $p$-value ≈ 0.0046 < 0.01, we reject $H_0$. There is evidence that the age group and homicide type are not independent. ■

## Example 11.12 The estimated $p$-value method

Recall Example 11.10, where we tested whether response ("very happy," "pretty happy," "not too happy") is independent of age group (Gen Nexters versus 26+). Here, we demonstrate the estimated $p$-value method for performing this same hypothesis test, using level of significance $\alpha = 0.05$.

### Solution

**Step 1** **State the hypotheses and check the conditions.**
The conditions were verified in Example 11.10.

$H_0$ : *Age group* and *response* are independent.

$H_a$ : *Age group* and *response* are not independent.

Reject $H_0$ if the $p$-value is less than 0.05.

**Step 2** **Find the test statistic $\chi^2_{\text{data}}$.** In Example 11.9, we found $\chi^2_{\text{data}} = 38.5192$.

**Step 3** **Estimate the $p$-value.** We have degrees of freedom = $(r - 1)(c - 1) = (3 - 1)(2 - 1) = 2$. We therefore examine the row for df = 2 in the $\chi^2$ table (see page 614). Note that $\chi^2_{\text{data}} = 38.5192$ would lie to the right of the rightmost entry in the row, $\chi^2 = 10.597$. Thus, the $p$-value is less than 0.005.

| Degrees of freedom | 0.10 | 0.05 | 0.025 | 0.01 | 0.005 |
|---|---|---|---|---|---|
| 1 | 2.706 | 3.841 | 5.024 | 6.635 | 7.879 |
| 2 | 4.605 | 5.991 | 7.378 | 9.210 | 10.597 |

$\chi^2_{\text{data}} \approx 38.5192$

***Step 4*** **State the conclusion and the interpretation.** The estimated $p$-value of less than 0.005 is less than $\alpha = 0.05$. We therefore reject $H_0$. There is evidence that *age group* is not independent of *response*.

## 3 Test for the Homogeneity of Proportions

Recall the two-sample $Z$ test for $p_1 - p_2$ from Chapter 10, where we compared the proportions of two independent populations. When we extend that hypothesis test to $k$ independent populations, the result uses a test statistic that follows a $\chi^2$ distribution. Just as the null hypothesis for the two-sample test assumed no difference between the population proportions, the null hypothesis for the $k$-sample test also assumes that all $k$ population proportions are equal. The alternative hypothesis states that not all the population proportions are equal. When performing the **test for the homogeneity of proportions**, we use the same steps as for the $\chi^2$ test for independence. The difference between the test for homogeneity of proportions and the test for independence (page 610) has to do with how the data are collected. If a single sample is taken and two variables are measured, then the test for independence is appropriate. If several ($k$) samples are taken and the sample proportion is measured for each sample, then the test for homogeneity of proportions is appropriate.

### Example 11.13 Airline on-time performance

Scott Olson/AFP/Getty Images

The Bureau of Transportation Statistics (**www.bts.gov**) reports on the proportion of airline passenger flights that are on time, for each major airline. The January–April 2007 statistics for the three busiest carriers are shown in Table 11.10. Test whether the population proportions of on-time flights are the same for the three airlines, using the $p$-value method, Minitab, and level of significance $\alpha = 0.05$.

Table 11.10 Observed on-time statistics for three major airlines, January–April 2007

| | Southwest Airlines | American Airlines | Skywest Airlines | Total |
|---|---|---|---|---|
| Number of on-time flights | 146,607 | 68,939 | 60,298 | 275,844 |
| Number of flights not on-time | 36,697 | 35,688 | 32,497 | 104,882 |
| Total flights | 183,304 | 104,627 | 92,795 | 380,726 |

#### What Results Might We Expect?

The observed sample proportions of on-time flights are as follows:

$$p_{\text{Southwest}} = \frac{146{,}607}{183{,}304} \approx 0.80 \quad p_{\text{American}} = \frac{68{,}939}{104{,}627} \approx 0.66 \quad p_{\text{Skywest}} = \frac{60{,}298}{92{,}795} \approx 0.65$$

The 80% on-time proportion of Southwest Airlines does seem to be somewhat higher than the on-time proportions of the other airlines. Thus, we would not be surprised if the hypothesis test found evidence that not all the population proportions were equal.

### Solution

The Minitab results are shown here. We use the same steps as for the $\chi^2$ test for independence.

```
Expected counts are printed below observed counts
Chi-Square contributions are printed below expected counts

        Southwest  American   Skywest   Total
    1      146607     68939     60298  275844
        132807.61  75804.46  67231.93
         1433.828   621.792   715.127

    2       36697     35688     32497  104882
         50496.39  28822.54  25563.07
         3771.027  1635.338  1880.815

Total      183304    104627     92795  380726

Chi-Sq = 10057.927, DF = 2, P-Value = 0.000
```

**Step 1** **State the hypotheses and the rejection rule. Check the conditions.**

$$H_0 : p_{\text{Southwest}} = p_{\text{American}} = p_{\text{Skywest}}$$

$$H_a : \text{Not all the proportions in } H_0 \text{ are equal.}$$

Reject $H_0$ if the $p$-value is less than 0.05.

None of the expected frequencies are less than either 1 or 5. Therefore, the conditions are met, and we may proceed with the hypothesis test.

**Step 2** **Find the test statistic $\chi^2_{\text{data}}$.** $\chi^2_{\text{data}}$ is shown as "Chi-Sq" = 10057.927. There are $r = 2$ rows and $c = 3$ columns, so the degrees of freedom are $(r - 1)(c - 1) = (2 - 1)(3 - 1) = 2$.

**Step 3** **Find the $p$-value.** Minitab provides the $p$-value, which is essentially 0.000.

**Step 4** **State the conclusion and the interpretation.** The $p$-value of 0.000 is less than $\alpha = 0.05$. We therefore reject $H_0$, as expected. There is evidence that not all population proportions of on-time flights are equal. ■

Gogo/Punchstock Images

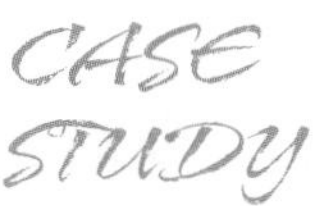


## Online Dating

We look at two tests for independence in this Case Study. The first examines whether the type of relationship reported by respondents depends on the gender of the respondent. The second investigates whether the self-reported physical appearance of online daters depends on the person's gender.

### Does the Reported Type of Relationship Depend on Gender?

As a preliminary to their analysis of online dating in their 2005 study, the Pew Internet and American Life Project examined whether single men and women differed with respect to their current relationships. The proportions are given in Table 11.11.

Table 11.11 Proportions reported in Pew study

| | Gender | |
|---|---|---|
| **Type of relationship** | **Single men** | **Single women** |
| In committed relationship | 30% | 23% |
| Not in committed relationship and not looking for partner | 42% | 65% |
| Not in committed relationship but looking for partner | 23% | 9% |
| Don't know/refused | 5% | 3% |

Unfortunately, the study did not provide sample sizes for this table, but it did provide the margins of error:

$E_{\text{men}} = 0.05$ for the single men and $E_{\text{women}} = 0.04$ for the single women

Using these margins of error, assuming a 95% confidence level, and recalling the formula for required sample size from Section 8.5, we estimate the sample sizes for the single men and single women as

$$n_{\text{men}} = \frac{0.5 \cdot Z_{\alpha/2}}{(E_{\text{men}})^2} = \frac{0.5 \cdot 1.96}{(0.05)^2} \approx 385 \qquad n_{\text{women}} = \frac{0.5 \cdot Z_{\alpha/2}}{(E_{\text{women}})^2} = \frac{0.5 \cdot 1.96}{(0.04)^2} \approx 601$$

Multiplying the proportions in Table 11.11 by these sample sizes, we derive the (estimated) observed frequencies shown in Table 11.12.

Table 11.12 Observed frequencies (estimated)

| | **Gender** | |
|---|---|---|
| **Type of relationship** | **Single men** | **Single women** |
| In committed relationship | 115 | 138 |
| Not in committed relationship and not looking for partner | 162 | 391 |
| Not in committed relationship but looking for partner | 89 | 54 |
| Don't know/refused | 19 | 18 |

We are interested in whether the type of relationship reported depends on the gender of the respondent. In other words, we will test whether the type of relationship is independent of gender. We will use the $p$-value method, with level of significance $\alpha = 0.05$, and we will follow the TI-83/84 instructions in the Step-by-Step Technology Guide at the end of this section (page 619) for the calculations.

## What Results Might We Expect?

Table 11.12 and Figure 11.22 indicate that the proportion of men who are "looking" is greater than the proportion of women who are "looking." Similarly, the proportion of women who are "not looking" is greater than for men. This is evidence that the type of relationship depends on gender and that we might expect to reject the null hypothesis of independence. ■

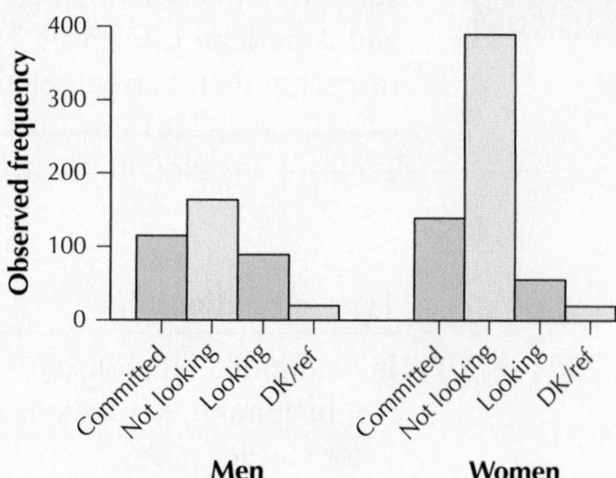


FIGURE 11.22

**Step 1** **State the hypotheses and the rejection rule. Check the conditions.**

$H_0$ : *Type of relationship* and *gender* are independent.
$H_a$ : *Type of relationship* and *gender* are not independent.

Reject $H_0$ if the $p$-value is less than 0.05.

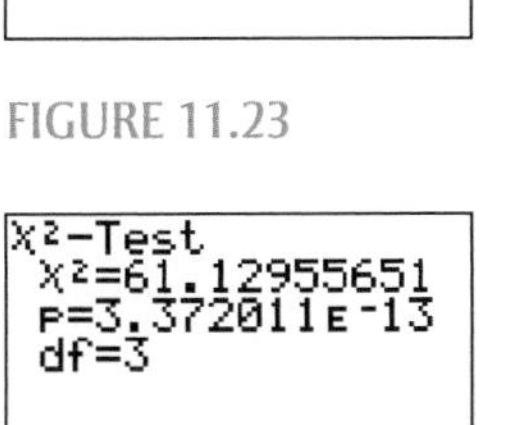


FIGURE 11.23

χ²-Test
χ²=61.12955651
p=3.372011E-13
df=3

FIGURE 11.24

Figure 11.23 shows the expected frequencies, none of which are less than 5. Thus, the conditions are met.

**Step 2** **Find $\chi^2_{\text{data}}$.** The TI-83/84 results in Figure 11.24 tell us

$$\chi^2_{\text{data}} = 61.12955651$$

**Step 3** **Find the $p$-value.** Figure 11.24 also gives us the $p$-value:

$$p\text{-value} = 3.372011\text{E-}13 \approx 0.0000000000003372011$$

**Step 4** **State the conclusion and the interpretation.** Since the $p$-value is less than $\alpha = 0.05$, we reject $H_0$, as we expected. There is evidence that the type of relationship reported in the study depends on the gender of the respondent. Since the $p$-value is so small, the significance of this result is not in doubt even though we estimated the observed frequencies.

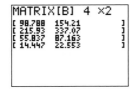

## Does Self-Reported Physical Appearance of Online Daters Depend on Gender?

A master's thesis from Massachusetts Institute of Technology examined the characteristics and behavior of online daters.[5] Table 11.13 contains the self-reported physical appearance and gender of 52,817 users of an online dating service.

Table 11.13 Gender and self-reported physical appearance

| | Physical Appearance | | | | |
|---|---|---|---|---|---|
| | **Very attractive** | **Attractive** | **Average** | **Prefer not to answer** | **Total** |
| **Female** | 3113 | 16,181 | 6093 | 3478 | 28,865 |
| **Male** | 1415 | 12,454 | 7274 | 2809 | 23,952 |
| Total | 4528 | 28,635 | 13,367 | 6287 | 52,817 |

Figure 11.25 is the clustered bar graph for the data. Are gender and self-reported physical appearance independent? We will test using the $p$-value method, with level of significance $\alpha = 0.01$, and Minitab.

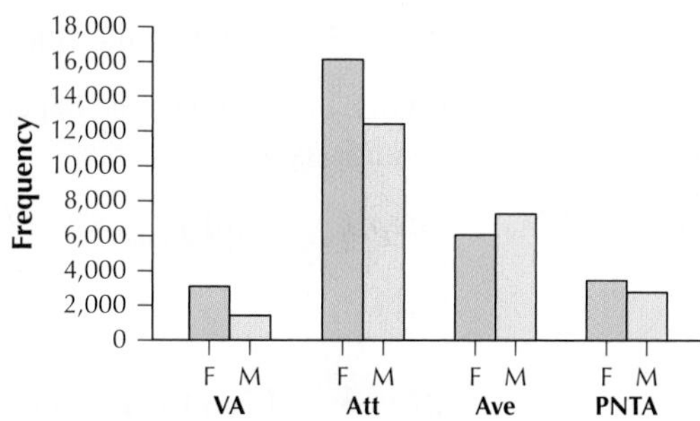


FIGURE 11.25

Figure 11.25 indicates that females have higher proportions of those self-reporting as either attractive or very attractive, while males have a higher proportion of those self-reporting as average. This is evidence that self-reported physical appearance does depend on gender and that we might expect to reject the null hypothesis of independence.

The hypotheses are

$H_0$ : *Self-reported physical appearance* and *gender* are independent.

$H_a$ : *Self-reported physical appearance* and *gender* are not independent.

We reject $H_0$ if the $p$-value is less than level of significance $\alpha = 0.01$.

The Minitab results in Figure 11.26 tell us

$$\chi^2_{\text{data}} = \text{"Chi-Sq"} = 847.702$$
$$p\text{-value} \approx 0$$

Figure 11.26 gives us the expected frequencies (highlighted in color), none of which are less than 5, allowing us to perform the hypothesis test. Since the $p$-value is less than $\alpha = 0.01$, we reject $H_0$, as we expected. There is evidence that the self-reported physical appearance depends on the gender of the online dater.

```
Expected counts are printed below observed counts
Chi-Square contributions are printed below expected counts

            VA        Att        Ave      PNTA  Total
  F       3113      16181       6093      3478  28865
       2474.60   15649.30   7305.19   3435.91
       164.698     18.065   201.147     0.516

  M       1415      12454       7274      2809  23952
       2053.40   12985.70   6061.81   2851.09
       198.480     21.770   242.406     0.621
Total     4528      28635      13367      6287  52817

Chi-Sq = 847.702, DF = 3, P-Value = 0.000
```

FIGURE 11.26 Minitab results.

## STEP-BY-STEP TECHNOLOGY GUIDE: Test for Independence or Test for the Homogeneity of Proportions

We demonstrate using Example 11.11 (pages 612–613).

### TI-83/84

#### Entering Matrix Data

***Step 1*** Press **2nd**, then **MATRIX**.
***Step 2*** Highlight **EDIT**, and press **ENTER**.
***Step 3*** Set the dimensions of **MATRIX[A]** (number of rows × number of columns). Table 11.9 has 2 rows and 3 columns, so enter **2**, press **ENTER**, enter **3**, and press **ENTER**.
***Step 4*** Press the down-arrow key. Enter the first number in the first cell, **31**, and press **ENTER**.
***Step 5*** Continue entering the data row by row until the matrix is complete (Figure 11.27).

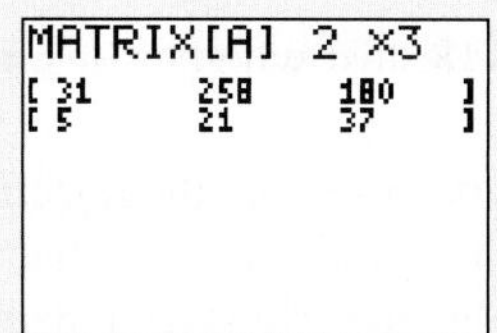


Figure 11.27

#### $\chi^2$ Test for Independence or Test for Homogeneity of Proportions

***Step 1*** Enter the data into **Matrix[A]**.
***Step 2*** Press **STAT**, highlight **TESTS**, select **C: $\chi^2$ Test**, and press **ENTER**.
***Step 3*** The expected frequencies are automatically generated and put into **MATRIX[B]**. Highlight **Calculate**, and press **ENTER**. The results are shown in Figure 11.20 in Example 11.11.
***Step 4*** To view the expected frequencies, press **2nd MATRIX**, highlight **EDIT**, choose **2** for **MATRIX[B]**, and press **ENTER**.

### EXCEL

#### $\chi^2$ Test for Independence or Test for Homogeneity of Proportions Using the WHFStat Macros

***Step 1*** Enter the data from Table 11.9, *including row and column totals,* in cells A1 to D3.
***Step 2*** Load the **WHFStat Macros**.
***Step 3*** Select **Add-Ins > Macros > Tables > Two Way Tables/Chi Squared Test**.
***Step 4*** Select cells A1 to D3 as the **Dataset Range**.
***Step 5*** Select **Chi-squared Test**, and click **OK**.

### MINITAB

#### $\chi^2$ Test for Independence or Test for Homogeneity of Proportions

***Step 1*** Enter the observed frequencies from Table 11.9 into the Minitab worksheet, as shown here.
***Step 2*** Click **Stat > Tables > Chi-Square Test**.
***Step 3*** Choose each of columns C1, C2, and C3 as the **Columns containing the table**. Then click **OK**. The results are shown in Figure 11.21 in Example 11.11.

| C1 | C2 | C3 |
|---|---|---|
| Age < 25 | Age 25 - 44 | Age > 44 |
| 31 | 258 | 180 |
| 5 | 21 | 37 |

## Section 11.2 SUMMARY

1 To determine whether two categorical variables are independent, using the data from a contingency table, we use a $\chi^2$ test for independence. The hypotheses take the form

$H_0$ : Variable *A* and Variable *B* are independent.

$H_a$ : Variable *A* and Variable *B* are not independent.

2 The $\chi^2$ test for independence is performed using the critical value method, the exact *p*-value method, or the estimated *p*-value method. The observed frequencies are compared with the expected frequencies on the assumption that $H_0$ is correct. Large differences lead to the rejection of the null hypothesis.

3 The *k*-sample test, called the test for the homogeneity of proportions, determines whether all *k* population proportions are equal. The result uses a test statistic that follows a $\chi^2$ distribution. The null hypothesis for the *k*-sample test assumes that all *k* population proportions are equal. The alternative hypothesis states that not all the population proportions are equal. When performing the test for the homogeneity of proportions, the same steps are used as for the $\chi^2$ test for independence.

## Section 11.2 EXERCISES

### CLARIFYING THE CONCEPTS

1. Explain what a contingency table is.
2. Explain in your own words what is meant by a test for independence.
3. What is the difference between the $\chi^2$ test for homogeneity of proportions and the two-sample $Z$ test for the difference in proportions from Chapter 10?
4. Explain how the expected frequencies are calculated without using the shortcut method.

### PRACTICING THE TECHNIQUES

For Exercises 5–10, the observed frequencies are provided in a contingency table of two categorical variables. Find the expected frequencies, on the assumption that the variables are independent.

5.

| | A1 | A2 |
|---|---|---|
| **B1** | 10 | 20 |
| **B2** | 12 | 18 |

6.

| | C1 | C2 |
|---|---|---|
| **D1** | 50 | 100 |
| **D2** | 60 | 90 |

7.

| | E1 | E2 | E3 |
|---|---|---|---|
| **F1** | 30 | 20 | 10 |
| **F2** | 35 | 24 | 8 |

8.

| | G1 | G2 |
|---|---|---|
| **H1** | 10 | 8 |
| **H2** | 8 | 10 |
| **H3** | 9 | 9 |

9.

| | I1 | I2 | I3 |
|---|---|---|---|
| **J1** | 100 | 90 | 105 |
| **J2** | 50 | 60 | 55 |
| **J3** | 25 | 15 | 20 |

10.

| | K1 | K2 | K3 | K4 |
|---|---|---|---|---|
| **L1** | 40 | 70 | 90 | 100 |
| **L2** | 20 | 40 | 60 | 70 |
| **L3** | 30 | 65 | 65 | 70 |

For Exercises 11–14, test whether or not the variables are independent.

- **a.** State the hypotheses.
- **b.** Verify that the conditions for performing the $\chi^2$ test for independence are met.
- **c.** Find $\chi^2_{\text{crit}}$ and state the rejection rule.
- **d.** Calculate $\chi^2_{\text{data}}$.
- **e.** Compare $\chi^2_{\text{data}}$ with $\chi^2_{\text{crit}}$. State the conclusion and the interpretation.

11. Exercise 5, level of significance $\alpha = 0.05$.
12. Exercise 7, level of significance $\alpha = 0.10$.
13. Exercise 9, level of significance $\alpha = 0.01$.
14. Exercise 9, level of significance $\alpha = 0.10$.

For Exercises 15–18, test whether or not the variables are independent.

- **a.** State the hypotheses and the rejection rule for the $p$-value method, and verify that the conditions for performing the $\chi^2$ test for independence are met.
- **b.** Find $\chi^2_{\text{data}}$.
- **c.** Estimate the $p$-value.
- **d.** Compare the $p$-value with $\alpha$. State the conclusion and the interpretation.

15. Exercise 6, level of significance $\alpha = 0.05$.
16. Exercise 8, level of significance $\alpha = 0.10$.
17. Exercise 10, level of significance $\alpha = 0.01$.
18. Exercise 10, level of significance $\alpha = 0.10$.

For Exercises 19–22, test whether or not the proportions of successes are the same for all populations.

- **a.** State the hypotheses.
- **b.** Calculate the expected frequencies and verify that the conditions for performing the $\chi^2$ test for homogeneity of proportions are met.
- **c.** Find $\chi^2_{\text{crit}}$ and state the rejection rule. Use level of significance $\alpha = 0.05$.
- **d.** Find $\chi^2_{\text{data}}$.
- **e.** Compare $\chi^2_{\text{data}}$ with $\chi^2_{\text{crit}}$. State the conclusion and the interpretation.

19.

| | Sample 1 | Sample 2 | Sample 3 |
|---|---|---|---|
| **Successes** | 10 | 20 | 30 |
| **Failures** | 20 | 45 | 62 |

20.

| | Sample 1 | Sample 2 | Sample 3 |
|---|---|---|---|
| **Successes** | 50 | 50 | 100 |
| **Failures** | 200 | 210 | 425 |

21.

| | Sample 1 | Sample 2 | Sample 3 | Sample 4 |
|---|---|---|---|---|
| **Successes** | 10 | 15 | 20 | 25 |
| **Failures** | 15 | 24 | 32 | 40 |

22.

| | Sample 1 | Sample 2 | Sample 3 | Sample 4 |
|---|---|---|---|---|
| **Successes** | 100 | 150 | 200 | 250 |
| **Failures** | 150 | 240 | 320 | 400 |

For Exercises 23–26, test whether or not the proportions of successes are the same for all populations.

**a.** State the rejection rule for the $p$-value method using level of significance $\alpha = 0.05$, calculate the expected frequencies, and verify that the conditions for performing the $\chi^2$ test for homogeneity of proportions are met.

**b.** Find $\chi^2_{\text{data}}$.

**c.** Estimate the $p$-value.

**d.** Compare the $p$-value with $\alpha$. State the conclusion and the interpretation.

23.

| | Sample 1 | Sample 2 | Sample 3 |
|---|---|---|---|
| **Successes** | 30 | 60 | 90 |
| **Failures** | 10 | 25 | 50 |

24.

| | Sample 1 | Sample 2 | Sample 3 |
|---|---|---|---|
| **Successes** | 100 | 120 | 140 |
| **Failures** | 20 | 25 | 30 |

25.

| | Sample 1 | Sample 2 | Sample 3 | Sample 4 |
|---|---|---|---|---|
| **Successes** | 10 | 12 | 24 | 32 |
| **Failures** | 6 | 10 | 15 | 30 |

26.

| | Sample 1 | Sample 2 | Sample 3 | Sample 4 |
|---|---|---|---|---|
| **Successes** | 100 | 200 | 300 | 400 |
| **Failures** | 30 | 70 | 150 | 300 |

## APPLYING THE CONCEPTS

**27. Conditioning Mice.** A psychologist is conducting research using white mice, brown mice, a classical conditioning stimulus, and an operant conditioning stimulus. The psychologist is interested in whether type of stimulus is independent of the type of mouse. One hundred mice were tested. The following table shows the number of each type of mice that completed their assigned task satisfactorily, given the type of stimulus. Test at level of significance $\alpha = 0.10$ whether type of stimulus and type of mouse are independent, using the estimated $p$-value method.

| | Type of Stimulus | | |
|---|---|---|---|
| **Type of mouse** | **Classical** | **Operant** | **Total** |
| White | 20 | 40 | 60 |
| Brown | 10 | 30 | 40 |
| Total | 30 | 70 | 100 |

**28. Information Technology Jobs.** In their report *The Digital Economy 2003,* the Economics and Statistics Administration of the U.S. Department of Commerce tracked the number of information technology–related jobs by year (1999 and 2002) and the level of education and training required (high, moderate, and low). Test whether year and level are independent, using the critical-value method and level of significance $\alpha = 0.05$.

| | High | Moderate | Low |
|---|---|---|---|
| **1999** | 3435 | 1652 | 1151 |
| **2002** | 3512 | 1411 | 1080 |

**29. Cable TV Content Restrictions.** A June 2004 *Chicago Tribune* Poll asked, "Should government restrict violence and sexual content that appears on cable TV, or should government not impose restrictions?" The responses were categorized by political affiliation. Test whether the population proportion favoring restriction is the same for all three groups, using level of significance $\alpha = 0.05$.

| | Restrict | Not restrict/don't know |
|---|---|---|
| Republicans | 59 | 41 |
| Independents | 52 | 48 |
| Democrats | 53 | 47 |

**30. History Course Preferences.** A history professor teaches three types of history courses: American history, world history, and European history. The professor would like to assess whether student preference for these courses differs among four ethnic groups. A random sample of 600 students resulted in the following table of preferences. Test using the critical-value method at level of significance $\alpha = 0.05$ whether type of course and ethnic group are independent.

| | Type of History Course | | | |
|---|---|---|---|---|
| **Ethnic group** | **American history** | **World history** | **European history** | **Total** |
| Caucasian | 100 | 40 | 60 | 200 |
| African American | 100 | 40 | 20 | 160 |
| Asian American | 50 | 60 | 30 | 140 |
| Hispanic | 50 | 20 | 30 | 100 |
| Total | 300 | 160 | 140 | 600 |

**31. Gender Gap on Campus?** Much has been written about the gender gap in the electoral process. Is there a gender gap on campus? A random sample of 500 students were selected from one campus, and their gender and political party preference were noted. Test using the $p$-value method at level of significance $\alpha = 0.01$ whether a gender gap exists on this campus (that is, whether or not gender and political party preference are independent).

| Gender | Political Party Preference | | | Total |
|---|---|---|---|---|
| | Democrat | Republican | Independent | |
| Female | 100 | 50 | 100 | 250 |
| Male | 50 | 100 | 100 | 250 |
| Total | 150 | 150 | 200 | 500 |

**32. Immigrant Origins and Preferences.** Does the state where immigrants wish to settle depend on where the immigrant is coming from? The U.S. Department of Homeland Security tracks the continent of origin and the desired state of settlement for immigrants. Some of the 2002 data are shown here, in thousands. Test using the critical-value method whether continent of origin and state of settlement are independent, using level of significance $\alpha = 0.01$.

| | California | Florida | New York |
|---|---|---|---|
| Europe | 24.0 | 9.8 | 23.2 |
| Asia | 112.6 | 9.0 | 31.3 |
| South America | 8.0 | 16.1 | 17.7 |

**33. Email, Phone, or in Person?** What is the most effective way to handle a task at work: by email, by phone, or in person? Well, you probably say, it depends on the task. The Pew Internet and American Life Project Email at Work Survey surveyed 1000 randomly selected work email users, who chose the following methods as the best for handling certain work tasks. Test whether the proportions who favor email differ between the two tasks, using level of significance $\alpha = 0.05$ and the estimated $p$-value method.

| Task | By email | By phone or in person |
|---|---|---|
| Edit or review documents | 670 | 330 |
| Arrange meetings or appointments | 630 | 370 |

**34. Using Graphs as Evidence for Your Conclusion.** Sick of spam (unsolicited broadcast email)? Do you get more spam at your work, school, or home email address? The Pew Internet and American Life Project Email at Work Survey examined the proportion of spam in email users' work and home email accounts. Using only the information in the clustered bar graph below, would you conclude that the proportion of those who report "a lot of spam" is the same for work email and personal email? Why?

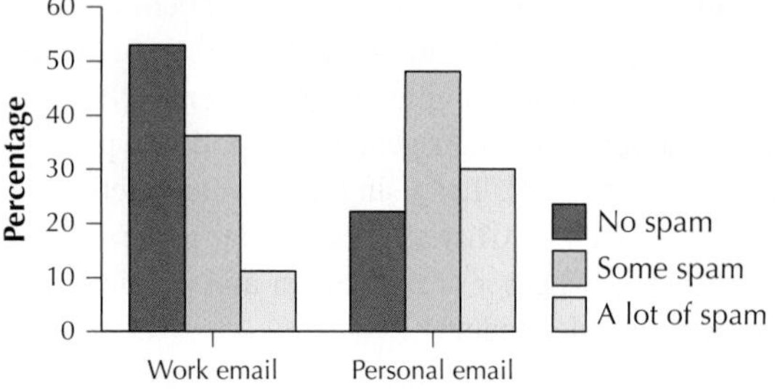


**35. Spam, Spam, Spam.** Continue your work from the previous exercise. The following contingency table shows the actual percentages in the graph above based on samples of size 100 for each of work email and personal email. Test whether the proportions who report "a lot of spam" are the same for work email and personal email, using the estimated $p$-value method and level of significance $\alpha = 0.01$. Does your conclusion agree with your conjecture in the previous exercise?

| | None | Some | A lot |
|---|---|---|---|
| Work email | 53% | 36% | 11% |
| Personal email | 22% | 48% | 30% |

**36. Gender Differences in Computer/Video/Online Gaming.** In June 2003, the Pew Internet and American Life Project released their data on the College Students Gaming Survey. Among the questions they asked 1720 randomly selected college students was "Which one of the following do you play the most: video games, computer games, or online games?" The results are summarized by gender in the following contingency table.

| | Video games | Computer games | Internet games |
|---|---|---|---|
| Male | 616 | 221 | 139 |
| Female | 198 | 372 | 174 |

**a.** Before you carry out the hypothesis test, *what result might you expect?* Look over the data set carefully to see whether you can detect significant differences between the levels of the variables. Then see whether your hypothesis test bears out your intuition.

**b.** Test whether *gender* and *game type* are independent, using the estimated $p$-value method and level of significance $\alpha = 0.01$.

**37. Online Dating.** A Pew Internet and American Life Project study reported that the proportion of urban residents who use online dating is 13%, while the proportion for suburban residents is 10% and the proportion for rural residents is 9%.[6] Test using level of significance $\alpha = 0.05$ whether there are differences among the population proportions of residents from the three categories who use online dating. Assume the study sample size was 1000. (*Hint:* The null hypothesis assumes that all proportions are equal.)

Use Minitab or Excel to perform the appropriate hypothesis test for each of Exercises 38–44 using the $p$-value method. Make sure you provide the hypotheses, the rejection rule, the test statistic, the $p$-value, the conclusion, and the interpretation.

**Goals of Middle School Students.** Open the **Goals** data set. The subjects are students in grades four, five, and six, from three schools districts in Michigan. The students were asked which of the following was most important to them: good grades, athletic ability, or popularity. Information about the students' age, gender, race, and grade was also gathered, as well as whether their school was in an urban, suburban, or rural setting.[7]

**38.** How many observations are in the data set? How many variables?

**39.** Comparing gender and goals.

**a.** Looking at the data, do you think that boys and girls at this age differ in what is most important to them: grades, popularity, or sports? In other words, do you think that the variables *gender* and *goals* are dependent or independent?

**b.** Perform the $\chi^2$ test for independence, using level of significance $\alpha = 0.05$.

**40.** Comparing gender and grade.

**a.** Looking at the data, do you think that the ratio of females to males differs significantly from grade to grade? In other words, do you think that the variables *gender* and *grade* are dependent or independent?

**b.** Perform the $\chi^2$ test for independence, using level of significance $\alpha = 0.05$.

**41.** Comparing goals and school setting.

**a.** Looking at the data, do you think that the setting of the school (urban, suburban, or rural) affects the goals of the students? Or do you think that it has no effect? In other words, do you think that the variables *urb_rur* and *goals* are independent or dependent?

**b.** Perform the $\chi^2$ test for independence, using level of significance $\alpha = 0.10$.

**42.** Comparing grades and goals.

**a.** One thing we know for sure is that, as students get older, they get more serious and grades get more important to them (don't they?). So we would expect that the variables *grade* and *goals* would be dependent, wouldn't we? Is this borne out by looking at the data?

**b.** Perform the $\chi^2$ test for independence, using level of significance $\alpha = 0.01$.

**43. 1970 Military Draft.** Is there evidence that the 1970 military draft, conducted at the height of the Vietnam War, was not truly random? For this exercise, birth dates were ranked from 1 (for the first date drawn) to 366 (the last date drawn). In 1970, only those young men with birth date rankings up to 195 were eventually drafted. Since 195 of the 366 dates were "drafted," the overall proportion of "drafted dates" is $195/366 \approx 0.5328$. Assuming the draft was truly random, we do not expect the proportion of "drafted dates" to vary significantly from month to month. In other words, the proportion of "drafted dates" should be about the same for each of the 12 months. We therefore define a multinomial random variable *drafted,* with the $k = 12$ months as categories. The monthly counts of dates not drafted and drafted are provided here. (For example, for April, 12 dates out of 30 were chosen to be drafted.) Test whether the proportions of "drafted dates" are equal for all months, using level of significance $\alpha = 0.01$.

| Month | Dates not drafted | Dates drafted | All |
|---|---|---|---|
| Jan. | 17 | 14 | 31 |
| Feb. | 16 | 13 | 29 |
| Mar. | 21 | 10 | 31 |
| Apr. | 18 | 12 | 30 |
| May | 17 | 14 | 31 |
| June | 16 | 14 | 30 |
| July | 13 | 18 | 31 |
| Aug. | 12 | 19 | 31 |
| Sept. | 11 | 19 | 30 |
| Oct. | 17 | 14 | 31 |
| Nov. | 8 | 22 | 30 |
| Dec. | 5 | 26 | 31 |
| All | 171 | 195 | 366 |

**44. 1971 Military Draft.** Criticism of the 1970 draft lottery led the U.S. Selective Service Bureau to focus on making sure that the 1971 draft lottery was truly random. Were their efforts successful? The results of the 1971 draft lottery are shown here. The Selective Service reports that all birth dates with a rank of 125 or less were chosen for the draft. Perform a $\chi^2$ test for homogeneity of proportions to determine whether the population proportion of "drafted dates" per month were all equal, using level of significance $\alpha = 0.10$.

| Month | Dates not drafted | Dates drafted |
|---|---|---|
| Jan. | 19 | 12 |
| Feb. | 19 | 9 |
| Mar. | 21 | 10 |
| Apr. | 21 | 9 |
| May | 22 | 9 |
| June | 21 | 9 |
| July | 19 | 12 |
| Aug. | 18 | 13 |
| Sept. | 23 | 7 |
| Oct. | 19 | 12 |
| Nov. | 16 | 14 |
| Dec. | 22 | 9 |

## *Try This in Class!* In-class activities to enhance your understanding of statistics

(*Note to instructor:* The small sample sizes involved in the following activities may not meet the assumptions for the $\chi^2$ test. You can alleviate this by taking a larger sample or by collapsing the categories with the smaller expected frequencies.)

### A $\chi^2$ GOODNESS OF FIT TEST FOR THE PROPORTIONS OF FRESHMEN, SOPHOMORES, JUNIORS, AND SENIORS AT YOUR COLLEGE

**1.** What do you think are the proportions of freshmen, sophomores, juniors, and seniors at your college? Write out the hypotheses for the test to determine whether these proportions are correct.

**2.** Use a sample of 25 students to test this hypothesis. What are the expected frequencies for each category? Do you have any data yet to evaluate your claim in (1)?

APPLET **3.** Each student should generate a sample of 25 students from the class, using either (a) technology, (b) the *Random Sample* applet, or (c) the method shown in the Chapter 1 Try This in Class!

**4.** For each student in your sample, ask which class he or she belongs to. Then add up the counts for each class. Which type of frequencies are these?

**5.** You will use a level of significance of $\alpha = 0.10$ for this test. What is the critical value $\chi^2_{\text{crit}}$ for this test? Will it be the same for each student?

**6.** Carry out the $\chi^2$ goodness of fit test using your sample data. Will the value of $\chi^2_{\text{data}}$ be the same for each student?

**7.** Record your name and your value for the test statistic $\chi^2_{\text{data}}$ calculated in (6) on a Post-It Note®.

**8.** Place your Post-It Note on the number line that the instructor prepared on the blackboard, noting the location of the cutoff value $\chi^2_{\text{crit}}$. We thus have a collection of values for the statistic $\chi^2_{\text{data}}$, representing an approximation of the sampling distribution of $\chi^2_{\text{data}}$. Can you make any comment on the shape of the distribution of $\chi^2_{\text{data}}$ values?

**9.** What proportion of students rejected the null hypothesis?

**10.** Reach a class consensus as to why most students either did or did not reject the null hypothesis.

## CHAPTER 11 FORMULAS AND VOCABULARY

### SECTION 11.1

- **Conditions for performing a goodness of fit test** (p. 598).
  - **a.** None of the expected frequencies is less than 1.
  - **b.** At most, 20% of the expected frequencies are less than 5.
- **$\chi^2$ Goodness of fit test** (p. 593). A hypothesis test used to ascertain whether a random variable follows a particular distribution. In a goodness of fit test, the hypotheses are

$H_0$ : The random variable follows a particular distribution.
$H_a$ : The random variable does not follow the distribution specified in $H_0$.

The observed frequencies are compared with the frequencies that we expect if we assume that $H_0$ is correct. Large differences lead to the rejection of the null hypothesis.

- **Multinomial random variable** (p. 591).
  - **a.** Each independent trial has $k$ possible outcomes, $k = 2, 3, 4, \ldots$.
  - **b.** The $i$th outcome (category) occurs with probability $p_i$, where $i = 1, 2, \ldots, k$.
  - **c.** $\sum_{i=1}^{k} p_i = 1$ (Law of Total Probability).
  - **d.** The expected frequency of the $i$th category is

$$\text{expected frequency}_i = E_i = n \cdot p_i$$

where $n$ is the number of trials, and $p_i$ is the population proportion for the $i$th category.

- **Test statistic for the goodness of fit test** (p. 594).

$$\chi^2_{\text{data}} = \sum \frac{(O_i - E_i)^2}{E_i}$$

where $O_i$ represents the observed frequency for category $i$, and $E_i$ represents the expected frequency for category $i$.

### SECTION 11.2

- **$\chi^2$ test for independence** (p. 607). The hypotheses take the form

$H_0$ : Variable $A$ and Variable $B$ are independent.
$H_a$ : Variable $A$ and Variable $B$ are not independent.

We compare the observed frequencies with the frequencies that we expect if we assume that $H_0$ is correct. Large differences lead to the rejection of the null hypothesis.

- **Conditions for performing both the test for independence and the test for the homogeneity of proportions** (p. 609).
  - **a.** None of the expected frequencies is less than 1.
  - **b.** At most, 20% of the expected frequencies are less than 5.
- **Test for the homogeneity of proportions** (p. 614). Compares the proportions of $k$ independent populations. The null hypothesis for the $k$-sample test assumes that all $k$ population proportions are equal. The alternative hypothesis states that not all the population proportions are equal.
- **Test statistic for both the test for independence and the test for the homogeneity of proportions** (p. 609).

$$\chi^2_{\text{data}} = \sum \frac{(O_i - E_i)^2}{E_i}$$

where $O_i$ represents the observed frequency for category $i$, and $E_i$ represents the expected frequency for category $i$.

# CHAPTER 11 REVIEW EXERCISES

## SECTION 11.1

For Exercises 1–6, perform the $\chi^2$ goodness of fit test.

1. **Truck-Hauled Trade.** In 2002, according to the U.S. Census Bureau, 32% of North American international truck-hauled trade (in dollars) went from the United States to Canada, 22% went from the United States to Mexico, 31% went from Canada to the United States, and 15% went from Mexico to the United States. Suppose that a survey this year showed that \$25 billion went from the United States to Canada, \$15 billion went from the United States to Mexico, \$20 billion went from Canada to the United States, and \$10 billion went from Mexico to the United States. Test whether the population proportions of truck-hauled trade have changed since 2002, using the $p$-value method and level of significance $\alpha = 0.05$.

2. **Death Row Inmates.** According to the report *Death Row, USA,* published by the National Association for the Advancement of Colored People, 42% of the people on death row in 2004 were black, 46% were white, 10% were Hispanic, and 2% were other. Suppose that a survey of randomly selected death row inmates this year found 1500 blacks, 1600 whites, 375 Hispanics, and 100 inmates otherwise classified. Test whether the population proportions have changed since 2004, using the critical value method and level of significance $\alpha = 0.10$.

3. **College Radio Preferences.** The college radio station would like to determine if the music preferences of college students have changed since 1995. In 1995, 40% of the students preferred rock music, 25% preferred rap and hip-hop, 15% preferred country music, 8% preferred blues, 7% preferred jazz, and 5% preferred classical music. A random sample of 200 university students this year showed that 60 preferred rock music, 60 preferred rap and hip-hop, 35 preferred country music, 12 preferred blues, 18 preferred jazz, and 15 preferred classical. Is there evidence at level of significance $\alpha = 0.05$ that the music preferences of college students have changed since 1995? Test using the critical value method.

4. **Alcohol Abuse and Dependence in College.** According to a 2002 report, 25% of college students had abused alcohol in the last 12 months, while a further 6% (not counted in the 25%) were alcohol dependent.[8] Suppose that a survey of 1000 randomly selected college students taken this year finds 275 who had abused alcohol in the last 12 months and a further 50 (not counted in the 275) who are alcohol dependent. Test whether the population proportions have changed since 2002, using the $p$-value method and level of significance $\alpha = 0.10$.

5. **Truly Random Lottery Drawing?** Have you ever wondered whether lottery drawings are truly random? For example, the histogram given here (see the second column) shows the frequencies of the third digit in the Maryland lottery's Pick 3 game (218 drawings from September 1989 to April 1990). In a Pick 3 game, you choose a three-digit number between 000 and 999, and if your number comes up, you win the cash prize. Notice that 1 appears as the third digit least of all the digits, and quite a bit less often than some of the other digits. Does the relative scarcity of 1s indicate that the system is flawed?

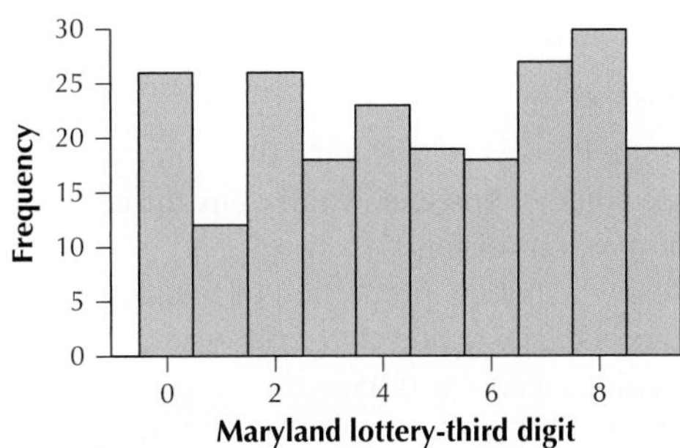


Frequency histogram of third digits in the Maryland lottery's Pick 3 game.

The relative frequency distribution of the third digit is shown in the following table. We would of course expect each digit to show up 10% of the time. Test whether the population proportions of digits are all 0.10, using level of significance $\alpha = 0.05$.

| Digit | Count | Percent |
|---|---|---|
| 0 | 26 | 11.93 |
| 1 | 12 | 5.50 |
| 2 | 26 | 11.93 |
| 3 | 18 | 8.26 |
| 4 | 23 | 10.55 |
| 5 | 19 | 8.72 |
| 6 | 18 | 8.26 |
| 7 | 27 | 12.39 |
| 8 | 30 | 13.76 |
| 9 | 19 | 8.72 |
| $N =$ | 218 | |

6. **Alternative Medicine Use.** A study conducted in 2006 examined the prevalence of alternative medicine usage by age group among persons with diabetes.[9] In the study, 5.7% of the subjects were aged 18–34 years, 20.7% were aged 35–49 years, 38.8% were aged 50–64 years, and 34.8% were age 65 or older. Suppose that a study conducted this year found that, of the 1000 randomly selected respondents with diabetes, 70 were 18–34 years old, 220 were 35–49 years old, 440 were 50–64 years old, and 270 were over age 65. Test using level of significance $\alpha = 0.05$ whether the proportions have changed since 2006.

## SECTION 11.2

7. **Grades and the SAT.** In their 2003 "Profile of College-Bound Seniors," the College Board provided the following

data on high school grade point average and gender for the students taking the SAT exam. We are interested in testing whether the proportion of females is the same across the six grade categories.

| | High School Grade Point Average | | | | | |
|---|---|---|---|---|---|---|
| **Gender** | **A+** | **A** | **A−** | **B** | **C** | **D–F** |
| Female | 60 | 62 | 59 | 53 | 43 | 43 |
| Male | 40 | 38 | 41 | 47 | 57 | 57 |

**a.** Before you perform any calculations, *what result might we expect*? Examine the table carefully to see whether you can identify any differences in the proportions of females.

**b.** Test whether the proportion of females is the same across the six grade categories, using level of significance $\alpha = 0.05$.

**8. Homicide Victims by Race.** The U.S. Bureau of Justice Statistics reported the following numbers of homicide victims, by gender and race, for the state of California in the year 2000. Test whether gender and race are independent, using level of significance $\alpha = 0.10$.

| Gender | White | Black | Other |
|---|---|---|---|
| Male | 1089 | 516 | 64 |
| Female | 297 | 85 | 23 |

**9. Pregnancy and HIV Testing.** A study examined the proportions of pregnant women in the United States who have had an HIV test in the past 12 months.[10] The proportions for the Northeast, Midwest, South, and West were 56.8%, 49.3%, 58.5%, and 50.2%. Test whether the population proportions of pregnant women who have had an HIV test in the past 12 months are the same across all four regions, using level of significance $\alpha = 0.01$. Assume that each sample size equals 1000.

**10. September 11 and Pearl Harbor.** The terrorist attacks on New York City and Washington, D.C., on September 11, 2001, were often compared to the Japanese attack on Pearl Harbor on December 7, 1941. In an NBC News Terrorism Poll, the following question was asked: Would you say that Tuesday's attacks are more serious than, equal to, or not as serious as the Japanese attack on Pearl Harbor? This poll was conducted on September 12, 2001, and the results are given in the following table (see the second column). Were there systematic differences in the way men and women responded to this question? In other words, are the variables *poll response* and *gender* independent? Perform the $\chi^2$ test for independence between *poll response* and *gender*, using the *p*-value method and level of significance $\alpha = 0.01$.

| | Gender | | |
|---|---|---|---|
| | **Male** | **Female** | **Total** |
| More serious | 200 | 212 | 412 |
| Equal | 70 | 84 | 154 |
| Not as serious | 23 | 6 | 29 |
| Not sure | 11 | 12 | 23 |
| Total | 304 | 314 | 618 |

**11. Radio-Listening Trends.** The Arbitron Corporation tracks radio-listening trends among demographic groups. In a 2003 survey, 240 teens aged 12–17 listened to pop contemporary hit radio stations while 170 teens listened to alternative radio stations. Also, 250 young adults aged 18–24 listened to pop contemporary hit radio stations while 260 young adults listened to alternative radio stations. Test whether age and radio station type are independent, using level of significance $\alpha = 0.05$.

**12. Happiness in Marriage.** The General Social Survey tracks trends in American society. The accompanying crosstabulation shows the responses to a question that asked people to characterize their feelings about being married. Test whether happiness in marriage is independent of gender, using level of significance $\alpha = 0.05$.

| | Happiness in Marriage | | | |
|---|---|---|---|---|
| **Respondents' gender** | **Very happy** | **Pretty happy** | **Not too happy** | **Total** |
| Male | 242 | 115 | 9 | 366 |
| Female | 257 | 149 | 17 | 423 |
| Total | 499 | 264 | 26 | 789 |

**13. The Digital Divide: Accounting for Income.** It is well known that a greater proportion of whites than blacks use the Internet. This is one aspect of what is known as the "digital divide." However, what if we control for income? That is, suppose that we consider only white, blacks, and Hispanics of a certain annual income range, say, more than $50,000. The Pew Internet and American Life Project conducted a survey in Spring 2002 in which the following proportions of respondents with incomes above $50,000 were found to be using the Internet. Test whether the digital divide exists after accounting for income. That is, test whether or not there is a significant difference in Internet use levels among the races. Use the estimated *p*-value method, with level of significance $\alpha = 0.05$. Assume each sample size equals 400.

| Whites | Blacks | Hispanics |
|---|---|---|
| 82% | 65% | 82% |

# CHAPTER 11 Quiz

### TRUE OR FALSE

**1.** True or false: In a goodness of fit test, large differences between the observed frequencies and the expected frequencies lead to rejection of the null hypothesis.

**2.** True or false: In a test for independence, the degrees of freedom equals $k - 1$.

**3.** True or false: In the test for the homogeneity of proportions, the alternative hypothesis states that all the population proportions are different.

### FILL IN THE BLANK

**4.** The conditions for performing a goodness of fit test are that none of the expected frequencies is less than ____, and, at most, 20% of the expected frequencies are less than ____.

**5.** In the test for the homogeneity of proportions, the null hypothesis states that all $k$ population proportions are _____.

**6.** The ________ ________ [two words] of the $i$th category is given by the formula $n \cdot p_i$, where $n$ is the number of trials, and $p_i$ is the population proportion for the $i$th category.

### SHORT ANSWER

**7.** Name the three methods for performing the $\chi^2$ goodness of fit test.

**8.** In the test for the homogeneity of proportions, which hypothesis states that not all population proportions are equal?

**9.** How does one calculate the degrees of freedom for the $\chi^2$ test for independence?

### CALCULATIONS AND INTERPRETATIONS

For Exercises 10–13, the alternative hypothesis takes the following form:

$H_a$ : The random variable does not follow the distribution specified in $H_0$.

For each exercise, do the following.

- **a.** Find the expected frequencies.
- **b.** Show that the conditions for performing the $\chi^2$ goodness of fit test are met.
- **c.** Find $\chi^2_{\text{crit}}$ for the $\chi^2$ distribution with the given degrees of freedom. Find the rejection rule.
- **d.** Find $\chi^2_{\text{data}}$.
- **e.** Compare $\chi^2_{\text{data}}$ with $\chi^2_{\text{crit}}$. State the conclusion and the interpretation.

**10.** $H_0: p_1 = 0.2, p_2 = 0.2, p_3 = 0.2, p_4 = 0.2, p_5 = 0.2$; $O_1 = 8, O_2 = 9, O_3 = 10, O_4 = 11, O_5 = 12$; $\alpha = 0.05$

**11.** $H_0: p_1 = 0.3, p_2 = 0.25, p_3 = 0.20, p_4 = 0.15, p_5 = 0.06$, $p_6 = 0.04$; $O_1 = 50, O_2 = 40, O_3 = 30, O_4 = 20, O_5 = 10$; $O_6 = 10$; $\alpha = 0.05$

**12.** $H_0: p_1 = 0.2, p_2 = 0.2, p_3 = 0.2, p_4 = 0.2, p_5 = 0.2$; $O_1 = 18, O_2 = 19, O_3 = 21, O_4 = 22, O_5 = 20$; $\alpha = 0.01$

**13.** $H_0: p_1 = 0.3, p_2 = 0.25, p_3 = 0.20, p_4 = 0.15, p_5 = 0.06$, $p_6 = 0.04$; $O_1 = 65, O_2 = 55, O_3 = 30, O_4 = 25, O_5 = 15$, $O_6 = 10$; $\alpha = 0.05$

**14. Illicit Drug Use Among Young People.** Monitoring the Future (**www.monitoringthefuture.org**), at the University of Michigan, is an "an ongoing study of the behaviors, attitudes, and values of American secondary school students, college students, and young adults." They reported the lifetime prevalence of the use of any illicit drug among 8th-graders, 10th-graders, and 12th-graders, as shown in the table.

| | 8th-graders | 10th-graders | 12th-graders |
|---|---|---|---|
| Have used an illicit drug | 3,655 | 6,527 | 7,461 |
| Have never used an illicit drug | 13,345 | 9,873 | 7,139 |

- **a.** Before you carry out the hypothesis test, *what result might you expect?* Look over the data set carefully to see whether you can detect significant differences between the levels of the variables. Then see whether your hypothesis test bears out your intuition.
- **b.** Test using level of significance $\alpha = 0.01$ for differences among the proportions of children in those grades who have ever used an illicit drug.

**15. The Digital Divide: Accounting for Education Level.** Refer to Exercise 13 on page 626. Suppose that we consider only whites, blacks, and Hispanics of a certain education level, say, with a bachelor's degree or more. The Pew Internet and American Life Project conducted a survey in Spring 2002 in which the following proportions of respondents with bachelor's degrees or more were found to be using the Internet. Test whether the digital divide exists after accounting for education level. That is, test whether or not there is a significant difference in Internet use levels among the races. Use the estimated $p$-value method, with level of significance $\alpha = 0.01$. Assume each sample size equals 400.

| Whites | Blacks | Hispanics |
|---|---|---|
| 83% | 76% | 87% |

**16. Sport Preference and Gender.** A student group would like to determine whether or not gender and participatory sport preference are independent on campus. A survey of 200 randomly selected college students showed the following sport preferences. Test at level of significance $\alpha = 0.05$ whether gender and sport preference are independent.

| | Sport Preference | | | |
|---|---|---|---|---|
| Gender | Basketball | Soccer | Swimming | Total |
| Female | 30 | 20 | 50 | 100 |
| Male | 50 | 30 | 20 | 100 |
| Total | 80 | 50 | 70 | 200 |

**17. Beef Cattle and Farm Size.** The National Agricultural Statistics Service publishes data on farm products in the United States.[11] The accompanying table shows the number of beef cattle on smaller-scale operations (farms having fewer than 50 head) for three states. Test whether the proportions of cattle on smaller farms are the same across all three states, using level of significance $\alpha = 0.05$.

| | Texas | Oklahoma | Pennsylvania |
|---|---|---|---|
| Beef cattle on smaller scale operations | 103,000 | 3,600 | 11,400 |
| Beef cattle on operations that are not smaller scale | 28,000 | 44,400 | 600 |

Photo Illustration by Chris Jackson/Getty Images

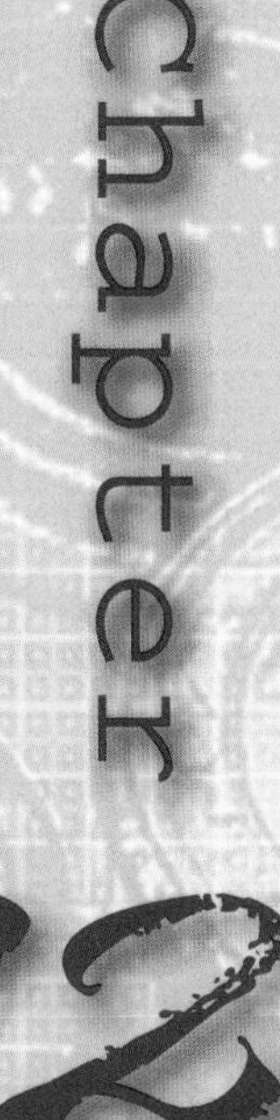

Chapter 12

# Analysis of Variance

CASE STUDY

## Professors on Facebook

To improve communication with students, many instructors have been turning to computer-mediated interaction via online social networks. Online social networks serve as virtual meeting spots, where instructors and students can share information and interests outside the classroom. Examples of online social networks include Facebook, MySpace, and Second Life. Do you think that the amount of information a professor posts about himself or herself, that is, *self-disclosure,* has an effect on the motivation, affective learning, classroom climate, or other important instructional issues? A recently published research paper asked precisely this question.[1] The researchers' primary research tool was *analysis of variance* (*ANOVA*), our topic here in Chapter 12. We investigate their results in Section 12.2 in the Case Study, Professors on Facebook.

OVERVIEW



## The Big Picture — Where we are coming from, and where we are headed...

In Chapter 11 we examined the goodness of fit of a categorical data set to the hypothesized distribution. We also learned how to test for the independence of two categorical variables and for the homogeneity of proportions from several populations.

Chapters 12 and 13 introduce you to perhaps the most powerful and widespread statistical procedures in the world. In this chapter, we compare the means of several populations to determine whether significant differences exist. Suppose you belong to a fraternity and have friends who belong to three different fraternities. You are interested in whether the mean grade point averages among the four fraternities differ or are all the same. How would you go about investigating this question?

This chapter introduces us to *analysis of variance,* a way to approach problems like this. The idea is to compare the population means of several different groups and determine whether significant differences exist between these means. But first we need to introduce a new distribution in Section 12.1, called the $F$ distribution. The $F$ distribution will help us perform hypothesis tests in Chapters 12 and 13. You may be relieved to hear that it is the last new distribution we shall meet in this text.

There is only one more chapter left in the text, Chapter 13, Regression Analysis. In that chapter we shall examine whether there is evidence for a relationship between two continuous variables.

# 12.1 Introduction to Analysis of Variance (ANOVA)

## Objectives: By the end of this section, I will be able to...

1. Apply the characteristics of the $F$ distribution to find the $F$ critical value
2. Explain that ANOVA works by comparing the variation between samples to the variation within the samples
3. Construct an ANOVA table, and compute and interpret the statistics contained in the table

In Chapter 11, we used the $\chi^2$ distribution to analyze categorical data. Here, in Chapter 12, we need to learn about a new distribution, the $F$ distribution, which will help us with the analytic methods we will learn in this chapter and in Chapter 13.

## 1 *F* Distribution

A. Barrington Brown / Photo Researchers.

The ***F* distribution** was named in honor of the "grandfather of statistics," Sir Ronald A. Fisher. Like the $\chi^2$ distribution, the *F* distribution is right-skewed, never takes negative values, and has an infinite number of different *F* curves (Figure 12.1). The shape of the curve depends on two different degrees of freedom.

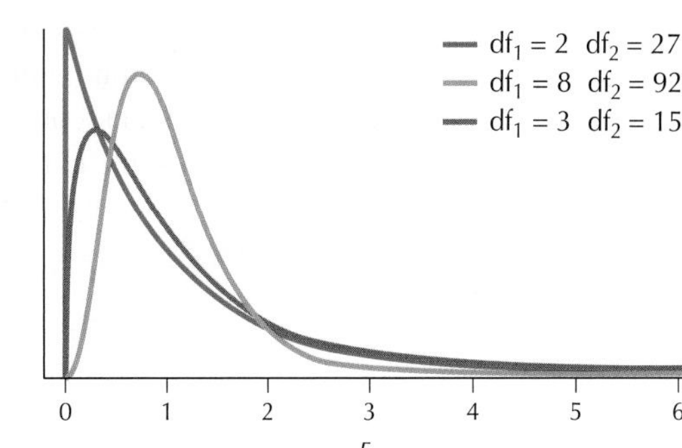


FIGURE 12.1
Shape of the *F* distribution for various degrees of freedom.

Note that the *F* distribution resembles the $\chi^2$ distribution. This is not surprising since the values of the *F* distribution represent ratios of two $\chi^2$ distributions. Moreover, the *F* distribution has two different degrees of freedom, which we shall call $df_1$ and $df_2$, derived from the degrees of freedom of the two $\chi^2$ distributions represented in the ratio. Often, $df_1$ is called the *numerator degrees of freedom,* and $df_2$ is called the *denominator degrees of freedom.*

**Properties of the *F* Curve**

1. The total area under the *F* curve equals 1.
2. The value of the *F* random variable is never negative, so the *F* curve starts at 0. However, it extends indefinitely to the right. The curve approaches but never quite meets the horizontal axis.
3. Because of the characteristics described in (2), the *F* curve is right-skewed.
4. There is a different *F* curve for each different pair of degrees of freedom, $df_1$ and $df_2$.

Since the *F* distribution is continuous, we can find probabilities associated with values of *F*, and vice versa, just as we did with the normal, *t*, and $\chi^2$ distributions. Just as for any continuous distribution, probability is represented by the area below the *F* curve above an interval.

To perform the hypothesis tests in Chapters 12 and 13, we need to find the critical values of an *F* distribution for a given $\alpha$. Similar to the $\chi^2$ procedures from Chapter 11, we are interested only in the area to the *right* of a particular value of *F*. For example, we need to find the value of an *F* distribution that has area $\alpha$ to the right of it. To find these *F* critical values, we shall work with the *F* tables (see Table F in the Appendix, pages T-13–T-16). The *F* tables are somewhat different from the other tables we have worked with so far.

When you are asked to find the critical value $F_{crit}$, you will be given $\alpha$, $df_1$, and $df_2$. To find $F_{crit}$:

- Look across the top of the *F* table until you find your $df_1$.
- Then go down that column until you see your $df_2$ on the left.
- For each $df_2$ on the left you will see a range of $\alpha$ values from 0.100 to 0.001. Choose the row next to $df_2$ that has your value of $\alpha$.
- The *F* value in that row and column is your value of $F_{crit}$.

## Example 12.1 Finding $F_{crit}$

Use the excerpt from the $F$ table in Figure 12.2 to find the $F$ critical value $F_{crit}$ when $\alpha = 0.05$, $df_1 = 2$, and $df_2 = 7$.

**Solution**

First we go across the top of the $F$ table until we get to $df_1 = 2$. Then we go down that column until we see $df_2 = 7$ on the left. Next to the 7 is a range of $\alpha$ values (0.10, 0.05, and so on). We choose the row with $\alpha = 0.05$. The $F$ value in that row and column is 4.74. Thus, $F_{crit} = 4.74$, as shown in Figure 12.2. ■

*F* distribution critical values

| | | $df_1$ | | |
|---|---|---|---|---|
| $df_2$ | **Area in right tail** | **1** | **2** | **3** |
| **1** | 0.100 | 39.86 | 49.59 | 53.59 |
| | 0.050 | 161.45 | 199.50 | 215.71 |
| | 0.025 | 647.79 | 799.50 | 864.16 |
| | 0.010 | 4052.20 | 4999.50 | 5403.40 |
| | 0.001 | 405284.00 | 500000.00 | 540379.00 |
| **2** | 0.100 | 8.53 | 9.00 | 9.16 |
| | 0.050 | 18.51 | 19.00 | 19.16 |
| | 0.025 | 38.51 | 39.00 | 39.17 |
| | 0.010 | 98.50 | 99.00 | 99.17 |
| | 0.001 | 998.50 | 999.00 | 999.17 |
| **3** | 0.100 | 5.54 | 5.46 | 5.39 |
| | 0.050 | 10.13 | 9.55 | 9.28 |
| | 0.025 | 17.44 | 16.04 | 15.44 |
| | 0.010 | 34.12 | 30.82 | 29.46 |
| | 0.001 | 167.03 | 148.50 | 141.11 |
| **4** | 0.100 | 4.54 | 4.32 | 4.19 |
| | 0.050 | 7.71 | 6.94 | 6.59 |
| | 0.025 | 12.22 | 10.65 | 9.98 |
| | 0.010 | 21.20 | 18.00 | 16.69 |
| | 0.001 | 74.14 | 61.25 | 56.18 |
| **5** | 0.100 | 4.06 | 3.78 | 3.62 |
| | 0.050 | 6.61 | 5.79 | 5.41 |
| | 0.025 | 10.01 | 8.43 | 7.76 |
| | 0.010 | 16.26 | 13.27 | 12.06 |
| | 0.001 | 47.18 | 37.12 | 33.20 |
| **6** | 0.100 | 3.78 | 3.46 | 3.29 |
| | 0.050 | 5.99 | 5.14 | 4.76 |
| | 0.025 | 8.81 | 7.26 | 6.60 |
| | 0.010 | 13.75 | 10.92 | 9.78 |
| | 0.001 | 35.51 | 27.00 | 23.70 |
| **7** | 0.100 | 3.59 | 3.26 | 3.07 |
| | 0.050 | 5.59 | 4.74 | 4.35 |
| | 0.025 | 8.07 | 6.54 | 5.89 |
| | 0.010 | 12.25 | 9.55 | 8.45 |
| | 0.001 | 29.25 | 21.69 | 18.77 |

FIGURE 12.2 *F* table.

## 2 How Analysis of Variance (ANOVA) Works

**Analysis of variance (ANOVA)** is an inferential method for testing whether the means of different populations are equal. Why is it called *analysis of variance* if we use it for testing *means*? The answer lies in the way ANOVA works—by comparing different variance estimates, as we shall see. But first, we learn conceptually how ANOVA works.

Suppose we are interested in determining whether there are significant differences in grade point averages (GPAs) among residents of three dormitories, A, B, and C. Table 12.1 displays three random samples of GPAs of ten residents from each dormitory.

Table 12.1 Sample GPAs from Dorms A, B, and C

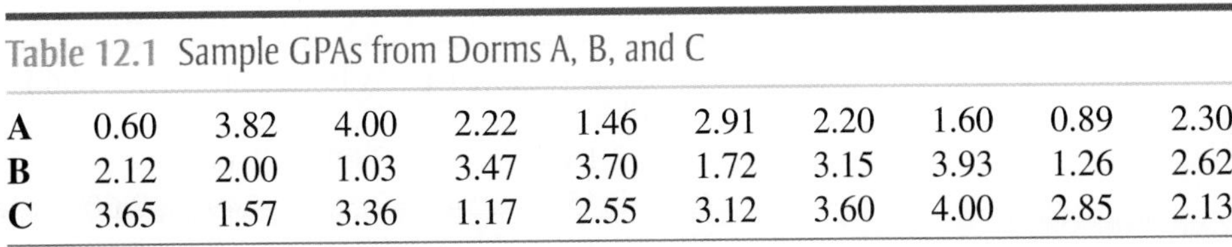

| | | | | | | | | | | |
|---|---|---|---|---|---|---|---|---|---|---|
| **A** | 0.60 | 3.82 | 4.00 | 2.22 | 1.46 | 2.91 | 2.20 | 1.60 | 0.89 | 2.30 |
| **B** | 2.12 | 2.00 | 1.03 | 3.47 | 3.70 | 1.72 | 3.15 | 3.93 | 1.26 | 2.62 |
| **C** | 3.65 | 1.57 | 3.36 | 1.17 | 2.55 | 3.12 | 3.60 | 4.00 | 2.85 | 2.13 |

The sample mean GPA for Dormitory A is

$$\bar{x}_A = \frac{0.60 + 3.82 + 4.00 + 2.22 + 1.46 + 2.91 + 2.20 + 1.60 + 0.89 + 2.30}{10} = 2.2$$

Similarly, we can find the sample mean GPAs for the other dormitories: $\bar{x}_B = 2.5$ and $\bar{x}_C = 2.8$. We note that the sample means are not equal. The question is, Are the population means equal? Let $\mu_A$, $\mu_B$, and $\mu_C$ represent the population mean GPAs for Dormitories A, B, and C, respectively. We are interested in the following hypotheses, where $\mu_i$ represents the population mean GPA for dormitory $i$:

$$H_0: \mu_A = \mu_B = \mu_C \quad \text{versus} \quad H_a: \text{not all the population means are equal}$$

*Note:* In analysis of variance, the null hypothesis *always* states that all the population means are equal and the alternative hypothesis *always* states that not all the population means are equal. Note that $H_a$ is *not* stating that the population means are all different. For $H_a$ to be true, it is sufficient for a single population mean to be different, even though all the other population means are equal.

Sufficient differences in the sample means would represent evidence that the population means were not equal. The question is, What represents "sufficiently" different? We need something to compare against, such as the *spread* of each sample. One measure of spread or variability is the *range*:

$$\text{range} = \text{max} - \text{min}$$

We have

$$\text{range (Dorm A)} = 4.00 - 0.60 = 3.40$$
$$\text{range (Dorm B)} = 3.93 - 1.03 = 2.90$$
$$\text{range (Dorm C)} = 4.00 - 1.17 = 2.83$$

These ranges are rather large spreads, and there is a considerable amount of overlap among the different dormitory GPAs, as shown in Figure 12.3.

Figure 12.3 shows the difference among the means for the three dorm GPAs compared with the spread of each dorm's GPAs, as measured by the range. The red triangles represent the sample means, $\bar{x}_A = 2.2$, $\bar{x}_B = 2.5$, and $\bar{x}_C = 2.8$. The spread of the sample means (shown by the red arrows) is much less than the spreads of the individual dorm GPAs (shown by the blue arrows). Thus, the sample means $\bar{x}_A = 2.2$, $\bar{x}_B = 2.5$, and $\bar{x}_C = 2.8$ are not sufficiently different when compared against the spread of the GPAs. This graph would therefore not provide evidence to reject the null hypothesis that the population mean GPAs are all equal.

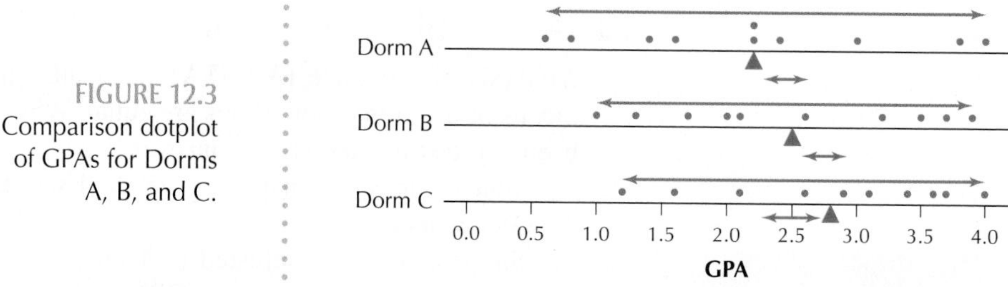


FIGURE 12.3
Comparison dotplot of GPAs for Dorms A, B, and C.

Now we make a similar comparison for the GPAs for Dormitories D, E, and F in Table 12.2.

Table 12.2 Sample GPAs from Dorms D, E, and F

| | | | | | | | | | | |
|---|---|---|---|---|---|---|---|---|---|---|
| **D** | 2.16 | 2.23 | 2.09 | 2.17 | 2.25 | 2.19 | 2.24 | 2.28 | 2.25 | 2.14 |
| **E** | 2.45 | 2.34 | 2.58 | 2.49 | 2.60 | 2.42 | 2.55 | 2.62 | 2.45 | 2.50 |
| **F** | 2.80 | 2.75 | 2.93 | 2.68 | 2.88 | 2.75 | 2.87 | 2.81 | 2.73 | 2.80 |

The sample mean GPAs for Dormitories D, E, and F are the *same* as those for Dormitories, A, B, and C, respectively: $\bar{x}_D = 2.2$, $\bar{x}_E = 2.5$, and $\bar{x}_F = 2.8$. Again we are interested in whether the population means are equal.

$$H_0 : \mu_D = \mu_E = \mu_F \quad \text{versus} \quad H_a : \text{not all the population means are equal}$$

Consider the comparison dotplot in Figure 12.4. There now seems to be better evidence for concluding that the three population means are not all equal. There is no overlap among the three samples because the spread *within* each dormitory is much smaller than for Dormitories A, B, and C.

$$\text{range (Dorm D)} = 2.28 - 2.09 = 0.19$$
$$\text{range (Dorm E)} = 2.62 - 2.34 = 0.28$$
$$\text{range (Dorm F)} = 2.93 - 2.68 = 0.25$$

Figure 12.4 shows the difference among the means for the three dorm GPAs compared with the range of each dorm's GPAs. The red triangles represent the sample means, $\bar{x}_D = 2.2$, $\bar{x}_E = 2.5$, and $\bar{x}_F = 2.8$. The spread of the sample means (red arrows) is much greater than the spreads of the individual dorm GPAs (blue arrows). Thus, the sample means $\bar{x}_D = 2.2$, $\bar{x}_E = 2.5$, and $\bar{x}_F = 2.8$ are sufficiently different when compared against the range of the GPAs. This graph would, therefore, provide some evidence to reject the null hypothesis that the population mean GPAs are all equal.

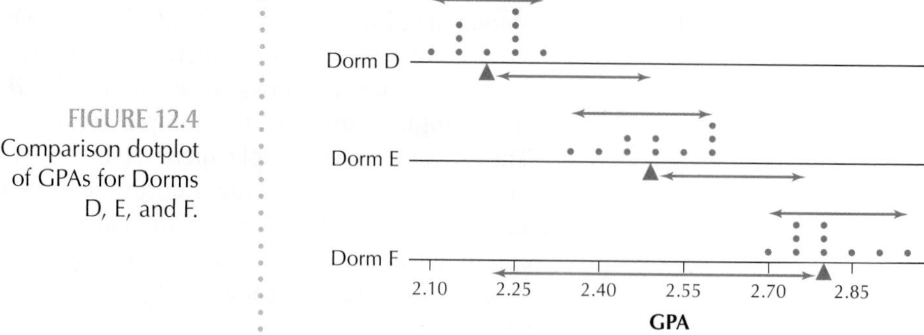


FIGURE 12.4
Comparison dotplot of GPAs for Dorms D, E, and F.

Note that we arrived at *opposite conclusions* for the two sets of dormitories, *even though the sample means of the first group are identical to the sample means of the second group.* Here is the key: compared to the large within-sample spread of Dormitories A, B, and C, the difference in sample means did not seem large. However, compared to the small within-sample spread of Dormitories D, E, and F, the difference in sample means did seem large. These are the types of comparisons that the ANOVA method makes.

Instead of using the range as the measure of spread, analysis of variance uses the standard deviation of the individual samples. Recall that samples with larger spread have larger standard deviations, just as they have larger range.

Developing Your Statistical Sense

### How Does Analysis of Variance Work?

The key to how analysis of variance works is the following comparison. Compare

**a.** the variability in the sample means—that is, how large the differences are between the sample means (indicated by the lengths of the red arrows in Figures 12.3 and 12.4)—with

**b.** the variability within each sample—that is, the within-sample spreads (indicated by the lengths of the blue arrows in Figures 12.3 and 12.4).

When **(a)** is much larger than **(b)**, this is evidence that the population means are not all equal and that we should reject the null hypothesis. Thus, our analysis depends on measuring variability. And hence the term *analysis of variance.* ■

Just as for hypothesis-testing procedures from previous chapters, analysis of variance can be performed only if certain conditions are satisfied.

**Requirements for Performing Analysis of Variance**

1. Each of the $k$ populations is normally distributed.
2. The variances ($\sigma^2$) of the populations are all equal.
3. The samples are independently drawn.

Our hypotheses for testing for the equality of the population mean GPA for Dormitories A, B, and C are

$$H_0 : \mu_A = \mu_B = \mu_C \quad \text{versus} \quad H_a : \text{not all the population means are equal}$$

Think about what $H_0$ is saying, in the light of the requirements specified. If $H_0$ is true, then all three dormitories would have the same population mean GPA: $\mu_A = \mu_B = \mu_C = \mu$, where we denote the hypothesized common mean as $\mu$. Further, requirement 1 states that each population is normally distributed and requirement 2 states that all the population variances are equal. Let's call this common variance $\sigma^2$. Putting all this together, the null hypothesis for performing analysis of variance assumes that the observations from each population come from the same distribution, and this distribution is normal with mean $\mu$ and variance $\sigma^2$.

Suppose we then take samples of size $n$ from each group. Fact 4 in Chapter 7 states that the sampling distribution of the sample mean $\bar{x}$ for a sample of size $n$ taken from a normal population with mean $\mu$ and standard deviation $\sigma$ (that is, variance $\sigma^2$) is also normal, with mean $\mu$ and standard deviation $\sigma/\sqrt{n}$ (that is, variance $\sigma^2/n$), as shown in Figure 12.5. Since each dormitory's GPA is assumed (under $H_0$) to come from the same sampling distribution, we would expect the sample means to be fairly close together.

On the other hand, when $H_0$ is not true, then not all the population means are equal (Figure 12.6). In this case, there is no sampling distribution common to all sample means. We might not expect the sample means to be close together. Note in Figure 12.6 that each distribution nevertheless has the same shape (normal) and spread (i.e., variance) because of the requirements.

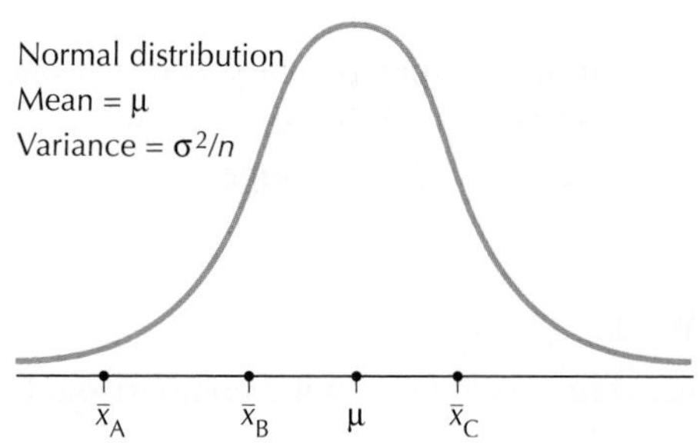


FIGURE 12.5 Common sampling distribution when $H_0$ is true.

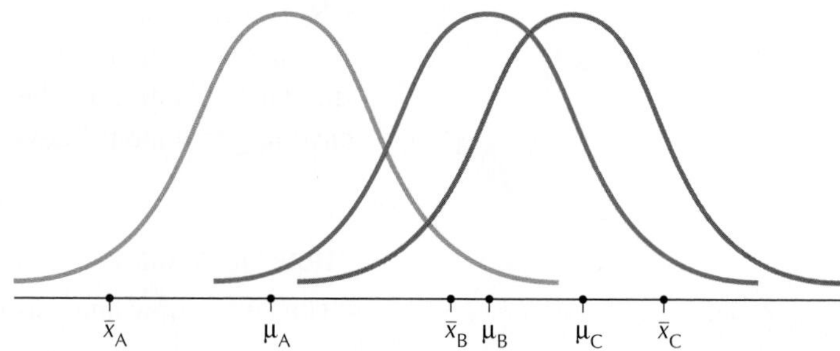


FIGURE 12.6 No common sampling distribution when $H_0$ is not true.

*Note:* Normal probability plots were introduced in Chapter 7.

We verify the normality requirement using normal probability plots for each sample. The equal variances requirement is met if the largest sample standard deviation is not larger than twice the smallest sample standard deviation.

**Procedure for Verifying the Requirements for Analysis of Variance**

***Step 1*** **Normality.** Check that the data from each group are normally distributed.

***Step 2*** **Equal Variances.** Compute the sample standard deviation for each group to verify that the largest standard deviation is not larger than twice the smallest standard deviation.

***Step 3*** **Independence.** Verify that the samples drawn from each group are independently drawn.

## Example 12.2 Verify the requirements for performing an analysis of variance

Verify the requirements for performing an analysis of variance using the hypotheses

$$H_0 : \mu_A = \mu_B = \mu_C \quad \text{versus} \quad H_a : \text{not all the population means are equal}$$

where $\mu_i$ represents the population mean GPA for Dormitory $i$, using data from Table 12.1.

### Solution

***Step 1*** **Normality.** To verify that each of the $k = 3$ populations is normally distributed, we examine normal probability plots of each sample, shown in Figure 12.7. Each plot indicates acceptable normality.

***Step 2*** **Equal Variances.** To find the standard deviation for Dorm A, we first find

$$\begin{aligned}\sum(x - \bar{x})^2 = {} & (0.60 - 2.2)^2 + (3.82 - 2.2)^2 + (4.00 - 2.2)^2 + (2.22 - 2.2)^2 \\ & + (1.46 - 2.2)^2 + (2.91 - 2.2)^2 + (2.20 - 2.2)^2 + (1.60 - 2.2)^2 \\ & + (0.89 - 2.2)^2 + (2.30 - 2.2)^2 = 11.5626\end{aligned}$$

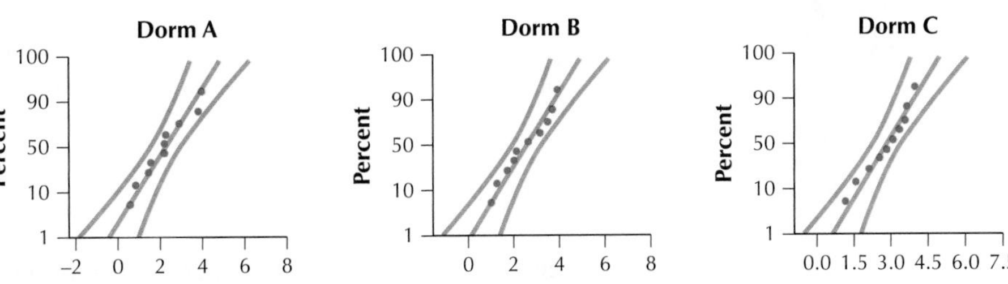


FIGURE 12.7 Normal probability plots.

Then

$$s_A = \sqrt{\frac{\sum(x - \bar{x})^2}{n - 1}} = \sqrt{\frac{11.5626}{10 - 1}} \approx 1.133460777$$

*Note:* We retain many decimal places when calculating $s_A$, $s_B$, and $s_C$ because these values are used to calculate other quantities later on.

We similarly find $s_B \approx 1.030857248$ and $s_C \approx 0.9370284$. The largest, $s_A \approx 1.133460777$, is not larger than twice the smallest, $s_C \approx 0.9370284$. Thus, the equal variance requirement is satisfied.

***Step 3*** **Independence.** Since the students are randomly sampled from each dormitory, with the selection of students in one dormitory not affecting the selection of students sampled from the other dormitories, the independence assumption is also validated. ■

Assuming that $H_0$ is true, we estimate the common population mean $\mu$ using the **overall sample mean, $\bar{\bar{x}}$**:

$$\bar{\bar{x}} = \frac{(n_1\bar{x}_1 + n_2\bar{x}_2 + \cdots + n_k\bar{x}_k)}{n_t}$$

where there are $k$ samples and $n_t$ is the "total sample size" (sum of the $k$ sample sizes). The overall sample mean $\bar{\bar{x}}$ is simply the mean of all the observations from all the samples. For the special case when all the sample sizes are equal, the overall sample mean $\bar{\bar{x}}$ is simply the mean of the sample means,

$$\bar{\bar{x}} = \frac{(\bar{x}_1 + \bar{x}_2 + \cdots + \bar{x}_k)}{k}$$

## Example 12.3 Calculating $\bar{\bar{x}}$

For the sample GPA data given in Table 12.1 for Dorms A, B, and C, calculate the overall sample mean, $\bar{\bar{x}}$.

### Solution

We have $k = 3$ dormitories, with sample mean GPAs $\bar{x}_A = 2.2$, $\bar{x}_B = 2.5$, $\bar{x}_C = 2.8$. Also, $n_A = n_B = n_C = 10$, and $n_t = 10 + 10 + 10 = 30$. Thus,

$$\bar{\bar{x}} = \frac{(10(2.2) + 10(2.5) + 10(2.8))}{30} = 2.5$$

Since all the sample sizes are equal, we can also calculate $\bar{\bar{x}}$ as follows:

$$\bar{\bar{x}} = \frac{(2.2 + 2.5 + 2.8)}{3} = 2.5$$

■

**What Does This Number Mean?**

$\bar{\bar{x}} = 2.5$ is the mean GPA for all 30 students from all three samples. We can use $\bar{\bar{x}}$ as our estimate of the common population mean $\mu$ assumed in $H_0$. ■

## 3 ANOVA Table

Recall that analysis of variance works by comparing the variability in the sample means to the variability within each sample. We use the following statistics to measure these variabilities.

The **mean square treatment (MSTR)** measures the variability in the sample means. MSTR is the sample variance of the sample means, weighted by sample size.

$$\text{MSTR} = \frac{\sum n_i(\bar{x}_i - \bar{\bar{x}})^2}{k - 1}$$

where $n_i$ and $\bar{x}_i$ are the sample size and mean of the $i$th sample, $\bar{\bar{x}}$ is the overall sample mean, and there are $k$ populations.

The **mean square error (MSE)** measures the variability within the samples. MSE is the mean of the sample variances, weighted by sample size.

$$\text{MSE} = \frac{\sum(n_i - 1)s_i^2}{n_t - k}$$

where $n_i$ and $s_i^2$ are the sample size and variance of the $i$th sample, $n_t$ is the total sample size, and there are $k$ populations.

The **test statistic for analysis of variance ($F_{\text{data}}$)** is

$$F_{\text{data}} = \frac{\text{MSTR}}{\text{MSE}}$$

$F_{\text{data}}$ measures the variability among the sample means, compared to the variability within the samples. $F_{\text{data}}$ follows an $F$ distribution with $\text{df}_1 = k - 1$ and $\text{df}_2 = n_t - k$, when the following requirements are met: (1) each of the $k$ populations is normally distributed, (2) the variances of the populations are all equal, and (3) the samples are independently drawn.

The term *mean square* represents a weighted mean of quantities that are squared. Each mean square itself consists of two parts, the *sum of squares* in the numerator and the *degrees of freedom* in the denominator. The numerator for MSTR is called the **sum of squares treatment (SSTR)**, and the numerator for MSE is called the **sum of squares error (SSE)**.

$$\text{MSTR} = \frac{\text{sum of squares treatment}}{\text{df}_1} = \frac{\text{SSTR}}{\text{df}_1} = \frac{\sum n_i(\bar{x}_i - \bar{\bar{x}})^2}{k - 1}$$

$$\text{MSE} = \frac{\text{sum of squares error}}{\text{df}_2} = \frac{\text{SSE}}{\text{df}_2} = \frac{\sum(n_i - 1)s_i^2}{n_t - k}$$

Finally, the **total sum of squares (SST)** is found by adding SSTR and SSE:

$$\text{SST} = \text{SSTR} + \text{SSE}$$

## Example 12.4 Calculating SSTR, SSE, SST, MSTR, MSE, and $F_{\text{data}}$

Using the summary statistics in Table 12.3 for the sample GPAs for Dorms A, B, and C, calculate the following: (a) SSTR, (b) SSE, (c) SST, (d) MSTR, (e) MSE, (f) $F_{\text{data}}$.

Table 12.3 Summary statistics for sample GPAs for Dorms A, B, and C

| | Dorm A | Dorm B | Dorm C |
|---|---|---|---|
| Mean | $\bar{x}_A = 2.2$ | $\bar{x}_B = 2.5$ | $\bar{x}_C = 2.8$ |
| Standard deviation | $s_A \approx 1.133460777$ | $s_B \approx 1.030857248$ | $s_C \approx 0.9370284$ |
| Sample size | $n_1 = 10$ | $n_2 = 10$ | $n_3 = 10$ |

### Solution

We have $k = 3$ dormitories, and total sample size $n_t = 10 + 10 + 10 = 30$. Thus,

**a.** $\text{SSTR} = \sum n_i(\bar{x}_i - \bar{\bar{x}})^2 = 10(2.2 - 2.5)^2 + 10(2.5 - 2.5)^2 + 10(2.8 - 2.5)^2$
$= 10[(-0.3)^2 + 0^2 + 0.3^2] = 1.8$

**b.** $\text{SSE} \approx (10 - 1)1.133460777^2 + (10 - 1)1.030857248^2 + (10 - 1)0.9370284^2$
$\approx 29.0288$

**c.** $\text{SST} = \text{SSTR} + \text{SSE} = 1.8 + 29.0288 = 30.8288$

**d.** $\text{MSTR} = \text{SSTR}/(k - 1) = \dfrac{1.8}{3 - 1} = 0.9$

**e.** $\text{MSE} = \text{SSE}/(n_t - k) = \dfrac{29.0288}{30 - 3} = 1.0751407407$

**f.** $F_{\text{data}} = \text{MSTR/MSE} = \dfrac{0.9}{1.0751407407} = 0.8370997079 \approx 0.84$ ■

What Do These Numbers Mean?

- **MSTR.** Since MSTR measures the variability among the sample means, the larger the MSTR, the greater the distance between the sample means.
- **MSE.** Since MSE measures the variability within the samples themselves, the larger the MSE, the larger the standard deviation of the $k$ samples.
- $\boldsymbol{F_{\text{data}}}$**.** Therefore, since $F_{\text{data}} = \text{MSTR/MSE}$, the larger the $F_{\text{data}}$, the greater the distance between the samples compared to the spread within the individual samples. ■

The ANOVA Table shown in Table 12.4 is a convenient way to display the various statistics calculated during an analysis of variance. Note that the quantities in the "Mean square" column equal the ratio of the two columns to its left.

Table 12.4 ANOVA table

| Source of variation | Sum of squares | Degrees of freedom | Mean square | $F$-test statistic |
|---|---|---|---|---|
| Treatment | SSTR | $\text{df}_1 = k - 1$ | $\text{MSTR} = \dfrac{\text{SSTR}}{k - 1}$ | $F_{\text{data}} = \dfrac{\text{MSTR}}{\text{MSE}}$ |
| Error | SSE | $\text{df}_2 = n_t - k$ | $\text{MSE} = \dfrac{\text{SSE}}{n_t - k}$ | |
| Total | SST | | | |

## Example 12.5 ANOVA table

Construct the ANOVA table for the sample GPA data for Dorms A, B, and C.

**Solution**

Using the statistics calculated in Example 12.4, we construct the ANOVA table as follows. For clarity, the results are rounded. ■

| Source of variation | Sum of squares | Degrees of freedom | Mean square | $F$-test statistic |
|---|---|---|---|---|
| Treatment | SSTR = 1.8 | $df_1 = 3 - 1 = 2$ | $MSTR = \frac{1.8}{2} = 0.9$ | $F_{data} = \frac{0.9}{1.075} \approx 0.84$ |
| Error | SSE = 29.0288 | $df_2 = 30 - 3 = 27$ | $MSE = \frac{29.0288}{27} \approx 1.075$ | |
| Total | SST = 30.8288 | | | |

We can use either MSTR or MSE to estimate the common population variance $\sigma^2$. MSE is always a good estimate of $\sigma^2$. However, MSTR is a good estimate of $\sigma^2$ only when $H_0$ is true. Suppose MSTR and MSE are close to each other. Since MSE is always a good estimate of $\sigma^2$, then there is evidence in favor of $H_0$, since MSTR is close to MSE here. If MSTR is close to MSE, then the ratio $F_{data}$ = MSTR/MSE is close to 1.0. Thus, if this ratio $F_{data}$ = MSTR/MSE is close to 1.0, then this is evidence in favor of $H_0$.

On the other hand, MSTR and MSE not close to each other is evidence against $H_0$, since MSTR *overestimates* $\sigma^2$ when $H_0$ is not true. If MSTR is much greater than MSE, then the ratio $F_{data}$ = MSTR/MSE is much greater than 1.0. Thus, if the ratio $F_{data}$ = MSTR/MSE is much greater than 1, this is evidence against $H_0$.

Is our value of $F_{data} \approx 0.84$ from Example 12.5 close to 1.0? It seems so. But to make sure, we must perform the analysis of variance hypothesis test. We learn how to do this in Section 12.2.

*Note:* The Step-by-Step Technology Guide for performing analysis of variance is provided at the end of Section 12.2 (pages 654–655).

## Section 12.1 SUMMARY

1 The $F$ distribution is right-skewed, never takes negative values, and has an infinite number of different $F$ curves, the shape of a curve depending on two different degrees of freedom, $df_1$ and $df_2$.

2 Analysis of variance is an inferential method for testing whether the means of different populations are equal. The null hypothesis always states that all the population means are equal, and the alternative hypothesis always states that not all the population means are equal. ANOVA works by comparing (a) the variability in the sample means and (b) the variability within each sample. If (a) is large *compared with* (b), this is evidence that the true means are not all equal and we should reject the null hypothesis.

3 The ANOVA table is a convenient way to display the values for mean square treatment, mean square error, and the other statistics calculated during an analysis of variance.

## Section 12.1 EXERCISES

**CLARIFYING THE CONCEPTS**

In Exercises 1–5, determine whether each statement is true or false. If the statement is false, restate the proposition correctly.

1. The value of the $F$ random variable is never negative.
2. The minimum value for the $F$ distribution is 1.0.
3. The $F$ curve is right-skewed.
4. There is an infinite number of $F$ curves.
5. The $F$ distribution has two different degrees of freedom, $df_1$ and $df_2$.

**6.** Does the overall sample mean always equal the mean of the sample means? Explain.

**7.** What does MSTR measure? What does MSE measure?

**8.** Why do we call the process *analysis of variance* when the hypotheses are testing whether there are differences in the means?

**9.** In your own words, explain how ANOVA works.

**10.** If $F_{\text{data}}$ = MSTR/MSE is close to 1.0, what does it mean in terms of the evidence for or against $H_0$? How about if $F_{\text{data}}$ = MSTR/MSE is much greater than 1.0?

**11.** What are the required conditions for performing an analysis of variance?

**12.** How do we assess whether the equal variances requirement is satisfied?

## PRACTICING THE TECHNIQUES

For Exercises 13–22, **(a)** find the value of the $F$ critical value $F_{\text{crit}}$ for the given $\alpha$, $\text{df}_1$, and $\text{df}_2$, and **(b)** draw the $F$ curve, showing the critical value on the number line and the area to the right of it equaling $\alpha$.

**13.** $\alpha = 0.05$, $\text{df}_1 = 1$, $\text{df}_2 = 7$

**14.** $\alpha = 0.10$, $\text{df}_1 = 1$, $\text{df}_2 = 7$

**15.** $\alpha = 0.01$, $\text{df}_1 = 2$, $\text{df}_2 = 15$

**16.** $\alpha = 0.05$, $\text{df}_1 = 2$, $\text{df}_2 = 15$

**17.** $\alpha = 0.10$, $\text{df}_1 = 2$, $\text{df}_2 = 15$

**18.** $\alpha = 0.01$, $\text{df}_1 = 3$, $\text{df}_2 = 16$

**19.** $\alpha = 0.05$, $\text{df}_1 = 3$, $\text{df}_2 = 16$

**20.** $\alpha = 0.10$, $\text{df}_1 = 3$, $\text{df}_2 = 16$

**21.** $\alpha = 0.05$, $\text{df}_1 = 4$, $\text{df}_2 = 16$

**22.** $\alpha = 0.10$, $\text{df}_1 = 4$, $\text{df}_2 = 16$

For Exercises 23–26, calculate the following measures.

- **a.** $\text{df}_1$ and $\text{df}_2$
- **b.** $\bar{\bar{x}}$
- **c.** SSTR
- **d.** SSE
- **e.** SST

**23.**

| Sample A | Sample B | Sample C |
|---|---|---|
| $\bar{x}_A = 10$ | $\bar{x}_B = 12$ | $\bar{x}_C = 8$ |
| $s_A = 1$ | $s_B = 1$ | $s_C = 1$ |
| $n_A = 5$ | $n_B = 5$ | $n_C = 5$ |

**24.**

| Sample A | Sample B | Sample C | Sample D |
|---|---|---|---|
| $\bar{x}_A = 10$ | $\bar{x}_B = 12$ | $\bar{x}_C = 8$ | $\bar{x}_D = 14$ |
| $s_A = 1$ | $s_B = 1$ | $s_C = 1$ | $s_D = 1$ |
| $n_A = 5$ | $n_B = 5$ | $n_C = 5$ | $n_D = 5$ |

**25.**

| Sample A | Sample B | Sample C | Sample D |
|---|---|---|---|
| $\bar{x}_A = 50$ | $\bar{x}_B = 75$ | $\bar{x}_C = 100$ | $\bar{x}_D = 125$ |
| $s_A = 5$ | $s_B = 4$ | $s_C = 6$ | $s_D = 5$ |
| $n_A = 100$ | $n_B = 150$ | $n_C = 200$ | $n_D = 250$ |

**26.**

| Sample A | Sample B | Sample C | Sample D |
|---|---|---|---|
| $\bar{x}_A = 0$ | $\bar{x}_B = 10$ | $\bar{x}_C = 20$ | $\bar{x}_D = 10$ |
| $s_A = 1.5$ | $s_B = 2.25$ | $s_C = 1.75$ | $s_D = 2.0$ |
| $n_A = 50$ | $n_B = 100$ | $n_C = 50$ | $n_D = 100$ |

In Exercises 27–30, refer to the exercises cited and calculate the following measures.

- **a.** MSTR
- **b.** MSE
- **c.** $F_{\text{data}}$

**27.** Exercise 23

**28.** Exercise 24

**29.** Exercise 25

**30.** Exercise 26

## APPLYING THE CONCEPTS

For Exercises 31–37 assume that the data are independently drawn random samples from normal populations.

- **a.** Verify the equal variances assumption.
- **b.** Calculate the following measures.
  - **i.** $\text{df}_1$ and $\text{df}_2$ **ii.** $\bar{\bar{x}}$ **iii.** SSTR **iv.** SSE
  - **v.** SST **vi.** MSTR **vii.** MSE **viii.** $F_{\text{data}}$
- **c.** Construct the ANOVA table.

**31. Online, Hybrid, and Traditional Classrooms.** A researcher randomly selected six students for each of three different treatment groups. The first group of students took elementary statistics online. The second group of students took the same course in the traditional in-class way. The third group of students took a hybrid course, which met once each week and also had an online component. The table shows the grade results. Researchers are interested in whether significant differences exist among the mean grades for the three groups.

| Online grades | Traditional grades | Hybrid grades |
|---|---|---|
| 70 | 75 | 95 |
| 75 | 75 | 60 |
| 60 | 95 | 90 |
| 90 | 60 | 75 |
| 85 | 60 | 85 |
| 50 | 80 | 75 |
| $\bar{x} = 71.6667$ | $\bar{x} = 74.1667$ | $\bar{x} = 80$ |
| $s = 15.0555$ | $s = 13.1972$ | $s = 12.6491$ |

**32. Store Sales.** The district sales manager would like to determine whether there are significant differences in the mean sales among the four franchise stores in her district. Sales (in thousands of dollars) were tracked over 5 days at each of the four stores. The resulting data are summarized in the following table.

| Store A sales | Store B sales | Store C sales | Store D sales |
|---|---|---|---|
| 10 | 20 | 3 | 30 |
| 15 | 20 | 7 | 25 |
| 10 | 25 | 5 | 30 |
| 20 | 15 | 10 | 35 |
| 20 | 20 | 4 | 30 |
| $\bar{x} = 15$ | $\bar{x} = 20$ | $\bar{x} = 5.8$ | $\bar{x} = 30$ |
| $s = 5$ | $s = 3.5355$ | $s = 2.7749$ | $s = 3.5355$ |

**33. ANOVA Can Be Applied to Two Populations.** Researchers are interested in whether the mean heart rates of women and men differ. The following table provides summary statistics of random samples of pulse rates drawn from groups of women and men. After solving **(a)–(c)** (see the instructions for Exercises 31–36), answer the following question. Which method of inference from an earlier chapter could we also use to solve this problem?

| | Females | Males |
|---|---|---|
| $n$ | 65 | 65 |
| $\bar{x}$ | 98.384 | 98.104 |
| $s$ | 0.743 | 0.699 |

**34. 1971 Draft Lottery.** Criticism of the non-random aspect of the 1970 draft lottery led the U.S. Selective Service System to focus on making sure that the 1971 draft lottery was truly random. The table shows the birth months, with the mean and standard deviation of the order of drawing in the draft lottery for each month.

| | N | Mean | Std. Deviation |
|---|---|---|---|
| Jan | 31 | 151.84 | 87.51 |
| Feb | 28 | 198.89 | 119.34 |
| Mar | 31 | 179.77 | 97.50 |
| Apr | 30 | 182.17 | 93.34 |
| May | 31 | 183.52 | 103.22 |
| Jun | 30 | 194.57 | 112.52 |
| Jul | 31 | 183.58 | 122.36 |
| Aug | 31 | 194.35 | 113.89 |
| Sep | 30 | 209.87 | 95.21 |
| Oct | 31 | 172.97 | 113.14 |
| Nov | 30 | 163.13 | 105.44 |
| Dec | 31 | 183.45 | 103.86 |

**35. Education and Religious Background.** The General Social Survey collected data on the number of years of education and the religious preference of the respondent. The summary statistics are shown here.

| | N | Mean | Std. Deviation |
|---|---|---|---|
| PROTESTANT | 1660 | 13.10 | 2.87 |
| CATHOLIC | 683 | 13.51 | 2.74 |
| JEWISH | 68 | 15.37 | 2.80 |
| NONE | 339 | 13.52 | 3.22 |
| OTHER | 141 | 14.46 | 3.18 |

**36. Do Taller People Have Lower Voices?** The heights of the members of the New York Choral Society were recorded.[2] Choral singers are roughly divided into four groups, from highest-pitched voices to lowest: sopranos, and altos (usually female), tenors and basses (usually male). The summary statistics are shown here.

| Variable | n | Mean | StDev |
|---|---|---|---|
| Sopranos | 36 | 64.250 | 1.873 |
| Altos | 35 | 64.886 | 2.795 |
| Tenors | 20 | 69.150 | 3.216 |
| Basses | 39 | 70.718 | 2.361 |

**37. The Full Moon and Emergency Room Visits.** Is there a difference in emergency room visits before, during, and after a full moon? A study looked at the admission rate (number of patients per day) to the emergency room of a Virginia mental health clinic over a series of 12 full moons.[3] The data are provided in the table.

| Before | During | After |
|---|---|---|
| 6.4 | 5 | 5.8 |
| 7.1 | 13 | 9.2 |
| 6.5 | 14 | 7.9 |
| 8.6 | 12 | 7.7 |
| 8.1 | 6 | 11.0 |
| 10.4 | 9 | 12.9 |
| 11.5 | 13 | 13.5 |
| 13.8 | 16 | 13.1 |
| 15.4 | 25 | 15.8 |
| 15.7 | 14 | 13.3 |
| 11.7 | 14 | 12.8 |
| 15.8 | 20 | 14.5 |

## 12.2 Performing Analysis of Variance

*Objectives:* **By the end of this section, I will be able to...**

1. Perform analysis of variance using the $p$-value method.
2. Perform analysis of variance using the estimated $p$-value method.
3. Perform analysis of variance using the critical value method.

## ANOVA: *p*-Value Method

In Section 12.1, we became acquainted with how analysis of variance works. In this section we learn how to perform analysis of variance.

**Analysis of Variance for the Equality of *k* Population Means: *p*-Value Method**
We have taken random samples from each of *k* populations and want to test whether the population means of the *k* populations are all equal.

Required conditions:

1. Each of the *k* populations is normally distributed.
2. The variances ($\sigma^2$) of the populations are all equal.
3. The samples are independently drawn.

***Step 1*** **State the hypotheses, and state the rejection rule.**

$$H_0: \mu_1 = \mu_2 = \cdots = \mu_k \quad \text{versus} \quad H_a: \text{not all the population means are equal}$$

where the $\mu$'s represent the population mean from each population. The rejection rule is *Reject $H_0$ if the p-value is less than $\alpha$.*

***Step 2*** **Calculate the value of the test statistic, $F_{\text{data}}$.**

$$F_{\text{data}} = \frac{\text{MSTR}}{\text{MSE}}$$

where

$$\text{MSTR} = \frac{\sum n_i(\bar{x}_i - \bar{\bar{x}})^2}{k-1} \quad \text{and} \quad \text{MSE} = \frac{\sum (n_i - 1)s_i^2}{n_t - k}$$

$F_{\text{data}}$ follows an $F$ distribution with $\text{df}_1 = k - 1$ and $\text{df}_2 = n_t - k$ if the required conditions are satisfied, where $n_t$ represents the total sample size.

***Step 3*** **Find the *p*-value.** *p*-Value = $P(F > F_{\text{data}})$, as shown in Figure 12.8. If you have access to technology, use it to find the *p*-value. Otherwise, estimate the *p*-value using the method shown in Example 12.9.

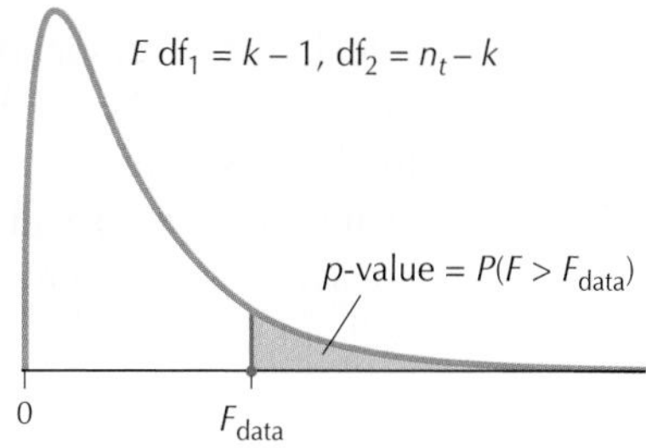


FIGURE 12.8 *p*-Value for the ANOVA *F* test.

***Step 4*** **State the conclusion and interpretation.** Compare the *p*-value with $\alpha$.

## Example 12.6 Finding the *p*-value using the TI-83/84

In Example 12.1, we were interested in determining whether the population mean GPAs for Dormitories A, B, and C were all equal.

$$H_0: \mu_A = \mu_B = \mu_C \quad \text{versus} \quad H_a: \text{not all the population means are equal}$$

In Example 12.4, we found that $F_{\text{data}} = 0.8370997079$, and we have $\text{df}_1 = k - 1 = 3 - 1 = 2$ and $\text{df}_2 = n_t - k = 30 - 3 = 27$. Use technology to find the *p*-value.

### Solution

We use the instructions provided in the Step-by-Step Technology Guide at the end of this section (pages 654–655). From Figures 12.9 and 12.10, we have

$$p\text{-value} = P(F > F_{\text{data}}) = P(F > 0.8370997079) = 0.4438929572 \approx 0.4439$$

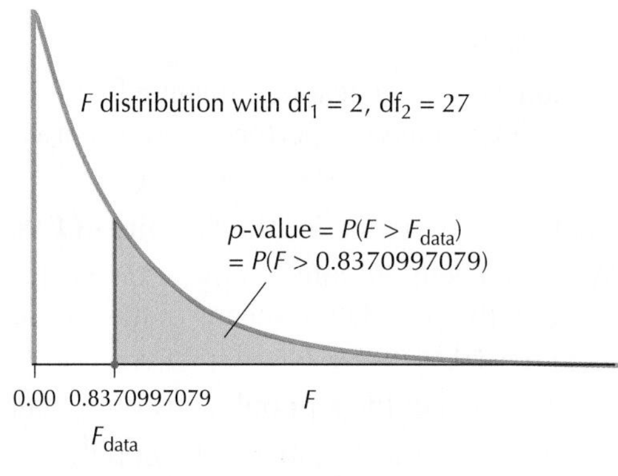


FIGURE 12.9 $p$-Value = $P(F > 0.8370997079)$.

```
Fcdf(0.837099707
9,1E99,2,27)
       .4438929572
```

FIGURE 12.10 TI-83/84 $p$-value.

When calculating the $p$-value for analysis of variance, always retain as many decimal places in the value of $F_{\text{data}}$ as you can. This will make the $p$-value as accurate as possible. Rounding $F_{\text{data}}$ too much will make the $p$-value less accurate.

The *Analysis of Variance* applet allows you to experiment with various values for the sample means and the sample variability in order to see how changes in these values affect $F_{\text{data}}$ and the $p$-value.

## Example 12.7 Analysis of variance using the $p$-value method

Test using the $p$-value method at $\alpha = 0.05$ whether the population mean GPAs differ among the students in dormitories A, B, and C.

### What Result Might We Expect?

Recall that the comparison dotplot in Figure 12.3 (page 634) showed a large amount of overlap in the GPAs among the three dormitories. The large ranges illustrate the large within-dormitory spread of the GPAs for these dorms. When compared against this large within-sample variability, the variability in sample means may not seem large. Therefore, we might expect that the null hypothesis of no difference will not be rejected. ■

### Solution

We already verified the requirements for performing the analysis of variance in Example 12.2.

**Step 1** **State the hypotheses, and state the rejection rule.** Define $\mu_i$.

$$H_0 : \mu_A = \mu_B = \mu_C \quad \text{versus} \quad H_a : \text{not all the population means are equal}$$

where $\mu_i$ represents the population mean GPA of students from dormitory $i$. The rejection rule is *Reject $H_0$ if the p-value is less than $\alpha$.*

**Step 2** **Calculate the value of the test statistic, $F_{\text{data}}$.** From Example 12.4, we have MSTR = 0.9, MSE = 1.0751407407, and

$$F_{\text{data}} = \frac{\text{MSTR}}{\text{MSE}} = \frac{0.9}{1.0751407407} = 0.8370997079$$

$F_{\text{data}}$ follows an $F$ distribution with $\text{df}_1 = k - 1 = 3 - 1 = 2$ and $\text{df}_2 = n_t - k = 30 - 3 = 27$.

**Step 3** **Find the $p$-value.** In Example 12.6, we found the $p$-value to be = $P(F > 0.8370997079) = 0.4438929572 \approx 0.4439$.

**Step 4** **State the conclusion and interpretation.** Compare the $p$-value with $\alpha$. Since the $p$-value of 0.4439 is not less than $\alpha = 0.05$, we do not reject $H_0$. There is not enough evidence to conclude that not all population mean GPAs are equal.

Note that this conclusion confirms our earlier intuition from the graphical evidence in Figure 12.3 that the population mean GPAs would not be found significantly different. ■

## Example 12.8 Performing ANOVA using technology

Alamy

Researchers from the Institute for Behavioral Genetics at the University of Colorado investigated the effect that the enzyme protein kinase C (PKC) has on anxiety in mice. PKC is involved in the messages sent along neural pathways in the brain. The genotype for a particular gene in a mouse (or a human) consists of two copies (alleles) of each chromosome, one each from the father and mother. There are two distinct alleles for PKC. The investigators in the study separated the mice into three groups. In Group 00, neither of the mice's alleles for PKC produced the enzyme. In Group 01 mice, one of the two alleles for PKC produced the enzyme and the other did not. In Group 11 mice, both PKC alleles produced the enzyme. To measure the anxiety in the mice, scientists measured the time (in seconds) the mice spent in the "open-ended" sections of an elevated maze. It was surmised that mice spending more time in open-ended sections exhibit decreased anxiety. The data are provided in Table 12.5. Use technology to test at $\alpha = 0.01$ whether the mean time spent in the open-ended sections of the maze differs among the three groups.

Table 12.5 Time spent in open-ended section of maze

| Group 00 | Group 01 | Group 11 |
|---|---|---|
| 15.8 | 5.2 | 10.6 |
| 16.5 | 8.7 | 6.4 |
| 37.7 | 0.0 | 2.7 |
| 28.7 | 22.2 | 11.8 |
| 5.8 | 5.5 | 0.4 |
| 13.7 | 8.4 | 13.9 |
| 19.2 | 17.2 | 0.0 |
| 2.5 | 11.9 | 16.5 |
| 14.4 | 7.6 | 9.2 |
| 25.7 | 10.4 | 14.5 |
| 26.9 | 7.7 | 11.1 |
| 21.7 | 13.4 | 3.5 |
| 15.2 | 2.2 | 8.0 |
| 26.5 | 9.5 | 20.7 |
| 20.5 | 0.0 | 0.0 |

### *What Result Might We Expect?*

Figure 12.11 shows a plot of the time in open-ended sections for the mice in the three groups. Note that the Group 01 and Group 11 mice spent on average about the same amount of time in the open-ended sections but that Group 00 spent on average somewhat more time in the open-ended sections. This would tend to suggest that the null hypothesis that all three population means are equal should be rejected. ■

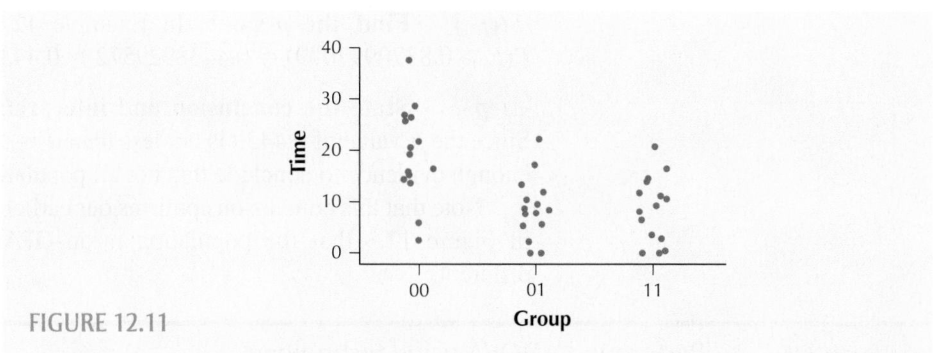


FIGURE 12.11

## Solution

We use the instructions provided in the Step-by-Step Technology Guide at the end of this section (pages 654–655). We first verify whether the requirements are met.

- The normal probability plots in Figure 12.12 indicate acceptable normality.
- The group standard deviations are $s_{00} \approx 9.0$, $s_{01} \approx 6.0$, and $s_{11} \approx 6.4$. Thus, the largest standard deviation is not greater than twice the smaller, which verifies the equal variances requirement.
- The selection of a mouse to a particular group did not affect the selection of mice to the other groups, so that the samples are independent.

Thus, we proceed with the ANOVA.

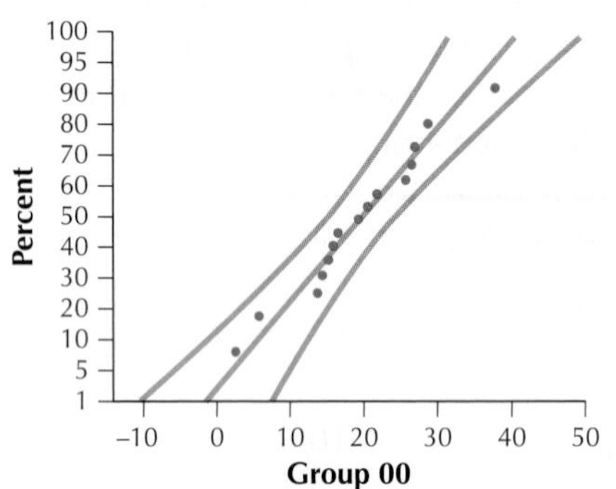


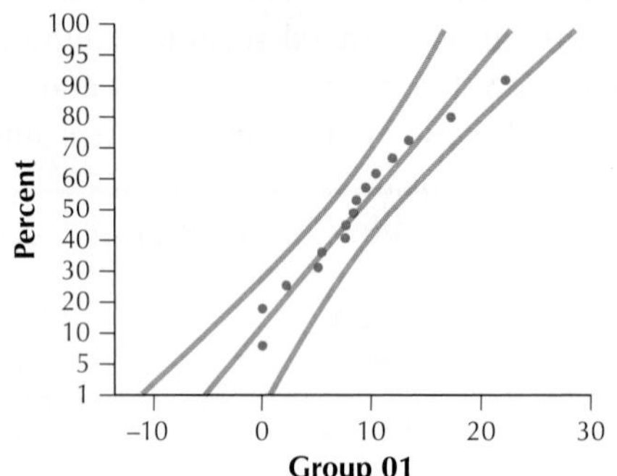


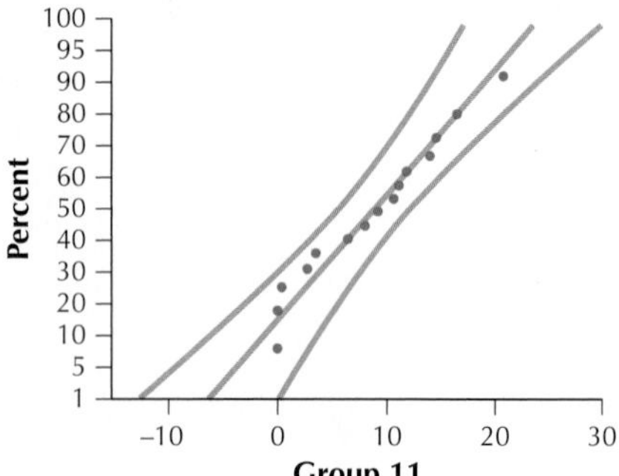


FIGURE 12.12 Normal probability plots.

Figure 12.13 contains the results from the TI-83/84, showing where each statistic corresponds to the ANOVA table structure in Table 12.4. We have $F_{\text{data}} = 10.906$, with a $p$-value of "1.5320224E-4" $= 0.00015320224$. Since this $p$-value is less than $\alpha = 0.01$, we reject $H_0$. There is evidence that the population mean times in the open-ended sections of the maze are not equal for all three groups.

```
One-way ANOVA
 F=10.90607167
 p=1.5320224E-4
 Factor
  df=2
  SS=1154.92044
↓ MS=577.460222

One-way ANOVA
↑ MS=577.460222
 Error
  df=42
  SS=2223.83733
  MS=52.9485079
 Sxp=7.27657254
```

| Source of variation | Sum of squares | Degrees of freedom | Mean square | $F$-test statistic |
|---|---|---|---|---|
| **Treatment** | SSTR ≈ 1154.92 | $df_1 = 2$ | MSTR ≈ 577.46 | $F_{\text{data}} \approx 10.906$ |
| **Error** | SSE ≈ 2223.84 | $df_2 = 42$ | MSE ≈ 52.95 | |
| **Total** | SST ≈ 3378.75 | | | |

FIGURE 12.13 Correspondence between TI-83/84 ANOVA output and the ANOVA table.

Figure 12.14 contains the Excel ANOVA results, and Figure 12.15 contains the Minitab ANOVA results. Values differ slightly due to rounding.

| ANOVA | | | | | | |
|---|---|---|---|---|---|---|
| *Source of Variation* | *SS* | *df* | *MS* | *F* | *P-value* | *F crit* |
| Between Groups | 1154.92 | 2 | 577.4602 | 10.90607 | 0.000153 | 3.219938 |
| Within Groups | 2223.84 | 42 | 52.94851 | | | |
| | | | | | | |
| Total | 3378.76 | 44 | | | | |

FIGURE 12.14 Excel ANOVA results.

| Source | DF | SS | MS | F | P |
|---|---|---|---|---|---|
| Group | 2 | 1154.9 | 577.5 | 10.91 | 0.000 |
| Error | 42 | 2223.8 | 52.9 | | |
| Total | 44 | 3378.8 | | | |

FIGURE 12.15 Minitab ANOVA results.

## 2 ANOVA: Estimated *p*-Value Method

If a computer or calculator is not available, we may still estimate the *p*-value for a given value of $F_{data}$, using the $F$ table. Suppose we are interested in testing whether the population mean recovery times of three surgical procedures are equal. Six patients for each procedure are randomly selected from last year's surgeries. Thus we have $k = 3$, $n_t = 6 + 6 + 6 = 18$, $df_1 = k - 1 = 3 - 1 = 2$, and $df_2 = n_t - k = 18 - 3 = 15$. When the results are in, suppose we find that $F_{data} = 3.2$. Consider Figure 12.16, excerpted from the $F$ table (see Table $F$ in the Appendix, pages T-13–T-16), and Figure 12.17.

| | Area in right tail | 1 | (2) | 3 |
|---|---|---|---|---|
| | 0.100 | 3.18 | 2.81 | 2.61 |
| | 0.050 | 4.75 | 3.89 | 3.49 |
| (12) | 0.025 | 6.55 | 5.10 | 4.47 |
| | 0.010 | 9.33 | 6.93 | 5.95 |
| | 0.001 | 18.64 | 12.97 | 10.80 |
| | 0.100 | 3.07 | 2.70 | 2.49 |
| | 0.050 | 4.54 | 3.68 | 3.29 |
| (15) | 0.025 | 6.20 | 4.77 | 4.15 |
| | 0.010 | 8.68 | 6.36 | 5.42 |
| | 0.001 | 16.59 | 11.34 | 9.34 |

FIGURE 12.16 Excerpt from the $F$ table.

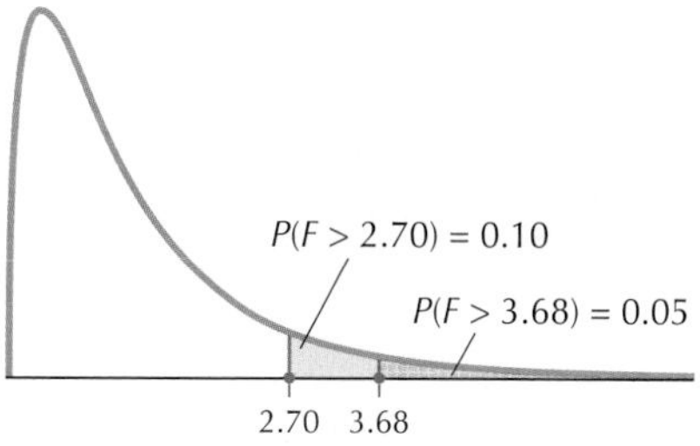

FIGURE 12.17

The $F$ critical values for $df_1 = 2$ and $df_2 = 15$ for various values of $\alpha$ are highlighted in Figure 12.16. Each $F$ value has the indicated area $\alpha$ to the right. For example, $F = 2.70$ has area 0.10 to the right of it, and $F = 3.68$ has area 0.05 to the right of it (see Figure 12.17). $F_{data} = 3.2$ falls between $F = 2.70$ and $F = 3.68$. Therefore, the area to the right of $F_{data} = 3.2$ lies between 0.10 and 0.05. By similar reasoning, for all values of $F$ less than 2.70, the *p*-value will be larger than 0.10. For all values of $F$ greater than 11.34 (see Figure 12.16), the *p*-value will be less than 0.001.

### Example 12.9 Performing analysis of variance using the estimated *p*-value method

The U.S. National Water and Climate Center tracks the percent capacity of lakes and reservoirs. Table 12.6 contains the percent capacity (how full the lake or reservoir is) on a particular day for a random sample of lakes and reservoirs in Arizona, California, and Colorado.[4] Test whether the population mean capacities are equal for all three states, using $\alpha = 0.05$.

Table 12.6 Percent capacity of lakes and reservoirs, with summary statistics

| Arizona | % | California | % | Colorado | % |
|---|---|---|---|---|---|
| Salt River Reservoir | 61 | Boca Reservoir | 80 | Black Hollow Lake | 35 |
| Lake Pleasant | 49 | Bridgeport Reservoir | 83 | Cobb Lake | 46 |
| Verde River Reservoir | 33 | Big Bear Lake | 77 | Fossil Creek | 55 |
| Show Low Lake | 51 | Loon Lake | 94 | Green Mountain Lake | 53 |
| | | El Capitán Reservoir | 84 | Point of Rocks | 46 |
| | | Lake Tahoe | 68 | | |
| $n_{AZ} = 4$ | | $n_{CA} = 6$ | | $n_{CO} = 5$ | |
| $\bar{x}_{AZ} = 48.5$ | | $\bar{x}_{CA} = 81$ | | $\bar{x}_{CO} = 47$ | |
| $s^2_{AZ} = 134.3333$ | | $s^2_{CA} = 73.6$ | | $s^2_{CO} = 61.5$ | |
| $s_{AZ} \approx 11.5902$ | | $s_{CA} \approx 8.5790$ | | $s_{CO} \approx 7.8422$ | |

### What Result Might We Expect?

The comparison dotplot in Figure 12.18 shows no overlap at all between the percent capacities in California and those of the other two states. Therefore, we might expect that the null hypothesis of no difference will be rejected.

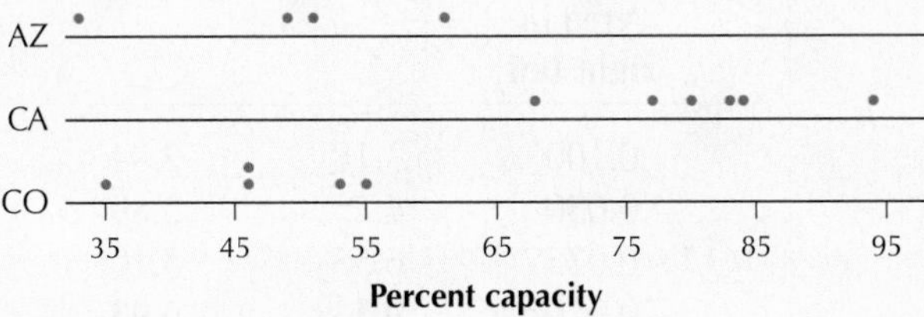


FIGURE 12.18 Comparison dotplots of percent capacity.

## Solution

First we verify whether the requirements are satisfied.

- The normal probability plots shown in Figure 12.19 indicate acceptable normality.
- Of the sample standard deviations in the table, the largest, $s_{AZ} \approx 11.5902$ is not larger than twice the smallest, $s_{CO} \approx 7.8422$. Thus, the equal variance requirement is satisfied.
- Since each lake or reservoir was randomly sampled from each state and the selection of lakes in a particular state did not affect the selection of lakes in the other states, the independence assumption is also validated.

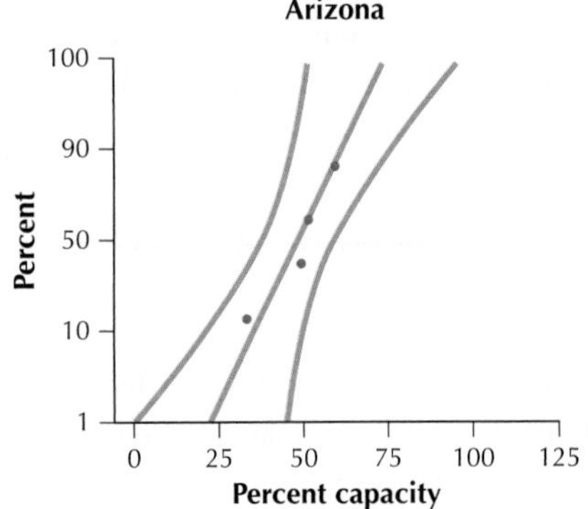


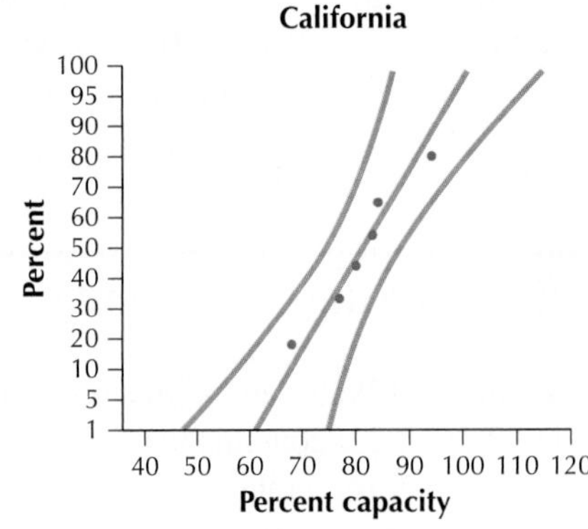


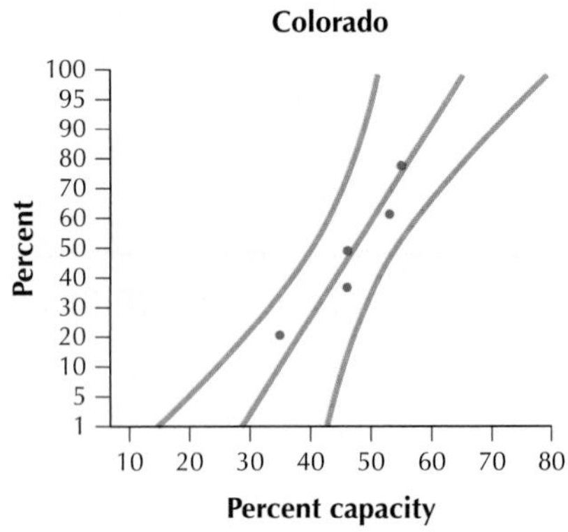


FIGURE 12.19

We therefore proceed with the ANOVA.

**Step 1** **State the hypotheses, and state the rejection rule.**

$$H_0: \mu_{AZ} = \mu_{CA} = \mu_{CO} \quad \text{versus} \quad H_a: \text{not all the population means are equal}$$

where $\mu_{AZ}$, $\mu_{CA}$, and $\mu_{CO}$ represent the population mean percent capacity of reservoirs and lakes in Arizona, California, and Colorado, respectively. Reject $H_0$ if the $p$-value is less than $\alpha = 0.05$.

**Step 2** **Calculate the value of the test statistic, $F_{\text{data}}$.**
First, we find $\bar{\bar{x}}$. The total sample size is $n_t = 4 + 6 + 5 = 15$, so

$$\bar{\bar{x}} = \frac{(n_{AZ}\bar{x}_{AZ} + n_{CA}\bar{x}_{CA} + n_{CO}\bar{x}_{CO})}{n_t} = \frac{(4(48.5) + 6(81) + 5(47))}{15} = 61$$

Then

$$\text{SSTR} = \sum n_i(\bar{x}_i - \bar{\bar{x}})^2 = 4(48.5 - 61)^2 + 6(81 - 61)^2 + 5(47 - 61)^2 = 4005$$

and

$$\text{MSTR} = \frac{\text{SSTR}}{k - 1} = \frac{4005}{3 - 1} = 2002.5$$

We also have

$$\text{SSE} = \sum (n_i - 1)s_i^2 = (4 - 1)134.3333 + (6 - 1)73.6 + (5 - 1)61.5 = 1017$$

and

$$\text{MSE} = \frac{\text{SSE}}{n_t - k} = \frac{1017}{15 - 3} = 84.75$$

which gives

$$F_{\text{data}} = \frac{\text{MSTR}}{\text{MSE}} = \frac{2002.5}{84.75} = 23.62831858 \approx 23.6$$

$F_{\text{data}}$ follows an $F$ distribution with $\text{df}_1 = k - 1 = 3 - 1 = 2$ and $\text{df}_2 = n_t - k = 15 - 3 = 12$.

| | Area in right tail | 1 | (2) |
|---|---|---|---|
| | 0.100 | 3.29 | 2.92 |
| | 0.050 | 4.96 | 4.10 |
| 10 | 0.025 | 6.94 | 5.46 |
| | 0.010 | 10.04 | 7.56 |
| | 0.001 | 21.04 | 14.91 |
| | 0.100 | 3.18 | 2.81 |
| | 0.050 | 4.75 | 3.89 |
| (12) | 0.025 | 6.55 | 5.10 |
| | 0.010 | 9.33 | 6.93 |
| | 0.001 | 18.64 | 12.97 |

FIGURE 12.20

**Step 3** **Estimate the $p$-value.**
We search the $F$ table excerpt in Figure 12.20 for $\text{df}_1 = 2$ and $\text{df}_2 = 12$. The $F_{\text{data}} \approx 23.6$ is larger than the largest value of $F$ shown for $\text{df}_1 = 2$ and $\text{df}_2 = 12$, $F = 12.97$. Therefore, the estimated $p$-value is smaller than the smallest displayed "area in right tail," 0.001. That is, the estimated $p$-value is smaller than 0.001.

**Step 4** Since the estimated $p$-value of "smaller than 0.001" is less than $\alpha = 0.05$, we reject $H_0$. There is evidence that not all the population mean percent capacities are equal among Arizona, California, and Colorado. Note that this conclusion confirms our earlier perception based on the graphical evidence in Figure 12.18. ■

## 3 ANOVA: Critical Value Method

ANOVA can also be performed using the critical value method, as we have done for other hypothesis tests.

**Analysis of Variance for the Equality of $k$ Population Means: Critical Value Method**

We have taken random samples from each of $k$ populations, and we want to test whether the population means of the $k$ populations are all equal.

Required conditions:

1. Each of the $k$ populations is normally distributed.
2. The variances ($\sigma^2$) of the populations are all equal.
3. The samples are independently drawn.

***Step 1*** **State the hypotheses.**

$H_0: \mu_1 = \mu_2 = \cdots = \mu_k$ versus $H_a$: not all the population means are equal

where the $\mu$'s represent the population mean from each population.

***Step 2*** **Find the critical value $F_{crit}$ and state the rejection rule.** The rejection rule is always: Reject $H_0$ if $F_{data} > F_{crit}$.

***Step 3*** **Calculate the value of the test statistic, $F_{data}$.**

$$F_{data} = \frac{\text{MSTR}}{\text{MSE}}$$

where

$$\text{MSTR} = \frac{\sum n_i(\bar{x}_i - \bar{\bar{x}})^2}{k - 1} \quad \text{and} \quad \text{MSE} = \frac{\sum (n_i - 1)s_i^2}{n_t - k}$$

$F_{data}$ follows an $F$ distribution with $df_1 = k - 1$ and $df_2 = n_t - k$ if the required conditions are satisfied, where $n_t$ represents the total sample size.

***Step 4*** **State the conclusion and interpretation.** Compare $F_{data}$ with $F_{crit}$.

## Example 12.10 Analysis of variance using the critical value method

A study compared the size of ants from different colonies.[5] Researchers measured the masses (in milligrams) of random samples of ants from three different colonies. The summary statistics are shown in Table 12.7. Test using the critical value method at $\alpha = 0.05$ whether the population mean sizes differ among the three ant colonies.

Table 12.7 Summary statistics for ant sizes from Colonies 1, 2, and 3

| | Colony 1 | Colony 2 | Colony 3 |
|---|---|---|---|
| Mean | $\bar{x}_1 = 89.66$ mg | $\bar{x}_2 = 99.99$ mg | $\bar{x}_3 = 92.16$ mg |
| Standard deviation | $s_1 = 29.00$ mg | $s_2 = 26.96$ mg | $s_3 = 31.40$ mg |
| Sample size | $n_1 = 130$ | $n_2 = 111$ | $n_3 = 120$ |

### Solution

First we verify whether the requirements are satisfied.

- The normal probability plots indicate acceptable normality (Figure 12.21).
- The largest sample standard deviation, $s_3 = 31.40$, is not larger than twice the smallest, $s_2 = 26.96$. Thus, the equal variance requirement is satisfied.
- Since the ants are randomly sampled from each colony, the independence assumption is also validated. We therefore proceed with the ANOVA.

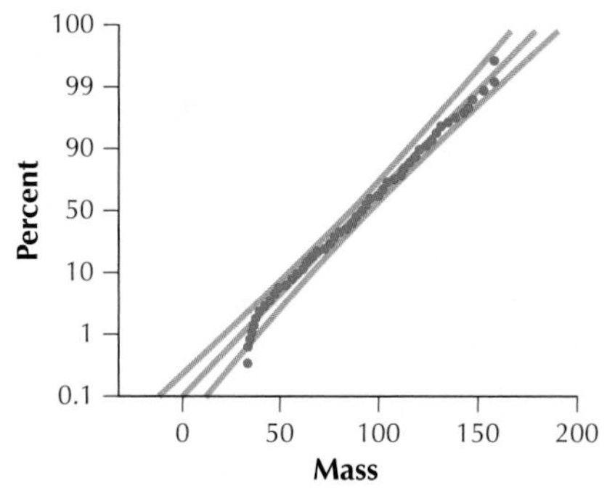


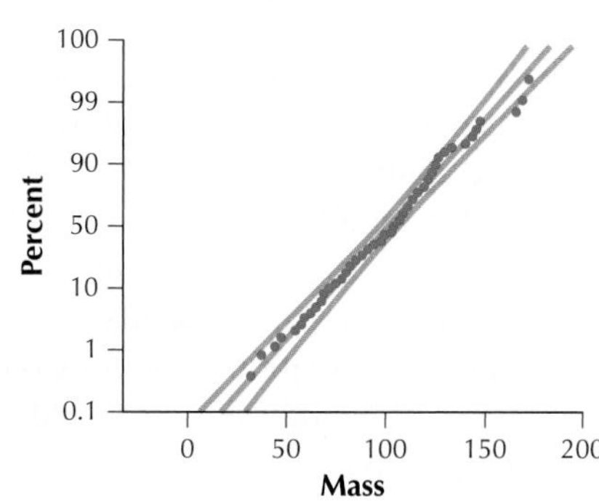


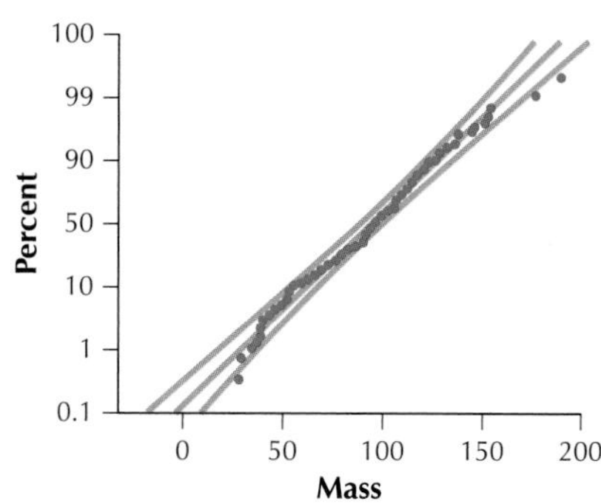


FIGURE 12.21

**Step 1** $H_0: \mu_1 = \mu_2 = \mu_3$ versus $H_a$: not all the population means are equal where $\mu_i$ represents a population mean mass of ants from each colony.

**Step 2** We have $k = 3$ colonies, and total sample size $n_t = 130 + 111 + 120 = 361$. Thus, $df_1 = 3 - 1 = 2$ and $df_2 = 361 - 3 = 358$. Referring to the $F$ tables in the Appendix (pages T-13–T-16), there are no entries for $df_2 = 358$, so we take the closest value, $df_2 = 200$. For $\alpha = 0.05$, $F_{crit} = 3.04$. We reject $H_0$ if $F_{data} > 3.04$.

**Step 3** **Calculate $F_{data}$.** Minitab yields the ANOVA provided here.

```
Source   DF      SS     MS     F       P
Colony    2    6798   3399  3.98   0.020
Error   358  305802    854
Total   360  312600
```

We see that $df_1 = 2$ and $df_2 = 358$, as we found earlier. The row labeled "Colony" represents the "treatment" row in the ANOVA table. Also, Minitab switches the "DF" and "SS" columns. Notice the relationships among the statistics, as shown in the ANOVA table below.

| Source | Sum of squares | Degrees of freedom | Mean square | $F$ |
|---|---|---|---|---|
| Treatment | SSTR = 6798 | $df_1 = 2$ | $MSTR = \frac{SSTR}{df_1} = \frac{6798}{2} = 3399$ | $F_{data} = \frac{MSTR}{MSE} = \frac{3399}{854} = 3.98$ |
| Error | SSE = 305,802 | $df_2 = 358$ | $MSE = \frac{SSE}{n_t - k} = \frac{305{,}802}{358} = 854$ | |
| Total | SST = SSTR + SSE = 312,600 | | | |

**Step 4** **State the conclusion and interpretation.** Compare $F_{data}$ with $F_{crit}$. In Step 2, we said that we would reject $H_0$ if $F_{data} > 3.04$. Since $F_{data} = 3.98 > 3.04$, we reject $H_0$. There is evidence that not all the population mean sizes are equal among the three ant colonies. ■

*Developing Your Statistical Sense*

### Do Not Draw the Wrong Conclusion

Note that the following interpretation is incorrect: "There is evidence that all the population mean sizes are different." Our conclusion is simply that the population means are not all equal, which does *not* imply that they are all different. For example, two of the three colony mean sizes might still be equal, and if the third mean size is different, we would still reject $H_0$.

Also, we cannot determine which colony has the larger population mean size, even though the sample mean of Colony 2 ($\bar{x}_2 = 99.99$ mg) is greater than the sample mean of Colony 1 ($\bar{x}_1 = 89.66$ mg). Since the original hypothesis stated that all population means are equal, the ANOVA results alone do not allow us to conclude formally that the population mean for Colony 2 is greater than that of Colony 1. To test these types of hypotheses, we would need to apply a type of analysis known as multiple comparisons.[6] ■

## What If Scenario ? Analysis of Variance

Consider Example 12.10, where the null hypothesis was rejected. However, what if there had been a systematic error due to an equipment malfunction. The malfunction resulted in each observation in Colony 2 needing to be adjusted downward by the same amount. However, after the adjustment, Colony 2 continues to retain the largest sample mean. Explain how and why this adjustment would affect the following measures. Would the statistic increase, decrease, or stay the same, or is there insufficient information to determine?

**a.** $n_t$
**b.** $k$
**c.** SSTR
**d.** SSE
**e.** SST
**f.** MSTR
**g.** MSE
**h.** $F_{\text{data}}$
**i.** $p$-value
**j.** Conclusion

### Solution

**a.** $n_t$ would stay the same. We are neither adding nor deleting data values. Instead, we are adjusting existing data values. So the total sample size remains the same.

**b.** $k$ would stay the same. The number of colonies is unaffected by the shift in the data values.

**c.** Recall that SSTR measures the variability among the sample means. Figure 12.22 shows the sample mean mass (size) for the three colonies, indicated by red triangles. If the sample mean for Colony 2 is decreasing (while still remaining the largest), then the overall distance between the various sample means is decreasing. That is, the variability in sample means would be decreasing. Thus, SSTR would be decreasing.

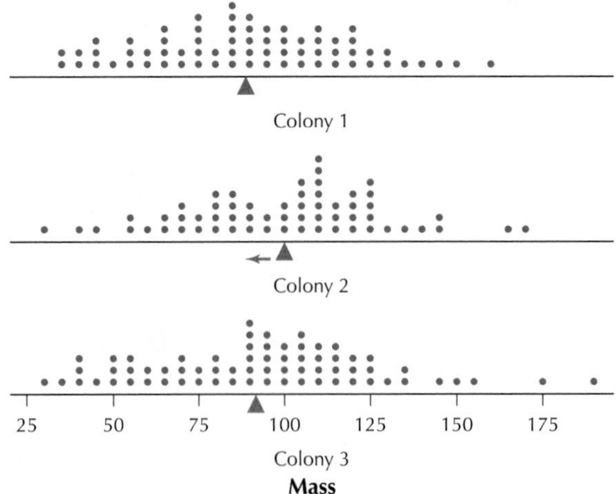


FIGURE 12.22

**d.** Recall that SSE measures the within-sample variability. Since each data value in the highest sample is being shifted by the same amount, the error would have no effect on the spread of this sample. Therefore, SSE stays the same.

**e.** Use **(c)** and **(d)**, as well as the identity SST = SSTR + SSE. If SSTR is decreasing and SSE stays the same, then SST would also decrease.

**f.** Use **(b)** and **(c)**, as well as the identity MSTR = SSTR/$df_1$. If SSTR is decreasing while $df_1 = k - 1$ stays the same, then MSTR would also be decreasing.

**g.** Use **(a)**, **(b)**, and **(d)**, as well as the identity MSE = SSE/$df_2$. If SSE and $df_2 = n_t - k$ stay the same, then MSE would also stay the same.

**h.** Use **(f)** and **(g)**, as well as the identity $F_{data}$ = MSTR/MSE. If MSTR decreases while MSE stays the same, then $F_{data}$ would also decrease.

**i.** For the $F$ test, the $p$-value represents the area to the right of $F_{data}$. If $F_{data}$ is decreasing, then the $p$-value must be increasing.

**j.** We are told that the original conclusion was to reject $H_0$, indicating that the $p$-value was less than $\alpha$. Now, from **(i)**, we know that the $p$-value is increasing; however, we do not know by how much. We don't know whether the $p$-value is now larger or smaller than $\alpha$. There is therefore insufficient information to determine whether or not to reject $H_0$.

Photo Illustration by Chris Jackson/Getty Images

## CASE STUDY | Professors on Facebook

A recent study investigated whether the amount of information a professor posts about himself or herself (that is, *self-disclosure*) on the online social network Facebook is related to student motivation.[7] A professor constructed three different Facebook sites, one offering low self-disclosure, one offering medium self-disclosure, and one offering high self-disclosure. For example, the low-disclosure site offered information only about her position at the university. The medium-disclosure site also showed the professor's favorite movies, books, and quotes. On the high-disclosure site, fictitious comments from "friends" were posted on "the Wall," highlighting social gatherings.

Study participants (students not enrolled in the professor's courses) were then randomly assigned to access and browse one of the three Facebook sites, develop an impression of the professor, and complete the research questionnaire. Student motivation was measured using a set of 16 items, and the sum of the 16 items was calculated to form the total motivation score. The items measured student interest, involvement, stimulation, level of excitement, and whether the student was inspired or challenged. Use technology to test at $\alpha = 0.05$ whether the population mean motivation scores are equal for the three types of Facebook pages, low, medium, and high self-disclosure.

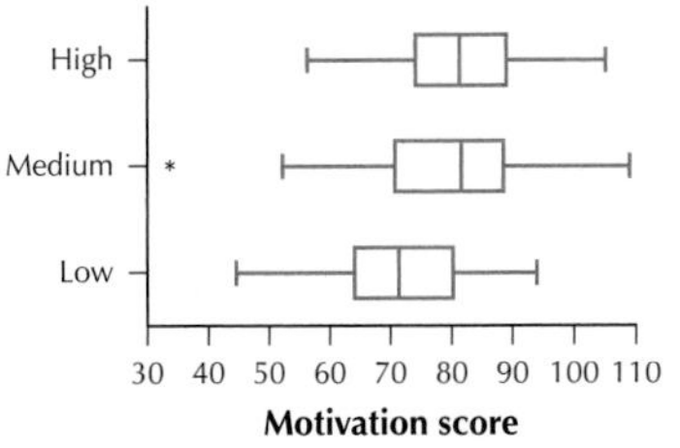


FIGURE 12.23

### What Result Might We Expect?

The comparison boxplot of the motivation scores is shown in Figure 12.23. The plot for the low-disclosure group seems to be somewhat lower than the other two plots. This may be considered mild evidence against the null hypothesis that all population mean motivation scores were equal. So it would not surprise us if the null hypothesis were rejected. ■

### Solution

First we verify whether the requirements are satisfied.

- The normal probability plots in Figure 12.24 indicate acceptable normality.

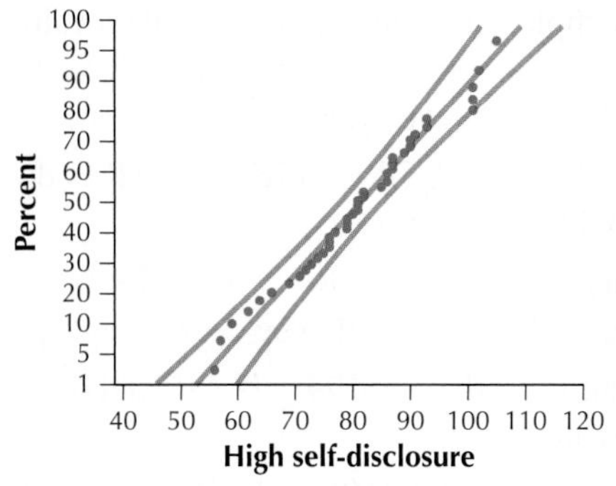


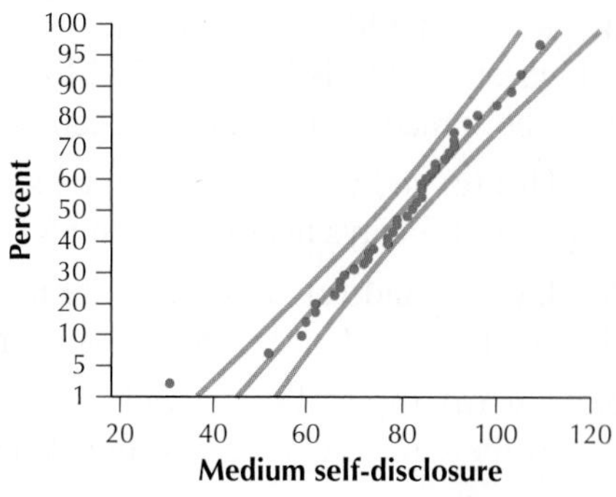


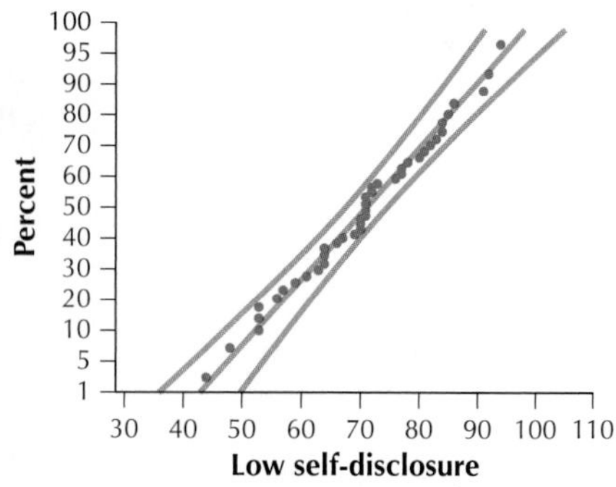


FIGURE 12.24

- The standard deviations are shown in the Minitab output in Figure 12.25. The largest, $s_{\text{medium}} = 14.70$, is not larger than twice the smallest, $s_{\text{low}} = 11.88$. Thus, the equal variance requirement is satisfied.

FIGURE 12.25
Minitab output for Facebook ANOVA.

```
Source   DF     SS     MS     F      P
Group     2   2712   1356  8.05  0.001
Error   127  21386    168
Total   129  24098

S = 12.98   R-Sq = 11.25%   R-Sq(adj) = 9.86%


                             Individual 95% CIs For Mean Based on
                             Pooled StDev
Level    N   Mean  StDev  -------+---------+---------+---------+--
High    43  81.09  12.12                         (-------*-------)
Medium  44  79.36  14.70                      (-------*------)
Low     43  70.63  11.88  (-------*-------)
                          -------+---------+---------+---------+--
                               70.0      75.0      80.0      85.0

Pooled StDev = 12.98
```

- Since the student participants were randomly selected for each level of self-disclosure, the independence assumption is also validated.

We therefore proceed with the ANOVA.

$$H_0: \mu_{\text{high}} = \mu_{\text{medium}} = \mu_{\text{low}} \quad \text{versus} \quad H_a: \text{not all the population means are equal}$$

where $\mu_i$ represents a population mean motivation score for each self-disclosure level. Reject $H_0$ if the $p$-value is less than $\alpha = 0.05$.

From Figure 12.25, we get $F_{\text{data}} = 8.05$, with an associated $p$-value of approximately 0.001. Since this $p$-value is less than $\alpha = 0.05$, we reject $H_0$. There is evidence that not all population mean motivation scores are equal across all levels of self-disclosure. Informally, we may observe that the sample mean motivation score was lowest for the Facebook Web site with low self-disclosure.

## STEP-BY-STEP TECHNOLOGY GUIDE: Analysis of Variance

### TI-83/84

**Performing ANOVA with Technology**

(Example 12.8, pages 645–647, is used to illustrate the procedure.)

***Step 1*** Enter the Group 00 data in **L1**, the Group 01 data in **L2**, and the Group 11 data in **L3**.

***Step 2*** Press **STAT**, highlight **TESTS**, select **"ANOVA("**, and press **ENTER**.

***Step 3*** On the home screen, enter **"L1, L2, L3)"** and press **ENTER** (Figure 12.26). The results are shown as in Figure 12.13 (page 646).

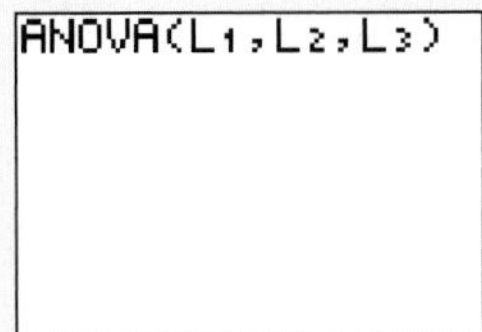


Figure 12.26

## Finding the *p*-Value for a Given $F_{data}$

(Example 12.6, pages 643–644, is used to illustrate the procedure.)

*p*-value = $P(F > F_{data}) = P(F > 0.8370997079)$, where $df_1 = 2$ and $df_2 = 27$.

***Step 1*** Press **2nd > DISTR**.
***Step 2*** Select **Fcdf(** and press **ENTER**.
***Step 3*** Enter **"0.8370997079, 1E99, 2, 27)"** and press **ENTER**. The results are shown in Figure 12.10 (page 644).

# EXCEL

## Performing ANOVA with Technology

(Example 12.8, pages 645–647, is used to illustrate the procedure.)

***Step 1*** Enter the Group 00 data in column A, the Group 01 data in column B, and the Group 11 data in column C.
***Step 2*** Click **Data > Data Analysis > Anova: Single Factor**, and click **OK**.
***Step 3*** Select the input range of the data by clicking and dragging over the data in columns A, B, and C. Then click **OK**. Excel produces the output shown in Figure 12.14 (page 647).

## Finding the *p*-Value for a Given $F_{data}$.

(Example 12.6, pages 643–644, is used to illustrate the procedure.)

*p* value = $P(F > F_{data}) = P(F > 0.8370997079)$, where $df_1 = 2$ and $df_2 = 27$.

***Step 1*** Select cell **A1**. Click the **Insert Function** icon $f_x$.
***Step 2*** For **Search for a Function,** type **FDIST** and click **OK**.
***Step 3*** For **X,** enter **0.8370997079,** for **Deg_freedom 1**, enter **2**, and for **Deg_freedom 2**, enter **27**. Then click **OK**. The cell now contains the *p*-value: 0.4438929572.

# MINITAB

## Performing ANOVA with Technology

(Example 12.8, pages 645–647, is used to illustrate the procedure.)

Minitab accepts data in two different forms for performing ANOVA, *stacked* or *unstacked*. *Unstacked* refers to the data of each group being in a separate column. *Stacked* merges each group's data together in a single column, with the group numbers in a different column.

## ANOVA (Stacked)

***Step 1*** Enter the **Time** data for all three groups in C1 and the values for the categorical variable **Group** in C2.
***Step 2*** Click on **Stat > ANOVA > One-Way**.
***Step 3*** Choose the quantitative variable **Time** as your **response** and the categorical variable **Group** as your **factor**. Then click **OK**. Minitab then displays the ANOVA results shown in Figure 12.15 (page 647).

## ANOVA (Unstacked)

***Step 1*** Enter the Group 00 data in **C1**, the Group 01 data in **C2**, and the Group 11 data in **C3**.
***Step 2*** Click **Stat > ANOVA > One-Way (Unstacked)**.
***Step 3*** For **Responses (in separate columns)**, select columns C1–C3 and click **OK**. Minitab then displays the ANOVA results shown in Figure 12.15 (page 647).

## Finding the *p*-Value for a Given $F_{data}$

(Example 12.6, pages 643–644, is used to illustrate the procedure.)

*p*-value = $P(F > F_{data}) = P(F > 0.8370997079)$, where $df_1 = 2$ and $df_2 = 27$.

***Step 1*** Click **Calc** > **Probability Distributions > F**.
***Step 2*** Select **Cumulative Probability**, enter **2** for **Numerator degrees of freedom** and **27** for **Denominator degrees of freedom**.
***Step 3*** Select **Input Constant**, enter **0.8370997079**, and click **OK**.
***Step 4*** Minitab then displays the cumulative probability $P(F < 0.8370997079) = 0.5561070428$. This cumulative probability represents the area to the left of 0.8370997079 (the unshaded area in Figure 12.9, page 644). Since the entire area under the curve equals 1, to get the *p*-value we need to subtract $P(F < 0.8370997079) = 0.5561070428$ from 1:

$$p\text{-value} = P(F > 0.8370997079) = 1 - P(F < 0.8370997079)$$
$$= 1 - 0.5561070428 = 0.4438929572$$

## Section 12.2 SUMMARY

1 ANOVA, which tests whether the population means of several groups are all equal, may be performed using the critical value method, the $p$-value method, or the estimated $p$-value method. Section 12.2 details the steps involved in carrying out the analysis of variance using each of these methods. In the $p$-value method, technology is found to calculate the exact $p$-value and compare it to $\alpha$.

2 In the estimated $p$-value method, the $p$-value is estimated using the $F$ table and comparing the estimated $p$-value to $\alpha$.

3 In the critical value method, $F_{data}$ is compared with $F_{crit}$. If $F_{data}$ is larger than $F_{crit}$, the null hypothesis is rejected.

## Section 12.2 EXERCISES

### CLARIFYING THE CONCEPTS

**1.** If we are unable to find the correct degrees of freedom in the $F$ tables (see the Appendix), what should we do?

**2.** Say we would like to use the $p$-value method but do not have a computer or calculator. Which method should we use?

**3.** A comparison dotplot of the SAT scores of three sororities shows no overlap at all between the groups. Does this represent evidence for or against the null hypothesis that all population means are equal?

**4.** True or false: If we reject the null hypothesis in an analysis of variance, then there is evidence that all the population mean sizes are different. If the statement is false, explain why it is false.

### PRACTICING THE TECHNIQUES

For Exercises 5–12, assume that the ANOVA assumptions are verified.

For the Exercises 5 and 6, test whether the population means differ, using the estimated $p$-value method and $\alpha = 0.05$.

**a.** State the hypotheses and the rejection rule.
**b.** Find $F_{data}$. (*Hint:* You already calculated $F_{data}$ in Exercises 27 and 28 in Section 12.1.)
**c.** Estimate the $p$-value.
**d.** Compare the $p$-value with $\alpha = 0.05$. State the conclusion and the interpretation.

**5.**

| Sample A | Sample B | Sample C |
|---|---|---|
| $\bar{x}_A = 10$ | $\bar{x}_B = 12$ | $\bar{x}_C = 8$ |
| $s_A = 1$ | $s_B = 1$ | $s_C = 1$ |
| $n_A = 5$ | $n_B = 5$ | $n_C = 5$ |

**6.**

| Sample A | Sample B | Sample C | Sample D |
|---|---|---|---|
| $\bar{x}_A = 10$ | $\bar{x}_B = 12$ | $\bar{x}_C = 8$ | $\bar{x}_D = 14$ |
| $s_A = 1$ | $s_B = 1$ | $s_C = 1$ | $s_D = 1$ |
| $n_A = 5$ | $n_B = 5$ | $n_C = 5$ | $n_D = 5$ |

For Exercises 7 and 8, we are interested in testing whether the population means differ, using the critical value method and $\alpha = 0.05$.

**a.** State the hypotheses.
**b.** Find $F_{crit}$ and the rejection rule.
**c.** Find $F_{data}$. (*Hint:* You already calculated $F_{data}$ in Exercises 29 and 30 in Section 12.1.)
**d.** Compare $F_{data}$ with $F_{crit}$. State the conclusion and the interpretation.

**7.**

| Sample A | Sample B | Sample C | Sample D |
|---|---|---|---|
| $\bar{x}_A = 50$ | $\bar{x}_B = 75$ | $\bar{x}_C = 100$ | $\bar{x}_D = 125$ |
| $s_A = 5$ | $s_B = 4$ | $s_C = 6$ | $s_D = 5$ |
| $n_A = 100$ | $n_B = 150$ | $n_C = 200$ | $n_D = 250$ |

**8.**

| Sample A | Sample B | Sample C | Sample D |
|---|---|---|---|
| $\bar{x}_A = 0$ | $\bar{x}_B = 10$ | $\bar{x}_C = 20$ | $\bar{x}_D = 10$ |
| $s_A = 1.5$ | $s_B = 2.25$ | $s_C = 1.75$ | $s_D = 2.0$ |
| $n_A = 50$ | $n_B = 100$ | $n_C = 50$ | $n_D = 100$ |

**9.** Part of an ANOVA table for an analysis of variance involving seven groups for a study follows. Each sample contained ten data values.

| Source of variation | Sum of squares | Degrees of freedom | Mean square | $F$ |
|---|---|---|---|---|
| Treatment | 120 | ___ | ___ | ___ |
| Error | 315 | ___ | ___ | |
| Total | ___ | | | |

**a.** Find all six missing values in the table and fill in the blanks.
**b.** Perform the appropriate hypothesis test using the estimated $p$-value method and $\alpha = 0.05$.

**10.** Part of an ANOVA table for an analysis of variance involving three groups follows. Each sample contained 6 data values.

| Source of variation | Sum of squares | Degrees of freedom | Mean square | $F$ |
|---|---|---|---|---|
| Treatment | ___ | ___ | ___ | ___ |
| Error | 90 | ___ | ___ | |
| Total | 150 | | | |

**a.** Find all six missing values in the table.
**b.** Perform the appropriate hypothesis test using the critical value method and $\alpha = 0.01$.

**11.** Part of an ANOVA table follows.

| Source of variation | Sum of squares | Degrees of freedom | Mean square | $F$ |
|---|---|---|---|---|
| Treatment | ___ | 4 | 10 | 1.0 |
| Error | ___ | ___ | ___ | |
| Total | 440 | | | |

**a.** Find all four missing values in the table and fill in the blanks.
**b.** Perform the appropriate hypothesis test using the critical value method and $\alpha = 0.10$.

**12.** Part of an ANOVA table follows.

| Source of variation | Sum of squares | Degrees of freedom | Mean square | $F$ |
|---|---|---|---|---|
| Treatment | ___ | 2 | ___ | 2.0 |
| Error | 480 | ___ | 24 | |
| Total | ___ | | | |

**a.** Find all four missing values in the table and fill in the blanks.
**b.** Perform the appropriate hypothesis test using the estimated $p$-value method and $\alpha = 0.05$.

## APPLYING THE CONCEPTS

For Exercises 13 and 14, the assumptions were verified and the statistics were calculated in the Section 12.1 exercises. Perform the appropriate analysis of variance using $\alpha = 0.05$ and either the $p$-value method or the estimated $p$-value method. For each ANOVA, provide the following:

**a.** Hypotheses and rejection rule. Clearly define the $\mu_i$.
**b.** $F_{data}$
**c.** $p$-value
**d.** Conclusion and interpretation

**13. Education and Religious Background.** The General Social Survey collected data on the number of years of education and the religious preference of the respondents. The summary statistics are shown here (see Exercise 35 in Section 12.1).

| | N | Mean | Std. Deviation |
|---|---|---|---|
| PROTESTANT | 1660 | 13.10 | 2.87 |
| CATHOLIC | 683 | 13.51 | 2.74 |
| JEWISH | 68 | 15.37 | 2.80 |
| NONE | 339 | 13.52 | 3.22 |
| OTHER | 141 | 14.46 | 3.18 |

**14. ANOVA can be applied to two populations.** Researchers are interested in whether the mean heart rates of women and men differ. The following table summarizes the data (see Exercise 33 in Section 12.1).

| | Females | Males |
|---|---|---|
| $n$ | 65 | 65 |
| $\bar{x}$ | 98.384 | 98.104 |
| $s$ | 0.743 | 0.699 |

For Exercises 15 and 16, the assumptions were verified and the statistics were calculated in the Section 12.1 exercises. Perform the appropriate analysis of variance using the critical value method and $\alpha = 0.05$. For each ANOVA, provide the following:

**a.** Hypotheses. Clearly define the $\mu_i$.
**b.** $F_{crit}$ and the rejection rule
**c.** $F_{data}$
**d.** Conclusion and interpretation

**15. The Full Moon and Emergency Room Visits.** Is there a difference in emergency room visits before, during, and after a full moon? A study looked at the admission rate (number of patients per day) to the emergency room of a Virginia mental health clinic over a series of 12 full moons.[8] The data are provided in the table. Is there evidence of a difference in emergency room visits before, during, and after the full moon (see Exercise 37 in Section 12.1)?

| Before | During | After |
|---|---|---|
| 6.4 | 5 | 5.8 |
| 7.1 | 13 | 9.2 |
| 6.5 | 14 | 7.9 |
| 8.6 | 12 | 7.7 |
| 8.1 | 6 | 11.0 |
| 10.4 | 9 | 12.9 |
| 11.5 | 13 | 13.5 |
| 13.8 | 16 | 13.1 |
| 15.4 | 25 | 15.8 |
| 15.7 | 14 | 13.3 |
| 11.7 | 14 | 12.8 |
| 15.8 | 20 | 14.5 |

**16. 1971 Draft Lottery.** Criticism of the non-random aspect of the 1970 draft lottery led the U.S. Selective Service System to focus on making sure the 1971 draft lottery was truly random. The descriptive statistics are shown here. Were their efforts successful (see Exercise 34 in Section 12.1)?

| | N | Mean | Std. Deviation |
|---|---|---|---|
| Jan | 31 | 151.84 | 87.51 |
| Feb | 28 | 198.89 | 119.34 |
| Mar | 31 | 179.77 | 97.50 |
| Apr | 30 | 182.17 | 93.34 |
| May | 31 | 183.52 | 103.22 |
| Jun | 30 | 194.57 | 112.52 |
| Jul | 31 | 183.58 | 122.36 |
| Aug | 31 | 194.35 | 113.89 |
| Sep | 30 | 209.87 | 95.21 |
| Oct | 31 | 172.97 | 113.14 |
| Nov | 30 | 163.13 | 105.44 |
| Dec | 31 | 183.45 | 103.86 |

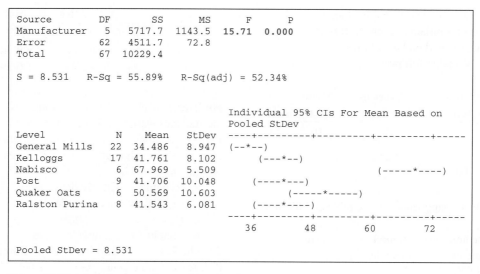

```
Source          DF       SS      MS      F      P
Manufacturer     5   5717.7  1143.5  15.71  0.000
Error           62   4511.7    72.8
Total           67  10229.4

S = 8.531   R-Sq = 55.89%   R-Sq(adj) = 52.34%

                                        Individual 95% CIs For Mean Based on
                                        Pooled StDev
Level            N    Mean   StDev   ----+---------+---------+---------+-----
General Mills   22  34.486   8.947   (--*--)
Kelloggs        17  41.761   8.102        (---*--)
Nabisco          6  67.969   5.509                          (-----*----)
Post             9  41.706  10.048       (----*---)
Quaker Oats      6  50.569  10.603             (-----*-----)
Ralston Purina   8  41.543   6.081       (----*----)
                                     ----+---------+---------+---------+-----
                                        36        48        60        72

Pooled StDev = 8.531
```

Figure 12.27 Minitab output for nutritional ratings ANOVA.

**Breakfast Cereals.** Use this information for Exercises 17–20. Does the nutritional rating of breakfast cereals vary by manufacturer? The data set **Breakfast Cereals 2** contains information on the nutritional rating and manufacturer for 68 breakfast cereals.[9] Assume that the cereals were selected randomly from the manufacturers so that the selection of a cereal from one manufacturer did not affect the selection of cereals from other manufacturers.

**17. What Result Might We Expect?** The comparison boxplot of the nutritional ratings is shown here. Assume that the normality condition is satisfied.

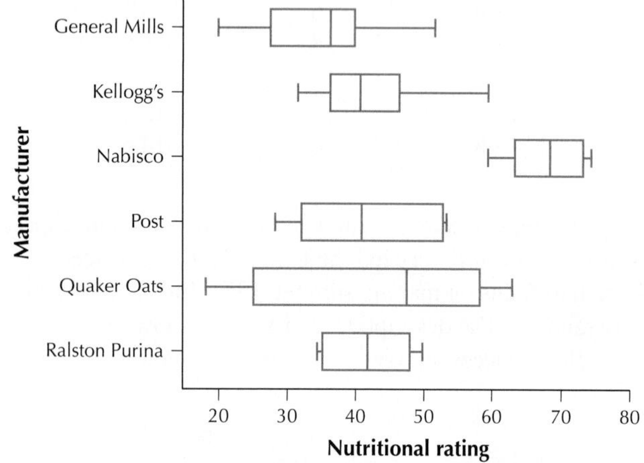


**a.** Explain whether the plot may be considered evidence for or against the null hypothesis that all population mean nutritional ratings were equal.

**b.** Would we expect the null hypothesis to be rejected or not to be rejected?

**18.** The largest standard deviation is, $s_{\text{Quaker Oats}} = 10.603$, while the smallest is $s_{\text{Nabisco}} = 5.509$. State whether the equal variance requirement is satisfied.

**19.** Explain why the independence assumption is validated.

**20.** The Minitab output in Figure 12.27 contains the ANOVA results. $F_{\text{data}}$ and the $p$-value are in boldface. Use the $p$-value method and $\alpha = 0.05$ to test whether all population mean nutritional ratings are equal across all manufacturers. Provide all steps.

**Online, Hybrid, and Traditional Classrooms.** Use this information for Exercises 21–23. A researcher randomly selected six students for each of three different treatment groups. The first group of students took elementary statistics online. The second group of students took the same course in a traditional class. The third group of students took a hybrid course which met once each week in person and also had an online component. The table shows the grade results. Researchers are interested in whether significant differences exist among the mean grades for the three groups. Assume that the assumptions are satisfied.

| Online | Traditional | Hybrid |
|---|---|---|
| 70 | 75 | 95 |
| 75 | 75 | 60 |
| 60 | 95 | 90 |
| 90 | 60 | 75 |
| 85 | 60 | 85 |
| 50 | 80 | 75 |
| $\bar{x} = 71.6667$ | $\bar{x} = 74.1667$ | $\bar{x} = 80$ |
| $s = 15.0555$ | $s = 13.1972$ | $s = 12.6491$ |

**21.** Perform the appropriate ANOVA using the critical value method at $\alpha = 0.01$.

WHAT IF? Refer to Exercise 21. Without redoing the calculations and without re-running the ANOVA, answer **(a)** and **(b)** for Exercises 22 and 23.

**a.** How would each of the following changes affect each statistic in the ANOVA table? Would it increase the statistic, decrease it, or leave it unaffected. Why?

**b.** How would each change affect the conclusion and why? (If there is insufficient information in a particular case to respond properly, indicate so.)

**22.** *What if* the 50 recorded for the 6th online student should really be 100.

**23.** *What if* each of the online scores is 10 points lower than indicated.

**24. Store Sales.** The district sales manager would like to determine whether there are significant differences in the mean sales among the four franchise stores in her district. Sales (in thousands of dollars) over 5 days at each of the four stores are summarized in the following table. Assume that the assumptions are satisfied.

| Store A sales | Store B sales | Store C sales | Store D sales |
|---|---|---|---|
| 10 | 20 | 3 | 30 |
| 15 | 20 | 7 | 25 |
| 10 | 25 | 5 | 30 |
| 20 | 15 | 10 | 35 |
| 20 | 20 | 4 | 30 |
| $\bar{x} = 15$ | $\bar{x} = 20$ | $\bar{x} = 5.8$ | $\bar{x} = 30$ |
| $s = 5$ | $s = 3.5355$ | $s = 2.7749$ | $s = 3.5355$ |

**a.** Perform the appropriate analysis of variance, using the $p$-value method and $\alpha = 0.05$.

**b.** Construct the ANOVA table.

WHAT IF? **25.** Refer to Exercise 24. *What if* the data are all wrong and all the stores actually have a sample mean sales of $30,000. Try to answer the following questions without touching your calculator.

**a.** Find the value of SSTR, MSTR, and $F_{\text{data}}$.

**b.** What would be the $p$-value of the ANOVA hypothesis test?

**c.** What would be the conclusion?

**Gas Mileage for European, Japanese, and American Cars.** Use this information for the Exercises 26–28. The following figure shows a comparison box plot of the vehicle mileage (in miles per gallon) for random samples of automobiles manufactured in Europe, Japan, and the United States. The summary statistics are provided. We are interested in testing using $\alpha = 0.01$ whether population mean gas mileage differs among automobiles from the three regions. Assume that the assumptions are satisfied.

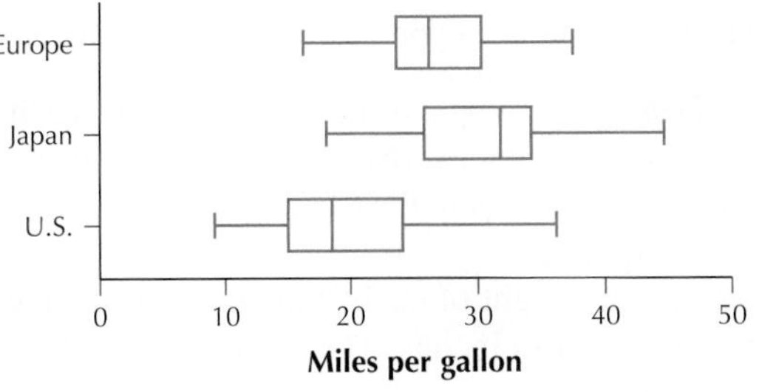


| MPG | Sample 1: Europe | Sample 2: Japan | Sample 3: USA |
|---|---|---|---|
| Sample mean | $\bar{x}_1 = 27.603$ | $\bar{x}_2 = 30.451$ | $\bar{x}_3 = 20.033$ |
| Sample standard deviation | $s_1 = 6.58$ | $s_2 = 6.09$ | $s_3 = 6.440$ |
| Sample size | $n_1 = 68$ | $n_2 = 79$ | $n_3 = 245$ |

**26. What Result Might We Expect?**

**a.** Based on the graphical evidence in the comparison boxplot, what might be the conclusion? Explain your reasoning.

**b.** Perform the ANOVA, using whichever method you prefer.

**c.** Is your intuition from part a supported?

**27. Confidence Intervals as Further Clues in ANOVA.** Refer to Exercise 26. Suppose we construct a confidence interval for each of the population means. If at least one confidence interval does not overlap the others, then it is evidence against the null hypothesis.

**a.** Use a $t$ interval from Section 8.2 to construct a 99% confidence interval for the population mean gas mileage of

**i.** European cars

**ii.** Japanese cars

**iii.** American cars

**b.** Is there one confidence interval from **(a)** that does not overlap the other two? If so, what does this mean in terms of the null hypothesis that all the population means are equal?

WHAT IF? **28.** Refer to the table of descriptive summaries of vehicle mileage. *What if* we discovered that we made a mistake in the data collection and that every Japanese vehicle tested actually had 1 mpg higher gas mileage than previously recorded. Explain how and why this change would affect the following measures—increase, decrease, or no change.

**a.** $n$

**b.** $k$

**c.** SSTR

**d.** SSE

**e.** SST

**f.** MSTR

**g.** MSE

**h.** $F_{\text{data}}$

**i.** $p$-value

**j.** Conclusion

**Do Taller People Have Lower Voices?** Use this information for Exercises 29 and 30. The heights of the members of the New York Choral Society were recorded.[10] Choral singers are roughly divided into four groups, from highest-pitched voices to lowest: sopranos, altos (usually female), tenors, and basses (usually male). Assume that the assumptions are satisfied.

**29. What Result Might We Expect?** The comparison boxplot of the heights is shown here. Is there evidence for differences in the population mean heights?

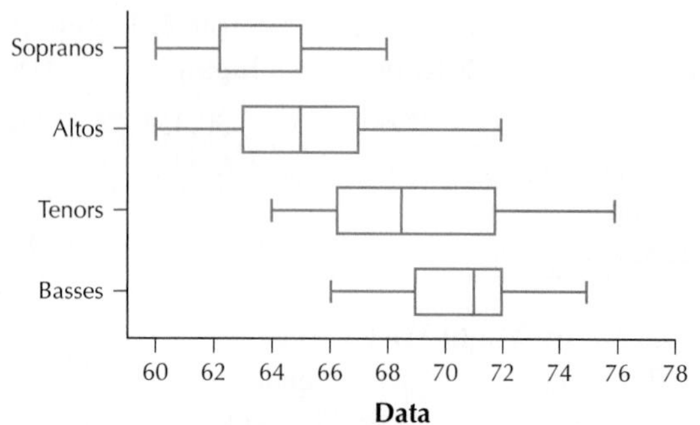


**30.** The summary statistics are shown here. Test at $\alpha = 0.01$ whether the population mean heights of the groups differ.

| Variable | n | Mean | StDev |
|---|---|---|---|
| Sopranos | 36 | 64.250 | 1.873 |
| Altos | 35 | 64.886 | 2.795 |
| Tenors | 20 | 69.150 | 3.216 |
| Basses | 39 | 70.718 | 2.361 |

Use technology to solve Exercises 31–36. Perform the appropriate hypothesis test using the *p*-value method. Provide the following measures.

- **a.** Hypotheses
- **b.** Rejection rule
- **c.** Test statistic $F_{\text{data}}$
- **d.** *p*-value
- **e.** Conclusion
- **f.** Interpretation
- **g.** ANOVA table

Open the **Crash** data set, which contains information about the severity of injuries sustained by crash dummies when the National Transportation Safety Board crashed automobiles into a wall at 35 miles per hour.

**31. Head Injuries and Vehicle Size.** The variable *head_inj* contains a measure of the severity of the head injury sustained by the dummies. The variable *size2* categorizes the type of vehicle, such as light, medium, heavy, pick-up truck, mpv (suv), and so on. The values of the variable *size2* are as follows: 1 = compact car, 2 = light car, 3 = medium car, 4 = heavy car, 5 = mini-compact car, 6 = van, 7 = pickup truck, and 8 = mpv (suv). Would you expect the population mean severity of head injuries suffered by the dummies to be the same across all the size categories? Use Excel or Minitab to perform the analysis of variance, using $\alpha = 0.05$. Comment on the results.

**32. Head Injuries and Vehicle Year.** The variable *year* indicates the model year of the vehicle (87–91). Would you expect the population mean severity of head injuries suffered by the dummies to be the same across all model years? Use Excel or Minitab to perform the analysis of variance, using $\alpha = 0.05$. Comment on the results.

**33. Chest Injury and Protection Device.** The variable *chest_in* is a measure of the severity of the chest injury suffered by the crash dummy. The variable *protect2* categorizes the type of protection that the vehicle offers: 1 = manual seat belts, 2 = motorized seat belts, 3 = passive seat belts, 4 = dual air bags. Would you expect the population mean severity of chest injuries suffered by the dummies to be the same for all categories of protection? Use Excel or Minitab to perform the analysis of variance, using $\alpha = 0.01$. Comment on the results.

**34.** Refer to Exercise 33. Now use $\alpha = 0.05$. How does this affect your conclusion?

APPLET Use the *Analysis of Variance* applet for Exercises 35 and 36.

**35.** Move the group means so that they are about the same by clicking and dragging the black dots so that they are about even horizontally.

- **a.** What happens to the value of $F$ ($F_{\text{data}}$)?
- **b.** Explain why this happens, using the concept of between-sample variability and the statistics SSTR, MSTR, and $F_{\text{data}}$.

**36.** Click **Reset.** Increase the **Within-Sample Variability.**

- **a.** What happens to the value of $F$ ($F_{\text{data}}$)?
- **b.** Explain why this change happens, using the concept of within-sample variability and the statistics SSE, MSE, and $F_{\text{data}}$.

## *Try This in Class!* In-class activities to enhance your understanding of statistics

### ANALYSIS OF VARIANCE FOR THE MEAN HEIGHT OF FRESHMEN, SOPHOMORES, JUNIORS, AND SENIORS

You are going to test whether there are differences in the population mean height of freshmen, sophomores, juniors, and seniors. If the frequency of any class is too small, either eliminate that class or combine it with another class. However, do not go below three classes.

1. State the appropriate hypotheses, and define your notation.

APPLET 2. Generate a sample of four students from each class, using either (a) technology, (b) the *Random Sample* applet, or (c) the method shown in the Chapter 1 Try This in Class Activities (page 29).

3. Find out the height of each student in the sample and express the height in inches. Then find the sample mean height for each class.

**4.** The sample means you calculated in Step 3 were not all the same. Isn't this sufficient evidence to reject the null hypothesis that the population means are all the same? Why or why not?
**5.** Using the level of significance of $\alpha = 0.10$, what is the critical value $F_{\text{crit}}$ for this test? Will it be the same for each student?
**6.** Carry out the analysis of variance using your sample data. Use $\alpha = 0.10$. Will the value of the test statistic $F_{\text{data}}$ be the same for each student?
**7.** Record your name and the value for the test statistic $F_{\text{data}}$ calculated in Step 6 on a Post-It Note.
**8.** Place your Post-It Note on the number line that the instructor has prepared on the board, noting the location of the cutoff value $F_{\text{crit}}$. You thus have a collection of the values for the statistic $F_{\text{data}}$. Can you make any comment on the shape of the distribution of $F_{\text{data}}$ values?
**9.** What proportion of students rejected the null hypothesis?
**10.** Reach a class consensus as to why most students either did or did not reject the null hypothesis.

## CHAPTER 12 FORMULAS AND VOCABULARY

### SECTION 12.1

- **Analysis of variance (ANOVA)** (p. 633). Inferential method for testing whether the means of different populations are equal. The null hypothesis always states that all the population means are equal, and the alternative hypothesis always states that not all the population means are equal. The requirements to perform analysis of variance are (1) each of the $k$ populations is normally distributed, (2) the variances ($\sigma^2$) of the populations are all equal, and (3) the samples are independently drawn.
- **$F_{\text{data}}$** (p. 638). Test statistic for performing an analysis of variance.

$$F_{\text{data}} = \frac{\text{MSTR}}{\text{MSE}}$$

$F_{\text{data}}$ measures the variability among the sample means compared to the variability within the samples. $F_{\text{data}}$ follows an $F$ distribution with $\text{df}_1 = k - 1$ and $\text{df}_2 = n_t - k$, when the requirements are met, where $n_t$ is the total sample size.
- ***F* distribution** (p. 631). Right-skewed distribution that never takes negative values and has an infinite number of different $F$ curves, the shape of the curve depending on degrees of freedom. The $F$ distribution has two different degrees of freedom, $\text{df}_1$ and $\text{df}_2$.
- **Mean square error (MSE)** (p. 638). Measures the variability within the samples.

$$\text{MSE} = \frac{\sum (n_i - 1)s_i^2}{n_t - k}$$

where $n_i$ and $s_i^2$ are the sample size and variance of the $i$th sample, $n_t$ is the total sample size, and there are $k$ populations.
- **Mean square treatment (MSTR)** (p. 638). Measures the variability in the sample means.

$$\text{MSTR} = \frac{\sum n_i(\bar{x}_i - \bar{\bar{x}})^2}{k - 1}$$

where $n_i$ and $\bar{x}_i$ are the sample size and mean of the $i$th sample, $\bar{\bar{x}}$ is the overall sample mean, and there are $k$ populations.
- **Overall sample mean, $\bar{\bar{x}}$** (p. 637). Mean of all the observations from all the samples:

$$\bar{\bar{x}} = \frac{(n_1\bar{x}_1 + n_2\bar{x}_2 + \cdots + n_k\bar{x}_k)}{n_t}$$

- **Sum of squares error (SSE)** (p. 638). Numerator for MSTE.

$$\text{SSE} = \sum (n_i - 1)s_i^2$$

- **Sum of squares treatment (SSTR)** (p. 638). Numerator for MSTR.

$$\text{SSTR} = \sum n_i(\bar{x}_i - \bar{\bar{x}})^2$$

- **Total sum of squares (SST)** (p. 638). Measure of the overall variability in the data across all samples.

$$\text{SST} = \text{SSTR} + \text{SSE}$$

### SECTION 12.2

- **$F_{\text{crit}}$** (p. 650). $F$ critical value, which separates the values of the test statistic $F_{\text{data}}$ for which the null hypothesis will be rejected from those for which the null hypothesis will not be rejected. It is found using $\text{df}_1 = k - 1$ and $\text{df}_2 = n_t - k$.
- **Hypotheses for analysis of variance** (p. 650).

$$H_0: \mu_1 = \mu_2 = \cdots = \mu_k \quad \text{versus}$$
$$H_a: \text{not all the population means are equal}$$

- ***p*-Value** (p. 643). $p\text{-value} = P(F > F_{\text{data}})$.

## CHAPTER 12 REVIEW EXERCISES

### SECTION 12.1

For the following data, assume that the ANOVA assumptions are met, and calculate the measures in Exercises 1–8.

| Sample A | Sample B | Sample C | Sample D |
|---|---|---|---|
| $\bar{x}_A = 0$ | $\bar{x}_B = 10$ | $\bar{x}_C = 20$ | $\bar{x}_D = 10$ |
| $s_A = 1.5$ | $s_B = 2.25$ | $s_C = 1.75$ | $s_D = 2.0$ |
| $n_A = 50$ | $n_B = 100$ | $n_C = 50$ | $n_D = 100$ |

1. $\text{df}_1$ and $\text{df}_2$
2. $\bar{\bar{x}}$
3. SSTR
4. SSE
5. SST
6. MSTR
7. MSE
8. $F_{\text{data}}$
9. Construct the ANOVA table for the statistics in Exercises 1–8.

### SECTION 12.2

For Exercises 10 and 11, assume that the ANOVA assumptions are met. Perform the appropriate analysis of variance using the critical method and $\alpha = 0.05$. Note that you will have to find the summary statistics yourself, so make use of a computer or calculator, if available. For each exercise, provide the following:

- **a.** Hypotheses
- **b.** $F$-critical value
- **c.** Rejection rule
- **d.** Test statistic
- **e.** Conclusion
- **f.** Interpretation
- **g.** ANOVA table

**10. Diversity Awareness Test Scores.** A corporate multicultural affairs officer would like to assess whether there are differences in the diversity awareness test scores for new employee training at three corporate locations, Branches A, B, and C. She randomly selects eight new employees from each of the three different branches. The following table summarizes the test score results. Test whether significant differences exist in the mean diversity awareness test scores for the three branches.

| Branch A | Branch B | Branch C |
|---|---|---|
| 50 | 80 | 85 |
| 90 | 90 | 95 |
| 60 | 95 | 90 |
| 90 | 60 | 75 |
| 85 | 60 | 85 |
| 50 | 80 | 75 |
| 60 | 70 | 65 |
| 70 | 75 | 80 |

**11. Differences in Medical Treatments.** A psychologist is interested in investigating whether differences in mean client improvement exist for three medical treatments. Seven clients undergoing each medical treatment were asked to rate their level of satisfaction on a scale of 0 to 100. The data are summarized in the following table.

| Medical treatment 1 | Medical treatment 2 | Medical treatment 3 |
|---|---|---|
| 75 | 75 | 100 |
| 100 | 100 | 100 |
| 0 | 25 | 50 |
| 50 | 75 | 90 |
| 50 | 50 | 75 |
| 40 | 75 | 75 |
| 25 | 60 | 90 |

For Exercises 12 and 13, assume that the ANOVA assumptions are met. Perform the appropriate analysis of variance using $\alpha = 0.05$ and either the $p$-value method or the estimated $p$-value method. For each ANOVA, provide the following measures.

- **a.** Hypotheses and the rejection rule. Clearly define the $\mu_i$.
- **b.** $F_{\text{data}}$
- **c.** $p$-value
- **d.** Conclusion and interpretation
- **e.** ANOVA table

**12. Customer Satisfaction.** The district sales manager of a local chain store would like to determine whether there are significant differences in the mean customer satisfaction among the four franchise stores in her district. Customer satisfaction data were gathered over 7 days at each of the four stores. The resulting data are summarized in Table 12.8.

**TABLE 12.8 Customer satisfaction in four stores**

| Store A | Store B | Store C | Store D |
|---|---|---|---|
| 50 | 60 | 25 | 75 |
| 40 | 45 | 30 | 60 |
| 60 | 70 | 50 | 80 |
| 60 | 70 | 30 | 90 |
| 50 | 60 | 40 | 70 |
| 45 | 65 | 25 | 85 |
| 55 | 70 | 45 | 95 |
| $\bar{x}_A = 51.43$ | $\bar{x}_B = 62.86$ | $\bar{x}_C = 35.00$ | $\bar{x}_D = 79.29$ |
| $s_A = 7.48$ | $s_B = 9.06$ | $s_C = 10.00$ | $s_D = 12.05$ |

**13. The Price of Apples.** A market researcher is interested in whether differences exist in mean price per pound for apples at local supermarket chains. Random samples of apples from each chain were taken, and their prices per pound (in dollars) are summarized in Table 12.9. Does it pay to comparison shop?

**TABLE 12.9 Apple prices**

| Supermarket A | Supermarket B | Supermarket C | Supermarket D | Supermarket E |
|---|---|---|---|---|
| 1.50 | 2.00 | 2.60 | 3.00 | 3.30 |
| 1.90 | 2.10 | 2.40 | 2.90 | 3.90 |
| 1.50 | 2.00 | 2.70 | 3.50 | 3.30 |
| 1.00 | 2.30 | 1.80 | 2.60 | 3.50 |
| 1.50 | 3.00 | 2.50 | 3.00 | 3.20 |
| 2.40 | | 2.90 | | |

# CHAPTER 12 *Quiz*

## TRUE OR FALSE

1. True or false: The $F$ curve is symmetric.
2. True or false: The total area under the $F$ curve equals 1.
3. True or false: If we reject the null hypothesis in an ANOVA, we conclude that there is evidence that all the population means are different.

## FILL IN THE BLANK

4. A weighted average (mean) of quantities that are squared is called a __________ __________ [two words].
5. The __________ __________ __________ [three words] measures the variability in the sample means.
6. The __________ __________ __________ [three words] measures the variability within the samples.

## SHORT ANSWER

7. What do we use for an estimate of the overall population mean?
8. In the ANOVA table, what is the relationship between the quantities in the mean square column with respect to the quantities in the sum of squares column and the degrees of freedom column?
9. For analysis of variance, the $p$-value represents the area to the right of what?

## CALCULATIONS AND INTERPRETATIONS

For Exercises 10–13, assume that the ANOVA assumptions are met and that the requirements have been met.

**a.** Calculate the following measures.

- **i.** $df_1$ and $df_2$
- **ii.** $\bar{\bar{x}}$
- **iii.** SSTR
- **iv.** SSE
- **v.** SST
- **vi.** MSTR
- **vii.** MSE
- **viii.** $F_{data}$

**b.** Construct the ANOVA table.

**10. Gas Mileage and Number of Cylinders.** When it comes to getting good gas mileage, does the number of cylinders in your engine make a difference? The following table provides the summary statistics regarding miles per gallon for 4-cylinder, 6-cylinder, and 8-cylinder cars.

| | 4 cylinders | 6 cylinders | 8 cylinders |
|---|---|---|---|
| $n$ | 199 | 83 | 103 |
| $\bar{x}$ | 29.3 | 20.0 | 15.0 |
| $s$ | 5.7 | 3.8 | 2.9 |

**11. Hours Worked and Marital Status.** The General Social Survey tracks demographic trends. Here we are interested in whether the mean number of hours worked differs by marital status. The summary statistics are shown here.

| | N | Mean | Std. Deviation |
|---|---|---|---|
| MARRIED | 964 | 42.76 | 14.08 |
| WIDOWED | 72 | 40.13 | 14.28 |
| DIVORCED | 342 | 43.69 | 13.93 |
| SEPARATED | 79 | 41.66 | 15.71 |
| NEVER MARRIED | 478 | 41.03 | 14.03 |

**12. Calories in Breakfast Cereals.** A dietary researcher is interested in whether differences exist in the mean number of calories in breakfast cereals made by different manufacturers. The summary statistics for the samples from three manufacturers appear in the following table.

| | Kellogg's | Quaker | Ralston Purina |
|---|---|---|---|
| $n$ | 23 | 8 | 8 |
| $\bar{x}$ | 109 | 95 | 115 |
| $s$ | 22 | 29 | 23 |

**13. GPA Differences by Class.** A researcher is interested in whether differences exist in the mean grade point averages of freshmen, sophomores, juniors, seniors, and graduate students at a local university. Random samples of students from each category are taken, and their grade point averages are summarized in Table 12.10.

For Exercises 14 and 15, assume that the ANOVA assumptions are met. Perform the appropriate analysis of variance using $\alpha = 0.05$ and either the $p$-value method or the estimated $p$-value method. Provide the following measures.

**a.** Hypotheses and rejection rule. Clearly define the $\mu_i$.

**b.** $F_{\text{data}}$

**c.** $p$-value

**d.** Conclusion and interpretation

**14.** Data in Exercise 10

**15.** Data in Exercise 11

For Exercises 16 and 17, assume that the ANOVA assumptions are met. Perform the appropriate analysis of variance using the critical value method and $\alpha = 0.05$. Provide the following measures.

**a.** Hypotheses. Clearly define the $\mu_i$.

**b.** $F_{\text{crit}}$ and rejection rule

**c.** $F_{\text{data}}$

**d.** Conclusion and interpretation

**16.** Data in Exercise 12

**17.** Data in Exercise 13

**TABLE 12.10 Grade point averages**

| Freshmen | Sophomores | Juniors | Seniors | Graduate students |
|---|---|---|---|---|
| 1.0 | 2.2 | 2.5 | 2.9 | 3.5 |
| 1.5 | 1.5 | 2.5 | 3.0 | 4.0 |
| 1.5 | 2.0 | 2.7 | 3.5 | 3.3 |
| 1.0 | 2.3 | 1.8 | 2.6 | 3.5 |
| 1.5 | 2.0 | 2.5 | 3.0 | 3.2 |
| 2.5 | | 3.0 | | |
| $\bar{x} = 1.5$ | $\bar{x} = 2.0$ | $\bar{x} = 2.5$ | $\bar{x} = 3.0$ | $\bar{x} = 3.5$ |
| $s = 0.5477$ | $s = 0.3082$ | $s = 0.3950$ | $s = 0.3240$ | $s = 0.3082$ |

Chapter 13

# Regression Analysis

CASE STUDY

## How Fair Is the Scoring in Scrabble?

Scrabble is a game in which the players choose letters randomly from a pool of letter tiles and take turns building words. Each letter tile is worth a specific number of points. For example, the letter A is worth 1 point, and the letter Z is worth 10 points. Do you think the way the various letters are valued is fair? Do you dread picking up the letter Z so much that you think it should be worth 25 points rather than 10? Do you think there are too many I's, perhaps, and not enough T's?

We can use *regression analysis,* the topic of this chapter, to help us find out why it often seems that there are too many I's and not enough T's. Regression can also help us understand why we would rather pick up an H tile rather than a V tile, even though each tile is worth 4 points. We can compare the distribution of letter points in Scrabble to the number of letter tiles in the game and to the frequency distribution of letters in the English language. To see more, check out this chapter's Case Study, How Fair Is the Scoring in Scrabble? in Section 13.3 (page 694).

# OVERVIEW



## The Big Picture

### Where we are coming from, and where we are headed...

In the last part of *Discovering Statistics,* we have been studying more advanced topics in inferential statistics. In Chapter 10 we learned how to compare means or proportions for two populations. That chapter was related in a way to Chapter 12, where we studied how to compare the means of several different populations. In Chapter 11, we investigated whether a relationship existed between two qualitative (categorical) variables.

Here, in Chapter 13, we learn about regression analysis. **Regression analysis** develops an equation that can describe the relationship between two quantitative variables, often for the purpose of prediction. We were introduced to some simple methods for investigating the relationship between two quantitative variables in Chapter 4. Look back at those sections now as a preparation for this last chapter.

In Section 13.1, we review some regression concepts and learn about the sum of squares error SSE and the standard error of the estimate $s$. In Section 13.2, we are introduced to the coefficient of determination $r^2$ and the relationship among the sums of squares. Finally, in Section 13.3 we examine the regression model, which allows us to perform inference regarding the slope of the regression line.

Where do we go from there? That's up to you. The world is drowning in data and needs people like you who understand the statistical techniques and methods for uncovering the knowledge locked in the data. This course in statistics has tried to provide you with these tools. Here's hoping that your new expertise as a data analyst will help you succeed in the Age of Information.

## 13.1 Introduction to Linear Regression

### Objectives: By the end of this section, I will be able to...

1 Recall the regression concepts covered in Chapter 4.

2 Explain prediction error, calculate SSE, and utilize the standard error $s$ as a measure of a typical prediction error.

### 1 Review of Regression Concepts

In Chapter 4, we learned about scatterplots, correlation, and the regression line. These are all methods for analyzing the relationship between two continuous variables. In this chapter, we examine regression analysis more closely and learn some inferential methods for regression. Example 13.1 provides a review of some of the material we learned earlier. Recall that the regression line approximates the relationship between two continuous variables and is described by the regression equation $\hat{y} = b_0 + b_1x$, where $b_0$ is the *y intercept* and $b_1$ is the slope, $x$ represents the *predictor variable*, $y$ represents the *response variable*, and $\hat{y}$ represents the *estimated or predicted y-value*.

## Example 13.1 Review of regression concepts

*Copyright David Young-Wolff/Photo Edit*

Table 13.1 shows the results for ten student subjects who were given a set of short-term memory tasks to perform within a certain amount of time. These tasks included memorizing nonsense words and random patterns. Later, the students were asked to repeat the words and patterns, and the students were scored according to the number of words and patterns memorized and the quality of the memory. Partially remembered words and patterns were given partial credit, so the score was a continuous variable.

Table 13.1 Results of short-term memory test

| Student | Time to memorize (in minutes) ($x$) | Short-term memory score ($y$) |
|---|---|---|
| 1 | 1 | 9 |
| 2 | 1 | 10 |
| 3 | 2 | 11 |
| 4 | 3 | 12 |
| 5 | 3 | 13 |
| 6 | 4 | 14 |
| 7 | 5 | 19 |
| 8 | 6 | 17 |
| 9 | 7 | 21 |
| 10 | 8 | 24 |

**a.** Draw the scatterplot and find the regression line.
**b.** Estimate the score for a subject given 3.5 minutes to study.

### Solution

**a.** Figure 13.1 displays the scatterplot of $y$ = score versus $x$ = time, together with the regression line $\hat{y} = 7 + 2x$, as calculated by Minitab.

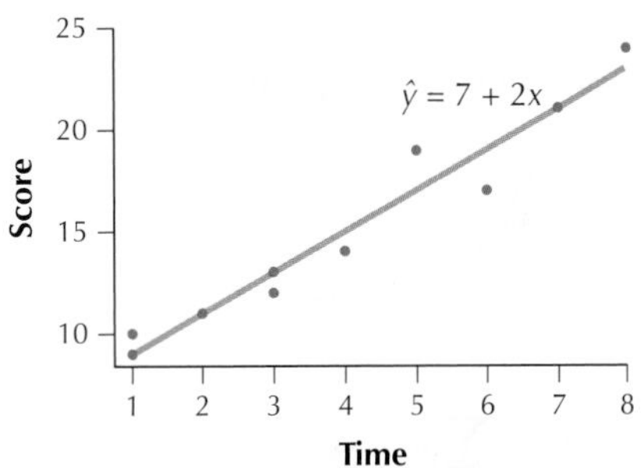


FIGURE 13.1 Scatterplot with regression line.

```
The regression equation is
Score = 7.00 + 2.00 Time
```

Minitab regression results (excerpt).

**b.** Substituting $x = 3.5$ in the equation $\hat{y} = 7 + 2x$ gives $\hat{y} = 7 + 2(3.5) = 14$. The estimated score when a subject is given 3.5 minutes to study is 14 points.

The *Correlation and Regression* applet allows you to insert your own data values and see how the regression line changes for different data values.

In Chapter 4, we introduced the definition formula and a computational formula for calculating the slope $b_1$ of the regression line with equation $\hat{y} = b_0 + b_1x$. The formula for the $y$ intercept in both cases is $b_0 = \bar{y} - (b_1 \cdot \bar{x})$.

**Formulas for Slope $b_1$**

Definition: $b_1 = \dfrac{\sum(x - \bar{x})(y - \bar{y})}{\sum(x - \bar{x})^2}$ Computational: $b_1 = \dfrac{\sum xy - \dfrac{(\sum x)(\sum y)}{n}}{\sum x^2 - \dfrac{(\sum x)^2}{n}}$

## Example 13.2 Calculating $b_0$ and $b_1$

Calculate $b_0$ and $b_1$ for the regression line in Example 13.1 using the computational formula for $b_1$.

### Solution

Construct Table 13.2.

Table 13.2 Calculation of the summary statistics $\sum x$, $\sum y$, $\sum xy$, $\sum x^2$, and $\sum y^2$

| Student | $x$ | $y$ | $xy$ | $x^2$ | $y^2$ |
|---|---|---|---|---|---|
| 1 | 1 | 9 | 9 | 1 | 81 |
| 2 | 1 | 10 | 10 | 1 | 100 |
| 3 | 2 | 11 | 22 | 4 | 121 |
| 4 | 3 | 12 | 36 | 9 | 144 |
| 5 | 3 | 13 | 39 | 9 | 169 |
| 6 | 4 | 14 | 56 | 16 | 196 |
| 7 | 5 | 19 | 95 | 25 | 361 |
| 8 | 6 | 17 | 102 | 36 | 289 |
| 9 | 7 | 21 | 147 | 49 | 441 |
| 10 | 8 | 24 | 192 | 64 | 576 |
| Summation | $\sum x = 40$ | $\sum y = 150$ | $\sum xy = 708$ | $\sum x^2 = 214$ | $\sum y^2 = 2478$ |

CAUTION The notation $\Sigma x^2$ means that we square each $x$-value and then add up the $x^2$-values. The notation $(\Sigma x)^2$ means that we add up the $x$-values first and square the resulting sum.

Using the summary statistics from Table 13.2,

$$b_1 = \frac{\sum xy - (\sum x)(\sum y)/n}{\sum x^2 - (\sum x)^2/n} = \frac{708 - (40)(150)/10}{214 - (40)^2/10} = \frac{108}{54} = 2$$

Now,

$$\bar{y} = \frac{\sum y}{n} = \frac{150}{10} = 15 \quad \text{and} \quad \bar{x} = \frac{\sum x}{n} = \frac{40}{10} = 4$$

giving

$$b_0 = \bar{y} - (b_1 \cdot \bar{x}) = 15 - (2)(4) = 7$$ ■

**What Do the Slope and y Intercept Mean?**

The slope $b_1$ is interpreted as the "estimated change in $y$ for a unit increase in $x$." The $y$ intercept $b_0$ is interpreted as the "estimated $y$ when $x$ equals 0." For example, here $b_1 = 2$, meaning that for every additional minute of study time, the estimated increase in score is 2 points. And $b_0 = 7$ means that the estimated score for a study time of $x = 0$ minutes is 7 points. Using the score $b_0 = 7$, however, represents **extrapolation**, since $x = 0$ lies outside the range of $x$. ■

**Extrapolation** consists of using the regression equation to make estimates or predictions based on x-values that are outside the range of the x-values in the data set.

Extrapolation should be avoided, if possible, because the relationship between the variables may no longer be linear outside the range of $x$. A regression line based solely on the available data (white background) and ignoring the hidden data (gray background) is shown in Figure 13.2. Since the regression line is based on incomplete data, in this case, predicting $y$ at the point $x = a$ resulted in a large difference between the predicted value $\hat{y}$ and the actual value $y$, called the *prediction error*, or *residual*.

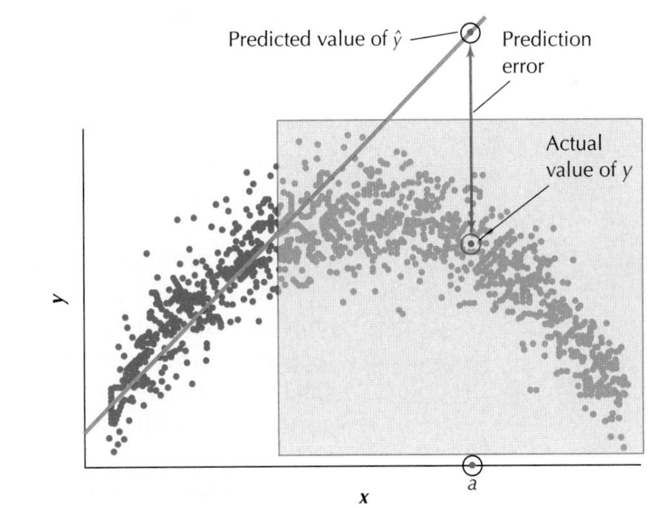

FIGURE 13.2 Dangers of extrapolation.

## Example 13.3 Extrapolation

Using the regression equation from Example 13.1, estimate the score for the following times. If the estimate represents extrapolation, indicate so.

**a.** 5 minutes

**b.** 20 minutes

### Solution

We evaluate the regression equation for $x = 5$ and $x = 20$. In Table 13.1, the smallest $x$ is 1 minute and the largest $x$ is 8 minutes, so estimates for any value of $x$ between 1 and 8 would not represent extrapolation.

**a.** $\hat{y} = 7 + 2(5) = 17$. $x = 5$ lies between $x = 1$ and $x = 8$, so the estimate does not represent extrapolation.

**b.** $\hat{y} = 7 + 2(20) = 47$. $x = 20$ does not lie between $x = 1$ and $x = 8$, so the estimate represents extrapolation. ■

## 2 Prediction Error, SSE, and Standard Error of the Estimate $s$

Figure 13.3 shows the regression line for the data in Example 13.1. The $(x, y)$ data values are represented by the blue dots, while the points whose $y$-values are the $\hat{y}$'s, the predicted values of $y$, lie exactly on the regression line. The student (7) who was

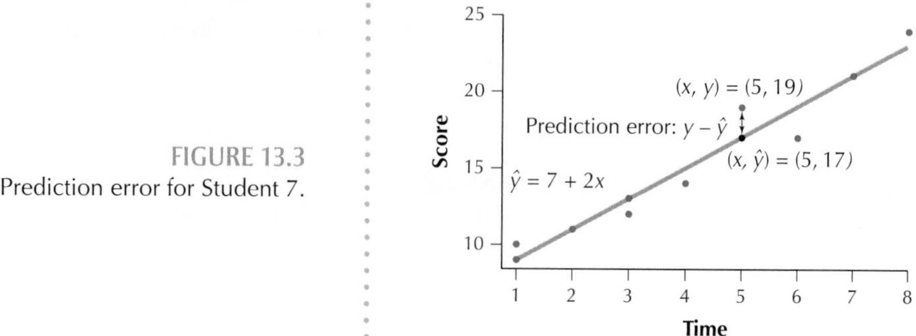


FIGURE 13.3
Prediction error for Student 7.

given $x = 5$ minutes obtained a short-term memory score of $y = 19$. However, the predicted score $\hat{y}$ for this student was somewhat lower than the actual score of $y = 19$. Using the regression equation, we can calculate this predicted score:

$$\hat{y} = 7 + 2x = 7 + 2(5) = 17$$

> The **prediction error** or **residual** *$(y - \hat{y})$* measures how far the predicted value $\hat{y}$ is from the actual value of $y$ observed in the data set.

This predicted score $\hat{y} = 17$ for $x = 5$ is given by the point $(x, \hat{y}) = (5, 17)$, which falls exactly on the regression line (black dot). Thus, the prediction generated by the regression equation is "off" by two points, giving us a prediction error of $y - \hat{y} = 19 - 17 = 2$ points. The vertical distance between each dot and the regression line represents the difference between the actual value of $y$ (19) and the predicted value $\hat{y}$ (17). The term for this difference $(y - \hat{y})$ is the **prediction error** or the **residual**.

The residual can be calculated for every observation in the data set. In the example for Student 7, the residual is $(y - \hat{y}) = (19 - 17) = 2$. If we wish to use the regression to make useful predictions, we would like to keep all our prediction errors small. To measure the prediction errors, we calculate the sum of squared prediction errors, or more simply, the **sum of squares error (SSE)**:

> **Sum of Squares Error (SSE)**
>
> $$\text{SSE} = \sum(y - \hat{y})^2 = \sum(\text{residual})^2 = \sum(\text{prediction error})^2$$

Since we want our prediction errors to be small, it follows that we want SSE to be as small as possible. The least-squares criterion states that the regression line will be the *line for which the SSE is minimized.* That is, out of all possible straight lines, the least-squares criterion chooses the line with the smallest SSE to be the regression line.

### Example 13.4 Calculating SSE

Calculate the sum of squares error (SSE) for Example 13.1.

**Solution**

Table 13.3 shows the $\hat{y}$-values and residuals for the data in Table 13.1. The sum of squares error is then found by squaring each residual and taking the sum. Thus, $\text{SSE} = \Sigma(y - \hat{y})^2 = 12$. Since we know that $\hat{y} = 7 + 2x$ is the regression line, according to the least-squares criterion, no other possible straight line would result in a smaller SSE. ■

Table 13.3 Calculation of the SSE for the short-term memory test example

| Student | Time ($x$) | Actual score ($y$) | Predicted score ($\hat{y} = 7 + 2x$) | Residual ($y - \hat{y}$) | (Residual)$^2$ ($(y - \hat{y})^2$) |
|---|---|---|---|---|---|
| 1 | 1 | 9 | 9 | 0 | 0 |
| 2 | 1 | 10 | 9 | 1 | 1 |
| 3 | 2 | 11 | 11 | 0 | 0 |
| 4 | 3 | 12 | 13 | −1 | 1 |
| 5 | 3 | 13 | 13 | 0 | 0 |
| 6 | 4 | 14 | 15 | −1 | 1 |
| 7 | 5 | 19 | 17 | 2 | 4 |
| 8 | 6 | 17 | 19 | −2 | 4 |
| 9 | 7 | 21 | 21 | 0 | 0 |
| 10 | 8 | 24 | 23 | 1 | 1 |
| | | | | | $\text{SSE} = \Sigma(y - \hat{y})^2 = 12$ |

A useful interpretive statistic is $s$, the **standard error of the estimate**. The formula for $s$ follows.

**Standard Error of the Estimate $s$**

$$s = \sqrt{\frac{\text{SSE}}{n - 2}}$$

The standard error of the estimate gives a measure of the typical residual. That is, $s$ is a measure of the size of the *typical prediction error*, the typical difference between the predicted value of $y$ and the actual observed value of $y$. If the typical prediction error is large, then the regression line may not be useful.

### Example 13.5 Calculating $s$, the standard error of the estimate

Calculate and interpret the standard error for Example 13.1.

**Solution**

SSE = 12 and $n = 10$, so

$$s = \sqrt{\frac{\text{SSE}}{n - 2}} = \sqrt{\frac{12}{8}} \approx 1.2247$$

*Note:* Here we are rounding $s = 1.2247$ for reporting purposes. However, we use $s$ for calculating other quantities later and do not round until the last calculation.

Thus, the typical error in prediction is 1.2247 points. In other words, if we know the amount of time ($x$) a given student spent memorizing, then our estimate of the student's score on the short-term memory test will typically differ from the student's actual score by only 1.2247 points.

## Section 13.1 SUMMARY

1 Linear regression analysis is used to approximate the relationship between two quantitative variables, using a straight line. For a given data set, this line is described by the regression equation $\hat{y} = b_0 + b_1 x$, where $\hat{y}$ is the estimated or predicted value of $y$ for the given value of $x$, $b_0$ is the $y$ intercept, and $b_1$ is the slope of the regression equation.

2 The prediction error, or residual, is the difference $(y - \hat{y})$, measuring how far off the predicted value of $y$ is from the actual value of $y$ observed in the data set. The sum of squared prediction errors is referred to as sum of squares error, $\text{SSE} = \Sigma(y - \hat{y})^2$. The standard error of the estimate, $s = \sqrt{\frac{\text{SSE}}{n - 2}}$, is an indicator of the precision of the estimates derived from the regression equation, since it provides a measure of the typical residual or prediction error.

## Section 13.1 EXERCISES

### CLARIFYING THE CONCEPTS

**1.** What is the meaning of $b_0$? How do we interpret the meaning of $b_1$?

**2.** Explain what a prediction error is. Do we want our prediction errors to be large or small?

**3.** How does the least-squares criterion choose the "best" line to approximate the relationship between the predictor variable and the response variable?

**4.** What does SSE measure? Would we want SSE to be large or small? Why?

**5.** What does $s$ measure?

**6.** Suppose we would like to use the predictor variable to estimate the value of the response variable. Would we want the standard error of the estimate $s$ to be large or small? Why?

### PRACTICING THE TECHNIQUES

For Exercises 7–12, follow these steps.

**a.** Draw a scatterplot.

**b.** Construct a table like Table 13.2 and calculate $\sum x$, $\sum y$, $\sum xy$, $\sum x^2$, and $\sum y^2$.

**c.** Calculate $b_0$ and $b_1$ using the computational formulas, and write the regression equation.

**7.**

| $x$ | $y$ |
|---|---|
| 1 | 15 |
| 2 | 20 |
| 3 | 20 |
| 4 | 25 |
| 5 | 25 |

**8.**

| $x$ | $y$ |
|---|---|
| 0 | 10 |
| 5 | 20 |
| 10 | 45 |
| 15 | 50 |
| 20 | 75 |

**9.**

| $x$ | $y$ |
|---|---|
| −5 | 0 |
| −4 | 8 |
| −3 | 8 |
| −2 | 16 |
| −1 | 16 |

**10.**

| $x$ | $y$ |
|---|---|
| −3 | −5 |
| −1 | −15 |
| 1 | −20 |
| 3 | −25 |
| 5 | −30 |

**11.**

| $x$ | $y$ |
|---|---|
| 10 | 100 |
| 20 | 95 |
| 30 | 85 |
| 40 | 85 |
| 50 | 80 |

**12.**

| $x$ | $y$ |
|---|---|
| 0 | 11 |
| 20 | 11 |
| 40 | 16 |
| 60 | 21 |
| 80 | 26 |

For Exercises 13–18, follow these steps.

**a.** Construct a table like Table 13.3, and calculate the following quantity for each observation:

**i.** $\hat{y}$ (the estimated $y$)

**ii.** $(y - \hat{y})$, the prediction error or residual

**iii.** $(y - \hat{y})^2$ the squared residual

**b.** Calculate SSE, the sum of squares error. What does this quantity measure?

**c.** Calculate and interpret $s = \sqrt{\text{SSE}/(n-2)}$, the standard error of the estimate.

**13.** Data in Exercise 7

**14.** Data in Exercise 8

**15.** Data in Exercise 9

**16.** Data in Exercise 10

**17.** Data in Exercise 11

**18.** Data in Exercise 12

### APPLYING THE CONCEPTS

For Exercises 19–27, follow these steps.

**a.** Construct the scatterplot.

**b.** Compute the regression equation.

**c.** Interpret the meaning of $b_0$ and $b_1$ in the context of the particular exercise.

**d.** Calculate $s = \sqrt{\text{SSE}/(n-2)}$, the standard error of the estimate, and interpret this statistic.

**19. Volume and Weight.** The table contains the volume ($x$, in cubic meters) and weight ($y$, in kilograms) of five randomly chosen packages shipped to a local village.

| Volume ($x$) | Weight ($y$) |
|---|---|
| 4 | 10 |
| 8 | 16 |
| 12 | 25 |
| 16 | 30 |
| 20 | 35 |

**20. Family Size and Pets.** Shown in the table (see next page) are the number of family members ($x$) in a random sample taken from a suburban neighborhood, along with the number of pets ($y$) belonging to each family.

| Family size ($x$) | Pets ($y$) |
|---|---|
| 2 | 1 |
| 3 | 2 |
| 4 | 2 |
| 5 | 3 |
| 6 | 3 |

**21. SAT Math and Verbal Scores.** The College Board reported the SAT I Verbal scores ($x$) and SAT I Math scores ($y$) for the five states with the best participation rates:

| State | SAT I Verbal ($x$) | SAT I Math ($y$) |
|---|---|---|
| New York | 497 | 510 |
| Connecticut | 515 | 515 |
| Massachusetts | 518 | 523 |
| New Jersey | 501 | 514 |
| New Hampshire | 522 | 521 |

**22. World Temperatures.** Listed in the table are the low ($x$) and high ($y$) temperatures for a particular day in 2006, measured in degrees Fahrenheit, for a random sample of cities worldwide.

| City | Low ($x$) | High ($y$) |
|---|---|---|
| Kolkata, India | 57 | 77 |
| London, England | 36 | 45 |
| Montreal, Quebec | 7 | 21 |
| Rome, Italy | 39 | 55 |
| San Juan, Puerto Rico | 70 | 83 |
| Shanghai, China | 34 | 45 |

**23. NCAA Power Ratings.** For the NCAA basketball power ratings, is there a relationship between a school's winning percentage and the power rating? The table shows each team's winning percentage ($x$) and power rating ($y$) for the 2004 NCAA Basketball Tournament, according to **www.teamrankings.com**.

| School | Win % ($x$) | Rating ($y$) |
|---|---|---|
| Duke | 0.838 | 96.020 |
| St. Joseph's | 0.938 | 95.493 |
| Connecticut | 0.846 | 95.478 |
| Oklahoma State | 0.882 | 95.320 |
| Pittsburgh | 0.857 | 94.541 |
| Georgia Tech | 0.737 | 93.091 |
| Stanford | 0.938 | 92.862 |
| Kentucky | 0.844 | 92.692 |
| Gonzaga | 0.903 | 92.609 |
| Mississippi State | 0.867 | 91.912 |

**24. Midterm Exams and Overall Grade.** Can you predict how you will do in a course based on the result of the midterm exam only? The midterm exam score and the overall grade were recorded for a random sample of 12 students in an elementary statistics course. The results are shown in the following table.

| Student | Midterm exam score ($x$) | Overall grade ($y$) |
|---|---|---|
| 1 | 50 | 65 |
| 2 | 90 | 80 |
| 3 | 70 | 75 |
| 4 | 80 | 75 |
| 5 | 60 | 45 |
| 6 | 90 | 95 |
| 7 | 90 | 85 |
| 8 | 80 | 80 |
| 9 | 70 | 65 |
| 10 | 70 | 70 |
| 11 | 60 | 65 |
| 12 | 50 | 55 |

**25. Teenage Birth Rate.** The National Center for Health Statistics publishes data on state birth rates. The table contains the overall birth rate and the teenage birth rate for ten randomly chosen states. The overall birth rate is defined by the NCHS as "live births per 1000 women," and the teenage birth rate is defined as "live births per 1000 women aged 15–19." We are interested in estimating teenage birth rate based on overall birth rate.

| State | Overall birth rate ($x$) | Teenage birth rate ($y$) |
|---|---|---|
| Alabama | 13.1 | 52.4 |
| Arizona | 16.3 | 60.1 |
| California | 15.2 | 39.5 |
| Florida | 12.5 | 42.4 |
| Georgia | 15.7 | 53.4 |
| New York | 13.0 | 26.9 |
| Ohio | 13.0 | 38.5 |
| Pennsylvania | 11.7 | 30.5 |
| Texas | 17.0 | 62.6 |
| Virginia | 13.9 | 35.2 |

**26. Cotton Prices.** The National Agricultural Statistics Service (**www.nass.usda.gov**) reports the prices for commodities produced in the United States. The table (see next page) contains the prices in dollars per pound reported by NASS for Georgia-grown cotton in 2005–2006. (No data are available for September of either year.) We are interested in predicting 2006 cotton prices based on 2005 prices.

| Month | 2005 prices ($x$) | 2006 prices ($y$) |
|---|---|---|
| Jan | 0.401 | 0.507 |
| Feb | 0.388 | 0.491 |
| Mar | 0.389 | 0.471 |
| Apr | 0.383 | 0.478 |
| May | 0.374 | 0.462 |
| Jun | 0.374 | 0.464 |
| Jul | 0.389 | 0.455 |
| Aug | 0.397 | 0.456 |
| Oct | 0.516 | 0.510 |
| Nov | 0.502 | 0.519 |
| Dec | 0.501 | 0.510 |

**27. Height and Weight.** The university medical unit is collecting data on the heights (in inches) and weights (in pounds) of the female students on campus. A random sample of ten female students showed the following heights and weights.

| Student | Height ($x$) | Weight ($y$) |
|---|---|---|
| 1 | 60 | 100 |
| 2 | 60 | 115 |
| 3 | 62 | 110 |
| 4 | 63 | 125 |
| 5 | 63 | 135 |
| 6 | 64 | 145 |
| 7 | 66 | 140 |
| 8 | 67 | 145 |
| 9 | 68 | 150 |
| 10 | 70 | 155 |

For Exercises 28–34, make the indicated estimation. If the estimate represents extrapolation, report the estimate and explain why it may not be appropriate.

**28. SAT Math and Verbal Scores.** Refer to Exercise 21. Estimate the SAT I Math score for someone with the following SAT I Verbal scores.

**a.** 500

**b.** 550

**29. World Temperatures.** Refer to Exercise 22. Estimate the high temperature for a city with the following low temperatures.

**a.** 5

**b.** 50

**30. NCAA Power Ratings.** Refer to Exercise 23. Estimate the power rating for a college with the following winning percentages.

**a.** 0.950

**b.** 0.900

**31. Midterm Exams and Overall Grade.** Refer to Exercise 24. Estimate the overall grade for a student with the following midterm exam scores.

**a.** 100

**b.** 75

**32. Teenage Birth Rate.** Refer to Exercise 25. Estimate the teenage birth rate for a state with the following overall birth rates.

**a.** 15.0

**b.** 20.0

**33. Cotton Prices.** Refer to Exercise 26. Estimate the 2006 cotton prices for a month from the following 2005 prices.

**a.** 0.50

**b.** 0.35

**34. Height and Weight.** Refer to Exercise 27. Estimate the weights of students with the following heights.

**a.** 72 inches

**b.** 62 inches

For Exercises 35–37, use Minitab or Excel.

**a.** Construct the scatterplot.

**b.** Compute the regression equation.

**c.** Calculate and interpret $s = \sqrt{\frac{SSE}{n-2}}$, the standard error of the estimate.

**35.** Open the **Darts** data set, which we used for the Chapter 3 Case Study. Use the Dow Jones Industrial Average ($x$) to estimate the pros' performance ($y$).

**36.** Open the **Nutrition** data set. Estimate the number of calories per gram ($y$) using the amount of fat per gram ($x$).

**37.** Open the **Pulse and Temp** data set. Estimate body temperature ($y$) using heart rate ($x$).

### CREATE YOUR OWN DATA SETS

Suppose we have a tiny data set with the following $(x, y)$ pairs.

| $x$ | $y$ |
|---|---|
| 1 | ? |
| 2 | ? |
| 3 | ? |

For Exercises 38–42, create a set of $y$-values that would fulfill each specification.

**38.** The slope of the line is positive.

**39.** The slope of the line is negative.

**40.** The slope of the line is 0.

**41.** The slope of the line is equal to 2.

**42.** The slope of the line is equal to $-3$.

APPLET Use the *Correlation and Regression* applet for Exercises 43–45.

**43.** In these applet exercises, use the "thermometer" above the graph (where it says "Sum of squares =") to help find the least-squares regression line interactively.

**a.** Select 5 points so that the correlation coefficient is about 0.8. Then select "Draw line."

**b.** Make your best guess about where the least-squares regression line should be, and draw the line there.

**44.** The blue section of the thermometer is a measure of the sum of squares error, the total squared vertical distance from the data points to the actual regression line. Recall that the least-squares regression line minimizes this distance. The green section of the thermometer tells you how much "extra" squared error you get from using the line you constructed in Exercise 43a.

**a.** Adjust the line you drew in Exercise 43a by clicking and dragging on the points until the green section of the thermometer has disappeared.

**b.** What does the disappearance of the green part tell you about the adjusted line you constructed?

**c.** Will the line now coincide with the least-squares regression line?

**45.** Verify that your adjusted line from Exercise 44 coincides with the least-squares regression line by selecting "Show least-squares line."

## 13.2 Coefficient of Determination $r^2$

**Objectives: By the end of this section, I will be able to...**

1 Describe how total variability, prediction error, and improvement are measured by SST, SSE, and SSR.

2 Explain the meaning of $r^2$ as a measure of the usefulness of the regression.

In Section 13.1, we reviewed regression analysis and introduced the concept of prediction error. Here in Section 13.2, we derive a statistic, the coefficient of determination $r^2$, which will provide an indication of the usefulness of the regression.

### 1 SST, SSR, and SSE

The coefficient of determination $r^2$ depends on the values of two new statistics, SST and SSR. To learn about SST and SSR, we continue developing the short-term memory scores example from Section 13.1. The regression line was calculated as $\hat{y} = 7 + 2x$, where $x$ is time and $y$ is score. We found that $\text{SSE} = \sum(y - \hat{y})^2 = 12$ and $s = \sqrt{\frac{\text{SSE}}{n-2}} = \sqrt{\frac{12}{8}} \approx 1.2247$. The least-squares criterion guarantees that this value of $\text{SSE} = 12$ is the smallest possible value for SSE, given the data in Table 13.1 (page 667). However, this guarantee in itself does not tell us that the regression is useful. For the regression to be useful, the prediction error (and therefore SSE) must be small. However, we cannot yet tell whether the value of $\text{SSE} = 12$ is indeed small, since we have nothing to compare it against.

Suppose for a moment that we wanted to estimate short-term memory scores without knowledge of the amount of time ($x$) for memorizing. Then the best estimate for $y$ is simply $\bar{y} = 15$, the mean of the sample of short-term memory test scores. We could use this value as an estimated score for each subject, regardless of the amount of time given to memorize (since that information is not available). The graph of $\bar{y} = 15$ is the horizontal line in Figure 13.4.

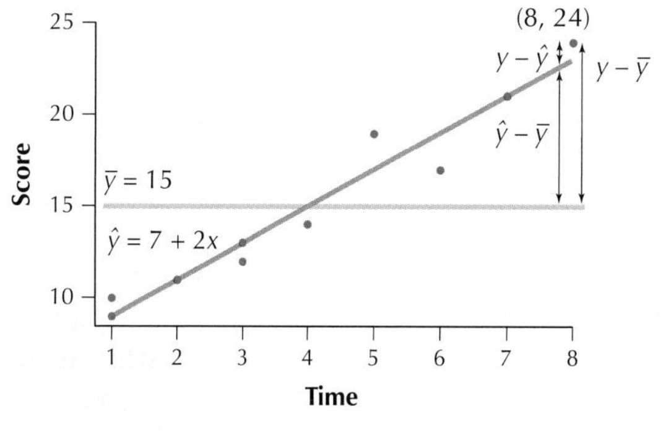

FIGURE 13.4
Comparing $(y - \hat{y})$ and $(y - \bar{y})$.

In general, the data points are closer to the regression line than they are to the horizontal line $\bar{y} = 15$, indicating that the errors in prediction are smaller when using the regression equation. Consider Student 10, who had a short-term memory score of $y = 24$ after memorizing for $x = 8$ minutes. Using $\bar{y} = 15$ as the estimate, the prediction error (residual) for Student 10 is

$$(y - \bar{y}) = 24 - 15 = 9$$

This error is shown in Figure 13.4 as the vertical distance $(y - \bar{y})$.

Suppose we found this value $(y - \bar{y})$ for every student in the data set and summed the squared $(y - \bar{y})$, just as we did for the $(y - \hat{y})$ when finding SSE. The resulting statistic is called the **total sum of squares (SST)** and is a measure of the total variability in the values of the $y$ variable:

$$\text{SST} = \sum(y - \bar{y})^2$$

Developing Your Statistical Sense

### Relationship Between SST and the Variance of the *y*'s

Note that SST ignores the presence of the $x$ information; it is simply a measure of the variability in $y$. Recall that the *variance* of a sample of $y$-values is given by $s^2 = \sum(y - \bar{y})^2/(n - 1)$. Thus, SST is proportional to the variance of the $y$'s and, as such, is a measure of the variability in the $y$ data. ■

## Example 13.6 Calculating SST

Calculate the total sum of squares (SST) for Example 13.1.

### Solution

Table 13.4 shows the values for $(y - \bar{y}) = (y - 15)$ for the data in Table 13.1. Thus, $\text{SST} = \sum(y - \bar{y})^2 = 228$. ■

Table 13.4 Calculation of SST

| Student | Score ($y$) | $(y - \bar{y})$ | $(y - \bar{y})^2$ |
|---|---|---|---|
| 1 | 9 | −6 | 36 |
| 2 | 10 | −5 | 25 |
| 3 | 11 | −4 | 16 |
| 4 | 12 | −3 | 9 |
| 5 | 13 | −2 | 4 |
| 6 | 14 | −1 | 1 |
| 7 | 19 | 4 | 16 |
| 8 | 17 | 2 | 4 |
| 9 | 21 | 6 | 36 |
| 10 | 24 | 9 | 81 |
| | | | $\text{SST} = \sum(y - \bar{y})^2 = 228$ |

Consider Figure 13.4 once again. For Student 10, note that the error in prediction when ignoring the $x$ data is $(y - \bar{y}) = 9$, while the error in prediction when using the regression equation is $(y - \hat{y}) = 1$. (Recall that $\hat{y} = 7 + 2(8) = 23$, since Student

10's time is $x = 8$.) The amount of improvement (diminishing the prediction error) is the difference between $\hat{y}$ and $\bar{y}$:

$$(\hat{y} - \bar{y}) = 23 - 15 = 8$$

Once again, we can find $(\hat{y} - \bar{y})$ for each observation in the data set, square them, and sum the squared results to obtain $\sum(\hat{y} - \bar{y})^2$. The resulting statistic is **SSR**, the **sum of squares regression**.

$$\text{SSR} = \sum(\hat{y} - \bar{y})^2$$

SSR measures the amount of *improvement* in the accuracy of our estimates when using the regression equation compared with relying only on the $y$-values and ignoring the $x$ information. Note in Figure 13.4 that the distance $(y - \bar{y})$ is the same as the sum of the distances $(\hat{y} - \bar{y})$ and $(y - \hat{y})$. It can be shown, using algebra, that the following also holds true.

**Relationship Among SST, SSR, and SSE**

$$\text{SST} = \text{SSR} + \text{SSE}$$

None of these sums of squares can ever be negative. If any two of these sums of squares are known, the third can be calculated as well, as shown in the following example.

## Example 13.7 Using SST and SSE to find SSR

Use SST and SSE to find the value of SSR for the data from Example 13.6.

**Solution**

From Example 13.4, we have SSE = 12, and from Example 13.6, we have SST = 228. That leaves us with just one unknown in the equation SST = SSR + SSE, so we can solve for the unknown SSR:

$$\text{SSR} = \text{SST} - \text{SSE} = 228 - 12 = 216$$

## 2 Calculating and Interpreting the Coefficient of Determination $r^2$

SSR represents the amount of variability in the response variable that is accounted for by the regression equation, that is, the linear relationship between the response variable $y$ and the predictor variable $x$. SSE represents the amount of variability in the response variable that is left unexplained after accounting for the relationship between $x$ and $y$ (including random error). Since we know that SST represents the sum of SSR and SSE, it makes sense to consider the *ratio* of SSR and SST, called the **coefficient of determination $r^2$**.

The **coefficient of determination $r^2$** = SSR/SST measures the goodness of fit of the regression equation to the data. We interpret $r^2$ as the proportion of the variability in $y$ that is accounted for by the linear relationship between $y$ and $x$. The values that $r^2$ can take are $0 \le r^2 \le 1$.

## Example 13.8 Finding the coefficient of determination $r^2$

Calculate and interpret the value of the coefficient of determination $r^2$ for the data in Example 13.1.

**Solution**

From Example 13.6 we have SST = 228, and from Example 13.7 we have SSR = 216. Hence,

$$r^2 = \frac{\text{SSR}}{\text{SST}} = \frac{216}{228} \approx 0.9474$$

Thus, 94.74% of the variability in the memory test score ($y$) is accounted for by the linear relationship between score ($y$) and the time given for study ($x$). ■

What Does This Number Mean?

What does the value of $r^2 \approx 0.9474$ mean? Consider that the memory test scores have a certain amount of variability: some scores are higher than others. In addition to the amount of time ($x$) given for memorizing, there may be several other factors that might account for variability in the scores, such as the memorizing ability of the students, how much sleep the students had, and so on. However, $r^2 \approx 0.9474$ indicates that 94.74% of this variability in memory scores ($y$) is explained by the single factor "amount of time given for study" ($x$). All other factors (such as amount of sleep) account for only 100% − 94.74% = 5.26% of the variability in the memory test scores. ■

Suppose that the regression equation was a perfect fit to the data, so that every observation lay exactly on the regression line. Since there would be no errors in prediction, SSE would equal 0, which would imply that

$$\text{SST} = \text{SSR} + 0 = \text{SSR}$$

Since SST = SSR, then

$$r^2 = \frac{\text{SSR}}{\text{SST}} = \frac{\text{SST}}{\text{SST}} = 1$$

Conversely, if SSR = 0, then *no improvement at all* is gained by using the regression equation. That is, the regression equation accounts for no variability at all, and $r^2 = 0/\text{SST} = 0$.

The closer the value of $r^2$ is to 1, the better the fit of the regression equation to the data set. A value near 1 indicates that the regression equation fits the data extremely well. A value near 0 indicates that the regression equation fits the data extremely poorly.

Here are the alternate computational formulas for finding SST and SSR. Note that the formula for SSR is the same as the computational formula for the slope $b_1$ of the regression equation, except that the numerator is squared.

**Computational Formulas for SST and SSR**

$$\text{SST} = \sum y^2 - \frac{\left(\sum y\right)^2}{n} \qquad \text{SSR} = \frac{\left[\sum xy - \frac{\left(\sum x\right)\left(\sum y\right)}{n}\right]^2}{\sum x^2 - \frac{\left(\sum x\right)^2}{n}}$$

## Example 13.9 Calculating SSR and SST using the computational formulas

Use the computational formulas to find SSR and SST for the data from Example 13.1.

**Solution**

Using the summary statistics from Table 13.2,

$$\text{SST} = \sum y^2 - \frac{\left(\sum y\right)^2}{n} = 2478 - \frac{(150)^2}{10} = 228$$

$$\text{SSR} = \frac{\left[\sum xy - \frac{\left(\sum x\right)\left(\sum y\right)}{n}\right]^2}{\sum x^2 - \frac{\left(\sum x\right)^2}{n}} = \frac{\left[708 - \frac{(40)(150)}{10}\right]^2}{214 - \frac{(40)^2}{10}} = \frac{[108]^2}{54} = 216$$

Then SSE = SST – SSR = 228 − 216 = 12. This value SSE = 12 agrees with the value we calculated earlier using Table 13.3.

Recall from Chapter 4 that the correlation coefficient $r$ is given by

$$r = \frac{\sum(x - \bar{x})(y - \bar{y})}{(n - 1)\, s_x s_y}$$

where $s_x$ and $s_y$ represent the sample standard deviation of the $x$ data and the $y$ data, respectively. The **correlation coefficient *r*** is a measure of the strength of the linear relationship between two continuous variables. We can express the correlation coefficient $r$ as

$$r = \pm\sqrt{r^2}$$

The correlation coefficient $r$ takes the same sign as the slope $b_1$. If the slope $b_1$ of the regression equation is positive, then $r = \sqrt{r^2}$; if the slope $b_1$ of the regression equation is negative, then $r = -\sqrt{r^2}$.

## Example 13.10 Calculate the correlation coefficient

Use $r^2$ to calculate the value of the correlation coefficient $r$ for the data in Example 13.1.

**Solution**

From Example 13.1, the slope $b_1 = 2$, which is positive and tells us that the sign of the correlation coefficient $r$ is positive. Hence,

$$r = \sqrt{r^2} = \sqrt{0.9474} \approx 0.9733$$

Thus, student scores on the short-term memory test are strongly positively correlated with the amount of time allowed for memorization.

## Section 13.2 SUMMARY

1 The total variability in the $y$ variable is measured by the total sum of squares, $\text{SST} = \Sigma(y - \bar{y})^2$, and may be divided into the sum of squares regression, $\text{SSR} = \Sigma(\hat{y} - \bar{y})^2$, and the sum of squares error, $\text{SSE} = \Sigma(y - \hat{y})^2$. SSR measures the amount of improvement in the accuracy of estimates when using the regression equation compared to ignoring the $x$ information.

2 The coefficient of determination, $r^2 = \text{SSR/SST}$, measures the goodness of fit of the regression equation as an approximation of the relationship between $x$ and $y$. Finally, the correlation coefficient $r$ may be expressed as $r = \pm\sqrt{r^2}$, taking the positive or negative sign of the slope $b_1$.

## Section 13.2 EXERCISES

### CLARIFYING THE CONCEPTS

1. What does SST measure?
2. Do the values of $x$ affect SST at all?
3. What does SSR measure?
4. Suppose we would like to use the predictor variable to estimate the value of the response variable. Would we want SSR to be large or small? Why?
5. What does it mean when $r^2$ is close to 1? What does it mean when $r^2$ is close to 0?
6. How do we find the correlation coefficient $r$ using the regression results?

### PRACTICING THE TECHNIQUES

For Exercises 7–12, follow steps **(a)–(e)**. Note that SSE was calculated in Section 13.1, Exercises 13–18.

**a.** Construct a table like Table 13.4, calculating $(y - \bar{y})$ and $(y - \bar{y})^2$ for each observation.
**b.** Calculate SST, the total sum of squares. What does this quantity measure?
**c.** Calculate SSR = SST − SSE. What does this quantity measure?
**d.** Calculate and interpret the coefficient of determination $r^2$.
**e.** Calculate and interpret the correlation coefficient $r$.

**7.**

| $x$ | $y$ |
|---|---|
| 1 | 15 |
| 2 | 20 |
| 3 | 20 |
| 4 | 25 |
| 5 | 25 |

**8.**

| $x$ | $y$ |
|---|---|
| 0 | 10 |
| 5 | 20 |
| 10 | 45 |
| 15 | 50 |
| 20 | 75 |

**9.**

| $x$ | $y$ |
|---|---|
| −5 | 0 |
| −4 | 8 |
| −3 | 8 |
| −2 | 16 |
| −1 | 16 |

**10.**

| $x$ | $y$ |
|---|---|
| −3 | −5 |
| −1 | −15 |
| 1 | −20 |
| 3 | −25 |
| 5 | −30 |

**11.**

| $x$ | $y$ |
|---|---|
| 10 | 100 |
| 20 | 95 |
| 30 | 85 |
| 40 | 85 |
| 50 | 80 |

**12.**

| $x$ | $y$ |
|---|---|
| 0 | 11 |
| 20 | 11 |
| 40 | 16 |
| 60 | 21 |
| 80 | 26 |

### APPLYING THE CONCEPTS

For Exercises 13–24, follow these steps.

**a.** Calculate and interpret the coefficient of determination $r^2$.
**b.** Calculate and interpret the correlation coefficient $r$.

**13. Volume and Weight.** The table shows the volume ($x$, in cubic meters) and weight ($y$, in kilograms) of five randomly chosen packages shipped to a local college.

| Volume ($x$) | Weight ($y$) |
|---|---|
| 4 | 10 |
| 8 | 16 |
| 12 | 25 |
| 16 | 30 |
| 20 | 35 |

**14. Family Size and Pets.** Shown in this table are the number of family members ($x$) in a random sample taken from a suburban neighborhood, along with the number of pets ($y$) belonging to each family.

| Family size ($x$) | Pets ($y$) |
|---|---|
| 2 | 1 |
| 3 | 2 |
| 4 | 2 |
| 5 | 3 |
| 6 | 3 |

**15. SAT Math and Verbal Scores.** The College Board reported the SAT I Verbal scores ($x$) and SAT I Math scores ($y$) for the five states with the best participation rates:

| State | Verbal ($x$) | Math ($y$) |
|---|---|---|
| NY | 497 | 510 |
| CT | 515 | 515 |
| MA | 518 | 523 |
| NJ | 501 | 514 |
| NH | 522 | 521 |

**16. World Temperatures.** Listed in the table are the low ($x$) and high ($y$) temperatures for a particular day in 2006, measured in degrees Fahrenheit, for a random sample of cities worldwide.

| City | Low ($x$) | High ($y$) |
|---|---|---|
| Kolkata, India | 57 | 77 |
| London, England | 36 | 45 |
| Montreal, Quebec | 7 | 21 |
| Rome, Italy | 39 | 55 |
| San Juan, Puerto Rico | 70 | 83 |
| Shanghai, China | 34 | 45 |

**17. NCAA Power Ratings.** The table shows each team's winning percentage ($x$) and power rating ($y$) for the 2004 NCAA Basketball Tournament, according to **www.teamrankings.com**.

| School | Win % ($x$) | Rating ($y$) |
|---|---|---|
| Duke | 0.838 | 96.020 |
| St. Joseph's | 0.938 | 95.493 |
| Connecticut | 0.846 | 95.478 |
| Oklahoma State | 0.882 | 95.320 |
| Pittsburgh | 0.857 | 94.541 |
| Georgia Tech | 0.737 | 93.091 |
| Stanford | 0.938 | 92.862 |
| Kentucky | 0.844 | 92.692 |
| Gonzaga | 0.903 | 92.609 |
| Mississippi State | 0.867 | 91.912 |

**18. Midterm Exams and Overall Grade.** The table shows the midterm exam scores ($x$) and the overall grade ($y$) for a random sample of 12 students in an elementary statistics course.

| Midterm exam ($x$) | Overall grade ($y$) |
|---|---|
| 50 | 65 |
| 90 | 80 |
| 70 | 75 |
| 80 | 75 |
| 60 | 45 |
| 90 | 95 |
| 90 | 85 |
| 80 | 80 |
| 70 | 65 |
| 70 | 70 |
| 60 | 65 |
| 50 | 55 |

**19. State of Maine Veterans.** The U.S. Department of Veterans Affairs reports demographic information on veterans. The table contains the number of veterans in each county in the state of Maine under age 65 and age 65 and over. We are interested in predicting the number of veterans 65 and over based on the number of veterans under 65.

| County | Veterans under 65 | Veterans 65 and over |
|---|---|---|
| Androscoggin | 7,246 | 3,744 |
| Aroostook | 4,750 | 3,509 |
| Cumberland | 15,223 | 9,629 |
| Franklin | 1,915 | 1,292 |
| Hancock | 3,463 | 2,561 |
| Kennebec | 8,134 | 4,901 |
| Knox | 2,614 | 2,277 |
| Lincoln | 2,327 | 2,122 |
| Oxford | 3,701 | 2,697 |
| Penobscot | 8,881 | 5,642 |
| Piscataquis | 1,387 | 889 |
| Sagadahoc | 3,222 | 1,554 |
| Somerset | 3,254 | 2,343 |
| Waldo | 2,626 | 1,407 |
| Washington | 2,528 | 1,745 |
| York | 13,631 | 7,848 |

**20. Used Car Prices.** Looking to buy a used car? Do you think you can predict the price of a used car based on how old it is? The table contains the price and the age in years of 14 previously owned automobiles of the same make and model.

| Car | Age ($x$) | Price ($y$) |
|---|---|---|
| 1 | 2 | $17,500 |
| 2 | 2 | 16,000 |
| 3 | 3 | 14,000 |
| 4 | 3 | 15,500 |
| 5 | 4 | 13,500 |
| 6 | 4 | 14,500 |
| 7 | 5 | 10,500 |
| 8 | 5 | 12,000 |
| 9 | 6 | 9,500 |
| 10 | 6 | 8,000 |
| 11 | 7 | 8,500 |
| 12 | 7 | 7,000 |
| 13 | 8 | 6,000 |
| 14 | 8 | 5,500 |

**21. GPA and Combined SAT Scores.** The college admissions office would like to determine whether there is a relationship between the combined SAT score and the grade point average of first-year students, using the data in the table.

| Student | Combined SAT score $(x)$ | Grade point average $(y)$ |
|---|---|---|
| 1 | 900 | 1.5 |
| 2 | 925 | 2.6 |
| 3 | 950 | 1.9 |
| 4 | 975 | 2.7 |
| 5 | 1000 | 2.0 |
| 6 | 1000 | 2.5 |
| 7 | 1025 | 3.0 |
| 8 | 1025 | 2.4 |
| 9 | 1050 | 2.1 |
| 10 | 1050 | 2.9 |
| 11 | 1075 | 2.7 |
| 12 | 1075 | 3.1 |
| 13 | 1100 | 3.0 |
| 14 | 1125 | 3.3 |
| 15 | 1150 | 3.2 |

**22. Batting Average and Runs Scored.** The table shows the top ten hitters in Major League Baseball for 2007. We are interested in estimating the number of runs scored $(y)$ using the player's batting average $(x)$.

| Player | Team | Batting average $(x)$ | Runs $(y)$ |
|---|---|---|---|
| M. Ordonez | Detroit Tigers | .363 | 117 |
| I. Suzuki | Seattle Mariners | .351 | 111 |
| P. Polanco | Detroit Tigers | .341 | 105 |
| M. Holliday | Colorado Rockies | .340 | 120 |
| J. Posada | New York Yankees | .338 | 91 |
| C. Jones | Atlanta Braves | .337 | 108 |
| D. Ortiz | Boston Red Sox | .332 | 116 |
| H. Ramirez | Florida Marlins | .332 | 125 |
| E. Renteria | Atlanta Braves | .332 | 87 |
| C. Utley | Philadelphia Phillies | .332 | 104 |

**23. Stock Prices.** Would you expect there to be a relationship between the price $(x)$ of a stock and its change $(y)$ in price on a particular day? The table provides stock price and stock price change for August 2, 2004, for a random sample of ten stocks listed on the New York Stock Exchange.

| Stock | Price $(x)$ | Change $(y)$ |
|---|---|---|
| Nortel Networks | 3.86 | +0.04 |
| Qwest Communications | 3.41 | −0.56 |
| Tyco International | 32.02 | +0.78 |
| Lucent Technologies | 3.04 | −0.03 |
| Vishay Intertechnology | 13.96 | −1.96 |
| Tenet Healthcare | 10.36 | −0.82 |
| Select Medical Group | 14.74 | +1.62 |
| Cox Communications | 33.19 | +0.03 |
| Verizon Communications | 39.15 | +0.46 |
| General Electric | 33.05 | −0.21 |

**24. Magma Hot Spot.** Earth scientists believe that there is a magma "hot spot" beneath the mid-Pacific, which has been generating volcanic islands for millions of years. As the Pacific plate inches toward Japan, the hot spot sends up another volcanic island in an apparently different spot, although the hotspot is stationary and it is the oceanic plate that is moving. Scientists have measured the age ($y$, in millions of years) of each volcanic island and the distance ($x$, in miles) from Kilauea to determine whether there is an association.

| Volcano | Distance $(x)$ | Age $(y)$ |
|---|---|---|
| Kilauea | 0 | 0.20 |
| Mauna Kea | 54 | 0.38 |
| Kohala | 100 | 0.43 |
| East Maui | 182 | 0.75 |
| Kahoolawe | 185 | 1.03 |
| West Maui | 221 | 1.32 |
| Lanai | 226 | 1.28 |
| East Molokai | 256 | 1.76 |
| West Molokai | 280 | 1.90 |
| Koolau | 339 | 2.60 |
| Waianae | 374 | 3.70 |
| Kauai | 519 | 5.10 |
| Niihau | 565 | 4.89 |
| Nihoa | 780 | 7.20 |
| [Unnamed] | 913 | 9.60 |
| Necker | 1058 | 10.30 |
| La Perouse | 1209 | 12.00 |
| Brooks Bank | 1256 | 13.00 |
| Gardner | 1435 | 12.30 |
| Laysan | 1818 | 19.90 |
| Northampton | 1841 | 26.60 |
| Pearl/Hermes | 2291 | 20.60 |
| Midway | 2432 | 27.70 |

Source: D. A. Clague and B. G. Dalrymple, "Tectonics, Geomorphology and Origin of the Hawaiian-Emperor Volcanic Chain," in E. L. Winterer, D. M. Hussong, and R. E. Decker (Eds.), *The Geology of North America,* Vol. N: *Eastern Pacific Ocean and Hawaii* (Geological Society of America 1989), pp. 188–217.

**25. Mammal Brain and Body Weight.** A study compared the body weight (in kilograms) and brain weight (in grams) for a random sample of mammals.[1] We are interested in estimating brain weight based on body weight.

| Body weight (kg) | Brain weight (g) |
|---|---|
| 52.16 | 440 |
| 60 | 81 |
| 27.66 | 115 |
| 85 | 325 |
| 36.33 | 119.5 |
| 100 | 157 |
| 35 | 56 |
| 62 | 1320 |
| 83 | 98.2 |
| 55.5 | 175 |

a. Construct a scatterplot of the data. Is there a particular mammal that is off by itself in the scatterplot? Which mammal do you think this might be, given the large brain weight for its body weight?
b. Use technology to perform the regression of brain weight versus body weight.
c. Omit the mammal you identified in (a) and perform the regression again. Describe how the following statistics have changed:
   i. $b_0$
   ii. $b_1$
   iii. $r^2$
   iv. $s$

WHAT IF? **26. Mammal Brain and Body Weight.** Refer to Exercise 25. *What if* we add 10 grams to each mammal's brain weight. Describe how and why this change will affect the following quantity. Will the quantity increase, decrease, or remain unchanged, or is there insufficient information to determine the effect?
a. $b_0$
b. $b_1$
c. $r^2$
d. $s$

For Exercises 27–29, use Minitab or Excel and follow steps **(a)–(e)**.
a. Construct the scatterplot.
b. Compute the regression equation.
c. Calculate and interpret the coefficient of determination $r^2$.
d. Calculate and interpret $s = \sqrt{\text{SSE}/(n-2)}$, the standard error of the estimate.
e. Calculate and interpret the correlation coefficient $r$.

**27.** Open the **Darts** data set, which we used for the Chapter 3 Case Study. We are interested in predicting the pros' performance ($y$) using the Dow Jones Industrial Average ($x$).

**28.** Open the **Nutrition** data set. We are interested in estimating the number of calories per gram ($y$) based on the amount of fat per gram ($x$).

**29.** Open the **Pulse and Temp** data set. We would like to predict body temperature ($y$) using heart rate ($x$).

## 13.3 Inference About the Slope of the Regression Line

*Objectives:* By the end of this section, I will be able to...

1. Explain and apply the regression model.
2. Construct and interpret the confidence interval for the slope of the regression line.
3. Perform and interpret the $t$ test for the relationship between the predictor variable and the response variable.

Thus far in this chapter, we have not discussed inferential statistics such as hypothesis tests or confidence intervals. In this section we learn about a hypothesis test and a confidence interval for the slope of the regression line.

### 1 Regression Model

Recall Example 13.1, where we examined time and score data from a sample of ten students. Based on the sample, we obtained the regression line $\hat{y} = 7 + 2x$. Suppose we examine a different sample of student times and scores. It is likely that this second sample will differ from the first, giving us a different regression line and different values for $b_0$ and $b_1$. In fact, for every different sample, $b_0$ and $b_1$ take different values since $b_0$ and $b_1$ are sample statistics. However, every sample comes from a population. Since we do not have data on all students, we are not able to find the characteristics of the population regression equation for time and score for all students. The $y$ intercept $\beta_0$ and slope $\beta_1$ of this "true" population regression equation are unknown population parameters, just as $\mu$ and $p$ are parameters in other contexts. The values of $\beta_0$ and $\beta_1$ are unknown, so we need to perform inference to learn about them.

The **regression model** may be used to approximate the relationship between the predictor variable $x$ and the response variable $y$ for the *entire population* of $(x, y)$ pairs.

**Regression Model**

The **population regression equation** is defined as

$$y = \beta_0 + \beta_1 x + \varepsilon$$

where $\beta_0$ is the $y$ intercept of the population regression line, $\beta_1$ is the slope of the population regression line, and $\varepsilon$ is the error term.

Students 1 and 2 in Table 13.1 both studied for the same $x = 1$ minute, but Student 1 had a score of $y = 9$ while Student 2 had a score of $y = 10$. Thus, any linear approximation of the true relationship between $x$ and $y$ will introduce a certain amount of error. This is why the error term $\varepsilon$ is needed.

We can estimate the population (true) regression equation $y = \beta_0 + \beta_1 x + \varepsilon$ by generating a regression equation from a sample. What if the value of $\beta_1$ was 0? Then the population regression equation would be

$$y = \beta_0 + (0)x + \varepsilon$$

that is,

$$y = \beta_0 + \varepsilon$$

Note that there is no longer any $x$ term in the population regression equation if $\beta_1 = 0$. In other words, *if $\beta_1$ equals 0, then there is no relationship between x and y.* If $\beta_1$ takes on any other value than 0, then there does exist a linear relationship between $x$ and $y$. This idea forms the basis for the inference we perform in this section.

## Regression Model Assumptions

The regression model operates under a set of four assumptions that must be valid in order to perform the inference in this section.

**Regression Model Assumptions**

1. **Zero mean assumption.** The error term $\varepsilon$ is a random variable, with a mean of 0. That is, the expected value of the random variable $\varepsilon$ is 0: $E(\varepsilon) = 0$.
2. **Constant variance assumption.** The variance of $\varepsilon$, which is denoted as $\sigma^2$, is the same regardless of the value of $x$.
3. **Independence assumption.** The values of $\varepsilon$ are independent of each other.
4. **Normality assumption.** The error term $\varepsilon$ is a normal random variable.

To summarize: the values of the error term $\varepsilon$ are independent normal random variables, with mean 0 and variance $\sigma^2$. The four assumptions lead to the following four behaviors of the response variable $y$.

**Behaviors of the Response Variable *y*, Based on the Regression Model Assumptions**

1. The zero mean assumption gives

$$E(y) = E(\beta_0 + \beta_1 x + \varepsilon) = E(\beta_0) + E(\beta_1 x) + E(\varepsilon) = \beta_0 + \beta_1 x$$

That is, for each value of $x$, the mean of the $y$-values lies on the *population regression line*:

$$E(y) = \beta_0 + \beta_1 x$$

2. From the constant variance assumption, we derive the variance of $y$:

$$\text{var}(y) = \text{var}(\beta_0 + \beta_1 x + \varepsilon) = \text{var}(\varepsilon) = \sigma^2$$

That is, the variance of the $y$-values is always constant, regardless of the value of $x$.

3. From the independence assumption we conclude that, for any particular value of $x$, the values of $y$ are independent as well.
4. From the normality assumption, we conclude that $y$ is also a normally distributed random variable. Why? We know that $y = \beta_0 + \beta_1 x + \varepsilon$. The only randomness in the quantity $\beta_0 + \beta_1 x + \varepsilon$ lies in the error term $\varepsilon$, since $\beta_0$ , $\beta_1$, and $x$ are each constants. Thus, since $\varepsilon$ is normal, so is $y$.

Behaviors 1 and 2 are derived using the following results from mathematical statistics theory.

- The expected value of a constant $a$ is equal to $a$. Thus, $E(\beta_0) = \beta_0$.
- The expected value of a constant times a data value is the constant times the data value. Thus, $E(\beta_1 x) = \beta_1 x$.
- The variance of a constant is 0. Thus $\text{var}(\beta_0) = 0$.
- The variance of a constant times a data value is 0. Thus $\text{var}(\beta_1 x) = 0$.

To summarize, for each value of $x$, the values of $y$ come from a normally distributed population with a mean on the population regression line $E(y) = \beta_0 + \beta_1 x$ and constant standard deviation $\sigma^2$. Figure 13.5 illustrates how $y$ is distributed for each

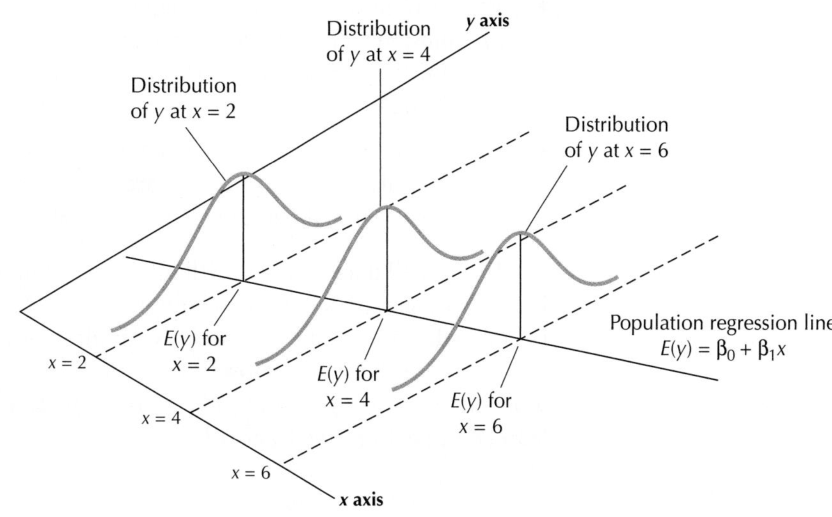


FIGURE 13.5 Illustrating the regression assumptions.

value of $x$. Note that each normal curve has the same shape, indicating constant variance for each $x$.

## Verifying the Regression Assumptions

To check the regression model assumptions, we construct two graphs:

1. Scatterplot of the residuals (prediction errors $y - \hat{y}$) against the fitted values (**fitted values** refers to the predicted values, $\hat{y}$)
2. Normal probability plot of the residuals

Figure 13.6 shows four types of patterns that might be observed in the residuals versus fitted values plots.

- Plot (a) is a "healthy" plot, displaying no noticeable patterns.
- In plot (b) we see a curve, which indicates a violation of the independence assumption. Independence implies that knowing the value of a particular $y$ does not help to predict the value of a different $y$. However, a curve suggests that knowing the value of a previous $y$ helps in knowing the value of the next $y$.

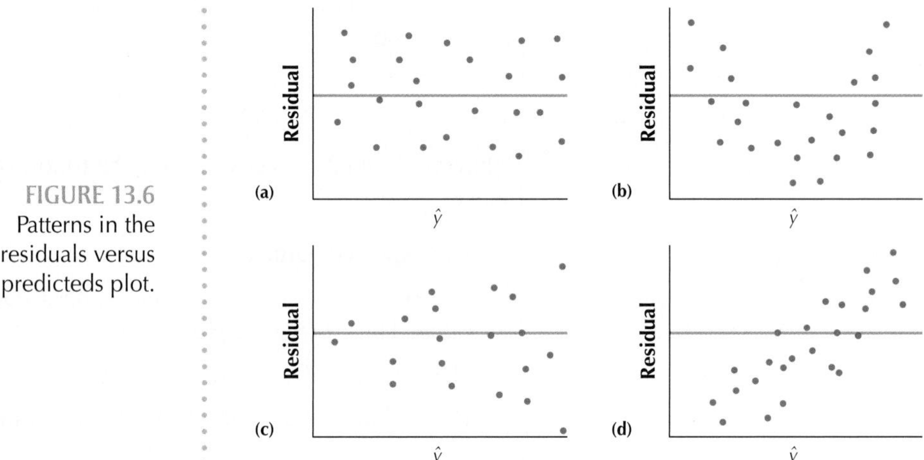


FIGURE 13.6 Patterns in the residuals versus predicteds plot.

- Plot (c) shows a "funnel" pattern, which contradicts the constant variance assumption. The residuals on the left are close together vertically (small variability), while the residuals on the right are far apart vertically (large variability).
- In plot (d) we see an increasing pattern, that violates the zero-mean assumption. The residuals on the left are all below the midline, so $E(y) < \beta_0 + \beta_1 x$, while the residuals on the right are all above the midline, so $E(y) > \beta_0 + \beta_1 x$.

With small data sets, it is difficult to ascertain whether or not patterns really exist. Be wary of seeing patterns where none exist. If one or more regression assumptions is violated, we should not proceed with inferential methods such as hypothesis tests or confidence intervals. However, even if one or more regression assumptions is violated, we can still report and interpret the descriptive regression statistics that we learned in Sections 13.1 and 13.2.

## Example 13.11 Verifying the regression assumptions

Verify the regression assumptions for the data in Example 13.1.

**Solution**

Figure 13.7 is a scatterplot of the residuals versus fitted values (predicted values) and Figure 13.8 contains a normal probability plot of the residuals for the data in Example 13.1.

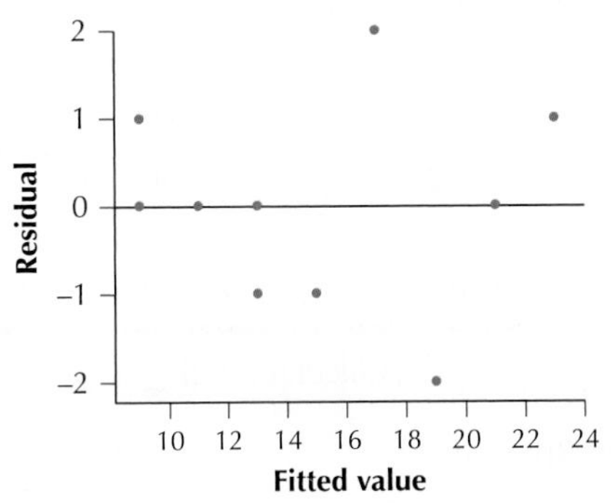


FIGURE 13.7 Residuals versus fitted values plot.

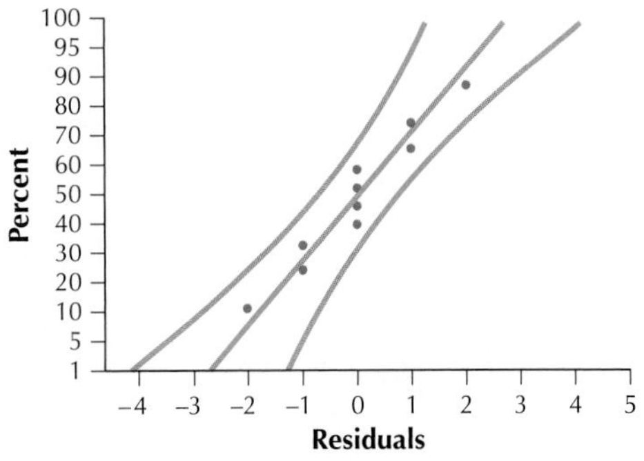

FIGURE 13.8 Normal probability plot of the residuals.

Figure 13.7 contains no strong evidence of the unhealthy patterns shown in Figure 13.6. Thus, the independence assumption, the constant variance assumption, and the zero-mean assumption are verified. Also, Figure 13.8 indicates no evidence of departures from normality in the residuals. Therefore we conclude that the regression assumptions are verified. ■

## 2 Confidence Interval for Slope $\beta_1$

The slope $b_1$ of the regression equation is a statistic that has a sampling distribution.

**Sampling Distribution of $b_1$**

When the regression assumptions are satisfied, then the quantity

$$t = \frac{b_1 - \beta_1}{s_{b_1}}$$

has approximately a $t$ distribution, with $n - 2$ degrees of freedom, where $n$ is the sample size and $b_1$ is the estimate of the slope $\beta_1$ of the population regression equation.

The **standard error $s_{b_1}$ of the statistic $b_1$** is

$$s_{b_1} = \frac{s}{\sqrt{\sum(x_i - \bar{x})^2}}$$

where $s$ represents the standard error of the estimate from Section 13.1. The standard error $s_{b_1}$ measures the size of the typical error in using $b_1$ to estimate $\beta_1$, and it is measured in the same units as the response variable $y$.

## Example 13.12 Calculating the standard error $s_{b_1}$

Calculate and interpret the standard error $s_{b_1}$ for the data in Example 13.1.

**Solution**

We know from Example 13.2 that $\bar{x} = 4$. Table 13.5 shows that

$$\sum(x - \bar{x})^2 = 54$$

In Example 13.5, we found that $s = \sqrt{\text{SSE}/(n - 2)} = \sqrt{12/8} \approx 1.2247$. Hence,

$$s_{b_1} = \frac{s}{\sqrt{\sum(x - \bar{x})^2}} = \frac{\sqrt{\frac{12}{8}}}{\sqrt{54}} = \frac{1}{6} \approx 0.1667$$

The size of the typical error in using $b_1$ to estimate $\beta_1$ is thus $s_{b_1} = 0.1667$ points. ■

Table 13.5 Calculation of $\sum(x - \bar{x})^2$

| Student | $x$ | $(x - \bar{x})^2$ |
|---|---|---|
| 1 | 1 | $(1 - 4)^2 = 9$ |
| 2 | 1 | $(1 - 4)^2 = 9$ |
| 3 | 2 | $(2 - 4)^2 = 4$ |
| 4 | 3 | $(3 - 4)^2 = 1$ |
| 5 | 3 | $(3 - 4)^2 = 1$ |
| 6 | 4 | $(4 - 4)^2 = 0$ |
| 7 | 5 | $(5 - 4)^2 = 1$ |
| 8 | 6 | $(6 - 4)^2 = 4$ |
| 9 | 7 | $(7 - 4)^2 = 9$ |
| 10 | 8 | $(8 - 4)^2 = 16$ |
| | | $\sum(x - \bar{x})^2 = 54$ |

Recall that in Chapter 8 we constructed a confidence interval estimate for a population parameter, consisting of an interval of numbers that contain the parameter with a certain confidence level. We can find a similar **confidence interval for $\beta_1$.**

**Confidence Interval for $\beta_1$**

When the regression assumptions are met, a $100(1 - \alpha)\%$ confidence interval for $\beta_1$ is given by

$$b_1 \pm t_{\alpha/2} \cdot s_{b_1}$$

where $n$ is the sample size and $b_1$ is the estimate of the true slope $\beta_1$, $s_{b_1} = \frac{s}{\sqrt{\sum(x - \bar{x})^2}}$ is the standard error of $b_1$, $s$ is the standard error of the estimate, and $t_{\alpha/2}$ is associated with the confidence level and has $n - 2$ degrees of freedom.

**Margin of Error $E$**

The margin of error for a $100(1 - \alpha)\%$ confidence interval for $\beta_1$ is given by

$$E = t_{\alpha/2} \cdot \text{standard error of } b_1$$
$$= t_{\alpha/2} \cdot s_{b_1} = t_{\alpha/2} \cdot \frac{s}{\sqrt{\sum(x - \bar{x})^2}}$$

Thus, the confidence interval for $\beta_1$ takes the form $b_1 \pm E$.

## Example 13.13 Confidence interval for $\beta_1$

Find a 95% confidence interval for the true slope $\beta_1$ of the relationship between time and score.

**Solution**

The regression assumptions were verified in Example 13.11. We have $b_1 = 2$ from Example 13.1, and we calculated $s_{b_1} = \frac{1}{6}$ in Example 13.12. From the $t$ table, we find that $t_{\alpha/2}$ for $n - 2 = 10 - 2 = 8$ degrees of freedom is $t_{\alpha/2} = 2.306$. Hence our margin of error $E$ is

$$E = t_{\alpha/2} \cdot s_{b_1} = (2.306)\left(\frac{1}{6}\right) \approx 0.3843$$

The 95% confidence interval for $\beta_1$ is then given by

$$b_1 \pm E = 2 \pm 0.3843 = (1.6157, 2.3843)$$

What Do These Numbers Mean?

The margin of error $E = 0.3843$ means that, when we repeatedly take samples from this population, 95% of the time the sample estimate $b_1$ will be within $E = 0.3843$ of the unknown value of the true slope $\beta_1$. In other words, we are 95% confident that the interval (1.6157, 2.3843) captures the true slope $\beta_1$ of the relationship between time and score. ■

## 3 Hypothesis Tests for Slope $\beta_1$

We next construct the hypothesis test to determine whether or not $\beta_1$ equals 0. The hypotheses are

$H_0 : \beta_1 = 0$ There is no linear relationship between $x$ and $y$.

$H_a : \beta_1 \neq 0$ There is a linear relationship between $x$ and $y$.

The hypothesis test is based on the same approximate $t$ distribution as the $t$ interval for $\beta_1$. Assuming $H_0$, ($\beta_1 = 0$), the test statistic is

$$t_{\text{data}} = \frac{b_1 - \beta_1}{s_{b_1}} = \frac{b_1 - 0}{s_{b_1}} = \frac{b_1}{s_{b_1}} = \frac{b_1}{\dfrac{s}{\sqrt{\sum (x - \bar{x})^2}}}$$

The **test statistic $t_{\text{data}}$** measures the difference between the slope $b_1$ and 0, expressed in terms of standard errors $s_{b_1}$. If $t_{\text{data}}$ is extreme (far from 0), then it represents evidence against the null hypothesis.

## Example 13.14 Calculating the test statistic $t_{\text{data}}$

For the data from Example 13.1, calculate the value of the test statistic $t_{\text{data}}$.

**Solution**

From Example 13.1, we have $b_1 = 2$, and from Example 13.12, we have $s_{b_1} = 1/6$. Therefore,

$$t_{\text{data}} = \frac{b_1}{s_{b_1}} = \frac{2}{(1/6)} = 12$$

We illustrate performing the $t$ test for the slope $\beta_1$ using the $p$-value method and the critical value method.

**Hypothesis Test for Slope $\beta_1$ : $p$-Value Method**

***Step 1*** **State the hypotheses, and state the rejection rule.**

$H_0 : \beta_1 = 0$ There is no linear relationship between $x$ and $y$.

$H_a : \beta_1 \neq 0$ There is a linear relationship between $x$ and $y$.

The rejection rule is *Reject $H_0$ if the p-value is less than $\alpha$.*

***Step 2*** **Calculate the value of the test statistic $t_{data}$.**

$$t_{data} = \frac{b_1}{s_{b_1}} = \frac{b_1}{\dfrac{s}{\sqrt{\sum (x - \bar{x})^2}}}$$

which follows an approximate $t$ distribution with $n - 2$ degrees of freedom.

***Step 3*** **Find the $p$-value. If you have access to technology, use it. Otherwise, estimate the $p$-value using the following formula.**

$$\begin{aligned} p\text{-value} &= P(t > |t_{data}|) + P(t < -|t_{data}|) \\ &= 2 \cdot P(t > |t_{data}|) \end{aligned}$$

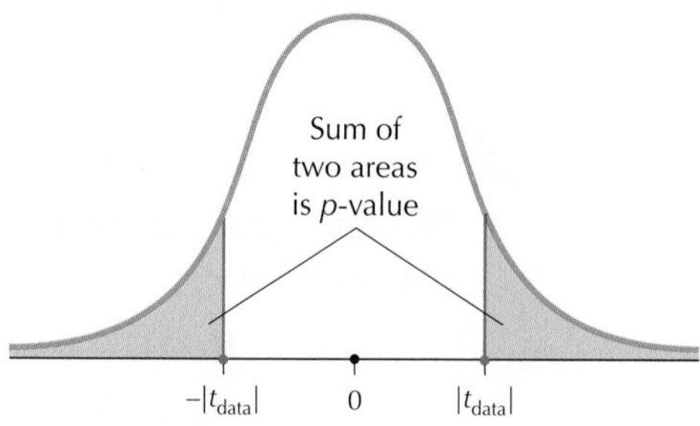


***Step 4*** **State the conclusion and interpretation.** Compare the $p$-value with $\alpha$.

## Example 13.15 Hypothesis test for the slope $\beta_1$ using the $p$-value method and technology

Test using the $p$-value method whether a linear relationship exists between time and score, using $\alpha = 0.01$.

### Solution

We use the instructions provided in the Step-by-Step Technology Guide at the end of this section. The regression assumptions were verified in Example 13.11.

**Step 1** $H_0 : \beta_1 = 0$ There is no linear relationship between time and score.
$H_a : \beta_1 \neq 0$ There is a linear relationship between time and score.

*Reject $H_0$ if the p-value is less than* 0.01.

**Step 2** From Example 13.14, we have $t_{data} = 12$.

**Step 3** Find the $p$-value. The regression results (including the $p$-value) for the TI-83/84, Excel, and Minitab are shown in Figures 13.9, 13.10, and 13.11. (Differing results are due to rounding.)

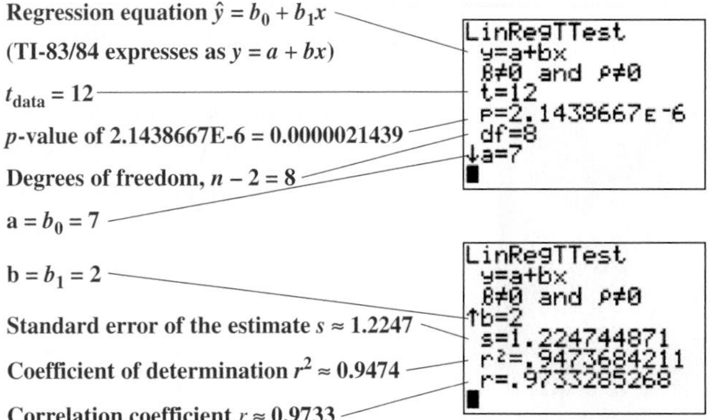


FIGURE 13.9 TI-83/84 regression results.

| Regression Statistics | | |
|---|---|---|
| Multiple R | 0.9733 | Correlation coefficient $r = 0.9733$ |
| R Square | 0.9474 | Coefficient of determination $r^2 = 0.9474$ |
| Adjusted R Square | 0.9408 | |
| Standard Error | 1.2247 | Standard error of the estimate $s = 1.2247$ |
| Observations | 10 | Sample size $n = 10$ |

| | Coefficients | Standard Error | t Stat | P-value |
|---|---|---|---|---|
| Intercept | 7 | 0.7710 | 9.0791 | 0.00001738 |
| X Variable - Time | 2 | 0.1667 | 12.0000 | 0.00000214 |

$b_0 = 7$
$b_1 = 2$
$t_{data} = 12$
p-value = 0.00000214

FIGURE 13.10 Excel regression results.

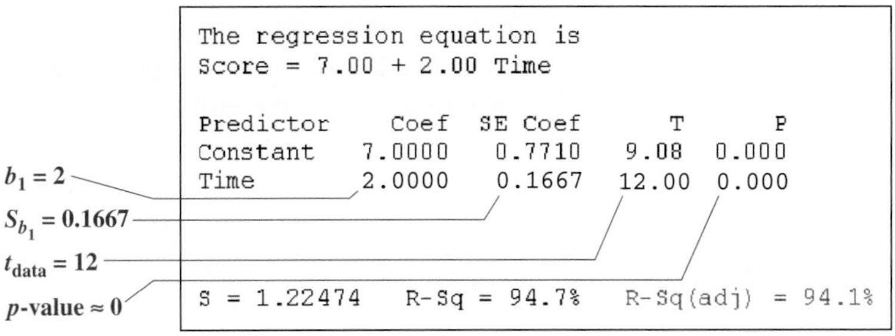

The regression equation is
Score = 7.00 + 2.00 Time
Predictor Coef SE Coef T P
Constant 7.0000 0.7710 9.08 0.000
Time 2.0000 0.1667 12.00 0.000
S = 1.22474 R-Sq = 94.7% R-Sq(adj) = 94.1%
$b_1 = 2$
$S_{b_1} = 0.1667$
$t_{data} = 12$
p-value ≈ 0


FIGURE 13.11 Minitab regression results.

***Step 4*** Since the *p*-value of about 0.000 is less than $\alpha = 0.01$, we reject $H_0$. There is evidence for a linear relationship between time and score. ■

**Hypothesis Test for Slope $\beta_1$: Critical Value Method**

***Step 1*** **State the hypotheses.**

$H_0: \beta_1 = 0$ There is no linear relationship between *x* and *y*.

$H_a: \beta_1 \neq 0$ There is a linear relationship between *x* and *y*.

***Step 2*** **Find the *t* critical value $t_{crit}$ and the rejection rule.**

To find $t_{crit}$, use the *t*-Distribution table (Table D in the Appendix, page T-11) for a two-tailed test and degrees of freedom $n - 2$. The rejection rule is *Reject $H_0$ if $t_{data} > t_{crit}$ or $t_{data} < -t_{crit}$.*

***Step 3*** **Calculate the value of the test statistic $t_{data}$.**

$$t_{data} = \frac{b_1}{s_{b_1}} = \frac{b_1}{\dfrac{s}{\sqrt{\sum (x - \bar{x})^2}}}$$

which follows an approximate *t* distribution with $n - 2$ degrees of freedom.

***Step 4*** **State the conclusion and interpretation.** Compare $t_{data}$ with $t_{crit}$.

## Example 13.16 Hypothesis test for the slope $\beta_1$ using the critical value method

AP Photo/Ed Reinke

The following table shows the winning percentage ($x$) and power rating ($y$) for ten teams in the 2004 NCAA Men's Basketball Tournament, according to **www.teamrankings.com**. Is there a linear relationship between winning percentage and power rating? Test using the critical value method using $\alpha = 0.05$.

| School | Win % ($x$) | Rating ($y$) |
|---|---|---|
| Duke | 0.838 | 96.020 |
| St. Joseph's | 0.938 | 95.493 |
| Connecticut | 0.846 | 95.478 |
| Oklahoma State | 0.882 | 95.320 |
| Pittsburgh | 0.857 | 94.541 |
| Georgia Tech | 0.737 | 93.091 |
| Stanford | 0.938 | 92.862 |
| Kentucky | 0.844 | 92.692 |
| Gonzaga | 0.903 | 92.609 |
| Mississippi State | 0.867 | 91.912 |

### What Results Might We Expect?

The scatterplot in Figure 13.12 indicates no evidence for a linear relationship between winning percentage and power rating. As winning percentage increases (as we move left to right), the power rating tends to remain unchanged. We might expect to *not reject* the null hypothesis that there is no linear relationship.

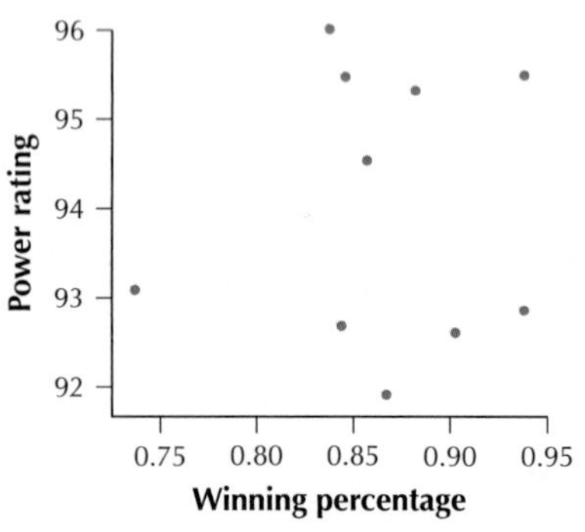

FIGURE 13.12 Scatterplot of power rating versus winning percentage.

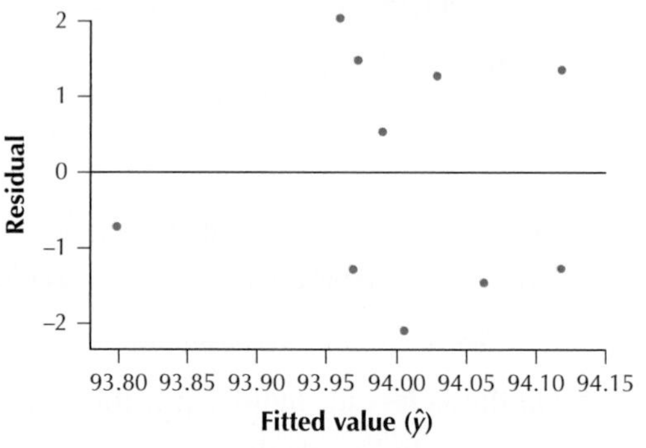

FIGURE 13.13 Plot of residuals versus fitted values.

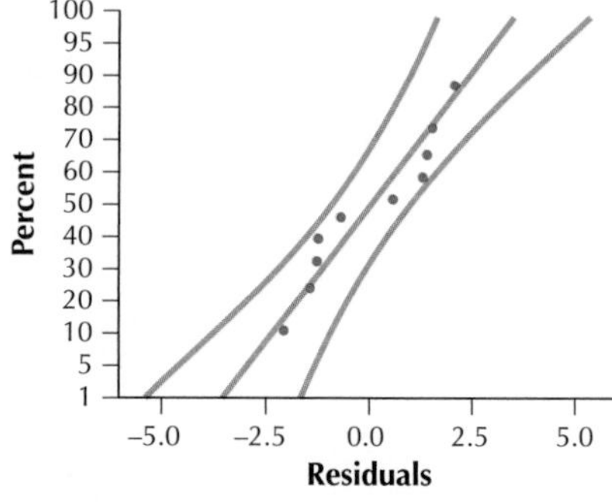

FIGURE 13.14 Normality plot of residuals.

There is no strong evidence in Figures 13.13 and 13.14 that the regression assumptions have been violated. We thus proceed with the hypothesis test.

**Step 1** $H_0: \beta_1 = 0$ There is no linear relationship between winning percentage and power rating.
$H_a: \beta_1 \neq 0$ There is a linear relationship between winning percentage and power rating.

**Step 2** We find $t_{\text{crit}}$ in the $t$ table for a two-tailed test with $\alpha = 0.05$ and df $= n - 2 = 10 - 2 = 8$: $t_{\text{crit}} = 2.306$. Reject $H_0$ if $t_{\text{data}} > 2.306$ or $t_{\text{data}} < -2.306$.

**Step 3** **Find $t_{\text{data}}$.** We first calculate $\bar{x} = \sum x/n = 8.65/10 = 0.865$ and $\sum(x - \bar{x})^2 = 0.030374$. The calculations are provided in the accompanying table. In Exercise 23 of

Section 13.1 (page 673), we found that $b_1 = 1.587$ and $s = 1.60388$. Thus,

$$s_{b_1} = \frac{s}{\sqrt{\sum (x-\bar{x})^2}} = \frac{1.60388}{\sqrt{0.030374}} \approx 9.2028$$

and

$$t_{\text{data}} = \frac{b_1}{s_{b_1}} = \frac{1.587}{9.2028} \approx 0.1724$$

**Calculating $\sum (x - \bar{x})^2$**

| School | $x$ | $(x - \bar{x})^2$ |
|---|---|---|
| Duke | 0.838 | $(0.838 - 0.865)^2 = 0.000729$ |
| St. Joseph's | 0.938 | $(0.938 - 0.865)^2 = 0.005329$ |
| Connecticut | 0.846 | $(0.846 - 0.865)^2 = 0.000361$ |
| Oklahoma State | 0.882 | $(0.882 - 0.865)^2 = 0.000289$ |
| Pittsburgh | 0.857 | $(0.857 - 0.865)^2 = 0.000064$ |
| Georgia Tech | 0.737 | $(0.737 - 0.865)^2 = 0.016384$ |
| Stanford | 0.938 | $(0.938 - 0.865)^2 = 0.005329$ |
| Kentucky | 0.844 | $(0.844 - 0.865)^2 = 0.000441$ |
| Gonzaga | 0.903 | $(0.903 - 0.865)^2 = 0.001444$ |
| Mississippi State | 0.867 | $(0.867 - 0.865)^2 = 0.000004$ |
| | $\sum x = 8.65$ | $\sum (x - \bar{x})^2 = 0.030374$ |

***Step 4*** Since $t_{\text{data}} = 0.1724$ is not greater than 2.306 and is not less than $-2.306$, we therefore do not reject $H_0$. There is insufficient evidence for a linear relationship between winning percentage and power rating. ■

Regression analysis can help us uncover unusual observations, such as outliers. An **outlier** is an observation that has a very large absolute residual $|y - \hat{y}|$.

## Example 13.17 Palm Beach outlier

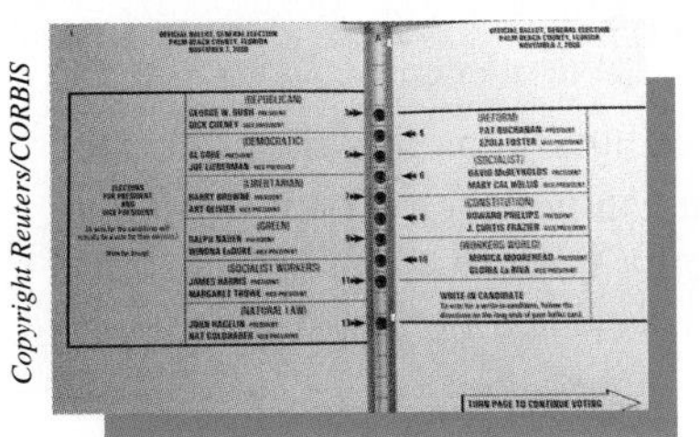


In the 2000 presidential election, many Palm Beach County, Florida, residents found the ballot confusing, and many voters chose Reform Party candidate Pat Buchanan (see the data set **Florida**[2]). The scatterplot in Figure 13.15 shows, county-by-county for all Florida's counties, the number of votes for Buchanan ($y$) versus the total number of county votes ($x$). There is a clear trend of a linear (straight line) relationship, except for one glaring exception, which is Palm Beach County. Determine whether this exception is an outlier.

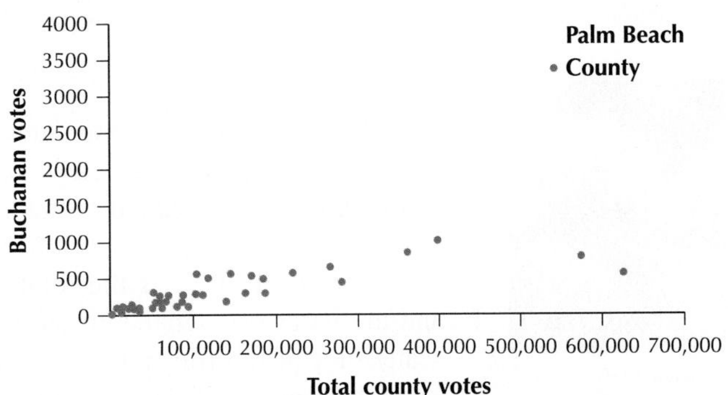


FIGURE 13.15 Scatterplot of Buchanan votes versus total county votes, 2000 presidential election in Florida.

### Solution

Consider the Minitab results excerpt in Figure 13.16 of the regression of Buchanan votes on total votes. The regression equation is given as

$$\hat{y} = b_0 + b_1x = 54.2 + 0.00232(\text{total votes})$$

```
The regression equation is
Buchanan = 54.2 + 0.00232 Total

Predictor        Coef      SE Coef     T       P
Constant        54.23        49.14  1.10   0.274
Total       0.0023229    0.0003104  7.48   0.000

S = 332.719    R-Sq = 46.3%    R-Sq(adj) = 45.5%

Unusual Observations
Obs   Total     Buchanan     Fit   SE Fit  Residual  St Resid
 50  433186       3411.0  1060.5    114.3    2350.5     7.52RX

R denotes an observation with a large standardized residual.
```

FIGURE 13.16 Excerpt from Minitab regression results.

Minitab also shows that observation 50 (Palm Beach County) is an "unusual observation," with the "R" denoting a large residual. In other words, Minitab identifies Palm Beach County as an outlier because it has a large absolute residual. The number of Buchanan votes ($y$) in Palm Beach County was 3411, while the predicted ("fit") number of votes for Buchanan was $\hat{y} = 1060.5$, giving a residual of $y - \hat{y} = 2350.5$. Is this large? To find out, we compare it to the standard error of the estimate $s$. For a typical Florida county, knowing the total votes in the county, we can predict the number of votes for Buchanan in that county to within about $s = 332.719$ votes, typically. Now, $2350.5/332.719 \approx 7.06$. Thus, the prediction error for Buchanan's vote in Palm Beach County lies more than seven standard errors above what was expected. Assuming normality of the residuals (regression assumption 4), the probability of observing this data point by chance would be (using Case 2 of Table 6.7 [page 298]):

$$P(Z > 7.06) = 0.0000000005$$

To compare, the probability of winning the jackpot (multimillions of dollars for your \$1 bet) in the Connecticut Lotto is 0.0000014. The Palm Beach County result is *28 times less likely* than winning the Connecticut Lotto. Clearly, there is something going on with the Palm Beach County data. Whatever went wrong there crucially affected the 2000 presidential election. ■

Copyright Kayte M. Deioma/Photo Edit

CASE STUDY

## How Fair Is the Scoring in Scrabble?

In this case study, we consider the frequency and point values of Scrabble tiles. Table 13.6 shows the relative frequency in the English language, the frequency (number of tiles) in Scrabble, and the point value in Scrabble.

First of all, what is the relationship between the tile frequencies in Scrabble and the letter frequencies in the English language? Figure 13.17 shows a Minitab scatterplot of the tile frequencies in Scrabble against the letter frequencies in the English language. There appears to be a positive relationship between the two variables. That is, as the English frequencies increase, game frequencies also tend to increase.

Table 13.6 Frequency in English, frequency in Scrabble, and Scrabble point value of the letters in the alphabet

| Letter | Rel. freq. in English language | Frequency in Scrabble | Point value in Scrabble | Letter | Rel. freq. in English language | Frequency in Scrabble | Point value in Scrabble |
|---|---|---|---|---|---|---|---|
| A | 0.073 | 9 | 1 | N | 0.078 | 6 | 1 |
| B | 0.009 | 2 | 3 | O | 0.074 | 8 | 1 |
| C | 0.030 | 2 | 3 | P | 0.027 | 2 | 3 |
| D | 0.044 | 4 | 2 | Q | 0.003 | 1 | 10 |
| E | 0.130 | 12 | 1 | R | 0.077 | 6 | 1 |
| F | 0.028 | 2 | 4 | S | 0.063 | 4 | 1 |
| G | 0.016 | 3 | 2 | T | 0.093 | 6 | 1 |
| H | 0.035 | 2 | 4 | U | 0.027 | 4 | 1 |
| I | 0.074 | 9 | 1 | V | 0.013 | 2 | 4 |
| J | 0.002 | 1 | 8 | W | 0.016 | 2 | 4 |
| K | 0.003 | 1 | 5 | X | 0.005 | 1 | 8 |
| L | 0.035 | 4 | 1 | Y | 0.019 | 2 | 4 |
| M | 0.025 | 2 | 3 | Z | 0.001 | 1 | 10 |

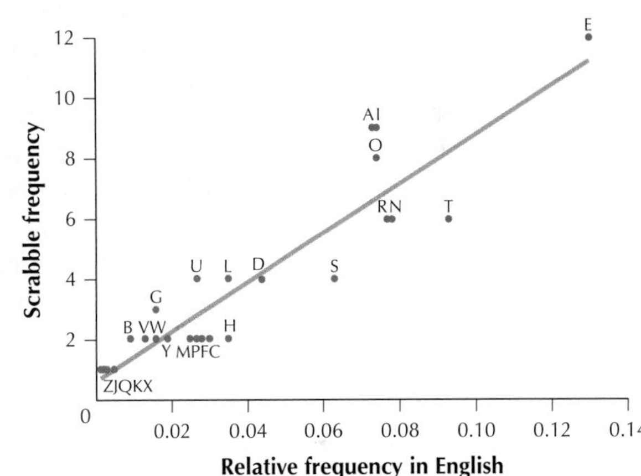


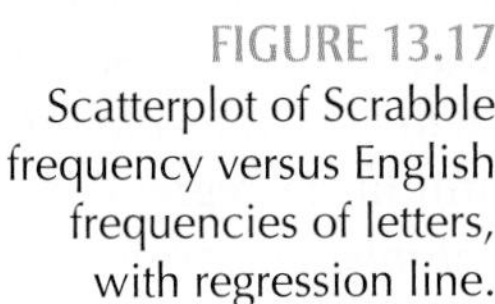
FIGURE 13.17 Scatterplot of Scrabble frequency versus English frequencies of letters, with regression line.

Note that the letters above the regression line occur "too frequently" in the game, whereas the letters below the line occur "not frequently enough." Playing typical English words during a game of Scrabble would tend to leave you with a rack of letters similar to those above the regression line. Note that S is one of the letters that is rarer in the game than in the language.

Figure 13.18 displays the MINITAB results from a regression of the tile frequencies against the English language relative frequencies. The regression equation is

$$\hat{y} = 0.636 + 81.5(\text{relative frequency in English}).$$

The slope is positive, which is not surprising, is it? The game wouldn't sell very well if there were more Z's than E's, would it?

Next we turn to the hypothesis test:

$H_0: \beta_1 = 0$ There is no linear relationship between Scrabble frequency and English relative frequency.

$H_a: \beta_1 \neq 0$ There is a linear relationship between Scrabble frequency and English relative frequency.

```
The regression equation is
Scrabble Freq = 0.636 + 81.5 Rel Freq Eng

Predictor         Coef  SE Coef      T      P
Constant        0.6362   0.3484   1.83  0.080
Rel Freq Eng    81.458    6.856  11.88  0.000


S = 1.16096   R-Sq = 85.5%   R-Sq(adj) = 84.9%


Analysis of Variance

Source          DF      SS      MS       F      P
Regression       1  190.27  190.27  141.17  0.000
Residual Error  24   32.35    1.35
Total           25  222.62
```

FIGURE 13.18 Minitab regression output.

The 0.000 in red represents the $p$-value for the $t$ test. Since this $p$-value is smaller than any $\alpha$, we reject the null hypothesis that there is no linear relationship between the game frequencies and the English frequencies. Does the model fit the data well? The coefficient of determination $r^2$ is 0.855, which is good, and the correlation coefficient $r$ equals 0.924, which indicates that the variables are positively correlated.

But the fit really could be better. Look at the value of $s$, the standard error of the estimate: $s = 1.16$. This means that, given the English language frequency of a letter, the estimate of the tile frequency will typically differ from the actual tile frequency by more than one tile.

Next, what is the relationship between the Scrabble point values and the English relative frequencies? Figure 13.19 shows a scatterplot of these two variables. The first thing you might notice about this relationship is that it is not linear. Therefore, it would not be appropriate to perform linear regression on this data set. In fact, this data set calls for the technique of *multiple regression,* which is beyond the scope of this book.

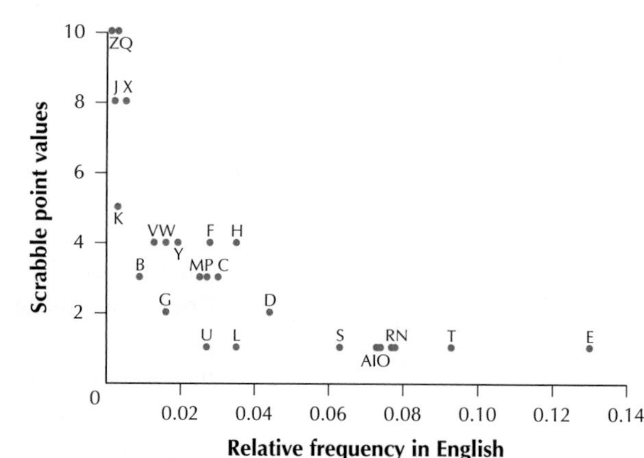


FIGURE 13.19 Scatterplot of Scrabble point values versus English frequencies. Linear regression would not be appropriate.

We can, nevertheless, make some descriptive remarks.

- What is a "good" Scrabble tile to pick up? In the best case, it would be a letter with high English frequency worth lots of Scrabble game points. Unfortunately, the two do not go together. The high-frequency letters like E and T have low point values, and the high point-value letters like Q and Z have low frequencies. But we can still make comparisons.

- Which would you rather pick up, a D or a G? It would seem that D would be preferable since it has the same point value as G with much higher English frequency.

- Which do you prefer between J and X? They are worth the same points, but X has a higher frequency in English, so it is easier to make words with it in the game. The letter H seems to have a good combination of high points and moderate frequency.

Now, if we could only get regression analysis to help us with that triple-word score!

## STEP-BY-STEP TECHNOLOGY GUIDE: Regression Analysis

Data from Example 13.15 (page 690) are used to illustrate the steps.

### TI-83/84

***Step 1*** Enter the **X (Time)** data in **L1** and the **Y (Score)** data in **L2**.
***Step 2*** Press **STAT**, highlight **CALC**, and press **8** (not 4!) to choose **LinReg(a+bx)**. On the home screen, the following command appears: **LinReg(a+bx)**.
***Step 3*** Press **ENTER**. The output shows **y = a +bx, a=7, b=2.** The TI-83/84 denotes the $y$ intercept $b_0$ as **a** and the slope $b_1$ as **b**. Thus the TI-83/84 is telling you that the estimated regression equation is $\hat{y} = 7 + 2x$.
***Step 4*** Now Press **STAT** again and press the right arrow key until **TESTS** is highlighted.
***Step 5*** Press the **down arrow** key until **E** is highlighted (for **LinRegTTest**).
***Step 6*** Press **ENTER**. The LinRegTTest menu appears.
***Step 7*** For **Xlist**, enter **L1** (or whichever list you entered the X data in).
***Step 8*** For **Ylist**, enter **L2** (or whichever list you entered the Y data in).
***Step 9*** For **Freq**, enter **1**, and for $\beta$ & $\rho$ highlight "$\neq 0$".
***Step 10*** Move the cursor over **Calculate**, make sure all your entries are correct, and press **ENTER**. The results are as shown in Figure 13.9 (page 691).

### EXCEL

***Step 1*** Enter the "Time" variable in column **A** and the "Score" variable in column **B**.
***Step 2*** Click on **Data > Data Analysis > Regression** and click **OK**.
***Step 3*** For **Input Y Range**, select cells **B1 – B10**. For **Input X Range**, select cells **A1 – A10**.
***Step 4*** If you would like to verify the regression assumptions, then select **Residual Plots** and **Normal Probability Plots**.
***Step 5*** Click **OK**. The results are as shown in Figure 13.10 (page 691).

### MINITAB

***Step 1*** Enter the "Time" variable in **C1** and the "Score" variable in **C2**.
***Step 2*** Click on **Stat > Regression > Regression**.
***Step 3*** Select "Score" as your **Response Variable** and "Time" as your **Predictor Variable**.
***Step 4*** If you would like to verify the regression assumptions, click the button labeled **Graphs** and select **Four in One**.
***Step 5*** Click **OK** twice. The results are as shown in Figure 13.11 (page 691).

## Section 13.3 SUMMARY

1 This section examines inferential methods for regression analysis. The regression model, or the (population) regression equation, is $y = \beta_0 + \beta_1 x + \varepsilon$, where $\beta_0$ is the $y$ intercept of the population regression line, $\beta_1$ is the slope of the population regression line, and $\varepsilon$ is the error term.

2 We can perform hypothesis tests or construct confidence intervals for the true value of the population regression slope $\beta_1$ since it is unknown. The $100(1 - \alpha)\%$ confidence interval for $\beta_1$ is $b_1 \pm t_{\alpha/2}\, s_{b_1}$, where $t_{\alpha/2}$ is based on $n - 2$ degrees of freedom.

3 The test statistic for the $t$ test about $\beta_1$ is given as $t_{\text{data}} = b_1/s_{b_1}$, where $s_{b_1}$ is the standard error of the slope $b_1$.

## Section 13.3 EXERCISES

### CLARIFYING THE CONCEPTS

1. What is the difference between the regression equation (calculated using the sample) and the population regression equation?
2. What are the four regression model assumptions?
3. How do we go about verifying the regression model assumptions?
4. What is the difference between $b_0$ and $b_1$ on the one hand and $\beta_0$ and $\beta_1$ on the other hand?
5. What does it mean for the relationship between $x$ and $y$ when $\beta_1$ equals 0?
6. What is an outlier?

### PRACTICING THE TECHNIQUES

For Exercises 7–12, follow these steps. Note that the predicted values and the residuals were calculated in Exercises 13–18 in Section 13.1.

a. Construct a scatterplot of the residuals versus the predicted values.
b. Use technology to construct a normal probability plot of the residuals.
c. Verify that the regression assumptions are valid.

7.

| $x$ | $y$ |
|---|---|
| 1 | 15 |
| 2 | 20 |
| 3 | 20 |
| 4 | 25 |
| 5 | 25 |

8.

| $x$ | $y$ |
|---|---|
| 0 | 10 |
| 5 | 20 |
| 10 | 45 |
| 15 | 50 |
| 20 | 75 |

9.

| $x$ | $y$ |
|---|---|
| −5 | 0 |
| −4 | 8 |
| −3 | 8 |
| −2 | 16 |
| −1 | 16 |

10.

| $x$ | $y$ |
|---|---|
| −3 | −5 |
| −1 | −15 |
| 1 | −20 |
| 3 | −25 |
| 5 | −30 |

11.

| $x$ | $y$ |
|---|---|
| 10 | 100 |
| 20 | 95 |
| 30 | 85 |
| 40 | 85 |
| 50 | 80 |

12.

| $x$ | $y$ |
|---|---|
| 0 | 11 |
| 20 | 11 |
| 40 | 16 |
| 60 | 21 |
| 80 | 26 |

For Exercises 13–18, follow these steps. Assume that the regression model assumptions are valid.

a. Find $s_{b_1}$. Explain what this number means.
b. Find $t_{\alpha/2}$ for a 95% confidence interval for $\beta_1$.
c. Construct a 95% confidence interval for $\beta_1$.

13. Data in Exercise 7, where $b_1 = 2.5$
14. Data in Exercise 8, where $b_1 = 3.2$
15. Data in Exercise 9, where $b_1 = 4.0$
16. Data in Exercise 10, where $b_1 = -3.0$
17. Data in Exercise 11, where $b_1 = -0.5$
18. Data in Exercise 12, where $b_1 = 0.2$

For Exercises 19–21, follow these steps. Assume that the regression model assumptions are valid.

a. Calculate $t_{\text{data}}$ using $s_{b_1}$ from Exercises 13–15.
b. Find $p$-value $= 2 \cdot P(t > |t_{\text{data}}|)$.
c. Perform the hypothesis test for the linear relationship between $x$ and $y$ using the $p$-value method and $\alpha = 0.05$.

19. Data in Exercise 7, where $b_1 = 2.5$
20. Data in Exercise 8, where $b_1 = 3.2$
21. Data in Exercise 9, where $b_1 = 4.0$

For Exercises 22–24, follow these steps. Assume that the regression model assumptions are valid.

a. Find $t_{\text{crit}}$ for a two-tailed test with $\alpha = 0.05$ and df $= n - 2$.
b. Calculate $t_{\text{data}}$ using $s_{b_1}$ from Exercises 16–18.
c. Perform the hypothesis test for the linear relationship between $x$ and $y$, using the critical value method, and $\alpha = 0.05$.

22. Data in Exercise 10, where $b_1 = -3.0$
23. Data in Exercise 11, where $b_1 = -0.5$
24. Data in Exercise 12, where $b_1 = 0.2$

### APPLYING THE CONCEPTS

For Exercises 25–32, follow steps **(a)**–**(d)**. Assume that the regression model assumptions are valid. The regression equations for Exercises 25–29 were calculated in Section 13.1, Exercises 19–23.

a. Find $s_{b_1}$. Explain what this number means.
b. Find $t_{\alpha/2}$ for a 95% confidence interval for $\beta_1$.
c. Construct a 95% confidence interval for $\beta_1$.
d. Interpret the interval.

25. **Volume and Weight.** The table contains the volume ($x$, in cubic meters) and weight ($y$, in kilograms) of five randomly chosen packages shipped to a local college.

| Volume ($x$) | Weight ($y$) |
|---|---|
| 4 | 10 |
| 8 | 16 |
| 12 | 25 |
| 16 | 30 |
| 20 | 35 |

**26. Family Size and Pets.** Shown in the table are the number of family members ($x$) in a random sample taken from a suburban neighborhood and the number of pets ($y$) belonging to each family.

| Family size ($x$) | Pets ($y$) |
|---|---|
| 2 | 1 |
| 3 | 2 |
| 4 | 2 |
| 5 | 3 |
| 6 | 3 |

**27. SAT Math and Verbal Scores.** The College Board reported the SAT I Verbal scores ($x$) and SAT I Math scores ($y$) for the five states with the best participation rates:

| State | Verbal ($x$) | Math ($y$) |
|---|---|---|
| NY | 497 | 510 |
| CT | 515 | 515 |
| MA | 518 | 523 |
| NJ | 501 | 514 |
| NH | 522 | 521 |

**28. World Temperatures.** Listed in the table are the low ($x$) and high ($y$) temperatures for a particular day in 2006, measured in degrees Fahrenheit, for a random sample of cities worldwide.

| City | Low ($x$) | High ($y$) |
|---|---|---|
| Kolkata, India | 57 | 77 |
| London, England | 36 | 45 |
| Montreal, Quebec | 7 | 21 |
| Rome, Italy | 39 | 55 |
| San Juan, Puerto Rico | 70 | 83 |
| Shanghai, China | 34 | 45 |

**29. NCAA Power Ratings.** The table shows the team's winning percentage ($x$) and power rating ($y$) for the 2004 NCAA Basketball Tournament, according to **www.teamrankings.com**.

| School | Win% ($x$) | Rating ($y$) |
|---|---|---|
| Duke | 0.838 | 96.020 |
| St. Joseph's | 0.938 | 95.493 |
| Connecticut | 0.846 | 95.478 |
| Oklahoma State | 0.882 | 95.320 |
| Pittsburgh | 0.857 | 94.541 |
| Georgia Tech | 0.737 | 93.091 |
| Stanford | 0.938 | 92.862 |
| Kentucky | 0.844 | 92.692 |
| Gonzaga | 0.903 | 92.609 |
| Mississippi State | 0.867 | 91.912 |

**30. Midterm Exams and Overall Grade.** The table shows the midterm exam scores ($x$) and the overall grade ($y$) for a random sample of 12 students in an elementary statistics course.

| Midterm exam ($x$) | Overall grade ($y$) |
|---|---|
| 50 | 65 |
| 90 | 80 |
| 70 | 75 |
| 80 | 75 |
| 60 | 45 |
| 90 | 95 |
| 90 | 85 |
| 80 | 80 |
| 70 | 65 |
| 70 | 70 |
| 60 | 65 |
| 50 | 55 |

**31. GPA and Combined SAT Scores.** The college admissions office would like to determine if there is a relationship between the combined SAT score and the grade point average of first-year students, using the data in the table.

| Student | Combined SAT score ($x$) | Grade point average ($y$) |
|---|---|---|
| 1 | 900 | 1.5 |
| 2 | 925 | 2.6 |
| 3 | 950 | 1.9 |
| 4 | 975 | 2.7 |
| 5 | 1000 | 2.0 |
| 6 | 1000 | 2.5 |
| 7 | 1025 | 3.0 |
| 8 | 1025 | 2.4 |
| 9 | 1050 | 2.1 |
| 10 | 1050 | 2.9 |
| 11 | 1075 | 2.7 |
| 12 | 1075 | 3.1 |
| 13 | 1100 | 3.0 |
| 14 | 1125 | 3.3 |
| 15 | 1150 | 3.2 |

**32. Stock Prices.** Would you expect there to be a relationship between the price ($x$) of a stock and its change ($y$) in price on a particular day? The table provides stock price and stock price change for August 2, 2004, for a random sample of ten stocks listed on the New York Stock Exchange.

| Stock | Price ($x$) | Change ($y$) |
|---|---|---|
| Nortel Networks | 3.86 | +0.04 |
| Qwest Communications | 3.41 | −0.56 |
| Tyco International | 32.02 | +0.78 |
| Lucent Technologies | 3.04 | −0.03 |
| Vishay Intertechnology | 13.96 | −1.96 |
| Tenet Healthcare | 10.36 | −0.82 |
| Select Medical Group | 14.74 | +1.62 |
| Cox Communications | 33.19 | +0.03 |
| Verizon Communications | 39.15 | +0.46 |
| General Electric | 33.05 | −0.21 |

**SAT Verbal and Math Scores.** Use this information for Exercises 33–38. The table shows the SAT I scores for the five states with the best participation rates, as reported by the College Board in 2006. We are interested whether a linear relationship exists between the SAT I Verbal score ($x$) and the SAT I Math score ($y$).

| State | SAT I Verbal | SAT I Math |
|---|---|---|
| New York | 497 | 510 |
| Connecticut | 515 | 515 |
| Massachusetts | 518 | 523 |
| New Jersey | 501 | 514 |
| New Hampshire | 522 | 521 |

**33. What Result Might We Expect?** Consider the accompanying scatterplot of math score versus verbal score. Is there evidence for or against the null hypothesis that no linear relationship exists? Explain.

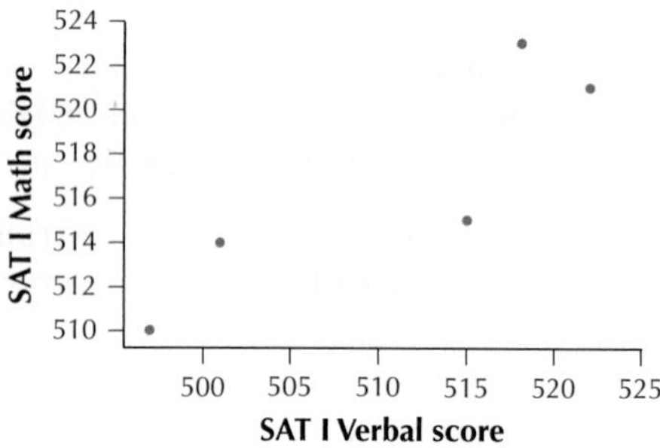


**34.** Consider Figures 13.20 and 13.21. Is there strong evidence that the regression assumptions are violated?

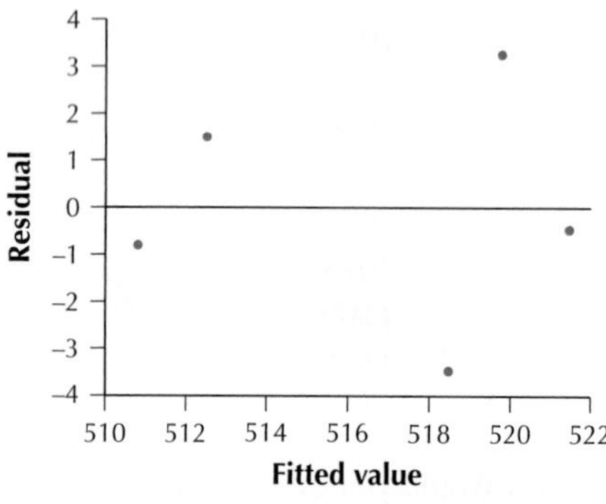


Figure 13.20 Plot of residuals versus fitted values.

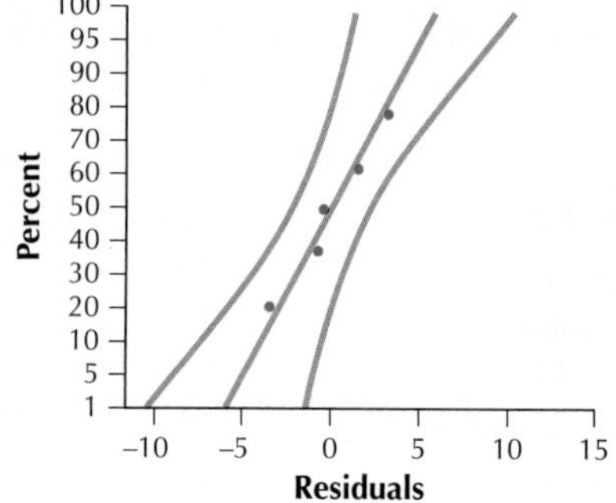


Figure 13.21 Normality plot of residuals.

**35.** Write out the hypotheses for testing whether or not a linear relationship exists between the verbal score and the math score.

**36.** Find the $t$ critical value $t_{\text{crit}}$ and the rejection rule. Use level of significance $\alpha = 0.05$.

**37.** Calculate the value of the test statistic $t_{\text{data}}$. Note that $\bar{x} = 510.6$ and $\Sigma(x - \bar{x})^2 = 481.2$. Also, from Exercise 21 of Section 13.1 (page 673), we found that $b_1 = 0.4264$ and $s = 2.92665$. (*Hint*: Use $s$ and $\Sigma(x - \bar{x})^2$ to find $s_{b_1}$. Then use $b_1$ and $s_{b_1}$ to find $t_{\text{data}}$.)

**38.** Compare $t_{\text{data}}$ with $t_{\text{crit}}$. State the conclusion and interpretation.

For Exercises 39–42, follow steps **(a)**–**(e)**.

**a.** Verify the regression model assumptions. (*Hint:* You can use either Excel or Minitab; see the Step-by-Step Technology Guide on page 697.)
**b.** Based on the confidence interval constructed in the indicated exercise, would you expect the hypothesis test to reject the null hypothesis that $\beta_1 = 0$?
**c.** Calculate $t_{\text{data}}$ using $s_{b_1}$ from the indicated exercise.
**d.** Find $p$-value $= 2 \cdot P(t > |t_{\text{data}}|)$.
**e.** Perform the hypothesis test for the linear relationship between $x$ and $y$ using the $p$-value method and $\alpha = 0.05$.

**39.** Data from Exercise 25; $b_1 = 1.6$
**40.** Data from Exercise 27; $b_1 = 0.4264$
**41.** Data from Exercise 29; $b_1 = 1.587$
**42.** Data from Exercise 31; $b_1 = 0.005569$

For Exercises 43–46, follow steps **(a)**–**(e)**.

**a.** Verify the regression model assumptions.
**b.** Based on the confidence interval constructed in the indicated exercise, would you expect the hypothesis test to reject the null hypothesis that $\beta_1 = 0$?
**c.** Find $t_{\text{crit}}$ for a two-tailed test with $\alpha = 0.05$ and df $= n - 2$.
**d.** Calculate $t_{\text{data}}$ using $s_{b_1}$ from the indicated exercise.
**e.** Perform the hypothesis test for the linear relationship between $x$ and $y$, using the critical value method and $\alpha = 0.05$.

**43.** Data from Exercise 26; $b_1 = 0.5$
**44.** Data from Exercise 28; $b_1 = 1.04658$
**45.** Data from Exercise 30; $b_1 = 0.7711$
**46.** Data from Exercise 32; $b_1 = 0.01944$

**47. Batting Average and Runs Scored.** The table shows the top ten hitters in Major League Baseball for 2007. We are interested in estimating the number of runs scored ($y$) using the player's batting average ($x$).

| Player | Team | Batting average ($x$) | Runs scored ($y$) |
|---|---|---|---|
| M. Ordonez | Detroit Tigers | .363 | 117 |
| I. Suzuki | Seattle Mariners | .351 | 111 |
| P. Polanco | Detroit Tigers | .341 | 105 |
| M. Holliday | Colorado Rockies | .340 | 120 |
| J. Posada | New York Yankees | .338 | 91 |
| C. Jones | Atlanta Braves | .337 | 108 |
| D. Ortiz | Boston Red Sox | .332 | 116 |
| H. Ramirez | Florida Marlins | .332 | 125 |
| E. Renteria | Atlanta Braves | .332 | 87 |
| C. Utley | Philadelphia Phillies | .332 | 104 |

a. What type of pattern do you see in the residuals versus predicted values plot?
b. Which regression assumption is violated?
c. Should we construct a confidence interval or perform a hypothesis test for the slope of the regression line?
d. Is it still appropriate to report the descriptive statistics we learned in Sections 13.1 and 13.2?

**48.** Suppose we have an estimated regression equation whose slope was not significant (that is, the null hypothesis was not rejected). *What if* we add five new data values to the original data set, and all five data values are identical, $(\bar{x}, \bar{y})$. How and why will this affect the following statistics? Will the statistic increase, decrease, or remain unchanged, or is there insufficient information to determine? (*Hint:* The data point $(\bar{x}, \bar{y})$ always lies on the estimated regression line.)

a. $n$
b. SSE
c. SST
d. SSR
e. MSE
f. MSR

**49.** Refer to Exercise 48. How and why will the change affect the following items?

a. $t_{\text{data}}$
b. $r^2$
c. $s$
d. $p$-value
e. Conclusion

**50.** Suppose a regression analysis of $y$ on $x$ was found to be significant (that is, the null hypothesis was rejected). *What if* we get ten new data values, all with different values of $x$ and *all of which can be found on the estimated regression line of the original model.* How and why will this change affect the following statistics? Will the statistic increase, decrease, or remain unchanged, or is there insufficient information to determine?

a. $n$
b. SSE
c. SST
d. SSR
e. MSE
f. MSR

**51.** Refer to Exercise 50. How and why will this change affect the following measures?

a. $t_{\text{data}}$
b. $r^2$
c. $s$
d. $p$-value
e. Conclusion

**52. Challenger Exercise.** Suppose a regression analysis of $y$ on $x$ was found to be significant (that is, the null hypothesis was rejected) and the slope $b_1 > 0$. Consider the observation (max $x$, $y$), which represents the $(x, y)$ data value for the maximum value of $x$ in the data set. Suppose the residual for (max $x$, $y$) is negative. *What if* we increase max $x$ by an arbitrary amount $c$ so that the new data value is (max $x + c$, $y$). (All other data values in the data set are unchanged.) How will this increase affect the following measures? Will they increase, decrease, or remain unchanged, or is there insufficient information to determine the effect?

a. $n$
b. SSE
c. SST
d. SSR
e. MSE
f. MSR
g. $F$

**53. Challenger Exercise.** Refer to Exercise 52. How and why will the change affect the following measures?

a. $t_{\text{data}}$
b. $r^2$
c. $s$
d. $p$-value
e. Conclusion

For Exercises 54–56, you can use Excel, Minitab, or some other statistical computing software to solve the following problems.

a. Verify the regression model assumptions.
b. Construct and interpret a 95% confidence interval for $\beta_1$.
c. Based on the confidence interval constructed in **(b)**, would you expect the hypothesis test to reject the null hypothesis that $\beta_1 = 0$?
d. Test at $\alpha = 0.05$ whether a linear relationship exists between $x$ and $y$.

**54.** Open the **Darts** data set, which we used for the Chapter 3 Case Study. Use the Dow Jones Industrial Average ($x$) to estimate the pros' performance ($y$).

**55.** Open the **Nutrition** data set. Estimate the number of calories per gram ($y$) using the amount of fat per gram ($x$).

**56.** Open the **Pulse and Temp** data set. Estimate body temperature ($y$) using heart rate ($x$).

**Crash Tests.** Use Excel, Minitab, or some other statistical computing software for Exercises 57–60. Open the **Crash** data set, which contains information about the severity of injuries sustained by crash dummies when the National Transportation Safety Board crashed automobiles into a wall at 35 miles per hour.

**57.** The variable *head_inj* contains a measure of the severity of the head injury sustained by the dummies. The variable *chest_in* is a measure of the severity of the chest injury suffered by the crash dummies.

a. Would you expect there to be a linear relationship between the severity of head injuries and chest injuries?
b. Construct a scatterplot of the *head_inj* against *chest_in*. Would you say that there is a positive relationship, a negative relationship, or no apparent linear relationship between the variables?

**c.** If we were to perform a regression analysis using these two variables, is it clear which of two variables we should label as the predictor and which we should label as the response? Explain.

**58.** Perform a regression of the head injury severity ($y$) on the chest injury severity ($x$).

**a.** What is the regression equation? Write it out in words and numbers so that a nonstatistician would understand it.

**b.** Find, define, and interpret the statistic that tells you how good a fit is the model for the data. (*Hint:* $r^2$.)

**c.** Find and interpret the correlation coefficient.

**d.** Perform the appropriate hypothesis test using the $p$-value method. Make sure you provide the hypotheses, the rejection rule, the test statistic, the $p$-value, the conclusion, and the interpretation. Use $\alpha = 0.01$.

**e.** Clearly interpret the meaning of the slope estimate $b_1$.

**f.** Construct and interpret a 99% confidence interval for the true slope $\beta_1$ of the relationship between severity of head injury and severity of chest injury. How does your confidence interval support your conclusion in **(d)**?

**59.** The variable *lleg_inj* contains a measure of the severity of the injury sustained by the dummies' left legs. The variable *weight* contains the weight of the vehicles.

**a.** Would you expect there to be a linear relationship between the severity of left leg injuries and the weight of the vehicles?

**b.** Construct a scatterplot of the *lleg_inj* against *weight*. Would you say that there is a positive relationship, a negative relationship, or no apparent linear relationship between the variables?

**c.** If we were to perform a regression analysis using these two variables, is it clear which of the two variables we should label as the predictor and which we should label as the response? Explain.

**60.** Perform a regression of the leg injury severity ($y$) on the vehicle weight ($x$).

**a.** What is the regression equation? Write it out in words and numbers so that a nonstatistician would understand it.

**b.** Find, define, and interpret the statistic that tells you how good a fit is the model for the data. (*Hint:* $r^2$.)

**c.** Find and interpret the correlation coefficient.

**d.** Is the relationship significant? Perform the appropriate hypothesis test using the $p$-value method. Make sure you provide the hypotheses, the rejection rule, the test statistic, the $p$-value, the conclusion, and the interpretation. Use a value of your own choosing for $\alpha$.

**e.** Construct and interpret a 95% confidence interval for the true slope $\beta_1$ of the relationship between vehicle weight and severity of left leg injury. How does your confidence interval support your conclusion in **(d)**?

## Try This in Class! In-class activities to enhance your understanding of statistics

### REGRESSION ACTIVITY FOR DISTANCE AND TIME TO CLASS

In this activity we discuss whether the amount of time it takes a student to get to class depends on how far away the student lives.

APPLET **1.** Generate a sample of 15 students from your class, using either (a) technology, (b) the *Random Sample* applet, or (c) the method shown in the Chapter 1 In-Class Activities. Each student's sample should be different.

**2.** Find out how long it takes each student in your sample to get to class and how far away the student lives. If the student lives on campus, get the data regarding the student's permanent residence (home mailing address).

**a.** Construct a scatterplot of the time in minutes against the distance in miles. Will everyone's scatterplot look exactly the same? Will everyone's scatterplot look somewhat similar?

**b.** Draw a straight line that you think best approximates the relationship between time and distance.

**c.** Estimate the slope and $y$ intercept of the line you drew in **(b)**.

**d.** Now calculate the regression equation using the data from your sample in Activity 1. Draw the regression line on your scatterplot.

**3.** Explore the regression results.

**a.** Clearly describe the meaning of the slope $b_1$.

**b.** Does the literal meaning of the $y$ intercept $b_0$ make sense? What is its literal meaning? What is wrong with interpreting it this way?

**c.** Are your estimates in Activity 2c close to the least-squares results in Activity 2d?

**d.** Are time and distance positively correlated? Compute the correlation coefficient.

**4.** Make a prediction, and find the prediction error.

**a.** Choose one student from your scatterplot and calculate the predicted time for the student. Then calculate the prediction error for the student.

**b.** What is the typical difference between the predicted amount of time and the actual amount of time? Do you want this statistic (*Hint:* $s$) to be large or small?

5. Work with the sampling distribution of $t_{\text{data}}$.
   **a.** Carry out the $t$ test to determine whether a linear relationship exists between time and distance using $\alpha = 0.05$.
   **b.** On a Post-It Note, record your name and the value for the test statistic $t_{\text{data}}$ calculated in **(a)**.
   **c.** Place your Post-It Note on the number line the instructor has prepared on the board, noting the location of the cutoff values $\pm\, t_{\text{crit}}$. You thus have a collection of the values for the statistic $t_{\text{data}}$. Can you make any comment on the shape of the distribution of $t_{\text{data}}$ values?
   **d.** Construct a 95% confidence interval for the true slope $\beta_1$. Does your confidence interval contain 0? What does this mean with respect to the linear relationship between time and distance?
   **e.** Come to a class consensus about the evidence for a linear relationship between the amount of time it takes a student to get to class and the distance the student travels.

## CHAPTER 13 FORMULAS AND VOCABULARY

### SECTION 13.1

- **Extrapolation** (p. 669) Using the regression equation to make estimates or predictions based on $x$-values that are outside the range of the $x$-values in the data set.
- **Prediction error or residual** $(y - \hat{y})$ (p. 670). Measure of how far the predicted value of $y$ is from the actual value of $y$ observed in the data set.
- **Regression analysis** (p. 666). Develops an equation that can describe the relationship between two quantitative variables, often for the purpose of prediction.
- **Regression equation** (p. 667). The equation of the regression line, given by $\hat{y} = b_0 + b_1x$, where: $\hat{y}$ is the estimated or predicted value of $y$ for the given value of $x$, $b_0$ is the $y$ intercept, and $b_1$ is the slope of the regression equation.

$$\text{Slope: } b_1 = \frac{\sum xy - \dfrac{\left(\sum x\right)\left(\sum y\right)}{n}}{\sum x^2 - \dfrac{\left(\sum x\right)^2}{n}}$$

$$y \text{ intercept: } b_0 = \bar{y} - b_1\bar{x}$$

- **Regression line** (p. 667). The straight line that best (according to the least-squares criterion) approximates the relationship between the predictor variable $x$ and the response variable $y$.
- **SSE, sum of squares error** (p. 670). Measure of how much error in prediction results from using the regression equation.

$$\text{SSE} = \sum (y - \hat{y})^2$$

- **Standard error of the estimate** $s$ (p. 671). Measure of the typical residual or prediction error. Thus, $s$ is an indicator of the precision of the estimates derived from the regression equation.

$$s = \sqrt{\text{MSE}} = \sqrt{\frac{\text{SSE}}{n-2}}$$

### SECTION 13.2

- **Coefficient of determination** $r^2$ (p. 677). Measures the goodness of fit of the regression equation as an approximation of the relationship between $x$ and $y$.

$$r^2 = \text{SSR/SST}$$

- **Correlation coefficient** $r$ (p. 679). Can be expressed as $r = \pm\sqrt{r^2}$, taking the positive or negative sign of the slope $b_1$.
- **SSR, sum of squares regression** (p. 677). Measures the amount of improvement in the accuracy of our estimates when using the regression equation as compared to ignoring the $x$ information. $\text{SSR} = \sum(\hat{y} - \bar{y})^2$.

$$\text{Computational formula:} \quad \text{SSR} = \frac{\left[\sum xy - \left(\sum x\right)\left(\sum y\right)/n\right]^2}{\sum x^2 - \left(\sum x\right)^2/n}$$

- **SST, total sum of squares** (p. 676). Measures the total variability in the values of the $y$ variable. $SST = \sum(y - \bar{y})^2$.

$$\text{Computational formula: SST} = \sum y^2 - \left(\sum y\right)^2/n$$

### SECTION 13.3

- **Confidence interval for the true slope $\beta_1$ of the regression line** (p. 688). $b_1 \pm t_{\alpha/2} \cdot s_{b_1}$, where $t_{\alpha/2}$ is based on $n - 2$ degrees of freedom.
- **Fitted values** (p. 686). The predicted values, $\hat{y}$.
- **Outlier** (p. 693). An observation that has a very large absolute residual.
- **Regression model, or regression equation** (p. 684). $y = \beta_0 + \beta_1x + \varepsilon$, where $\beta_0$ is the $y$ intercept of the population regression line, $\beta_1$ is the slope of the population regression line, and $\varepsilon$ is the error term.
- **Standard error of the $t$ statistic** (p. 687)

$$s_{b_1} = \frac{s}{\sqrt{\sum(x_i - \bar{x})^2}}$$

- **Test statistic $t_{\text{data}}$** (p. 689)

$$t_{\text{data}} = b_1/s_{b_1}$$

## CHAPTER 13 REVIEW EXERCISES

### SECTION 13.1

For Exercises 1–4, follow steps **(a)–(f)**.

**a.** Construct a scatterplot of the *y* variable against the *x* variable.

**b.** Describe the relationship between the variables ("As the *x* variable increases, the *y* variable tends to increase, decrease, or stay the same.")

**c.** Compute the regression equation.

**d.** Clearly interpret the meaning of the value you obtained for the slope $b_1$.

**e.** Explain the literal meaning of the *y* intercept $b_0$. Does it make sense in this data set? Why or why not?

**f.** Calculate *s*, the standard error of the estimate. What does this number mean?

**1. Education and Earnings.** The U.S. Census Bureau reports the mean annual earnings of American citizens according to the number of years of education. Here is an excerpt from the 2003 findings. We are interested in the relationship between earnings (*y*, in thousands of dollars) and years of education (*x*).

| Education (*x*) | Annual earnings (*y*) |
|---|---|
| 8 | 18.6 |
| 10 | 18.9 |
| 12 | 27.3 |
| 13 | 29.7 |
| 14 | 34.2 |
| 16 | 51.2 |
| 18 | 60.4 |

**2. High School GPA and College GPA.** The college admissions office would like to determine if there is a relationship between the high school grade point average and the first-year college grade point average of first-year college students, using the data in the following table.

| Student | High school GPA (*x*) | First-year college GPA (*y*) |
|---|---|---|
| 1 | 2.4 | 2.6 |
| 2 | 2.5 | 1.9 |
| 3 | 2.9 | 2.7 |
| 4 | 2.7 | 2.5 |
| 5 | 3.0 | 2.4 |
| 6 | 3.5 | 2.9 |
| 7 | 3.0 | 2.7 |
| 8 | 3.6 | 3.1 |
| 9 | 3.4 | 3.0 |
| 10 | 3.9 | 3.3 |

**3. NBA Points Leaders.** The table lists the 2003–2004 National Basketball Association leaders in points per game, as reported by nba.com. Would you expect there to be a linear relationship between field goals (*x*) and free throws (*y*)?

| Player | Field goals (*x*) | Free throws (*y*) |
|---|---|---|
| 1 Tracy McGrady (Orlando Magic) | 653 | 398 |
| 2 Predrag Stojakovic (Sacramento Kings) | 665 | 394 |
| 3 Kevin Garnett (Minnesota Timberwolves) | 804 | 368 |
| 4 Kobe Bryant (Los Angeles Lakers) | 516 | 454 |
| 5 Paul Pierce (Boston Celtics) | 602 | 517 |
| 6 Baron Davis (New Orleans Hornets) | 554 | 237 |
| 7 Vince Carter (Toronto Raptors) | 608 | 336 |
| 8 Tim Duncan (San Antonio Spurs) | 592 | 352 |
| 9 Dirk Nowitzki (Dallas Mavericks) | 605 | 371 |
| 10 Michael Redd (Milwaukee Bucks) | 633 | 383 |

**4. Used Cars: Price vs Age.** Do you think you can predict the price of a used car based on how old it is? The table shows the price (in thousands of dollars) and the age (in years) of 10 previously owned vehicles of the same make and model.

| Car | Age (*x*) | Price (*y*) |
|---|---|---|
| 1 | 1 | 18.0 |
| 2 | 2 | 16.0 |
| 3 | 3 | 15.5 |
| 4 | 4 | 13.5 |
| 5 | 4 | 14.5 |
| 6 | 5 | 10.5 |
| 7 | 5 | 12.0 |
| 8 | 6 | 9.5 |
| 9 | 7 | 8.5 |
| 10 | 8 | 7.0 |

### SECTION 13.2

For Exercises 5–8, follow these steps.

**a.** Calculate the three sums of squares SSR, SST, and SSE.

**b.** Calculate $r^2$. Comment on how useful the *x* variable is in predicting the *y* variable.

**c.** Find the value of the correlation coefficient. Are the *x* variable and the *y* variable positively correlated, negatively correlated, or uncorrelated?

5. Data in Exercise 1

6. Data in Exercise 2
7. Data in Exercise 3
8. Data in Exercise 4

**SECTION 13.3**

For Exercises 9–12, follow these steps.

**a.** Find $s_{b_1}$. Explain what this number means.
**b.** Find $t_{\alpha/2}$ for a 95% confidence interval for $\beta_1$.
**c.** Construct and interpret a 95% confidence interval for $\beta_1$.

9. Data in Exercise 1
10. Data in Exercise 2
11. Data in Exercise 3
12. Data in Exercise 4

For Exercises 13 and 14, follow steps **(a)**–**(e)**.

**a.** Verify the regression model assumptions.
**b.** Based on the confidence interval constructed in **(c)** of Exercises 9 and 10, would you expect the hypothesis test to reject the null hypothesis that $\beta_1 = 0$?
**c.** Calculate $t_{data}$.
**d.** Find $p$-value $= 2 \cdot P(t > |t_{data}|)$.
**e.** Perform the hypothesis test for the linear relationship between $x$ and $y$ using the $p$-value method and $\alpha = 0.05$.

13. Data in Exercise 1
14. Data in Exercise 2

For Exercises 15 and 16, follow steps **(a)**–**(d)**.

**a.** Based on the confidence interval constructed in **(c)** of Exercises 11 and 12, would you expect the hypothesis test to reject the null hypothesis that $\beta_1 = 0$?
**b.** Find $t_{crit}$ for a two-tailed test with $\alpha = 0.05$ and df $= n - 2$.
**c.** Calculate $t_{data}$.
**d.** Perform the hypothesis test for the linear relationship between $x$ and $y$ using the critical value method and $\alpha = 0.05$.

15. Data in Exercise 3
16. Data in Exercise 4

## CHAPTER 13 Quiz

**TRUE OR FALSE**

1. True or false: The difference $(y - \hat{y})$, which measures how far the predicted value of $y$ is from the actual value of $y$ observed in the data set, is called the sum of squares error.
2. True or false: The standard error of the estimate $s$ provides a measure of the typical residual or prediction error.
3. True or false: SSR measures the total variability in the values of the $y$ variable

**FILL IN THE BLANK**

4. We may consider $r^2$ to be the proportion of the variability in ___________ that is explained by the relationship between ___________ and ___________.
5. The standard error of the estimate $s$ is a measure of the typical difference between ___________ and ___________.
6. The ___________ ___________ ___________ [3 words] measures the goodness of fit of the regression equation.

**SHORT ANSWER**

7. What is the range of possible values of $r^2$?
8. What is the relationship between the three sums of squares?
9. The critical value $t_{cirt}$ is based on how many degrees of freedom?

**CALCULATIONS AND INTERPRETATIONS**

For Exercises 10–12, follow these steps.

**a.** Construct a scatterplot of the $y$ variable against the $x$ variable.
**b.** Describe the relationship between the variables ("As the $x$ variable increases, the $y$ variable tends to increase, decrease, or stay the same.")
**c.** Compute the regression equation.

10. **Men's Heights and Weights.** The university medical unit is collecting data on the heights and weights of the male students on campus. A random sample of ten male students showed the following heights (in inches) and weights (in pounds).

| Student | Height ($x$) | Weight ($y$) |
|---|---|---|
| 1 | 66 | 150 |
| 2 | 68 | 145 |
| 3 | 69 | 160 |
| 4 | 70 | 165 |
| 5 | 70 | 165 |
| 6 | 71 | 180 |
| 7 | 72 | 175 |
| 8 | 72 | 180 |
| 9 | 73 | 195 |
| 10 | 75 | 210 |

11. **Ratio Accounting Grades.** An accounting professor is trying to predict the performance of her students in the second semester of the introductory accounting course by their performance in the first semester. The first-semester grade and second-semester grade were recorded for a random sample of 12 students taking the two-semester course at a local college. The results are shown in the table (see next page).

| Student | First-semester grade ($x$) | Second-semester grade ($y$) |
|---|---|---|
| 1 | 75 | 65 |
| 2 | 80 | 90 |
| 3 | 50 | 75 |
| 4 | 70 | 60 |
| 5 | 90 | 80 |
| 6 | 75 | 80 |
| 7 | 50 | 60 |
| 8 | 95 | 90 |
| 9 | 80 | 75 |
| 10 | 50 | 40 |
| 11 | 60 | 55 |
| 12 | 75 | 70 |

**12. Popular First Names.** The U.S. Census Bureau tracks the popularity of first names in the United States. Here is a table of the most popular first names for males as of 2003. The variable "Pct" indicates the percentage of males born with that name. For example, the proportion of males named John is 3.318% = 0.03318. Let "Rank" be the $x$ variable.

| Name | Rank ($x$) | Pct ($y$) |
|---|---|---|
| James | 1 | 3.318 |
| John | 2 | 3.271 |
| Robert | 3 | 3.143 |
| Michael | 4 | 2.629 |
| William | 5 | 2.451 |
| David | 6 | 2.363 |
| Richard | 7 | 1.703 |
| Charles | 8 | 1.523 |
| Paul | 13 | 0.948 |
| Joseph | 9 | 1.404 |
| Mark | 14 | 0.938 |
| Thomas | 10 | 1.380 |
| Donald | 15 | 0.931 |
| Christopher | 11 | 1.035 |
| George | 16 | 0.927 |
| Daniel | 12 | 0.974 |
| Kenneth | 17 | 0.826 |
| Steven | 18 | 0.780 |
| Edward | 19 | 0.779 |
| Brian | 20 | 0.736 |

For Exercises 13–15, follow these steps.

**a.** Clearly interpret the meaning of the value you obtained for the slope $b_1$.

**b.** Explain the literal meaning of the $y$ intercept $b_0$. Does it make sense in this data set? Why or why not?

**c.** Calculate $s$, the standard error of the estimate. What does this number mean?

**13.** Data in Exercise 10

**14.** Data in Exercise 11

**15.** Data in Exercise 12

For Exercises 16–18, follow these steps.

**a.** Calculate the three sums of squares SSR, SST, and SSE.

**b.** Calculate $r^2$. Comment on how useful the $x$ variable is in predicting the $y$ variable.

**c.** Find the value of the correlation coefficient. Are the $x$ variable and the $y$ variable positively correlated, negatively correlated, or uncorrelated?

**16.** Data in Exercise 10

**17.** Data in Exercise 11

**18.** Data in Exercise 12

For Exercises 19–21, follow these steps.

**a.** Find $s_{b_1}$. Explain what this number means.

**b.** Find $t_{\alpha/2}$ for a 95% confidence interval for $\beta_1$.

**c.** Construct and interpret a 95% confidence interval for $\beta_1$.

**19.** Data in Exercise 10

**20.** Data in Exercise 11

**21.** Data in Exercise 12

For Exercises 22 and 23, perform the hypothesis test for the linear relationship between $x$ and $y$ using the $p$-value method and $\alpha = 0.05$.

**22.** Data in Exercise 10

**23.** Data in Exercise 11

**24.** For the data in Exercise 12, perform the hypothesis test for the linear relationship between $x$ and $y$ using the critical value method and $\alpha = 0.05$.

# Answers to Odd-Numbered Exercises and Chapter Quizzes

## CHAPTER 1

### Section 1.1

**1, 3, 5.** Answers will vary.
**7.** Note the large differences in the comparative heights of the rectangles that measure responses of sadness, anger and disbelief.
**9.** Yes; the point corresponding to Palm Beach County votes lies significantly above the rest of the data and outside the otherwise linear trend.
**11.** About 3500 votes.

### Section 1.2

**1.** Answers will vary.
**3.** False.
**5.** A *qualitative variable* is usually classified into categories; a *quantitative variable* takes on numerical values.
**7.** True.
**9.** A *statistic* is a characteristic of a sample.
**11.** Statistical inference.
**13.** Answers will vary. It may be impossible or too expensive to sample the entire population, for instance.
**15.** **(a)** quantitative **(b)** interval
**17.** **(a)** quantitative **(b)** ratio
**19.** **(a)** quantitative **(b)** ratio
**21.** **(a)** qualitative **(b)** ordinal
**23.** **(a)** qualitative **(b)** nominal
**25.** **(a)** qualitative **(b)** nominal
**27.** **(a)** quantitative **(b)** ratio
**29.** Population: all home sales in Tarrant County, Texas; sample: 100 home sales selected.
**31.** Population: all 4-H clubs in Utah; sample: ten selected 4-H clubs.
**33.** Population: all students at Portland Community College; sample: 50 selected Portland Community College students.
**35.** Descriptive statistics; the variable describes a sample.
**37.** Statistical inference; the sample was used to draw a conclusion about the entire population.
**39.** Descriptive statistics; the variable describes a sample.
**41.** Descriptive statistics; the variable describes a sample.
**43.** Sample (out of a total of 11 species)
**45.** **(a)** Elements are the companies. **(b)** Employees and Industry. **(c)** Employees **(d)** Industry **(e)** It has 1,755 employees and is a health services industry.
**47.** **(a)** Elements are the types of crime. **(b)** 2005 Total, per 100,000 People, National per 100,000 People, and Compared to National Average. **(c)** 2005 Total, per 100,000 People, and National per 100,000 People. **(d)** Compared to National Average **(e)** In 2005, there were 55 automobile thefts in Stillwater, Oklahoma, at a rate of 134.1 per 100,000 people. This was better than the national rate of 526.5 thefts per 100,000 people.
**49.** **(a)** Elements are the hospitals. **(b)** Beds, City, and Zip. **(c)** Beds **(d)** City and Zip **(e)** It has 72 beds and is in the city of Kosciusko with a zip code of 39090.
**51.** Four: ERA, HR, BB, SO; quantitative.
**53.** Earned-run average: 4.41; home runs given up:170; walks given up: 496; strikeouts made: 1019.
**55.** Births, Average Maternal Age; quantitative.
**57.** **(a)** Elements are the types of music that *all* radio listeners listened to in the winter of 2004; we can only list Mainstream AC, Hot AC, Mod AC, and Soft AC—other elements are not given. **(b)** Percentage of listeners for each type; quantitative. **(c)** The proportion is (% Mainstream AC ÷ % AC) *100% = (7.8/14)100% = 55.71%.
**59.** They compared the average lifetime of a sample of their own light bulb to the reported average lifetimes of other current models of light bulbs.
**61.** **(a)** Years. **(b)** Deaths; quantitative **(c)** No; years were not selected randomly.

### Section 1.3

**1.** Answers will vary.
**3.** Answers will vary; could have chosen a random sample of houses and apartments and surveying the people door to door, for instance.
**5, 7.** Answers will vary.
**9.** Observational study—observes association between explanatory and response variables; experimental study—investigates how varying the explanatory variable affects the response by placing subjects into treatment and control groups.
**11.** Stratified sampling
**13.** Systematic sampling
**15.** Random sampling
**17.** Leading question
**19.** Observational study; explanatory variable: whether or not the company gives large bonuses to their CEO's; response variable: stock price.
**21.** Answers will vary; samples are different because each was randomly selected.
**23.** Target population: all high schools in New England; potential population: all high schools in greater Boston. Potential population is not a random sample of all high schools in New England; drop out rate for target population may differ from potential population.
**25.** Desired response type is open to interpretation: preference or yes/no.
**27.** No; the conclusion leaves out the details of the survey question
**29.** **(a)** No, different random samples may contain different largest employers. **(b)** Answers will vary. **(c)** No; see answer **(a)**. **(d)** Answers will vary.
**31.** **(a)** No, different random samples may contain different lowest stock prices. **(b)** Answers will vary. **(c)** No; see answer **(a)**. **(d)** Answers will vary.
**33.** Predictor variable: patient diet, Mediterranean or Western; response variable: risk for a second heart attack.
**35.** **(a)** Type of bracelet worn, placebo or ionized **(b)** Wearing the ionized bracelet. **(c)** Measure of pain.
**37.** Answers will vary.
**39.** **(a)** Answers will vary. **(b)** No. **(c)** Answers will vary. **(d)** We cannot tell in advance.

### Chapter 1 Review

**1.** **(a)** The ten elements are the countries. **(b)** Continent, Climate, Water Use (Per Capita Gallons per Day), and Main Use. **(c)** It is on the continent of North America, it has a temperate climate, its water use is 1268 gallons per capita per day, and its main use of water is industry. **(d)** Iraq **(e)** "Per person."

**3.** United States and Canada. **(a)** No; the United States uses 1565 gallons per capita per day, Canada uses 1268 gallons per capita per day. **(b)** No; the United States has more farming and industry than Canada so it uses more water.

**5. (a)** Turn on all one million light bulbs, leave them all on until they burn out, compute the average of all bulb lifetimes. **(b)** Use the average lifetime of a sample to estimate the average lifetime of the population.

**7. (a)** All statistics students **(b)** The students in the statistics class who were selected for the sample **(c)** Left-handed or not; qualitative **(d)** No; not likely to be very far away from the population proportion since enrollment in a specific statistics class is not dependent on being left-handed or not.

**9. (a)** Observational study **(b)** It would be impractical to randomly reassign people to a statistics class after classes have started.

**11. (a)** Randomization **(b)** Without randomization, results may be biased.

**13.** No; there may be other factors that determine a child's cognitive skills.

## Chapter 1 Quiz

**1.** False
**2.** True
**3.** False
**4.** collecting
**5.** observation
**6.** equal chance
**7.** sample
**8.** Observational study
**9.** Convenience sampling
**10.** Experimental study
**11.** Predictor variable: drug given to an elderly patient with Alzheimer's, new or placebo; response variable: whether or not the patient's Alzheimer's symptoms are reduced.
**12.** The elements are the three sectors: Hardware, Software, Communications.
**13.** Gross domestic product in each of the years from 1996 through 2003; all variables are quantitative.
**14.** 2001
**15.** Predictor variable: form of pesticide used, new or traditional; response variable: the amount of insect damage to crops.
**16.** Experimental study.
**17.** Treatment: new pesticide; control: traditional pesticide.
**18.** Different people have different interpretations of the words often, occasionally, sometimes, and seldom.
**19.** No; it would have been too time-consuming and expensive.
**20.** Statistics, since they most likely came from a sample.
**21.** Answers will vary. According to the survey, 60.6% of all African Americans living in the Northeast prefer Kool cigarettes.

# CHAPTER 2

## Section 2.1

**1.** We use graphical and tabular form to summarize data in order to organize it in a format where we can better assess the information. If we just report the raw data, it may be extremely difficult to extract the information contained in the data.

**3.** True.

**5.**

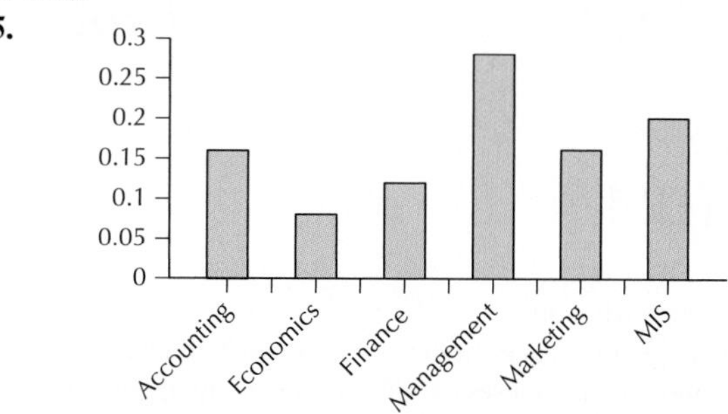


**7. (a)** and **(b)**

| Climate | Frequency | Relative frequency |
|---|---|---|
| Arid | 5 | 0.50 |
| Temperate | 4 | 0.40 |
| Tropical | 1 | 0.10 |

**(c)**

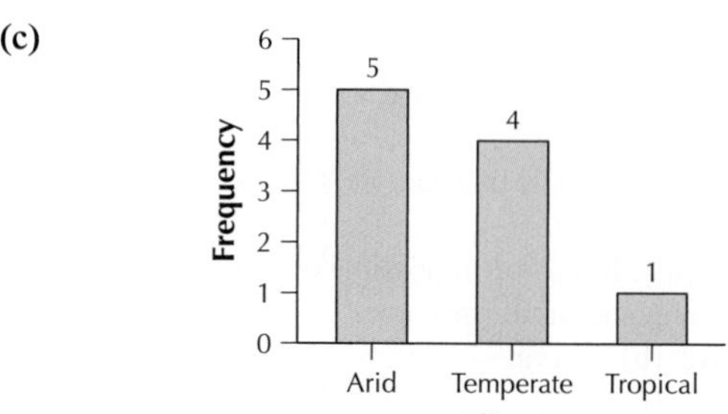


**(d)**

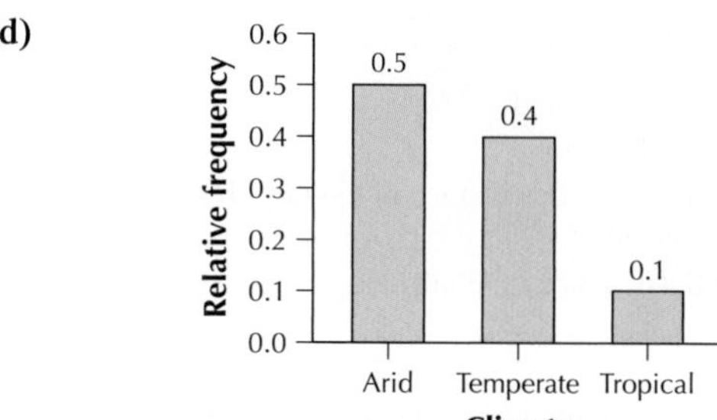


**(e)**

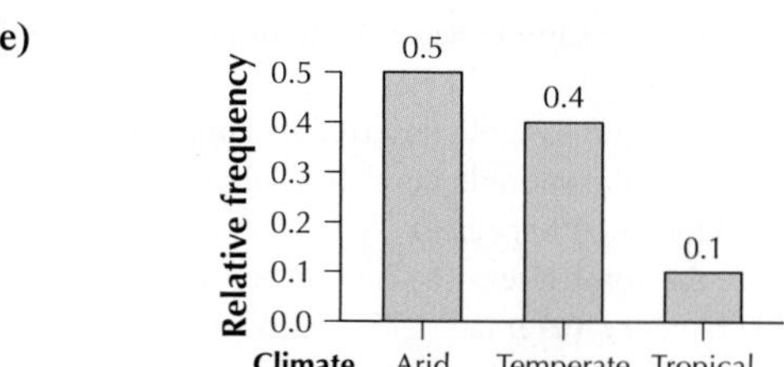


**(f)**

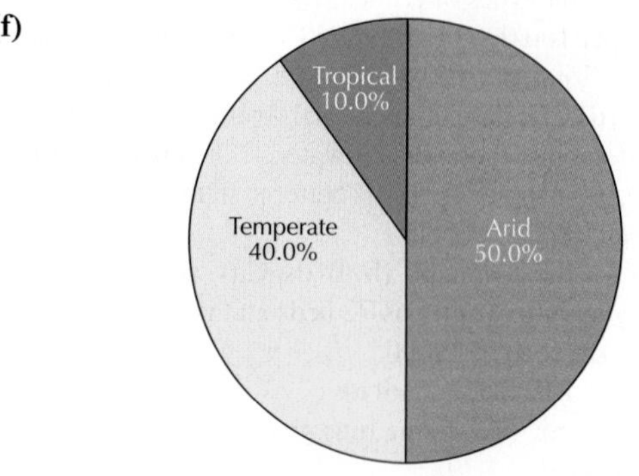

**9. (a)**

| End-use sector | Relative frequency |
|---|---|
| Residential | 0.2049 |
| Commercial | 0.1746 |
| Industrial | 0.2931 |
| Transportation | 0.3274 |

**(b)**

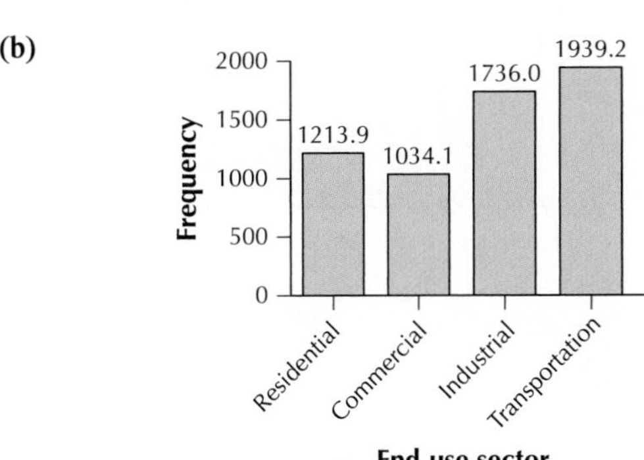


**(c)**

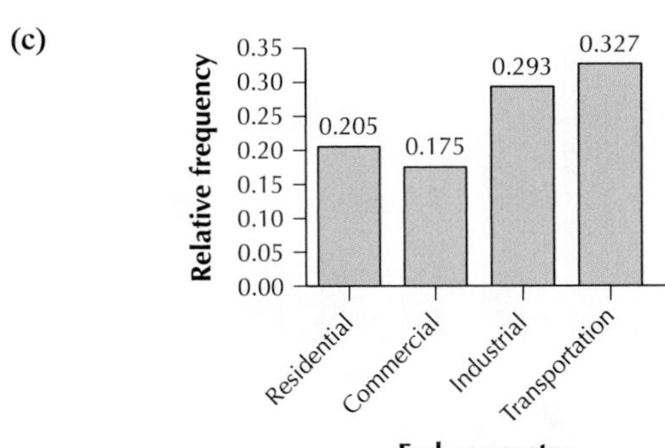


**(d)**

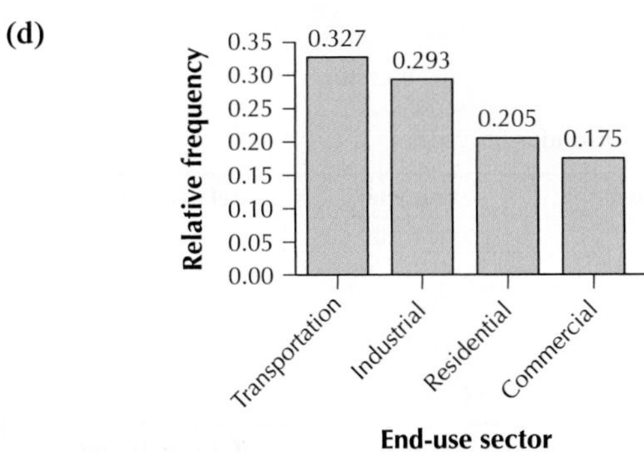


**(e)** *Note:* In the chart, the relative frequencies are expressed as percentages.

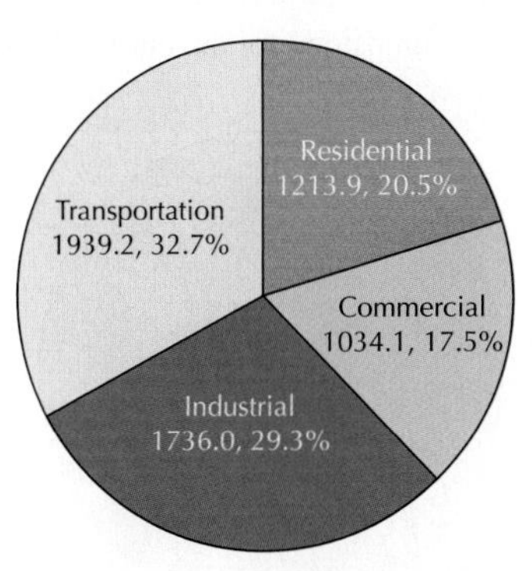


**11. (a) and (b)**

| Type | Frequency | Relative frequency |
|---|---|---|
| F | 5 | 0.5 |
| NF | 5 | 0.5 |

**(c)**

**(d)**

**(e)**

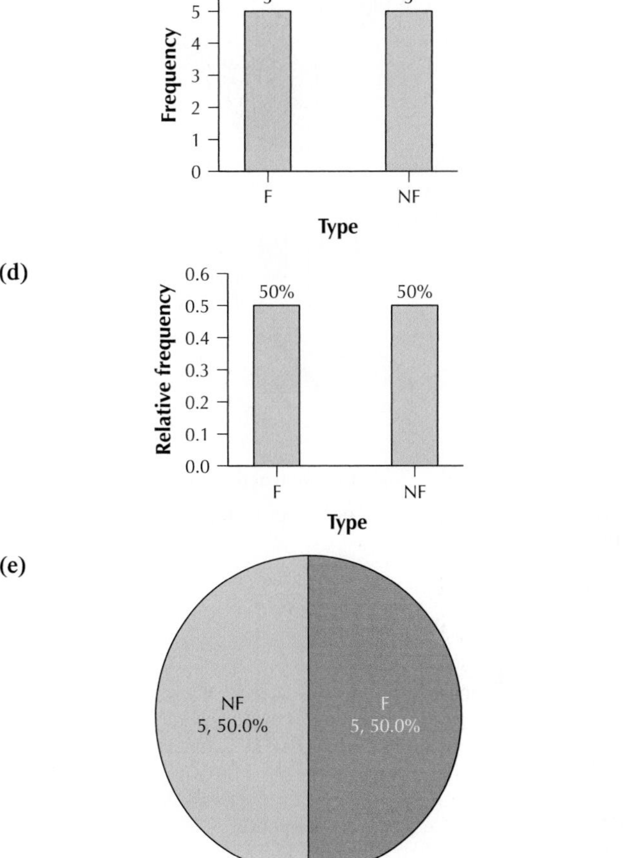


**13. (a) and (b)**

| Gender | Frequency | Relative frequency |
|---|---|---|
| Female | 5 | 0.5 |
| Male | 5 | 0.5 |

**(c)**

**(d)**

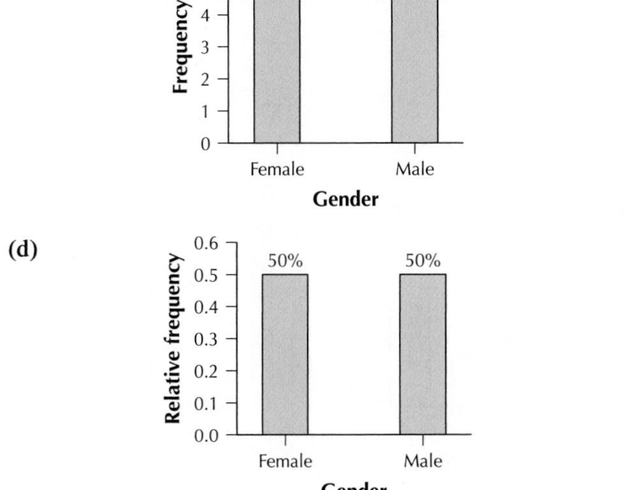

**(e)**

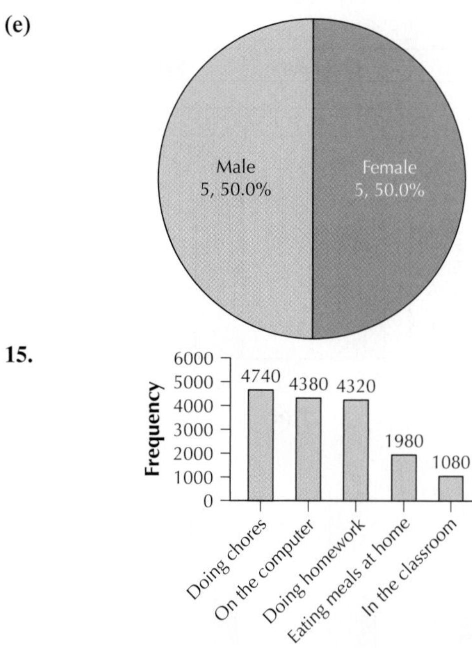


**15.**

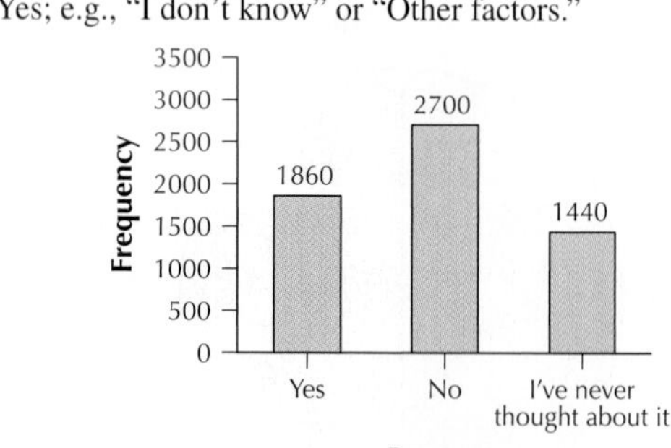


**17.** Yes; e.g., "I don't know" or "Other factors."

**19.**



**21. (a)**

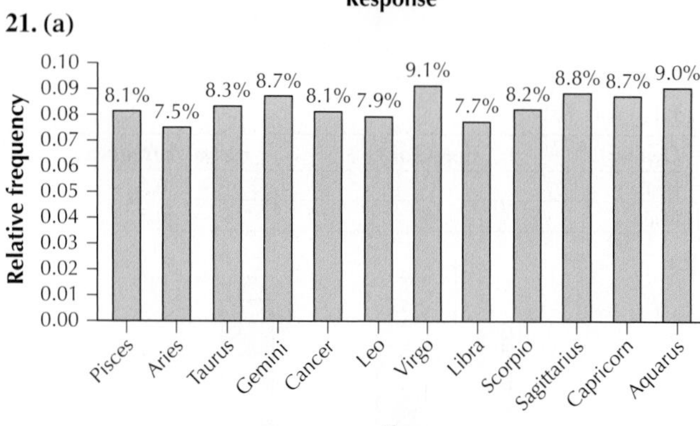


**(b)**

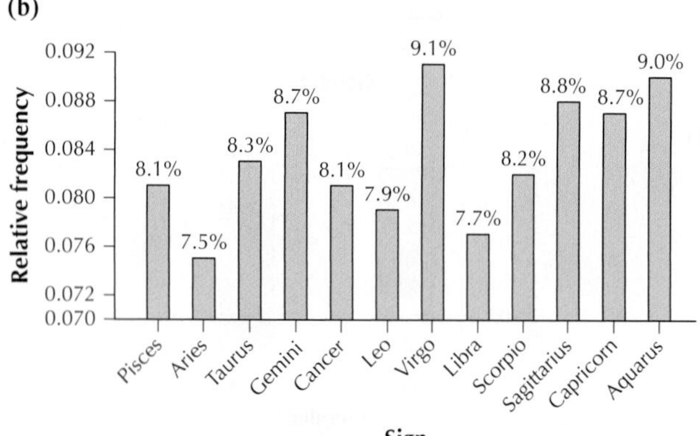


The graph in **(b)** uses an adjusted scale, which is misleading. Use this graph to magnify the small variability in percentages.

**23.** Categories are qualitative; interpretation will vary.

**25.** Yes

**27. (a)**

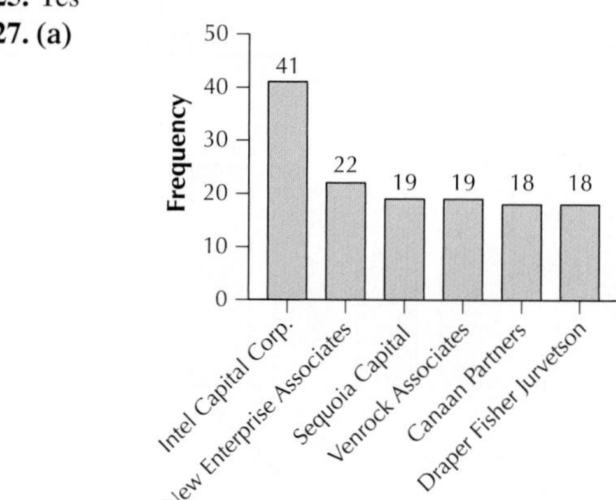


**(b)**

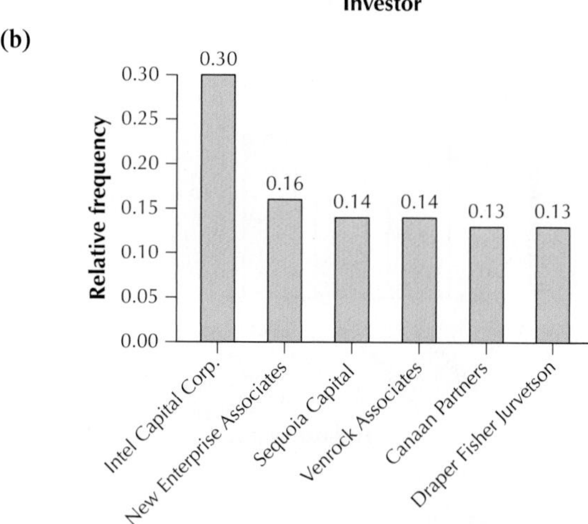


**29.** Distribution for under 65 years

| Type of insurance | Frequency | Relative frequency |
|---|---|---|
| Private | 126,800,000 | 0.6967 |
| Medicaid | 13,700,000 | 0.0753 |
| Other | 5,700,000 | 0.0313 |
| Uninsured | 35,800,000 | 0.1967 |

**31.** Distribution for over 65 years

| Type of insurance | Frequency | Relative frequency |
|---|---|---|
| Private | 20,800,000 | 0.5977 |
| Medicaid | 11,700,000 | 0.3362 |
| Other | 2,100,000 | 0.0603 |
| Uninsured | 200,000 | 0.0057 |

**33.** The proportion of uninsured for the under 65 group is greater than those 65 and over.

**35.**

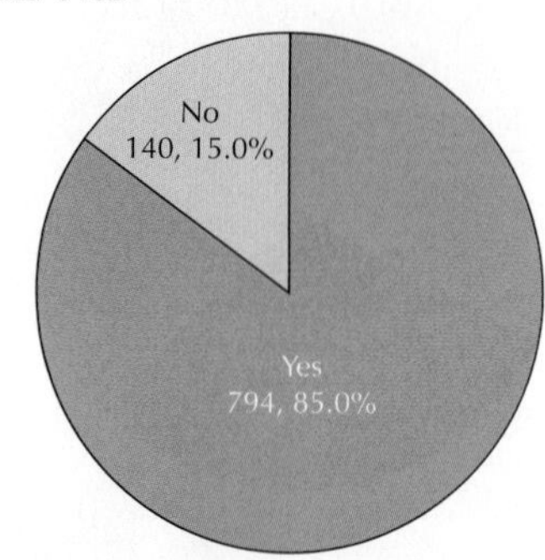

**37.**

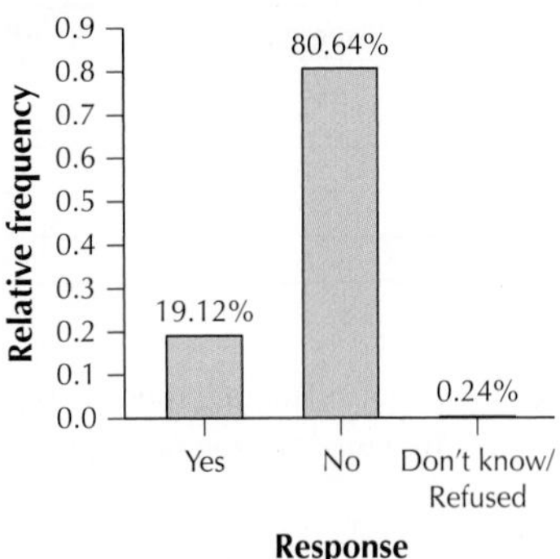


**39.**

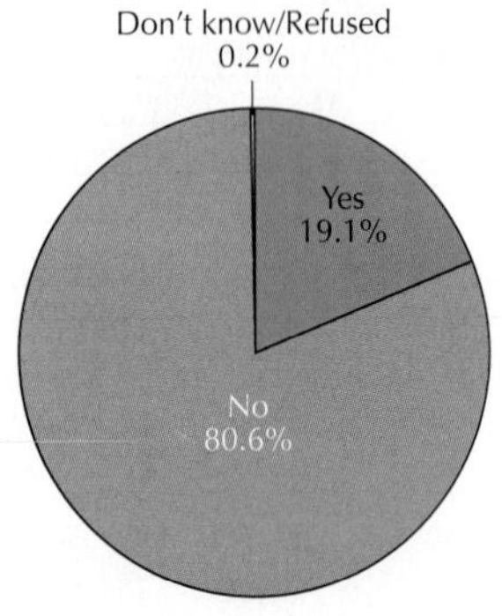


**41.**

| Food Store | Frequency | Relative frequency |
|---|---|---|
| Supermarkets | 73,357 | 0.3695 |
| Convenience food stores | 30,748 | 0.1549 |
| Convenience food/gasoline | 23,035 | 0.1160 |
| Delicatessens | 6,123 | 0.0308 |
| Meat and fish market | 8,941 | 0.0450 |
| Retail bakeries | 20,418 | 0.1029 |
| Fruit and vegetable markets | 2,971 | 0.0150 |
| Candy, nuts, confectionery | 5,029 | 0.0253 |
| Dairy products store | 2,340 | 0.0118 |
| Other food stores | 25,552 | 0.1287 |

**43.**

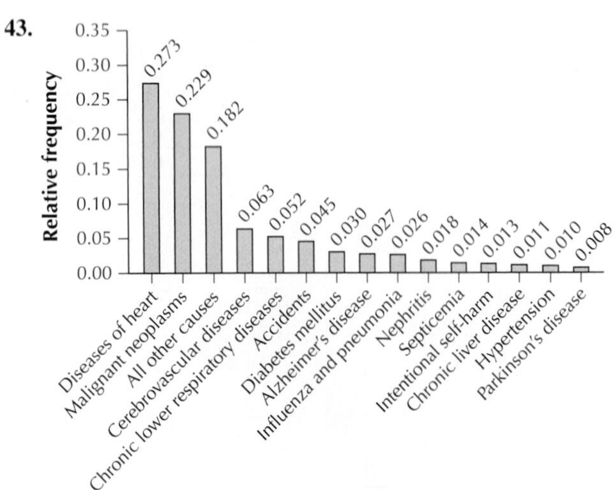


**45.** Combined categories present the clearest picture here.

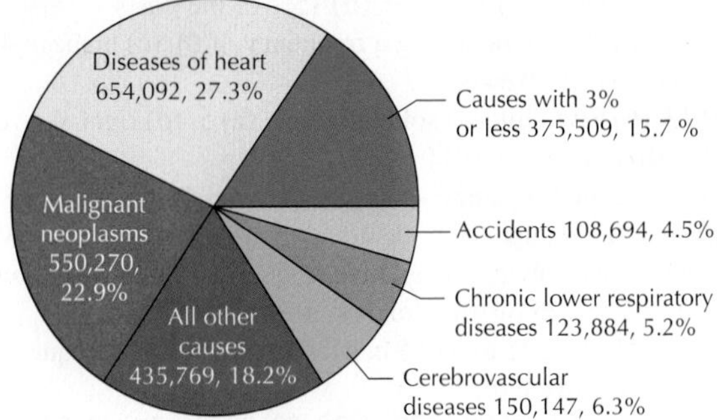


**47.**

| Gender | Frequency | Relative frequency |
|---|---|---|
| Boy | 26 | 0.5909 |
| Girl | 18 | 0.4091 |

**49.**

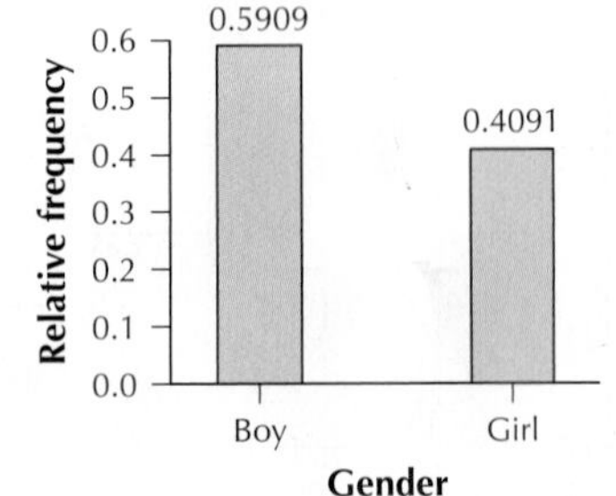


No, just change the scale of the vertical axis to represent frequencies instead of relative frequencies.

**51. (a)**

```
Tally for Discrete Variables: GENDER

GENDER  Percent
   boy    47.49
  girl    52.51
```

**(b)**

```
Tally for Discrete Variables: GOALS

  GOALS  Percent
 Grades    51.67
Popular    29.50
 Sports    18.83
```

**53.** Qualitative: State; Quantitative: Tot_hhld, Fam_tpc, Fam_mpc, Fam_fpc, Nfm_tpc, Nfm_lpc, and Ave_size.

**55. (a)** and **(b)**

| Class | Frequency | Relative frequency |
|---|---|---|
| Freshman | 5 | 0.25 |
| Sophomore | 5 | 0.25 |
| Junior | 5 | 0.25 |
| Senior | 5 | 0.25 |

**57.** Answers will vary.

## Section 2.2

**1.** Both: frequency distribution, relative frequency distribution; quantitative data only: histograms, frequency polygons, stem-and-leaf displays, dotplot.
**3.** Between 5 and 20
**5.** Answers will vary.

**7.** Answers will vary.
**9. (a)** 4 **(b)** 1 and 6 **(c)** 15 times **(d)** 15% of the times
**11. (a)** 46 **(b)** 33 (not including a frequency of 0) **(c)** highest: 49: lowest: 33. **(d)** left-skewed
**13. (a)** 13 **(b)** Five children sold only one **(c)** 5 **(d)** right-skewed
**15. (a)** 8 **(b)** 0.08 **(c)** 6 **(d)** 0.06
**17. (a)** Divide the frequency values by the total frequency—classes not affected **(b)** change the scale along the relative frequency (vertical) axis by multiplying the relative frequency values by the total frequency—shape of distribution not affected **(c)** 19
**19. (a)** 0 **(b)** 0 **(c)** \$25 to \$27.5 has the largest relative frequency, 4/19 = 0.2105. **(d)** 3 **(e)** 0
**21.** Data set: 23 24 25 26 27 28 28 29 30 31 31 32 32 32 33 35 36 37 39 40
**23.** Histogram with five classes

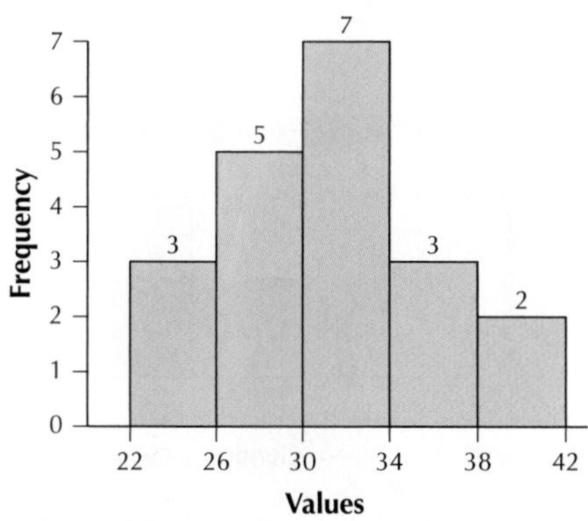


**25. (a)** 15 **(b)** 37.5 **(c)** 52.5 **(d)** 67.5 to 82.5 **(e)** 22.5 to 37.5
**27. (a)** 2000 **(b)** 1000 **(c)** 1000 to 3000 **(d)** 17,000 to 19,000.
**29. (a)** and **(b)**

| Continent | Frequency | Relative frequency |
|---|---|---|
| Africa | 1 | 0.1 |
| Asia | 5 | 0.5 |
| Europe | 1 | 0.1 |
| North America | 2 | 0.2 |
| South America | 1 | 0.1 |

**(c)** Cannot construct because the variable is qualitative. **(d)** Cannot construct because the variable is qualitative. **(e)** We cannot construct because the variable is qualitative.
**31. (a)** Highest: 20–39; lowest: 0–19, 40–59, and 120–139 (not including a frequency of zero). **(b)** Right-skewed **(c)** No
**33.** Stem-and-leaf plot for the number of beds.

| | |
|---|---|
| 1 | 5 |
| 2 | 55 |
| 3 | 04 |
| 4 | 9 |
| 5 | |
| 6 | 7 |
| 7 | 2 |
| 8 | |
| 9 | |
| 10 | |
| 11 | |
| 12 | |
| 13 | 4 |

**35.**

| Classes | Frequency | Relative frequency |
|---|---|---|
| 900–1099 | 4 | 0.4 |
| 1100–1299 | 2 | 0.2 |
| 1300–1499 | 0 | 0.0 |
| 1500–1699 | 2 | 0.2 |
| 1700–1899 | 2 | 0.2 |

**37.** Vertical axis of frequency histogram represents frequency; vertical axis of relative frequency histogram represents relative frequency.

**39.**

| Classes | Frequency | Relative frequency |
|---|---|---|
| 550–599 | 1 | 1/12 ≈ 0.0833 |
| 600–649 | 1 | 1/12 ≈ 0.0833 |
| 650–699 | 1 | 1/12 ≈ 0.0833 |
| 700–749 | 0 | 0/12 = 0.0000 |
| 750–799 | 3 | 3/12 = 0.2500 |
| 800–849 | 3 | 3/12 = 0.2500 |
| 850–899 | 1 | 1/12 ≈ 0.0833 |
| 900–949 | 2 | 2/12 ≈ 0.1667 |

**41.**

550 600 650 700 750 800 850 900
**Arrests**

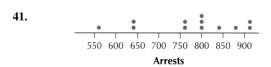

**43.** It is difficult to determine the most common number of shutouts since we have intervals of values.

| Interval | Frequency | Relative frequency |
|---|---|---|
| 2–3 | 1 | 1/14 = 0.071 |
| 4–5 | 1 | 1/14 = 0.071 |
| 6–7 | 2 | 2/14 = 0.143 |
| 8–9 | 8 | 8/14 = 0.571 |
| 10–11 | 0 | 0 |
| 12–13 | 2 | 2/14 = 0.143 |

**45. (a)** 10/14 = 0.7143 **(b)** 0 **(c)** 2/14 = 0.1429 **(d)** 12/14 = 0.8571 **(e)** 2/14 = 0.1429
**47.** 62%; middle.
**49. (a)** and **(b)**

| Exam scores | Frequency | Relative frequency |
|---|---|---|
| 50–59 | 2 | 0.10 |
| 60–69 | 4 | 0.20 |
| 70–79 | 6 | 0.30 |
| 80–89 | 5 | 0.25 |
| 90–99 | 3 | 0.15 |

**(c)**

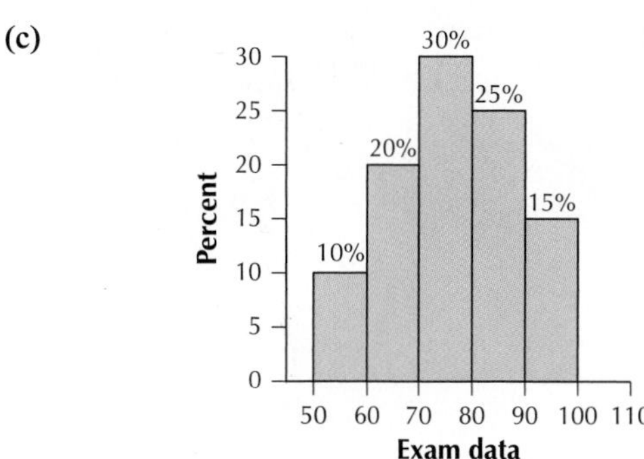

**51.**

| | Dotplot | Histogram | Stem-and-leaf |
|---|---|---|---|
| **(a)** Symmetry and skewness | Appropriate to use for small ranges of data | Appropriate to use | Appropriate to use for small ranges of data |
| **(b)** Construct using pencil and paper | Easily done for small ranges of data | Easily done for small ranges of data | Easily done for small ranges of data |
| **(c)** Presentation in front of non-statisticians | Appropriate | Appropriate | May not be appropriate |
| **(d)** Maximum leeway for presentation | Appropriate | Appropriate | Appropriate |

**53.** 51; 8; State; tot_hhld, fam_tpc, fam_mpc, fam_fpc, nfm_tpc, nfm_lpc, ave_size.
**55.** Yes, California (10,381,206 households); histogram indicates that this is unusually large.
**57.** Yes; District of Columbia (2.26); Hawaii (3.01) and Utah (3.15); histogram indicates that these are unusually large average households.
**59.** 961; 22
**61.** Yes; fats and oils.
**63.** One whole cheesecake (2053 grams of cholesterol).
**65.** **(a)** 2 **(b)** 4.00, 4.30
**67.** **(a)** Increases **(b)** Decreases
**69.** **(a)** 6 **(b)** 57 **(c)** Split stems
**71.** Answers will vary.

## Section 2.3

**1.** A frequency distribution gives the frequency counts for each class (grouped or ungrouped). A cumulative frequency distribution gives the number of values which are less than or equal to the upper limit of a given class for grouped data or it gives the number of values which are less than or equal to a given number for ungrouped data.
**3.** Ogive.
**5.** Time series data.
**7.** **(a)**, **(b)**, and **(c)**

| Value | Frequency | Cumulative frequency | Relative frequency | Cumulative relative frequency |
|---|---|---|---|---|
| 1 | 13 | 13 | 0.13 | 0.13 |
| 2 | 20 | 33 | 0.20 | 0.33 |
| 3 | 15 | 48 | 0.15 | 0.48 |
| 4 | 24 | 72 | 0.24 | 0.72 |
| 5 | 15 | 87 | 0.15 | 0.87 |
| 6 | 13 | 100 | 0.13 | 1.00 |

**9.** **(a)–(c)**

| Stock prices | Frequency | Cumulative frequency | Relative frequency | Cumulative relative frequency |
|---|---|---|---|---|
| 5.00–7.49 | 1 | 1 | 0.0526 | 0.0526 |
| 7.50–9.99 | 1 | 2 | 0.0526 | 0.1052 |
| 10.00–12.49 | 2 | 4 | 0.1053 | 0.2105 |
| 12.50–14.99 | 1 | 5 | 0.0526 | 0.2631 |
| 15.00–17.49 | 2 | 7 | 0.1053 | 0.3684 |
| 17.50–19.99 | 0 | 7 | 0.0000 | 0.3684 |
| 20.00–22.49 | 3 | 10 | 0.1579 | 0.5263 |
| 22.50–24.99 | 3 | 13 | 0.1579 | 0.6842 |
| 25.00–27.49 | 4 | 17 | 0.2105 | 0.8947 |
| 27.50–29.99 | 2 | 19 | 0.1053 | 1.0000 |

**11.** **(a)**

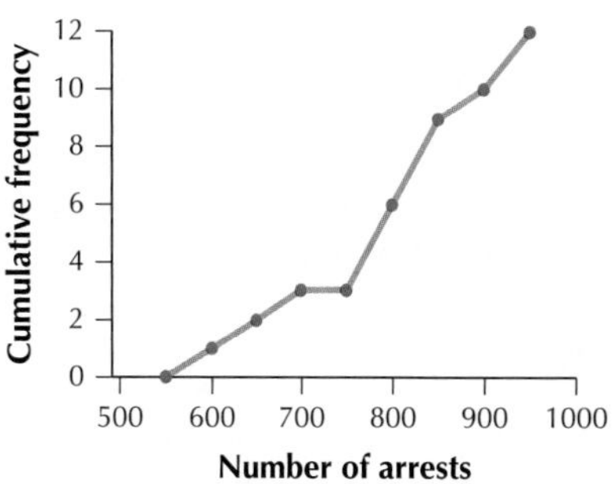


**(b)**

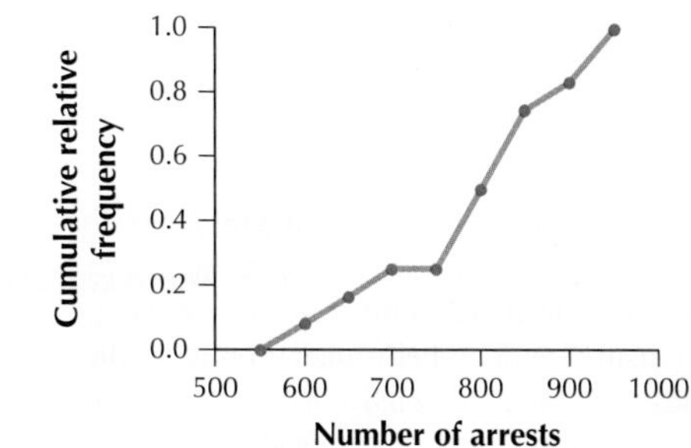


**13.** **(a)** 0.8 **(b)** 2.39 **(c)** 1.99
**15.** Divide the frequency values along the *y* axis by 367; will not affect the points or the line segments along the graph.
**17.** **(a)** Approximately 45% **(b)** Approximately 75% **(c)** Approximately 30%
**19.** **(a)** 1.5 to 3 (not inclusive); nine states **(b)** 10.5 to 12 (not inclusive); California **(c)** Georgia, Pennsylvania, Michigan, and South Dakota.

**21.**

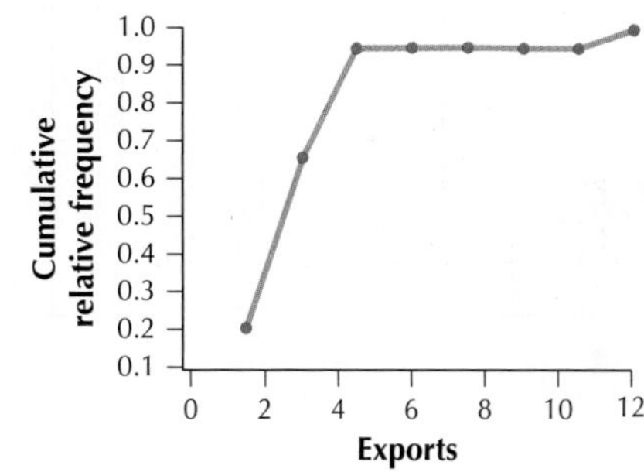


**23.** **(a)** 5.26% **(b)** 94.74%
**25.** **(a)** 82.45% **(b)** 17.55%

**27.** **(a)**

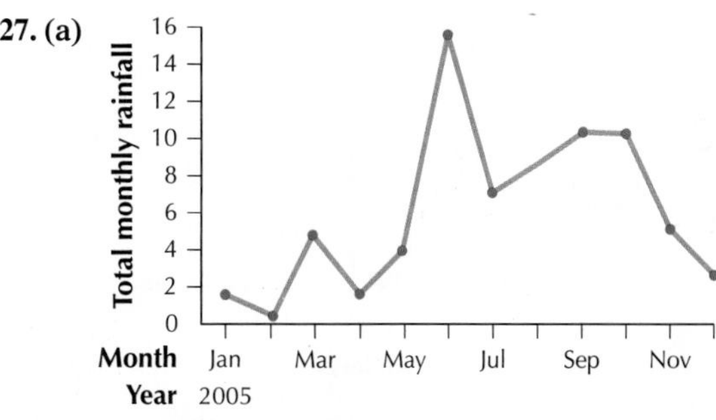


**(b)** It is wetter in the summer than in winter.

**29. (a)**

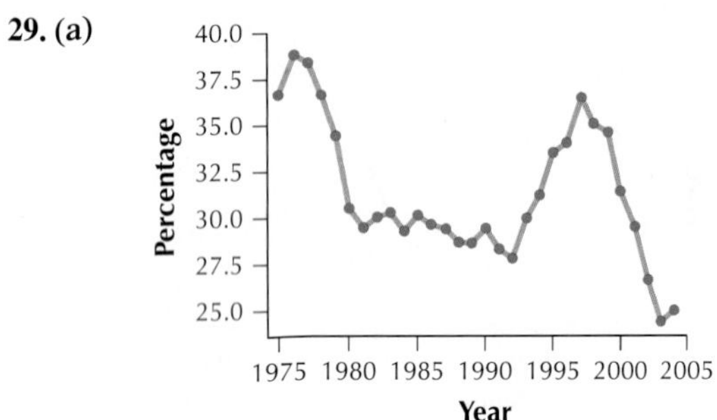


**(b)** Downward until 1992, back upward until 1997, then again downward.

## Section 2.4

**1.** Answers will vary.
**3.** Figure 2.31 visually reinforces the magnitude of the differences.
**5.** Table 2.26 gives the actual number of cars stolen.
**7. (a)** Biased distortion or embellishment; omitting the zero on the relevant scales; inaccuracy in relative lengths of bars in a bar chart. **(b)** A Pareto chart or pie chart can be used.
**9. (a)** The percentage (or number) of doctors who are devoted to family practice is decreasing **(b)** inaccuracy in relative lengths of bars in the bar chart **(c)** One may use a bar chart.
**11. (a)** Omitting the zero on the vertical scale.

**(b)**

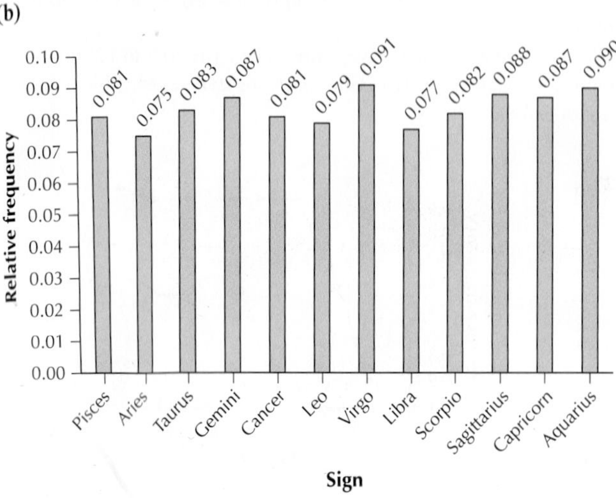


**13. (a)**

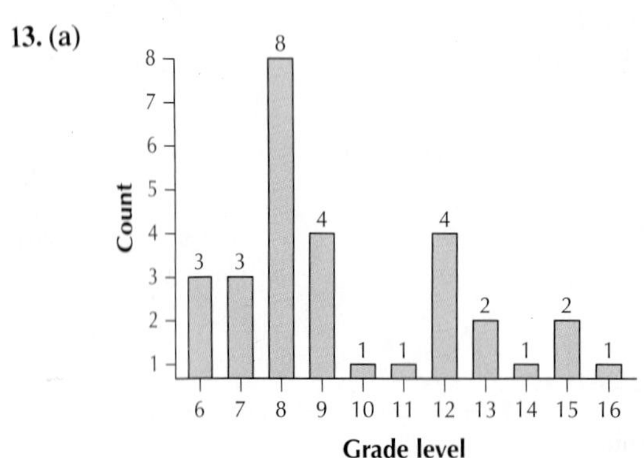


**(b)** Manipulating the scale, omitting the zero on the vertical scale.

**(c)**

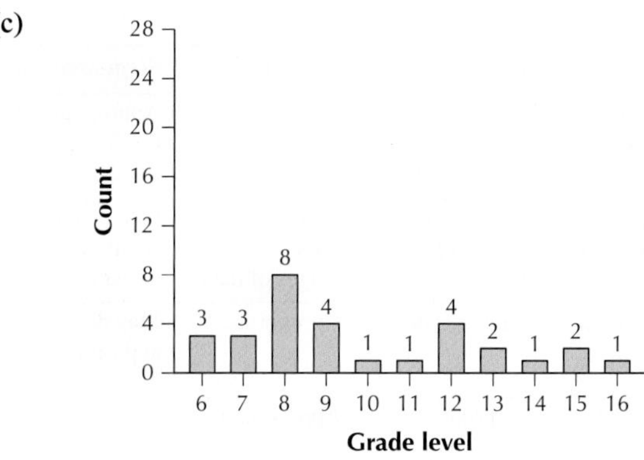


**(d)** Manipulating the scale.

**(e)**

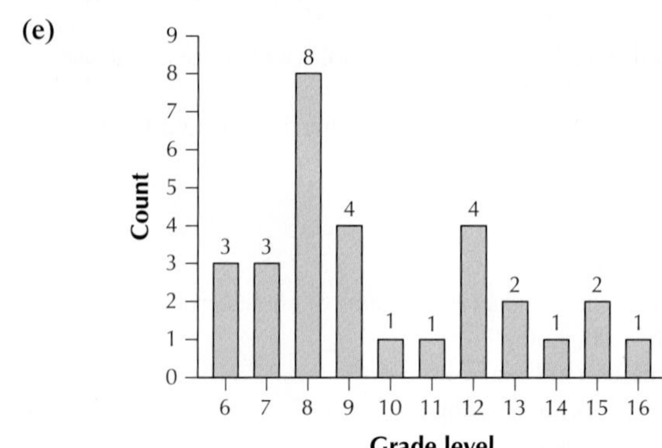


**15, 17.** Answers will vary.

## Chapter 2 Review

**1.** No, because the variable is categorical.
**3.** The relative frequencies are expressed as percentages.

**(a)** and **(b)**

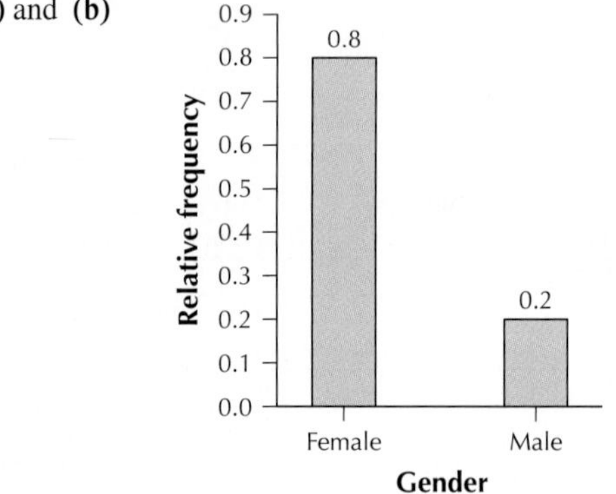


**(c)** and **(d)**

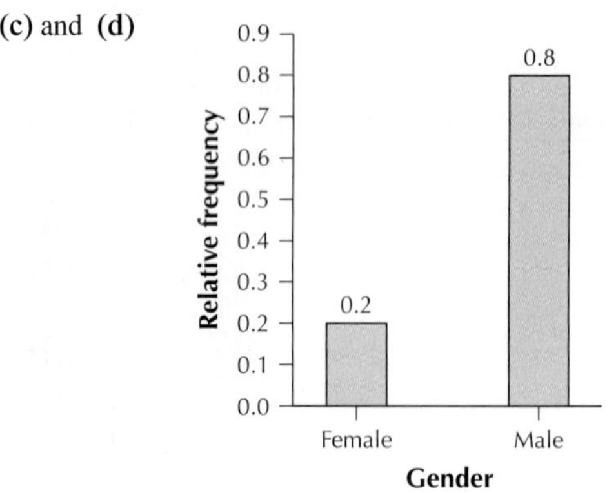

**(e)**

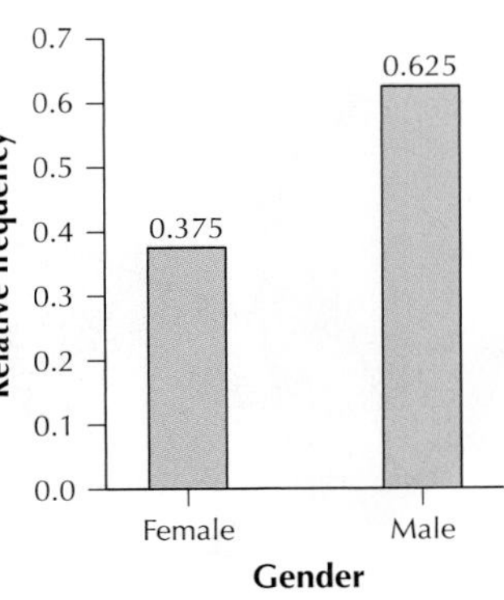


**(f)**

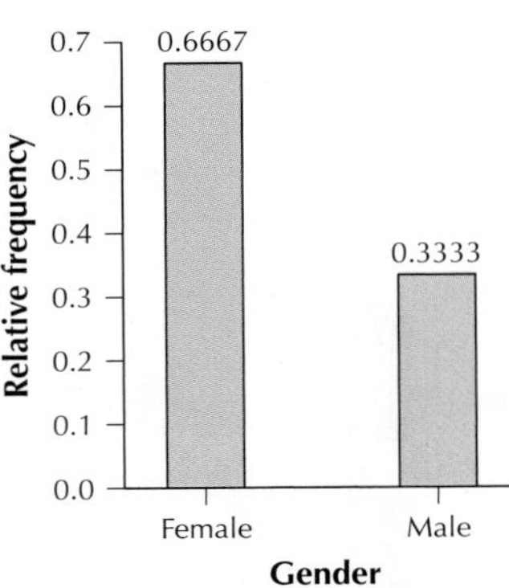


**5.** No, because we only have the frequencies for the categories inexpensive, moderately priced, and expensive and not the original data.

**7.** No, we would need the total number of observations.

**9.** Around a value of 62.

**11.** **(b)** tending to be right-skewed

**13.** Shape would not be affected; The scale would move 0.5 units to the right

**15.** **(a)** and **(b)**

| Interval | Frequency | Cumulative frequency | Relative frequency | Cumulative relative frequency |
|---|---|---|---|---|
| 0–19 | 1 | 1 | $1/9 \approx 0.1111$ | 0.1111 |
| 20–39 | 4 | 5 | $4/9 \approx 0.4444$ | 0.5555 |
| 40–59 | 1 | 6 | $1/9 \approx 0.1111$ | 0.6666 |
| 60–79 | 2 | 8 | $2/9 \approx 0.2222$ | 0.8888 |
| 80–99 | 0 | 8 | $0/9 = 0.0000$ | 0.8888 |
| 100–119 | 0 | 8 | $0/9 = 0.0000$ | 0.8888 |
| 120–139 | 1 | 9 | $1/9 \approx 0.1111$ | $0.9999 \approx 1$ |

**17.**

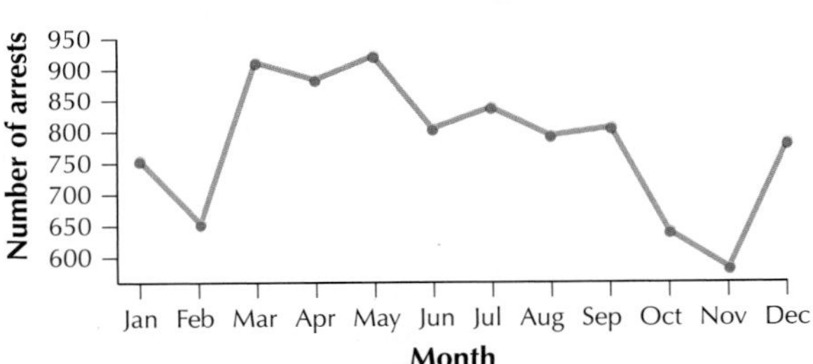


**19.** **(a)**

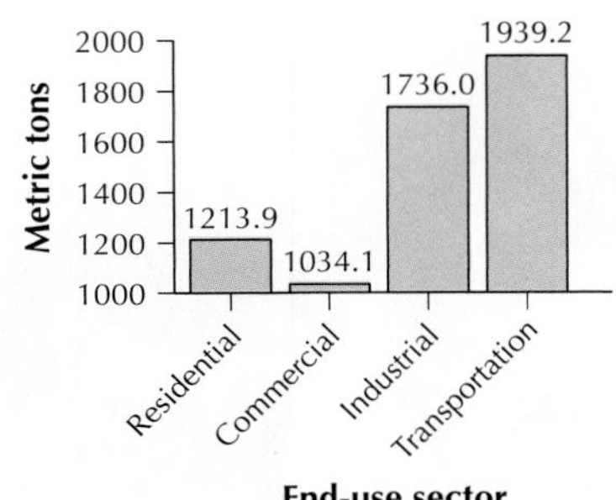


**(b)** Manipulating the scale.

**(c)**

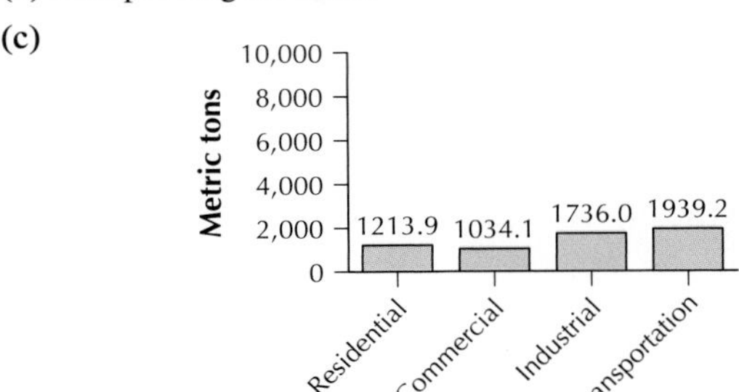


**(d)** Manipulating the scale.

**(e)**

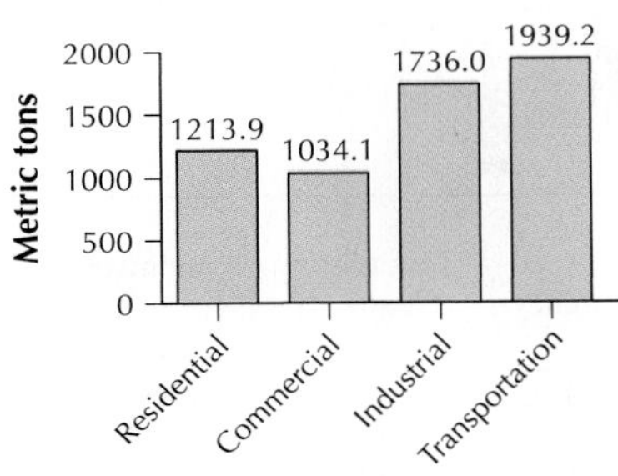


**21.** **(a)**

**Tally for Discrete Variables: SCHOOL**

| SCHOOL | Percent |
|---|---|
| Brent El | 14.02 |
| Brent Mid | 17.57 |
| Brown Mid | 10.88 |
| Elm | 4.39 |
| Main | 14.23 |
| Portage | 12.76 |
| Ridge | 10.04 |

**(b)** Brent Middle School, Elm

**(c)**

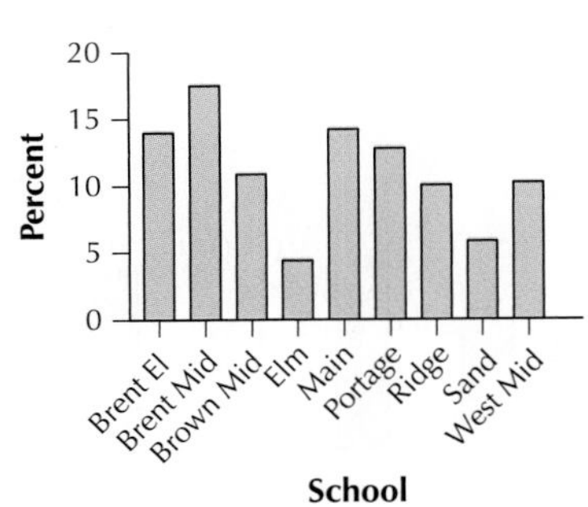

Percent within all data

**(d)**

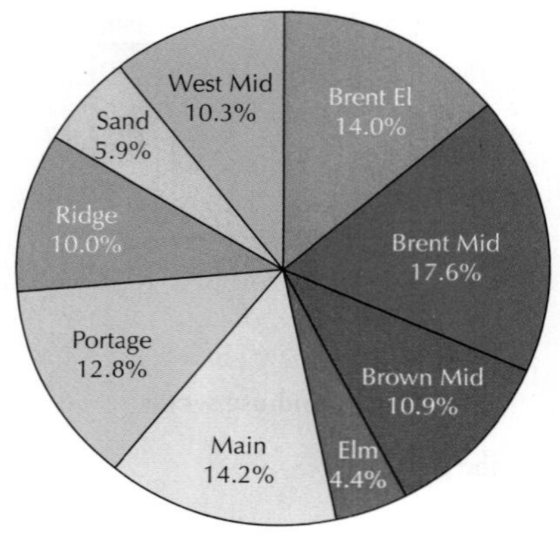


**(e)** Answers will vary.

## Chapter 2 Quiz

**1.** False
**2.** True
**3.** True
**4.** sample size.
**5.** bar chart.
**6.** frequency distribution
**7.** 1
**8.** Symmetric
**9.** Right skewed

**10–13.**

| Vowels | Frequency | Cumulative frequency | Relative frequency | Cumulative relative frequency |
|---|---|---|---|---|
| a | 73 | 73 | 0.1931 | 0.1931 |
| e | 130 | 203 | 0.3439 | 0.5370 |
| i | 74 | 277 | 0.1958 | 0.7328 |
| o | 74 | 351 | 0.1958 | 0.9286 |
| u | 27 | 378 | 0.0714 | 1.0000 |

**14.**

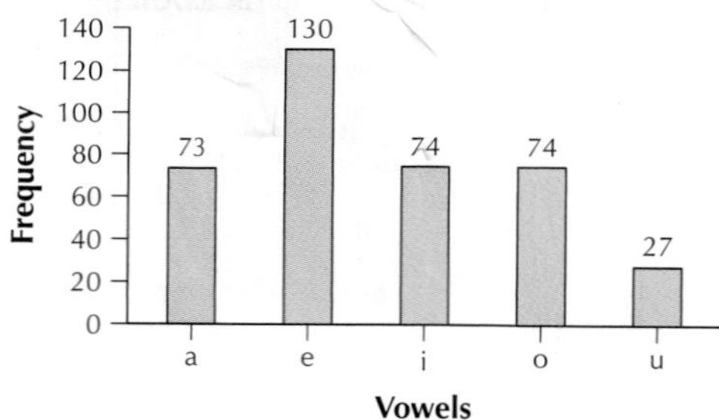


**15.**

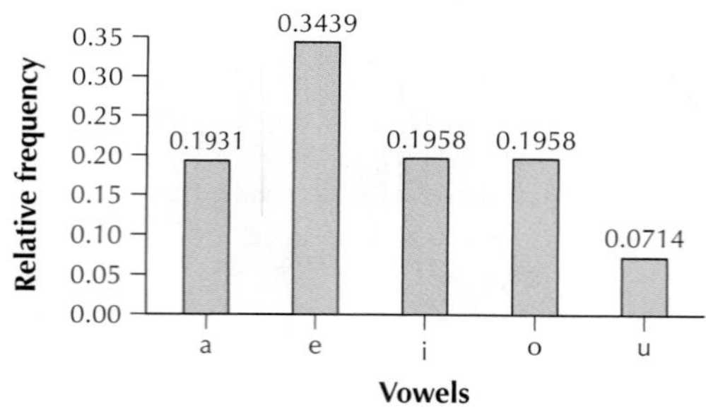


**16.**

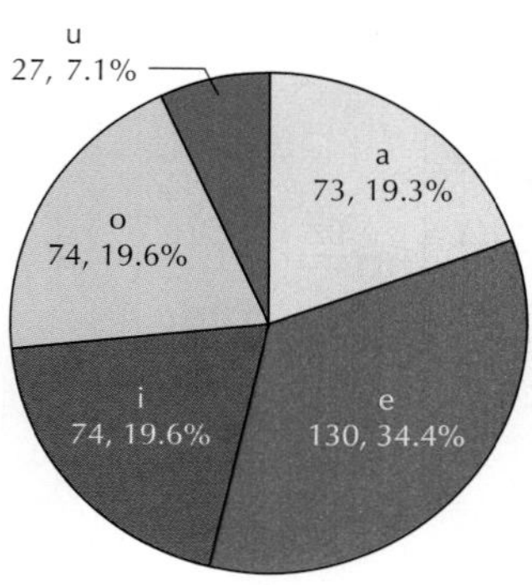


**17, 18.** Can't construct because variable is qualitative.

**19.**

| Response | Frequency | Relative frequency |
|---|---|---|
| More serious than Pearl Harbor | 412 | 0.6667 |
| Equal to Pearl Harbor | 154 | 0.2492 |
| Not as serious as Pearl Harbor | 29 | 0.0469 |
| Not sure | 23 | 0.0372 |

**20.**

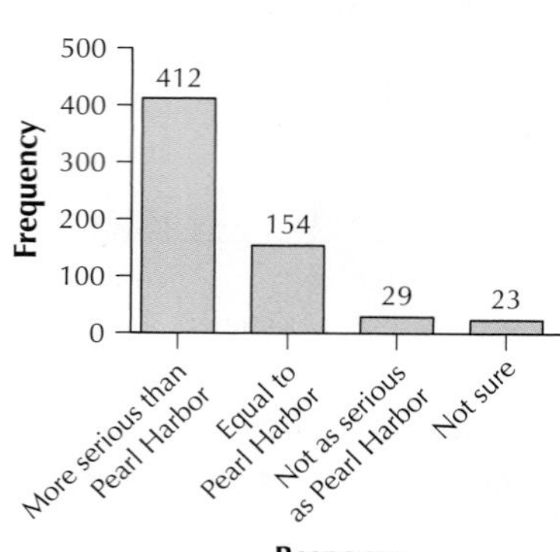


**21.**

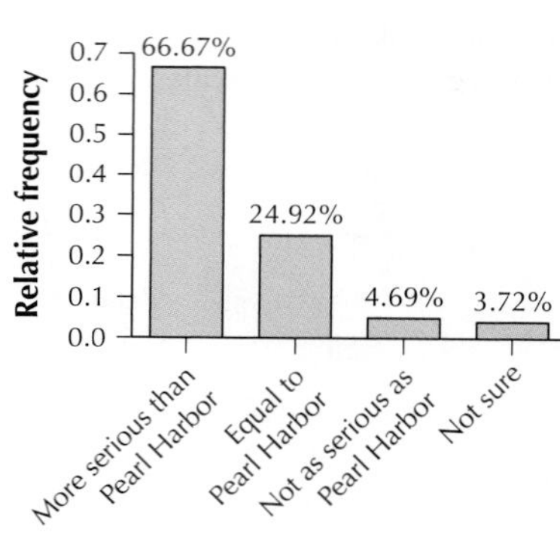


Multiply the values along the *y* scale by the sample size to get the frequencies; bars would not need to be redrawn.

**22.**

| Type of music | Relative frequency | Frequency |
|---|---|---|
| Hip-hop/rap | 0.27 | 1620 |
| Pop | 0.23 | 1380 |
| Rock/punk | 0.17 | 1020 |
| Alternative | 0.07 | 420 |
| Christian/gospel | 0.06 | 360 |
| R & B | 0.06 | 360 |
| Country | 0.05 | 300 |
| Techno/house | 0.04 | 240 |
| Jazz | 0.01 | 60 |
| Other | 0.04 | 240 |

**23.**

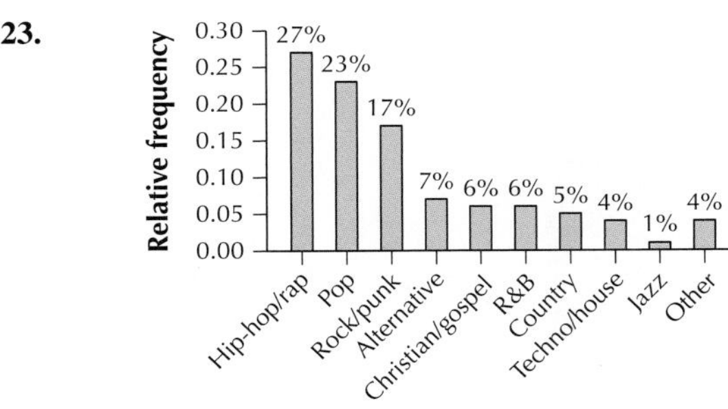

**24.**

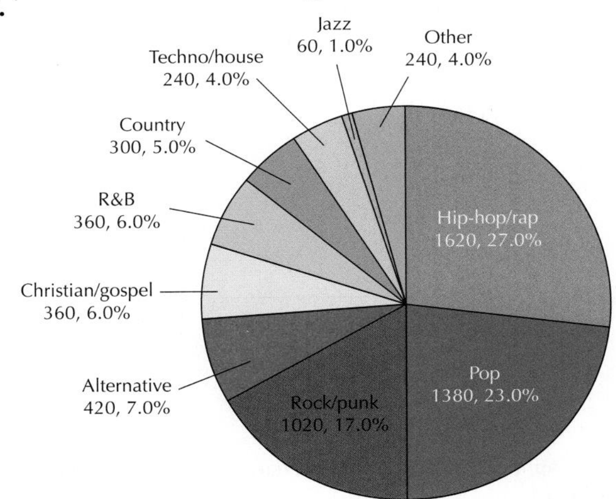


**25.** Height of the bars are the same; frequency chart has frequency plotted on $y$ axis, relative frequency chart has relative frequency plotted on $y$ axis.

# CHAPTER 3

## Section 3.1

**1.** A value that locates the center of the data set.
**3.** Because the mean depends in part on the sum of all data values, an outlier will skew the mean (pull it in one direction or another). Since the median simply depends on position in an ordered list, it is not sensitive to outliers.
**5. (a)** 22.5 **(b)** 20 **(c)** 20
**7. (a)** 16.6 **(b)** 17 **(c)** 5 and 15
**9. (a)** 7 **(b)** 7.5 **(c)** 4 and 9
**11. (a)** 33.4 **(b)** 33.5 **(c)** 35
**13. (a)** 507.38 **(b)** 503 **(c)** 501
**15. (a)** English; no. **(b)** No, the data are qualitative. **(c)** Economics does not occur with the highest frequency.
**17. (a)** \$119.84 **(b)** \$119.84 = \$23.968 × 5 **(c)** If a set of values is multiplied by a positive constant then the mean for this new data set will equal the mean of the original data set times the constant.
**19. (a)** 2 **(b)** Mean = 3.5; Median = 3.
**21. (a)** Mean < Median < Mode **(b)** Mode < Median < Mean **(c)** Mean = Median = Mode.
**23.** The mean is larger than the median, which implies the distribution is positively skewed.
**25.** The mean, median and mode will all be halved. Each statistic will be affected equally.
**27. (a)** 73 **(b)** 76 **(c)** Females; yes
**29. (a)** Decrease **(b)** Unchanged **(c)** Unchanged
**31.** \$2,400,000; 11.0462%.
**33.** No; not representative of all 50 states because sample values are from northeastern states and from the states that have the highest participation rates.
**35. (a)** 98.1 **(b)** 98.4 **(c)** Females; yes
**37. (a)** Decrease **(b)** Unchanged **(c)** Unchanged
**39, 41, 43.** Answers will vary.

## Section 3.2

**1.** Deviation for a data value gives the distance the value is from the mean.
**3.** Less variability will lead to more precise estimates, higher confidence in conclusions, and greater discerning power.
**5.** 6
**7.** 25
**9.** 9.08
**11.** 41.67
**13.** 24 years
**15.** Variance = 49.5 mpg squared; Standard deviation = 7.04 mpg.
**17.** Variance = 4.336 years squared; Standard deviation = 2.082 years. The standard deviation is more easily understood because it has the same units as the variable *age*.
**19.** The Short Tracks variable has the greater variability.
**21.** Super Speedway variable: 24, Road Courses: 8; Super Speedway has greater variability.
**23.** Zooplankton: 6.86, phytoplankton: 9.96 **(a)** phytoplankton **(b)** phytoplankton
**25.** Mean number of cases sold = 676.6 million.

| Number of cases (in millions) | Deviations (in millions) |
|---|---|
| 1929 | 1252.4 |
| 1385 | 708.4 |
| 811 | 134.4 |
| 541 | −135.6 |
| 537 | −139.6 |
| 536 | −140.6 |
| 530 | −146.6 |
| 220 | −456.6 |
| 180 | −496.6 |
| 97 | −579.6 |

**27. (a)** It would reduce the values of the range, variance and standard deviations. **(b)** Range: 1288, variance: 151,973, standard deviation: 390
**29.** Colony A = 480.04; Colony B = 694.49. **(a)** Colony B **(b)** Colony B; yes **(c)** Colony B, because it has the larger variance.
**31.** Range for SAT I Verbal = 25; range for SAT I Math = 24. Variance for SAT I Verbal = 90.55; variance for SAT I Math = 84.84. Standard deviation for SAT I Verbal = 9.52; standard deviation for SAT I Math = 9.21. These values support the judgment in the previous exercise since they are similar in magnitude.
**33. (a)** 6 inches **(b)** 2 inches
**35. (a)** Increase the range, variance and standard deviation. **(b)** New range = 10. New variance = 15.2. New standard deviation = 3.9. Yes, the judgment in **(a)** was supported.
**37.** About 300.
**39.** Standard deviation = 276.2. **(a)** 24 **(b)** The frequency counts for the syllables typically differ from the mean of 396.8 by only 276.2.
**41. (a)** They are the same. **(b)** 8.11 **(c)** Slight difference of 0.11. **(d)** 5.875 **(e)** They are the same. **(f)** Female; yes **(g)** Females
**43. (a)** Males: 98.105; females: 98.394; very close. **(b)** Males: maximum = 99.5, minimum = 96.3; range = 3.2. Females: maximum = 100.8, minimum = 96.4, range = 4.4; they are the same. **(c)** Females: 0.743; Males: 0.699. **(d)** Females.
**45. (a)** Answers will vary. **(b)** June
**47. (a)** Remain the same. **(b)** Increase. **(c)** Increase
**49, 51.** Answers will vary.

## Section 3.3

**1.** These formulas will provide only estimates because we will not know the exact data values.
**3.** Population mean
**5.** 69
**7.** 3.2
**9.** 2.3446
**11.**

| Class limits | Midpoints |
|---|---|
| 10–12.49 | 11.25 |
| 12.50–14.99 | 13.75 |
| 15–17.49 | 16.25 |
| 17.50–19.99 | 18.75 |

**13.** 14.2857
**15.** 450
**17.** Estimated standard deviation = 7.310096; estimated variance = 53.4375.
**19. (a)**

| Age | Frequency | Midpoints |
|---|---|---|
| 0–4.99 | 63,422 | 2.5 |
| 5–17.99 | 240,629 | 11.5 |
| 18–64.99 | 540,949 | 41.5 |

**(b)** Estimated mean = 30.0298 years **(c)** Estimated standard deviation = 15.455909 years; estimated variance = 238.8851 years squared
**21. (a)**

| Dollar value | Housing units (frequency) | Midpoints |
|---|---|---|
| \$0–\$49,999 | 5,430 | 25,000 |
| \$50,000–\$99,000 | 90,605 | 75,000 |
| \$100,000–\$149,999 | 90,620 | 125,000 |
| \$150,000–\$199,999 | 54,295 | 175,000 |
| \$200,000–\$299,000 | 34,835 | 250,000 |
| \$300,000–\$499,999 | 15,770 | 400,000 |
| \$500,000–\$999,999 | 5,595 | 750,000 |

**(b)** Estimated mean = \$158,079.25 **(c)** Estimated standard deviation = \$116,223.13; estimated variance = 13,507,815,862.70 (dollars squared)
**23.** 82.8%
**25.** 93.67 milligrams
**27.** If $w_i = 1$ for all $i$ then the weighted mean formula will be equivalent to the formula for the sample mean.

## Section 3.4

**1, 3.** Answers will vary.
**5.** It is possible for the 1st percentile to equal the 99th percentile if all of the data values are the same.
**7. (a)** 20 **(b)** 24 **(c)** 18
**9. (a)** 0.21 **(b)** –0.78 **(c)** −0.64
**11.** No; *z* score is 1.92 and $-2 < 1.92 < 2$.
**13. (a)** 105 **(b)** 130
**15. (a)** –1.65 **(b)** 0.93
**17. (a)** 4,200,000 **(b)** 7,100,000
**19. (a)** 1.11 **(b)** 0.6033
**21. (a)**

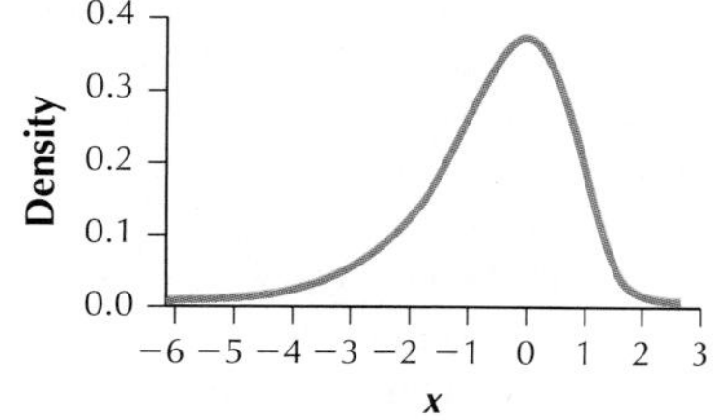

**(b)**

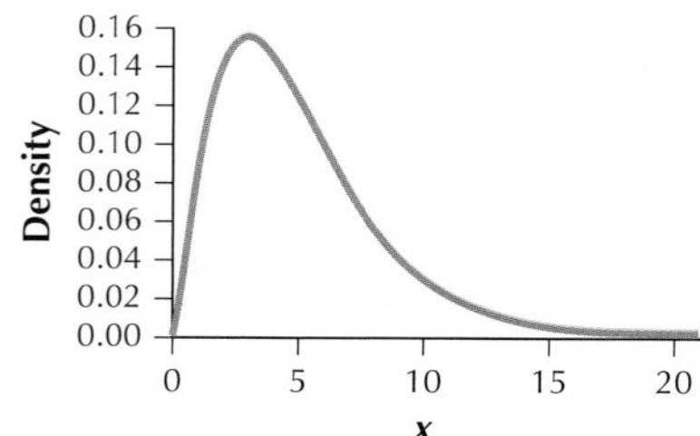


**(c)** 503
**23. (a)** \$48,422 **(b)** \$78,312 **(c)** \$47,550 **(d)** \$82,406
**25. (a)** Right-skewed **(b)** Greater than, since the distribution is right-skewed.

**(c)**

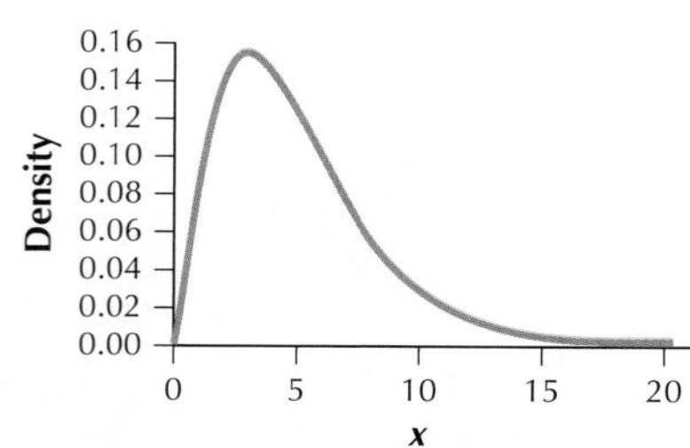


## Section 3.5

**1.** May be applied to any continuous data set; value of *k* must be greater than 1.
**3.** The interval between the mean minus two standard deviations and the mean plus two standard deviations.
**5.** True
**7.** Yes.
**9. (a)** At least 75% **(b)** At least 88.9%
**11. (a)** At least 75% **(b)** At least 93.75%
**13. (a)** Between 80 and 120 **(b)** 120 **(c)** 80
**15. (a)** At least 75% **(b)** At least 88.9% **(c)** Can't find since $k = 1$.
**17. (a)** About 95% **(b)** About 99.7% **(c)** About 68%
**19. (a)** (i) 102,000,000 (ii) He is very truthful. **(b)** The day corresponding to the 25th percentile. **(c)** At least 75%
**21. (a), (b),** and **(c)** cannot be done since the distribution is unknown and $k = 1$.
**23. (a)** About 95% **(b)** About 2.5%
**25.** July: about 2.5%; January: between 16% and 50%.
**27. (a)** Cannot be done since $k = 1$. **(b)** At least 75% **(c)** At most 25%
**29.** Answers in Exercise 28 are more precise; the distribution is bell-shaped.
**31.** Lower-performing school *z*-score 0.8, so 4.0 GPA is not considered unusual; higher-performing school *z*-score 2.8, so a GPA of 4.0 is considered moderately unusual.
**33. (a)** Answers will vary. Mean: 2.8251 mg/g. **(b)** Not bell-shaped **(c)** Since distribution is not bell-shaped, the Empirical Rule should not be used.

## Section 3.6

**1.** We need statistics that can summarize a data set while being less sensitive to outliers; mean and the standard deviation might not be very representative of center and spread of the data because of sensitivity to outliers.
**3.** The 3rd quartile is the 75th percentile; the 3rd percentile is the data value at which 3% of all other values is less than or equal to.

**5.** **(a)** All numbers between Q1 and Q2 would be the same; median line and line for Q1 would be the same. **(b)** All numbers in the data set would have to be the same. The box plot would be a single line. **(c)** Right-skewed with a few values much larger than the rest; median line of box plot closer to the line for Q3 than the line for Q1. **(d)** Left-skewed with a few values much smaller than the rest; median line of box plot closer to the line for Q1 than the line for Q3. **(e)** Can't do. The 50th percentile of a data set can never be smaller than the 25th percentile.
**(f)** The numbers between Q1 and Q3 can't all be the same.
**7.** False
**9.** **(a)** 28 **(b)** 33.5 **(c)** 35
**11.** Min = 25, Q1 = 28; median = 33.5, Q3 = 35; max = 49.
**13.**

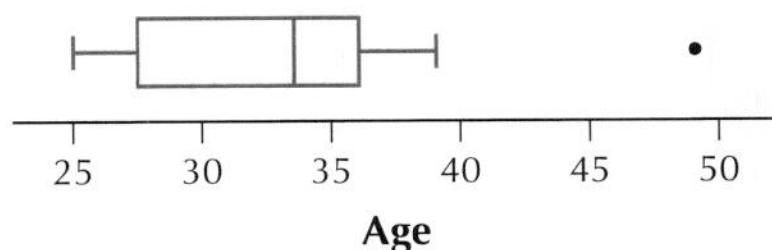


**15.** 17.5
**17.** Q1 − 1.5 * IQR = 41.75 and Q3 + 1.5 * IQR = 111.75. There are no values in the data set that lie outside of this interval so there are no outliers.
**19.** For WMU: min = 60, Q1 = 70, median = 70, Q3 = 70, max = 75. For NCU: min = 66, Q1 = 67, median = 70, Q3 = 70, max = 72; probably not.
**21.** The boxplots look so unusual because each data set only has 5 numbers.
**23.** IQR = 24.1, which is the spread of the middle 50% of the data.
**25.**

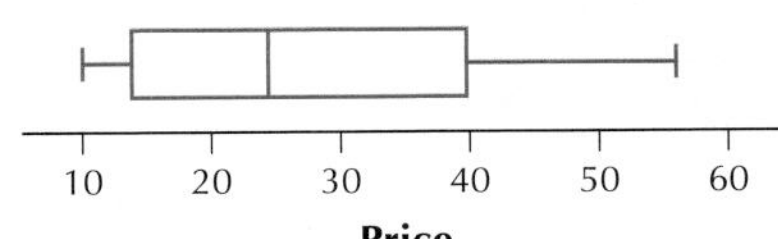


**27.** IQR = 0.22, which is the spread of the middle 50% of the data.
**29.**

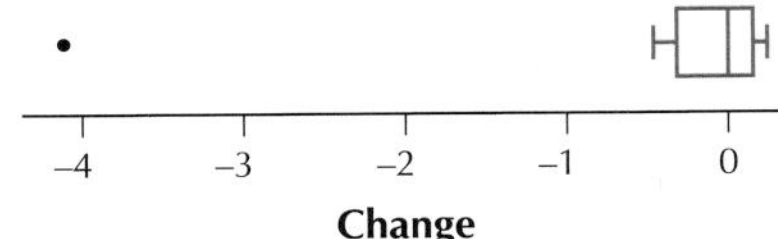


**31.** IQR = 10, which is the spread of the middle 50% of the data.
**33.**

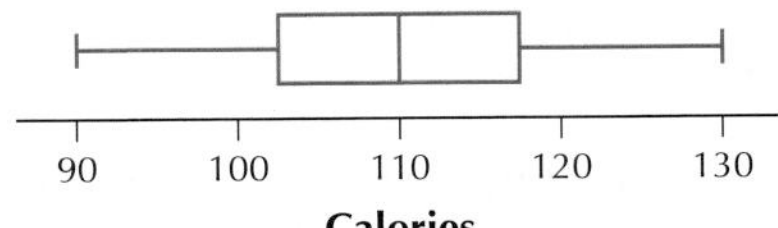


**35.** The $z$-score is 1.79, so Basic 4 is not an outlier. Q1 − 1.5 IQR = 90 and Q3 + 1.5 IQR = 130. Since Basic 4 has 130 calories per serving, it is an outlier by the robust method.
**37.** IQR = 4,300,000, which is the spread of the middle 50% of the data.
**39.**

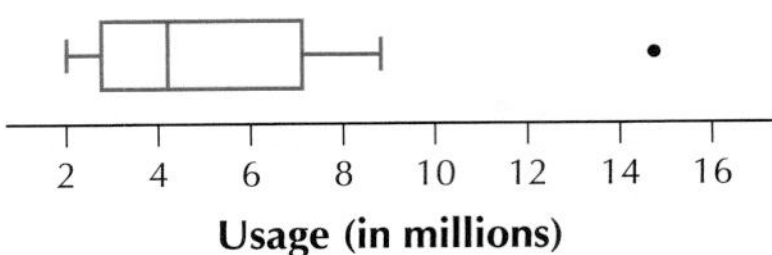


**41.** The $z$-score is 2.87, so echinacea usage is moderately unusual but not an outlier. The robust method indicates that echinacea usage is an outlier.
**43.** IQR = $7850, which is the spread of the middle 50% of the data.
**45.**

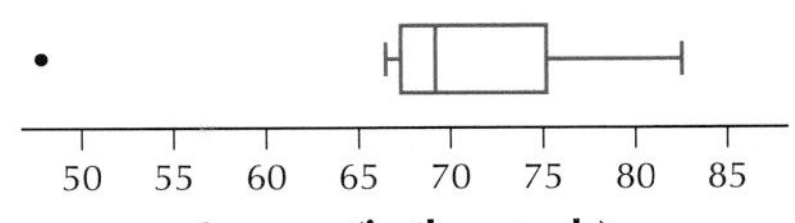


**47.** The $z$-score is −2.87, so the median income for West Virginia is moderately unusual but not an outlier. The robust method indicates the median income is an outlier.
**49.** Mean = 1.784 mg, standard deviation = 3.138 mg, min = 0.000 mg, Q1 = 0.300 mg, median = 0.800 mg, Q3 = 1.700 mg, max = 37.600 mg. Range = 37.600 mg − 0.000 mg = 37.600 mg. IQR = 1.700 mg − 0.300 mg = 1.400 mg
**51.** The boxplot is very right-skewed.

## Chapter 3 Review

**1.** Mean number of cases = 676,600, median number of cases = 536,500. The distribution should be right-skewed.
**3.** The median since there are outlying values in the data set and the data set is right-skewed.
**5.** **(a)** The mode, since the value with the largest frequency is unaffected by the deletion of values less than 60 calories. **(b)** The median, since the center of the data will be shifted slightly to the right. **(c)** The mean, since its computation uses all the values.
**7.** The mean, median, and mode will all be doubled.
**9.** **(a)** Range = 15; standard deviation = 5.48. **(b)** Adding a positive constant to each value in a data set will not change the value of the original range or standard deviation. **(c)** (i) Stayed the same. (ii) Yes.
**11.** **(a)** Increase because the frequency is a large, outlying value relative to the rest of the data. **(b)** Range with "the" = 7206. Standard deviation with "the" = 2101. The above values bear out the intuition that the variability will increase.
**13.** Estimated mean = 3.8042. Estimated standard deviation = 0.7903.
**15.** Estimated mean = 41.2055. Estimated standard deviation = 10.7061.
**17.** 1.78545 mg/g.
**19.** −0.41
**21.** −0.55
**23.** **(a)** 166 **(b)** Zero **(c)** It is impossible to hit a negative number of triples. **(d)** No, there will not be batters in the low end of the triples distribution; see answer **(c)**.
**25.** Between 2.5% and 16%.
**27.** Since we don't know the distribution we can't find the 50th percentile.
**29.** The 36th percentile lies between 65 and 70 mph.
**31.** Min = 1.2292, Q1 = 1.45295, median = 1.78545, Q3 = 2.86575, max = 12.5294.
**33.** **(a)** The $z$-score for the cholesterol level of the yolk of a raw egg is 3.63, so it is an outlier. **(b)** Outliers are 12.5294 and 6.3000; $z$-score method identified only 12.5294 as an outlier.

**35.** IQR = 23, which is the spread of the middle 50% of the data set.
**37. (a)** No outliers **(b)** No outliers **(c)** Yes

## Chapter 3 Quiz

**1.** True
**2.** False
**3.** False
**4.** False
**5.** True
**6.** 2
**7.** outlier
**8.** center
**9.** mean
**10.** average
**11.** deviation
**12.** squared deviations
**13.** median, Q2—second quartile
**14.** robust measures
**15.** mode
**16.** Zero
**17.** Class midpoint
**18. (a)** Left-skewed **(b)** Right-skewed **(c)** Symmetric unimodal
**19. (a)** Mean = 87,453 **(b)** Median = 98,008
**20. (a)** Range = 86,910 **(b)** Standard deviation = 33,857
**21.** Typically, the number of passengers differs from the mean of 87,453 by 33,857.
**22.** Estimated mean = 332.5; estimated standard deviation = 73.101.
**23.** Estimated mean = 61.6527; estimated standard deviation = 18.4518.
**24. (a)** 0 **(b)** 1 **(c)** 2 **(d)** −1 **(e)** −2
**25. (a)** 1.5 **(b)** −1 **(c)** 1 **(d)** −1.5 **(e)** 0
**26. (a)** About the 2.5th percentile. **(b)** (i) At least 75% (ii) Can't be done since $k = 1$.
**27. (a)** 60 **(b)** between 34% and 81.5% **(c)** No, furthermore we must assume that one of the values of $k$ is less than 1. **(d)** Between 2.5% and 16%.
**28. (a)** 501.5 **(b)** 512 **(c)** 518
**29.** IQR = 16.5
**30.** Min = 499, Q1 = 501.5, median = 512, Q3 = 518, max = 523.
**31.** Q1 − 1.5 * IQR = 476.75 and Q3 + 1.5 * IQR = 542.75. All the SAT scores lie between 476.75 and 542.75 so there are no outliers.
**32.**

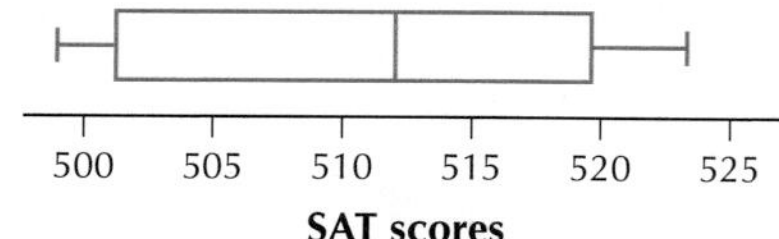


# CHAPTER 4

## Section 4.1

**1.** Answers will vary.
**3.** The headings of the columns and the rows of a crosstabulation are categories, not numbers. We could recode the variables in order to use crosstabulation by dividing the numbers up into classes, as we did for grouped frequency distributions. Then we could code each class as a category.

**5.**

| | Beverages | Food | Household/ personal | Tobacco | All |
|---|---|---|---|---|---|
| Glass | 2 | 0 | 0 | 0 | 2 |
| Metal | 0 | 0 | 2 | 0 | 2 |
| Paper | 0 | 0 | 1 | 3 | 4 |
| Plastic | 1 | 2 | 1 | 0 | 4 |
| All | 3 | 2 | 4 | 3 | 12 |

**7.**

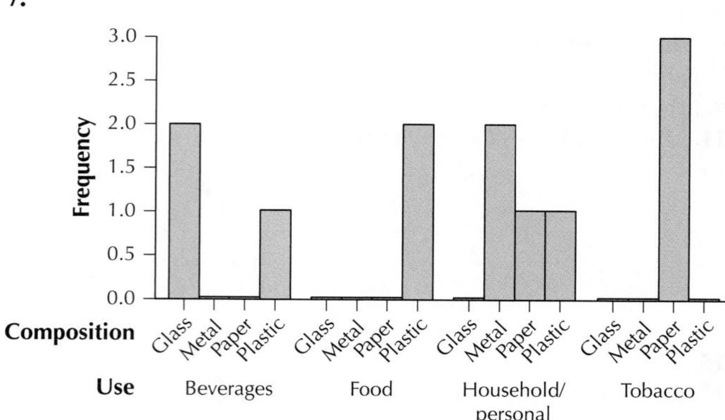


**9.**

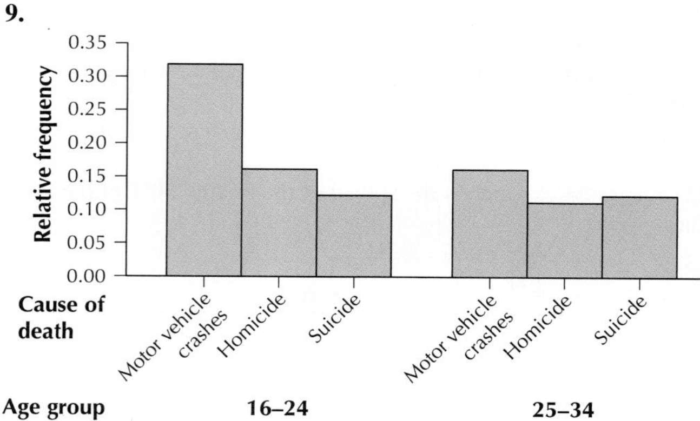


**11. (a)**

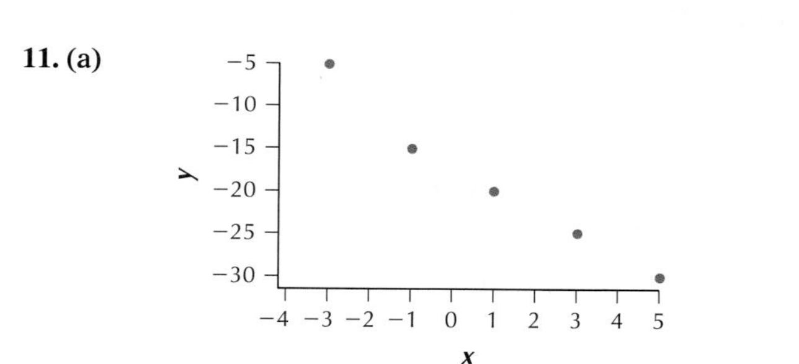


**(b)** Negative relationship. **(c)** As $x$ increases, $y$ decreases. Smaller values of $x$ are associated with larger values of $y$; larger values of $x$ are associated with smaller values of $y$.

**13. (a)**

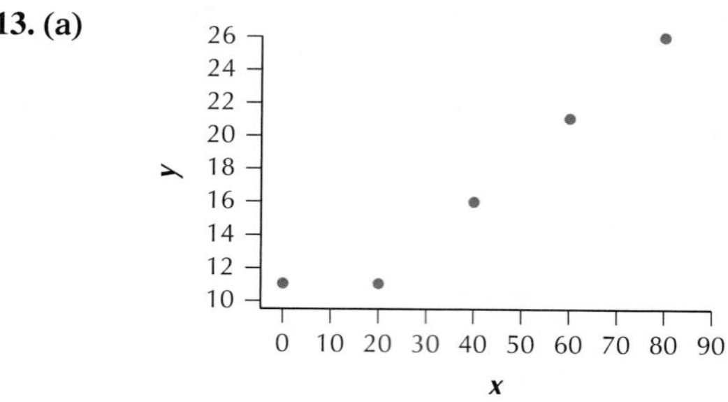

**(b)** Positive relationship. **(c)** As $x$ increases, $y$ increases. Smaller values of $x$ are associated with smaller values of $y$; larger values of $x$ are associated with larger values of $y$.

**15. (a)**

| | Freshman | Junior | Senior | Sophomore | All |
|---|---|---|---|---|---|
| Art | 0 | 1 | 0 | 1 | 2 |
| Business | 0 | 1 | 0 | 2 | 3 |
| Communication | 1 | 0 | 0 | 2 | 3 |
| Economics | 0 | 0 | 2 | 0 | 2 |
| English | 4 | 0 | 0 | 1 | 5 |
| History | 0 | 3 | 0 | 0 | 3 |
| Philosophy | 0 | 0 | 1 | 0 | 1 |
| Political Science | 0 | 0 | 1 | 0 | 1 |
| Psychology | 0 | 2 | 0 | 3 | 5 |
| All | 5 | 7 | 4 | 9 | 25 |

No apparent relationship; no patterns

**(b)**

| Tally for Discrete Variables: Major | |
|---|---|
| **Major** | **Count** |
| Art | 2 |
| Business | 3 |
| Communication | 3 |
| Economics | 2 |
| English | 5 |
| History | 3 |
| Philosophy | 1 |
| Political Science | 1 |
| Psychology | 5 |
| N = | 25 |

| Tally for Discrete Variables: Class | |
|---|---|
| **Class** | **Count** |
| Freshman | 5 |
| Junior | 7 |
| Senior | 4 |
| Sophomore | 9 |
| N = | 25 |

**17.**

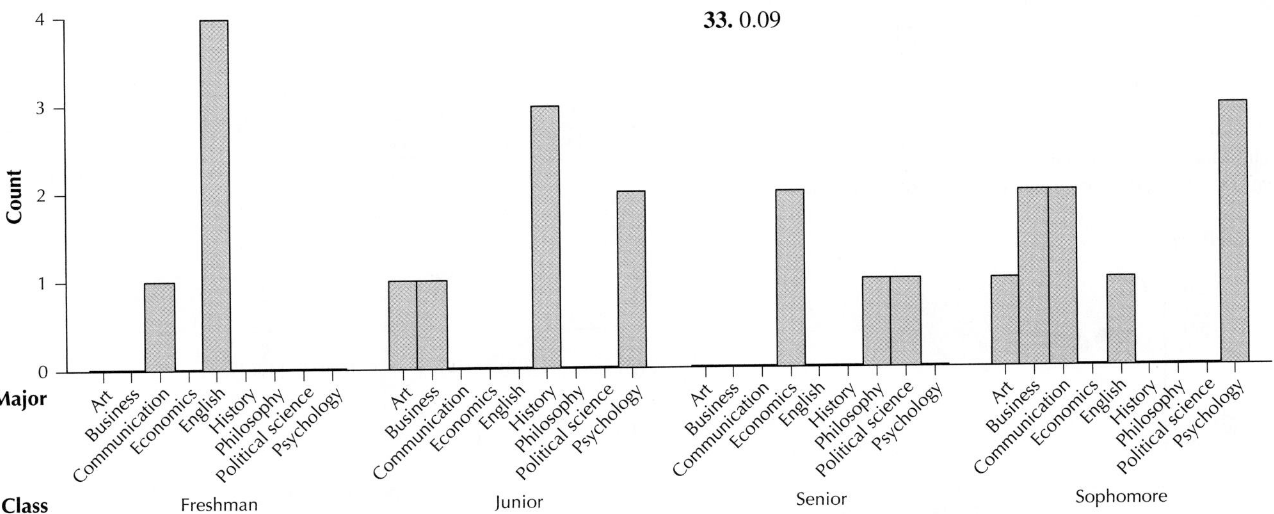


**19.** Missing values are in red.

| "How much do you enjoy shopping?" | Gender | | |
|---|---|---|---|
| | **Male** | **Female** | **Total** |
| **A lot** | 388 | 950 | 1338 |
| **Some** | 528 | 673 | 1255 |
| **Only a little** | 662 | 497 | 1159 |
| **Not at all** | 497 | 220 | 717 |
| **Don't know/refused** | 20 | 25 | 45 |
| **Total** | 2149 | 2365 | 4514 |

**21. (a)** Women **(b)** Women **(c)** Men **(d)** Men

**23. (a)** .7100 **(b)** .5363 **(c)** .4288 **(d)** .3068

**25. (a)**

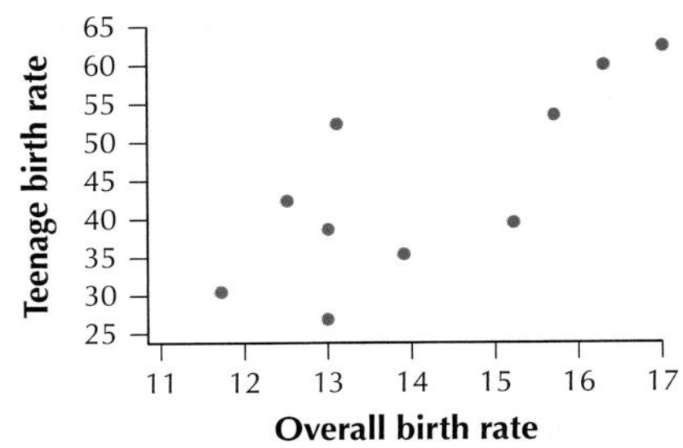


**(b)** The relationship between overall birth rate and teenage birth rate appears to be positive.

**27. (a)**

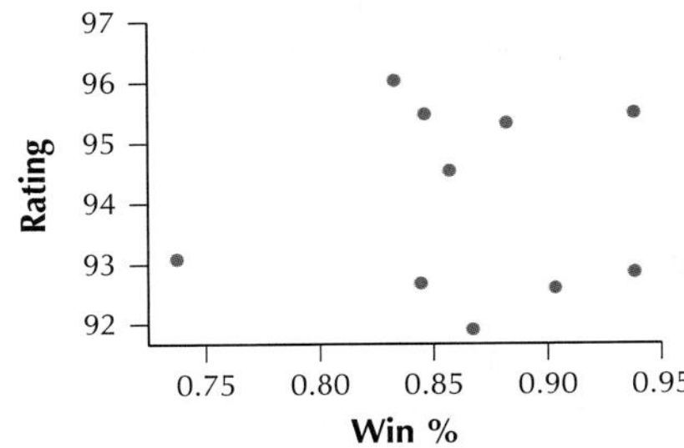


**(b)** There appears to be no relationship between a team's winning percentage and a team's power rating.

**29.** No; the scatterplot indicates that there is no relationship between a player's age and a player's batting average.

**31.** There appears to be a positive relationship between the number of words and the number of sentences in the *New Yorker* ads.

**33.** 0.09

**35.**

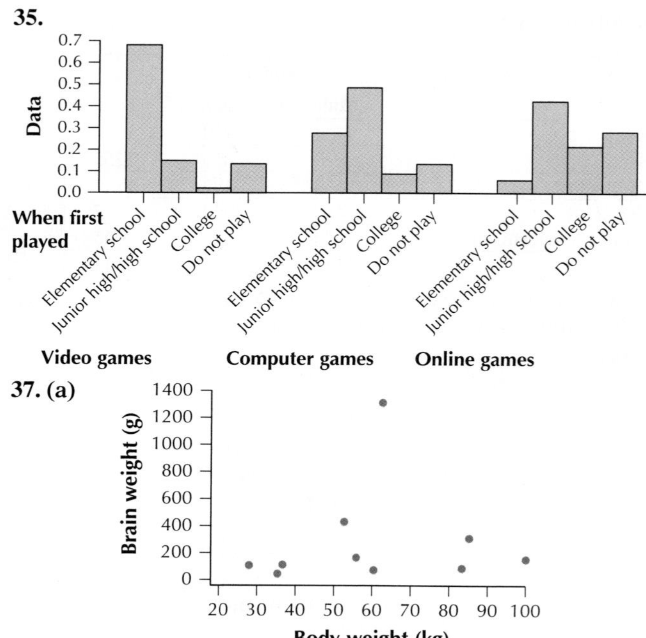


**37. (a)**

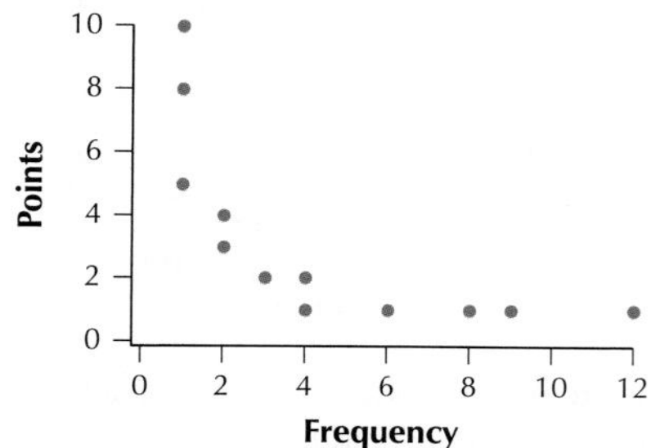


**(b)** There appears to be no relationship between body weight and brain weight. **(c)** Yes **(d)** Humans

**39.** There seems to be a negative relationship between the points, but it is not linear.



**41.** Missing values are in red.

**Visited Memorial Web Site**

| Student Status | Yes | No | Total |
|---|---|---|---|
| Full time | 45 | 171 | 216 |
| Part time | 53 | 189 | 242 |
| No | 371 | 1629 | 2000 |
| Total | 469 | 1989 | 2458 |

**43.** People who weren't students; full-time students

**45.** The relationship is not apparent.

**47.** Low values of $x$ and high values of $x$ are both associated with the values of $y$.

**49.** Answers will vary.

**51. (a)**

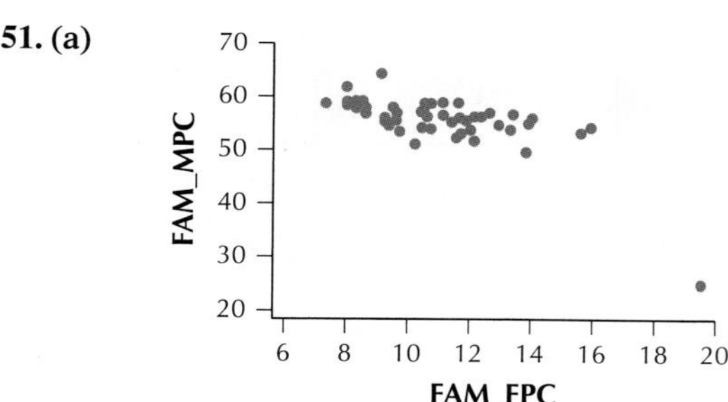


**(b)** There is a negative relationship between the variables.

**53. (a)**

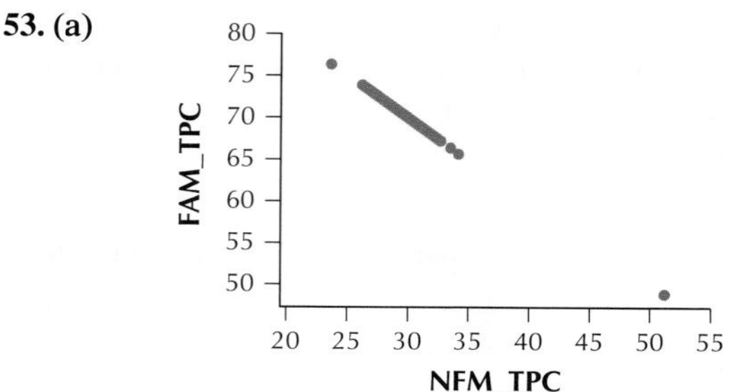


**(b)** Negative. The higher the percentage of nonfamily households, the lower the percentage of family households. **(c)** It is almost a perfect straight line. Two variables usually don't have this perfect of a linear relationship.

## Section 4.2

**1.** Scatterplot

**3.** Between –1 and 1, inclusive

**5.** i

**7.** ii

**9.** iv

**11. (a)** (6,8), (7,7), (8,7), (9,8), (10,8), (11,8), (12,7), (13,8), (14,8), (15,7) **(b)** Minitab: Correlations: $x$, $y$; Pearson correlation of $x$ and $y = 0.000$. TI-84: $r = 0$

**13. (a)** (1,11), (2,9), (3,7), (4,7), (5,6), (6,6), (7,4), (8,3), (9,3), (10,1) **(b)** Minitab: Correlations: $x$, $y$; Pearson correlation of $x$ and $y = -0.978$. TI-84: $r = -0.9781316853$

**15.**

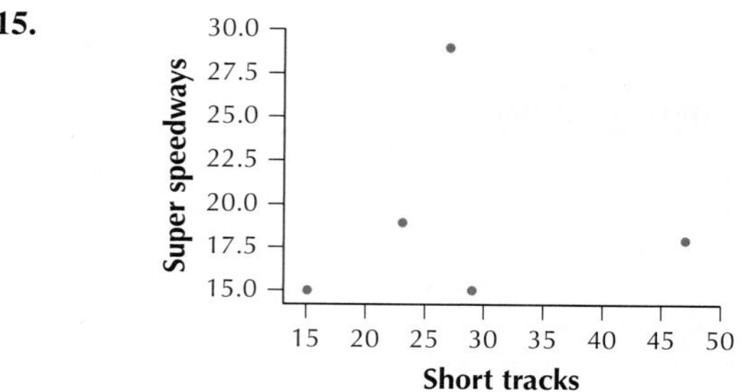


**17. (a)** $\bar{x} = 28.2, \bar{y} = 19.2$

**(b)**

| $x$ | $y$ | $x - \bar{x}$ | $(x - \bar{x})^2$ | $y - \bar{y}$ | $(y - \bar{y})^2$ | $(x - \bar{x})(y - \bar{y})$ |
|---|---|---|---|---|---|---|
| 47 | 18 | 18.8 | 353.44 | −1.2 | 1.44 | −22.56 |
| 27 | 29 | −1.2 | 1.44 | 9.8 | 96.04 | −11.76 |
| 15 | 15 | −13.2 | 174.24 | −4.2 | 17.64 | 55.44 |
| 29 | 15 | 0.8 | 0.64 | −4.2 | 17.64 | −3.36 |
| 23 | 19 | −5.2 | 27.04 | −.2 | .04 | 1.04 |
| | | | $\Sigma(x - \bar{x})^2 = 556.8$ | | $\Sigma(y - \bar{y})^2 = 132.8$ | $\Sigma(x - \bar{x})(y - \bar{y}) = 18.8$ |

**(c)** $s_x = 11.79830496$; $s_y = 5.761944116$

**(d)** $r = \dfrac{\sum(x - \bar{x})(y - \bar{y})}{(n-1)s_x s_y} = \dfrac{18.8}{(5-1)(11.79830496)(5.761944116)}$

$= 0.0691367879$

**(e)** Minitab: Correlations: Short Tracks, Super Speedways; Pearson correlation of Short Tracks and Super Speedways = 0.069. TI-84: $r = 0.691367879$.

**19.** Positively correlated

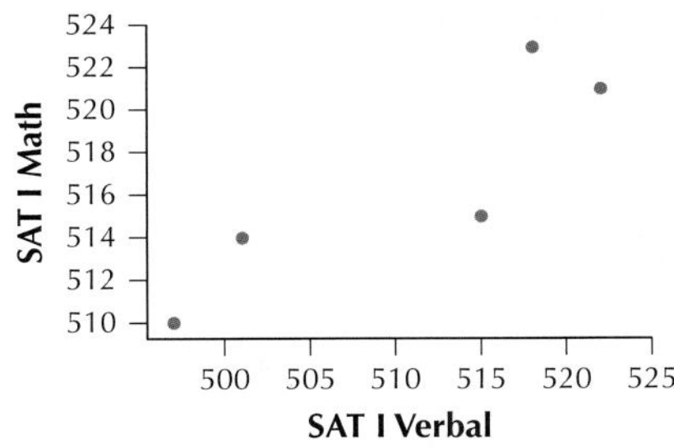


**21.** SAT I Verbal scores and SAT I Math scores are positively correlated. As the SAT I Verbal score increases, the SAT I Math score increases. Yes.
**23.** Minitab: Correlations: Body Weight (kg), Brain Weight (g); Pearson correlation of body weight (kg) and brain weight (g) = 0.099. TI-84: $r = 0.0992032294$
**25.** Positive
**27.** The number of calories in a serving of breakfast cereal and the amount of sugar in a serving of breakfast cereal are positively correlated. As the amount of sugar in a serving of breakfast cereal increases the number of calories in a serving of the breakfast cereal tends to increase.
**29.** A player's age and a player's batting average are not correlated. As a player's age increases the player's batting average tends to remain unchanged.
**31.** Positive; positive
**33. (a)** 18–65 group **(b)** No
**35. (a)** $\bar{x} = 9.16666667$; $\bar{y} = 5.5$; $s_x = 9.432214303$; $s_y = 4.679743583$; $r = -0.7453498716$. Minitab: Correlations: Hip-Hop CDs owned, Country CDs owned; Pearson correlation of Hip-Hop CDs owned and Country CDs owned = −0.745. TI-84: $r = -0.7453498716$. **(b)** Yes **(c)** The variables number of Hip-Hop CDs owned and number of Country CDs owned are negatively correlated. As the number of Country CDs owned increases, the number of Hip-Hop CDs owned decreases.
**37. (a)** Yes **(b)** The variables birth weight and time born are not correlated. As the time born increases, the birth weight tends to remain the same.
**39, 41.** Answers will vary.
**43. (a)** Positively correlated **(b)** Negatively correlated **(c)** Not correlated
**45.** Answers will vary.

## Section 4.3

**1.** To approximate the relationship between two numerical variables using the regression line and the regression equation.
**3.** We can find the predicted value of $y$ by plugging a given value of $x$ into the regression equation and simplifying.
**5.** Extrapolation is the process of making predictions based on $x$-values that are beyond the range of the $x$-values in our data set.
**7. (a)** $b_1 = 0.0337643678$ **(b)** $b_0 = 18.24784483$ **(c)** $\hat{y} = b_0 + b_1x = 8.24784483 + 0.0337643678\,x$. The estimated value of the number of Super Speedways won is 18.24784483 plus 0.0337643678 times the number of Short Tracks won.
**9. (a)** $\cong 19$ **(b)** $x = 50$ is not in the range of the $x$-values of the data set. **(c)** $\cong 19$
**11. (a)** The slope $b_1 = 0.4264339152$ means that the estimated SAT I Math Score increases by 0.4264339152 points for every increase of one point in the SAT I Verbal Score. **(b)** The $y$-intercept $b_0 = 298.8628429$ means that the estimated SAT I Math Score is 298.8628429 when the SAT I Verbal Score is 0. **(c)** The SAT I Verbal Score can't be 0, so this situation will never happen.
**13. (a)** $\hat{y} = 193.5 + 1.595x$ **(b)** The estimated brain weight of a mammal is 193.5 grams plus 1.595 times the body weight.
**15. (a)** 273.25 grams **(b)** 353 grams **(c)** The value $x = 200$ kg is outside the range of the $x$-values of the data set, which is between 27.66 kg and 100 kg.
**17.** Curved
**19. (a)** 10.5 **(b)** In a state with 0 households, 10.5% of the households are headed by women. Since all states have households, the value $x = 0$ would not occur. **(c)** This estimate would be considered extrapolation since the value of $x = 0$ is outside the range of $x$-values in the data set. **(d)** 0.000000282 **(e)** For each increase of one household, the percentage of households headed by women increases by 0.000000282. **(f)** (Percentage of households headed by women) = 10.5 + 0.000000282 (Total number of households). The estimated percentage of households headed by women equals 10.5 plus 0.000000282 times the total number of households. **(g)** Positive since the slope is positive.
**21. (a)** 12.474% **(b)** $x = 100{,}000$ is not in the range of the $x$-values in the data set.
**23. (a)** 1.032 **(b)** The estimated percent change in the stock prices selected by the Darts increases by 1.032% for each unit percent increase in the Dow Jones Industrial Average. **(c)** Darts = −2.49 + 1.032 DJIA. The estimated percent change in the stock prices selected by the Darts is equal to −2.49 plus 1.032 times the percent change in the Dow Jones Industrial Average. **(d)** Positive since the slope of the regression line is positive.
**25. (a)** 20.214% **(b)** −12.81% **(c)** $x = -22$ is not in the range of the $x$-values of the data set.
**27.** The slope of the regression line will decrease from 0.9865 to 0.7027 or less because the low temperature of 10 degrees is below all the other low temperatures and the high temperature of 50 degrees or more is higher than the other $y$-values corresponding to smaller $x$-values .
**29, 31, 33.** Answers will vary.

## Chapter 4 Review

**1.** 0.6612
**3.** 0.0246
**5.** Answers will vary. May have clustered bar graph by happiness of marriage or clustered bar graph by sex.
**7. (a)** The relationship in each of the scatterplots in Exercise 6 is positive. **(b)** The $y$ variable increases as the $x$ variable increases. **(c)** Low values of $x$ are associated with low values of $y$ and high values of $x$ are associated with high values of $y$.
**9.** Zero
**11.** Positive
**13. (a)** 216 **(b)** 192 **(c)** Since all the points lie on a straight line, this $x$ variable is extremely useful in predicting this $y$ variable.
**15.** $b_0 \cong 0$

## Chapter 4 Quiz

**1.** True
**2.** False
**3.** False
**4.** two-way tables; contingency tables
**5.** estimate
**6.** 0
**7.** unit
**8.** clustered bar graphs
**9.** extrapolation
**10.** negative

**11.** Missing values are in red.

| | Exciting | Routine | Dull | Total |
|---|---|---|---|---|
| **Protestant** | 264 | 326 | 33 | 623 |
| **Catholic** | 107 | 128 | 6 | 241 |
| **Jewish** | 12 | 8 | 0 | 20 |
| **None** | 38 | 29 | 2 | 69 |
| **Other** | 9 | 5 | 0 | 14 |
| **Total** | 430 | 496 | 41 | 967 |

**12.**

| | Exciting | Routine | Dull |
|---|---|---|---|
| **Protestant** | 61.40% | 65.73% | 80.49% |
| **Catholic** | 24.88% | 25.81% | 14.63% |
| **Jewish** | 2.79% | 1.61% | 0% |
| **None** | 8.84% | 5.85% | 4.88% |
| **Other** | 2.09% | 1.01% | 0% |
| **Total** | 100.00% | 100.01% | 100.00% |

**13.** Protestant
**14.** Protestant
**15.** Protestant
**16.**

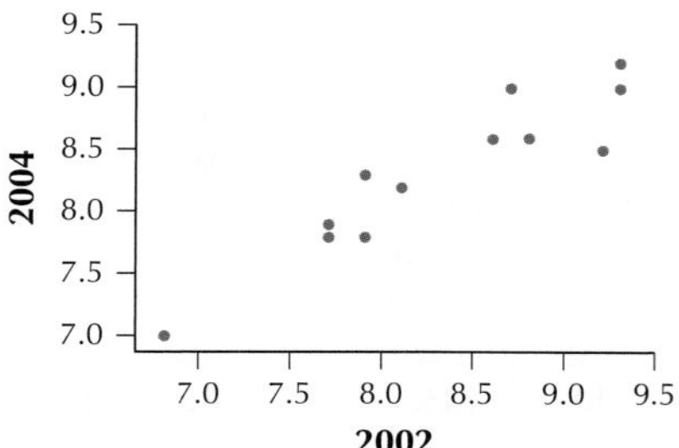


**17.** Positive
**18.** As the percent of violent crimes committed in 2002 increases, the percent of violent crimes committed in 2004 increases.
**19. (a)**

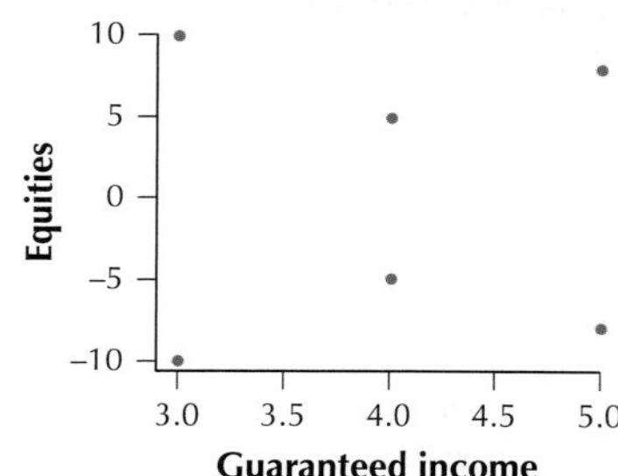


**(b)** No apparent relationship **(c)** Near zero; near zero
**20. (a)** $\bar{x} = 4$; $\bar{y} = 0$; $s_x = 0.894427191$; $s_y = 8.694826048$;

$$r = \frac{\sum(x - \bar{x})(y - \bar{y})}{(n - 1)s_x s_y} = \frac{0}{(6 - 1)(0.894427191)(8.694826048)} = 0$$

**(b)** Yes **(c)** There is no apparent relationship between the net price change for equities and the net price change for guaranteed income investments. As the net price change for guaranteed income investments increases, the net price change for equities remains the same.
**21. (a)** 0 **(b)** 0; all the estimated equities net price changes are 0. **(c)** Not very useful since all of the estimates are 0.
**22. (a)** $y$ intercept $b_1 = 97.1$ **(b)** A breakfast cereal with 0 grams of fat will have 97.1 calories per serving. Yes, because there are other ingredients in cereal besides fat and they have calories. **(c)** No, since $x = 0$ is an $x$-value in the data set. **(d)** The number of calories in a serving of breakfast cereal increases by 9.65 calories for every increase of 1 gram of fat. **(e)** Calories = 97.1 + 9.65 fat. The estimated number of calories in a serving of breakfast cereal equals 97.1 plus 9.65 times the number of grams of fat. **(f)** Positive, since the slope of the regression line is positive.
**23. (a)** Cereal A has 2 (9.65) = 19.3 more calories per serving than Cereal B has. **(b)** Cereal C has 3 (9.65) = 28.95 fewer calories per serving than Cereal D has.
**24. (a)** 135.7 **(b)** The value of $x = 10$ is outside the range of $x$ values in the data set.

# CHAPTER 5

## Section 5.1

**1.** Answers will vary; chance, likelihood.
**3.** Answers will vary.
**5.** The experiment has equally likely outcomes.
**7.** We consider all available information, tempered by our experience and intuition, and then assign a probability value that expresses our estimate of the likelihood that the outcome will occur.
**9.** First find out how many students are at your college and find out how many of them like hip-hop music. Then calculate the relative frequency of students who like hip-hop music.
Use the relative frequency method.
**11.** No, probability for females is greater than 1.
**13.** No, sum of probabilities is greater than 1.
**15.** No, sum of probabilities is less than 1.
**17.** 1/2
**19.** 1/3
**21.** 0
**23.** 1/4
**25.** 1/2
**27.** {HHH, HHT, HTH, THH, TTH, THT, HTT, TTT}
**29.** Choose a branch for the first toss, then choose a branch for the second toss, then choose a branch for the third toss. This will give one of the possible outcomes. If we follow all possible choices of branches for each toss, we get all the outcomes in the sample space.
**31.** outcome; event; event; event; event; event
**33. (a)** a sum of 7 **(b)** 1/6
**35. (a)** 0.33 **(b)** 0.67 **(c)** Relative frequency method
**37. (a)** 0.025 **(b)** Relative frequency method **(c)** The events "cutting class" and "not cutting class" are not equally likely.
**39. (a)**

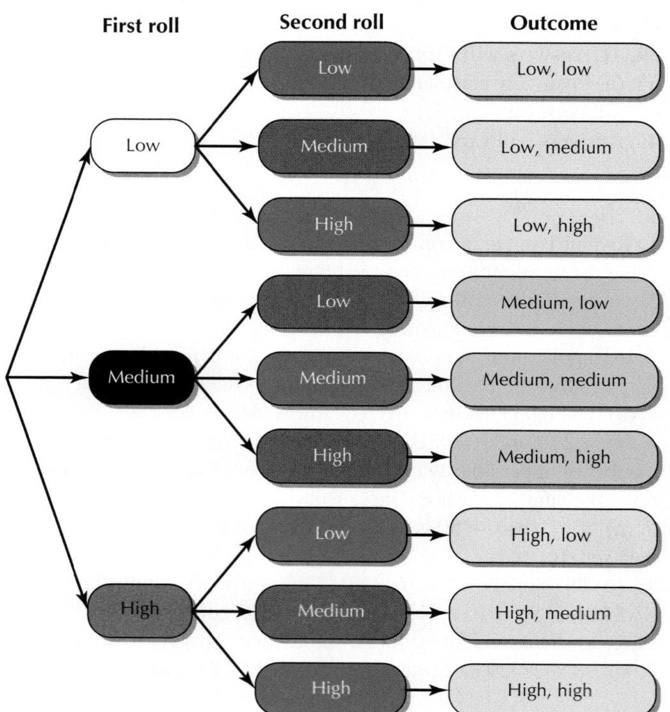

**(b)** {Low and Low, Low and Medium, Low and High, Medium and Low, Medium and Medium, Medium and High, High and Low, High and Medium, High and High} **(c)** 1/9; Classical

**41. (a)**

| | Frequency | Relative frequency |
|---|---|---|
| **Girls** | 18 | 18/44 = 0.4091 |
| **Boys** | 26 | 26/44 = 0.5909 |
| **Total** | 44 | 44/44 = 1.0000 |

**(b)**

| Outcome | Probability |
|---|---|
| Girl | 18/44 = 0.4091 |
| Boy | 26/44 = 0.5909 |

**(c)** Both $P$ (Girl) = 18/44 = 0.4091 and $P$ (Boy) = 26/44 = 0.5909 are between 0 and 1. $P$ (Girl) + $P$ (Boy) = 18/44 + 26/44 = 0.4091 + 0.5909 = 44/44 = 1.0000.
**43.** Answers will vary.
**45. (a)** E, Z, yes **(b)** consonant **(c)** 5/26 **(d)** 1/2 **(e)** Half of the most popular letters are vowels, but only 5/26 of all of the letters are vowels. **(f)** T, N, R, S, D
**47. (a)** 5/18 **(b)** 13/18 **(c)** $1.39

**49. (a)**

| Type of music | Probability |
|---|---|
| Hip-hop/rap | 0.27 |
| Pop | 0.23 |
| Rock/punk | 0.17 |
| Alternative | 0.07 |
| Christian/gospel | 0.06 |
| R&B | 0.06 |
| Country | 0.05 |
| Techno/house | 0.04 |
| Jazz | 0.01 |
| Other | 0.04 |

**(b)** All the probabilities are between 0 and 1 and the sum of the probabilities is 1. **(c)** Yes **(d)** Answers will vary. **(e)** As the sample size increases the relative frequencies approach the probabilities.
**51. (a)–(d)** Answers will vary.

## Section 5.2

**1.** Two events are mutually exclusive if they have no outcomes in common.
**3.** Yes
**5.** Answers will vary; 1 – $P$ (will rain); they are complementary events.
**7.** 25/36
**9.** 0
**11.** 25/36
**13.** 4/13
**15.** 1/2
**17.** 51/52
**19.** 1/2
**21.** 9/52
**23.** 3/52
**25.** 7/8
**27.** 5/8
**29.** 3/8
**31.** 3/4
**33.** 1/8
**35. (a)** 1/3 **(b)** 2/3
**37. (a)** 1/10 **(b)** 1/2
**39. (a)** 1/9 **(b)** 1/18 **(c)** 1/6
**41. (a)** 1/2 **(b)** 7/10 **(c)** 9/10
**43. (a)** 1/2 **(b)** 7/13 **(c)** 11/26 **(d)** 0 **(e)** 3/4
**45. (a)** 1/4 **(b)** 9/12 = 3/4; 1 − 1/4 = 3/4
**47. (a)** 0.85 **(b)** 0.50 **(c)** 0.65 **(d)** 0
**49. (a)** 1/3 **(b)** Relative frequency method **(c)** 2/3
**51. (a)** 0.57 **(b)** 0.55
**53. (a)** 0.4535 **(b)** 0.1190 **(c)** 0.0532 **(d)** 0.5193
**55. (a)** 0 **(b)** 0.0528

## Section 5.3

**1. (a)** Yes. **(b)** The probability of winning the football game depends on whether or not the star quarterback can play in the game.
**3.** For $P(A \mid B)$, we assume that the event $B$ has occurred, and now need to find the probability of event $A$, given event $B$. On the other hand, for $P(A \cap B)$, we do not assume that event $B$ has occurred, and instead need to determine the probability that both events occurred.
**5.** Answers will vary.
**7. (a)** Independent; sampling with replacement **(b)** Dependent; sampling without replacement
**9.** 0.24
**11.** 0.4
**13.** 0.1
**15.** 0.2
**17.** 0.1667
**19.** 1/8
**21.** 1/32
**23.** 0.2
**25.** 0.5
**27.** 0.8
**29.** 0.7
**31.** Yes. $P(C \text{ and } D) = 0.21 = (0.7)(0.3) = P(C)P(D)$, $P(C \mid D) = 0.7 = P(C)$, and $P(D \mid C) = 0.3 = P(D)$.
**33.** $P(E \text{ and } F)$
**35.** 1/6
**37.** 1/2
**39.** They are independent.
**41. (a)** 1/6 **(b)** 1/3 **(c)** $P$ (rolling a 6) = 1/6; $P$ (rolling a 6 | odd number) = 0
**43. (a)** 1/3 **(b)** 1/10 **(c)** 1/6 **(d)** 1/15
**45.** No; $P(C) = 1/2 \neq 5/9 = P(C \mid F)$, $P(C) = 1/2 \neq 5/12 = P(C \mid M)$, $P(F \text{ and } C) = 1/3 \neq 3/10 = P(F)P(C)$, and $P(M \text{ and } C) = 1/6 \neq 1/5 = P(M)P(C)$.
**47. (a)** 16/49 **(b)** 2/7
**49. (a)** 0.000361 **(b)** 0.962361 **(c)** 0.9085 **(d)** 0.0559
**51. (a)** 0.055 **(b)** 0.3025 **(c)** 0.55 **(d)** 0.55 **(e)** 0.003025 **(f)** It is not possible since we are given only the percent of female students that are business majors, not the percent of all students that are business majors.
**53. (a)** 3/5 **(b)** 2/5 **(c)** 1/3 **(d)** 5/9 **(e)** 1/3
**55. (a)** 0.3430 **(b)** 0.3236
**57.** No, $P$ (more serious than Pearl Harbor | female) = 0.6752 ≠ 0.6667 = $P$ (more serious than Pearl Harbor) and $P$ (more serious than Pearl Harbor | male) = 0.6579 ≠ 0.6667 = $P$ (more serious than Pearl Harbor).

## Section 5.4

**1.** Tree diagram
**3.** In a permutation, order is important. In a combination, order is not important.
**5.** Answers will vary.
**7.** 720
**9.** 1
**11.** 1
**13.** 210
**15.** 6720
**17.** 100

**19.** 93,326,215,443,944,152,681,699,238,856,266,700,490,715, 968,264,381,621,468,592,963,895,217,599,993,229,915,608,941, 463,976,156,518,286,253,697,920,827,223,758,251,185,210,916, 864,000,000,000,000,000,000,000,000

**21.** 35

**23.** 165

**25.** 11

**27.** 1

**29.** ${}_7C_3 = \frac{7!}{3!\cdot 4!} = \frac{7!}{4!\cdot 3!} = {}_7C_4$

**31.** {Amy, Bob, Chris}, {Amy, Chris, Bob}, {Bob, Amy, Chris}, {Bob, Chris, Amy}, {Chris, Amy, Bob}, {Chris, Bob, Amy}, {Amy, Bob, Danielle}, {Amy, Danielle, Bob}, {Bob, Amy, Danielle}, {Bob, Danielle, Amy}, {Danielle, Amy, Bob}, {Danielle, Bob, Amy}, {Amy, Chris, Danielle}, {Amy, Danielle, Chris}, {Chris, Amy, Danielle}, {Chris, Danielle, Amy}, {Danielle, Amy, Chris}, {Danielle, Chris, Amy}, {Bob, Chris, Danielle}, {Bob, Danielle, Chris}, {Chris, Bob, Danielle}, {Chris, Danielle, Bob}, {Danielle, Bob, Chris}, {Danielle, Chris, Bob}. ${}_4P_3 = 24$

**33.** {Amy, Bob, Chris}, {Amy, Chris, Bob}, {Chris, Amy, Bob}, {Chris, Bob, Amy}, {Bob, Amy, Chris}, and {Bob, Chris, Amy} are all different permutations but the same combination.

**35.** $r!$

**37.** (a)

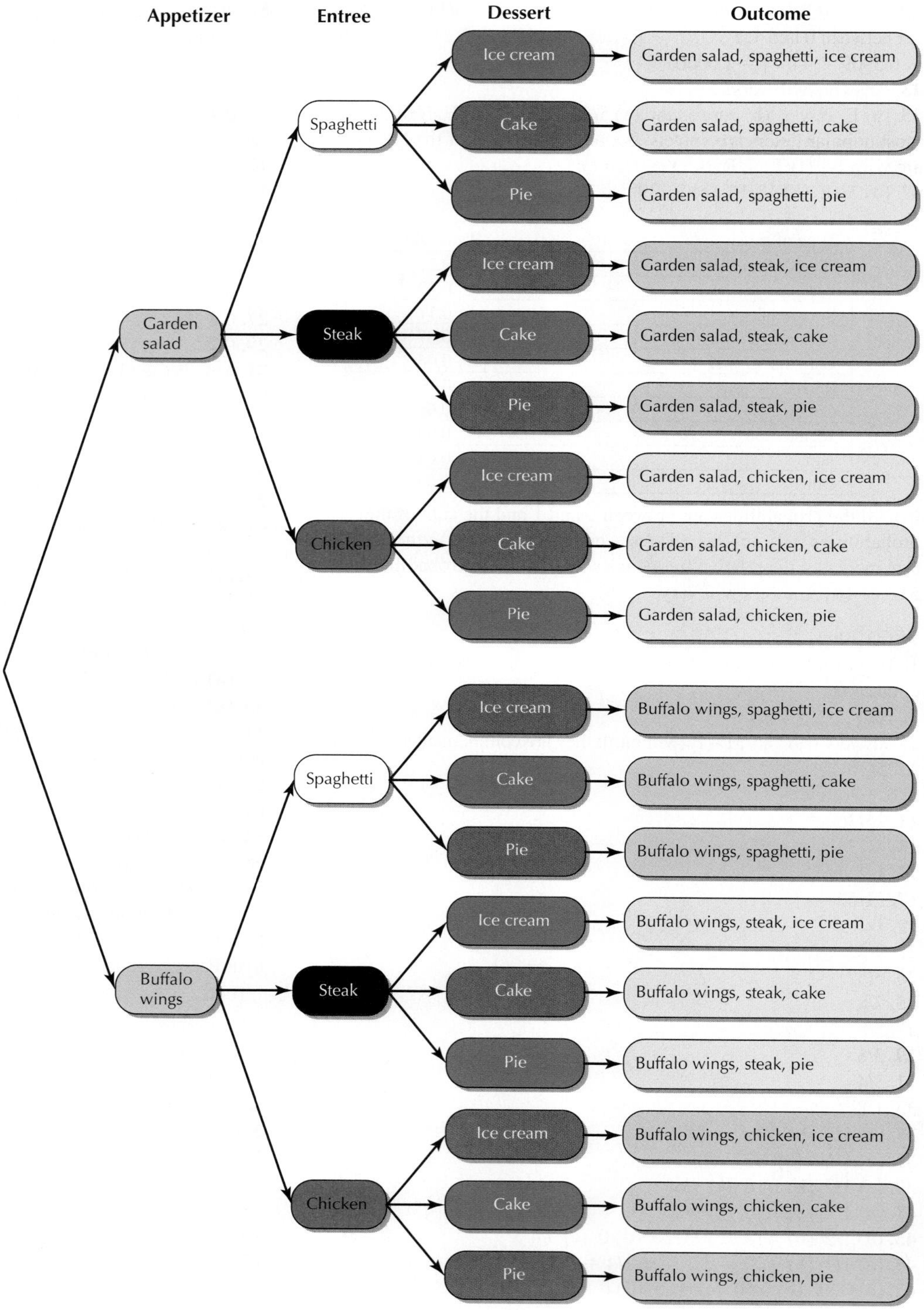

Appetizer
Entree
Dessert
Outcome
Garden salad
Spaghetti
Ice cream
Garden salad, spaghetti, ice cream
Cake
Garden salad, spaghetti, cake
Pie
Garden salad, spaghetti, pie
Steak
Ice cream
Garden salad, steak, ice cream
Cake
Garden salad, steak, cake
Pie
Garden salad, steak, pie
Chicken
Ice cream
Garden salad, chicken, ice cream
Cake
Garden salad, chicken, cake
Pie
Garden salad, chicken, pie
Buffalo wings
Spaghetti
Ice cream
Buffalo wings, spaghetti, ice cream
Cake
Buffalo wings, spaghetti, cake
Pie
Buffalo wings, spaghetti, pie
Steak
Ice cream
Buffalo wings, steak, ice cream
Cake
Buffalo wings, steak, cake
Pie
Buffalo wings, steak, pie
Chicken
Ice cream
Buffalo wings, chicken, ice cream
Cake
Buffalo wings, chicken, cake
Pie
Buffalo wings, chicken, pie

**(b)** 18
**39.** 362,880
**41.** 120
**43.** 720
**45.** 10
**47.** 300
**49.** 20
**51.** 184,756
**53.** 6720
**55.** **(a)** Combination; order is unimportant. **(b)** 260 **(c)** 0.0640
**57.** **(a)** 0.0265 **(b)** 0.0013 **(c)** 0.2652 **(d)** 0.4419

## Chapter 5 Review

**1.** 3/8
**3.** 0
**5.** 1/2
**7.** **(a)** 0.213 **(b)** 0.656 **(c)** 0
**9.** **(a)** Yes, since no student in the sample repeated twelfth grade. **(b)** 0.0290 **(c)** 0.0150
**11.** **(a)** 0.9269 **(b)** 0 **(c)** 1
**13.** **(a)** 0.3729 **(b)** 0.6271
**15.** **(a)** 1/2 **(b)** 1/3
**17.** **(a)** 5/18 **(b)** 5/12
**19.** 34,650
**21.** **(a)** Combination; order is not important. **(b)** 3 **(c)** 0.0196

## Chapter 5 Quiz

**1.** False
**2.** True
**3.** True
**4.** 0, 1
**5.** or, and
**6.** 0.5
**7.** permutation
**8.** 1
**9.** With replacement
**10.** Intersection of *A* and *B*
**11.** Multiplication Rule for Counting
**12.** 1/36
**13.** 1/36
**14.** 5/18
**15.** 1/6
**16.** 1/9
**17.** 8/9
**18.** 5/18
**19.** 1/18
**20.** 1/3
**21.** 0.2
**22.** 0.2125
**23.** **(a)** 1/4 **(b)** 3/13 **(c)** 1/13 **(d)** 1/2 **(e)** 1/52 **(f)** 1/26
**24.** **(a)** 0.3830 **(b)** 0.1654 **(c)** 0.1075 **(d)** 0.4409
**25.** **(a)** 0.5361 **(b)** 0.4639 **(c)** 0.0330
**26.** **(a)** 0.0215 **(b)** 0.0114
**27.** **(a)** 0.0402 **(b)** 0.0246
**28.** No, *P* (Not too happily married) = 0.0330 ≠ 0.0402 = *P* (Not too happily married | Female) and *P* (Not too happily married) = 0.0330 ≠ 0.0246 = *P* (Not too happily married | Male)
**29.** 4
**30.** **(a)** Permutation; the order in which the numbers are selected is important. **(b)** 6840 **(c)** 1/6840

# CHAPTER 6

## Section 6.1

**1.** Answers will vary.
**3.** Discrete: takes finite or a countable number of values that can be graphed as separate points on the number line; continuous takes infinitely many values that form an interval on the number line.
**5.** Discrete
**7.** Continuous
**9.** Discrete
**11.** {0, 1, 2, 3, 4, 5, 6, 7, 8, 9, 10, 11, 12, 13, 14, 15}
**13.** {0, 1, 2, 3, 4}
**15.** No, the probabilities don't add up to 1.
**17.** No, $P(X = 1)$ is negative.
**19.** **(a)** 0.30 **(b)** 0.06 **(c)** 0
**21.** **(a)** 2/3 **(b)** Gaining $10,000 **(c)** 1/3
**23.** Mean: $\mu = 3$; $Var(X) = 0.5$; $SD(X) = 0.7071$; yes
**25.** Mean: $\mu = 3.1$; $Var(X) = 1.29$; $SD(X) = 1.1358$; yes
**27.** About 5.4 games
**29.** **(a)** $\mu = 5.38$ games **(b)** Estimate is 0.02 games more than the actual mean.
**31.** **(a)**

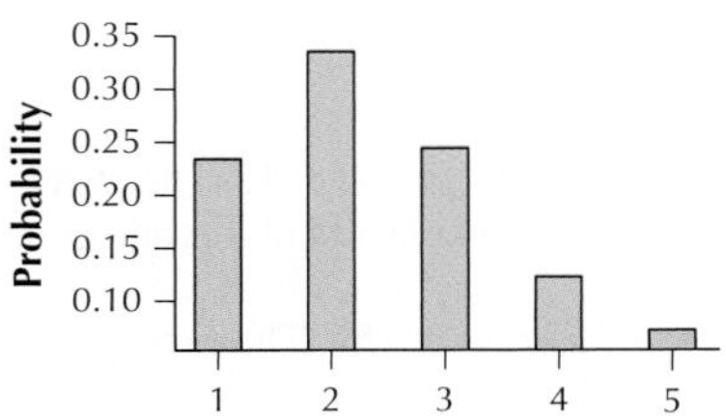


**(b)** 2 classes; 5 classes **(c)** $\mu = E(X) = 2.458$ classes **(d)** The "typical" number of courses taught by faculty at all degree-granting institutions of higher learning in the United States is 2.458.
**33.** **(a)** $SD(X) = 1.1689$ classes
**(b)** *Z*-score is 2.17; moderately unusual.
**35.** **(a)**

Probability
0.45
0.40
0.35
0.30
0.25
0.20
0.15
0.10
0.0 0.5 1.0 1.5 2.0 2.5 3.0
x = Number of vehicles

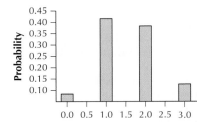

**(b)** 1 vehicle; 0 vehicles **(c)** $\mu = 1.547$ vehicle **(d)** The "typical" number of vehicles owned by residents of Florida is 1.547.
**37.** **(a)** $SD(X) = 0.8098$ **(b)** *Z*-score is −1.91; not unusual.
**39.** The mean is about 7.

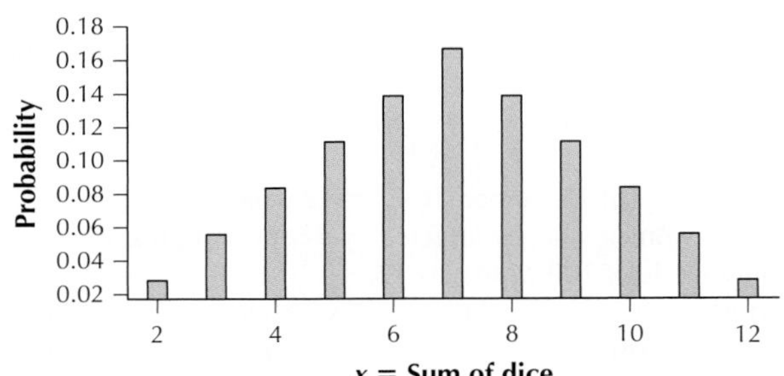


**41.** $SD(X) = 2.4152$.

**43.** $Z$-score is $-2.07$; moderately unusual. By symmetry, so is $X = 12$.
**45.** **(a)** Increases by $k$ **(b)** Remains unchanged

## Section 6.2

**1.** Not binomial; the events "Person A comes to party" and "Person B comes to party" may not be independent.
**3.** Not binomial; more than two different ethnicities.
**5.** Not binomial; more than two possible ages.
**7.** Not binomial; more than two possible total number of spots.
**9.** Binomial; $n = 3$, $X$ = number of even numbers, $p = 1/2$, $1 - p = 1/2$.
**11.** Binomial; $n = 2$, $X$ = number of games won, $p = 0.25$, $1 - p = 0.75$
**13.** $1/8 = 0.125$
**15.** $3/8 = 0.375$
**17.** $7/8 = 0.875$
**19.** 0.2461
**21.** 0.1216
**23.** $\mu = 5$, $\sigma^2 = 2.5$, $\sigma = 1.5811$. The expected number of heads in 10 tosses of a fair coin is 5.
**25.** $\mu = 2$, $\sigma^2 = 1.8$, $\sigma = 1.3416$. The expected number of people who are left-handed in a random sample of 20 people is 2.
**27.** 0.0741
**29.** 0.9614
**31.** $\mu = 0.765$, $\sigma^2 = 0.7306$, $\sigma = 0.8547$. The expected number of students who are from Canada in a random sample of 17 students is 0.765.
**33.** $\mu = 2.5$, $\sigma^2 = 1.25$, $\sigma = 1.1180$. The expected number of cars obeying the speed limit in a random sample of 5 cars on the Interstate is 2.5.
**35.** 0.6231
**37.** 0.1701
**39.** $\mu = 20$, $\sigma^2 = 10$, $\sigma = 3.1623$. The expected number of heads observed in 40 tosses of a fair coin is 20.
**41.** $\mu = 3$, $\sigma^2 = 2.55$, $\sigma = 1.5969$. The expected number of people who will support an Independent candidate for president in the next election in a random sample of 20 people is 3.
**43.** **(a)** 0.2581 **(b)** 0.6489 **(c)** 3 sophomores
**45.** **(a)** $\mu = 3$; $\sigma^2 = 2.25$; $\sigma = 1.5$. The expected number of sophomores in a random sample of 12 statistics students is 3. **(b)** $Z$-score is –2; moderately unusual.
**47.** **(a)** 0.1171 **(b)** 0.2447 **(c)** 8 women
**49.** **(a)** $\mu = 8$ women; $\sigma^2 = 4.8$; $\sigma = 2.1909$. The expected number of women in a random sample of 20 students from management courses is 8. **(b)** $Z$-score is 1.83; not unusual.
**51.** **(a)** 0.2496 **(b)** 9 women **(c)** $\mu = 8.7$ women. The expected number of women who reported that "Music I Like" is the biggest factor in deciding which radio station to listen to in a random sample of 10 women is 8.7.
**53.** **(a)** $\sigma^2 = 1.131$; $\sigma = 1.0635$. **(b)** $Z$-score is $-6.30$; unusual.
**55.** **(a)** 0.0795 **(b)** 0.2414 **(c)** 61 women **(d)** $\mu = 61$ women. The expected number of women with access to the Internet in a random sample of 100 women is 61.
**57.** **(a)** $\sigma^2 = 23.79$; $\sigma = 4.8775$ **(b)** $Z$-score is $-2.46$; moderately unusual.
**59.** **(a)** 0.3337 **(b)** 12 women **(c)** $\mu = 12$ women. The expected number of women who are affected by a depressive disorder in a random sample of 100 women is 12.
**61.** **(a)** $\sigma^2 = 10.56$; $\sigma = 3.2496$ **(b)** $Z$-score is $-0.62$; not unusual.
**63.** **(a)** 0.0311 **(b)** 68 music downloaders **(c)** $\mu = 68$ music downloaders. The expected number of music downloaders who say that they are using paid services in a sample of 400 music downloaders is 68.
**65.** **(a)** $\sigma^2 = 56.44$; $\sigma = 7.5127$ **(b)** $Z$-score is 2.93; moderately unusual.

## Section 6.3

**1.** The probability that $X$ equals some particular value is zero.
**3.** Area under the normal distribution curve above an interval.
**5.** 0
**7.** 0.5
**9.** Greater than 0.5. Since $X = 4285$ is greater than the mean of 3285 and the area to the left of $\mu = 3285$ is 0.5, the area to the left of $X = 4285$ is greater than the area to the left of $X = 3285$.
**11.** About 0.997.
**13.** $A$ has mean 10; $B$ has mean 25. The peak of a normal curve is at the mean; from the graphs we see that the mean of $A$ is less than the mean of $B$.
**15.** $\mu = 0$, $\sigma = 1$
**17.** $\mu = -10$, $\sigma = 10$
**19.** $\mu = 100$, $\sigma = 12.5$
**21.** $\mu = 0$, $\sigma = 2$
**23.** **(a)**

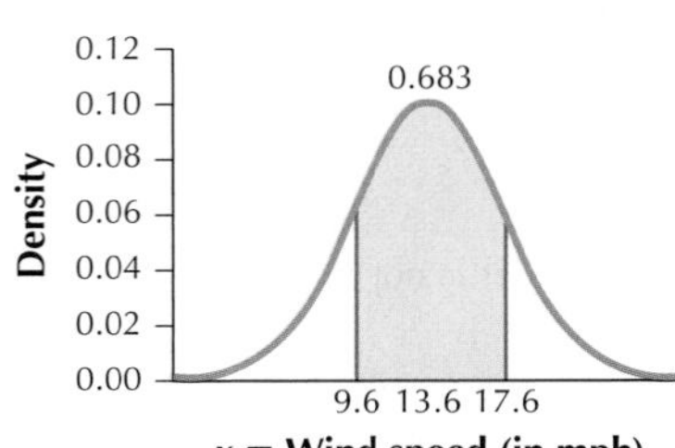


**(b)** About 0.68
**25.** **(a)**

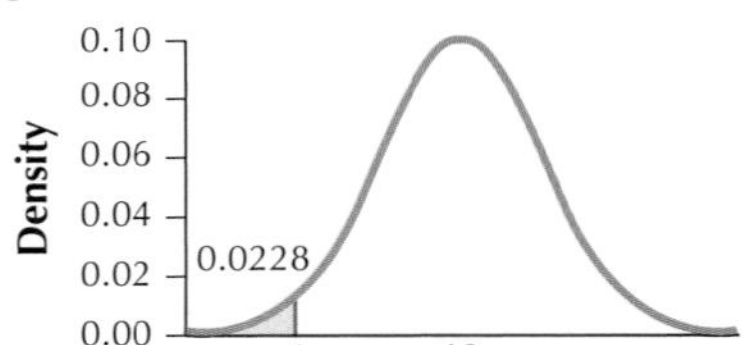


**(b)** About 0.025
**27.** **(a)**

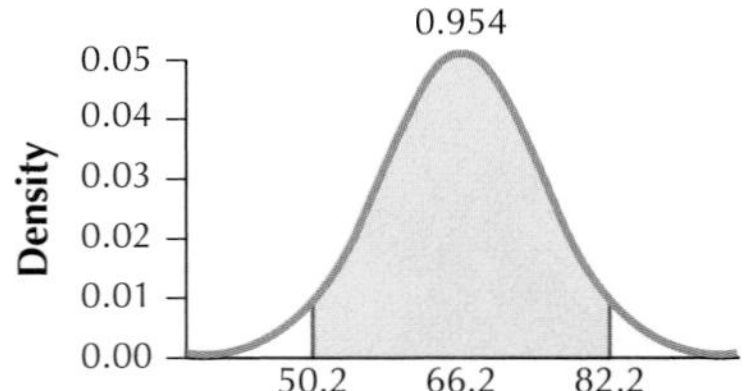


**(b)** About 0.95
**29.** About 0.135
**31.** About 0.025
**33.** There is no minimum value of $X$ for a normal distribution, but birth weights can't be less than 0.

## Section 6.4

**1.** $\mu = 0$
**3.** True
**5.** **(a)** Greater than 0.5 **(b)** 0.9750 **(c)** 0.9750 is greater than 0.5
**7.** **(a)** Less than 0.5 **(b)** 0.4821 **(c)** 0.4821 is less than 0.5
**9.** **(a)** Less than 0.5 **(b)** 0.0179 **(c)** 0.0179 is less than 0.5

**11. (a)** Greater than 0.5 **(b)** 0.8020 **(c)** 0.8020 is greater than 0.5
**13. (a)** Less than 0.5 **(b)** 0.1832 **(c)** 0.1832 is less than 0.5
**15. (a)** Greater than 0.5 **(b)** 0.9641 **(c)** 0.9641 is greater than 0.5
**17. (a)**

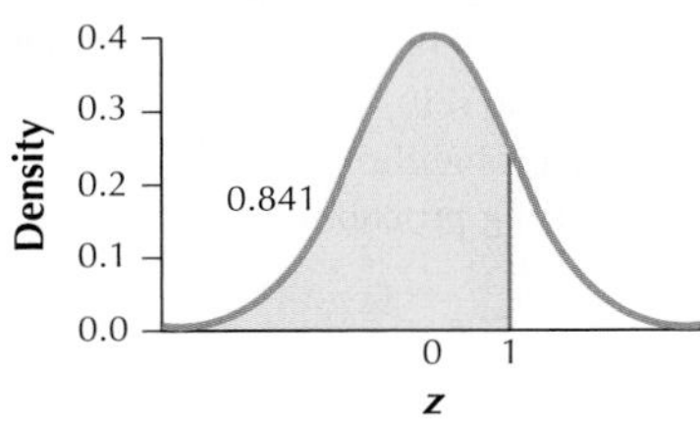


**(b)** Greater than 0.5 **(c)** 0.8413 **(d)** 0.8413 is greater than 0.5
**19. (a)**

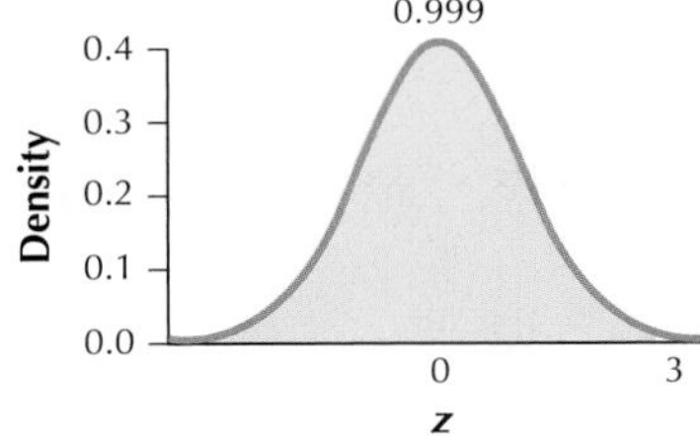


**(b)** Greater than 0.5 **(c)** 0.9987 **(d)** 0.9987 is greater than 0.5

**21. (a)**

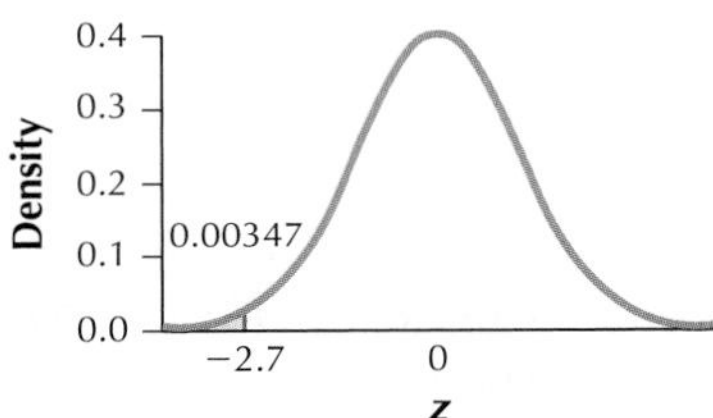


**(b)** Less than 0.5 **(c)** 0.0035 **(d)** 0.0035 is less than 0.5

**23. (a)**

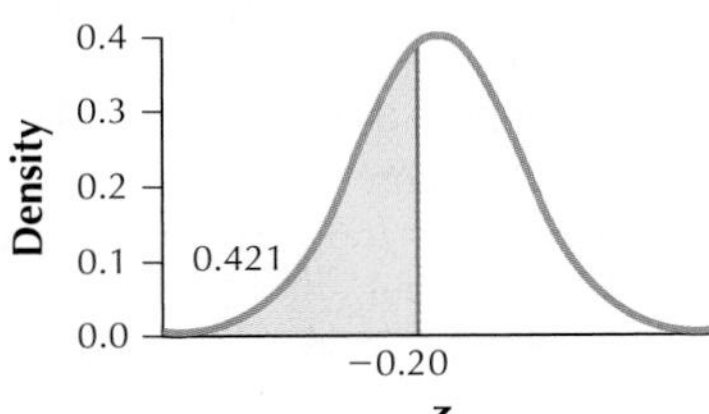


**(b)** Less than 0.5 **(c)** 0.4207 **(d)** 0.4207 is less than 0.5

**25. (a)**

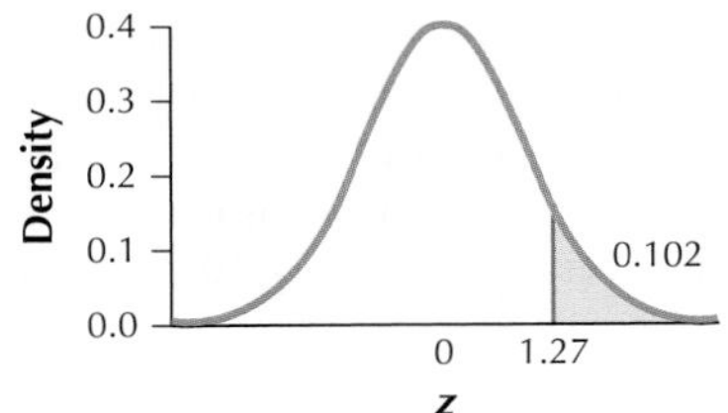


**(b)** Less than 0.5 **(c)** 0.1020 **(d)** 0.1020 is less than 0.5

**27. (a)**

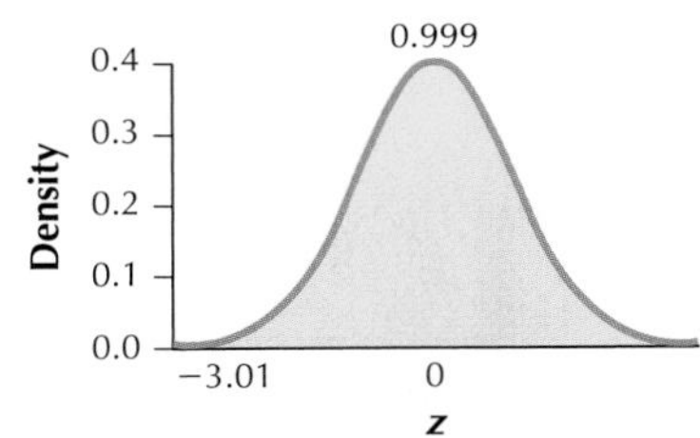


**(b)** Greater than 0.5 **(c)** 0.9987 **(d)** 0.9987 is greater than 0.5

**29. (a)**

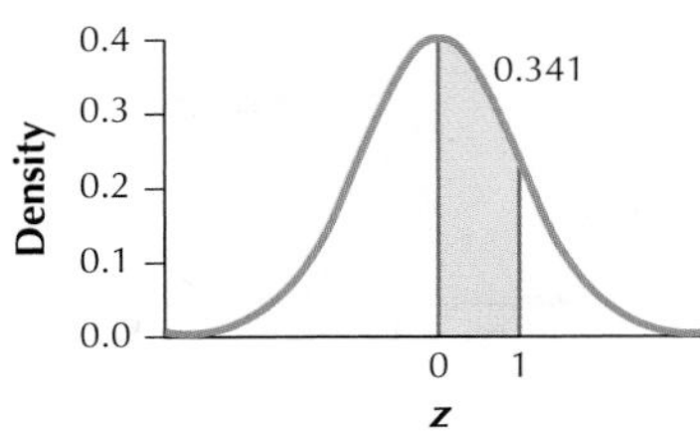


**(b)** Less than 0.5 **(c)** 0.3413 **(d)** 0.3413 is less than 0.5

**31. (a)**

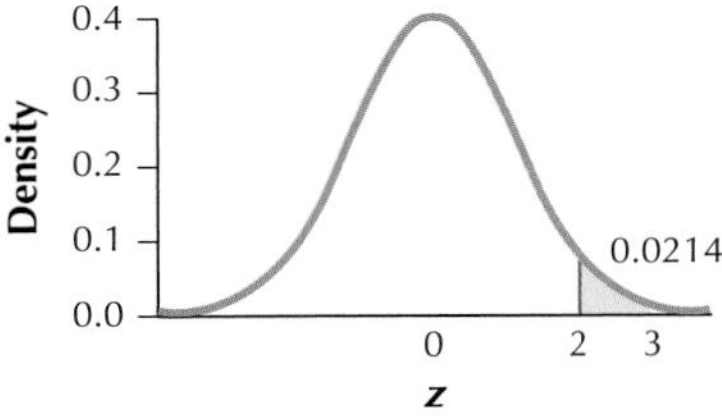


**(b)** Less than 0.5 **(c)** 0.0214 **(d)** 0.0214 is less than 0.5

**33. (a)**

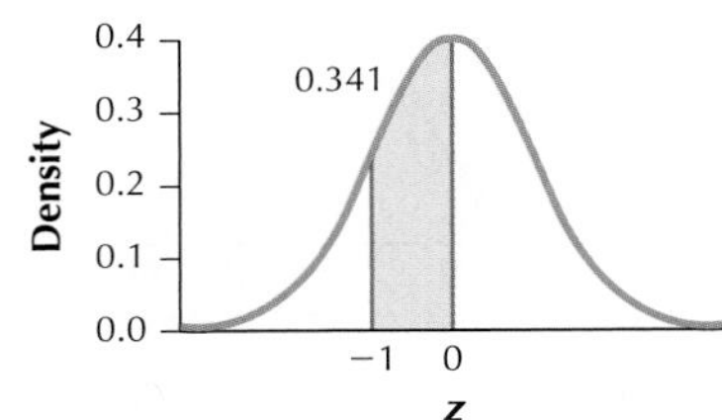


**(b)** Less than 0.5 **(c)** 0.3413 **(d)** 0.3413 is less than 0.5
**35. (a)**

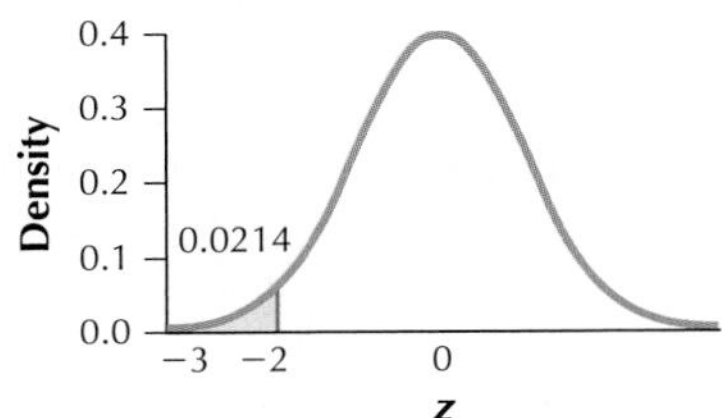


**(b)** Less than 0.5 **(c)** 0.0214 **(d)** 0.0214 is less than 0.5
**37. (a)**

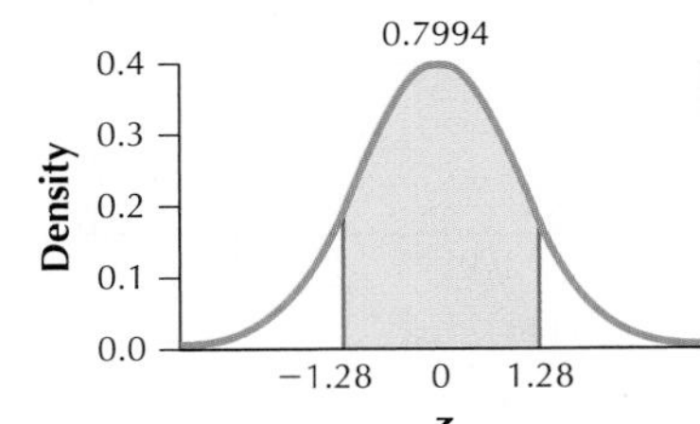


**(b)** Greater than 0.5 **(c)** 0.7994 **(d)** 0.7994 is greater than 0.5

**39. (a)**

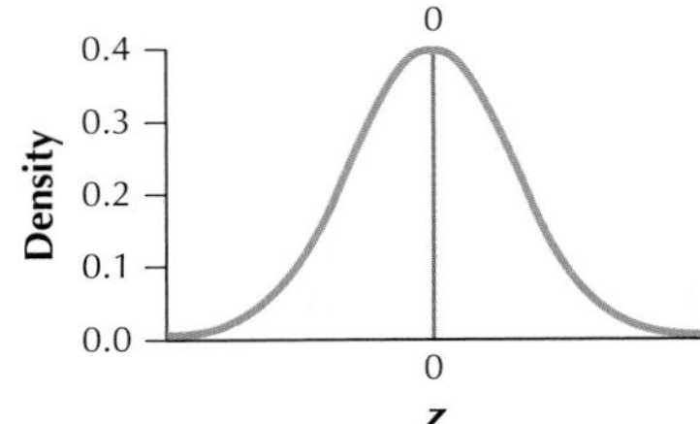

**(b)** 0 **(c)** 0 **(d)** the estimate equals the actual value

**41. (a)**

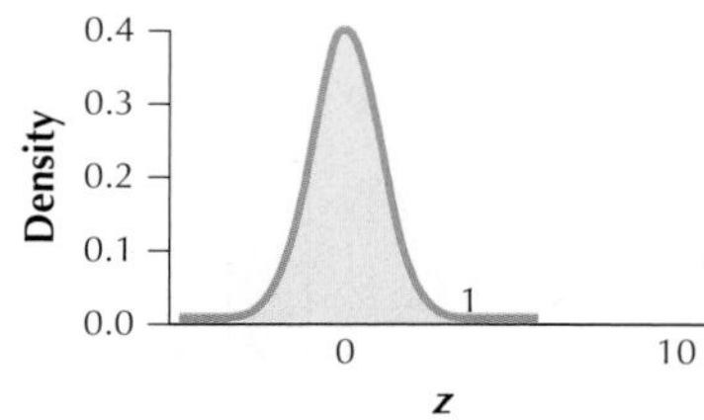


**(b)** 1 **(c)** 1 **(d)** the estimate equals the actual value

**43. (a)**

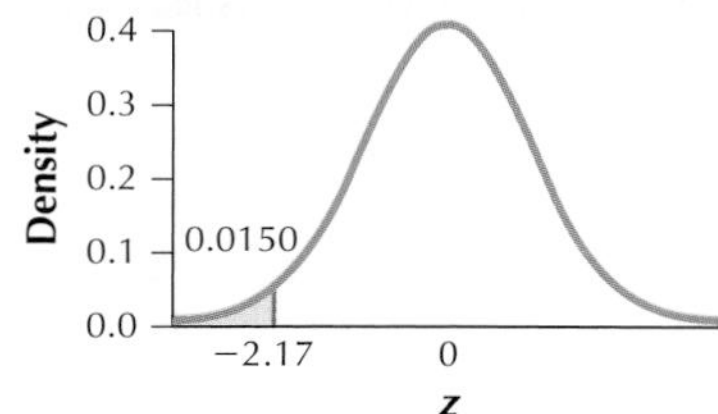


**(b)** Less than 0.5 **(c)** 0.0150 **(d)** 0.0150 is less than 0.5

**45. (a)**

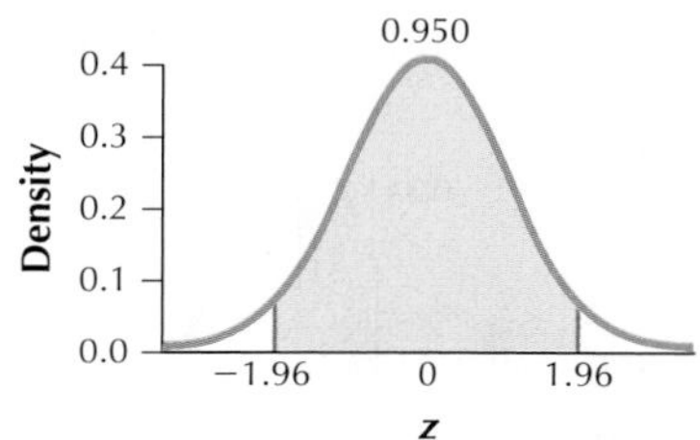


**(b)** Greater than 0.5 **(c)** 0.9500 **(d)** 0.9500 is greater than 0.5

**47. (a)**

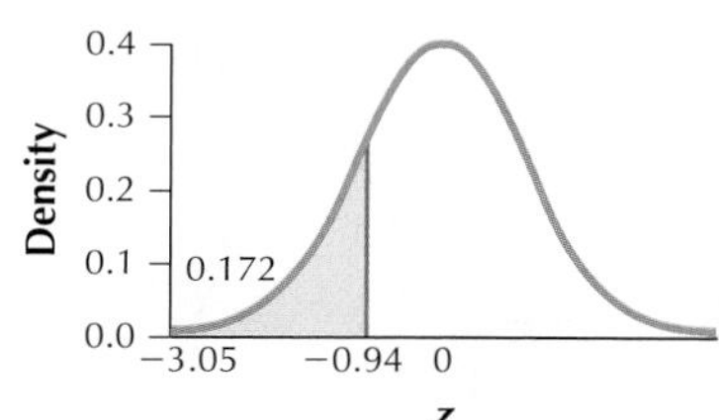


**(b)** Less than 0.5 **(c)** 0.1725 **(d)** 0.1725 is less than 0.5

**49. (a)**

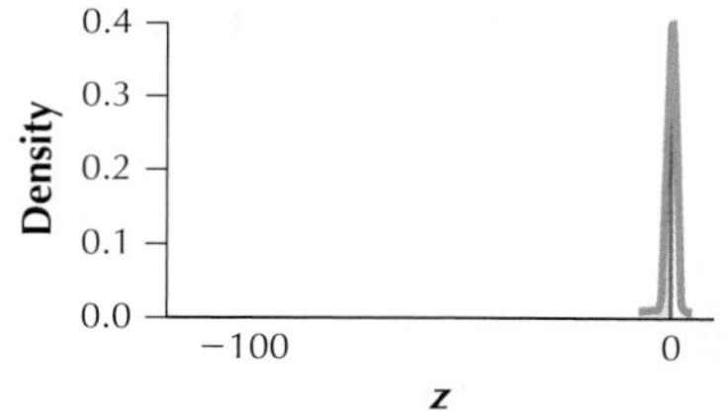


**(b)** About 0.5 **(c)** 0.5000 **(d)** The estimate equals the actual value.

**51.** Less than 0; $Z = -0.43$; $-0.43$ is less than 0

**53.** Less than 0; $Z = -0.45$; $-0.45$ is less than 0

**55.** Less than 0; $Z = -0.87$; $-0.87$ is less than 0

**57.** Less than 0; $Z = -2.03$; $-2.03$ is less than 0

**59.** $Z = 0$

**61.** $Z = 2.58$

**63.** The area between $Z = -2$ and $Z = 2$ is 0.9544. By the Empirical Rule, the area between $Z = -2$ and $Z = 2$ is about 0.95.

**65. (a)** 0.0668 **(b)** 0.9332 **(c)** 0.8664

**67.** $Z = -2.58$ and $Z = 2.58$.

**69.** $-0.67$; 0; 0.67

## Section 6.5

**1.** To standardize things means to make them all the same, uniform, or equivalent. To standardize a normal random variable $X$, we transform $X$ into the standard normal random variable $Z$ using the formula $Z = \frac{X - \mu}{\sigma}$. We do this so that we can use the standard normal table to find the probabilities.

**3.** You would get the wrong probability.

**5.** 0.5

**7.** 0.8413

**9.** 0.0062

**11.** 0.9332

**13.** 0.8400

**15.** 0.0049

**17.** $X = 86.45$

**19.** $X = 82.8$

**21.** $X = 50.4$ and $X = 89.6$.

**23.** Appropriate

**25.** Not appropriate

**27.** Not appropriate

**29.** 0.1272

**31.** 0.4364

**33.** 0.4364

**35.** 0.3616

**37.** 0.0992

**39.** 0.6772

**41.** 0.0853

**43.** 0.0992

**45. (a)** 0.1359 **(b)** 0.0228

**47. (a)** 0.1423 **(b)** 0.1423 **(c)** 26.67% **(d)** $X = 27.6$ mph **(e)** $Z$-score is –2.27; moderately unusual

**49. (a)** 0.3085 **(b)** 0.0668 **(c)** 0.0928 **(d)** $X = 14.70$ million viewers **(e)** $Z$-score is 3; unusual

**51. (a)** 0.9772 **(b)** 4.75% **(c)** 0.1574 **(d)** $X = 134.95$ years **(e)** $Z$-score is 3.33; unusual

**53. (a)** 0.6915 **(b)** 0.0228 **(c)** 0.8413 **(d)** $X = 6.34$ million tobacco-related deaths **(e)** $Z$-score is 1.5; not unusual

**55. (a)** 0.0287 **(b)** 0.0764 **(c)** 0.2667 **(d)** $-0.36$ yards and 23.16 yards **(e)** $Z$-score is 3.93; unusual

**57. (a)** 0.0062 **(b)** 0.0228 **(c)** 0.7506 **(d)** 0.2963 **(e)** 0.7734 **(f)** \$24,880 **(g)** \$26,640 and \$33,360

**59. (a)** $-1.04$ calories per gram **(b)** No, food cannot have negative calories per gram. **(c)** Yes **(d)** The distribution of the number of calories per gram is right-skewed, so it is not normal.

**61. (a)** 0.0026 **(b)** 0.2313 **(c)** 80.40 and 119.60

**63. (a)** 0.7157 **(b)** 0.0548 **(c)** 0.0003 **(d)** 0.0951

**65. (a)** 0.1350 **(b)** 0.5675 **(c)** 0.4325 **(d)** 0.5675 **(e)** 0.4325

**67. (a)** No **(b)** The normal distribution is not a good approximation to the binomial distribution ($n \cdot p = 2 < 10$, so not appropriate.

## Chapter 6 Review

**1. (a)** 0.95 **(b)** 2 games **(c)** 0.60

**3. (a)** $\mu = \$1500$, $Var(X) = 60{,}250{,}000$, $SD(X) = \$7762.09$

**(b)**

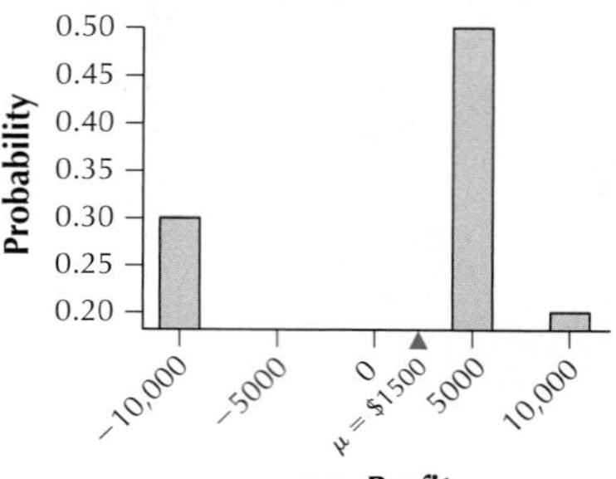

(c) This value for the mean makes sense as the point where the distribution balances.

**5. (a)**

| $X$ | 2 | 3 | 4 | 5 | 7 | 11 |
|---|---|---|---|---|---|---|
| $P(X)$ | 7/13 | 2/13 | 1/13 | 1/13 | 1/13 | 1/13 |

**(b)**

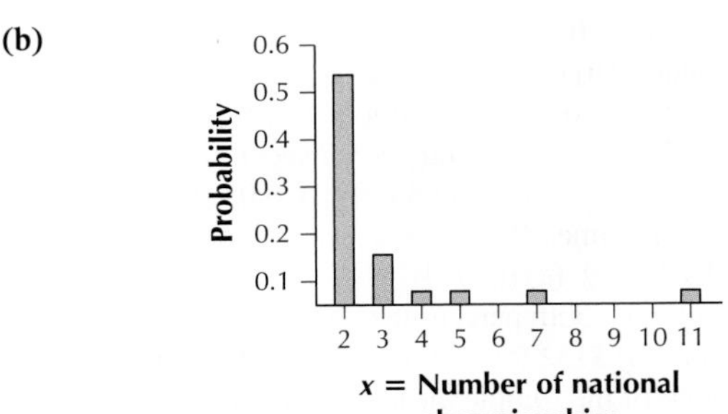


**(c)** 2 national championships; **(d)** $\mu = 3.62$ national championships
**7. (a)** $SD(X) = 7.7200$ **(b)** Z-score is 1.84; not unusual
**9. (a)** 0.0036 **(b)** 0.9964 **(c)** 0.1236
**11.** $\mu = 1.6$, $Var(X) = 1.472$, $SD(X) = 1.2133$. The expected number of pregnancies in which gestational diabetes occurs in a random sample of 20 pregnancies is 1.6.
**13.** $\mu = 5.25$, $Var(X) = 3.4125$, $SD(X) = 1.8473$. The expected number of Americans who said that the price of gasoline was the news story they followed more closely than any other news story in a random sample of 15 Americans is 5.25.
**15. (a)** 0.1057 **(b)** 0.1880
**17. (a)** $\mu = 17$ eighth graders, $Var(X) = 14.11$, $SD(X) = 3.7563$. The expected number of eighth graders who have consumed alcohol in the last month in a random sample of 100 eighth graders is 17. **(b)** Z-score is 2.66; moderately unusual
**19. (a)** 0.1079 **(b)** 0.5742
**21. (a)** $\mu = 16.3$ Baptists, $Var(X) = 13.6431$, $SD(X) = 3.6937$. The expected number of Baptists in a random sample of 100 church members is 16.3.
**(b)** Z-score is 2.36; moderately unusual
**23. (a)** 0.0853 **(b)** 0.5477
**25. (a)** $\mu = 32$ deaths, $Var(X) = 21.76$, $SD(X) = 4.6648$. The expected number of deaths due to motor vehicle accidents in a random sample of 100 deaths of people aged 16–24 is 32.
**(b)** Z-score is $-1.50$; not unusual
**27.** 0.5
**29.** Less than 0.5. Since the area to the right of the mean $\mu = 106$ mm is 0.5 and $X = 110$ mm is greater than the mean $\mu = 106$ mm, the area to the right of $X = 110$ mm is less than the area to the right of the mean $\mu = 106$.
**31.** About 0.68
**33.** About 0.997

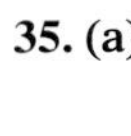

**35. (a)**

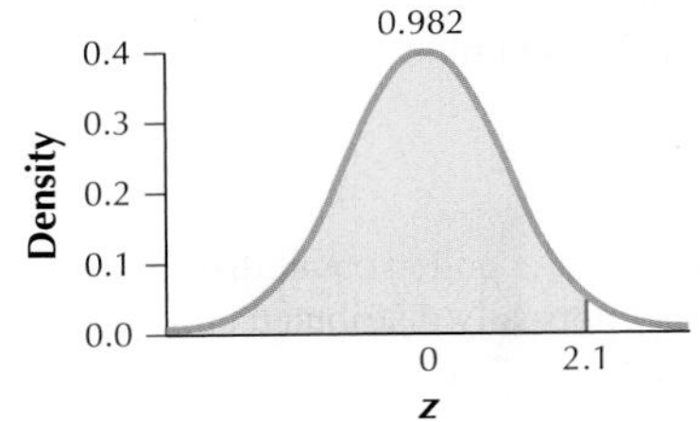


**(b)** Greater than 0.5 **(c)** 0.9821 **(d)** 0.9821 is greater than 0.5.

**37. (a)**

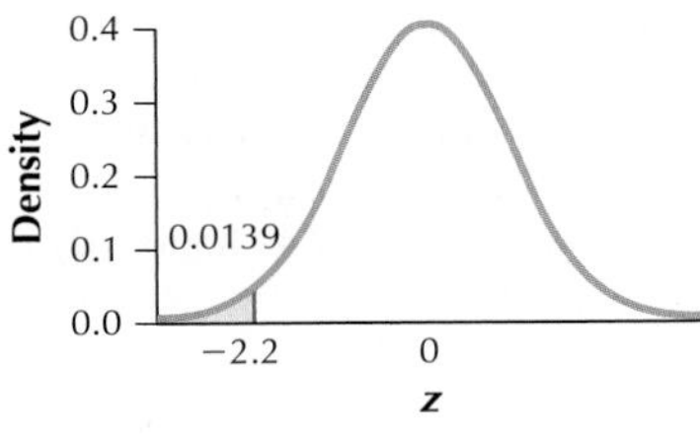


**(b)** Less than 0.5 **(c)** 0.0139 **(d)** 0.0139 is less than 0.5.

**39. (a)**

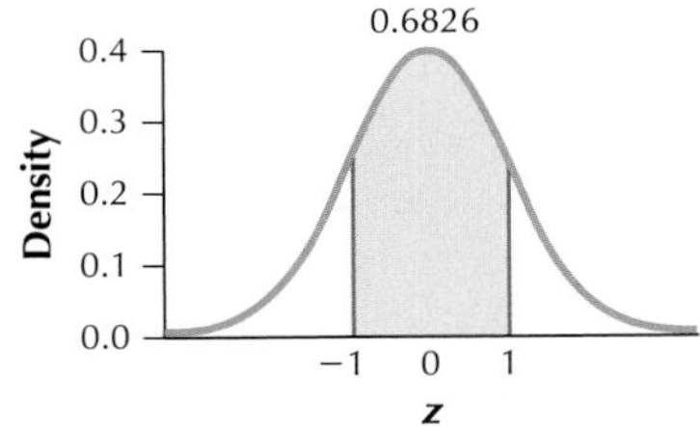


**(b)** Greater than 0.5 **(c)** 0.6826 **(d)** 0.6826 is greater than 0.5.

**41. (a)**

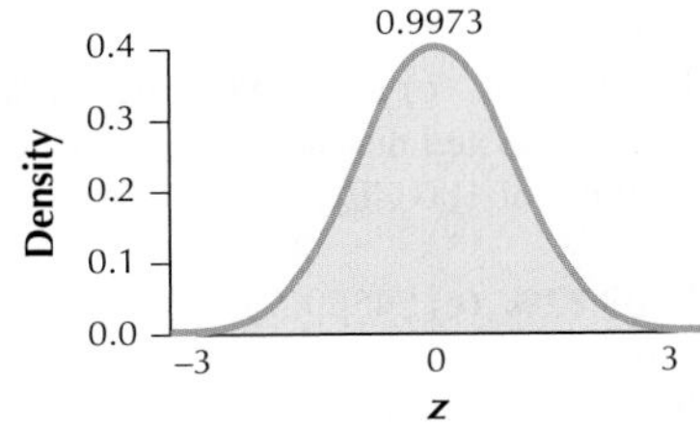


**(b)** Greater than 0.5 **(c)** 0.9973 **(d)** 0.9973 is greater than 0.5.

**43. (a)**

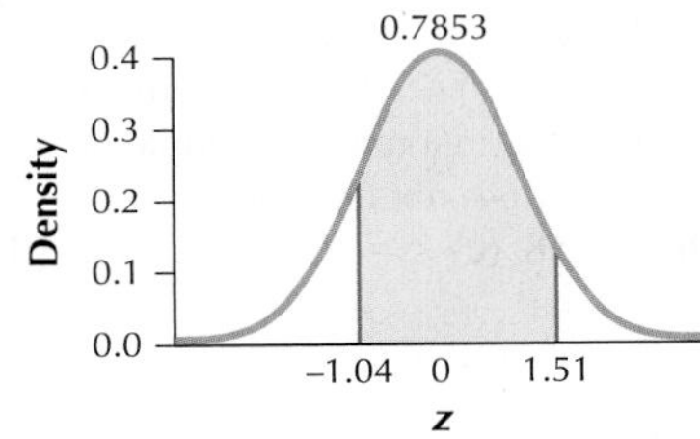


**(b)** Greater than 0.5 **(c)** 0.7853 **(d)** 0.7853 is greater than 0.5.

**45. (a)** and **(b)**

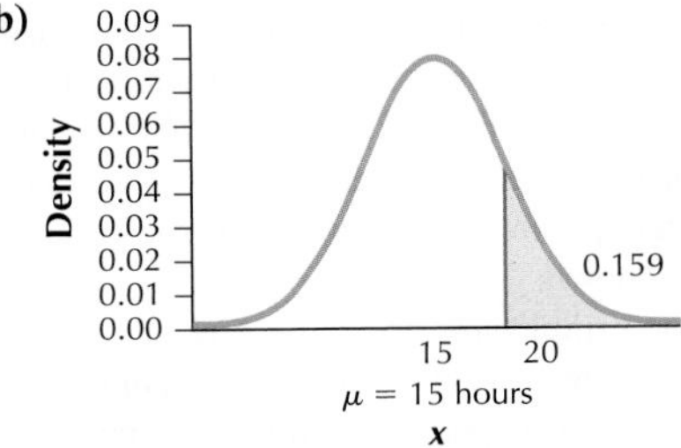


**(c)** The probability that a randomly selected reader of the magazine *Computer Gaming World* plays computer games for more than 20 hours a week is 0.1587. **(d)** 0.8413 **(e)** 0.5, 0.5
**47. (a)** 0.0736 **(b)** 0.3821 **(c)** 0.3085 **(d)** 0.6915 **(e)** 0.6179 **(f)** 0.3557
**49. (a)** 0.0228 **(b)** 2.28% **(c)** 0.3413 **(d)** $X = \$7.33$ per hour **(e)** Z-score is 1; not unusual
**51. (a)** 0.1660 **(b)** 38.97% **(c)** 0.2434 **(d)** $X = \$15{,}857$ **(e)** Z-score is 2.84; moderately unusual
**53. (a)** Since the mean and median of the population of incomes of white Americans living alone are not equal, then the assumption that

the distribution of incomes of white Americans living alone is normal is incorrect. **(b)** All our results in the previous exercise are incorrect.
**55. (a)** 0.1587 **(b)** 0.2018 **(c)** $X = 17$ **(d)** Z-score is $-3$; unusual. The producer will have to file a report.
**57. (a)** 0.4562 **(b)** 0.0993 **(c)** $X = 9.49$ hours **(d)** Z-score is $-2.94$; moderately unusual

## Chapter 6 Quiz

**1.** True
**2.** False
**3.** False
**4.** 0.5
**5.** 0
**6.** 0
**7.** discrete
**8.** binomial
**9.** $\mu = 0, \sigma = 1$
**10. (a)**

| X = Amount won | 0 | 5 |
|---|---|---|
| P(X) | 13/18 | 5/18 |

**(b)** \$1.39 **(c)** \$1.39
**11. (a)** 0.0962 **(b)** 19 CEOs **(c)** $\mu = 19$ CEOs, $Var(X) = 15.39$, $SD(X) = 3.9230$. The expected number of CEOs who drive luxury cars in a random sample of 100 CEOs is 19. **(d)** Z-score is 5.35; unusual
**12. (a)** 0.1003 **(b)** 33.22% **(c)** \$4329.50 **(d)** Z-score is $-2.05$; moderately unusual
**13. (a)** 0.0257 **(b)** 0.0074 **(c)** For cats, $\mu = 314$ households, $Var(X) = 215.404$, $SD(X) = 14.6766$. The expected number of American households that have a cat in a random sample of 1000 households is 314. For dogs, $\mu = 343$ households, $Var(X) = 225.351$, $SD(X) = 15.0117$. The expected number of American households that have a dog in a random sample of 1000 households is 343. **(d)** Z-score is $-1.64$; not unusual
**14. (a)** 0.4207 **(b)** 0.2358 **(c)** $X = 50{,}800$ visitors **(d)** Z-score is $-2.3$; moderately unusual

# CHAPTER 7

## Section 7.1

**1.** *Sampling error* is the distance between the point estimate and its target parameter. It measures how far the point estimate misses the actual target parameter. In the real world, the only time we know the actual value of the sampling error is when we know the true value of the target parameter. Usually, the true value of the target parameter is unknown.
**3.** The larger the sample size, the more precise the analysis is. However, there comes a point somewhere around $n = 80$ when the trade-off between higher precision and higher cost becomes less beneficial.
**5.** The standard deviation of the sampling distribution for $\bar{x}$ is the population standard deviation divided by the square root of the sample size.
**7.** 5
**9.** 5
**11.** 0.05
**13.** $\mu_{\bar{x}} = 100, \sigma_{\bar{x}} = 4$
**15.** $\mu_{\bar{x}} = 0, \sigma_{\bar{x}} = 3.3333$
**17.** $\mu_{\bar{x}} = -10, \sigma_{\bar{x}} = 0.5$
**19.** Normal with mean $\mu = 10$ and standard deviation $\sigma_{\bar{x}} = 1$
**21.** 0.1587
**23.** 11.96
**25.** 0.95
**27. (a)** $\bar{x} = 3021$ pounds **(b)** Probably not; different samples may have different average weights. **(c)** Not likely; since samples are subsets of the population, they are not perfect representations of the population.
**29. (a)** 2930.34 pounds **(b)** 90.66 pounds
**31. (a)** 627.13 pounds **(b)** 607 pounds **(c)** 20.13 pounds
**33. (a)** $\mu_{\bar{x}} = 1.7$ seconds, $\sigma_{\bar{x}} = 0.0667$ seconds **(b)** $\mu_{\bar{x}} = 1.7$ seconds, $\sigma_{\bar{x}} = 0.05$ seconds **(c)** $\mu_{\bar{x}} = 1.7$ seconds, $\sigma_{\bar{x}} = 0.04$ seconds **(d)** $\mu_{\bar{x}} = 1.7$ seconds, $\sigma_{\bar{x}} = 0.0333$ seconds
**35. (a)** $\mu_{\bar{x}}$ remains the same. **(b)** $\sigma_{\bar{x}}$ decreases.
**37. (a)** 0.1056 **(b)** 0.1359 **(c)** 0.4013
**39. (a)** \$100 million, the 50th percentile equals the mean for a normal distribution **(b)** \$132.90 million **(c)** \$67.10 million **(d)** The middle 90% of the sample mean IPO amounts.
**41. (a)** 0.1587 **(b)** 0.9974 **(c)** 0.1587
**43. (a)** 404.20 **(b)** 325.80 **(c)** The middle 95% of sample mean carbon dioxide concentrations of samples of size $n = 25$.
**45. (a)** ${}_5C_2 = 10$ samples of size $n = 2$ **(b)** $\mu = 64.12$. No; we don't usually know all values from the population. **(c)** $\sigma = 2.9991$. No; we don't usually know all values from the population.
**47. (a–b)** About 64.1
**49. (a)** 3.5 **(b)** 0.4270
**51. (a)** \$4 **(b)** \$2 **(c)** The sampling distribution of the $\bar{x}$ is normal with a mean of $\mu_{\bar{x}} = \$4$ and a standard deviation of $\sigma_{\bar{x}} = \$2$ by Fact 4.
**53. (a–b)** Unchanged **(c)** Decrease in $\sigma$ will result in a decrease in $P(X < 0)$. **(d)** Unchanged **(e)** Decrease
**55.** 0.4483
**57. (a)** 0.3015 **(b)** Sample means are less variable than individual observations, so 500 is more standard deviations below $\mu_{\bar{x}}$ than below $\mu$.
**59. (a–b)** Increase **(c)** Decrease
**61.** The sampling error decreases as the sample size increases.
**63.** The sampling error decreases as the sample size increases.

## Section 7.2

**1.** Symmetric; answers will vary.
**3.** The variability of the sampling distribution becomes smaller as the sample size becomes larger.
**5.** Case 1
**7.** Case 1
**9.** Case 2
**11.** Case 3
**13. (a)** 516 **(b)** 38.6667 **(c)** Normal
**15. (a)** \$60,000 **(b)** \$2500 **(c)** Normal
**17. (a)** 80 **(b)** 1 **(c)** Approximately normal
**19. (a)** 50 miles per gallon **(b)** 1.5 miles per gallon **(c)** Unknown
**21.** Not possible since the exchange rate for US \$100 in euros for 2007 is not normally distributed and the sample size of $n = 9$ is small ($n < 30$). The sampling distribution of $\bar{x}$ is unknown by Case 3.
**23.** 0.0918
**25.** 0.9544
**27.** Not possible since the pollen count distribution for Los Angeles in September is not normally distributed and the sample size of $n = 16$ is small ($n < 30$). The sampling distribution of $x$ is unknown by Case 3.
**29.** 579.61
**31.** \$3.00

**33.** \$56,800
**35.** 81.645
**37.** Not possible since the pollen count distribution for Los Angeles in September is not normally distributed and the sample size of $n = 16$ is small ($n < 30$). Case 3 applies and the sampling distribution of $\bar{x}$ is unknown.
**39. (a)** No, the distribution of serum cholesterol levels in Americans in 2005 is unknown. **(b)** Yes; 0.0918
**41. (a)** Yes; 0.0808 **(b)** No, since the sample size of $n = 4$ is small ($n < 30$).
**43. (a)** No, since the distribution of the salaries of all new sociology professors is right-skewed, it is not normal. **(b)** No, since the sample size of $n = 16$ is small ($n < 30$).
**45. (a)** No. Since the distribution of the salaries of all new sociology professors is right-skewed, it is not normal. **(b)** Yes; 0.0113
**47.** $P(\bar{x} > 50) = 0.1292$
**49.** For $x = 30$
**51. (a)** No, since the sampling distribution of the changes in SAT scores are not normally distributed. **(b)** Yes; 0
**53. (a)** Smaller. Since $0 - \mu$ is negative, an increase in the sample size will result in a decrease in $\sigma_{\bar{x}}$, which will result in a decrease in $(0 - \mu)/\sigma_{\bar{x}}$. Since $P(\bar{X} < 0) = P(Z < (0 - \mu)/\sigma_{\bar{x}})$, an increase in $n$ will result in a decrease in $P(\bar{X} < 0)$. **(b)** 0
**55. (a)** Decrease **(b)** Increase **(c)** Increase in the 2.5th percentile and decrease in the 97.5th percentile.
**57. (a)** Increase since $n$ is in the denominator. **(b)** Smaller. Since $0 - \mu_{\bar{x}}$ is negative, a decrease in $n$ results in an increase in $\sigma_{\bar{x}}$. Since $P(\bar{X} > 0) = P(Z > \frac{0 - \mu_{\bar{x}}}{\sigma_{\bar{x}}})$, a decrease in $n$ results in a decrease in $P(\bar{X} > 0)$.
**59. (a)** 0.1067 grams **(b)** 1.067 grams **(c)** About 0.997
**61. (a)** 0.0002 **(b)** 0.0002, 0.9998 **(c)** The value found in the original case study in the text favors the Master of the Mint.
**63.** $\mu = 127.78800654$ grams

## Section 7.3

**1.** Example 7.17 shows that a sample of size $n = 30$ may not be enough for the sampling distribution of $\hat{p}$ to be approximately normal.
**3.** Answers will vary.
**5. (a)** Decreasing **(b)** As $n$ increases, $\sigma_{\hat{p}} = \sqrt{\frac{p(1-p)}{n}}$ decreases.
**7. (a)** 0.5 **(b)** 0.05 **(c)** Approximately normal
**9. (a)** 0.01 **(b)** 0.0099 **(c)** Unknown
**11. (a)** 0.9 **(b)** 0.0474 **(c)** Unknown
**13.** 10
**15.** 50
**17.** 500

**19.**

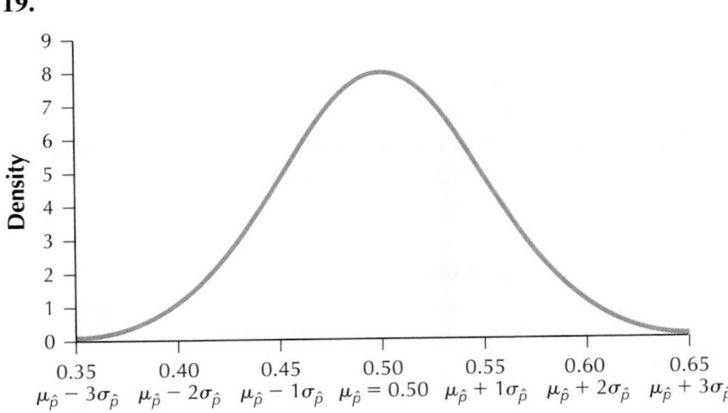


**21.**

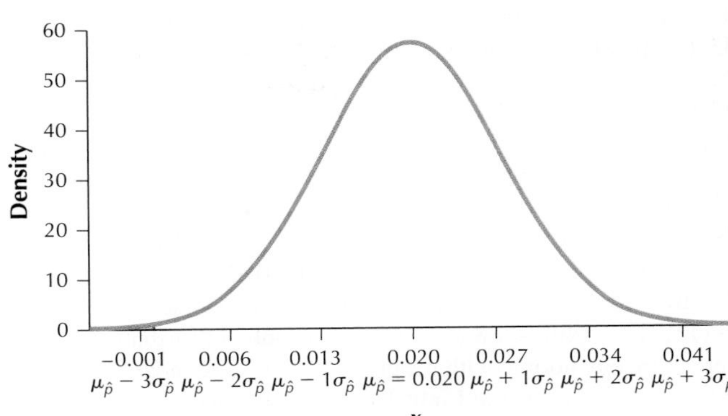


**23.**

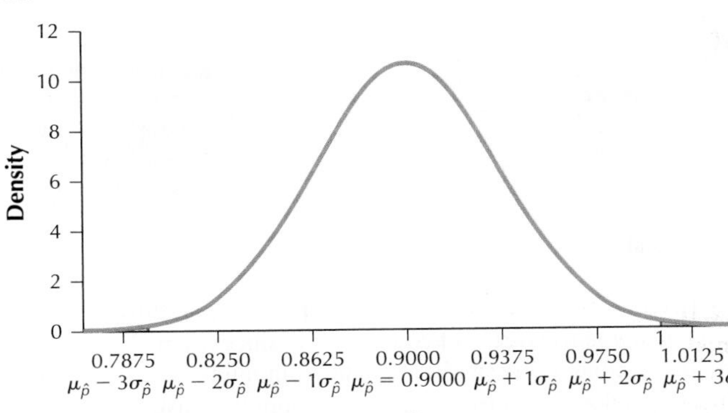


**25.**

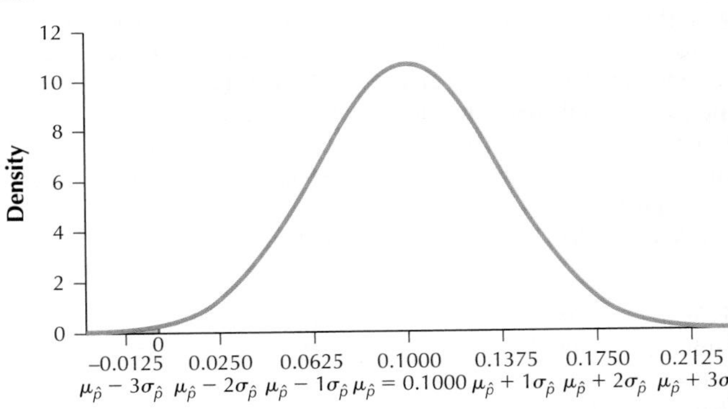


**27.**

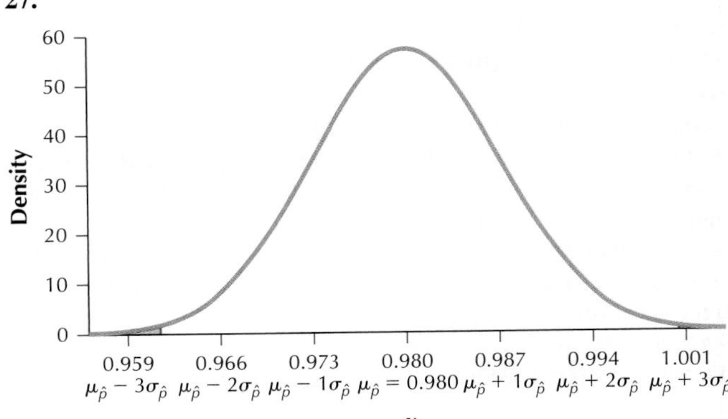

**29.** 0.1587
**31.** Not possible; sampling distribution of $\hat{p}$ is unknown.
**33.** Not possible; sampling distribution of $\hat{p}$ is unknown.
**35.** 0.564
**37.** 0.962
**39.** 0.052
**41.** **(a)** $\mu_{\hat{p}} = 0.25$ **(b)** $\sigma_{\hat{p}} = 0.0433$ **(c)** Approximately normal
**43.** **(a)** 0.1251 **(b)** 0.321
**45.** **(a)** $np = (100)(0.61) = 61 \geq 5$ and $n(1 - p) = 100(1 - 0.61) = 39 \geq 5$. **(b)** $\mu_{\hat{p}} = 0.61$ and $\sigma_{\hat{p}} = 0.0488$. **(c)** 0.4207
**47.** **(a)** $n^* = 13$ **(b)** $n^*p = (13)(0.40) = 5.2 \geq 5$ and $n^*(1 - p) = (13)(1 - 0.40) = 7.8 \geq 5$. **(c)** The sampling distribution of $\hat{p}$ is approximately normal with mean $\mu_{\hat{p}} = 0.40$ and standard deviation $\sigma_{\hat{p}} = 0.1359$. The Central Limit Theorem for Proportions allows us to say this. **(d)** $\mu_{\hat{p}} = 0.40$ and $\sigma_{\hat{p}} = 0.1095$. **(e)** 0.4286
**49.** **(a)** An increase in the sample size $n$ will result in a decrease of $\sigma_{\hat{p}}$. **(b)** Any value greater than 20
**51.**

| Response | Frequency | Relative frequency (frequency/total) |
|---|---|---|
| Many people | 2058 | 2058/4395 = 0.47 |
| Just a few people | 1806 | 1806/4395 = 0.41 |
| Hardly any people | 485 | 485/4395 = 0.11 |
| No one/None | 46 | 46/4395 = 0.01 |
| **Total** | **4395** | |

**53.** **(a)** $\mu_{\hat{p}} = 0.01$ and $\sigma_{\hat{p}} = 0.0044$. **(b)** Smaller because $p$ is so small and $n$ is so large. **(c)** Not possible, since a sample size of 200 respondents is smaller than the minimum sample size of 500 respondents required to produce a sampling distribution of $\hat{p}$ that is approximately normal. The sampling distribution of $\hat{p}$ is unknown.
**55.** **(a)** Increase **(b)** Decrease **(c)** Increase **(d)** Increase of $p$ up to 0.5 will result in an increase in $\sigma_{\hat{p}}$. An increase of $p$ to a value greater than 0.5 will result in a decrease in $\sigma_{\hat{p}}$ **(e)** An increase of $p$ to a value less than or equal to 0.99 will result in an increase in $n^*$, an increase in $p$ to $p = 0.99$ will result in the same value of $n^*$, and an increase in $p$ to a value between 0.99 and 1 exclusive will result in a decrease $n^*$.
**57.** **(a–c)** Unchanged **(d)** Decrease **(e)** Unchanged

## Chapter 7 Review

**1.** 0
**3.** 0.1
**5.** $\mu_{\bar{x}} = 2$ and $\sigma_{\bar{x}} = 0.0833$.
**7.** $\mu_{\bar{x}} = 50$ and $\sigma_{\bar{x}} = 10$.
**9.** 0.1056
**11.** 0.7888
**13.** **(a)** 2562 pounds **(b)** Statistic; parameter **(c)** Underestimate; see answer **15(d)**
**15.** **(a)** 1467 pounds **(b)** Sampling error **(c)** Underestimate **(d)** The true range is the largest weight in the population minus the smallest weight in the population. Since a sample is unlikely to contain both the car with the largest weight in the population and the car with the smallest weight in the population, it will be most likely to contain a car lighter than the heaviest car or a car heavier than the lightest car (or both). **(e)** Yes
**17.** Case 2
**19.** **(a)** 50 beats per second **(b)** 1 beat per second **(c)** Approximately normal
**21.** 0.0013
**23.** 48.72 beats per minute
**25.** **(a)** 47.18 years **(b)** 40.82 years **(c)** 40.82 years and 47.18 years **(d)**

0.950
Density
0.25, 0.20, 0.15, 0.10, 0.05, 0.00
40.82 2.5th percentile, 44, 47.18 97.5th percentile
x

**27.** **(a)** 0.1 **(b)** 0.0424 **(c)** Approximately normal
**29.** $n^* = 50$
**31.**

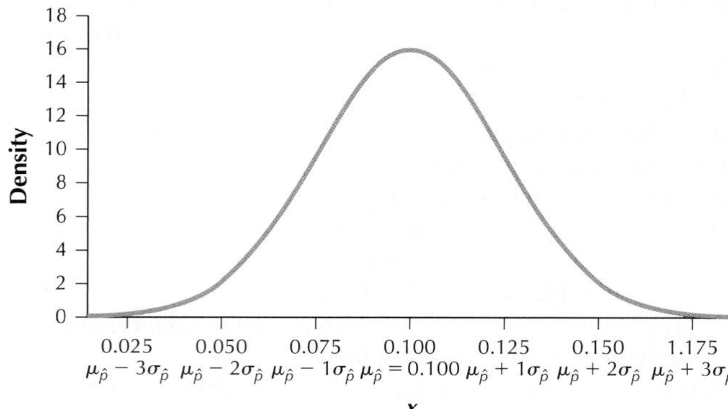


**33.** 0.6808
**35.** 0.0162
**37.** **(a)** We have $np = (120)(0.11) = 13.2 \geq 5$ and $n(1 - p) = (120)(1 - 0.11) = 106.8 \geq 5$, so the sampling distribution of $\hat{p}$ is approximately normal. **(b)** $\mu_{\hat{p}} = 0.11$ and $\sigma_{\hat{p}} = 0.0286$. **(c)** 0.1762

## Chapter 7 Quiz

**1.** True
**2.** False
**3.** False
**4.** Sampling error
**5.** 30
**6.** Approximately normal
**7.** Normal probability plot
**8.** No
**9.** $np \geq 5$ and $n(1 - p) \geq 5$
**10.** **(a)** 0.1587 **(b)** 0.9500 **(c)** 0.1056
**11.** **(a)** 45.15 grams **(b)** 34.85 grams **(c)** 34.85 grams and 45.15 grams
**12.** **(a)** 0.0228 **(b)** 0.0228 **(c)** 0.9544
**13.** **(a)** 68.77 inches **(b)** 67.23 inches **(c)** 67.23 and 68.77 inches
**14.** Case 3
**15.** Case 1
**16.** **(a)** 100 **(b)** 3 **(c)** Unknown
**17.** **(a)** 100 **(b)** 3 **(c)** Normal
**18.** Not possible, since scores on a psychological test are not normally distributed and the sample size of $n = 25$ is small ($n < 30$). Case 3 applies, so the sampling distribution of $\bar{x}$ is unknown.
**19.** 0.8185
**20.** Not possible, since scores on a psychological test are not normally distributed and the sample size of $n = 25$ is small ($n < 30$). Case 3 applies, so the sampling distribution of $\bar{x}$ is unknown.
**21.** 100
**22.** **(a)** No, the distribution of $X$ is unknown. **(b)** No, the sample size of $n = 15$ is small ($n < 30$).

**23.** **(a)** No, the distribution of $X$ is unknown. **(b)** Yes; 0.0016
**24.** **(a)** 0.99 **(b)** 0.0099 **(c)** Unknown
**25.** **(a)** 0.99 **(b)** 0.0044 **(c)** Approximately normal.
**26.** $n^* = 100$
**27.** $n^* = 500$
**28.**

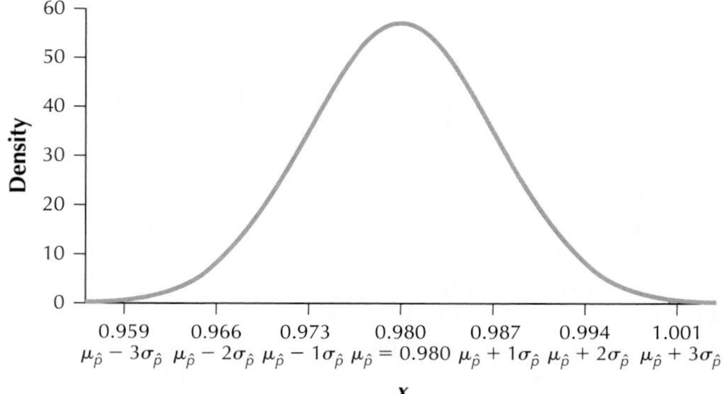

**29.**

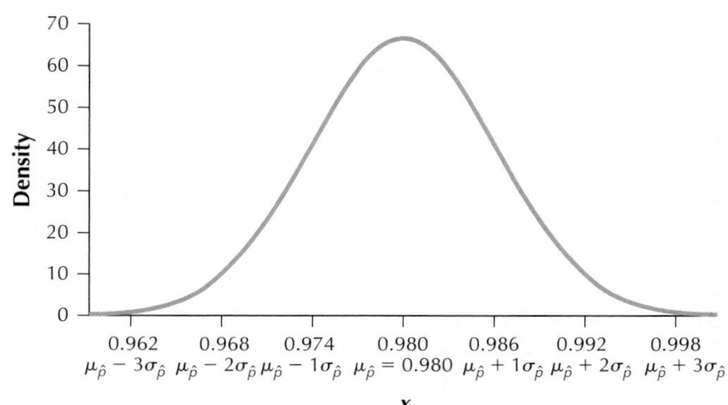


**30.** Not possible. Sampling distribution of $\hat{p}$ is unknown.
**31.** 0.6845
**32.** 0.98
**33.** 0.98
**34.** **(a)** We have $n\,p = (100)\,(0.12) = 12 \geq 5$ and $n\,(1 - p) = (100)\,(1 - 0.12) = 88 \geq 5$, so the sampling distribution of $\hat{p}$ is approximately normal. **(b)** $\mu_{\hat{p}} = 0.12$ and $\sigma_{\hat{p}} = 0.0325$. **(c)** 0.7324
**35.** **(a)**We have $n\,p = (400)\,(0.17) = 68 \geq 5$ and $n\,(1 - p) = (400)\,(1 - 0.17) = 332 \geq 5$, so the sampling distribution of $\hat{p}$ is approximately normal. **(b)** $\mu_{\hat{p}} = 0.17$ and $\sigma_{\hat{p}} = 0.0188$. **(c)** 0

# CHAPTER 8

## Section 8.1

**1.** A *confidence interval estimate* of a parameter consists of an interval of numbers generated by a point estimate, together with an associated *confidence level* specifying the probability that the interval contains the parameter.
**3.** **(a)** $Z_{\alpha/2}$ increases. **(b)** Since the confidence level is $(1 - \alpha) \times 100\%$, as the confidence level increases, $1 - \alpha$ increases. Thus $\alpha$ and $\alpha/2$ will decrease. Since $\alpha/2$ is the area underneath the standard normal curve to the right of $Z_{\alpha/2}$, a decrease in $\alpha/2$ will result in an increase in $Z_{\alpha/2}$.
**5.** Shorter, tighter confidence intervals are better, since the maximum difference between the sample mean and the true mean is reduced. Confidence intervals that are too wide are ineffectual and often useless.
**7.** A higher confidence level means that we can be more confident that our interval will contain the true value of the parameter. But a higher confidence level for the same sample data will result in a wider, less precise confidence interval.
**9.** **(a)** No **(b)** Yes **(c)** Yes
**11.** $Z_{\alpha/2} = 2.576$
**13.** $Z_{\alpha/2} = 2.576$
**15.** $Z_{\alpha/2} = 1.645$
**17.** $Z_{\alpha/2} = 1.645$
**19.** **(a)** Case 2 applies and we can use the $Z$ interval. **(b)** 0.5 **(c)** $Z_{\alpha/2} = 1.96$ **(d)** $E = 0.98$. We can estimate $\mu$ to within 0.98 with 95% confidence. **(e)** (9.02, 10.98). We are 95% confident that the true mean $\mu$ lies between 9.02 and 10.98.
**21.** **(a)** Case 1 applies, and we can use the $Z$ interval. **(b)** 0.5 **(c)** $Z_{\alpha/2} = 1.96$ **(d)** $E = 0.98$. We can estimate $\mu$ to within 0.98 with 95% confidence. **(e)** (34.02, 35.98). We are 95% confident that the true mean $\mu$ lies between 34.02 and 35.98.
**23.** **(a)** Case 2 applies, and we can use the $Z$ interval. **(b)** 2 **(c)** $Z_{\alpha/2} = 1.96$ **(d)** $E = 3.92$. We can estimate $\mu$ to within 3.92 with 95% confidence. **(e)** (96.08, 103.92). We are 95% confident that the true mean $\mu$ lies between 96.08 and 103.92.
**25.** **(a)** (9.342, 10.658). We are 90% confident that the true mean $\mu$ lies between 9.342 and 10.658. **(b)** (9.216, 10.784). We are 95% confident that the true mean $\mu$ lies between 9.216 and 10.784. **(c)** (8.9696, 11.0304). We are 99% confident that the true mean $\mu$ lies between 8.9696 and 11.0304. **(d)** The confidence interval for a given sample size becomes wider as the confidence level increases. **(e)** Case 1
**27.** **(a)** (86.08, 93.92). We are 95% confident that the true mean $\mu$ lies between 86.08 and 93.92. **(b)** (88.04, 91.96). We are 95% confident that the true mean $\mu$ lies between 88.04 and 91.96. **(c)** (89.02, 90.98). We are 95% confident that the true mean $\mu$ lies between 89.02 and 90.98. **(d)** For a given confidence level, as the sample size increases the width of the confidence interval decreases. **(e)** Case 1
**29.** **(a)** 7 ounces **(b)** 0.17 ounces **(c)** $Z_{\alpha/2} = 1.96$ **(d)** $E = 0.33$ ounces. We can estimate $\mu$, the mean amount of soda dispensed, to within 0.33 ounces with 95% confidence. **(e)** (6.67, 7.33). We are 95% confident that the true mean amount of soda dispensed $\mu$ lies between 6.67 ounces and 7.33 ounces.
**31.** **(a)** 2600 people **(b)** 141.42 people **(c)** $Z_{\alpha/2} = 2.576$ **(d)** $E = 364.30$ people. We can estimate $\mu$, the mean number of members in a Roman Catholic church parish, to within 364.30 people with 99% confidence. **(e)** (2235.70, 2964.30). We are 99% confident that the true mean number of members in a Roman Catholic church parish $\mu$ lies between 2235.70 people and 2964.30 people.
**33.** **(a)** Neither Case 1 nor Case 2 apply, so we should not use a $Z$ interval. **(b)** 19.5 seconds **(c)** $Z_{\alpha/2} = 1.96$ **(d)** $E = 38.22$ seconds. We can estimate $\mu$, the mean length of time that boys remain engaged with a science exhibit at a museum, to within 38.22 seconds with 95% confidence. **(e)** (68.78, 145.22). We are 95% confident that the true mean length of time that boys remain engaged with a science exhibit at a museum $\mu$ lies between 68.78 seconds and 145.22 seconds.
**35.** **(a)** $E = 1.05$ miles. We can estimate $\mu$, the mean commuting distance, to within 1.05 miles with 95% confidence. **(b)** (8.88,

11.00). We are 95% confident that the true mean commuting distance $\mu$ lies between 8.86 miles and 11.00 miles.

**37. (a)** $E = 3.70$ ng/g. We can estimate $\mu$, the mean concentration of the herbicide dicamba in Iowa homes, to within 3.70 ng/g with 95% confidence. **(b)** (176.30, 183.70). We are 95% confident that the true mean concentration of the herbicide dicamba in Iowa homes $\mu$ lies between 176.30 ng/g and 183.70 ng/g.

**39. (a)** $E = 0.74$ points. We can estimate $\mu$, the mean increase in IQ points for all children after listening to a Mozart piano sonata for about 10 minutes, to within 0.74 points with 90% confidence. **(b)** (7.76, 9.24). We are 90% confident that the true mean increase in IQ points in all children after listening to a Mozart piano sonata for about 10 minutes $\mu$ lies between 7.76 points and 9.24 points.

**41. (a)** $Z_{\alpha/2} = 1.96$ **(b)** 144 steps **(c)** 132 steps. We can estimate the true mean number of fewer steps for each additional hour of television viewing within 132 steps with 95% confidence. **(d)** $673.5\sqrt{n}$

**43. (a)** Since $\sigma$ is a population characteristic, it stays constant and is unaffected by a decrease in confidence level. **(b)** The quantity $\sigma/\sqrt{n}$ is unaffected by a decrease in confidence level. **(c)** A decrease in the confidence level will result in a decrease in $Z_{\alpha/2}$. The width of the confidence interval $= 2 \cdot E = 2 \cdot Z_{\alpha/2}(\sigma/\sqrt{n})$. Thus a decrease in $Z_{\alpha/2}$ will result in a decrease in the width of the confidence interval. **(d)** The quantity $\bar{x}$ depends only on the sample taken, so it will remain unaffected by a decrease in the confidence level. **(e)** A decrease in the confidence level will result in a decrease in $Z_{\alpha/2}$. Since the margin of error is $E = Z_{\alpha/2}(\sigma/\sqrt{n})$, a decrease in $Z_{\alpha/2}$ will result in a decrease in the margin of error.

**45. (a)** 6199 small firms **(b)** See the histogram in **(e)**.
**(c)**

Percent

Number of small firms per metropolitan area

Since the majority of the points lie outside of the curved lines, the normality assumption is not valid.

**(d)** (3188.95, 9209.05). We are 95% confident that the average number of small firms per metropolitan area lies between 3188.95 and 9209.05.

**(e)**

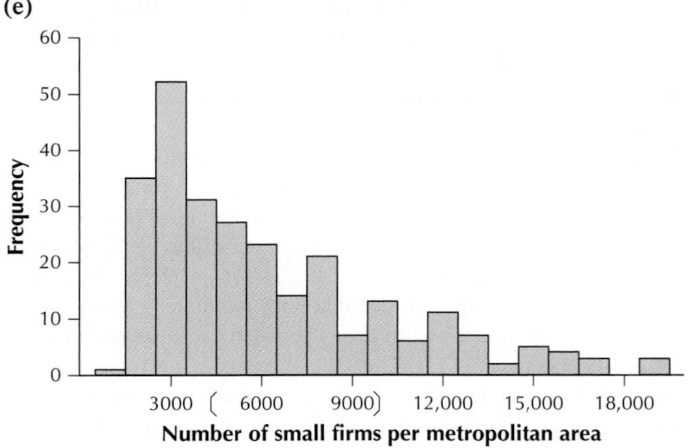


**47.** Answers will vary.

**49.** Answers will vary.

## Section 8.2

**1.** In most real-world problems, the population standard deviation $\sigma$ is unknown, so we can't use the $Z$ interval.

**3.** The $t$ curve approaches closer and closer to the $Z$ curve.

**5. (a)** $t_{\alpha/2} = 1.833$ **(b)** $t_{\alpha/2} = 2.262$ **(c)** $t_{\alpha/2} = 3.250$

**7. (a)** The value of $t_{\alpha/2}$ increases as the confidence level increases. **(b)** The larger the value of $1 - \alpha$, the larger the value of $t_{\alpha/2}$ will have to be in order to have an area of $1 - \alpha$ between $-t_{\alpha/2}$ and $t_{\alpha/2}$. $t_{\alpha/2} = 1.833$ for a 90% confidence interval with 9 degrees of freedom; $t_{\alpha/2} = 2.262$ for a 95% confidence interval with 9 degrees of freedom; $t_{\alpha/2} = 3.250$ for a 99% confidence interval with 9 degrees of freedom.

**9.** Conservative: $t_{\alpha/2} = 2.009$. Interpolation: $t_{\alpha/2} = 2.0054$.

**11.** Conservative: $t_{\alpha/2} = 1.684$. Interpolation: $t_{\alpha/2} = 1.680$.

**13. (a)** $t_{\alpha/2} = 2.064$ **(b)** $E = 2.064$. **(c)** (7.936, 12.064)

**15. (a)** $t_{\alpha/2} = 3.182$ **(b)** $E = 9.546$. **(c)** (40.454, 59.546)

**17. (a)** $t_{\alpha/2} = 1.860$ **(b)** $E = 3.720$. **(c)** $(-23.720, -16.280)$

**19. (a)** $t_{\alpha/2} = 1.987$ **(b)** $E = 1.987$ **(c)** (98.013, 101.987)

**21. (a)** $t_{\alpha/2} = 2.660$ **(b)** $E = 2.660$ **(c)** (32.340, 37.660)

**23. (a)** $t_{\alpha/2} = 1.664$ **(b)** $E = 1.1093$ **(c)** (–21.1093, –18.8907)

**25.** Since the distribution of the population is unknown, Case 1 does not apply. Since the sample size of $n = 25$ is small ($n < 30$), Case 2 does not apply. Thus we cannot construct the indicated confidence interval.

**27.** Case 1 (9.3387, 10.6613)

**29.** Since the distribution of the population is unknown, Case 1 does not apply. Since the sample size of $n = 16$ is small ($n < 30$), Case 2 does not apply. Thus we cannot construct the indicated confidence interval.

**31.** Case 2 (47.97, 52.03)

**33. (a)** $t_{\alpha/2} = 1.994$ **(b)** $E = \$6.91$. We can estimate $\mu$, the true mean revenue collected from all parking meters, within \$6.91 with 95% confidence. **(c)** (113.09, 126.91). We are 95% confident that the true mean revenue collected from all parking meters $\mu$ lies between \$113.09 and \$126.91.

**35. (a)** See the graph. All the data points lie between the curved lines. In fact all the points lie close to the center line. Thus the distribution appears to be normal. **(b)** $t_{\alpha/2} = 2.015$ **(c)** $E = 3.276$ miles per gallon. We can estimate $\mu$, the true mean city gas mileage for hybrid cars, within 3.276 miles per gallon with 90% confidence. **(d)** (27.057, 33.609). We are 90% confident that the true mean city gas mileage for hybrid cars $\mu$ lies between 27.057 miles per gallon and 33.609 miles per gallon.

**37. (a)**

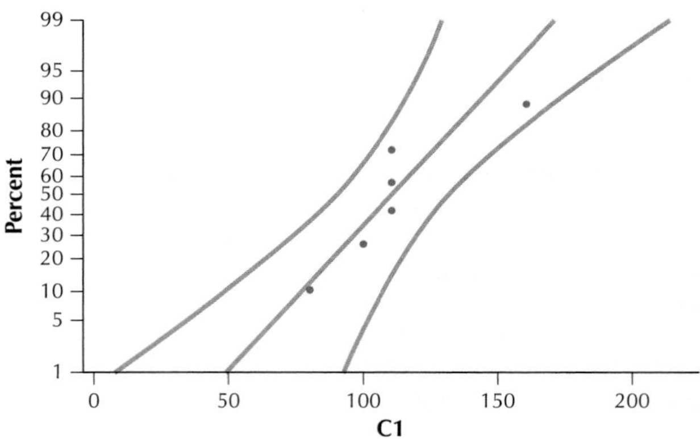

**(b)** Yes, the points do not appear to lie in a straight line. **(c)** Since the data do not appear to be normal, Case 1 does not apply. Since the sample size of $n = 12$ is small ($n < 30$), Case 2 does not apply. Thus a $t$ interval cannot be used.

**39 (a)** $t_{\alpha/2} = 2.045$ **(b)** $E = 1.1003$ miles. We can estimate $\mu$, the true mean commuting distance, within 1.1003 miles with 95% confidence. **(c)** (8.3330, 11.0336). We are 95% confident that the true mean commuting distance lies between 8.3330 miles and 11.0336 miles.

**41. (a)** $t$ interval: (29.6671, 54.3329). $Z$ interval = (32.2248, 51.7752). We are 99% confident that the population mean weekly number of HIV-related deaths in all large American cities lies between 29.6671 and 54.3329. **(b)** Lower the confidence level or increase the sample size.

**43. (a)** The histogram indicates that the data are left-skewed and therefore not normally distributed. The normal probability plot indicates that the data points all lie between the curved bounds but do not all lie along a straight line, so the normality assumption may not be valid. **(b)** $n = 10$, $\bar{x} = -0.306$, $s = 0.3435$ **(c)** $t_{\alpha/2} = 2.262$ **(d)** $(-0.5517, -0.0603)$ **(e)** $E = 0.2457$. We can estimate $\mu$, the true mean net change of all NYSE stocks, within 0.2457 with 95% confidence.

**45. (a)** Except for one outlier on the left, the data appear to be normally distributed. **(b)** (28.9902, 37.7098). We are 95% confident that the true mean rate of minority mortgage lending refusals for all banks lies between 28.9902% and 37.7098%.

**47. (a)** Since the margin of error is $E = t_{\alpha/2}(s/\sqrt{n})$, a decrease in the confidence level will result in a decrease in $t_{\alpha/2}$, which will result in a decrease in the margin of error. **(b)** Since the width of the confidence interval is 2 $E$, a decrease in the confidence level will result in a decrease in $E$, which will result in a decrease in the width of the confidence interval.

**49. (a)** An increase in the sample size will result in a decrease in $t_{\alpha/2}$. **(b)** Since the margin of error is $E = t_{\alpha/2}(s/\sqrt{n})$ and the sample size $n$ occurs in the denominator, an increase in the sample size will result in a decrease in $t_{\alpha/2}$ and a decrease in the margin of error. **(c)** Since the width of the confidence interval is 2 $E$, an increase in the sample size will result in a decrease in $E$, which will result in a decrease in the width of the confidence interval.

## Section 8.3

**1.** No, unless there is some reason to suspect that the value of $p$ has changed.

**3. (a)** $n\hat{p} = 25 \cdot 0.5 = 12.5 \geq 5$ and $n(1 - \hat{p}) = 25 \cdot (1 - 0.5) = 12.5 \geq 5$. **(b)** $Z_{\alpha/2} = 1.96$ **(c)** 0.196

**5. (a)** $n\hat{p} = 100 \cdot 0.95 = 95 \geq 5$ and $n(1 - \hat{p}) = 100 \cdot (1 - 0.95) = 5 \geq 5$. **(b)** $Z_{\alpha/2} = 2.576$ **(c)** 0.0561

**7. (a)** $n\hat{p} = 64 \cdot 0.4 = 25.6 \geq 5$ and $n(1 - \hat{p}) = 64 \cdot (1 - 0.4) = 38.4 \geq 5$. **(b)** $Z_{\alpha/2} = 1.96$ **(c)** 0.1200

**9. (a)** $n\hat{p} = 49 \cdot 0.3 = 14.7 \geq 5$ and $n(1 - \hat{p}) = 49 \cdot (1 - 0.3) = 34.3 \geq 5$. **(b)** $Z_{\alpha/2} = 1.645$ **(c)** 0.1077

**11. (a)** (0.304, 0.696)

**(b)**

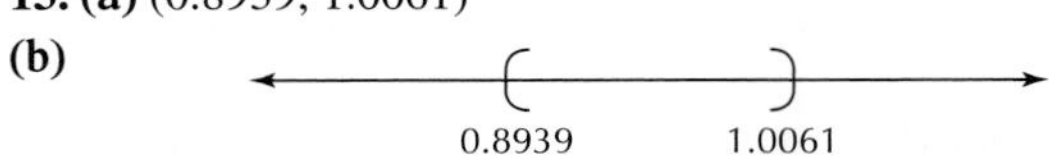


**13. (a)** (0.8939, 1.0061)

**(b)**

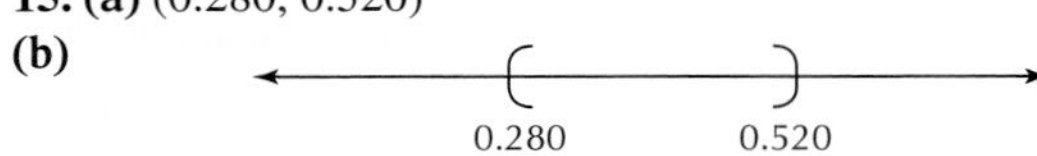


**15. (a)** (0.280, 0.520)

**(b)**

0.280 0.520

**17. (a)** (0.1923, 0.4077)

**(b)**

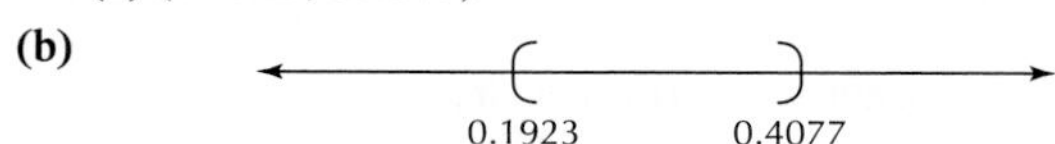


**19. (a)** $Z_{\alpha/2} = 1.96$ **(b)** Yes **(c)** 0.0535 **(d)** (0.0465, 0.1535)

**21. (a)** $Z_{\alpha/2} = 1.645$ **(b)** Yes **(c)** 0.2056 **(d)** (0.2944, 0.7056)

**23. (a)** $Z_{\alpha/2} = 1.96$ **(b)** No **(c)–(d)** Not applicable.

**25. (a)** $Z_{\alpha/2} = 1.645$ **(b)** Yes **(c)** 0.0979 **(d)** (0.0521, 0.2479)

**27. (a)** 0.3099 **(b)** 0.098 **(c)** 0.0310 **(d)** 0.0098

**29. (a)** Since the margin of error is $E = Z_{\alpha/2} \cdot \sqrt{\hat{p}(1 - \hat{p})/n}$, an increase in the sample size while $\hat{p}$ remains constant results in a decrease in the margin of error. **(b)** Since the width of the confidence interval is 2 $E$, an increase in the sample size while $\hat{p}$ remains constant results in a decrease in the width of the confidence interval.

**31. (a)** $Z_{\alpha/2} = 2.576$ **(b)** Yes **(c)** $E = 0.0518$. We can estimate $p$, the true proportion of workers who would be more likely to accept a global assignment, to within 0.0518 with 99% confidence. **(d)** (0.6179, 0.7215). We are 99% confident that the true proportion of workers who would be more likely to accept a global assignment lies between 0.6179 and 0.7215.

**33. (a)** $Z_{\alpha/2} = 1.96$ **(b)** Yes **(c)** $E = 0.0627$. We can estimate $p$, the true proportion of minorities who oppose race-conscious college admissions, within 0.0627 with 95% confidence. **(d)** (0.5773, 0.7027). We are 95% confident that the true proportion of minorities who oppose race-conscious college admissions lies between 0.5773 and 0.7027.

**35. (a)** $Z_{\alpha/2} = 1.96$ **(b)** Yes **(c)** $E = 0.0304$. We can estimate $p$, the true proportion of NASCAR racing attendees who own a pickup truck, within 0.0304 with 95% confidence. **(d)** (0.3696, 0.4304). We are 95% confident that the true proportion of NASCAR racing attendees who own a pickup truck lies between 0.3696 and 0.4304.

**37. (a)** Yes. **(b)** $E = 0.1001$. We can estimate $p$, the true proportion of all antiterrorist incidents that occurred since 2003 that took place in the Middle East, within 0.1001 with 90% confidence. **(c)** (0.2332, 0.4334). Thus we are 90% confident that the true proportion of all anti-American terrorist incidents since 2003 that took place in the Middle East lies between 0.2332 and 0.4334.

**39. (a)** Yes. **(b)** $E = 0.0353$. We can estimate $p$, the true proportion of all white twelfth-graders who are using Ecstasy, within 0.0353 with 99% confidence. **(c)** (0.0037, 0.0743). We are 99% confident that the true proportion of all white twelfth-graders who are using Ecstasy lies between 0.0037 and 0.0743.

**41. (a)** $E = 0.1313$. We can estimate $p$, the true proportion of all high-volume stocks to post an increase in price, within 0.1313 with 95% confidence. **(b)** (0.5422, 0.8048). We are 95% confident that the true proportion of all high-volume stocks to post an increase in price lies between 0.5422 and 0.8048.

**43.** 0.0303

**45. (a)** Decrease **(b)** Unchanged **(c)** Unchanged

**47.** We have $\hat{p} = x/n = 482/1005 = 0.4796$. We have $n\hat{p} = 1005(0.4796) = 481.998 \geq 5$ and $n(1 - \hat{p}) = 1005(1 - 0.48796) = 523.002 \geq 5$. Thus we can use the $Z$ interval for $p$.

**49. (a)** Decrease **(b)** Decrease **(c)** Decrease

**51. (a)** (0.4537, 0.5055) **(b)** We can estimate $p$, the true proportion of all studies sponsored by drug companies that have outcomes favoring the drug company, within 0.0406 with 90% confidence. **(c)** We have $n\hat{p} = 40(0.975) = 39 \geq 5$ but $n(1 - \hat{p}) = 40(1 - 0.975) = 1 < 5$. Thus we cannot use the $Z$ interval for $p$.

**53. (a)** (0.7198, 0.8694) **(b)** (i) Decrease in $Z_{\alpha/2}$ from 1.96 to 1.645. (ii) Decrease in the margin of error from 0.0748 to 0.0628.

(iii) Decrease in the width of the confidence interval from 0.1496 to 0.1256.

**55.** **(a)** Increase **(b)** Unchanged **(c)** Unchanged

## Section 8.4

**1.** The population must be normal.

**3.** To use this method, the distribution has to be symmetric and the $\chi^2$ curve is not symmetric.

**5.** True

**7.** True

**9.** $\chi^2_{1-\alpha/2} = \chi^2_{0.95} = 13.848$ and $\chi^2_{\alpha/2} = \chi^2_{0.05} = 36.415$.

**11.** $\chi^2_{1-\alpha/2} = \chi^2_{0.995} = 9.886$ and $\chi^2_{\alpha/2} = \chi^2_{0.005} = 45.559$.

**13.** $\chi^2_{1-\alpha/2} = \chi^2_{0.975} = 5.629$ and $\chi^2_{\alpha/2} = \chi^2_{0.025} = 26.119$.

**15.** For a given sample size $\chi^2_{1-\alpha/2}$, decreases and $\chi^2_{\alpha/2}$ increases as the confidence level increases.

**17.** Lower bound = 6.59, upper bound = 17.33

**19.** Lower bound = 5.27, upper bound = 24.28

**21.** Lower bound = 2.47, upper bound = 4.40

**23.** As the confidence level increases but the sample size stays the same, the lower bound for the confidence interval for $\sigma^2$ decreases and the upper bound for the confidence interval for $\sigma^2$ increases.

**25.** Lower bound = 4.73, upper bound = 33.33

**27.** Lower bound = 5.78, upper bound = 21.33

**29.** Lower bound = 2.32, upper bound = 4.99

**31.** As the sample size increases but the confidence level stays the same, the lower bound of a confidence interval for $\sigma^2$ increases and the upper bound of a confidence interval for $\sigma^2$ decreases.

**33.** **(a)** $\chi^2_{1-\alpha/2} = \chi^2_{0.975} = 0.484$ and $\chi^2_{\alpha/2} = \chi^2_{0.025} = 11.143$. **(b)** Lower bound = 154.82, upper bound = 3564.46. We are 95% confident that the population variance $\sigma^2$ lies between 154.82 prisoners squared and 3564.46 prisoners squared. **(c)** Lower bound = 12.44, upper bound = 59.70. We are 95% confident that the population standard deviation $\sigma$ lies between 12.44 prisoners and 59.70 prisoners.

**35.** **(a)** Prisoners squared **(b)** No **(c)** Prisoners **(d)** Yes

**37.** **(a)** $\chi^2_{1-\alpha/2} = \chi^2_{0.975} = 8.907$ and $\chi^2_{\alpha/2} = \chi^2_{0.025} = 32.852$. **(b)** Lower bound = 0.95, upper bound = 3.49. We are 95% confident that the population variance $\sigma^2$ lies between 0.95 centimeters squared and 3.49 centimeters squared. **(c)** Lower bound = 0.97, upper bound =1.87. We are 95% confident that the population standard deviation $\sigma$ lies between 0.97 centimeters and 1.87 centimeters. **(d)** The interpretation in (c) is easier to understand since it is in the same units as our data.

**39.** **(a)** $\chi^2_{1-\alpha/2} = \chi^2_{0.995} = 1.735$ and $\chi^2_{\alpha/2} = \chi^2_{0.005} = 23.589$. **(b)** Lower bound =1.85, upper bound = 6.81. We are 99% confident that the population standard deviation $\sigma$ lies between 1.85 trillion BTUs and 6.81 trillion BTUs.

## Section 8.5

**1.** The higher the confidence level, the wider the confidence interval.

**3.** As the margin of error is increased, $n$ decreases, and as the margin of error is decreased, $n$ increases.

**5.** 1

**7.** 5

**9.** 5

**11.** 11

**13.** 2

**15.** 17

**17.** 1068

**19.** 17,074

**21.** **(a)** The sample size required is increased by a factor of approximately 4. **(b)** The point of diminishing returns is somewhere between a margin of error of 0.005 and 0.010.

**23.** 385

**25.** 5

**27.** 752

**29.** 1844

**31.** 897

**33.** 897

**35.** 1844

**37.** **(a)** 3 **(b)** 62

**39.** **(a)** 271 **(b)** 6766

**41.** 1770

**43.** **(a)** 3 **(b–c)** Answers will vary.

**45.** **(a)** 35 **(b)** 865 **(c)** 35 **(d)** 865 **(e)** The value of the mean does not play any role in calculating the required sample size.

**47.** **(a)** 97 **(b)** 9604 days ≈ 26.31 years.

**49.** 98,345 days ≈ 269.44 years. Neither the analyst nor the boss will live long enough to collect a sample of this size.

## Chapter 8 Review

**1.** **(a)** Case 1 **(b)** $\sigma/\sqrt{n} = 2$ **(c)** $Z_{\alpha/2} = 1.96$ **(d)** $E = 3.92$. We can estimate $\mu$ to within 3.92 with 95% confidence. **(e)** (46.08, 53.92). We are 95% confident that the true mean $\mu$ lies between 46.08 and 53.92.

**3.** **(a)** Case 2 **(b)** $\sigma/\sqrt{n} = 1$ **(c)** $Z_{\alpha/2} = 1.96$ **(d)** $E = 1.96$. We can estimate $\mu$ to within 1.96 with 95% confidence. **(e)** (48.04, 51.96). We are 95% confident that the true mean $\mu$ lies between 48.04 and 51.96.

**5.** **(a)** (6.4156, 7.5844) **(b)** (6.2320, 7.7680) **(c)** The width of the confidence interval increases as the confidence level increases.

**7.** **(a)** \$45,000 **(b)** $\sigma/\sqrt{n} = \$1200$ **(c)** $Z_{\alpha/2} = 2.576$ **(d)** $E =$ \$3091.20. We can estimate $\mu$, the mean income of all working women in California to within \$3091.20 with 99% confidence. **(e)** (41,908.80, 48,091.20). We are 99% confident that the true mean income of all working women in California $\mu$ lies between \$41,908.80 and \$48,091.20.

**9.** Since the mean income of the population of all working women in California is a non-negative real number, any interval that contains the non-negative real numbers will be a 100% confidence interval for $\mu$.

**11.** Case 1; the confidence interval is (20.289, 23.711).

**13.** **(a)** (2,162.65, 2,621.85) **(b)** (2,052.56, 2,731.94) **(c)** The interval in (a) is more precise than the interval in (b) but the interval in (b) has higher confidence of containing $\mu$.

**15.** **(a)** $t_{\alpha/2} = 1.667$ **(b)** $E = \$5.77$. We can estimate $\mu$, the true mean amount of money collected by all parking meters, within \$5.77 with 90% confidence. **(c)** (114.23, 125.77). We are 90% confident that the true mean amount of money collected by all parking meters lies between \$114.23 and \$125.77.

**17.** **(a)** A 90% confidence interval will be shorter than a 99% confidence interval and therefore the estimate will be more precise. But we are more confident that a 99% confidence interval will contain the true value of the mean than we are for a 90% confidence interval. **(b)** 90% confidence level: $E = 0.9305$ colors 99% confidence level: $E = 1.5095$ colors. The margin of error increases as the confidence level increases.

**19.** **(a)** Unchanged. **(b)** Decrease **(c)** Increase from 5 to 6 **(d)** Decrease

**21.** **(a)** $Z_{\alpha/2} = 1.96$ **(b)** $E = 0.098$. We can estimate $p$, the true population proportion, within $E = 0.098$ with 95% confidence. **(c)** No **(d)** We have $n\hat{p} = 36(0.1) = 3.6 < 5$, so we cannot use the $Z$ interval for $p$.

**23.** **(a)** $Z_{\alpha/2} = 1.96$ **(b)** $E = 0.0195$. We can estimate $p$, the true population proportion, within 0.0195 with 95% confidence. **(c)** No **(d)** We have $n\hat{p} = 100(0.99) = 99 \geq 5$ but $n(1 - p) = 100(1 - 0.99) = 1 < 5$. Thus we cannot use the $Z$ interval for $p$.
**25.** **(a)** Yes **(b)** $E = 0.1140$. We can estimate $p$, the true population proportion, within $E = 0.1140$ with 95% confidence. **(c)** (0.1693, 0.3973). We are 95% confident that the true proportion of all anti-American terrorist incidents occurring since 2003 that took place in the Middle East lies between 0.1693 and 0.3973.
**27.** **(a)** We have $n\hat{p} = 1000(0.89) = 890 \geq 5$ and $n(1 - \hat{p}) = 1000(1 - 0.89) = 110 \geq 5$. Thus we can use the $Z$ interval for $p$. **(b)** The margin of error is $E = 0.0255$. We can estimate $p$, the true proportion of all erasures that resulted in a correct answer at the prestigious school in Fairfield, CT, within $E = 0.0255$ with 99% confidence. **(c)** (0.8645, 0.9155) **(d)** Since 0.70 does not lie in our confidence interval and none of the values below 0.70 lie in our confidence interval, we are 99% confident that no number less than or equal to 0.70 is a plausible value for $p$. All numbers less than or equal to 0.70 lie below our confidence interval. This is an indication that tampering may have occurred.
**29.** Lower bound = 65.786, upper bound = 170.159
**31.** Lower bound = 8.383, upper bound = 12.482
**33.** Lower bound = 7.620, upper bound = 14.268
**35.** 1
**37.** 1
**39.** 752
**41.** 385
**43.** 5
**45.** **(a)** 482 **(b)** 683 **(c)** 1,180

## Chapter 8 Quiz

**1.** False
**2.** True
**3.** True
**4.** 0
**5.** 4
**6.** less
**7.** $\alpha$ is a probability.
**8.** Either the population is normal or the sample size is large ($n \geq 30$).
**9.** The first method for finding the required sample size if $p$ is unknown is when prior information about $p$ is available. Use the formula $n = \hat{p}(1 - \hat{p})(Z_{\alpha/2}/E)^2$ where $\hat{p}$ is the sample proportion of successes available from an earlier sample. The second method for finding the required sample size if $p$ is unknown is when no prior information about $p$ is available. Use the formula $n = (0.5 \cdot (Z_{\alpha/2})/E))^2$. The second method will always deliver the larger sample size.
**10.** **(a)** \$30,500 **(b)** \$428.57 **(c)** $Z_{\alpha/2} = 1.645$ **(d)** $E = \$705$. We can estimate $\mu$, the mean cost of a college education, to within \$705 with 90% confidence. **(e)** (29,795, 31,205). We are 90% confident that the true mean cost of a college education lies between \$29,795 and \$31,205.
**11.** **(a)** 30 pounds **(b)** $Z_{\alpha/2} = 1.645$ **(c)** $E = 49.35$ pounds. We can estimate $\mu$, the mean femur load number in a frontal crash for the passenger in a 2005 Ford Equinox SUV, within 49.35 pounds with 90% confidence. **(d)** (953.65, 1052.35). We are 90% confident that the true mean femur load number in a frontal crash for the passenger in a 2005 Ford Equinox SUV lies between 953.65 pounds and 1052.35 pounds.
**12.** **(a)** (55.15, 58.85). We are 90% confident that the true mean day count of wolves in Isle Royale lies between 55.15 wolves and 58.85 wolves. **(b)** Since we can't decrease the confidence level the only way we can decrease the margin of error is to increase the sample size.
**13.** **(a)** (9.30, 9.70). We are 95% confident that the true mean fog index for all the books of this horror author lies between 9.30 and 9.70. **(b)** (9.234, 9.776); less precise than the 95% confidence interval in (a). **(c)** The 99% confidence interval is wider and less precise than the 95% confidence interval, but we are more confident that the 99% confidence contains the true value of the mean. **(d)** Since $\sigma$ is unknown, it would not be appropriate to use a $Z$ interval.
**14.** **(a)** Since all the points lie close to the center line, the normality assumption appears valid. Since $\sigma$ is unknown, Case 1 applies and we can use the $t$ interval. **(b)** (72.5763, 78.6237). We are 90% confident that the true mean heart rate of all women $\mu$ lies between 72.5763 beats per minute and 78.6237 beats per minute.
**15.** **(a)** $E = 0.0142$. We can estimate $p$, the true proportion of all Americans who attended a religious service in response to the attacks on the World Trade Center and the Pentagon, within 0.0142 with 95% confidence. **(b)** (0.2513, 0.2797). We are 95% confident that the true proportion of all Americans who attended a religious service in response to the attacks on the World Trade Center and the Pentagon lies between 0.2513 and 0.2797.
**16.** **(a)** $E = 0.0386$. We can estimate $p$, the true proportion of all Québecois who favor independence for the Province of Quebec, within 0.0386 with 99% confidence. **(b)** (0.3014, 0.3786)
**17.** **(a)** Yes, Amanda's and Bethany's delivery times are normally distributed. **(b)** (1.863, 4.944). We are 95% confident that the population standard deviation of Amanda's delivery times lies between 1.8623 minutes and 4.941 minutes. **(c)** (6.083, 16.147). We are 95% confident that the population standard deviation of Bethany's delivery times lies between 6.083 minutes and 16.147 minutes. **(d)** The confidence interval for the population standard deviation of Bethany's delivery times is wider than the confidence interval for the population standard deviation of Amanda's delivery times. This shows that there is strong statistical evidence that Bethany's delivery times are more highly variable than Amanda's .
**18.** **(a)** 363 **(b)** 626
**19.** **(a)** 554 **(b)** 2,213
**20.** **(a)** 17 **(b)** Since the margin of error ($E = 0.20$) is so large, it won't be very precise. **(c)** 752

# CHAPTER 9

## Section 9.1

**1.** Answers will vary.
**3.** The alternate hypothesis represents an alternative claim about the value of the parameter.
**5.**

| Form | Null hypothesis | Alternative hypothesis |
|---|---|---|
| 1 | $H_0: \mu \leq \mu_0$ vs. | $H_a: \mu > \mu_0$ |
| 2 | $H_0: \mu \geq \mu_0$ vs. | $H_a: \mu < \mu_0$ |
| 3 | $H_0: \mu = \mu_0$ vs. | $H_a: \mu \neq \mu_0$ |

**7.** **(1)** Finding the defendant guilty when in reality he did not commit the crime (a Type I error) and **(2)** finding the defendant not guilty when in reality he did commit the crime (a Type II error).
**9.** $H_0: \mu \leq 10$ vs. $H_a: \mu > 10$
**11.** $H_0: \mu = 0$ vs. $H_a: \mu \neq 0$
**13.** $H_0: \mu = 4.0$ vs. $H_a: \mu \neq 4.0$
**15.** $H_0: \mu = 36$ vs. $H_a: \mu \neq 36$
**17.** **(a)** $H_0: \mu \leq 43.9$ vs. $H_a: \mu > 43.9$ **(b)** A Type II error was made.

**19. (a)** $H_0: \mu \leq \mu_0$ vs. $H_a: \mu > \mu_0$ **(b)** A Type I error was made.
**21. (a)** $H_0: \mu \leq 3$ vs. $H_a: \mu > 3$ **(b)** No error was made.
**23. (a)** $H_0: \mu = 339.1$ vs. $H_a: \mu \neq 339.1$ **(b)** (1) the mean number of fatal injury collisions is different than 339.1 per year when the population mean number of fatal injury collisions is actually different than 339.1 per year and (2) the average number of fatal injury collisions is equal to 339.1 per year when in actuality the population mean number of fatal injury collisions is equal to 339.1 per year. **(c)** The mean number of fatal injury collisions is different than 339.1 per year when the population mean number of fatal injury collisions is actually equal to 339.1 per year. **(d)** The mean number of fatal injury collisions is equal to 339.1 per year when actually the population mean number of fatal injury collisions is not equal to 339.1 per year.
**25. (a)** $H_0: \mu = 175$ vs. $H_a: \mu \neq 175$ **(b)** (1) the mean height of Americans this year has changed from 175 centimeters when it actually is different from 175 centimeters and (2) the mean height of Americans has not changed from 175 centimeters when it actually is equal to 175 centimeters. **(c)** The mean height of Americans this year has changed from 175 centimeters when it actually is equal to 175 centimeters. **(d)** the mean height of Americans has not changed from 175 centimeters when it actually is not equal to 175 centimeters.
**27. (a)** $H_0: \mu \leq 52{,}200$ vs. $H_a: \mu > 52{,}200$ **(b)** (1) the mean salary of college graduates is actually greater than \$52,200 when it actually is greater than \$52,200 and (2) the mean salary of college graduates is less than or equal to \$52,200 when it actually is less than or equal to \$52,200. **(c)** The mean salary of college graduates is greater than \$52,200 when it actually is less than or equal to \$52,200. **(d)** The mean salary of college graduates is less than or equal to \$52,200 when it is actually greater than \$52,200.

## Section 9.2

**1.** Answers will vary.
**3.** The $p$-value is small when it is less than $\alpha$.
**5.** In Example 9.9 the $p$-value $= 0.0668$. If $\alpha = 0.10$, then the $p$-value $< \alpha$, so we would reject the null hypothesis. We would interpret our conclusion thus: "There is sufficient evidence that the population mean user rating for a Dell XPS 410 computer is less than 7.2."
**7.** $Z_{\text{data}} = -6$
**9.** $Z_{\text{data}} = 6$
**11.** $Z_{\text{data}} = -4$
**13. (a)** Since the $p$-value $< \alpha$, we reject the null hypothesis. **(b)** There is evidence that the population mean is less than $\mu_0$.
**15. (a)** Since the $p$-value $\geq \alpha$, we do not reject the null hypothesis. **(b)** There is insufficient evidence that the population mean is less than $\mu_0$.
**17. (a)** Since the $p$-value $< \alpha$, we reject the null hypothesis. **(b)** There is evidence that the population mean is less than $\mu_0$.
**19. (a)**

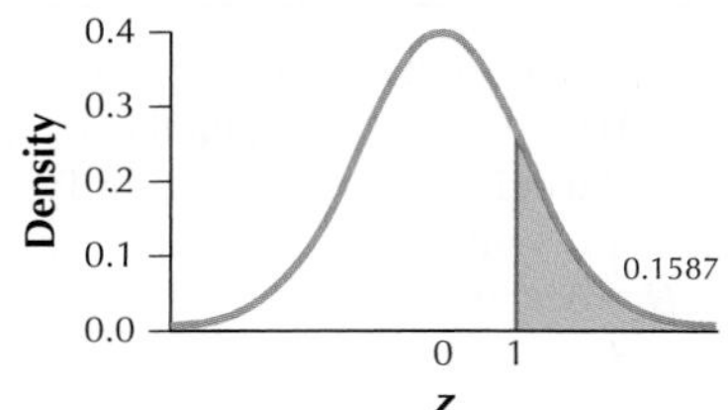


**(b)** $p$-value $= 0.1587$

**21. (a)**

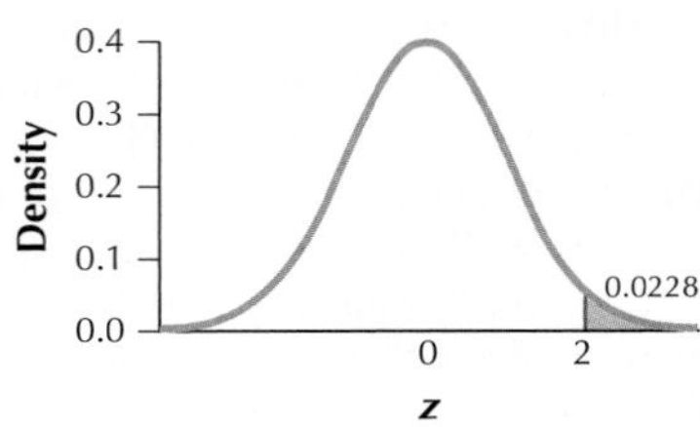


**(b)** $p$-value $= 0.0228$
**23. (a)**

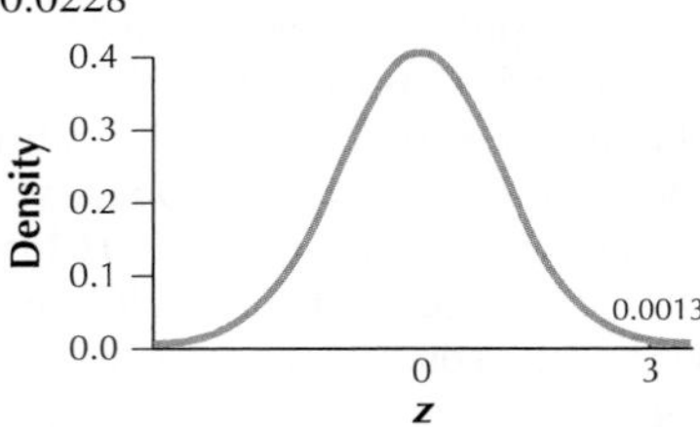


**(b)** $p$-value $= 0.0013$
**25. (a)** $H_0: \mu \leq 100$ vs. $H_a: \mu > 10.0$. Reject $H_0$ if the $p$-value $<$ 0.05. **(b)** $Z_{\text{data}} = 2$ **(c)** $p$-value $= 0.0228$.

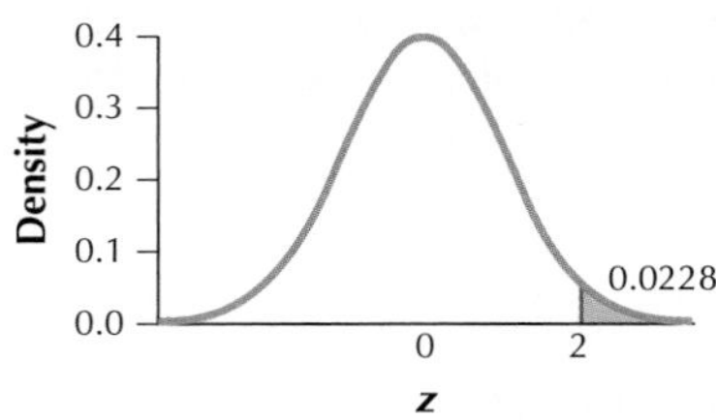


**(d)** Since the $p$-value $< \alpha$, reject $H_0$. There is evidence that the population mean is greater than 100.
**27. (a)** $H_0: \mu = 20$ vs. $H_a: \mu \neq 20$. Reject $H_0$ if the $p$-value $<$ 0.01. **(b)** $Z_{\text{data}} = 9.8$ **(c)** $p$-value $\approx 0$.

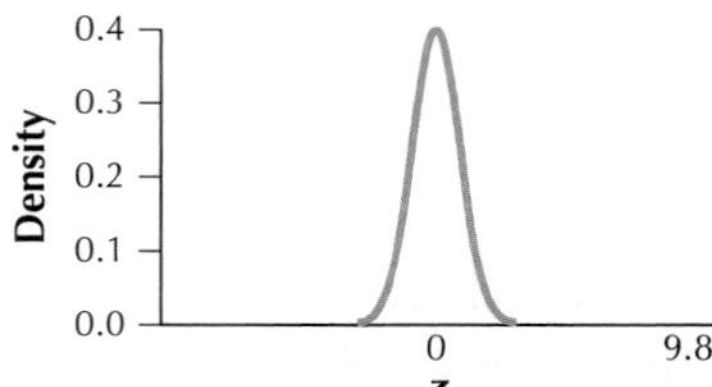


**(d)** Since the $p$-value $< \alpha$, reject $H_0$. There is evidence that the population mean is different than 20.
**29.** Very strong evidence against the null hypothesis
**31.** No evidence against the null hypothesis
**33.** Extremely strong evidence against the null hypothesis
**35.** No evidence against the null hypothesis
**37. (a)** $H_0: \mu \leq 1{,}600{,}000{,}000$ vs. $H_a: \mu > 1{,}600{,}000{,}000$. Reject $H_0$ if the $p$-value $< 0.05$. **(b)** $-1.2$ **(c)** $p$-value $= 0.8849$ **(d)** Since the $p$-value $\geq \alpha$, do not reject $H_0$. There is insufficient evidence that the population daily mean number of shares traded is greater than 1.6 billion shares.
**39. (a)** $H_0: \mu \leq 2$ vs. $H_a: \mu > 2$. Reject $H_0$ if the $p$-value $< 0.05$. **(b)** 24 **(c)** $p$-value $\approx 0$ **(d)** Since the $p$-value $< \alpha$, reject $H_0$. There is evidence that the population mean temperature increase in California is greater than 2 degrees.
**41. (a)** $H_0: \mu \leq 47.2$ vs. $H_a: \mu > 47.2$. Reject $H_0$ if the $p$-value $<$ 0.01. **(b)** 16.60 **(c)** $p$-value $\approx 0$. **(d)** Since the $p$-value $< \alpha$, reject $H_0$. There is evidence that the population mean DDT level in the breast milk of Hispanic women in the Yakima valley is greater than 47.2 parts per billion.

**43. (a)** $H_0: \mu \ge 3$ vs. $H_a: \mu < 3$. Reject $H_0$ if the $p$-value $< 0.01$. **(b)** $-13.5$ **(c)** $p$-value $\approx 0$ **(d)** Since the $p$-value $< \alpha$, reject $H_0$. There is evidence that the population mean time hybrid cars take to recoup their initial cost is less than 3 years.
**45. (a)** Yes, the normal probability plot indicates acceptable normality. **(b)** $H_0: \mu \ge 210$ vs. $H_a: \mu < 210$. Reject $H_0$ if $p$-value $< 0.01$. $Z_{data} = -1.69$. $p$-value $= 0.0455$. Since $p$-value $\ge \alpha$, do not reject $H_0$. There is insufficient evidence that the population mean sodium content per serving of breakfast cereal is less than 210 grams.
**47. (a)** $H_0: \mu \ge 210$ vs. $H_a: \mu < 210$. Reject $H_0$ if $p$-value $<$ 0.05. $Z_{data} = -1.69$. $p$-value $= 0.0455$. Since $p$-value $< \alpha$, reject $H_0$. There is evidence that the population mean sodium content per serving of breakfast cereal is less than 210 grams. **(b)** The data have not changed; α was increased to a value greater than the $p$-value. **(c)** Report the $p$-value and assess the strength of the evidence against the null hypothesis; obtain more data.
**49. (a)** Yes, the normal probability plot indicates acceptable normality. **(b)** $H_0: \mu \ge 78$ vs. $H_a: \mu < 78$. Reject $H_0$ if $p$-value $<$ 0.05. $Z_{data} = -1.03$. $p$-value $= 0.1515$. Since $p$-value $\ge \alpha$, do not reject $H_0$. There is insufficient evidence that the population mean heart rate for all women is less than 78 beats per minute. **(c)** $H_0$ : $\mu = 78$ vs. $H_a: \mu \ne 78$. Reject $H_0$ if $p$-value $< 0.05$. $Z_{data} = -1.03$. $p$-value $= 0.303$. Since $p$-value $\ge 0.05$, do not reject $H_0$. There is insufficient evidence that the population mean heart rate for all women is different than 78 beats per minute.
**51. (a)** $H_0: \mu \ge 3.15$ vs. $H_a: \mu < 3.15$. Reject $H_0$ if the $p$-value $< 0.05$. **(b)** $-1.5$. **(c)** $p$-value $= 0.0668$ **(d)** Since the $p$-value $\ge$ 0.05, do not reject $H_0$. There is insufficient evidence that the population mean family size in America is less than 3.15 persons.
**53. (a)** 0.0668 **(b)** The assumption $\sigma = 1$ person seems valid. This value would have been obtained from the sample standard deviation. **(c)** Type II error; Type I error **(d)** This headline is not supported by the data and our hypothesis test.
**55. (a)** $p$-value $= 0.0853$ **(b)** Since $p$-value $\ge 0.05$, we would not reject $H_0$.
**57. (a)** $p$-value $= 0.0384$ **(b)** Since $p$-value $< 0.05$, we would reject $H_0$.

## Section 9.3

**1.** Threshold for significance.
**3.** Threshold for significance.
**5.** Probability.
**7.** Not directly related.
**9.** Directly related.
**11.** Not directly related.
**13. (a)** $-1.28$ **(b)** Reject $H_0$ if $Z_{data} < -1.28$
**(c)**

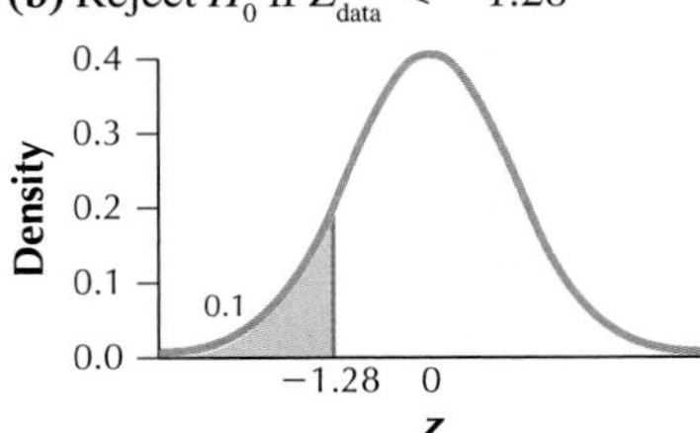


**15. (a)** $-2.33$ **(b)** Reject $H_0$ if $Z_{data} < -2.33$
**(c)**

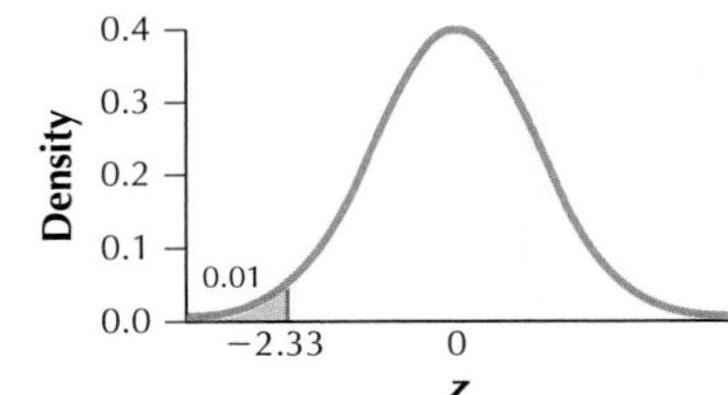


**17. (a)** $-1.645$ **(b)** Reject $H_0$ if $Z_{data} < -1.645$
**(c)**

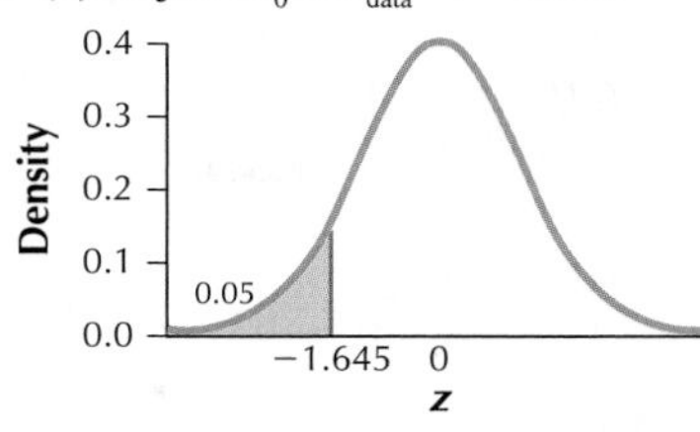


**19. (a)** 2.58 **(b)** Reject $H_0$ if $Z_{data} < -2.58$ or $Z_{data} > 2.58$.
**(c)**

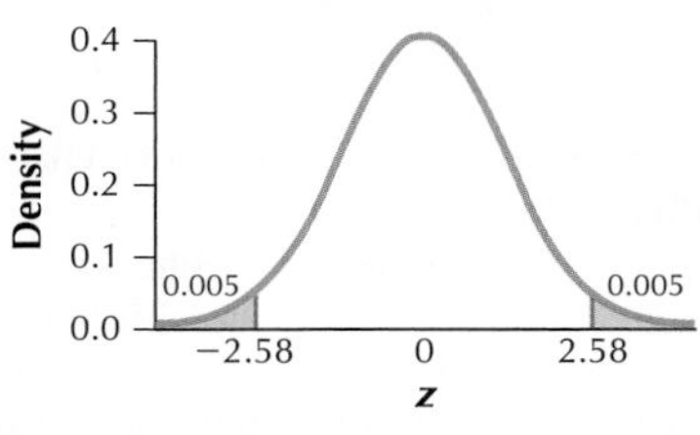


**21. (a)** 1.645 **(b)** Reject $H_0$ if $Z_{data} > 1.645$
**(c)**

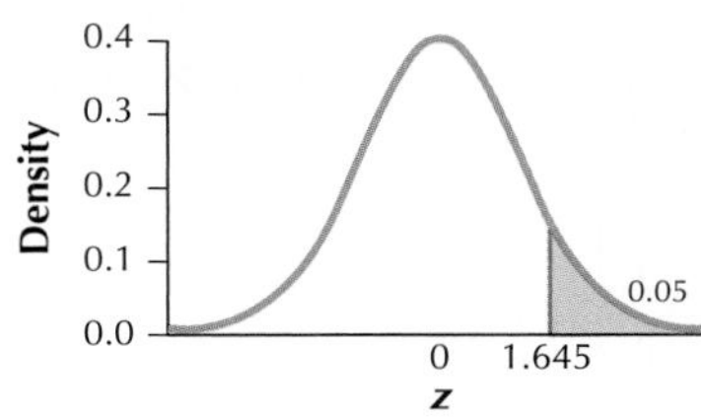


**(d)** $Z_{data} > 1.645$, so we reject $H_0$. **(e)** There is evidence that the population mean is greater than $\mu_0$.
**23. (a)** 1.645 **(b)** Reject $H_0$ if $Z_{data} > 1.645$
**(c)**

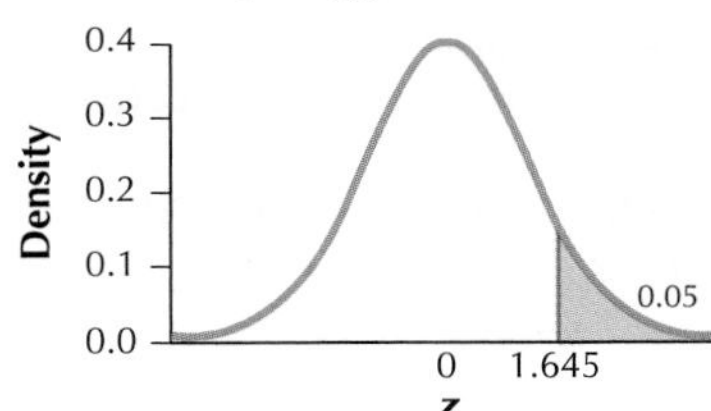


**(d)** $Z_{data} \le 1.645$, so we do not reject $H_0$. **(e)** There is insufficient evidence that the population mean is greater than $\mu_0$.
**25. (a)** $-1.28$ **(b)** Reject $H_0$ if $Z_{data} < -1.28$
**(c)**

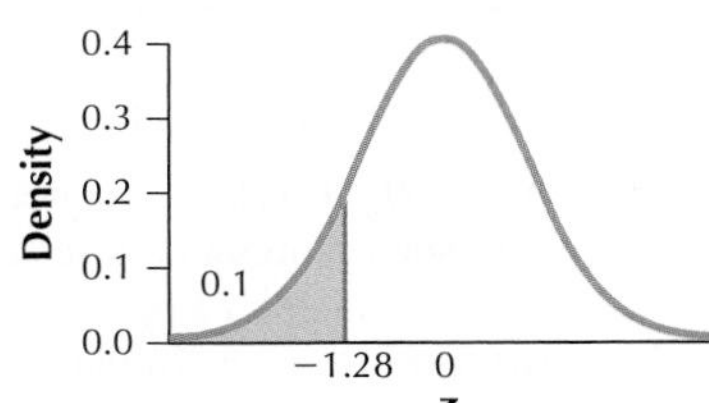


**(d)** Since $Z_{data} < -1.28$, we reject $H_0$. **(e)** There is evidence that the population mean is less than $\mu_0$.
**27. (a)** $H_0: \mu \le 100$ vs. $H_a: \mu > 100$ **(b)** $Z_{crit} = 1.645$. Reject $H_0$ if $Z_{data} > 1.645$. **(c)** $Z_{data} = 2$

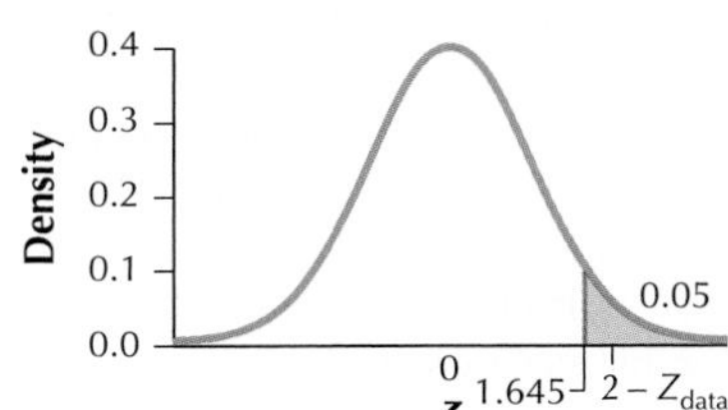

**(d)** Since $Z_{data} > 1.645$, we reject $H_0$. There is evidence that the population mean is greater than 100.

**29. (a)** $H_0: \mu = 20$ vs. $H_a: \mu \neq 20$ **(b)** $Z_{crit} = 2.58$. Reject $H_0$ if $Z_{data} < -2.58$ or $Z_{data} > 2.58$. **(c)** $Z_{data} = 9.8$

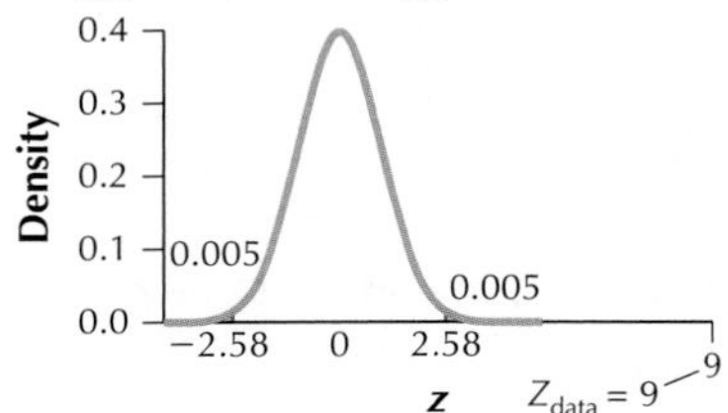


**(d)** Since $Z_{data} > 2.58$ we reject $H_0$. There is evidence that the population mean is different than 20.

**31. (a)** Since $\mu_0 = -3$ does not lie in the confidence interval, we reject $H_0$. **(b)** Since $\mu_0 = -2$ lies in the confidence interval we do not reject $H_0$. **(c)** Since $\mu_0 = 0$ lies in the confidence interval, we do not reject $H_0$. **(d)** Since $\mu_0 = 5$ lies in the confidence interval, we do not reject $H_0$. **(e)** Since $\mu_0 = 7$ does not lie in the confidence interval, we reject $H_0$.

**33. (a)** $H_0: \mu \leq 3.24$ vs. $H_a: \mu > 3.24$ **(b)** $Z_{crit} = 1.645$; Reject $H_0$ if $Z_{data} > 1.645$ **(c)** $Z_{data} = 6.4$

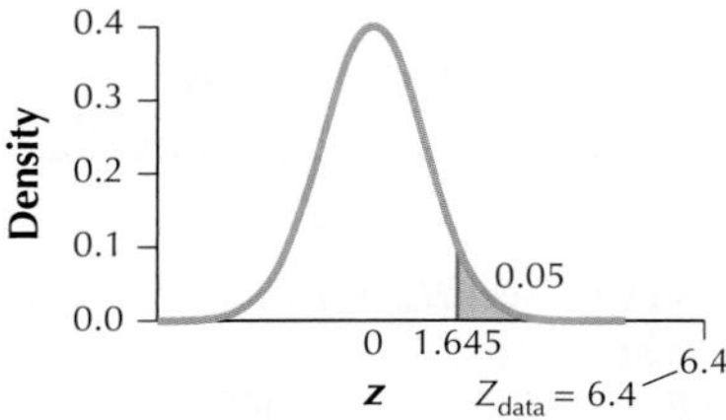


**(d)** Since $Z_{data} > 1.645$ we reject $H_0$. There is evidence that the population mean price of a gallon of milk this year is greater than \$3.24 per gallon.

**35. (a)** $H_0: \mu \leq 50{,}000$ vs. $H_a: \mu > 50{,}000$ **(b)** $Z_{crit} = 2.33$. Reject $H_0$ if $Z_{data} > 2.33$. **(c)** $Z_{data} = 15.75$

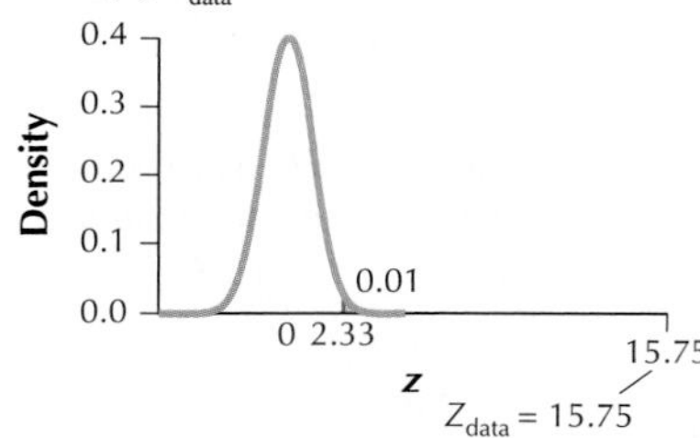


**(d)** Since $Z_{data} > 2.33$, we reject $H_0$. There is evidence that the population mean salary of assistant professors in science is greater than \$50,000.

**37. (a)** (23,160, 24,840) **(b)** (i) Since $\mu_0 = 24{,}000$ lies in the confidence interval, we do not reject $H_0$. (ii) Since $\mu_0 = 23{,}000$ does not lie in the confidence interval, we reject $H_0$. (iii) Since $\mu_0 = 23{,}200$ lies in the confidence interval, we do not reject $H_0$. (iv) Since $\mu_0 = 25{,}000$ does not lie in the confidence interval, we reject $H_0$.

**39. (a)–(c)** Unchanged **(d)** Value of $Z_{data}$ greater than 1.67 **(e)** Unchanged

**41.** The histogram indicates that the data are extremely right-skewed and therefore not normally distributed. Thus Case 1 does not apply. Since the sample size of $n = 16$ is small ($n < 30$), Case 2 does not apply. Thus it is not appropriate to apply the $Z$ test.

**43. (a)** Since the sample size of $n = 100$ is large ($n \geq 30$), Case 2 applies, so it is appropriate to apply the $Z$ test. **(b)** Even though the sample mean $\bar{x} = 6.2$ cents per mile is greater than the hypothesized mean $\mu_0 = 5.9$ cents per mile, this is not enough by itself to reject the null hypothesis. It also depends on the variability of the data and on α. **(c)** $\sigma_{\bar{x}} = 0.15$ cents per mile. **(d)** Since $\bar{x} = 6.2$ cents per mile is 2 standard errors above $\mu_0 = 5.9$ cents per mile, it is mildly extreme.

**45. (a)** $H_0: \mu \leq 5.9$ vs. $H_a: \mu > 5.9$ **(b)** $Z_{crit} = 1.645$. Reject $H_0$ if $Z_{data} > 1.645$ **(c)** $Z_{data} = 2$ **(d)** Since $Z_{data} > 1.645$ we reject $H_0$. There is evidence that the population mean cost of operating an automobile in the United States is greater than 5.9 cents per mile.

## Section 9.4

**1.** The population standard deviation $\sigma$ is known.

**3.** Case 1: The population is normal. Case 2: The sample size is large ($n \geq 30$).

**5.** $t_{crit} = -1.363$

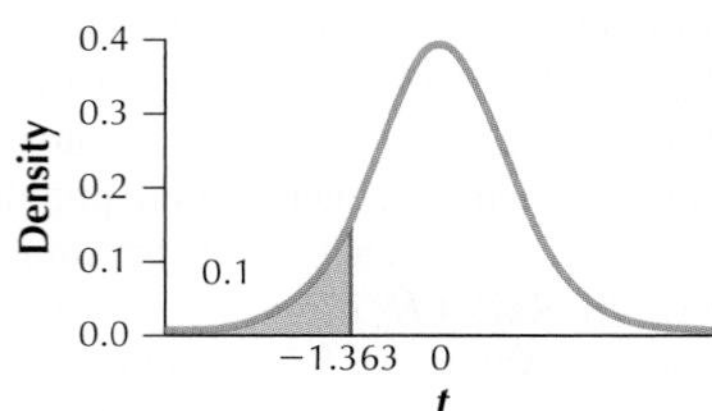


**7.** $t_{crit} = -2.718$

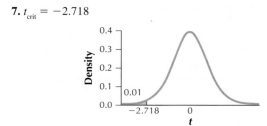

**9.** $t_{crit} = 1.740$

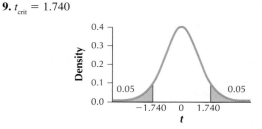

**11.** $t_{crit} = 2.898$

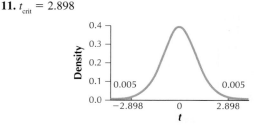

**13.** $t_{data} = -3$

**15.** $t_{data} = -2.67$

**17.** $t_{data} = 3.6$

**19.** $0.01 < p\text{-value} < 0.025$

**21.** $0.05 < p\text{-value} < 0.10$

**23.** $0.01 < p\text{-value} < 0.025$

**25.** Solid evidence

**27.** Mild evidence

**29.** Solid evidence

**31.** **(a)** $H_0: \mu = 0$ vs. $H_a: \mu \neq 0$. Reject $H_0$ if the $p$-value $< 0.05$. **(b)** $t_{data} = 6$ **(c)** $p$-value $= 0.0003233933213$ **(d)** Since the $p$-value $< 0.05$, we reject $H_0$. There is evidence that the population mean is different than 0.
**33.** **(a)** $H_0: \mu \geq 28$ vs. $H_a: \mu < 28$. Reject $H_0$ if the $p$-value $<$ 0.05. **(b)** $t_{data} = -1$ **(c)** $p$-value $= 0.1598742373$. **(d)** Since the $p$-value $\geq 0.05$, we do not reject $H_0$. There is insufficient evidence that the population mean is less than 28.
**35.** **(a)** $H_0: \mu \leq 100$ vs. $H_a: \mu > 100$ **(b)** $t_{crit} = 2.492$. Reject $H_0$ if $t_{data} > 2.492$.

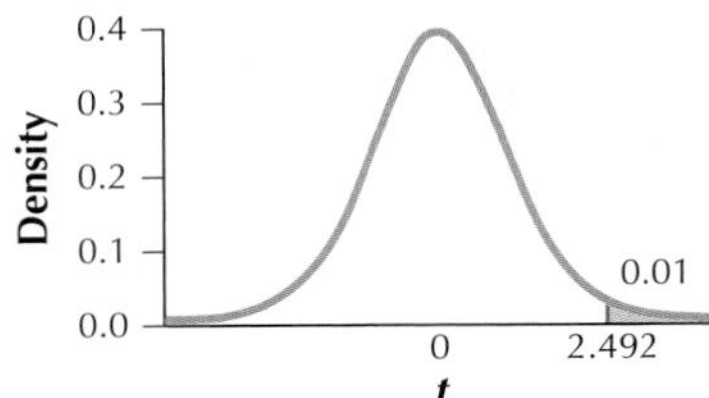


**(c)** 2 **(d)** Since $t_{data} \leq 2.492$, we do not reject $H_0$. There is insufficient evidence that the population mean is greater than 100.
**37.** **(a)** $H_0: \mu = 9$ vs. $H_a: \mu \neq 9$ **(b)** $t_{crit} = 1.690$. Reject $H_0$ if $t_{data} < -1.690$ or $t_{data} > 1.690$.

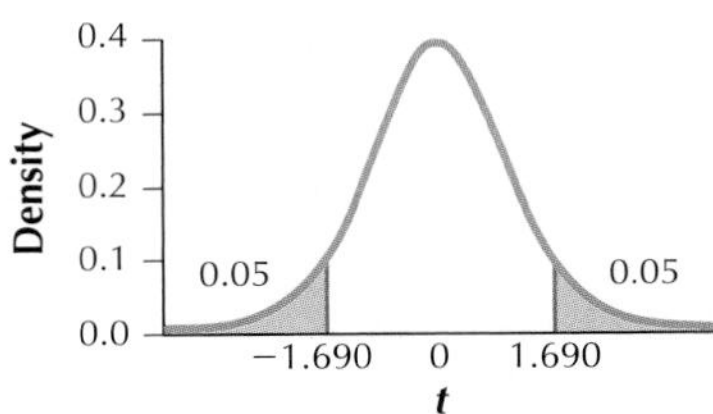


**(c)** 2 **(d)** Since $t_{data} > 1.690$, we reject $H_0$. There is evidence that the population mean is different than 9.
**39.** **(a)** 2.2723 **(b)** The sample mean number of home runs $\bar{x} =$ 13.8571 lies 0.94 standard errors below the hypothesized mean number of home runs $\mu_0 = 16$.
**41.** **(a)** 55.7832 **(b)** The sample mean math journal price $\bar{x} =$ \$506.82 lies 1.91 standard errors above the hypothesized mean math journal price $\mu_0 = \$400$.
**43.** **(a)** No, since the sample size of $n = 13$ is small ($n < 30$). **(b)** No **(c)** Since neither Case 1 nor Case 2 applies, we cannot proceed with the $t$ test.
**45.** **(a)** $H_0: \mu \leq 60$ vs. $H_a: \mu > 60$ **(b)** $t_{crit} = 1.771$. Reject $H_0$ if $t_{data} > 1.771$ **(c)** $t_{data} = 2.50$ **(d)** Since $t_{data} > 1.771$ we reject $H_0$. There is evidence that the population mean response time is greater than 60 milliseconds.
**47.** **(a)** Yes. Since the prices per bottle for Obsession by Calvin Klein are normally distributed, Case 1 applies. **(b)** No, a sample mean that is greater than the hypothesized mean is not enough to reject $H_0$. **(c)** 3.87 **(d)** $H_0: \mu \leq 46.42$ vs. $H_a: \mu > 46.62$. Reject $H_0$ if the $p$-value $< 0.01$. $t_{data} = 3.87$. $p$-value $< 0.005$. Since the $p$-value $< 0.01$, we reject $H_0$. There is evidence that the population mean price per bottle of Obsession is greater than \$46.42.
**49.** **(a)** Decrease **(b)** Increase **(c–d)** Decrease **(e)** Unchanged **(f)** Stronger **(g)** Unchanged
**51.** **(a)** $H_0: \mu \leq 30$ vs. $H_a: \mu > 30$. Reject $H_0$ if $p$-value $< 0.01$. $t_{data} = 3.54$. $p$-value $= 0.0031570524$. Since $p$-value $< 0.01$, we reject $H_0$. There is evidence that the population mean gas mileage is greater than 30 mpg. **(b)** $H_0: \mu \leq 30$ vs. $H_a: \mu > 30$. Reject $H_0$ if $p$-value $< 0.001$. $t_{data} = 3.54$. $p$-value $= 0.0031570524$. Since the $p$-value $\geq 0.001$, we do not reject $H_0$. There is insufficient evidence that the population mean gas mileage is greater than 30 mpg. **(c)** We could turn to a direct assessment of the strength of the evidence against $H_0$ or we could obtain more data. **(d)** Since $0.001 < p$-value $\leq 0.01$, there is very strong evidence against the null hypothesis that the population mean gas mileage is less than or equal to 30 mpg. This does not change for any value of $\alpha$ we use.
**53.** **(a)** Yes. Since all the points lie within the curved lines, the normal probability plot indicates acceptable normality. Case 1 applies. **(b)** $n = 10$, $\bar{x} = \$2538.92$, $s = \$404.75$ **(c)** $s_{\bar{x}} = \$404.75/\sqrt{10} =$ \$127.99 **(d)** $t_{data} = 2.09$. Therefore, the sample mean fees and tuition at community colleges nationwide $\bar{x} = \$2538.92$ is 2.09 standard errors above the hypothesized mean fees and tuition at community colleges nationwide $\mu_0 = \$2272$. This is mildly extreme. **(e)** It says, "Test of mu = 2272 vs. not = 2272" across the top of the printout. The researchers wanted to test the hypotheses "$H_0: \mu \leq 2272$ vs. $H_a: \mu > 2272$."
**55.** **(a)** Since the $p$-value for the two-tailed test is twice the $p$-value for the one-tailed test, it is possible to conclude that there is insufficient evidence that the population mean cost has changed but there is evidence that the population mean cost has increased if $\alpha$ is between the two $p$-values. **(b)** Since $0.01 < p$-value $\leq 0.05$, there is solid evidence against the null hypothesis that the population mean tuition and fees at community colleges is less than or equal to \$2272. **(c)** Since the test was done only for community colleges, not all colleges, our data and the hypothesis test do not support the headline.
**57.** $H_0: \mu \leq 26$ vs. $H_a: \mu > 26$. $t_{crit} = 1.290$: Reject $H_0$ if $t_{data} >$ 1.290. $t_{data} = 1.57$. Since $t_{data} > 1.290$ reject $H_0$. There is evidence that the population mean age at first marriage is greater than 26 years.
**59.** **(a)** $p$-value $= 0.1184412114$ **(b)** Since $0.10 < p$-value $\leq$ 0.15, there is slight evidence against the null hypothesis that the population mean marriage is 26 years. **(c)** 0.10. Since the $p$-value is greater than 0.10, it is closer to 0.10 than it is to 0.05. **(d)** This headline is not supported by the data and the hypothesis test.
**61.** **(a)** 254 observations, 7 variables **(b)** $\bar{x} = 23{,}901$, $s = 88{,}421$. The distribution is right-skewed. **(c)** Since the $p$-value $< 0.05$, we reject $H_0$. There is evidence that the population mean total occupied housing units for these counties is different than 40,000.

## Section 9.5

**1.** $\hat{p}$ is the sample proportion and $p$ is population proportion.
**3.** Answers will vary.
**5.** Between 0 and 1 inclusive: $0 \leq p_0 \leq 1$
**7.** 0.0707
**9.** 0.05
**11.** 0.0158
**13.** As the sample size increases, $\sigma_{\hat{p}}$ decreases.
**15.** 1.44
**17.** 4.33
**19.** −4.47
**21.** 0
**23.** 4.47
**25.** **(a)** We have $np_0 = 100(0.4) = 40 \geq 5$ and $n(1 - p_0) =$ $100\,(1 - 0.4) = 60 \geq 5$, so we can use the $Z$ test for proportions. **(b)** $H_0: p \leq 0.4$ vs. $H_a: p > 0.4$. Reject $H_0$ if the $p$-value $< 0.05$. **(c)** 0.82 **(d)** $p$-value $= 0.2061$ **(e)** Since the $p$-value $\geq 0.05$, we do not reject $H_0$. There is insufficient evidence that the population proportion is greater than 0.4.
**27.** **(a)** We have $np_0 = 900(0.5) = 450 \geq 5$ and $n(1 - p_0) =$ $900(1 - 0.5) = 450 \geq 5$, so we may use the $Z$ test for proportions.

**(b)** $H_0 : p = 0.5$ vs. $H_a : p \neq 0.5$. Reject $H_0$ if the $p$-value $< 0.05$. **(c)** 1.67 **(d)** $p$-value $= 0.095$ **(e)** Since the $p$-value $\geq 0.05$, we do not reject $H_0$. There is insufficient evidence that the population proportion is not equal to 0.5.

**29. (a)** We have $np_0 = 225(0.5) = 112.5 \geq 5$ and $n(1 - p_0) = 225(1 - 0.5) = 112.5 \geq 5$, so we can use the $Z$ test for proportions. **(b)** $H_0 : p \geq 0.5$ vs. $H_a : p < 0.5$ **(c)** $Z_{crit} = -1.645$. Reject $H_0$ if $Z_{data} < -1.645$. **(d)** $-1.67$ **(e)** Since $Z_{data} < -1.645$, we reject $H_0$. There is evidence that the population proportion is less than 0.5.

**31. (a)** We have $np_0 = 400(0.6) = 240 \geq 5$ and $n(1 - p_0) = 400(1 - 0.6) = 160 \geq 5$, so we can use the $Z$ test for proportions. **(b)** $H_0 : p \leq 0.6$ vs. $H_a : p > 0.6$ **(c)** $Z_{crit} = 1.645$. Reject $H_0$ if $Z_{data} > 1.645$. **(d)** 2.04 **(e)** Since $Z_{data} > 1.645$, we reject $H_0$. There is evidence that the population proportion is greater than 0.6.

**33.** $np_0 = 1000(0.368) = 368 \geq 5$ and $n(1 - p_0) = 1000(1 - 0.368) = 632 \geq 5$, so we may use the $Z$ test for proportions. $H_0 : p \leq 0.368$ vs. $H_a : p > 0.368$. Reject $H_0$ if $p$-value $< 0.05$. $Z_{data} = 0.79$. $p$-value $= 0.2148$. Since $p$-value $\geq 0.05$, we do not reject $H_0$. There is insufficient evidence that the population proportion of births to unmarried women is greater than 0.368.

**35.** $np_0 = 900\ (0.048) = 43.2 \geq 5$ and $n(1 - p_0) = 900\ (1 - 0.048) = 856.8 \geq 5$, so we may use the $Z$-test for proportions. $H_0 : p \leq 0.048$ vs. $H_a : p > 0.048$. Reject $H_0$ if $p$-value $< 0.01$. $Z_{data} = 1.68$. $p$-value $= 0.0465$. Since $p$-value $\geq 0.01$, we do not reject $H_0$. There is insufficient evidence that the population proportion of persons 12 or older that have used a prescription pain reliever non-medically is greater than 0.048.

**37.** $np_0 = 1000(0.07) = 70 \geq 5$ and $n(1 - p_0) = 1000(1 - 0.07) = 930 \geq 5$, so we may use the $Z$-test for proportions. $H_0 : p = 0.07$ vs. $H_a : p \neq 0.07$. Reject $H_0$ if $p$-value $< 0.10$. $Z_{data} = 1.24$. $p$-value $= 0.2150$. Since $p$-value $\geq 0.10$, we do not reject $H_0$. There is insufficient evidence that the population proportion of hospitalizations of 18- to 44-year-old American women for affective disorders is not equal to 0.07.

**39. (a)** Yes. We have $np_0 = 100(0.153) = 15.3 \geq 5$ and $n(1 - p_0) = 100(1 - 0.153) = 84.7 \geq 5$. **(b)** If the null hypothesis is correct, the sampling distribution of $\hat{p}$ will be centered at $p_0 = 0.153$. From (a), the shape of the sampling distribution of $\hat{p}$ will be approximately normal. **(c)** $\sigma_{\hat{p}} = 0.0360$. This is the standard deviation of the sampling distribution of $\hat{p}$ if $H_0$ is true. It is a measure of the variability of the sampling distribution of $\hat{p}$ if $H_0$ is true. **(d)** 2.14; mildly extreme **(e)** Right tail

**41. (a)** $H_0 : p = 0.153$ vs. $H_a : p \neq 0.153$ **(b)** Reject $H_0$ if the $P$-value $< 0.01$. **(c)** 2.14 **(d)** $p$-value $= 0.0324$ **(e)** Since $0.01 < p$-value $\leq 0.05$, there is solid evidence against the null hypothesis. **(f)** Since the $p$-value $\geq 0.01$, we do not reject $H_0$. There is insufficient evidence that the population proportion of Hispanic families that had a household income of at least \$75,000 is not equal to 0.153.

**43. (a)–(c)** Increase **(d)–(e)** Unchanged

**45.** $H_0 : p \geq 0.11$ vs. $H_a : p < 0.11$. $Z_{crit} = -1.645$. Reject $H_0$ if $Z_{data} < -1.645$. $Z_{data} = -1.60$. Since $Z_{data} \geq -1.645$, we do not reject $H_0$. There is insufficient evidence that the population proportion of children age 6 and under exposed to ETS at home on a regular basis is less than 0.11.

**47. (a)** Since the $p$-value $< 0.10$, we reject $H_0$. There is evidence that the population proportion of children age 6 and under exposed to environmental tobacco smoke at home on a regular basis is less than 0.11. **(b)** The difference is because we changed the value of $\alpha$ and not because we used different methods for the two different hypothesis tests. **(c)** Since $0.05 < p$-value $\leq 0.10$, there is mild evidence against the null hypothesis that the population proportion of children age 6 and under exposed to environmental tobacco smoke at home on a regular basis is greater than or equal to 0.11.

**49. (a)** Decrease **(b)** Increase by a factor of $\sqrt{2}$ **(c)** Decrease **(d)** Unchanged **(e)** The conclusion will now be to reject $H_0$.

## Section 9.6

**1.** Answers will vary.

**3.** No, $\sigma$ will never be less than 0.

**5.** Answers will vary.

**7.** $H_0 : \sigma \leq 10$ vs. $H_a : \sigma > 10$

**9.** $H_0 : \sigma = 3$ vs. $H_a : \sigma \neq 3$

**11.** $\chi^2_{data} = 60$

**13.** $\chi^2_{data} = 10.417$

**15.** $\chi^2_{data} = 6.125$

**17. (a)**

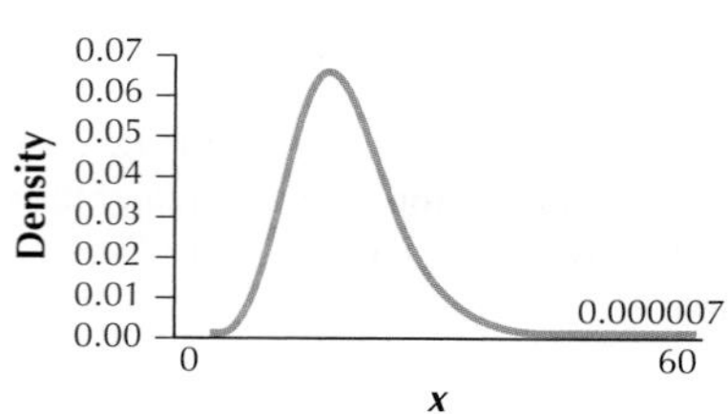


**(b)** $p$-value $= 7.121750863 \times 10^{-6}$ **(c)** Since the $p$-value $< 0.05$, we reject $H_0$. There is evidence that the population standard deviation is greater than 1.

**19. (a)**

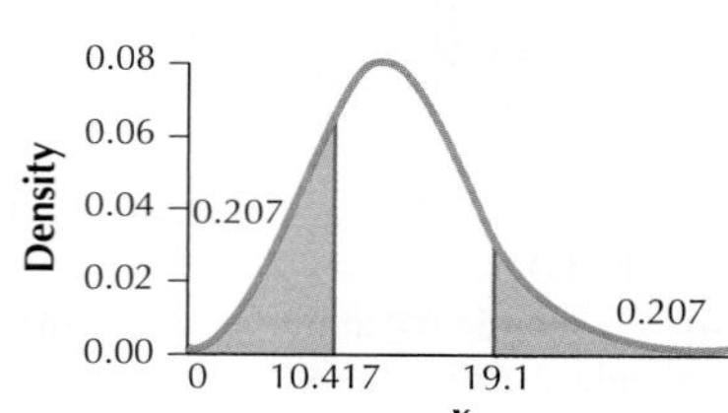


**(b)** $p$-value $= 0.4145552434$ **(c)** Since the $p$-value $\geq 0.05$, we do not reject $H_0$. There is insufficient evidence that the population standard deviation is different than 3.

**21. (a)**

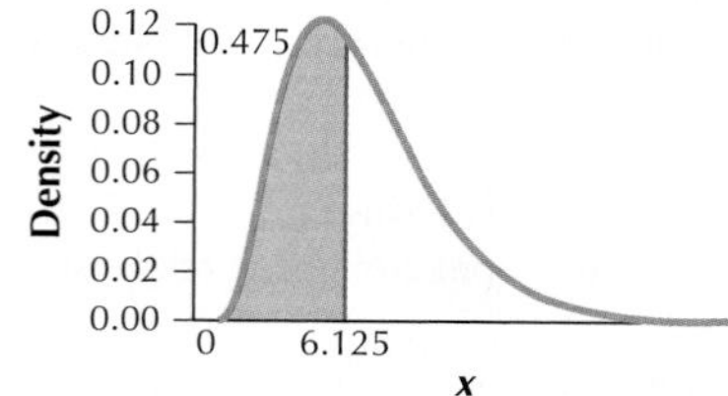


**(b)** $p$-value $= 0.4747679539$ **(c)** Since the $p$-value $\geq 0.05$, we do not reject $H_0$. There is insufficient evidence that the population standard deviation is less than 20.

**23.** $\chi^2_{\alpha} = \chi^2_{0.05} = 31.410$

**25.** $\chi^2_{\alpha/2} = \chi^2_{0.025} = 27.488$ and $\chi^2_{1-\alpha/2} = \chi^2_{0.975} = 6.262$

**27.** $\chi^2_{1-\alpha} = \chi^2_{0.90} = 2.833$

**29. (a)** Reject $H_0$ if $\chi^2_{data} > 31.41$ **(b)** Since $\chi^2_{data} > 31.410$, we reject $H_0$. There is evidence that the population standard deviation is greater than 1.

**31. (a)** Reject $H_0$ if $\chi^2_{data} < 6.262$ or $\chi^2_{data} > 27.488$. **(b)** Since $\chi^2_{data}$ is not less than 6.262 and $\chi^2_{data}$ is not greater than 27.488, we do not reject $H_0$. There is insufficient evidence that the population standard deviation is different than 3.

**33. (a)** Reject $H_0$ if $\chi^2_{data} < 2.833$. **(b)** Since $\chi^2_{data} \geq 2.833$, we do not reject $H_0$. There is insufficient evidence that the population standard deviation is less than 20.
**35. (a)** $H_0: \sigma \leq 36.5$ vs. $H_a: \sigma > 36.5$. Reject $H_0$ if the $p$-value $< 0.01$. **(b)** 982.75 **(c)** $p$-value $\approx 0$ **(d)** Since the $p$-value $< 0.01$, we reject $H_0$. **(e)** There is evidence that the population standard deviation of DDT level in the breast milk of Hispanic women in the Yakima valley is greater than 36.5 parts per billion.
**37. (a)** $H_0: \sigma = 30{,}000$ vs. $H_a: \sigma \neq 30{,}000$. Reject $H_0$ if the $p$-value $< 0.05$. **(b)** 0.00001497113333 **(c)** $p$-value $\approx 0$ **(d)** Since the $p$-value $< 0.05$, we reject $H_0$. **(e)** There is evidence that the population standard deviation of union membership differs from 30,000.
**39.** *Test using the p-value method:* $H_0: \sigma \leq 50$ vs. $H_a: \sigma > 50$. Reject $H_0$ if $p$-value $< 0.05$. $\chi^2_{data} = 104$. $p$-value $= 0.3721497012$. Since $p$-value $\geq 0.05$, we do not reject $H_0$. There is insufficient evidence that the population standard deviation of test scores for boys is greater than 50 points. *Test using the critical value method:* $H_0: \sigma \leq 50$ vs. $H_a: \sigma > 50$. $\chi^2_\alpha = \chi^2_{0.05} = 124.342$. Reject $H_0$ if $\chi^2_{data} > 124.342$. $\chi^2_{data} = 104$. Since $\chi^2_{data} \leq 124.342$, we do not reject $H_0$. There is insufficient evidence that the population standard deviation of test scores for boys is greater than 50 points.

## Chapter 9 Review

**1.** $H_0: \mu \geq 12$ vs. $H_a: \mu < 12$
**3.** $H_0: \mu \geq 0$ vs. $H_a: \mu < 0$
**5. (a)** $H_0: \mu \geq 202.7$ vs. $H_a: \mu < 202.7$ **(b)** We conclude that (1) the population mean number of speeding-related fatalities is less than 202.7 when it actually is and (2) the mean number of speeding-related fatalities is greater than or equal to 202.7 when it actually is. **(c)** The population mean number of speeding-related fatalities is less than 202.7 when it actually is greater than or equal to 202.7. **(d)** The population mean number of speeding-related fatalities is greater than or equal to 202.7 when it actually is less than 202.7.
**7.** $-1$
**9.** $-10$
**11. (a)** $H_0: \mu \geq -10$ vs. $H_a: \mu < -10$. Reject $H_0$ if the $p$-value $< 0.01$. **(b)** $-5$ **(c)** $p$-value $= 2.87105 \times 10^{-7}$

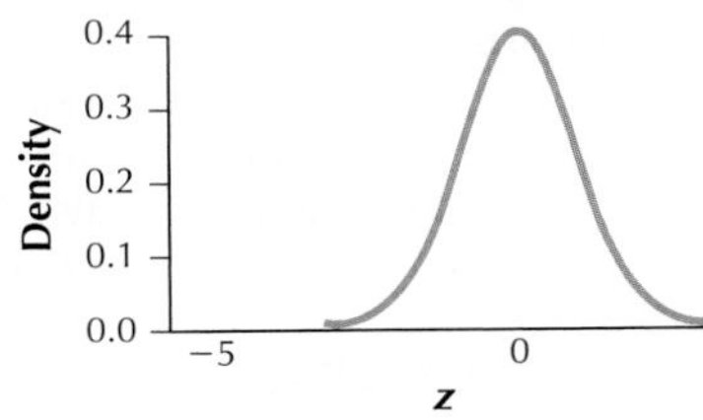


**(d)** Since the $p$-value $< 0.01$, reject $H_0$. There is evidence that the population mean is less than $-10$.
**13. (a)** 2.58 **(b)** Reject $H_0$ if $Z_{data} < -2.58$ or $Z_{data} > 2.58$.
**(c)**

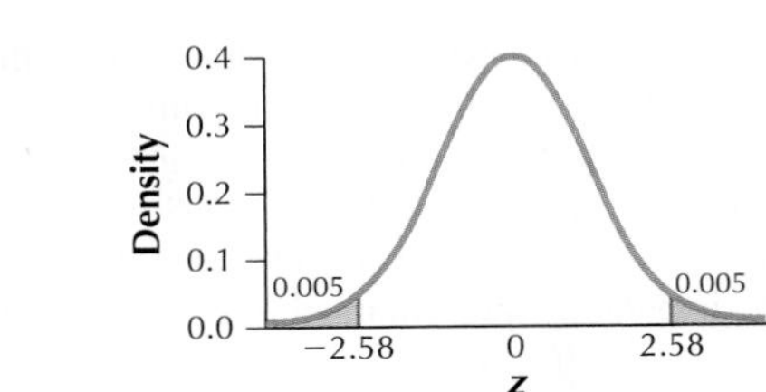


**(d)** Since $Z_{data}$ is not less than $-2.58$ and $Z_{data}$ is not greater than 2.58, we do not reject $H_0$. There is insufficient evidence that the population mean is different than $\mu_0$.
**15. (a)** 1.645 **(b)** Reject $H_0$ if $Z_{data} > 1.645$.
**(c)**

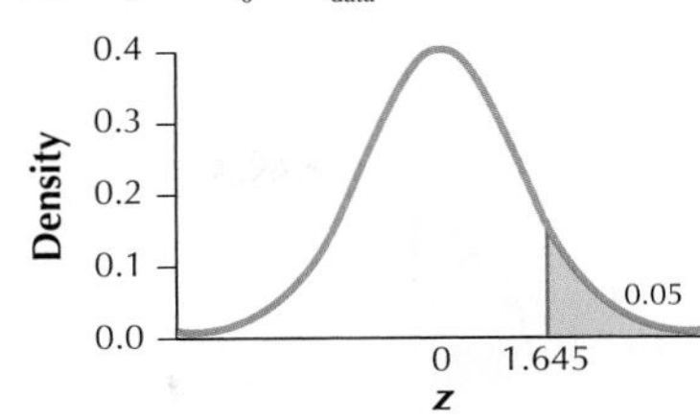


**(d)** Since $Z_{data} \leq 1.645$, we do not reject $H_0$. There is insufficient evidence that the population mean is greater than $\mu_0$.
**17. (a)** $H_0: \mu \leq 52{,}200$ vs. $H_a: \mu > 52{,}200$ **(b)** $Z_{crit} = 1.28$. Reject $H_0$ if $Z_{data} > 1.28$. **(c)** $Z_{data} = 9.33$

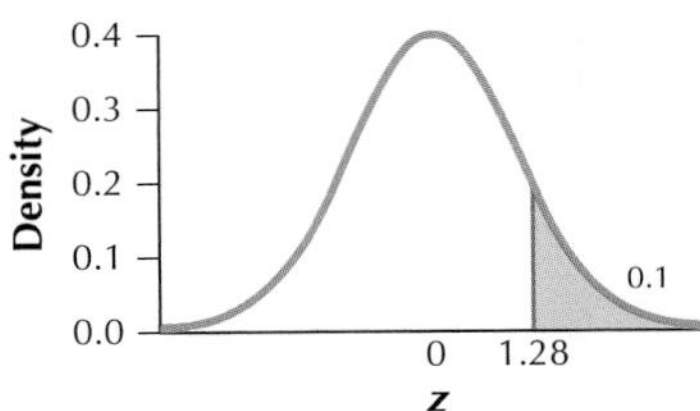


**(d)** Since $Z_{data} > 1.28$, we reject $H_0$. There is evidence that the population mean salary of college graduates is greater than \$52,200.
**19.** $t_{crit} = 1.415$

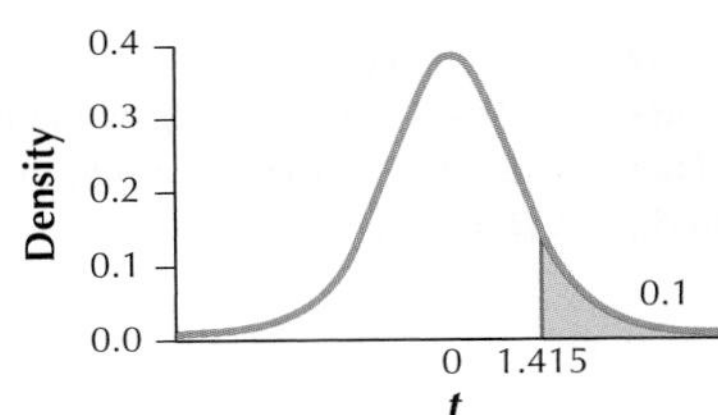


**21.** $t_{crit} = 2.998$

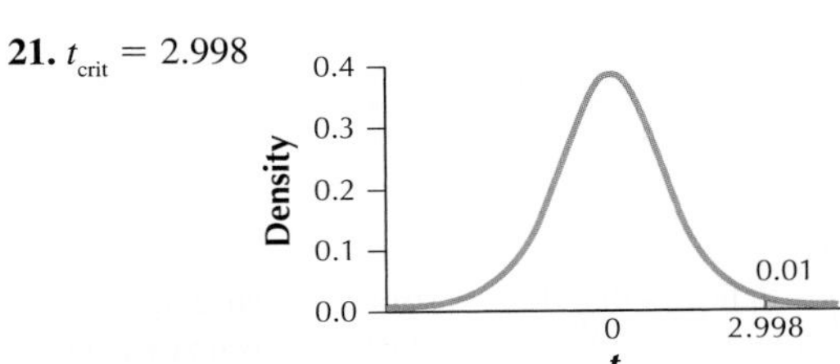


**23.** $H_0: \mu = 9$ vs. $H_a: \mu \neq 9$. $t_{crit} = 1.753$. Reject $H_0$ if $t_{data} < -1.753$ or $t_{data} > 1.753$. $t_{data} = 1.33$. Since $t_{data}$ is not less than $-1.753$ and $t_{data}$ is not greater than 1.753, we do not reject $H_0$. There is insufficient evidence that the population mean is different than 9.
**25. (a)** We have $np_0 = 144(0.7) = 100.8 \geq 5$ and $n(1 - p_0) = 144(1 - 0.7) = 43.2 \geq 5$. **(b)** $H_0: p = 0.7$ vs. $H_a: p \neq 0.7$. Reject $H_0$ if the $p$-value $< 0.05$. **(c)** $Z_{data} = 1.67$ **(d)** $p$-value $= 0.095$ **(e)** Since the $p$-value $\geq 0.05$, we do not reject $H_0$. There is insufficient evidence that the population proportion is different than 0.7.
**27. (a)** We have $np_0 = 1000(0.8) = 800 \geq 5$ and $n(1 - p_0) = 1000(1 - 0.8) = 200 \geq 5$. **(b)** $H_0: p \leq 0.8$ vs. $H_a: p > 0.8$ **(c)** $Z_{crit} = 1.28$. Reject $H_0$ if $Z_{data} > 1.28$. **(d)** $Z_{data} = 2.37$ **(e)** Since $Z_{data} > 1.28$, we reject $H_0$. There is evidence that the population proportion is greater than 0.8.
**29. (a)** We have $np_0 = 100(0.4) = 40 \geq 5$ and $n(1 - p_0) = 100(1 - 0.4) = 60 \geq 5$. **(b)** $H_0: p = 0.4$ vs. $H_a: p \neq 0.4$ **(c)** $Z_{crit} = 2.58$. Reject $H_0$ if $Z_{data} < -2.58$ or $Z_{data} > 2.58$. **(d)** $Z_{data} = 3.06$ **(e)** Since $Z_{data} > 2.58$, we reject $H_0$. There is evidence that the population proportion is not equal to 0.4.

**31. (a)** $H_0: \sigma \geq 35$ vs. $H_a: \sigma < 35$. Reject $H_0$ if the $p$-value $< \alpha$. **(b)** 6.857 **(c)** $p$-value $= 0.5560805474$

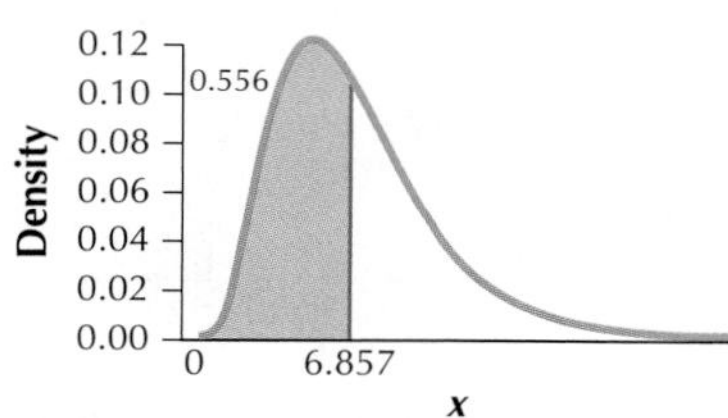


**(d)** Since the $p$-value $\geq 0.05$, we do not reject $H_0$. There is insufficient evidence that the population standard deviation is less than 35.

**33. (a)** $H_0: \sigma \leq 6$ vs. $H_a: \sigma > 6$ **(b)** $\chi^2_\alpha = \chi^2_{0.05} = 30.144$. Reject $H_0$ if $\chi^2_{data} > 30.144$ **(c)** $\chi^2_{data} = 42.75$

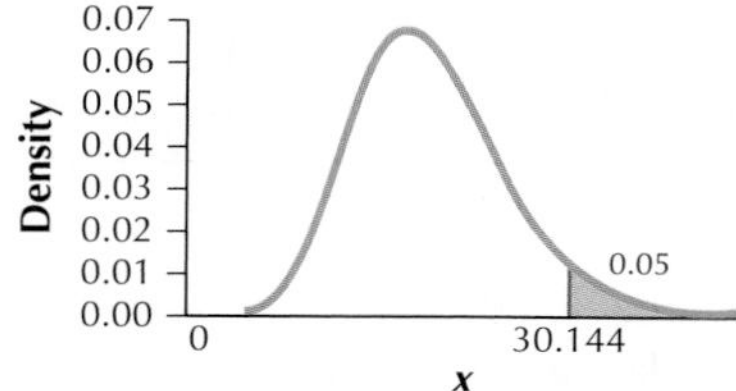


**(d)** Since $\chi^2_{data} > 30.144$, we reject $H_0$. There is evidence that the population standard deviation is greater than 6.

**35.** $H_0: \sigma = 50$ vs. $H_a: \sigma \neq 50$. Reject $H_0$ if the $p$-value $< 0.01$. $\chi^2_{data} = 0.690$. $p$-value $= 0.094887$. Since the $p$-value $\geq 0.01$, we do not reject $H_0$. There is insufficient evidence that the population standard deviation differs from 50.

## Chapter 9 Quiz

**1.** False
**2.** True
**3.** True
**4.** I
**5.** small
**6.** $\alpha$
**7.** $n\,p_0 \geq 5$ and $n\,(1 - p_0) \geq 5$
**8.** A small $p$-value indicates that there is strong evidence against the null hypothesis. A large $p$-value indicates that there is no evidence against the null hypothesis.
**9.** No
**10. (a)** $H_0: \mu = 47$ vs. $H_a: \mu \neq 47$. Reject $H_0$ if the $p$-value $< 0.10$. **(b)** 5.83 **(c)** $p$-value $= 5.560346071 \times 10^{-9}$ **(d)** Since the $p$-value $< 0.10$, we reject $H_0$. There is evidence that the population mean prison sentence length for felons convicted on drug offenses is different than 47 months. **(e)** Since the $p$-value $\leq 0.001$, there is extremely strong evidence against the null hypothesis that the population mean prison sentence length for felons is 47 months.
**11. (a)** $H_0: \mu \geq 1.14$ vs. $H_a: \mu < 1.14$ **(b)** $-1.645$ **(c)** Reject $H_0$ if $Z_{data} < -1.645$. **(d)** $-1.68$ **(e)** Since $Z_{data} < -1.645$, we reject $H_0$. There is evidence that the population mean fee charged by banks when withdrawing funds from an ATM machine not owned by your bank is less than \$1.14.
**12.** The conclusion remains the same since the $p$-value method and the critical-value method are equivalent.
**13.** Type I error, Type II error
**14. (a)** No, since the population standard deviation is not known. **(b)** $H_0: \mu = 32$ vs. $H_a: \mu \neq 32$ **(c)** 1.690 **(d)** Reject $H_0$ if $t_{data} < -1.690$ or $t_{data} > 1.690$. **(e)** 1.80 **(f)** Since $t_{data} > 1.690$, we reject $H_0$. There is evidence that the population mean years of potential life lost in alcohol-related fatal automobile accidents is different than 32 years.
**15. (a)** $0.05 < p\text{-value} \leq 0.10$ **(b)** There is mild evidence against the null hypothesis that the population mean years of potential life lost in alcohol-related fatal automobile accidents is 32 years.
**16. (a)** Yes. We have $n\,p_0 = 400\,(0.127) = 50.8 \geq 5$ and $n\,(1 - p_0) = 400\,(1 - 0.127) = 349.2 \geq 5$. **(b)** $\mu_{\hat{p}} = p = 0.127$, approximately normal **(c)** $\sigma_{\hat{p}} = 0.0166$. It is a measure of the variability of the distribution of the sample proportions if $H_0$ is true. **(d)** 0.93. No, since $-2 < 0.93 < 2$. **(e)** Near the center
**17. (a)** $H_0: p \leq 0.127$ vs. $H_a: p > 0.127$ **(b)** Reject $H_0$ if the $p$-value $< 0.05$. **(c)** 0.93 **(d)** $p$-value $= 0.1762$ **(e)** Since $0.15 <$ $p$-value, there is no evidence against the null hypothesis that the population proportion of preterm births is less than or equal to 0.127 **(f)** Since the $p$-value $\geq 0.05$, we do not reject $H_0$. There is insufficient evidence that the population proportion of preterm births is greater than 0.127.
**18.** $H_0: \sigma \geq 0.25$ vs. $H_a: \sigma < 0.25$. $\chi^2_{1-\alpha} = \chi^2_{0.90} = 4.168$. Reject $H_0$ if $\chi^2_{data} < 4.168$. $\chi^2_{data} = 16.992$. Since $\chi^2_{data} \geq 4.168$, we do not reject $H_0$. There is insufficient evidence that the population standard deviation of net price change is less than 25 cents.

# CHAPTER 10

## Section 10.1

**1.** When the subjects selected for the first sample do not determine the subjects in the second sample
**3.** Matched pairs or paired samples
**5.** Since both samples of games were based on the same players, this is an example of dependent sampling.
**7.** Since the same students are taking both tests, this is an example of dependent sampling.
**9. (a)** $\bar{x}_d = 1.1$, $s_d = 0.9618$ **(b)** $(-0.0940, 2.294)$
**11. (a)** $\bar{x}_d = -7.1429$, $s_d = 2.6726$ **(b)** $(-9.6127, -4.6711)$
**13.** Since the $p$-value $< 0.05$, we reject $H_0$. There is evidence that the population mean of the differences is greater than 0.
**15.** Since the $p$-value $\geq 0.05$, we do not reject $H_0$. There is insufficient evidence that the population mean of the differences is greater than 0.
**17.** Since $t_{data} \leq -1.943$, we reject $H_0$. There is evidence that the population mean of the differences is different than 0.
**19. (a)** $\bar{x}_d = 753.75$, $s_d = 986.1658$ **(b)** $(-815.2398, 2322.7398)$
**21.** There is insufficient evidence that 2007 models are on average more expensive.
**23. (a)** $\bar{x}_d = -28{,}825$, $s_d = 16{,}806.2241$ **(b)** $(-48{,}597.5227, -9{,}052.4773)$
**25.** Since $\mu_d = 0$ does not lie in the confidence interval, we reject $H_0$. There is evidence that the population mean difference between the 1st quarter 2007 median price and the 1st quarter 2008 median price differs from 0.
**27. (a)** $(1.9618, 23.5048)$ **(b)** Since 0 is not in the interval, there is evidence that the 2003 science test scores for fourth-graders are higher than the 1995 science test scores for fourth-graders.
**29. (a)** $-6.0333$ collisions per year **(b)** $(-10.0998, -1.9668)$ **(c)** Since 0 is not in the interval, there is evidence that the improvements lowered the population mean number of collisions per year.
**31.** *Critical value method:* $H_0: \mu_d \geq 0$ vs. $H_a: \mu_d < 0$. $t_{crit} = 3.365$. Reject $H_0$ if $t_{data} < -3.365$. $t_{data} = -4.843$. Since $t_{data} < -3.365$, we reject $H_0$. There is evidence that the population mean number of goals scored per game has increased.
**33.** The conclusion would still be to reject $H_0$.

## Section 10.2

**1.** Case 1: The two populations are normally distributed. Case 2: The sample sizes are large (at least 30).
**3.** It measures the size of the typical error in using $\bar{x}_1 - \bar{x}_2$ to estimate $\mu_1 - \mu_2$.
**5. (a)** Since both sample sizes are large ($n_1 \geq 30$ and $n_2 \geq 30$), it is appropriate to construct a 95% confidence interval for $\mu_1 - \mu_2$. **(b)** 2 **(c)** 0.4714 **(d)** 0.9569 **(e)** (1.0431, 2.9569). We are 95% confident that the interval captures the difference in population means.
**7. (a)** Since both sample sizes are large ($n_1 \geq 30$ and $n_2 \geq 30$), it is appropriate to construct a 95% confidence interval for $\mu_1 - \mu_2$. **(b)** −5 **(c)** 0.9832 **(d)** 2.011 **(e)** (−7.011, −2.989). We are 95% confident that the interval captures the difference in population means.
**9. (a)** Since both sample sizes are large ($n_1 \geq 30$ and $n_2 \geq 30$), it is appropriate to construct a 95% confidence interval for $\mu_1 - \mu_2$. **(b)** −1 **(c)** 0.4013 **(d)** 0.811 **(e)** (−1.811, −0.189). We are 95% confident that the interval captures the difference in population means.
**11. (a)** $H_0: \mu_1 = \mu_2$ vs. $H_a: \mu_1 \neq \mu_2$ **(b)** $t_{crit} = 1.690$. Reject $H_0$ if $t_{data} < -1.690$ or $t_{data} > 1.690$. **(c)** $t_{data} = 4.243$. The difference of the sample means $\bar{x}_1 - \bar{x}_2 = 2$ lies 4.243 standard errors above the difference of the hypothesized means $\mu_1 - \mu_2 = 0$. **(d)** Since $t_{data} > 1.690$, we reject $H_0$. There is evidence that the population mean for Population 1 is different than the population mean for Population 2.
**13. (a)** $H_0: \mu_1 = \mu_2$ vs. $H_a: \mu_1 \neq \mu_2$ **(b)** $t_{crit} = 2.756$. Reject $H_0$ if $t_{data} < -2.756$ or $t_{data} > 2.756$. **(c)** $t_{data} = -5.085$. The difference of the sample means $\bar{x}_1 - \bar{x}_2 = -5$ lies 5.085 standard errors below the difference of the hypothesized means $\mu_1 - \mu_2 = 0$. **(d)** Since $t_{data} < -2.756$, we reject $H_0$. There is evidence that the population mean for Population 1 is different than the population mean for Population 2.
**15. (a)** $H_0: \mu_1 \leq \mu_2$ vs. $H_a: \mu_1 > \mu_2$ **(b)** $t_{crit} = 2.423$. Reject $H_0$ if $t_{data} > 2.423$. **(c)** $t_{data} = -2.492$. The difference of the sample means $\bar{x}_1 - \bar{x}_2 = -1$ lies 2.492 standard errors below the difference of the hypothesized means $\mu_1 - \mu_2 = 0$. **(d)** Since $t_{data} \leq 2.423$, we do not reject $H_0$. There is insufficient evidence that the population mean for Population 1 is greater than the population mean for Population 2.
**17.** Since neither population is normal, Case 1 does not apply. Since both sample sizes are small ($n_1 < 30$ and $n_2 < 30$), Case 2 does not apply. Therefore, it is not appropriate to perform two-sample *t* inference.
**19. (a)** Since both sample sizes are large ($n_1 \geq 30$ and $n_2 \geq 30$), Case 2 applies. Thus it is appropriate to apply two-sample *t* inference. **(b)** −1192 **(c)** 1195.4657 **(d)** 2,426.795 **(e)** (−3,618.795, 1,234.795). We are 95% confident that the interval captures the difference of the population mean incomes for Sacramento County and Los Angeles County, California.
**21. (a)** $t_{crit} = 2.030$ **(b)** $t_{data} = -0.997$. The difference in the sample mean incomes for Sacramento County and Los Angeles County, California, $\bar{x}_1 - \bar{x}_2 = -1192$, lies 0.997 standard errors below the difference in the hypothesized mean incomes for Sacramento County and Los Angeles County, California, $\mu_1 - \mu_2 = 0$. **(c)** Since $t_{data} \geq -2.030$ and $t_{data} \leq 2.030$, we do not reject $H_0$. There is insufficient evidence that the population mean income for Sacramento County, California, differs from the population mean income for Los Angeles County, California. **(d)** Yes, since it is a two-tailed test.
**23. (a)** 1.6 **(b)** 0.9718 **(c)** 2.647 **(d)** (−1.047, 4.247). We are 95% confident that the interval captures the difference in the population mean number of children per teacher in the towns of Cupertino, California, and Santa Rosa, California.
**25. (a)** $t_{crit} = 2.724$ **(b)** $t_{data} = 1.646$. The difference in the sample mean number of children per teacher in the towns of Cupertino, California, and Santa Rosa, California, $\bar{x}_1 - \bar{x}_2 = 1.6$, lies 1.646 standard errors above the difference in the hypothesized mean number of children per teacher in the towns of Cupertino, California, and Santa Rosa, California, $\mu_1 - \mu_2 = 0$. **(c)** Since $t_{data} \geq -2.724$ and $t_{data} \leq 2.724$, we do not reject $H_0$. There is insufficient evidence that the population mean number of children per teacher in the town of Cupertino, California, differs from the population mean number of children per teacher in the town of Santa Rosa, California.
**27.** $H_0: \mu_1 = \mu_2$ vs. $H_1: \mu_1 \neq \mu_2$. $t_{crit} = 2.201$. Reject $H_0$ if $t_{data} < -2.201$ or if $t_{data} > 2.201$. $t_{data} = -5.829$. Since $t_{data} < -2.201$, we reject $H_0$. There is sufficient evidence that the difference between mean property taxes in Ohio and North Carolina is significant.
**29. (a)** (−74.6747, −5.3253). We are 95% confident that the interval captures the difference in the population mean number of daily visitors to Windvale Park and Cranebrook Park. **(b)** $H_0: \mu_1 \geq \mu_2$ vs. $H_a: \mu_1 < \mu_2$. Reject $H_0$ if *p*-value $< 0.05$. $t_{data} = -2.359$. *p*-value $= 0.0126370393$. Since the *p*-value $< 0.05$, we reject $H_0$. There is evidence that the population mean number of daily visitors to Windvale Park is less than the population mean number of daily visitors to Cranebrook Park. **(c)** No, since confidence intervals can be used only for two-tailed tests and this is a one-tailed test.
**31. (a)** $\bar{x}_1 - \bar{x}_2 = 8$ **(b)** (−12.699, 28.699) **(c)** $H_0: \mu_1 = \mu_2$ vs. $H_a: \mu_1 \neq \mu_2$. Reject $H_0$ if *p*-value $< 0.01$. $t_{data} = 1.017$. *p*-value $= 0.3116324129$. Since the *p*-value $\geq 0.01$, we do not reject $H_0$. There is insufficient evidence that the mean coached SAT score improvement is different than the mean noncoached SAT score improvement.
**33. (a)** $\bar{x}_1 - \bar{x}_2 = -498.5238$ **(b)** (−791.006, −206.042). We are 90% confident that the interval captures the difference in the population mean daily calorie intake of children in the inner city and the suburbs. **(c)** $H_0: \mu_1 \geq \mu_2$ vs. $H_a: \mu_1 < \mu_2$. Reject $H_0$ if *p*-value $< 0.10$. $t_{data} = -3.434$. *p*-value $= 0.0092775929$. Since the *p*-value $< 0.10$, we reject $H_0$. There is evidence that the population mean daily calorie intake of inner city children is less than that of children from the suburbs. **(d)** There is very strong evidence against the null hypothesis.
**35. (a)** (−3.161, 0.561). We are 99% confident that the interval captures the difference of the population mean number of hours worked in 2003 and 1994. **(b)** $H_0: \mu_1 \geq \mu_2$ vs. $H_a: \mu_1 < \mu_2$. Reject $H_0$ if *p*-value $< 0.01$. $t_{data} = -1.838$. *p*-value $= 0.0342678723$. Since the *p*-value $\geq 0.01$, we do not reject $H_0$. There is insufficient evidence that the population mean number of hours worked in 2003 is less than the population mean number of hours worked in 1994. **(c)** Answers will vary.
**37. (a)** This will result in an increase in $\bar{x}_1 - \bar{x}_2$, but $\bar{x}_1 - \bar{x}_2$ will still be negative. **(b)** This will result in an increase in $t_{data}$, but $t_{data}$ will still be negative. **(c)** Increase in *p*-value **(d)** Unchanged
**39. (a)** $\bar{x}_1 = 31{,}363$, $s_1 = 135{,}270$ **(b)** $\bar{x}_2 = 18{,}305$, $s_2 = 260{,}938$ **(c)** $H_0: \mu_1 = \mu_2$ vs. $H_a: \mu_1 \neq \mu_2$. Reject $H_0$ if the *p*-value $< 0.05$. $t_{data} = 1.259$; *p*-value $= 0.2084026891$. Since the *p*-value $\geq 0.05$, we do not reject $H_0$. There is insufficient evidence that the population mean number of residents in towns and cities in California is different than the population mean number of residents is towns and cities in New York.

## Section 10.3

**1.** $\hat{p}_1$ and $\hat{p}_2$
**3.** No, because the confidence interval is not based on the assumption that $p_1 = p_2$.
**5.** $Z_{data}$ measures the standardized distance between sample proportions. Extreme values of $Z_{data}$ indicate evidence against the null hypothesis.
**7. (a)** $x_1 = 80 \geq 5$, $n_1 - x_1 = 20 \geq 5$, $x_2 = 30 \geq 5$, and $n_2 - x_2 = 10 \geq 5$, so it is appropriate. **(b)** 0.05 **(c)** 0.0793. The typical error in estimating the unknown quantity $p_1 - p_2$ is 0.0793. **(d)** 0.1554. The point estimate $\hat{p}_1 - \hat{p}_2$ will lie within $E = 0.1554$ of the difference in population proportions $p_1 - p_2$ 95% of the time. **(e)** $(-0.1054, 0.2054)$. We are 95% confident that the difference in population proportions lies between $-0.1054$ and 0.2054.
**9. (a)** $x_1 = 60 \geq 5$, $n_1 - x_1 = 140 \geq 5$, $x_2 = 40 \geq 5$, and $n_2 - x_2 = 210 \geq 5$, so it is appropriate. **(b)** 0.14 **(c)** 0.0398. The typical error in estimating the unknown quantity $p_1 - p_2$ is 0.0398. **(d)** 0.078. The point estimate $\hat{p}_1 - \hat{p}_2$ will lie within $E = 0.078$ of the difference in population proportions $p_1 - p_2$ 95% of the time. **(e)** (0.062, 0.218). We are 95% confident that the difference in population proportions lies between 0.062 and 0.218.
**11. (a)** $x_1 = 490 \geq 5$, $n_1 - x_1 = 510 \geq 5$, $x_2 = 620 \geq 5$, and $n_2 - x_2 = 380 \geq 5$, so it is appropriate. **(b)** $-0.13$ **(c)** 0.0220. The typical error in estimating the unknown quantity $p_1 - p_2$ is 0.0220. **(d)** 0.0431. The point estimate $\hat{p}_1 - \hat{p}_2$ will lie within $E = 0.0431$ of the difference in population proportions $p_1 - p_2$ 95% of the time. **(e)** $(-0.1731, -0.0869)$. We are 95% confident that the difference in population proportions lies between $-0.1731$ and $-0.0869$.
**13. (a)** $H_0: p_1 = p_2$ vs. $H_a: p_1 \neq p_2$; $Z_{crit} = 1.645$. Reject $H_0$ if $Z_{data} < -1.645$ or $Z_{data} > 1.645$. **(b)** 0.7857 **(c)** 0.65 **(d)** Since $Z_{data} \geq -1.645$ and $Z_{data} \leq 1.645$, we do not reject $H_0$. There is insufficient evidence that the population proportion from Population 1 is different than the population proportion from Population 2.
**15. (a)** $H_0: p_1 \leq p_2$ vs. $H_a: p_1 > p_2$; $Z_{crit} = 1.645$. Reject $H_0$ if $Z_{data} > 1.645$. **(b)** 0.2222 **(c)** 3.55 **(d)** Since $Z_{data} > 1.645$, we reject $H_0$. There is evidence that the population proportion from Population 1 is greater than the population proportion from Population 2.
**17. (a)** $H_0: p_1 \leq p_2$ vs. $H_a: p_1 > p_2$. Reject $H_0$ if the $p$-value $< 0.10$. **(b)** 0.5625 **(c)** 3.56 **(d)** 0.000185467351 **(e)** Since the $p$-value $< 0.10$, we reject $H_0$. There is evidence that the population proportion from Population 1 is greater than the population proportion from Population 2.
**19. (a)** $H_0: p_1 = p_2$ vs. $H_a: p_1 \neq p_2$. Reject $H_0$ if the $p$-value $< 0.05$. **(b)** 0.7982 **(c)** $-1.28$ **(d)** 0.2005452669 **(e)** Since the $p$-value $\geq 0.05$, we do not reject $H_0$. There is insufficient evidence that the population proportion from Population 1 is different than the population proportion from Population 2.
**21. (a)** $x_1 = 360.38 \geq 5$, $n_1 - x_1 = 126.62 \geq 5$, $x_2 = 404.21 \geq 5$, and $n_2 - x_2 = 82.79 \geq 5$, so it is appropriate. **(b)** $-0.09$ **(c)** 0.0262. The typical error in estimating the unknown difference in the population proportion of teenage boys who post their photo on their online profile and the population proportion of teenage girls who do so $p_1 - p_2$ is 0.0262. **(d)** 0.0514. The point estimate of the difference in the population proportion of teenage boys who post their photo on their online profile and the population proportion of teenage girls who do so will lie within $E = 0.0514$ of the difference in population proportions $p_1 - p_2$ 95% of the time. **(e)** $(-0.1414, -0.0386)$. We are 95% confident that the difference in the population proportion of teenage boys who post their photo on their online profile and the population proportion of teenage girls who do so lies between $-0.1414$ and $-0.0386$.
**23. (a)** $H_0: p_1 = p_2$ vs. $H_a: p_1 \neq p_2$ where $p_1$ = the population proportion of teenage boys who posted their photo online and $p_2$ = the population proportion of teenage girls who posted their photo online. **(b)** Zero does not fall within the confidence interval. **(c)** Since 0 does not fall within the confidence interval, we reject $H_0$. There is evidence that the population proportion of teenage boys who post their photo on their online profile differs from the population proportion of teenage girls who do so.
**25. (a)** 0.12 **(b)** 0.3622 **(c)** $H_0: p_1 = p_2$ vs. $H_a: p_1 \neq p_2$. Reject $H_0$ if $p$-value $< 0.05$. $Z_{data} = 2.65$. $p$-value $= 0.008$. Since $p$-value $< 0.05$, we reject $H_0$. There is evidence of a generation gap in the perception of same-sex marriage.
**27. (a)** $(-0.4827, -0.1099)$ **(b)** $H_0: p_1 = p_2$ vs. $H_a: p_1 \neq p_2$. Since 0 does not fall in the confidence interval, we reject $H_0$. There is evidence that there is a difference in the population proportions of women who had fetal cells.
**29.** *Critical value method:* $H_0: p_1 \leq p_2$ vs. $H_a: p_1 > p_2$. $Z_{crit} = 1.645$. Reject $H_0$ if $Z_{data} > 1.645$. $\hat{p}_{pooled} = 0.042$, $Z_{data} = 4.11$. Since $Z_{data} > 1.645$, we reject $H_0$. There is evidence that the population proportion of minority-owned businesses in Michigan exceeds that in Minnesota.
**31. (a)** $x_1 = 920 \geq 5$, $n_1 - x_1 = 80 \geq 5$, $x_2 = 870 \geq 5$, and $n_2 - x_2 = 130 \geq 5$, so it is appropriate. **(b)** 0.05 **(c)** 0.0137. The typical error in estimating the unknown difference in the population proportion of 18- to 24-year-old males who listen to the radio each week and the population proportion of males 65 years and older who listen to the radio each week $p_1 - p_2$ is 0.0137. **(d)** 0.0269. The point estimate of the difference in the population proportion of 18- to 24-year-old males who listen to the radio each week and the population proportion of males 65 years and older who listen to the radio each week will lie within $E = 0.0269$ of the difference in population proportions $p_1 - p_2$ 95% of the time. **(e)** (0.0231, 0.0769). We are 95% confident that the difference in the population proportion of 18- to 24-year-old males who listen to the radio each week and the population proportion of males 65 years and older who listen to the radio each week lies between 0.0231 and 0.0769.
**33. (a)** It will remain 0.05. **(b)** Increase in $s_{\hat{p}_1 - \hat{p}_2}$. This will result in a confidence interval that is less precise. **(c)** Increase in the width of the confidence interval **(d)** Unchanged **(e)** Increase in the $p$-value to 0.2502 **(f)** The conclusion would change from "Reject $H_0$" to "Do not reject $H_0$."

## Chapter 10 Review

**1. (a)** $\bar{x}_d = -2.6875$, $s_d = 1.6146$ **(b)** $(-4.0376, -1.3374)$
**3.** $H_0: \mu_d \geq 0$ vs. $H_a: \mu_d < 0$. Reject $H_0$ if $p$-value $< 0.05$. $t_{data} = -4.708$. $p$-value $= 0.0010939869$. Since the $p$-value $< 0.05$, we reject $H_0$. There is evidence that the population mean of the differences is less than 0.
**5.** (4.111, 74.889)
**7.** Since both sample sizes are large ($n_1 \geq 30$ and $n_2 \geq 30$), it is appropriate to construct a 95% confidence interval for $\mu_1 - \mu_2$.
**9.** 0.0028
**11.** (0.094, 0.106). We are 95% confident that the interval captures the difference in population means.
**13. (a)** $n_1^* = 32$, $n_2^* = 125$ **(b)** 0.12 **(c)** (0.0096, 0.2304) **(d)** $H_0: p_1 = p_2$ vs. $H_a: p_1 \neq p_2$. Reject $H_0$ if $p$-value $< 0.10$. $\hat{p}_{pooled} = 0.0637$, $Z_{data} = 2.48$. $p$-value $= 0.013138259$. Since the $p$-value $< 0.10$, we reject $H_0$. There is evidence that the population proportion packet loss from Asian Web sites is different than the population proportion packet loss from North American Web sites.

**15. (a)** $-0.0197$ **(b)** $(-0.0716, 0.0322)$ **(c)** Since the $p$-value $\geq 0.01$, we do not reject $H_0$. There is insufficient evidence that the population proportion of new mothers in Florida who took their babies in for a checkup within one week of delivery is different from the proportion of new mothers in North Carolina who took their babies in for a checkup within one week of delivery.

## Chapter 10 Quiz

**1.** True
**2.** True
**3.** False
**4.** normal; large (greater than or equal to 30)
**5.** margin of error
**6.** $\bar{x}_d$
**7.** $\mu_1 - \mu_2$
**8.** $s_{\bar{x}_1-\bar{x}_2}$
**9.** No difference
**10. (a)** $\bar{x}_d = 14$ **(b)** $(6.6680, 21.3320)$ **(c)** Since 0 does not lie in the confidence interval we reject $H_0$. There is evidence that the population mean difference in the number of cigarettes smoked before and after attending Butt-Enders is different than 0.
**11. (a)** $-\$15{,}000$ **(b)** $(-21{,}913.75, -8{,}086.25)$. We are 90% confident that the interval captures the difference of the population mean income of Suburb A and the population mean income of Suburb B. **(c)** $(-23{,}304.69, -6{,}695.31)$. We are 95% confident that the interval captures the difference of the population mean income of Suburb A and the population mean income of Suburb B. **(d)** $H_0: \mu_1 \geq \mu_2$ vs. $H_a: \mu_1 < \mu_2$. $t_{\text{crit}} = 1.306$. Reject $H_0$ if $t_{\text{data}} < -1.306$. $t_{\text{data}} = -3.667$. Since $t_{\text{data}} < -1.306$, we reject $H_0$. There is evidence that the population mean income of Suburb A is less than the population mean income of Suburb B. **(e)** $H_0: \mu_1 = \mu_2$ vs. $H_a: \mu_1 \neq \mu_2$. $t_{\text{crit}} = 2.030$. Reject $H_0$ if $t_{\text{data}} < -2.030$ or $t_{\text{data}} > 2.030$. $t_{\text{data}} = -3.667$. Since $t_{\text{data}} < -2.030$, we reject $H_0$. There is evidence that the population mean income of Suburb A is different than the population mean income of Suburb B. **(f)** Since 0 does not lie in the confidence interval in **(c)**, we would reject $H_0$. There is evidence that the population mean income of Suburb A is different than the population mean income of Suburb B. We could have used the confidence interval in **(c)** to perform the hypothesis test in **(e)**.
**12. (a)** Yes, it is appropriate to apply two-sample $t$ inference. Since both sample sizes are large ($n_1 \geq 30$ and $n_2 \geq 30$), Case 2 applies. **(b)** 10 **(c)** 3.9051 **(d)** 7.7594 **(e)** $(2.2406, 17.7594)$. We are 95% confident that the interval captures the difference of the population mean number of bottles processed by the updated machine and the mean number of bottles processed by the non-updated machine.
**13. (a)** 1.662 **(b)** 2.561 **(c)** Since $t_{\text{data}} > 1.662$, we reject $H_0$. There is evidence that the population mean number of bottles processed by the updated machine is greater than the mean number of bottles processed by the non-updated machine. **(d)** Since confidence intervals can be used only to perform two-tailed tests and the hypothesis test in (c) is a one-tailed test, the confidence interval in Exercise 12(e) cannot be used to perform the hypothesis test in (c).
**14. (a)** $-\$5{,}782$ **(b)** $H_0: \mu_1 = \mu_2$ vs. $H_a: \mu_1 \neq \mu_2$. $t_{\text{crit}} = 1.662$. Reject $H_0$ if $t_{\text{data}} < -1.662$ or $t_{\text{data}} > 1.662$. $t_{\text{data}} = -6.129$. Since $t_{\text{data}} < -1.662$, we reject $H_0$. There is evidence that the population mean income of people 18 to 24 years old who never married is different than the population mean income of people 18 to 24 years old who are married. **(c)** No, the conclusion of the two-tailed hypothesis test for $\alpha = 0.10$ is "Reject $H_0$."
**15. (a)** 0.006 **(b)** $\hat{p}_{\text{pooled}} = 0.382$ **(c)** $H_0: p_1 \geq p_2; H_a: p_1 < p_2$. Reject $H_0$ if the $p$-value $< 0.05$. $Z_{\text{data}} = 0.28$; $p$-value $= 0.3897$. Since the $p$-value $\geq 0.05$, we do not reject $H_0$. There is insufficient evidence that the population proportion of 18- to 20-year-olds who used an illicit drug decreased from 2004 to 2005.
**16.** $H_0: p_1 \leq p_2$ vs. $H_a: p_1 > p_2$. $Z_{\text{crit}} = 1.645$. Reject $H_0$ if $Z_{\text{data}} > 1.645$. $\hat{p}_1 - \hat{p}_2 = 0.051$, $\hat{p}_{\text{pooled}} = 0.7155$, $Z_{\text{data}} = 2.53$. Since $Z_{\text{data}} > 1.645$, we reject $H_0$. There is evidence that the population proportion of people aged 35 to 44 who were married was lower in 2000 than in 1990.

# CHAPTER 11

## Section 11.1

**1.** (1) Each independent trial of the experiment has $k$ possible outcomes, $k = 2, 3, \ldots$ (2) The $i$th outcome (category) occurs with probability $p_i$, where $i = 1, 2, \ldots, k$ (3) $\sum_{i=1}^{k} p_i = 1$.
**3.** It is the long-run mean of that random variable after an arbitrarily large number of trials.
**5.** Multinomial
**7.** Multinomial
**9. (a)** $p$-value $> 0.10$
**(b)**



**11. (a)** $0.005 < p$-value $< 0.01$
**(b)**

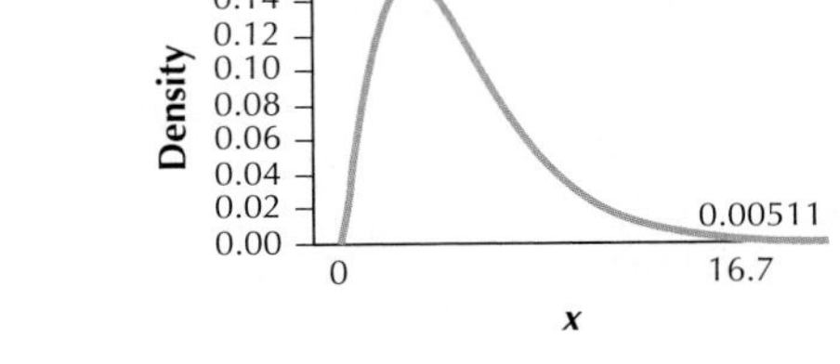


**13. (a)** $E_1 = 50, E_2 = 25, E_3 = 25$ **(b)** Conditions are met.
**15. (a)** $E_1 = 90, E_2 = 5, E_3 = 4, E_4 = 1$ **(b)** Conditions are not met.
**17.** 0.667
**19.** 7.333
**21.** 17.667
**23. (a)** $E_1 = 40, E_2 = 30, E_3 = 30$; conditions are met. **(b)** $\chi^2_{\text{crit}} = \chi^2_{0.05} = 5.991$. Reject $H_0$ if $\chi^2_{\text{data}} > 5.991$. **(c)** 4.167 **(d)** Since $\chi^2_{\text{data}} \leq 5.991$, we do not reject $H_0$. There is insufficient evidence that the random variable does not follow the distribution specified in $H_0$.
**25. (a)** $E_1 = 80, E_2 = 70, E_3 = 20, E_4 = 20, E_5 = 10$; conditions are met. **(b)** $\chi^2_{\text{crit}} = \chi^2_{0.10} = 7.779$. Reject $H_0$ if $\chi^2_{\text{data}} > 7.779$. **(c)** 6.607 **(d)** Since $\chi^2_{\text{data}} \leq 7.779$, we do not reject $H_0$. There is insufficient evidence that the random variable does not follow the distribution specified in $H_0$.
**27. (a)** Reject $H_0$ if the $p$-value $< 0.05$. $E_1 = 50, E_2 = 50$; conditions are met. **(b)** 4 **(c)** $0.025 < p$-value $< 0.05$ **(d)** Since the $p$-value $< 0.05$, we reject $H_0$. There is evidence that the random variable does not follow the distribution specified in $H_0$.
**29. (a)** Reject $H_0$ if the $p$-value $< 0.10$. $E_1 = 100, E_2 = 50, E_3 = 30, E_4 = 20$; conditions are met. **(b)** 6.083 **(c)** $p$-value $> 0.10$
**(d)** Since the $p$-value $\geq 0.10$, we do not reject $H_0$. There is insufficient evidence that the random variable does not follow the distribution specified in $H_0$.

**31.** Since $\chi^2_{data} \le 9.488$, we do not reject $H_0$. There is insufficient evidence that the distribution of education levels has changed since 2005.

**33.** $H_0: p_{phip} = 0.30, p_{mm} = 0.556, p_{other} = 0.144$. $H_a$ : The random variable does not follow the distribution specified in $H_0$. $E_{phip} = 300, E_{mm} = 556, E_{other} = 144$. Since none of the expected frequencies is less than 1 and none of the expected frequencies is less than 5, the conditions for performing the $\chi^2$ goodness of fit test are met. $\chi^2_{crit} = \chi^2_{0.05} = 5.991$. Reject $H_0$ if $\chi^2_{data} > 5.991$. $\chi^2_{data} = 14.224$. Since $\chi^2_{data} > 5.991$, we reject $H_0$. There is evidence that the population proportions of minority patients who suffered spinal cord injuries, who had a private health insurance provider, Medicare, Medicaid, or other arrangements, have changed.

**35.** $H_0: p_{males} = 0.4641403733, p_{females} = 0.5358596267$. $H_a$ : The random variable does not follow the distribution specified in $H_0$. $E_{males} = 1129.25, E_{females} = 1303.75$. Since none of the expected frequencies is less than 1 and none of the expected frequencies is less than 5, the conditions for performing the $\chi^2$ goodness of fit test are met. $\chi^2_{crit} = \chi^2_{0.05} = 3.841$. Reject $H_0$ if $\chi^2_{data} > 3.841$. $\chi^2_{data} = 17.278$. Since $\chi^2_{data} > 3.841$, we reject $H_0$. There is evidence that the population proportions of males and females living in college dormitories have changed.

**37.** $H_0: p_{Hertz} = 0.28, p_{Avis} = 0.214, p_{National} = 0.121, p_{Budget} = 0.105, p_{Alamo} = 0.105, p_{Dollar} = 0.085, p_{Others} = 0.09$. $H_a$ : The random variable does not follow the distribution specified in $H_0$. $E_{Hertz} = 28, E_{Avis} = 21.4, E_{National} = 12.1, E_{Budget} = 10.5, E_{Alamo} = 10.5, E_{Dollar} = 8.5, E_{Others} = 9$. Since none of the expected frequencies is less than 1 and none of the expected frequencies is less than 5, the conditions for performing the $\chi^2$ goodness of fit test are met. $\chi^2_{crit} = \chi^2_{0.05} = 12.592$. Reject $H_0$ if $\chi^2_{data} > 12.592$. $\chi^2_{data} = 0.2058$. Since $\chi^2_{data} \le 9.488$, we do not reject $H_0$. There is insufficient evidence that the population proportions of market shares held by the largest car rental agencies have changed.

**39.** $H_0: p_{every} = 0.31, p_{almost} = 0.12, p_{onceortwice} = 0.14, p_{few} = 0.24, p_{never} = 0.19$. $H_a$ : The random variable does not follow the distribution specified in $H_0$. $E_{every} = 31, E_{almost} = 12, E_{onceortwice} = 14, E_{few} = 24, E_{never} = 19$. Since none of the expected frequencies is less than 1 and none of the expected frequencies is less than 5, the conditions for performing the $\chi^2$ goodness of fit test are met. $\chi^2_{crit} = \chi^2_{0.10} = 7.779$. Reject $H_0$ if $\chi^2_{data} > 7.779$. $\chi^2_{data} = 0.531$. Since $\chi^2_{data} \le 7.779$, we do not reject $H_0$. There is insufficient evidence that the population proportions of how often Americans attend religious services have changed.

**41.** $H_0: p_{believe} = 0.78, p_{notsure} = 0.12, p_{didnotbelieve} = 0.10$. $H_a$ : The random variable does not follow the distribution specified in $H_0$. $E_{believe} = 780, E_{notsure} = 120, E_{didnotbelieve} = 100$. Since none of the expected frequencies is less than 1 and none of the expected frequencies is less than 5, the conditions for performing the $\chi^2$ goodness of fit test are met. $\chi^2_{crit} = \chi^2_{0.05} = 5.991$. Reject $H_0$ if $\chi^2_{data} > 5.991$. $\chi^2_{data} = 11.885$. Since $\chi^2_{data} > 5.991$, we reject $H_0$. There is evidence that the population proportions of people who either believe, are not sure, or don't believe in angels have changed.

**43.** $H_0: p_{personnel} = 0.24, p_{operations} = 0.28, p_{other} = 0.48$. $H_a$ : The random variable does not follow the distribution specified in $H_0$. $E_{personnel} = 91.056, E_{operations} = 106.232, E_{other} = 182.112$. Since none of the expected frequencies is less than 1 and none of the expected frequencies is less than 5, the conditions for performing the $\chi^2$ goodness of fit test are met. $\chi^2_{crit} = \chi^2_{0.05} = 5.991$. Reject $H_0$ if $\chi^2_{data} > 5.991$. $\chi^2_{data} = 3.7072$. Since $\chi^2_{data} \le 5.991$, we do not reject $H_0$. There is insufficient evidence that the population proportions of budget items have changed from 1985 to 2004.

## Section 11.2

**1.** Tabular summary of the relationship between two categorical variables

**3.** The two-sample *Z* test for the difference in proportions from Chapter 10 is for comparing proportions of two independent populations, and the $\chi^2$ test for homogeneity of proportions is for comparing proportions of *k* independent populations.

**5.**

| | A1 | A2 | Total |
|---|---|---|---|
| B1 | 11 | 19 | 30 |
| B2 | 11 | 19 | 30 |
| Total | 22 | 38 | 60 |

**7.**

| | E1 | E2 | E3 | Total |
|---|---|---|---|---|
| F1 | 30.71 | 20.79 | 8.50 | 60 |
| F2 | 34.29 | 23.21 | 9.50 | 67 |
| Total | 65 | 44 | 18 | 127 |

**9.**

| | I1 | I2 | I3 | Total |
|---|---|---|---|---|
| J1 | 99.2788 | 93.6058 | 102.1154 | 295 |
| J2 | 55.5288 | 52.3558 | 57.1154 | 165 |
| J3 | 20.1923 | 19.0385 | 20.7692 | 60 |
| Total | 174.9999 | 165.0001 | 180 | 520 |

**11. (a)** $H_0$ : Variable A and Variable B are independent. $H_a$ : Variable A and Variable B are not independent.

**(b)**

| | A1 | A2 | Total |
|---|---|---|---|
| B1 | 11 | 19 | 30 |
| B2 | 11 | 19 | 30 |
| Total | 22 | 38 | 60 |

Since none of the expected frequencies is less than 1 and none of the expected frequencies is less than 5, the conditions for performing the $\chi^2$ test for independence are met. **(c)** 3.841. Reject $H_0$ if $\chi^2_{data} > 3.841$. **(d)** 0.2871 **(e)** Since $\chi^2_{data} \le 3.841$, we do not reject $H_0$. There is insufficient evidence that variable A and variable B are not independent.

**13. (a)** $H_0$: Variable I and Variable J are independent. $H_a$ : Variable I and Variable J are not independent.

**(b)**

| | I1 | I2 | I3 | Total |
|---|---|---|---|---|
| J1 | 99.2788 | 93.6058 | 102.1154 | 295 |
| J2 | 55.5288 | 52.3558 | 57.1154 | 165 |
| J3 | 20.1923 | 19.0385 | 20.7692 | 60 |
| Total | 174.9999 | 165.0001 | 180 | 520 |

Since none of the expected frequencies is less than 1 and none of the expected frequencies is less than 5, the conditions for performing the $\chi^2$ test for independence are met. **(c)** 13.277. Reject $H_0$ if $\chi^2_{data} > 13.277$. **(d)** 4.000 **(e)** Since $\chi^2_{data} \le 13.277$, we do not reject $H_0$. There is insufficient evidence that variable I and variable J are not independent.

**15. (a)** $H_0$ : Variable C and Variable D are independent. $H_a$: Variable C and Variable D are not independent. Reject $H_0$ if the *p*-value $< 0.05$.

| | C1 | C2 | Total |
|---|---|---|---|
| D1 | 55 | 95 | 150 |
| D2 | 55 | 95 | 150 |
| Total | 110 | 190 | 300 |

Since none of the expected frequencies is less than 1 and none of the expected frequencies is less than 5, the conditions for performing the $\chi^2$ test for independence are met. **(b)** 1.4354 **(c)** *p*-value $> 0.10$ **(d)** Since the *p*-value $\ge 0.05$, we do not reject $H_0$. There is insufficient evidence that variable C and variable D are not independent.

**17. (a)** $H_0$ : Variable K and Variable L are independent. $H_0$ : Variable K and Variable L are not independent. Reject $H_0$ if $p$-value $< 0.01$.

| | K1 | K2 | K3 | K4 | Total |
|---|---|---|---|---|---|
| **L1** | 37.5 | 72.92 | 89.58 | 100 | 300 |
| **L2** | 23.75 | 46.18 | 56.74 | 63.33 | 190 |
| **L3** | 28.75 | 55.90 | 68.68 | 76.67 | 230 |
| **Total** | 90 | 175 | 215 | 240 | 720 |

Since none of the expected frequencies is less than 1 and none of the expected frequencies is less than 5, the conditions for performing the $\chi^2$ test for independence are met. **(b)** 4.9077 **(c)** $p$-value $> 0.10$ **(d)** Since $p$-value $\geq 0.01$, we do not reject $H_0$. There is insufficient evidence that variable K and variable L are not independent.

**19. (a)** $H_0 : p_1 = p_2 = p_3$. $H_a$ : Not all the proportions in $H_0$ are equal.

**(b)**

| | Sample 1 | Sample 2 | Sample 3 | Total |
|---|---|---|---|---|
| **Successes** | 9.63 | 20.86 | 29.52 | 60.01 |
| **Failures** | 20.37 | 44.14 | 62.48 | 126.99 |
| **Total** | 30 | 65 | 92 | 187 |

Since none of the expected frequencies is less than 1 and none of the expected frequencies is less than 5, the conditions for performing the $\chi^2$ test for homogeneity of proportions are met. **(c)** 5.991. Reject $H_0$ if $\chi^2_{\text{data}} > 5.991$. **(d)** 0.0846 **(e)** Since $\chi^2_{\text{data}} \leq 5.991$, we do not reject $H_0$. There is insufficient evidence that not all the proportions in $H_0$ are equal.

**21. (a)** $H_0 : p_1 = p_2 = p_3 = p_4$. $H_a$ : Not all the proportions in $H_0$ are equal.

**(b)**

| | Sample 1 | Sample 2 | Sample 3 | Sample 4 | Total |
|---|---|---|---|---|---|
| **Successes** | 9.67 | 15.08 | 20.11 | 25.14 | 70 |
| **Failures** | 15.33 | 23.92 | 31.89 | 39.86 | 111 |
| **Total** | 25 | 39 | 52 | 65 | 181 |

Since none of the expected frequencies is less than 1 and none of the expected frequencies is less than 5, the conditions for performing the $\chi^2$ test for homogeneity of proportions are met. **(c)** 7.815. Reject $H_0$ if $\chi^2_{\text{data}} > 7.815$. **(d)** 0.0213 **(e)** Since $\chi^2_{\text{data}} \leq 7.815$, we do not reject $H_0$. There is insufficient evidence that not all the proportions in $H_0$ are equal.

**23. (a)** $H_0 : p_1 = p_2 = p_3$. $H_a$ : Not all the proportions in $H_0$ are equal. Reject $H_0$ if the $p$-value $< 0.05$.

| | Sample 1 | Sample 2 | Sample 3 | Total |
|---|---|---|---|---|
| **Successes** | 27.17 | 57.74 | 95.09 | 180 |
| **Failures** | 12.83 | 27.26 | 44.91 | 85 |
| **Total** | 40 | 85 | 140 | 265 |

Since none of the expected frequencies is less than 1 and none of the expected frequencies is less than 5 the conditions for performing the $\chi^2$ test for homogeneity of proportions are met. **(b)** 2.0442 **(c)** $p$-value $> 0.10$ **(d)** Since the $p$-value $\geq 0.05$, we do not reject $H_0$. There is insufficient evidence that not all the proportions in $H_0$ are equal.

**25. (a)** $H_0 : p_1 = p_2 = p_3 = p_4$. $H_a$ : Not all the proportions in $H_0$ are equal. Reject $H_0$ if the $p$-value $< 0.05$.

| | Sample 1 | Sample 2 | Sample 3 | Sample 4 | Total |
|---|---|---|---|---|---|
| **Successes** | 8.98 | 12.35 | 21.88 | 34.79 | 78 |
| **Failures** | 7.02 | 9.65 | 17.12 | 27.21 | 61 |
| **Total** | 16 | 22 | 39 | 62 | 139 |

Since none of the expected frequencies is less than 1 and none of the expected frequencies is less than 5, the conditions for performing the $\chi^2$ test for homogeneity of proportions are met. **(b)** 1.264 **(c)** $p$-value $> 0.10$ **(d)** Since the $p$-value $\geq 0.05$, we do not reject $H_0$. There is insufficient evidence that not all the proportions in $H_0$ are equal.

**27.** $H_0$ : *Type of stimulus* and *type of mouse* are independent. $H_a$ : *Type of stimulus* and *type of mouse* are not independent. Reject $H_0$ if $p$-value $< 0.05$. Since none of the expected frequencies is less than 1 and none of the expected frequencies is less than 5, the conditions for performing the $\chi^2$ test for independence are met. $\chi^2_{\text{data}} = 0.7937$. $p$-value $> 0.10$. Since $p$-value $\geq 0.10$, we do not reject $H_0$. There is insufficient evidence that *type of stimulus* and *type of mouse* are not independent.

**29.** $H_0 : p_{\text{Republicans}} = p_{\text{Independents}} = p_{\text{Democrats}}$. $H_a$ : Not all the proportions in $H_0$ are equal. Since none of the expected frequencies is less than 1 and none of the expected frequencies is less than 5, the conditions for performing the $\chi^2$ test for homogeneity of proportions are met. $\chi^2_{\text{crit}} = \chi^2_{0.05} = 5.991$. Reject $H_0$ if $\chi^2_{\text{data}} > 5.991$. $\chi^2_{\text{data}} = 1.1568$. Since $\chi^2_{\text{data}} \leq 5.991$, we do not reject $H_0$. There is insufficient evidence that the population proportion favoring restriction is not the same for all three groups.

**31.** $H_0$ : *Gender* and *political party preference* are independent. $H_a$ : *Gender* and *political party preference* are not independent. Reject $H_0$ if $p$-value $< 0.01$. Since none of the expected frequencies is less than 1 and none of the expected frequencies is less than 5, the conditions for performing the $\chi^2$ test for independence are met. $\chi^2_{\text{data}} = 33.333$. $p$-value $< 0.005$. Since $p$-value $< 0.01$, we reject $H_0$.There is evidence that *gender* and *political party preference* are not independent.

**33.** $H_0 : p_{\text{Edit}} = p_{\text{Arrange}}$. $H_a$ : Not all the proportions in $H_0$ are equal. Reject $H_0$ if $p$-value $< 0.05$. Since none of the expected frequencies is less than 1 and none of the expected frequencies is less than 5, the conditions for performing the $\chi^2$ test for homogeneity of proportions are met. $\chi^2_{\text{data}} = 3.5165$. $0.05 < p$-value $< 0.10$. Since $p$-value $\geq 0.05$, we do not reject $H_0$. There is insufficient evidence that the proportions who favor email differ between the two tasks.

**35.** $H_0 : p_{\text{Work}} = p_{\text{Personal}}$. $H_a$ : Not all the proportions in $H_0$ are equal. Reject $H_0$ if $p$-value $< 0.01$. Since none of the expected frequencies is less than 1 and none of the expected frequencies is less than 5, the conditions for performing the $\chi^2$ test for homogeneity of proportions are met. $\chi^2_{\text{data}} = 23.3325$. $p$-value $< 0.005$. Since $p$-value $< 0.01$, we reject $H_0$. There is evidence that the population proportions who report "a lot of spam" are not the same for work email and personal email. Yes.

**37.** $H_0 : p_{\text{Urban}} = p_{\text{Suburban}} = p_{\text{Rural}}$. $H_a$ : Not all the proportions in $H_0$ are equal. Reject $H_0$ if $p$-value $< 0.05$. Since none of the expected frequencies is less than 1 and none of the expected frequencies is less than 5, the conditions for performing the $\chi^2$ test for homogeneity of proportions are met. $\chi^2_{\text{data}} = 9.095$. $0.01 < p$-value $< 0.025$. Since $p$-value $< 0.05$, we reject $H_0$. There is evidence that the population proportions of residents from the three categories who use online dating are not all the same.

**39. (a)** Dependent **(b)** Since the $p$-value $\approx 0$, $p$-value $< 0.05$. Thus we reject $H_0$. There is evidence that *gender* and *goals* are not independent.

**41. (a)** Dependent **(b)** Since the $p$-value 0.001, $p$-value $< 0.10$. Thus we reject $H_0$. There is evidence that *urb_rural* and *goals* are not independent.

**43.** $H_0 : p_{\text{Jan}} = p_{\text{Feb}} = p_{\text{Mar}} = p_{\text{Apr}} = p_{\text{May}} = p_{\text{June}} = p_{\text{July}} = p_{\text{Aug}} = p_{\text{Sept}} = p_{\text{Oct}} = p_{\text{Nov}} = p_{\text{Dec}}$. $H_a$ : Not all the proportions in $H_0$ are equal. Reject $H_0$ if $p$-value $< 0.01$. $\chi^2_{\text{data}} = 30.2571$. $p$-value $< 0.005$. Since $p$-value $< 0.01$, we reject $H_0$. There is evidence that the population proportion of "drafted dates" are not equal for all months.

## Chapter 11 Review

**1.** $H_0: p_{USCan} = 0.32, p_{USMex} = 0.22, p_{CanUS} = 0.31, p_{MexUS} = 0.15$. $H_a$: The random variable does not follow the distribution specified in $H_0$. $E_{USCan}$ = \$22.4 billion, $E_{USMex}$ = \$15.4 billion, $E_{CanUS}$ = \$21.7 billion, $p_{MexUS}$ = \$10.5 billion. Since none of the expected frequencies is less than 1 and none of the expected frequencies is less than 5, the conditions for performing the $\chi^2$ goodness of fit test are met. Reject $H_0$ if $p$-value $< 0.05$. $\chi^2_{data} = 0.469$. $0.90 < p\text{-value} < 0.95$. Since $p$-value $\geq$ 0.05, we do not reject $H_0$. There is insufficient evidence that the population proportions of truck-hauled trade have changed since 2002.

**3.** $H_0: p_{rock} = 0.40, p_{rap/hiphop} = 0.25, p_{country} = 0.15, p_{blues} = 0.08, p_{jazz} = 0.07, p_{classical} = 0.05$. $H_a$: The random variable does not follow the distribution specified in $H_0$. $E_{rock} = 80, E_{rap/hiphop} = 50, E_{country} = 30, E_{blues} = 16, E_{jazz} = 14, E_{classical} = 10$. Since none of the expected frequencies is less than 1 and none of the expected frequencies is less than 5, the conditions for performing the $\chi^2$ goodness of fit test are met. $\chi^2_{crit} = \chi^2_{0.05} = 11.071$. Reject $H_0$ if $\chi^2_{data} > 11.071$. $\chi^2_{data} = 12.4762$. Since $\chi^2_{data} > 11.071$, we reject $H_0$. There is evidence that the music preferences of college students have changed since 1995.

**5.** $H_0: p_0 = 0.10, p_1 = 0.10, p_2 = 0.10, p_3 = 0.10, p_4 = 0.10, p_5 = 0.10, p_6 = 0.10, p_7 = 0.10, p_8 = 0.10, p_9 = 0.10$. $H_a$: Not all the proportions in $H_0$ are equal. Reject $H_0$ if $p$-value $< 0.05$. $E_0 = 21.8, E_1 = 21.8, E_2 = 21.8, E_3 = 21.8, E_4 = 21.8, E_5 = 21.8, E_6 = 21.8, E_7 = 21.8, E_8 = 21.8, E_9 = 21.8$. Since none of the expected frequencies is less than 1 and none of the expected frequencies is less than 5, the conditions for performing the $\chi^2$ goodness of fit test are met. Reject $H_0$ if $p$-value $< 0.05$. $\chi^2_{data} = 12.4587$. $p$-value = 0.18866727. Since $p$-value $\geq 0.05$, we do not reject $H_0$. There is insufficient evidence that the population proportions of digits are not all 0.10.

**7.** **(a)** A higher proportion of the females with high GPAs take the SAT exam than the proportion of the females with lower GPAs. **(b)** $H_0: p_{A+} = p_A = p_{A-} = p_B = p_C = p_{D/F}$. $H_a$: Not all the proportions in $H_0$ are equal. Reject $H_0$ if $p$-value $< 0.05$. Since none of the expected frequencies is less than 1 and none of the expected frequencies is less than 5, the conditions for performing the $\chi^2$ test for homogeneity of proportions are met. $\chi^2_{data} = 15.0850$. $0.01 < p\text{-value} < 0.025$. Since $p$-value $< 0.05$, we reject $H_0$. There is evidence that the proportion of females is not all across the six grade categories.

**9.** $H_0: p_{Northeast} = p_{Midwest} = p_{South} = p_{West}$. $H_a$: Not all the proportions in $H_0$ are equal. Reject $H_0$ if $p$-value $< 0.01$. Since none of the expected frequencies is less than 1 and none of the expected frequencies is less than 5, the conditions for performing the $\chi^2$ test for homogeneity of proportions are met. $\chi^2_{data} = 25.846$. $p$-value $<$ 0.005. Since $p$-value $< 0.01$, we reject $H_0$. There is evidence that the population proportions of pregnant women who have had an HIV test in the past 12 months are not the same across all four regions.

**11.** $H_0$: Age and radio station type are independent. $H_a$: Age and radio station type are not independent. Reject $H_0$ if $p$-value $< 0.05$. Since none of the expected frequencies is less than 1 and none of the expected frequencies is less than 5, the conditions for performing the $\chi^2$ test for independence are met. $\chi^2_{data} = 8.269$. $p$-value $<$ 0.005. Since $p$-value $< 0.05$, we reject $H_0$.There is evidence that age and radio station type are not independent.

**13.** $H_0: p_{Whites} = p_{Blacks} = p_{Hispanics}$. $H_a$: Not all the proportions in $H_0$ are equal. Reject $H_0$ if $p$-value $< 0.05$. Since none of the expected frequencies is less than 1 and none of the expected frequencies is less than 5, the conditions for performing the $\chi^2$ test for homogeneity of proportions are met. $\chi^2_{data} = 42.658$. $p$-value $< 0.005$. Since $p$-value $< 0.05$, we reject $H_0$. There is evidence that Internet use levels is not the same for all races.

## Chapter 11 Quiz

**1.** True
**2.** False
**3.** False
**4.** 1, 5
**5.** equal
**6.** expected frequency
**7.** **(a)** Critical value method, **(b)** exact $p$-value method, or **(c)** estimated $p$-value method
**8.** $H_a$, the alternative hypothesis
**9.** Degrees of freedom = $(r - 1)(c - 1)$, where $r$ = the number of categories in the row variable and $c$ = the number of categories in the column variable.

**10.** **(a)** $E_1 = 10, E_2 = 10, E_3 = 10, E_4 = 10, E_5 = 10$ **(b)** Conditions are met. **(c)** 9.488. Reject $H_0$ if $\chi^2_{data} > 9.488$. **(d)** $\chi^2_{data} = 1$ **(e)** Since $\chi^2_{data} \leq 9.488$, we do not reject $H_0$. There is insufficient evidence that the random variable does not follow the distribution specified in $H_0$.

**11.** **(a)** $E_1 = 48, E_2 = 40, E_3 = 32, E_4 = 24, E_5 = 9.6, E_6 = 6.4$. **(b)** Conditions are met. **(c)** 11.071. Reject $H_0$ if $\chi^2_{data} > 11.071$. **(d)** $\chi^2_{data} = 2.917$ **(e)** Since $\chi^2_{data} \leq 11.071$, we do not reject $H_0$. There is sufficient evidence that the random variable does not follow the distribution specified in $H_0$.

**12.** **(a)** $E_1 = 20, E_2 = 20, E_3 = 20, E_4 = 20, E_5 = 20$ **(b)** Conditions are met. **(c)** 13.277. Reject $H_a$ if $\chi^2_{data} > 13.277$. **(d)** 0.5 **(e)** Since $\chi^2_{data} \leq 13.277$, we do not reject $H_0$. There is insufficient evidence that the random variable does not follow the distribution specified in $H_0$.

**13.** **(a)** $E_1 = 60, E_2 = 50, E_3 = 40, E_4 = 30, E_5 = 12, E_6 = 8$ **(b)** Conditions are met. **(c)** 11.071. Reject $H_a$ if $\chi^2_{data} > 11.071$. **(d)** 5.5 **(e)** Since $\chi^2_{data} \leq 11.071$, we do not reject $H_0$. There is insufficient evidence that the random variable does not follow the distribution specified in $H_0$.

**14.** **(a)** The higher the grade level, the higher the proportion of students who have used an illicit drug. **(b)** $H_0: p_{8thgraders} = p_{10thgraders} = p_{12thgraders}$. $H_a$: Not all the proportions in $H_0$ are equal. Reject $H_0$ if $p$-value $< 0.01$. Since none of the expected frequencies is less than 1 and none of the expected frequencies is less than 5, the conditions for performing the $\chi^2$ test for homogeneity of proportions are met. $\chi^2_{data} = 3060.14226$. $p$-value $< 0.005$. Since $p$-value $< 0.01$, we reject $H_0$. There is evidence that the proportions of children in those grades that have ever used an illicit drug are not all the same.

**15.** $H_0: p_{Whites} = p_{Blacks} = p_{Hispanics}$. $H_a$: Not all the proportions in $H_0$ are equal. Reject $H_0$ if $p$-value $< 0.01$. Since none of the expected frequencies is less than 1 and none of the expected frequencies is less than 5, the conditions for performing the $\chi^2$ test for homogeneity of proportions are met. $\chi^2_{data} = 16.802$. $p$-value $< 0.005$. Since $p$-value $< 0.01$, we reject $H_0$. There is evidence that Internet use levels among the races are not the same.

**16.** $H_0$: *Gender* and *sport preference* are independent. $H_a$: *Gender* and *sport preference* are not independent. Reject $H_0$ if $p$-value $<$ 0.05. Since none of the expected frequencies is less than 1 and none of the expected frequencies is less than 5, the conditions for performing the $\chi^2$ test for independence are met. $\chi^2_{data} = 19.857$. $p$-value $< 0.005$. Since $p$-value $< 0.05$, we reject $H_0$. There is evidence that *gender* and *sport preference* are not independent.

**17.** $H_0: p_{Texas} = p_{Oklahoma} = p_{Pennsylvania}$. $H_a$: Not all the proportions in $H_0$ are equal. Reject $H_0$ if $p$-value $< 0.05$. Since none of the expected frequencies is less than 1 and none of the expected frequencies is less than 5, the conditions for performing the $\chi^2$ test for homogeneity of proportions are met. $\chi^2_{data} = 81{,}246.70827$. $p$-value $< 0.005$. Since $p$-value $< 0.01$, we reject $H_0$. There is evidence that the population proportions of cattle on smaller farms are not the same across all three states.

# Chapter 12

## Section 12.1

**1.** True
**3.** True
**5.** True
**7.** MSTR measures the variability in the sample means. MSE measures the variability within the samples.
**9.** Answers will vary.
**11.** Each of the $k$ populations is normally distributed, the variances of the populations are all equal, and the samples are independently drawn.
**13. (a)** $F_{crit} = 5.59$
**(b)**

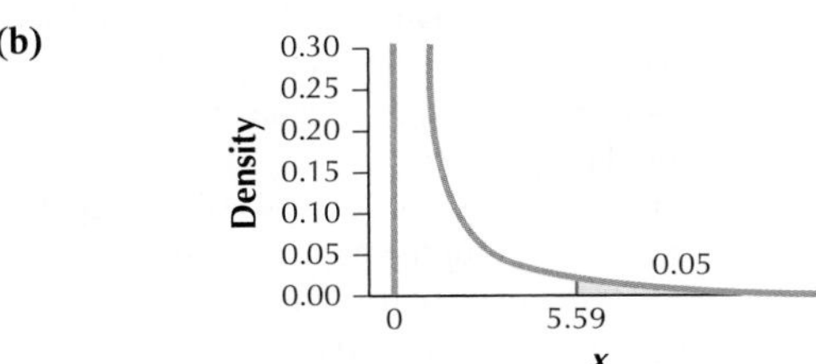


**15. (a)** $F_{crit} = 6.36$
**(b)**

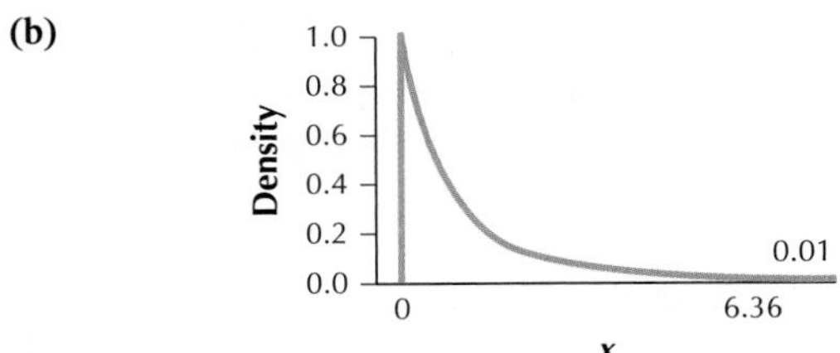


**17. (a)** $F_{crit} = 2.70$
**(b)**

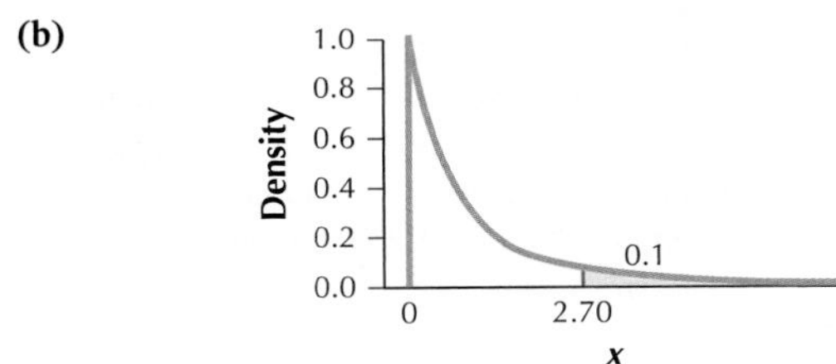


**19. (a)** $F_{crit} = 3.24$
**(b)**

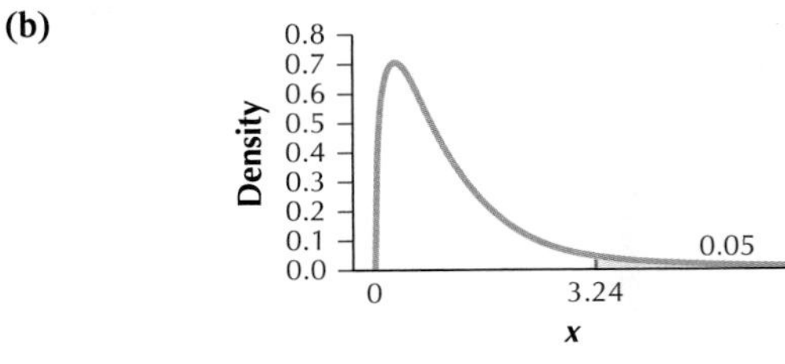


**21. (a)** $F_{crit} = 3.01$
**(b)**



**23. (a)** $df_1 = 2$, $df_2 = 12$ **(b)** 10 **(c)** 40 **(d)** 12 **(e)** 52
**25. (a)** $df_1 = 3$, $df_2 = 696$ **(b)** 96.42857143 **(c)** 491,071.4286 **(d)** 18,248 **(e)** 509,319.4286
**27. (a)** 20 **(b)** 1 **(c)** 20
**29. (a)** 32.14285714 **(b)** 26.2183908 **(c)** 1.225966055
**31. (a)** $S_{Online} = 15.0555 < 25.2982 = 2\,S_{Hybrid}$ **(b)** (i) $df_1 = 2$, $df_2 = 15$ (ii) 75.2778 (iii) 219.4426 (iv) 2804.1695 (v) 3023.6121 (vi) 109.7213 (vii) 186.944633 (viii) 0.5869
**(c)**

| Source of variation | Sum of squares | Degrees of freedom | Mean square | *F*-test statistic |
|---|---|---|---|---|
| **Treatment** | SSTR = 219.4426 | $df_1 = 2$ | MSTR = 109.7213 | $F_{data} =$ 0.5869 |
| **Error** | SSE = 2804.1695 | $df_2 = 15$ | MSE = 186.944633 | |
| **Total** | SST = 3023.6121 | | | |

**33. (a)** $S_{female} = 0.743 < 1.398 = 2\,S_{male}$ **(b)** (i) $df_1 = 1$, $df_2 = 128$ (ii) 98.244 (iii) 2.548 (iv) 66.6016 (v) 69.1496 (vi) 2.548 (vii) 0.520325 (viii) 4.896939413

**(c)**

| Source of variation | Sum of squares | Degrees of freedom | Mean square | *F*-test statistic |
|---|---|---|---|---|
| **Treatment** | SSTR = 2.548 | $df_1 = 1$ | MSTR = 2.548 | $F_{data} =$ 4.896939413 |
| **Error** | SSE = 66.6016 | $df_2 = 128$ | MSE = 0.520325 | |
| **Total** | SST = 69.1496 | | | |

**35. (a)** $S_{none} = 3.22 < 5.48 = 2\,S_{Catholic}$ **(b)** (i) $df_1 = 4$, $df_2 = 2886$ (ii) 13.36583535 (iii) 581.5002576 (iv) 24,230.7355 (v) 24,812.23576 (vi) 145.3750644 (vii) 8.395958247 (viii) 17.31488654
**(c)**

| Source of variation | Sum of squares | Degrees of freedom | Mean square | *F*-test statistic |
|---|---|---|---|---|
| **Treatment** | SSTR = 581.5002576 | $df_1 = 4$ | MSTR = 145.3750644 | $F_{data} =$ 17.31488654 |
| **Error** | SSE = 24,230.7355 | $df_2 = 2886$ | MSE = 8.395958247 | |
| **Total** | SST = 24,812.23576 | | | |

**37. (a)** $S_{During} = 5.5014 < 6.2188 = 2\,S_{After}$ **(b)** (i) $df_1 = 2$, $df_2 = 33$ (ii) 11.9306 (iii) 41.5146 (iv) 583.4119 (v) 624.9265 (vi) 20.7573 (vii) 17.6791 (viii) 1.1741

**(c)**

| Source of variation | Sum of squares | Degrees of freedom | Mean square | *F*-test statistic |
|---|---|---|---|---|
| **Treatment** | SSTR = 41.5146 | $df_1 = 2$ | MSTR = 20.7573 | $F_{data} =$ 1.1741 |
| **Error** | SSE = 583.4119 | $df_2 = 33$ | MSE = 17.6791 | |
| **Total** | SST = 624.9265 | | | |

## Section 12.2

**1.** Use the closest value.
**3.** Against
**5. (a)** $H_0: \mu_1 = \mu_2 = \mu_3$. $H_a$: Not all the population means are equal. Reject $H_0$ if the $p$-value $< 0.05$. **(b)** 20 **(c)** $p$-value $< 0.001$ **(d)** Since the $p$-value $< 0.05$, we reject $H_0$. There is evidence that not all the population means are equal.
**7. (a)** $H_0: \mu_A = \mu_B = \mu_C = \mu_D$. $H_a$: Not all the population means are equal. **(b)** $F_{crit} = 2.61$. Reject $H_0$ if $F_{data} > 2.61$. **(c)** 1.225966055 **(d)** Since $F_{data} \leq 2.61$, we do not reject $H_0$. There is insufficient evidence that not all the population means are equal.
**9. (a)** Missing values are in red.

| Source of variation | Sum of squares | Degrees of freedom | Mean square | *F*-test statistic |
|---|---|---|---|---|
| **Treatment** | SSTR = 120 | $df_1 = 6$ | MSTR = 20 | $F_{data} = 44$ |
| **Error** | SSE = 315 | $df_2 = 693$ | MSE = 0.4545454545 | |
| **Total** | SST = 435 | | | |

**(b)** $H_0: \mu_1 = \mu_2 = \mu_3 = \mu_4 = \mu_5 = \mu_6 = \mu_7$. $H_a$: Not all the population means are equal. Reject $H_0$ if the $p$-value $< 0.05$. $F_{data} = 44$; $p$-value $< 0.001$. Since the $p$-value $< 0.05$, we reject $H_0$. There is evidence that not all the population means are equal.
**11. (a)** Missing values are in red.

| Source of variation | Sum of squares | Degrees of freedom | Mean square | *F*-test statistic |
|---|---|---|---|---|
| **Treatment** | SSTR = 40 | $df_1 = 4$ | MSTR = 10 | $F_{data} = 1.0$ |
| **Error** | SSE = 400 | $df_2 = 40$ | MSE = 10 | |
| **Total** | SST = 440 | | | |

**(b)** $H_0: \mu_1 = \mu_2 = \mu_3 = \mu_4 = \mu_5$. $H_a$: Not all the population means are equal. $F_{crit} = 2.06$. Reject $H_0$ if $F_{data} > 2.06$. $F_{data} = 1.0$. Since $F_{data} \leq 2.06$, we do not reject $H_0$. There is insufficient evidence that not all the population means are equal.
**13. (a)** $H_0: \mu_{Protestant} = \mu_{Catholic} = \mu_{Jewish} = \mu_{none} = \mu_{other}$. $H_a$: Not all the population means are equal. $\mu_{Protestant}$ = the population mean number of years of education of people whose religious preference is Protestant. $\mu_{Catholic}$ = the population mean number of years of education of people whose religious preference is Catholic. $\mu_{Jewish}$ = the population mean number of years of education of people whose religious preference is Jewish. $\mu_{none}$ = the population mean number of years of education of people whose religious preference is None. $\mu_{other}$ = the population mean number of years of education of people whose religious preference is Other. Reject $H_0$ if the $p$-value $<$ 0.05. **(b)** 17.31488654 **(c)** $p$-value $< 0.001$ **(d)** Since the $p$-value $< 0.05$, we reject $H_0$. There is evidence that not all the population mean numbers of years of education are equal.
**15. (a)** $H_0: \mu_{before} = \mu_{during} = \mu_{after}$. $H_a$: Not all the population means are equal. $\mu_{before}$ = the population mean number of emergency room visits before a full moon. $\mu_{during}$ = the population mean number of emergency room visits during a full moon. $\mu_{after}$ = the population mean number of emergency room visits after a full moon. **(b)** $F_{crit} = 3.39$. Reject $H_0$ if $F_{data} > 3.39$. **(c)** 1.174110786 **(d)** Since $F_{data} \leq 3.39$, we do not reject $H_0$. There is insufficient evidence that not all the population mean numbers of emergency room visits are equal.
**17. (a)** Since one of the boxplots overlaps only 2 of the 5 other boxplots, the comparison boxplot of the nutritional ratings may be considered as evidence against the null hypothesis that all population mean nutritional ratings were equal. **(b)** Rejected
**19.** Yes, since the cereals were selected randomly from the manufacturers so that the selection of a cereal from one manufacturer did not affect the selection of cereals from other manufacturers.
**21.** $H_0: \mu_{online} = \mu_{traditional} = \mu_{hybrid}$. $H_a$: Not all the population means are equal. $F_{crit} = 6.36$. Reject $H_0$ if $F_{data} > 6.36$. $F_{data} =$ 0.5886918576. Since $F_{data} \leq 6.36$, we do not reject $H_0$. There is insufficient evidence that not all the population mean numbers of emergency room visits are equal.
**23. (a)** (i) $n_t$ would stay the same. (ii) $k$ would stay the same. (iii) *SSTR* would increase. (iv) *SSE* stays the same. (v) *SST* will increase. (vi) *MSTR* will increase. (vii) *MSE* will stay the same. (vii) $F_{data}$ will increase. **(b)** Since the original $F_{data}$ was less than 6.36, our conclusion was "Do not reject $H_0$." From (vii) we know that $F_{data}$ is increased. The new value for $F_{data}$ is 2.346449094. This is still less than 6.36, so our conclusion would still be "Do not reject $H_0$."
**25. (a)** $SSTR = 0$; $MSTR = 0$; $F_{data} = 0$. **(b)** $p$-value $= 1$. **(c)** Since the $p$-value $\geq 0.05$, we would not reject $H_0$.
**27. (a)** (i) (25.4805, 29.7255) (ii) (28.6366, 32.2654) (iii) (18.9526, 21.1134) **(b)** The confidence interval for the population mean gas mileage of American cars does not overlap the other two confidence intervals. This is evidence against the null hypothesis that all the population means are equal.
**29.** No evidence
**31. (a)** $H_0: \mu_1 = \mu_2 = \mu_3 = \mu_4 = \mu_5 = \mu_6 = \mu_7 = \mu_8$. $H_a$: Not all the population means are equal. **(b)** Reject $H_0$ if the $p$-value $<$ 0.05. **(c)** 8.27 **(d)** $p$-value $\approx 0$ **(e)** Since the $p$-value $< 0.05$, we reject $H_0$. **(f)** There is evidence that not all the population means are equal.

**(g)**

```
Source   DF        SS        MS      F      P
SIZE2     7  10879518   1554217   8.27  0.000
Error   330  62003922    187891
Total   337  72883441
```

**33. (a)** $H_0: \mu_1 = \mu_2 = \mu_3 = \mu_4$. $H_a$: Not all the population means are equal. **(b)** Reject $H_0$ if the $p$-value $< 0.01$. **(c)** 3.12 **(d)** $p$-value $= 0.026$ **(e)** Since the $p$-value $\geq 0.01$, we do not reject $H_0$. **(f)** There is insufficient evidence that not all the population means are equal.

**(g)**

```
Source      DF       SS     MS     F      P
PROTECT2     3    844.4  281.5  3.12  0.026
Error      334  30168.7   90.3
Total      337  31013.1
```

**35. (a, b)** Decreases

## Chapter 12 Review

**1.** $df_1 = 3$, $df_2 = 296$
**3.** 10,000
**5.** 11,157.5
**7.** 3.910472973
**9.**

| Source of variation | Sum of squares | Degrees of freedom | Mean square | *F*-test statistic |
|---|---|---|---|---|
| **Treatment** | SSTR = 10,000 | $df_1 = 3$ | MSTR = 3333.3333 | $F_{data} =$ 852.4117985 |
| **Error** | SSE = 1157.5 | $df_2 = 296$ | MSE = 3.910472973 | |
| **Total** | SST = 11,157.5 | | | |

**11. (a)** $H_0 : \mu_1 = \mu_2 = \mu_3$. $H_a$ : Not all the population means are equal. $\mu_1$ = population mean level of satisfaction for Medical Treatment 1. $\mu_2$ = population mean level of satisfaction for Medical Treatment 2. $\mu_3$ = population mean level of satisfaction for Medical Treatment 3. **(b)** $F_{\text{crit}} = 3.49$ **(c)** Reject $H_0$ if $F_{\text{data}} > 3.49$. **(d)** $F_{\text{data}} = 3.19$ **(e)** Since $F_{\text{data}} \leq 3.49$, we do not reject $H_0$. **(f)** There is insufficient evidence that not all of the population means are equal.

**(g)**

| Source | DF | SS | MS | F | P |
|---|---|---|---|---|---|
| Factor | 2 | 4114 | 2057 | 3.19 | 0.065 |
| Error | 18 | 11600 | 644 | | |
| Total | 20 | 15714 | | | |

**13. (a)** $H_0 : \mu_A = \mu_B = \mu_C = \mu_D = \mu_E$. $H_a$ : Not all the population means are equal. $\mu_A$ = the population mean price per pound (in dollars) of apples from Supermarket A. $\mu_B$ = the population mean price per pound (in dollars) of apples from Supermarket B. $\mu_C$ = the population mean price per pound (in dollars) of apples from Supermarket C. $\mu_D$ = the population mean price per pound (in dollars) of apples from Supermarket D. $\mu_E$ = the population mean price per pound (in dollars) of apples from Supermarket E. Reject $H_0$ if the $p$-value $< 0.05$. **(b)** $F_{\text{data}} = 17.53$ **(c)** $p$-value $\approx 0$ **(d)** Since the $p$-value $< 0.05$, we reject $H_0$. There is evidence that not all the population means are equal.

**(e)**

| Source | DF | SS | MS | F | P |
|---|---|---|---|---|---|
| Factor | 4 | 10.395 | 2.599 | 17.53 | 0.000 |
| Error | 22 | 3.262 | 0.148 | | |
| Total | 26 | 13.656 | | | |

## Chapter 12 Quiz

**1.** False
**2.** True
**3.** False
**4.** Mean square
**5.** Mean square treatment
**6.** Mean square error
**7.** $\bar{\bar{x}}$
**8.** Mean square equals the sum of squares divided by the degrees of freedom.
**9.** $F_{\text{data}}$
**10. (a)** (i) $\text{df}_1 = 2$, $\text{df}_2 = 382$ (ii) 23.46935065 (iii) 15,152.49833 (iv) 8474.92 (v) 23,627.41833 (vi) 7,576.249165 (vii) 22.1856445 (viii) 341.4931564
**(b)**

| Source of variation | Sum of squares | Degrees of freedom | Mean square | F-test statistic |
|---|---|---|---|---|
| **Treatment** | SSTR = 15,152.49833 | $\text{df}_1 = 2$ | MSTR = 7576.249165 | $F_{\text{data}}$ = 341.4931564 |
| **Error** | SSE = 8474.92 | $\text{df}_2 = 382$ | MSE = 22.1856445 | |
| **Total** | SST = 23,627.41833 | | | |

**11. (a)** (i) $\text{df}_1 = 4$, $\text{df}_2 = 1931$ (ii) 42.3542 (iii) 2001.432669 (iv) 384,702.6296 (v) 386,704.0623 (vi) 500.3581674 (vii) 199.3277874 (viii) 2.510227871
**(b)**

| Source of variation | Sum of squares | Degrees of freedom | Mean square | F-test statistic |
|---|---|---|---|---|
| **Treatment** | SSTR = 2001.432669 | $\text{df}_1 = 4$ | MSTR = 500.3581674 | $F_{\text{data}}$ = 2.510227871 |
| **Error** | SSE = 384,702.6296 | $\text{df}_2 = 1930$ | MSE = 199.3277874 | |
| **Total** | SST = 386,704.0623 | | | |

**12. (a)** (i) $\text{df}_1 = 2$, $\text{df}_2 = 36$ (ii) 107.3589744 (iii) 1750.974359 (iv) 20,238 (v) 21,988.97436 (vi) 875.4871795 (vii) 562.1666667 (viii) 1.557344523
**(b)**

| Source of variation | Sum of squares | Degrees of freedom | Mean square | F-test statistic |
|---|---|---|---|---|
| **Treatment** | SSTR = 1750.974359 | $\text{df}_1 = 2$ | MSTR = 875.4871795 | $F_{\text{data}}$ = 1.557344523 |
| **Error** | SSE = 20,238 | $\text{df}_2 = 36$ | MSE = 562.1666667 | |
| **Total** | SST = 21,988.97436 | | | |

**13. (a)** (i) $\text{df}_1 = 4$, $\text{df}_2 = 22$ (ii) $\bar{\bar{x}} = 2.462962963$ (iii) 13.46296296 (iv) 3.45980337 (v) 16.92276633 (vi) 3.365740741 (vii) 0.1572637895 (viii) 21.40187993
**(b)**

| Source of variation | Sum of squares | Degrees of freedom | Mean square | F-test statistic |
|---|---|---|---|---|
| **Treatment** | SSTR = 13.46296296 | $\text{df}_1 = 4$ | MSTR = 3.365740741 | $F_{\text{data}}$ = 21.40187993 |
| **Error** | SSE = 3.45980337 | $\text{df}_2 = 22$ | MSE = 0.1572637895 | |
| **Total** | SST = 16.92276633 | | | |

**14. (a)** $H_0 : \mu_4 = \mu_6 = \mu_8$. $H_a$ : Not all the population means are equal. $\mu_4$ = population mean miles per gallon for 4-cylinder cars. $\mu_6$ = population mean miles per gallon for 6-cylinder cars. $\mu_8$ = population mean miles per gallon for 8-cylinder cars. Reject $H_0$ if the $p$-value $< 0.05$. **(b)** 341.4931564 **(c)** $p$-value $< 0.001$ **(d)** Since the $p$-value $< 0.05$, we reject $H_0$. There is evidence that not all the population mean gas mileages are equal.
**15. (a)** $H_0 : \mu_{\text{married}} = \mu_{\text{widowed}} = \mu_{\text{divorced}} = \mu_{\text{separated}} = \mu_{\text{nevermarried}}$. $H_a$ : Not all the population means are equal. $\mu_{\text{married}}$ = population mean number of hours worked by people who are married people. $\mu_{\text{widowed}}$ = population mean number of hours worked by people who are widowed. $\mu_{\text{divorced}}$ = population mean number of hours worked by people who are divorced. $\mu_{\text{separated}}$ = population mean number of hours worked by people who are separated. $\mu_{\text{nevermarried}}$ = population mean number of hours worked by people who have never been married. Reject $H_0$ if the $p$-value $< 0.05$. **(b)** 2.510227871 **(c)** $0.025 < p$-value $< 0.05$ **(d)** Since the $p$-value $< 0.05$, we reject $H_0$. There is evidence that not all the population mean numbers of hours worked are equal.
**16. (a)** $H_0 : \mu_{\text{Kelloggs}} = \mu_{\text{Quaker}} = \mu_{\text{RalstonPurina}}$. $H_a$ : Not all the population means are equal. $\mu_{\text{Kelloggs}}$ = population mean number of calories per serving in breakfast cereals made by Kellogg's. $\mu_{\text{Quaker}}$ = population mean number of calories per serving in breakfast cereals made by Quaker. $\mu_{\text{RalstonPurina}}$ = population mean number of calories per serving in breakfast cereals made by Ralston Purina. **(b)** $F_{\text{crit}} = 3.18$.

Reject $H_0$ if $F_{data} > 3.18$. **(c)** 1.557344523 **(d)** Since $F_{data} \le 3.18$, we do not reject $H_0$. There is insufficient evidence that not all the population mean numbers of calories per serving are equal.
**17. (a)** $H_0 : \mu_{freshmen} = \mu_{sophomores} = \mu_{juniors} = \mu_{seniors} = \mu_{graduate\ students}$. $H_a$ : Not all the population means are equal. $\mu_{freshmen}$ = population mean GPA of freshmen at this university. $\mu_{sophomores}$ = population mean GPA of sophomores at this university. $\mu_{juniors}$ = population mean GPA of juniors at this university. $\mu_{seniors}$ = population mean GPA of seniors at this university. $\mu_{graduate\ students}$ = population mean GPA of graduate students at this university. **(b)** $F_{crit} = 2.87$. Reject $H_0$ if $F_{data} > 2.87$. **(c)** 21.40182689 **(d)** Since $F_{data} > 2.87$, we reject $H_0$. There is evidence that not all the population mean GPAs are equal.

# CHAPTER 13

## Section 13.1

**1.** The $y$ intercept $b_0$ is interpreted as "the estimated $y$ when $x$ equals zero." The slope $b_1$ is interpreted as "the estimated change in $y$ for a unit increase in $x$."
**3.** Out of all possible straight lines, the least-squares criterion chooses the line with the smallest SSE.
**5.** The standard error of the estimate $s$ is a measure of the size of the typical difference between the predicted value of $y$ and the observed value of $y$.
**7. (a)**

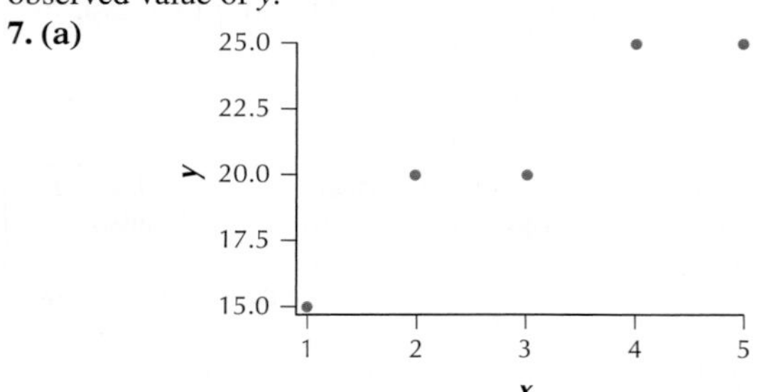


**(b)**

| $x$ | $y$ | $xy$ | $x^2$ | $y^2$ |
|---|---|---|---|---|
| 1 | 15 | 15 | 1 | 225 |
| 2 | 20 | 40 | 4 | 400 |
| 3 | 20 | 60 | 9 | 400 |
| 4 | 25 | 100 | 16 | 625 |
| 5 | 25 | 125 | 25 | 625 |
| $\Sigma x = 15$ | $\Sigma y = 105$ | $\Sigma xy = 340$ | $\Sigma x^2 = 55$ | $\Sigma y^2 = 2275$ |

**(c)** $b_0 = 13.5$, $b_1 = 2.5$, $\hat{y} = 13.5 + 2.5\,x$
**9. (a)**

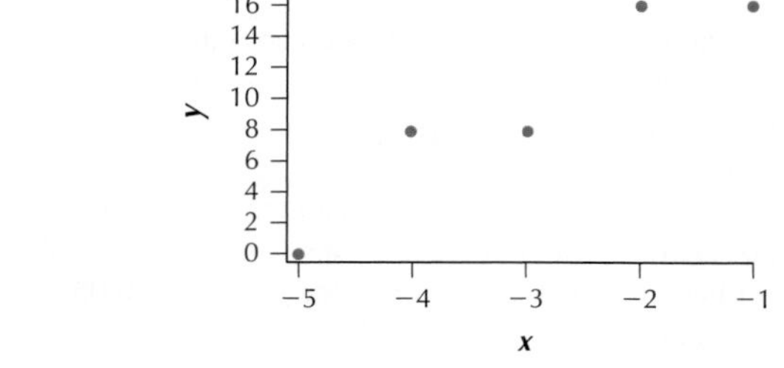


**(b)**

| $x$ | $y$ | $xy$ | $x^2$ | $y^2$ |
|---|---|---|---|---|
| −5 | 0 | 0 | 25 | 0 |
| −4 | 8 | −32 | 16 | 64 |
| −3 | 8 | −24 | 9 | 64 |
| −2 | 16 | −32 | 4 | 256 |
| −1 | 16 | −16 | 1 | 256 |
| $\Sigma x = -15$ | $\Sigma y = 48$ | $\Sigma xy = 104$ | $\Sigma x^2 = 55$ | $\Sigma y^2 = 640$ |

**(c)** $b_0 = 21.6$, $b_1 = 4$, $\hat{y} = 21.6 + 4\,x$
**11. (a)**

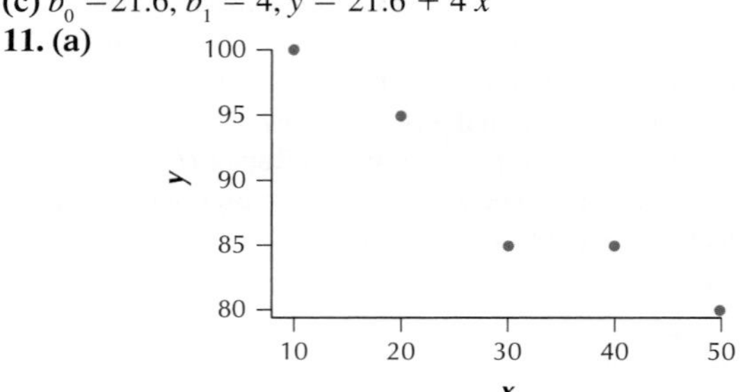


**(b)**

| $x$ | $y$ | $xy$ | $x^2$ | $y^2$ |
|---|---|---|---|---|
| 10 | 100 | 1000 | 100 | 10,000 |
| 20 | 95 | 1900 | 400 | 9,025 |
| 30 | 85 | 2550 | 900 | 7,225 |
| 40 | 85 | 3400 | 1600 | 7,225 |
| 50 | 80 | 4000 | 2500 | 6,400 |
| $\Sigma x = 150$ | $\Sigma y = 445$ | $\Sigma xy = 12,850$ | $\Sigma x^2 = 5,500$ | $\Sigma y^2 = 39,875$ |

**(c)** $b_0 = 104$, $b_1 = -0.5$, $\hat{y} = 104 - 0.5x$
**13. (a)**

| $x$ | $y$ | Predicted value $\hat{y} = 13.5 + 2.5\,x$ | Residual $(y - \hat{y})$ | (Residual)² $(y - \hat{y})^2$ |
|---|---|---|---|---|
| 1 | 15 | 16 | −1 | 1 |
| 2 | 20 | 18.5 | 1.5 | 2.25 |
| 3 | 20 | 21 | −1 | 1 |
| 4 | 25 | 23.5 | 1.5 | 2.25 |
| 5 | 25 | 26 | −1 | 1 |

**(b)** SSE = 7.5. The quantity SSE measures the sum of the squared distances between the predicted value $\hat{y}$ and the actual value $y$.
**(c)** $s = 1.5811$. The typical error in prediction is 1.5811.
**15. (a)**

| $x$ | $y$ | Predicted value $\hat{y} = 21.6 + 4\,x$ | Residual $(y - \hat{y})$ | (Residual)² $(y - \hat{y})^2$ |
|---|---|---|---|---|
| −5 | 0 | 1.6 | −1.6 | 2.56 |
| −4 | 8 | 5.6 | 2.4 | 5.76 |
| −3 | 8 | 9.6 | −1.6 | 2.56 |
| −2 | 16 | 13.6 | 2.4 | 5.76 |
| −1 | 16 | 17.6 | −1.6 | 2.56 |

**(b)** SSE = 19.2. The quantity SSE measures the sum of the squared distances between the predicted value $\hat{y}$ and the actual value $y$.
**(c)** $s = 2.5298$. The typical error in prediction is 2.5298.
**17. (a)**

| $x$ | $y$ | Predicted value $\hat{y} = 104 - 0.5\,x$ | Residual $(y - \hat{y})$ | (Residual)² $(y - \hat{y})^2$ |
|---|---|---|---|---|
| 10 | 100 | 99 | 1 | 1 |
| 20 | 95 | 94 | 1 | 1 |
| 30 | 85 | 89 | −4 | 16 |
| 40 | 85 | 84 | 1 | 1 |
| 50 | 80 | 79 | 1 | 1 |

**(b)** SSE = 20. The quantity SSE measures the sum of the squared distances between the predicted value $\hat{y}$ and the actual value $y$.
**(c)** $s = 2.5820$. The typical error in prediction is 2.5820.
**19. (a)**

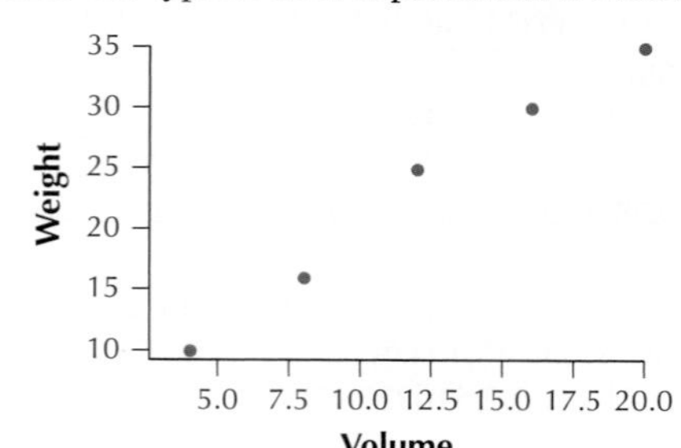

**(b)** $\hat{y} = 4 + 1.6x$ **(c)** $b_0 = 4$ means that the estimated weight of a package that has a volume of 0 cubic meters is 4 kg. $b_1 = 1.6$ means that for every additional cubic meter of volume the weight of the package is increased by 1.6 kg. **(d)** $s = 1.3166$ kg. If we know the volume ($x$) of the package, then our estimate of the weight of the package will typically differ from the actual weight of the package by 1.3166 kg.

**21. (a)**

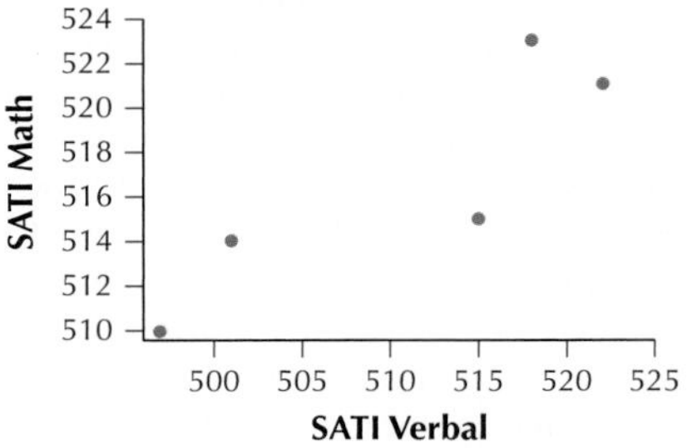

**(b)** $\hat{y} = 298.86 + 0.4264x$ **(c)** $b_0 = 298.86$ means that the estimated SAT I Math score for a student that has a SAT I Verbal score of 0 is 298.86. $b_1 = 0.4264$ means that for every additional point in the SAT I Verbal score the SAT I Math score is increased by 0.4264 points. **(d)** $s = 2.9266$ points. If we know the SAT I Verbal score ($x$), then our estimate of the SAT I Math score will typically differ from the actual SAT I Math score by 2.9266 points.

**23. (a)**

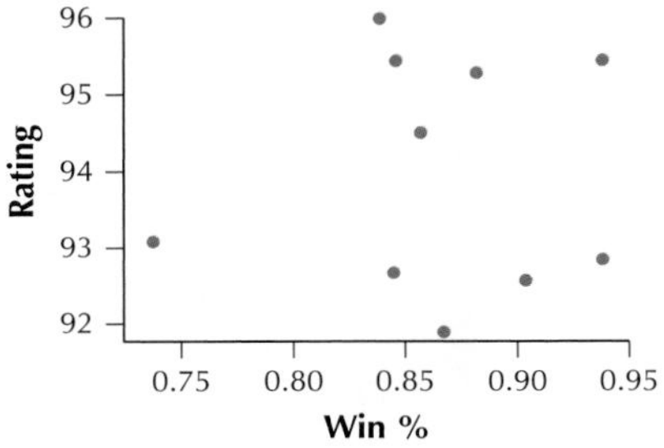

**(b)** $\hat{y} = 92.629 + 1.587x$ **(c)** $b_0 = 92.629$ means that the estimated rating for a team that has an 0% win percentage is 92.629. $b_1 = 1.587$ means that for every additional percentage point of wins, the rating increases by 1.587 percentage points. **(d)** $s = 1.60388$. If we know the percentage of wins, then our estimate of the rating will typically differ from the actual rating by 1.60388.

**25. (a)**

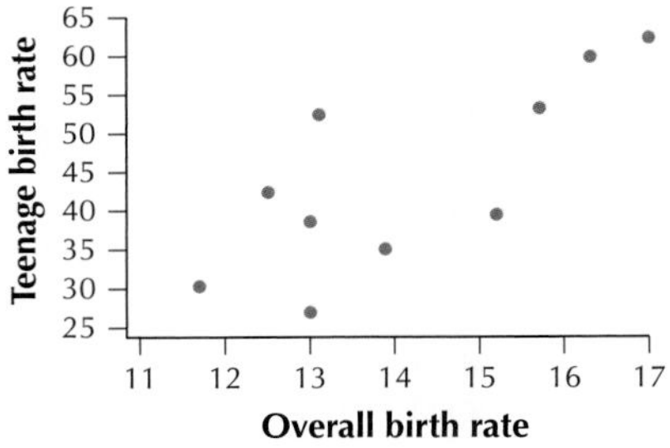

**(b)** $\hat{y} = -30.7993 + 5.3005x$ **(c)** $b_0 = -30.7993$ means that the estimated teenage birth rate for a state with an overall birth rate ($x$) of 0 is $-30.7993$. $b_1 = 5.3005$ means that for every additional birth per 1000 women in the overall birth rate the teenage birth rate increases by 5.3005 births per 1000 women. **(d)** $s = 8.36903$. If we know the overall birth rate, then our estimate of the teenage birth rate will typically differ from the actual teenage birth rate by 8.36903.

**27. (a)**

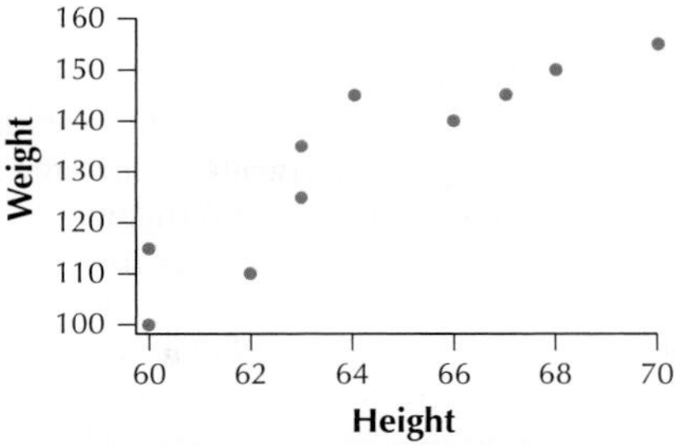

**(b)** $\hat{y} = -188.5553 + 4.9853x$ **(c)** $b_0 = -188.5553$ means that the estimated weight ($x$) for a female student whose height is 0 inches is $-188.5553$ pounds. $b_1 = 4.9853$ means that for every additional inch in height of a female student her weight increases by 4.9853 pounds. **(d)** $s = 8.45930$ pounds. If we know the female student's height, then the predicted weight will differ from the actual weight rate by 8.45930 pounds.

**29. (a)** $\hat{y} = 17.1799$°F. Since $x = 5$ does not lie between 7 and 70, this estimate represents extrapolation. It may not be appropriate to use extrapolation because the relationship between the variables may no longer be linear outside the range of $x$. **(b)** $\hat{y} = 64.2758$°F

**31. (a)** $\hat{y} = 93.0986$. Since $x = 100$ does not lie between 50 and 90, this estimate represents extrapolation. It may not be appropriate to use extrapolation because the relationship between the variables may no longer be linear outside of the range of $x$. **(b)** $\hat{y} = 73.8204$

**33. (a)** $\hat{y} = 0.5114$ dollars per pound **(b)** $\hat{y} = 0.4602$ dollars per pound. Since $x = 0.35$ does not lie between 0.374 and 0.516, this estimate represents extrapolation. It may not be appropriate to use extrapolation because the relationship between the variables may no longer be linear outside of the range of $x$.

**35. (a)**

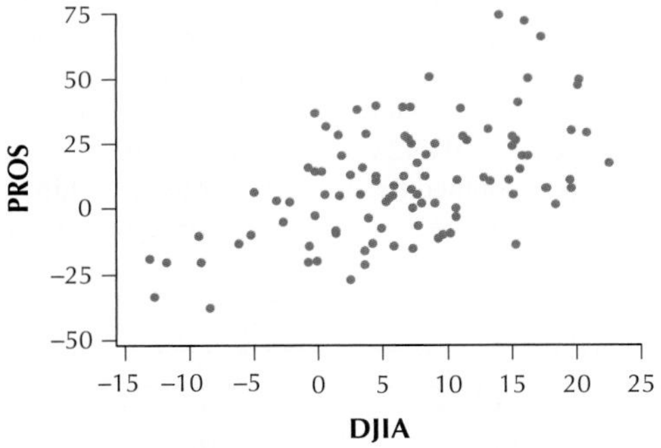

**(b)** $\hat{y} = 0.832 + 1.4890x$ **(c)** $s = 18.8545$. If we know the Dow Jones Industrial Average, then our estimate of the percent increase or decrease in the stock portfolio chosen by the pros will typically differ from the actual percent increase or decrease by 18.8545.

**37. (a)**

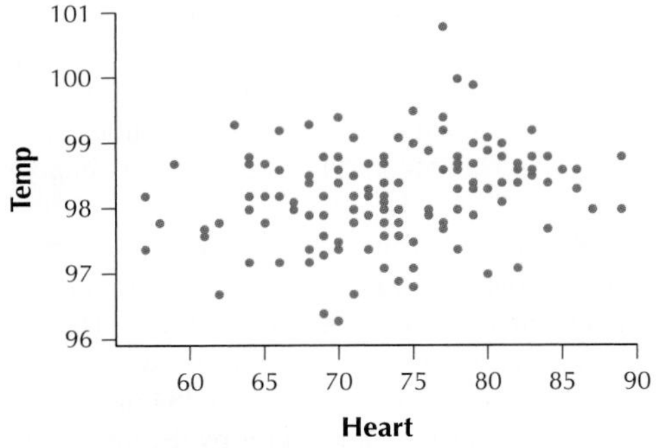

**(b)** $\hat{y} = 0.963071 + 0.026328x$ **(c)** $s = 0.711986$. If we know a person's heart rate, then our estimate of the person's temperature will typically differ from the person's actual temperature by 0.711986°F.

**39, 41, 43, 45.** Answers will vary.

## Section 13.2

**1.** Measure of the variability in $y$
**3.** It measures the amount of improvement in the accuracy of our estimate when using the regression equation compared with relying only on the $y$-values and ignoring the $x$ information.
**5.** A value of $r^2$ close to 1 indicates that the regression equation fits the data extremely well. A value of $r^2$ close to 0 indicates that the regression equation fits the data extremely poorly.
**7. (a)**

| $y$ | $y - \bar{y}$ | $(y - \bar{y})^2$ |
|---|---|---|
| 1 | −6 | 36 |
| 2 | −1 | 1 |
| 3 | −1 | 1 |
| 4 | 4 | 16 |
| 5 | 4 | 16 |

**(b)** $SST = 70$. This quantity measures the variability in $y$. **(c)** $SSR = 62.5$. This quantity measures the amount of improvement in the accuracy of our estimates when using the regression equation compared with relying only on the $y$-values and ignoring the $x$ information. **(d)** $r^2 = 0.8928571429$. Thus 0.8928571429 of the variability in $y$ is accounted for by the linear relationship between $y$ and $x$. **(e)** $r = 0.9449111825$. Thus $y$ and $x$ are strongly positively correlated.
**9. (a)**

| $y$ | $y - \bar{y}$ | $(y - \bar{y})^2$ |
|---|---|---|
| 0 | −9.6 | 92.16 |
| 8 | −1.6 | 2.56 |
| 8 | −1.6 | 2.56 |
| 16 | 6.4 | 40.96 |
| 16 | 6.4 | 40.96 |

**(b)** $SST = 179.2$. This quantity measures the variability in $y$. **(c)** $SSR = 160$. This quantity measures the amount of improvement in the accuracy of our estimates when using the regression equation compared with relying only on the $y$-values and ignoring the $x$ information. **(d)** $r^2 = 0.8928571429$. Thus 0.8928571429 of the variability in $y$ is accounted for by the linear relationship between $y$ and $x$. **(e)** $r = 0.9449111825$. Thus $y$ and $x$ are strongly positively correlated.
**11. (a)**

| $y$ | $y - \bar{y}$ | $(y - \bar{y})^2$ |
|---|---|---|
| 100 | 11 | 121 |
| 95 | 6 | 36 |
| 85 | −4 | 16 |
| 85 | −4 | 16 |
| 80 | 9 | 81 |

**(b)** $SST = 270$. This quantity measures the variability in $y$. **(c)** $SSR = 250$. This quantity measures the amount of improvement in the accuracy of our estimates when using the regression equation compared with relying only on the $y$-values and ignoring the $x$ information. **(d)** $r^2 = 0.9259259259$. Thus 0.9259259259 of the variability in $y$ is accounted for by the linear relationship between $y$ and $x$. **(e)** $r = -0.9622504486$. Thus $y$ and $x$ are strongly negatively correlated.
**13. (a)** $r^2 = 0.987463838$. Thus 0.987463838 of the variability in weight ($y$) is accounted for by the linear relationship between weight ($y$) and volume ($x$). **(b)** $r = 0.9937121505$. Thus weight ($y$) and volume ($x$) are strongly positively correlated.
**15. (a)** $r^2 = 0.7730056484$. Thus 0.7730056484 of the variability in SAT I Math scores ($y$) is accounted for by the linear relationship between SAT I Math scores ($y$) and SAT I Verbal scores ($x$). **(b)** $r = 0.8792073978$. Thus SAT I Math scores ($y$) and SAT I Verbal scores ($x$) are strongly positively correlated.
**17. (a)** $r^2 = 0.003701587$. Thus 0.003701587 of the variability in the rating ($y$) is accounted for by the linear relationship between rating ($y$) and win percentage ($x$). **(b)** $r = 0.0608406689$. Thus rating ($y$) and win percentage ($x$) are not correlated.
**19. (a)** $r^2 = 0.9738656406$. Thus 0.9738656406 of the variability in the number of veterans 65 and over ($y$) is accounted for by the linear relationship between the number of veterans 65 and over ($y$) and the number of veterans under 65 ($x$). **(b)** $r = 0.9868463105$. Thus the number of veterans 65 and over ($y$) and the number of veterans under 65 ($x$) are strongly positively correlated.
**21. (a)** $r^2 = 0.5768158096$. Thus 0.5768158096 of the variability in GPAs ($y$) is accounted for by the linear relationship between GPAs ($y$) and the combined SAT scores ($x$). **(b)** $r = 0.75948391$. Thus GPAs ($y$) and the combined SAT scores ($x$) are strongly positively correlated.
**23. (a)** $r^2 = 0.0830501863$. Thus 0.0830501863 of the variability in the change in stock price ($y$) is accounted for by the linear relationship between the change in stock price ($y$) and the stock price ($x$). **(b)** $r = 0.2881842923$. Thus change in stock price ($y$) and the stock price ($x$) are not correlated.
**25. (a)**



Yes, humans.
**(b)**

**Regression Analysis: Brain weight versus Body weight**

```
The regression equation is
Brain weight = 194 + 1.59 Body weight


Predictor       Coef   SE Coef      T      P
Constant       193.5     360.7   0.54  0.606
Body weight    1.595     5.656   0.28  0.785


S = 402.732    R-Sq = 1.0%    R-Sq(adj) = 0.0%


Analysis of Variance

Source          DF        SS       MS     F      P
Regression       1     12896    12896  0.08  0.785
Residual Error   8   1297544   162193
Total            9   1310441


Unusual Observations

       Body    Brain
Obs  weight   weight  Fit  SE Fit  Residual  St Resid
  8      62     1320  292     128      1028     2.69R

R denotes an observation with a large standardized
residual.
```

**(c)** (i) $b_0$ decreased from 193.5 to 110.6. (ii) $b_1$ decreased from 1.595 to 1.068. (iii) $r^2$ increased from 0.010 to 0.045. (iv) $s$ decreased from 402.732 to 132.469.

**27. (a)**

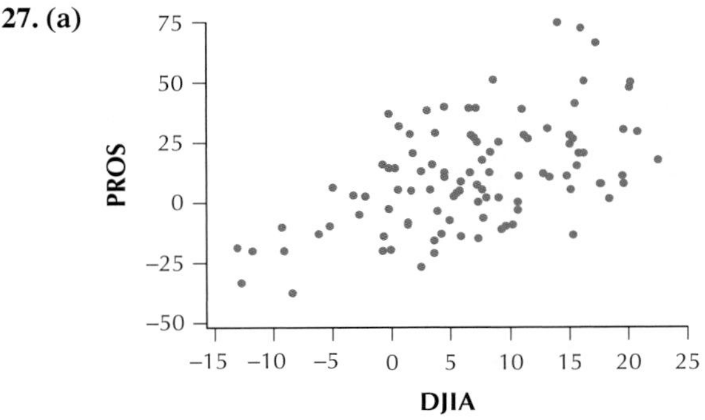


**(b)** $\hat{y} = 0.832 + 1.4890x$ **(c)** $r^2 = 0.289$. Thus 0.289 of the variability in the performance of the stock portfolios chosen by the pros ($y$) is accounted for by the linear relationship between the performance of the stock portfolios chosen by the pros ($y$) and the Dow Jones Industrial Average ($x$). **(d)** $s = 18.8545$. If we know the Dow Jones Industrial Average, then our estimate of the performance of the stock portfolio chosen by the pros will typically differ from the actual performance by 18.8545. **(e)** $r = 0.5379$. Thus the performance of the stock portfolios chosen by the pros ($y$) and the Dow Jones Industrial Average ($x$) are mildly positively correlated.

**29. (a)**

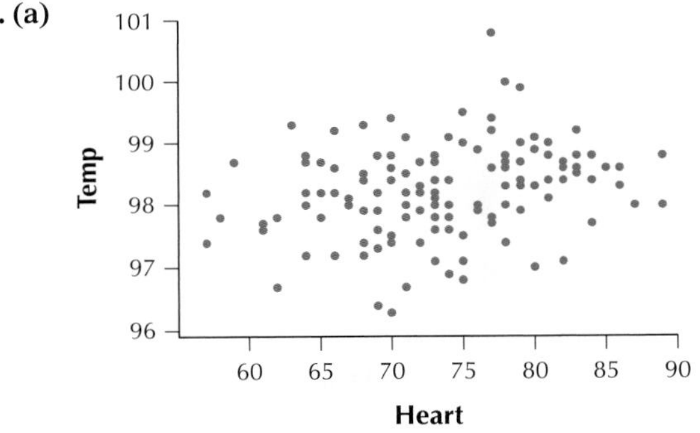


**(b)** $\hat{y} = 96.3071 + 0.026328x$ **(c)** $r^2 = 0.064$. Thus 0.064 of the variability in people's body temperatures ($y$) is accounted for by the linear relationship between people's body temperatures ($x$) and people's heart rates ($x$) **(d)** $s = 0.711986$. If we know a person's heart rate, then our estimate of the person's body temperature will typically differ from the person's actual body temperature by 0.711986°F. **(e)** $r = 0.253$. Thus a person's body temperature ($y$) and heart rate ($x$) are not correlated.

## Section 13.3

**1.** The regression equation is calculated from a sample and is valid only for values of $x$ in the range of the sample data. The population regression equation may be used to approximate the relationship between the predictor variable $x$ and the response variable $y$ for the entire population of $(x, y)$ pairs.
**3.** We construct a scatterplot of the residuals against the fitted values and a normal probability plot of the residuals. We must make sure that the scatterplot contains no strong evidence of any unhealthy patterns and that the normal probability plot indicates no evidence of departures from normality in residuals.
**5.** There is no relationship between $x$ and $y$.
**7. (a–b)** See Student Solutions Manual. **(c)** The scatterplot of the residuals contains an unhealthy pattern, so the regression assumptions are not verified.
**9. (a–b)** See Student Solutions Manual. **(c)** The scatterplot of the residuals contains an unhealthy pattern, so the regression assumptions are not verified.
**11. (a–b)** See Student Solutions Manual. **(c)** The scatterplot of the residuals contains an unhealthy pattern, so the regression assumptions are not verified.
**13. (a)** $s_{b_2} = 0.5$. The typical error in using $b_1$ to estimate $\beta_1$ is $s_{b_2} = 0.5$. **(b)** $t_{\alpha/2} = 3.182$ **(c)** (0.909, 4.091)
**15. (a)** $s_{b_2} = 0.8$. The typical error in using $b_1$ to estimate $\beta_1$ is $s_{b_2} = 0.8$. **(b)** $t_{\alpha/2} = 3.182$ **(c)** (1.4544, 6.5456)
**17. (a)** $s_{b_2} = 0.08165$. The typical error in using $b_1$ to estimate $\beta_1$ is $s_{b_2} = 0.08165$. **(b)** $t_{\alpha/2} = 3.182$ **(c)** (−0.7598, −0.2402)
**19. (a)** $t_{data} = 5$ **(b)** $p$-value $= 0.0153924381$ **(c)** $H_0 : \beta_1 = 0$: There is no relationship between $x$ and $y$. $H_a : \beta_1 \neq 0$: There is a linear relationship between $x$ and $y$. Reject $H_0$ if the $p$-value $< 0.05$. Since the $p$-value $< 0.05$, we reject $H_0$. There is evidence for a linear relationship between $x$ and $y$.
**21. (a)** $t_{data} = 5$ **(b)** $p$-value $= 0.0153924381$ **(c)** $H_0 : \beta_1 = 0$: There is no relationship between $x$ and $y$. $H_a : \beta_1 \neq 0$: There is a linear relationship between $x$ and $y$. Reject $H_0$ if the $p$-value $< 0.05$. Since the $p$-value $< 0.05$, we reject $H_0$. There is evidence for a linear relationship between $x$ and $y$.
**23. (a)** $t_{crit} = 3.182$ **(b)** $t_{data} = -6.1237$ **(c)** $H_0 : \beta_1 = 0$: There is no relationship between $x$ and $y$. $H_a : \beta_1 \neq 0$: There is a linear relationship between $x$ and $y$. Reject $H_0$ if $t_{data} < -3.182$ or $t_{data} > 3.182$. Since $t_{data} < -3.182$, we reject $H_0$. There is evidence for a linear relationship between $x$ and $y$.
**25. (a)** $s_{b_2} = 0.1041$. The typical error in using $b_1$ to estimate $\beta_1$ is $s_{b_2} = 0.1041$. **(b)** $t_{\alpha/2} = 3.182$ **(c)** (1.2688, 1.9312) **(d)** We are 95% confident that the interval (1.2688, 1.9312) captures the population slope $\beta_1$ of the relationship between volume and weight.
**27. (a)** $s_{b_2} = 0.1334$. The typical error in using $b_1$ to estimate $\beta_1$ is $s_{b_2} = 0.1334$. **(b)** $t_{\alpha/2} = 3.182$ **(c)** (0.0019, 0.8509) **(d)** We are 95% confident that the interval (0.0019, 0.8509) captures the population slope $\beta_1$ of the relationship between SAT I Verbal and SAT I Math.
**29. (a)** $s_{b_2} = 9.203$. The typical error in using $b_1$ to estimate $\beta_1$ is $s_{b_2} = 9.203$. **(b)** $t_{\alpha/2} = 2.306$ **(c)** (−19.6355, 22.8087) **(d)** We are 95% confident that the interval (−19.6355, 22.8087) captures the population slope $\beta_1$ of the relationship between Win% and Rating.
**31. (a)** $s_{b_2} = 0.0013$ The typical error in using $b_1$ to estimate $\beta_1$ is $s_{b_2} = 0.0013$. **(b)** $t_{\alpha/2} = 2.160$ **(c)** (0.0028, 0.0084) **(d)** We are 95% confident that the interval (0.0028, 0.0084) captures the population slope $\beta_1$ of the relationship between Combined SAT Score and Grade Point Average.
**33.** Against; the points appear to lie near a line with a positive slope.
**35.** $H_0 : \beta_1 = 0$: There is no relationship between SAT I Verbal ($x$) and SAT I Math ($y$). $H_a : \beta_1 \neq 0$: There is a linear relationship between SAT I Verbal ($x$) and SAT I Math ($y$).
**37.** $t_{data} = 3.1960$
**39. (a)** See Student Solutions Manual. The scatterplot of the residuals contains an unhealthy pattern, so the regression assumptions are not verified. **(b)** Yes **(c)** $t_{data} = 15.370$ **(d)** $p$-value $= 0.001$ **(e)** $H_0 : \beta_1 = 0$: There is no relationship between volume ($x$) and weight ($y$). $H_a : \beta_1 \neq 0$: There is a linear relationship between

volume ($x$) and weight ($y$). Reject $H_0$ if the $p$-value $< 0.05$. Since the $p$-value $< 0.05$, we reject $H_0$. There is evidence for a linear relationship between volume ($x$) and weight ($y$).

**41. (a)** See Student Solutions Manual. The scatterplot of the residuals contains no strong evidence of unhealthy patterns and the normal probability plot indicates no evidence of departures from normality in the residuals. Therefore we conclude that the regression assumptions are verified. **(b)** No **(c)** $t_{data} = 0.1724$ **(d)** $p$-value $= 0.867$ **(e)** $H_0 : \beta_1 = 0$: There is no relationship between Win% ($x$) and Rating ($y$). $H_a : \beta_1 \neq 0$: There is a linear relationship between Win% ($x$) and Rating ($y$). Reject $H_0$ if the $p$-value $< 0.05$. Since the $p$-value $\geq 0.05$, we do not reject $H_0$. There is insufficient evidence for a linear relationship between Win% ($x$) and Rating ($y$).

**43. (a)** See Student Solutions Manual. The scatterplot of the residuals contains an unhealthy pattern, so the regression assumptions are not verified. **(b)** Yes **(c)** $t_{crit} = 3.182$ **(d)** $t_{data} = 5$ **(e)** $H_0 : \beta_1 = 0$: There is no relationship between Family Size ($x$) and Pets ($y$). $H_a : \beta_1 \neq 0$: There is a linear relationship between Family Size ($x$) and Pets ($y$). Reject $H_0$ if $t_{data} < -3.182$ or $t_{data} > 3.182$. Since $t_{data} > 3.182$, we reject $H_0$. There is evidence for a linear relationship between Family Size ($x$) and Pets ($y$).

**45. (a)** See Student Solutions Manual. The scatterplot of the residuals contains no strong evidence of unhealthy patterns and the normal probability plot indicates no evidence of departures from normality in the residuals. Therefore we conclude that the regression assumptions are verified. **(b)** Yes **(c)** $t_{crit} = 2.228$ **(d)** $t_{data} = 4.847$ **(e)** $H_0 : \beta_1 = 0$: There is no relationship between Midterm Exam ($x$) and Overall Grade ($y$). $H_a : \beta_1 \neq 0$: There is a linear relationship between Midterm Exam ($x$) and Overall Grade ($y$). Reject $H_0$ if $t_{data} < -2.228$ or $t_{data} > 2.228$. Since $t_{data} > 2.228$, we reject $H_0$. There is evidence for a linear relationship between Midterm Exam ($x$) and Overall Grade ($y$).

**47. (a)** See Student Solutions Manual. The residuals vs. predicted values plot shows a funnel pattern. **(b)** The funnel pattern in the residuals vs. predicted values plot violates the constant variance assumption. **(c)** No, because one of the regression assumptions is violated. **(d)** Yes

**49. (a)** $t_{data}$ increases if $b_1$ is positive and decreases if $b_1$ is negative. **(b)** $r^2$ remains the same. **(c)** $s$ decreases. **(d)** $p$-value decreases. **(e)** Since we don't know what the new $p$-value will be, we don't know if the $p$-value will decrease enough to change the conclusion from "Do not reject $H_0$" to "Reject $H_0$."

**51. (a)** $t_{data}$ increases if $b_1$ is positive and decreases if $b_1$ is negative. **(b)** $r^2$ increases. **(c)** $s$ decreases. **(d)** $p$-value decreases. **(e)** Unchanged.

**53. (a–b)** Decrease **(c–d)** Increase **(e)** Depends on the new $p$-value.

**55. (a)** The scatterplot of the residuals contains evidence of an unhealthy pattern and the normal probability plot indicates evidence of departures from normality in the residuals. Therefore we conclude that the regression assumptions are not verified. **(b)** (13.5483, 14.6201). We are 95% confident that the interval (13.5483, 14.6201) captures the population slope $\beta_1$ of the relationship between fat per gram and calories per gram. **(c)** Yes **(d)** $H_0 : \beta_1 = 0$: There is no relationship between fat per gram ($x$) and calories per gram ($y$). $H_a : \beta_1 \neq 0$: There is a linear relationship between fat per gram ($x$) and calories per gram ($y$). Reject $H_0$ if the $p$-value $< 0.05$. $t_{data} = 52.15$. $p$-value $\approx 0$. Since the $p$-value $< 0.05$, we reject $H_0$. There is evidence for a linear relationship between fat per gram ($x$) and calories per gram ($y$).

**57. (a)** No **(b)** Positive relationship **(c)** Unclear

**59. (a)** No **(b)** No apparent relationship between the variables **(c)** The weight of vehicles is the predictor variable and the severity of the leg injuries should be the response variable.

## Chapter 13 Review

**1. (a)**

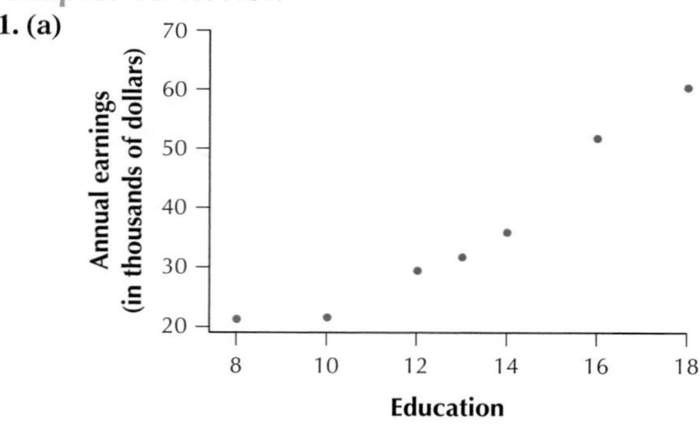


**(b)** As the number of years of education increases, the annual earnings increase. **(c)** $\hat{y} = -23.763 + 4.4686x$ **(d)** $b_1 = 4.4686$ means that for every additional year of education the annual earnings are increased by \$4468.60 **(e)** $b_0 = -23.763$ means that the estimated annual earnings of a person who has 0 years of education is $-$\$23,763. This does not make sense in this data set because a person can't earn a negative amount of money. **(f)** $s = 4.95712$. If we know the number of years of education ($x$) a person has, then our estimate of the person's annual earnings will typically differ from the person's actual earnings by \$4957.12.

**3. (a)**

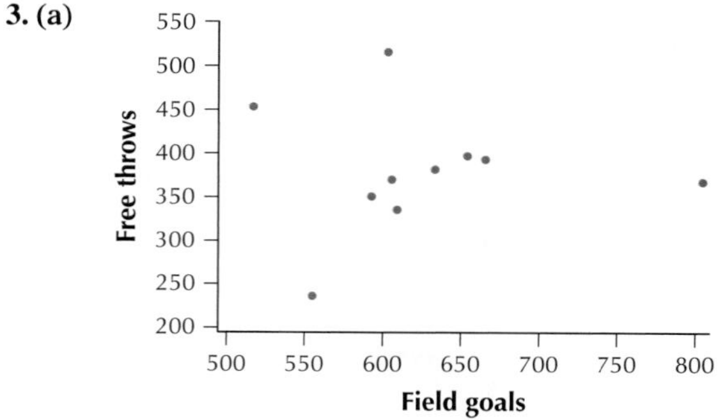


**(b)** As the number of field goals increases, the number of free throws stays the same. **(c)** $\hat{y} = 383.9 - 0.0047x$ **(d)** $b_1 = -0.0047$ means that for every additional field goal the number of free throws decreases by 0.0047. **(e)** $b_0 = 383.9$ means that the estimated number of free throws for a player who has 0 field goals is 383.9. This makes sense in this data set because it is possible for a person to have 384 free throws and 0 field goals. **(f)** $s = 77.5861$. If we know the number of field goals ($x$) that a player has then our estimate of the player's free throws typically differs from the player's actual number of free throws by 77.5861.

**5. (a)** SSR $= 1397.8$, SSE $= 122.9$, SST $= 1520.6$ **(b)** $r^2 = 0.9192013859$. Since $r^2$ is close to 1, the regression equation fits the data extremely well. **(c)** $r = 0.9587499079$. Positively correlated.

**7. (a)** SSR $= 1$, SSE $= 48{,}157$, SST $= 48{,}158$ **(b)** $r^2 = 0.00002495406$. Since $r^2$ is close to 0, the regression equation fits the data extremely poorly. **(c)** $r = -0.0049954037$. Not correlated.

**9. (a)** $s_{b_1} = 0.5925$. The typical error in using $b_1$ to estimate $\beta_1$ is $s_{b_1} = 0.5925$. **(b)** $t_{\alpha/2} = 2.571$ **(c)** (2.9453, 5.9919). We are 95% confident that the interval (2.9453, 5.9919) captures the population slope $\beta_1$ of the relationship between Education and Annual Earnings.

**11. (a)** $s_{b_1} = 0.3349$. The typical error in using $b_1$ to estimate $\beta_1$ is $s_{b_1} = 0.3349$. **(b)** $t_{\alpha/2} = 2.306$ **(c)** $(-0.7770, 0.7676)$. We are 95% confident that the interval $(-0.7770, 0.7676)$ captures the population slope $\beta_1$ of the relationship between Field Goals and Free Throws.
**13. (a)** The scatterplot of the residuals contains an unhealthy pattern, so the regression assumptions are not verified. **(b)** Yes **(c)** $t_{\text{data}} = 7.54$ **(d)** $p$-value $= 0.001$ **(e)** $H_0 : \beta_1 = 0$: There is no relationship between Education ($x$) and Annual Earnings ($y$). $H_a : \beta_1 \neq 0$: There is a linear relationship between Education ($x$) and Annual Earnings ($y$). Reject $H_0$ if the $p$-value $< 0.05$. Since the $p$-value $< 0.05$, we reject $H_0$. There is evidence for a linear relationship between Education ($x$) and Annual Earnings ($y$).
**15. (a)** Since 0 is in the confidence interval, we would expect to not reject the null hypothesis that $\beta_1 = 0$. **(b)** $t_{\text{crit}} = 2.306$ **(c)** $t_{\text{data}} = -0.014$ **(d)** $H_0 : \beta_1 = 0$: There is no relationship between Field Goals ($x$) and Free Throws ($y$). $H_a : \beta_1 \neq 0$: There is a linear relationship between Field Goals ($x$) and Free Throws ($y$). Reject $H_0$ if $t_{\text{data}} < -2.306$ or $t_{\text{data}} > 2.306$. Since $t_{\text{data}}$ is not less than $-2.306$ and $t_{\text{data}}$ is not greater than 2.306, we do not reject $H_0$. There is insufficient evidence for a linear relationship between Field Goals ($x$) and Free Throws ($y$).

## Chapter 13 Quiz

**1.** False
**2.** True
**3.** False.
**4.** $y$, $y$, $x$
**5.** predicted value of $y$, actual observed value of $y$
**6.** coefficient of determination
**7.** between 0 and 1 inclusive
**8.** SST = SSR + SSE
**9.** $n - 2$
**10. (a)**

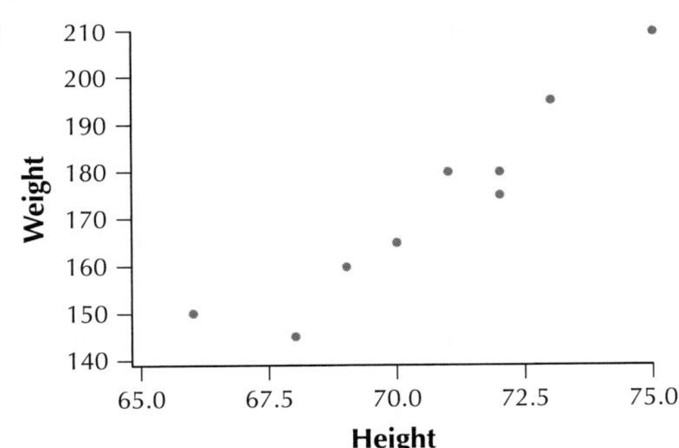


**(b)** As the height of a person increases, the weight of the person increases. **(c)** $\hat{y} = -341.80 + 7.2848x$

**11. (a)**

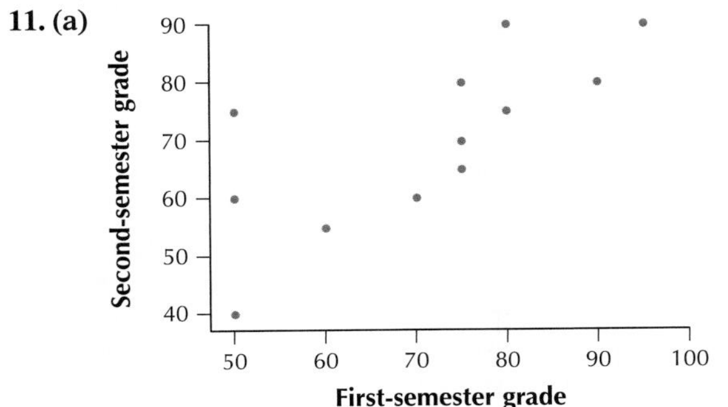


**(b)** As the first-semester grade increases, the second-semester grade increases. **(c)** $\hat{y} = 20.12 + 0.7042x$

**12. (a)**

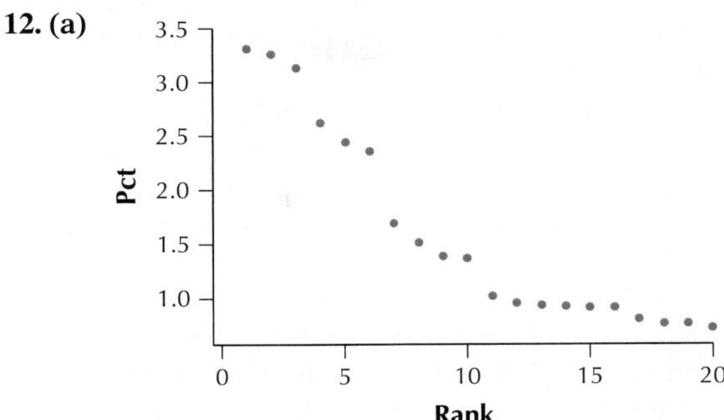


**(b)** As the rank of the boy's name increases, the percent of boys with that name decreases. **(c)** $\hat{y} = 3.1016 - 0.14272\,x$
**13. (a)** $b_1 = 7.2848$ means that for every additional inch in height the weight of the person increases by 7.2848 pounds. **(b)** $b_0 = -341.80$ means that the estimated weight of a person who is 0 inches tall is $-341.80$ pounds. This does not make sense in this data set because it is impossible for a person to weigh a negative number of pounds and it is impossible for a person to be 0 inches tall. **(c)** $s = 6.68208$. If we know the height ($x$) of a person then our estimate of the person's weight typically differs from the person's actual weight by 6.68208 pounds.
**14. (a)** $b_1 = 0.7042$ means that for every additional point of the first-semester grade, the second-semester grade increases by 0.7042 points. **(b)** $b_0 = 20.12$ means that the estimated grade of a student with a grade of 0 in the first semester is 20.12. This does not make sense in this data set because a student with a grade of 0 in the first semester of accounting would not be eligible to take the second semester of accounting. **(c)** $s = 10.5587$. If we know the grade the student earned in the first semester of accounting ($x$) then our estimate of the student's grade in the second semester of accounting typically differs from the person's actual grade by 10.5587 points.
**15. (a)** $b_1 = -0.14272$ means that for every increase of 1 in rank of a boy's name the percent of boys with that name decreases by 0.14272. **(b)** $b_0 = 3.1016$ means that the estimated percent of boys with a name that has a rank of 0 is 3.1016%. This does not make sense in this data set because a name can't have a rank of 0. **(c)** $s = 0.350698$. If we know the rank of a boy's name ($x$) then our estimate of the percent of boys with that name typically differs from the actual percent of boys with that name by 0.350698%.
**16. (a)** SSR = 3205.3, SSE = 357.2, SST = 3562.5 **(b)** $r^2 = 0.8997327756$. Since $r^2$ is close to 1 the regression equation fits the data extremely well. **(c)** $r = 0.948542448$; positively correlated
**17. (a)** SSR = 1285.1, SSE = 1114.9, SST = 2400.0 **(b)** $r^2 = 0.53544702572$. Since $r^2$ is close to 1 the regression equation fits the data extremely well. **(c)** $r = 0.7317583325$; positively correlated
**18. (a)** SSR = 13.546, SSE = 2.214, SST = 15.760 **(b)** $r^2 = 0.8595299263$. Since $r^2$ is close to 1 the regression equation fits the data extremely well. **(c)** $r = -0.9271083681$; negatively correlated.
**19. (a)** $s_{b_2} = 0.8598$. The typical error in using $b_1$ to estimate $\beta_1$ is $s_{b_2} = 0.8598$. **(b)** $t_{\alpha/2} = 2.306$ **(c)** $(5.3021, 9.2675)$. We are 95% confident that the interval $(5.3021, 9.2675)$ captures the population slope $\beta_1$ of the relationship between Weight and Height.

**20. (a)** $s_{b_2} = 0.2074$. The typical error in using $b_1$ to estimate $\beta_1$ is $s_{b_2} = 0.2074$. **(b)** $t_{\alpha/2} = 2.228$ **(c)** (0.2421, 1.1663). We are 95% confident that the interval (0.2421, 1.1663) captures the population slope $\beta_1$ of the relationship between First-Semester Grade and Second-Semester Grade.
**21. (a)** $s_{b_2} = 0.01360$. The typical error in using $b_1$ to estimate $\beta_1$ is $s_{b_2} = 0.01360$. **(b)** $t_{\alpha/2} = 2.101$ **(c)** $(-0.1713, -0.1141)$. We are 95% confident that the interval $(-0.1713, -0.1141)$ captures the population slope $\beta_1$ of the relationship between Rank and Pct.
**22.** $H_0 : \beta_1 = 0$: There is no relationship between Height ($x$) and Weight ($y$). $H_a : \beta_1 \neq 0$: There is a linear relationship between Height ($x$) and Weight ($y$). Reject $H_0$ if the $p$-value $< 0.05$. $t_{\text{data}} = 8.47$, $p$-value $\approx 0$. Since the $p$-value $< 0.05$, we reject $H_0$. There is evidence for a linear relationship between Height ($x$) and Weight ($y$).
**23.** $H_0 : \beta_1 = 0$: There is no relationship between First-Semester Grade ($x$) and Second-Semester Grade ($y$). $H_a : \beta_1 \neq 0$: There is a linear relationship between First-Semester Grade ($x$) and Second-Semester Grade ($y$). Reject $H_0$ if the $p$-value $< 0.05$. $t_{\text{data}} = 3.40$, $p$-value $= 0.007$. Since the $p$-value $< 0.05$, we reject $H_0$. There is evidence for a linear relationship between First-Semester Grade ($x$) and Second-Semester Grade ($y$).
**24.** $H_0 : \beta_1 = 0$: There is no relationship between Rank ($x$) and Pct ($y$). $H_a : \beta_1 \neq 0$: There is a linear relationship between Rank ($x$) and Pct ($y$). $t_{\text{crit}} = 2.101$. Reject $H_0$ if $t_{\text{data}} < -2.101$ or $t_{\text{data}} > 2.101$. $t_{\text{data}} = -10.49$. Since $t_{\text{data}} < -2.101$, we reject $H_0$. There is evidence for a linear relationship between Rank ($x$) and Pct ($y$).

# Tables Appendix

## Table A Random numbers

| | | | | | | | | | | | | | |
|---|---|---|---|---|---|---|---|---|---|---|---|---|---|
| 10480 | 15011 | 01536 | 02011 | 81647 | 91646 | 67179 | 14194 | 62590 | 36207 | 20969 | 99570 | 91291 | 90700 |
| 22368 | 46573 | 25595 | 85393 | 30995 | 89198 | 27982 | 53402 | 93965 | 34095 | 52666 | 19174 | 39615 | 99505 |
| 24130 | 48360 | 22527 | 97265 | 76393 | 64809 | 15179 | 24830 | 49340 | 32081 | 30680 | 19655 | 63348 | 58629 |
| 42167 | 93093 | 06243 | 61680 | 07856 | 16376 | 39440 | 53537 | 71341 | 57004 | 00849 | 74917 | 97758 | 16379 |
| 37570 | 39975 | 81837 | 16656 | 06121 | 91782 | 60468 | 81305 | 49684 | 60672 | 14110 | 06927 | 01263 | 54613 |
| 77921 | 06907 | 11008 | 42751 | 27756 | 53498 | 18602 | 70659 | 90655 | 15053 | 21916 | 81825 | 44394 | 42880 |
| 99562 | 72905 | 56420 | 69994 | 98872 | 31016 | 71194 | 18738 | 44013 | 48840 | 63213 | 21069 | 10634 | 12952 |
| 96301 | 91977 | 05463 | 07972 | 18876 | 20922 | 94595 | 56869 | 69014 | 60045 | 18425 | 84903 | 42508 | 32307 |
| 89579 | 14342 | 63661 | 10281 | 17453 | 18103 | 57740 | 84378 | 25331 | 12566 | 58678 | 44947 | 05584 | 56941 |
| 85475 | 36857 | 43342 | 53988 | 53060 | 59533 | 38867 | 62300 | 08158 | 17983 | 16439 | 11458 | 18593 | 64952 |
| 28918 | 69578 | 88231 | 33276 | 70997 | 79936 | 56865 | 05859 | 90106 | 31595 | 01547 | 85590 | 91610 | 78188 |
| 63553 | 40961 | 48235 | 03427 | 49626 | 69445 | 18663 | 72695 | 52180 | 20847 | 12234 | 90511 | 33703 | 90322 |
| 09429 | 93969 | 52636 | 92737 | 88974 | 33488 | 36320 | 17617 | 30015 | 08272 | 84115 | 27156 | 30613 | 74952 |
| 10365 | 61129 | 87529 | 85689 | 48237 | 52267 | 67689 | 93394 | 01511 | 26358 | 85104 | 20285 | 29975 | 89868 |
| 07119 | 97336 | 71048 | 08178 | 77233 | 13916 | 47564 | 81056 | 97735 | 85977 | 29372 | 74461 | 28551 | 90707 |
| 51085 | 12765 | 51821 | 51259 | 77452 | 16308 | 60756 | 92144 | 49442 | 53900 | 70960 | 63990 | 75601 | 40719 |
| 02368 | 21382 | 52404 | 60268 | 89368 | 19885 | 55322 | 44819 | 01188 | 65255 | 64835 | 44919 | 05944 | 55157 |
| 01011 | 54092 | 33362 | 94904 | 31273 | 04146 | 18594 | 29852 | 71585 | 85030 | 51132 | 01915 | 92747 | 64951 |
| 52162 | 53916 | 46369 | 58586 | 23216 | 14513 | 83149 | 98736 | 23495 | 64350 | 94738 | 17752 | 35156 | 35749 |
| 07056 | 97628 | 33787 | 09998 | 42698 | 06691 | 76988 | 13602 | 51851 | 46104 | 88916 | 19509 | 25625 | 58104 |
| 48663 | 91245 | 85828 | 14346 | 09172 | 30168 | 90229 | 04734 | 59193 | 22178 | 30421 | 61666 | 99904 | 32812 |
| 54164 | 58492 | 22421 | 74103 | 47070 | 25306 | 76468 | 26384 | 58151 | 06646 | 21524 | 15227 | 96909 | 44592 |
| 32639 | 32363 | 05597 | 24200 | 13363 | 38005 | 94342 | 28728 | 35806 | 06912 | 17012 | 64161 | 18296 | 22851 |
| 29334 | 27001 | 87637 | 87308 | 58731 | 00256 | 45834 | 15398 | 46557 | 41135 | 10367 | 07684 | 36188 | 18510 |
| 02488 | 33062 | 28834 | 07351 | 19731 | 92420 | 60952 | 61280 | 50001 | 67658 | 32586 | 86679 | 50720 | 94953 |
| 81525 | 72295 | 04839 | 96423 | 24878 | 82651 | 66566 | 14778 | 76797 | 14780 | 13300 | 87074 | 79666 | 95725 |
| 29676 | 20591 | 68086 | 26432 | 46901 | 20849 | 89768 | 81536 | 86645 | 12659 | 92259 | 57102 | 80428 | 25280 |
| 00742 | 57392 | 39064 | 66432 | 84673 | 40027 | 32832 | 61362 | 98947 | 96067 | 64760 | 64584 | 96096 | 98253 |
| 05366 | 04213 | 25669 | 26422 | 44407 | 44048 | 37937 | 63904 | 45766 | 66134 | 75470 | 66520 | 34693 | 90449 |
| 91921 | 26418 | 64117 | 94305 | 26766 | 25940 | 39972 | 22209 | 71500 | 64568 | 91402 | 42416 | 07844 | 69618 |
| 00582 | 04711 | 87917 | 77341 | 42206 | 35126 | 74087 | 99547 | 81817 | 42607 | 43808 | 76655 | 62028 | 76630 |
| 00725 | 69884 | 62797 | 56170 | 86324 | 88072 | 76222 | 36086 | 84637 | 93161 | 76038 | 65855 | 77919 | 88006 |
| 69011 | 65797 | 95876 | 55293 | 18988 | 27354 | 26575 | 08625 | 40801 | 59920 | 29841 | 80150 | 12777 | 48501 |
| 25976 | 57948 | 29888 | 88604 | 67917 | 48708 | 18912 | 82271 | 65424 | 69774 | 33611 | 54262 | 85963 | 03547 |
| 09763 | 83473 | 73577 | 12908 | 30883 | 18317 | 28290 | 35797 | 05998 | 41688 | 34952 | 37888 | 38917 | 88050 |
| 91567 | 42595 | 27958 | 30134 | 04024 | 86385 | 29880 | 99730 | 55536 | 84855 | 29080 | 09250 | 79656 | 73211 |
| 17955 | 56349 | 90999 | 49127 | 20044 | 59931 | 06115 | 20542 | 18059 | 02008 | 73708 | 83517 | 36103 | 42791 |
| 46503 | 18584 | 18845 | 49618 | 02304 | 51038 | 20655 | 58727 | 28168 | 15475 | 56942 | 53389 | 20562 | 87338 |
| 92157 | 89634 | 94824 | 78171 | 84610 | 82834 | 09922 | 25417 | 44137 | 48413 | 25555 | 21246 | 35509 | 20468 |
| 14577 | 62765 | 35605 | 81263 | 39667 | 47358 | 56873 | 56307 | 61607 | 49518 | 89656 | 20103 | 77490 | 18062 |
| 98427 | 07523 | 33362 | 64270 | 01638 | 92477 | 66969 | 98420 | 04880 | 45585 | 46565 | 04102 | 46880 | 45709 |
| 34914 | 63976 | 88720 | 82765 | 34476 | 17032 | 87589 | 40836 | 32427 | 70002 | 70663 | 88863 | 77775 | 69348 |
| 70060 | 28277 | 39475 | 46473 | 23219 | 53416 | 94970 | 25832 | 69975 | 94884 | 19661 | 72828 | 00102 | 66794 |
| 53976 | 54914 | 06990 | 67245 | 68350 | 82948 | 11398 | 42878 | 80287 | 88267 | 47363 | 46634 | 06541 | 97809 |
| 76072 | 29515 | 40980 | 07391 | 58745 | 25774 | 22987 | 80059 | 39911 | 96189 | 41151 | 14222 | 60697 | 59583 |
| 90725 | 52210 | 83974 | 29992 | 65831 | 38857 | 50490 | 83765 | 55657 | 14361 | 31720 | 57375 | 56228 | 41546 |
| 64364 | 67412 | 33339 | 31926 | 14883 | 24413 | 59744 | 92351 | 97473 | 89286 | 35931 | 04110 | 23726 | 51900 |
| 08962 | 00358 | 31662 | 25388 | 61642 | 34072 | 81249 | 35648 | 56891 | 69352 | 48373 | 45578 | 78547 | 81788 |
| 95012 | 68379 | 93526 | 70765 | 10593 | 04542 | 76463 | 54328 | 02349 | 17247 | 28865 | 14777 | 62730 | 92277 |
| 15664 | 10493 | 20492 | 38391 | 91132 | 21999 | 59516 | 81652 | 27195 | 48223 | 46751 | 22923 | 32261 | 85653 |

Table B Binomial distribution

| | | *p* | | | | | | | | |
|---|---|---|---|---|---|---|---|---|---|---|
| *n* | *X* | **0.10** | **0.15** | **0.20** | **0.25** | **0.30** | **0.35** | **0.40** | **0.45** | **0.50** |
| **2** | 0 | 0.8100 | 0.7225 | 0.6400 | 0.5625 | 0.4900 | 0.4225 | 0.3600 | 0.3025 | 0.2500 |
| | 1 | 0.1800 | 0.2550 | 0.3200 | 0.3750 | 0.4200 | 0.4550 | 0.4800 | 0.4950 | 0.5000 |
| | 2 | 0.0100 | 0.0225 | 0.0400 | 0.0625 | 0.0900 | 0.1225 | 0.1600 | 0.2025 | 0.2500 |
| **3** | 0 | 0.7290 | 0.6141 | 0.5120 | 0.4219 | 0.3430 | 0.2746 | 0.2160 | 0.1664 | 0.1250 |
| | 1 | 0.2430 | 0.3251 | 0.3840 | 0.4219 | 0.4410 | 0.4436 | 0.4320 | 0.4084 | 0.3750 |
| | 2 | 0.0270 | 0.0574 | 0.0960 | 0.1406 | 0.1890 | 0.2389 | 0.2880 | 0.3341 | 0.3750 |
| | 3 | 0.0010 | 0.0034 | 0.0080 | 0.0156 | 0.0270 | 0.0429 | 0.0640 | 0.0911 | 0.1250 |
| **4** | 0 | 0.6561 | 0.5220 | 0.4096 | 0.3164 | 0.2401 | 0.1785 | 0.1296 | 0.0915 | 0.0625 |
| | 1 | 0.2916 | 0.3685 | 0.4096 | 0.4219 | 0.4116 | 0.3845 | 0.3456 | 0.2995 | 0.2500 |
| | 2 | 0.0486 | 0.0975 | 0.1536 | 0.2109 | 0.2646 | 0.3105 | 0.3456 | 0.3675 | 0.3750 |
| | 3 | 0.0036 | 0.0115 | 0.0256 | 0.0469 | 0.0756 | 0.1115 | 0.1536 | 0.2005 | 0.2500 |
| | 4 | 0.0001 | 0.0005 | 0.0016 | 0.0039 | 0.0081 | 0.0150 | 0.0256 | 0.0410 | 0.0625 |
| **5** | 0 | 0.5905 | 0.4437 | 0.3277 | 0.2373 | 0.1681 | 0.1160 | 0.0778 | 0.0503 | 0.0312 |
| | 1 | 0.3280 | 0.3915 | 0.4096 | 0.3955 | 0.3602 | 0.3124 | 0.2592 | 0.2059 | 0.1562 |
| | 2 | 0.0729 | 0.1382 | 0.2048 | 0.2637 | 0.3087 | 0.3364 | 0.3456 | 0.3369 | 0.3125 |
| | 3 | 0.0081 | 0.0244 | 0.0512 | 0.0879 | 0.1323 | 0.1811 | 0.2304 | 0.2757 | 0.3125 |
| | 4 | 0.0004 | 0.0022 | 0.0064 | 0.0146 | 0.0284 | 0.0488 | 0.0768 | 0.1128 | 0.1562 |
| | 5 | | 0.0001 | 0.0003 | 0.0010 | 0.0024 | 0.0053 | 0.0102 | 0.0185 | 0.0312 |
| **6** | 0 | 0.5314 | 0.3771 | 0.2621 | 0.1780 | 0.1176 | 0.0754 | 0.0467 | 0.0277 | 0.0156 |
| | 1 | 0.3543 | 0.3993 | 0.3932 | 0.3560 | 0.3025 | 0.2437 | 0.1866 | 0.1359 | 0.0938 |
| | 2 | 0.0984 | 0.1762 | 0.2458 | 0.2966 | 0.3241 | 0.3280 | 0.3110 | 0.2780 | 0.2344 |
| | 3 | 0.0146 | 0.0415 | 0.0819 | 0.1318 | 0.1852 | 0.2355 | 0.2765 | 0.3032 | 0.3125 |
| | 4 | 0.0012 | 0.0055 | 0.0154 | 0.0330 | 0.0595 | 0.0951 | 0.1382 | 0.1861 | 0.2344 |
| | 5 | 0.0001 | 0.0004 | 0.0015 | 0.0044 | 0.0102 | 0.0205 | 0.0369 | 0.0609 | 0.0938 |
| | 6 | | | 0.0001 | 0.0002 | 0.0007 | 0.0018 | 0.0041 | 0.0083 | 0.0156 |
| **7** | 0 | 0.4783 | 0.3206 | 0.2097 | 0.1335 | 0.0824 | 0.0490 | 0.0280 | 0.0152 | 0.0078 |
| | 1 | 0.3720 | 0.3960 | 0.3670 | 0.3115 | 0.2471 | 0.1848 | 0.1306 | 0.0872 | 0.0547 |
| | 2 | 0.1240 | 0.2097 | 0.2753 | 0.3115 | 0.3177 | 0.2985 | 0.2613 | 0.2140 | 0.1641 |
| | 3 | 0.0230 | 0.0617 | 0.1147 | 0.1730 | 0.2269 | 0.2679 | 0.2903 | 0.2918 | 0.2734 |
| | 4 | 0.0026 | 0.0109 | 0.0287 | 0.0577 | 0.0972 | 0.1442 | 0.1935 | 0.2388 | 0.2734 |
| | 5 | 0.0002 | 0.0012 | 0.0043 | 0.0115 | 0.0250 | 0.0466 | 0.0774 | 0.1172 | 0.1641 |
| | 6 | | 0.0001 | 0.0004 | 0.0013 | 0.0036 | 0.0084 | 0.0172 | 0.0320 | 0.0547 |
| | 7 | | | | 0.0001 | 0.0002 | 0.0006 | 0.0016 | 0.0037 | 0.0078 |
| **8** | 0 | 0.4305 | 0.2725 | 0.1678 | 0.1001 | 0.0576 | 0.0319 | 0.0168 | 0.0084 | 0.0039 |
| | 1 | 0.3826 | 0.3847 | 0.3355 | 0.2670 | 0.1977 | 0.1373 | 0.0896 | 0.0548 | 0.0312 |
| | 2 | 0.1488 | 0.2376 | 0.2936 | 0.3115 | 0.2965 | 0.2587 | 0.2090 | 0.1569 | 0.1094 |
| | 3 | 0.0331 | 0.0839 | 0.1468 | 0.2076 | 0.2541 | 0.2786 | 0.2787 | 0.2568 | 0.2188 |
| | 4 | 0.0046 | 0.0185 | 0.0459 | 0.0865 | 0.1361 | 0.1875 | 0.2322 | 0.2627 | 0.2734 |
| | 5 | 0.0004 | 0.0026 | 0.0092 | 0.0231 | 0.0467 | 0.0808 | 0.1239 | 0.1719 | 0.2188 |
| | 6 | | 0.0002 | 0.0011 | 0.0038 | 0.0100 | 0.0217 | 0.0413 | 0.0703 | 0.1094 |
| | 7 | | | 0.0001 | 0.0004 | 0.0012 | 0.0033 | 0.0079 | 0.0164 | 0.0313 |
| | 8 | | | | | 0.0001 | 0.0002 | 0.0007 | 0.0017 | 0.0039 |

Note: Blank entries indicate a binomial probability of less than 0.00005.

*(Continued)*

**Table B** Binomial distribution *(continued)*

| | | *p* | | | | | | | | |
|---|---|---|---|---|---|---|---|---|---|---|
| *n* | *X* | **0.10** | **0.15** | **0.20** | **0.25** | **0.30** | **0.35** | **0.40** | **0.45** | **0.50** |
| **9** | 0 | 0.3874 | 0.2316 | 0.1342 | 0.0751 | 0.0404 | 0.0207 | 0.0101 | 0.0046 | 0.0020 |
| | 1 | 0.3874 | 0.3679 | 0.3020 | 0.2253 | 0.1556 | 0.1004 | 0.0605 | 0.0339 | 0.0176 |
| | 2 | 0.1722 | 0.2597 | 0.3020 | 0.3003 | 0.2668 | 0.2162 | 0.1612 | 0.1110 | 0.0703 |
| | 3 | 0.0446 | 0.1069 | 0.1762 | 0.2336 | 0.2668 | 0.2716 | 0.2508 | 0.2119 | 0.1641 |
| | 4 | 0.0074 | 0.0283 | 0.0661 | 0.1168 | 0.1715 | 0.2194 | 0.2508 | 0.2600 | 0.2461 |
| | 5 | 0.0008 | 0.0050 | 0.0165 | 0.0389 | 0.0735 | 0.1181 | 0.1672 | 0.2128 | 0.2461 |
| | 6 | 0.0001 | 0.0006 | 0.0028 | 0.0087 | 0.0210 | 0.0424 | 0.0743 | 0.1160 | 0.1641 |
| | 7 | | | 0.0003 | 0.0012 | 0.0039 | 0.0098 | 0.0212 | 0.0407 | 0.0703 |
| | 8 | | | | 0.0001 | 0.0004 | 0.0013 | 0.0035 | 0.0083 | 0.0176 |
| | 9 | | | | | | 0.0001 | 0.0003 | 0.0008 | 0.0020 |
| **10** | 0 | 0.3487 | 0.1969 | 0.1074 | 0.0563 | 0.0282 | 0.0135 | 0.0060 | 0.0025 | 0.0010 |
| | 1 | 0.3874 | 0.3474 | 0.2684 | 0.1877 | 0.1211 | 0.0725 | 0.0403 | 0.0207 | 0.0098 |
| | 2 | 0.1937 | 0.2759 | 0.3020 | 0.2816 | 0.2335 | 0.1757 | 0.1209 | 0.0763 | 0.0439 |
| | 3 | 0.0574 | 0.1298 | 0.2013 | 0.2503 | 0.2668 | 0.2522 | 0.2150 | 0.1665 | 0.1172 |
| | 4 | 0.0112 | 0.0401 | 0.0881 | 0.1460 | 0.2001 | 0.2377 | 0.2508 | 0.2384 | 0.2051 |
| | 5 | 0.0015 | 0.0085 | 0.0264 | 0.0584 | 0.1029 | 0.1536 | 0.2007 | 0.2340 | 0.2461 |
| | 6 | 0.0001 | 0.0012 | 0.0055 | 0.0162 | 0.0368 | 0.0689 | 0.1115 | 0.1596 | 0.2051 |
| | 7 | | 0.0001 | 0.0008 | 0.0031 | 0.0090 | 0.0212 | 0.0425 | 0.0746 | 0.1172 |
| | 8 | | | 0.0001 | 0.0004 | 0.0014 | 0.0043 | 0.0106 | 0.0229 | 0.0439 |
| | 9 | | | | | 0.0001 | 0.0005 | 0.0016 | 0.0042 | 0.0098 |
| | 10 | | | | | | | 0.0001 | 0.0003 | 0.0010 |
| **12** | 0 | 0.2824 | 0.1422 | 0.0687 | 0.0317 | 0.0138 | 0.0057 | 0.0022 | 0.0008 | 0.0002 |
| | 1 | 0.3766 | 0.3012 | 0.2062 | 0.1267 | 0.0712 | 0.0368 | 0.0174 | 0.0075 | 0.0029 |
| | 2 | 0.2301 | 0.2924 | 0.2835 | 0.2323 | 0.1678 | 0.1088 | 0.0639 | 0.0339 | 0.0161 |
| | 3 | 0.0853 | 0.1720 | 0.2362 | 0.2581 | 0.2397 | 0.1954 | 0.1419 | 0.0923 | 0.0537 |
| | 4 | 0.0213 | 0.0683 | 0.1329 | 0.1936 | 0.2311 | 0.2367 | 0.2128 | 0.1700 | 0.1208 |
| | 5 | 0.0038 | 0.0193 | 0.0532 | 0.1032 | 0.1585 | 0.2039 | 0.2270 | 0.2225 | 0.1934 |
| | 6 | 0.0005 | 0.0040 | 0.0155 | 0.0401 | 0.0792 | 0.1281 | 0.1766 | 0.2124 | 0.2256 |
| | 7 | | 0.0006 | 0.0033 | 0.0115 | 0.0291 | 0.0591 | 0.1009 | 0.1489 | 0.1934 |
| | 8 | | 0.0001 | 0.0005 | 0.0024 | 0.0078 | 0.0199 | 0.0420 | 0.0762 | 0.1208 |
| | 9 | | | 0.0001 | 0.0004 | 0.0015 | 0.0048 | 0.0125 | 0.0277 | 0.0537 |
| | 10 | | | | | 0.0002 | 0.0008 | 0.0025 | 0.0068 | 0.0161 |
| | 11 | | | | | | 0.0001 | 0.0003 | 0.0010 | 0.0029 |
| | 12 | | | | | | | | 0.0001 | 0.0002 |
| **15** | 0 | 0.2059 | 0.0874 | 0.0352 | 0.0134 | 0.0047 | 0.0016 | 0.0005 | 0.0001 | |
| | 1 | 0.3432 | 0.2312 | 0.1319 | 0.0668 | 0.0305 | 0.0126 | 0.0047 | 0.0016 | 0.0005 |
| | 2 | 0.2669 | 0.2856 | 0.2309 | 0.1559 | 0.0916 | 0.0476 | 0.0219 | 0.0090 | 0.0032 |
| | 3 | 0.1285 | 0.2184 | 0.2501 | 0.2252 | 0.1700 | 0.1110 | 0.0634 | 0.0318 | 0.0139 |
| | 4 | 0.0428 | 0.1156 | 0.1876 | 0.2252 | 0.2186 | 0.1792 | 0.1268 | 0.0780 | 0.0417 |
| | 5 | 0.0105 | 0.0449 | 0.1032 | 0.1651 | 0.2061 | 0.2123 | 0.1859 | 0.1404 | 0.0916 |
| | 6 | 0.0019 | 0.0132 | 0.0430 | 0.0917 | 0.1472 | 0.1906 | 0.2066 | 0.1914 | 0.1527 |
| | 7 | 0.0003 | 0.0030 | 0.0138 | 0.0393 | 0.0811 | 0.1319 | 0.1771 | 0.2013 | 0.1964 |
| | 8 | | 0.0005 | 0.0035 | 0.0131 | 0.0348 | 0.0710 | 0.1181 | 0.1647 | 0.1964 |
| | 9 | | 0.0001 | 0.0007 | 0.0034 | 0.0016 | 0.0298 | 0.0612 | 0.1048 | 0.1527 |
| | 10 | | | 0.0001 | 0.0007 | 0.0030 | 0.0096 | 0.0245 | 0.0515 | 0.0916 |
| | 11 | | | | 0.0001 | 0.0006 | 0.0024 | 0.0074 | 0.0191 | 0.0417 |
| | 12 | | | | | 0.0001 | 0.0004 | 0.0016 | 0.0052 | 0.0139 |
| | 13 | | | | | | 0.0001 | 0.0003 | 0.0010 | 0.0032 |
| | 14 | | | | | | | | 0.0001 | 0.0005 |
| | 15 | | | | | | | | | |

Note: Blank entries indicate a binomial probability of less than 0.00005.

Table B Binomial distribution (*continued*)

| | | *p* | | | | | | | | |
|---|---|---|---|---|---|---|---|---|---|---|
| *n* | *X* | **0.10** | **0.15** | **0.20** | **0.25** | **0.30** | **0.35** | **0.40** | **0.45** | **0.50** |
| **18** | 0 | 0.1501 | 0.0536 | 0.0180 | 0.0056 | 0.0016 | 0.0004 | 0.0001 | | |
| | 1 | 0.3002 | 0.1704 | 0.0811 | 0.0338 | 0.0126 | 0.0042 | 0.0012 | 0.0003 | 0.0001 |
| | 2 | 0.2835 | 0.2556 | 0.1723 | 0.0958 | 0.0458 | 0.0190 | 0.0069 | 0.0022 | 0.0006 |
| | 3 | 0.1680 | 0.2406 | 0.2297 | 0.1704 | 0.1046 | 0.0547 | 0.0246 | 0.0095 | 0.0031 |
| | 4 | 0.0700 | 0.1592 | 0.2153 | 0.2130 | 0.1681 | 0.1104 | 0.0614 | 0.0291 | 0.0117 |
| | 5 | 0.0218 | 0.0787 | 0.1507 | 0.1988 | 0.2017 | 0.1664 | 0.1146 | 0.0666 | 0.0327 |
| | 6 | 0.0052 | 0.0301 | 0.0816 | 0.1436 | 0.1873 | 0.1941 | 0.1655 | 0.1181 | 0.0708 |
| | 7 | 0.0010 | 0.0091 | 0.0350 | 0.0820 | 0.1376 | 0.1792 | 0.1892 | 0.1657 | 0.1214 |
| | 8 | 0.0002 | 0.0022 | 0.0120 | 0.0376 | 0.0811 | 0.1327 | 0.1734 | 0.1864 | 0.1669 |
| | 9 | | 0.0004 | 0.0033 | 0.0139 | 0.0386 | 0.0794 | 0.1284 | 0.1694 | 0.1855 |
| | 10 | | 0.0001 | 0.0008 | 0.0042 | 0.0149 | 0.0385 | 0.0771 | 0.1248 | 0.1669 |
| | 11 | | | 0.0001 | 0.0010 | 0.0046 | 0.0151 | 0.0374 | 0.0742 | 0.1214 |
| | 12 | | | | 0.0002 | 0.0012 | 0.0047 | 0.0145 | 0.0354 | 0.0708 |
| | 13 | | | | | 0.0002 | 0.0012 | 0.0045 | 0.0134 | 0.0327 |
| | 14 | | | | | | 0.0002 | 0.0011 | 0.0039 | 0.0117 |
| | 15 | | | | | | | 0.0002 | 0.0009 | 0.0031 |
| | 16 | | | | | | | | 0.0001 | 0.0006 |
| | 17 | | | | | | | | | 0.0001 |
| | 18 | | | | | | | | | |
| **20** | 0 | 0.1216 | 0.0388 | 0.0115 | 0.0032 | 0.0008 | 0.0002 | | | |
| | 1 | 0.2702 | 0.1368 | 0.0576 | 0.0211 | 0.0068 | 0.0020 | 0.0005 | 0.0001 | |
| | 2 | 0.2852 | 0.2293 | 0.1369 | 0.0669 | 0.0278 | 0.0100 | 0.0031 | 0.0008 | 0.0002 |
| | 3 | 0.1901 | 0.2428 | 0.2054 | 0.1339 | 0.0716 | 0.0323 | 0.0123 | 0.0040 | 0.0011 |
| | 4 | 0.0898 | 0.1821 | 0.2182 | 0.1897 | 0.1304 | 0.0738 | 0.0350 | 0.0139 | 0.0046 |
| | 5 | 0.0319 | 0.1028 | 0.1746 | 0.2023 | 0.1789 | 0.1272 | 0.0746 | 0.0365 | 0.0148 |
| | 6 | 0.0089 | 0.0454 | 0.1091 | 0.1686 | 0.1916 | 0.1712 | 0.1244 | 0.0746 | 0.0370 |
| | 7 | 0.0020 | 0.0160 | 0.0545 | 0.1124 | 0.1643 | 0.1844 | 0.1659 | 0.1221 | 0.0739 |
| | 8 | 0.0004 | 0.0046 | 0.0222 | 0.0609 | 0.1144 | 0.1614 | 0.1797 | 0.1623 | 0.1201 |
| | 9 | 0.0001 | 0.0011 | 0.0074 | 0.0271 | 0.0654 | 0.1158 | 0.1597 | 0.1771 | 0.1602 |
| | 10 | | 0.0002 | 0.0020 | 0.0099 | 0.0308 | 0.0686 | 0.1171 | 0.1593 | 0.1762 |
| | 11 | | | 0.0005 | 0.0030 | 0.0120 | 0.0336 | 0.0710 | 0.1185 | 0.1602 |
| | 12 | | | 0.0001 | 0.0008 | 0.0039 | 0.0136 | 0.0355 | 0.0727 | 0.1201 |
| | 13 | | | | 0.0002 | 0.0010 | 0.0045 | 0.0146 | 0.0366 | 0.0739 |
| | 14 | | | | | 0.0002 | 0.0012 | 0.0049 | 0.0150 | 0.0370 |
| | 15 | | | | | | 0.0003 | 0.0013 | 0.0049 | 0.0148 |
| | 16 | | | | | | | 0.0003 | 0.0013 | 0.0046 |
| | 17 | | | | | | | | 0.0002 | 0.0011 |
| | 18 | | | | | | | | | 0.0002 |
| | 19 | | | | | | | | | |
| | 20 | | | | | | | | | |

Note: Blank entries indicate a binomial probability of less than 0.00005.

*(Continued)*

## Table B Binomial distribution (*continued*)

| | | *p* | | | | | | | | |
|---|---|---|---|---|---|---|---|---|---|---|
| *n* | *X* | **0.55** | **0.60** | **0.65** | **0.70** | **0.75** | **0.80** | **0.85** | **0.90** | **0.95** |
| **2** | 0 | 0.2025 | 0.1600 | 0.1225 | 0.0900 | 0.0625 | 0.0400 | 0.0225 | 0.0100 | 0.0025 |
| | 1 | 0.4950 | 0.4800 | 0.4550 | 0.4200 | 0.3750 | 0.3200 | 0.2550 | 0.1800 | 0.0950 |
| | 2 | 0.3025 | 0.3600 | 0.4225 | 0.4900 | 0.5625 | 0.6400 | 0.7225 | 0.8100 | 0.9025 |
| **3** | 0 | 0.0911 | 0.0640 | 0.0429 | 0.0270 | 0.0156 | 0.0080 | 0.0034 | 0.0010 | 0.0001 |
| | 1 | 0.3341 | 0.2880 | 0.2389 | 0.1890 | 0.1406 | 0.0960 | 0.0574 | 0.0270 | 0.0071 |
| | 2 | 0.4084 | 0.4320 | 0.4436 | 0.4410 | 0.4219 | 0.3840 | 0.3251 | 0.2430 | 0.1354 |
| | 3 | 0.1664 | 0.2160 | 0.2746 | 0.3430 | 0.4219 | 0.5120 | 0.6141 | 0.7290 | 0.8574 |
| **4** | 0 | 0.0410 | 0.0256 | 0.0150 | 0.0081 | 0.0039 | 0.0016 | 0.0005 | 0.0001 | |
| | 1 | 0.2005 | 0.1536 | 0.1115 | 0.0756 | 0.0469 | 0.0256 | 0.0115 | 0.0036 | 0.0005 |
| | 2 | 0.3675 | 0.3456 | 0.3105 | 0.2646 | 0.2109 | 0.1536 | 0.0975 | 0.0486 | 0.0135 |
| | 3 | 0.2995 | 0.3456 | 0.3845 | 0.4116 | 0.4219 | 0.4096 | 0.3685 | 0.2916 | 0.1715 |
| | 4 | 0.0915 | 0.1296 | 0.1785 | 0.2401 | 0.3164 | 0.4096 | 0.5220 | 0.6561 | 0.8145 |
| **5** | 0 | 0.0185 | 0.0102 | 0.0053 | 0.0024 | 0.0010 | 0.0003 | 0.0001 | | |
| | 1 | 0.1128 | 0.0768 | 0.0488 | 0.0284 | 0.0146 | 0.0064 | 0.0022 | 0.0005 | |
| | 2 | 0.2757 | 0.2304 | 0.1811 | 0.1323 | 0.0879 | 0.0512 | 0.0244 | 0.0081 | 0.0011 |
| | 3 | 0.3369 | 0.3456 | 0.3364 | 0.3087 | 0.2637 | 0.2048 | 0.1382 | 0.0729 | 0.0214 |
| | 4 | 0.2059 | 0.2592 | 0.3124 | 0.3601 | 0.3955 | 0.4096 | 0.3915 | 0.3281 | 0.2036 |
| | 5 | 0.0503 | 0.0778 | 0.1160 | 0.1681 | 0.2373 | 0.3277 | 0.4437 | 0.5905 | 0.7738 |
| **6** | 0 | 0.0083 | 0.0041 | 0.0018 | 0.0007 | 0.0002 | 0.0001 | | | |
| | 1 | 0.0609 | 0.0369 | 0.0205 | 0.0102 | 0.0044 | 0.0015 | 0.0004 | 0.0001 | |
| | 2 | 0.1861 | 0.1382 | 0.0951 | 0.0595 | 0.0330 | 0.0154 | 0.0055 | 0.0012 | 0.0001 |
| | 3 | 0.3032 | 0.2765 | 0.2355 | 0.1852 | 0.1318 | 0.0819 | 0.0415 | 0.0146 | 0.0021 |
| | 4 | 0.2780 | 0.3110 | 0.3280 | 0.3241 | 0.2966 | 0.2458 | 0.1762 | 0.0984 | 0.0305 |
| | 5 | 0.1359 | 0.1866 | 0.2437 | 0.3025 | 0.3560 | 0.3932 | 0.3993 | 0.3543 | 0.2321 |
| | 6 | 0.0277 | 0.0467 | 0.0754 | 0.1176 | 0.1780 | 0.2621 | 0.3771 | 0.5314 | 0.7351 |
| **7** | 0 | 0.0037 | 0.0016 | 0.0006 | 0.0002 | 0.0001 | | | | |
| | 1 | 0.0320 | 0.0172 | 0.0084 | 0.0036 | 0.0013 | 0.0004 | 0.0001 | | |
| | 2 | 0.1172 | 0.0774 | 0.0466 | 0.0250 | 0.0115 | 0.0043 | 0.0012 | 0.0002 | |
| | 3 | 0.2388 | 0.1935 | 0.1442 | 0.0972 | 0.0577 | 0.0287 | 0.0109 | 0.0026 | 0.0002 |
| | 4 | 0.2918 | 0.2903 | 0.2679 | 0.2269 | 0.1730 | 0.1147 | 0.0617 | 0.0230 | 0.0036 |
| | 5 | 0.2140 | 0.2613 | 0.2985 | 0.3177 | 0.3115 | 0.2753 | 0.2097 | 0.1240 | 0.0406 |
| | 6 | 0.0872 | 0.1306 | 0.1848 | 0.2471 | 0.3115 | 0.3670 | 0.3960 | 0.3720 | 0.2573 |
| | 7 | 0.0152 | 0.0280 | 0.0490 | 0.0824 | 0.1335 | 0.2097 | 0.3206 | 0.4783 | 0.6983 |
| **8** | 0 | 0.0017 | 0.0007 | 0.0002 | 0.0001 | | | | | |
| | 1 | 0.0164 | 0.0079 | 0.0033 | 0.0012 | 0.0004 | 0.0001 | | | |
| | 2 | 0.0703 | 0.0413 | 0.0217 | 0.0100 | 0.0038 | 0.0011 | 0.0002 | | |
| | 3 | 0.1719 | 0.1239 | 0.0808 | 0.0467 | 0.0231 | 0.0092 | 0.0026 | 0.0004 | |
| | 4 | 0.2627 | 0.2322 | 0.1875 | 0.1361 | 0.0865 | 0.0459 | 0.0185 | 0.0046 | 0.0004 |
| | 5 | 0.2568 | 0.2787 | 0.2786 | 0.2541 | 0.2076 | 0.1468 | 0.0839 | 0.0331 | 0.0054 |
| | 6 | 0.1569 | 0.2090 | 0.2587 | 0.2965 | 0.3115 | 0.2936 | 0.2376 | 0.1488 | 0.0515 |
| | 7 | 0.0548 | 0.0896 | 0.1373 | 0.1977 | 0.2670 | 0.3355 | 0.3847 | 0.3826 | 0.2793 |
| | 8 | 0.0084 | 0.0168 | 0.0319 | 0.0576 | 0.1001 | 0.1678 | 0.2725 | 0.4305 | 0.6634 |

Note: Blank entries indicate a binomial probability of less than 0.00005.

Table B Binomial distribution (*continued*)

| | | *p* | | | | | | | | |
|---|---|---|---|---|---|---|---|---|---|---|
| *n* | *X* | **0.55** | **0.60** | **0.65** | **0.70** | **0.75** | **0.80** | **0.85** | **0.90** | **0.95** |
| **9** | 0 | 0.0008 | 0.0003 | 0.0001 | | | | | | |
| | 1 | 0.0083 | 0.0035 | 0.0013 | 0.0004 | 0.0001 | | | | |
| | 2 | 0.0407 | 0.0212 | 0.0098 | 0.0039 | 0.0012 | 0.0003 | | | |
| | 3 | 0.1160 | 0.0743 | 0.0424 | 0.0210 | 0.0087 | 0.0028 | 0.0006 | 0.0001 | |
| | 4 | 0.2128 | 0.1672 | 0.1181 | 0.0735 | 0.0389 | 0.0165 | 0.0050 | 0.0008 | |
| | 5 | 0.2600 | 0.2508 | 0.2194 | 0.1715 | 0.1168 | 0.0661 | 0.0283 | 0.0074 | 0.0006 |
| | 6 | 0.2119 | 0.2508 | 0.2716 | 0.2668 | 0.2336 | 0.1762 | 0.1069 | 0.0446 | 0.0077 |
| | 7 | 0.1110 | 0.1612 | 0.2162 | 0.2668 | 0.3003 | 0.3020 | 0.2597 | 0.1722 | 0.0629 |
| | 8 | 0.0339 | 0.0605 | 0.1004 | 0.1556 | 0.2253 | 0.3020 | 0.3679 | 0.3874 | 0.2985 |
| | 9 | 0.0046 | 0.0101 | 0.0207 | 0.0404 | 0.0751 | 0.1342 | 0.2316 | 0.3874 | 0.6302 |
| **10** | 0 | 0.0003 | 0.0001 | | | | | | | |
| | 1 | 0.0042 | 0.0016 | 0.0005 | 0.0001 | | | | | |
| | 2 | 0.0229 | 0.0106 | 0.0043 | 0.0014 | 0.0004 | 0.0001 | | | |
| | 3 | 0.0746 | 0.0425 | 0.0212 | 0.0090 | 0.0031 | 0.0008 | 0.0001 | | |
| | 4 | 0.1596 | 0.1115 | 0.0689 | 0.0368 | 0.0162 | 0.0055 | 0.0012 | 0.0001 | |
| | 5 | 0.2340 | 0.2007 | 0.1536 | 0.1029 | 0.0584 | 0.0264 | 0.0085 | 0.0015 | 0.0001 |
| | 6 | 0.2384 | 0.2508 | 0.2377 | 0.2001 | 0.1460 | 0.0881 | 0.0401 | 0.0112 | 0.0010 |
| | 7 | 0.1665 | 0.2150 | 0.2522 | 0.2668 | 0.2503 | 0.2013 | 0.1298 | 0.0574 | 0.0105 |
| | 8 | 0.0763 | 0.1209 | 0.1757 | 0.2335 | 0.2816 | 0.3020 | 0.2759 | 0.1937 | 0.0746 |
| | 9 | 0.0207 | 0.0403 | 0.0725 | 0.1211 | 0.1877 | 0.2684 | 0.3474 | 0.3874 | 0.3151 |
| | 10 | 0.0025 | 0.0060 | 0.0135 | 0.0282 | 0.0563 | 0.1074 | 0.1969 | 0.3487 | 0.5987 |
| **12** | 0 | 0.0001 | | | | | | | | |
| | 1 | 0.0010 | 0.0003 | 0.0001 | | | | | | |
| | 2 | 0.0068 | 0.0025 | 0.0008 | 0.0002 | | | | | |
| | 3 | 0.0277 | 0.0125 | 0.0048 | 0.0015 | 0.0004 | 0.0001 | | | |
| | 4 | 0.0762 | 0.0420 | 0.0199 | 0.0078 | 0.0024 | 0.0005 | 0.0001 | | |
| | 5 | 0.1489 | 0.1009 | 0.0591 | 0.0291 | 0.0115 | 0.0033 | 0.0006 | | |
| | 6 | 0.2124 | 0.1766 | 0.1281 | 0.0792 | 0.0401 | 0.0155 | 0.0040 | 0.0005 | |
| | 7 | 0.2225 | 0.2270 | 0.2039 | 0.1585 | 0.1032 | 0.0532 | 0.0193 | 0.0038 | 0.0002 |
| | 8 | 0.1700 | 0.2128 | 0.2367 | 0.2311 | 0.1936 | 0.1329 | 0.0683 | 0.0213 | 0.0021 |
| | 9 | 0.0923 | 0.1419 | 0.1954 | 0.2397 | 0.2581 | 0.2362 | 0.1720 | 0.0852 | 0.0173 |
| | 10 | 0.0339 | 0.0639 | 0.1088 | 0.1678 | 0.2323 | 0.2835 | 0.2924 | 0.2301 | 0.0988 |
| | 11 | 0.0075 | 0.0174 | 0.0368 | 0.0712 | 0.1267 | 0.2062 | 0.3012 | 0.3766 | 0.3413 |
| | 12 | 0.0008 | 0.0022 | 0.0057 | 0.0138 | 0.0317 | 0.0687 | 0.1422 | 0.2824 | 0.5404 |
| **15** | 0 | | | | | | | | | |
| | 1 | 0.0001 | | | | | | | | |
| | 2 | 0.0010 | 0.0003 | 0.0001 | | | | | | |
| | 3 | 0.0052 | 0.0016 | 0.0004 | 0.0001 | | | | | |
| | 4 | 0.0191 | 0.0074 | 0.0024 | 0.0006 | 0.0001 | | | | |
| | 5 | 0.0515 | 0.0245 | 0.0096 | 0.0030 | 0.0007 | 0.0001 | | | |
| | 6 | 0.1048 | 0.0612 | 0.0298 | 0.0116 | 0.0034 | 0.0007 | 0.0001 | | |
| | 7 | 0.1647 | 0.1181 | 0.0710 | 0.0348 | 0.0131 | 0.0035 | 0.0005 | | |
| | 8 | 0.2013 | 0.1771 | 0.1319 | 0.0811 | 0.0393 | 0.0138 | 0.0030 | 0.0003 | |
| | 9 | 0.1914 | 0.2066 | 0.1906 | 0.1472 | 0.0917 | 0.0430 | 0.0132 | 0.0019 | |
| | 10 | 0.1404 | 0.1859 | 0.2123 | 0.2061 | 0.1651 | 0.1032 | 0.0449 | 0.0105 | 0.0006 |
| | 11 | 0.0780 | 0.1268 | 0.1792 | 0.2186 | 0.2252 | 0.1876 | 0.1156 | 0.0428 | 0.0049 |

Note: Blank entries indicate a binomial probability of less than 0.00005.

*(Continued)*

**Table B** Binomial distribution *(continued)*

| | | *p* | | | | | | | | |
|---|---|---|---|---|---|---|---|---|---|---|
| *n* | *X* | **0.55** | **0.60** | **0.65** | **0.70** | **0.75** | **0.80** | **0.85** | **0.90** | **0.95** |
| | 12 | 0.0318 | 0.0634 | 0.1110 | 0.1700 | 0.2252 | 0.2501 | 0.2184 | 0.1285 | 0.0307 |
| | 13 | 0.0090 | 0.0219 | 0.0476 | 0.0916 | 0.1559 | 0.2309 | 0.2856 | 0.2669 | 0.1348 |
| | 14 | 0.0016 | 0.0047 | 0.0126 | 0.0305 | 0.0668 | 0.1319 | 0.2312 | 0.3432 | 0.3658 |
| | 15 | 0.0001 | 0.0005 | 0.0016 | 0.0047 | 0.0134 | 0.0352 | 0.0874 | 0.2059 | 0.4633 |
| **18** | 0 | | | | | | | | | |
| | 1 | | | | | | | | | |
| | 2 | 0.0001 | | | | | | | | |
| | 3 | 0.0009 | 0.0002 | | | | | | | |
| | 4 | 0.0039 | 0.0011 | 0.0002 | | | | | | |
| | 5 | 0.0134 | 0.0045 | 0.0012 | 0.0002 | | | | | |
| | 6 | 0.0354 | 0.0145 | 0.0047 | 0.0012 | 0.0002 | | | | |
| | 7 | 0.0742 | 0.0374 | 0.0151 | 0.0046 | 0.0010 | 0.0001 | | | |
| | 8 | 0.1248 | 0.0771 | 0.0385 | 0.0149 | 0.0042 | 0.0008 | 0.0001 | | |
| | 9 | 0.1694 | 0.1284 | 0.0794 | 0.0386 | 0.0139 | 0.0033 | 0.0004 | | |
| | 10 | 0.1864 | 0.1734 | 0.1327 | 0.0811 | 0.0376 | 0.0120 | 0.0022 | 0.0002 | |
| | 11 | 0.1657 | 0.1892 | 0.1792 | 0.1376 | 0.0820 | 0.0350 | 0.0091 | 0.0010 | |
| | 12 | 0.1181 | 0.1655 | 0.1941 | 0.1873 | 0.1436 | 0.0816 | 0.0301 | 0.0052 | 0.0002 |
| | 13 | 0.0666 | 0.1146 | 0.1664 | 0.2017 | 0.1988 | 0.1507 | 0.0787 | 0.0218 | 0.0014 |
| | 14 | 0.0291 | 0.0614 | 0.1104 | 0.1681 | 0.2130 | 0.2153 | 0.1592 | 0.0700 | 0.0093 |
| | 15 | 0.0095 | 0.0246 | 0.0547 | 0.1046 | 0.1704 | 0.2297 | 0.2406 | 0.1680 | 0.0473 |
| | 16 | 0.0022 | 0.0069 | 0.0190 | 0.0458 | 0.0958 | 0.1723 | 0.2556 | 0.2835 | 0.1683 |
| | 17 | 0.0003 | 0.0012 | 0.0042 | 0.0126 | 0.0338 | 0.0811 | 0.1704 | 0.3002 | 0.3763 |
| | 18 | | 0.0001 | 0.0004 | 0.0016 | 0.0056 | 0.0180 | 0.0536 | 0.1501 | 0.3972 |
| **20** | 0 | | | | | | | | | |
| | 1 | | | | | | | | | |
| | 2 | | | | | | | | | |
| | 3 | 0.0002 | | | | | | | | |
| | 4 | 0.0013 | 0.0003 | | | | | | | |
| | 5 | 0.0049 | 0.0013 | 0.0003 | | | | | | |
| | 6 | 0.0150 | 0.0049 | 0.0012 | 0.0002 | | | | | |
| | 7 | 0.0366 | 0.0146 | 0.0045 | 0.0010 | 0.0002 | | | | |
| | 8 | 0.0727 | 0.0355 | 0.0136 | 0.0039 | 0.0008 | 0.0001 | | | |
| | 9 | 0.1185 | 0.0710 | 0.0336 | 0.0120 | 0.0030 | 0.0005 | | | |
| | 10 | 0.1593 | 0.1171 | 0.0686 | 0.0308 | 0.0099 | 0.0020 | 0.0002 | | |
| | 11 | 0.1771 | 0.1597 | 0.1158 | 0.0654 | 0.0271 | 0.0074 | 0.0011 | 0.0001 | |
| | 12 | 0.1623 | 0.1797 | 0.1614 | 0.1144 | 0.0609 | 0.0222 | 0.0046 | 0.0004 | |
| | 13 | 0.1221 | 0.1659 | 0.1844 | 0.1643 | 0.1124 | 0.0545 | 0.0160 | 0.0020 | |
| | 14 | 0.0746 | 0.1244 | 0.1712 | 0.1916 | 0.1686 | 0.1091 | 0.0454 | 0.0089 | 0.0003 |
| | 15 | 0.0365 | 0.0746 | 0.1272 | 0.1789 | 0.2023 | 0.1746 | 0.1028 | 0.0319 | 0.0022 |
| | 16 | 0.0139 | 0.0350 | 0.0738 | 0.1304 | 0.1897 | 0.2182 | 0.1821 | 0.0898 | 0.0133 |
| | 17 | 0.0040 | 0.0123 | 0.0323 | 0.0716 | 0.1339 | 0.2054 | 0.2428 | 0.1901 | 0.0596 |
| | 18 | 0.0008 | 0.0031 | 0.0100 | 0.0278 | 0.0669 | 0.1369 | 0.2293 | 0.2852 | 0.1887 |
| | 19 | 0.0001 | 0.0005 | 0.0020 | 0.0068 | 0.0211 | 0.0576 | 0.1368 | 0.2702 | 0.3774 |
| | 20 | | | 0.0002 | 0.0008 | 0.0032 | 0.0115 | 0.0388 | 0.1216 | 0.3585 |

Note: Blank entries indicate a binomial probability of less than 0.00005.

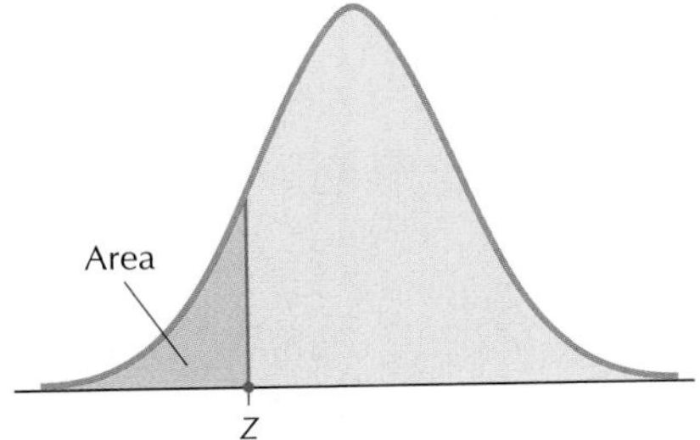


Table C Standard normal distribution

| z | 0.00 | 0.01 | 0.02 | 0.03 | 0.04 | 0.05 | 0.06 | 0.07 | 0.08 | 0.09 |
|---|---|---|---|---|---|---|---|---|---|---|
| **–3.4** | 0.0003 | 0.0003 | 0.0003 | 0.0003 | 0.0003 | 0.0003 | 0.0003 | 0.0003 | 0.0003 | 0.0002 |
| **–3.3** | 0.0005 | 0.0005 | 0.0005 | 0.0004 | 0.0004 | 0.0004 | 0.0004 | 0.0004 | 0.0004 | 0.0003 |
| **–3.2** | 0.0007 | 0.0007 | 0.0006 | 0.0006 | 0.0006 | 0.0006 | 0.0006 | 0.0005 | 0.0005 | 0.0005 |
| **–3.1** | 0.0010 | 0.0009 | 0.0009 | 0.0009 | 0.0008 | 0.0008 | 0.0008 | 0.0008 | 0.0007 | 0.0007 |
| **–3.0** | 0.0013 | 0.0013 | 0.0013 | 0.0012 | 0.0012 | 0.0011 | 0.0011 | 0.0011 | 0.0010 | 0.0010 |
| **–2.9** | 0.0019 | 0.0018 | 0.0018 | 0.0017 | 0.0016 | 0.0016 | 0.0015 | 0.0015 | 0.0014 | 0.0014 |
| **–2.8** | 0.0026 | 0.0025 | 0.0024 | 0.0023 | 0.0023 | 0.0022 | 0.0021 | 0.0021 | 0.0020 | 0.0019 |
| **–2.7** | 0.0035 | 0.0034 | 0.0033 | 0.0032 | 0.0031 | 0.0030 | 0.0029 | 0.0028 | 0.0027 | 0.0026 |
| **–2.6** | 0.0047 | 0.0045 | 0.0044 | 0.0043 | 0.0041 | 0.0040 | 0.0039 | 0.0038 | 0.0037 | 0.0036 |
| **–2.5** | 0.0062 | 0.0060 | 0.0059 | 0.0057 | 0.0055 | 0.0054 | 0.0052 | 0.0051 | 0.0049 | 0.0048 |
| **–2.4** | 0.0082 | 0.0080 | 0.0078 | 0.0075 | 0.0073 | 0.0071 | 0.0069 | 0.0068 | 0.0066 | 0.0064 |
| **–2.3** | 0.0107 | 0.0104 | 0.0102 | 0.0099 | 0.0096 | 0.0094 | 0.0091 | 0.0089 | 0.0087 | 0.0084 |
| **–2.2** | 0.0139 | 0.0136 | 0.0132 | 0.0129 | 0.0125 | 0.0122 | 0.0119 | 0.0116 | 0.0113 | 0.0110 |
| **–2.1** | 0.0179 | 0.0174 | 0.0170 | 0.0166 | 0.0162 | 0.0158 | 0.0154 | 0.0150 | 0.0146 | 0.0143 |
| **–2.0** | 0.0228 | 0.0222 | 0.0217 | 0.0212 | 0.0207 | 0.0202 | 0.0197 | 0.0192 | 0.0188 | 0.0183 |
| **–1.9** | 0.0287 | 0.0281 | 0.0274 | 0.0268 | 0.0262 | 0.0256 | 0.0250 | 0.0244 | 0.0239 | 0.0233 |
| **–1.8** | 0.0359 | 0.0351 | 0.0344 | 0.0336 | 0.0329 | 0.0322 | 0.0314 | 0.0307 | 0.0301 | 0.0294 |
| **–1.7** | 0.0446 | 0.0436 | 0.0427 | 0.0418 | 0.0409 | 0.0401 | 0.0392 | 0.0384 | 0.0375 | 0.0367 |
| **–1.6** | 0.0548 | 0.0537 | 0.0526 | 0.0516 | 0.0505 | 0.0495 | 0.0485 | 0.0475 | 0.0465 | 0.0455 |
| **–1.5** | 0.0668 | 0.0655 | 0.0643 | 0.0630 | 0.0618 | 0.0606 | 0.0594 | 0.0582 | 0.0571 | 0.0559 |
| **–1.4** | 0.0808 | 0.0793 | 0.0778 | 0.0764 | 0.0749 | 0.0735 | 0.0721 | 0.0708 | 0.0694 | 0.0681 |
| **–1.3** | 0.0968 | 0.0951 | 0.0934 | 0.0918 | 0.0901 | 0.0885 | 0.0869 | 0.0853 | 0.0838 | 0.0823 |
| **–1.2** | 0.1151 | 0.1131 | 0.1112 | 0.1093 | 0.1075 | 0.1056 | 0.1038 | 0.1020 | 0.1003 | 0.0985 |
| **–1.1** | 0.1357 | 0.1335 | 0.1314 | 0.1292 | 0.1271 | 0.1251 | 0.1230 | 0.1210 | 0.1190 | 0.1170 |
| **–1.0** | 0.1587 | 0.1562 | 0.1539 | 0.1515 | 0.1492 | 0.1469 | 0.1446 | 0.1423 | 0.1401 | 0.1379 |
| **–0.9** | 0.1841 | 0.1814 | 0.1788 | 0.1762 | 0.1736 | 0.1711 | 0.1685 | 0.1660 | 0.1635 | 0.1611 |
| **–0.8** | 0.2119 | 0.2090 | 0.2061 | 0.2033 | 0.2005 | 0.1977 | 0.1949 | 0.1922 | 0.1894 | 0.1867 |
| **–0.7** | 0.2420 | 0.2389 | 0.2358 | 0.2327 | 0.2296 | 0.2266 | 0.2236 | 0.2206 | 0.2177 | 0.2148 |
| **–0.6** | 0.2743 | 0.2709 | 0.2676 | 0.2643 | 0.2611 | 0.2578 | 0.2546 | 0.2514 | 0.2483 | 0.2451 |
| **–0.5** | 0.3085 | 0.3050 | 0.3015 | 0.2981 | 0.2946 | 0.2912 | 0.2877 | 0.2843 | 0.2810 | 0.2776 |
| **–0.4** | 0.3446 | 0.3409 | 0.3372 | 0.3336 | 0.3300 | 0.3264 | 0.3228 | 0.3192 | 0.3156 | 0.3121 |
| **–0.3** | 0.3821 | 0.3783 | 0.3745 | 0.3707 | 0.3669 | 0.3632 | 0.3594 | 0.3557 | 0.3520 | 0.3483 |
| **–0.2** | 0.4207 | 0.4168 | 0.4129 | 0.4090 | 0.4052 | 0.4013 | 0.3974 | 0.3936 | 0.3897 | 0.3859 |
| **–0.1** | 0.4602 | 0.4562 | 0.4522 | 0.4483 | 0.4443 | 0.4404 | 0.4364 | 0.4325 | 0.4286 | 0.4247 |
| **–0.0** | 0.5000 | 0.4960 | 0.4920 | 0.4880 | 0.4840 | 0.4801 | 0.4761 | 0.4721 | 0.4681 | 0.4641 |

*(Continued)*

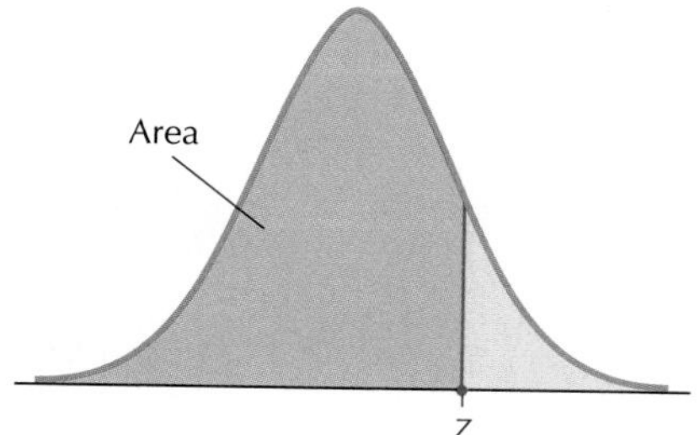


**Table C** Standard normal distribution (*continued*)

| *z* | 0.00 | 0.01 | 0.02 | 0.03 | 0.04 | 0.05 | 0.06 | 0.07 | 0.08 | 0.09 |
|---|---|---|---|---|---|---|---|---|---|---|
| **0.0** | 0.5000 | 0.5040 | 0.5080 | 0.5120 | 0.5160 | 0.5199 | 0.5239 | 0.5279 | 0.5319 | 0.5359 |
| **0.1** | 0.5398 | 0.5438 | 0.5478 | 0.5517 | 0.5557 | 0.5596 | 0.5636 | 0.5675 | 0.5714 | 0.5753 |
| **0.2** | 0.5793 | 0.5832 | 0.5871 | 0.5910 | 0.5948 | 0.5987 | 0.6026 | 0.6064 | 0.6103 | 0.6141 |
| **0.3** | 0.6179 | 0.6217 | 0.6255 | 0.6293 | 0.6331 | 0.6368 | 0.6406 | 0.6443 | 0.6480 | 0.6517 |
| **0.4** | 0.6554 | 0.6591 | 0.6628 | 0.6664 | 0.6700 | 0.6736 | 0.6772 | 0.6808 | 0.6844 | 0.6879 |
| **0.5** | 0.6915 | 0.6950 | 0.6985 | 0.7019 | 0.7054 | 0.7088 | 0.7123 | 0.7157 | 0.7190 | 0.7224 |
| **0.6** | 0.7257 | 0.7291 | 0.7324 | 0.7357 | 0.7389 | 0.7422 | 0.7454 | 0.7486 | 0.7517 | 0.7549 |
| **0.7** | 0.7580 | 0.7611 | 0.7642 | 0.7673 | 0.7704 | 0.7734 | 0.7764 | 0.7794 | 0.7823 | 0.7852 |
| **0.8** | 0.7881 | 0.7910 | 0.7939 | 0.7967 | 0.7995 | 0.8023 | 0.8051 | 0.8078 | 0.8106 | 0.8133 |
| **0.9** | 0.8159 | 0.8186 | 0.8212 | 0.8238 | 0.8264 | 0.8289 | 0.8315 | 0.8340 | 0.8365 | 0.8389 |
| **1.0** | 0.8413 | 0.8438 | 0.8461 | 0.8485 | 0.8508 | 0.8531 | 0.8554 | 0.8577 | 0.8599 | 0.8621 |
| **1.1** | 0.8643 | 0.8665 | 0.8686 | 0.8708 | 0.8729 | 0.8749 | 0.8770 | 0.8790 | 0.8810 | 0.8830 |
| **1.2** | 0.8849 | 0.8869 | 0.8888 | 0.8907 | 0.8925 | 0.8944 | 0.8962 | 0.8980 | 0.8997 | 0.9015 |
| **1.3** | 0.9032 | 0.9049 | 0.9066 | 0.9082 | 0.9099 | 0.9115 | 0.9131 | 0.9147 | 0.9162 | 0.9177 |
| **1.4** | 0.9192 | 0.9207 | 0.9222 | 0.9236 | 0.9251 | 0.9265 | 0.9279 | 0.9292 | 0.9306 | 0.9319 |
| **1.5** | 0.9332 | 0.9345 | 0.9357 | 0.9370 | 0.9382 | 0.9394 | 0.9406 | 0.9418 | 0.9429 | 0.9441 |
| **1.6** | 0.9452 | 0.9463 | 0.9474 | 0.9484 | 0.9495 | 0.9505 | 0.9515 | 0.9525 | 0.9535 | 0.9545 |
| **1.7** | 0.9554 | 0.9564 | 0.9573 | 0.9582 | 0.9591 | 0.9599 | 0.9608 | 0.9616 | 0.9625 | 0.9633 |
| **1.8** | 0.9641 | 0.9649 | 0.9656 | 0.9664 | 0.9671 | 0.9678 | 0.9686 | 0.9693 | 0.9699 | 0.9706 |
| **1.9** | 0.9713 | 0.9719 | 0.9726 | 0.9732 | 0.9738 | 0.9744 | 0.9750 | 0.9756 | 0.9761 | 0.9767 |
| **2.0** | 0.9772 | 0.9778 | 0.9783 | 0.9788 | 0.9793 | 0.9798 | 0.9803 | 0.9808 | 0.9812 | 0.9817 |
| **2.1** | 0.9821 | 0.9826 | 0.9830 | 0.9834 | 0.9838 | 0.9842 | 0.9846 | 0.9850 | 0.9854 | 0.9857 |
| **2.2** | 0.9861 | 0.9864 | 0.9868 | 0.9871 | 0.9875 | 0.9878 | 0.9881 | 0.9884 | 0.9887 | 0.9890 |
| **2.3** | 0.9893 | 0.9896 | 0.9898 | 0.9901 | 0.9904 | 0.9906 | 0.9909 | 0.9911 | 0.9913 | 0.9916 |
| **2.4** | 0.9918 | 0.9920 | 0.9922 | 0.9925 | 0.9927 | 0.9929 | 0.9931 | 0.9932 | 0.9934 | 0.9936 |
| **2.5** | 0.9938 | 0.9940 | 0.9941 | 0.9943 | 0.9945 | 0.9946 | 0.9948 | 0.9949 | 0.9951 | 0.9952 |
| **2.6** | 0.9953 | 0.9955 | 0.9956 | 0.9957 | 0.9959 | 0.9960 | 0.9961 | 0.9962 | 0.9963 | 0.9964 |
| **2.7** | 0.9965 | 0.9966 | 0.9967 | 0.9968 | 0.9969 | 0.9970 | 0.9971 | 0.9972 | 0.9973 | 0.9974 |
| **2.8** | 0.9974 | 0.9975 | 0.9976 | 0.9977 | 0.9977 | 0.9978 | 0.9979 | 0.9979 | 0.9980 | 0.9981 |
| **2.9** | 0.9981 | 0.9982 | 0.9982 | 0.9983 | 0.9984 | 0.9984 | 0.9985 | 0.9985 | 0.9986 | 0.9986 |
| **3.0** | 0.9987 | 0.9987 | 0.9987 | 0.9988 | 0.9988 | 0.9989 | 0.9989 | 0.9989 | 0.9990 | 0.9990 |
| **3.1** | 0.9990 | 0.9991 | 0.9991 | 0.9991 | 0.9992 | 0.9992 | 0.9992 | 0.9992 | 0.9993 | 0.9993 |
| **3.2** | 0.9993 | 0.9993 | 0.9994 | 0.9994 | 0.9994 | 0.9994 | 0.9994 | 0.9995 | 0.9995 | 0.9995 |
| **3.3** | 0.9995 | 0.9995 | 0.9995 | 0.9996 | 0.9996 | 0.9996 | 0.9996 | 0.9996 | 0.9996 | 0.9997 |
| **3.4** | 0.9997 | 0.9997 | 0.9997 | 0.9997 | 0.9997 | 0.9997 | 0.9997 | 0.9997 | 0.9997 | 0.9998 |

Table D *t*-Distribution

| | Confidence level | | | | |
|---|---|---|---|---|---|
| | 80% | 90% | 95% | 98% | 99% |
| | Area in one tail | | | | |
| | 0.10 | 0.05 | 0.025 | 0.01 | 0.005 |
| | Area in two tails | | | | |
| **df** | 0.20 | 0.10 | 0.05 | 0.02 | 0.01 |
| **1** | 3.078 | 6.314 | 12.706 | 31.821 | 63.657 |
| **2** | 1.886 | 2.920 | 4.303 | 6.965 | 9.925 |
| **3** | 1.638 | 2.353 | 3.182 | 4.541 | 5.841 |
| **4** | 1.533 | 2.132 | 2.776 | 3.747 | 4.604 |
| **5** | 1.476 | 2.015 | 2.571 | 3.365 | 4.032 |
| **6** | 1.440 | 1.943 | 2.447 | 3.143 | 3.707 |
| **7** | 1.415 | 1.895 | 2.365 | 2.998 | 3.499 |
| **8** | 1.397 | 1.860 | 2.306 | 2.896 | 3.355 |
| **9** | 1.383 | 1.833 | 2.262 | 2.821 | 3.250 |
| **10** | 1.372 | 1.812 | 2.228 | 2.764 | 3.169 |
| **11** | 1.363 | 1.796 | 2.201 | 2.718 | 3.106 |
| **12** | 1.356 | 1.782 | 2.179 | 2.681 | 3.055 |
| **13** | 1.350 | 1.771 | 2.160 | 2.650 | 3.012 |
| **14** | 1.345 | 1.761 | 2.145 | 2.624 | 2.977 |
| **15** | 1.341 | 1.753 | 2.131 | 2.602 | 2.947 |
| **16** | 1.337 | 1.746 | 2.120 | 2.583 | 2.921 |
| **17** | 1.333 | 1.740 | 2.110 | 2.567 | 2.898 |
| **18** | 1.330 | 1.734 | 2.101 | 2.552 | 2.878 |
| **19** | 1.328 | 1.729 | 2.093 | 2.539 | 2.861 |
| **20** | 1.325 | 1.725 | 2.086 | 2.528 | 2.845 |
| **21** | 1.323 | 1.721 | 2.080 | 2.518 | 2.831 |
| **22** | 1.321 | 1.717 | 2.074 | 2.508 | 2.819 |
| **23** | 1.319 | 1.714 | 2.069 | 2.500 | 2.807 |
| **24** | 1.318 | 1.711 | 2.064 | 2.492 | 2.797 |
| **25** | 1.316 | 1.708 | 2.060 | 2.485 | 2.787 |
| **26** | 1.315 | 1.706 | 2.056 | 2.479 | 2.779 |
| **27** | 1.314 | 1.703 | 2.052 | 2.473 | 2.771 |
| **28** | 1.313 | 1.701 | 2.048 | 2.467 | 2.763 |
| **29** | 1.311 | 1.699 | 2.045 | 2.462 | 2.756 |
| **30** | 1.310 | 1.697 | 2.042 | 2.457 | 2.750 |
| **31** | 1.309 | 1.696 | 2.040 | 2.453 | 2.744 |
| **32** | 1.309 | 1.694 | 2.037 | 2.449 | 2.738 |
| **33** | 1.308 | 1.692 | 2.035 | 2.445 | 2.733 |
| **34** | 1.307 | 1.691 | 2.032 | 2.441 | 2.728 |
| **35** | 1.306 | 1.690 | 2.030 | 2.438 | 2.724 |
| **36** | 1.306 | 1.688 | 2.028 | 2.435 | 2.719 |
| **37** | 1.305 | 1.687 | 2.026 | 2.431 | 2.715 |
| **38** | 1.304 | 1.686 | 2.024 | 2.429 | 2.712 |
| **39** | 1.304 | 1.685 | 2.023 | 2.426 | 2.708 |
| **40** | 1.303 | 1.684 | 2.021 | 2.423 | 2.704 |
| **50** | 1.299 | 1.676 | 2.009 | 2.403 | 2.678 |
| **60** | 1.296 | 1.671 | 2.000 | 2.390 | 2.660 |
| **70** | 1.294 | 1.667 | 1.994 | 2.381 | 2.648 |
| **80** | 1.292 | 1.664 | 1.990 | 2.374 | 2.639 |
| **90** | 1.291 | 1.662 | 1.987 | 2.368 | 2.632 |
| **100** | 1.290 | 1.660 | 1.984 | 2.364 | 2.626 |
| **1000** | 1.282 | 1.646 | 1.962 | 2.330 | 2.581 |
| $z$ | 1.282 | 1.645 | 1.960 | 2.326 | 2.576 |

## Table E Chi-square ($\chi^2$) distribution

| Degrees of freedom | Area to the right of critical value | | | | | | | | | |
|---|---|---|---|---|---|---|---|---|---|---|
| | **0.995** | **0.99** | **0.975** | **0.95** | **0.90** | **0.10** | **0.05** | **0.025** | **0.01** | **0.005** |
| **1** | — | — | 0.001 | 0.004 | 0.016 | 2.706 | 3.841 | 5.024 | 6.635 | 7.879 |
| **2** | 0.010 | 0.020 | 0.051 | 0.103 | 0.211 | 4.605 | 5.991 | 7.378 | 9.210 | 10.597 |
| **3** | 0.072 | 0.115 | 0.216 | 0.352 | 0.584 | 6.251 | 7.815 | 9.348 | 11.345 | 12.838 |
| **4** | 0.207 | 0.297 | 0.484 | 0.711 | 1.064 | 7.779 | 9.488 | 11.143 | 13.277 | 14.860 |
| **5** | 0.412 | 0.554 | 0.831 | 1.145 | 1.610 | 9.236 | 11.071 | 12.833 | 15.086 | 16.750 |
| **6** | 0.676 | 0.872 | 1.237 | 1.635 | 2.204 | 10.645 | 12.592 | 14.449 | 16.812 | 18.548 |
| **7** | 0.989 | 1.239 | 1.690 | 2.167 | 2.833 | 12.017 | 14.067 | 16.013 | 18.475 | 20.278 |
| **8** | 1.344 | 1.646 | 2.180 | 2.733 | 3.490 | 13.362 | 15.507 | 17.535 | 20.090 | 21.955 |
| **9** | 1.735 | 2.088 | 2.700 | 3.325 | 4.168 | 14.684 | 16.919 | 19.023 | 21.666 | 23.589 |
| **10** | 2.156 | 2.558 | 3.247 | 3.940 | 4.865 | 15.987 | 18.307 | 20.483 | 23.209 | 25.188 |
| **11** | 2.603 | 3.053 | 3.816 | 4.575 | 5.578 | 17.275 | 19.675 | 21.920 | 24.725 | 26.757 |
| **12** | 3.074 | 3.571 | 4.404 | 5.226 | 6.304 | 18.549 | 21.026 | 23.337 | 26.217 | 28.299 |
| **13** | 3.565 | 4.107 | 5.009 | 5.892 | 7.042 | 19.812 | 22.362 | 24.736 | 27.688 | 29.819 |
| **14** | 4.075 | 4.660 | 5.629 | 6.571 | 7.790 | 21.064 | 23.685 | 26.119 | 29.141 | 31.319 |
| **15** | 4.601 | 5.229 | 6.262 | 7.261 | 8.547 | 22.307 | 24.996 | 27.488 | 30.578 | 32.801 |
| **16** | 5.142 | 5.812 | 6.908 | 7.962 | 9.312 | 23.542 | 26.296 | 28.845 | 32.000 | 34.267 |
| **17** | 5.697 | 6.408 | 7.564 | 8.672 | 10.085 | 24.769 | 27.587 | 30.191 | 33.409 | 35.718 |
| **18** | 6.265 | 7.015 | 8.231 | 9.390 | 10.865 | 25.989 | 28.869 | 31.526 | 34.805 | 37.156 |
| **19** | 6.844 | 7.633 | 8.907 | 10.117 | 11.651 | 27.204 | 30.144 | 32.852 | 36.191 | 38.582 |
| **20** | 7.434 | 8.260 | 9.591 | 10.851 | 12.443 | 28.412 | 31.410 | 34.170 | 37.566 | 39.997 |
| **21** | 8.034 | 8.897 | 10.283 | 11.591 | 13.240 | 29.615 | 32.671 | 35.479 | 38.932 | 41.401 |
| **22** | 8.643 | 9.542 | 10.982 | 12.338 | 14.042 | 30.813 | 33.924 | 36.781 | 40.289 | 42.796 |
| **23** | 9.260 | 10.196 | 11.689 | 13.091 | 14.848 | 32.007 | 35.172 | 38.076 | 41.638 | 44.181 |
| **24** | 9.886 | 10.856 | 12.401 | 13.848 | 15.659 | 33.196 | 36.415 | 39.364 | 42.980 | 45.559 |
| **25** | 10.520 | 11.524 | 13.120 | 14.611 | 16.473 | 34.382 | 37.652 | 40.646 | 44.314 | 46.928 |
| **26** | 11.160 | 12.198 | 13.844 | 15.379 | 17.292 | 35.563 | 38.885 | 41.923 | 45.642 | 48.290 |
| **27** | 11.808 | 12.879 | 14.573 | 16.151 | 18.114 | 36.741 | 40.113 | 43.194 | 46.963 | 49.645 |
| **28** | 12.461 | 13.565 | 15.308 | 16.928 | 18.939 | 37.916 | 41.337 | 44.461 | 48.278 | 50.993 |
| **29** | 13.121 | 14.257 | 16.047 | 17.708 | 19.768 | 39.087 | 42.557 | 45.722 | 49.588 | 52.336 |
| **30** | 13.787 | 14.954 | 16.791 | 18.493 | 20.599 | 40.256 | 43.773 | 46.979 | 50.892 | 53.672 |
| **40** | 20.707 | 22.164 | 24.433 | 26.509 | 29.051 | 51.805 | 55.758 | 59.342 | 63.691 | 66.766 |
| **50** | 27.991 | 29.707 | 32.357 | 34.764 | 37.689 | 63.167 | 67.505 | 71.420 | 76.154 | 79.490 |
| **60** | 35.534 | 37.485 | 40.482 | 43.188 | 46.459 | 74.397 | 79.082 | 83.298 | 88.379 | 91.952 |
| **70** | 43.275 | 45.442 | 48.758 | 51.739 | 55.329 | 85.527 | 90.531 | 95.023 | 100.425 | 104.215 |
| **80** | 51.172 | 53.540 | 57.153 | 60.391 | 64.278 | 96.578 | 101.879 | 106.629 | 112.329 | 116.321 |
| **90** | 59.196 | 61.754 | 65.647 | 69.126 | 73.291 | 107.565 | 113.145 | 118.136 | 124.116 | 128.299 |
| **100** | 67.328 | 70.065 | 74.222 | 77.929 | 82.358 | 118.498 | 124.342 | 129.561 | 135.807 | 140.169 |

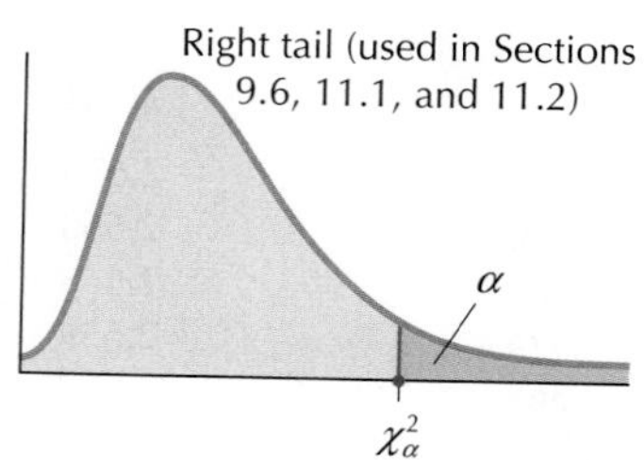


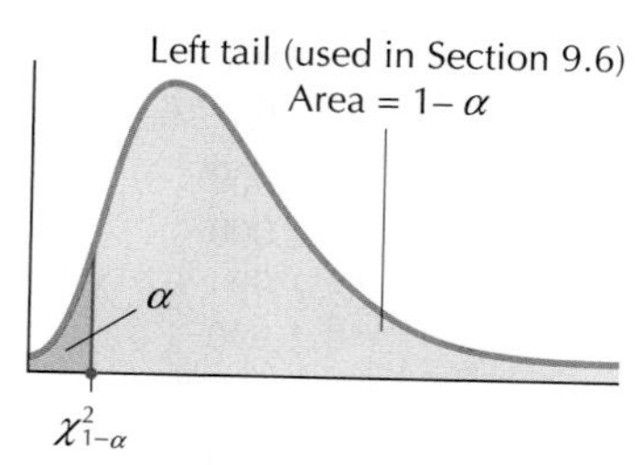


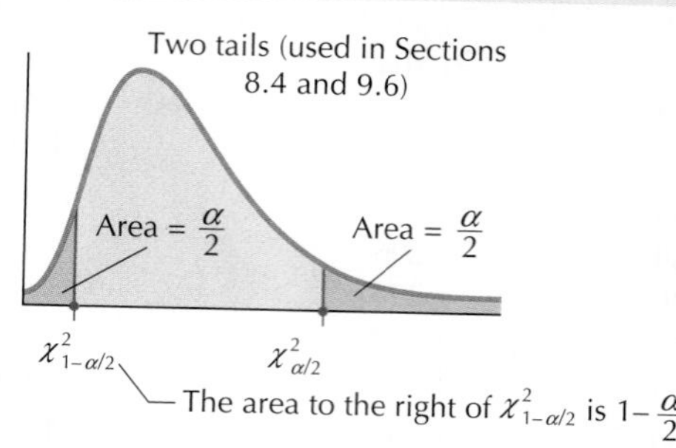

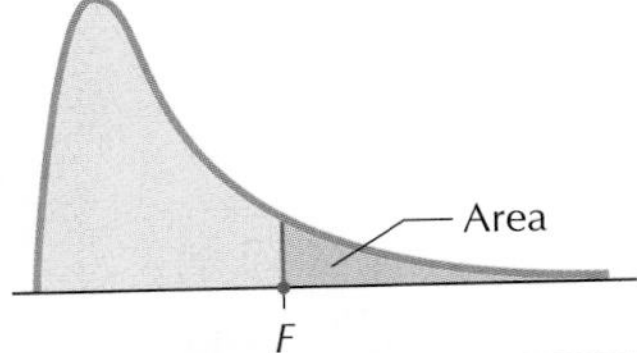


Table F *F*-Distribution critical values

| $df_2$ | Area in right tail | $df_1$ = 1 | 2 | 3 | 4 | 5 | 6 | 7 | 8 |
|---|---|---|---|---|---|---|---|---|---|
| 1 | 0.100 | 39.86 | 49.59 | 53.59 | 55.83 | 57.24 | 58.20 | 58.91 | 59.44 |
| | 0.050 | 161.45 | 199.50 | 215.71 | 224.58 | 230.16 | 233.99 | 236.77 | 238.88 |
| | 0.025 | 647.79 | 799.50 | 864.16 | 899.58 | 921.85 | 937.11 | 948.22 | 956.66 |
| | 0.010 | 4052.20 | 4999.50 | 5403.40 | 5624.60 | 5763.60 | 5859.00 | 5928.40 | 5981.10 |
| | 0.001 | 405284.00 | 500000.00 | 540379.00 | 562500.00 | 576405.00 | 585937.00 | 592873.00 | 598144.00 |
| 2 | 0.100 | 8.53 | 9.00 | 9.16 | 9.24 | 9.29 | 9.33 | 9.35 | 9.37 |
| | 0.050 | 18.51 | 19.00 | 19.16 | 19.25 | 19.30 | 19.33 | 19.35 | 19.37 |
| | 0.025 | 38.51 | 39.00 | 39.17 | 39.25 | 39.30 | 39.33 | 39.36 | 39.37 |
| | 0.010 | 98.50 | 99.00 | 99.17 | 99.25 | 99.30 | 99.33 | 99.36 | 99.37 |
| | 0.001 | 998.50 | 999.00 | 999.17 | 999.25 | 999.30 | 999.33 | 999.36 | 999.37 |
| 3 | 0.100 | 5.54 | 5.46 | 5.39 | 5.34 | 5.31 | 5.28 | 5.27 | 5.25 |
| | 0.050 | 10.13 | 9.55 | 9.28 | 9.12 | 9.01 | 8.94 | 8.89 | 8.85 |
| | 0.025 | 17.44 | 16.04 | 15.44 | 15.10 | 14.88 | 14.73 | 14.62 | 14.54 |
| | 0.010 | 34.12 | 30.82 | 29.46 | 28.71 | 28.24 | 27.91 | 27.67 | 27.49 |
| | 0.001 | 167.03 | 148.50 | 141.11 | 137.10 | 134.58 | 132.85 | 131.58 | 130.62 |
| 4 | 0.100 | 4.54 | 4.32 | 4.19 | 4.11 | 4.05 | 4.01 | 3.98 | 3.95 |
| | 0.050 | 7.71 | 6.94 | 6.59 | 6.39 | 6.26 | 6.16 | 6.09 | 6.04 |
| | 0.025 | 12.22 | 10.65 | 9.98 | 9.60 | 9.36 | 9.20 | 9.07 | 8.98 |
| | 0.010 | 21.20 | 18.00 | 16.69 | 15.98 | 15.52 | 15.21 | 14.98 | 14.80 |
| | 0.001 | 74.14 | 61.25 | 56.18 | 53.44 | 51.71 | 50.53 | 49.66 | 49.00 |
| 5 | 0.100 | 4.06 | 3.78 | 3.62 | 3.52 | 3.45 | 3.40 | 3.37 | 3.34 |
| | 0.050 | 6.61 | 5.79 | 5.41 | 5.19 | 5.05 | 4.95 | 4.88 | 4.82 |
| | 0.025 | 10.01 | 8.43 | 7.76 | 7.39 | 7.15 | 6.98 | 6.85 | 6.76 |
| | 0.010 | 16.26 | 13.27 | 12.06 | 11.39 | 10.97 | 10.67 | 10.46 | 10.29 |
| | 0.001 | 47.18 | 37.12 | 33.20 | 31.09 | 29.75 | 28.83 | 28.16 | 27.65 |
| 6 | 0.100 | 3.78 | 3.46 | 3.29 | 3.18 | 3.11 | 3.05 | 3.01 | 2.98 |
| | 0.050 | 5.99 | 5.14 | 4.76 | 4.53 | 4.39 | 4.28 | 4.21 | 4.15 |
| | 0.025 | 8.81 | 7.26 | 6.60 | 6.23 | 5.99 | 5.82 | 5.70 | 5.60 |
| | 0.010 | 13.75 | 10.92 | 9.78 | 9.15 | 8.75 | 8.47 | 8.26 | 8.10 |
| | 0.001 | 35.51 | 27.00 | 23.70 | 21.92 | 20.80 | 20.03 | 19.46 | 19.03 |
| 7 | 0.100 | 3.59 | 3.26 | 3.07 | 2.96 | 2.88 | 2.83 | 2.78 | 2.75 |
| | 0.050 | 5.59 | 4.74 | 4.35 | 4.12 | 3.97 | 3.87 | 3.79 | 3.73 |
| | 0.025 | 8.07 | 6.54 | 5.89 | 5.52 | 5.29 | 5.12 | 4.99 | 4.90 |
| | 0.010 | 12.25 | 9.55 | 8.45 | 7.85 | 7.46 | 7.19 | 6.99 | 6.84 |
| | 0.001 | 29.25 | 21.69 | 18.77 | 17.20 | 16.21 | 15.52 | 15.02 | 14.63 |
| 8 | 0.100 | 3.46 | 3.11 | 2.92 | 2.81 | 2.73 | 2.67 | 2.62 | 2.59 |
| | 0.050 | 5.32 | 4.46 | 4.07 | 3.84 | 3.69 | 3.58 | 3.50 | 3.44 |
| | 0.025 | 7.57 | 6.06 | 5.42 | 5.05 | 4.82 | 4.65 | 4.53 | 4.43 |
| | 0.010 | 11.26 | 8.65 | 7.59 | 7.01 | 6.63 | 6.37 | 6.18 | 6.03 |
| | 0.001 | 25.41 | 18.49 | 15.83 | 14.39 | 13.48 | 12.86 | 12.40 | 12.05 |

*(Continued)*

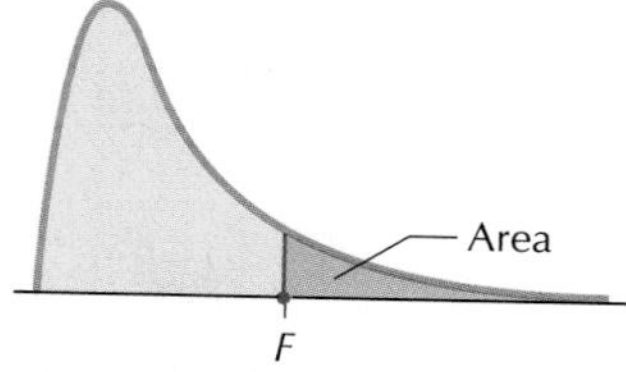


Table F *F*-Distribution critical values (*continued*)

| $df_2$ | Area in right tail | $df_1$ 9 | 10 | 15 | 20 | 30 | 60 | 120 | 1000 |
|---|---|---|---|---|---|---|---|---|---|
| 1 | 0.100 | 59.86 | 60.19 | 61.22 | 61.74 | 62.26 | 62.79 | 63.06 | 63.30 |
| | 0.050 | 240.54 | 241.88 | 245.95 | 248.01 | 250.10 | 252.20 | 253.25 | 254.19 |
| | 0.025 | 963.28 | 968.63 | 984.87 | 993.10 | 1001.4 | 1009.8 | 1014.0 | 1017.7 |
| | 0.010 | 6022.5 | 6055.8 | 6157.3 | 6208.7 | 6260.6 | 6313.0 | 6339.4 | 6362.7 |
| | 0.001 | 602284.0 | 605621.0 | 615764.0 | 620908.0 | 626099.0 | 631337.0 | 633972.0 | 636301.0 |
| 2 | 0.100 | 9.38 | 9.39 | 9.42 | 9.44 | 9.16 | 9.47 | 9.48 | 9.49 |
| | 0.050 | 19.38 | 19.40 | 19.43 | 19.45 | 19.46 | 19.48 | 19.49 | 19.49 |
| | 0.025 | 39.39 | 39.40 | 39.43 | 39.45 | 39.46 | 39.48 | 39.49 | 39.50 |
| | 0.010 | 99.39 | 99.40 | 99.43 | 99.45 | 99.47 | 99.48 | 99.49 | 99.50 |
| | 0.001 | 999.39 | 999.40 | 999.43 | 999.45 | 999.47 | 999.48 | 999.49 | 999.50 |
| 3 | 0.100 | 5.24 | 5.23 | 5.20 | 5.18 | 5.17 | 5.15 | 5.14 | 5.13 |
| | 0.050 | 8.81 | 8.79 | 8.70 | 8.66 | 8.62 | 8.57 | 8.55 | 8.53 |
| | 0.025 | 14.47 | 14.42 | 14.25 | 14.17 | 14.08 | 13.99 | 13.95 | 13.91 |
| | 0.010 | 27.35 | 27.23 | 26.87 | 26.69 | 26.50 | 26.32 | 26.22 | 26.14 |
| | 0.001 | 129.86 | 129.25 | 127.37 | 126.42 | 125.45 | 124.47 | 123.97 | 123.53 |
| 4 | 0.100 | 3.94 | 3.92 | 3.87 | 3.84 | 3.82 | 3.79 | 3.78 | 3.76 |
| | 0.050 | 6.00 | 5.96 | 5.86 | 5.80 | 5.75 | 5.69 | 5.66 | 5.63 |
| | 0.025 | 8.90 | 8.84 | 8.66 | 8.56 | 8.46 | 8.36 | 8.31 | 8.26 |
| | 0.010 | 14.66 | 14.55 | 14.20 | 14.02 | 13.84 | 13.65 | 13.56 | 13.47 |
| | 0.001 | 48.47 | 48.05 | 46.76 | 46.10 | 45.43 | 44.75 | 44.40 | 44.09 |
| 5 | 0.100 | 3.32 | 3.30 | 3.24 | 3.21 | 3.17 | 3.14 | 3.12 | 3.11 |
| | 0.050 | 4.77 | 4.74 | 4.62 | 4.56 | 4.50 | 4.43 | 4.40 | 4.37 |
| | 0.025 | 6.68 | 6.62 | 6.43 | 6.33 | 6.23 | 6.12 | 6.07 | 6.02 |
| | 0.010 | 10.16 | 10.05 | 9.72 | 9.55 | 9.38 | 9.20 | 9.11 | 9.03 |
| | 0.001 | 27.24 | 26.92 | 25.91 | 25.39 | 24.87 | 24.33 | 24.06 | 23.82 |
| 6 | 0.100 | 2.96 | 2.94 | 2.87 | 2.84 | 2.80 | 2.76 | 2.74 | 2.72 |
| | 0.050 | 4.10 | 4.06 | 3.94 | 3.87 | 3.81 | 3.74 | 3.70 | 3.67 |
| | 0.025 | 5.52 | 5.46 | 5.27 | 5.17 | 5.07 | 4.96 | 4.90 | 4.86 |
| | 0.010 | 7.98 | 7.87 | 7.56 | 7.40 | 7.23 | 7.06 | 6.97 | 6.89 |
| | 0.001 | 18.69 | 18.41 | 17.56 | 17.12 | 16.67 | 16.21 | 15.98 | 15.77 |
| 7 | 0.100 | 2.72 | 2.70 | 2.63 | 2.59 | 2.56 | 2.51 | 2.49 | 2.47 |
| | 0.050 | 3.68 | 3.64 | 3.51 | 3.44 | 3.38 | 3.30 | 3.27 | 3.23 |
| | 0.025 | 4.82 | 4.76 | 4.57 | 4.47 | 4.36 | 4.25 | 4.20 | 4.15 |
| | 0.010 | 6.72 | 6.62 | 6.31 | 6.16 | 5.99 | 5.82 | 5.74 | 5.66 |
| | 0.001 | 14.33 | 14.08 | 13.32 | 12.93 | 12.53 | 12.12 | 11.91 | 11.72 |
| 8 | 0.100 | 2.56 | 2.54 | 2.46 | 2.42 | 2.38 | 2.34 | 2.32 | 2.30 |
| | 0.050 | 3.39 | 3.35 | 3.22 | 3.15 | 3.08 | 3.01 | 2.97 | 2.93 |
| | 0.025 | 4.36 | 4.30 | 4.10 | 4.00 | 3.89 | 3.78 | 3.73 | 3.68 |
| | 0.010 | 5.91 | 5.81 | 5.52 | 5.36 | 5.20 | 5.03 | 4.95 | 4.87 |
| | 0.001 | 11.77 | 11.54 | 10.84 | 10.48 | 10.11 | 9.73 | 9.53 | 9.36 |

Table F *F*-Distribution critical values (*continued*)

| $df_2$ | Area in right tail | $df_1$ = 1 | 2 | 3 | 4 | 5 | 6 | 7 | 8 | 9 | 10 |
|---|---|---|---|---|---|---|---|---|---|---|---|
| 9 | 0.100 | 3.36 | 3.01 | 2.81 | 2.69 | 2.61 | 2.55 | 2.51 | 2.47 | 2.44 | 2.42 |
| | 0.050 | 5.12 | 4.26 | 3.86 | 3.63 | 3.48 | 3.37 | 3.29 | 3.23 | 3.18 | 3.14 |
| | 0.025 | 7.21 | 5.71 | 5.08 | 4.72 | 4.48 | 4.32 | 4.20 | 4.10 | 4.03 | 3.96 |
| | 0.010 | 10.56 | 8.02 | 6.99 | 6.42 | 6.06 | 5.80 | 5.61 | 5.47 | 5.35 | 5.26 |
| | 0.001 | 22.86 | 16.39 | 13.90 | 12.56 | 11.71 | 11.13 | 10.70 | 10.37 | 10.11 | 9.89 |
| 10 | 0.100 | 3.29 | 2.92 | 2.73 | 2.61 | 2.52 | 2.46 | 2.41 | 2.38 | 2.35 | 2.32 |
| | 0.050 | 4.96 | 4.10 | 3.71 | 3.48 | 3.33 | 3.22 | 3.14 | 3.07 | 3.02 | 2.98 |
| | 0.025 | 6.94 | 5.46 | 4.83 | 4.47 | 4.24 | 4.07 | 3.95 | 3.85 | 3.78 | 3.72 |
| | 0.010 | 10.04 | 7.56 | 6.55 | 5.99 | 5.64 | 5.39 | 5.20 | 5.06 | 4.94 | 4.85 |
| | 0.001 | 21.04 | 14.91 | 12.55 | 11.28 | 10.48 | 9.93 | 9.52 | 9.20 | 8.96 | 8.75 |
| 12 | 0.100 | 3.18 | 2.81 | 2.61 | 2.48 | 2.39 | 2.33 | 2.28 | 2.24 | 2.21 | 2.19 |
| | 0.050 | 4.75 | 3.89 | 3.49 | 3.26 | 3.11 | 3.00 | 2.91 | 2.85 | 2.80 | 2.75 |
| | 0.025 | 6.55 | 5.10 | 4.47 | 4.12 | 3.89 | 3.73 | 3.61 | 3.51 | 3.44 | 3.37 |
| | 0.010 | 9.33 | 6.93 | 5.95 | 5.41 | 5.06 | 4.82 | 4.64 | 4.50 | 4.39 | 4.30 |
| | 0.001 | 18.64 | 12.97 | 10.80 | 9.63 | 8.89 | 8.38 | 8.00 | 7.71 | 7.48 | 7.29 |
| 15 | 0.100 | 3.07 | 2.70 | 2.49 | 2.36 | 2.27 | 2.21 | 2.16 | 2.12 | 2.09 | 2.06 |
| | 0.050 | 4.54 | 3.68 | 3.29 | 3.06 | 2.90 | 2.79 | 2.71 | 2.64 | 2.59 | 2.54 |
| | 0.025 | 6.20 | 4.77 | 4.15 | 3.80 | 3.58 | 3.41 | 3.29 | 3.20 | 3.12 | 3.06 |
| | 0.010 | 8.68 | 6.36 | 5.42 | 4.89 | 4.56 | 4.32 | 4.14 | 4.00 | 3.89 | 3.80 |
| | 0.001 | 16.59 | 11.34 | 9.34 | 8.25 | 7.57 | 7.09 | 6.74 | 6.47 | 6.26 | 6.08 |
| 20 | 0.100 | 2.97 | 2.59 | 2.38 | 2.25 | 2.16 | 2.09 | 2.04 | 2.00 | 1.96 | 1.94 |
| | 0.050 | 4.35 | 3.49 | 3.10 | 2.87 | 2.71 | 2.60 | 2.51 | 2.45 | 2.39 | 2.35 |
| | 0.025 | 5.87 | 4.46 | 3.86 | 3.51 | 3.29 | 3.13 | 3.01 | 2.91 | 2.84 | 2.77 |
| | 0.010 | 8.10 | 5.85 | 4.94 | 4.43 | 4.10 | 3.87 | 3.70 | 3.56 | 3.46 | 3.37 |
| | 0.001 | 14.82 | 9.95 | 8.10 | 7.10 | 6.46 | 6.02 | 5.69 | 5.44 | 5.24 | 5.08 |
| 25 | 0.100 | 2.92 | 2.53 | 2.32 | 2.18 | 2.09 | 2.02 | 1.97 | 1.93 | 1.89 | 1.87 |
| | 0.050 | 4.24 | 3.39 | 2.99 | 2.76 | 2.60 | 2.49 | 2.40 | 2.34 | 2.28 | 2.24 |
| | 0.025 | 5.69 | 4.29 | 3.69 | 3.35 | 3.13 | 2.97 | 2.85 | 2.75 | 2.68 | 2.61 |
| | 0.010 | 7.77 | 5.57 | 4.68 | 4.18 | 3.85 | 3.63 | 3.46 | 3.32 | 3.22 | 3.13 |
| | 0.001 | 13.88 | 9.22 | 7.45 | 6.49 | 5.89 | 5.46 | 5.15 | 4.91 | 4.71 | 4.56 |
| 50 | 0.100 | 2.81 | 2.41 | 2.20 | 2.06 | 1.97 | 1.90 | 1.84 | 1.80 | 1.76 | 1.73 |
| | 0.050 | 4.03 | 3.18 | 2.79 | 2.56 | 2.40 | 2.29 | 2.20 | 2.13 | 2.07 | 2.03 |
| | 0.025 | 5.34 | 3.97 | 3.39 | 3.05 | 2.83 | 2.67 | 2.55 | 2.46 | 2.38 | 2.32 |
| | 0.010 | 7.17 | 5.06 | 4.20 | 3.72 | 3.41 | 3.19 | 3.02 | 2.89 | 2.78 | 2.70 |
| | 0.001 | 12.22 | 7.96 | 6.34 | 5.46 | 4.90 | 4.51 | 4.22 | 4.00 | 3.82 | 3.67 |
| 100 | 0.100 | 2.76 | 2.36 | 2.14 | 2.00 | 1.91 | 1.83 | 1.78 | 1.73 | 1.69 | 1.66 |
| | 0.050 | 3.94 | 3.09 | 2.70 | 2.46 | 2.31 | 2.19 | 2.10 | 2.03 | 1.97 | 1.93 |
| | 0.025 | 5.18 | 3.83 | 3.25 | 2.92 | 2.70 | 2.54 | 2.42 | 2.32 | 2.24 | 2.18 |
| | 0.010 | 6.90 | 4.82 | 3.98 | 3.51 | 3.21 | 2.99 | 2.82 | 2.69 | 2.59 | 2.50 |
| | 0.001 | 11.50 | 7.41 | 5.86 | 5.02 | 4.48 | 4.11 | 3.83 | 3.61 | 3.44 | 3.30 |
| 200 | 0.100 | 2.73 | 2.33 | 2.11 | 1.97 | 1.88 | 1.80 | 1.75 | 1.70 | 1.66 | 1.63 |
| | 0.050 | 3.89 | 3.04 | 2.65 | 2.42 | 2.26 | 2.14 | 2.06 | 1.98 | 1.93 | 1.88 |
| | 0.025 | 5.10 | 3.76 | 3.18 | 2.85 | 2.63 | 2.47 | 2.35 | 2.26 | 2.18 | 2.11 |
| | 0.010 | 6.76 | 4.71 | 3.88 | 3.41 | 3.11 | 2.89 | 2.73 | 2.60 | 2.50 | 2.41 |
| | 0.001 | 11.15 | 7.15 | 5.63 | 4.81 | 4.29 | 3.92 | 3.65 | 3.43 | 3.26 | 3.12 |
| 1000 | 0.100 | 2.71 | 2.31 | 2.09 | 1.95 | 1.85 | 1.78 | 1.72 | 1.68 | 1.64 | 1.61 |
| | 0.050 | 3.85 | 3.00 | 2.61 | 2.38 | 2.22 | 2.11 | 2.02 | 1.95 | 1.89 | 1.84 |
| | 0.025 | 5.04 | 3.70 | 3.13 | 2.80 | 2.58 | 2.42 | 2.30 | 2.20 | 2.13 | 2.06 |
| | 0.010 | 6.66 | 4.63 | 3.80 | 3.34 | 3.04 | 2.82 | 2.66 | 2.53 | 2.43 | 2.34 |
| | 0.001 | 10.89 | 6.96 | 5.46 | 4.65 | 4.14 | 3.78 | 3.51 | 3.30 | 3.13 | 2.99 |

(*Continued*)

Table F *F*-Distribution critical values (*continued*)

| $df_2$ | Area in right tail | $df_1$ 12 | 15 | 20 | 25 | 30 | 40 | 50 | 60 | 120 | 1000 |
|---|---|---|---|---|---|---|---|---|---|---|---|
| 9 | 0.100 | 2.38 | 2.34 | 2.30 | 2.27 | 2.25 | 2.23 | 2.22 | 2.21 | 2.18 | 2.16 |
| | 0.050 | 3.07 | 3.01 | 2.94 | 2.89 | 2.86 | 2.83 | 2.80 | 2.79 | 2.75 | 2.71 |
| | 0.025 | 3.87 | 3.77 | 3.67 | 3.60 | 3.56 | 3.51 | 3.47 | 3.45 | 3.39 | 3.34 |
| | 0.010 | 5.11 | 4.96 | 4.81 | 4.71 | 4.65 | 4.57 | 4.52 | 4.48 | 4.40 | 4.32 |
| | 0.001 | 9.57 | 9.24 | 8.90 | 8.69 | 8.55 | 8.37 | 8.26 | 8.19 | 8.00 | 7.84 |
| 10 | 0.100 | 2.28 | 2.24 | 2.20 | 2.17 | 2.16 | 2.13 | 2.12 | 2.11 | 2.08 | 2.06 |
| | 0.050 | 2.91 | 2.85 | 2.77 | 2.73 | 2.70 | 2.66 | 2.64 | 2.62 | 2.58 | 2.54 |
| | 0.025 | 3.62 | 3.52 | 3.42 | 3.35 | 3.31 | 3.26 | 3.22 | 3.20 | 3.14 | 3.09 |
| | 0.010 | 4.71 | 4.56 | 4.41 | 4.31 | 4.25 | 4.17 | 4.12 | 4.08 | 4.00 | 3.92 |
| | 0.001 | 8.45 | 8.13 | 7.80 | 7.60 | 7.47 | 7.30 | 7.19 | 7.12 | 6.94 | 6.78 |
| 12 | 0.100 | 2.15 | 2.10 | 2.06 | 2.03 | 2.01 | 1.99 | 1.97 | 1.96 | 1.93 | 1.91 |
| | 0.050 | 2.69 | 2.62 | 2.54 | 2.50 | 2.47 | 2.43 | 2.40 | 2.38 | 2.34 | 2.30 |
| | 0.025 | 3.28 | 3.18 | 3.07 | 3.01 | 2.96 | 2.91 | 2.87 | 2.85 | 2.79 | 2.73 |
| | 0.010 | 4.16 | 4.01 | 3.86 | 3.76 | 3.70 | 3.62 | 3.57 | 3.54 | 3.45 | 3.37 |
| | 0.001 | 7.00 | 6.71 | 6.40 | 6.22 | 6.09 | 5.93 | 5.83 | 5.76 | 5.59 | 5.44 |
| 15 | 0.100 | 2.02 | 1.97 | 1.92 | 1.89 | 1.87 | 1.85 | 1.83 | 1.82 | 1.79 | 1.76 |
| | 0.050 | 2.48 | 2.40 | 2.33 | 2.28 | 2.25 | 2.20 | 2.18 | 2.16 | 2.11 | 2.07 |
| | 0.025 | 2.96 | 2.86 | 2.76 | 2.69 | 2.64 | 2.59 | 2.55 | 2.52 | 2.46 | 2.40 |
| | 0.010 | 3.67 | 3.52 | 3.37 | 3.28 | 3.21 | 3.13 | 3.08 | 3.05 | 2.96 | 2.88 |
| | 0.001 | 5.81 | 5.54 | 5.25 | 5.07 | 4.95 | 4.80 | 4.70 | 4.64 | 4.47 | 4.33 |
| 20 | 0.100 | 1.89 | 1.84 | 1.79 | 1.76 | 1.74 | 1.71 | 1.69 | 1.68 | 1.64 | 1.61 |
| | 0.050 | 2.28 | 2.20 | 2.12 | 2.07 | 2.04 | 1.99 | 1.97 | 1.95 | 1.90 | 1.85 |
| | 0.025 | 2.68 | 2.57 | 2.46 | 2.40 | 2.35 | 2.29 | 2.25 | 2.22 | 2.16 | 2.09 |
| | 0.010 | 3.23 | 3.09 | 2.94 | 2.84 | 2.78 | 2.69 | 2.64 | 2.61 | 2.52 | 2.43 |
| | 0.001 | 4.82 | 4.56 | 4.29 | 4.12 | 4.00 | 3.86 | 3.77 | 3.70 | 3.54 | 3.40 |
| 25 | 0.100 | 1.82 | 1.77 | 1.72 | 1.68 | 1.66 | 1.63 | 1.61 | 1.59 | 1.56 | 1.52 |
| | 0.050 | 2.16 | 2.09 | 2.01 | 1.96 | 1.92 | 1.87 | 1.84 | 1.82 | 1.77 | 1.72 |
| | 0.025 | 2.51 | 2.41 | 2.30 | 2.23 | 2.18 | 2.12 | 2.08 | 2.05 | 1.98 | 1.91 |
| | 0.010 | 2.99 | 2.85 | 2.70 | 2.60 | 2.54 | 2.45 | 2.40 | 2.36 | 2.27 | 2.18 |
| | 0.001 | 4.31 | 4.06 | 3.79 | 3.63 | 3.52 | 3.37 | 3.28 | 3.22 | 3.06 | 2.91 |
| 50 | 0.100 | 1.68 | 1.63 | 1.57 | 1.53 | 1.50 | 1.46 | 1.44 | 1.42 | 1.38 | 1.33 |
| | 0.050 | 1.95 | 1.87 | 1.78 | 1.73 | 1.69 | 1.63 | 1.60 | 1.58 | 1.51 | 1.45 |
| | 0.025 | 2.22 | 2.11 | 1.99 | 1.92 | 1.87 | 1.80 | 1.75 | 1.72 | 1.64 | 1.56 |
| | 0.010 | 2.56 | 2.42 | 2.27 | 2.17 | 2.10 | 2.01 | 1.95 | 1.91 | 1.80 | 1.70 |
| | 0.001 | 3.44 | 3.20 | 2.95 | 2.79 | 2.68 | 2.53 | 2.44 | 2.38 | 2.21 | 2.05 |
| 100 | 0.100 | 1.61 | 1.56 | 1.49 | 1.45 | 1.42 | 1.38 | 1.35 | 1.34 | 1.28 | 1.22 |
| | 0.050 | 1.85 | 1.77 | 1.68 | 1.62 | 1.57 | 1.52 | 1.48 | 1.45 | 1.38 | 1.30 |
| | 0.025 | 2.08 | 1.97 | 1.85 | 1.77 | 1.71 | 1.64 | 1.59 | 1.56 | 1.46 | 1.36 |
| | 0.010 | 2.37 | 2.22 | 2.07 | 1.97 | 1.89 | 1.80 | 1.74 | 1.69 | 1.57 | 1.45 |
| | 0.001 | 3.07 | 2.84 | 2.59 | 2.43 | 2.32 | 2.17 | 2.08 | 2.01 | 1.83 | 1.64 |
| 200 | 0.100 | 1.58 | 1.52 | 1.46 | 1.41 | 1.38 | 1.34 | 1.31 | 1.29 | 1.23 | 1.16 |
| | 0.050 | 1.80 | 1.72 | 1.62 | 1.56 | 1.52 | 1.46 | 1.41 | 1.39 | 1.30 | 1.21 |
| | 0.025 | 2.01 | 1.90 | 1.78 | 1.70 | 1.64 | 1.56 | 1.51 | 1.47 | 1.37 | 1.25 |
| | 0.010 | 2.27 | 2.13 | 1.97 | 1.87 | 1.79 | 1.69 | 1.63 | 1.58 | 1.45 | 1.30 |
| | 0.001 | 2.90 | 2.67 | 2.42 | 2.26 | 2.15 | 2.00 | 1.90 | 1.83 | 1.64 | 1.43 |
| 1000 | 0.100 | 1.55 | 1.49 | 1.43 | 1.38 | 1.35 | 1.30 | 1.27 | 1.25 | 1.38 | 1.08 |
| | 0.050 | 1.76 | 1.68 | 1.58 | 1.52 | 1.47 | 1.41 | 1.36 | 1.31 | 1.24 | 1.11 |
| | 0.025 | 1.96 | 1.85 | 1.72 | 1.64 | 1.58 | 1.50 | 1.45 | 1.41 | 1.29 | 1.13 |
| | 0.010 | 2.20 | 2.06 | 1.90 | 1.79 | 1.72 | 1.61 | 1.54 | 1.50 | 1.35 | 1.16 |
| | 0.001 | 2.77 | 2.54 | 2.30 | 2.14 | 2.02 | 1.87 | 1.77 | 1.69 | 1.49 | 1.22 |

# Notes and Data Sources

## Chapter 1

**1.** As reported at **www.cnn.com/2001/ALLPOLITICS/03/11/palmbeach.recount/**.

**2.** T. J. Scanlon, R. N. Luben, F. L. Scanlon, and N. Singleton, "Is Friday the 13th bad for your health?" *British Medical Journal* 307 (December 1993).

**3.** Pew Internet and American Life Project, "Cyberbullying and online teens," June 2007, **www.pewinternet.org**.

**4.** U.S. Census Bureau, *The Population Profile of the United States: 2000*, **www.consensus.gov/population/www/pop-profile/profile2000**.

**5.** National Agricultural Statistics Service.

**6.** Iain McGregor and Wayne Hall, "MDMA (Ecstasy) neurotoxicity: assessing and communicating the risks," *Lancet* 355 (9217, May 20, 2000): 1818–21.

**7.** U.S. Department of Health and Human Services, *The Health Consequences of Involuntary Exposure to Tobacco Smoke: A Report of the Surgeon General—Executive Summary*, U.S. Department of Health and Human Services, Centers for Disease Control and Prevention, Coordinating Center for Health Promotion, National Center for Chronic Disease Prevention and Health Promotion, Office on Smoking and Health, 2006.

**8.** Michel de Lorgeril, Patricia Salen, Jean-Louis Martin, Isabelle Monjaud, Jacques Delaye, and Nicole Mamelle, "Mediterranean diet, traditional risk factors, and the rate of cardiovascular complications after myocardial infarction, final report of the Lyon Diet Heart Study," *Circulation: Journal of the American Heart Association* 99 (1999): 779–85. The American Heart Association (**www.americanheart.org**) identifies the following characteristics as common to most Mediterranean diets. There is a "high consumption of fruits, vegetables, bread and other cereals, potatoes, beans, nuts and seeds. Olive oil is an important monounsaturated fat source. Dairy products, fish and poultry are consumed in low to moderate amounts, and little red meat is eaten."

**9.** R. L. Bratton et al., "Effect of 'ionized' wrist bracelets on musculoskeletal pain: a randomized, double-blind, placebo-controlled trial," *Mayo Clinic Proceedings* 77 (2002):1164–68.

## Chapter 2

**1.** National Health Interview Survey, U.S. Centers for Disease Control and Prevention, 2005.

**2.** M. A. Chase and G. M. Dummer, "The role of sports as a social determinant for children," *Research Quarterly for Exercise and Sport* 63 (1992): 418–24.

**3.** U.S. Bureau of Labor Statistics.

**4.** U.S. Energy Information Administration.

**5.** U.S. Centers for Disease Control and Prevention.

**6.** See Note 2.

## Chapter 3

**1.** U.S. Centers for Disease Control and Prevention, U.S. Department of Health and Human Services, **www.cdc.gov/flu/avian/index.htm**.

**2.** For more on clickstream analysis, see Zdravko Markov and Daniel Larose, *Data Mining the Web: Uncovering Patterns in Web Content, Structure, and Usage* (John Wiley and Sons, 2007).

**3.** U.S. Census Bureau.

**4.** P. A. Mackowiak, S. S. Wasserman, and M. M. Levine, "A critical appraisal of 98.6 degrees F, the upper limit of the normal body temperature, and other legacies of Carl Reinhold August Wunderlich," *Journal of the American Medical Association* 268 (1992): 1578–80.

**5.** See Note 4.

**6.** Michael Brett and Charles Goldman, "A meta-analysis of the freshwater trophic cascade," *Proceedings of the National Academy of Sciences* 93 (July 1996).

**7.** Dr. Peter Nonacs, "Foraging habits of thatch ants," Department of Statistics, University of California at Los Angeles and the Sierra Nevada Aquatic Research Laboratory, **www.stat.ucla.edu/datasets/**.

**8.** Children's Bureau, Administration for Children and Families, U.S. Department of Health and Human Services.

**9.** National Water and Climate Center, U.S. Department of Agriculture.

**10.** See Note 7.

**11.** B. S. Glenn et al., "Changes in systolic blood pressure associated with lead in blood and bone," *Epidemiology* 17 (September 2006).

**12.** National Center for Education Statistics, 2005.

**13.** M. Donald Thomas and William L. Bainbridge, "Grade inflation: the current fraud," *Effective School Research*, January 1997.

**14.** National Center for Health Statistics, *Health*, United States, 2006.

## Chapter 4

**1.** Roper Center, University of Connecticut.

**2.** T. Allison and D. V. Cicchetti, "Sleep in mammals: ecological and constitutional correlates," *Science*, 194 (1976): 732–34. Web reference: Quantitative Environmental Learning Project, Joseph Hull and Greg Langkamp, Seattle Central Community College, **www.seattlecentral.org/qelp/sets/017/017.html**.

**3.** United Nations Educational, Scientific, and Cultural Organization (UNESCO).

**4.** *Crime in the United States, 2004*, **www.fbi.gov**.

## Chapter 5

**1.** Amanda Lenhart et al., *Writing, Technology, and Teens*, Pew Internet and American Life Project, December 2007.

**2.** U.S. Census Bureau, 2004 American Community Survey.

**3.** Andrew Rocco Tresolini Fiore, "Romantic regressions: an analysis of behavior in online dating systems," Master's thesis, Massachusetts Institute of Technology, 2004.

**4.** *Percentage Baseball* (MIT Press, 1966).

**5.** Washington Initiative (**greaterwashington.org**).

**6.** Bureau of Labor Statistics.

**7.** *Profile of Hired Farmworkers, A 2008 Update/ERR-60*, Economic Research Service/USDA.

## Chapter 6

**1.** U.S. National Center for Education Statistics. The category "5 or more" has been changed to "5" for this exercise.

**2.** U.S. Census Bureau. The category "9 or more" has been changed to "9" for this exercise.

**3.** U.S. Census Bureau. The category "3 or more" has been changed to "3" for this exercise.

**4.** Gunter Hitsch, Ali Hortacsu, and Dan Ariely, "What makes you click: an empirical analysis of online dating;" available online at **www.aeaweb.org/annual_mtg_papers/2006/0106_0800_0502.pdf**.

**5.** D. L. Olds, C. R. Henderson Jr., R. Tatelbaum, et al., "Improving the delivery of prenatal care and outcomes of pregnancy: a randomized trial of nurse home visitation," *Pediatrics* 77 (1986): 16–28.

**6.** M. Donald Thomas and William L. Bainbridge, "Grade inflation: the current fraud," *Effective School Research*, January 1997.

**7.** Lynn Unruh and Myron Fottler, "Patient turnover and nursing staff adequacy," *Health Services Research*, April 2006.

**8.** See Note 6.

**9.** Harvard School of Public Health, survey of 5046 adults in hurricane high-risk areas, June–July 2007.

**10.** The Associated Press/Ipsos Poll actually contacted 1000 adults in June 2007.

**11.** L. D. Johnston, P. M. O'Malley, J. G. Bachman, and J. E. Schulenberg, *Monitoring the Future: National Results on Adolescent Drug Use; Overview of Key Findings, 2006*, NIH Publication no. 07-6202 (Bethesda, MD: National Institute on Drug Abuse, 2007).

**12.** The National Survey on Environmental Management of Asthma and Children's Exposure to Environmental Tobacco Smoke (NSEMA/CEE), U.S. Environmental Protection Agency, 2004.

**13.** Barry Kosmin, Egon Mayer, and Ariela Keysar, Graduate Center of the City University of New York, 2001.

**14.** Barbara Alving et al., "Trends in blood pressure among children and adolescents," *Journal of the American Medical Association* 291 (May 2004): 2107–13.

**15.** Phillida Bunkle and John Lepper, "Women's participation in gambling: whose reality? A public health issue," paper presented to the European Association for the Study of Gambling Conference, Barcelona, Spain, October 2002.

## Chapter 7

**1.** "The magical number seven, plus or minus two: some limits on our capacity for processing information," *Psychological Review* 63 (1956), 81–97.

**2.** A small business is defined by the SBA as having fewer than 20 employees.

**3.** The notation "±" always indicates two numbers. For example, $\mu_{\hat{p}} \pm \sigma_{\hat{p}}$ indicates $\mu_{\hat{p}} + \sigma_{\hat{p}}$ and $\mu_{\hat{p}} - \sigma_{\hat{p}}$.

**4.** Murray Mittleman et al., "Determinants of myocardial onset study," *Circulation: Journal of the American Heart Association*, June 1999.

## Chapter 8

**1.** Adapted from A. Johnson, "Results from analyzing metals in 1999 Spokane River fish and crayfish samples," Quantitative Environmental Learning Project, Washington State Department of Ecology report 00-03-017, **www.seattlecentral.edu/qelp/sets/021/021.html**.

**2.** Kevin Crowley et al., "Parents explain more often to boys than girls during shared scientific thinking," *Psychological Science* 12 (3, May 2001): 258–61.

**3.** George Miller, "The magical number seven, plus or minus two: some limits on our capacity for processing information," *Psychological Review* 63 (1956): 81–97.

**4.** Mary H. Ward et al., "Proximity to crops and residential exposure to agricultural herbicides in Iowa," *Environmental Health Perspectives* 114 (6, June 2006): 893–97.

**5.** Irene Yen et al., "Perceived neighborhood problems and quality of life, physical functioning, and depressive symptoms among adults with asthma," *American Journal of Public Health*, 96 (5, May 2006): 873–79.

**6.** Gary Bennett et al., "Television viewing and pedometer-determined physical activity among multiethnic residents of low-income housing," *American Journal of Public Health* 96 (9, September 2006), 1681–85.

**7.** Mary C. Meyer, "Wider shoes for wider feet?" *Journal of Statistics Education* 14 (1, 2006), **www.amstat.org/publications/jse/v14n1/datasets.meyer.html**.

**8.** Clive Lipchin, "Report from the field: perceptions of water use in the Arava Valley of Israel and Jordan," *Journal of the International Institute, University of Michigan* (Summer 2004). A cubic meter of water is equivalent to 264 gallons.

**9.** Fact 2 in Section 7.1 stated that the standard deviation of the sampling distribution of the sample mean $\bar{x}$ is $\sigma_{\bar{x}} = \sigma/\sqrt{n}$. This quantity was used in Section 8.1 to calculate the $Z$ interval for the population mean when $\sigma$ is known. However, here in Section 8.2, we no longer assume that $\sigma$ is known. When $\sigma$ is not known, we use the following quantity to estimate the standard deviation of the sampling distribution of the sample mean: $s_{\bar{x}} = s/\sqrt{n}$. The term $s_{\bar{x}}$ is called the *standard error of the sample mean*. (Some textbooks use the term *standard error* to refer to the true value of the standard deviation of a statistic, for example, $\sigma/\sqrt{n}$ for the sample mean. In that case, the value of the standard error is unknown and not observed. However, the use of the term *standard error* as an observed value, deriving from a sample, is widespread, both in the methodological literature and in statistical software.)

**10.** Robert J. Pianta et al., "Teaching: opportunities to learn in America's elementary classroom," *Science* 315 (March 30, 2007): 1795–96.

**11.** Community College Survey of Student Engagement (CCSSE), 2007, **www.ccsse.org**. The survey reported that 178 of 307 (57.98045603%) students worked with classmates outside class to prepare a group assignment during the current academic year. The sample results in Example 8.16 (174 of 300, or 58%) were chosen for ease of calculation.

**12.** Christopher Reynolds, "Prey tell," *American Demographics* 25 (8, October 2003): 48.

**13.** "Survey: most support a diverse college campus, but not admissions by race alone," *Black Issues in Higher Education,* 20 (10, July, 2003): 9.

**14.** *MDMA (Ecstacy) Abuse*, National Institute on Drug Abuse Research Report Series, U.S. Department of Health and Human Services, February 2006.

**15.** Mildred Cho and Lisa Bero, "The quality of drug studies published in symposium proceedings," *Annals of Internal Medicine*, 124 (5, March 1996): 485–89.

**16.** See Note 7.

**17.** Manchester (CT) *Journal-Inquirer*, June 9, 2008, p. 1.

**18.** See Note 2.

**19.** Karen Avenoso, "Erasing the passed," *Boston Globe*, July 14, 1996. (Thanks to *Chance News* **www.dartmouth.edu/~chance/**).

**20.** See Note 3.

**21.** See Note 12.

## Chapter 9

**1.** Y. Kvach and C. A. Stepien, "The invasive round goby *Apollonia melanostoma* (Actinopterygii: Gobiidae)—a new intermediate host of the trematode *Neochasmus umbellus* (Trematoda: Cryptogonimidae) in Lake Erie, Ohio, USA," *Journal of Applied Ichthyology*, 24 (1, February 2008), 103–05.

**2.** NASA, "Climate data shows California has been heating up," *Science Daily* 31 (March 2007). **www.sciencedaily.com/releases/2007/03/070330221144.htm**. (Accessed June 5, 2008.)

**3.** Press release, August 23, 2007: "Consumers report eating at home more in the wake of high gas prices," NPD Group, Inc., 900 West Shore Road, Port Washington, NY 11050.

**4.** "When it comes to height, Americans no longer stand tallest," *Research News*, The Ohio State University, **researchnews.osu.edu/**.

**5.** "Genomics of cardiovascular development, adaptation, and remodeling," NHLBI Program for Genomic Applications, Harvard Medical School, **www.cardiogenomics.org**. (Accessed April 2007.)

**6.** K. Marien, A. Conseur, and M. Sanderson, "The effect of fish consumption on DDT and DDE levels in breast milk among Hispanic immigrants," *Journal of Human Lactation* 14 (3, 1998): 237–42.

**7.** C. J. Earle, L. B. Brubaker, and G. Segura, International Tree Ring Data Base, NOAA/NGDC Paleoclimatology Program, Boulder, CO.

**8.** See Note 2.

**9.** "Online shoppers will pay extra to protect privacy, Carnegie Mellon study shows," *ScienceDaily*, **www.sciencedaily.com/releases/2007/06/070606235316.htm**. (Accessed January 20, 2008).

**10.** *Digital Transactions News*, September, 2007.

**11.** G. M. Darling, J. A. Johns, P. I. McCloud, and S. R. Davis, "Estrogen and progestin compared with simvastatin for hypercholesterolemia in postmenopausal women," *New England Journal of Medicine* 337 (9, 1997), 595–601.

**12.** **health.usnews.com/sections/health/west-hospitals**.

**13.** Caroline Davis, Elizabeth Blackmore, Deborah Katzman, and John Fox, "Anorexia nervosa case study," paper presented at Statistical Society of Canada Annual Conference, Montreal, 2004. We have reversed the research question from that of the original case study.

**14.** Courtesy American Heritage Center, University of Wyoming.

**15.** Data courtesy of OzDASL (Australian Data and Story Library) at **statsci.org**. The original source is Cara Dubois, ed., *Lowie's Selected Papers in Anthropology* (University of California Press, 1960).

**16.** Mary Madden and Amanda Lenhart "Online dating," Pew Internet and American Life Project, 2006.

**17.** Barry Kosmin and Egon Mayer, "Principal investigators," American Religious Identification Survey, Graduate Center, City University of New York.

**18.** Brady Hamilton, Joyce Martin, and Stephanie Ventura, "Births: preliminary data for 2005," *National Vital Statistics Reports* 55 (11), U.S. Department of Health and Human Services.

**19.** "Patterns and trends in nonmedical prescription pain reliever use: 2002 to 2005," in *NSDUH Report*, Substance Abuse and Mental Health Services Administration, April 6, 2007.

**20.** Jeff Humphries, "The multicultural economy: minority buying power in the new century," Selig Center for Economic Growth, Terry College of Business, University of Georgia, 2006.

**21.** "Trends in the prevalence of alcohol use among eighth graders: Monitoring the Future Study, 1991–2003," NIAAA, National Institutes of Health.

**22.** "Fact sheet: National Survey on Environmental Management of Asthma and Children's Exposure to Environmental Tobacco Smoke," U.S. Environmental Protection Agency, May 17, 2005.

**23.** Energy Information Administration, "Annual electric generator report," Form EIA-906.

**24.** U.S. Census Bureau.

**25.** See Note 6.

**26.** See Note 7.

**27.** U.S. Bureau of Labor Statistics.

**28.** Mary C. Meyer, "Wider shoes for wider feet?" *Journal of Statistics Education*, 14 (1, 2006).

**29.** Steve Strand, Ian Deary, and Pauline Smith, "Sex differences in cognitive abilities test scores: a UK national picture," *British Journal of Educational Psychology* 76 (2006): 463–80.

**30.** Siobhan Banks and David Dinges, "Behavioral and physiological consequences of sleep restriction," *Journal of Clinical Sleep Medicine*, 15 (2007): 519–28.

**31.** See the Infoplease Web site: **www.infoplease.com/ipa/A0934556.html**.

**32.** "A nation online: entering the broadband age," Economics and Statistics Administration, U.S. Department of Commerce.

**33.** U.S. Bureau of Justice Statistics.

**34.** Joyce A. Martin et al., "Births: final data for 2005," *National Vital Statistics Reports*, 56 (6, December 5, 2007).

## Chapter 10

**1.** Kelley, H. H., "The warm-cold variable in first impression of persons," *Journal of Personality* 18 (1950): 431–39.

**2.** A. Towler and R. L. Dipboye, "The effect of instructor reputation and need for cognition on student behavior," poster presented at American Psychological Society conference, May 1998.

**3.** K. J. Thomas et al., "Randomized controlled trial of a short course of traditional acupuncture compared with usual care for persistent nonspecific low back pain," *British Medical Journal* 23 (September 2006).

**4.** Karin Olson and John Hanson, "Using reiki to manage pain," *Cancer Prevention and Control* 1 (2, 1997): 108–13.

**5.** "Highway safety projects—before and after study update," *Measures, Markers, and Mileposts*, Washington State Department of Transportation, December 2005.

**6.** P. A. Mackowiak, S. S. Wasserman, and M. M. Levine, "A critical appraisal of 98.6 degrees F, the upper limit of the normal body temperature, and other legacies of Carl Reinhold August Wunderlich," *Journal of the American Medical Association* 268 (1992):1578–80.

**7.** George W. Snedecor and William G. Cochran, *Statistical Methods*, 8th Ed. (Iowa State University Press, 1989).

**8.** D. L. Olds, C. R. Henderson Jr, R. Tatelbaum et al., "Improving the delivery of prenatal care and outcomes of pregnancy: a randomized trial of nurse home visitation," *Pediatrics* 77 (1986): 16–28.

**9.** Amanda Lenhart and Mary Madden, "Teens, privacy, and online social networks: how teens manage their online identities and personal information in the age of MySpace," Pew Internet and American Life Project, April 2007.

**10.** See Note 9.

**11.** Vijayakrishna K. Gadi et al., "Case-control study of fetal microchimerism and breast cancer," *PLoS one* 3 (March 5, 2008). (plos one, doi; 10:1371/journal.pone.0001706).

**12.** R. L. Bratton et al., "Effect of 'ionized' wrist bracelets on musculoskeletal pain: a randomized, double-blind, placebo-controlled trial," *Mayo Clinic Proceedings* 77 (2002): 1164–68.

## Chapter 11

**1.** Mary Madden and Amanda Lenhart, *Online Dating*, Pew Internet and American Life Project, 2005.

**2.** U.S. Department of Education, National Center for Education Statistics, Adult Education Survey of the 2005 National Household Education Surveys Program.

**3.** Derek M. Burnett et al., "Impact of minority status following traumatic spinal cord injury," *NeuroRehabilitation* 17 (2002): 187–94.

**4.** Pew Research Center for the People and the Press, *How Young People View Their Lives, Futures, and Politics: A Portrait of "Generation Next"* (Washington, D.C., 2007).

**5.** Andrew Rocco Tresolini Fiore, "Romantic regressions: an analysis of behavior in online dating systems," Master's thesis, Program in Media Arts and Sciences, Massachusetts Institute of Technology, 2004.

**6.** See Note 1.

**7.** M. A. Chase and G. M. Dummer, "The role of sports as a social determinant for children," *Research Quarterly for Exercise and Sport* 63 (1992): 418–24.

**8.** J. R. Knight, H. Wechsler, M. Kuo, M. Seibring, E. R. Weitzman, and M. Schuckit, "Alcohol abuse and dependence among U.S. college students," *Journal of Studies on Alcohol* 63, (3, 2002): 263–70.

**9.** Donald Garrow and Leonard Egede, "National patterns and correlates of complementary and alternative medicine use in adults with diabetes," *Journal of Alternative and Complementary Medicine* 12 (2006): 895–902.

**10.** J. E. Anderson and S. Sansom, "HIV testing in a national sample of pregnant US women: Who is not getting tested?" *AIDS Care* 19 (March 2007): 375–80.

**11.** National Agricultural Statistics Service, *Agricultural Statistics*, **www.usda.gov/nass** 2006.

## Chapter 12

**1.** Joseph Maze, Richard Murphy, and Cheri Simonds, "I'll see you on Facebook: the effects of computer-mediated teacher self-disclosure on student motivation, affective learning, and classroom climate," *Communication Edition* 56 (2007): 1–17.

**2.** William S. Cleveland, *Visualizing Data* (Hobart Press, 1993).

**3.** S. Blackman and D. Catalina, "The Moon and the Emergency Room," *Perceptual and Motor Skills* 37 (1973): 624–26.

**4.** U.S. National Water Climate and Climate Center, U.S. Department of Agriculture.

**5.** Peter Novacs, "Foraging Habits of Thatch Ants," **www.stat.ucla.edu/datasets/**.

**6.** See, for example, David S. Moore, George P. McCabe, and Bruce A. Craig, *Introduction to the Practice of Statistics*, 6th ed. (W. H. Freeman and Company, 2009), Section 2.2.

**7.** See Note 1.

**8.** See Note 3.

**9.** The data set is adapted from the **CEREALS** data set from the Data and Story Library, **ib.stat.cmu.edu/DASL**.

**10.** See Note 2.

## Chapter 13

**1.** T. Allison and D. V. Cicchetti, "Sleep in Mammals: ecological and Constitutional Correlates," *Science* 194 (1976): 732–34. Web reference: Quantitative Environmental Learning Project, Joseph Hull and Greg Langkamp, Seattle Central Community College, **www.seattlecentral.org/qelp/sets/017/017.html**.

**2.** Results certified by the Florida Department of State. Data set available at the National Organization for Research and Computing at the University of Chicago: **www2.norc.org/fl/results/index.html**.

# Index

**boldface** indicates a definition *italics* indicates a figure *t* indicates a table